Handbook of Sample Preparation

Handbook of Sample Preparation

Edited by

Janusz Pawliszyn

Heather L. Lord

WILEY

A John Wiley & Sons, Inc., Publication

Published by John Wiley & Sons, Inc., Hoboken, New Jersey.
Published simultaneously in Canada.

Library of Congress Cataloging-in-Publication Data:

Handbook of sample preparation / edited by Janusz Pawliszyn.
p. cm.
Includes index.
ISBN 978-0-470-09934-6 (cloth)
1. Sample preparation (Chemistry) 2. Extraction (Chemistry) 3. Chemistry, Analytic. I. Pawliszyn, Janusz.
QD75.4.S24H36 2011
543′.19—dc22

2010010828

10 9 8 7 6 5 4 3 2 1

Contents

Preface

Sample preparation is a critical part of the analytical process and should be part of any analytical chemistry teaching curriculum. Often though, it is either not mentioned or glossed over during graduate or undergraduate analytical courses. The primary reason for this situation is that sample preparation is not typically considered a separate part of analytical science with unique challenges to be considered, but rather a set or routine process conducted without much consideration during analytical method development. The result has been that "advances" in sample preparation in the past three decades (since the introduction of solid-phase extraction [SPE]) have primarily revolved around repackaging or repurposing existing technologies.

The main difficulty in recognizing the scientific principles of sample preparation is that the fundamentals of extraction, involving natural and frequently complex samples, are much less developed and understood compared with physicochemically simpler systems used in the separation and quantification steps of the analytical process, such as chromatography and mass spectrometry. This situation creates an impression that rational design and optimization of extraction systems is not possible. Therefore, the development of sample preparation procedures is frequently considered to be more of an "art" rather than a "science."

Given its significance in the overall success of analysis, advances in the science of sample preparation hold the promise of providing important gains in analytical method development. Until quite recently, sample preparation has been based on very simple "low-tech" approaches such as sample–solvent or sample–headspace partitioning, while underlying more scientifically challenging problems associated with the sample matrix have been ignored. This situation is presently changing with the introduction of nontraditional technologies, which address the need for solvent-free alternatives, automation, and miniaturization. These approaches are frequently simpler to operate but more difficult to optimize, requiring more fundamental knowledge by the analytical chemist not only about equilibrium conditions, but, more importantly, about the kinetics of mass transfer in the extraction systems. For some years, we have been actively involved in teaching the fundamental aspects of modern sample preparation technology to practitioners of analytical chemistry, mainly industrial chemists. We recognized a need to provide the fundamental background, not only to assist users, but also to help educators in developing their undergraduate and graduate programs. Designing teaching programs to address the new developments in extraction technologies is challenging as the scientific literature's emphasis is mostly placed on the differences between techniques rather than on their common features, which would facilitate general understanding.

The present trend in analytical instrumentation is toward miniaturization and portability. These developments will eventually enable the attainment of a major goal of the analytical chemist: to perform the analysis at the place where a sample is located, rather than the current practice of moving the sample to a laboratory. This new approach will reduce errors and time associated with sample transport and storage and thus result in more accurate, precise, and faster data. Simplification of sample preparation technologies and their integration with sampling and introduction of extracted components to analytical instrumentation are both challenges to and opportunities for the contemporary analytical chemist. The design of easy-to-use but powerful sample preparation tools will have a profound effect on the future of analytical methodology.

The purpose of this book is to address the needs and challenges outlined above in a single resource that provides practical information on the use of a wide variety of sample preparation strategies. Leading scientists in this area have contributed chapters on modern aspects of liquid, solid-phase, and membrane extractions with and without derivatization as well as the challenges associated with different types of matrices. In the first chapter on the Theory of Extraction, an attempt has been made to outline common features among extraction technologies. The following chapters are dedicated to different extraction technologies and applications for different types of matrices, and focus on the impact of new technologies on the science of sample preparation. Many authors emphasize the fact that extraction technologies should not be considered in isolation but should be well integrated with the steps of sampling and the introduction to analytical instrumentation. This is particularly important when implementing the analytical technology directly on-site.

This book is not intended to provide a comprehensive review of the topic of sample preparation, but rather to be a first step toward a unified treatment of analytical sample preparation technologies. It is hoped that it will be helpful for learning more about sample preparation and for identifying the commonalities, as well as for encouraging an interest in it by outlining the present practice of the technology and by indicating research opportunities.

JANUSZ PAWLISZYN and HEATHER L. LORD
University of Waterloo
Waterloo, ON, Canada

Contributors

Rocio Aguilar-Martínez, Department of Analytical Chemistry, University Complutense of Madrid, Ciudad Universitaria, 28040 Madrid, Spain

Ian J. Allan, Contaminants in the Marine Environment Section, Norsk Institutt for Vannforskning (NIVA), Gaustadalleen 21, NO-0349 Olso, Norway

Teresa Artuso, School of Biological Sciences and Applied Chemistry, Seneca College, 70 The Pond Road, North York, ON, M3J 3M6, Canada

Jacqueline M.R. Bélanger, Environmental Science and Technology Centre, Environment Canada (ret.), Ottawa, ON, K1A 0H3, Canada

Janine Brümmer, School of Biological Sciences, University of Portsmouth, King Henry I Street, Portsmouth, PO1 2DY, Hampshire, UK

Andrew Chan, Chemical Sensors Group, Department of Chemical and Physical Sciences, University of Toronto Mississauga, 3359 Mississauga Rd. North, Mississauga, ON, L5L 1C6, Canada

Ray E. Clement, Ontario Ministry of the Environment, Laboratory Services Branch, 125 Resources Road, Etobicoke, ON, M9P 3V6, Canada

John R. Dean, Biomolecular and Biomedical Research Centre, School of Life Sciences, Northumbria University, Newcastle upon Tyne, NE1 8ST, UK

Richard Greenwood, School of Biological Sciences, University of Portsmouth, King Henry I Street, Portsmouth, PO1 2DY, Hampshire, UK

Chunyan Hao, Ontario Ministry of the Environment, Laboratory Services Branch, 125 Resources Road, Etobicoke, ON, M9P 3V6, Canada

Toh Ming Hii, Department of Chemistry, National University of Singapore, 3 Science Drive 3, 119260, Republic of Singapore

Henryk H. Jeleń, Faculty of Food Science and Nutrition, Poznań University of Life Sciences, Wojska Polskiego 31, 60-624 Poznań, Poland

Jan Åke Jönsson, Department of Analytical Chemistry, Lund University, Getingevägen 60, Lund, 22100, Sweden

Hiroyuki Kataoka, School of Pharmacy, Shujitsu University, Nishigawara, Okayama 703-8516, Japan

Raimo A. Ketola, Centre for Drug Research, Faculty of Pharmacy, University of Helsinki, P.O. Box 56 (Viikinkaari 5E), Helsinki, FI-00014, Finland

Jesper Knutsson, Water Environment Technology, Chalmers University of Technology, Sven Hultins gata 8, SE-412 96, Göteborg, Sweden

Tapio Kotiaho, Laboratory of Analytical Chemistry, Department of Chemistry and Division of Pharmaceutical Chemistry, Faculty of Pharmacy, University of Helsinki, A.I. Virtasen aukio 1 (P.O. Box 55), Helsinki, FI-00014, Finland

Byron Kratochvil, Department of Chemistry, University of Alberta, 11227 Saskatchewan Drive, Edmonton, AB, T6G 2G2, Canada

Jeremy J. Kroon, Department of Chemistry and Biochemistry, South Dakota State University, Brookings, SD, 57007

Ulrich J. Krull, Chemical Sensors Group, Department of Chemical and Physical Sciences, University of Toronto Mississauga, 3359 Mississauga Rd. North, Mississauga, ON, L5L 1C6, Canada

Frants R. Lauritsen, Department of Pharmaceutics and Analytical Chemistry, Faculty of Pharmaceutical Sciences, Copenhagen University, Universitetsparken 2, Copenhagen, 2100, Denmark

Hian Kee Lee, Department of Chemistry, National University of Singapore, 3 Science Drive 3, 119260, Republic of Singapore

Heather L. Lord, Department of Chemistry, University of Waterloo, 200 University Avenue West, Waterloo, ON, N2L 3G1, Canada

Susan M. Lunte, Department of Pharmaceutical Chemistry and Ralph N. Adams Institute for Bioanalytical Chemistry, University of Kansas, 132 Simons Biosciences Research Laboratories, 2095 Constant Avenue, Lawrence, KS, 66047

Renli Ma, Biomolecular and Biomedical Research Centre, School of Applied Sciences, Northumbria University, Newcastle upon Tyne, NE1 8ST, UK

Ronald E. Majors, Agilent Technologies Inc., 2850 Centreville Road, Wilmington, DE, 19808

Abdul Malik, Department of Chemistry, University of South Florida, 4202 E. Fowler Avenue, CHE 205, Tampa, FL 33620

Antonio Martin Esteban, Dpto. Medio Ambiente, INIA, Carretera de A Coruña Km 7.5, 28040 Madrid, Spain

Graham A. Mills, School of Pharmacy and Biomedical Sciences, University of Portsmouth, White Swan Road, Portsmouth, PO1 2DT, Hampshire, UK

Pradyot Nandi, University of Colorado Denver, 2100 N. Ursula Street, Unit 329, Aurora, CO, 80045

Gangfeng Ouyang, School of Chemistry and Chemical Engineering, Sun Yat-sen University, 135 Xingang Street West, Guangzhou, 510275, China

J.R. Jocelyn Paré, Your World, 94A Newton Street, Moncton, NB, E1E 3A4, Canada

Janusz Pawliszyn, Department of Chemistry, University of Waterloo, 200 University Avenue West, Waterloo, ON N2L 3G1, Canada

Stig Pedersen-Bjergaard, School of Pharmacy, University of Oslo, P.O. Box 1068, Blindern, Oslo, N-0316, Norway

Zelda E. Penton, Varian Inc. (ret.); Email: z.penton@comcast.net

Knut Einar Rasmussen, School of Pharmacy, University of Oslo, P.O. Box 1068, Blindern, Oslo, N-0316, Norway

Douglas E. Raynie, Department of Chemistry and Biochemistry, South Dakota State University, Brookings, SD 57007

Sanja Risticevic, Department of Chemistry, University of Waterloo, 200 University Avenue West, Waterloo, ON N2L 3G1, Canada

Jack Rosenfeld, Department of Pathology and Molecular Medicine, McMaster University (Professor Emeritus), 14 Huntingwood Avenue, Unit 10, Hamilton, ON, L8H 6X3, Canada

Keita Saito, School of Pharmacy, Shujitsu University, 1-6-1, Nishigawara, Okayama 703-8516, Japan

Scott Segro, Department of Chemistry, University of South Florida, 4202 E. Fowler Avenue, CHE 205, Tampa, FL 33620

Börje Sellergren, INFU, Technische Universität Dortmund, Otto Hahn Strasse 6, 44221 Dortmund, Germany

Roger M. Smith, Department of Chemistry, Loughborough University, Loughborough, Leics, LE11 3TU, UK

Branislav Vrana, Research Centre for Toxic Compounds in the Environment (RECETOX), Masaryk University, Kamenice 126/3, 625 00 Brno, Czech Republic

Dajana Vuckovic, Centre for Cellular and Biomolecular Research, 160 College Street, Room 940, Toronto, ON, M5S 3E1, Canada

Atsushi Yokoyama, Criminal Investigation Laboratory, Okayama Prefectural Police Headquarters, Tonda-cho, Okayama 700-0816, Japan

Part I

Fundamental Extraction Techniques

Chapter 1

Theory of Extraction

Janusz Pawliszyn

1.1. PERSPECTIVE ON SAMPLE PREPARATION

Over the last two decades, active research on sample preparation has been fueled by interest in elimination of organic solvent from environmental analysis, and rapid analysis of combinatorial chemistry and biological samples requiring high-level automation with robots that are able to process multiwell plates containing an ever-increasing number of samples. These new developments resulted in miniaturization of the extraction process, leading to new micro-configurations and solvent-free approaches. Fundamental understanding of extraction principles has advanced in parallel with the development of new technologies. This progress has been very important in the development of novel approaches resulting in new trends in sample preparation, for example, microextraction, miniaturization, and integration of the sampling and separation and/or quantification steps of the analytical process.

The fundamentals of the sampling and sample preparation processes are substantially different from those related to chromatographic separations or other traditional disciplines of analytical chemistry. Sampling and sample preparation frequently resemble engineering approaches on a smaller scale. The sample preparation step typically consists of extraction of components of interest from the sample matrix. This procedure can vary in degree of selectivity, speed, and convenience, depending on the approach and conditions used, as well as on geometric configurations of the extraction phase. Optimization of this process enhances overall analytical performance. Proper design of the extraction devices and procedures facilitates rapid and convenient on-site implementation, integration with separation and quantification steps, and/or automation. The key to rational choice, optimization, and design is an understanding of fundamental principles governing mass transfer of analytes in multiphase systems. There is a tendency to divide extraction techniques according to random criteria. The objectives of this chapter are to emphasize common principles among different extraction techniques, to describe a unified theoretical treatment, and to discuss future research opportunities in integration and miniaturization trends.

1.1.1. Steps in the Analytical Process

The analytical procedure for complex samples consists of several steps typically including sampling, sample preparation, separation, quantification, statistical evaluation, and decision making (Fig. 1.1). Each step is critical for obtaining correct and informative results. The sampling step includes deciding where to get samples that properly define the object or problem being characterized, and choosing a method to obtain samples in the right amount. The objective of the sample preparation step is to isolate the components of interest from a sample matrix, because most analytical instruments cannot handle the matrix directly. Sample preparation involves extraction procedures and can also include "cleanup" procedures for very complex, "dirty" samples. This step must also bring the analytes to a concentration level suitable for detection, and therefore, sample preparation methods typically include enrichment. During the separation step of the analytical process, the isolated complex mixture containing the target analytes is divided into its constituents, typically by means of a chromatographic or an electrophoretic technique, which are subsequently identified and quantified. The identification can be based on retention time or migration time combined with selective detection, for example, mass spectrometry (MS). Statistical evaluation of the results provides an estimate of the concentration of the target compound in the sample being analyzed. The resulting data are used to make appropriate decisions, which might include a move to take another sample for further investigation of the object or problem.

It is important to note, as emphasized in Figure 1.1, that analytical steps follow one after another, and a subsequent step cannot begin until the preceding one has been

Handbook of Sample Preparation, Edited by Janusz Pawliszyn and Heather L. Lord

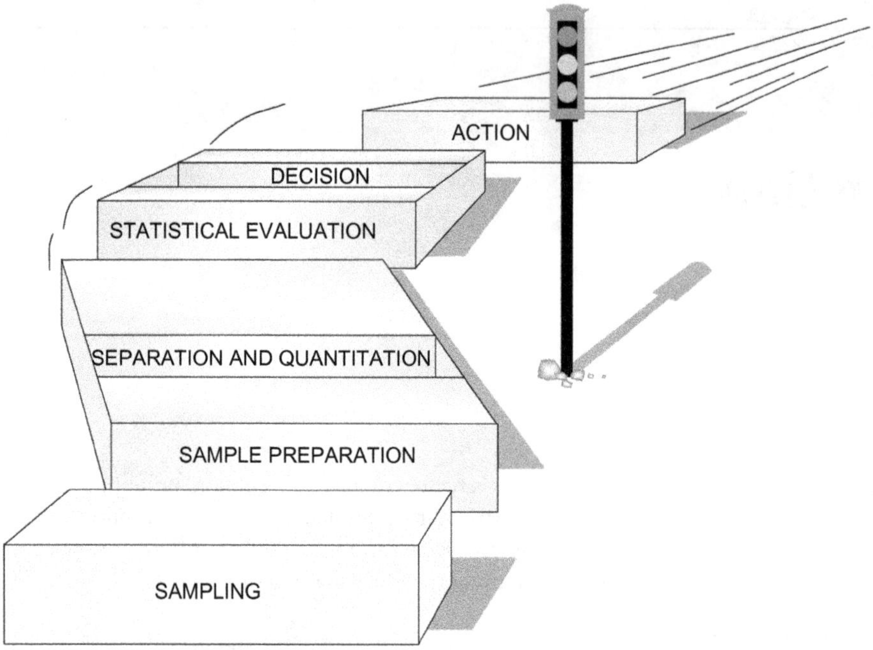

Figure 1.1. Steps in analytical process. Copyright Wiley-VCH, 1997. Reprinted with permission.

completed. Therefore, the slowest step determines the overall speed of the analytical process, and improving the speed of a single step may not necessarily result in an increase in throughput. To increase throughput, all steps need to be considered. Also, errors performed in any preceding step, including sampling, will result in the overall poor performance of the procedure.

1.1.2. Sample Preparation as Part of the Analytical Process

There have been major breakthroughs in the development of improved instrumentation, which involve miniaturization of analytical devices and hyphenation of different steps into one system. It is recognized that an ideal instrument will perform all the analytical steps with minimal human intervention, preferably directly on the site where an investigated system is located rather than moving the sample to laboratory, as is a common practice at the present time. This approach will eliminate errors and reduce the time associated with sample transport and storage, and therefore, result in faster analysis and more accurate, precise data. Although such a device has not yet been built, today's sophisticated instruments, such as the gas chromatography–mass spectrometry (GC-MS) or liquid chromatography–mass spectrometry (LC-MS), can separate and quantify complex mixtures and automatically apply chemometric methods to statistically evaluate results. It is much more difficult to hyphenate sampling and sample preparation steps, primarily because the current state-of-the-art sample preparation techniques employ multistep procedures involving organic solvents. This characteristic makes it difficult to develop a method that integrates sampling and sample preparation with separation methods, for the purpose of automation. The result is that over 80% of analysis time is currently spent on sampling and sample preparation steps for complex samples.

One of the reasons for slow progress in the area of sample preparation is that the fundamentals of extraction involving natural, frequently complex samples are much less developed and understood compared with physicochemically simpler systems used in separation and quantification steps such as chromatography and MS. This situation creates an impression that rational design and optimization of extraction systems is not possible. Therefore, the development of sample preparation procedures is frequently considered to be an "art" and not a "science."

1.1.3. Classification of Extraction Techniques

Figure 1.2 provides a classification of extraction techniques and unifies the fundamental principles behind the different extraction approaches. In principle, exhaustive extraction approaches do not require calibration, because most analytes are transferred to the extraction phase by employing overwhelming volumes of it. In practice, however, confirmation of satisfactory recoveries is implemented in the method by using surrogate standards. To reduce the amount of solvents and time required to accomplish exhaustive removal, batch equilibrium techniques (e.g., liquid–liquid extractions [LLEs]) are frequently replaced by flow-through techniques. For example, a sorbent bed can be packed with extraction phase dispersed on a supporting material; when the sample is passed through, the analytes in the sample are retained on the bed. Large volumes of sample can be passed through a small cartridge, and the flow through the well-packed bed facilitates efficient mass transfer. The extraction procedure

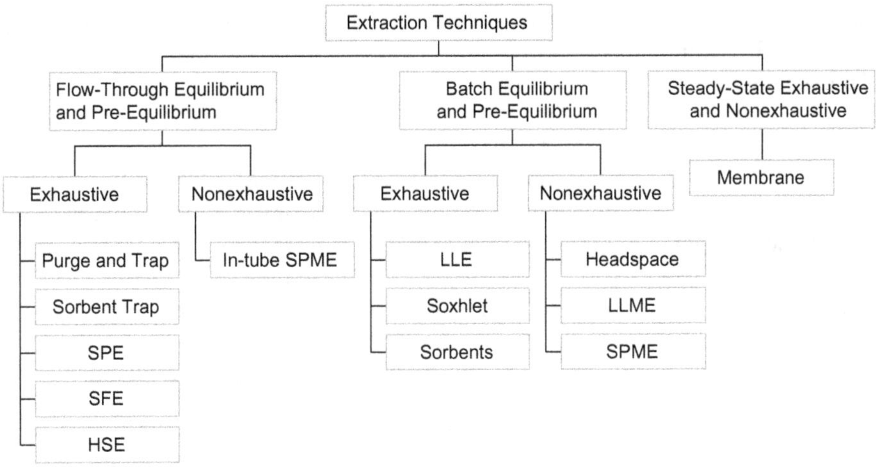

Figure 1.2. General classification of extraction techniques. Copyright Wiley-VCH, 1997. Reprinted with permission.

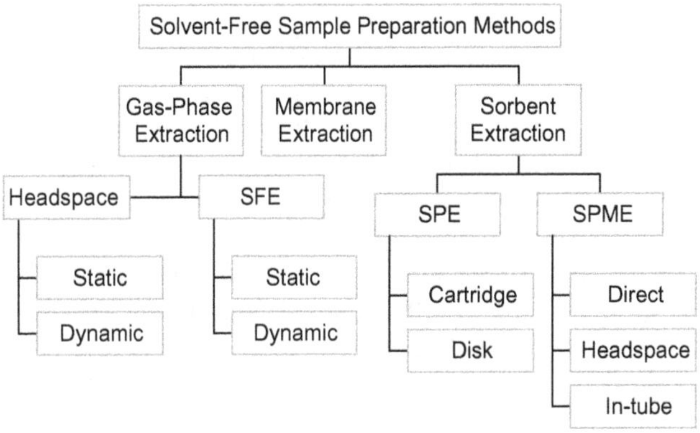

Figure 1.3. Classification of solvent-free extraction techniques. Copyright Wiley-VCH, 1997. Reprinted with permission.

is followed by desorption of analytes into a small volume of solvent, resulting in substantial enrichment and concentration of the analytes. This strategy is used in sorbent-trap techniques and in solid-phase extraction (SPE).[1] Alternatively, the sample (typically, a solid) can be packed in the bed, and the extraction phase can be used to remove and transport the analytes to the collection point. In supercritical fluid extraction (SFE), compressed gas is used to wash analytes from the sample matrix; an inert gas at atmospheric pressure performs the same function in purge-and-trap methods. In dynamic solvent extraction, for example, in a Soxhlet apparatus, the solvent continuously removes the analytes from the matrix at the boiling point of the solvent. In more recent pressurized fluid extraction (PFE) techniques, smaller volumes of organic solvent or even water are used to achieve greater enrichment at the same time as extraction, because of the increased solvent capacity and elution strength at high temperatures and pressures.[2]

Alternatively, nonexhaustive approaches can be designed on the basis of the principles of equilibrium, pre-equilibrium, and permeation techniques.[3] Although equilibrium nonexhaustive techniques are fundamentally analogous to equilibrium-exhaustive techniques, the capacity of the extraction phase is smaller and is usually insufficient to remove most of the analytes from the sample matrix. This is because of the use of a small volume of the extracting phase relative to the sample volume, such as is employed in microextraction (solvent microextraction[4] or solid-phase microextraction [SPME][5]), or in the cases of a low sample matrix–extraction phase distribution constant, as is typically encountered in gaseous headspace techniques.[6] Pre-equilibrium conditions are accomplished by breaking the contact between the extraction phase and the sample matrix before equilibrium with the extracting phase has been reached. Although the devices used are frequently identical with those of microextraction systems, shorter extraction times are employed. The pre-equilibrium approach is conceptually similar to the flow injection analysis (FIA) approach,[7] in which quantification is performed in a dynamic system and system equilibrium is not required to obtain acceptable levels of sensitivity, reproducibility, and accuracy. In permeation techniques, for example, membrane extraction,[8] continuous steady-state transport of analytes through the extraction phase is accomplished by simultaneous re-extraction of analytes. Membrane extraction can be made exhaustive by designing appropriate membrane modules and optimizing the sample and stripping flow conditions,[9] or it can be optimized for throughput and sensitivity in nonexhaustive, open-bed extraction.[10]

In addition to classification of methods based on more fundamental principles as discussed above, it is also instructive to divide techniques according to particular characteristics. For example, recently there is a trend toward solvent-free techniques (Fig. 1.3).[11] This is an important direction, not only because it addresses health and pollution prevention issues, but also because such approaches tend to be easier to implement for on-site monitoring in field conditions. This direction has generated a lot of interest and research opportunities recently, and it is expected to continue to be a very active area in the near future. The most promising solventless techniques are headspace, membrane, and sorbent approaches. SFE is able to selectively remove semivolatile and nonvolatile trace components from solid matrices, but field implementation of this technology is very difficult since it uses heavy and inconvenient components. However, new developments in the technology, such as miniaturized fluid delivery systems,[12] will aid on-site implementations of this technology.

1.2. FUNDAMENTALS

As the preceding discussion and Figure 1.2 indicate, there is a fundamental similarity among extraction techniques used in the sample preparation process. In all techniques, the extraction phase is in contact with the sample matrix, and analytes are transported between the phases. For exhaustive techniques, the phase ratio is higher and geometries are more restrictive to ensure the quantitative transfer of analytes compared with nonexhaustive approaches. The thermodynamics of the process are defined by the extraction phase–sample matrix distribution constant. It is instructive to consider in more detail the kinetics of processes occurring at the extraction phase–sample matrix interface since this defines the time of the analytical procedure. In many cases, the analytes are re-extracted from the extraction phase, but this step is not discussed here since this process is analogous and much simpler in principle compared with removing analytes from a more complex sample matrix. The main objective of this chapter is to outline the common fundamental principles among various extraction techniques to facilitate a better understanding of selection criteria for appropriate techniques, device geometries, and operational conditions.

1.2.1. Thermodynamics

The fundamental thermodynamic principle common to all chemical extraction techniques involves the distribution of analyte between the sample matrix and the extraction phase.

1.2.1.1. Distribution constant. When a liquid is used as the extraction medium, then the distribution constant, K_{es},

$$K_{es} = a_e / a_s = C_e / C_s, \tag{1.1}$$

defines the equilibrium conditions and ultimate enrichment factors achievable in the technique, where a_e and a_s are the activities of analytes in the extraction phase and matrix, correspondingly, and can be approximated by the appropriate concentrations. Figure 1.4 shows the schematic example of the extraction system for LLE. For solid extraction phase

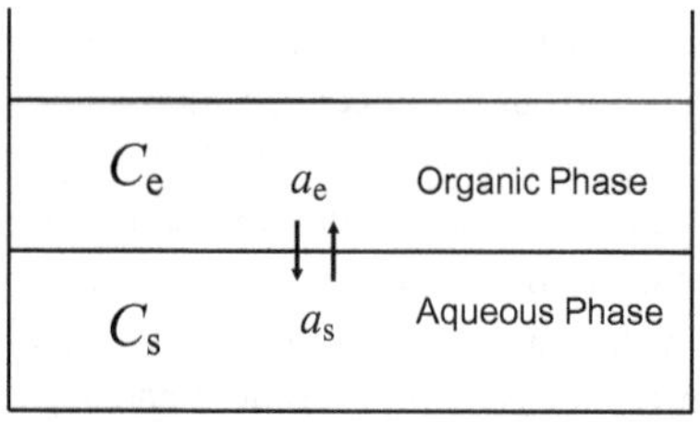

Figure 1.4. Partitioning between aqueous sample matrix and organic extraction phase. Copyright Wiley-VCH, 1997. Reprinted with permission.

adsorption, equilibria can be explained by the following equation:

$$K_{es}^{s} = S_e / C_s, \tag{1.2}$$

where S_e is the solid extraction phase surface concentration of adsorbed analytes. The relationship above is similar to Equation 1.1, except that the extraction phase concentration is replaced with surface concentration. The S_e term in the numerator indicates that the sorbent surface area available for adsorption must also be considered. This complicates the calibration at equilibrium conditions because of displacement effects and the nonlinear adsorption isotherm.[13] These equations can be used to calculate the amount of analyte in the extraction phase at equilibrium conditions.[4] For example, for equilibrium liquid microextraction techniques and large samples, including direct extraction from the investigated system, the appropriate expression for the amount of analyte, n, is very simple,

$$n = K_{es} V_e C_s, \tag{1.3}$$

where K_{es} is the extraction phase–sample matrix distribution constant, V_e is the volume of the extraction phase, and C_s is the concentration of the sample. This equation is valid when the amount of analytes extracted is insignificant compared with the amount of analytes present in a sample (large V_s and/or small K_{es}), resulting in negligible depletion of analyte concentration in the original sample. In Equation 1.3, K_{es} and V_e determine the sensitivity of the microextraction method, whereas K_{es} determines its selectivity. The sample volume can be neglected, thus integrating sampling and extraction without the need for a separate sampling procedure, as discussed in more detail later. The nondepletion property of the small dimensions typically associated with microextraction systems results in minimum disturbance of the investigated system, facilitating convenient speciation, investigation of multiphase distribution equilibria, and repeated sampling from the same system to follow a process of interest.

When significant depletion occurs, the sample volume, V_s, has some impact on the amount extracted and, therefore, on sensitivity.[14] This effect can be calculated using the following equation:

$$n = \frac{K_{es} V_e C_0 V_s}{K_{es} V_e + V_s}. \tag{1.4}$$

In heterogeneous samples (headspace, immiscible liquids, and solids), the components of the sample partition in the multiphase system and are less available for extraction. This effect depends on analyte affinities and capacities of the competing phases and can be calculated if appropriate volumes and distribution constants are known. The distribution constants are dependent on various parameters including temperature, pressure, and sample matrix conditions such as pH, salt, and organic component concentration. All these parameters need to be optimized for maximum transfer of analytes to the extraction phase during the method development process. In practice, however, kinetic factors defined by the dissociation constants, diffusion coefficient, and agitation conditions frequently determine the amounts of extracted analytes from complex samples since the overall rates are slow, and therefore, extraction amounts for time-limited experiments do not reach equilibrium values.

1.2.1.2. Matrix effects. Two potential complications are typically observed when extracting analytes from complex matrices. One is associated with competition among different phases for the analyte and the other with the fouling of the extraction phase, because of the adsorption of macromolecules such as proteins and humic materials at the interface. The components of heterogeneous samples (including headspace, immiscible liquids, and solids) partition in the multiphase system and are less available for extraction. This effect depends on analyte affinity and the volume of the competing phases and can be estimated if appropriate volumes and distribution constants are known. The mass of an analyte extracted by an extraction phase in contact with a multiphase sample matrix can be calculated using the following equation:

$$n = \frac{K_{es}V_e C_0 V_s}{K_{es}V_e + \sum_{i=1}^{i=m} K_{is}V_i + V_s}, \tag{1.5}$$

where $K_{is} = C_i^{\infty}/C_s^{\infty}$ is the distribution constant of the analyte between the ith phase and the matrix of interest.[15] Equation 1.5 simplifies to Equation 1.4 if there are no competing phases in the sample matrix.

The typical approach used to reduce fouling of the extraction phase involves the introduction of a barrier between the sample matrix and the extraction phase to restrict transport of high-molecular-weight interferences (Fig. 1.5). For example, the extraction phase can be surrounded by a porous membrane with pores smaller than the size of the interfering macromolecules (Fig. 1.5a), for example, use of a dialysis membrane with the appropriate molecular weight cutoff. This approach is conceptually similar to membrane dialysis

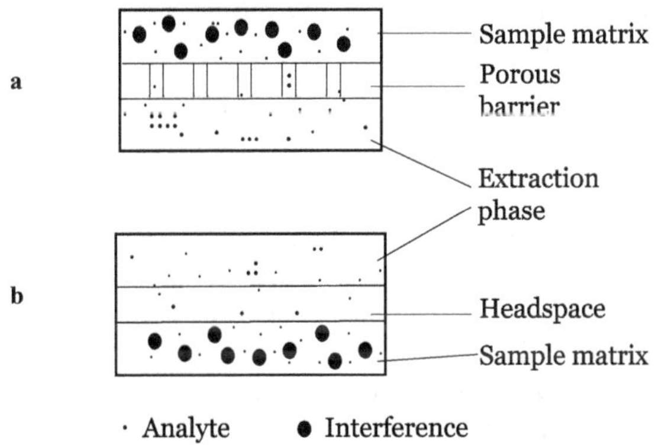

Figure 1.5. Integrated cleanup and extraction using selective barrier approaches based on size exclusion with a porous membrane (a) and based on volatility with a headspace gap (b). Copyright Wiley-VCH, 1997. Reprinted with permission.

from complex matrices, in which the porous membrane is used to prevent large molecules from entering the dialyzed solution.[16] Membrane separation has also been used to protect SPME fibers from humic material.[17] More recently, hollow fiber membranes have been used in solvent microextraction, both to support the small volume of solvent and to eliminate interferences when extracting biological fluids.[18] This concept has been further explored by integrating the protective structure and the extraction phase in individual sorbent particles, resulting in restricted access material (RAM).[19] The chemical nature of the small inner pore surface of the particles is hydrophobic, facilitating extraction of small target analytes, whereas the outer surface is hydrophilic, thus preventing adsorption of excluded large proteins. In practice, fouling of the hydrophobic interface occurs to a large extent only when the interfering macromolecules are hydrophobic in nature.

A gap made of gas is also a very effective separation barrier (Fig. 1.5b). Analytes must be transported through the gaseous barrier to reach the coating, thus resulting in exclusion of nonvolatile components of the matrix. This approach is practically implemented by placing the extraction phase in the headspace above the sample; it results in a technique such as headspace SPME, which is suitable for extraction of complex aqueous and solid matrices.[20] The major limitation of this approach is that the rates of extraction are low for poorly volatile or polar analytes, because of their small Henry's law constants. In addition, sensitivity for highly volatile compounds can suffer, because these analytes have high affinity for the gas phase, where they are concentrated. The effect of the headspace on the amount of analytes extracted and, therefore, on sensitivity can be calculated using Equation 1.5, which indicates that reducing its gaseous volume minimizes the effect.

Extraction at elevated temperatures enhances Henry's law constants by increasing the concentrations of the analytes in the headspace; this results in rapid extraction by the extraction phase. The coating/sample distribution coefficient also decreases with increasing temperature; however, this results in diminution of the equilibrium amount of the analyte extracted. To prevent this loss of sensitivity, the extraction phase can be cooled simultaneously with sample heating. This "cold finger" effect results in increased accumulation of the volatilized analytes on the extraction phase. This additional enhancement in the sample matrix–extraction phase distribution constant associated with the temperature gap present in the system can be described by the equation,[21]

$$K_T = K_0 \frac{T_s}{T_e} \exp\left[\frac{C_p}{R}\left(\frac{\Delta T}{T_e} + \ln\frac{T_e}{T_s}\right)\right], \tag{1.6}$$

where $K_T = C_e(T_e)/C_s(T_s)$ is the distribution constant of the analyte between cold extraction phase on the fiber having temperature T_e and hot headspace at temperature T_s; C_p is the constant-pressure heat capacity of the analyte;

$\Delta T = T_s - T_e$; and K_0 is the coating/headspace distribution constant of the analyte when both coating and headspace are at temperature T_e. Because of enhancement of the sample matrix–extraction phase distribution constant, quantitative extraction of many analytes,[22] including volatile compounds, is possible with this method.[23]

1.2.1.3. Characteristics of the extraction phase. The properties of the extraction phase should be carefully optimized, because they determine the selectivity and reliability of the method. These properties include both bulk physicochemical properties, for example, polarity, and physical properties, for example, thermal stability and chemical inertness. Solvents and liquid polymeric phases, for example, polydimethylsiloxane (PDMS),[24] are very popular because they have wide linear dynamic ranges associated with linear absorption isotherms. They also facilitate "gentle" sample preparation, because chemisorption and catalytic properties, frequently associated with solid surfaces, are absent. No loss or modification of the analyte occurs during extraction and/or desorption. Despite these attractive properties of liquid extraction media, solid phases are frequently used because of their superior selectivity and extraction efficiency for some groups of compounds. For example, carbon-based sorbents are effective for extraction of volatile analytes.

The development of selective extraction materials often parallels that of the corresponding selective chemical sensors.[25] Similar manufacturing approaches and structures similar to those of sensor surfaces have been implemented as extraction phases. For example, phases with specific properties, such as molecularly imprinted polymers[26] and immobilized antibodies,[27] have recently been developed for extraction. These types of sorbents rely on differences between bulk properties of the extraction phase, and the highly specific molecular recognition centers dissolved in it to facilitate high-selectivity extraction with minimum nonspecific adsorption.[28] In addition, chemically tuneable properties of the extraction phase can be controlled during the preparation procedure. For example, polypyrrole has been used successfully for a range of applications ranging from ion exchange extraction to hydrophobic extraction based on selective interaction between the polymer and the target analytes.[29] In addition, tuneable properties of the polymer, for example, the oxidation/reduction equilibrium in conductive polypyrrole, can be explored to control adsorption and desorption.[30]

Demands on the specificity of extraction phases are typically less stringent than for sensor surfaces, because a powerful separation and quantification technique, for example, GC-MS or LC-MS, is usually used after extraction, facilitating accurate identification of the analyte. More demand is placed, however, on the thermal stability and chemical inertness of the extraction phase, because the extraction materials are frequently exposed to high temperatures and different solvents during extraction and introduction to the analytical separation instruments. New coating chemistries, for example, the sol-gel polymerization approach, have been developed to address these needs.[31]

To optimize sensitivity, the choice of the extraction phase is frequently based on its affinity toward the target analyte. In practice, however, kinetic factors defined by dissociation constants, diffusion coefficients, and agitation conditions frequently determine the amounts of analytes extracted from complex samples. Because overall extraction rates are slow, the amounts of analytes extracted during experiments of limited duration do not reach equilibrium values.

1.2.2. Kinetics

1.2.2.1. LLE. It is instructive to consider a simple case of static extraction of water with organic solvent, as illustrated in Figure 1.6 to consider the effects of different parameters on extraction kinetics. An appropriate equation showing concentration profiles in each of the phases can be obtained by solving Fick's second law differential equation for appropriate boundary conditions:

$$\frac{\partial C(x,t)}{\partial t} = D\frac{\partial^2 C(x,t)}{\partial x^2}. \tag{1.7}$$

If no convection is present in the system, the distribution constant is defined by Equation 1.1, and the two phases are placed in contact with each other at $t = 0$, then, the solution can be found using a Laplace transform approach for aqueous sample phase ($x < 0$) to be

$$C_s(x,t) = C_0\frac{\frac{z}{K_{es}} + erf\left(z\sqrt{tD_s}\right)}{1+\frac{z}{K_{es}}}. \tag{1.8a}$$

For organic extraction phase ($x > 0$),

$$C_e(x,t) = C_0\frac{z\left\{1 - erf\left(x/z\sqrt{tD_e}\right)\right\}}{1+\frac{z}{K_{es}}}, \tag{1.8b}$$

where C_0 is the initial concentration of the analyte in the aqueous phase, D_e and D_s are the diffusion of analyte in the extraction phase and in the sample, respectively; $z = D_e/D_s$; and K_{es} is an appropriate distribution constant defined by Equation 1.1. The solution to the above equation is shown graphically in Figure 1.6 for several extraction times when diffusion of analytes in aqueous and organic phases is 10^{-5} cm^2/s. Figure 1.6 illustrates that the concentration gradient is decreasing and extending deeper into both phases as a function of time. The flux of analytes is decreasing proportionally with decrease in the gradient. The concentration effect of analytes at the boundary on the organic side compared with the bulk aqueous concentration is not observed at the beginning of extraction due to the drop of the concentration on the aqueous side. Therefore, a decrease in boundary layer thickness and the diffusion length by agitation of one or both phases increases the rate of extraction dramati-

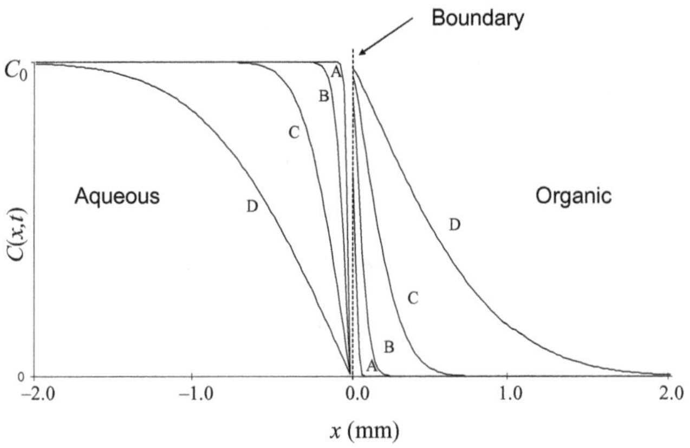

Figure 1.6. Concentration profiles at the interface between infinite volume sample and extraction phases for analyte characterized by identical diffusion coefficient in aqueous and organic phase (10^{-5} cm^2/s). The profiles correspond to A, 1 s; B, 10 s; C, 100 s; D, 1000 s after merging both phases. Copyright Wiley-VCH, 1997. Reprinted with permission.

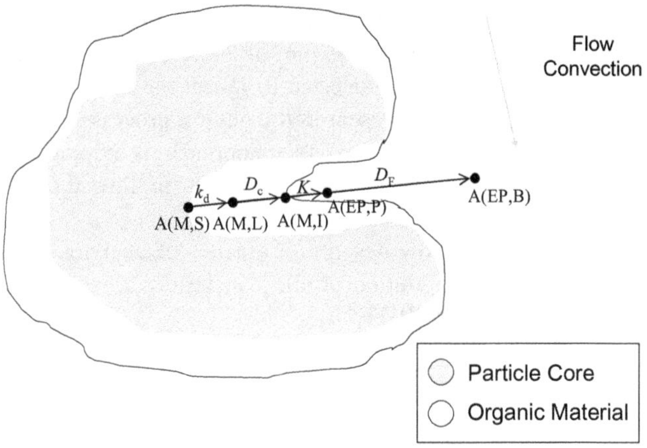

Figure 1.7. Processes involved in the extraction of heterogeneous samples containing porous solid particles. The symbols/terms in the figure are discussed in the text. Copyright Wiley-VCH, 1997. Reprinted with permission.

cally. The effects of agitation can be calculated using the boundary layer model described later. One other way to improve mass transfer is to use thin films of sample matrix and/or extraction phase to decrease the diffusion length. In addition, a combination of agitation of the sample and use of a thin extraction phase also facilitate shorter extraction times. If the extraction media and matrix phases are of different states of matter, then it is more critical to overall extraction kinetics to agitate or use the thin-film format of the phase, since it is characterized by a smaller diffusion coefficient. For example, when extracting gas or liquid samples with PDMS, it is critical to disperse the extraction phase as a thin film so the equilibrium between the phases may be rapidly reached.

1.2.2.2. Extraction of solids. The most challenging extractions occur when a solid is present as a part of the sample matrix. This case can be considered as the most general example of extraction since it involves a number of fundamental processes occurring during extraction. If we assume that a matrix particle consists of an organic layer on an impermeable but porous core and the analyte is adsorbed onto the pore surface, the extraction process can be modeled by considering several basic steps as shown in Figure 1.7. To remove the analyte from the extraction vessel, the compound must first be desorbed from the surface (A(M,S), Fig. 1.7); then, it must diffuse through the organic part of the matrix (A(M,L)) to reach the matrix/fluid interface (A(M,I)). At this point, the analyte must be solvated by the extraction phase (A(EP,P)), and then it must diffuse through the static phase present inside the pore to reach the portion of the extraction phase influenced by convection, to be transported through the interstitial pores of the matrix, and eventually reach the bulk of the extraction phase (A(EP,B)). The simplest way to design a kinetic model for this problem is to adopt equations developed by engineers to investigate mass transport through porous media.[32,33]

For the purpose of this discussion, we consider the efficient and frequently applied experimental arrangement for removing solid-bound semivolatile analytes, involving the use of a piece of stainless steel tubing as the extraction vessel. The sample is typically placed inside the tubing and a linear flow restrictor is attached to maintain the pressure at the end of the vessel. During the process, the extraction phase continuously removes analytes from the matrix, which are then transferred to the collection vessel after the expansion of the fluid. This leaching process is very similar to chromatographic elution with packed columns, particularly to the frontal method. The main difference is that in sample preparation, analytes are dispersed in the matrix at the beginning of the experiment, while in chromatographic frontal analysis, a long plug is introduced into the column at the initial stage of the separation process. The principal objective of the extraction is to remove analytes from the vessel in the shortest period of time, requiring elution conditions under which the analytes are unretained. In chromatography, on the other hand, the ultimate goal is to separate components of the sample, which requires retention of analytes in the column. Another major difference is that the packing matrix is usually well characterized in chromatography, but in sample preparation, it is often unknown.

One way to develop a mathematical model for this extraction approach is to establish the mass balance equation for the system after careful consideration of the individual mass transfer steps occurring during the extraction process (see Fig. 1.7) and specific boundary conditions.[34] Extensive investigations on similar topics have already been conducted by engineers who have studied the mass transfer in porous media[12] and chromatographers.[13] In these studies, the relationship between various matrix parameters and flow conditions on the elution profile were described mathematically and verified experimentally. In chromatography, this relationship is usually described as contributions from each of the mass transfer steps to the height equivalent to a theoretical plate (HETP). The overall performance of the system can be defined as the sum of the relevant individual components judiciously selected to reflect the most significant individual steps present in the elution process. For the purpose of this discussion, this approach is adopted to develop a model for extraction kinetics in flow-through techniques.

The effect of slow desorption kinetics of analytes from the matrix on the elution profile can be described as the contribution to the HETP,[8] h_{RK},

$$h_{RK} = \frac{2k\,u_e}{(1+k)^2\,(1+k_o)\,k_d}, \tag{1.9}$$

where k is the partition ratio; k_d is the dissociation rate constant of the analyte–matrix complex of reversible process; k_o is the ratio of the intraparticulate void volume to the interstitial void space and is expressed as

$$k_o = \frac{\varepsilon_i(1-\varepsilon_e)}{\varepsilon_e}, \tag{1.10}$$

where ε_i is intraparticulate porosity and ε_e is interstitial porosity; and u_e is the interstitial linear extraction phase velocity expressed as

$$u_e = u(1+k_o), \tag{1.11}$$

where $u = L/t_0$ is the chromatographic linear velocity; L is the length of the extraction vessel; and t_0 is the time required to remove one void volume of the extraction phase from the vessel. Chromatographic and interstitial linear velocities are identical if matrix particles have low porosity. This analysis can be extended to elution through a matrix having multiple adsorption sites characterized by different dissociation rate constants by using the approach described by Giddings.[35]

The diffusion of the analyte in the liquid or swollen solid part of the matrix is important when polymeric materials are extracted, or the matrix has substantial organic content. Its contribution can be expressed as h_{DC},

$$h_{DC} = \frac{2}{3}\frac{k}{(k+1)^2}\frac{d_c^2}{D_s}u_e, \tag{1.12}$$

where d_c is the thickness of the matrix component permeable to analyte and D_s is the diffusion coefficient of the analyte in the sample matrix.

The analytes migrate in and out of a pore structure of the matrix during the elution. This can be described as resistance to mass transfer in the fluid associated with the porous nature of the environmental matrices, which gives rise to the following HETP component, h_{DP},

$$h_{DP} = \frac{\theta(k_o+k+k\,k_o)^2\,d_p^2 u_e}{30\,k_o\,(1+k_o)^2\,(1+k)^2\,D_p}, \tag{1.13}$$

where θ is tortuosity factor for the porous particle, d_p is the diameter of the particulate matter, and D_p is the diffusion coefficient of the analyte in the material filling the pores, which, in most practical cases, will be an extraction phase, and therefore, $D_p = D_e$, where D_e is the diffusion coefficient of the analyte in the extraction phase. This contribution can be quite important considering the relatively large particle size (about 1 mm) of environmental matrices and becomes particularly important when the pores are filled with dense organic material, such as humic matter rather than the extraction phase.

In the flowing bulk of the fluid, an analyte experiences resistance to mass transfer associated with eddy diffusion (random paths of the analytes through the vessel filled with the particles), which is given by h_{ED},

$$h_{ED} = 2\lambda d_p, \tag{1.14}$$

where λ is a structural parameter and is close to 1 for spherical matrix particles. This contribution to band broadening is the most important factor in high-performance liquid chromatography (HPLC) separations, and it is expected to remain significant in extractions because matrices typically have large particle sizes.

In addition, we should also consider analyte diffusion along the axis of the vessel (longitudinal diffusion), which can be defined as h_{LD},

$$h_{LD} = \frac{\gamma_M D_e}{u_e}, \tag{1.15}$$

where γ_M is the obstruction factor that characterizes the structure of the matrix. The contribution of this component is expected to be small. The analyte concentration profile generated during the experiment as a function of time $C(x,t)$ can be represented using the equation that describes the dispersion of a plug of finite width:[9]

$$\frac{C(x,t)}{C_0} = \frac{1}{2}\left\{ERF\left(\frac{\frac{L}{2}-x-\frac{ut}{1+k}}{\sigma\sqrt{2}}\right)+ERF\left(\frac{\frac{L}{2}+x+\frac{ut}{1+k}}{\sigma\sqrt{2}}\right)\right\}, \tag{1.16}$$

where L is the length of the vessel; C_0 is the initial concentration of analyte in the extraction vessel; and σ is the mean square root dispersion of the band expressed as

$$\sigma = \sqrt{H\,t\frac{u}{1+k}}, \tag{1.17}$$

where H is equivalent to the HETP in chromatographic systems and is a sum of the contributions discussed above, $H = h_{RK} + h_{DC} + h_{DP} + h_{ED} + h_{LD}$. The mass of analyte eluted from the vessel during a given extraction time t can be calculated from the following equation:

$$\frac{m(t)}{m_0} = \frac{\int_{-\infty}^{-\frac{1}{2}L} C(x,t)\,dx}{C_s L}, \tag{1.18}$$

where $m(t)$ is the extracted mass of analyte and m_0 is the total amount of analyte in the vessel at the beginning of the experiment. We will refer to this function as the "time elution profile" emphasizing the similarity of the extraction process in this simple case to chromatographic elution.

*1.2.2.3. **Convolution model of extraction.*** The above discussion applies only to the situation when the analytes are initially present in a fluid phase, which, in flow-through techniques, corresponds to elution of uniform spikes from the extraction vessel, or when weakly adsorbed native analytes are removed from an organic-poor matrix such as sand. In other words, the above relationships are suitable for systems in which the partitioning equilibrium between the matrix and extraction fluid is reached quickly compared with the fluid flow. They are also suitable to model static/dynamic extractions, under good solubility conditions ($k = 0$), in which the sample is initially exposed to the static extraction phase (vessel is capped) for a time required to achieve an equilibrium condition prior to elution by fluid flow. If dynamic extraction is performed from the beginning of extraction, then in the majority of practical cases, the system is not expected to achieve the initial equilibrium conditions. This is because of slow mass transport between the matrix and the fluid (e.g., slow desorption kinetics or slow diffusion in the matrix). The expected relationship between amount of analyte removed from the vessel versus time can be obtained in this case by convoluting the function describing the rate of mass transfer between the phases $F(\tau)$ with the elution time profile $m/m_0(t)$ derived above (Eq. 1.18):[36]

$$\int_{\tau=0}^{\tau=t} \frac{m(t-\tau)}{m_0} F(\tau)\,d\tau. \tag{1.19}$$

The resulting function describes a process where elution and mass transfer between the phases occur simultaneously. In this discussion, we will refer to this function as the "extraction time profile" to emphasize the point that in a majority of extraction cases, these two processes are expected to be combined. $F(\tau)$ describes the kinetics of the process, which defines the release rate of analyte from the sample matrix and can include, for example, the matrix–analyte complex dissociation rate constant, the diffusion coefficient, the time constant that describes the swelling of the matrix that will facilitate the removal of the analyte, or a combination of the above. Detailed discussion, graphical representations, and applications of this model to describe and/or investigate processes in SFE have been described in detail elsewhere.[37,38]

The conclusion above can be stated in a more general way. Convolution among functions describing individual processes occurring during the extraction describes the overall extraction process and represents a unified way to describe the kinetics of these complex processes. The exact mathematical solution to the convolution integral is frequently difficult to obtain, but graphical representation of the solution can be calculated using Fourier transform or numerical approaches. Frequently, it is possible to incorporate mathematical functions that describe a combination of the unit processes. In the example of the flow-through system discussed above, the elution function describes the effect of porosity and analyte affinity toward the extraction matrix on the extraction rate. It should be emphasized that the convolution approach considers all processes equivalently. In practice, however, a small number and frequently just one unit process controls the overall rate of extraction so the equations can be simplified by considering this fact.

Determination of the limiting step is not possible exclusively by qualitative agreement with the mathematical model since the effect on recovery of most of the unit processes has an exponential decay nature. To properly recognize them, quantitative agreement and/or effect of extraction parameters need to be examined. Identification of the limiting process provides valuable insight on the most effective approach to optimization of extraction.

1.3. OPTIMIZATION OF THE EXTRACTION PROCESS

A fundamental understanding of the process leads to better strategies for optimization of performance. In heterogeneous samples, for example, the release of solid-bound analytes from the sample matrix, through a reversal of chemisorption or inclusion, frequently controls the extraction rate. By recognizing this fact, extraction parameters can be changed to increase the extraction rates. For example, dissociation of the chemisorbed analytes can be accomplished either by using high temperature or application of catalysts. Recognition of this fact led to the development of high-temperature SFE,[39] followed by the evolution of both the hot solvent extraction approach[40] and microwave extraction, with more selective energy focusing at the sample matrix–extraction phase interface.[41] There is also an indication that milder conditions can be applied by taking advantage of the catalytic properties of the extraction phase or additives.[42] However, to realize this opportunity, more research needs to be performed to gain insight about the nature of interactions between analytes and matrices. Benefits are not only improved speed, but also selectivity resulting from application of appropriate conditions. This strategy of simultaneous extraction and cleanup has been applied successfully to a very difficult case of extraction of polychlorinated dibenzo-*p*-dioxins from fly ash.[43]

If the extraction rate is controlled by mass transport of analytes in the pores of the matrix, then the process can be successfully enhanced by application of sonic and microwave energy, which induce convection even in the small dimensions of the pore. Frequently, diffusion through the whole or portion of sample matrix containing natural or synthetic polymeric material controls the extraction rate.[44] In this case, swelling the matrix and increasing temperature result in increased diffusion coefficients and, therefore, increased extraction rates.

1.3.1. Flow-Through Techniques

For homogeneous samples and flowing fluid extraction phase, the description of the extraction process is much simpler and can be based directly on the chromatographic theory for liquid stationary phases. Let us consider another case of the flow-through system where the extraction phase is dispersed as a thin layer inside the extraction bed, and the sample flows through the cartridge. The bed can be constructed of a piece of fused-silica capillary, internally coated with a thin film of extracting phase[45] (a piece of open tubular capillary GC column; in-tube SPME[46]), or the bed may be packed with extracting phase dispersed on an inert supporting material (SPE cartridge). In these geometric arrangements, the concentration profile along the x-axis, of the tubing containing the extracting phase as a function of time t, can be described by adopting the expression for dispersion of a concentration front:

$$C(x,t) = 0.5C_s\left(1 - erf\frac{x - \frac{u_s t}{1+k}}{\sigma\sqrt{2}}\right), \tag{1.20}$$

where u_s is linear velocity of the sample through the tube, k is the partition ratio defined as

$$k = K_{es}\frac{V_e}{V_v}, \tag{1.21}$$

where K_{es} is the extraction phase–sample matrix distribution constant, V_e is the volume of the extracting phase, and V_v is the void volume of the tubing containing the extracting phase. σ is the mean square root dispersion of the front defined as

$$\sigma = \sqrt{Ht\frac{u_s}{1+k}}, \tag{1.22}$$

where H is equivalent to the HETP in chromatographic systems. This can be calculated as a sum of individual contributions to the front dispersion. These contributions are dependent on the particular geometry of the extracting system, as discussed previously.

Figure 1.8 illustrates the normalized concentration profiles produced in the bed during extraction.[25] Full breakthrough is obtained for the rightmost curve, which corresponds to the breakthrough volume of the sample matrix. The time required to pass this required volume through this extraction system corresponds to the equilibration time of the compound with the bed.

Equation 1.20 and Figure 1.8 indicate that the front of analyte migrates through the capillary/bed with speed proportional to the linear velocity of the sample, and inversely related to the partition ratio. For in-tube SPME and short capillaries with a small dispersion, the minimum extraction time at equilibrium conditions can be assumed to be similar to the time required for the center of the band to reach the end of the capillary,

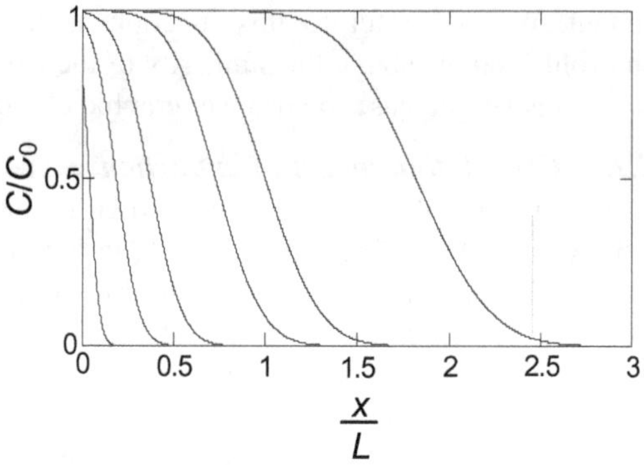

Figure 1.8. Normalized concentration profiles for in-tube SPME calculated using the equation discussed in the text. Copyright Wiley-VCH, 1997. Reprinted with permission.

$$t_e = \frac{L\left(1 + K_{es}\frac{V_e}{V_v}\right)}{u_s}, \tag{1.23}$$

where L is the length of the capillary holding the extraction phase. For packed bed extractors typically used in SPE techniques, analogous equations can be developed. In that case, the calculated time corresponds to the maximum extraction time before breakthrough occurs. As expected, the extraction time is proportional to the length of the capillary and inversely proportional to the linear flow rate of the sample. Extraction time also increases with an increase in the extraction phase–sample distribution constant and with the volume of the extracting phase, but decreases with an increase of the void volume of the capillary.

1.3.2. Batch Techniques

Coupling equations for systems involving convection caused by flow through a tube, such as the discussion above, are frequently not available for other means of agitation and other geometric configurations. In these cases, the most successful approach is to consider the boundary layer formed at the interface between the sample matrix and the extraction phase. Independent of the agitation level, fluid contacting the extraction phase surface is always stationary, and as the distance from the surface increases, the fluid movement gradually increases until it corresponds to bulk flow in the sample. To model mass transport, the gradation in fluid motion and convection of molecules in the space surrounding the extraction phase surface can be simplified as a zone of a defined thickness in which no convection occurs, and perfect agitation occurs in the bulk of the fluid everywhere else. This static layer zone is called the Prandtl boundary layer (Fig. 1.9).[47]

1.3.3. Boundary Layer Model

A precise understanding of the definition and thickness of the boundary layer in this sense is useful. The thickness of the boundary layer (δ) is determined by both the rate of

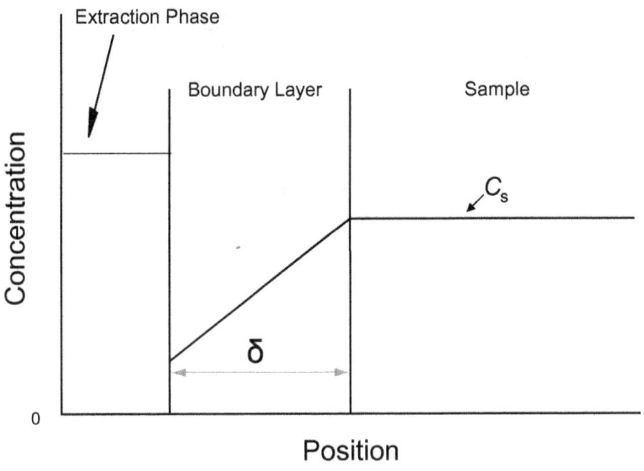

Figure 1.9. Boundary layer model. Copyright Wiley-VCH, 1997. Reprinted with permission.

convection (agitation) in the sample and an analyte's diffusion coefficient. Thus, in the same extraction process, the boundary layer thickness will be different for different analytes. Strictly speaking, the boundary layer is a region where analyte flux is progressively more dependent on analyte diffusion and less on convection, as the extraction phase is approached. For convenience, however, analyte flux in the bulk of the sample (outside of the boundary layer) is assumed to be controlled by convection, whereas analyte flux within the boundary layer is assumed to be controlled by diffusion. δ is defined as the position where this transition occurs, or the point at which convection toward the extraction phase is equal to diffusion away. At this point, analyte flux from δ toward the extraction phase (diffusion controlled) is equal to the analyte flux from the bulk of the sample toward δ, controlled by convection.

In many cases, when the extraction phase is dispersed well to form a thin coating, the diffusion of analytes through the boundary layer controls the extraction rate. The equilibration time, t_e, can be estimated as the time required to extract 95% of the equilibrium amount and calculated for these cases from the equation below:[5]

$$t_e = B\frac{\delta b K_{es}}{D_s}, \tag{1.24}$$

where b is the extraction phase thickness; D_s is the analyte's diffusion coefficient in the sample matrix; K_{es} is the analyte's distribution constant between the extraction phase and the sample matrix; and B is a geometric factor referring to the geometry of the supporting material upon which the extraction phase is dispersed on. The boundary layer thickness can be calculated for given convection conditions using engineering principles, and it is discussed in more detail later. Equation 1.24 can be used to predict equilibration times when the extraction rate is controlled by diffusion in the boundary layer, which is valid for thin extraction phase coatings ($b < 200\,\mu m$) and high distribution constants ($K_{es} > 100$).

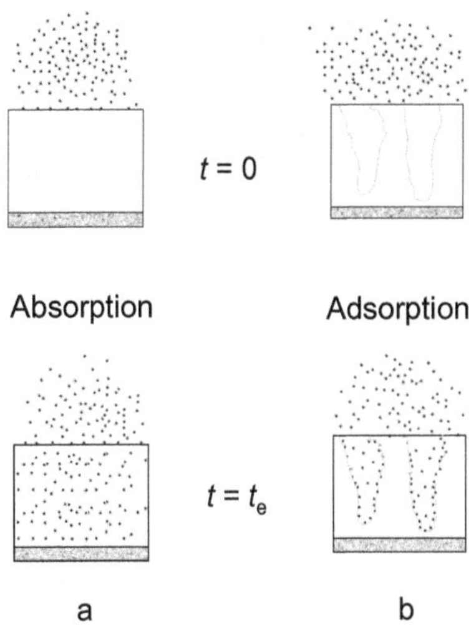

Figure 1.10. Extraction using absorptive (a) and adsorptive (b) extraction phases immediately after exposure of the phase to the sample ($t = 0$) and after completion of the extraction ($t = t_e$). Copyright Wiley-VCH, 1997. Reprinted with permission.

1.3.4. Solid versus Liquid Sorbents

There is a substantial difference in performance between liquid and solid coatings (Fig. 1.10). In the case of liquid coatings, the analytes partition into the extraction phase, where the molecules are solvated by the coating molecules. The diffusion coefficient in the liquid coating allows the molecules to penetrate the whole volume of the coating within a reasonable extraction time if the coating is thin (Fig. 1.10a). In the case of solid sorbents (Fig. 1.10b), the coating has a well-defined crystalline or amorphous structure, which, if dense, substantially reduces the diffusion coefficients within the structure. Therefore, within the experimental time, sorption occurs only on the porous surface of the coating (Fig. 1.10b). During extraction by solid phase, compounds with poor affinity toward the phase are frequently displaced at longer extraction times by analytes characterized by stronger binding, or those present in the sample at high concentrations. This effect is associated with the fact that there is only a limited surface area available for adsorption. If this area is substantially occupied, then a competition effect occurs[6] and the equilibrium amount extracted can vary with concentrations of both the target and other analytes. On the other hand, in the case of extraction with liquid phases, partitioning between the sample matrix and extraction phase occurs. In this case, equilibrium extraction amounts vary only if the bulk coating properties are modified by the extracted components, which only occurs when the amount extracted is a substantial portion (a few percent) of the extraction phase. This is rarely observed, since extraction/enrichment techniques are typically used to determine trace contamination samples; however, it cannot be neglected as a possible cause of nonlinearity when quantifying very complex matrices.

1.3.5. Diffusion-Based Calibration

The only way to overcome the fundamental limitation of porous coatings, as suggested in Figure 1.10, is to use an extraction time much less than the equilibrium time, so that the total amount of analytes accumulated onto the porous coating is substantially below the saturation value. At saturation, all surfaces available for adsorption are occupied. When performing experiments with pre-equilibrium short extraction times, it is critical to precisely control extraction times and convection conditions since they determine the thickness of the diffusion layer. One way of eliminating the need for compensation of differences in convection is to normalize (use consistent) agitation conditions. For example, the use of a stirring means at a well-defined rotation rate in the laboratory, or fans for field air monitoring, can ensure consistent convection.[48,49] The short-time exposure measurement described above has an advantage associated with the fact that the rate of extraction is defined by diffusivity of analytes through the boundary layer of the sample matrix and corresponding diffusion coefficients, rather than by distribution constants. This situation is illustrated in Figure 1.11 for cylindrical geometry of the extraction phase dispersed on the supporting rod.

The analyte concentration in the bulk of the matrix can be considered constant when a short sampling time is used and there is a constant supply of an analyte via convection. These assumptions are true for most cases of sampling, where the volume of sample is much greater then than the volume of the interface, and the extraction process does not affect the bulk sample concentration. In addition, the solid coating can be treated as a perfect sink. The adsorption binding is frequently instantaneous and essentially irreversible. The analyte concentration on the coating surface is far from saturation and can be assumed to be negligible for short sampling times and relatively low analyte concentrations in a typical sample. The analyte concentration profile can be assumed to be linear from C_s to C_0. In addition, the initial analyte concentration on the coating surface, C_0, can be assumed to be equal to zero when extraction begins. Diffusion of analytes inside the pores of a solid coating controls mass transfer from the outer to the inner surface of the coating.

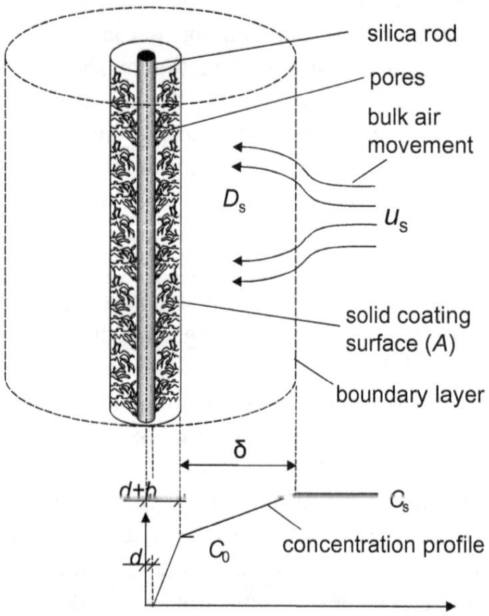

Figure 1.11. Schematic of the diffusion-based calibration model. The symbols/terms are defined in the text. Copyright Wiley-VCH, 1997. Reprinted with permission.

The function describing the mass of extracted analyte with sampling time can be derived,[50] which results in the following equation:

$$n(t) = \frac{B_1 A D_s}{\delta} \int_0^t C_s(t)\,dt, \tag{1.25}$$

where n is the mass of extracted analyte over sampling time (t); D_s is the gas-phase molecular diffusion coefficient; A is the surface area of the sorbent; δ is the thickness of the boundary layer surrounding the extraction phase; B_1 is a geometric factor; and C_s is analyte concentration in the bulk of the sample. It can be assumed that the analyte concentration is constant for very short sampling times, and therefore, Equation 1.25 can be further reduced to

$$n(t) = \frac{B_1 D_s A}{\delta} C_s t, \tag{1.26}$$

where t is the sampling time.[51]

It can be seen from Equation 1.26 that the amount of extracted mass is proportional to the sampling time t, D_s for each analyte and bulk sample concentration, and inversely proportional to δ. This is consistent with the fact that an analyte with a greater D_s will cross the interface and reach the surface of the fiber coating faster. Values of D_s for each analyte can be found in the literature or estimated from physicochemical properties.[25] This relationship allows for quantitative analysis. Equation 1.26 can be rearranged to estimate the analyte concentration in the sample for rapid sampling with solid sorbents:

$$C_s = \frac{n\delta}{B_1 D_s A t}. \tag{1.27}$$

The amount of extracted analyte (n) can be estimated from the detector response.

The thickness of the boundary layer (δ) is a function of sampling conditions. The most important factors affecting δ are the geometric configuration of the extraction phase, sample velocity, temperature, and D_s for each analyte. The effective thickness of the boundary layer can be estimated for the coated fiber geometry (Fig. 1.11) using Equation 1.28, adapted from the heat transfer theory:

$$\delta = 9.52 \frac{d}{R_e^{0.62} S_e^{0.38}}, \tag{1.28}$$

where R_e is the Reynolds number $= 2u_s d/\nu$; u_s is the linear sample velocity; ν is the kinematic viscosity of matrix; S_c is the Schmidt number $= \nu/D_s$; and d is the fiber diameter. The effective thickness of the boundary layer in Equation 1.28 is a surrogate (or average) estimate and does not take into account changes of the thickness that may occur when the flow separates and/or a wake is formed. Equation 1.28 indicates that the thickness of the boundary layer will decrease with an increase of the linear sample velocity (Fig. 1.11). Similarly, when sample temperature (T_g) increases, the kinematic viscosity also increases. Since the kinematic viscosity term is present in the denominator of R_e and in the numerator of S_c, the overall effect on δ is small. A reduction of the boundary layer and an increase of the mass transfer rate for an analyte can be achieved in at least two ways, that is, by increasing the sample velocity and by increasing the sample temperature. However, the temperature increase will reduce the solid sorbent efficiency. As a result, the sorbent coating may not be able to adsorb all molecules reaching the surface, and therefore, may stop behaving as a zero sink for all analytes.

1.3.6. Calibrants in the Extraction Phase

Internal standardization and standard addition are important calibration approaches that are very effective when quantifying target analytes in complex matrices. They compensate for additional capacity or activity of the sample matrix. However, such approaches require delivery of the standard. This is incompatible in some sampling situations, such as on-site or *in vivo* investigations. The standard in the extraction phase approach is not practical for conventional exhaustive extraction techniques, since the extraction parameters are designed to facilitate the complete removal of the analytes from the matrix. However, in microextraction, a substantial portion of the analytes remains in the matrix during the extraction and after equilibrium is reached. This suggests that the standard could be added to the investigated system together with the extraction phase. This property of the microextraction techniques has been explored to integrate addition of the calibrant with the rest of automated high-throughput analysis.[52] In addition, two calibration methods, kinetic calibration with standard[5,3,53] and standard-free kinetic calibration, were proposed.[54]

1.3.6.1. Kinetic calibration with standard in the extraction phase. Chen and Pawliszyn demonstrated the symmetry of absorption and desorption in the SPME liquid fiber coating, and a new calibration method, kinetic calibration, was proposed.[53] This kinetic calibration method, uses the desorption of the standards, which are preloaded in the extraction phase, to calibrate the extraction of the analytes.

For field sampling, the desorption of standard from an extraction phase can be described by

$$\frac{Q}{q_0} = \exp(-at), \tag{1.29}$$

where q_0 is the amount of pre-added standard in the extraction phase and Q is the amount of the standard remaining in the extraction phase after exposure of the extraction phase to the sample matrix for the sampling time, t.

Ai proposed a theoretical model based on a diffusion-controlled mass transfer process to describe the entire kinetic process of SPME:[55,56]

$$n = [1-\exp(-at)]\frac{K_{fs}V_fV_s}{K_{fs}V_f + V_s}C_0, \tag{1.30}$$

where a is a rate constant that is dependent on the extraction phase, headspace, and sample volumes; the mass transfer coefficients; the distribution coefficients; and the surface area of the extraction phase.

Equation 1.30 can be changed to

$$\frac{n}{n_e} = 1-\exp(-at), \tag{1.31}$$

where n_e is the amount of the extracted analyte at equilibrium. When the constant a has the same value for the absorption of target analytes and the desorption of preloaded standards, the sum of Q/q_0 and n/n_e should be 1 at any desorption/absorption time:[53]

$$\frac{n}{n_e} + \frac{Q}{q_0} = 1. \tag{1.32}$$

Then, the initial concentrations of target analytes in the sample, C_0, can be calculated with Equation 1.33:[57,58]

$$C_0 = \frac{q_0 n}{K_{es}V_e(q_0 - Q)}, \tag{1.33}$$

where V_e is the volume of the extraction phase. K_{es} is the distribution coefficient of the analyte between the extraction phase and the sample.

The change of environmental variables will affect the extraction of the analyte and the desorption of the preloaded standard simultaneously; therefore, the effect of environmental factors, such as biofouling, temperature, or turbulence, can be calibrated with this approach. The feasibility of this technique for time-weighted average (TWA) water sampling was demonstrated by both theoretical derivations and field trials.[59]

This technique is a pre-equilibrium method and can be used for the entire sampling period. The concentration determined before the sampling reaches equilibrium is a TWA concentration because the desorption of the preloaded standard calibrated the extraction of the analytes and the extraction is an integrative process. If the sampling reaches equilibrium, the determined data are the concentrations of the analytes in the sample at the time the samplers were retrieved.

The standard in the extraction phase technique makes it possible to use a simple PDMS rod or PDMS membrane as a passive sampler to obtain the TWA concentrations of target analytes in a sampling environment. Both PDMS-rod and PDMS-membrane samplers are simple and easy to deploy and retrieve. They have large sampling rates, and the sensitivity is much higher than the fiber-retracted SPME device since the samplers are in direct contact with the sample matrix.[60] The concept of calibrants in the extraction phase has been extended to determine the concentrations of target analytes directly in the veins of animals, indicating that this approach is useful for *in vivo* studies as well.[61] Experiments demonstrated that this calibration corrected for the sample matrix effects and minimized the displacement effects due to the use of pre-equilibrium extraction. The pharmacokinetic profiles of diazepam, nordiazepam, and oxazepam obtained by kinetic calibration based on deuterated standards are quite similar to those determined by standard calibration method.[62] The applications of this technique for quantitative analysis of liquid-phase microextraction (LPME) were also reported.[55]

Deuterated compounds are expensive and sometimes not available. Zhou et al. proposed an alternative calibration method, which employs the target analytes as the internal standards by the means of dominant desorption.[63] Dominant pre-equilibrium desorption not only offers a shorter sample preparation time but also provides time constants for the purpose of quantitative analysis. This kinetic calibration method was successfully applied to on-site sampling of polycyclic aromatic hydrocarbons (PAHs) in a flow-through system and *in vivo* direct pesticide sampling in the leaves of a jade plant.[64]

Using kinetic calibration with standard in the extraction phase method, the samplers require preloading of a certain amount of standards, either deuterated compounds or target analytes. Zhao et al. reported several standard loading approaches, which include (1) headspace extraction of the standard dissolved in a solvent or pumping oil, (2) headspace extraction of pure standard in a vial, (3) direct extraction in a standard solution, and (4) direct transfer of the standard solution from the syringe to the fiber.[64]

The existing SPME kinetic calibration technique, using desorption of preloaded standards to calibrate extraction of the analytes requires that the physicochemical properties of the standard be similar to those of the analyte, which limits the application of the technique. Recently, a new method, termed the one-calibrant technique, which uses only one standard to calibrate all extracted analytes, was proposed.[65] The theoretical considerations were validated in a flow-through system, using PDMS SPME fibers as passive samplers. The newly proposed one-calibrant technique makes the SPME kinetic calibration method more convenient and more applicable.

1.3.6.2. Standard-free kinetic calibration. Kinetic calibration with standard in the extraction phase can be used for both grab sampling and long-term monitoring. For fast on-site or *in vivo* analysis, preloading standards is inconvenient Also, this calibration method may not work in some fast sampling situations because the loss of the standard will be too small to detect. Recently, a standard-free kinetic calibration method was proposed for fast on-site and *in vivo* analysis.[54] With this calibration method, all analytes can be directly calibrated with only two samplings.

Equilibrium extraction results in the highest sensitivity in SPME because the amount of analyte extracted with the fiber coating is maximized when equilibrium is reached. If sensitivity is not a major concern in analysis, reduction of the extraction time is desirable. When the extraction conditions are kept constant, for example, fast sampling, Equation 1.34 can be used for the calculation of n_e, the amount of analyte extracted at equilibrium:

$$\frac{t_2}{t_1}\ln\left(1-\frac{n_1}{n_e}\right)=\ln\left(1-\frac{n_2}{n_e}\right), \tag{1.34}$$

where n_1 and n_2 are the amount of analyte extracted at sampling times t_1 and t_2, respectively. Then, the concentration of the analyte in the sample can be calculated with Equations 1.3 or 1.4.

The feasibility of this calibration method was validated in a standard aqueous solution flow-through system and a standard gas flow-through system. Using this standard-free kinetic calibration method, the sampling time can be markedly shortened. In the reported study, typical sampling times for the extraction of PAHs in a water environment, which typically range from 2–24 h (equilibrium), can be shortened to 2–5 min, and sampling times for benzene, toluene, ethylbenzene, and xylene (BTEX) in air can be shortened from 5–10 min to 5–10 s.[54]

This calibration method can be used for the entire sampling period, without considering whether the system reaches equilibrium. This aspect of the technique is desirable for systems when the equilibrium time is not known, and particularly useful for instances when a number of compounds are measured simultaneously. The method is unsuitable for long-term monitoring of pollutants in the environment since the method requires that the sampling rate remains constant and the determined concentration is therefore representative of a spot sampling.

1.3.7. Headspace Extraction

Equations 1.24 and 1.26 indicate that the use of the headspace above the sample as an intermediate phase might be an interesting approach to accelerate the extraction of analytes characterized by high Henry constants. When a thin extraction phase is used, the initial extraction rate, and hence, the extraction time, is controlled by the diffusion of analytes through the boundary layer present in the sample matrix. The presence of a gaseous headspace facilitates rapid transport into the extraction phase because of the high diffusion coefficients. To increase the transport from the sample matrix into the headspace, the system can be designed to produce a large sample/headspace interface. This can be accomplished by using large diameter vials with good agitation, purge, or even spray systems. At room temperature, only volatile analytes are transported through the headspace. For low volatility compounds, heating of the sample is a good approach, if loss in magnitude of the distribution constant is reasonable. The ultimate approach is to heat the sample and cool the extraction phase at the same time. Heating of the sample not only increases the Henry constant but also induces convection of the headspace due to density gradients associated with temperature gradients present in the system, resulting in higher mass transport rates. On the other hand, cooling the sorbent increases its capacity. Collection of analytes can be performed in the same vial[66] or can be separated in space, as in the purge-and-trap technique. In the heating–cooling experiments, both kinetic and thermodynamic factors are addressed simultaneously. Headspace approaches are also interesting since adverse affects associated with the presence of solid, oily, or high-molecular-weight interferences, which can cause fouling of the extraction phase, are eliminated.

1.3.8. Passive TWA Sampling

Consideration of different arrangements of the extraction phase is always beneficial in order to select the most appropriate geometry for a given application. For example, extension of the boundary layer by a protective shield that restricts convection will result in TWA measurement of analyte concentration (Eq. 1.24). Various diffusive samplers have been developed based on this principle. For example, when the extracting phase in an SPME device is not exposed directly to the sample, but is contained in a protective tubing (needle) without any flow of the sample through it (Fig. 1.12), the diffusive transfer of analytes occurs through the static sample (gas phase or other matrix) trapped in the needle. The system consists of an externally coated fiber with the extraction phase withdrawn into the needle (Fig. 1.12b). This geometric arrangement represents a very powerful method, capable of generating a response proportional to the integral of the analyte concentration over time and space (when the needle is moved through space).[67] In this case, the only mechanism of analyte transport to the extracting phase is diffusion through the matrix contained in the needle. During this process, a linear concentration profile (shown in Fig. 1.12b) is established in the tubing between the small needle opening, characterized by surface area A and the distance Z between the needle opening and the position of the extracting phase. The amount of analyte extracted, dn, during time interval, dt, can be calculated by considering Fick's first law of diffusion:[5]

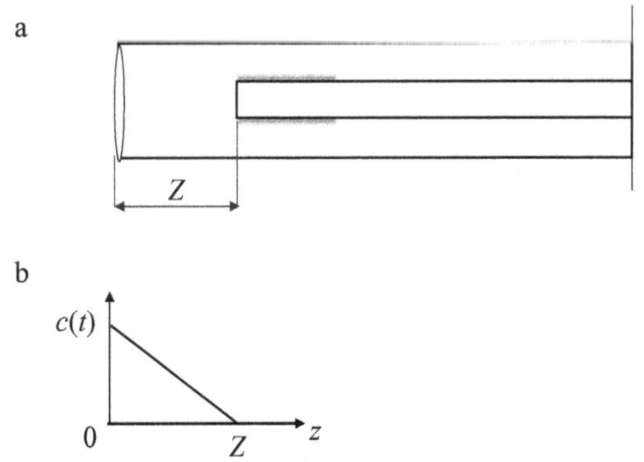

Figure 1.12. SPME/TWA approaches based on in-needle fiber. Copyright Wiley-VCH, 1997. Reprinted with permission.

$$dn = AD_{\mathrm{m}}\frac{dc}{dz}dt = AD_{\mathrm{m}}\frac{\Delta C(t)}{Z}dt, \qquad (1.35)$$

where $\Delta C(t)/Z$ is an expression of the gradient established in the needle between the needle opening and the position of the extracting phase, Z; $\Delta C(t) = C(t) - C_Z$, where $C(t)$ is a time-dependent concentration of analyte in the sample in the vicinity of the needle opening; and C_Z is the concentration of the analyte in the gas phase in the vicinity of the coating. If C_Z is close to zero for a high extraction phase/matrix distribution constant capacity, then, $\Delta C(t) = C(t)$. The concentration of analyte at the coating position in the needle, C_Z, will increase with integration time, but it will be kept low compared with the sample concentration because of the presence of the extraction phase. Therefore, the accumulated amount over time can be calculated as

$$n = D_{\mathrm{m}}\frac{A}{Z}\int C_{\mathrm{s}}(t)dx. \qquad (1.36)$$

As expected, the extracted amount of analyte is proportional to the integral of the sample concentration over time, the diffusion coefficient of analyte in the matrix filling the needle, D_{m}, in the area of the needle opening, A, and inversely proportional to the distance of the coating position with respect to the needle opening, Z. It should be emphasized that Equations 1.27 and 1.28 are valid only in a situation where the amount of analyte extracted onto the sorbent is a small fraction (below %RSD of the measurement, typically 5%) of the equilibrium amount with respect to the lowest concentration in the sample. To extend integration times, the coating can be placed further into the needle (larger Z), the opening of the needle can be reduced by placing an additional orifice (smaller A), or a higher capacity sorbent can be used.[68] The first two solutions will result in low measurement sensitivity. An increase of sorbent capacity presents a more attractive opportunity and can be achieved by either increasing the volume of the coating or the affinity of coating toward the analyte. An increase of the coating volume will require an increase of the device size. Therefore, the optimum approach to increased integration time is to use sorbents characterized by large coating/gas distribution constants. If the matrix filling the needle is different than the sample matrix, then, an appropriate diffusion coefficient should be used, as discussed below in the case of membrane extraction.

1.3.9. Extraction Combined with Derivatization

The capacity of the extraction phase for analytes that are difficult to extract, such as polar or ionic species, is frequently enhanced by introducing a derivatization step. The objective of this approach is frequently not only to convert the native analytes into less polar derivatives that are extracted more efficiently, but also to label them for better detection and/or chromatographic separation. The most interesting implementation of this approach is simultaneous extraction/derivatization. In this technique, the derivatization reagent is present in the extraction phase during the extraction. The major advantage of this approach is that two steps are integrated into one. There are two limiting cases describing the combination between extraction and derivatization. The first occurs when mass transfer to the fiber is slow compared with the reaction rate. In this case, Equation 1.25, as discussed above, describes the accumulation rate of analytes, assuming that the derivative is trapped in the extraction phase. In the second limiting case, the situation is reversed in that the reaction rate is slow compared with the transport of analytes to the extraction phase. In other words, at any time during the extraction, the extraction phase is at equilibrium with the analyte remaining in a well-agitated sample, resulting in a uniform reaction rate throughout the coating. This is a typical case for thinly dispersed extraction phase, since the equilibration time for well-agitated conditions is very short compared with a typical reaction rate constant. The accumulation rate of the product in the extraction phase n/t can then be defined by

$$n = V_e k_r K_{es} \int C_s(t)\,dt, \tag{1.37}$$

where C_s is the initial concentration of the analyte in the sample and k_r is chemical reaction rate constant. In other words, when the sample volume is large, such as in direct sampling in the field, the reaction and accumulation of the analyte in the extraction phase proceeds with the same rate as long as the reagent is present in an excess amount. It is worth noting that the rate is also proportional to the extraction phase–sample matrix distribution constant. If the concentration varies during the accumulation, the collected amount corresponds to the integral over concentration and time, as discussed above in the case of TWA sampling. For a limited sample volume, however, the concentration of the analyte in the sample phase decreases with time as it is partitioned into the coating and converted to trapped product, resulting in a gradual decrease of the rate. The time required to exhaustively extract analytes from a limited volume can be estimated using experimental parameters.[5]

1.3.10. Membrane Extraction Techniques

For continuous monitoring applications, membrane extraction is an attractive approach. Permeation through a membrane is a specific extraction process, where the sorption into and desorption out of the extraction phase occurs simultaneously. The sample (donor phase) is in contact with one side of the membrane where extraction into membrane material occurs, while permeated analytes are removed by the stripping phase (acceptor) on the other side. For membrane extraction with good flow (agitation) conditions at both acceptor and donor sites and efficient stripping, the rate of mass transport through the membrane is controlled by the diffusion of analytes through the membrane material. The concentration gradient, which facilitates transport across the membrane, is formed by the difference in analyte concentration between the sample side ($K_{es}C_s$) and the stripping phase, which is close to zero for high flow rates of stripping phase (Fig. 1.13). The mass transfer rate through the membrane, n/t, can be estimated at steady-state conditions using the following equation:

$$n/t = B_2 A D_e K_{es} C_s / b, \tag{1.38}$$

where A is the surface area of the membrane; D_e is the diffusion coefficient in the membrane material; K_{es} is the membrane material/sample matrix distribution constant; b is the thickness of the membrane; and B_2 is a geometric factor defined by the shape of the membrane. The permeation rate through the membrane is proportional to both the diffusion coefficient (D_e) and the distribution constant (K_{es}), and inversely proportional to b. D_e determines the rate of analyte migration through the membrane, and K_{es} determines the magnitude of the concentration gradient generated in the membrane.[69] This information can be used to calibrate the extraction process a priori if these parameters are obtained from tables or experimental data.[70]

The concentration of the unknown can be calculated by converting Equation 1.38:

$$C_s = \frac{bn}{B_2 A D_e K_{es} t}. \tag{1.39}$$

The membrane material/sample matrix distribution constant K_{es} determines the sensitivity of membrane extraction (Eq. 1.39), indicating that the membrane, although a physical barrier, is also a concentrating medium, analogous to the extraction phase in other configurations. However, the concentration in the stripping gas phase is lower compared with the sample since the gradient needs to exist in the membrane

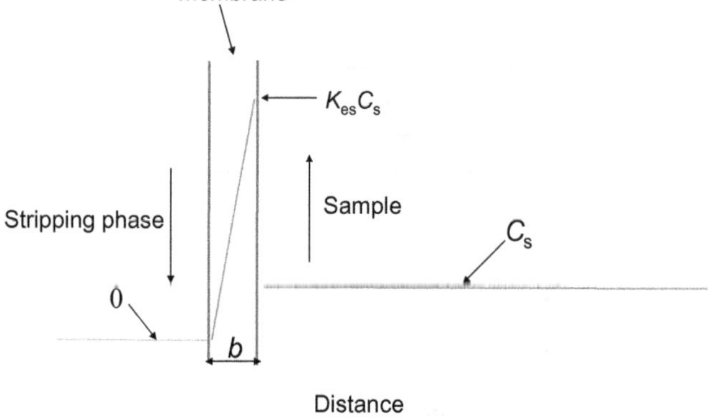

Figure 1.13. Membrane extraction at good sample agitation and stripping conditions. The symbols/terms are defined in the text. Copyright Wiley-VCH, 1997. Reprinted with permission.

to generate diffusive mass transfer through the membrane material (Fig. 1.13). Therefore, incorporating a sorbent after the membrane will allow concentration of the extract and, therefore, sensitive analysis. When the membrane is in direct contact with the aqueous phase, the mass transfer through the boundary layer surrounding the membrane can contribute to the overall mass transfer in the system. Therefore, for analytes characterized by high Henry constants, it is important to consider a headspace membrane extraction arrangement.

Quantification is a weak point in membrane extraction. Addressing this limitation represents a unique opportunity to facilitate wider use of membrane extraction techniques in designing total analysis systems (TASs), in particular, the systems based on micromachining technologies: micro-TAS. The strength of the membrane extraction approach is that it constitutes a simple selective contact between the sample and the instrument. Therefore, it can be used with portable instrumentation. However, it is impossible to relate correctly the amount of target analyte transferred through the membrane to its concentration in the matrix since both concentration in the matrix and the mass transfer conditions affect the extracted amount. Therefore, it is critical to characterize the mass transfer conditions to facilitate correct quantification. To address this challenge, a new technique for calibration in membrane extraction processes has been proposed by adding an analytically noninterfering internal calibrant in the receiving phase (stripping phase).[71] The membrane extraction with a sorbent interface (MESI) system was used to evaluate this approach. During the membrane extraction process, the internal calibrant present in the carrier gas, which acts as a stripping phase, and the target analyte present in the sample matrix will permeate simultaneously through the membrane in opposite directions. The changes in accumulated amounts (relative loss) of internal calibrant can be used as means of calibration to correct the variations of extraction rate due to variation in environmental factors, such as sample velocity and membrane temperature, which determine the extraction conditions. Thus, this approach allows for more accurate estimates of the concentrations of the target analytes at various sampling or monitoring conditions during field analysis. This approach was validated, and the results indicated the advantages of the proposed approach, in particular, for on-site analysis.[72]

1.4. SUMMARY: SIGNIFICANCE OF FUNDAMENTAL DEVELOPMENTS

1.4.1. Optimization

Whenever a new type of complex sample is considered, a small research project is frequently conducted to find the optimum extraction conditions enabling the most efficient and most complete release of native analytes from the matrix, and their partitioning into the extraction phase. Typically, an empirical approach is used and several parameters are varied, for example, the chemical properties of the extraction phase or types of additive/reagent used. Better understanding of the analyte–matrix interaction will facilitate a more rational choice of extraction conditions on the basis of models, when the characteristics of the analyte and the matrix are known. With solid matrices, however, greater research effort is required to reach the level of fundamental knowledge necessary for practical implementation of such an approach.

1.4.2. Obtaining More Information about the Investigated System and Analyte

The objective of exhaustive extraction techniques is to remove all the analytes to the extraction phase. In that process, information about the nature of interaction and proportion of different species of analytes are lost. On the other hand, in headspace, microextraction, and membrane approaches, the amount of analytes extracted is related to the free concentration of analytes, which is in equilibrium with that of the bound analytes, or of different species.[73] Therefore, it is possible to obtain information not only about the biologically available portion of analyte, but also about binding and speciation of analytes in the matrix. Therefore, depending on the required information, different extraction techniques should be chosen. Further development of experimental and data analysis procedures will allow more convenient deconvolution of the required information.

1.4.3. Calibration

Reliance on physicochemical constants in calibration results in rapid and cost-effective procedures, but might seem unconventional or even uncomfortable to some researchers. As theory indicates, however, these constants define the extraction process, and there is an opportunity to take advantage of this. Physicochemical constants can be frequently estimated from simple experiments, or calculated by considering the molecular structures of analytes, extraction phase, and matrix, and this adds to the attractiveness of this approach. For equilibrium microextraction techniques, the extraction phase–sample matrix distribution constant is used to quantify the concentration of analytes in the sample matrix (Eq. 1.3). For extraction approaches controlled by mass transfer, calibration can be based on the diffusion coefficient in the sample matrix for constant extraction time under well-defined convection conditions (Eqs. 1.26 and 1.36). When derivatization reaction kinetics control the rate of extraction, the rate constant can provide a means of calibration. Occasionally, for example, in membrane extraction, a combination of constants defines the rate of extraction and can be used for calibration also. The major argument against using this approach is that physicochemical constants are affected by many experimental conditions, for example, temperature and the properties of the matrix. The impact of temperature change, however, can be compensated for by monitoring it and using correction factors, and allowing for direct calibration for simple matrices.[74] For more complex matrices, internal standard or standard addition calibrations routinely applied in exhaustive techniques to monitor recoveries can be used to compensate for matrix variations.[75] In future research, correlations between distribution constants and simple measurements, such as turbidity and pH, might be found to account for matrix variations for a given type of matrix and, therefore, eliminate the need for internal calibration.

One can draw several parallels between developments and applications of extraction techniques with electrochemical methods. For instance, the coulometric technique corresponds to total or exhaustive extraction methods. Although the most precise, this technique is not used frequently because of the time required to complete it. Equilibrium potentiometric techniques are more frequently used (pH electrode), particularly when the sample is a simple mixture and/or the selectivity of the membrane in an ion-selective electrode is sufficient to quantify the target analyte in complex matrices. The equilibrium microextraction approach has further advantages in selectivity, because the extraction is coupled with separation and/or specific detection (e.g., MS), which enables identification and quantification of many components simultaneously. The advantage of electrochemical methods is a short response time, because of the low capacities of electrodes. Design of microsystems with cylindrical geometry facilitates rapid extraction, as in microelectrodes.[76] Some electrochemical methods, for example, amperometry, are based on mass transport through the boundary layer, as in pre-equilibrium extraction techniques (e.g., TWA and diffusion based). Similarly, extraction calibration based on diffusion coefficients can be accomplished as long as the agitation conditions are constant, the extraction times are short, and the coating has high affinity for the analytes.

1.4.4. On-Site Implementation

The advantages of nonexhaustive extraction are its fundamental simplicity and fewer geometric restrictions. This facilitates several interesting on-site implementations by integrating sampling and sample preparation. For instance, sample introduction to miniaturized analytical field instrumentation can be more convenient. More information about the investigated system can also be obtained. For example, it is possible to speciate and to determine the distribution of analytes in multiphase systems, because the extraction process does not disturb the equilibrium naturally present in the systems. Different forms of an analyte are therefore extracted and quantified according to their corresponding distribution constants and/or diffusion coefficients.

Simplification of sample preparation technologies and their integration with sampling and/or separation/quantification steps in the analytical processes are significant challenges and opportunities for the contemporary analytical chemist. These developments will eventually enable attainment of a major goal of the analytical chemist—to perform analysis at the place where the sample is taken, rather than moving the sample to a laboratory, as is traditional. The on-site analysis approach reduces errors and the time associated with sample transport and storage, resulting in faster analysis and more accurate and precise data. The trend in analytical instrumentation is toward miniaturization, which results in portability and on-site compatibility. Simplification and miniaturization of sampling and sample preparation is a logical next step.

1.4.5. Miniaturization and Integration

Practical integration of sample preparation with the rest of the analytical process has been accomplished in several ways. The concept of FIA has facilitated the performance of sequential sample preparation processes and quantification in a single device with the help of a flowing stream.[77] These devices can be made very small by using capillary flowing systems integrated with, for example, a small semiconductor light emission and detection devices, which use fiber optics,[78] and can be implemented on-site, for example, in combination with single solvent drop detection.[79] Further miniaturization of FIA technology results in a whole sample preparation process being performed in the body of a single valve ("lab-in-a-valve").[80]

The application of micromachining technology to the construction of highly integrated analytical systems (μTAS or "lab-on-a-chip") has recently resulted in sample preparation being performed in machined microchannels.[81] μTAS enables effective coupling of separation/detection processes with sample preparation similar to capillary-based devices but can, potentially, be mass-produced at much lower cost.

Integration of sample preparation in the microdevices with the other steps of the analytical process can be accomplished in two fundamentally different ways. First, analogous to FIA, sample preparation may be performed directly in the capillaries/microchannels in the flowing systems. This approach would typically use flow-through sample preparation techniques (Fig. 1.2). These devices are expected to be structurally complex and relatively large, because they must incorporate valves to control flows, pumps, or high voltage supply; sample; reagent ports; and detection components.[82] In addition, because of the high surface area/volume ratio, there is a possibility of sample losses and carryover in a complex channel network. Integration of the sampling step with this complex system might be a challenge. It can be addressed by using moisture-repellent sorbents, electromigration focusing mechanisms, membranes, and solvent microextraction when the mobile phase in the separation technique is a solvent. For example, attempts are being made to integrate capillary electrophoresis (CE) with sampling/concentration.[83] Membrane sampling interfaced to an investigated system could facilitate sampling of aqueous media, as is currently performed in microdialysis systems coupled to condensed phase separations,[84] or MESI coupled to microgas chromatography.[85] Recent developments in polymer manufacturing of microfluidic systems, including PDMS,[86] will facilitate these approaches, because this material is an excellent extraction phase. Because the overall size of the fully integrated device is expected to be relatively large, there will always be some restrictions on the dimensions of the object that can be investigated. The most significant limitation of this approach, however, is expected to be the cost, because unique configurations will be required for each specific application. This restriction will make the approach cost-competitive only for very popular applications, when mass production is justified.

The alternative approach involves integration of sampling with sample preparation only, by performing extraction and sample processing directly in the sampling device followed by the on-site introduction to a portable microseparation/quantification instrument. The extraction process can be made very selective for target analytes, limiting disturbance of the system investigated. If the sampling/sample preparation device is small enough, it can deliver the prepared samples directly into the separation channel/capillary of the separation/quantification microdevice. For example, in-microneedle and on-fiber microsampling devices could enable such a method, because processing reagents can be either drawn into the needle or delivered onto the fiber by dipping or by use of a spray.[87] Prepared analytes can subsequently be introduced to the microdevice for separation and quantification. Because sample preparation is performed directly in the sampling system, external to the separation/quantification device, restrictions applied to one device will not have to be arbitrarily applied to the other. Low-cost generic microseparation/detection devices can be used as long as they are designed to accommodate a specific configuration of sampler. The major limitation of this approach is in monitoring and parallel analysis applications for which separate miniaturized automated systems will be required to control the device to perform sample preparation and, occasionally, sampling. It is, however, sometimes possible to prepare an extraction phase, which already contains the required reagents before sampling.[88] In this approach to on-site analysis, optimization of the design of the sampling/sample preparation systems is conducted independently. Much smaller and flexible devices are expected, compared with the previously described fully integrated single microdevice. Several of the sample preparation technologies listed in Figure 1.2, including batch extraction techniques such as coated microfibers, thin films, or packed microneedles, can be explored for this application.

1.4.6. *In Vivo* Analysis

The sampling procedures in the integrated on-site microdevices described above are a significant departure from conventional "sampling" techniques, in which a portion of the system under study is removed from its natural environment and the compounds of interest are extracted and analyzed in a laboratory environment. There are two main motivations for exploring these types of configuration. The first is the desire to study chemical processes in association with the normal biochemical milieu of a living system; the second is the lack of availability, or the impracticality, frequently associated with size, of removing suitable samples for study from the living system. New approaches of an externally coated extraction phase on a microfiber mounted in a syringe-like device, packed microneedles, or online sampling from a membrane interface seem to be logical targets for the development of such tools. As with any microextraction or membrane technique, compounds of interest are not exhaustively removed from the investigated system. On the contrary, conditions can be devised in which only a small proportion of the total compounds are removed and none of the matrix is removed, thus avoiding disturbance of the normal balance of chemical components. Second, because it is either a syringe-like device that can be physically removed from the laboratory environment for sampling or an integrated micromembrane system, it is suitable for monitoring of a living system in its natural environment, rather than trying to move the living system to an unnatural laboratory environment. Microdialysis systems are already used in animal studies[89] and MESI has been used in breath monitoring.[72,90] The coated microfiber approach has recently been used in drug metabolism studies—the components of interest were extracted directly from a peripheral vein of an animal.[91] To further improve the capability of SPME for *in vivo* sampling, new specific coatings, for example, affinity phases, should be developed for a range of important target analytes. The ultimate goal is to remove only those compounds required to characterize the system investigated and none of the matrix, using molecular recognition approaches, as is frequently performed in sensor arrays.[92] This specific direct extraction approach is critical to minimizing interference with the operation of the investigated system. For example, removal of neurotransmitters from the synaptic cleft results in elimination of the signal coming down the nerves and/or depletion the presynaptic stores of the transmitter. In addition, specific nonequilibrium direct extraction might facilitate sampling at the speed of biological processes. The extraction can be limited to a small number of molecules and combined with on-probe amplification approaches, and/or single molecular detection schemes, facilitating investigation of analytes present in the system at low copy number.

Frequently Used Symbols and Abbreviations

A	area of the extraction phase
B, B_1, B_2	geometric factors
b	thickness of the extraction phase
C_0	initial concentration of the analyte in the sample
C_e	concentration of analyte in extraction phase
C_s	concentration of the analyte in the sample

d	radius of the fiber core
d_c	thickness of the matrix component permeable to analyte
D_e	diffusion coefficient of the analyte in the extraction phase
δ	boundary layer thickness
d_p	diameter of solid particulate matter
D_s	diffusion coefficient of analytes in the sample matrix
ε_e	interstitial porosity
ε_i	intraparticulate porosity
H	height equivalent to theoretical plate (HETP)
k	partition ratio
k_d	dissociation rate constant
K_{es}	extraction phase–sample matrix distribution constant ($K_{es} = C_e/C_s$)
k_o	the ratio of the intraparticulate void volume to the interstitial void space
k_r	chemical reaction rate constant
K_{es}^{s}	sample matrix–solid extraction phase distribution constant ($K_{ses} = S_e/C_s$)
L	length of the extraction phase
n	amount of analyte extracted onto the extraction phase
PDMS	polydimethylsiloxane
S_e	surface concentration of the analyte adsorbed on solid extraction phase
SFE	supercritical fluid extraction
SPE	solid-phase extraction
SPME	solid-phase microextraction
t_0	time required to remove one void volume of the extraction phase
t_e	equilibration time
u	chromatographic linear velocity
u_e	interstitial linear extraction phase velocity
V_e	volume of the extraction phase
V_v	void volume
V_s	sample volume
Z	distance between the sample and the extracting phase
z	D_e/D_s

REFERENCES

1. Thurman, E.; Mills, M. *Solid Phase Extraction*. New York: John Wiley; 1998.
2. Dean, J. *Extraction Methods for Environmental Analysis*. New York: John Wiley; 1998.
3. Handley, A., Ed. *Extraction Methods in Organic Analysis*. Sheffield, UK: Sheffield Academic Press; 1999.
4. Cantwell, F.; Losier, M. Liquid–liquid extraction. In *Sampling and Sample Preparation for Field and Laboratory*, Pawliszyn, J., Ed. Amsterdam: Elsevier; 2002.
5. Pawliszyn, J. *Solid Phase Microextraction*. New York: Wiley-VCH; 1997.
6. Ioffe, B.; Vitenberg, A. *Headspace Analysis and Related Methods in Gas Chromatography*. New York: John Wiley; 1984.
7. Ruzicka, J.; Hansen, E. *Flow Injection Analysis*, 1st ed. New York: Wiley; 1981.
8. Stern, S. *Membrane Separation Technology*. Amsterdam: Elsevier; 1995.
9. Pratt, K.; Pawliszyn, J. Gas extraction dynamics of volatile organic species from water with a hollow fibre membrane. *Anal. Chem.* **1992**, *64*, 2101–2106.
10. Yang, M.; Adams, M.; Pawliszyn, J. Kinetic model of membrane extraction with a sorbent interface (MESI). *Anal. Chem.* **1996**, *68*, 2782–2789.
11. Pawliszyn, J. New directions in sample preparation for analysis of organic compounds. *Trends Anal. Chem.* **1995**, *14*, 113–122.
12. Adams, M.; Otu, E.; Kozliner, M.; Szubra, J.; Pawliszyn, J. Portable thermal pump for supercritical fluid delivery. *Anal. Chem.* **1995**, *67*, 212–219.
13. Gorecki, T.; Yu, X.; Pawliszyn, J. Theory of analyte extraction by selected porous polymer SPME fibres. *Analyst* **1999**, *124*, 643–649.
14. Gorecki, T.; Pawliszyn, J. The effect of sample volume on quantitative analysis by SPME. Part I: Theoretical considerations. *Analyst* **1997**, *122*, 1079–1086.
15. Pawliszyn, J. *Solid Phase Microextraction, Theory and Practice*. New York: Wiley; **1997**; pp. 44–47.
16. Mulder, M. *Basic Principles of Membrane Technology*. Dordrecht: Kluwer; 1991.
17. Zhang, Z.; Poerschmann, J.; Pawliszyn, J. Direct solid phase microextraction of complex aqueous samples with hollow fibre membrane protection. *Anal. Commun.* **1996**, *33*, 129–131.
18. Rasmussen, K.E.; Pedersen-Bjergaard, S; Krogh, M; Grefslie Ugland, H.; Grønhaug, T. Development of a simple in-vial liquid-phase microextraction device for drug analysis compatible with capillary gas chromatography, capillary electrophoresis and high-performance liquid chromatography. *J. Chromatogr. A* **2000**, *873*, 3–11.
19. Boos, K.; Grimm, C-H. High-performance liquid chromatography integrated solid-phase extraction in bioanalysis using restricted access precolumn packings. *Trends Anal. Chem.* **1999**, *18*, 175–180.
20. Zhang, Z.; Pawliszyn, J. Headspace solid phase microextraction. *Anal. Chem.* **1993**, *65*, 1843–1852.
21. Zhang, Z.; Pawliszyn, J. Quantitative extraction using an internally cooled solid phase microextraction device. *Anal. Chem.* **1995**, *67*, 34–43.
22. Ghiasvand, A; Hosseinzadeh, S.; Pawliszyn, J. New cold-fiber headspace SPME device for quantitative extraction of PAH in sediment. *J. Chromatogr. A* **2006**, *1124*, 35–42.
23. Chen, Y.; Begnaud, F.; Chaintreau, A.; Pawliszyn, J. Analysis of flavor and perfume using an internally cooled coated fiber device. *J. Sep. Sci.* **2007**, *30*, 1037–1043.
24. Louch, D.; Motlagh, S.; Pawliszyn, J. Extraction dynamics of organic compounds from water using liquid-coated fused silica fibres. *Anal. Chem.* **1992**, *64*, 1187–1199.
25. Eggins, B. *Chemical Sensors and Biosensors*, 2nd ed. New York: Wiley-VCH; 2002.
26. Sellegren, B., Ed. *Molecularly Imprinted Polymers—Man-Made Mimics of Antibodies and Their Applications in Analytical Chemistry*. Amsterdam: Elsevier; 2001.
27. Pichon, V.; Bouzige, M.; Miege, C.; Hennion, M.-C. Immunosorbents: Natural molecular recognition materials for sample preparation of complex environmental matrices. *Trends Anal. Chem.* **1999**, *18*, 219–235.
28. Li, S.; Weber, S. Determination of barbiturates by solid-phase microextraction and capillary electrophoresis. *Anal. Chem.* **1997**, *69*, 1217–1222.
29. Wu, J.; Pawliszyn, J. Preparation and applications of polypyrrole films in SPME. *J. Chromatogr.* **2001**, *909*, 37–52.
30. Wu, J.; Mullett, W.; Pawliszyn, J. Electrochemically controlled solid-phase microextraction (SPME) based on conductive polypyrrole films. *Anal. Chem.* **2002**, *74*, 4855–4859.

31. Chong, S.-L.; Wang, D-X.; Hayes, J.; Wilhite, B.; Malik, A. Sol-gel column technology for single-step deactivation, coating, and stationary-phase immobilization in high-resolution capillary gas chromatography. *Anal. Chem.* **1997**, *69*, 4566–4576.

32. Dullien, F.A.L. *Porous Media*. San Diego, CA: Academic Press; 1992.

33. Horváth, C.; Lin, H.-J. Band spreading in liquid chromatography, general plate height equation and a method for the evaluation of the individual plate height contributions. *J. Chromatogr.* **1978**, *149*, 43–70.

34. Crank, J. *Mathematics of Diffusion*. Oxford: Clarendon Press; 1989.

35. Giddings, J.C. Kinetic origin of tailing in chromatography. *Anal. Chem.* **1963**, *35*, 1999–2002.

36. Cadzow, J.A.; van Landingham, H.F. *Signals, Systems, and Transforms*. Englewood Cliffs, NJ: Prince Hall; 1985.

37. Pawliszyn, J. Kinetic model of supercritical fluid extraction. *J. Chromatogr. Sci.* **1993**, *31*, 31–37.

38. Langenfeld, J.; Hawthorne, S.; Miller, D.; Pawliszyn, J. Kinetic study of supercritical fluid extraction of organic contaminants from heterogeneous environmental samples with carbon dioxide and elevated temperatures. *Anal. Chem.* **1995**, *67*, 1727–1736.

39. Langenfeld, J.; Hawthorne, S.; Miller, D.; Pawliszyn, J. Effects of temperature and pressure on supercritical fluid extraction efficiencies of polycyclic aromatic hydrocarbons and polychlorinated biphenyls. *Anal. Chem.* **1993**, *65*, 338–344.

40. Richter, B.E.; Jones, B.A.; Ezzell, J.L.; Porter, N.L.; Avdalovic, N.; Pohl, C. Accelerated solvent extraction: A technique for sample preparation. *Anal. Chem.* **1996**, *68*, 1033–1039.

41. Pare, J.R.J.; Belanger, J.M.R.; Li, K.; Stafford, S.S. Microwave-assisted process (MAP): Application to the headspace analysis of VOCs in water. *J. Microcolumn Sep.* **1995**, *7*, 37–40.

42. Alexandrou, N.; Pawliszyn, J. Supercritical fluid extraction for the rapid determination of polychlorinated dibenzo-p-dioxins and dibenzofurans in municipal incinerator fly ash. *Anal. Chem.* **1989**, *61*, 2770–2776.

43. Miao, Z.; Zhang, Z.; Pawliszyn, J. Supercritical fluid extraction and clean-up with temperature fractionation: Application to determination of polychlorinated dibenzo-p-dioxins. *J. Microcolumn Sep.* **1994**, *6*, 459–465.

44. Bartle, K.D.; Boddington, T.; Clifford, A.A.; Cotton, N. Supercritical fluid extraction and chromatography for the determination of oligomers in poly(ethylene terephthalate) films. *Anal. Chem.* **1991**, *63*, 2371–2377.

45. Eisert, R.; Pawliszyn, J. Automated in-tube solid phase microextraction coupled to high-performance liquid chromatography. *Anal. Chem.* **1997**, *69*, 3140–3147.

46. Eisert, R.; Pawliszyn, J. New trends in solid phase microextraction. *Crit. Rev. Anal. Chem.* **1997**, *27*, 103–135.

47. Young, A.D. *Boundary Layers*. Oxford: BSP Professional Books; 1989.

48. Augusto, F.; Koziel, J.; Pawliszyn, J. Design and validation of portable SPME devices for rapid field sampling and diffusion-based calibration. *Anal. Chem.* **2001**, *73*, 481–486.

49. Sukola, K.; Koziel, J.; Augusto, F.; Pawliszyn, J. Diffusion-based calibration for SPME analysis of aqueous samples. *Anal. Chem.* **2001**, *73*, 13–18.

50. Carslaw, H.S.; Jaeger, J.C. *Conduction of Heat in Solids*. Oxford: Clarendon Press; 1986.

51. Koziel, J.; Jia, M.; Pawliszyn, J. Rapid air sampling with porous SPME fibres. *Anal. Chem.* **2000**, *72*, 5178–5186.

52. Chen, Y.; O'Reilly, J.; Wang, Y.; Pawliszyn, J. Standards in the extraction phase, a new approach to calibration of microextraction processes. *Analyst* **2004**, *129*, 702–703.

53. Chen, Y.; Pawliszyn, J. Kinetics and the on-site application of standards in a solid-phase microextraction fiber. *Anal. Chem.* **2004**, *76*, 5807–5815.

54. Ouyang, G.; Cai, J.; Zhang, X.; Li, H.; Pawliszyn, J. Standard-free kinetic calibration for rapid on-site analysis by solid-phase microextraction. *J. Sep. Sci.* **2008**, *31*, 1167–1172.

55. Ai, J. Solid phase microextraction for quantitative analysis in non-equilibrium situations. *Anal. Chem.* **1997**, *69*, 1230–1236.

56. Ai, J. Headspace solid phase microextraction. Dynamics and quantitative analysis before reaching a partition equilibrium. *Anal. Chem.* **1997**, *69*, 3260–3266.

57. Ouyang, G.; Zhao, W.; Pawliszyn, J. Kinetic calibration for automated headspace liquid-phase microextraction. *Anal. Chem.* **2005**, *77*, 8122–8128.

58. Zhao, W.; Ouyang, G.; Alaee, M.; Pawliszyn, J. On-rod standardization technique for time-weighted average water sampling with PDMS rod. *J. Chromatogr. A* **2006**, *1124*, 112–120.

59. Bragg, L.; Qin, Z.; Alaee, M.; Pawliszyn, J. Field sampling with a polydimethylsiloxane thin-film. *J. Chromatogr. Sci.* **2006**, *44*, 317–323.

60. Ouyang, G.; Zhao, W.; Bragg, L.; Qin, Z.; Alaee, M.; Pawliszyn, J. Time-weighted average water sampling in Lake Ontario with solid-phase microextraction passive samplers. *Environ. Sci. Technol.* **2007**, *41*, 4026–4031.

61. Musteata, F.M.; Musteata, M.L.; Pawliszyn, J. Fast in-vivo microextraction: A new tool for clinical analysis. *Clin. Chem.* **2006**, *52*, 708–715.

62. Zhang, X.; Es-Haghi, A.; Musteata, F.M.; Ouyang, G.; Pawliszyn, J. Quantitative in-vivo microsampling for pharmacokinetic studies based on an integrated SPME system. *Anal. Chem.* **2007**, *79*, 4507–4513.

63. Zhou, S.N.; Zhao, W.; Pawliszyn, J. Kinetic calibration using dominant pre-equilibrium desorption for on-site and in vivo sampling by solid-phase microextraction. *Anal. Chem.* **2008**, *80*, 481–490.

64. Zhao, W.; Ouyang, G.; Pawliszyn, J. Preparation and application of in-fibre internal standardization SPME. *Analyst* **2007**, *132*, 256–261.

65. Ouyang, G.; Cui, S.; Qin, Z.; Pawliszyn, J. One-calibrant kinetic calibration for on-site water sampling with solid-phase microextraction. *Anal. Chem.* **2009**, *81*, 5629–5636.

66. Chen, Y.; Pawliszyn, J. Miniaturization and automation of an internally cooled coated fiber device. *Anal. Chem.* **2006**, *78*, 5222–5226.

67. Chai, M.; Pawliszyn, J. Analysis of environmental air samples by solid phase microextraction and gas chromatography-ion trap mass spectrometry. *Environ. Sci. Technol.* **1995**, *29*, 693–701.

68. Chen, Y.; Pawliszyn, J. Solid-phase microextraction field sampler. *Anal. Chem.* **2004**, *76*, 6823–6828.

69. Luo, Y.; Adams, M.; Pawliszyn, J. Kinetic study of membrane extraction with a sorbent interface for air analysis. *Anal. Chem.* **1998**, *70*, 248–254.

70. Luo, Y.; Pawliszyn, J. Calibration of membrane extraction for air analysis. *Anal. Chem.* **2000**, *72*, 1064–1071.

71. Liu, X; Pawliszyn, J. Internal calibrant in the stripping gas. An approach to calibration of MESI. *Anal. Chem.* **2006**, *78*, 3001–3009.

72. Ma, W.; Liu, X.; Pawliszyn, J. Analysis of breath with microextraction techniques and continuous monitoring carbon dioxide concentration. *Anal. Bioanal. Chem.* **2006**, *385*, 1398–1408.

73. Musteata, F.M.; Pawliszyn, J. Study of ligand-receptor binding using SPME: Investigation of receptor, free and total ligand concentrations. *J. Proteome Res.* **2005**, *4*, 789–800.

74. Martos, P.; Pawliszyn, J. Calibration of solid phase microextraction for air analyses based on physical chemical properties of the coating. *Anal. Chem.* **1997**, *69*, 206–215.

75. Grote, C.; Levsen, K. The application of SPME in water analysis. In *Applications of Solid Phase Microextraction*, Pawliszyn, J., Ed. Cambridge, UK: RSC; 1999.

76. Heinze, J. Ultramicroelectrodes in electrochemistry. *Angew. Chem. Int. Ed. Engl.* **1993**, *32*, 1268–1288.

77. Fang, Z-L. *Flow Injection Separation and Preconcentration*. Weinheim: VCH; 1993.

78. Pawliszyn, J. Properties and applications of the concentration gradient sensor to detection of flowing samples. *Anal. Chem.* **1986**, *58*, 3207–3215.

79. Liu, H.; Dasgupta, P. A renewable liquid droplet as a sampler and a windowless optical cell. Automated sensor for gaseous chlorine. *Anal. Chem.* **1995**, *67*, 4221–4228.

80. Wu, C-H.; Scampavia, L.; Ruzicka, J. Microsequential injection: Anion separations using 'lab-on-valve' coupled with capillary electrophoresis. *Analyst* **2002**, *127*, 898–905 and references therein.

81. Greenwood, P.; Greenway, G. Sample manipulation in micro total analytical systems. *Trends Anal. Chem.* **2002**, *21*, 726–740.

82. Huang, Y.; Mather, E.; Bell, J. MEMS-based sample preparation for molecular diagnostics. *Anal. Bioanal. Chem.* **2002**, *372*, 49–65.

83. Zhu, L.Y.; Tu, C.H.; Lee, H.K. Liquid-phase microextraction of phenolic compounds combined with on-line preconcentration by field-amplified sample injection at low pH in micellar electrokinetic chromatography. *Anal. Chem.* **2001**, *73*, 5655–5660.

84. Blakely, R.; Wages, S.; Justice, J. Jr.; Herndon, J.; Neil, D. Neuroleptics increase striatal catecholamine metabolites but not ascorbic acid in dialyzed perfusate. *Brain Res.* **1984**, *308*, 1–12.

85. Segal, A.; Gorecki, T.; Mussche, P.; Lips, J.; Pawliszyn, J. Development of membrane extraction with a sorbent interface—Micro GC for field analysis. *J. Chromatogr. A* **2000**, *873*, 13–27.

86. Ng, J.; Stroock, A.; Whitesides, G. Components for integrated poly(dimethylsiloxane) microfluidic systems. *Electrophoresis* **2002**, *23*, 3461–3473.

87. Pawliszyn, J. Solid phase microextraction. In *Sampling and Sample Preparation for Field and Laboratory*, Pawliszyn, J., Ed. Amsterdam: Elsevier; 2002.

88. Koziel, J.; Noah, J.; Pawliszyn, J. Field sampling and determination of formaldehyde in indoor air with SPME and on-fiber derivatisation. *Environ. Sci. Technol.* **2001**, *35*, 1481–1486.

89. Song, Y.; Lunte, C. Calibration methods for microdialysis sampling in vivo: Muscle and adipose tissue. *Anal. Chim. Acta* **1999**, *400*, 143–152.

90. Lord, H.; Yu, W.; Segal, A.; Pawliszyn, J. Breath analysis and monitoring by membrane extraction with sorbent interface. *Anal. Chem.* **2002**, *74*, 5650–5657.

91. Lord, H.; Grant, R.; Incledon, B; Walles, M.; Pawliszyn, J. Development and evaluation of a solid phase microextraction probe for in-vivo pharmacokinetic studies. *Anal. Chem.* **2003**, *75*, 5103–5115.

92. Michael, K.; Taylor, L.; Schultz, S.; Walt, D. Randomly ordered addressable high-density optical sensor arrays. *Anal. Chem.* **1998**, *70*, 1242–1248.

Chapter 2

Headspace Gas Chromatography

Zelda E. Penton

2.1. OVERVIEW

2.1.1. Introduction

Headspace gas chromatography (GC) is a sample preparation method for determining volatile compounds in solid and liquid samples. The technique has existed since the late 1950s[1] and is still actively used. The popularity of headspace analysis is due to its simplicity and the fact that it is a very clean method of introducing volatile analytes into a GC—the injector system and column should require virtually no maintenance.

A typical headspace analysis involves a liquid or solid sample in a sealed vial containing a gas phase. A portion of the gas phase is removed and introduced into a GC. The analytes are volatile compounds that are dissolved in a water-based matrix or in an organic solvent with a high boiling point (>150°C). The matrix can also be a solid or semisolid material. The analytes are normally present at concentrations ranging from parts per billion to low percentage levels.

Some common applications of the headspace technique are determination of blood alcohol, residual solvents in pharmaceuticals, and monomers in polymers. However, there are many other applications of headspace GC. Some of these will be discussed in Section 2.6.

2.1.2. Static versus Dynamic Headspace

The term "headspace GC" is commonly used to describe static headspace but is sometimes used to refer to dynamic headspace or purge and trap. In the static headspace technique (Fig. 2.1), the sample is placed in a vial, which is sealed and usually heated. The volatile analytes begin to partition between the sample and the gas phase. Eventually, an equilibration state is reached. At this point, an aliquot of the gas phase is transferred to the GC.

With dynamic headspace (Fig. 2.2), an inert gas is bubbled through the sample, and the volatiles are transferred to an absorbent trap. The trap is heated and the volatiles are released or desorbed and transferred to the GC. Finally, the trap is heated to a higher temperature than was used during the desorption step, to remove residual analytes and moisture, and the system is ready for another sampling.

Purge and trap is generally more sensitive than static headspace, since the former technique should theoretically remove all of the analytes from the matrix, while the latter is limited to removal of a single aliquot of the gas phase. However, a purge-and-trap system requires more maintenance and is subject to problems such as foaming of the sample. The two techniques are compared in Table 2.1. This chapter will focus on static headspace sampling.

2.2. THEORY OF CONVENTIONAL STATIC HEADSPACE

2.2.1. Partition Coefficient and Phase Ratio (β)

In order to develop and optimize headspace methods, the analyst should understand the effect of two parameters on the sensitivity that will be achieved in a headspace analysis. These are the partition coefficient and the phase ratio. These parameters are discussed below.

A sample that is to be analyzed with the headspace technique is usually collected from the source and refrigerated in a vial with no gas phase over the sample. The sample is transferred to a headspace vial just prior to analysis. These vials normally hold a total volume of 5–22 mL and are sealed with a septum and crimped metal cap. The sample is placed in the vial, leaving a significant amount of headspace over the sample; then, the vial is sealed and heated. The analytes (volatile compounds) begin to move into the gas phase above the sample until a state of

Handbook of Sample Preparation, Edited by Janusz Pawliszyn and Heather L. Lord

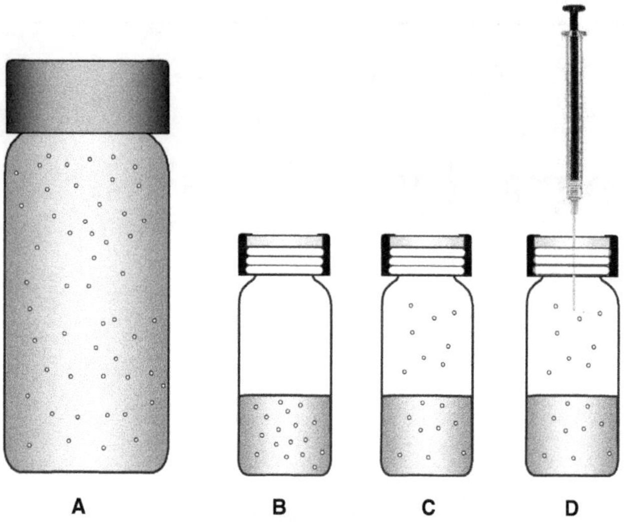

Figure 2.1. Static headspace sampling. The sample is collected in a container with no space over the liquid so that volatiles are not lost (A). An aliquot of the sample is placed in a headspace vial that is sealed with a septum and a crimp-top vial (B). The volatiles in the liquid phase enter the gas phase until a state of equilibrium is reached (C). An aliquot of the gas phase is removed with a gastight syringe (D) and injected into the GC.

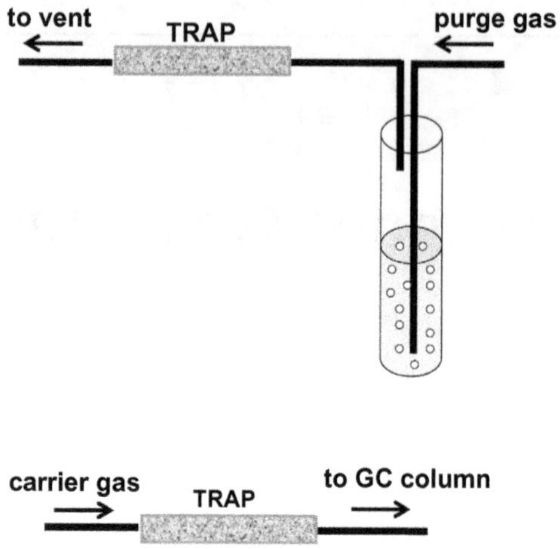

Figure 2.2. Dynamic headspace sampling or purge and trap. The top diagram is a schematic of the purge mode. The purge gas (helium or nitrogen) passes through the sample, extracting the volatiles, which pass onto an adsorbent trap. The lower diagram shows the desorption mode where the gas flow is reversed and the trap is heated. The desorbed gas sweeps the analytes out of the trap to the GC column for separation. Finally, the trap is heated to a higher temperature to clean out residual analytes and water.

Table 2.1. Comparison of Static and Dynamic Headspace

	Static Headspace	Dynamic Headspace (Purge and Trap)
Concentration range	Parts per billion to low percent levels	Parts per trillion to parts per million
Hardware required	Ranges from a simple gastight syringe to a sophisticated dedicated instrument	Complex dedicated instrument
Automation	Can be manual with a gastight syringe but several automated systems can handle over 100 samples	Purge-and-trap autosamplers can handle a maximum of ~70 samples
Quantitation	Highly matrix dependent since there is only partial recovery of the analyte	Less matrix dependent since most of the analyte is removed
Ease of maintenance	Little maintenance required	Labor intensive, traps must be cleaned, foaming a problem
Common applications	Blood alcohol Residual solvents in pharmaceuticals Monomers in polymers Flavors in foods Volatiles in drinking and wastewater (Europe), soil (United States)	Monomers in polymers Flavors in foods Volatiles in drinking and wastewater and soil (United States)

equilibrium is reached. At this point, the ratio of the analyte concentrations in the gas phase and in the liquid or solid phase is a constant. The constant is defined as the partition coefficient:

$$\text{Partition coefficient} \quad K = \frac{C_L}{C_G} \tag{2.1}$$

The ratio (β) of the gas phase to the liquid phase is

$$\beta = \frac{V_G}{V_L}. \tag{2.2}$$

The equilibration process is illustrated in Figure 2.3 where

C_L is the concentration in the liquid phase after equilibration,
C_G is the concentration in the gas phase after equilibration,
C_O is the original concentration of the analyte in the sample,
V_L is the volume of the liquid sample,
V_G is the volume of the gas phase, and
V is $V_L + V_G$ (total volume of the headspace vial).

Equation 2.3, the relationship of the partition coefficient (K) and the phase ratio (β) to headspace sensitivity (C_G) can be calculated from Equations 2.1 and 2.2:

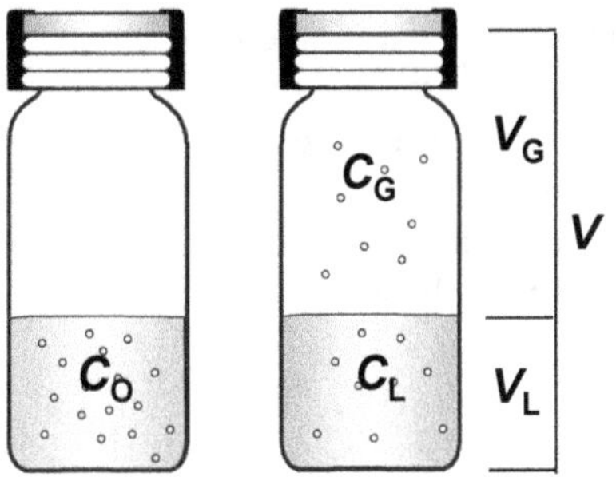

Figure 2.3. Showing the equilibrium process in static headspace. In the vial on the left, the analytes are at the concentration that they were in the original sample. On the right, the analytes have reached equilibrium between the gas and liquid phases.

$$C_G = \frac{C_O}{K+\beta}. \tag{2.3}$$

Equation 2.3 is based on the assumption that the solution is dilute. It is possible to draw several conclusions from Equation 2.3:

- The volume of the headspace vial has no effect on sensitivity.
- As K increases, the sensitivity is reduced and β becomes less significant.
- As K decreases, β becomes more significant; a small value of β will tend to increase sensitivity.

Figure 2.4 illustrates these conclusions by plotting headspace sensitivity versus volume of liquid sample in a 10-mL vial for three compounds with partition coefficients of 4, 40, and 400. These values roughly correspond to the partition coefficients of benzene, ethyl acetate, and isopropanol in water at 50°C. Note that the headspace sensitivity of the compound with a low partition coefficient increases dramatically as the relative volume of the gas phase decreases (smaller β). The compound with a partition coefficient of 400 (comparable to isopropanol) shows almost no change in sensitivity as β decreases.

2.2.2. Parameters Affecting the Partition Coefficient

Some parameters affecting the partition coefficient of a compound in a given matrix are solubility and temperature. As the solubility of an analyte increases in a given matrix, the partition coefficient increases and the headspace sensitivity is reduced. In this situation, altering the matrix will have a significant effect on the partition coefficient. Thus, addition of salt to water will reduce the solubility of polar compounds in the water and increase the sensitivity of headspace analysis. Nonpolar compounds with a low partition coefficient in water will not be affected significantly by the addition of salt (Fig. 2.5).

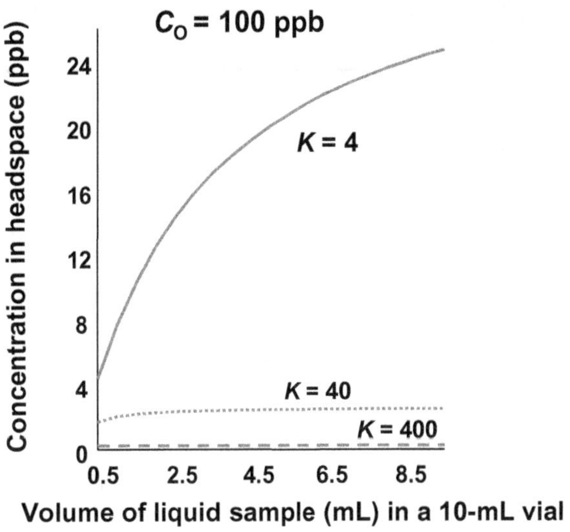

Figure 2.4. Hypothetical curves calculated using Equation 2.3 for three different analytes with partition coefficients of 4, 40, and 400. For all of the compounds, the original concentration of the analyte in the liquid sample was 100 ppb. The x-axis represents volumes of liquid sample in a 10-mL vial ranging from 0.5 to 9.5 mL. The y-axis is the concentration of the analyte in the gas phase for each volume of liquid sample. Note that the volume of the sample has virtually no effect on the headspace concentration when the partition coefficient is 400, but the sample volume has a significant effect when the partition coefficient is 4.

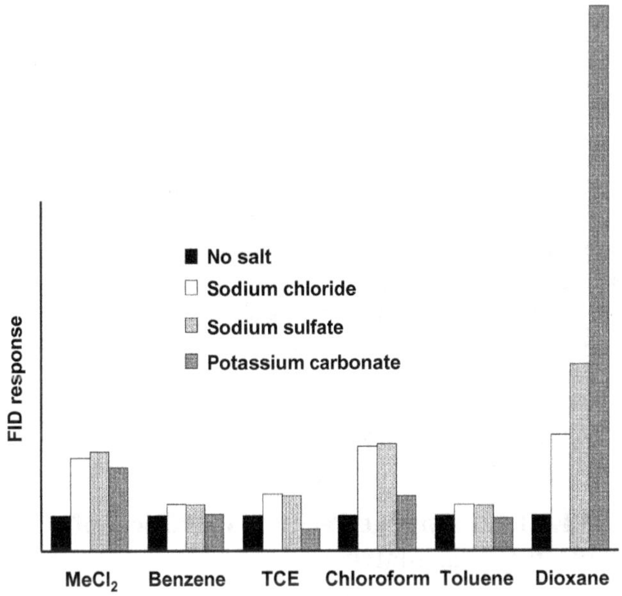

Figure 2.5. Effect of saturation with salt on increasing the headspace sensitivity of various compounds in water. Note that "salting out" has the most significant effect on dioxane, the polar compound, where the response was enhanced 23-fold (off-scale) when saturating with potassium carbonate. TCE, 1,1,2-trichloroethylene.

Table 2.2. Air–Water Partition Coefficients of 10 Selected Compounds versus Temperature

	Partition Coefficient (K)		
Compound	40°C	60°C	80°C
Dioxane	1618	642	288
Ethanol	1355	511	216
Ethyl acetate	62.4	29.3	17.5
Benzene	Not available	2.27	1.66
Toluene	2.82	1.77	1.27
o-Xylene	2.44	1.31	0.99
Dichloromethane	5.65	3.31	2.07
1,1,1-Trichloromethane	1.65	1.47	1.18
Tetrachloroethylene	1.48	1.27	0.87
Cyclohexane	0.08	0.05	0.02

The partition coefficient is also affected by temperature:

$$\ln K = \frac{A}{RT} - B, \tag{2.4}$$

where

K is the partition coefficient,
R is the ideal gas constant,
A and B are thermodynamic constants, and
T is the absolute temperature in kelvin (K).

Table 2.2 contains partition coefficients of some organic compounds in water at different temperatures. Generally, the more soluble an analyte is in a given matrix, the more the partition coefficient will vary with temperature. Compare the changes in the partition coefficients of 1,1,1-trichloromethane and dioxane in water as the temperature increases.

During a headspace analysis, the temperature of the samples normally must be kept constant to assure excellent precision. There are a few exceptions, however, when the partition coefficient varies only slightly with temperature or an internal standard is present that shows the same variation in partition coefficient as the analyte.[2]

2.3. OPTIMIZATION OF HEADSPACE RESPONSE

2.3.1. Decreasing the Partition Coefficient of Analytes in the Matrix

As discussed above, a typical approach to decreasing the partition coefficient is to increase the temperature of the sample. However, problems can arise when a headspace vial is heated to a temperature approaching the boiling point of the matrix. These include leaks due to increasing pressure in the vial and decomposition of the sample.

Altering the matrix to reduce the solubility of the analytes can also decrease the partition coefficient. "Salting out," as mentioned above, is an effective method for driving polar analytes from biological fluids and soil. It is important to saturate the sample with salt to maximize the effect of the salt and also to avoid variations in salt concentrations from sample to sample, which will affect the relative responses of the analytes.

It is sometimes useful to change the pH of the matrix to maximize response—that is, lower the pH to convert volatile acids to the molecular state or raise the pH to determine amines in the sample.

2.3.2. Mixing and Equilibration Time

Most commercially available headspace systems have the ability to mix the sample during the equilibration period. Mixing tends to reduce the time required to achieve equilibration.[3]

This brings up the question of sampling the headspace before reaching equilibrium. The main advantage of sampling the headspace, after the analytes have reached full equilibrium between the matrix and the gas phase, is to achieve maximum sensitivity and precision. In many laboratories, speed of analysis is more important than maximizing sensitivity and precision. An example of an application where the GC run time is much shorter than the time necessary to reach equilibrium is the determination of a residual monomer in a polymer. In the case of volatiles in a solid, equilibrium may take hours or even days to achieve. The only practical solution may be to sample the headspace before equilibrium is attained, and to heat all of the samples for the same length of time prior to analysis, to assure comparability.

When headspace GC is used in thermodynamic studies such as the determination of activity coefficients,[4–6] it is important to sample after equilibrium has been reached.

2.3.3. Injection Volume

In most headspace GC analyses, fused-silica columns with inner diameters ranging from 0.25 to 0.53 mm are utilized. A film thickness of 0.5 µm or greater is often selected to increase the capacity of the GC column. Most columns have a limited ability to focus very volatile compounds such as freons at temperatures 35°C and above. Therefore, a smaller injection volume is desirable to achieve narrow early-eluting peaks. The default injection volume in many commercial headspace systems is usually 1 mL. This volume is often too large for optimum chromatographic results. When developing a headspace method, it is a good idea to use as small an injection volume as is consistent with the sensitivity requirements (Fig. 2.6). If it is necessary to inject volumes of 1 mL or greater, a cryogenic oven or a cold trap should be used to focus the analytes on the column.

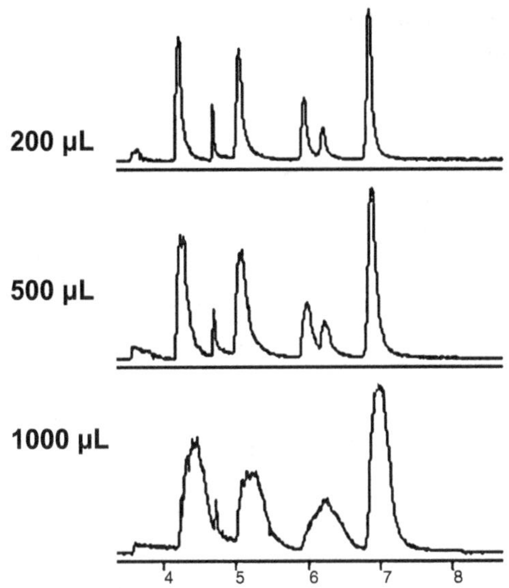

Figure 2.6. GC-MS chromatograms of volatile halogenated hydrocarbons in the headspace over water. Note that the chromatography improves as the injection volume is reduced. Although, a high-capacity, fused-silica column was used to help focus the volatile compounds, a cryogenic trap would have been useful for this sample.

2.3.4. Vial Size

As discussed in Section 2.2.1, the sensitivity is determined by the partition coefficient and the phase ratio and is independent of the vial size. However, there are advantages to large vials. It is easier to add solid samples to a 20-mL vial than to a 2-mL vial, and if a relatively large sample is added to a vial, it is more likely to be representative of the material to be analyzed.

2.3.5. Headspace Trapping

In recent years, the major headspace system manufacturers have introduced an optional trapping capability for headspace samplers that results in a lower detection limit as compared with traditional static headspace. This option is free of the drawbacks of the purge-and-trap technique mentioned in Table 2.1.

During the headspace sampling, a valve isolates the GC column from the carrier gas that is flowing through the headspace system.* The headspace vapor is transferred to a cool trap, where the analytes are retained. Next, the trap is purged with carrier gas to remove water vapor, and then, it is rapidly heated to desorb the analytes. At this point, the valve switches to end the isolation of the column from the headspace system, and the analytes are directed into the GC column. Analyte detection limits are enhanced not only as a result of the multiple samplings, but also because the rapid

*There may be an auxiliary flow of carrier gas through the GC column, depending on the headspace system.

Table 2.3. Detection Limit of Solvents Determined with Conventional Static Headspace and Static Headspace with Trapping on a Teledyne Tekmar HT3TM Headspace System

	Detection Limit (ppb)	
Solvent	Without Trapping (Loop)	With Trapping
Benzene	210	1.32
Carbon tetrachloride	650	1.54
1,2-Dichloroethane	440	1.32
1,1,1-Trichloroethane	1150	1.41
N,N-dimethylacetamide	8460	3.07
N,N-dimethylformamide	7640	2.06
1,4-Dioxane	6890	2.51
Hexane	8030	2.14
Methanol	5620	70
Pyridine	8860	4.07

Partial list of data taken from Cox, J. Application Note: Residual solvents by HT3TM headspace in reference to USP 467 with a comparison of static versus dynamic headspace analysis. www.teledynetekmar.com. Printed with permission of Teledyne Tekmar Co.

desorption and transfer of analytes from the trap can yield better focusing on the head of the GC column as compared with the transfer of analytes from a 1-mL sample loop or syringe. The resulting peaks from trap desorption are taller and narrower, giving a better signal-to-noise ratio.

Table 2.3 compares detection limits with conventional static headspace and static headspace with trapping for several solvents in water. When using headspace trapping, the user should remember that different trapping materials are recommended for different analytes and the appropriate* trap should be installed.

2.4. QUANTITATION IN HEADSPACE

2.4.1. Quantitation Techniques

Headspace GC differs from most other analytical chemistry techniques in that recovery is almost never near 100%—sometimes it is well under 1%! In spite of this, it is possible to obtain extremely reproducible and accurate results with headspace. The main difficulty in achieving good quantitation with headspace is in matching the matrix of the samples and standards. The significance of the partition coefficient in determining headspace response was described previously (Section 2.2.1). Slight differences in the matrix, such as content of organic material in soil samples (Fig. 2.7) or protein or fat in biological fluids, and differences in ionic strength and pH will alter the matrix, thereby affecting headspace response. Some approaches to overcoming the matrix effect are discussed below.

*Chromatography suppliers such as Supelco (division of Sigma-Aldrich) sell prepared traps and provide information in their catalogs on the performance of the packings for retaining different analytes.

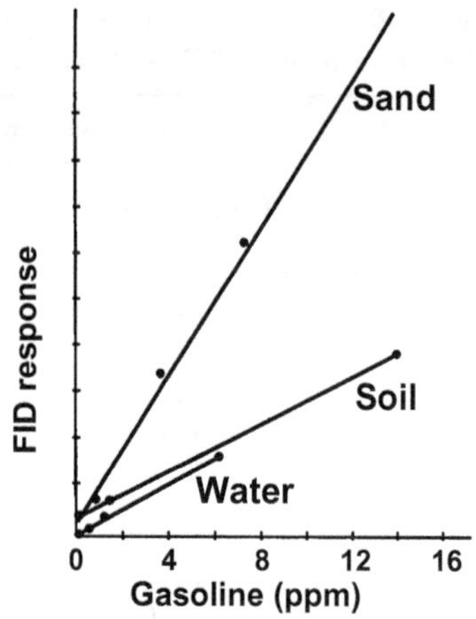

Figure 2.7. An illustration of the matrix effect on recovery of gasoline from sand, soil, and water by headspace GC.

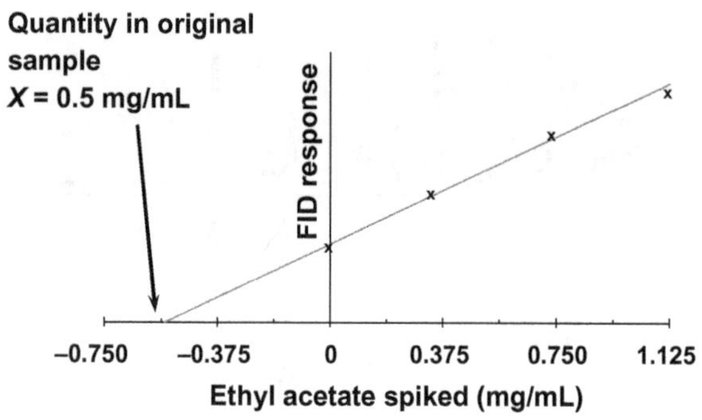

Figure 2.8. "Standard additions" quantitation in the headspace analysis of ethyl acetate in an industrial mixture. The detector responses were determined for the unspiked sample and for the sample spiked with increasing quantities of ethyl acetate. The resulting curve was back-extrapolated to the *x*-axis to determine the quantity of ethyl acetate in the original sample.

2.4.1.1. Using a blank that exactly matches the sample matrix. This method could be used for the analysis of organic volatiles in drinking water. It is not difficult to prepare a solution of water that does not contain measurable volatile organics. The water is spiked with volatile organics of known concentration, and a calibration curve is prepared.

2.4.1.2. Dilute the samples and standards with a matrix modifier. This method is used in determining blood alcohol (Section 2.6.1). It is also the basis of the United States Environmental Protection Agency (USEPA) Method 5021[7] for the analysis of volatile organics in soil (Section 2.6.3). The matrix modifier is often a solution of water saturated with salt. For maximum effect, the sample and standards should be diluted as much as possible while retaining a high enough concentration of the analytes to meet the requirements of the analysis. In the blood alcohol analysis described later,[2] three volumes of matrix modifier were added to one volume of the samples and standards.

2.4.1.3. Spike the sample with known increments of analyte. This technique, known as "standard additions," is useful when it is not possible to find a blank matrix and the sensitivity requirements of the analysis do not permit dilution with a matrix modifier. An example would be the determination of a trace flavor compound in wine such as linalool. Three to five aliquots of wine samples are spiked with increasing quantities of linalool. The spiked samples and an unspiked sample are analyzed. The quantity spiked is plotted versus the response, and the resulting calibration curve is extrapolated to the *x*-intercept to determine the original quantity of the analyte in the sample. Figure 2.8 illustrates the use of standard additions for the determination of ethyl acetate in an industrial solution. When using this calibration method, the analyst should verify that the analyte is not bonded to components in the matrix; for this reason, standard additions are recommended for liquid rather than for solid samples.

2.4.1.4. Use the multiple headspace extraction (MHE) technique. In a typical headspace analysis, the vial is sampled only once and then discarded. After one sampling some of the analyte is removed from the headspace, and after a new equilibrium occurs, there is less analyte in the gas phase. Therefore, unlike the situation with liquid injection, only the first injection represents the original sample.

There is a headspace quantitation technique that utilizes multiple sampling from a single vial. MHE was proposed in 1970 by Suzuki et al.[8] and modified by Kolb and Ettre.[9] With MHE, the sample is sealed in a headspace vial and sampled repeatedly at equal time intervals. It is assumed that the concentration of volatiles, under these conditions, will decay exponentially (Fig. 2.9). If an infinite number of extractions are carried out, the volatiles will be completely removed from the vial. The total area count of the peaks corresponding to the analyte should be equal to the sum of the areas from each individual extraction.

MHE is normally used for determining volatiles in solid samples. With solids, a blank matrix is usually not available, and spiking the solid with the analyte and then sampling often does not result in valid results, because the spiked analyte is released from the solid material more readily than the analyte that is present from the manufacturing process.

Kolb and Ettre[10] derived a simplified method of quantitation for MHE. The equations for MHE are shown below; the calculations were illustrated by the authors with an analysis of ethylene oxide in surgical silk sutures. The silk sample was placed in a headspace vial and sealed. A known quantity of ethylene oxide was injected into an empty headspace vial for calibration. Kolb showed that when an analyte is sampled several times from a headspace vial, the total area from all of the samplings can be represented by Equation 2.5:

$$\sum A_i = \frac{A_1}{1-e^k}, \tag{2.5}$$

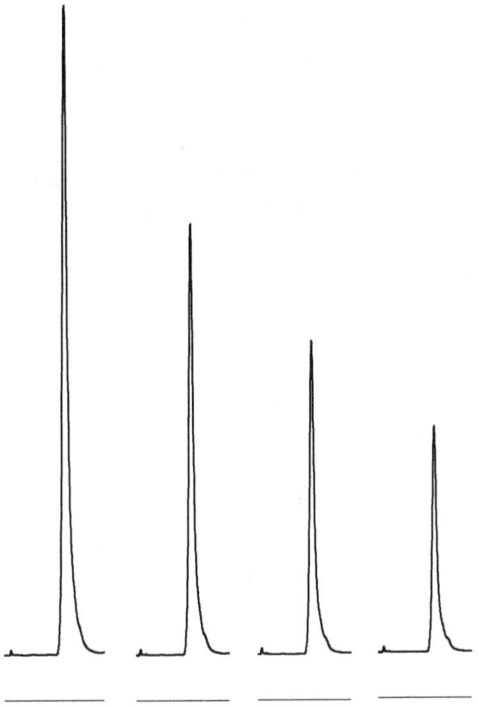

Figure 2.9. FID chromatograms of vinyl chloride injected four times from the same vial in a multiple headspace extraction analysis. The vial contained a polyvinyl chloride polymer.

where

ΣA_i is the total area count from an infinite number of samplings,
A_1 is the area from the first sampling, and
k is the slope of the plot obtained by plotting the natural logarithm of the area counts versus the number of the sampling.

In order to validate the method for a particular analysis, it is necessary to demonstrate a linear response when the natural logorithm of the area counts is plotted against the number of the sampling. A nonlinear response may be due to trapping of the analyte in the interstitial spaces of a solid sample. Linear plots for MHE samplings of vinyl chloride in a polymer are shown in Figure 2.10.

After validating the method by demonstrating a linear response, a simplified form of Equation 2.5 can be used, which requires only two samplings:

$$\sum A_i = \frac{A_1^2}{(A_1 - A_2)}, \tag{2.6}$$

where

ΣA_i is the total area count,
A_1 is the area from the first sampling, and
A_2 is the area from the second sampling.

While results would be expected to be more accurate with more than two samplings, Equation 2.6 is practical for routine analysis. The total area of the analyte in the calibration sample would also be determined using Equation 2.6. Finally, the weight of the analyte would be calculated with Equation 2.7:

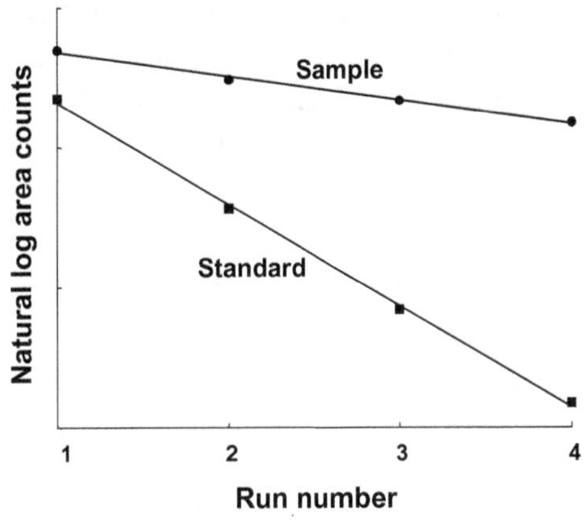

Figure 2.10. Validation curves for determining vinyl chloride in polyvinyl chloride with multiple headspace extraction (MHE). If a solid sample binds the analyte chemically or traps the analyte so that the rate of release into the headspace varies after each sampling, then the curves will not be linear and MHE cannot be used.

$$W_{smp} = \frac{(A_{smp})(W_{std})}{A_{std}}, \tag{2.7}$$

where

W_{smp} is the weight of the analyte in the sample,
A_{smp} is the total area of the analyte in the sample as calculated in Equation 2.6,
W_{std} is the weight of the analyte in the standards vial, and
A_{std} is the total area of the analyte in the standard vial as calculated in Equation 2.6.

MHE is useful only if the partition coefficient between the matrix and the analyte is small, so that a substantial quantity is removed at each extraction. MHE can also be used for liquid samples, but generally, the quantitation methods described in Sections 2.4.1.1–2.4.1.3 are simpler.

2.4.1.5. Use the "full evaporative technique" (FET). FET[11] is an approach to eliminating the matrix effect in headspace GC. With this technique, there is no equilibrium between the sample and the headspace. A very small liquid sample (a few microliters) is injected into an empty headspace vial, and the vial is heated until the entire sample is in the vapor state. Then, a sample of the gas phase in the vial is injected into the GC. Since there is no equilibrium between two phases, there is no matrix effect. Standards are prepared by injecting the appropriate quantity of pure or diluted analyte into empty headspace vials.

A drawback to the FET is the sensitivity or limit of detection (LOD) that one can obtain with compounds that have a low partition coefficient. For example, with dioxane in water at 40°C ($K = 1618$), there is almost no change in LOD with FET if one compares the LOD from a 10-μL sample in a 20-mL headspace vial to the LOD with conventional headspace (10-mL sample in a 20-mL vial). However, with *o*-xylene in water at 40°C ($K = 2.44$), the LOD with conventional headspace is lower by a factor of over 1000 as compared with the FET.

2.4.2. Internal Standards in Headspace Quantitation

The use of internal standards in headspace is not as common as in liquid injection. As is the case in liquid injection, if there is a complex mixture of analytes, the internal standards may coelute with some of the compounds of interest. It can also be difficult to find a compound that has a partition coefficient that closely matches several analytes. In a simple analysis such as blood alcohol determination, an internal standard is normally used.

2.4.3. Problems with Quantitation

The most frequent cause of problems with quantitation in headspace is due to loss of volatiles during sample handling. This includes losses during preparation of standards. It is important to fill the container to the top when collecting samples, to store samples in the refrigerator, and to analyze them as soon as possible. Standards should be prepared and stored with equal care. Occasionally, analysts spike solvents with levels of standards that exceed the solubility of these compounds in the matrix. This is another error that must be avoided to obtain valid results.

Other problems such as low response or no response are usually due to improper sealing of the vials. Some vials may leak only at higher temperatures.

With automated systems, where the sample is removed from the vial by puncturing the septum with a needle, the needle position may be set too high, so that the septum is not fully punctured.

2.5. HEADSPACE HARDWARE

This section contains a brief description of the various types of apparatus that are commercially available.

2.5.1. Manual Headspace

The simplest method of headspace analysis is to sample the vials with a manual gastight syringe. The vials may be unheated or heated in a temperature bath. Manual headspace is practical if the lab does not have many samples and the samples contain volatile compounds with low partition coefficients. This type of sample does not require precise temperature control, and there is no danger of condensation in an unheated syringe. Manual headspace is an inexpensive way for a laboratory to evaluate the headspace technique for a particular application.

2.5.2. Automated Headspace

Dedicated headspace systems should have the following features:

- constant heating time for each sample;
- heating of the next sample in a sequence during the GC run of the current sample, so that there is no wasted time between runs;
- an inert sample path;
- variable injection volume;
- mixing;
- automated method optimization (the ability to download different methods to study the effect of varying parameters such as sample temperature or equilibration time); and
- ability to handle a large number of samples, if required.

2.5.2.1. Syringe-based systems. A syringe-based automated system can be as simple as a modified liquid autosampler that has been equipped with a gastight syringe.[2,12] These systems perform well for some applications, but the sample vials and the syringe are unheated; therefore, they are inadequate for analyzing solid samples or relatively high-boiling analytes such as diesel fuels. There are also dedicated headspace systems that use a heated syringe and have all of the desirable features listed above.

2.5.2.2. Valve and loop-based systems. Figure 2.11 shows the schematics of a valve and loop-based headspace autosampler that is equipped with an optional trap. The system has an eight-port valve; two ports are plugged off when the trap option is not installed. There is also a 1-mL* sample loop and a transfer line that connects to the GC.

The steps in introducing the sample are the following:

- The vial is heated for the length of time specified by the user, and then, the vial is pierced with a needle and pressurized with an inert gas.
- The valve is turned so that the flow of gas changes direction and a portion of the headspace flows into the sample loop.
- The valve is turned again so that the gas in the sample loop is flushed through the transfer line and into the GC.
- The system returns to the standby position where clean inert gas flushes the sample loop and the transfer line.

When the trapping option is used, the sample flows into the trap, rather than the sample loop. The subsequent steps were described in Section 2.3.5.

2.5.2.3. Pressure-balanced headspace sampling. Pressure-balanced headspace sampling was invented and patented by PerkinElmer, Inc. Systems using this type of headspace sampling are available only from this company.

*With some of these systems, the sample loop is accessible and can be easily changed to a different size.

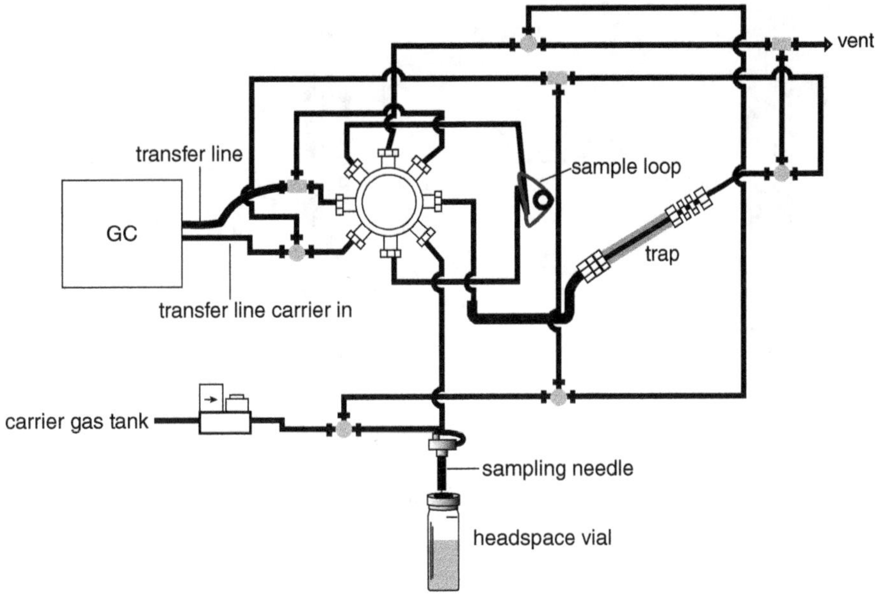

Figure 2.11. Schematic of a valve-loop automated headspace system with optional trap. When the system is purchased without the trapping option, two of the ports in the eight-port valve are plugged off. The valve, sample loop, and tubing that lead to the needle that penetrates the sample vial, and the transfer line that leads to the GC, are heated. The sequences of events for the loop mode and the trap mode are described in the text (Sections 2.5.2.2 and 2.3.5). Drawing printed with permission of Teledyne Tekmar Co.

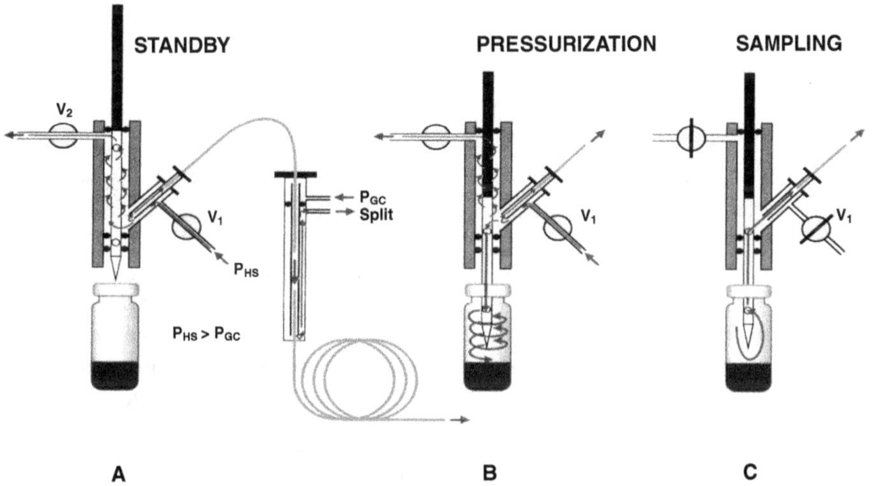

Figure 2.12. Schematic of the PerkinElmer pressure-balanced headspace system. The sequence of events is described in the text (Section 2.5.2.3). Drawing printed with permission of PerkinElmer, Inc.

The technique is quite different from other headspace sampling methods, in that the quantity of headspace gas that is introduced into the GC column is controlled by the pressure in the headspace vial and the sampling time, rather than by the volume as in a loop or syringe-based system.

Pressure-balanced headspace sampling can best be understood by referring to the schematic in Figure 2.12A, where the system is in standby while the sample in the vial reaches equilibrium. After reaching equilibrium, a rotating needle penetrates the headspace vial (Fig. 2.12B) and pressurizes it with carrier gas until the pressure is equal to the carrier gas inlet pressure of the column. In the sampling stage, Figure 2.12C, the solenoid valve closes, interrupting the flow of carrier gas, and the pressurized headspace in the vial flows through the column for the specified injection time. This technique has been modified over the years[10] so that the GC injector can maintain its original source of carrier gas flow and the headspace system has an independent source of carrier gas. It is available with a trapping option (Fig. 2.13).

All three types of dedicated headspace apparatus described here should perform adequately for most applications. Table 2.4 summarizes some of the advantages and disadvantages of these systems.

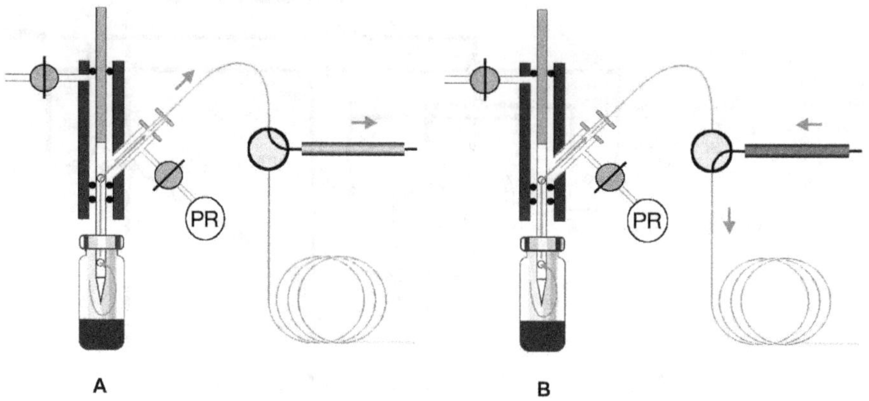

Figure 2.13. Schematic of the PerkinElmer pressure-balanced headspace system with the optional trap. Drawing printed with permission of PerkinElmer, Inc. The sample is pressurized, and the excess pressure in the vial passes through the cooled trap where the analytes in the headspace sample are absorbed. This cycle can be repeated up to a total of four times. A purge of dry carrier gas follows to remove moisture from the trap. The trap is rapidly heated, and the analyte are released from the trap and backflushed onto the GC column.

Table 2.4. Summary of Advantage and Potential Disadvantages with Dedicated Automated Headspace Systems

Automated Headspace Technique	Advantages	Disadvantages
Syringe based	Inert sample path	Gastight syringes have Teflon-tipped plungers, which can leak (once they are used at a given temperature, the syringe should not be used at a lower temperature)
	GC injector accessible for troubleshooting in headspace and for liquid injection	Occasional needle bending
	Simple to vary injector volume	
Valve and loop	GC injector usually accessible for troubleshooting in headspace and for liquid injection	Activity in the sample path was a problem but has been reduced with the introduction of inert materials
	Usually easy to move from one GC to another	Ability to change sample loop (and sample volume) is limited in some systems
Pressure balanced	Inert sample path	GC injector usually not accessible
	Simple system with minimum of moving parts	Troubleshooting difficult
		Works best with a column that requires a high back pressure

All of the systems can be equipped with optional trapping of the headspace to increase headspace sensitivity.

2.6. HEADSPACE APPLICATIONS

There are literally hundreds of headspace GC applications in the literature. This section will briefly cover the most common applications and the scope of applications where headspace is used. Some recent interesting applications will also be described.

2.6.1. Volatiles in Biological Fluids

Determination of ethanol in the blood of suspected drunk drivers is one of the most important applications of headspace GC. An internal standard is almost always used—most frequently *n*-propanol—but other compounds including *tert*-butanol, acetonitrile, methyl ethyl ketone, and isoamyl alcohol have also been reported.[13] There is a significant matrix effect when comparing the response of ethanol in blood to the response in water or urine. To minimize differences from sample to sample and from samples to standards, Penton[2] diluted the samples and standards fourfold with a solution that was saturated with sodium chloride and also contained *n*-propanol internal standard.

In addition to ethanol, there are headspace methods for numerous other solvents in biological fluids. Sharp[14] developed a method for screening of more than 40 organic compounds in biological matrices, using a two-column system with flame ionization detector (FID) detection and mass spectrometry (MS) confirmation. Matrix effects were eliminated with a 15-fold dilution. Frequent findings in samples were ethanol, toluene, and ethyl ether. Russo and Campanella[15] used headspace to determine benzene at a level of 10 ng/g in small tissue samples (50–70 mg).

2.6.2. Pharmaceutical and Other Industrial Applications

2.6.2.1. Residual solvents in pharmaceuticals. Another major application for headspace is the determination of residual solvents in pharmaceuticals. Methodology has been published by the International Conference of Harmonization (ICH) of Technical Requirements for Registration of Pharmaceuticals for Human Use and the United States Pharmacopeia (USP).

A new version* of the USP Method 467 for determining residual solvents in pharmaceuticals[16] was implemented on July 1, 2008. Solvents are divided into three classes depending on their toxicity: Class 1 (solvents to be avoided), Class 2 (solvents to be limited), and Class 3 (solvents with low toxic potential). Method 467 describes detailed procedures for determining Classes 1 and 2 solvents by static headspace. The maximum levels of the solvents are shown in Table 2.5. There are two similar procedures described: one for water-soluble and one for water-insoluble pharmaceuticals. The procedures are summarized as follows: the pharmaceutical compound is dissolved in water or *N,N*-dimethylformamide,† depending on whether it is water soluble or not. Next, corresponding standards of the Class 1 and Class 2 compounds are prepared by diluting purchased USP standards‡ in the same solvent or diluent as the pharmaceutical compound. If a headspace analysis of the pharmaceutical compound yields a peak with a retention time corresponding to one of the solvents in the standards, and if that peak has a larger area count than the standard, the entire test must be repeated with a second GC column of greater polarity. If the second test confirms the presence of the residual solvent at a higher level than was found in the standard, then a third headspace test must be performed for quantitation of the residual solvent.

The use of high-boiling diluents such as dimethylformamide and dimethyl sulfoxide to determine residual solvents in water-insoluble pharmaceuticals cannot be avoided, but detection limits for most residual solvents are considerably higher than when water is used. Hong and Altorfer[17] compared water as a dissolution medium with *N,N*-dimethylformamide, *N,N*-dimethylacetamide, and 1,3-dimethyl-2-imidazolidinone. They compared the headspace responses and precision of 14 compounds in the four dissolution media and concluded that while most of the compounds exhibited the best headspace response in water, 1,3-dimethyl-2-imidazolidinone was a better dissolution medium for hydrophobic drugs than dimethylformamide or dimethylacetamide. Later, the same authors[18] developed a micro-method for this analysis. The sample size was reduced from 100–200 mg to 5–30 mg. The equilibration time was reduced from 45–60 min to 5–10 min.

*This new USP method is more in accord with the ICH method than the previous one.

†Dimethyl sulfoxide may be used instead of *N,N*-dimethylformamide for water-insoluble compounds.

‡Standards may be purchased directly from the USP (http://www.usp.org/hottopics/residualSolvents.html) or from chromatography supply companies.

Table 2.5. Limits on Residual Solvents in Pharmaceuticals (as of 2008)

	Solvent	Limit (ppm)
Class 1	Benzene	2
	Carbon tetrachloride	4
	1,2-Dichloroethane	5
	1,1-Dichloroethene	8
	1,1,1-Trichloroethane	1500
Class 2	Acetonitrile	410
	Chlorobenzene	360
	Chloroform	60
	Cyclohexane	3880
	1,2-Dichlorethene	1870
	1,2-Dimethoxyethane	100
	N,N-dimethylacetamide	1090
	N,N-dimethylformamide	880
	1,4-Dioxane	380
	2-Ethoxyethanol[a]	160
	Ethylene glycol[a]	620
	Formamide[a]	220
	Hexane	290
	Methanol	3000
	2-Methoxyethanol[a]	50
	Methybutylketone	50
	Methylcyclohexane	1180
	Methylene chloride	600
	N-methylpyrrolidone[a]	530
	Nitromethane	50
	Pyridine	200
	Sulfolane[a]	160
	Tetrahydrofuran	720
	Tetralin	100
	Toluene	890
	Trichloroethylene	80
	Xylene	2170

Source: United States Pharmacopeia Method 467.

[a] These compounds cannot be determined with static headspace and should be determined with other appropriate validated methods.

2.6.2.2. Food and packaging. The food and packaging industries use headspace GC to characterize flavors,[19–21] to look for decomposition products,[22] and to detect compounds that contribute to an "off" taste from packaging materials.[23,24] Kaipainen et al.[25] looked at the decomposition of sugars with headspace and compared GC-MS detection with an electronic nose. Schulz et al.[26] determined volatile constituents in spirits, using headspace trapping and found extraction yields 35–55 times higher than those seen with conventional static headspace.

Pattern recognition techniques are important in the food industry; with this technique, individual compounds are not identified; the analyst looks for changes in a headspace chromatogram as a quality assurance technique.

2.6.2.3. Volatiles in high-boiling oils. Headspace GC is used for determining not only volatile organic solvents but also permanent gases. Jalbert et al.[27] measured dissolved gases (H_2, N_2, O_2, CO, CO_2, methane, ethane, ethylene, acetylene, propane) and traces of water in mineral insulating oils. The authors separated the analytes with a two-channel system. The first channel consisted of two columns for separating the permanent gases; the second channel consisted of a column with an ethylene glycol phase to separate the water from the other analytes. A helium photoionization pulsed discharge detector was used in this work.

Levermore et al.[28] published an interesting study in which they identified certain "signature" compounds in engine oil that indicated when it was time to change the oil. Some of these compounds were identified by GC-MS as acetaldehyde, acetone, butanal, benzaldehyde, and acetic and benzoic acids.

2.6.3. Environmental Methods

Volatile organic compounds in drinking water and wastewater are determined with static headspace or purge and trap, depending on the country. In Europe, static headspace methods tend to dominate, while most of the methods in the United States use purge and trap. USEPA Method 5021[7] is a static headspace method for determining volatile organic compounds in soils and other solid matrices. This method covers a wide range of aromatics and halogenated hydrocarbons. A matrix-modifying solution is used to minimize the differences between samples and to decrease the partition coefficients of the analytes. This consists of water, saturated with sodium chloride and adjusted to a pH of 2.0 with phosphoric acid. The matrix modifier is spiked with volatiles to prepare calibration standards. Unspiked matrix modifier is mixed with the soil samples as a diluent to minimize matrix differences (10-mL matrix modifier to 2 g of soil).

Barani et al.[29] used static headspace with trapping to determine halogenated and aromatic volatile organic compounds in groundwater, mineral water, and drinking water. The concentration range was 1–10,000 ng/L. All of the required detection limits specified in the Italian laws were met, except for 1,2,3-trichloropropane, which had an LOD of 8 ng/L, rather than the required 1 ng/L. Linearity and precision were good for all compounds tested.

One would not normally think of headspace as a technique for determining pesticides, but Royer et al.[30] proposed an automated headspace method for determining dithiocarbamates in plant matrices. The method was suggested as a replacement for the spectrophotometric method, European Norm EN 12396-1 (1996), and the manual headspace method, EN 12396-2 (1999). The new method offered several advantages including greater sample throughput and the need for fewer reagents. The authors mixed 2 g of plant matrix with 6 mL of a stannous chloride/HCl solution and 4 mL of a diethanolamine solution in a headspace vial; the vial was then heated to 90°C. The dithiocarbamates released carbon disulfide, which was determined in the headspace. Recoveries from potato, onion, and currants varied from 85% to 103%. The detection limit was 0.05 mg/kg.

2.7. HEADSPACE GC AS A TOOL FOR DETERMINING THERMODYNAMIC CONSTANTS

Headspace GC is more than an analytical tool; it has been used for determining thermodynamic constants including partition coefficients,[31–33] activity coefficients,[4–6] and Henry's constant.[34,35] An example follows: Kolb et al.[32] determined the partition coefficients of 12 compounds in water at temperatures ranging from 40 to 80°C with the "vapor phase calibration" technique. For calibration, a known quantity of analyte was injected into an empty vial and allowed to completely vaporize. An aliquot of the vapor phase was injected into the GC, and the area count of the resulting peak was measured. Then, a known quantity of the analyte was injected into a vial containing the matrix, and the vial was kept at a given temperature until equilibration was reached between the matrix and the headspace. After equilibration occurred, a portion of the headspace, equal to the volume injected in the calibration step, was injected into the GC. The area count of the resulting peak was compared with the area count that was obtained in the calibration step, and the partition coefficient was calculated. Some of the data from this work are shown in Table 2.2. There was excellent agreement with partition coefficients derived from other sources.

2.8. CONCLUSION

In this era of rapid technological changes, headspace GC has survived for several decades as a useful analytical tool for determining volatiles in liquids and solids. The potential user can determine if the technique is feasible for a given analytical problem by examining a vast body of literature and by a few simple experiments using GC, some vials, and a gastight syringe.

ACKNOWLEDGMENTS

Tom Hartlein and Janet Gleason of Teledyne Tekmar Co. and Allesandro Baldi and Giuilia Orsanigo of PerkinElmer were very helpful in providing information for this chapter. This article was originally published in *Sampling and Sample Preparation for Field and Laboratory*, Vol. 37, Penton, Z.E., Headspace gas chromatography, 279–296, Copyright Elsevier (2002) and is reprinted with permission from Elsevier. The chapter has been updated since the original publication.

REFERENCES

1. Ioffe, B.V.; Vitenberg, A.G. *Headspace Analysis and Related Methods in Gas Chromatography*. New York: Wiley-Interscience; 1984.

2. Penton, Z. Headspace measurement of ethanol in blood with a modified autosampler. *Clin. Chem.* **1985**, *31*, 439–441.

3. Penton, Z. Optimization of conditions in static headspace GC. *J. High Resolut. Chromatogr.* **1992**, *15*, 834–836.

4. Asprion, N.; Hasse, H; Maurer, G. Limiting activity coefficients in alcohol-containing organic solutions from headspace gas chromatography. *J. Chem. Eng. Data* **1998**, *43*, 74–80.

5. Castells, C.B.; Eikens, D.I.; Carr, P.W. Headspace gas chromatographic measurements of limiting activity coefficients of eleven alkanes in organic solvents at 25°C 1. *J. Chem. Eng. Data* **2000**, *45*, 369–375.

6. Castells, C.B.; Eikens, D.I.; Carr, P.W. Headspace gas chromatographic measurements of limiting activity coefficients of eleven alkanes in organic solvents at 25°C 2. Accuracy and precision. *J. Chem. Eng. Data* **2000**, *45*, 376–381.

7. Test Methods for Evaluating Solid Waste Physical/Chemical Methods, SW-846, Vol. IB, Chap. 4, Sec 4.2.1. Cincinnati, OH: U.S. Environmental Protection Agency; **1996**.

8. Suzuki, M.; Tsuge, S.; Takeuchi, T. Gas chromatographic estimation of occluded solvents in adhesive tape by periodic introduction method. *Anal. Chem.* **1970**, *42*, 1705–1708.

9. Kolb, B.; Ettre, L. Theory and practice of multiple headspace extraction. *Chromatographia* **1991**, *32*, 505–513.

10. Kolb, B.; Ettre, L. *Static Headspace Gas Chromatography: Theory and Practice*. New York: Wiley-VCH; 1997.

11. Markelov, M.; Guzowski, J. Matrix independent headspace gas chromatographic analysis—The full evaporation technique. *Anal. Chim. Acta* **1993**, *276*, 235–245.

12. Penton, Z. Applications of a GC autosampler modified for headspace sampling. *J. High Resolut. Chromatogr.* **1994**, *17*, 647–650.

13. Tagliaro, F.; Lubli, G.; Ghielmi, S.; Franchi, D.; Marigo, M. Chromatographic methods for blood alcohol determination. *J. Chromatogr. B* **1992**, *580*, 161–190.

14. Sharp, M.E. Technical note: A comprehensive screen for volatile organic compounds in biological fluids. *J. Anal. Toxicol.* **2001**, *25*, 631–636.

15. Russo, M.V.; Campanella, L. Static headspace analysis by GC-MS (in SIM mode) to determine the benzene in human tissues. *Anal. Lett.* **2001**, *34*, 883–891.

16. <467> Residual Solvents. *USP 32/NF 27, United States Pharmacopeial Convention*. Rockville, MD; **2009**; pp. 163–175.

17. Hong, L.; Altorfer, H.R. A comparison study of sample dissolution media in headspace analysis of organic volatile impurities in pharmaceuticals. *Pharm. Acta Helv.* **1997**, *72*, 95–104.

18. Hong, L.; Altorfer, H. A micro-sized headspace GC technique for determination of organic volatile impurities in water-insoluble pharmaceuticals. *Chromatographia* **2001**, *53*, 76–80.

19. Alonso, L.; Fraga, M.J. Simple and rapid analysis for quantitation of the most important volatile flavor compounds in yogurt by headspace gas chromatography-mass spectrometry. *J. Chromatogr. Sci.* **2001**, *39*, 297–300.

20. Fernandez-Garcia, E. Use of headspace sampling in the quantitative analysis of artisanal Spanish cheese aroma. *J. Agric. Food Chem.* **1996**, *44*, 1833–1839.

21. Mathieu, F.; Malosse, C.; Cain, A.; Frerot, B. Comparative headspace analysis of fresh red coffee berries from different cultivated varieties of coffee trees. *J. High Resolut. Chromatogr.* **1996**, *19*, 298–300.

22. Medina, I.; Satue-Gracia, M.T.; Frankel, E.N. Static headspace gas chromatographic analyses to determine oxidation of fish muscle lipids during thermal processing. *J. Am. Oil Chem. Soc.* **1999**, *76*, 231–236.

23. Du, L.; Xu, Q.; Lin, A. Determination of organic residues in plastic packing bags by gas chromatography with large-bore capillary columns. *Fenxi Ceshi Xuebao* **1999**, *18*, 59–60, 61.

24. Forsgren, G.; Frisell, H.; Ericsson, B. Taint- and odor-related quality monitoring of two food packaging board products using gas chromatography, gas sensors and sensory analysis. Nord. *Pulp Pap. Res. J.* **1999**, *14*, 5–16.

25. Kaipainen, A.; Ylisuutari, S.; Lucas, Q.; Moy, L. A new approach to odor detection. Comparison of thermal desorption GC-MS and electronic nose. Two techniques for the analysis of headspace aroma profiles of sugar. *Int. Sugar J.* **1997**, *99*, 403–408.

26. Schulz, K.; Dressler, J.; Sohnius E.; Lachenmeier, D. Determination of volatile constituents in spirits using headspace trap technology. *J. Chromatogr. A* **2007**, *1145*, 204–209.

27. Jalbert, J.; Gilbert, R.; Tétreault, P. Simultaneous determination of dissolved gases and moisture in mineral insulating oils by static headspace gas chromatography with helium ionization pulsed discharge detection. *Anal. Chem.* **2001**, *73*, 3382–3391.

28. Levermore, D.M.; Josowicz, M.; Rees, W.S.; Janata, J. Headspace analysis of engine oil by gas chromatography/mass spectroscopy. *Anal. Chem.* **2001**, *73*, 1361–1365.

29. Barani, F.; Dell'Amico, N.; Griffone, L.; Santoro, M.; Tarabella, C. Determination of volatile organic compounds by headspace trap. *J. Chromatogr. Sci.* **2006**, *44*, 625–630.

30. Royer, A.; Ménand, M.; Grimault, A.; Communal, P.Y. Development of automated headspace gas chromatography determination of dithiocarbamates in plant matrixes. *J. Agric. Food Chem.* **2001**, *49*, 2152–2158.

31. Cummins, T.M.; Robbins, G.A.; Henebry, B.J.; Goad, R.C.; Gilbert, E.J.; Miller, M.E.; Stuart, J.D. A water extraction, static headspace sampling, gas chromatographic method to determine MTBE in heating oil and diesel fuel. *Environ. Sci. Technol.* **2001**, *35*, 1202–1208.

32. Kolb, B.; Welter, C.; Bichler, C. Determination of partition coefficients by automatic equilibrium headspace gas chromatography by vapor phase calibration. *Chromatographia* **1992**, *34*, 235–240.

33. Chaintreau, A.; Grade, A.; Munoz-Box, R. Determination of partition coefficients and quantitation of headspace volatile components. *Anal. Chem.* **1995**, *67*, 3300–3304.

34. Chai, X.S.; Zhu, J.Y. Simultaneous measurements of solute concentration and Henry's constant using multiple headspace extraction gas chromatography. *Anal. Chem.* **1998**, *70*, 3481–3487.

35. Peng, J.; Wan, A. Measurement of Henry's constants of high-volatility organic compounds using a headspace autosampler. *Environ. Sci. Technol.* **1997**, *31*, 2998–3003.

Chapter 3

Liquid–Liquid Extraction in Environmental Analysis

Toh Ming Hii and Hian Kee Lee

3.1. INTRODUCTION

Liquid–liquid extraction (LLE), also known as solvent extraction, is the most common technique used for the extraction of organic analytes from aqueous samples. Historically, LLE was the first sample preparation technique used in analytical chemistry,[1] being generally utilized to extract and therefore isolate an analyte from a complex mixture to simplify its analysis. Its use is not confined to the analytical field, however, since synthetic chemists usually use the procedure to isolate newly synthesized substances from aqueous solutions.[2] As perhaps the earliest separation, or fractionation, method for organic compounds, LLE has become a powerful sample preparation technique over the last 50 years[3] and continues to be widely used today, despite its limitations (see Section 3.10).

To date, LLE has been developed and applied very broadly in diverse fields, such as for studying equilibria of inorganic and organic complexes; for separations in analytical chemistry; for large-scale inorganic, organic, pharmaceutical, and biochemical industrial separation processes; and for industrial waste treatment.[4] This chapter focuses mainly on the application of LLE to sample preparation involving organic analytes. The theory, extraction modes, applications to environmental samples, and future trends of the procedure are discussed.

3.2. THEORY OF LLE

The theory of LLE is based on the fact that if an analyte is to be extracted from an aqueous sample into a water-immiscible organic solvent, and the two phases are mixed together, the analyte is distributed between the two. Two terms are used to describe the distribution of an analyte between two immiscible solvents, that is, the distribution ratio (D) and distribution constant (K_D).[5,6]

According to the definition by the International Union of Pure and Applied Chemistry, the distribution ratio is the ratio of the total analytical concentration of a solute (analyte) in the organic phase (regardless of its chemical form) to its total analytical concentration in the aqueous phase, usually measured at equilibrium (Eq. 3.1). The distribution constant is the ratio of the concentration of a substance in a single definite form A, in the organic phase (C_o) to its concentration in the same form in the aqueous phase (C_{aq}) at equilibrium (Eq. 3.2).[7,8] The distribution ratio is identical to the distribution constant when there is no chemical dissociation involved:[5,6]

$$D = \frac{\text{Concentration of } A \text{ in all chemical forms in the organic phase}}{\text{Concentration of } A \text{ in all chemical forms in the aqueous phase}} \tag{3.1}$$

$$K_D = \frac{C_o}{C_{aq}}. \tag{3.2}$$

Figure 3.1 illustrates the principles of LLE. A separatory funnel (usually pear shaped) is often used for LLE; it contains two layers of liquids, one is generally water and the other is generally a water-immiscible organic solvent. In the case considered, the organic solvent is less dense than water; it separates from the water and rises to the top as a distinct upper layer. LLE can be performed by vigorously shaking the two immiscible liquids together in the funnel. The separatory funnel may need to be vented to release the organic solvent vapor. This shaking–venting operation is repeated several times before allowing the two phases to separate for a period of time. The analyte A, which is initially present in only one of the two liquids, will distribute between the two phases. Typically, a neutral and nonpolar solute (analyte) tends to be more soluble in an organic solvent than a polar solvent such as water.[9] When the distribution achieves equilibrium, the concentration of analyte in the organic and

Handbook of Sample Preparation, Edited by Janusz Pawliszyn and Heather L. Lord

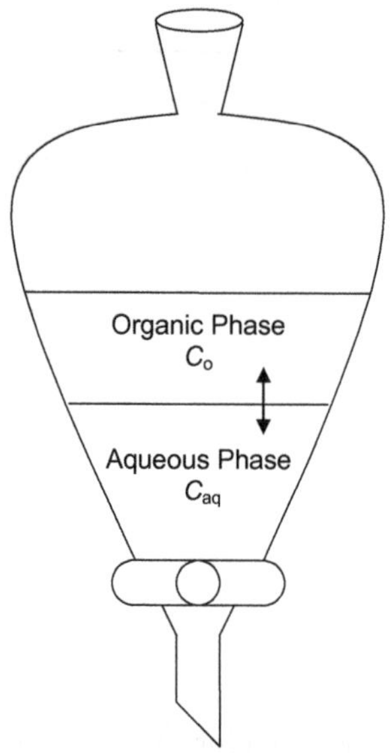

Figure 3.1. Discontinuous liquid–liquid extraction between two immiscible phases in a separatory funnel.

aqueous layers are C_o and C_{aq}, respectively, and the ratio of the concentration remains constant as predicted by the distribution constant (Eq. 3.2).

The distribution constant is useful because it allows us to calculate the concentration of an analyte remaining in a solution after a certain number of extractions, and also provides a guide as to the most efficient way to perform an extractive separation.[10] The efficiency of the extraction can be determined by the fraction of analyte extracted (E), which is often expressed in percentage terms (Eq. 3.3):[5,6,11]

$$E = \frac{C_o V_o}{C_o V_o + C_{aq} V_{aq}}, \tag{3.3}$$

where C_o and C_{aq} are the concentration of the analytes in the organic and aqueous phases, respectively, and V_o and V_{aq} are the volumes of the organic and aqueous phases, respectively.

By dividing the numerator and denominator by $C_o V_o$,

$$E = \frac{1}{1 + \left(\frac{C_{aq} V_{aq}}{C_o V_o} \right)}. \tag{3.4}$$

By substituting Equation 3.2,

$$E = \frac{1}{1 + \frac{1}{K_D V}} = \frac{1}{\left(\frac{K_D V + 1}{K_D V} \right)} = \frac{K_D V}{K_D V + 1}, \tag{3.5}$$

where V is the volume ratio of organic phase to aqueous phase, that is, V_o/V_{aq}.

From Equation 3.5, it is seen that a single-stage extraction will not provide high extraction efficiency. Only with a large K_D, that is, >10, and volume ratio within the practical range, that is, $0.1 < V < 10$, would extraction efficiency be >99%.[5,6]

Better efficiency can be achieved by conducting successive multiple extractions using a smaller volume of organic solvent for each stage.[11] This means that the total volume of the solvent should be equally divided into several smaller portions to perform LLE separately on the single sample solution, after which the extracts are combined. The amount of analytes extracted after successive multiple extractions (U) is given in Equation 3.6:[5,6,11]

$$U = 1 - \left[\frac{1}{K_D V + 1} \right]^n, \tag{3.6}$$

where n is the number of extractions. For instance, if $K_D = 2$ for an analyte and the volume of the two phases are equal ($V = 1$), then four extractions ($n = 4$) may be carried out to achieve >98% extraction efficiency. By using the entire volume of solvent for a single extraction, such a high extraction efficiency cannot be achieved.

3.3. EXTRACTION MODES

LLE can be carried out in four different ways, that is, discontinuous extraction, continuous extraction, countercurrent extraction, and online extraction.[12] Discontinuous extraction is the most traditional extraction method, which can be carried out in a single- or multiple-step extraction where equilibrium is established between two immiscible phases (Fig. 3.1). In discontinuous LLE, the equilibrium process is influenced by several factors, such as pH adjustment to avoid ionization of acids or bases, formation of ion pairs with ionizable analytes, formation of hydrophobic complexes with metal ions, and reduction of analyte solubility by adding a neutral salt.[5,6]

When selecting an extraction solvent, three factors need to be considered, that is, miscibility, density, and solubility of the solvent. The solvent miscibility chart[13] as depicted in Figure 3.2 may be used in choosing suitable immiscible solvent pairs for LLE. Despite that water and organic solvents may be considered immiscible, nevertheless, they have some solubility in each other and will become mutually saturated when mixed together, even though they form two visibly distinct phases. If the solubility of water in a solvent is greater than the solubility of that solvent in water, then a disposal problem may arise due to the aqueous phase being saturated with the organic solvent.[14]

Moreover, the selection of aqueous and organic pairs of solvents also depends on the nature of the target analytes. For instance, if the target analyte is to be analyzed by high-performance liquid chromatography (HPLC), then it is best

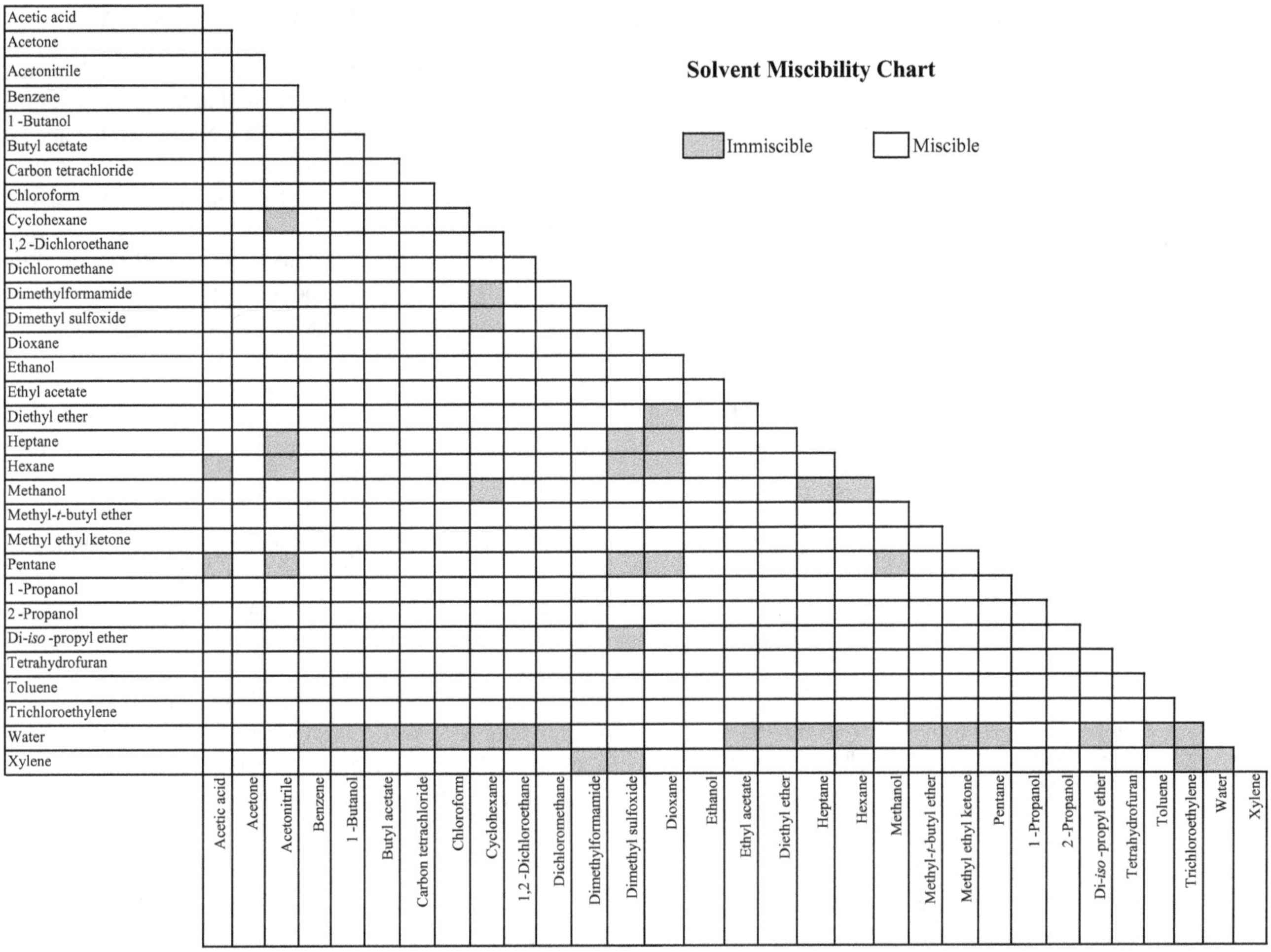

Figure 3.2. Solvent miscibility chart (modified from Phenomenex[13]).

isolated in the aqueous phase, whereas if the target analyte is to be analyzed by gas chromatography (GC), then the final extract should be the organic solvent.[5,6]

Continuous extraction is used when the distribution constant, K_D, is very small or when the sample volume is large.[12] Figure 3.3 shows the extractors used for continuous LLE. In Figure 3.3a, the extraction solvent, for example, dichloromethane, is denser than the liquid sample. Fresh organic solvent is boiled from the flask and condensed into the extraction vessel. Dense droplets of organic solvent are continuously falling through the liquid sample and extracting the analytes. The organic solvent is then pushed through the return tube and overflows into the extraction solvent reservoir. However, if the extraction solvent is less dense than the liquid sample, for example, hexane, which is more often the case (Fig. 3.3b), the organic solvent will fall through a funnel into the bottom of the extraction vessel before rising through the liquid sample and in doing so, extracting the analytes. Eventually, all analytes are slowly extracted from the liquid sample into the extraction solvent reservoir. After completion of the extraction and allowing sufficient time for the solvent to cool, solvent evaporation, for example, under reduced pressue in rotary evaporator, or solvent blowdown, is performed to concentrate the target analytes before chromatographic analysis.[9,11,15]

In continuous extraction, the solvent is recycled and its volume remains constant. The extraction process usually takes between 18 and 24 h but provides enrichment factors up to 10^5, which is an essential requirement for trace analysis.[5] Continuous LLE is usually restricted to semivolatile and nonvolatile analytes because the volatile analytes may easily be recycled back to extraction vessel with the organic solvent or be lost through the condenser, consequently leading to a decrease in the extraction efficiency. However, continuous LLE can be carried out unattended and is less labor intensive than discontinuous LLE.[11]

Countercurrent extraction is carried out when complex samples with analytes of similar distribution ratios are to be extracted.[12] A single extraction usually does not provide sufficient efficiency for complex samples. Some interfering solutes will also be extracted with the analytes, but some analytes will be left behind in the original solvent as well. So, multiple extractions are needed for both extracting solvent, that is, to remove interfering solutes that were extracted with the analytes and to completely extract analytes that were not extracted by the previous extraction.

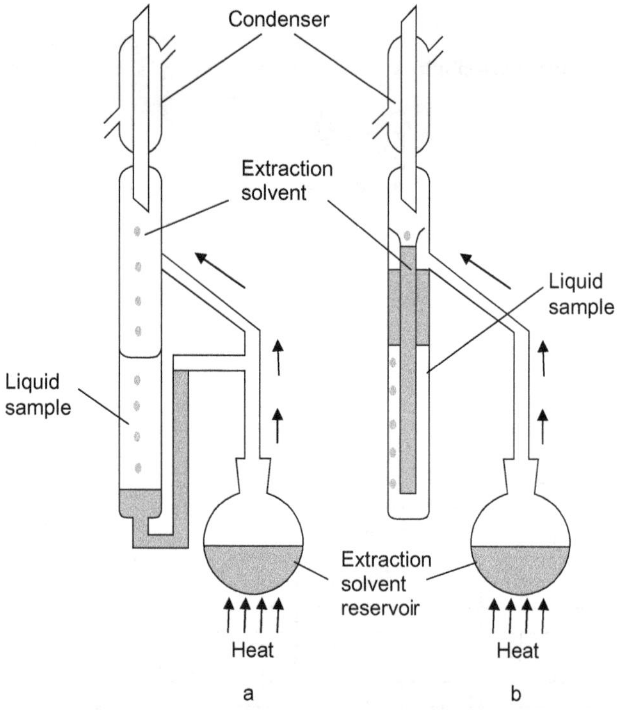

Figure 3.3. Continuous liquid–liquid extraction apparatus used when the extraction solvent is (a) denser than the liquid sample and (b) lighter than the liquid sample.

Countercurrent extraction is based on the exposure of separated aqueous samples with fresh, separate portions of organic solvent in an experimental arrangement known as the countercurrent distribution or the Craig apparatus.[16] The extracting solvent, after first being in contact with the original solution, is moved to second separatory funnel in which there is a fresh original solution, while the fresh extracting solvent is brought into the first funnel and the extraction is carried out. Then, the extracting solvent from the second funnel is moved to a third funnel containing a fresh original solution; the extracting solvent from the first is moved to the second funnel, and the fresh extracting solvent is introduced into the first funnel. The extraction process continues in this manner until the desired extraction efficiency is achieved.[17]

Online LLE is a dynamic process that allows the extraction of low-volume samples and reduces organic solvent consumption. However, it is not popular due to the complexity of the instrumentation.[12] A combination of online LLE with GC has been used for the determination of volatile organic trace compounds in aqueous sample,[18] halocarbons in seawater,[19] organochlorine pesticides (OCPs) in groundwater,[20] and pesticide intermediates in water.[21] Besides these, LLE-GC has also been used for the simultaneous extraction and derivatization of organic acids.[22] The different methods and benefits associated with the online combination of sample preparation with GC, including LLE-GC, have been discussed in the literature.[23–25]

3.4. LLE IN ENVIRONMENTAL ANALYSIS

LLE has been an important technique in trace analysis and remains so. When analytes are present at low concentrations in a water sample, LLE is used to preconcentrate the analyte from a large volume of water into a much smaller volume. Furthermore, LLE can be used to clean up the analyte from matrix interferences.[5]

As referred to earlier, the selectivity and efficiency of the LLE process is mainly dependent on the choice of the two immiscible solvents.[5,6] The extraction efficiency depends on the distribution constant, pH, ionic strength, the solvent-to-aqueous phase ratio, and the number of successive extractions.[26] There is no "universal solvent," and the choice of a suitable solvent selection must be made individually for a particular separation problem.[4]

LLE has been used mainly in the chromatographic analysis of environmental samples, such as pesticides, herbicides, polychlorinated biphenyls (PCBs), polycyclic aromatic hydrocarbons (PAHs), endocrine-disrupting chemicals (EDCs), *N*-nitrosamines, organic acids, phthalate esters, and surfactants. In this review, we focus particularly on the LLE of these compounds. Selected applications of environmental sample analysis in which LLE is used are summarized in Table 3.1, which also provides, at a glance, details of the extraction solvents and the chromatographic techniques used, as well as the analytical figures of merit achieved.

3.5. PESTICIDES

Pesticides are used extensively in agriculture for protecting crops during their growth and storage. For instance, in the United States alone, over 800 pesticide active ingredients are formulated in 21,000 different commercial products.[27] The use of these chemicals to counter human disease vectors such as mosquitoes is also substantial. As a consequence of all these uses, various environmental compartments, for example, water, soil, sediment, and biota, are frequently contaminated with pesticides. Pesticides also possess potential environmental hazards to human and animals through their existence and concentration in the food chain.[28] Studies have shown that less than 0.3% of the amount of pesticides applied actually reaches the target pests, and the rest has the potential to move into other environmental compartments, thus leading to the exposure of nontarget organisms.[29,30] Among the synthetic pesticides, carbamates, OCPs, and, to a lesser extent, organophosphorus pesticides (OPPs) exhibit significant persistence and stability in the environment.[31]

3.5.1. Carbamate Pesticides

Carbamate pesticides are relatively more labile compared with OCPs but are generally more persistent than OPPs.[32] Carbamates are usually of low solubility in water but are soluble in the more polar organic solvents, such as acetone, ethanol, and methanol. They are also moderately soluble in medium polarity solvents, such as benzene, chloroform, dichloromethane, 1,2-dichloroethane, toluene, and xylene; they are generally poorly soluble in nonpolar solvents, such as hexane and petroleum ether.[33,34]

The method used for the extraction of carbamate pesticides from environmental samples depends on their polarity and the sample matrix involved. LLE is commonly used in the United States Environmental Protection Agency (USEPA) methods (e.g., Methods 632 and 8318) for the preconcentration of pesticides in liquid samples. Usually, mixed solvents are used for fine adjustment of the solvent strength for the extraction of various carbamate pesticides from aqueous environmental samples.[32]

3.5.2. OCPs

The widespread use of OCPs in developed as well as developing countries was based on the fact that these are extremely stable compounds with low solubility in water and high solubility in organic solvent. They are highly toxic to insects and, at that time, were considered to have low toxicity to humans. However, in the wake of the concern over their persistence and endocrine-disrupting properties, these substances were subsequently banned or had their production and use restricted.[35] For example, dichlorodiphenyltrichloroethane (DDT) was banned in the United States in 1972, although it is still used in many developing countries. Exposure to DDT can affect the central nervous system and can cause cardiac and respiratory failure.[36] DDT exposure also caused the thinning of birds' eggshells (by interfering with calcium metabolism) that led to the drastic decline in bird population in the late 1950s.[37]

The amenability of OCPs to organic solvents permits the favorable use of LLE for trace analysis of these compounds in environmental samples. For example, USEPA Method 508 describes the use of methyl chloride for the extraction of OCPs in water.[38] In a report, a microscale version of LLE, termed micro-LLE by its authors, was developed to enable the extracts be analyzed directly by GC-electron capture detector (ECD) without additional cleanup steps.[39] On the other hand, a normal-scale, continuous-flow liquid–liquid extractor called Goulden large-scale extractor (GLSE) was used for large-volume sample preconcentration (up to 112 L); this compares to the conventional-scale 1-L extractor.[40–42]

3.5.3. OPPs

During the past 30 years, there has been a trend toward the replacement of the persistent and bioaccumulative OCPs, with less toxic and shorter-lifetime compounds for agricultural activities. OPPs have played a significant role here. In the United States alone, there are about 200 different OPPs available on the market, which comprise 45% of registered pesticides. The popularity of OPPs has been due to their higher degradability in the environment. Nevertheless, after having been used extensively for more than 40 years, OPPs are now widely found in groundwater, and surface and drinking water. Concern about this contamination continues to guide the monitoring of the presence of these compounds in these matrices.[43,44]

Most OPPs are slightly water soluble, but they are freely soluble in alcohol, esters, ethers, ketones, and aromatic hydrocarbons.[45] In general, pesticides are usually present in the environment at very low concentration levels (at the parts per billion level). Thus, it is important that at least a 1000-fold enrichment (preconcentration) is required to increase the sensitivity of chromatographic determinations.[43]

The most common organic solvents used for the extraction of OPPs are dichloromethane, ethyl acetate, and a mixture of dichloromethane and hexane. Ethyl acetate is the favored solvent for extraction of the more polar OPPs,[46] whereas for the least polar OPPs, dichloromethane–hexane is a better choice.[43]

3.5.4. Herbicides

Herbicides also form a large group of chemical pesticides extensively used in agriculture to control weeds in cereal, vegetable, and fruit tree crops.[47] For instance, a 2004 article estimated that the worldwide consumption of pesticides was 2 million metric tons,[48] with more than half of the total consumption being herbicides.[49] This amount has been increasing steadily during the last two decades. The most widely used herbicides include triazines, chloroacetanilides, phenoxyalkanoic acids, and phenylurea herbicides (PUHs).[50] Due to their polar nature and water solubility, herbicides and their degradation products (which sometimes may be more toxic and persistent than the parent compounds[51]) are widely dispersed in the environment via leaching and runoff processes.[52,53] Thus, they contaminate surface water, groundwater, and drinking water.

LLE with traditional solvents, such as chloroform, dichloromethane, ether diethyl oxide, hexane, and octanol, is still used for the extraction of herbicides from water samples.[50] Among the solvents used, dichloromethane is the most favorable, due to the herbicides having a broad range of polarity.[54] LLE, however, cannot be used to extract the polar degradation products of herbicides.[50]

3.6. PCBs

PCBs are another class of widespread environmental pollutants. They are mixtures of up to 209 chemical compounds, known individually as congeners, which were previously used widely as dielectric fluids in transformers and capacitors; as heat-transfer and hydraulic fluids; as plasticizers in paints, adhesives, sealants, and plastics; in the formulation of lubricating and cutting oils; and in many other industrial applications.[55–58] The widespread industrial use of PCBs and improper disposal practices over the decades up to the time they were banned have led to their ubiquity in the environment. As a consequence, PCBs have been found all over the world, including in the air, water, sediments, fish, wildlife, and even human adipose tissue, milk, and serum.[58,59]

The USEPA banned most of the use of PCBs in 1997 because of their potential health effects; for example, these compounds may act as cancer initiators and cause reproductive failure in animals.[56] Most countries of the world have followed suit. However, these previously released persistent pollutants are still being (re)cycled in the environment from soil to air and back to soil again. Furthermore, PCBs can also be produced unintentionally as by-products in some chemical processes, water chlorination, and thermal degradation of chlorinated organic compounds.[57]

PCBs are hydrophobic and lipophilic compounds that can generally be extracted from samples using nonpolar solvents. The most widely used solvents in water extraction include dichloromethane, isooctane, hexane, pentane, or a mixture of these solvents. It should be noted that the extraction solvent can be a source of contamination in LLE. To avoid this, all solvents used in the pesticides extraction must be at least of pesticide grade.[56] Additionally, blank analyses must be conducted regularly.[56]

3.7. PAHs

PAHs are a group of fused aromatic ring hydrocarbons that are derived from endogenous and anthropogenic sources. They usually arise from incomplete combustion or high-temperature pyrolytic processes (e.g., coke production) involving organic materials. It has been estimated that 230,000 t of PAHs enter the global environment annually from petroleum spills and seeps, direct discharges from industrial or domestic sources, aerial transport, and biosynthesis.[60,61] Due to their persistence, PAHs usually accumulate in soil, sediments, surface water, and atmosphere, as well as organisms. They have been included in the European Union (EU) and USEPA lists of priority pollutants[61,62] due to their carcinogenic, mutagenic, and, it is also suspected, endocrine-disrupting properties.[63–65]

Two USEPA methods are available for the LLE of PAHs from water and wastewater samples, that is, Methods 610[66] and 3510C.[67] The extraction solvent specified is dichloromethane for GC analysis. This solvent is exchanged with acetonitrile for HPLC analysis.[61]

3.8. EDCs

Recently, there has been increasing concern and alarm about a significant number of natural and synthetic chemicals that are present in the environment, interfering with the functioning of endocrine systems, thus affecting reproduction and development in wildlife and human beings.[68] These chemicals are described as EDCs or, in some instances, hormonally active agents. The list of suspected or known EDCs include pesticides, organohalogens, alkyphenols, heavy metals, organotins, phthalates, natural hormones, pharmaceuticals, phytoestrogens, phenols, and aromatic hydrocarbons,[69] and therefore encompasses many of the pollutants and contaminants already encountered in the preceeding paragraphs. EDCs are released from various sources such as domestic sewage, agricultural activities, animal waste, industrial wastes, mining activity, and landfills. As a consequence, their presence is clearly evident in the environment, industrial products, household items, and, indeed, also the food we consume.[70,71]

LLE is frequently applied to isolate EDCs in aqueous samples, and the most commonly used organic solvents are dichloromethane and hexane.[68] However, extraction efficiencies for various EDCs can vary due to the formation of emulsions,[72,73] influence of sample pH,[74,75] and sample ionic strength.[76] For some EDCs like organotins, extraction and derivatization (since GC is the analytical method of choice) can be performed simultaneously.[77–79]

3.9. *N*-NITROSO COMPOUNDS, ORGANIC ACIDS, PHTHALATES, AND OTHER CONTAMINANTS

Among the *N*-nitroso compounds, *N*-nitrosamines have received significant attention in the etiology of human cancer due to their mutagenic, carcinogenic, and teratogenic properties.[80] Humans are exposed to *N*-nitroso compounds from different sources such as food, water, occupational environments, tobacco, and cosmetics.[81] One of the *N*-nitroso compounds, *N*-nitrosodimethylamine (NDMA) is very soluble and volatile; thus, it is very likely to be found in surface water, seawater, wastewater, and drinking water.[82] They are present in the last mentioned matrix because of the use of chloramination, which is increasingly being used to replace chlorination for disinfection purposes.

Dichloromethane is the most popular extraction solvent used in the LLE of *N*-nitrosamines. In the extraction process, sodium chloride is usually added to break up any emulsion. However, it is also recommended that continuous LLE (6 h or overnight) is used for the extraction of wastewater due to the formation of emulsions during shaking of the separatory funnel when used in the conventional way.[83–85]

Organic acids, being hydrocarbon based, have structures that vary from aliphatic to aromatic, saturated to unsaturated, and straight chain to branched. Short-chain organic acids play an important part in the metabolism of living organisms; therefore, they have most commonly been found in environmental matrices. Short-chain organic acids are released into the environment through photochemical and biochemical degradation of anthropogenic and natural organic material, anthropogenic emissions, and excretion by microorganisms, plants, and animals. This class of compounds is present in all environment compartments ranging from the low microgram per liter to a few hundred milligram per liter concentration levels.[86] The LLE of organic acids is challenging, and the recovery achieved by this extraction technique is usually low and irreproducible due to their hydrophilic properties.[87,88] However, the use of tri-*n*-octylphosphine oxide with methyl *tert*-butyl ether as extraction solvent reportedly showed good recovery when compared with that with diethyl ether.[89]

Phthalate is the general term for dialkyl or alkyl aryl esters of 1,2-benzenedicarboxylic acid. The main application of phthalate esters is as plasticizers for the production of polyvinylchloride in the plastic manufacturing industry.[90] The annual global production of phthalate esters in 1990s was about 4 million metric tons.[91] Phthalate esters can be released into the environment during their production and incorporation into plastic materials. Some phthalate esters have been shown to cause liver and kidney damage.[90] They also possess endocrine-disruption potential.[90]

The most commonly used procedures for the extraction of phthalate esters from water samples are LLE and solid-phase extraction (SPE). There are numerous official methods specified for phthalate esters that involve LLE, such as USEPA Methods 506, 525, 606, 625, 1625, 8061, 8270, and 8410. The most frequently used solvents are dichloromethane and hexane.[90]

Surfactants have both a hydrophilic and a hydrophobic group in their amphiphilic structures. Based on the hydrophilic part of the structure, surfactants can be divided into anionic, nonionic, and cationic types. They are found in laundry detergents, cleaning agents, and personal care products. Currently, linear alkylbenzene sulfonates are the most widely used surfactant class in detergents. The latter causes environmental pollution if they remain in the treated water before it is released from sewage treatment plants.[92]

Direct LLE of surfactants from aqueous samples normally represents a challenging task. This is due to the tendency of surfactants to concentrate at the phase boundaries, thus leading to the formation of emulsions. This problem can be solved by adding suitable counterions to form lipophilic ion pairs during the extraction[93,94] (Table 3.1).

Table 3.1. List of Studies on the Determination of Environmental Contaminants and Pollutants in Various Types of Samples in which Liquid–Liquid Extraction Is Involved

Analytes	Matrix	Extraction Solvent	Chromatographic Determination	Analytical Performance	Reference
Carbamate pesticides	Aqueous solution	Dichloromethane	LC-MS	Rec = 71% RSD = 11%	95
Carbamate pesticides	Soil	Ethyl acetate	GC-NPD	Rec = 89–109% RSD = 2–10% LOD = 0.001–0.02 μg/g	96
Carbamate pesticides	Aqueous solution	Dichloromethane	GC-NPD	Rec = 94.5% RSD = 3.13% LOD = 33 μg/L	97
Carbamate pesticides	River and well water	Methyl chloride	GC-ECD	Rec = 82–102% LOD = 2.5–10 ppb	98
Carbamate pesticides	Water	Hexane	GC-MS	Rec = 97% LOD = 0.11 μg/L	99
Carbamate pesticides	Air particulates	Isooctane	GC-MS	Rec = 70–110% RSD < 15% LOD = 0.05–3 μg/device	100
Carbamate pesticides	Soil	Acetonitrile–water	GC-MS	Rec = 81–131% LOD = 5 ppb	101
Carbamate pesticides	Water	Dichloromethane	GC-MS	Rec = 83–127% RSD = 2.6–22.6% LOD = 0.014–0.18 ng/mL	102
Carbamate pesticides	Water	Ethyl acetate	GC-FID	RSD = 2.15–3.92% LOD = 0.2–0.4 mg/L	103
OCPs	Water	Dichloromethane	GC-ECD	EPA Method 508 Rec = 47–148% RSD = 3–43% LOD = 0.0015–2.2 μg/L	38

Table 3.1. (*Continued*)

Analytes	Matrix	Extraction Solvent	Chromatographic Determination	Analytical Performance	Reference
Organohalide pesticides, PCBs	Water	Hexane	GC-MS	EPA Method 505 Rec = 37–191% RSD = 0.4–29.8% LOD = 0.002–15 µg/L	104
Pesticides	Natural and treated water	NA	NA	Review	105
OCPs	Drinking water	Toluene	GC-ECD and NPD	Rec > 50% RSD = 5.2–7.9%	39
OCPs, OPPs, PCBs	Raw and finished drinking water	Dichloromethane–hexane	GC-ECD	Rec = 85–116% RSD = 5–21% LOD = 1.6–104.1 ng/L	106
OPPs	Water	Dichloromethane	GC-NPD	EPA Method 507 Rec = 48–216% RSD = 3–30% LOD = 0.014–2.8 µg/L	107
OPPs	Water	Dichloromethane	GC-FPD and NPD	EPA Method 8141B Rec = 0–129% RSD = 0–37% LOD = 0.04–0.80 µg/L	108
Pesticides and degradation products	Shallow groundwater	Dichloromethane	LC-MS	LOD = 0.002–0.1 µg/L	109
Chloroacetanilide herbicides	Soil	Acetonitrile–water	LC-MS	Rec = 90–120% RSD < 15% LOD = 0.3–0.7 µg/kg	110
Phenoxyacid herbicides	Soil	Potassium hydroxide (0.5 M)	LC-MS/MS	Rec = 75–91% RSD = 4–10% LOD = 0.3 µg/kg	111
Sulfonylurea herbicides	Soil	Acetonitrile–water, trichloromethane	LC–MS/MS	LOD = 0.05 ppb RSD < 12%	112
PCBs	Seawater	Hexane	GC-MS and ECD	Rec > 75% RSD < 15%	113
PCBs	Lake water	Dichloromethane	GC-ECD	Rec = 75%	114
PCBs	Stormwater	Hexane–dichloromethane	GC	Rec = 62–70% LOD = 0.11–0.24 ng/L	115
PCBs	Seawater	Pentane	GC-ECD	Rec = 66.5–97.3% RSD = 7.2–29.9% LOD = 15 pg/L	116
PCBs	Fish tissue	Chloroform–methanol, dichloromethane–hexane	GC-ECD	NA	117
PAHs and PCBs	Bivalve tissue	Dichloromethane–hexane	GC-MS/MS	NA	118
PCBs	Human serum	Hexane	GC-ECD	NA	119
PCBs	Cormorant egg	Hexane–petroleum ether	GC-MS	NA	120
PCBs	Human serum	Diethyl ether–hexane	GC-MS	Rec = 113% RSD = 16% LOD – 0.01–0.08 µg/L	121
PCBs	Fish tissue	Hexane–MTBE	GC-ECD	Rec = 85% RSD = 4%	122
PAHs	Bulk precipitation and surface water	Hexane	HPLC-FLD	Rec = 80–120%	123

Table 3.1. *(Continued)*

Analytes	Matrix	Extraction Solvent	Chromatographic Determination	Analytical Performance	Reference
PAHs and PCBs	Seawater	Hexane	HPLC-FLD, GC-MS	NA	124
PAHs	Surface runoff	Dichloromethane	GC-MS	Rec = 11.02–123.99% RSD = 0.07–3%	125
Chlorophenoxy acid herbicides	Water	Hexane	GC-MS	RSD = 8–15% LOD = 10–60 ng/L	126
Urea pesticides	Water	NA	NA	Review	127
Alkylphenols	Wastewater	Dichloromethane	HPLC, GC-MS	LOD = 2–6 μg/L LOD = 0.001–0.02 μg/L	72
Alkylphenols	Wastewater	Hexane	GC-MS, LC-MS	Rec = 74–87% RSD < 6% LOD = 0.4–6 ng/L	73
Phenols	Seawater and springwater	Dichloromethane	GC-MS	RSD = 1.5–8.6% LOD = 0.6 ng/mL	74
Phenols	Water	Dichloromethane	GC-MS	Rec = 59.5–88.6% RSD = 1.6–10.3% LOD = 1 ng/mL	75
Phenols	Wastewater	Trichloromethane	GC-MS	RSD = 5.54–6.3% LOD = 0.006–0.13 μg/L	76
NDMA	Drinking water and fruit drink	Dichloromethane	GC-TEA, GC-MS	LOD = 15 pg/g LOD = 1 pg/g Rec = 74–105%	82
NDMA	Wastewater and natural water	Dichloromethane	GC-MS/MS	LOD = 0.09 μg/L	83
NDMA	Water	Dichloromethane	GC-MS	Rec = 85–115% LOD = 0.003 pg/μL	84
NDMA	Drinking and groundwater	Dichloromethane	HPLC-CLND	Rec = 71% LOD = 2 ng/L	85
Organic acids	Smog	TOPO in MTBE	GC-ITMS	NA	128
Organic acids	Drinking water	Diethyl ether	GC-MS and ECD	NA	129
Organic acids	Seawater	TOPO in MTBE	GC-ECD	LOD = 3 nM	89
Phthalate esters	Leachates	Ether	GC-MS	RSD < 20% LOD = 1 μg/L	130
Phthalate esters	River, well, and tap water	Hexane–ethyl acetate	GC-FID	Rec = 95–97% LOD = 0.03 μg/L	131
Phthalate esters	Surface water	Hexane	LC-UV	NA	132
Cationic surfactants	Sewage influent, sewage effluent, and river water	Trichloromethane	LC-MS	Rec = 95–97% RSD = 4–8%	93
Linear alkylbenzene sulfonates	Domestic waste	Trichloromethane	GC-FID	NA	94

CLND, chemiluminescent nitrogen detector; ECD, electron capture detector; FID, flame ionization detector; FLD, fluorescence detector; FPD, flame photometric detector; GC, gas chromatography; HPLC, high-performance liquid chromatography; ITMS, ion trap mass spectrometry; LC, liquid chromatography; LOD, limit of detection; MS, mass spectrometry; MS/MS, tandem MS; MTBE, methyl-t-butyl ether; NA, not applicable; NDMA, N-nitrosodimethylamine; NPD, nitrogen–phosphorus detector; OCPs, organochlorine pesticides; OPPs, organophosphorus pesticides; PCBs, polychlorinated biphenyls; Rec, recovery; RSD, relative standard deviation; TEA, thermal energy analyzer; TOPO, tri-n-octylphosphine oxide; UV, UV detector.

3.10. FUTURE TRENDS

The conventional LLE approach (Fig. 3.1) is still in common use due to the wide availability of pure solvents, simplicity of the extraction process and apparatus, and its extensive implementation in official methods, such as USEPA methodologies and EU standard methods for nonvolatile and semivolatile analytes in environmental samples.[5,6,12] Yet, it is fundamentally an environmentally unfriendly procedure, particularly with respect to the large volumes of solvents consumed and the subsequent generation of waste. Apart from this, there are several other disadvantages, that is, the extended time needed to perform the extraction, formation of emulsions resulting in analytes losses, extensive use of glassware for cleanup operations that can contaminate the sample, magnification of solvent impurities, need of sample preconcentration prior to analysis, evaporative losses of analytes, poor repeatability, and loss of sensitivity as a consequence of the injection of only a fraction of the extracted compounds.[35] Although continuous extraction is possible, it appears to be only sparingly used; thus, in general, LLE also suffers from poor sample throughput and is labor intensive.

Yet, although, miniaturization, simplification, and automation in sample preparation (of which extraction is a critical part) have been the major trends in modern analytical chemistry during the last few years,[12,133] LLE has remained a commonly used technique. Currently, there are extensive efforts to miniaturize sample preparation techniques to replace LLE; these popular alternatives in sample preparation include solid-phase microextraction (see Chapter 5) and stir-bar sorptive extraction, which are commercially available solventless approaches, and liquid-phase microextraction (see Chapter 7), in its various formats, which uses only microliter volumes of solvent. These are equilibrium-based, not exhaustive extraction methods, so absolute analyte recoveries are generally low. The advent of GC injection approaches has helped to overcome some of the limitations of this type of extraction. For example, large-volume injection GC is a powerful tool as it allows the introduction of up to few hundreds microliters of extract while maintaining good chromatographic characteristics.[133,134] SPE, introduced commercially in the mid-1970s, as perhaps the first of the miniaturized extraction approaches, will, in all likelihood, continue to be important for processing large aqueous volumes. It seems likely that these more environmentally benign microscale or miniaturized procedures will replace LLE one day, but until that day arrives, LLE will continue to be a mainstay procedure in many analytical laboratories. An indication of the enormous application database of LLE and its common usage is provided by the fact that modern advanced extraction techniques are still benchmarked against it, surely a tribute to its longevity and important contribution to the practice of analytical chemistry, particularly in the environmental analytical field.

REFERENCES

1. Loconto, P.R. *Trace Environmental Quantitative Analysis: Principles, Techniques and Applications*, 2nd ed. Boca Raton, FL: CRC Press; 2006.

2. Higson, S.P.J. *Analytical Chemistry*. New York: Oxford University Press; 2004.

3. Aguilar, M.; Cortina, J.L.; Maria, A. Analytical applications of solvent extraction. In *Solvent Extraction Principles and Practice*, Rydberg, J.; Cox, M.; Musikas, C.; Choppin, G.R., Eds., 2nd ed. New York: Marcel Dekker; 2004; pp. 549–594.

4. Cox, M.; Rydberg, J. Introduction to solvent extraction. In *Solvent Extraction Principles and Practice*, Rydberg, J.; Cox, M.; Musikas, C.; Choppin, G.R., Eds., 2nd ed. New York: Marcel Dekker; 2004; pp. 1–24.

5. Dean, J.R. *Methods for Environmental Trace Analysis*. England: Wiley; 2003.

6. Dean, J.R. *Extraction Methods for Environmental Analysis*. England: Wiley; 1998.

7. Rice, N.M.; Irving, H.M.N.H.; Leonard, M.A. Nomenclature for liquid-liquid distribution (solvent extraction). *Pure Appl. Chem.* **1993**, *65*, 2373–2396.

8. Berthod, A.; Carda-Broch, S. Determination of liquid-liquid partition coefficients by separation methods. *J. Chromatogr. A* **2004**, *1037*, 3–14.

9. Harris, D.C. *Quantitative Chemical Analysis*, 7th ed. New York: W.H. Freeman; 2007.

10. Skoog, D.A.; West, D.M.; Holler, F.J.; Crouch, S.R. *Fundamentals of Analytical Chemistry*, 8th ed. Belmont, CA: Brooks Cole; 2004.

11. Kebbekus, B.B.; Mitra, S. *Environmental Chemical Analysis*. London: Thomson Science; 1998.

12. Domini, C.E.; Hristozov, D.; Almagro, B.; Roman, I.P.; Prats, S.; Canals, A. Sample preparation for chromatographic analysis of environmental samples. In *Chromatographic Analysis of the Environment*, Nollet, L.M.L., Ed., 3rd ed. Boca Raton, FL: CRC Press; 2006; pp. 31–132.

13. Phenomenex Chromatography Catalog 2007/2008, Torrance, CA; 2007, p. 336.

14. Wells, M.J.M. Principles of extraction and the extraction of semivolatile organics from liquids. In *Sample Preparation Techniques in Analytical Chemistry*, Mitra, S., Ed. Hoboken, NJ: Wiley; 2003; pp. 37–138.

15. Beesley, T.E.; Buglio, B.; Scott, R.P.W. *Quantitative Chromatographic Analysis*. New York: Marcel Dekker; 2001.

16. Rossi, D.T.; Miller, K.G. Sample preparation methods for the analysis of pharmaceutical materials. In *Handbook of Isolation and Characterization of Impurities in Pharmaceuticals*, Ahuja, S.; Alsante, K.M., Eds. San Diego, CA: Academic Press; 2003; pp. 177–179.

17. Kenkel, J. *Analytical Chemistry for Technicians*, 3rd ed. Boca Raton, FL: CRC Press; 2003.

18. Roeraade, J. Automated monitoring of organic trace components in water. I. Continuous flow extraction together with on-line capillary gas chromatography. *J. Chromatogr.* **1985**, *330*, 263–274.

19. Fogelqvist, E.; Krysell, M.; Danielsson, L.-G. On-line liquid-liquid extraction in a segmented flow directly coupled to on-column injection into a gas chromatography. *Anal. Chem.* **1986**, *58*, 1516–1520.

20. Goosens, E.C.; Bunschoten, R.G.; Engelen, V.; de Jong, D.; van den Berg, J.H.M. Determination of hexachlorocyclohexanes in ground water by coupled liquid-liquid extraction and capillary gas chromatography. *J. High Resolut. Chromatogr.* **1990**, *13*, 438–442.

21. Goosens, E.C.; de Jong, D.; de Jong, G.J.; Brinkman, U.A.T. A continuous liquid-liquid extraction system coupled on-line with capillary gas chromatography for environmental and ecotoxicological analysis. *J. High Resolut. Chromatogr.* **1997**, *20*, 325–332.

22. Goosens, E.C.; Broekman, M.H.; Wolters, M.H.; Strijker, R.E.; de Jong, D.; de Jong, G.J.; Brinkman, U.A.T. A continuous two-phase reaction system coupled on-line with capillary chromatography for the determination of polar solutes in water. *J. High Resolut. Chromatogr.* **1992**, *15*, 242–248.

23. Hyotylainen, T.; Riekkola, M.L. On-line combinations of pressurized hot-water extraction and microporous membrane liquid-liquid extraction with chromatography. *LC-GC Eur.* **2002**, *15*, 298–302.

24. Vreuls, J.J.; Louter, A.J.H.; Brinkman, U.A.T. On-line combination of aqueous-sample preparation and capillary gas chromatography. *J. Chromatogr. A* **1999**, *856*, 279–314.

25. Barcelo, D.; Hennion, M.-C. On-line sample handling strategies for the trace-level determination of pesticides and their degradation products in environmental waters. *Anal. Chim. Acta* **1995**, *318*, 1–14.

26. Culea, M.; Gocan, S. Organophosphates. In *Handbook of Water Analysis*, Nollet, L.M.L., Ed. New York: Marcel Dekker; 2000; pp. 571–608.

27. Barr, D.B.; Needham, L.L. Analytical methods for biological monitoring of exposure to pesticides: A review. *J. Chromatogr. B* **2002**, *778*, 5–29.

28. Landis, W.G.; Yu, M.-H. *Introduction to Environmental Toxicology: Impacts of Chemicals upon Ecological System*, 3rd ed. Boca Raton, FL: CRC Press; 2004.

29. Pimentel, D. Amounts of pesticides reaching target pests: Environmental impacts and ethics. *J. Agric. Environ. Ethics* **1995**, *8*, 17–29.

30. Sabik, H.; Jeannot, R.; Rondeau, B. Multiresidue methods using solid-phase extraction techniques for monitoring priority pesticides, including triazines and degradation products, in ground and surface waters. *J. Chromatogr. A* **2000**, *885*, 217–236.

31. van der Hoff, G.R.; van Zooneen, P. Trace analysis of pesticides by gas chromatography. *J. Chromatogr. A* **1999**, *843*, 301–322.

32. Tribaldo, E.B. Chromatographic determination of carbamate pesticides in environmental samples. In *Chromatographic Analysis of the Environment*, Nollet, L.M.L., Ed., 3rd ed. Boca Raton, FL: CRC Press; 2006; pp. 889–934.

33. Tribaldo, E.B. Residue analysis of carbamate pesticides in water. In *Handbook of Water Analysis*, Nollet, L.M.L., Ed. New York: Marcel Dekker; 2000; pp. 537–570.

34. Ripley, B.D.; Chau, A.S.Y. Carbamate pesticides. In *Analysis of Pesticides in Water, Vol. 3, Nitrogen-Containing Pesticides*, Robinson, J.W.; Chau, A.S.Y.; Afghan, B.K., Eds. Boca Raton, FL: CRC Press; 1982; pp. 1–182.

35. Maione, M.; Mangani, F. Chromatographic analysis of insecticides chlorinated compounds in water and soil. In *Chromatographic Analysis of the Environment*, Nollet, L.M.L., Ed., 3rd ed. Boca Raton, FL: CRC Press; 2006; pp. 803–840.

36. Spear, R. Recognized and exposure to pesticides. In *Handbook of Pesticide Toxicology*, Hayes, W.J. Jr.; Laws, E.R. Jr., Eds. New York: Academic Press; 1991; pp. 20–27.

37. Nakamaru, M.; Iwasa, Y.; Nakanishi, J. Extinction risk to bird populations caused by DDT exposure. *Chemosphere* **2003**, *53*, 377–387.

38. US Environmental Protection Agency. Method 508: Determination of Chlorinated Pesticides in Water by Gas Chromatography with an Electron Capture Detector, **1995**.

39. Zapf, A.; Heyer, R.; Stan, H.J. Rapid micro liquid-liquid extraction method for trace analysis of organic contaminants in drinking water. *J. Chromatogr. A* **1995**, *694*, 453–461.

40. Foster, G.D.; Gates, P.M.; Foreman, W.T.; McKenzie, S.W.; Rinella, F.A. Determination of dissolved-phase pesticides in surface water from the Yakima River Basin, Washington, using the Goulden large-sample extractor and gas chromatography/mass spectrometry. *Environ. Sci. Technol.* **1993**, *27*, 1911–1917.

41. Foreman, W.T.; Gates, P.M. Matrix-enhanced degradation of p,p′-DDT during gas chromatographic analysis: A consideration. *Environ. Sci. Technol.* **1997**, *31*, 905–910.

42. Headley, J.V.; Dickson, L.C.; Swyngedouw, C.; Crosley, B.; Whitley, G. Evaluation of the Goulden large-sample extractor for acidic compounds in natural waters. *Environ. Toxicol. Chem.* **1996**, *15*, 1937–1944.

43. Jeannot, R.; Dagnac, T. Organophosphorus compounds in water, soils, waste and air. In *Chromatographic Analysis of the Environment*, Nollet, L.M.L., Ed., 3rd ed. Boca Raton, FL: CRC Press; 2006; pp. 841–888.

44. Hernandez, F.; Serrano, R.; Pitarch, E.; Lopez, F.J. Automated sample clean-up procedure for organophosphorus pesticides in several aquatic organisms using normal phase liquid chromatography. *Anal. Chim. Acta* **1998**, *374*, 215–229.

45. Barcelo, D.; Hennion, M.-C. *Determination of Pesticides and Their Degradation Products in Water*. Amsterdam: Elsevier; 1997.

46. Mol, H.G.J.; van Dam, R.C.J.; Steijger, O.M. Determination of polar organophosphorus pesticides in vegetables and fruits using liquid chromatography with tandem mass spectrometry: Selection of extraction solvent. *J. Chromatogr. A* **2003**, *1015*, 119–127.

47. Bogialli, S.; Curini, R.; Corcia, A.D.; Nazzari, M. Analysis of urea derivative herbicides in water and soil. In *Chromatographic Analysis of the Environment*, Nollet, L.M.L., Ed., 3rd ed. Boca Raton, FL: CRC Press; 2006; pp. 935–976.

48. Gupta, P.K. Pesticide exposure—Indian scene. *Toxicology* **2004**, *198*, 83–90.

49. Gupta, P.K. Toxicity of herbicides. In *Veterinary Toxicology: Basic and Clinical Principles*, Gupta, R.C., Ed. New York: Academic Press; 2007; pp. 567–586.

50. Dagnac, T.; Jeannot, R. Herbicide residues in the environment. In *Chromatographic Analysis of the Environment*, Nollet, L.M.L., Ed., 3rd ed. Boca Raton, FL: CRC Press; 2006; pp. 977–1026.

51. Crespin, M.A.; Gallego, M.; Valcarcel, M. Study of the degradation of the herbicides 2,4-D and MCPA at different depths in contaminated agricultural soil. *Environ. Sci. Technol.* **2001**, *35*, 4265–4270.

52. Kloppel, H.; Kordel, W.; Stein, B. Herbicide transport by surface runoff and herbicide retention in a filter strip—Rainfall and runoff simulation studies. *Chemosphere* **1997**, *35*, 129–141.

53. Felding, G.; Sorensen, J.B.; Mogensen, B.B.; Hansen, A.C. Phenoxyalkanoic acid herbicides in run-off. *Sci. Total Environ.* **1995**, *175*, 207–218.

54. Baltussen, E.; Snijders, H.; Janssen, H.-G.; Sandra, P.; Cramers, C.A. Determination of phenylurea herbicides in water samples using on-line sorptive preconcentration and high-performance liquid chromatography with UV or electrospray mass spectrometric detection. *J. Chromatogr. A* **1998**, *802*, 285–295.

55. Moret, I.; Piazza, R.; Benedetti, M.; Gambaro, A.; Barbante, C.; Cescon, P. Determination of polychlorobiphenyls in Venice Lagoon sediments. *Chemosphere* **2001**, *43*, 559–565.

56. Ceccarini, A.; Giannarelli, S. Polychlorobiphenyls. In *Chromatographic Analysis of the Environment*, Nollet, L.M.L., Ed., 3rd ed. Boca Raton, FL: CRC Press; 2006; pp. 667–710.

57. Erickson, M.D. *Analytical Chemistry of PCBs*, 2nd ed. Boca Raton: CRC Press; 1997.

58. Safe, S. Toxicology, structure-function relationship, and human and environmental health impacts of polychlorinated biphenyls: Progress and problems. *Environ. Health Perspect.* **1993**, *100*, 259–268.

59. Fuoco, R.; Ceccarini, A. Polychlorobiphenyls in Antarctic matrices. In *Environmental Contamination in Antarctica—A Challenge to Analytical Chemistry*, Caroli, S.; Cescon, P.; Walton, D.W.H., Eds. Oxford: Elsevier; 2001; pp. 237–274.

60. Williamson, K.S.; Petty, J.D.; Huckins, J.N.; Lebo, J.A.; Kaiser, E.M. HPLC-PFD determination of priority pollutant PAHs in water, sediment, and semipermeable membrane devices. *Chemosphere* **2002**, *49*, 703–715.

61. McGowin, A.E. Polycyclic aromatic hydrocarbons. In *Chromatographic Analysis of the Environment*, Nollet, L.M.L., Ed., 3rd ed. Boca Raton, FL: CRC Press; 2006; pp. 555–616.

62. Wise, S.A.; Sander, L.C.; May, W.E. Determination of polycyclic aromatic hydrocarbons by liquid chromatography. *J. Chromatogr.* **1993**, *642*, 329–349.

63. Brindle, I.D.; Li, X.-F. Investigation into the factors affecting performance in the determination of polycyclic aromatic hydrocarbons using capillary gas chromatography-mass spectrometry with splitless injection. *J. Chromatogr.* **1990**, *498*, 11–24.

64. Thompson, D.; Jolley, D.; Maher, W. Determination of polycyclic aromatic hydrocarbons in oyster tissues by high-performance liquid chromatography with ultraviolet and fluorescence detection. *Microchem. J.* **1993**, *47*, 351–362.

65. Tran, D.Q.; Ide, C.F.; McLachlan, J.A.; Arnold, S.F. The anti-estrogenic activity of selected polynuclear aromatic hydrocarbons in yeast expressing human estrogen receptor. *Biochem. Biophys. Res. Commun.* **1996**, *229*, 102–108.

66. US Environmental Protection Agency. Method 610: Polynuclear Aromatic Hydrocarbons; 1994.

67. US Environmental Protection Agency. Method 3510C: Separatory Funnel Liquid-Liquid Extraction; 1996.

68. Ying, G.-G. Chromatographic analysis of endocrine disrupting chemicals in the environmental. In *Chromatographic Analysis of the Environment*, Nollet, L.M.L., Ed., 3rd ed. Boca Raton, FL: CRC Press; 2006; pp. 1241–1267.

69. Ying, G.-G.; Kookana, R.S. Endocrine disruption: An Australian perspective. *AWA Water J.* **2002**, *29*, 42–45.

70. Ying, G.-G.; Williams, B.; Kookana, R. Environmental fate of alkylphenols and alkylphenol ethoxylates—A review. *Environ. Int.* **2002**, *28*, 215–226.

71. Ying, G.-G.; Kookana, R.S.; Ru, Y.-J. Occurrence and fate of hormone steroids in the environment. *Environ. Int.* **2002**, *28*, 545–551.

72. Rudel, R.A.; Melly, S.J.; Geno, P.W.; Sun, G.; Brody, J.G. Identification of alkylphenols and other estrogenic phenolic compounds in wastewater, septage, groundwater on Cape Cod, Massachusetts. *Environ. Sci. Technol.* **1998**, *32*, 861–869.

73. Espejo, R.; Valter, K.; Simona, M.; Janin, Y.; Arrizabalaga, P. Determination of nineteen 4-alkylphenol endocrine disrupters in Geneva municipal sewage wastewater. *J. Chromatogr. A* **2002**, *976*, 335–343.

74. del Olmo, M.; Gonzalez-Casado, A.; Navas, N.A.; Vilchez, J.L. Determination of bisphenol A (BPA) in water by gas chromatography-mass spectrometry. *Anal. Chim. Acta* **1997**, *346*, 87–92.

75. Helaleh, M.I.H.; Takabayashi, Y.; Fujii, S.; Korenaga, T. Gas chromatographic-mass spectrometric method for separation and detection of endocrine disruptors from environmental water samples. *Anal. Chim. Acta* **2001**, *428*, 227–234.

76. Vilchez, J.L.; Zafra, A.; Gonzalez-Casado, A.; Hontoria, E.; del Olmo, M. Determination of trace amounts of bisphenol F, bisphenol A and their diglycidyl ethers in wastewater by gas chromatography-mass spectrometry. *Anal. Chim. Acta* **2001**, *431*, 31–40.

77. Bancon-Montigny, C.; Lespes, G.; Potin-Gautier, M. Optimisation of the storage of natural freshwaters before organotin speciation. *Water Res.* **2001**, *35*, 224–232.

78. Bancon-Montigny, C.; Lespes, G.; Potin-Gautier, M. Improved routine speciation of organotin compounds in environmental samples by pulsed flame photometric detection. *J. Chromatogr. A* **2000**, *896*, 149–158.

79. Tolosa, I.; Readman, J.W. Simultaneous analysis of the antifouling agents: Tributyltin, triphenyltin and IRGAROL 1051 in marine water samples. *Anal. Chim. Acta* **1996**, *335*, 267–274.

80. Perera, A.M.A. N-nitrosamines. In *Chromatographic Analysis of the Environment*, Nollet, L.M.L., Ed., 3rd ed. Boca Raton, FL: CRC Press; 2006; pp. 419–452.

81. Preussmann, R.; Eisenbrand, G. N-nitroso carcinogens in the environment. In *Chemical Carcinogens; ACS Monograph No.182*, Searle, C.E., Ed. Washington, DC: American Chemical Society; 1984; pp. 829–844.

82. Sen, N.P.; Baddoo, P.A.; Weber, D.; Boyle, M. A sensitive and specific method for the determination of N-nitrosodimethylamine in drinking water and fruit drinks. *Int. J. Environ. Anal. Chem.* **1994**, *56*, 149–163.

83. Mitch, W.A.; Gerecke, A.C.; Sedlak, D.L. A N-nitrosodimethylamine (NDMA) precursor analysis for chlorination of water and wastewater. *Water Res.* **2003**, *37*, 3733–3741.

84. Raksit, A.; Johri, S. Determination of N-nitrosodimethylamine in environmental aqueous samples by isotope-dilution GC/MS-SIM. *J. AOAC Int.* **2001**, *84*, 1413–1419.

85. Tomkins, B.A.; Griest, W.H.; Higgins, C.E. Determination of N-nitrosodimethylamine at part-per-trillion levels in drinking waters and contaminated groundwaters. *Anal. Chem.* **1995**, *67*, 4387–4395.

86. Peldszus, S. Organic acids. In *Chromatographic Analysis of the Environment*, Nollet, L.M.L., Ed., 3rd ed. Boca Raton, FL: CRC Press; 2006; pp. 453–512.

87. Manni, G.; Caron, F. Calibration and determination of volatile fatty acids in waste leachates by gas chromatography. *J. Chromatogr. A* **1995**, *690*, 237–242.

88. Richard, J.J.; Chriswell, C.D.; Fritz, J.S. Concentration and determination of organic acids in complex aqueous samples. *J. Chromatogr.* **1980**, *199*, 143–148.

89. Vairavamurthy, A.; Andreae, M.O.; Brooks, J.M. Determination of acrylic acid in aqueous samples by electron capture gas chromatography after extraction with tri-n-octylphosphine oxide and derivatization with pentafluorobenzyl bromide. *Anal. Chem.* **1986**, *58*, 2684–2687.

90. Llompart, M.; Garcia-Jares, C.; Landin, P. Phthalate esters. In *Chromatographic Analysis of the Environment*, Nollet, L.M.L., Ed., 3rd ed. Boca Raton, FL: CRC Press; 2006; pp. 1103–1154.

91. Lin, Z.-P.; Ikonomou, M.G.; Jing, H.; Mackintosh, C.; Gobas, F.A.P.C. Determination of phthalate ester congeners and mixtures by LC/ESI-MS in sediments and biota of an urbanized marine inlet. *Environ. Sci. Technol.* **2003**, *37*, 2100–2108.

92. Thiele, B. Surfactants. In *Chromatographic Analysis of the Environment*, Nollet, L.M.L., Ed., 3rd ed. Boca Raton, FL: CRC Press; 2006; pp. 1173–1198.

93. Radke, M.; Behrends, T.; Forster, J.; Herrmann, R. Analysis of cationic surfactants by microbore high-performance liquid chromatography-electrospray mass spectrometry. *Anal. Chem.* **1999**, *71*, 5362–5366.

94. Takada, H.; Ishiwatari, R. Biodegradation experiments of linear alkylbenzenes (LABs): Isomeric composition of C_{12} LABs as an indicator of the degree of LAB degradation in the aquatic environment. *Environ. Sci. Technol.* **1990**, *24*, 86–91.

95. Bellar, T.A.; Budde, W.L. Determination of nonvolatile organic compounds in aqueous environmental samples using liquid chromatography/mass spectrometry. *Anal. Chem.* **1988**, *60*, 2076–2083.

96. Sanchez-Brunete, C.; Perez, R.A.; Miguel, E.; Tadeo, J.L. Multiresidue herbicide analysis in soil samples by means of extraction in small columns and gas chromatography with nitrogen-phosphorus and mass spectrometry detection. *J. Chromatogr. A* **1998**, *823*, 17–24.

97. Ling, C.F.; Perez-Melian, G.; Jimenez-Conde, F.; Revilla, E. Determination of carbofuran in a nutrient solution by GLC/NPD and HPLC. *Chromatographia* **1990**, *30*, 421–423.

98. Nagasawa, K.; Uchiyama, H.; Ogamo, A.; Shinozuka, T. Gas chromatographic determination of microamounts of carbaryl and 1-naphthol in natural water as sources of water supplies. *J. Chromatogr.* **1977**, *144*, 77–84.

99. Muino, M.A.F.; Gandara, J.S.; Lozano, J.S. Simultaneous determination of pentachlorophenol and carbaryl in water. *Chromatographia* **1991**, *32*, 238–240.

100. Coldwell, M.R.; Pengelly, I.; Rimmer, D.A. Determination of dithiocarbamate pesticides in occupational hygiene sampling devices using the isooctane method and comparison with an automatic thermal desorption (ATD) method. *J. Chromatogr. A* **2003**, *984*, 81–88.

101. Papadopoulou-Mourkidou, E.; Patsias, J.; Kotopoulou, A. Determination of pesticides in soils by gas chromatography-ion trap mass spectrometry. *J. AOAC Int.* **1997**, *80*, 447–454.

102. Okumura, T.; Imamura, K.; Nishikawa, Y. Determination of carbamate pesticides in environmental samples as their trifluoroacetyl or methyl derivatives by using gas chromatography-mass spectrometry. *Analyst* **1995**, *120*, 2675–2681.

103. Ballesteros, E.; Gallego, M.; Valcarcel, M. Automatic determination of N-methylcarbamate pesticides by using a liquid-liquid extractor derivatization module coupled on-line to a gas chromatography equipped with a flame ionization detector. *J. Chromatogr.* **1993**, *633*, 169–176.

104. US Environmental Protection Agency. Method 505: Analysis of Organohalide Pesticides and Commercial Polychlorinated Biphenyl (PCB) Products in Water by Microextraction and Gas Chromatography; 1995.

105. Biziuk, M.; Przyjazny, A.; Czerwinski, J.; Wiergowski, M. Occurrence and determination of pesticides in natural and treated waters. *J. Chromatogr. A* **1996**, *754*, 103–123.

106. Fernandez, M.J.; Garcia, C.; Garcia-Villanova, R.J.; Gomez, J.A. Evaluation of liquid-solid extraction with a new sorbent and liquid-liquid extraction for multiresidue pesticides. Determination in raw and finished drinking waters. *J. Agric. Food Chem.* **1996**, *44*, 1790–1795.

107. US Environmental Protection Agency. Method 507: Determination of Nitrogen and Phosphorus-Containing Pesticides in Water by Gas Chromatography with a Nitrogen-Phosphorus Detector; 1995.

108. US Environmental Protection Agency. Method 8141B: Organophosphorus Compounds by Gas Chromatography; 1998.

109. Spliid, N.H.; Koppen, B. Occurrence of pesticides in Danish shallow ground water. *Chemosphere* **1998**, *37*, 1307–1316.

110. Dagnac, T.; Jeannot, R.; Mouvet, C.; Baran, N. Determination of oxanilic and sulfonic acid metabolites of acetochlor in soils by liquid chromatography-electrospray ionisation mass spectrometry. *J. Chromatogr. A* **2002**, *957*, 69–77.

111. Pozo, O.; Pitarch, E.; Sancho, J.V.; Hernandez, F. Determination of the herbicide 4-chloro-2-methylphenoxyacetic acid and its main metabolite, 4-chloro-2-methylphenol in water and soil by liquid chromatography-electrospray tandem mass spectrometry. *J. Chromatogr. A* **2001**, *923*, 75–85.

112. Li, L.Y.T.; Campbell, D.A.; Bennett, P.K.; Henion, J. Acceptance criteria for ultratrace HPLC-tandem mass spectrometry: Quantitative and qualitative determination of sulfonylurea herbicides in soil. *Anal. Chem.* **1996**, *68*, 3397–3404.

113. Fuoco, R.; Colombini, M.P. Polychlorobiphenyls in the environment: Analytical procedures and data evaluation. *Microchem. J.* **1995**, *51*, 106–121.

114. Pearson, R.F.; Hornbuckle, K.C.; Eisenreich, S.J.; Swackhamer, D.L. PCBs in Lake Michigan water revisited. *Environ. Sci. Technol.* **1996**, *30*, 1429–1436.

115. Rossi, L.; de Alencastro, L.; Kupper, T.; Tarradellas, J. Urban stormwater contamination by polychlorinated biphenyls (PCBs) and its importance for urban water system in Switzerland. *Sci. Total Environ.* **2004**, *322*, 179–189.

116. Kelly, A.G.; Cruz, I.; Wells, D.E. Polychlorobiphenyls and persistent organochlorine pesticides in sea water at the pg l^{-1} level. Sampling apparatus and analytical methodology. *Anal. Chim. Acta* **1993**, *276*, 3–13.

117. Drouillard, K.G.; Hagen, H.; Haffner, G.D. Evaluation of chloroform/methanol and dichloromethane/hexane extractable lipids as surrogate measures of sample partition capacity for organochlorines in fish tissues. *Chemosphere* **2004**, *55*, 395–400.

118. Fournier, M.; Pellerin, J.; Lebeuf, M.; Brousseau, P.; Morin, Y.; Cyr, D. Effects of exposure of *Mya arenaria* and *Mactromeris polynyma* to contaminated marine sediments on phagocytic activity of hemocytes. *Aquat. Toxicol.* **2002**, *59*, 83–92.

119. Ahmed, M.T.; Loutfy, N.; Shiekh, E.E. Residue levels of DDE and PCBs in the blood serum of women in the Port Said region of Egypt. *J. Hazard. Mater.* **2002**, *A89*, 41–48.

120. Konstantinou, I.K.; Goutner, V.; Albanis, T.A. The incidence of polychlorinated biphenyl and organochlorine pesticide residues in the eggs of the cormorant (*Phalacrocorax carbo sinensis*): An evaluation of the situation in four Greek wetlands of international importance. *Sci. Total Environ.* **2000**, *257*, 61–79.

121. Kontsas, H.; Pekari, K. Determination of polychlorinated biphenyls in serum using gas chromatography-mass spectrometry with negative chemical ionization for exposure estimation. *J. Chromatogr. B* **2003**, *791*, 117–125.

122. Olsson, A.; Vitinsh, M.; Plikshs, M.; Bergman, A. Halogenated environmental contaminants in perch (*Perca fluviatilis*) from Latvian coastal areas. *Sci. Total Environ.* **1999**, *239*, 19–30.

123. Manoli, E.; Samara, C.; Konstantinou, I.; Albanis, T. Polycyclic aromatic hydrocarbons in the bulk precipitation and surface waters of Northern Greece. *Chemosphere* **2000**, *41*, 1845–1855.

124. Telli-Karakoc, F.; Tolun, L.; Henkelmann, B.; Klimm, C.; Okay, O.; Schramm, K.-W. Polycyclic aromatic hydrocarbons (PAHs) and polychlorinated biphenyls (PCBs) distributions in the Bay of Marmara sea: Izmit Bay. *Environ. Pollut.* **2002**, *119*, 383–397.

125. Ogunfowokan, A.O.; Asubiojo, O.I.; Fatoki, O.S. Isolation and determination of polycyclic aromatic hydrocarbons in surface runoff and sediments. *Water Air Soil Pollut.* **2003**, *147*, 245–261.

126. Catalina, M.I.; Dalluge, J.; Vreuls, R.J.J.; Brinkman, U.A.T. Determination of chlorophenoxy acid herbicides in water by in situ esterification followed by in-vial liquid-liquid extraction combined with large-volume on-column injection and gas chromatography-mass spectrometry. *J. Chromatogr. A* **2000**, *877*, 153–166.

127. Berrada, H.; Font, G.; Molto, J.C. Determination of urea pesticide residues in vegetable, soil and water samples. *Crit. Rev. Anal. Chem.* **2003**, *33*, 19–41.

128. Chien, C.-J.; Charles, M.J.; Sexton, K.G.; Jeffries, H.E. Analysis of airborne carboxylic acids and phenols as their pentafluorobenzyl derivatives: Gas chromatography/ion trap mass spectrometry with a novel chemical ionization reagent, PFBOH. *Environ. Sci. Technol.* **1998**, *32*, 299–309.

129. Xiong, F.; Croue, J.-P.; Legube, B. Long-term ozone consumption by aquatic fulvic acids acting as precursors of radical chain reactions. *Environ. Sci. Technol.* **1992**, *26*, 1059–1064.

130. Jonsson, S.; Ejlertsson, J.; Ledin, A.; Mersiowsky, I.; Svensson, B.H. Mono- and diesters from o-phthalic acid in leachates from different European landfills. *Water Res.* **2003**, *37*, 609–617.

131. Hashizume, K.; Nanya, J.; Toda, C.; Yasui, T.; Nagano, H.; Kojima, N. Phthalate esters detected in various water samples and biodegradation of phthalates by microbes isolated from river water. *Bio. Pharm. Bull.* **2002**, *25*, 209–214.

132. Ruminski, J.K.; Dejewska, B.; Wojtanis, J. Environmental research, part I, investigation on dioctyl phthalate (DEHP) pollution in soil and surface water near Wabrzezno (Torun District). *Pol. J. Environ. Stud.* **1995**, *4*, 65–69.

133. Lopez, F.J.; Beltran, J.; Forcada, M.; Hernandez, F. Comparison of simplified methods for pesticide residue analysis. Use of large-volume injection in capillary gas chromatography. *J. Chromatogr. A* **1998**, *823*, 25–33.

134. Hoh, E.; Mastovska, K. Large volume injection techniques in capillary gas chromatography. *J. Chromatogr. A* **2008**, *1186*, 2–15.

Chapter 4

Solid-Phase Extraction

Ronald E. Majors

4.1. BASICS OF SOLID-PHASE EXTRACTION (SPE)

SPE is a widely used sample preparation technique with principles similar to those of high-performance liquid chromatography (HPLC). It is used for the selective sorption and concentration of analytes or interferences from complex matrices and precedes measurement by chromatography and other analytical techniques. The application of SPE spans a wide variety of sample types and market segments. It is often used in the investigation of drugs in biological fluids, trace organics in environmental samples (e.g., water, soil, sludge), and toxins in foodstuffs. The sample is usually in a gaseous or liquid form, although there are subsets of SPE (e.g., matrix solid-phase dispersion [MSPD], see Section 4.14.4) where solid samples can be handled directly. SPE is most often used as a replacement for liquid–liquid extraction (LLE).

4.2. COMPARISON TO OTHER TECHNOLOGIES

4.2.1. SPE versus LLE

SPE can be used in a similar fashion as LLE but has a number of advantages over LLE:

- more complete extraction of the analyte,
- more efficient separation of interferences from analytes,
- reduced organic solvent consumption,
- no emulsion formation,
- easier collection of analytes,
- more convenient manual procedures,
- removal of particulates, and
- more easily automated.

LLE procedures that require several successive extractions to recover +99% of the analyte often can be replaced by one-step SPE methods. Because SPE is a more efficient separation process than LLE, it is easier to obtain a higher recovery of the analyte. With SPE, it is also possible to obtain the complete removal of interferences from the analyte fraction. Reversed-phase (RP)-SPE procedures are the most popular because only small amounts of organic solvent are required, while maintaining a higher concentration of analyte. Analyte purity may be better with SPE particularly when a large volume of the LLE organic phase containing trace analytes must be concentrated by evaporation to dryness. With SPE, the amount of organic solvent used in the elution step is quite small (in the microliter to milliliter range), and thus, when concentration is required, any solvent impurities would be reduced proportionally. There is no need for phase separation as required for LLE, so the total analyte fraction is easily collected in SPE, eliminating errors associated with variable or inaccurately measured extract volumes. As a side benefit, larger particulates are trapped by the SPE cartridge and do not pass through into the analyte fraction. Thus, SPE also serves as a filtration step.

On the other hand, there are some disadvantages of SPE versus LLE. These include the irreversible adsorption of some analytes on the SPE packing, frit or housing materials, and potential variability of some HPLC packings, and there is more complex method development needed to optimize the SPE experiment. Irreversible adsorption may occur when very basic ionizable analytes at trace levels interact with residual silanols on the surface of a bonded silica SPE packing or small amounts of other reactive groups may be present on other SPE phases. The surface area of an LLE device (e.g., separatory funnel) is quite small (and less active) compared with an SPE cartridge, so that irreversible binding of analyte (with lower recoveries) is less likely with LLE versus SPE. The batch-to-batch variability of SPE packings was formerly a bigger problem than it is today. Initiatives to improve production quality have led to major improvements in cartridge reproducibility. Still, similar to HPLC columns, SPE devices of the same phase

Handbook of Sample Preparation, Edited by Janusz Pawliszyn and Heather L. Lord

from different manufacturers may vary in bonding chemistry, carbon loading, residual silanols, variations in polymeric composition, and other factors that may not allow easy transfer from one brand to another. The solvents used in LLE are usually pure and well defined, so that LLE separations are quite reproducible from laboratory to laboratory. Method development in SPE involves four distinct steps (see Section 4.6), while for LLE, the selection of the two immiscible solvents, their relative volumes, and appropriate buffered pH (for ionizable analytes) is performed quite easily.

4.2.2. SPE versus HPLC

Since SPE and HPLC have similar retention modes, it is useful to compare and contrast the two techniques (see Table 4.1). Although several different forms (devices) are commercially available (see Section 4.4), in its simplest and most popular format, SPE uses a small, plastic, disposable column or cartridge, often a polypropylene medical syringe barrel packed with 0.1- to 1.0-g sorbent, which is contained by polyethylene frits. In further discussions, we will refer to a "cartridge" as the SPE device, but this "generic" term could refer to any one of several formats (see Section 4.4). Although many sorbent phases are available (see Table 4.2), for purposes of comparison, we will assume that the sorbent is an RP material (e.g., ODS-silica) that resembles RP chromatography (RPC) in its separation characteristics. Although silica-gel-based, bonded-phase packings were introduced first, polymeric sorbents have become available in recent years and have been gaining in popularity.

For SPE, the particle size typically is larger than that in HPLC. Because of shorter packed bed lengths, larger particles, and less well-packed beds, SPE cartridges are much less efficient than an HPLC column. For cost reasons, irregularly shaped, type-A-silica packings are usually used in SPE. Type A silicas are older silica gel materials that have higher metal content than "modern" high-purity silicas. The higher metal content generally gives rise to more acidic surface silanols, which may lead to irreversible adsorption and even ion exchange interactions with basic analytes. Modern, higher-purity silicas are referred to as type B silicas. Recently, spherical silicas for SPE have come on the market but have not strongly impacted the sale of the most popular products. Polymeric sorbents generally are spherical but more expensive than silica-based packings. Some SPE disks do, however, use more expensive spherical SPE packings with particle diameters in the 8- to 12-μm range. Overall, the principles of separation, selection of conditions, and method development are similar for both SPE and HPLC, except that SPE uses a series of isocratic steps during retention and elution of the analyte. SPE is generally referred to as a "digital chromatography" process since the analyte is either on (K_A very large) (see Eq. 4.1 in Section 4.6)* or off the cartridge (K_A very small). In contrast, HPLC is a continuous, elution process where analytes are separated by subtle changes in resolution in an analog manner as the analytes move down the column using an isocratic or gradient mobile phase.

Table 4.1. Comparison of HPLC and SPE

Factor	HPLC	SPE
Format	Stainless-steel column	Plastic cartridge
Particle size (μm)	1.5–5	40–60
Particle shape	Spherical	Irregular
Plates/column	20–25,000	<100
Separation	Continuous elution	Digital elution
Cost of column	\$300–\$400	\$1.00–\$2.00
Operation	Reusable	Used once
Separation modes	Many	Many
Cost of operation	Moderate-high	Low
Cost of equipment	High	Low

One major difference between SPE and HPLC is that the SPE cartridge is generally used once and discarded, since potential interferences may remain on the cartridge, whereas HPLC columns are used many times. The equipment to perform SPE is low cost (see Section 4.5), while commercial HPLC systems can cost tens of thousands of dollars.

4.3. USES OF SPE

SPE is primarily used for two main purposes in sample preparation: (1) removal of sample interferences and (2) concentration or trace enrichment of the analyte.

4.3.1. Interference Removal

Interferences that overlap analyte bands in a chromatographic separation or in the spectral output of a spectroscopic investigation complicate method development and can adversely affect assay results. In some cases, especially for complex samples (e.g., natural products, environmental samples), a large number of interferences in the original sample can make it almost impossible to separate these from one or more analyte bands by means of a single HPLC separation or to perform spectroscopy on the original sample. SPE can be used to reduce or eliminate those interferences prior to analysis. Some samples contain components, such as hydrophobic substances (e.g., fats, oils, greases), proteins, polymeric materials, or particulates that can either plug or deactivate the HPLC column or that can lodge in the injection port or head of a gas chromatography (GC) column. These interferences can often be removed by SPE and therefore prolong column lifetime.

4.3.2. Analyte Concentration

SPE can often be used to concentrate a trace component. If an SPE phase can be selected so that $K_A \gg 1$ for the analyte,

*The K term here is defined as shown in Equation 4.1.

Table 4.2. Various SPE Phases and Conditions

Mechanism of Separation	Typical Phases	Structure(s)	Analyte Type	Loading Solvent	Eluting Solvent[a]
Normal phase (adsorption)	Silica, Alumina, Florisil	–SiOH, AlOH, Mg_2SiO_3	Slightly to moderately polar	Small ε, for example, hexane, $CHCl_3$/hexane	Large ε, for example, methanol, ethanol
Normal phase (polar bonded phase)	Cyano, amino, diol	–CN, $–NH_2$, –CH(OH)–CH(OH)–	Moderately to strongly polar	Small ε, for example, hexane	Large ε, for example, methanol, ethanol
Reversed phase (nonpolar bonded phase—strongly hydrophobic)	Octadecylsiloxane, Octylsiloxane, PS-DVB, DVB (polymeric)	$(–CH_2–)_{17}CH_3$, $(–CH_2–)_7CH_3$, PS-DVB, DVB	Hydrophobic (strongly nonpolar)	High P', for example, H_2O, MeOH/H_2O, ACN/H_2O	Intermediate P', for example, MeOH, ACN
Reversed phase (nonpolar bonded phase—intermediate hydrophobicity)	Cyclohexyl, phenyl, diphenyl	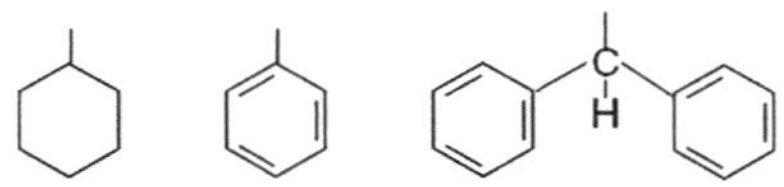	Moderately nonpolar	High P', for example, H_2O, MeOH/H_2O, ACN/H_2O	Intermediate P', for example, MeOH, ACN
Reversed phase (nonpolar bonded phase—low hydrophobicity)	Butyl, ethyl, methyl	$(–CH_2–)_3CH_3$, $–C_2H_5$, $–CH_3$	Slightly polar to moderately nonpolar	High P', for example, H_2O	Intermediate P', for example, MeOH, ACN
Polymeric reversed phase (hydrophobic–hydrophilic balanced)	Polyamide, poly[*n*-vinylpyrrolidone-divinylbenzene(DVB)], methacrylate-DVB	Various polymers	Acidic, basic, neutral	Water or buffer	Intermediate P', for example, MeOH, ACN
Anion exchange (weak)	Amino, 1^0, 2^0-amino	$(–CH_2–)_3NH_2$, $(–CH2–)_3NHCH_2CH_2NH_2$	Ionic (ionizable), acidic	Water or buffer ($pH = pK_a + 2$)	A. Buffer ($pH = pK_a – 2$) B. pH value where sorbent or analyte is neutral C. Buffer with high ionic strength

(Continued)

Table 4.2. (*Continued*)

Mechanism of Separation	Typical Phases	Structure(s)	Analyte Type	Loading Solvent	Eluting Solvent[a]
Anion exchange (strong)	Quaternary amine	$(-CH_2-)_3N^+(CH_3)_3$	Ionic (ionizable), acidic	Water or buffer ($pH = pK_a + 2$)	A. Buffer ($pH = pK_a - 2$) B. pH value where analyte is neutral C. Buffer with high ionic strength
Cation exchange (weak)	Carboxylic acid	$(-CH_2-)_3COOH$	Ionic (ionizable), basic	Water or buffer ($pH = pK_a - 2$)	A. Buffer ($pH = pK_a + 2$) B. pH where sorbent or analyte is neutral C. Buffer with high ionic strength[a]
Cation exchange (strong)	Alkyl sulfonic acid, aromatic sulfonic acid	$(-CH_2-)_3SO_3H$, 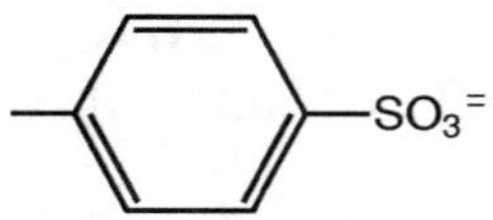	Ionic (ionizable), basic	Water or buffer ($pH = pK_a - 2$)	A. Buffer ($pH = pK_a + 2$) B. pH value where analyte is neutral C. Buffer with high ionic strength[a]

[a]For ion exchange, three possible elution conditions exist: A, buffer 2 units above (acids) or below (bases) pK_a of analyte; B, pH where either analyte or sorbent (weak exchangers) is neutral; C, high ionic strength. P' represents polarity (octanol–water partition coefficient). ε, strength of a solvent in the normal phase mode of chromatography or solid-phase extraction.

a relatively large volume of sample (e.g., several milliliters up to several hundred milliliters), can be applied before the analyte saturates and breaks through the cartridge. An increase in analyte concentration can then be achieved, if the cartridge is eluted with a small volume of strong solvent ($K_A < 1$). This process is called *trace enrichment* and is often used to concentrate sub-part per billion concentrations of organic pollutants such as polynuclear aromatic hydrocarbons or pesticides from environmental water samples using an RP-SPE cartridge. A strong solvent (e.g., acetonitrile [ACN] or methanol [MeOH]) elutes these analytes from the cartridge in a small volume, which also saves on evaporation time. Once evaporation removes the noncompatible solvent, the sample can then be redissolved in a solvent compatible with the subsequent HPLC or GC separation. Alternatively, the eluted sample can be diluted using a suitable injection solvent, but the concentration of analyte will be decreased.

To a lesser extent, SPE is also used for desalting, solvent exchange, and sample storage and transport. RP-SPE can be used to desalt samples, particularly biological samples that may contain high salt concentrations. It is often used prior to ion exchange chromatography, where a low-ionic-strength sample is desirable. Conditions of pH and % organic are selected to retain the analyte initially, which allows the soluble inorganic (and possibly organic) salts to be washed from the cartridge with water. The analyte can then be eluted (salt free) with organic solvent.[1]

Solvent exchange is sometimes required when the analytical technique is incompatible with an organic solvent used for sample cleanup. The SPE phase is selected to retain the analyte of interest from a volatile organic solvent. By passage of nitrogen or air through the cartridge, the organic solvent can be removed by evaporation. Then, a suitable compatible solvent is used to remove the sorbed analyte, hopefully in a small volume, for further processing.

Sample storage and transport is a little known phenomenon that can be performed using SPE devices. Sometimes when an analyte is retained in a sorbed state, it is stabilized relative to being in solution or exposed to the atmosphere in an open container. Thus, a sample collected in the field can be stored for later analysis or even shipped by mail to a remote laboratory without fear of loss by volatilization or oxidation. Of course, each particular situation should be investigated independently.

4.4. SPE FORMATS

Several SPE devices are in popular use (Fig. 4.1). As stated earlier, the *SPE cartridge* is, by far, the most widely used. A typical SPE disposable cartridge (syringe-barrel format) is depicted in Figure 4.1a. The syringe barrel is usually medical-grade polypropylene, fitted with a Luer tip, so that a needle can be affixed to direct the effluent to a small container or vial. For special applications where a high degree of inertness is required, cartridge bodies of polytetrafluoroethylene (PTFE) or serological-grade glass are available. The frits that hold the particle bed in the cartridge are made of polyethylene, PTFE, or stainless steel with a porosity of 10–20 μm, and thus offer little flow resistance. SPE cartridges may vary in design to fit an automated instrument or robotic system. SPE cartridges are relatively inexpensive and generally are used a single time and discarded to avoid sample cross-contamination.

To accommodate a wide range of SPE applications, sorbent-packed cartridges are available with packing weights of 35 mg to 2 g per cartridge, as well as with reservoir volumes (the volume above the packing in the cartridge) of 0.5–10 mL. For very large samples, "mega" cartridges contain up to 10 g of packing with a 70-mL reservoir. Cartridges with a larger amount of packing should be used for dirty samples that may overload a low-capacity cartridge. However, cartridges containing 100 mg of packing or less are preferred for relatively clean liquid samples, where cartridge capacity is not an issue, as well as for trace analysis and small sample volumes. The amount of solvent used in SPE is roughly proportional to bed mass, so smaller cartridge weights generally save solvent.

Due to the higher surface areas of polymeric SPE packings, less packing is needed than for silica-based particles. Weights of polymeric packing in the 5- to 60-mg range can accommodate smaller volumes of sample and solvent. In most cases, it is desirable to collect the analyte in the smallest possible volume, which means that the SPE cartridge generally should also be as small as possible.

The second oldest SPE configuration is the disk (Fig. 4.1b). *SPE disks* combine the advantages of membranes (see below) and SPE. In their appearance, the disks closely resemble membrane filters: they are flat, usually ≤1 mm thick, and 4–90 mm in diameter. The physical construction of the SPE disks differs from membrane filters. One popular type is the flexible or expanded PTFE network (0.5 mm thick) membranes filled with silica- or resin-based packings (e.g., Empore, 3M, St. Paul, MN). Here, the packing is embedded in the PTFE fibrils and comprised of about 60–90% of the total weight of the membrane. These disks are sold individually, and since they are flexible, they must be supported in a reusable filter holder. Others are sold preloaded in disposable holders or cartridges with Luer fittings for easy connection to syringes.

Rigid fiberglass disks containing packing material are also available (Agilent Technologies, Santa Clara, CA). These disks are more rigid than the PTFE membranes and can be used at higher flow rates. Depending on the disk size, packing amounts vary from 15 to 56 mg. Low-bed-mass, rigid fiberglass disks with packing mass down to 1.5 mg are useful for pretreating small clinical samples (e.g., plasma or serum[2,3]). Their reduced sorbent mass and small volume reduce solvent consumption (and any related sample contamination by solvent impurities). An advantage of this type of disk is an absence of frits that are a possible further source of contamination.

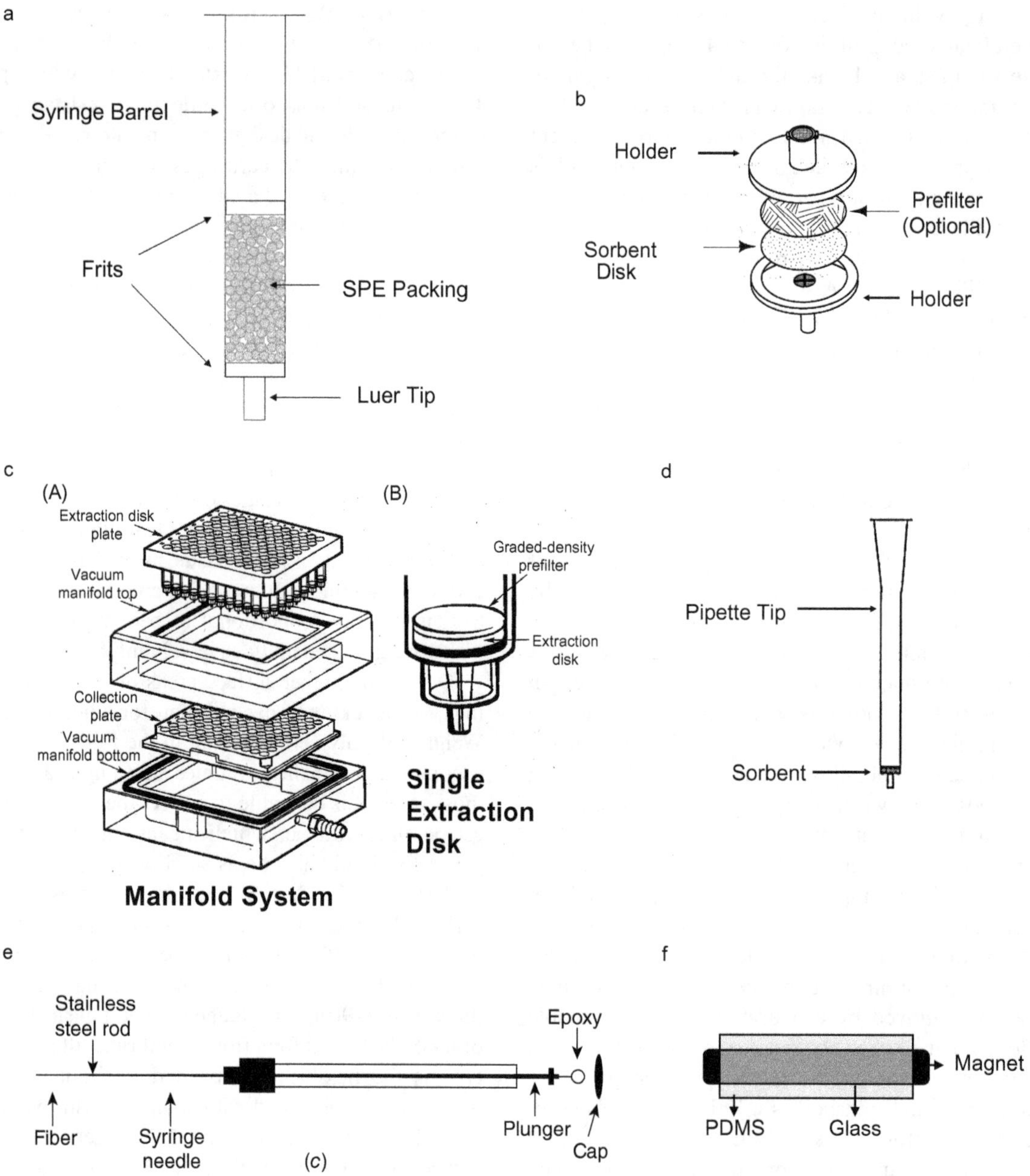

Figure 4.1. SPE devices. (a) Typical syringe barrel cartridge design. (b) Typical SPE disk configuration. (c) Schematic diagram of a 96-well SPE extraction plate system (3M Corp.). (d) SPE pipette tip. (e) SPME syringe assembly. (f) Coated stir bar.

SPE disks and cartridges differ mainly in their length/diameter ratio (L/d): disks have $L/d < 1$ and cartridges have $L/d > 1$. Compared with SPE cartridges, this characteristic of the disk enables higher flow rates and faster extraction. Even at higher flow rates, the linear velocity is fairly low (0.10–0.30 cm/s) due to the high cross-sectional areas of disks. "Dirty" water or water-containing particulates, such as wastewater, can plug the porous disks, just as in the case of cartridges. In either case, a prefilter should be used prior to the SPE treatment. Some disk products come with a built-in prefilter. Since packing is embedded in the disks, channeling, which can cause uneven flow through poorly packed cartridges, is not a problem with disks. Due to the thinness of the disk (typically 0.5–2 mm); however, compounds with low k values tend to have lower breakthrough volumes than for SPE cartridges. Note that disk cartridges are also available to compete with low-bed-mass cartridge formats.

SPE disks are especially useful for environmental applications, such as the analysis of trace organics in surface water, which require a large sample volume to obtain the necessary sensitivity. The United States Environmental Protection Agency (USEPA) has approved SPE technology as an alternative for large-volume LLE methods[4] in the preparation of water samples for HPLC analysis. Examples of environmental methods include procedures for phenols,[5] pesticides and polychlorinated biphenyls (PCBs),[6] and nitrosoamines in drinking water.[7]

The move toward miniaturization and high throughput in analytical laboratories has prompted the development of new configurations of SPE devices. The most popular device, especially in the pharmaceutical laboratory, is the *96-well SPE plate* (Fig. 4.1c). The 96-well plate is well suited for automation and the SPE processing of a large number of small samples. In this format, 96 flow-through SPE "wells" of 0.5–2.0 mL volume contain small masses of packing (usually <100 mg) contained by small frits or embedded into individual disks. The plate is analogous to an 8 × 12 array of miniaturized SPE cartridges. The plates can be handled by robotic instrumentation to completely automate the SPE process. Both vacuum manifolds and pressurized systems can be used for the movement of liquid samples through the plates. Dedicated 96-port automation equipment such as xyz liquid-handling systems can perform all of the necessary steps in the SPE experiment. In addition, 96-position evaporation stations can automatically evaporate the elution solvent from collection plates. The evaporated samples then can be reconstituted with a suitable injection solvent and injected directly from the 96-well collection plate using a modern autosampler capable of handling such plates.

The *micropipette tip* (MPT) format (Fig. 4.1d) permits the handling of sub-microliter amounts of sample, such as biological fluids. SPE has been performed with various packings that are placed in loosely in the pipette tip (contained with frits), or embedded in, or coated on the internal walls of the tip. With the tips where coatings are on or embedded in the internal surface of the tip, liquid samples can be drawn up and expelled without undue pressure drop or plugging. Sample is drawn into the tip, where it interacts with the SPE packing. Next, the tip is rinsed with wash solvent to enhance cleanup; then, it is eluted with a strong solvent, just as with SPE cartridges. Many popular SPE modes have been adapted to MPTs, including RP, ion exchange, hydrophobic interaction, hydrophilic interaction, immobilized-metal affinity, and affinity chromatography. MPTs have mainly been used for purification, concentration, and selective isolation (e.g., affinity, metal chelation) of proteins and peptides and are an essential tool for matrix-assisted laser desorption ionization–mass spectrometry (MALDI) and for other advanced mass spectrometry (MS) techniques. One of the main advantages of MPTs is that they can be used with micropipettors or in liquid-handling automation designed for picking up/discarding the MPTs. Solvent flow can be bidirectional in the tips so that each of the SPE steps can be performed with a single micropipette. Compared with 96-well plates (either the SPE plate or the collection plate) that can accidentally be overfilled with solvent or sample with possible cross-contamination, pipette tips are used singly, and thus, carryover and cross-contamination cannot occur.

Polymer-coated fibers are used for solid-phase microextraction (SPME). As depicted in Figure 4.1e, in this design, a fused-silica fiber affixed to a syringe plunger is coated with a polymeric stationary phase, such as a polydimethylsiloxane (PDMS) or a polyacrylate.[8] The fiber is dipped into the solution or placed in the headspace of the sample to be analyzed, and analytes diffuse to and partition into the coating as a function of their distribution coefficients. Usually, stirring is used to enhance convective diffusion. Once equilibrium is achieved, the fiber is removed from the solution and placed in the injection port of a gas chromatograph for thermal desorption of sorbed analytes. For HPLC, solutes are displaced by solvent in a special adaptation of an HPLC valve or directly into an autosampler vial where analytes are dissolved from the fiber with a strong solvent. An entire chapter is devoted to SPME (see Chapter 5), and no additional information shall be provided here.

Stir-bar sorptive extraction (SBSE)[9] is similar in concept to the use of coated fibers, but the greatly increased sorbent volume allows for greater mass sensitivity. The stir bar, with a polymeric sorptive coating (Fig. 4.1f), is placed in an aqueous liquid, and the solution is stirred while analyte/matrix partitioning takes place. After equilibrium, the stir bar is removed, dried to remove traces of water, and then transferred to a special device where the analytes are displaced into the GC or HPLC column. Both coated-fiber and stir-bar devices are more popular in GC, where thermal desorption is more efficient in volatilizing sorbed analytes into the gas phase than solvent desorption in the liquid phase.

4.5. SPE APPARATUS

The equipment needed for SPE can be very simple to very complicated automated systems (Fig. 4.2). With the relatively large particle size of SPE packings and the short beds, gravity can be used as the driving force, but flow through the cartridge with "real" viscous samples can be quite slow. An interlocking syringe can be used to manually push solvent or sample through the cartridge (Fig. 4.2a), but this approach can fail for samples that are viscous or which contain particulates, so vacuum- or pressure-driven flow is preferred. For example, a simple vacuum flask fitted with a rubber stopper can handle one cartridge at a time (Fig. 4.2b). However, when several samples must be processed simultaneously, a vacuum-manifold system for processing multiple cartridges at a time is recommended (Fig. 4.2d). These manifolds are constructed of chemically resistant, vacuum-safe glass and can have up to 24 positions for holding SPE cartridges. A removable, height-adjustable rack is located inside the vacuum manifold to hold various sizes of test tubes or vials for eluant collection. In some units, a vacuum bleed valve, a flow-control stopcock valve for each cartridge, and a vacuum gauge are incorporated to allow better control of the solvent flow. Positive-pressure manifold systems are also available that provide individual flow control for each of the cartridges. As the degree of sophistication increases, so does the price of the apparatus.

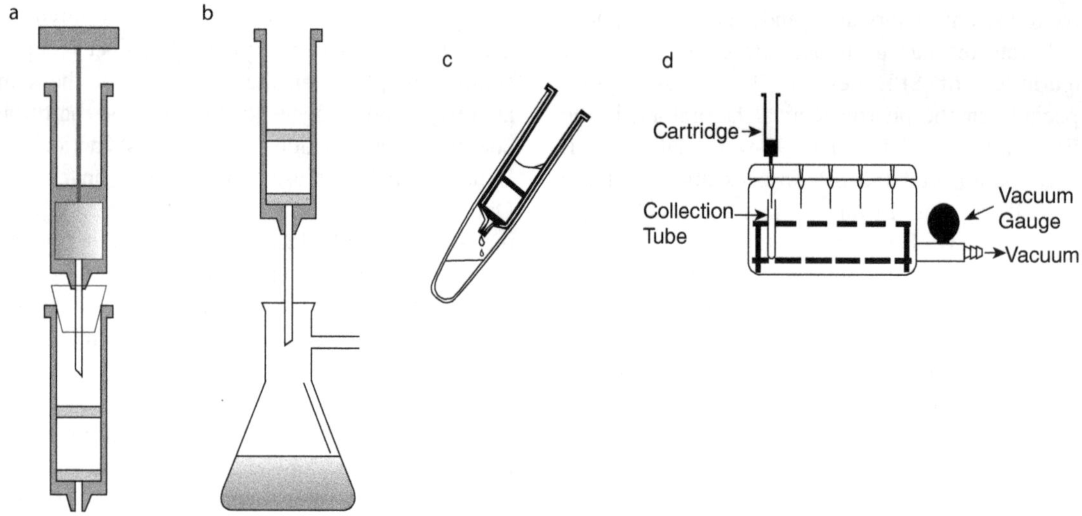

Figure 4.2. SPE apparatus: (a) pressurization, (b) vacuum, (c) centrifugation, and (d) vacuum manifold.

Centrifugation (Fig. 4.2c) is used less commonly to drive liquid through the cartridge.

For 96-well plates, vacuum manifold processing stations are also available with similar controls of the cartridge version. These units are constructed of rigid polypropylene and provide a small footprint. Generally, they are supplied with a disposable reservoir tray to collect conditioning and wash solvents. A collection plate with 96 wells is used to collect sample effluent. Spacer inserts allow the use of collection plates of differing heights to ensure that the exit of each well has maximum penetration into the collection plates to eliminate possible well-to-well contamination.

Regardless of the method used to create flow through the SPE cartridge or other SPE device, the flow rate should allow sufficient time for the sample to contact the packing. There is a kinetic factor that must allow analytes to diffuse into the packing to interact with the stationary phase and then to diffuse out of the packing for the process to continue. Ion exchange mechanisms have an additional kinetic factor, so even slower flow rates should be used in ion exchange SPE. Similar to HPLC, higher flow rates also decrease separation efficiency. For typical SPE applications, flow rates ≤10 mL/min[10] for a cartridge and ≤50 mL/min for a 90-mm disk[11] are recommended.

When the number of samples increases to the point where SPE sample preparation becomes the "bottleneck," automation of the entire process becomes attractive. There are three basic approaches to SPE automation. The simplest and least expensive are dedicated SPE systems that perform the four basic steps (see Section 4.6). Many of these systems are designed to use standard SPE cartridges, sometimes with rimless barrels. SPE workstations for 96-well plates are available, some of which can be interfaced to 96-well plate autosamplers for unattended operation for high-throughput laboratories.

Modified xyz liquid-handling systems perform liquid-handling functions such as dilution, mixing, and internal standard addition. These units have been adapted to perform SPE, and they too may be interfaced to an HPLC injection valve. Pipette tips are often the preferred format since the xyz liquid-handling systems are often adapted to accommodate them. Some can handle 96-well plates as well.

Robotic systems are the most versatile technique to perform sample preparation functions. Although a robot can be interfaced to devices that perform each of the steps of the SPE procedure, it is more time- and cost-effective to interface the robot to a dedicated SPE workstation. The robot serves to move sample containers to and from the SPE workstation, as well as to and from other sample preparation devices (e.g., balances, mixers, dilutors, autosamplers). Commercial robotics systems are available with different capabilities from many manufacturers (e.g., Beckman-Coulter, Gilson, Hamilton, Tomtec).

4.6. THEORY AND PRACTICE OF SPE

SPE draws its theoretical foundations from HPLC and pertinent equations describing analyte retention, and elution are described elsewhere.[12] However, the practice of SPE is performed differently. There are two common modes of SPE: (1) the analyte adsorption mode sometimes referred to as the "bind–elute" mode and (2) the matrix adsorption mode (interference removal mode). Figure 4.3 depicts the four steps of the SPE bind–elute process:

1. conditioning of the packing,
2. sample application (loading),
3. washing the packing (removal of interferences), and
4. recovery of the analyte.

The initial step 1, conditioning the dry packing prior to the loading of the sample, is also performed in HPLC by continuously running and equilibrating solvent through the column. However, in SPE, the conditioning step is mainly

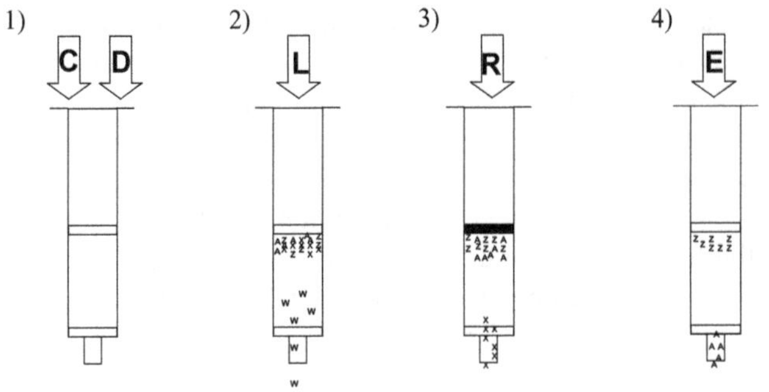

Figure 4.3. SPE steps. Sample components: A, analyte(s) of interest; W, weakly retained, undesired matrix component(s); X, intermediate undesired matrix component(s); Z, strongly retained, undesired matrix component(s). Solvents: C, conditioning solvent; D, equilibration solvent; L, loading solvent; R, rinsing solvent; E, eluting solvent.

used to wet the adsorbent or the bonded phase on the sorbent so that it is fully solvated in order to receive and interact with the sample. In step 2, sample application is performed by loading a solubilized sample and matrix onto the SPE cartridge. The distribution coefficient, K_A, for an analyte A and matrix/interference components W, X, and Z is defined in Equation 4.1:

$$K_A = (A)_{\text{sorbent}} / (A)_{\text{sample solution}}, \tag{4.1}$$

where $(A)_{\text{sorbent}}$ is the concentration of A in the SPE sorbent and $(A)_{\text{sample solution}}$ is the concentration of A in the solubilized sample. In SPE, the analyte(s), matrix components, and interferences are all solubilized in the loading solvent. The goal of the normal practice of SPE is to maximize the transfer of analyte A away from the sample matrix/interferences (components W, X, and Z) that should either remain in solution or be more strongly retained by the SPE sorbent. The loading solvent is selected such that the analyte is retained on the sorbent (high $K_A >> 1$) and weakly retained matrix/interference components (low K_W where K_W represents the distribution coefficient for a typical weakly retained compound W in Fig. 4.3, step 2) pass through the sorbent. An intermediate retained matrix/interference component (component X) will require fine-tuning with further solvents to separate them from the analyte of interest. Other more strongly retained components/interferences ($K_Z >>> 1$) remain on the sorbent.

Step 3 is the rinsing (washing) step where the elution power of the solvent is adjusted so that the intermediate component (X) is removed from the cartridge without breakthrough of the analyte of interest A. It is permissible for analyte A to move down the cartridge, but it should not elute. Hopefully, if the proper solvent composition is selected, strongly retained component Z remains at the top of the cartridge bed. Finally, step 4 is the elution step. Here, a solvent with preferential solubility for the analyte is chosen to elute it in the smallest possible elution volume. Also, it is preferable for the solvent used in this step to be volatile so it can be easily removed in order for the separated analyte to be redissolved in a compatible solvent for the analytical step that follows.

Let us explore the four steps in more detail since understanding these steps is important for the successful development of an SPE method. In this discussion, we assume RP-SPE using a nonpolar bonded phase on a silica gel backbone. We will also assume that we desire the initial retention of the analyte; for other phases and more detail for specific products, consult the manufacturer's literature.

4.6.1. Conditioning of the Sorbent (Step 1, Fig. 4.3 [1])

Performed prior to the addition of sample, the dry bonded silica packing in the cartridge is "conditioned" by passage of a few bed volumes of conditioning solvent C, typically methanol (MeOH) or acetonitrile (ACN). The role of the conditioning step is twofold: (1) it removes any impurities that may have been collected while the cartridge was exposed to the laboratory environment, or was initially present in the cartridge as supplied by the manufacturer, and (2) it allows the sorbent to be solvated. Solvation is important, since RP silica-based packings (especially C_8, C_{18}, or phenyl) that have been allowed to dry out, especially after the conditioning step, often exhibit a considerable decrease in sample retention. In addition, varying states of packing dryness lead to nonreproducible analyte recoveries. On the other hand, polymeric packings with a balance of hydrophobic–hydrophilic surface character can dry out slightly and still maintain their performance.

Once the packing is conditioned, the excess conditioning solvent can be removed by a flow of air through the cartridge until the solvent no longer drips from the bottom of the cartridge. This step is not depicted in Figure 4.3 (1). However, the airflow should not be continued past this point, as this can dry the packing and adversely affect analysis reproducibility (especially with silica-based SPE sorbents; polymeric packings are more forgiving). Alternatively, a few column volumes of water (solvent D in Fig. 4.3) can be passed through the SPE cartridge to flush out the excess organic conditioning solvent and ready the cartridge for the introduction of an aqueous sample. Do not allow too much time (e.g., >5 min) between this water-conditioning step and the sample-loading step, because if the packing sits in water too long, the solvating solvent on the bonded phase may slowly partition into the water, thereby "dewetting" the packing.

4.6.2. Sample-Loading Step (Step 2, Fig. 4.3 [2])

This step in the SPE procedure involves sample application (loading); the sample, dissolved in a weak aqueous solvent (water or buffer with ≤10% organic), is added to the cartridge causing strong retention of the analyte. The sample for SPE can be applied with a pipette or syringe, or, for large volumes (e.g., greater than 50 mL of environmental water), the sample can be pumped into the cartridge. The sample mass and cartridge sizes must be matched so as not to overload the cartridge. As a rule of thumb, a typical capacity for an adsorption mechanism would be 2–3 mg/g of packing. Remember that the capacity of the cartridge must be sufficient to handle the analytes, matrix, and interferences, all of which may be retained during the loading step. The sample solution should be passed through the cartridge without allowing the cartridge to dry out. Compared with HPLC, the flow rate is not precisely controlled in SPE, but it can be adjusted by varying the vacuum and the delivery rate from a syringe or pump, or by varying the pressure in a pressurized manifold system. Flow rates of 2–4 mL/min are usually acceptable. Too fast of a flow rate may not allow equilibrium to take place and therefore allowing some (or all) of the analyte to pass unretained. Too slow of a flow rate may greatly increase the time of the SPE experiment.

4.6.3. Rinsing (Washing) of the Cartridge (Step 3, Fig. 4.3 [3])

This step involves the removal of matrix and/or interferences by washing the cartridge with a solvent R, chosen for its intermediate strength. The wash-solvent strength and volume should be carefully chosen, as too large a volume and/or excessive solvent strength may result in partial elution of the analyte. Optimally, the wash step is discontinued just before analyte begins to leave the cartridge. In this way, interferences that are more weakly retained than the analyte are washed from the cartridge, but no loss of analyte occurs. Sometimes, water or buffered water is used as the wash solvent in RP-SPE, but this may not provide a maximum removal of interferences. A small, controlled amount of organic solvent may be added to the wash solution (solvent R) to aid in the removal of more hydrophobic interferences, but care must be taken that the analyte of interest is not removed at the same time. Because of the potential variability of the SPE separation from cartridge to cartridge, there must be some safety margin in the optimum wash-solvent strength and volume used to remove interferences from the cartridge. The primary goal is to collect 100% of the analyte in the elution step; otherwise, poor and variable recoveries will result.

4.6.4. Analyte Elution Step (Step 4, Fig. 4.3 [4])

The final step is the elution and collection of the analyte fraction. For optimum detection sensitivity, the goal should be to collect the analyte in as small a volume as possible. The elution solvent E should be quite strong, so that $K_A \approx 0$ for the analyte band during elution. Alternatively, the use of a weaker solvent E that still provides elution of the analyte (e.g., $K_A \approx 1$) will minimize the elution of more strongly retained matrix components/interferences that are preferentially left on the cartridge. Evaporation to dryness is often required, since the elution solvent E may be too strong or too large of a volume to be the sample solvent for the HPLC separation or too polar for a GC column. For this reason, choose a solvent E that is relatively volatile; otherwise, an excessive time for evaporation may be required.

A variant of the analyte retention (bind–elute) mode is the fractionation mode where one retains and then sequentially elutes different classes of analyte(s). This mode is similar in concept to gradient elution liquid chromatography (LC) using step gradients.

An alternative approach to practice SPE is to adjust the loading solvent such that the matrix and interferences are retained while allowing the analyte(s) of interest to pass through the cartridge unretained (schematic not shown here). Earlier, we referred to this mode as the matrix adsorption or interference removal mode. Since the analyte passes through in its original sample concentration, this option does not provide for any concentration of the analyte in the collected fraction. It is also not possible to separate the analyte from more weakly retained interferences that will also pass through the cartridge with the analyte(s). Therefore, this SPE mode usually provides "dirtier" analyte fractions, whereas the bind–elute procedure of Figure 4.3 allows the separation of analyte from both weakly and strongly retained sample components. For this reason, the matrix adsorption procedure is used much less often and will not be discussed further.

4.7. SPE METHOD DEVELOPMENT

In order to have a rugged and robust SPE method, a systematic approach should be used to ensure that one reaches the goals of maximum analyte purity, recovery, and reproducibility. Each of the four steps of SPE must be optimized during the process of method development. Figure 4.4 gives a generic approach to addressing each of the steps that should be performed in successful method development. In SPE, one must consider the following interactions: analyte ↔ sorbent; matrix ↔ sorbent; analyte ↔ matrix. Of course, the interaction of the solvent with all of these individual entities is also of major importance. By choosing the appropriate mode (e.g., adsorption, RP, ion exchange, etc.); sample capacity and retention characteristics of the sorbent; choice of the solvents, buffers, pH, and so on, used in the various steps; the flow rate; temperature; and other experimental parameters, one can reach the goals of SPE.

4.7.1. Research the Problem

SPE has been around a long time, and there are thousands of publications in the literature, textbooks, Internet, applications bibliographies, and manufacturer's websites. There is a good chance that your analyte (or type of analyte) and matrix have already been addressed. You can save yourself

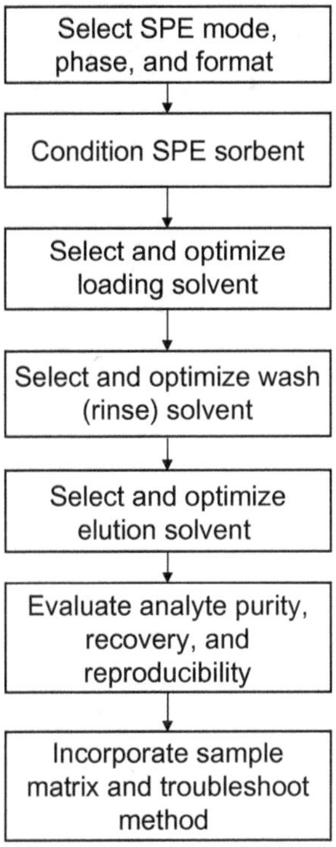

Figure 4.4. Systematic SPE method development.

a great deal of time to start with the SPE conditions already worked out.

4.7.2. Characterize the Analyte

In many cases, you know the structure and some of the chemical and physical properties of the analyte. The analyte's functional groups and polarity and other chemical properties can dictate the SPE mode that will allow maximum retention. Parameters, such as its solubility in various solvents, stability, pK_a, and log P (octanol/water partition coefficient), can help to maximize (or minimize) retention on the selected stationary phase. The concentration of analyte will have a profound effect on the experimental approach used to isolate it. Concentrations in the parts per billion or lower present challenges in recovery, purity, and reproducibility, while high concentrations make method development much easier.

4.7.3. Characterize the Sample Matrix

Knowledge of the sample matrix is important in selecting the appropriate loading, washing, and elution conditions. If the matrix is quite different than your analyte of interest, it may make method development easier. Possible compounds that have similar functional groups or pK_a to your analyte can make method development more difficult. Knowledge of the matrix solubility characteristics, stability, pH, ionic strength, and so on, can give one an idea of mode/sorbent selection.

4.7.4. Develop or Apply Effective HPLC or GC Method

Obviously, one needs an analytical method to monitor the progress of method development to determine purity, recovery, and reproducibility. Sometimes, because of their similarity, the development of a successful analytical method for HPLC gives one an idea of the mode and retention characteristics that can be employed for SPE.

4.7.5. Select and Test Sorbents/Loading and Elution Solvents

Once the mode is selected, the sorbent that provides maximum analyte retention should be determined. There are tens of different sorbents available for all modes. Fine-tuning the SPE cleanup can be further enhanced by the use of "mixed-mode" SPE phases (see Section 4.10.1). For example, a phase that includes both ion exchange and RP characteristics can be used to advantage in the cleanup of ionizable analytes. Also, an SPE cleanup that is "orthogonal" to the analytical column (i.e., has different selectivity or sorption characteristics) is likely to result in less overlap of analyte peaks by interferences.

A study of loading solvents on this sorbent to maximize retention yet yield high recovery should be performed. Finally, this is the time to determine a good elution solvent and volume to ensure complete elution.

4.7.6. Identify Optimum Wash (Rinse) Solvent

The most effective wash solvent should be found that will not elute the sorbed analyte yet allow interferences to be removed. Adjusting the pH of the wash solvent can be an effective way to moderate the retention and/or release of the analyte (e.g., acidic analytes will be more retained at low pH, and less retained at high pH).

4.7.7. Test Blank and Fortified Matrix

Once the method is optimized for analyte recovery, purity, and reproducibility, a fortified matrix as well as a blank should be run using the optimized wash and elution solvents.

4.7.8. Test Real Samples and Fortified Samples

The method should now be tested with real and spiked samples to see if analyte purity, recovery, and reproducibility are affected. If so, further refinements in the method have to be made.

A method development flowchart for SPE is depicted in Figure 4.5. This flowchart guides one through the main steps in developing an SPE method. For very complex samples, additional sample preparation protocols may be required

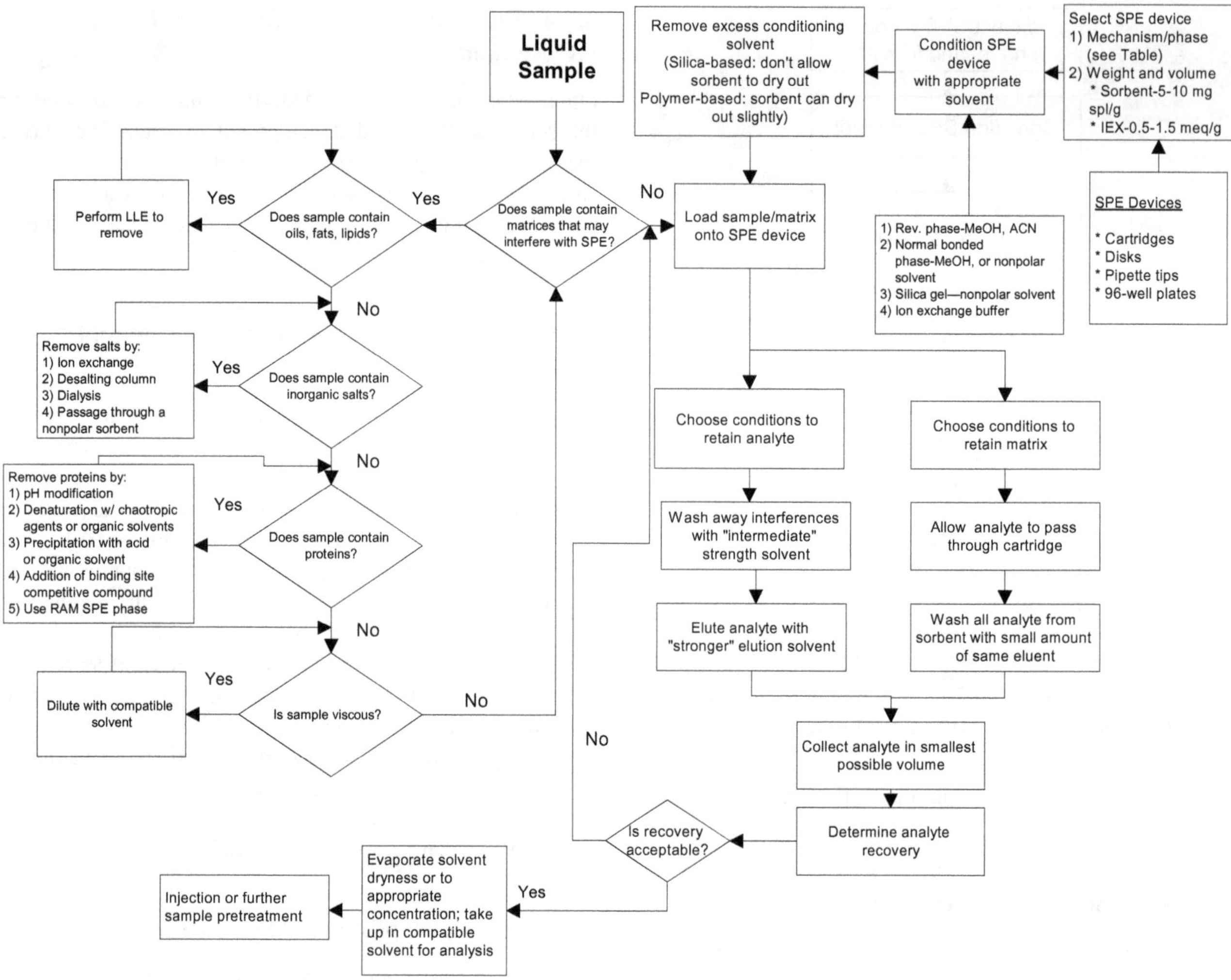

Figure 4.5. Method development flowchart for SPE.

prior to or after the SPE cleanup. However, often an optimized SPE method provides a sufficiently clean sample for direct analysis. The high selectivity and sensitivity of tandem mass spectrometry (MS/MS) has allowed simpler sample preparation protocols to be used.

Once an SPE method is developed, validation may be required. Some of the variables that should be considered when validating a method are depicted in Table 4.3. Figure 4.6 shows one of the more important investigations that should be conducted—the effect of flow rate and recovery in SPE, especially for the loading, washing, and elution steps. There is a kinetic element in sorption. The analyte needs time to diffuse into the porous SPE support and interact with the active functional group. If the flow rate is too fast, there is insufficient time for this equilibrium to occur. Too slow of a flow rate will just waste time. The flow rate for ion exchange SPE is even more important since ion exchange kinetics can be rather slow. Typically, a flow rate of 5 mL/min or less for adsorption and normal phase cartridges and 1–2 mL/min for ion exchange phases is sufficient for a typical cartridge format (e.g., 200 mg/3 mL).

Another important parameter to be investigated is sample breakthrough. The sorbents used in SPE have a finite sample capacity. Placing too large of a sample mass or volume on the SPE device can result in loss of analyte resulting in low recovery. The "sample" in this case is the total contribution of the analyte, the interferences, and the matrix since all components of the sample use up sorbent active sites. To determine the breakthrough, a sample solution is pumped through the SPE device at a slow flow rate (1–2 mL/min) while monitoring the baseline, say, with a UV detector or other detectors that can measure all sample components. Figure 4.7 shows that the sample will eventually saturate the stationary phase and elute from the exit of the cartridge. The breakthrough volume (V_B) represents the point where the breakthrough begins. V_M represents the maximum volume of sample that can be placed onto the SPE device. By knowing the concentration of the sample and the flow rate, one can calculate the mass of sample that can be handled by a particular weight of sorbent. Ensure that this mass is not exceeded; otherwise, analyte loss may occur and the recovery will be decreased.

Table 4.3. Variables to Consider in SPE Method Validation

Test Required	Experimental Parameter(s) to be Investigated
Sorbent	Phase selection, weight, cartridge format and size, check different lots
Conditioning solvent	Solvent strength (weak or strong solvent), contact time, volume
Loading solvent	Type, volume, % organic, pH, ionic strength, flow rate, breakthrough volume, drying time, analyte recovery/loss, interference/matrix removal
Wash solvent	Type, volume, % organic, pH, ionic strength, flow rate, analyte recovery/loss, drying time, interference/matrix removal
Elution solvent	Type, volatility, strength, volume, flow rate, pH, ionic strength, interference/matrix removal, analyte recovery/loss
Analyte and matrix stability	Tested in each step of method
Sample/matrix loadability	Different analyte concentrations
Detectability	Limits of detection (LOD), limits of quantitation (LOQ)
Method linearity and range	Tested over expected concentration range of analyte, test as function of matrix loading

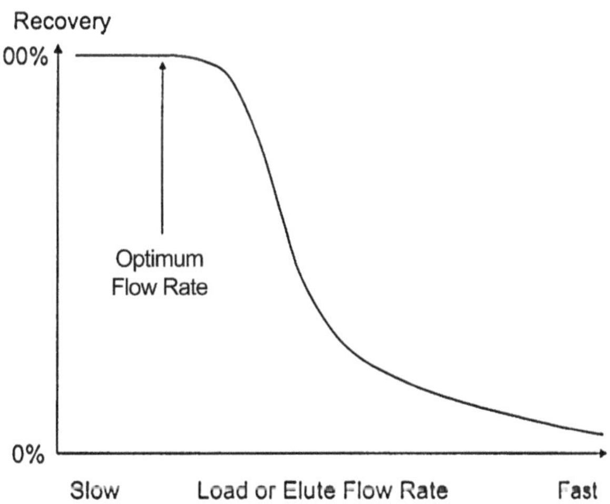

Figure 4.6. Relationship between sample recovery and flow rate in SPE.

4.8. SPE PHASES

Because SPE represents a low-efficiency adaptation of HPLC, many packings used in HPLC are also available for SPE. Table 4.2 lists the more popular SPE packings and the analyte types for which they are suited. Bonded silicas are used most often, but other inorganic and polymeric materials are commercially available. Compared with HPLC, as pointed out earlier, the SPE packing materials are usually of

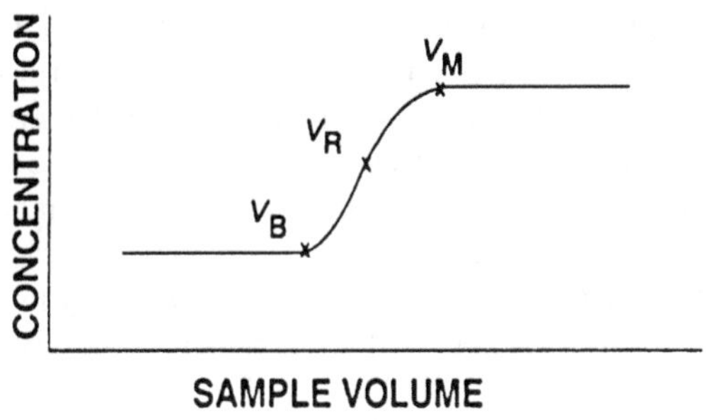

Figure 4.7. Breakthrough curve for SPE device. V_B, breakthrough volume; V_R, retention volume; V_M, maximum sampling volume.

larger particle size and often irregularly shaped. In addition to the generic packings shown in Table 4.2, specialty packings are available for the isolation of drugs of abuse in urine and plasma,[13,14] oil and grease from wastewater,[15,16] catecholamines from plasma,[17] and many other popular assays. Florisil (activated magnesium silicate) and alumina are most often used for the cleanup of pesticide residues for GC and GC-MS analysis; many published methods exist[18] for the isolation of pesticides using Florisil. The use of graphitized carbon for SPE has been increasing, especially for the selective removal of pigments such as carotenoids and chlorophyll from plant extracts. For instructions on the use of specific SPE products, consult the manufacturer's literature or one of the textbooks on the subject.[19–23]

Due to the nature of electrostatic interactions, ion exchange can be a powerful and selective SPE technique for ionic and ionizable compounds. Cation exchange packings are used for protonated bases and other cations, while anion exchange packings are used for ionized acids and other anions. Ion exchange packings come in two forms: "strong" and "weak"; strong ion exchangers are normally preferred if strong retention of the analyte is the main objective. Ionization (and thus retention) of weak ion exchangers is a function of pH. The choice of pH is a compromise between maintaining the ionic character of the stationary phase, while ensuring that the ionic analyte is remains in an ionic state. Thus, pH becomes a powerful variable for both optimizing retention and releasing the analyte from a weak ion exchanger. Weak ion exchangers often find use for compounds that are ionized at most pH values, such as alkyl sulfonates (by weak anion exchange) or tetraalkylammonium salts (by weak cation exchange).

SPE cartridge packings are generally of lower quality and cost than corresponding high-purity HPLC packings, and this contributes to the problem of batch-to-batch retention variability. Whereas high-purity, type-B silica-based column packings are preferred in RPC, RP-SPE packings will generally be more "acidic" (type A); their silanol interactions will tend to be more pronounced and more variable from lot to lot. However, because SPE is usually practiced as an "on–off" (digital) technique, small differences in retention should be less important than in HPLC, where small differences in selectivity can be more important.

Compared with silica-based SPE packings, polymeric packings have several advantages: (1) higher surface area (thus higher capacity); (2) better wettability; (3) tolerance to partial drying after the conditioning step without affecting recovery and reproducibility; (4) an absence of silanols (less chance of irreversible adsorption of highly basic compounds); and (5) a wide pH range (more flexibility in adjusting conditions). However, they are more expensive, and, compared with silica SPE, there are fewer stationary phases available but they are growing strongly due to their advantages. When packed into 96-well plates, due to their higher capacity, only a few tens of milligrams are required for most assays. By miniaturizing the entire process, sample and solvent requirements for the various SPE steps are decreased proportionally.

4.9. AN EXAMPLE OF SPE METHOD DEVELOPMENT: THE ISOLATION OF *N*-ACYL HOMOSERINE LACTONES (AHLs) IN GROWTH MEDIA

AHLs are autoinducers that mediate bacterial cell–cell communication, and the determination of their presence at trace amounts in culture medium is important. Direct injection of the media into an HPLC column will quickly block it. LLE has been used to remove AHLs, but large amounts of organic solvent are required. SPE was developed as an alternative technique prior to LC-MS/MS.[24] The structures of the AHLs of interest are shown in Figure 4.8a. After the development of an RP HPLC method using a C18 sub-two micron column, a study was done to select an optimum SPE phase. Seventeen sorbents (normal phase, RP, ion exchange) were tried. The presence of various functional groups permitted the use of an RP mechanism (hydrophobic groups), normal phase (polar functional groups), or an ion exchange mechanism (the secondary amine functionality). Some preliminary experiments revealed that cation exchangers could not be used for discriminating the AHLs since a fraction of the loaded compounds would not elute and recovery was low. As the normal phase sorbents could not discriminate the various carbon numbers and recovery was dependent on loaded amount, they were dismissed. Since an RP mechanism was used for the HPLC analytical method and the AHLs could be discriminated by carbon number, seven RP-SPE sorbents with different properties were further investigated. Preliminary SPE experiments revealed that

a

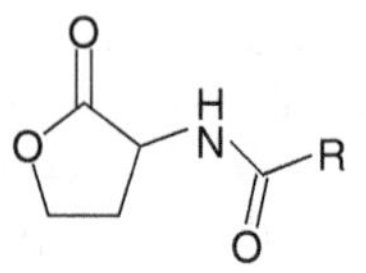

R	Name	Abbreviation	Log *P**
$—C_2H_4—CH_3$	*N*-butanoyl-homoserime lactone	C_4-AHL	0.03
$—C_4H_8—CH_3$	*N*-hexanoyl-homoserime lactone	C_6-AHL	1.02
$—C_5H_{10}—CH_3$	*N*-heptanoyl-homoserime lactone	C_7-AHL	1.94
$—C_6H_{12}—CH_3$	*N*-octanoyl-homoserime lactone	C_8-AHL	1.97
$—C_8H_{16}—CH_3$	*N*-decanoyl-homoserime lactone	C_{10}-AHL	2.96
$—C_{10}H_{20}—CH_3$	*N*-dodecanoyl-homoserime lactone	C_{12}-AHL	4.02
$—C_{12}H_{24}—CH_3$	*N*-tetradecanoyl-homoserime lactone	C_{14}-AHL	5.09

b

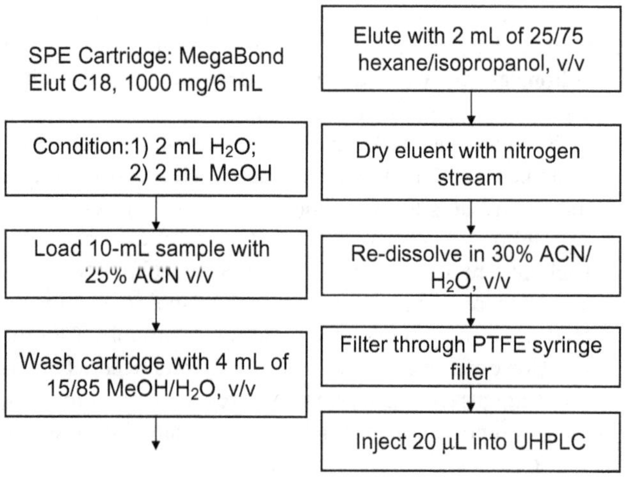

Figure 4.8. Example SPE method. (a) Chemical structures of AHLs. *Log *P* is *n*-octanol/water partition coefficient expressing the analyte's hydrophobicity. (b) SPE method flowchart. UHPLC, ultra high pressure liquid chromatography.

breakthrough occurred for three of the seven, so final experiments were performed on the remaining four: (A) Chromabond HR-P (Macherey-Nagel, Duren, Germany), an end-capped C18, 200 mg/3 mL cartridge; (B) Octadecyl Polar Plus (Mallinkrodt/Baker, St. Louis, MO), a C18 phase, 2000 mg/6 mL cartridge; (C) MegaBond Elut C18 (Varian), an end-capped C18, 1000 mg/6 mL cartridge; and (D) Bond Elut PPL (Varian), a modified polystyrene-divinylbenzene (PS-DVB), 200 mg/3 mL cartridge.

Acetonitrile was added to spiked samples at volume ratios between 10% and 40% (v/v) to reduce the matrix effect. A 25% (v/v) volume ratio of acetonitrile added to the aqueous sample gave the greatest reduction of the matrix effect. A study of wash solvent was undertaken. After conditioning the sorbents with 2 mL of water followed by 2 mL methanol, a 2 mL sample of aqueous stock solution of the seven AHLs under investigation was loaded onto the cartridge. Water and water–methanol mixtures were tested as wash solvents. No significant decrease in recoveries were noted until the methanol content was lower than 15% (v/v) so the wash solvent was chosen to be 15:85 methanol : water. The addition of methanol to the wash solvent was useful to remove highly polar solutes present in liquid culture media extracts.

For elution (step 4), three solvents (acetonitrile, methanol, and isopropanol) were investigated. Best recoveries were noted with isopropanol, and it was used for further investigation. It was noted that with an increase in the alkanoyl side-chain length of the AHLs, recoveries were decreased on SPE phases B and D. The end-capped C18 sorbents gave the best overall recovery. However, after the isopropanol elution, sorbent A gave more interferences than sorbent C. So for final experiments, the MegaBond Elut C18 was used. In order to increase the recoveries of the long-chain AHLs (C12-AHL and C14-AHL), the polarity of the eluting solvent (isopropanol) was modified by the addition of hexane, toluene, and tetrahydrofuran at various volume ratios. Maximum recovery of the long-chain AHLs was determined to be with a 25:75, v/v hexane : isopropanol ratio.

The final SPE method is depicted in Figure 4.8b. In order to increase the sensitivity, a separate study revealed that 10 mL of sample could be loaded onto MegaBond Elut C18 cartridge (1000 mg/6 mL) with no loss of analyte. For spiked media, the relative standard deviations (RSDs) were found to range from 1% to 4%, quite acceptable for concentrations of 10 mg/L for the AHLs. Overall, recoveries of the AHLs of Figure 4.8a were all higher than 90%.

4.10. SPECIAL TOPICS IN SPE

4.10.1. Multimodal and Mixed-Phase Extractions

Most SPE methods use a single separation mode (e.g., RP) and a single SPE device (e.g., cartridge). However, when more than one type of analyte is of interest, or if additional selectivity is required for the removal of interferences, multimodal SPE can prove useful. Multimodal SPE refers to the intentional use of two (or more) sequential separation modes or cartridges (e.g., RP and ion exchange). Experimentally, there are two approaches to multimodal SPE. In the serial approach, two (or more) SPE cartridges are connected in series. Thus, for the separate isolation of acids, strong bases, and neutrals, an anion and cation exchange cartridge could be connected in series. By adjusting the sample and wash solvent to pH 7, both the acids and bases will be fully ionized. As a result, the acids will be retained on the anion exchange cartridge, the bases will be retained on the cation exchange column, and the neutrals will pass through both columns (separated from acids and bases). The acids and bases can then be separately collected from each cartridge. The experimental setup for this form of multimodal SPE is shown in Figure 4.9.

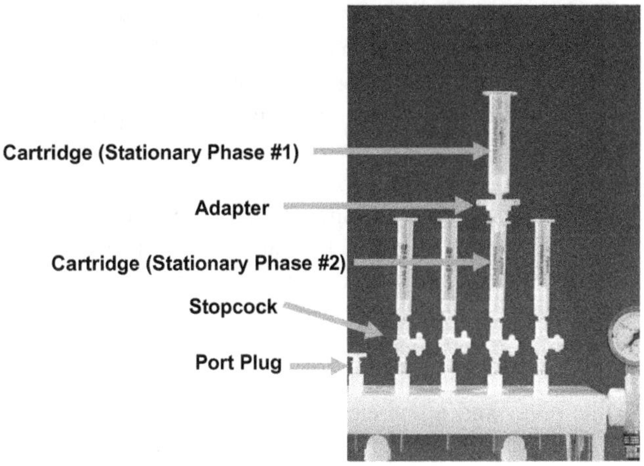

Figure 4.9. Experimental setup for multimodal SPE for cleanup of complex samples.

A second approach to multimodal SPE uses mixed phases. Here a single cartridge might possess two (or more) functional groups to retain multiple species, or to provide a unique selectivity. The multiple functional groups can be present on a single particle or can be mixed phases where some proportion of two separate packings are blended. One popular application of multimodal SPE is the isolation of drugs of abuse and other pharmaceuticals from biological fluids.[25] Still another version of multimodal SPE is the use of layered packings,[26] where two (or more) different packings are used to isolate differing molecular species.

4.10.2. Restricted Access Materials (RAMs)

RAMs are a special class of SPE packings used for the direct injection of biological fluids such as plasma, serum, or blood. Unlike SPE cartridges, RAMs are actually HPLC columns that incorporate sample preparation and—unlike SPE cartridges—are characterized by high plate numbers and reusability. RAMs are most often selected for the analysis of low-molecular-weight drugs, their impurities, and metabolites.[27–29] Many variations of these packings have been described: (1) internal-surface RPs, (2) shielded hydrophobic phases, (3) semipermeable surfaces, (4) dual-zone phases, and (5) mixed functional phases. See Reference 27 for a description and tabulation of commercial products.

Dual-mode porous packings are used in the most popular RAM columns. These packings are used typically for the analysis of drugs in blood, because proteins present in these samples will accumulate on an RPC column—leading to its failure after a few injections. The packing consists of small-pore particles with a C_8 or C_{18} layer covering the inside of the pores and a nonretentive hydrophilic layer covering the exterior of the particle. Proteins are unable to access the pores because of their large size and are unretained by the particle exterior; consequently, they pass through the column with $k \approx 0$. Small-molecule analytes can enter the pores and are retained sufficiently to elute after the proteins (often using gradient elution). Because proteins are not retained on these columns, a column can be used for a considerable number of samples. However, some care must be exercised in the choice of mobile-phase pH and organic solvent; otherwise, protein precipitation can occur—with precipitation on the column.

Alternatively, a RAM column can be connected via a switching valve to a conventional RPC column. The valve is initially positioned for elution of the RAM column to waste, and the sample is injected; after the proteins leave the column, the valve is switched to connect the two columns. Analytes are then eluted from the RAM column and enter the RPC column for further processing, usually by means of gradient elution. The RPC column also has a longer lifetime, as plasma proteins never contact this column.

4.10.3. Molecularly Imprinted Polymers (MIPs)

MIPs are among the most selective phases used in SPE, being designed for enhanced retention of a specific analyte. An MIP is a stable polymer with recognition sites that are adapted to the three-dimensional shape and functionalities of an analyte of interest (much like antibody binding). The most common approach involves noncovalent imprinting; this MIP synthesis is shown schematically in Figure 4.10. An analyte is used as a template and is chemically coupled with a monomer (most often methacrylic acid or methacrylate). After polymerization, the bound analyte is extracted to yield a selective binding site (receptor). The selective interactions between the analyte and the MIP include hydrogen bonding, ionic, and/or hydrophobic interactions. The action of an MIP is based on a "lock and key" fit, where a selective receptor or cavity on the surface of a polymer perfectly fits the analyte that was used to prepare the MIP. The concept is similar to immunoaffinity (IA) SPE phases, but obtaining a suitable antibody for these IA sorbents can be very time-consuming. Chapter 23 of this volume covers the basics of MIP technology, while review articles[30–33] and

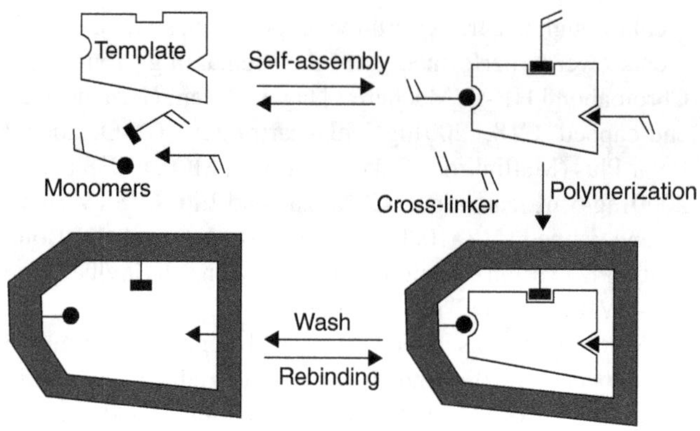

Figure 4.10. Synthetic route for molecularly imprinted polymers (MIPs).

a book[34] provide detailed information on the use and potential of MIPs in SPE.

Incomplete removal of the template analyte from the MIP during its preparation is one of its main problems. This residual template analyte frequently bleeds, resulting in baseline drift and interference with the assay of the desired analyte—especially for low analyte concentrations. This problem is most often addressed by using a structural analog to the template molecule as the template analyte. For example, one can employ a brominated version of the template rather than the chlorinated version. The imprint is very similar in size and therefore will be able to recognize the analyte of interest. Under actual use, the small amount of template bleed should not interfere with the chromatography, assuming that the brominated and chlorinated forms can be separated on the analytical column. There may be some swelling or shrinkage of the MIP with a change in solvent, which can modify the size of the receptor and reduce the retention of the target analyte. A major disadvantage of the MIP approach is that each sorbent must be custom made, either by the user or an outside supplier. Because of the high cost of synthesizing MIPs, their use is restricted to high-volume assays or when there is no other way to perform sample cleanup. Recently, several off-the-shelf MIPs have been commercialized, mainly for high-volume assays.

4.10.4. Immunoaffinity (IA) Extraction

IA packings are based on antibodies that are attached to a particle. As in the case of MIPs, the analyte is retained by a receptor that is highly complementary, so as to provide a "lock and key" fit. As a result, IA packings are quite specific for an individual analyte and are used for the selective extraction and concentration of individual compounds or classes of compounds from the sample—often in a one-step process. Antibodies for large biomolecules are readily available and have been used for many years in immunology and medical research (affinity chromatography). Because antibodies for small molecules are more difficult to obtain, the development of small-molecule IA extraction is more recent and less developed. Some excellent review articles describe IA extractions in more detail.[35–38]

As long as an antibody can be prepared, the numbers of IA packings can be almost unlimited. However, a great deal of time and effort is required in their production, so their use is restricted for the same reasons as for MIP packings. Nevertheless, several commercial IA packings have become available. Class-specific packings are available for a variety of pharmaceutical, food, and environmental applications.[39–40] An example of the successful use of an IA phase is the IA removal of high-abundance proteins from human blood plasma.[41–45] When molecular biologists and biochemists are looking for biomarkers that may be present in blood at parts per billion or lower concentrations, the highly abundant ("housekeeping") proteins will obscure these tiny amounts of biomarkers. So antibodies specific for human serum albumin, IgG, and up to 18 other high-abundance proteins are bonded to a polymeric media and used in a flow-through system. The high-abundance proteins are specifically removed allowing the low-abundance proteins to pass through the column. These low-abundance proteins are collected and trace enriched for further LC-MS/MS investigations.

4.10.5. Quick, Easy, Cheap, Effective, Rugged, and Safe (QuEChERS) Extraction and Dispersive SPE

QuEChERS is an extraction technique for the sample preparation of multiresidue, multiclass pesticides in fruits and vegetables.[46] The technique uses simple plasticware and a minimal amount of organic solvent. A QuEChERS method flowchart is depicted in Figure 4.11. QuEChERS has two main steps: (1) an extraction step and (2) a dispersive SPE step. First, the addition of a water-miscible, hydrophilic solvent such as acetonitrile or acetone to a homogenized portion of a sample allows the extraction of pesticides into the solvent. Subsequent addition of salt (e.g., $MgSO_4$ + NaCl) leads to separation of the organic solvent from water associated with the sample and promotes extraction of pesticides into the organic solvent ("salting out" effect). The internal standard is added next, and after shaking and centrifugation, an aliquot of the organic phase is subjected to further cleanup using dispersive SPE. Dispersive SPE involves the addition of small amounts of bulk SPE packing (e.g., C18, graphitized carbon, amino) to the extract for the purpose of removing interferences from the organic solvent. After centrifugation, the supernatant is sampled and analyzed. The analytical technique is usually LC- or GC-MS (or MS/MS) that provides sufficient sensitivity and selectivity to analyze trace amounts of pollutants in a wide variety of sample matrices.

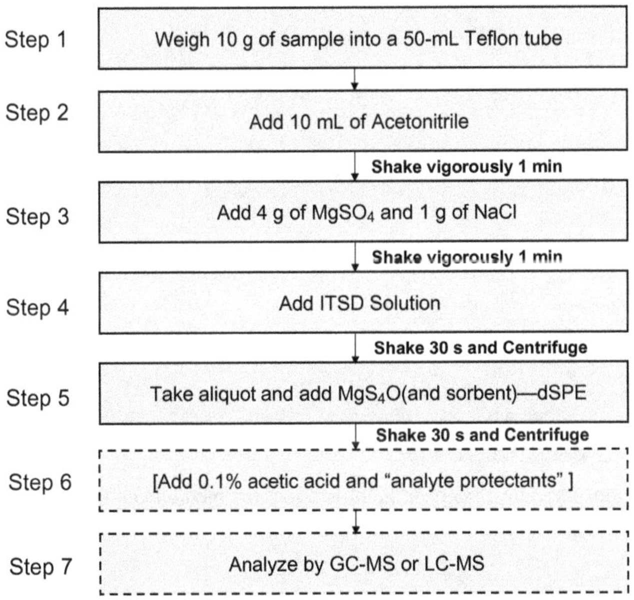

Figure 4.11. Flow diagram of the QuEChERS process. ITSD, internal standard; dSPE, dispersive solid-phase extraction.

QuEChERS has been found particularly useful for screening the food supply for multiple pesticides. Official QuEChERS methods from the American Association of Official Analytical Chemists (AOAC)[47,48] and the European EN 15662: 2007[49] are now available. QuEChERS has been investigated for hundreds of pesticides in a variety of fruit and vegetable matrices;[50–52] analyte recoveries (for concentrations of 100 ng/g) generally range from 70% to 110%, with a variability of less than 10%. The technique has been extended to new matrices such as meat products, as well as other analytes such as antibiotics and other drugs.[53–57]

To illustrate the application of QuEChERS, 16 representative pesticides were extracted from apples and analyzed by LC-MS/MS.[58] The pesticide mix was chosen to be representative of various classes of interest. Using the AOAC Method 2007.01, pesticides were spiked at three different levels (10, 50, and 200 ng/g) into a finely comminuted organically grown apple sample. The sample was prepared according to the AOAC method.[47] The AOAC method employs an acetonitrile extraction using $MgSO_4$ + NaOAc/HOAc salt addition, which is buffered. After centrifugation, the extract was treated with anhydrous $MgSO_4$ + primary–secondary amine (PSA) bonded phase sorbent, which further cleaned it.

After further centrifugation, the extract was ready for LC-electrospray ionization (ESI)-MS/MS analysis. The powerful selectivity of this technique using the multiple reaction monitoring (MRM) mode allows one to optimize the selectivity for each compound. The blank apple extract was very clean (not shown), and the chromatogram of the 5 ng/g fortified apple extract shown in Figure 4.12 allowed the identification and quantitation of all 16 pesticides, even when some were chromatographically unresolved. The recoveries averaged 97.5% and RSDs averaged 4.5%, which were quite acceptable for such low levels of pesticides. For more information, consult the original reference.[58]

a

Peak Number	Pesticide	Category
1	Methamidophos	Organophosphate
2	Acephate	Organophosphate
3	Pymetrozine	Pyridine
4	Carbendazim	Benzimidazole
5	Imidacloprid	Neonicotinoid
6	Thiabendazole	Benzimidazole
7	Dichlorvos	Organophosphate
8	Propoxur	Carbamate
9	Thiophanate-methyl	Benzimidazole
10	Carbaryl	Carbamate
11	Ethoprophos	Organophosphate
12	Penconazole	Triazole
13	Cyprodinil	Anilinopyrimidine
14	Dichlofluanid	Sulfamid
15	Kresoxim-methyl	Strobilurin
16	Tolyfluanid	Sulfamide

b

Instrument Condition

HPLC Conditions

Column: Agilent Eclipse Phenyl-Hexyl 150 mm x 3.0 mm. 3.5 μm (p/n 959963-312)

Flow rate: 0.3 ml/min

Column temperature: 30 °C

Injection volume: 10 μL

Mobile Phase: A: 5mM NH_4OAc, pH 5.0 in 20:80 Me OH/H_2O
B: 5mM NH_4OAc, pH 5.0 in ACN

Needle wash: 1:1:1:1 ACN/MeOH/IPA/H_2O (0.2% FA)

Gradient:

Time	% B	Flow rate (mL/min)
0	20	0.3
0.5	20	0.3
8.0	100	0.3
10.0	100	0.3
10.01	20	0.5
12.0	100	0.5
13.0	STOP	

Post run: 4 min

Total cycle time: 17 min

MS conditions

Positive mode

Gas Temperature: 350 °C

Gas Flow: 10 L/min

Nebulizer: 40 psi

Capillary: 4000 V

c

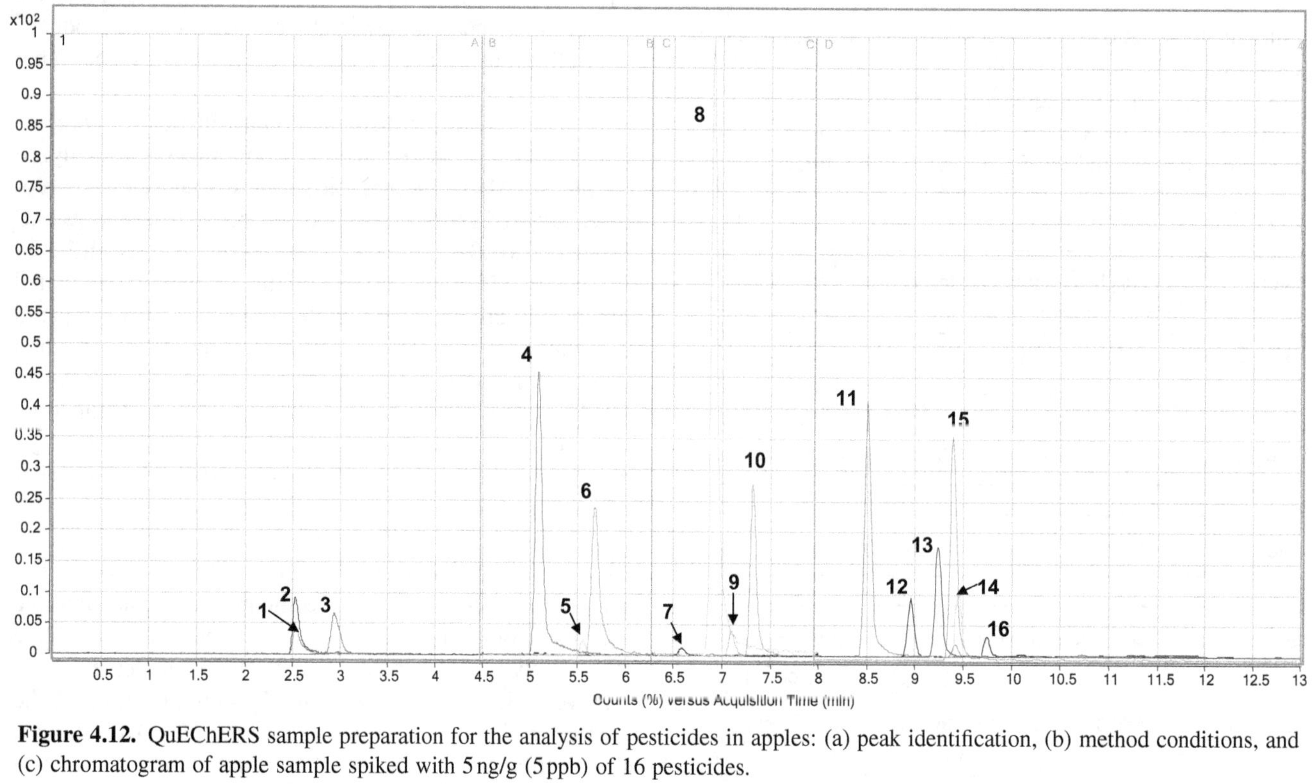

Figure 4.12. QuEChERS sample preparation for the analysis of pesticides in apples: (a) peak identification, (b) method conditions, and (c) chromatogram of apple sample spiked with 5 ng/g (5 ppb) of 16 pesticides.

4.11. CLASS-SPECIFIC SPE CARTRIDGES

Over the years, specialty phases have been introduced that are compound or class specific. While MIP and IA packings can provide extreme specificity or selectivity, specialty packings with special functional groups can selectively interact with certain compound classes.

4.11.1. Phenylboronic Acid (PBA) Phase

Under basic conditions, immobilized PBA selectively binds analytes that possess vicinol diols (e.g., sugars and catechols). Other compounds that are also selectively retained include alpha-hydroxy acids, aromatic *o*-hydroxy acids and amides, and amino alcohol-containing compounds. Covalent bonds between packing and analyte are formed, allowing interfering compounds to be washed from the packing with a variety of different solvents. Once washed, the covalent bonds can be broken by washing the phase with an acidic buffer/solvent that hydrolyzes the covalent bonds. A popular application of the PBA phases is the isolation of catecholamines in biological fluids.[59]

4.11.2. Crown Ethers as SPE Phases

The unusual ability of the crown ethers to selectively recognize cations has led to the development of unique SPE phases that display excellent selectivity for a class of metal ions or even a single metal ion in the presence of other ion orders of magnitude higher in concentration.[60] In particular, the 3M Rad Disks combine the Empore PTFE disk technology with the crown ether specificity to provide sorbents that are extremely selective for radium, technetium, and strontium. The Rad Disks have found applications for the selective removal of radiochemicals from aqueous solution. Trace levels of Technetium 99 can be determined in the presence of many other ions that may be present in drinking water, surface water, and groundwater. Nitrate levels up to 1000 mg/L do not interfere.

4.11.3. Ion Removal by SPE

In many applications, especially in ion chromatography, high concentrations of ionic components from the sample matrix can be undesirable interferences. The use of ion exchanger resins with specific functionalities can be used to remove these ionic species. For example, a strong cation exchanger in the barium form will selectively remove high concentrations of sulfate from aqueous solution. The same can be said for a cation ion exchanger in silver form for the removal of chloride ion. Chelating ion exchangers can remove transition metals.

For example, salt brine contains large amounts of sodium chloride, and if one is interested in the analysis of trace amounts of other anions, chloride ion can dominate the chromatogram. Figure 4.13 shows an example of how an

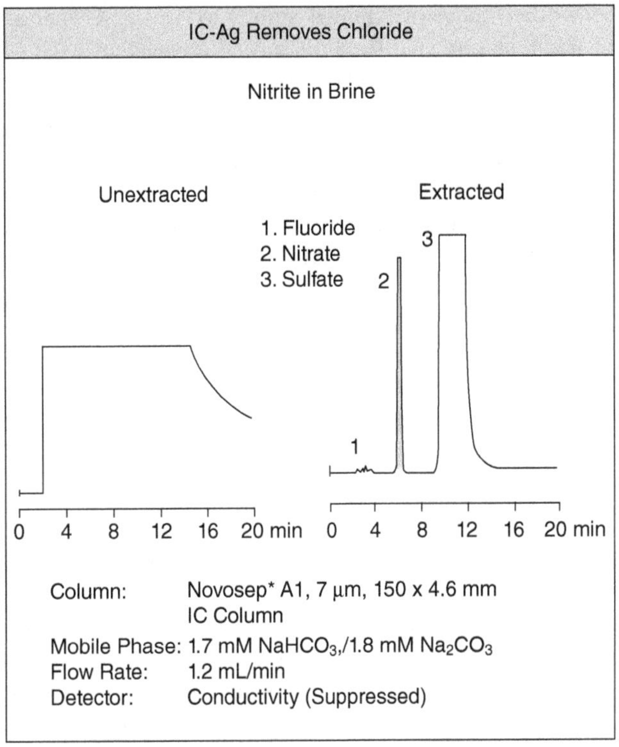

Figure 4.13. Chloride removal using IC-Ag (figure courtesy of Grace Davison Discovery Sciences, Deerfield, IL).

Extract-Clean SPE IC-Ag cartridge (Grace Alltech, Columbia, MD) can help in this situation. The left chromatogram shows how chloride dominates the ion chromatogram, while after the removal of the chloride ion, a much cleaner chromatogram is observed.

4.12. COLUMN SWITCHING

Closely related to SPE, online column switching can be a powerful technique for sample preparation as well as for two-dimensional separations. For sample preparation, a portion of the chromatogram from an initial column (column 1, the enrichment column) is selectively transferred to a second column (column 2, the analytical column) for further separation. Column switching for sample preparation is used for

- removal of "column killers" prior to column 2,
- removal of late eluters prior to column 2,
- removal of interferences that can overlap analyte bands in column 2, and
- trace enrichment.

The achievement of one or more of these goals often results in increased sample throughput due to the presentation of a cleaner sample to the column. Unwanted interferences can be directed to waste or backflushed to prevent their transfer to the HPLC column. The goal of column switching is separation of the analyte from interfering compounds by the initial column, that is, the same goal as for SPE.

While column switching is similar to the HPLC analysis of fractions provided by SPE, several advantages exist (d_p is the diameter of the sorbent particle):

- SPE cartridges are used only once and discarded; the initial column in column switching is used repeatedly, which is more cost-efficient.
- The initial column has a higher efficiency (e.g., 5 μm d_p) compared with an SPE cartridge (e.g., 40 μm d_p), making it easier to divert an entire peak or a major portion of a peak to column 2.
- Less sample loss occurs with column switching since it is a closed system.
- Many valve configurations are possible for heart cutting, backflushing, diverting contaminants directly to waste, and so on.

Of course, there are a few disadvantages associated with column switching. First, the system requirements are more complex than SPE with valves, tubing, precise timing requirements, and electronic interfaces required. Samples must be particulate free since the frits on column 1 will retain them and eventually block the column or, at least, experience a pressure buildup. Strongly retained sample components and matrix interferences can build up on the initial column requiring replacement or periodic cleaning.

4.13. PROTEIN PRECIPITATION VERSUS SPE FOR DRUGS IN BIOLOGICAL FLUIDS

One of the important application areas of SPE is the isolation of drugs and drug metabolites in biological fluids (i.e., urine, plasma, cerebral spinal fluid). These low-molecular-weight compounds can be selectively removed from biological fluids using well-established screening protocols. Recently, in the pharmaceutical laboratory, there has been a movement toward the use of protein precipitation rather than SPE for this task. The widespread use of LC-MS/MS has allowed a reduction in the purity requirement for drug/metabolite isolation. The selectivity and sensitivity of MS/MS provides excellent results on samples that are only partially clean, provided that little or no ion suppression is present.

To illustrate a comparison of techniques, the measurement of the drug amitriptyline in plasma was investigated using three different sample preparation protocols: protein precipitation, SPE on a neutral polymeric sorbent, and a cation ion exchange sorbent.[61] Protein precipitation, also referred to as protein "crashing," is used to remove proteins that may interfere with the separation by fouling the HPLC column. These experiments are frequently performed in 96-well crash plates, which either contain 96-wells of 1 or 2 mL volume or can be a flow-through 96-well filtration plate. Only a small amount (50 μL) of plasma is required, which is important for small animal studies. The protein is precipitated by the addition of a small volume of acetonitrile, acid, or salt. After agitation, the precipitated protein forms a bead that can be removed by centrifugation or by filtering using a vacuum or positive pressure. The supernatant is then collected, evaporated to dryness, reconstituted in a compatible solvent to 75 μL, and injected into an RP HPLC column connected to an MS/MS system. The resulting chromatogram (chromatographic conditions not shown), shown in Figure 4.14a, is quite messy due to the large number of plasma components that were not removed by simple protein precipitation. In addition, the presence of these matrix compo-

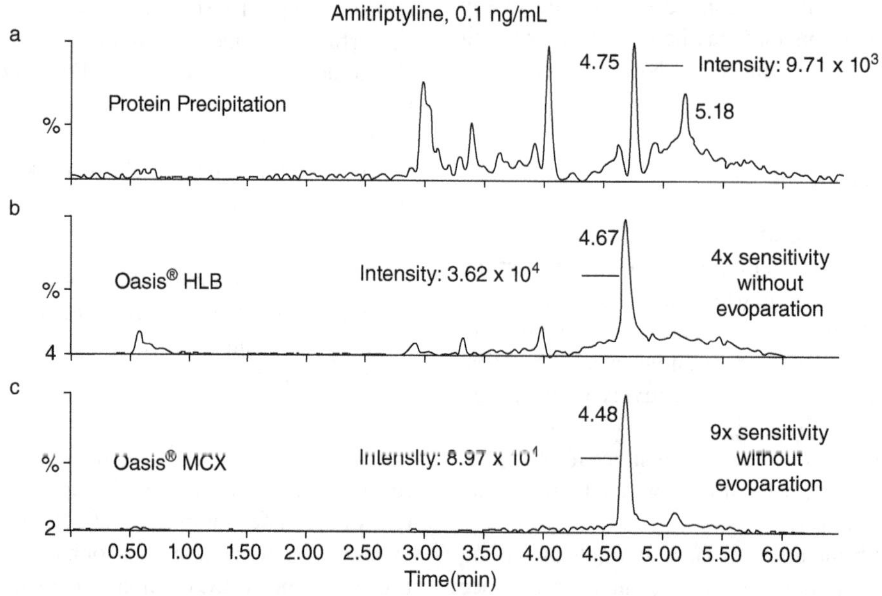

Figure 4.14. Comparison of protein precipitation and SPE cleanup of plasma using the 96-well plate format—Oasis μElution Plate. LC/MS runs using water-ammonia gradient on Xterra MS column (courtesy of Waters).

nents in the injected samples also resulted in ion suppression with resulting loss of signal as evidenced by the measured intensity of the amitryptyline peak shown at 4.75 min.

For the SPE experiments, 96-well flow-through SPE plates were used that were packed with Oasis (Waters, Milford, MA) *m*-divinylbenzene-*N*-vinylpyrrolidone copolymers. The Oasis HLB is a neutral polymer, while the MCX is the same base polymer but derivatized with sulfonic groups, making it a strong cation exchange packing. Because of the high capacity of the two polymeric packings, only 5 mg is packed into each well. This reduced amount of material cuts down on the amount of sample and solvents used for the experiments. The SPE protocol for the Oasis HLB plate involved the conditioning (MeOH), equilibrating (water), loading (50 µL of rat plasma spiked with 0.1 ng/mL with amitryptyline diluted with 50 µL of water with an added internal standard), washing (5% MeOH in water), and eluting (ACN : IPA 40:60 v/v containing 2% formic acid) steps. The SPE protocol for the Oasis MCX plate had the same conditioning, equilibration, and loading steps, but the wash step (wash #1: water with 2% formic acid; wash #2: MeOH) and the elution step (ACN : IPA 40:60 v/v containing 5% ammonia) were different. Both filtrates were diluted to 75 µL so that they could be compared directly with the protein precipitation experiment. Figure 4.14b shows the chromatogram observed with the Oasis HLB plate, while Figure 4.14c depicts the chromatogram for the Oasis MCX cation exchange method. Note that in both cases, the signal intensity observed for the amitryptyline was significantly higher than for the protein precipitation cleanup (Fig. 4.13a). In fact, the peak intensity of amitryptyline for the HLB cleanup was 4× the sensitivity of the crash experiment, while the MCX gave 9× the sensitivity, even with the evaporation step left out for both SPE experiments. The higher intensities were undoubtedly due to the reduction of ion suppression. In addition, the chromatograms were cleaner with interferences substantially reduced compared with protein precipitation. The selectivity of the cation exchange cleanup was much better than the neutral resin cleanup. Only compounds that showed a positive charge were retained, and the rinse step removed many of those compounds that showed up in the HLB extract.

This example showed that SPE is still a useful cleanup method even though protein precipitation is simpler. Despite this advantage, workers still prefer to use protein crashing since there is virtually no method development time required, relying on the sensitivity and selectivity of the MS/MS system to provide a quick answer.

4.14. TECHNOLOGIES RELATED TO SPE

4.14.1. SPME/Single-Drop Microextraction (SDME)

The simplicity and low cost of SPME, developed by Arthur and Pawliszyn,[62] in 1990, has made it into a popular sampling and sample preparation technique for GC and, to a lesser extent, for LC. In SPME, a fiber coated with a polymeric stationary phase is placed into a solution or in the headspace of the sample, and analytes diffuse and/or are moved by convection into the stationary phase. The concentrated analytes are transferred to the chromatography column by thermal desorption (GC) or liquid extraction (LC). The popularity of the technique has spurred the development of similar technologies such as SBSE (see Chapter 5 of this book).

One such technology termed "SDME" describes a configuration where a droplet of solvent contained at the end of a PTFE rod or syringe needle replaces the coated fiber. The analytes diffuse into this droplet in a similar manner as into the SPME fiber. The original work first described by Cantwell and Jeannot.[63] A schematic of the SDME experiment is shown in Figure 4.15. When equilibrium is achieved, the microdrop is retracted into the syringe and its volume injected into an injector manually or the entire experiment can be performed with a programmable autosampler.

One advantage of SDME is that the extracted sample contained in a small volume (1 or 2 µL) of organic solvent could potentially be transferred directly into an HPLC injector. Interfacing SPME to HPLC is more difficult since the procedure involves a complex instrumental arrangement, and the rate of dissolution of many analytes from the SPME fiber is quite slow, resulting in the initial band spreading during the displacement to the HPLC column.

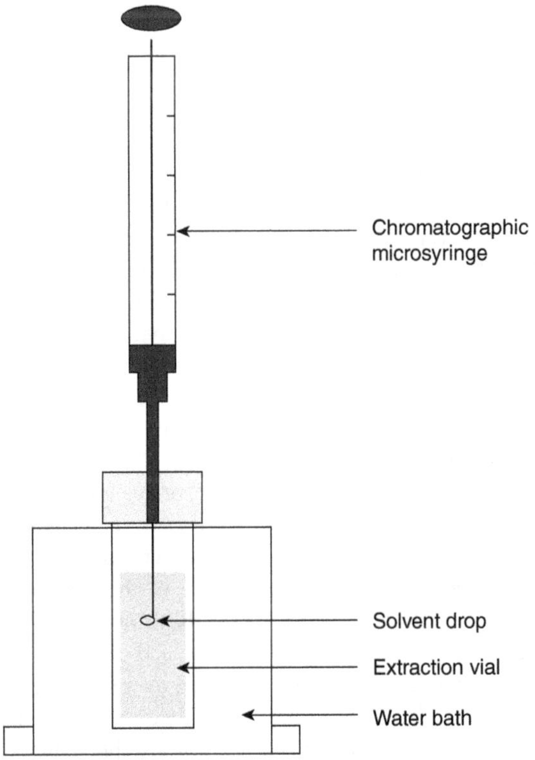

Figure 4.15. Schematic of the experimental setup for single-drop microextraction.

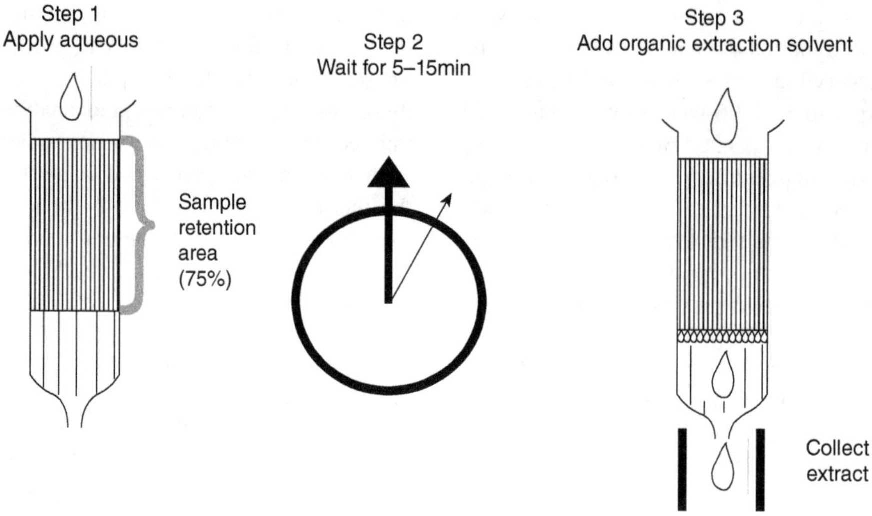

Figure 4.16. Schematic representation of supported liquid–liquid extraction. (courtesy of Biotage, Charlottesville, VA).

A recent publication described single-drop liquid-phase microextraction followed by HPLC for the analysis of hypercins in deproteinated plasma and urine.[64] Rather than attempting to inject the droplet directly into an HPLC injector, the authors transferred the droplet to a microvial and diluted the sample to 30 µL with an aqueous compatible solvent (methanol) for injection into an RP HPLC column. With some additional work on a better transfer system, it would be possible to perform online extractions and injection. Sometimes, the use of a hollow fiber filled with a small volume of organic solvent is more useful in containing the solvent droplet. This procedure is referred to as liquid-phase microextraction.

4.14.2. Solid-Supported LLE

Instead of using a separatory funnel to perform LLE, one can immobilize one liquid phase in an inert medium packed into a polypropylene tube and percolate the other immiscible liquid phase through the immobilized liquid in a manner similar to chromatography (Fig. 4.16). The most frequently used inert material is high-purity diatomaceous earth with a high surface area and high capacity for aqueous adsorption. The process is termed solid-supported LLE or supported liquid extraction (SLE) and is a popular alternative to the classical LLE experiment. In practice, the aqueous phase, which could be diluted plasma, urine, or even milk, is coated onto the diatomaceous earth and allowed to disperse for a period of time, usually a few minutes. The aqueous sample spreads over the hydrophilic surface of the diatomaceous earth in a very thin layer. Next, the immiscible organic solvent is added to the top of the tube and comes in contact with the aqueous layer finely dispersed over the high surface area packing. Rapid extraction of analyte occurs during this intimate contact between the two immiscible phases. The solvent moves through the packing by gravity flow or by use of a gentle vacuum.

The tubes used in SLE resemble SPE cartridges and their volumes can range from 0.3 to 300 mL. Some suppliers provide prebuffered pH 4.5 and 9.0 cartridges for extracting acidic and basic substances, respectively. For example, at low pH, acids will be in their unionized form and thus will be extractable from the immobilized aqueous phase. At high pH, amines will be in their neutral form and thereby be extracted into the organic phase. It is possible to add salt to the aqueous sample so that a "salting out" effect occurs thereby leading to better extraction efficiency of certain analytes. The SLE tubes can also be used to remove small amounts of water from organic samples.

Since there is no vigorous shaking as in conventional LLE, there is no possibility of emulsion formation. Since the packed tubes are considered disposable, there is no glassware to be cleaned after use. The entire process is amenable to automation and packed 96-well plates with several hundred milligrams of packing in each well are readily available to perform this task. The 96-well plates are suitable for extraction of 150–200 µL of aqueous sample and thus miniaturize the conventional LLE experiment as well. Examples of commercial products that perform SLE are Varian's Hydromax, Biotage's Isolute HM-N (Charlottesville, VA) and Merck's Extrelut (Darmstadt, Germany).

4.14.3. Immobilized Liquid Extraction (ILE)

The ILE process involves a device coated with a thin layer of sorptive elastomeric polymer such as PDMS that acts as the extraction medium. A targeted compound is extracted by partitioning into the device's coating from an aqueous solution, the interferences washed away, and the analyte desorbed (back extracted) in a small volume of organic solvent. The process is very reminiscent of SPME where a coating is applied to a solid fused-silica tube. Instead, the extraction is performed in a closed system such as the vial where the coating is applied to the cap, on the inner walls of a sealed

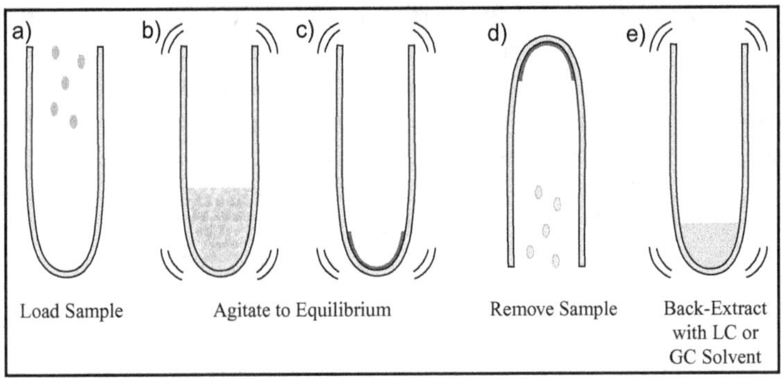

Figure 4.17. ILE well plate procedure. Steps are explained in text (figure courtesy of ILE, Ferndale, CA).

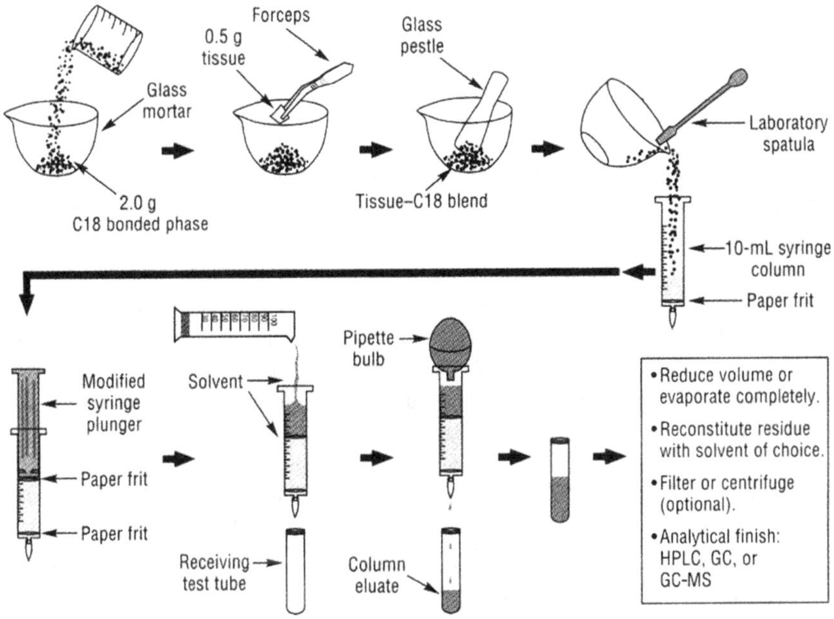

Figure 4.18. Flow diagram of matrix solid-phase dispersion process. Adapted from Reference 74.

96-well plate, or on the inner flow path of an MPT. In the vial version, aqueous sample is loaded (step 1, Fig. 4.17a). The sample is agitated while in contact with the extracting surface (Fig. 4.17b) until analytes have reached equilibrium (Fig. 4.17c) across the aqueous matrix and the extracting layer. The depleted sample solution is then removed (Fig. 4.17d). An optional wash step can follow to selectively remove interferences and matrix components. Finally, the analyte of interest is back extracted (eluted) from the immobilized phase with an appropriate solvent until equilibrium is obtained (Fig. 4.16e). The sample can be directly injected into the chromatograph or blown down and taken up in a more compatible solvent.

4.14.4. MSPD

MSPD is a sample preparation technique for solid (e.g., animal or plant tissue, milk, vegetables) and viscous samples.[65] It utilizes bonded phase solid supports, usually silica based, as both an abrasive to produce a disruption of sample architecture and as a bound "solvent" to assist in complete sample disruption during the process of blending the sample. Figure 4.18 provides a simplified description of the overall process. With this blending, the sample disperses itself over the surface of the bonded phase/support material providing a new mixed phase for conducting analyte isolation from a variety of sample matrices. The blending does not have to be performed with rigorous grinding by the mortar and pestle, but gentle stirring is sufficient for most matrices. This dispersed sample provides more access to solvents and reagents used for analyte isolation. After the disruption process, the blended sample matrix and its distribution onto the bonded phase/support are transferred to a syringe barrel column fitted with a frit. A second frit is placed at the top of the packed bed. Various solvents, usually of progressive stronger polarity depending on the analytes to be eluted, are by HPLC or GC, but often a second cleanup step is required. In some cases, the effluent from the MSPD column is directed to a conventional SPE cartridge where further purification takes place.

The MSPD technique has found widespread applications, especially in the food industry, such as for the sample preparation of pesticides in fruit,[66] baby food[67] and fruit juices,[68] veterinary drugs in tissue,[69] antibacterial residues in foodstuff,[70] isoflavones in clover leaves,[71] polyaromatic hydrocarbons in bivalves,[72] and antibiotics in milk.[73,74] The advantages of the method are its simplicity, good extraction efficiencies with reasonable recoveries, solvent savings over traditional extraction methods, and wide application range; automation of the technique is quite cumbersome.

4.15. FUTURE DIRECTIONS IN SPE

The SPE technique, in its various forms, is still one of the most widely used sample preparation techniques in the analytical laboratory.[75] The technique has been around for over three decades and shows no signs of slowing up. As a sample concentration and cleanup technique, it rivals any other sample preparation technique currently available. With market drivers such as the increase in complex samples in the food safety, natural product, and forensic and environmental markets, effective cleanup techniques will continue to be required in the future. In line with the "green" analytical chemistry movement, SPE minimizes the amount of organic solvent used in the laboratory. With over 100 different stationary phases commercially available, selectivity choices are widespread. As detection techniques such as MS/MS in GC and LC continue to be more selective and more sensitive, ion suppression effects caused by coeluting interferences and undesirable matrix compounds will still need the extraction power of SPE to remove these troublesome compounds.

The use of MS/MS as a detection principle will allow more simplified methods of sample preparation. Already some SPE procedures for drugs in biological fluids have given way to simple protein crashing methods where the sensitivity and selectivity of MS (Section 4.13) is exploited. QuEChERS with dispersive SPE with LC- or GC-MS/MS analysis (Section 4.10.5) is another technique where simple salting out extraction is used to isolate trace pesticides from fruit and vegetable samples. It is anticipated that this trend will continue in the future, as these powerful MS systems become in common use. Of course, one will always have to investigate the effects of any impurities that may be coextracted, and more rigorous SPE procedures may still be needed where ion suppression affect quantitation and sensitivity.

With the trend in miniaturization of analytical processes, SPE has also seen the reduction of bed masses to accommodate smaller samples now encountered. Besides their application in automated SPE systems, the advent of SPE MPTs (Section 4.4) and 96-well SPE plates with the reduction in bed mass has allowed a parallel reduction in solvent usage and overall time. For example, in step 4 (Section 4.6) of the bind–elute SPE process, a minimization in the bed mass allows a corresponding reduction in elution solvent volume, which leads to a shortening of the time of evaporation and reconstitution. Furthermore, with the increased use of polymers as SPE sorbents, the higher capacity has also allowed a further reduction in bed mass compared with the silica-based SPE particles. The emerging lab-on-a-chip and microfluidics instrumentation, incorporating online SPE trapping columns, allow the use of nanoliter volumes of sample and solvent. In the future, such systems may allow the entire SPE-chromatographic separation–detection to be performed on just a few nanoliters of total solvent, a truly "green" analytical technique.

With the advantages of polymeric sorbents in SPE over silica-based SPE sorbents, the use of these more rugged materials will continue to grow in the future. Polymeric sorbents allow the generation of "generic" methods that make method development simpler, so that instead of starting from "scratch," users can try the generic method first and then "tweak" the experimental parameters to optimize the final procedure. Some of the polymeric SPE products have hydrophilic surfaces and do not have to be thoroughly conditioned before sample loading. The ability to dry out such polymeric phases without analyte recovery loss brings some practical advantages to the user.

Specialized, application-specific SPE sorbents have always been around finding focused "niche" applications areas and such phases will continue to be developed as the need arises. Phases such as immobilized PBA for catecholamines (Section 4.11.1), crown ethers for radionuclide metals (Section 4.11.2), graphitized carbon black for chlorophyll,[75] and titania for phospholipids[19,76,77] have proven to be useful for selective isolations. The MIPs (Section 4.10.3) and IA phases (Section 4.10.4) are among the most selective and further developments may generate additional commercial products in this area. Of course, these specialized phases will need a sufficient market volume to justify investment from commercial entities. Selective phases are particularly important when the chemist only has a less selective detector such as UV or refractive index for LC or a flame ionization detector for GC available for use.

Although polymeric and silica-based SPE porous particles are still among the most widely used, some of the technology associated with chromatography may work its way into the sample prep domain. For example, monoliths are getting to be more accepted in chromatography and should become more important in SPE. Their higher permeability should allow the use with higher-viscosity liquids, and their good mass transfer should allow somewhat higher flow rates, thereby speeding up cleanup times.

Although not often used at present, SPE should be able to be moved closer to the source (sampling) point. Some polymeric sorbents do not need preconditioning, so samples could be collected on a cartridge or disk at the source (e.g., lake, waste site, polluted air, etc.) and thereby not requiring transport of the entire sample to the lab for analysis. For example, a liter of water could be run through the sorbent, trace organics concentrated on the sorbent, and the SPE cartridge transported back to the lab or even placed in the mail. The analysis can then be completed in a more convenient laboratory environment. Manufacturers could provide a convenient kit to perform such on-site sampling.

For high-throughput laboratories where complex samples demand some degree of sample preparation, automation of the sample preparation including SPE relieves some of the drudgery of manually handling large numbers of samples. In addition, automation frequently provides a great level of reproducibility since automated instruments do not tire out and can precisely deliver reagents, weigh samples, and other error-prone activities associated with manual procedures. A number of dedicated SPE devices are currently available, and xyz liquid-handling systems have been modified to perform SPE operations including 96-well plates, MPTs, disks, and cartridges. Full robotics systems that often mimic human operations are less frequently employed in the automation laboratories than several years ago. There is a movement to incorporating sample preparation closer to the analytical instrument so that there is a seamless integration of sample prep and analysis. Thus, chromatography autosamplers sometimes have basic sample preparation features built in, and everything is controlled by the same workstation. So automation of SPE still has a bright future, and with miniaturization becoming more popular, smaller automated systems should be possible.

REFERENCES

1. Bäckström, D.; Moberg, M.; Sjöberg, P.J.R.; Bergquist, J.; Danielsson, R. Multivariate comparison between peptide mass fingerprints obtained by liquid chromatography-electrospray ionization-mass spectrometry with different trypsin digestion procedures. *J. Chromatogr. A* **2007**, *1171*, 69–79.

2. Kushnir, M.M.; Crossett, J.; Brown, P.I.; Urry, F.M. Analysis of gabapentin in serum and plasma by solid-phase extraction and gas chromatography-mass spectrometry for therapeutic drug monitoring. *J. Anal. Toxicol.* **1999**, *23*, 1–6.

3. Sadjadi, S.; Gerhards, R. Minimizing SPE elution volumes. *LC-GC Eur.* **2005**, Special Supplement, 52.

4. Eichelberger, J.W.; Behymer, T.D.; Budde, W.L. *Determination of Organic Compounds in Drinking Water by Liquid-Solid Extraction and Capillary Column Gas Chromatography/Mass Spectrometry*, Federal Register Vol. 59, No. 232, Office of Research and Development, U.S. Environmental Protection Agency, Cincinnati, OH.

5. Sirvent, G.; Sanchez, J.M. Preconcentration and determination of priority pollutant phenols in waters at trace levels using a polymeric solid-phase extraction cartridge. *J. Sep. Sci.* **2004**, *27*, 1524–1530.

6. Russo, M.V. Diol Sep-Pak cartridges for enrichment of polychlorobiphenyls and chlorinated pesticides from water samples; determination by GC-ECD. *Chromatographia* **2000**, *52*, 93–98.

7. Planas, C.; Palacios, O.; Ventura, F.; Rivera, J.; Caixach, J. Analysis of nitrosamines in water by automated SPE and isotope dilution GC/HRMS. *Talanta* **2008**, *76*, 906–913.

8. Arthur, C.L.; Potter, D.W.; Buchholz, K.D.; Motlagh, S.; Pawliszyn, J. Solid-phase microextraction for the direct analysis of water: Theory and practice. *LCGC* **1992**, *10*, 656–661.

9. Baltussen, E.; Sandra, P.; David, F.; Cramers, C.A. Stir bar sorptive extraction (SBSE), a novel extraction technique for aqueous samples: Theory and principles. *J. Microcol. Sep.* **1999**, *11*, 737–747.

10. Pichon, V.; Hennion, M.-C. Determination of pesticides in environmental waters by automated online trace-enrichment and liquid chromatography. *J. Chromatogr. A* **1994**, *665*, 269–281.

11. Dirksen, T.A.; Price, S.M.; Mary, S.J. Solid-phase disk extraction of particulate-containing water samples. *Am. Lab.* **1993**, *25*, 24.

12. Snyder, L.R.; Kirkland, J.J.; Dolan, J.W. *Introduction to Modern Liquid Chromatography*, 3rd ed. New York: John Wiley & Sons; 2009.

13. Platoff, G.E. Jr.; Gere, J.A. Solid phase extraction of abused drugs from urine. *Forensic Sci. Rev.* **1991**, *3*, 117–133.

14. Chen, X.-H.; Franke, J.-P.; Ensing, K.; Wijsbeek, J.; De Zeeuw, R.A. Semi-automated solid-phase extraction procedure for drug screening in biological fluids using the ASPEC system in combination with Clean Screen DAU columns. *J. Chromatogr. B* **1993**, *613*, 289–294.

15. Stefkovich, J., *The "Greening" of Oil and Grease Analysis*, http://www.laboratoryequipment.com/article-the-greening-of-oil-and-grease.aspx, March 11, 2010.

16. Wells, M.J.M.; Ferguson, D.M.; Green, J.C. Determination of oil and grease in wastewater by solid-phase extraction. *Analyst*, **1995**, *120*, 1715–1721.

17. Dixit, V.; Dixit, V.M. Sample preparation for the analysis of catecholamines and their metabolites in human urine. *J. Liq. Chromatogr.* **1991**, *14*, 2779–2800.

18. Hsu, R.-C.; Biggs, I.; Saini, N.K. Solid-phase extraction cleanup of halogenated organic pesticides. *J. Agric. Food Chem.* **1991**, *39*, 1658–1611.

19. Simpson, N.J.K., Ed. *Solid-Phase Extraction: Principles, Techniques, and Applications*. New York: Marcel Dekker; 2000.

20. Thurman, E.M.; Mills, M.S. *Solid-Phase Extraction: Principles and Practice*. New York: John Wiley & Sons; 1998.

21. Fritz, J.S. *Analytical Solid-Phase Extraction*. New York: Wiley-VCH; 1999.

22. Simpson, N.; van Horne, K.C. *Sorbent Extraction Technology Handbook*. Harbor City, CA: Varian Sample Preparation Products; 1993.

23. Telepchak, M.J. *Forensic and Clinical Applications of Solid Phase Extraction*. Totowa, NJ: Humana Press; 2004.

24. Li, X.; Fekete, A.; Englmann, M.; Götz, C.; Rothballer, M.; Frommberger, M.; Buddrus, K.; Fekete, J.; Cai, C.; Schröeder, P.; Hartmann, A.; Chen, G.; Schmitt-Kopplin, P. Development and application of a method for the analysis of N-acylhomoserine lactones by solid-phase extraction and ultra high pressure liquid chromatography. *J. Chromatogr. A* **2006**, *1134*, 186–193.

25. Rook, E.J.; Hillebrand, M.J.X.; Rosing, H.; van Ree, J.M.; Beijnen, J.H. The quantitative analysis of heroin, methadone and their metabolites and the simultaneous detection of cocaine, acetylcodeine and their metabolites in human plasma by high-performance liquid chromatography coupled with tandem mass spectrometry. *J. Chromatogr. B* **2005**, *824*, 213–221.

26. Raisglid, M.; Burke, M.F. Solid phase extraction using layered, mixed, and single sorbent phases. *Pittsburgh Conference*, Abstract No. 653, Atlanta, GA; 1997.

27. Boos, K.-S.; Rudolphi, A. The use of restricted-access media in HPLC, part I. Classification and review. *LCGC* **1997**, *15*, 602–611.

28. Rudolphi, A.; Boos, K.-S. The use of restricted-access media in HPLC. Part II. Applications. *LCGC* **1997**, *15*, 814–823.

29. Cassiano, N.M.; Lima, V.V.; Oliveira, R.V.; de Pietra, A.C.; Cass, Q.B. Development of restricted-access media supports and their application to the direct analysis of biological fluid samples via high-performance liquid chromatography. *Anal. Bioanal. Chem.* **2006**, *384*, 1462–1469.

30. Dmitrienko, S.G.; Irkha, V.V.; Kuznetsova, A.Y.; Zolotov, Y.A. Use of molecular imprinted polymers for the separation and preconcentration of organic compounds. *J. Anal. Chem.* **2004**, *59*, 808–817.

31. Baggiani, C.; Anfossi, L.; Giovannoli, C. Molecular imprinted polymers: Useful tools for pharmaceutical analysis. *Curr. Pharm. Anal.* **2006**, *2*, 219–247.

32. Mahony, J.O.; Nolan, K.; Smyth, M.R.; Mizaikoff, B. Molecularly imprinted polymers—Potential and challenges in analytical chemistry. *Anal. Chim. Acta* **2005**, *534*, 31–39.

33. Cormack, P.A.G.; Elorza, A.Z. Molecularly imprinted polymers: Synthesis and characterization *J. Chromatogr. B* **2004**, *804*, 173–182.

34. Piletsky, S.; Turner, A. *Molecular Imprinting of Polymers*. Austin, TX: Landes Bioscience; 2006.

35. Hennion, M.-C.; Pichon, V. Immuno-based sample preparation for trace analysis. *J. Chromatogr. A* **2003**, *1000*, 29–52.

36. Pichon, V.; Delaunary-Bertoncini, N.; Hennion, M.-C. Immunosorbents in sample preparation. In *Comprehensive Analytical Chemistry*, Pawliszyn, J., Ed., Vol. *37*. Amsterdam: Elsevier Science and Technology; 2002; pp. 1081–1100.

37. Delaunay, N.; Pichon, V.; Hennion, M.-C. Immunoaffinity solid-phase extraction for the trace-analysis of low-molecular-mass analytes in complex sample matrices. *J. Chromatogr. B* **2000**, *745*, 15–37.

38. Stevenson, D. Immuno-affinity solid-phase extraction. *J. Chromatogr. B* **2000**, *745*, 39–48.

39. Delaunay-Bertoncini, N.; Pichon, V.; Hennion, M.-C. Comparison of immunoextraction sorbents prepared from monoclonal and polyclonal anti-isoproturon antibodies and optimization of the appropriate monoclonal antibody-based sorbent for environmental and biological applications. *Chromatographia* **2001**, *53*(Suppl.), S224–S230.

40. Venture, A.F. Immunoaffinity Columns and Cartridges, Brochure 522, Grace Davison Discovery Sciences, Deerfield, IL, June 2008.

41. Majors, R.E. Advanced topics in solid-phase extraction: Chemistries. *LCGC North Am.* **2006**, *25*, 16–32.

42. Zhang, K.; Zolotarjova, N.; Nicol, G.; Martosella, J.; Yang, L.-S.; Szafranski, C.; Bailey, J.; Boyes, B. Agilent multiple affinity removal system for the depletion of high-abundant proteins from human serum—A new technology from Agilent. Agilent Application Note 5988-9813EN; 2003.

43. Brand, J.; Haslberger, T.; Zolg, W.; Pestlin, G.; Palme, S. Depletion efficiency and recovery of trace markers from a multiparameter immunodepletion column. *Proteomics* **2006**, *6*, 3236–3242.

44. Shen, Z.; Want, E.J.; Chen, W.; Keating, W.; Nussbaumer, W.; Moore, R.; Gentle, T.M.; Siuzdak, G. Sepsis plasma protein profiling with immunodepletion, three-dimensional liquid chromatography tandem mass spectrometry, and spectrum counting. *J. Proteome Res.* **2006**, *5*, 3154–3160.

45. Fang, X.; Huang, L.; Feitelson, J.S.; Zhang, W.-W. Affinity separation: Divide and conquer the proteome. *Drug Discov. Today Technol.* **2004**, *1*, 141–148.

46. Anastassiades, M.; Lehotay, S.J.; Stajnbaher, D.; Schenck, F.J. Fast and easy multiresidue method employing acetonitrile extraction/partitioning and "dispersive solid-phase extraction" for the determination of pesticide residues in produce. *J. AOAC Int.* **2003**, *86*, 412–431.

47. Lehotay, S.J.; Mastovská, K.; Lightfield, A. Use of buffering and other means to improve results of problematic pesticides in a fast and easy method for residue analysis of fruits and vegetables. *J. AOAC Int.* **2005**, *88*, 615–629.

48. Lehotay, S.J. Determination of pesticide residues in foods by acetonitrile extraction and partitioning with magnesium sulfate: Collaborative study. *J. AOAC Int.* **2007**, *90*, 485–520.

49. European Committee for Standardization/Technical Committee CEN/TC 275. Foods of plant origin: Determination of pesticide residues using GC-MS and/or LC-MS/MS following acetonitrile extraction/partitioning and cleanup by dispersive SPE-QuEChERS method. Brussels: European Committee for Standardization; 2007.

50. Mastovska, K.; Lehotay, S.J.; Anastassiades, M. Combination of analyte protectants to overcome matrix effects in routine GC analysis of pesticide residues in food matrixes. *Anal. Chem.* **2005**, *77*, 8129–8137.

51. Anastassiades, M. 2004. http://www.quechers.com/docs/quechers_recov.pdf.

52. Lehotay, S.J.; de Kok, A.; Hiemstra, M.; Van Bodegraven, P. Validation of a fast and easy method for the determination of residues from 229 pesticides in fruits and vegetables using gas and liquid chromatography and mass spectrometric detection. *J. AOAC Int.* **2005**, *88*, 595–614.

53. Lee, J.-M.; Park, J.-W.; Jang, G.-C.; Hwang, K.-J. Comparative study of pesticide multi-residue extraction in tobacco for gas chromatography–triple quadruple mass spectrometry. *J. Chromatogr. A*, **2008**, *1187*, 25–33.

54. Martin, S.H.; Pinto, C.G.; Pavon, J.L.P.; Cordero, B.M. Determination of trihalomethanes in soil matrices by simplified, quick, easy, cheap, effective, rugged and safe extraction and fast gas chromatography with electron capture detection. *J. Chromatogr. A*, **2010**, *1217*, 4883–4889.

55. Mastovska, K.; Lehotay, S.J. Rapid sample preparation method for LC-MS/MS or GC-MS analysis of acrylamide in various food matrices. *J. Agric. Food Chem.* **2006**, *54*, 7001–7008.

56. Plossl, F.; Giera, M.; Bracher, F. Multiresidue analytical method using dispersive solid-phase extraction and gas chromatography/ion trap mass spectrometry to determine pharmaceuticals in whole blood. *J. Chromatogr. A*, **2010**, *1217*, 4612–4622.

57. Stubbings, G.; Bigwood, T. The development and validation of a multiclass liquid chromatography tandem mass spectrometry (LC-MS/MS) procedure for the determination of veterinary drug residues in animal tissue using a QuEChERS (QUick, Easy, CHeap, Effective, Rugged and Safe) approach. *Anal. Chim. Acta* **2009**, *637*, 68–78.

58. Zhao, L.; Schultz, D.; Stevens, J. *Analysis of Pesticide Residues in Apples using Agilent SampliQ QuEChERS Kit by LC-MS/MS Detection*. Santa Clara, CA: Agilent Technologies; **2009**.

59. Talwar, D.; Williamson, C.; McLaughlin, A.; Gill, A.; O'Reilly, D.S.J. Extraction and separation of urinary catecholamines as their diphenyl boronate complexes using C18 solid-phase extraction sorbent and high-performance liquid chromatography. *J. Chromatogr. B* **2002**, *769*, 341–349.

60. Bradshaw, J.S.; Izatt, R.M. Crown ethers: The search for selective ion ligating agents. *Acc. Chem. Res.* **1997**, *30*, 338–345.

61. http://www.waters.com.

62. Arthur, C.L.; Pawliszyn, J. Solid phase microextraction with thermal desorption fused silica optical fibres. *Anal. Chem.* **1990**, *62*, 2145–2148.

63. Jeannot, M.A.; Cantwell, F.F. Solvent microextraction into a single drop. *Anal. Chem.* **1996**, *68*, 2236–2240.

64. Gioti, E.M.; Skalkos, D.C.; Fiamegos, Y.C.; Stalikas, C.D. Single-drop liquid-phase microextraction for the determination of hypericin, pseudohypericin and hyperforin in biological fluids by high performance liquid chromatography. *J. Chromatogr. A* **2005**, *1093*, 1–10.

65. Barker, S.A.; Long, A.R.; Short, C.R. Isolation of drug residues from tissues by solid phase dispersion. *J. Chromatogr.* **1989**, *475*, 353–361.

66. Ramos, J.J.; Gonzalez, M.J.; Ramos, L. Comparison of gas chromatography-based approaches after fast miniaturised sample preparation for the monitoring of selected pesticide classes in fruits. *J. Chromatogr. A* **2009**, *1216*, 7307–7313.

67. Georgakopoulos, P.; Mylona, A.; Athanasopoulos, P.; Drosinos, E.H.; Skandamis, P.N. Evaluation of cost-effective methods in the pesticide residue analysis of non-fatty baby foods. *Food Chem.* **2009**, *115*, 1164–1169.

68. Radisic, M.; Grujic, S.; Vasiljevic, T.; Lausevic, M. Determination of selected pesticides in fruit juices by matrix solid-phase dispersion and liquid chromatography-tandem mass spectrometry. *Food Chem.* **2009**, *113*, 712–719.

69. Wang, S.; Mu, H.; Bai, Y.; Zhang, Y.; Liu, H. Multiresidue determination of fluoroquinolones, organophosphorus and N-methyl carbamates simultaneously in porcine tissue using MSPD and HPLC-DAD. *J. Chromatogr. B* **2009**, *877*, 2961–2966.

70. Marazuela, M.D.; Bogialli, S. A review of novel strategies of sample preparation for the determination of antibacterial residues in foodstuffs

using liquid chromatography-based analytical methods. *Anal. Chim. Acta* **2009**, *645*, 5–17.

71. Visnevschi-Necrasov, T.; Cunha, S.C.; Nunes, E.; Oliveira, M.B.P.P. Optimization of matrix solid-phase dispersion extraction method for the analysis of isoflavones in *Trifolium pratense*. *J. Chromatogr. A* **2009**, *1216*, 3720–3724.

72. Campins-Falco, P.; Verdu-Andres, J.; Sevillano-Cabeza, A.; Molins-Legua, C.; Herraez-Hernandez, R. New micromethod combining miniaturized matrix solid-phase dispersion and in-tube in-valve solid-phase microextraction for estimating polycyclic aromatic hydrocarbons in bivalves. *J. Chromatogr. A* **2008**, *1211*, 13–21.

73. Mamani, M.C.V.; Reyes, F.G.R.; Rath, S. Multiresidue determination of tetracyclines, sulphonamides and chloramphenicol in bovine milk using HPLC-DAD. *Food Chem.* **2009**, *117*, 545–552.

74. Barker, S.A. Matrix solid-phase dispersion. In *Solid-Phase Extraction: Principles, Techniques, and Applications*, Simpson, N., Ed., New York: Marcel-Dekker; 2000; pp. 361–380.

75. Majors, R.E. Trends in sample preparation. *LCGC North Am.* **2002**, *20*, 1098–1113.

76. Gonzalvez, A.; Preinerstorfer, B.; Lindner, W. Selective enrichment of phosphatidylchlolines from food and biological matrices using metal oxides as solid-phase extraction materials prior to analysis by HPLC-ESI-MS/MS. *Anal. Bioanal. Chem.*, **2010**, *396*, 2965–2975.

77. Calvano, C.D.; Jensen, O.N.; Zambonin, C.G. Selective extraction of phospholipids from dairy products by micro-solid phase extraction based on titanium dioxide microcolumns followed by MALDI-TOF-MS analysis. *Anal. Bioanal. Chem.* **2009**, *394*, 1453–1461.

Chapter 5

Solid-Phase Microextraction

Sanja Risticevic, Dajana Vuckovic, and Janusz Pawliszyn

5.1. HISTORICAL PERSPECTIVE AND DEVELOPMENT OF FIBER SOLID-PHASE MICROEXTRACTION (SPME)

SPME was developed to address the need for rapid sample preparation procedures both in the laboratory and on-site where the investigated system is located. During the preliminary studies conducted in our laboratory, the need for rapid sample preparation techniques was recognized to retain the time efficiency advantages made possible by using high-speed analytical instrumentation and, at the same time, to eliminate some of the drawbacks of conventional sample preparation techniques, namely the extensive consumption of organic solvents, large sample amount, and long sample preparation time requirements.[1] In the initial work on SPME, sections of fused-silica optical fibers, both uncoated and coated with liquid and solid polymeric phases, were dipped into an aqueous sample containing test analytes and then placed in a gas chromatography (GC) injector.[2] Despite their basic nature, the early experiments provided very important preliminary data that confirmed the usefulness of this simple approach, since both polar and nonpolar chemical species were extracted rapidly and reproducibly from aqueous samples.

The development of the technique accelerated rapidly with the implementation of coated fibers incorporated into a microsyringe, resulting in the first SPME device based on the Hamilton™ 7000 series microsyringe (Hamilton Co., Reno, NV).[3] The metal rod, which serves as the piston in a microsyringe, is replaced with stainless steel microtubing having an inside diameter slightly larger than the outside diameter of the fused-silica rod. Typically, the first 5 mm of the coating is removed from a 1.5-cm-long fiber, which is then inserted into the microtubing. High-temperature epoxy glue is used to mount the fiber permanently. Sample injection is performed in a manner similar to standard syringe injection. Movement of the plunger exposes the fiber during extraction and desorption, and protects the fiber in the needle during storage and penetration of the septa. In fact, SPME devices do not need expensive syringes like the Hamilton syringes. A useful device can be built from a short piece of stainless steel microtubing (to hold the fiber), another piece of larger tubing (to work as a "needle"), and a septum (to seal the connection between the microtubing and the "needle").

The design in Figure 5.1 is the basic building block of a commercial SPME device introduced by Supelco (Bellefonte, PA) in 1993. The device is similar in operation principle to the custom-made device originally described.[3] An additional improvement in the commercial device is the ability to adjust the depth of the fiber with respect to the end of the needle, which allows the control of the exposure depth in the injector and extraction vial. The commercial device also incorporates such useful features as color marking of the fiber assemblies to distinguish among various coating types.

Another simple SPME construction is based on a piece of internally coated tubing.[4] This tubing can be mounted inside a needle or it can constitute the "needle" of a syringe itself.[5] Elimination of mechanical movement of a plunger of a syringe can be accomplished by sealing the tubing at one end and installing a microheater. The expansion of air caused by temperature increase allows the removal of desorbed analytes from the extraction phase located inside the tubing. A coated tubing approach is useful in the design of passive sampling devices, since in this case, the extraction rate is limited by the diffusion of analytes into the needle.[6] In addition, active sampling is possible by heating and cooling of air contained in the upper part of the tubing, which causes the movement of liquid or gaseous samples in and out of the tubing, facilitating mass transport of analytes from the sample to the coating.

The in-tube concept was also expanded to facilitate the automation of sample preparation for high-performance liquid chromatography (HPLC) as will be described later in Section 5.6.1.2. In this in-tube SPME approach, the sample components are extracted by the coating located on the inner surface of the hollow tubing, and, after the extraction is

Handbook of Sample Preparation, Edited by Janusz Pawliszyn and Heather L. Lord

completed, the analytes are washed into the HPLC column using the mobile phase or solvent. This concept is very similar to solid-phase extraction (SPE).[7] However, the fundamental difference between the two techniques is that SPE relies on exhaustive extraction, while in-tube SPME relies on equilibrium.

The extraction coating has also been used on other elements of the analytical system. In addition to widely used fiber and in-tube geometry, SPME devices can take different configurations such as the coated interior of vessels,[8] the coated exterior of magnetic stirring bars,[9] or even pieces of polydimethylsiloxane (PDMS) tubes and thin membranes.[10–12] The main reason for developing these alternative approaches is to enhance sensitivity by using a larger volume of the extraction phase and to improve the kinetics of the mass transfer between sample and the extraction phase by increasing the surface to volume ratio of the extraction phase. The main disadvantage of these approaches, however, is the loss of the convenience associated with a syringe configuration, in particular for the introduction of the sample into the analytical instrument.

Although, to date, SPME devices have been used mainly in laboratory applications, recent research has been directed toward remote monitoring, particularly for clinical, field environmental, and industrial hygiene applications. The operating principles of such devices are analogous to those of the devices described above, but modifications are made for greater convenience in given applications. An example is shown in Figure 5.2: adding a tube with a small opening to cover the needle of the SPME syringe results in a useful device for breath analysis in a noninvasive clinical application.[13] This design can be further improved by adding two one-way valves, mounted at the mouthpiece and on the exit aperture, but the concept of operation remains the same.[14] The two valves allow entrapment of alveolar air and prevent accidental inhalation of it. The degree of dilution of alveolar air with environmental air can be measured by monitoring the carbon dioxide concentration.[15]

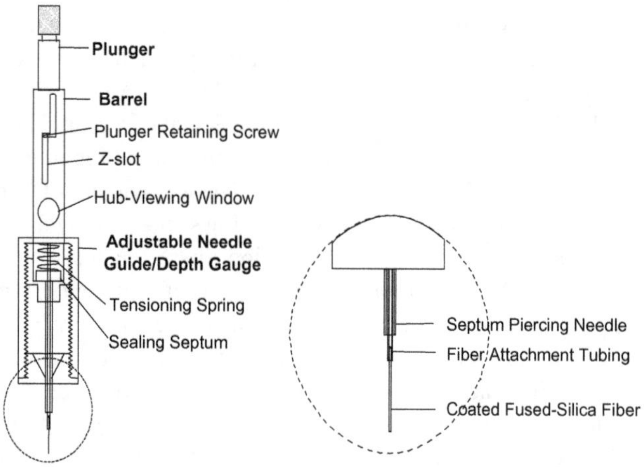

Figure 5.1. Design of the first commercial SPME device made by Supelco.

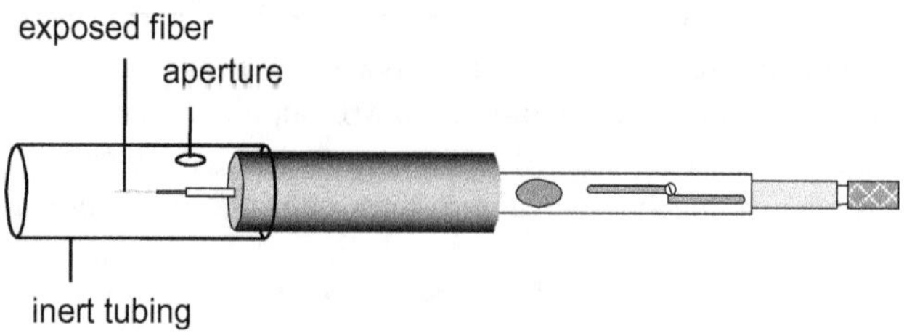

Figure 5.2. SPME device modified for breath analysis.

5.2. THEORY OF SPME

Theory has been developed to understand the principal processes involved in SPME by applying fundamental principles of thermodynamics and kinetics and to provide direction when developing SPME methods.

The most widely used form of SPME sampling consists of exposing a thin polymeric coating directly to the sample matrix for a predetermined amount of time (Figure 5.3).[16] Once the fiber is exposed to the sample matrix, the transport of analytes from the matrix to the coating begins immediately. SPME extraction is considered to be complete when the analyte concentration has reached distribution equilibrium between the sample matrix and the fiber coating, which means that past the equilibrium time point, the amount of analyte remains constant and does not change with further increase in extraction time.[16]

The distribution constant of the analyte between the fiber coating and the sample matrix is defined by the following equation:

$$K_{fs} = C_f^{\infty} / C_s^{\infty} \tag{5.1}$$

where C_f^{∞} and C_s^{∞} are the analyte equilibrium concentrations in the fiber coating and sample matrix, respectively. The distribution constant defines the enrichment factors achievable by the use of the technique and reflects the chemical composition of the extraction phase. Extraction conditions that affect K_{fs} are temperature, salt addition, pH, and organic solvent content (see Section 5.5.3).

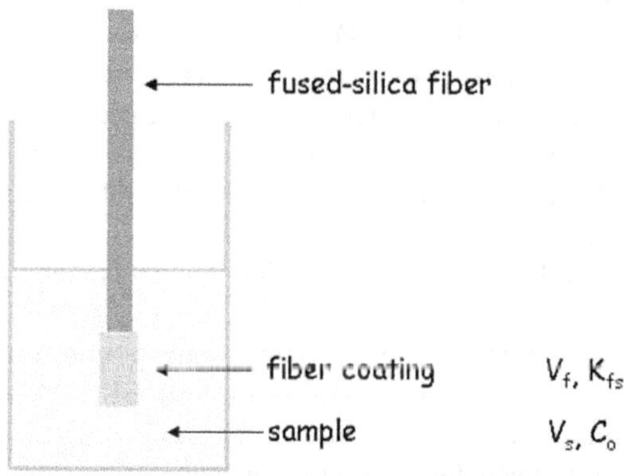

Figure 5.3. Graphical illustration of the SPME process. K_{fs}, fiber coating/sample matrix distribution constant; V_f, volume of the fiber coating; V_s, sample volume; C_o, initial concentration of a given analyte in the sample.

The number of moles of analyte extracted (n) by the coating at equilibrium can be expressed by Equation 5.2:

$$n = \frac{K_{fs}V_fV_sC_o}{K_{fs}V_f + V_s}. \qquad (5.2)$$

While thermodynamic principles may be used to predict the amount of analyte extracted under equilibrium conditions, kinetic theory determines the extraction rate and indicates strategies to increase the speed of extraction. Some level of agitation is required to facilitate rapid extraction and transport analytes from the bulk of the solution to the vicinity of the fiber. This is especially critical for aqueous samples since fluid contacting the fiber's surface is always stationary independent of the agitation level, and as the distance from the fiber surface increases, the fluid movement increases as well until it corresponds to the bulk flow in the sample.[16] In order to model the mass transport from the sample matrix to the fiber coating, a zone referred to as the Prandtl boundary layer is usually considered as a region where analyte flux is progressively more dependent on analyte diffusion and less on agitation, as the extraction phase is approached.[16] The thickness of the boundary layer is determined by the rate of agitation, the reason for which agitation affects the length of equilibration time and mass of analyte extracted in pre-equilibrium conditions.

5.3. RATIONAL DESIGN OF SPME DEVICES

Theoretical principles of SPME allow the development of devices that offer improved extraction efficiency.

5.3.1. Cold Fiber SPME

Extraction temperature is an important SPME parameter to be considered. At elevated temperatures, native analytes can effectively dissociate from the matrix and move into the headspace (HS) for rapid extraction by the fiber coating. However, the fiber coating/sample matrix distribution constant also decreases with an increase in sample temperature, resulting in a reduction of the equilibrium amount of analyte extracted, as described in Section 5.5.3.11. To prevent loss of sensitivity, the coating can be cooled simultaneously with sample heating. This idea was implemented in the design of the internally cooled SPME device.

In this device, a fused-silica tube is sealed and coated at one end (outer surface of the capillary). Liquid carbon dioxide is delivered via an inner capillary to the coated end of the outer capillary, resulting in a coating temperature lower than that of the sample. This "cold finger" effect results in accumulation of the volatilized analytes at the tip of the fiber.

The internally cooled fiber approach can be used effectively to increase the sensitivity and extraction speed of SPME. Quantitative extraction of many analytes, including volatiles, is possible with this method.[17] Complete recoveries are possible with the internal cooling approach, even for

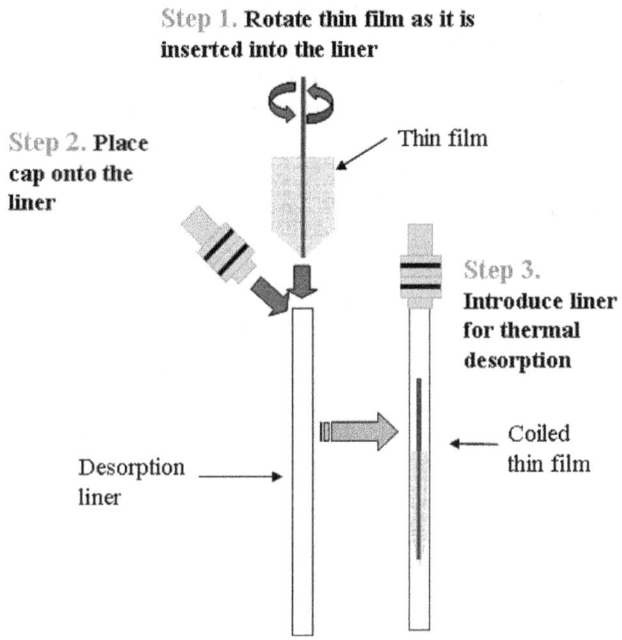

Figure 5.4. Introduction of a thin-film microextraction device into a GC injector.

volatile organic compounds, such as substituted benzenes, in a range of matrices. An alternative approach to cooling the fiber uses a Peltier cooler.[18]

5.3.2. High Surface Area Samplers (Thin-Film Microextraction [TFME])

The extraction rate after the exposure of the SPME device to the sample is proportional to the contact surface area between the sorbent and the sample. In order to increase the mass uptake rates and therefore the analytical sensitivity, large surface area sorbent geometries can be used. For example, a section of PDMS extraction phase can be cut into a house-like shape and fixed between a length of steel wire bent in half to secure the thin film in place.

In this case, a high surface area-to-volume ratio is obtained, resulting in very high accumulation rates. This approach is particularly beneficial for hydrophobic semivolatile components characterized by very high distribution constants.[1] To facilitate convenient introduction to the analytical instrument, the membrane attached to the holding rod/wire can be rolled around the rod by coiling while rotating during insertion into the GC liner, which is subsequently introduced into the GC for thermal desorption. This process is described in Figure 5.4 (refer to Section 5.5.2.4 for the details regarding the automation of procedure).

5.4. ON-SITE SAMPLERS

5.4.1. Field Samplers

An important feature of a field sampling device is the ability to preserve extracted analytes in the coating. The simplest practical way to accomplish this goal is to seal the end of the needle with a piece of septum. Additionally, cooling extends the storage time. Silicone-based septum material, however, may cause losses of analytes from the fiber, and therefore, it can be replaced by a suitable size sealing cap made from Teflon. Another more appropriate approach is to use metal-to-metal (or solid polymer) seals constructed based on a "leaf" closure.[19]

Muller described several equilibrium-based SPME field sampling devices and discussed the typical losses of volatiles and contamination of the fiber during storage and transport.[20] The percentage of compound retained in the coating was evaluated for the four devices, for different storage times and temperatures, and for different fiber coatings. The PDMS fiber demonstrated the lowest ability to retain these compounds.

5.4.2. Needle Trap (NT) Devices

As an alternative to the coated fiber in the needle SPME device, the sorbent can fill the whole diameter of the needle, forming an NT device.[21] Convenient NT field sampling devices have been designed in the form of badges and pens for both spot sampling and time-weighted average (TWA) sampling.[22]

There are two ways to immobilize the sorbent. One technique is to use a side hole needle and the other is to use an open tubular needle with the sorbent particles glued together and to the surface. The side hole needle device is easier to prepare, but it is more costly because the side hole needle is more expensive. Two types of sorbents were used to build NT devices: a single layer and a segmented sorbent with progression of the sorbent strength from the tip to the inside of the needle. The segmented sorbent allows convenient desorption because in needle operation, the flow of the desorption system is naturally reversed compared with the extraction. Different desorption gas delivery techniques have been applied with NT such as (1) the syringe piston delivery of a well-defined volume of the desorption gas, (2) divergence of the gas flow after needle placement in the injector (the technique used primarily with the automated system available from PAS Technology, Magdala, Germany), and, finally, (3) the design of a narrow neck insert/liner in combination with the NT-containing hole above the sorbent position (Fig. 5.5A) to facilitate convenient desorption.

In contrast to SPME, NT is an exhaustive technique, but it can work very effectively in combination with fiber SPME device to better characterize the sample directly on-site. Figure 5.5B illustrates GC chromatograms obtained for SPME and NT of mosquito coil smoke.[23] This figure indicates that SPME does not extract volatile components very well because of low distribution constant values, but performs very well for mid-range compounds, which exist as free analytes in air. Heavy (less volatile) components are

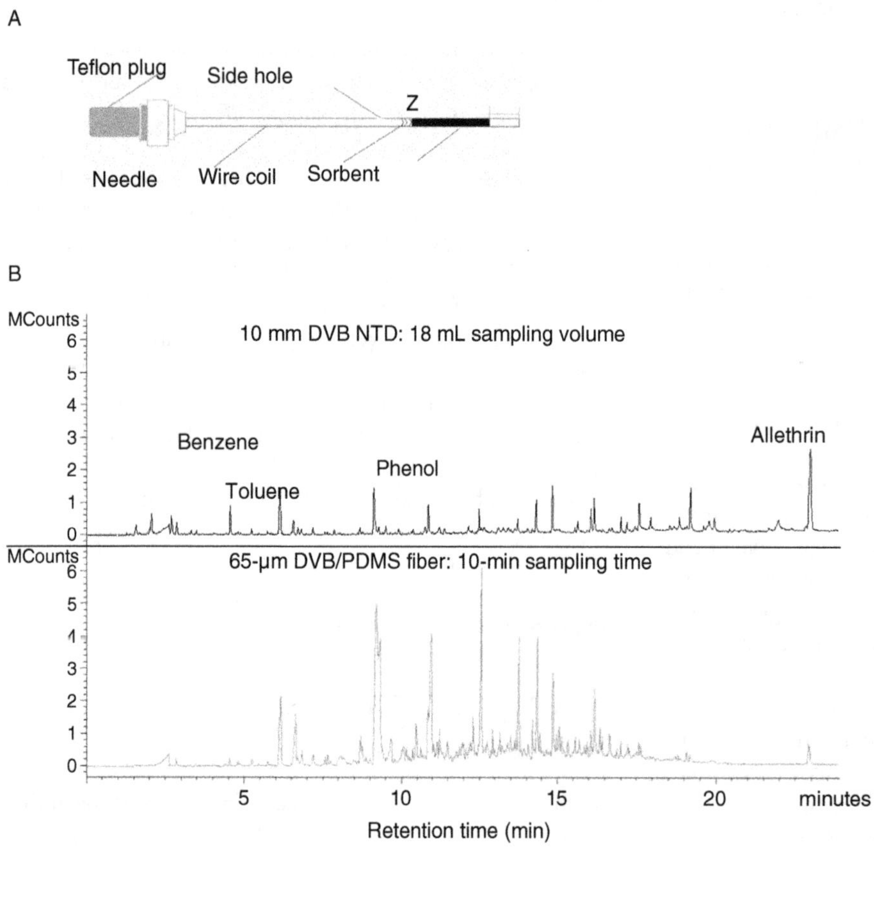

Figure 5.5. Needle trap (NT) device. (A) Schematic of an NT device with the side hole to facilitate convenient on-site desorption. (B) GC-MS chromatograms for SPME fiber and NT device after sampling from mosquito coil smoke. Reprinted with permission from reference 22. Copyright 2008 American Chemical Society.

particle bound and are not available for extraction by the fiber. However, they are well extracted using NT. Therefore, the combination of two needle-based devices (NT and SPME) provides a good characterization of complex air samples. SPME can be used to concentrate freely available analytes, which can cause immediate impact, while NT measures the total amount of analytes in the sample. The combination of the two approaches can therefore provide useful information as to the extent of binding of the analytes to the particulate matter in air. This measurement can be performed directly and conveniently on-site using a single portable analytical instrument.

5.4.3. TWA Devices

The commercially available SPME device (with the fiber retracted in the needle) was used as a TWA diffusive sampler by Martos and Pawliszyn.[24] Airborne (industrial) formaldehyde concentrations were measured with the device using on-fiber derivatization, and the results were compared with those produced by active air sampling using the NIOSH-2541 method.[25] The device was also used for TWA air sampling of normal alkanes from C_5 to C_{15}.[26,27] Field sampling trials were performed at a house, an apartment, and a school. The results were compared with those of active sampling through charcoal tubes using the NIOSH-1550 method.

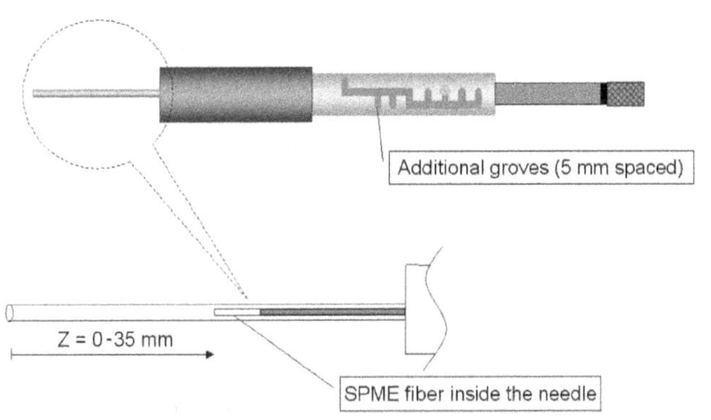

Figure 5.6. SPME TWA passive air sampling device: additional groves allow retraction of the fiber inside the needle for the desired length. Reprinted from reference 28 with permission from the American Chemical Society. Copyright 2003.

Later, Chen and Pawliszyn used the fiber-retracted SPME device to determine the TWA concentrations of volatile organic contaminants (VOCs) in air (Fig. 5.6)[28,29] and demonstrated that the face velocity of air across the needle opening does not affect sampling, due to the extremely small inner diameter of the fiber needle.

Ouyang et al. extended the applications of this type of SPME device to TWA passive water sampling (Fig. 5.7).[30] A Hamilton 500-μL gastight syringe was modified as a TWA water sampling device, to ensure all of the air in the SPME needle could be replaced with water. In addition, a removable needle was designed to avoid the effect of the adsorption of the target analytes on the outside wall of needle. This field TWA water sampling device was used to monitor polyaromatic hydrocarbons (PAHs) in Hamilton Harbour and Laurel Creek, Canada.[30,31]

5.4.4. *In Vivo* SPME Samplers

In vivo research is better suited than *in vitro* research for observing a dynamic effect. SPME is one of the most promising techniques for *in vivo* sampling and subsequent analysis. According to SPME theory, when $K_{fs} << V_s$ (Eq. 5.2), the amount extracted by SPME is independent of sample volume; therefore, it is not necessary to define a specific sample size for the analysis, which is very desirable for

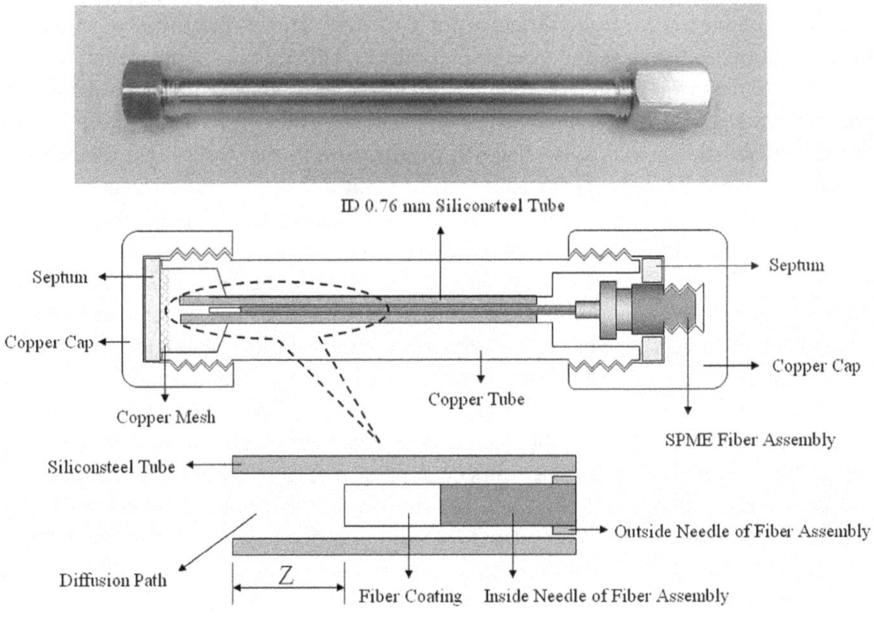

Figure 5.7. Schematic diagram and graphic of the SPME TWA field water sampler. Reprinted from reference 30 with permission from Elsevier. Copyright 2007.

in vivo and on-site sampling. Additionally, SPME directly extracts a small fraction of free analyte, so a negligible depletion of the free fraction is achieved after extraction. This depletion will be compensated by dynamic systems containing both bound and unbound analyte and the system will not be disturbed significantly. Finally, the technology is easy to miniaturize and thus suitable for both small living systems and microanalytical instruments. These advantages make SPME a promising tool for direct assay of *in vivo* chemical concentrations.

Highly or moderately volatile compounds are usually determined in the HS mode using existing commercial SPME devices. For HS sampling, the investigated individuals, plant parts, or insects, are enclosed in special chambers that are usually designed to minimize interferences.[32–41] Studies involving human exhudates have also been conducted in a similar manner.[42–44] When using HS sampling, any extraction phase is suitable, as long as it is robust and innocuous, but the use of PDMS and PDMS/divinylbenzene (DVB) coatings predominates in most applications.

In the case of polar, nonvolatile analytes, only direct extraction is feasible. The fiber coating is either carefully rubbed against the sample[32–34] or inserted with a special *in vivo* device.[45,46] The development of biocompatible extraction phases for SPME has led to significant advances in bioanalysis: all sample preparation steps can be combined into a single one, even for complex biological samples such as whole blood or plasma. Furthermore, biocompatible devices permit direct extraction of target analytes from the flowing blood of living organisms.

The biocompatible *in vivo* device, which will soon become commercially available from Supelco, Inc., consists of an SPME fiber that is housed inside a hypodermic needle. The main features of SPME coating used in this device are that it is (1) biocompatible in order to prevent adverse reactions during sampling and to minimize fouling of SPME coating surface by the adhesion of proteins present in biological fluid, (2) immobilized on flexible metal alloy to render it sufficiently robust for *in vivo* use, and (3) compatible with most commonly used solvents in order to facilitate desorption. A similar device without the hypodermic needle will also be introduced for *in vitro* analysis of biological fluids.

A fast *in vivo* microextraction technique that has the potential to replace current methods of analysis based on drawing blood was recently reported.[45,47,48] *In vivo* SPME was used for direct extraction of drugs from the flowing blood of beagle dogs, over a period of 8 h in order to construct pharmacokinetic profiles for diazepam and its metabolites. The drugs extracted on the probes were subsequently quantified by liquid chromatography (LC)–tandem mass spectrometry (MS/MS). Two calibration strategies—external and standard on the fiber—were employed to correlate the amount extracted with the *in vivo* concentration. The microextraction technique was validated by comparison with conventional plasma analysis, with a correlation factor of 0.99.[46] In addition to total concentrations, the method was very useful for simultaneously determining free drug concentrations.

5.5. HYPHENATION OF SPME AND GC

5.5.1. SPME-GC Interface

The analytical instrument used most frequently with SPME has been the gas chromatograph. Only the extracted analytes are introduced into the instrument, since the extracting phase is nonvolatile. Thus, there is no need for complex injectors designed to deal with large amounts of solvent vapor, and injector design can be simplified for use with SPME.

Standard GC injectors, such as split/splitless, can be applied to SPME as long as a narrow-bore (reduced inside diameter) insert with an inside diameter close to the outside diameter of the needle is used. Liners (0.75 mm i.d.) are commonly employed. The narrow-bore inserts are required to increase the linear flow around the fiber coating, resulting in efficient removal of extracted analytes. In order to increase the carrier gas flow rates around the fiber coating and accelerate the transfer of analytes onto the column, the split flow should be deactivated as well during SPME injection, which in turn allows for optimum sensitivity and desorption efficiency. In fact, SPME injection does not introduce liquid solvent into the system; therefore, large-volume inserts/liners typically used to accommodate the expansion of the evaporated solvent and split valve opening employed to ensure the removal of remaining solvent vapors are not required. Under these conditions, the desorption of analytes from the fiber is very rapid, not only because the coatings are thin but because high injector temperatures produce a dramatic decrease in the coating/gas distribution constant and an increase in the diffusion coefficients. In many cases, the speed of desorption is limited by the time required to introduce the fiber into the heated zone.

In summary, when attempting to optimize SPME desorption efficiency for GC applications, the following parameters should be carefully considered: (1) high carrier gas flow rate around the fiber is ensured by using low-volume/narrow-bore liners and deactivating split flow during desorption of analytes; (2) decrease in the coating/gas distribution constant and increase in the diffusion coefficient are achieved by using high injector temperatures; and (3) complete removal of analytes from the fiber coating is ensured by optimizing desorption time at the previously determined optimum desorption temperature. The latter parameter is verified by carryover tests, which are conducted by reinjecting a fiber. After the extraction of target analytes from the sample matrix and desorption for a particular desorption time, the device is sealed with a septum or Teflon sealing cap and reinjected upon a GC ready signal. Ensuring an optimum desorption time minimizes memory effects.

One way to obtain sharper injection zones and faster separation times is to use rapid injection autosampling devices, which will be discussed in Section 5.5.2. An alternative solution is to use a dedicated injector, which should be cold during needle introduction but which heats up very rapidly after exposure of the fiber to the carrier gas stream. A schematic of such an injector has been described.[49] This type of injector has been able to accomplish the separation of 28 volatile organic compounds listed in the United States Environmental Protection Agency (USEPA) Method 624 in 150 s with reproducibility better than 5% relative standard deviation (RSD) for most analytes.[50]

The fiber can also be designed to contain the heating element.[50,51] In this case, no injector is necessary. The modified fiber can be introduced directly into the front of the column, and analytes can be desorbed rapidly by heating with a capacitive discharge current after the fiber has been exposed from the needle.

5.5.2. SPME-GC Automation and High-Throughput Approaches

Even though the development and commercialization of the original fiber-SPME device took place in 1989 and 1993, respectively, and were followed by continuous developments of new SPME devices, the original commercial fiber-SPME device still continues to be the most widely used form of this technique.[52] The fiber-SPME device consists of a fiber holder originally developed for the performance of manual SPME processes and the fiber assembly (Fig. 5.1) with a built-in fused-silica fiber inside a needle coated with a thin layer of polymeric stationary phase, which enables the extraction of the analytes by concentrating them during absorption or adsorption processes.[53] The similarity of the fiber-SPME device with the traditional GC syringe for liquid injection resulted in significant breakthroughs in the area of SPME-GC automation, especially since the hyphenation of traditional sampling and sample preparation methods with separation/detection systems is extremely difficult.[16] In addition, numerous efforts toward the development of high-speed separation and detection systems have simultaneously created demand for the availability of compatible sample preparation/introduction methods so that the speed of the overall analytical process is increased.[54]

5.5.2.1. SPME-GC autosamplers. The automation of SPME-GC analysis was first reported by Arthur et al. in 1992.[55] The fiber-adaptable Varian model 8100 syringe (Varian, Palo Alto, CA) autosampler was adjusted to control the following SPME automation requirements: (1) the movement of the plunger to protect the fiber/fiber coating inside the needle during vial/injector septa penetration, (2) the depth of the needle/fiber coating during the sampling in the vial to account for different sample volumes, (3) the plunger movement to expose/retract the fiber coating from/into the needle before/after sampling, (4) the timing procedures for extraction from the vial and desorption in the injector, and (5) the depth of the fiber in the injector during desorption so that temperature gradient effects in the GC injector port can be compensated for. Magnetic stirring was used for agitation, which was achieved by setting up a microstirrer in close proximity to the vial being sampled. Based on this initial work, the first SPME-GC autosampler (model 8200, Varian) was commercialized in 1993.[52] Even though this early model was characterized by static sampling and unavailability of temperature control during the extraction processes, it was capable of initiating the extraction of the next sample in the sequence while the previously extracted sample underwent the GC analysis, thus greatly contributing to the throughput of analysis. Later versions launched in 1996 and 1998 allowed for the vibration of the fiber to agitate the sample and temperature control ability, respectively.

CTC Analytics (Zwingen, Switzerland) introduced an SPME-compatible CombiPAL™ autosampler in 1999.[52] This autosampler has provided additional advantages such as stirring provided by rotation of an agitator tray, full temperature control of individual samples (sample trays that can be heated or cooled), and a fiber-conditioning station for external fiber cleaning procedures. After the commercialization of the CombiPAL autosampler, several additional versions of GC autosamplers having SPME compatibility were launched. Gerstel GmbH (Mulheim an der Ruhr, Germany) recently introduced their MultiPurpose Sampler (MPS 2) as a new generation autosampler having a compatibility of automating SPME-GC procedures. This autosampler offers both orbital agitation and a stirring mechanism during incubation and extraction processes as well as Gerstel Maestro software, which is user-friendly and features flexible programming for the development of SPME methods. Thermo Fisher Scientific (Milan, Italy) commercialized their version of an autosampler having an SPME-GC automation capability (TriPlus autosampler, Fig. 5.8). The TriPlus autosampler is significantly modified compared with the previous two versions due to the utilization of a rocking agitation mechanism instead of an orbital agitation and the possibility to close the agitator tray during extraction. Lastly, PAS Technology introduced its CONCEPT autosampler for SPME-GC applications. This autosampler incorporates the use of both the orbital and magnetic stirring agitation mechanisms.

As can be seen, the commercially available autosamplers differ in terms of sample agitation methods by employing (1) fiber vibration (Varian autosampler), (2) tray rotation (CombiPAL, MPS 2, and CONCEPT), and (3) tray rocking motion (TriPlus).[56] On the other hand, magnetic stirring is the most common technique for manual fiber-SPME operations. Therefore, fiber vibration, tray rotation, and tray rocking motion eliminate the need for a stirring bar, which results in faster sample preparation times and eliminates the risk of sample contamination since a foreign object is not added to the sample. The fiber vibration technique proved to be more efficient than magnetic stirring in small 2-mL vials; however, it was not particularly effective in larger vials, and it does place stress on the needle of SPME fiber assembly.[56] Stress is also placed on the fiber needle through

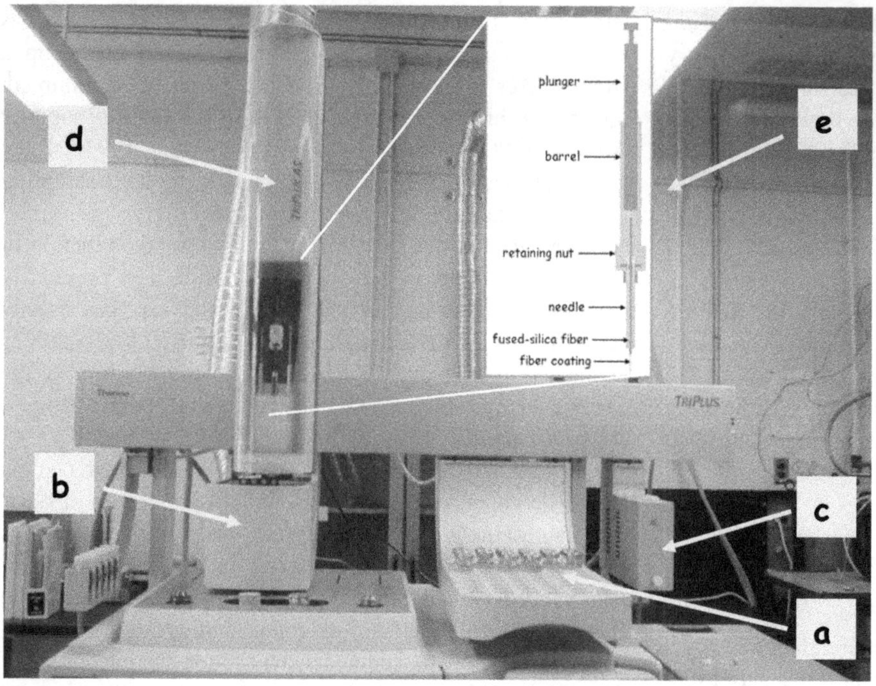

Figure 5.8. The graphical illustration of commercially available TriPlus autosampler (Thermo Fisher Scientific) for the performance of automated SPME processes. (a) Sample tray; (b) sample preparation station/agitator tray; (c) fiber-conditioning station/needle heater; (d) sample preparation/injection arm; (e) schematic of automated SPME fiber holder.

the agitating action of rotating and rocking tray mechanisms, which is the reason agitation is restricted to speeds ranging from 250 to 750 rpm with the CombiPAL, MPS 2, and CONCEPT systems. Therefore, for the commercial SPME fiber assemblies having 24-gauge needle size, stirring speed often must be significantly lower than 750 rpm to avoid breakage of the needle due to stress. This problem has been partially overcome with the development of fiber assemblies with 23-gauge needle size for automated processes, as they can withstand higher agitation speeds. However, it has been reported previously that the lifetime of silica-based fiber assemblies for automated procedure is limited to 100 extraction/desorption cycles, which is an additional reason that maximum agitation speeds of 500 rpm are strongly recommended.[54] These stirring speed restrictions may result in slower equilibration times and a lower amount of analyte extracted in pre-equilibrium conditions than would otherwise be obtained, since agitation speed is an important parameter that affects the extraction efficiency of SPME processes (Section 5.5.3.4). Similarly, the type of agitation mechanism itself is an important SPME efficiency controlling parameter due to the effects it might have on the speed of equilibrium extraction, sensitivity of pre-equilibrium SPME analysis, and ability to reduce the thickness of the boundary layer.[16]

As opposed to the manual procedure, automated SPME processes offer the following advantages: (1) repeatable timing events, which are critical for the achievement of satisfactory reproducibility under pre-equilibrium conditions and (2) high sample throughput associated with performance of all SPME steps in an automated fashion and the sample preparation of the next sample in the sequence while the previously prepared sample is still undergoing GC analysis. All these features contribute to time-effective SPME method development and optimization, high-throughput SPME-GC applications, and minimum analyst supervision of the operating system.

*5.5.2.2. **Automated fiber SPME-GC procedures.*** After the development of appropriate methods (specifications for agitator temperature, agitation speed, incubation time, sampling depth in the vial, extraction time, depth of fiber coating exposure in the injector, desorption time, fiber cleaning temperature, and fiber cleaning time) and loading of sample trays with samples to be analyzed, all of the aforementioned autosamplers are capable of automating the following sequence of SPME processes: (1) the sample vial is first transported by the sample preparation/injection arm (Fig. 5.8d) from a specified tray position (Fig. 5.8a) to the agitator (Fig. 5.8b); (2) extraction process is enabled by moving the sample preparation/injection arm equipped with automated fiber assembly holder (Fig. 5.8e) to the agitator; (3) desorption process is initiated by moving the sample preparation/injection arm to the injection port; and (4) the fiber cleaning procedure is initiated prior to the next extraction cycle by moving the sample preparation/injection arm to the fiber-conditioning station/needle heater (Fig. 5.8c).

In addition to the performance of the standard SPME procedures described above, the flexibility of currently available technology allows the automation of more complicated SPME procedures such as those involving internal standard loading and derivatization.[56]

A new calibration approach, recently introduced in SPME, involves preloading of an internal standard onto the fiber coating prior to extraction, instead of internal standard spiking directly into the sample.[54] This calibration method was developed to address the following needs during the preparation of complex sample matrices: (1) compensation for matrix activity effects; (2) elimination of internal standard spiking step into the sample matrix, which in turn saves sample preparation time and increases analysis throughput; (3) control of convection conditions in on-site determinations; and (4) internal standard delivery to the sample together with the extraction phase, which allows the determination of chemical concentrations directly *in vivo* and on-site.[54] An automated internal standard loading procedure involves the performance of the following steps: (1) internal standard preloading, (2) simultaneous internal standard desorption into the sample matrix and target analyte extraction from the sample matrix by exposure of preloaded fiber coating to the sample, and (3) desorption of extracted target analyte and remaining preloaded internal standard.

The first step of this calibration process involves automated internal standard preloading onto the fiber coating by performing one of the following approaches. It should be kept in mind that the optimum standard loading procedure is largely dependent on the physicochemical properties of the compound: (1) HS extraction of the standard dissolved in nonvolatile pump oil, (2) HS extraction of pure standard from a vial, and (3) direct extraction from the vial containing a standard dissolved in aqueous solution/phosphate-buffered saline (PBS) buffer solvent.[57,58] The second step of this calibration approach involves the exposure of the standard previously loaded on the fiber coating to the sample matrix in an automated fashion to allow for simultaneous symmetric desorption of the internal standard from the coating into the sample and extraction of the target analyte from the sample into the coating. The third step involves the desorption of the remaining internal standard and extracted target analyte into an analytical instrument. When short-time pre-equilibrium extraction conditions are employed, the amount of analyte extracted is related to the extraction time, and subsequent quantification is then performed on the basis of the timed accumulation of the analytes into the fiber coating.[54] Additionally, since desorption and extraction processes occur in symmetric fashion, the information from the desorption process is directly used for estimating the initial concentration of the target analyte in the sample matrix. This internal standardization SPME approach saves sample preparation time (elimination of standard spiking step), ensures high-throughput analysis (short pre-equilibrium times are employed), and allows *in vivo* and on-site sampling and can be completely automated (by placing standard loading vial into a different position in agitator tray as compared with the sample containing vial and ensuring automated loading, desorption, and extraction processes). So far, this approach has been successfully applied in the SPME analyses of wine, milk, on-site water, *in vivo* plant, and *in vivo* animal systems.[45,57–63]

In addition to the complete automation of in-fiber standard loading SPME procedures, automated in-fiber (or more accurately in-coating) derivatization methods have also been reported.[56] Similar to standard loading procedures, the derivatization agent can be loaded onto the fiber coating prior to the extraction process by placing the derivatization agent loading vial in a separate agitator tray station as compared with the vial containing target analytes. Due to the additional complexity and time costs typically encountered in derivatization methods, the throughput and reproducibility advantages of automation are even more appreciated here than for simpler methodology. Automated in-fiber derivatization methods have been employed in various applications such as, for example, the analysis of aldehydes with pentafluorobenzyl hydroxylamine (PFBHA) derivatization and carboxylic acids with pyrenyldiazomethane (PDAM) derivatization and many other studies.[56,64,65]

5.5.2.3. Development of more robust SPME fiber assemblies. It was emphasized previously that commercially available fused-silica fiber assemblies possess lifetimes of about 100 extraction/desorption cycles, which is perfectly acceptable for routine analyses and/or sample sequences containing up to 100 samples. However, having in mind that sample throughput has become the most important factor that is considered when choosing an analytical method, these fiber assemblies may not be suitable in high-throughput applications requiring a large number of samples to be analyzed. This aspect is particularly crucial in the area of food analysis as well as metabolomics studies since several hundreds or thousands of classification and characterization studies typically need to be performed to verify food authenticity or change in metabolic profiles.[56] Therefore, such characterization studies require robust and reliable fiber assemblies especially to avoid breakage of the fused-silica assembly anytime during the automated sequence run and hence minimize user intervention. The durability aspect of SPME fibers has been addressed by Supelco with the introduction of new-generation metal (super elastic) fibers. The fiber needle, plunger, and fiber core of these new fiber assemblies are made of a special type of inert, flexible alloy that makes the assemblies more robust and able to perform several hundreds of successive analyses. The development of these assemblies offered several advantages namely, (1) automation compatibility together with the possibility to use higher agitation speeds and longer agitation periods without compromising the fiber lifetime, (2) bevelled design of the needle allows easier penetration of vial and injector septa, and (3) improved robustness and durability characteristics.

These fibers were recently evaluated in studies that reported excellent durability (600 extraction/desorption cycles performed with negligible loss in extraction efficiency) and satisfactory results for both single-fiber and interfiber reproducibility.[54,66]

However, in any fiber SPME-GC application, septum coring may be an issue when traditional GC septa are employed, due to the wider 23- or 24-gauge needle of an SPME fiber assembly, as compared with the 26-gauge needles used for standard liquid injection. This drawback becomes even more significant when using automated methods and metal fiber assemblies, the latter condition due to the blunt design of the needle of the SPME fiber assembly. Septum coring may result in loss of sensitivity (due to the adsorption of desorbed analytes onto the septa pieces), decrease in sample throughput (due to the septa replacements and subsequent liner cleaning procedures), and eventual breakage of the fiber assembly. An efficient way toward resolving this problem is to employ septumless injection devices such as the high-pressure Merlin Microseal from Merlin Instrument (Half Moon Bay, CA), the septumless head (SLH) on the Gerstel programmed temperature vaporizer (PTV), and similar septumless sealing systems from Varian and other manufacturers.[54,56]

A critical maintenance procedure when using metal fibers for SPME-GC automation is to change the driving belt in the autosampler on a yearly basis and perform regular alignment checks between the needle position and the hole in the injector nut.

5.5.2.4. Automation of SPME formats other than fiber SPME. An automated cold fiber SPME device was introduced by Chen and Pawliszyn.[18] The development of the automated device consisted of miniaturizing the technology by accommodating the additional parts of the device in an 18-gauge needle. The cold fiber device was mounted on the CombiPAL autosampler with only small changes to the GC system required, such as enlarging the injector nut hole and support. This methodology was used for the automated determination of chloroanisoles in cork, volatile, and semivolatile constituents in tropical fruit samples and polycyclic aromatic hydrocarbons in sediment.[67–69] However, the cold fiber device could perform only a limited number of injections before the GC septum needed to be replaced, otherwise significant pressure, septum coring, and leak problems would be observed. These problems were partially overcome with the use of a specially adjusted SLH on the PTV manufactured by Gerstel, Inc. The system is currently being tested toward the improvement of automated cold fiber procedures.

In recent years, significant breakthroughs were accomplished in the automation of TFME as well. These were achieved with the aid of the automated tube exchange system (ATEX) on the MPS 2 autosampler and the thermal desorption system (TDS) mounted on the GC system via the CIS-4 cooled injection system inlet (Gerstel, Inc.). GC liners/tubes equipped with thin films (see Section 5.3.2.) would be introduced into the GC for thermal desorption of analytes. One substantial disadvantage of this approach is that thin films have to be manually placed into the glass tubes, which makes the technique only partially automated. Nevertheless, the technique was used in a number of environmental studies.[70]

With regard to the automation of other SPME formats, it is worth commenting that currently, the CONCEPT GC autosampler (PAS Technology) is being tested at the University of Waterloo laboratory for its capability of automating NT procedures.[22]

Overall, the percentage of SPME-GC analyses conducted in an automated fashion is certain to increase in the coming years, and, with this trend, the use of this sample preparation method will continue to grow.

5.5.3. Method Development for SPME-GC Applications

Even though the SPME sample preparation procedure is very simple and involves the performance of only two steps, namely, (1) partitioning of analytes between the polymeric coating and the sample matrix and (2) desorption of concentrated extracts into an analytical instrument, several parameters typically need to be optimized in order to provide highly sensitive analytical methods. The theory of the technique has been addressed in Section 5.2 and comprehensively described elsewhere,[16,53,56,71] and it allows for a more accurate optimization of the parameters discussed in the next sections of this document and displayed in Figure 5.9.

5.5.3.1. Step 1—optimization of the fiber coating. The first step in SPME method development involves the selection of the appropriate fiber coating. As described elsewhere,[16,56] SPME extraction efficiency is dependent on the fiber coating/sample matrix distribution constant,[72] a characteristic parameter that describes the properties of a coating and its selectivity toward the target analyte versus other matrix components. The only current manufacturer of commercially available SPME fibers is Supelco, and it offers fibers coated with various single polymer or mixed polymer materials differing in polarity, thickness, and length. The mechanisms of analyte extraction with single polymer (liquid) and mixed polymer (solid) phases are quite different. When using liquid coatings, the analytes are extracted via the process of absorption, since the diffusion coefficient of the analyte in these coatings is high enough for the solute molecules to penetrate the entire volume of the coating within a reasonable extraction time.[56] On the other hand, for solid sorbents, which can be either amorphous or have a well-defined crystalline structure, the analyte diffusion coefficients are substantially reduced, and the extraction occurs via the process of adsorption on the coating surface.[71] One advantage of solid coatings is that shorter extraction times can be achieved. However, when using mixed polymer phases, smaller dynamic ranges and displacement effects are quite likely to occur at long extraction times and in high-concentration samples. In these circumstances, displacement of analytes with weaker affinities takes place since there is a limited number of sites available for adsorption, and deviation from linearity may occur as the sorbent becomes saturated.[16,56]

Similar to the choice of the chromatographic stationary phases, fiber coating affinity toward target analytes is deter-

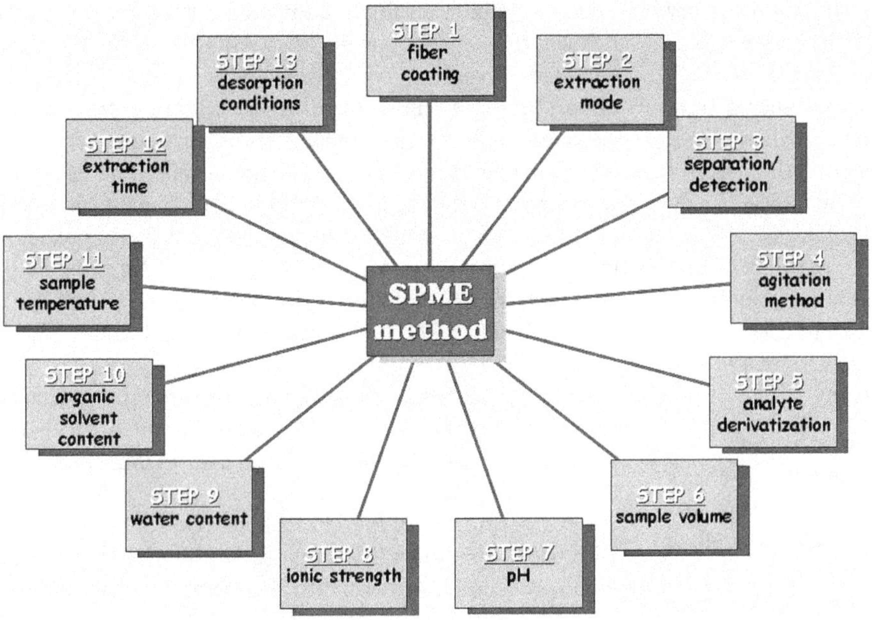

Figure 5.9. Steps involved in general SPME method development.

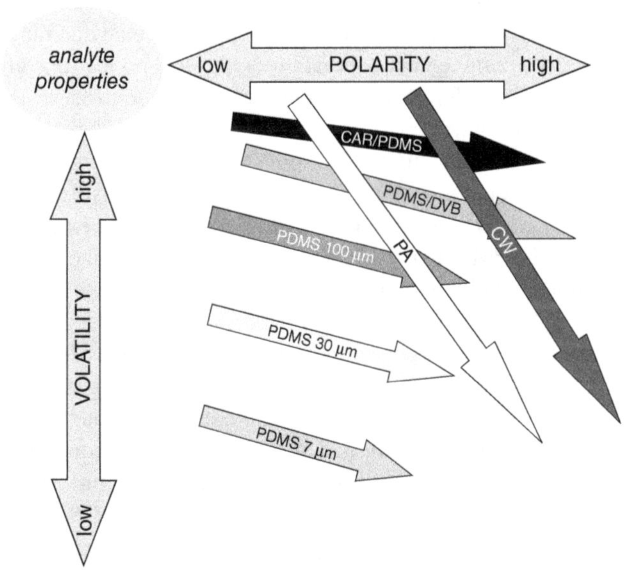

Figure 5.10. SPME fiber coating selection guide.

Table 5.1. Suitability of DI-SPME and HS-SPME Extraction Modes

Sampling Mode	Analyte Properties	Sample Matrix
DI-SPME	Low-to-medium volatility	Gas
	High-to-medium polarity	Simple liquid
HS-SPME	High-to-medium volatility	Simple and complex liquid
	Low-to-medium polarity	Solid

mined by the coating polarity, and it depends on the principle of "like dissolves like." Liquid coatings such as the nonpolar PDMS and polar polyacrylate (PA) provide high sensitivity for the extraction of apolar and polar compounds, respectively.[53,73] Mixed-phase coatings are suitable for the extraction of volatile low molecular mass and polar analytes.[53] The polyethylene glycol (PEG) fiber coating was recently commercialized by Supelco for the extraction of polar compounds. The suitability of different commercially available fiber coatings for the extraction of analytes having various physicochemical properties is illustrated in Figure 5.10. One important consideration worth commenting on is that SPME method sensitivity can be increased by increasing the coating thickness;[74] however, the use of thicker coatings results in longer equilibration times, just as the higher distribution constant does.[72]

5.5.3.2. Step 2—optimization of the extraction mode. The two most commonly used SPME extraction modes are direct immersion (DI-SPME) and HS (HS-SPME).[53] DI-SPME is more suitable for gaseous and simple liquid sample matrices. On the other hand, in order to protect the fiber coating from damage by high molecular mass and other nonvolatile interferences present in the sample matrix, HS-SPME is suggested and even considered as the only possibility for extraction of analytes from complex liquid and solid samples. In addition, while HS extraction is more suitable for the extraction of analytes of high-to-medium volatility and low-to-medium polarity, DI extraction is utilized for the extraction of compounds with low-to-medium volatility and high-to-medium polarity. These characteristics are summarized in Table 5.1.

In those circumstances in which HS-SPME mode is utilized, it is important to consider the effect of the HS volume in a vial, since extraction efficiency increases with the decreasing HS volume.[75]

5.5.3.3. Step 3—selection of the separation and detection systems. Even though this particular section is exclusively associated with hyphenation of SPME and GC, it is worth noting that significant accomplishments have been obtained with regard to combination of SPME and LC, as is discussed in Section 5.6. With regard to the use of detection systems in combination with SPME-GC, it is worth noting that due to its solvent-free nature, SPME can be interfaced to both conventional (flame ionization detector [FID], nitrogen phosphorus detector [NPD], etc.) and mass selective detection systems. The advantages associated with the ideal combination of SPME and mass spectrometry (MS) were recently summarized by Vas and Vékey.[73]

5.5.3.4. Step 4—selection of the agitation method. As was emphasized previously (Section 5.2), agitation is a very important parameter in SPME. Agitation mechanism and speed affect equilibration time and the mass of analyte extracted in pre-equilibrium conditions. In other words, the more efficient the agitation is, the shorter the equilibration times and the higher the amount of analyte extracted in pre-equilibrium conditions.[71] If perfect agitation is achieved, the time required to reach the equilibrium would only depend on the geometry of the extraction phase and the diffusion coefficients of analyte in the sample matrix.[56] Different agitation methods used in combination with SPME are shown in Table 5.2 and have been described elsewhere.[6,16,56,76]

It has been observed that increasing air speed enhances extraction when performing extraction from gaseous samples using short sampling times (pre-equilibrium).[77] The data indicate that the air speed or air bulk movement significantly affects the VOC mass transfer process from bulk air to the fiber. The VOC mass loading on the fiber increases as the air velocity increases from 0 to 5 cm/s. No further change was observed as the air speed was increased from 5 to 20 cm/s. This indicates that the thickness of the boundary layer between the fiber and air is diminished as the air speed increases, which results in an increase of the mass transfer rate and mass loading on the fiber. After the thickness of the boundary layer has been decreased to a certain degree, uptake becomes limited by diffusion within the pores of the polymer, and no further increase in extraction is seen with increasing air speed. In practice, when using SPME for field sampling, it would be helpful to use a fan to move air across the fiber during sampling, to eliminate possible imprecision in extraction due to variations in air speed. An example of an agitation device for field air sampling, consisting of a modified hair dryer fan with a mounting for the SPME device, has been described.[78]

Efficient agitation for aqueous TFME sampling can be achieved using an electric handheld drill, which is capable of holding both the SPME fiber and PDMS thin film (Fig. 5.11).[19] The drill chuck can be tightened onto the colored plastic hub of the SPME fiber to keep it secure. The wire used to secure PDMS thin films can also be attached to the drill similar to a drill bit. Glass sampling vials are held immobile using ring stands and clamps.[78,79]

5.5.3.5. Step 5—analyte derivatization. Derivatization is used in SPME-GC analyses, especially since a wide range of derivatizing agents compatible with GC applications are readily available.[72] For example, derivatization process can be used to enhance extraction efficiency (since derivatives typically possess higher distribution constants toward the coating), separation efficiency, and detection sensitivity. In addition, if analytes are too weakly volatile to be extracted with HS-SPME, derivatization is often used to convert them to more volatile analogs.[80]

The following three derivatization strategies can be used with SPME: (1) pre-extraction derivatization, (2) post-extraction derivatization, and (3) simultaneous extraction and derivatization.[72] In pre-extraction derivatization, derivatizing agent is manually added to the vial containing target analytes, and following the derivatization reaction, SPME

Table 5.2. The Performance Characteristics of Different Agitation Methods Used in Combination with SPME

Agitation Method	Advantages	Disadvantages
Static	Simple; good performance for gaseous phase	Limited to volatile analytes; limited to HS-SPME
Magnetic stirring	Simple equipment; good performance	Stirring bar required; heating of the stirring plate might occur above the optimum sample temperature
Intrusive stirring	Good performance	Difficulty to seal the sample
Vortex/moving vial	Good performance; stirring bar not required	Stress placed on fiber and needle
Fiber movement	Good performance for small sample volumes; stirring bar not required	Stress placed on fiber and needle; limited to small vial sizes
Flow through	Good performance at rapid flows	Constant flows required; possible cross-contamination
Sonication	Short extraction times	Sample heating occurs

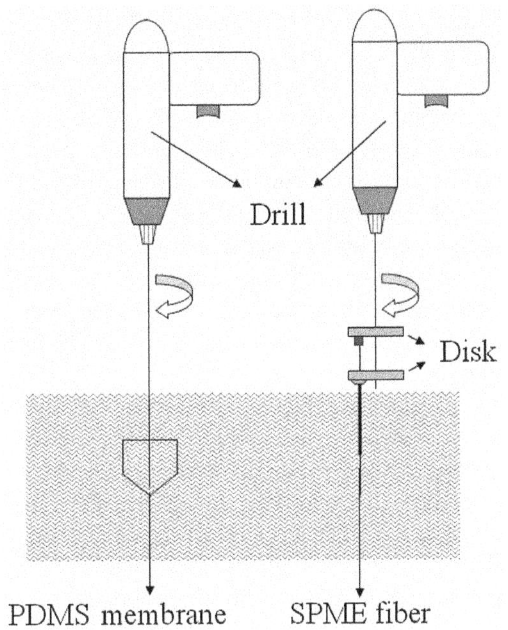

Figure 5.11. SPME field water sampling devices using a rotated fiber and a rotated membrane. Reprinted from reference 19 with permission from Elsevier. Copyright 2006.

extraction and desorption of derivatized compounds take place.[16] Post-extraction derivatization involves SPME extraction of underivatized target analytes, which should have a reasonably high affinity toward the fiber coating or reasonably high distribution constant values. Subsequently, SPME fiber containing underivatized analogs is exposed to the vial containing the derivatization agent so that on-fiber derivatization of extracted compounds can occur.[16] The third derivatization strategy, namely, simultaneous extraction and derivatization first involves the loading of derivatizing agent onto the fiber coating, which is then exposed to the sample matrix to allow simultaneous derivatization and extraction processes on the fiber. While the first and third approach can lead to the improvement in the extraction efficiency, chromatographic behavior, and detection properties, the second approach can only improve the chromatographic behavior and detection properties and requires suitable extraction phases.[72]

5.5.3.6. Step 6—optimization of sample volume. From Equation 5.2, it is clear that the amount of analyte extracted increases with the sample size, and this effect is more prominent for the compounds having higher K_{fs} values.[16] However, laboratory arrangements for testing the effect of sample volume are often limited by the size of sample vials (1.5- to 20-mL vials typically used in combination with SPME) and by the availability of sample.[72] For such in-laboratory experiments, the amount of analyte extracted usually increases with the sample volume.

However, for most on-site and *in vivo* applications, the sample volume becomes much larger than the product of K_{fs} and V_f, and this effect leads to the simplification of Equation 5.2 to the one by which the amount of analyte extracted by the fiber coating is dependent only on K_{fs}, V_f, and C_o and not on sample volume. In practice, this simplification implies that SPME fiber can be exposed to any reasonably sized living or environmental system, without the need to collect a defined sample volume prior to the analysis.

5.5.3.7. Step 7—optimization of sample pH. When using SPME, it is important to ensure that the analytes are fully converted into neutral forms, since SPME can extract only the undissociated/neutral species of analytes of interest.[81] This requirement and subsequent increase in sensitivity is usually achieved by pH adjustment. Accordingly, low pH values improve the extraction efficiency for acidic compounds, whereas high pH values improve the extraction of basic compounds. For those compounds that possess both of these functionalities, a pH providing optimal extraction efficiency must be determined empirically. For nearly neutral analytes, the extraction efficiency is not affected to a significant degree by pH variations.[56]

When adjusting sample pH, HS-SPME is the preferred extraction mode since direct contact of the fiber coating with sample can lead to deterioration of fiber coating at very low or high pH levels ($<$pH 2 or $>$pH 10).

5.5.3.8. Step 8—optimization of ionic strength. When a soluble salt is added to a sample, the sample matrix/fiber coating distribution constant potentially increases (depending on the target analyte of interest), which in turn leads to improvement in method sensitivity.[56] Aqueous solubilities of most organic compounds decrease in the presence of salt, due to the salting-out effect that causes the analyte molecules to pass more readily from the sample matrix to the fiber coating and from the sample matrix to the HS.[72] On the other hand, for compounds whose aqueous solubilities do not decrease in the presence of salt, the addition of salt may decrease the amount of analyte extracted by decreasing the activity coefficients and thus the distribution constants of the analytes. In SPME-GC analysis, the most commonly used salts are NaCl, Na_2SO_4, K_2CO_3, and NH_2SO_4.[80,81] It is important to note that salt addition can increase or decrease the amount of analyte extracted, depending on the target analyte of interest and salt concentration, and, to date, the effect of salt addition on SPME extraction efficiency has been determined only by experiment.[56] This effect has been demonstrated for the extraction of organophosphorus pesticides from fruit juices where 15% (w/v) sodium sulfate amount was selected as optimum.[82] Another important consideration is that, in some cases, an increase in ionic strength improves the extraction efficiency for both the target analytes and the interfering compounds, which is highly undesirable, especially when mixed-phase SPME coatings are employed. Furthermore, an HS-SPME extraction mode is preferred when salt is added because fiber coatings are prone to deterioration during agitation by DI-SPME.

5.5.3.9. Step 9—optimization of water content. Since many real-life samples are complex, matrix effects can have a significant impact on target analyte recovery. In fact, the effect of interfering sample matrix components on the SPME extraction efficiency may result in the following phenomena: (1) adsorption of target analytes onto the interfering components instead of the fiber coating, (2) decreased diffusion coefficients of target analytes, and (3) low recovery of target analytes.[72] In order to compensate for these matrix activity effects, water has proven to be a very effective additive, by decreasing the matrix effects with a simple dilution step.[71] An example of the effect of water dilution factor on the relative recovery of pesticides from fruit juices has been reported.[83]

5.5.3.10. Step 10—effects of organic solvent content. The presence of an organic solvent in an aqueous sample matrix may decrease the distribution constant; therefore, the amounts of organic solvents should be kept at a minimum.[16] Typically, for optimum extraction efficiencies, organic solvent amount should not exceed 1%, since above this threshold, the properties of water and distribution constant values change substantially.

5.5.3.11. Step 11—optimization of sample temperature. Sample/extraction temperature is another crucial parameter that needs to be optimized in order to obtain good sensitivity in SPME methods. The following two phenomena occur upon increasing sample temperature: (1) increased sample temperature causes an increase in HS capacity (when HS-SPME mode is used) and an increase in the analyte diffusion coefficient, which in turn results in increased extraction rate, increased rate of mass transfer from the sample matrix into the fiber coating, and shorter equilibration times; and (ii) increased sample temperature leads to a decrease in the sample matrix/fiber coating distribution constant, which leads to decreased method sensitivity and analyte recovery at equilibrium.

The effect of temperature on the extraction time profile and speed of analysis is illustrated in Figure 5.12 for the

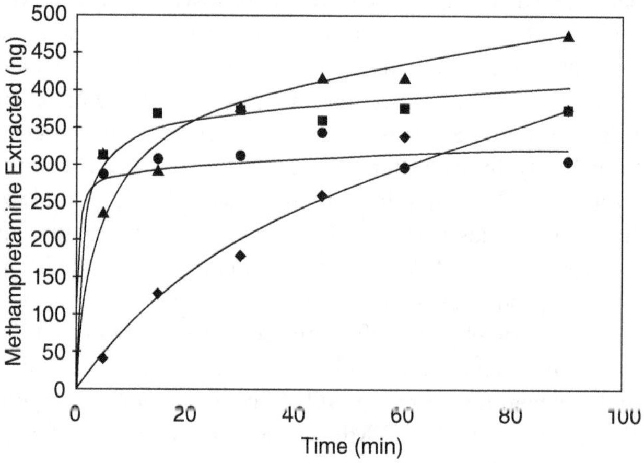

Figure 5.12. The effect of sample temperature on analyte recovery and speed of analysis. Legend: (◆), 22°C; (▲), 40°C; (■), 60°C; (●), 73°C. Reprinted from reference 71 with permission from Elsevier. Copyright 2000.

SPME extraction of methamphetamine.[71] Accordingly, the lowest extraction temperature results in the highest amount of analyte extracted at equilibrium and simultaneously the longest equilibration times. On the other hand, the highest temperature results in lower analyte recovery at equilibrium and shortest equilibration times. While the amount of analyte extracted at equilibrium conditions will be lower at high extraction temperatures and higher at low extraction temperatures, it is important to emphasize that in pre-equilibrium conditions, the amount of analyte extracted will still be higher at higher temperatures than what it would be at lower temperatures.

The two opposite effects derived from kinetic and thermodynamic viewpoints and taking place when the extraction temperature is raised should be taken into account and optimized based on the objective of analysis. For example, if the main objective of analysis is focused on method sensitivity under equilibrium conditions, then lower extraction temperatures should be used. On the other hand, if the sensitivity aspect for a particular application of interest is not as crucial as the speed of analysis and time required to reach equilibrium, higher sample temperatures should be employed. Alternatively, since higher sample temperature leads to higher amount of analyte extracted in pre-equilibrium conditions, pre-equilibrium extraction accompanied by high sample temperatures can be utilized, provided that constant extraction timing procedures are ensured from one sample onto the next.

Finally, since an increase in sample temperature affects both the diffusion coefficient and distribution constant, the optimum sample temperature condition will depend on the physicochemical properties of target analytes of interest. This effect has been described for the extraction of volatile and semivolatile aromatic hydrocarbons in olive oil samples.[84] Consequently, higher extraction temperatures favored the uptake of higher-molecular-weight and less volatile compounds, due to the increase in diffusion coefficients and consequent improvement of the mass transfer processes. On the other hand, the uptake of more volatile analytes decreased after 40 and 60°C (depending on the volatility) as a result of the distribution constant decrease since extraction by SPME is an exothermic process.

5.5.3.12. Step 12—optimization of extraction time. The choice of a suitable extraction time will very much depend on the objectives of analysis, since the optimum extraction time is always a compromise between the length, sensitivity, and repeatability of the analytical method.[72] Accordingly, if sample throughput is the major objective of analysis and if equilibration times are extremely long (which is usually the case in DI-SPME from aqueous samples applications), pre-equilibrium conditions should be employed. On the other hand, optimum sensitivity is achieved under equilibrium conditions, and such conditions should be employed when method sensitivity is the major objective of analysis.[56] Similarly, if the method repeatability is the analysis objective, equilibrium conditions should be used, unless perfectly repeatable timing events are guaranteed under pre-equilibrium conditions (e.g., with the use of an autosampler). The longer the extraction times and the less steep the extraction profile curve slope, the smaller the relative errors under pre-equilibrium conditions. Therefore, when the selected extraction time is in the steep area of extraction time profile curve, a small error in timing may result in much higher errors in amount of analyte extracted by the coating as compared with the extraction times selected closer to or after the equilibration time point.

5.5.3.13. Step 13—optimization of desorption conditions. For the procedure regarding the optimization of desorption conditions for SPME-GC applications, please refer to Section 5.5.1.

5.6. HYPHENATION OF SPME AND LC

5.6.1. SPME-LC Interfaces

Many analytes of interest, such as drugs, pesticides, peptides, and hormones, tend to be nonvolatile and/or polar and are thus more amenable for analysis using LC rather than GC. SPME can be coupled to LC using four different strategies: (1) manual interface commercially available from Supelco or lab-made interface with similar or reduced internal volume, (2) automated in-tube SPME, (3) manual offline desorption, and (4) high-throughput automated offline desorption using the Concept 96 autosampler. The main advantages and disadvantages of these SPME-LC coupling strategies are summarized in Table 5.3. For a detailed description, the reader is referred to recent review.[85]

5.6.1.1. Manual SPME interface for HPLC. The commercially available, manual SPME interface consists of

Table 5.3. Summary of Advantages and Disadvantages of various SPME-LC Coupling Strategies

SPME-LC Coupling Mode	Advantages	Disadvantages
Manual direct SPME-LC interface	High sensitivity since all of the analyte extracted is introduced into LC instrument No solvent front peak if dynamic desorption is used Applicable to complex samples or samples containing particulates	Coating may be damaged or stripped in the interface Possible leaks and loss of sample Labor intensive Low throughput and no automation Peak broadening Possible carryover problems due to slow desorption kinetics
In-tube SPME	Automated Reduced total analysis time Increased throughput Better precision and accuracy Decreased handling of biohazardous sample Good sensitivity (same or better than for direct manual interface) Less likelihood of carryover	Limited to particulate-free samples Sample preparation takes up instrument time Not applicable to *in vivo* and on-site analysis
Manual offline desorption	Does not require additional instrumentation Wide choice of desorption solvents available Physical damage to fiber is not likely Ability to carry out desorption of multiple fibers in parallel, reducing total analysis time	Offline Lack of automation Possible loss of analytes due to adsorption to walls of vial or insert used for desorption Poorer sensitivity because all of analyte is not injected
Concept 96 robotic station	Full automation Highest sample throughput Wide choice of desorption solvents available Suitable for use with pre-equilibrium SPME with no loss of precision Applicable to complex samples such as whole blood with no sample pretreatment required	Poorer sensitivity because all of analyte is not injected Applicable to nonvolatile analytes only Possible loss of analytes due to adsorption to walls of 96-well plates Cost

a desorption chamber and standard six-port injection valve, and allows for static and/or dynamic desorption of the fiber into mobile phase or another suitable desorption solvent. Alternatively, an SPME desorption chamber can also be easily constructed using stainless steel three-way tee and appropriate ferrules and tubing and incorporated in place of the injection loop.[86,87] Briefly, the SPME fiber is desorbed directly in the LC interface, and the desorbed analytes are swept onto the analytical column using flowing mobile phase. Each sample is processed serially and all steps of extraction, desorption, and sample introduction are performed manually. The main advantage of this type of coupling is high sensitivity, because all of the analytes desorbed from the fiber are injected directly in the LC system (similar to thermal desorption in GC).

5.6.1.2. Automated in-tube SPME. In-tube SPME provides a fully automated approach for coupling SPME to LC using commercial HPLC autosampler and chromatographic software.[88] In this configuration of SPME, the extraction phase is internally coated on the inside of a capillary column. For example, a short section of GC capillary column can be used for this purpose. The extraction is performed by repeated aspiration of the sample from the vial into the tube using an appropriate autosampler program as shown in Table 5.4 or by pumping sample through the extraction capillary for a defined time. Desorption is typically performed by directing mobile phase flow through the extraction capillary. In-tube SPME was successfully implemented on a variety of commercial HPLC autosamplers[89,90] and applied to a number of environmental, clinical, forensic, and food applications as reviewed by Kataoka.[91,92]

Table 5.4. Typical Autosampler Program for In-Tube SPME

Autosampler Step
Divert mobile phase flow away from the capillary (capillary bypass where capillary is under ambient pressure)
Aspirate/dispense cycle(s) of one or more conditioning solvents
Aspirate/dispense cycles of sample
Wash injection needle tip
Allow mobile phase through the capillary to desorb the analytes (main pass where capillary is under high pressure)

5.6.1.3. Manual and automated offline desorption. Offline desorption is the simplest and most flexible way to couple SPME to LC analysis. If performed manually using HPLC vials, small-volume inserts, or multiwell plates, this approach requires no additional instrumentation and offers maximum flexibility, because a wide variety of solvents and/or agitation methods can be used in order to achieve effective analyte desorption from the SPME fiber. Offline desorption can be performed either (1) using minimum volume of desorption solvent required to completely immerse the SPME extraction phase (direct desorption method) or (2) using large volume of desorption solvent followed by evaporation to dryness and reconstitution in minimal amount of solvent (evaporation/reconstitution method). Clearly, evaporation/reconstitution approach provides better sensitivity, while direct desorption approach provides higher sample throughput.

Very recently, the full automation of offline desorption in 96-well plate format was achieved using Concept 96 commercial robotic station (PAS Technology).[93] The main components of the system include a 96-fiber (or 96-thin-film device), three robotic arms, three orbital agitators, one wash station, and 96-well nitrogen blowdown device (Fig. 5.13A). The original device was designed using fiber geometry,[93] but the final commercial product uses thin-film geometry (TFME).[94] The main reason for adopting thin-film geometry (Fig. 5.13B) was to further increase surface area and volume of extraction phase, which results in improved analytical sensitivity and faster extraction rates as discussed in Section 5.3.2. Coatings used to date with this system include silica-sorbent-based coatings (5-μm silica particles coated with C18 or C16-amide),[93] carbon tape,[95] and Empore membrane disks.[96]

The Concept 96 can perform steps such as preconditioning, extraction, rinsing, solvent desorption, agitation, addition of internal standard, solvent evaporation, and solvent reconstitution in a fully automated fashion using the Concept software. The only steps of the analysis that are

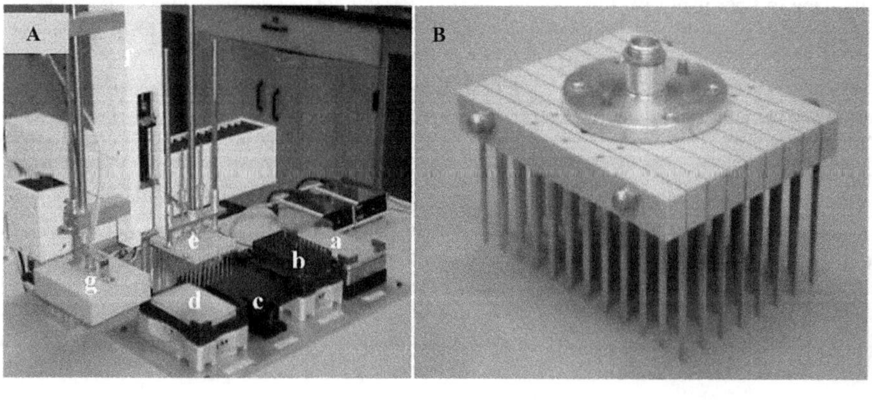

Figure 5.13. (A) Photo of Concept 96 automated SPME station with main components labeled (a) preconditioning station, (b) extraction station, (c) wash station, (d) desorption station, (e) 96-thin-film device, (f) syringe arm, and (g) 96-well nitrogen blowdown device, and (B) close-up of the 96-thin-film device.

not automated are (1) the dispensing of sample, desorption solvent and preconditioning solutions in multiwell plates and (2) the transfering of the completed plate to the HPLC autosampler for the analysis. These steps can be performed manually by the user or using appropriate existing commercially available robotic stations and/or plate feeders. The system can be further customized based on the user's exact requirements. Initial prototype designs of the Concept 96 also included an HPLC injection port, which allowed the direct interfacing of the unit with LC-MS/MS instrumentation for online operation.[93] However, significantly greater sample throughput is achievable if the unit is operated as a sample preparation station, with a separate conventional HPLC autosampler used to perform LC injection from multiwell plates.

The use of the Concept 96 for automated offline desorption offers significant advantages over manual desorption because it improves method precision and drastically increases sample throughput. In fact, sample throughput of >1000 samples per day is possible, as demonstrated by example applications of the analysis of benzodiazepines in whole blood[93] or Ochratoxin A in urine.[95] The system can also be used for automated ligand–receptor binding studies, which are important in drug discovery.[97] The automation of the desorption step of *in vivo* SPME is currently underway in our laboratory on the basis of the recently proposed parallel desorption device.[48]

5.6.2. Advantages of Automated High-Throughput SPME-LC

The automation of SPME/TFME in multiwell plate format allows significant reduction in analyst time and significant increase in sample throughput. In addition, automated SPME/TFME also improves method precision because of more accurate timing of extraction and desorption steps and excellent reproducibility in fiber/thin-film positioning within the wells. This allows the use of pre-equilibrium SPME extraction times without deterioration of method precision as is typically reported when manual SPME with pre-equilibrium extraction times is employed (Section 5.5.3.12). In comparison to automated in-tube SPME, the Concept 96 system is able to handle more complex samples (e.g., whole blood) without the need to perform any sample pretreatment steps and provides increased sample throughput due to multiplexing.

Automated SPME/TFME also offers some unique advantages in comparison to other sample preparation methods commonly employed such as automated SPE and liquid–liquid extraction (LLE). SPME can handle direct extraction from complex matrices such as whole blood or tissue homogenates because the extraction is performed in an open-bed format. This reduces the total number of steps required for sample preparation, which results in significant time savings, decreases the potential for sample contamination, and minimizes sample losses. Furthermore, SPME can be used to determine both free and total concentration of analyte from the analysis of a single sample as long as two appropriate calibration curves are constructed.[47,48,98–100] Detailed discussion of calibration theory is beyond the scope of this chapter, and the interested reader is referred to a recent review article.[101] In contrast, traditional methods generally rely on the disruption of the binding between analyte and proteins and thus can provide only total concentration. This means that a separate ultrafiltration or dialysis experiment must be carried out to determine free drug concentration. In terms of speed, automated SPME/TFME is faster than online SPE procedures, which usually require about 2–10 min per sample.[102,103] Its current speed is comparable to automated multiwell SPE and LLE methods typically requiring about 60 min for the preparation of 96 samples, but this speed can be further improved with additional research into the development of thinner coatings and coatings with high affinity for target analytes.

The availability of Concept 96 enables the use of SPME-LC in high-throughput applications for the first time, so this type of application can be expected to grow in the future.

5.6.3. Method Development for SPME-LC Applications

Although method development for SPME-LC methods is generally similar to method development for SPME-GC applications (Section 5.5.3), a few important differences will be discussed briefly in this section.[104] In particular, the analyst is faced with a choice of interfacing strategies, and proper optimization of desorption conditions is of critical importance for developing a successful SPME-LC method.

5.6.3.1. Selection of SPME-LC interfacing strategy. Table 5.3 summarizes the advantages and disadvantages of available SPME-LC interfaces to help in the selection of the most appropriate strategy for a given application.

5.6.3.2. Selection of SPME coatings for LC applications. Section 5.5.3.1 describes commercially available coatings and their suitability for the extraction of various analytes. Although these coatings were designed for use with GC applications in combination with thermal desorption, historically, these types of coatings were also employed for the development of SPME-LC applications. However, problems with significant swelling of the coating and poor extraction efficiency for very polar, nonvolatile, and/or large molecular weight analytes were encountered. A new line of SPME coatings suitable for use with LC applications was recently developed by Supelco to address such limitations.[105] These new coatings are immobilized on the metal fiber core and consist of a mixture of proprietary biocompatible binder and various types of coated silica (octadecyl, polar embedded, and cyano) particles. They are available in two formats: (1) regular assemblies for *in vitro* use and (2) *in vivo* assemblies described in Section 5.4.4. These coatings exhibit excellent chemical and mechanical stability, thus permitting their use in common LC solvents and high agitation speeds without breakage of fiber or inadvertent loss of coating.[105] Overall, C18 coatings performed the best for the analysis of compounds such as drugs and steroids, with RP-amide C16 (RPA) coatings showing slightly better extraction efficiency toward more polar compounds due to presence of an embedded amide group.

5.6.3.3. Optimization of desorption conditions. The optimization of desorption conditions is a crucial part of the SPME-LC method development. During method development, the following desorption parameters should be considered: (1) composition of desorption solvent, (2) volume of desorption solvent, (3) desorption time, and (4) the evaluation of carryover. Complete desorption of the analyte from the fiber is more difficult to achieve in SPME-LC applications than in SPME-GC applications, since the kinetics of the desorption process in liquid phase are significantly slower than in gas phase. The slow kinetics of desorption can result in three main problems: (1) band broadening for manual interface and in-tube SPME, (2) analyte carryover due to incomplete desorption, and (3) need for longer desorption times, so these potential issues should be thoroughly evaluated during desorption optimization experiments.

5.6.3.4. Fiber preconditioning. The presence of organic solvents from preconditioning or fiber cleaning may affect the extraction efficiency of the fiber. This effect should be evaluated by comparing extraction results from a fiber that was not preconditioned with results for the extraction where the fiber was appropriately preconditioned. For some recently introduced fibers, such as Supelco biocompatible HPLC fibers based on C18-coated silica particles, a preconditioning step is necessary to wet the C18 chains and obtain the best extraction efficiency.[105] Similarly, Lord et al. found approximately 2.5× fold increase in extraction efficiency of polypyrrole fibers if the probes were preconditioned by rinsing with methanol for 15 s, followed by rinsing with water for 15 s.[87]

5.6.3.5. Fiber rinsing. In many instances, a short fiber rinsing (2–60 s) step should be introduced after extraction. Addition of this step prior to desorption may improve the reproducibility of the analysis by removing any analyte loosely attached to the surface of the fiber by surface tension.[47,87] The incorporation of a short rinsing step is also recommended for automated SPME applications involving complex, heterogeneous matrices and for *in vivo* SPME methods in order to remove any particulate matter or cells from the surface of the fiber prior to the desorption.[93,95] If such a rinsing step is used, it is important to (1) choose a weak solvent, (2) keep rinsing times short to ensure the analyte is not accidentally desorbed from the fiber, and (3) perform this step immediately after extraction for best results.

5.7. CONCLUDING REMARKS

The development of SPME has addressed some of the major drawbacks of conventional sample preparation methods since the technique provides the following advantages: short sample preparation times; small sample volume requirements; liquid, gaseous, and solid sample matrix sampling; and solvent consumption elimination. In addition, with recent trends toward the development of improved SPME devices, high-throughput approaches, and on-site/*in vivo* sampling, the use of this technique will continue to grow.

REFERENCES

1. Pawliszyn, J.; Liu, S. Sample introduction for capillary gas chromatography with laser desorption and optical fibers. *Anal. Chem.* **1987**, *59*, 1475–1478.

2. Belardi, R.G.; Pawliszyn, J. The application of chemically modified fused silica fibers in the extraction of organics from water matrix samples and their rapid transfer to capillary columns. *Water Qual. Res. J. Can.* **1989**, *24*, 179–191.

3. Arthur, C.L.; Pawliszyn, J. Solid phase microextraction with thermal desorption using fused silica optical fiber. *Anal. Chem.* **1990**, *62*, 2145–2148.

4. Pawliszyn, J. Method & device for solid phase microextraction & desorption, PCT, International Patent Publication Number WO 91/15745 & national counterparts; 1991.

5. McComb, M.; Giller, E.; Gesser, H.D. Abs 528. *78th Canadian Society for Chemistry Conference & Exhibition*, University of Guelph, Guelph, ON; 1995.

6. Chai, M.; Pawliszyn, J. Analysis of environmental air samples by solid-phase microextraction and gas chromatography/ion trap mass spectrometry. *Environ. Sci. Technol.* **1995**, *29*, 693–701.

7. Baltussen, E.; David, F.; Sandra, P.; Janssen, H.-G.; Cramers, C. Sorption tubes packed with polydimethylsiloxane: A new and promising technique for the preconcentration of volatiles and semi-volatiles from air and gaseous samples. *J. High Resolut. Chromatogr.* **1998**, *21*, 332–340.

8. Nickerson, M.A. Sample screening & preparation within a collection vessel. U.S. Patent No. 5,827,944; 1998.

9. Baltussen, E.; Sandra, P.; David, F.; Janssen, H.-G.; Cramers, C. Study into the equilibrium mechanism between water and poly(dimethylsiloxane) for very apolar solutes: Adsorption or sorption? *Anal. Chem.* **1999**, *71*, 5213–5216.

10. Bruheim, I.; Liu, X.; Pawliszyn, J. Thin-film microextraction. *Anal. Chem.* **2003**, *75*, 1002–1010.

11. Zhao, W.; Ouyang, G.; Alaee, M.; Pawliszyn, J. On-rod standardization technique for time-weighted average water sampling with a polydimethylsiloxane rod. *J. Chromatogr. A* **2006**, *1124*, 112–120.

12. Popp, P.; Moeder, M. *ExTech'2001*, Barcelona, Spain, September; 2001

13. Grote, C.; Pawliszyn, J. Solid-phase microextraction for the analysis of human breath. *Anal. Chem.* **1997**, *69*, 587–596.

14. Lord, H.; Yu, W.; Segal, A.; Pawliszyn, J. Breath analysis and monitoring by membrane extraction with sorbent interface. *Anal. Chem.* **2002**, *74*, 5650–5657.

15. Ma, W.; Liu, X.; Pawliszyn, J. Analysis of human breath with micro extraction techniques and continuous monitoring of carbon dioxide concentration. *Anal. Bioanal. Chem.* **2006**, *385*, 1398–1404.

16. Pawliszyn, J. *Solid Phase Microextraction: Theory & Practice*. New York: Wiley-VCH; 1997.

17. Zhang, Z.; Pawliszyn, J. Quantitative extraction using an internally cooled solid phase microextraction device. *Anal. Chem.* **1995**, *67*, 34–43.

18. Chen, Y.; Pawliszyn, J. Miniaturization and automation of an internally cooled coated fiber device. *Anal. Chem.* **2006**, *78*, 5222–5226.

19. Ouyang, G.; Pawliszyn, J. Recent developments in SPME for on-site analysis and monitoring. *Trends Anal. Chem.* **2006**, *25*, 692–703.

20. Muller, L. Field analysis by SPME. In *Applications of Solid Phase Microextraction*, Pawliszyn, J., Ed. Cambridge, UK: RSC Chromatography Monographs—The Royal Society of Chemistry; 1999.

21. Wang, A.-P.; Fang, F.; Pawliszyn, J. Sampling and determination of volatile organic compounds with needle trap devices. *J. Chromatogr. A* **2005**, *1072*, 127–135.

22. Gong, Y.; Eom, I.-Y.; Lou, D.-W.; Hein, D.; Pawliszyn, J. Development and application of a needle trap device for time-weighted average diffusive sampling. *Anal. Chem.* **2008**, *80*, 7275–7282.

23. Eom, I.-Y.; Niri, V.; Pawliszyn, J. Development of a syringe pump assisted dynamic headspace sampling technique for needle trap device. *J. Chromatogr. A* **2008**, *1196–1197*, 10–14.

24. Martos, P.A.; Pawliszyn, J. Time-weighted average sampling with solid-phase microextraction device: Implications for enhanced personal exposure monitoring to airborne pollutants. *Anal. Chem.* **1999**, *71*, 1513–1520.

25. Koziel, J.; Noah, J.; Pawliszyn, J. Field sampling and determination of formaldehyde in indoor air with solid-phase microextraction and on-fiber derivatization. *Environ. Sci. Technol.* **2001**, *35*, 1481–1486.

26. Khaled, A.; Pawliszyn, J. Time-weighted average sampling of volatile and semi-volatile airborne organic compounds by the solid-phase microextraction device. *J. Chromatogr. A* **2000**, *892*, 455–467.

27. Koziel, J.; Jia, M.; Khaled, A.; Noah, J.; Pawliszyn, J. Field air analysis with SPME device. *Anal. Chim. Acta* **1999**, *400*, 153–162.

28. Chen, Y.; Pawliszyn, J. Time-weighted average passive sampling with a solid-phase microextraction device. *Anal. Chem.* **2003**, *75*, 2004–2010.

29. Chen, Y.; Pawliszyn, J. Solid-phase microextraction field sampler. *Anal. Chem.* **2004**, *76*, 6823–6828.

30. Ouyang, G.; Zhao, W.; Alaee, M.; Pawliszyn, J. Time-weighted average water sampling with a diffusion-based solid-phase microextraction device. *J. Chromatogr. A* **2007**, *1138*, 42–46.

31. Ouyang, G.; Zhao, W.; Bragg, L.; Qin, Z.; Alaee, M.; Pawliszyn, J. Time-weighted average water sampling in Lake Ontario with solid-phase microextraction passive samplers. *Environ. Sci. Technol.* **2007**, *41*, 4026–4031.

32. Said, I.; Gaertner, C.; Renou, M.; Rivault, C. Perception of cuticular hydrocarbons by the olfactory organs in *Periplaneta americana* (L.) (Insecta: Dictyoptera). *J. Insect Physiol.* **2005**, *51*, 1384–1389.

33. Tentschert, J.; Bestmann, H.J.; Heinze, J. Cuticular compounds of workers and queens in two Leptothorax ant species—A comparison of results obtained by solvent extraction, solid sampling, and SPME. *Chemoecology* **2002**, *12*, 15–21.

34. Tentschert, J.; Kolmer, K.; Holldobler, B.; Bestmann, H.J.; Delabie, J.H.C.; Heinze, J. Chemical profiles, division of labor and social status in *Pachycondyla queens* (Hymenoptera: Formicidae). *Naturwissenschaften* **2001**, *88*, 175–178.

35. Peeters, C.; Monnin, T.; Malosse, C. Cuticular hydrocarbons correlated with reproductive status in a queenless ant. *Proc. R. Soc. B Lond. Biol. Sci.* **1999**, *266*, 1323–1327.

36. Monnin, T.; Malosse, C.; Peeters, C. Solid-phase microextraction and cuticular hydrocarbon differences related to reproductive activity in queenless ant *Dinoponera quadriceps*. *J. Chem. Ecol.* **1998**, *24*, 473–490.

37. Gilley, D.C.; De Grandi-Hoffman, G.; Hooper, J.E. Volatile compounds emitted by live European honey bee (*Apis mellifera* L.) queens. *J. Insect Physiol.* **2006**, *52*, 520–527.

38. Djozan, D.; Baheri, T.; Farshbaf, R.; Azhari, S. Investigation of solid-phase microextraction efficiency using pencil lead fiber for in vitro and in vivo sampling of defensive volatiles from insect's scent gland. *Anal. Chim. Acta* **2005**, *554*, 197–201.

39. Anderbrant, O.; Östrand, F.; Bergström, G.; Wassgren, A.-B.; Auger-Rozenberg, M.-A.; Geri, C.; Hedenström, E.; Högberg, H.-E.; Herz, A.; Heitland, W. Release of sex pheromone and its precursors in the pine sawfly *Diprion pini* (Hym., Diprionidae). *Chemoecology* **2005**, *15*, 147–151.

40. Rochat, D.; Ramirez-Lucas, P.; Malosse, C.; Aldana, R.; Kakul, T.; Morin, J.P. Role of solid-phase microextraction in the identification of highly volatile pheromones of two Rhinoceros beetles *Scapanes australis* and *Strategus aloeus* (Coleoptera, Scarabaeidae, Dynastinae). *J. Chromatogr. A* **2000**, *885*, 433–444.

41. Chen, X.; Nakamuta, K.; Nakanishi, T.; Nakashima, T.; Tokoro, M.; Mochizuki, F.; Fukumoto, T. Female sex pheromone of a carpenter moth, *Cossus insularis* (Lepidoptera: Cossidae). *J. Chem. Ecol.* **2006**, *32*, 669–679.

42. Pionnier, E.; Chabanet, C.; Mioche, L.; Le Quere, J.-L.; Salles, C. 1. In vivo aroma release during eating of a model cheese: relationships with oral parameters. *J. Agric. Food Chem.* **2004**, *52*, 557–564.

43. Pionnier, E.; Semon, E.; Chabanet, C.; Salles, C. Evaluation of the solid phase microextraction (SPME) technique for the analysis of human breath during eating. *Sci. Aliment.* **2005**, *25*, 193–206.

44. Zhang, Z.-M.; Cai, J.-J.; Ruan, G.-H.; Li, G.-K. The study of fingerprint characteristics of the emanations from human arm skin using the original sampling system by SPME-GC/MS. *J. Chromatogr. B* **2005**, *822*, 244–252.

45. Musteata, F.M.; Musteata, M.L.; Pawliszyn, J. Fast in vivo microextraction: A new tool for clinical analysis. *Clin. Chem.* **2006**, *52*, 708–715.

46. Musteata, F.M.; Pawliszyn, J. In vivo sampling with solid phase microextraction. *J. Biochem. Biophys. Methods* **2007**, *70*, 181–193.

47. Es-haghi, A.; Zhang, X.; Musteata, F.M.; Bagheri, H.; Pawliszyn, J. Evaluation of bio-compatible poly(ethylene glycol)-based solid-phase microextraction fiber for in vivo pharmacokinetic studies of diazepam in dogs. *Analyst* **2007**, *132*, 672–678.

48. Zhang, X.; Es-haghi, A.; Musteata, F.M.; Ouyang, G.; Pawliszyn, J. Quantitative in vivo microsampling for pharmacokinetic studies based on an integrated solid-phase microextraction system. *Anal. Chem.* **2007**, *79*, 4507–4513.

49. Gorecki, T.; Pawliszyn, J. Solid phase microextraction/isothermal GC for rapid analysis of complex organic samples. *J. High Resolut. Chromatogr.* **1995**, *18*, 161–166.

50. Gorecki, T.; Pawliszyn, J. Sample introduction approaches for solid phase microextraction/rapid GC. *Anal. Chem.* **1995**, *67*, 3265–3274.

51. Gorecki, T.; Pawliszyn, J. Field-portable solid-phase microextraction/fast GC system for trace analysis. *Field Anal. Chem. Technol.* **1997**, *1*, 277–284.

52. O'Reilly, J.; Wang, Q.; Setkova, L.; Hutchinson, J.P.; Chen, Y.; Lord, H.L.; Linton, C.M.; Pawliszyn, J. Automation of solid-phase microextraction. *J. Sep. Sci.* **2005**, *28*, 2010–2022.

53. Kataoka, H.; Lord, H.L.; Pawliszyn, J. Applications of solid-phase microextraction in food analysis. *J. Chromatogr. A* **2000**, *880*, 35–62.

54. Risticevic, S.; Chen, Y.; Kudlejova, L.; Vatinno, R.; Baltensperger, B.; Stuff, J.R.; Hein, D.; Pawliszyn, J. Protocol for the development of automated high-throughput SPME-GC methods for the analysis of volatile and semivolatile constituents in wine samples. *Nat. Protoc.* **2010**, *5*, 162–176.

55. Arthur, C.L.; Killam, L.M.; Buchholz, K.D.; Pawliszyn, J.; Berg, J.R. Automation and optimization of solid-phase microextraction. *Anal. Chem.* **1992**, *64*, 1960–1966.

56. Pawliszyn, J. *Handbook of Solid Phase Microextraction*. Waterloo, Canada: University of Waterloo; 2007.

57. Zhao, W.; Ouyang, G.; Pawliszyn, J. Preparation and application of in-fibre internal standardization solid-phase microextraction. *Analyst* **2007**, *132*, 256–261.

58. Zhou, S.N.; Zhao, W.; Pawliszyn, J. Kinetic calibration using dominant pre-equilibrium desorption for on-site and in vivo sampling by solid-phase microextraction. *Anal. Chem.* **2008**, *80*, 481–490.

59. Chen, Y.; O'Reilly, J.; Wang, Y.; Pawliszyn, J. Standards in the extraction phase, a new approach to calibration of microextraction processes. *Analyst* **2004**, *129*, 702–703.

60. Wang, Y.; O'Reilly, J.; Chen, Y.; Pawliszyn, J. Equilibrium in-fibre standardisation technique for solid-phase microextraction. *J. Chromatogr. A* **2005**, *1072*, 13–17.

61. Chen, Y.; Pawliszyn, J. Kinetics and the on-site application of standards in a solid-phase microextraction fiber. *Anal. Chem.* **2004**, *76*, 5807–5815.

62. Zhou, S.N.; Zhang, X.; Ouyang, G.; Es-haghi, A.; Pawliszyn, J. On-fiber standardization technique for solid-coated solid-phase microextraction. *Anal. Chem.* **2007**, *79*, 1221–1230.

63. Zhou, S.N.; Oakes, K.D.; Servos, M.R.; Pawliszyn, J. Application of solid-phase microextraction for in vivo laboratory and field sampling of pharmaceuticals in fish. *Environ. Sci. Technol.* **2008**, *42*, 6073–6079.

64. Wang, Q.; O'Reilly, J.; Pawliszyn, J. Determination of low-molecular mass aldehydes by automated headspace solid-phase microextraction with in-fibre derivatisation. *J. Chromatogr. A* **2005**, *1071*, 147–154.

65. Sporkert, F.; Pragst, F.; Hubner, S.; Mills, G. Determination of low-molecular mass aldehydes by automated headspace solid-phase microextraction with in-fibre derivatisation. *J. Chromatogr. B* **2002**, *772*, 45–51.

66. Setkova, L.; Risticevic, S.; Linton, C.M.; Ouyang, G.; Bragg, L.M.; Pawliszyn, J. Solid-phase microextraction–gas chromatography–time-of-flight mass spectrometry utilized for the evaluation of the new-generation super elastic fiber assemblies. *Anal. Chim. Acta* **2007**, *581*, 221–231.

67. Carasek, E.; Cudjoe, E.; Pawliszyn, J. Fast and sensitive method to determine chloroanisoles in cork using an internally cooled solid-phase microextraction fiber. *J. Chromatogr. A* **2007**, *1138*, 10–17.

68. Carasek, E.; Pawliszyn, J. Screening of tropical fruit volatile compounds using solid-phase microextraction (SPME) fibers and internally cooled SPME fiber. *J. Agric. Food Chem.* **2006**, *54*, 8688–8696.

69. Ghiasvand, A.R.; Hosseinzadeh, S.; Pawliszyn, J. New cold-fiber headspace solid-phase microextraction device for quantitative extraction of polycyclic aromatic hydrocarbons in sediment. *J. Chromatogr. A* **2006**, *1124*, 35–42.

70. Risticevic, S.; Niri, V.H.; Vuckovic, D.; Pawliszyn, J. Recent developments in solid-phase microextraction. *Anal. Bioanal. Chem.* **2009**, *393*, 781–795.

71. Lord, H.; Pawliszyn, J. Evolution of solid-phase microextraction technology. *J. Chromatogr. A* **2000**, *885*, 153–193.

72. Risticevic, S.; Lord, H.; Górecki, T.; Arthur, C.L.; Pawliszyn, J. Protocol for solid phase microextraction method development. *Nat. Protoc.* **2010**, *5*, 122–139.

73. Vas, G.; Vékey, K. Solid-phase microextraction: A powerful sample preparation tool prior to mass spectrometric analysis. *J. Mass Spectrom.* **2004**, *39*, 233–254.

74. Tsoutsi, C.; Konstantinou, I.; Hela, D.; Albanis, T. Screening method for organophosphorus insecticides and their metabolites in olive oil samples based on headspace solid-phase microextraction coupled with gas chromatography. *Anal. Chim. Acta* **2006**, *573-574*, 216–222.

75. Mestres, M.; Busto, O.; Guasch, J. Headspace solid-phase microextraction analysis of volatile sulphides and disulphides in wine aroma. *J. Chromatogr. A* **1998**, *808*, 211–218.

76. Motlagh, S.; Pawliszyn, J. On-line monitoring of flowing samples using solid phase microextraction-gas chromatography. *Anal. Chim. Acta* **1993**, *284*, 265–273.

77. Augusto, F.; Koziel, J.; Pawliszyn, J. Design and validation of portable SPME devices for rapid field air sampling and diffusion-based calibration. *Anal. Chem.* **2001**, *73*, 481–486.

78. Ouyang, G.; Pawliszyn, J. SPME in environmental analysis. *Anal. Bioanal. Chem.* **2006**, *386*, 1059–1073.

79. Qin, Z.; Bragg, L.; Ouyang, G.; Pawliszyn, J. Comparison of thin-film microextraction and stir bar sorptive extraction for the analysis of polycyclic aromatic hydrocarbons in aqueous samples with controlled agitation conditions. *J. Chromatogr. A* **2008**, *1196-1197*, 89–95.

80. Mills, G.A.; Walker, V. Headspace solid-phase microextraction procedures for gas chromatographic analysis of biological fluids and materials. *J. Chromatogr. A* **2000**, *902*, 267–287.

81. Prosen, H.; Zupančič-Kralj, L. Solid-phase microextraction. *Trends Anal. Chem.* **1999**, *18*, 272–282.

82. Lambropoulou, D.A.; Albanis, T.A. Headspace solid phase microextraction applied to the analysis of organophosphorus insecticides in strawberry and cherry juices. *J. Agric. Food Chem.* **2002**, *50*, 3359–3365.

83. Zambonin, C.G.; Quinto, M.; Vietro, N.D.; Palmisano, F. Solid-phase microextraction–gas chromatography mass spectrometry: A fast and simple screening method for the assessment of organophosphorus pesticides residues in wine and fruit juices. *Food Chem.* **2004**, *86*, 269–274.

84. Vichi, S.; Pizzale, L.; Conte, L.S.; Buxaderas, S.; López-Tamames, E. Simultaneous determination of volatile and semi-volatile aromatic hydrocarbons in virgin olive oil by headspace solid-phase microextraction coupled to gas chromatography/mass spectrometry. *J. Chromatogr. A* **2005**, *1090*, 146–154.

85. Lord, H.L. Strategies for interfacing solid-phase microextraction with liquid chromatography. *J. Chromatogr. A* **2007**, *1152*, 2–13.

86. Chen, J.; Pawliszyn, J.B. Solid phase microextraction coupled to high-performance liquid chromatography. *Anal. Chem.* **1995**, *67*, 2530–2533.

87. Lord, H.L.; Grant, R.P.; Walles, M.; Incledon, B.; Fahie, B.; Pawliszyn, J.B. Development and evaluation of a solid-phase microextraction probe for in vivo pharmacokinetic studies. *Anal. Chem.* **2003**, *75*, 5103–5115.

88. Eisert, R.; Pawliszyn, J. Automated in-tube solid-phase microextraction coupled to high-performance liquid chromatography. *Anal. Chem.* **1997**, *69*, 3140–3147.

89. Gou, Y.; Eisert, R.; Pawliszyn, J. Automated in-tube solid-phase microextraction-high-performance liquid chromatography for carbamate pesticide analysis. *J. Chromatogr. A* **2000**, *873*, 137–147.

90. Gou, Y.; Tragas, C.; Lord, H.; Pawliszyn, J. On-line coupling of in-tube solid phase microextraction (SPME) to HPLC for analysis of carbamates in water samples: Comparison of two commercially available autosamplers. *J. Microcolumn Sep.* **2000**, *12*, 125–134.

91. Kataoka, H. Automated sample preparation using in-tube solid-phase microextraction and its application—A review. *Anal. Bioanal. Chem.* **2002**, *373*, 31–45.

92. Kataoka, H. New trends in sample preparation for clinical and pharmaceutical analysis, TrAC. *Trends Anal. Chem.* **2003**, *22*, 232–244.

93. Vuckovic, D.; Cudjoe, E.; Hein, D.; Pawliszyn, J. Automation of solid-phase microextraction in high-throughput format and applications to drug analysis. *Anal. Chem.* **2008**, *80*, 6870–6880.

94. Cudjoe, E.; Vuckovic, D.; Hein, D.; Pawliszyn, J. Investigation of the effect of the extraction phase geometry on the performance of automated solid-phase microextraction. *Anal. Chem.* **2009**, *81*, 4226–4232.

95. Vatinno, R.; Vuckovic, D.; Zambonin, C.G.; Pawliszyn, J. Automated high-throughput method using solid-phase microextraction-liquid chromatography-tandem mass spectrometry for the determination of ochratoxin A in human urine. *J. Chromatogr. A* **2008**, *1201*, 215–221.

96. Cudjoe, E.; Pawliszyn, J. A new approach to the application of solid phase extraction disks with LC-MS/MS for the analysis of drugs on a 96-well plate format. *J. Pharm. Biomed. Anal.* **2009**, *50*, 556–562.

97. Vuckovic, D.; Pawliszyn, J. Automated study of ligand-receptor binding using solid-phase microextraction. *J. Pharm. Biomed. Anal.* **2009**, *50*, 550–555.

98. Musteata, F.M.; Pawliszyn, J. Determination of free concentration of paclitaxel in liposome formulations. *J. Pharm. Pharm. Sci.* **2006**, *9*, 231–237.

99. Musteata, F.M.; Pawliszyn, J. Study of ligand-receptor binding using SPME: Investigation of receptor, free, and total ligand concentrations. *J. Proteome Res.* **2005**, *4*, 789–800.

100. Musteata, F.M.; Pawliszyn, J. Assay of stability, free and total concentration of chlorhexidine in saliva by solid phase microextraction. *J. Pharm. Biomed. Anal.* **2005**, *37*, 1015–1024.

101. Ouyang, G.; Pawliszyn, J. A critical review in calibration methods for solid-phase microextraction. *Anal. Chim. Acta* **2008**, *627*, 184–197.

102. Xu, R.N.; Fan, L.; Rieser, M.J.; El-Shourbagy, T.A. Recent advances in high-throughput quantitative bioanalysis by LC-MS/MS. *J. Pharm. Biomed. Anal.* **2007**, *44*, 342–355.

103. Mullett, W.M. Determination of drugs in biological fluids by direct injection of samples for liquid-chromatographic analysis. *J. Biochem. Biophys. Methods* **2007**, *70*, 263–273.

104. Kudlejova, L.; Risticevic, S.; Vuckovic, D. Solid-phase microextraction method development. In *Handbook of Solid-Phase Microextraction*, Pawliszyn, J., Ed. Beijing, Ching: Chemical Industry Press; 2009.

105. Vuckovic, D.; Shirey, R.; Chen, Y.; Sidisky, L.; Aurand, C.; Stenerson, K.; Pawliszyn, J. In vitro evaluation of new biocompatible coatings for solid-phase microextraction: Implications for drug analysis and in vivo sampling applications. *Anal. Chim. Acta* **2009**, *638*, 175–185.

Chapter 6

Microdialysis Sampling as a Sample Preparation Method

Pradyot Nandi and Susan M. Lunte

6.1. INTRODUCTION

In 1862, Thomas Graham coined the term dialysis to describe the process of separation of substances in solution by means of their unequal diffusion through a semipermeable membrane.[1] He subsequently employed this method to extract urea from urine—a process that ultimately provided the foundation for kidney dialysis. Since that time, dialysis has been used for many other applications including isolation of proteins, purification of water, and, more recently, as a sample preparation tool in analytical chemistry.[2,3]

Microdialysis is a specific form of dialysis that employs small-diameter hollow fiber membranes for sampling. The first application of microdialysis was for sampling neurotransmitters in the brain. Studies by Bito in 1966 and Delago in 1972 using dialysis to measure the content of the extracellular fluid (ECF) in the brain led to the development of microdialysis sampling by Ungerstedt in 1974.[4,5] Ungerstedt realized that the small hollow fibers used for kidney dialysis could be implanted in the brain to monitor neurotransmitter concentrations in the ECF.[6,7]

Modern microdialysis sampling is accomplished using a probe that contains a semipermeable membrane that is placed in the sample or physiological region of interest (Fig. 6.1).[8] A solution of ionic composition and pH similar to that of the exterior environment is pumped through the probe by a syringe pump, producing a concentration gradient of the analyte between the perfusate and the surrounding medium. This causes analytes to diffuse into the probe based on the steepness of the concentration gradient. The solution coming from the probe, defined as the dialysate, is then delivered by the syringe pump to a fraction collector or an appropriate analytical system.

In 1974, Ungerstedt used microdialysis to measure the release of dopamine into the ECF of the brain,[6,9] and this is still the most popular application of microdialysis.[10,11] In this case, the dialysis probe acts as a synthetic blood vessel that is capable of both recovery and delivery of compounds in the external environment. Because there is no net fluid loss, it is possible to continuously monitor neurotransmitters and other substances in the extracellular space over long periods of time. In addition, if a compound is placed into the perfusate, it will diffuse out of the probe into the ECF, making it possible to study the region-specific metabolism of compounds in living tissue.

Although the majority of applications of microdialysis continue to be in the brain, this technique has expanded into many other areas, including drug metabolism and distribution,[11] pharmacokinetics,[12–14] process monitoring,[15] environmental monitoring,[16] and sample preparation for mass spectrometry (MS).[16–18]

6.2. THE MICRODIALYSIS SAMPLING PROCESS

Microdialysis sampling is a dynamic process. Typical flow rates range from 1 to 5 µL/min. At these flow velocities, there is not enough time for complete equilibration between the perfusate and the surrounding environment to occur. Hence, the concentration of the analyte in the perfusate reflects only a percentage of the total amount of analyte that is present in the extracellular space or sample. Recovery is the term used to describe the ratio of the dialysate concentration to the actual tissue concentration and is defined by the overall mass transport of the analyte across the probe. The percent recovery can be calculated using the following equation:

$$\%R = C_d / C_s \times 100, \qquad (6.1)$$

where C_d is the concentration in the dialysis membrane and C_s is the concentration in the sample being interrogated.

Handbook of Sample Preparation, Edited by Janusz Pawliszyn and Heather L. Lord

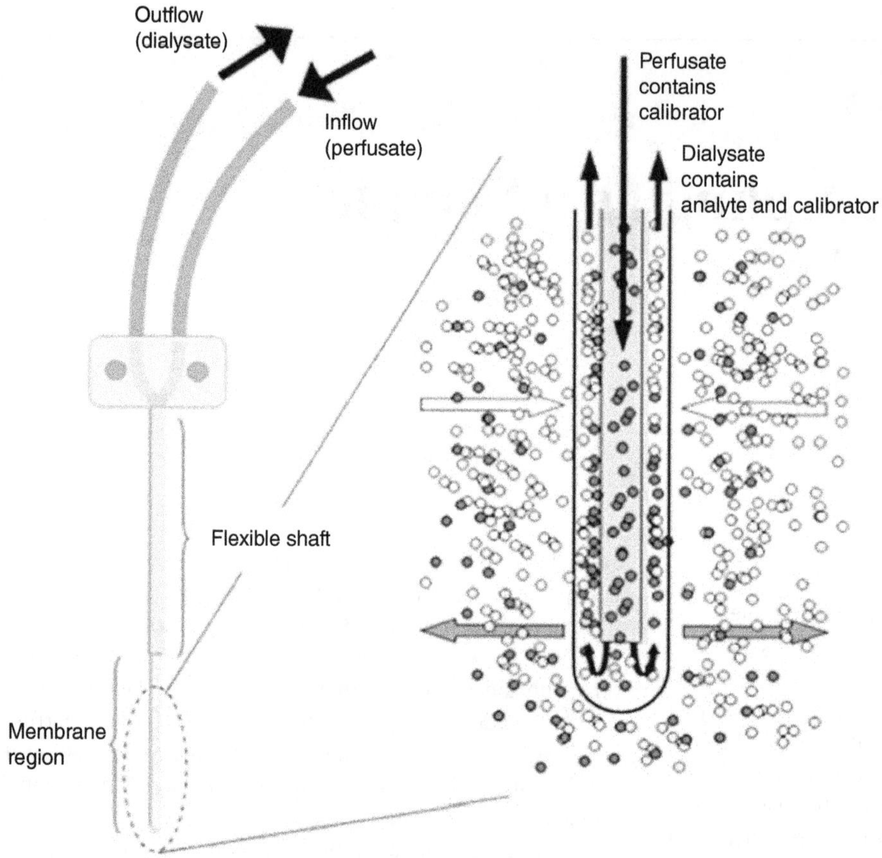

Figure 6.1. A concentric microdialysis probe. The black circles indicate the calibrator compound. The analyte from the extracellular space diffuses inside the probe and some of the calibration compound diffuses out of the probe based on concentration gradient. Reprinted from reference 8 with permission from Springer Science and Business Media. Copyright 2007.

Recovery in microdialysis is a function of the concentration gradient that is produced across the dialysis membrane. The magnitude of this concentration gradient is dependent on a number of parameters, including the chemical composition, length, thickness, and molecular weight cutoff (MWCO) of the membrane that is employed; the probe design and dimensions; the composition of the perfusate as well as the outer medium; the diffusion coefficient of the analyte of interest; the flow rate that is used for sampling; and whether the sampling is performed under hydrodynamic or static conditions.

More specifically, recovery can be described by the following equation:

$$R = 1 - \exp(-1/(Q_d(R_d + R_m + R_e))), \quad (6.2)$$

where Q_d is the volumetric flow rate of the perfusate, and R_d, R_m, and R_e are the resistances of dialysate, membrane, and extracellular space, respectively, to the mass transfer of analyte(s). R_e includes factors such as the rate of diffusion of the analyte of interest through the sample matrix as well as metabolism and other chemical or enzymatic reactions that the analyte may undergo during the sampling process. The membrane resistance, R_m, takes into account the probe-related factors including probe dimensions and MWCO, as well as the diffusion coefficient of the analyte across the probe membrane. R_d takes into consideration the diffusion of the analyte in the microdialysate.[19,20] This can be affected by the molecular weight and size of the compound of interest as well as the composition of the perfusate.

It is generally assumed that $R_m >> R_e$, which implies that recovery information obtained for a probe *in vitro* can be applied to *in vivo* experiments. However, this is not always the case because diffusion of the analyte from distant extracellular space to the area around the probe can be slow and is dependent on the diffusion coefficient of the analyte(s) in the extracellular matrix. For large peptides or proteins, this slow diffusion process can affect the recovery. In particular, peptides can undergo metabolism and other kinetic processes in the brain before reaching the probe membrane. Other considerations include determining the effects of loss of tissue integrity due to surgery and changes in blood flow and immunological reactions following insertion of the probe.

6.2.1. Dialysis Membranes

The first microdialysis membranes consisted of the polymer fibers that are used in kidney dialysis machines. Today, there are several different membrane materials to choose from, including cellulose acetate, polyacrylonitrile (PAN) and polycarbonate-polyether (PCE), polyethersulfone (PES), and cuprophan (CUP). Selection of the appropriate membrane for the application of interest is dependent on the target analyte and sample matrix.

Characteristics of the membrane that affect recovery include charge, thickness, and length, as well as pore size and shape. Zhao et al. investigated the effect of membrane type on the *in vitro* analyte recovery and delivery for 15 small molecule analytes varying in pI and hydrophobicity.[21] They found that the recoveries of anionic compounds using a PAN membrane were much lower than those obtained for neutral and basic compounds. Recovery decreased with increasing membrane thickness for all compounds. Although all the analytes that were investigated had molecular weights less than 500 amu, the MWCO of the membrane was shown to affect the recovery of some of the compounds. Lastly, as would be expected, recovery increased with increasing membrane length.

In a later study, Snyder compared the recoveries and apparent diffusion coefficients across the membrane for 10 different compounds having molecular weights between 94 and 1355 Da and exhibiting different log P values.[22] It was found that for some compounds, recovery was dramatically affected by the type of membrane used. For example, vitamin B_{12} yielded relative recovery values greater than 20% at 0.5 μL/min using CUP and PAN membranes but only a 1% relative recovery with the polycarbonate membranes. It was also found that ultrafiltration did not affect analyte relative recovery with either polycarbonate (PC) or PAN membranes. Lastly, the presence of 4% bovine serum albumin or 0.3% fibrinogen in the sample matrix did not significantly alter the membrane diffusion coefficients for most of the membrane–analyte combinations.

Most commercially available microdialysis probes posses MWCOs between 20,000 and 60,000 Da. With probes of this type, recoveries decrease dramatically for compounds with molecular weight greater than 10,000 Da. More recently, probes with a 100 kDa MWCO have become available with the goal of sampling larger biomolecules *in vivo*. Stenken's group investigated the mass transport properties of fluorescein-5-isothiocyanate (FITC)-labeled dextrans ranging from 10 to 70 kDa using a 100 kDa commercially available probe. They found that as the molecules increased in size, the membrane became a significant barrier to mass transport, most likely due to hindered diffusion.[23]

6.2.2. Probe Designs

There are several different types of commercially available microdialysis probes (Fig. 6.2). The concentric cannula design (Fig. 6.2A) is most commonly employed for neurochemical studies. These probes are composed of stainless steel and can be implanted into the specific brain region of interest using a guide cannula. A typical probe used for rat brain studies is 15 mm long with an outer diameter between 200 and 500 μm. The dialysis membrane is located at the end of the concentric cannula and is usually from 1 to 4 mm in length. Probes are also available for brain sampling in mice. The mouse probes employ shorter shafts (7–12 mm) and have outer diameters between 220 and 380 μm. Cannula-style probes have also been modified for bioreactor monitoring and environmental sampling.[24–26] One example is a unique tunable probe developed by Laurell and Buttler that can be used to optimize recovery of analyte in bioreactors.[27]

For sampling soft tissue, bioreactors, and environmental samples, a linear probe is often employed (Fig. 6.2B). These probes consist of a microdialysis membrane (4–10 mm) placed between two pieces of flexible tubing. Soft tissues such as skin, liver, and heart are more homogeneous than brain tissue, and therefore, spatial resolution is not as important. Thus, the surface area of these probes is generally larger than that of the brain probe. Linear probes have also been employed for sampling water and soil.[24] For soil sampling, the probe membranes are protected from mechanical stress by a polyvinylchloride (PVC) sleeve containing 300-μm holes.[16,28]

For blood sampling, a flexible probe has been developed (Fig. 6.2C). This type of probe was first described in 1992 by Telting-Diaz et al.[29] Its design is similar to that of the brain probe and consists of two pieces of fused-silica tubing attached to the dialysis membrane. The flexible probe can bend when the animal moves, minimizing any damage to blood vessels. In general, analyte recoveries from blood are higher than from the stagnant interstitial fluid due to better mass transport.[19]

A shunt probe has also been developed for sampling from moving fluids *in vivo* or *in vitro* (Fig. 6.2D).[30] This probe has been used to sample bile in awake, freely moving animals. For these experiments, the bile duct is cannulated and bile is sampled by a second dialysis membrane filled with fluid of similar ionic composition running in the opposite direction. This probe can also be used for desalting protein samples prior to introduction into a mass spectrometer.[31]

In cases of extremely fast sampling, diffusion across the dialysis membrane can be the limiting factor for temporal resolution.[32] To circumvent these problems and make it possible to sample larger molecules as well as obtain higher recoveries of small molecules, alternative sampling approaches have been developed.[33] For example, Shippy's group has developed a low-flow, push–pull perfusion device.[34] The push–pull perfusion probe consists of a 27-gauge stainless steel cannula that is perfused with saline. A fused-silica capillary acts as the inner cannula and is used to withdraw fluid from the extracellular space; the fluid is then transferred to the analysis system. Flow rates using this method are from 10 to 50 nL/min, and analyte recovery is 100%.

The selection of the dialysis probe and membrane for a particular application will be dictated by the target analytes and sample matrix. There are several commercial probes available (e.g., CMA, Solna, Sweden; Bioanalytical Systems, West Lafayette, IN). In addition, it is possible to have probes custom-made either commercially or in-house.[25,29,35–37] Accessories such as guide cannulas, tubing connectors, and

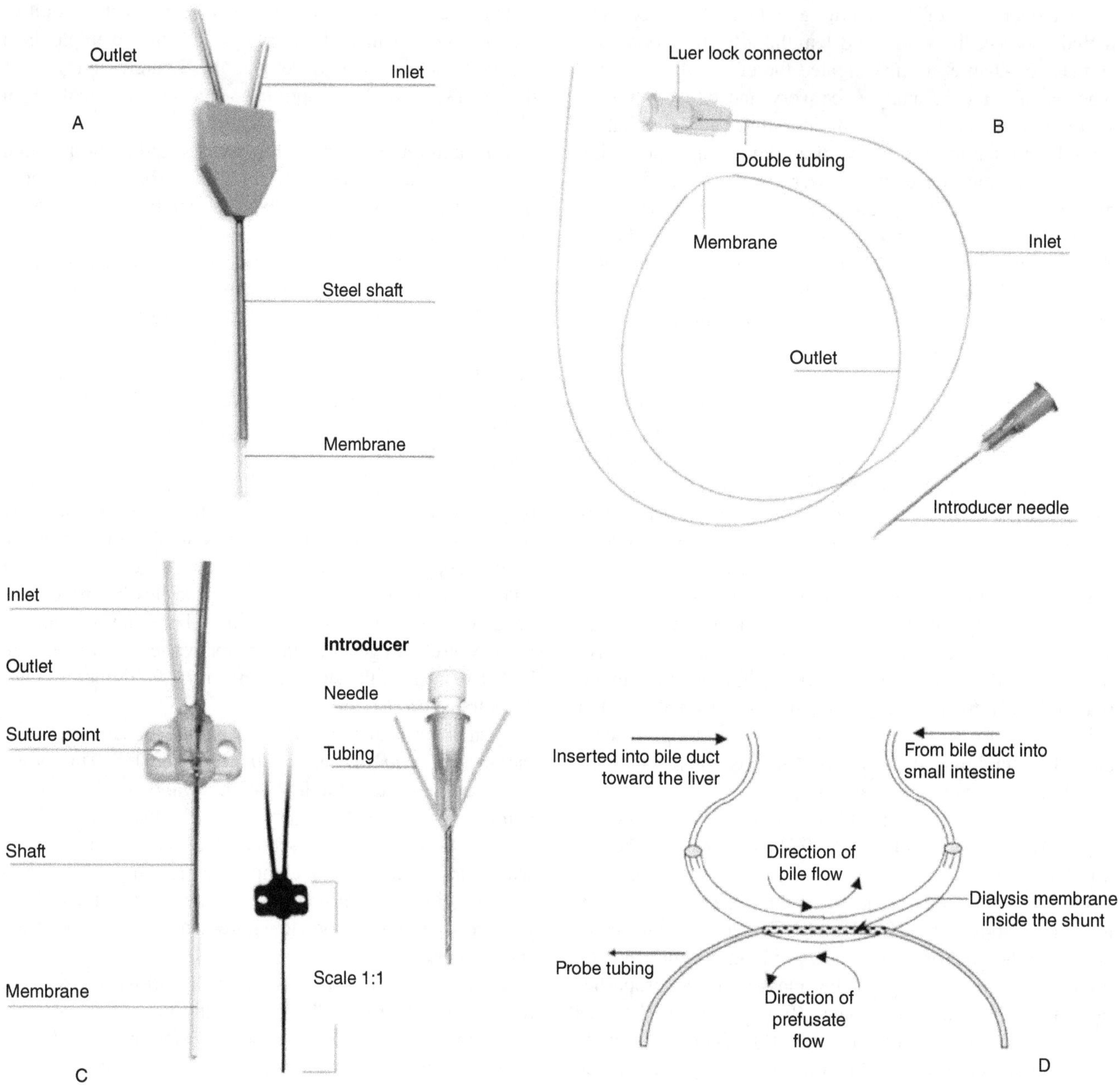

Figure 6.2. Microdialysis probes. (A) Concentric brain probe or cannula probe. (B) Linear probe. (C) Flexible probe. (D) Shunt probe. Images courtesy of CMA Microdialysis, CMA, Solna, Sweden.

swivels must be compatible with probe dimensions and the type of study that is being undertaken. Dialysis membranes should always be conditioned with perfusion fluid prior to commencing the sampling process.

6.3. QUANTITATION

Quantitation in microdialysis sampling can be a controversial subject, and there have been several excellent reviews published in recent years concerning the pros and cons of different approaches.[8,19,38–43] For many *in vivo* and *in vitro* applications, however, enough information can be obtained by monitoring the percent change of analyte with respect to the basal experiment concentration by using the untreated animal (or sample) as its own control. An example of this type of experiment is monitoring a rapid change in neurotransmitters in the brain due to the administration of a drug known to affect a neurochemical pathway.[19] During such experiments, it is generally assumed that that the probe is performing consistently throughout the duration of the experiment. However, this assumption should always be tested as there are cases where uptake or metabolism can change analyte recovery during the course of the experiment.[44,45]

If it is necessary to obtain more accurate *in vivo* concentration values, then quantitation has to be carefully considered. In addition to optimizing probe membrane type and size, location of sampling probe, and flow rate, the probe must also be calibrated *in vivo*. This can be difficult because there are many *in vivo* processes such as metabolism, elimination in the blood, and uptake that can influence the concentration gradient near the dialysis probe. In particular, getting accurate values for the concentration of endogenous compounds (e.g., dopamine) is difficult using microdialysis.

One technique that has been employed as the gold standard for determining *in vivo* concentrations of compounds using microdialysis is the no-net-flux (NNF) method.[46,47] This procedure can be used to determine steady-state concentrations of a drug or endogenous compound in tissue. The protocol involves adding a known concentration of analyte to the perfusate. This concentration is then varied over a range that is both higher and lower than the expected extracellular concentration of the exogenous or endogenous compound. The concentration of the analyte in the dialysate is measured and considered to be at the "NNF" condition when there is no exchange of analyte between extracellular space and perfusate. This point is considered to be the ECF concentration. A graph showing the calculation of the endogenous concentrations of dopamine using this method is shown in Fig. 6.3.[46]

The NNF method becomes more complicated if dynamic changes in the concentration of an endogenous or exogenous compound are to be monitored.[44] Administration of a drug or other substance to an animal can produce changes in the extracellular environment around the probe along with the pharmacological action of interest. This change in the extracellular environment can change the recovery of the analyte through the probe and, hence, the accuracy of the estimation of the extracellular concentration.

In 1993, Olson and Justice described a procedure for obtaining extracellular concentrations of endogenous neurotransmitters such as dopamine under transient conditions.[44] This method, termed dynamic no-net-flux (DNNF) uses groups of subjects instead of a single animal to measure the analyte concentration and recovery as a function of time. In their experiment, three groups of four rats were used to measure the effect of cocaine on dopamine release. The perfusate for each of the three groups contained 0, 10, and 40 nM dopamine, respectively. The change in dopamine concentration in the three groups of rats following cocaine administration was plotted versus time. The NNF plots (dopamine in/dopamine out) for each time point could then be employed to calculate the extracellular concentration of dopamine. These investigators found that the extracellular dopamine increased but recovery decreased after cocaine administration (Fig. 6.4).[44] This is consistent with the microdialysis recovery model that says that any process generating a sink for the compound of interest will affect probe recovery. Processes include synthesis or metabolism, release, and uptake as well as transport of the substance under investigation in and out of the blood vessels.

A less tedious method of probe calibration is called retrodialysis or reverse dialysis and is particularly applicable for *in vivo* experiments involving exogenous compounds such as drugs.[48,49] The underlying assumption of this method is that diffusion across the probe membrane is quantitatively equal in both directions. This method uses an internal standard whose physical, biological, and pharmacokinetic properties closely resemble those of the compound under scrutiny. Such a compound is added at a known concentration to the perfusate, and its rate of disappearance is calculated from the equation

$$\text{Recovery}\,(\%) = 100 - \left(100 \times C_{\text{dialysate}} / C_{\text{perfusate}}\right). \quad (6.3)$$

Calibration should ideally be performed before an experiment when there is no analyte present in the tissue in order to maintain the membrane concentration gradient during the experiment. Also, the probe should be perfused thoroughly with the physiological solution to remove the drug delivered

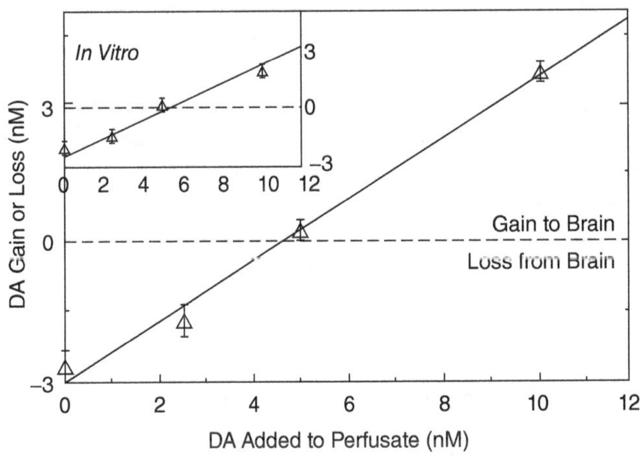

Figure 6.3. Extracellular concentration and *in vivo* recovery of dopamine (DA) in the nucleus accumbens. The zero point on the *y*-axis represents a steady state where there is neither net gain nor net loss of DA across the dialysis membrane. At this point, the concentration of DA added to the perfusate equals the extracellular concentration of DA. Reprinted from reference 46 with permission from Elsevier. Copyright 1993.

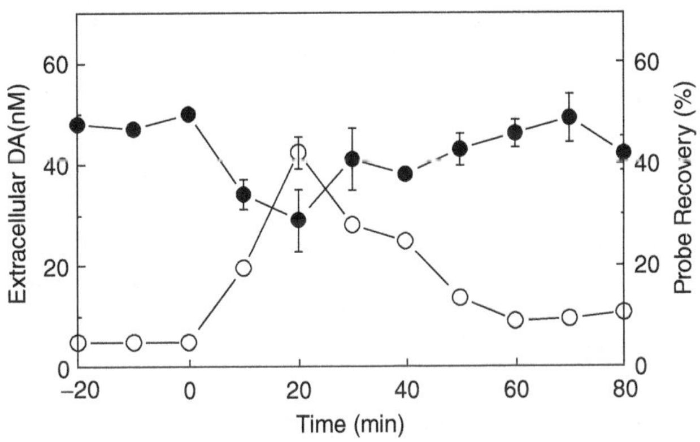

Figure 6.4. Time course of extracellular dopamine (DA) (open circles) and probe recovery (filled circles) in the nucleus accumbens following cocaine administration. Reprinted from reference 44 with permission from the American Chemical Society. Copyright 1993.

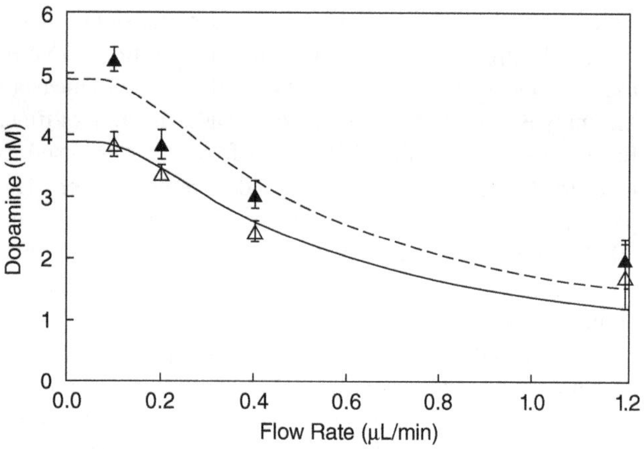

Figure 6.5. Variation of recovery with changes in perfusion flow rate. Recovery values at 100 nL/min are close to 100%. Reprinted from reference 46 with permission from Elsevier. Copyright 1993.

to the tissue during retrodialysis before commencing the pharmacological experiment.[12,38]

Song and Lunte compared retrodialysis with NNF for exogenous compounds both *in vivo* and *in vitro* in several different tissues. Using acetaminophen and caffeine as model compounds, they found no difference in the numbers obtained using recovery, delivery, or NNF experiments *in vitro*. The delivery (retrodialysis) method of calibration was then compared with that of NNF *in vivo*. No difference was found between the two approaches in muscle tissue for either acetaminophen or caffeine. However, in the brain tissue, the extraction efficiency determined by delivery was higher for caffeine and lower for acetaminophen than the value obtained using NNF. For caffeine, this was determined to be due to saturable active transport across the blood–brain barrier. This active transport resulted in the extraction efficiency being dependent on the concentration of caffeine in the brain.[50]

Menacherry et al. reported that at flow rates of less than 100 nL/min, the recovery of the analyte across the probe is essentially 100% as shown in Figure 6.5.[20,46] This makes it possible to quantitate analytes in the ECF without an internal or external reference. However, sampling at these low flow rates can generate significant analytical challenges regarding the manipulation and analysis of submicroliter samples. It also generally obviates application of this method to awake, freely moving animals because the swivels and tubing used in these studies have large dead volumes. Recent advances in using capillary electrophoresis both online and offline have made it possible to analyze these small volumes in near real time.[51]

Recently, Westerink's group developed a microdialysis probe for general use that samples at a flow rate of less than 200 nL/min and employs a makeup flow generated within the probe to produce sample at a rate of 1–2 µL/min. A diagram of the probe is shown in Figure 6.6.[52] This approach takes advantage of the quantitative recovery obtained at very low flow rates combined with the need for larger flow rates (volumes) for analysis and awake animal studies.[52] After correcting for dilution in the probe, an accurate determination of the *in vivo* concentration of drugs and neurotransmitters can be obtained.

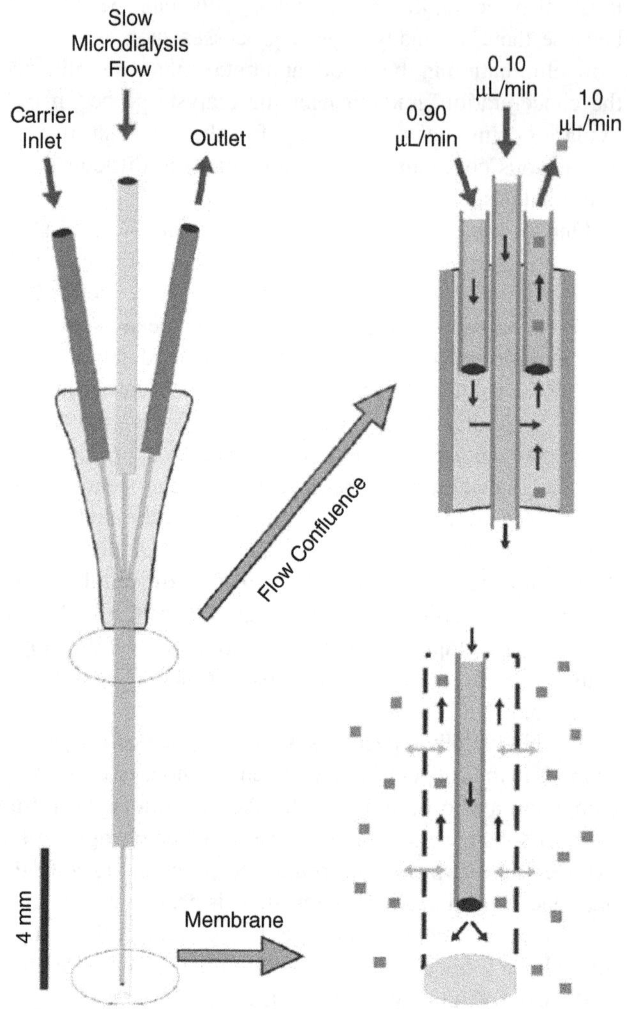

Figure 6.6. Diagram of a MetaQuant probe. The probe allows sampling at very low flow rates (0.1–0.5 µL/min) for near quantitative recovery. A makeup flow generates enough volume for the analytical method. Reprinted from reference 52 with permission from Elsevier. Copyright 2009.

Another semiquantitative approach has been described by the Pawliszyn group. This method employs two probes in close proximity with one probe perfused at a flow rate that is one-half that of the other.[53] The use of two probes makes it possible to calculate a concentration correction factor based on experimental data that can be used to approximate the initial analyte concentration in the sample matrix. This approach was first evaluated using an agarose gel spiked with pesticide as a model system and was later employed to determine the approximate concentrations of pesticides in the leaves of jade plants. Table 6.1 lists the different calibration methods that have been employed for quantitative microdialysis sampling.

Table 6.1. Calibration Methods

Method	Description of Process	Comments	Reference
In vitro calibration	Place probe in a solution of similar composition to the extracellular environment being interrogated; 37° with stirring	Gives an estimate that could be two to three times off; best for *in vitro* applications, to compare probes or if only a relative change is being measured	7
No-net-flux (NNF)	Estimate the extracellular concentration and then perfuse the probe with analyte at concentrations above and below plot C_{out} versus C_{in} to determine the concentration in the ECF; use linear regression to calculate recovery and C_{ECF}	Gives most accurate value for the concentration of the analyte in the extracellular fluid; experiments are long and require several hours to a day; animal serves as its own control	47
Dynamic no-net-flux (DNNF)	Several animals are perfused with the compound of interest over the expected concentration range; changes in the concentration of the compound are measured over time (as a function of some stimulus); the actual concentration can be determined by producing an NNF calibration plot a the different concentrations over the different animals	Can be used to correct for changes in recovery due to changes in the extracellular environment over time; uptake, metabolism, or synthesis requires a large number of animals	44,46
Retrodialysis or internal standard method	Include a compound of similar structure in the perfusate to account for changes in recovery during the experiment (assume delivery = recovery)	Compound must not have a physiological affect on the tissue around the probe or the recovery of the compound being investigated	48,49,54–56
In vivo delivery	Delivery of the substance of interest through the probe prior to the experiment followed by a washout period prior to administration of the drug; assume delivery = recovery; to check for changes in probe over the course of the experiment, perform the delivery experiment again at the end		9,21,36
Ultraslow flow method (extrapolation to zero flow)	At flow rates approaching 100 nL/min, recovery is close to 100%; therefore, the perfusate is a reflection of the ECF concentration		20
Mathematical models	Mass transport models of diffusion of analyte through tissue and probe membrane		9,42,57,58

6.4. ADDITIVES FOR IMPROVING RECOVERY

One approach to increase the recovery of the analyte of interest using microdialysis is to add a substance to the perfusate that has a strong affinity for the compound of interest.[54,55] This increases the overall flux of the analytes into the probe because the analyte concentration in proximity to the probe membrane will be close to zero, driving mass transport into the probe (Fig. 6.7).[55]

Several compounds that have been investigated as affinity agents are listed in Table 6.2. In 1999, Stenken's group evaluated the use of cyclodextrin as an additive to increase the recovery of hydrophobic drugs.[56] Ibuprofen was used as a model compound, and extraction efficiencies were compared *in vitro* with those obtained for antipyrine. The phenyl ring on ibuprofen forms a strong complex with cyclodextrin, while antipyrine exhibits no interaction. Recoveries ranged from 10% to over 100% for ibuprofen, depending on probe type, cyclodextrin concentration, and flow rate.

This approach was further investigated for the recovery of tricyclic antidepressants and structurally related analogs[57] using a PCE membrane. β-Cyclodextrin and hydroxypropyl β-cyclodextrin were investigated as additives. Enhancements varied from 1.5- to up to almost 10-fold, depending on the tricyclic compound investigated. CUP and AN-69 membranes generated recoveries about half of those obtained for the polycarbonate membranes. The use of cyclodextrins as an additive for the analysis of enkephalins has also been investigated.[58] However, recoveries were increased by less than a factor of two using this method.

Antibodies have been investigated for enhancing the recovery of peptides in microdialysis sampling. In this case, antibodies are immobilized on microspheres, which are then included in the perfusate.[59] This strategy has been investigated by Stenken's group for the analysis of cytokines,[59] neuropeptides,[58] and endocrine hormones *in vitro* using probes containing 100 kDa MWCO polysulfone microdialysis membranes. Although the antibodies were found to significantly enhance the recovery of these peptides *in vitro* (3- to 20-fold), one drawback of this approach for *in vivo* monitoring is the potential saturation of all the available binding sites on the microspheres, which would limit the overall dynamic range of the assay.

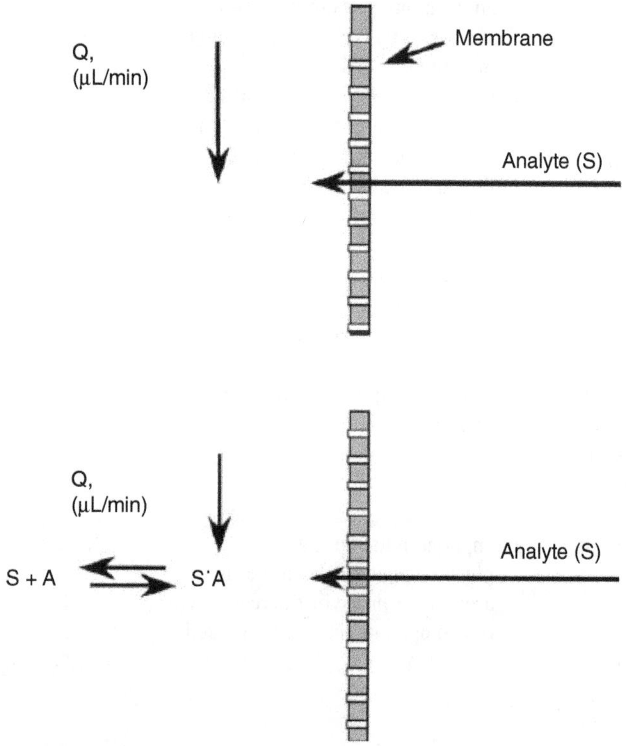

Figure 6.7. A schematic illustrating the trapping process within the perfusion fluid. The affinity agent (A) binds with the targeted analyte (S) while the fluid is passing through at a flow rate denoted by "Q." Once bound, this causes the free concentration of the analyte to be reduced at the membrane wall, thus driving additional mass transport. Reprinted from reference 55 with permission from Elsevier. Copyright 2006.

Due to the high expense and relatively low loading capacity of antibodies, the use of heparin has also been investigated as a potential agent to increase the recovery of cytokines.[54] Heparin has been shown to bind many of the human cytokines *in vivo*, and it is an inexpensive and soluble reagent that can be added at fairly high concentrations to the perfusate. The relative recoveries for IL-6, IL-7, MCP-1, and TNF-α were all improved with the addition of heparin. Heparin also did not interfere with the enzyme-linked immunosorbent assay (ELISA) of the peptides following collection.

Additives have also been used to increase the recovery of metal ions through the probe. Torto's group investigated a number of chelating agents to determine their effect on the recovery of metals from microdialysis samples.[60–62] The inclusion of 8-hydroxyquinoline (8-HQ) was found to improve the recovery of several metal ions.[63] Mogopodi and Torto investigated the effect of addition of poly-L-aspartic acid and poly-L-histidine on the recovery of Cr^{3+}, Cu^{2+}, Ni^{2+}, and Pb^{2+}.[61] The extraction of copper was enhanced by both additives, while only polyaspartic acid improved the recovery for Pb^{2+}. The recoveries for Cr^{3+} and Ni^{2+} were not significantly improved. However, later it was found that a cocktail consisting of 20% (w/v) poly-L-histidine, 0.032% (w/v) poly-L-aspartic acid, and 1 mM 8-HQ yielded recoveries for all four ions of 80–90% using an *in situ* tunable microdialysis probe made with a polysulfone membrane having an MWCO of 30 kDa.[60] More recently, it has been shown that the addition of 0.05% (w/v) humic acid to the

Table 6.2. Additives to Enhance Recovery

Additive	Analyte	Enhancement	Reference
Cyclodextrin	Tricyclics	2- to 2.5-fold	62
	Ibuprofen	1.5–2.0	61
	Enkephalins	1.4- to 1.9-fold	63
	Antipyrine	No enhancement	61
Antibodies	Cytokines	4-to 12-fold	64,65
	Enkephalins	2.5-fold	63
	Endocrine peptides	3- to 20-fold	60
BSA-heparin	TNF-α	Threefold	60
	IL-4, IL-6 cytokines		59
Poly-L-aspartic acid	Cu, Pb	64–90%	66
Poly-L-histidine	Cu	66%	
8-Hydroxyquinoline	Cr, Pb	20%	67
Poly-L-aspartic acid, poly-L-histidine, 8-hydroxyquinoline	Cu, Ni, Pb, Cr	80–90%	66
Humic acid	Cu, Ni	100%	68
EDTA	Cr, Cu, Ni, and Pb	100%	66

perfusate can lead to 100% recoveries for Cu^{2+} and Ni^{2+}. This approach has great potential as a tool for monitoring changes in metal ion concentrations in the environment, including plants, water, and soil samples.

6.5. TEMPORAL RESOLUTION

Microdialysis is a dynamic sampling technique with time-dependent concentration information embedded in the dialysate. Temporal resolution can be defined as the smallest increment of time over which the change in a dynamic process can be observed. The temporal resolution that can be obtained in microdialysis is dependent on the concentration of the analyte of interest in the external medium, the recovery across the probe, and the flow rate of the perfusate, as well as the volume requirements and sensitivity of the analytical method that is being employed. As an example, if sampling is being performed at a flow rate of 1 μL/min, a temporal resolution of 1 min requires the ability to measure the amount of analyte contained in a 1-μL sample of dialysate. The reproducible analysis of such small-volume aqueous samples has many challenges, including sample loss due to evaporation and surface tension. One approach to overcome these drawbacks is to couple microdialysis online to the analytical system.[18,51]

For offline analysis, analyte recovery and the sensitivity and volume requirements of the analytical method are the key factors that determine temporal resolution. Recovery of analyte can be improved through the choice of an appropriate probe design and membrane type, optimization of the flow rate, and/or the use of additives in the perfusate. The analytical method must be able to analyze sample volumes that provide the temporal resolution needed for the experiment. Liquid chromatography (LC) typically requires from 5 to 20 μL for analysis yielding temporal resolutions between 5 and 20 min. The sensitivity and selectivity of the analytical method are also important. A more sensitive method will make it possible to measure lower concentrations of analytes and, hence, smaller-volume samples. Selectivity minimizes the number of interferences that must be eliminated in the analysis, leading to shorter analysis times.

In some cases, it is possible to analyze small-volume samples offline. One such example is when a fluorescence derivatization step is incorporated into the assay to monitor the substances of interest. Under these conditions, it is possible to use smaller-sample volumes and then dilute them with reagent and buffer to the volume needed for the analytical system.[64,65] For example, it is possible to detect picomolar to nanomolar (final) concentrations of primary amines using LC or capillary electrophoresis following derivatization with a fluorescent reagent. Therefore, if the analytes of interest are in the micromolar range, the sample can be diluted 100- to 1000-fold and still be detectable. Hernandez used this approach to monitor extracellular glutamate with 1-s temporal resolution. He collected nanoliter volumes of dialysate in a capillary tube and derivatized them offline with a fluorescent derivatizing agent.[64] In these cases, analyte diffusion within the tubing and transport across the dialysis membrane become the limiting factors in temporal resolution.

One of the main considerations for both offline and online analysis is the choice of an analytical method. The first question that needs to be addressed is whether the device or the method is sensitive enough to actually detect the quantity of analyte(s) present in the small volume of sample generated by very fast sampling. LC with fluorescence or electrochemical detection is a popular method for the analysis of amino acids and catecholamines in microdialysis samples due to their high sensitivity and selectivity. LC coupled to MS is commonly employed for peptides, drugs, and environmental samples.

An additional consideration with online systems is that the temporal resolution of the technique will also be dependent on the overall analysis time. If the analysis time is longer than the duration of the event being measured, then the change will appear digital (or instantaneous) and appear in the next injection. On the other hand, in those cases where the analysis step is much faster than the event being measured, it is possible to detect the change in concentration as a function of time. In this case, the rise time is defined as the time required for the signal to increase from 10% to 90% of maximum intensity.[66] For very fast analyses, the rise time becomes dependent on the rate of diffusion of analyte across the probe membrane. However, in most cases, it is the dead volume in the system, the injection method, and the flow rate of the dialysate that determine how fast a concentration change can be measured with an online system.

As mentioned in the previous section, higher analyte recovery can be achieved by using very slow flow rates. At flow rates of 100 nL/min, analyte recoveries of 100% have been reported.[20] However, the increase in concentration of the analyte in the perfusate is offset by the extremely small sample volumes that are generated per unit time. In this case, an analytical method that is both sensitive and capable of analyzing submicroliter sample volumes is necessary if 1-min temporal resolution is to be obtained. This is one of the driving forces for the use of capillary and microchip electrophoresis for the analysis of microdialysis samples.[18,51]

For most online separation-based microdialysis systems, analysis is performed on substances such as amino acid neurotransmitters whose *in vivo* concentrations are relatively high and well within the detection limits of laser-induced fluorescence (LIF) detection. In such cases, the factor that dictates temporal resolution is the time that is required to separate the compounds so that serial analyses can be performed without overlapping the analysis peaks from two different runs. In other words, the difference in injection time between run 1 and run 2 should ideally be the separation time for run 1 so that as soon as run 1 ends, run 2 begins, and this cycle can be repeated to get uninterrupted *in vivo* data. If the analysis time can be made very short (a few seconds), such a mechanism leads to high temporal resolution.

As discussed in the previous section, mass transfer processes are important parameters influencing recovery from a microdialysis probe. The probe samples analytes that are present in the extracellular matrix surrounding the membrane. However, many times the analyte needs to diffuse from the regions farther away from the probe surface to the surrounding matrix in order to be sampled. For slow diffusing analytes, the recovery will be low, indicating that one will have to wait longer to obtain enough mass to be detected. In addition, diffusion of analyte can occur within the microdialysate due to Taylor dispersion while it is being pumped to the analysis system, compromising the temporal integrity of the zones and thereby diffusing zones of analyte that have different concentrations. Such phenomena can be mitigated by increasing the flow rate or reducing the length and inner diameter of the tubing.

6.6. APPLICATIONS

6.6.1. Brain Sampling

The most popular application of microdialysis is sampling of the ECF in the brain. This is evident from the numerous books and review papers on this topic, including a recent book by Westerink and Cremers[10] and the highly cited monograph by Robinson and Justice.[7] A comprehensive review of all the applications of microdialysis to brain sampling is beyond the scope of this chapter. In this section, we will discuss only sampling issues directly related to brain tissue.

Brain microdialysis is almost always accomplished using a cannula probe, the size of which depends on the size of the brain of the animal being sampled. Probes for rats and mice microdialysis studies are commercially available from a number of vendors. Probes for use in other animals can be custom-made either by a vendor or in-house. A significant advantage of microdialysis for neurochemical studies is that it can be performed on awake, freely moving animals. This makes it possible to correlate the concentrations of drugs and/or neurotransmitters in the ECF of the brain with behavior. Because of the small size and relatively noninvasive nature of the microdialysis probes, it is also possible to have multiple probes in a single animal. It is therefore possible to measure blood, brain, and tissue concentrations of drugs or endogenous substances simultaneously. If the animal is awake, these measurements can be correlated temporally with behavior.

A very nice example of the use of multiple probes in a single animal is shown in Figure 6.8.[67] In this experiment, the rat was given an i.v. injection of methylphenidate (Ritalin). Brain and blood sampling were accomplished using a concentric cannula probe and a flexible probe, respectively. Dialysates were collected offline, and the concentrations of methylphenidate and dopamine in both the brain and blood were determined using LC with electrochemical detection. In this manner, it was possible to measure the transport of Ritalin across the blood–brain

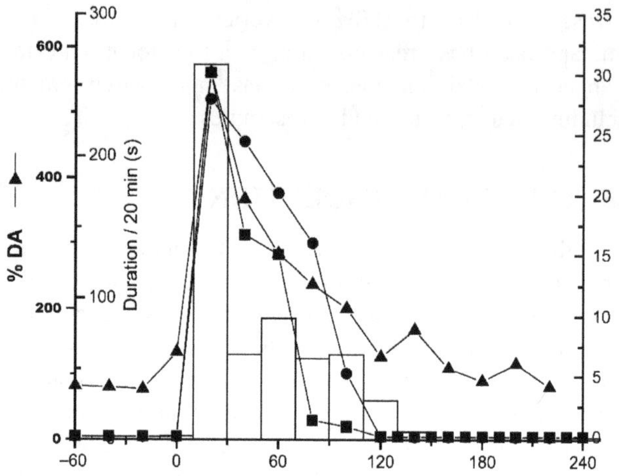

Figure 6.8. Sampling from multiple probes in a single animal. The rat was given an i.v. dose of methylphenidate (MPD). Blood concentrations of Ritalin were measured using a flexible probe. Brain concentrations of the drug and dopamine (DA) were monitored using a cannula probe. Activity was measured using the Rat Turn apparatus. Reprinted from reference 67 with permission from Elsevier. Copyright 2002.

barrier as well as its effect on catecholamine release. Lastly, by using a "Rat Turn," the extracellular concentration of these substances could be directly correlated with the overall activity level of the rat.

A major concern with microdialysis sampling in the brain is changes in the environment around the probe during the sampling process. If the microdialysis studies will be performed over a fairly long period of time, tissue damage associated with probe implantation and the potential for an immune response must be taken into consideration. Fibrosis or gliosis has been reported following several days of probe implantation.[12] Grabb et al. compared the effects of acute (2–4 h) and chronic (24 h) implantation of microdialysis probes in the brain tissue. Inflammation, hemorrhage, and edema in the area around the microdialysis probe were observed 24 h after implantation. The development of fibrin-like polymer was also observed around the probe. Both of these factors can adversely affect the recovery of the probe. Extracellular edema increases the diffusional distance between the probe and the ECF, and the fibrin-like polymer generated via gliosis can create a physical barrier between the dialysis probe and the ECF surrounding the cells.[68] In a separate study, based on local cerebral blood flow (LCBF) and local cerebral glucose metabolism (LCGM), Benveniste et al. recommended a 24-h period of recovery after probe implantation.[69]

The integrity of the blood brain barrier (BBB) following probe implantation has also been an important and controversial issue in brain microdialysis.[70] Studies conducted using autoradiography with ^{14}C-AIB (which does not cross the BBB under normal conditions) as well as transport characteristics of hydrophilic and moderately lipophilic drugs (following i.v. injection) post surgery indicate that the BBB integrity is maintained overall.[71,72] However, other studies have shown a significant effect of probe implantation on

BBB permeability using Cr-[51]-ethylenediaminetetraacetic acid (EDTA) transport.[73]

Michael's group has used a combination of microdialysis and *in vivo* voltammetry to investigate the effects of the microdialysis probe on measured dopamine concentrations in the brain tissue as well as the integrity of the blood–brain barrier. The microelectrode used for the voltammetry studies is only 7 μm in diameter and is therefore substantially smaller than the brain probe cannula (280 μm). The studies show that the probe does cause injury to the tissue and generates a diffusional barrier for the transport of dopamine to the probe.[74–78] An estimate of the extent of "injury" based on the combined studies is approximately 1 mm from the probe membrane. This increase in diffusion distance explains the difference in response times for release of dopamine by microdialysis in comparison to *in vivo* voltammetry studies.

For the interested reader, other interesting examples of applications of microdialysis in the brain are as follows: monitoring oxidative metabolism in the brain,[79] understanding addiction,[80] central nervous system (CNS) disorders,[81] stress,[82] behavior,[83] proteomics,[84] brain trauma,[85] and stroke,[12] as well as many others. There are also many different analytical methods that have been employed for the determination of neurotransmitters and neuropeptides in the brain,[11] including LC with electrochemical detection, capillary and microchip electrophoresis,[18,51] and MS.[84,86–88] Readers are referred to the book by Westerlink and Cremers[10] or the recent reviews cited above for further information on applications and analytical methods related to brain sampling. Table 6.3 lists some recent applications and useful reviews on this topic.

6.6.2. Other Tissues

In addition to the brain, microdialysis probes can be implanted in virtually any organ or peripheral tissue of the body; this is demonstrated in the literature by a wide variety of studies. Microdialysis sampling from the extracellular space of tissues and organs is an attractive alternative to animal sacrifice or tissue biopsy as this technique is minimally invasive. A series of publications by Lonnroth's group emphasizes the fact that the microdialysis catheter induces minimal "tissue trauma" based on the adenosine concentration in the dialysate.[89,90]

Table 6.3. Neurochemical Applications

Topic	Reference
General references	7,10,11,51,96–98
Calibration	9,20,41,44–46,48,50,52,54,55,84,85
Neurocritical care applications	88,92
Analytical methodology	32,43,51,52,71,95,99–103
Amino acids	72,104–108
Catecholamine	46,82–84
Peptides and protein	91,109–112
Oxidative stress	86,89,113
Blood–brain barrier	39,56,75,79–81,114,115
Nucleotides	116
Behavior	74,90,117
Early applications	4,5,76,78
Acetylcholine	108,118,119
Lactate, glucose, and ascorbate	86,120–122

An additional advantage of microdialysis is that the concentration of the analyte can be monitored in the tissue of interest in an awake or even a freely moving animal. The most popular application of microdialysis for tissue sampling is pharmacokinetic studies. In this case, microdialysis is used to monitor the concentration of the drug (and/or metabolite) in the tissue or organ of interest over time. Compared with traditional pharmacokinetics, where the data are collected at each time point, microdialysis data reflect an average concentration over the time period in which sampling is done, and this needs to be taken into consideration for area under the curve (AUC) calculations.[91] However, the time period of sampling can be reduced (fraction of a second, see section on high temporal resolution) such that the concentration data collected during this time period will be very close to the data collected by manual sampling at a given time point. However, achieving such small collection times depends on the sensitivity of the detector as well as the capability of the analysis system to perform fast injections.

The first successful pharmacokinetic study using microdialysis was performed by Craig Lunte's group in 1990. Continuous blood sampling was achieved using a flexible probe following the administration of acetaminophen.[37,92,93] The pharmacokinetics of aspirin were also investigated by microdialysis compared with manual blood sampling.[94] In assessing pharmacokinetic parameters using microdialysis, it must be taken into account that protein-bound drug cannot cross the membrane. Therefore, the concentration obtained with microdialysis experiments is representative of only the free fraction of the drug.[95] Later, the same group demonstrated the use of a linear probe for transdermal studies.[35,96] This same probe design was then widely utilized for studies of tissue, including liver,[97] tumor,[36] and muscle.[98]

Since the first use of microdialysis for pharmacokinetic studies in the early 1990s, a large amount of literature has been produced concerning its use in pharmacokinetic and drug metabolism studies in a wide variety of organs and tissues. The following sections attempt to give the reader some examples of recent applications of this technique in this area. A list of some useful references is provided in Table 6.4.

6.6.2.1. Ocular (eye) sampling. Linear probes have been employed to sample from the eye chamber. In a recent study, Hosoya et al. investigated the *in vivo* transport of anionic drugs to characterize the organic anion transporters (OATs) in the rat retina; in this study, the microdialysis probe was placed in the vitreous chamber of the eye.[99] The

Table 6.4. Pharmacokinetic Studies by Microdialysis

Region of Interest	Probe Type	Analyte	Reference
Rat brain	Metaquant brain probe	Amphetamine	115
Rat brain	CMA/11 cylindrical brain probe	(–)-Stepholidine	132
Rat brain	Brain probe	Tetramethylpyrazine	133
Rabbit ocular anterior chamber	MD-2005 BAS linear probe	(–)-Satropane	134
Human skeletal muscle and subcutaneous adipose tissue	CMA 60 microdialysis catheters	Cefuroxime, Ceftobiprole, Telithromycin	135–137
Rat dermis	BAS linear probe	Cephalexin	138
Rat lung		Voriconazole	139
Rat jugular vein, hind leg muscle	Custom linear probes for vein, CMA 20 for muscle	Clenbuterol	140
Rat jugular vein, rat dermis	Linear probes	Oxymatrine and metabolites	141
Rat kidney cortex	CMA 20 probe	Voriconazole	142
Human brain tumor (glioma)	Custom cylindrical probe (see reference)	Methotrexate	114

Mitra group also performed several studies concerned with the investigation of corneal drug delivery and pharmacokinetics as well as characterization of transporters present in the eye. In all of these studies, a linear probe was implanted in the anterior chamber of the eye for sampling.[100] Several additional reports in the literature demonstrate the popularity of ocular microdialysis sampling for pharmacokinetic studies.[101–103]

6.6.2.2. Muscle sampling. Microdialysis sampling in muscle has been used to monitor both endogenous and exogenous compounds. There are several endogenous compounds that are potential biomarkers in muscle, for example, glycerol for lipolysis and lactate for glycolysis.[104] Lonnroth et al. reported no major damage to various tissues in terms of inflammation or bleeding, which was assessed by histological examination.[105,106] In a recent publication, Flodgren et al. monitored the concentration of norepinephrine in active muscle as an effect of low-load exercise and mental load.[107] In another human study, the distribution of cefuroxime in morbidly obese patients was measured in interstitial space of soft tissues.[108] The concentrations of bradykinin and kallidin in muscle were compared in female individuals at rest, during a 20-min repetitive low-force exercise, and following recovery.[109] Hamrin et al. found that glucose concentration of interstitial fluid in insulin-resistant skeletal muscle is markedly decreased for several hours following a single exercise session.[110]

The concentration-time profile of clarithromycin in soft tissue interstitial space was investigated in another report where microdialysis probes were inserted in the thigh muscle of healthy male volunteers.[111] A number of muscle microdialysis studies have also been performed in rats. The concentration of acetaminophen in muscle was determined following topical administration and compared with that of intramuscular administration.[112] Awake rat studies were performed by Marchand et al. in which they investigated the distribution of imipenem in muscle ECF.[113]

6.6.2.3. Dermal microdialysis. Dermal microdialysis is a sampling technique that allows continuous monitoring of endogenous and exogenous compounds with minimal tissue trauma.[8] Figure 6.9[114] shows the placement of a small perfused membrane within the dermis.[114] Dermal microdialysis has been employed for a number of studies aimed at investigating the distribution of drugs in subcutaneous tissue following topical application.[112,115,116] In other publications, dermal microdialysis was used to deliver compounds and measure changes in the concentration of endogenous analytes.[117]

Derendorf et al. studied the use of microdialysis for monitoring the concentration of antibiotics in the peripheral tissue.[104,118,119] Microdialysis has also been applied to the lungs,[120] kidneys,[121] liver,[122] pancreas,[123] blood,[124] and inner ear.[125,126] By 2007, microdialysis had evolved enough as a tool for pharmacokinetic and drug metabolism studies that the United States Food and Drug Administration (FDA) sponsored a workshop on the role of microdialysis in the drug evaluation process.[8]

6.6.2.4. Environmental sampling. A completely different application of microdialysis is monitoring pollutants in the environment. Microdialysis has many advantages for environmental samples.[127] In water and soil samples, humic acid and other large molecules (proteins, etc.) are excluded from the membrane, eliminating tedious sample preparation steps that are often necessary for environmental sample analysis. Both concentric cannulas and linear probes have been employed in these studies. Due to the size and homogeneity of most environmental samples (compared with a rat or mouse brain), it is possible to use big probes with larger membranes leading to high (usually close to 100%) analyte recoveries. This makes quantitation simpler in these studies and places less of a burden on the sensitivity of the analytical methods. Most laboratory studies of environmental samples can be performed under hydrodynamic conditions, so mass transport to the probe is enhanced compared with tissue

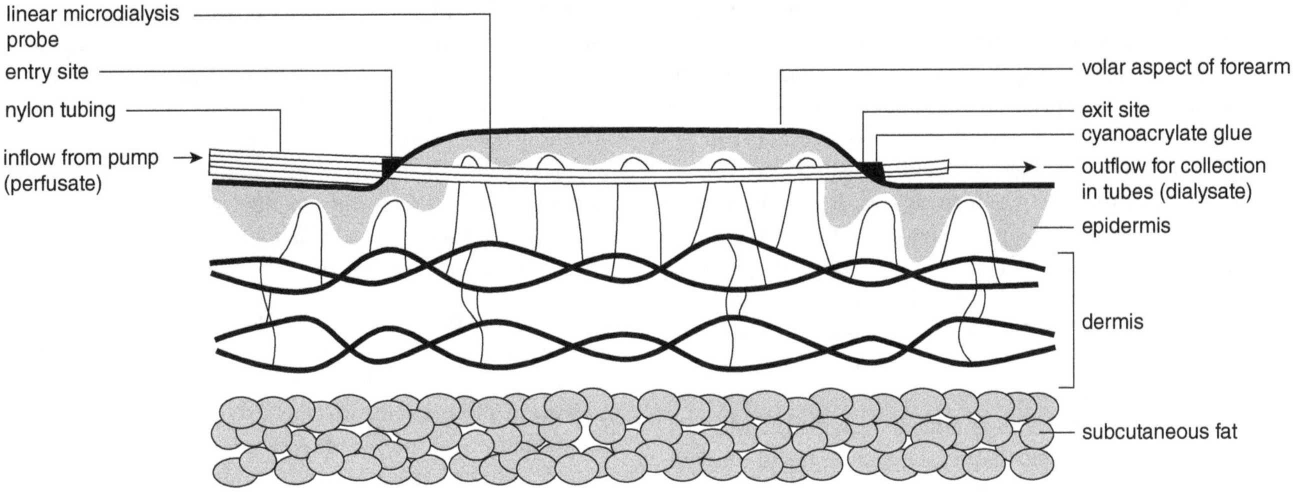

Figure 6.9. Pictorial representation of the process of dermal microdialysis. Reprinted from reference 114 with permission from Elsevier. Copyright 2009.

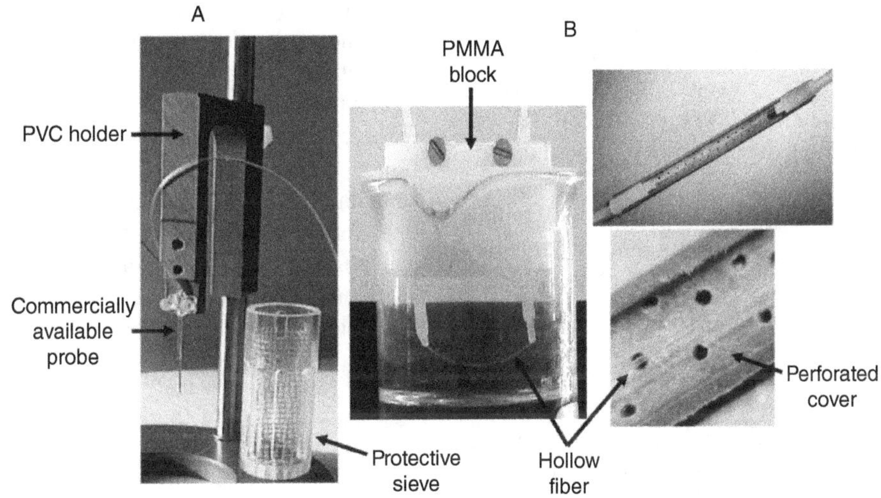

Figure 6.10. Magnified view of the various microdialysis probes assessed for environmental assays. (A) Commercial CMA/20 needle-type microdialyzer with a height-adjustable holder. Prior to accommodation into real-life samples, the microdialysis probe is housed within the protective sieve shown in the photograph. (B) Dedicated capillary-type microdialyzer. The perforated cover also illustrated was designed to prevent mechanical stress on the membrane surface after insertion into the soil layer. PMMA, poly(methyl methacrylate). Reprinted from reference 25 with permission from the American Chemical Society. Copyright 2004.

sampling. Since only a small amount of analyte is removed from the sample, the chemical equilibrium occurring in the sample is also maintained. Applications of microdialysis to environmental analysis include the determination of metals and pesticides in fresh waters, industrial effluents, and plants and soil samples.[16,127]

Microdialysis was used to determine aniline and 2-chloroaniline in industrial wastewater using an online system employing reversed-phase LC.[128] The effluent from a beverage processing plant was monitored for sugars and inorganic anions, including fluoride, nitrate, chloride, and phosphate, along with the metal ions Zn^{2+} and Ni^{2+}, using microdialysis.[129]

The fate of small molecules in soil has also been investigated using this technique. Miro's group used microdialysis coupled online with potentiometric detection for the determination of chloride in soil samples.[130] Later, Sulyok et al. developed a microdialysis sampling approach to monitor the presence of low-molecular-weight organic ions in soil.[28] A special linear probe was employed that allowed high recovery and enhanced ruggedness.[25] This probe consisted of a single cellulose-regenerated hollow fiber that was protected from damage by a PVC tube perforated with 300-μm holes (Fig. 6.10).[25] The dialysates were analyzed offline using conductivity detection.

Torto et al. used microdialysis to measure the amounts of Cr, Cd, Cu, Ni, and Pb in tomatoes grown from sewage sludge.[63] The results were compared with those obtained through traditional acid digestion and dissolution techniques. Although the amount of dialyzable metal in the tomato was significantly lower (about 10%) than the amount of total metal obtained with conventional methods, this approach has great promise for monitoring the fate of metals in tomatoes over time. Later, they used microdialysis to measure the metals in suspensions (slurries) generated from plants grown in a copper and nickel mineralized site.[131] The correlation between the concentrations obtained via microdialysis and those obtained via acid digestion and atomic absorption spectrometry was investigated. It was found that the ratios were constant, with values of 0.014 and 0.044 for copper and nickel, respectively. The results suggest that microdialysis sampling could potentially be used as an alternative to acid digestion for the determination of metals in plant suspensions by atomic absorption spectrometry.

As mentioned earlier, Zhou et al. employed simultaneous dual-probe microdialysis in combination with LC-MS/MS to determine the pesticide distributions in leaves of a living jade plant.[53] Two 10-mm concentric probes were inserted into the leaves of the plants close to one another with one probe perfused at half the flow rate of the other. Using an experimentally derived constant, it was possible to approximate the concentration of the pesticides in the plant tissues.

A particularly attractive feature of microdialysis for environmental sampling is the ability to perform online analysis. Flow-through probes can then be coupled directly to the analytical instrumentation, making near real-time monitoring or in-field measurements possible. Miro's group has developed a lab-on-a-valve system for continuous online sample preparation and analysis of environmental samples that can incorporate microdialysis as one of the sample preparation steps.[132]

6.6.2.5. Dissolution testing. Dissolution is an important test used in the pharmaceutical industry to determine the potential bioavailability of drugs. Normally, the drug is placed in a dissolution apparatus and aliquots are removed at distinct time points to monitor the amount of drug that is dissolved. Dissolution time profiles can be particularly important for sustained release pharmaceuticals. Microdialysis has several advantages as a monitoring system for dissolution studies.[133] First, it is possible to continuously monitor the concentration of drug in solution without having to remove aliquots from the apparatus thereby disturbing the equilibrium.[134,135] Second, by coupling microdialysis online to the analytical system, it is possible to obtain the dissolution profile in near real time. Lastly, using automated valving systems, it is possible to monitor multiple dissolution apparatus simultaneously.

Shah et al. described an online microdialysis LC-UV system using a loop probe suspended in the dissolution medium to monitor the dissolution of acetaminophen and Sulfatrim tablets and compared these to the profiles obtained with manual sampling.[135] The dissolution was performed in 0.1 M HCl, but the use of microdialysis sampling allowed the use of a buffered perfusate to make the sample compatible with the online LC-based analysis system. The profiles obtained using microdialysis were similar to those obtained via manual sampling.

Microdialysis has also been used to investigate the dissolution of sustained-release pharmaceuticals such as Accutrim.[134] Dash et al. used microdialysis sampling to investigate the dissolution of implantable drug delivery systems using ciprofloxacin microcapsules in polylactic acid (PLA) and poly(lactic-glycolic) acid (PLGA) as a model system.[136] Later, Fang's group developed a stopped-flow microdialysis sampling system for multivessel dissolution testing with high temporal resolution.[136,137]

6.6.2.6. Bioreactor monitoring. Microdialysis has many advantages for monitoring bioprocesses.[15,138] These include the ability to monitor small molecules (such as glucose) in the presence of cells and extracellular proteins without additional sample preparation steps. The probes can be sterilized and inserted directly into the bioreactor for continuous monitoring without fluid loss. Since microdialysis is a "generic" sampling technique, samples can be monitored either online or offline, depending on the requirements of the overall system (Fig. 6.11).[15]

Some of the important issues that have to be addressed with bioreactors are probe membrane type and geometry.[26,27] The type of membrane used can dramatically affect the recovery of analytes from the complex bioreactor sample.[24,139] Laurell and Buttler developed a probe with an adjustable cannula that can be used to maximize the recovery of analytes for bioreactor studies.[27] Membranes must also be able to withstand the higher temperatures that are employed in some bioreactors, and the composition of the perfusate can dramatically affect analyte recovery.

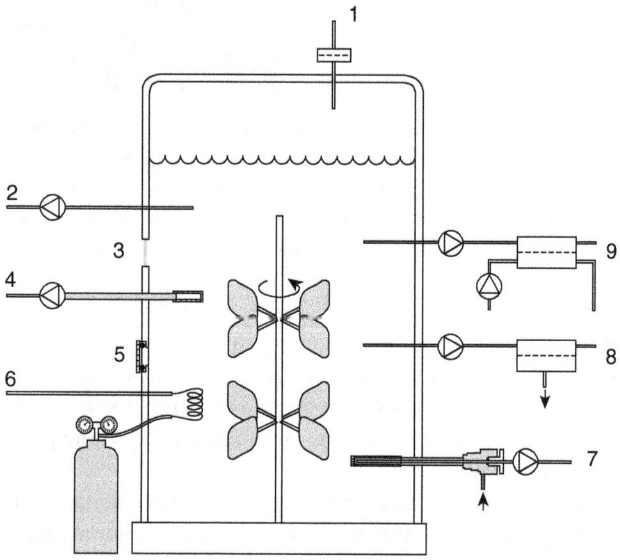

Figure 6.11. A schematic representation of a bioreactor with different sampling or monitoring devices (1) headspace analysis; (2) total bioprocess sampling; (3) radiation window for spectrophotometric techniques; (4) filtration device/probe; (5) planar probe; (6) gas- or temperature-sensing device; (7) microdialysis probe; (8) dialysis unit; and (9) filtration unit. Reprinted from reference 15 with permission from Elsevier. Copyright 1998.

Several groups have used microdialysis sampling to monitor enzyme reactions. Torto et al. extensively investigated the use of microdialysis for monitoring the enzymatic hydrolysis of starch.[140] The solution can be stirred and the membrane excludes the enzyme, making it possible to measure the reactant and the products using the analytical method of choice. Modi and LaCourse used microdialysis sampling coupled to high-performance liquid chromatography (HPLC)-UV to monitor carbohydrate enzymatic reactions. One application was the identification of complex carbohydrates present in willow bark tea by adding β-galactosidase to the sample.[141] Huyhn et al. used online microdialysis coupled to microchip electrophoresis to monitor the reaction of fluorescein-mono-β-D-galactopyranoside (FMG) with β-galacotsidase *in vitro* with a temporal resolution of approximately 15 s shown in Figure 6.12.[142]

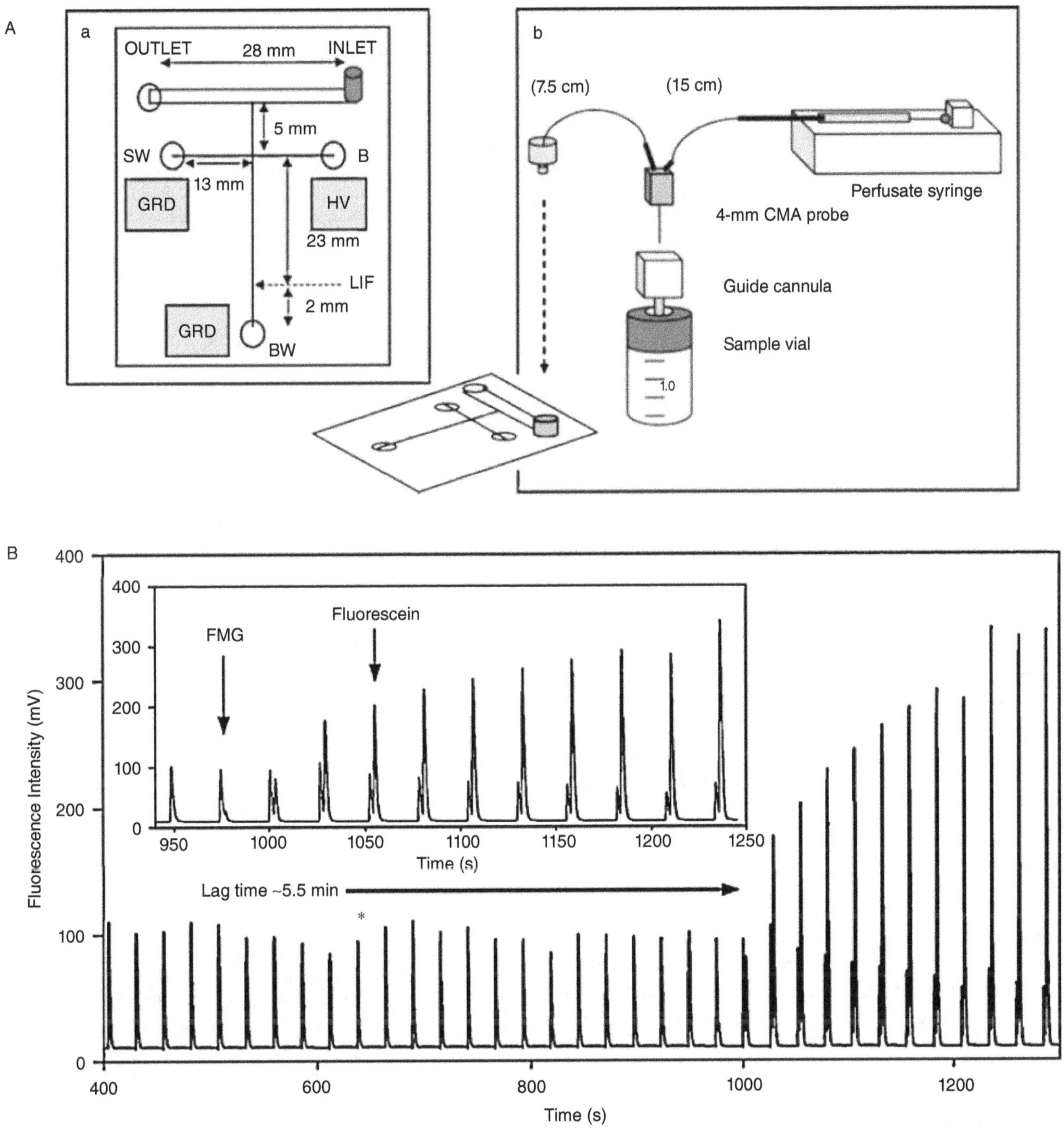

Figure 6.12. Online *in vitro* microdialysis microchip. (A) Chip design and setup. (B) *In vitro* enzyme assay of β-galactosidase using fluorescein-mono-β-D-galactopyranoside (FMG) as a substrate. Reprinted from reference 142 with permission from the American Chemical Society. Copyright 2004.

6.6.2.7. Other applications. Microdialysis has been used to desalt protein solutions prior to mass spectrometric analysis.[31,143–146] Affinity dialysis in a microfluidic format has been used as a sample preparation method prior to electrospray ionization (ESI) MS for the trace analysis of food residues as well as for high-throughput drug assays.[147] Since microdialysis measures only the non-protein-bound fraction, it has also been used to determine the degree of protein binding of drugs.[148,149]

6.7. SUMMARY

Microdialysis is a powerful sampling tool that makes it possible to continually monitor analytes in an external environment in near real time. The membrane excludes high-molecular-weight molecules in the external matrix minimizing the amount of additional sample preparation that needs to be performed. Recovery is an important parameter in microdialysis and is dependent on the membrane size and type, probe design, and the perfusion flow rate. Recovery can be significantly enhanced by using additives to the perfusate or very low flow rates. Quantitation in microdialysis can be difficult. There are several methods for estimating the actual concentration of analytes in the external medium. The method that is chosen depends on the external medium being sampled, the information that is needed, and whether the recovery changes over time due to changes in the external environment (e.g., due to uptake or metabolism). Temporal resolution for microdialysis sampling is dependent on the recovery, the sample volume requirements, and sensitivity of the analytical method. Online methods can permit near real-time monitoring, but the temporal resolution is then dependent on the analysis time as well.

There are many applications of microdialysis sampling. Although it is most commonly used for brain sampling, this technique can also be used extensively in pharmacokinetic/pharmacodynamic studies and drug metabolism. More recently, microdialysis can be applied in bioreactor monitoring and environmental studies. Future directions include the development of microdialysis coupled to microfabricated online monitoring systems for clinical and environmental applications.

ACKNOWLEDGMENTS

Support for this work was provided in part by a research grant from the National Institutes of Health R01 (NS042929-04). Support in the form of a predoctoral fellowship from the American Heart Association for Pradyot Nandi is gratefully acknowledged. The authors would also like to thank Nancy Harmony for her assistance in the preparation of this manuscript.

REFERENCES

1. Graham, T. On liquid diffusion applied to analysis. *J. Chem. Soc.* **1862**, *15*, 216–270.

2. Luque de Castro, M.D. Membrane-based separation techniques: Dialysis, gas diffusion and pervaporation. *Compr. Anal. Chem.* **2008**, *54*, 203–234.

3. Luque de Castro, M.D.; Priego Capote, F.; Sanchez Avila, N. Is dialysis alive as a membrane-based separation technique? *Trends Anal. Chem.* **2008**, *27*, 315–326.

4. Bito, L.Z.; Davson, H.; Levin, E.; Murray, M.; Snider, N. Concentration of free amino acids and other electrolytes in cerebrospinal fluid, in vivo dialyzate of brain, and blood plasma of the dog. *J. Neurochem.* **1966**, *13*, 1057–1067.

5. Delgado, J.M.; DeFeudis, F.V.; Roth, R.H.; Ryugo, D.K.; Mitruka, B.M. Dialytrode for long term intracerebral perfusion in awake monkeys. *Arch. Int. Pharmacodyn. Ther.* **1972**, *198*, 9–21.

6. Ungerstedt, U.; Pycock, C. Functional correlates of dopamine neurotransmission. *Bull. Schweiz. Akad. Med. Wiss.* **1974**, *30*, 44–55.

7. Robinson, T.E.; Justice, J.B. Jr. *Techniques in the Behavioral and Neural Sciences, Vol. 7: Microdialysis in the Neurosciences*. Amsterdam: Elsevier; 1991.

8. Chaurasia, C.S.; Mueller, M.; Bashaw, E.D.; Benfeldt, E.; Bolinder, J.; Bullock, R.; Bungay, P.M.; DeLange, E.C.M.; Derendorf, H.; Elmquist, W.F.; Hammarlund-Udenaes, M.; Joukhadar, C.; Kellogg, D.L. Jr.; Lunte, C.E.; Nordstrom, C.H.; Rollema, H.; Sawchuk, R.J.; Cheung, B.W.Y.; Shah, V.P.; Stahle, L.; Ungerstedt, U.; Welty, D.F.; Yeo, H. AAPS-FDA Workshop White Paper: Microdialysis principles, application and regulatory perspectives. *Pharm. Res.* **2007**, *24*, 1014–1025.

9. Lindefors, N.; Amberg, G.; Ungerstedt, U. Intracerebral microdialysis: I. Experimental studies of diffusion kinetics. *J. Pharmacol. Methods* **1989**, *22*, 141–156.

10. Westerink, B.; Cremers, T.I.F.H. *Handbook of Microdialysis*, Vol. 16. Amsterdam: Academic Press; 2007.

11. Perry, M.; Li, Q.; Kennedy, R.T. Review of recent advances in analytical techniques for the determination of neurotransmitters. *Anal. Chim. Acta* **2009**, *653*, 1–22.

12. Brunner, M.; Derendorf, H. Clinical microdialysis: Current applications and potential use in drug development. *Trends Anal. Chem.* **2006**, *25*, 674–680.

13. Hocht, C.; Opezzo, J.A.W.; Bramuglia, G.F.; Taira, C.A. Application of microdialysis for pharmacokinetic-pharmacodynamic modelling. *Expert Opin. Drug Discov.* **2006**, *1*, 289–301.

14. Zhou, Q.; Gallo, J.M. In vivo microdialysis for PK and PD studies of anticancer drugs. *AAPS J.* **2005**, *7*, E659–E667.

15. Torto, N.; Laurell, T.; Gorton, L.; Marko-Varga, G. Recent trends in the application of microdialysis in bioprocesses. *Anal. Chim. Acta* **1998**, *374*, 111–135.

16. Miro, M.; Jimoh, M.; Frenzel, W. A novel dynamic approach for automatic microsampling and continuous monitoring of metal ion release from soils exploiting a dedicated flow-through microdialyser. *Anal. Bioanal. Chem.* **2005**, *382*, 396–404.

17. Hansen, D.K.; Davies, M.I.; Lunte, S.M.; Lunte, C.E. Pharmacokinetic and metabolism studies using microdialysis sampling. *J. Pharm. Sci.* **1999**, *88*, 14–27.

18. Nandi, P.; Lunte, S.M. Recent trends in microdialysis sampling integrated with conventional and microanalytical systems for monitoring biological events: A review. *Anal. Chim. Acta* **2009**, *651*, 1–14.

19. Stenken, J.A. Methods and issues in microdialysis calibration. *Anal. Chim. Acta* **1999**, *379*, 337–357.

20. Menacherry, S.; Hubert, W.; Justice, J.B. Jr. In vivo calibration of microdialysis probes for exogenous compounds. *Anal. Chem.* **1992**, *64*, 577–583.

21. Zhao, Y.; Liang, X.; Lunte, C.E. Comparison of recovery and delivery in vitro for calibration of microdialysis probes. *Anal. Chim. Acta* **1995**, *316*, 403–410.

22. Snyder, K.L.; Nathan, C.E.; Yee, A.; Stenken, J.A. Diffusion and calibration properties of microdialysis sampling membranes in biological media. *Analyst* **2001**, *126*, 1261–1268.

23. Wang, X.; Stenken, J.A. Microdialysis sampling membrane performance during in vitro macromolecule collection. *Anal. Chem.* **2006**, *78*, 6026–6034.

24. Torto, N.; Bang, J.; Richardson, S.; Nilsson, G.S.; Gorton, L.; Laurell, T.; Marko-Varga, G. Optimal membrane choice for microdialysis sampling of oligosaccharides. *J. Chromatogr. A* **1998**, *806*, 265–278.

25. Miro, M.; Frenzel, W. Implantable flow-through capillary-type microdialyzers for continuous in situ monitoring of environmentally relevant parameters. *Anal. Chem.* **2004**, *76*, 5974–5981.

26. Wisniewski, N.; Torto, N. Optimization of microdialysis sampling recovery by varying inner cannula geometry. *Analyst* **2002**, *127*, 1129–1134.

27. Laurell, T.; Buttler, T. A microdialysis probe offering arbitrary membrane length and an in situ tunable relative recovery. *Anal. Methods Instrum.* **1995**, *2*, 197–201.

28. Sulyok, M.; Miro, M.; Stingeder, G.; Koellensperger, G. The potential of flow-through microdialysis for probing low-molecular weight organic anions in rhizosphere soil solution. *Anal. Chim. Acta* **2005**, *546*, 1–10.

29. Telting-Diaz, M.; Scott, D.O.; Lunte, C.E. Intravenous microdialysis sampling in awake, freely-moving rats. *Anal. Chem.* **1992**, *64*, 806–810.

30. Huff, J.K.; Heppert, K.E.; Davies, M.I. The microdialysis shunt probe: Profile of analytes in rats with erratic bile flow or rapid changes in analyte concentration in the bile. *Curr. Sep.* **1999**, *18*, 85–90.

31. Wu, Q.; Liu, C.; Smith, R.D. Online microdialysis desalting for electrospray ionization-mass spectrometry of proteins and peptides. *Rapid Commun. Mass Spectrom.* **1996**, *10*, 835–838.

32. Lada, M.W.; Vickroy, T.W.; Kennedy, R.T. High temporal resolution monitoring of glutamate and aspartate in vivo using microdialysis online with capillary electrophoresis with laser-induced fluorescence detection. *Anal. Chem.* **1997**, *69*, 4560–4565.

33. Schneiderheinze, J.M.; Hogan, B.L. Selective in vivo and in vitro sampling of proteins using miniature ultrafiltration sampling probes. *Anal. Chem.* **1996**, *68*, 3758–3762.

34. Kottegoda, S.; Shaik, I.; Shippy, S.A. Demonstration of low flow push-pull perfusion. *J. Neurosci. Methods* **2002**, *121*, 93–101.

35. Ault, J.M.; Lunte, C.E.; Meltzer, N.M.; Riley, C.M. Microdialysis sampling for the investigation of dermal drug transport. *Pharm. Res.* **1992**, *9*, 1256–1261.

36. Palsmeier, R.K.; Lunte, C.E. Microdialysis sampling in tumor and muscle: Study of the disposition of 3-amino-1,2,4-benzotriazine-1,4-di-N-oxide (SR 4233). *Life Sci.* **1994**, *55*, 815–825.

37. Scott, D.O.; Lunte, C.E. In vivo microdialysis sampling in the bile, blood, and liver of rats to study the disposition of phenol. *Pharm. Res.* **1993**, *10*, 335–342.

38. Cano-Cebrian, M.J.; Zornoza, T.; Polache, A.; Granero, L. Quantitative in vivo microdialysis in pharmacokinetic studies: Some reminders. *Curr. Drug Metab.* **2005**, *6*, 83–90.

39. Dai, H.; Elmquist, W.F. Drug transport studies using quantitative microdialysis. *Methods Mol. Med.* **2003**, *89*, 249–264.

40. Gonzales, R.A.; Tang, A.; Robinson, D.L. Quantitative microdialysis for in vivo studies of pharmacodynamics. In *Methods in Alcohol-Related Neuroscience Research*, Liu, Y. and Lovinger, M.D., Eds., Boca Raton, FL: CRC Press; 2002; pp. 287–317.

41. Peters, J.L.; Yang, H.; Michael, A.C. Quantitative aspects of brain microdialysis. *Anal. Chim. Acta* **2000**, *412*, 1–12.

42. Kehr, J. A survey on quantitative microdialysis: Theoretical models and practical implications. *J. Neurosci. Methods* **1993**, *48*, 251–261.

43. Davies, M.I.; Cooper, J.D.; Desmond, S.S.; Lunte, C.E.; Lunte, S.M. Analytical considerations for microdialysis sampling. *Adv. Drug Deliv. Rev.* **2000**, *45*, 169–188.

44. Olson, R.J.; Justice, J.B. Jr. Quantitative microdialysis under transient conditions. *Anal. Chem.* **1993**, *65*, 1017–1022.

45. Cosford, R.J.O.; Vinson, A.P.; Kukoyi, S.; Justice, J.B. Jr. Quantitative microdialysis of serotonin and norepinephrine: Pharmacological influences on in vivo extraction fraction. *J. Neurosci. Methods* **1996**, *68*, 39–47.

46. Justice, J.B. Jr. Quantitative microdialysis of neurotransmitters. *J. Neurosci. Methods* **1993**, *48*, 263–276.

47. Lonnroth, P.; Jansson, P.A.; Smith, U. A microdialysis method allowing characterization of intercellular water space in humans. *Am. J. Physiol.* **1987**, *253*, E228–E231.

48. Scheller, D.; Kolb, J. The internal reference technique in microdialysis: A practical approach to monitoring dialysis efficiency and to calculating tissue concentration from dialyzate samples. *J. Neurosci. Methods* **1991**, *40*, 31–38.

49. Yokel, R.A.; Allen, D.D.; Burgio, D.E.; McNamara, P.J. Antipyrine as a dialyzable reference to correct differences in efficiency among and within sampling devices during in vivo microdialysis. *J. Pharmacol. Toxicol. Methods* **1992**, *27*, 135–142.

50. Song, Y.; Lunte, C.E. Comparison of calibration by delivery versus no net flux for quantitative in vivo microdialysis sampling. *Anal. Chim. Acta* **1999**, *379*, 251–262.

51. Schultz, K.N.; Kennedy, R.T. Time-resolved microdialysis for in vivo neurochemical measurements and other applications. *Annu. Rev. Anal. Chem.* **2008**, *1*, 627–661.

52. Cremers, T.I.F.H.; de Vries, M.G.; Huinink, K.D.; van Loon, J.P.; van der Hart, M.; Ebert, B.; Westerink, B.H.C.; De Lange, E.C.M. Quantitative microdialysis using modified ultraslow microdialysis: Direct rapid and reliable determination of free brain concentrations with the MetaQuant technique. *J. Neurosci. Methods* **2009**, *178*, 249–254.

53. Zhou, S.N.; Oakes, K.D.; Servos, M.R.; Pawliszyn, J. Use of simultaneous dual-probe microdialysis for the determination of pesticide residues in a jade plant (*Crassula ovata*). *Analyst* **2009**, *134*, 748–754.

54. Wang, Y.; Stenken, J.A. Affinity-based microdialysis sampling using heparin for in vitro collection of human cytokines. *Anal. Chim. Acta* **2009**, *651*, 105–111.

55. Duo, J.; Fletcher, H.; Stenken, J.A. Natural and synthetic affinity agents as microdialysis sampling mass transport enhancers: Current progress and future perspectives. *Biosens. Bioelectron.* **2006**, *22*, 449–457.

56. Khramov, A.N.; Stenken, J.A. Enhanced microdialysis extraction efficiency of ibuprofen in vitro by facilitated transport with beta-cyclodextrin. *Anal. Chem.* **1999**, *71*, 1257–1264.

57. Khramov, A.N.; Stenken, J.A. Enhanced microdialysis recovery of some tricyclic antidepressants and structurally related drugs by cyclodextrin-mediated transport. *Analyst* **1999**, *124*, 1027–1033.

58. Fletcher, H.J.; Stenken, J.A. An in vitro comparison of microdialysis relative recovery of Met- and Leu-enkephalin using cyclodextrins and antibodies as affinity agents. *Anal. Chim. Acta* **2008**, *620*, 170–175.

59. Ao, X.; Sellati, T.J.; Stenken, J.A. Enhanced microdialysis relative recovery of inflammatory cytokines using antibody-coated microspheres analyzed by flow cytometry. *Anal. Chem.* **2004**, *76*, 3777–3784.

60. Mogopodi, D.; Torto, N. Maximizing metal ions flux across a microdialysis membrane by incorporating poly-L-aspartic acid, poly-L-histidine, 8-hydroxyquinoline and ethylenediaminetetraacetic acid in the perfusion liquid. *Anal. Chim. Acta* **2005**, *534*, 239–246.

61. Mogopodi, D.; Torto, N. Enhancing microdialysis recovery of metal ions by incorporating poly-L-aspartic acid and poly-L-histidine in the perfusion liquid. *Anal. Chim. Acta* **2003**, *482*, 91–97.

62. Torto, N.; Mogopodi, D. Opportunities in microdialysis sampling of metal ions. *Trends Anal. Chem.* **2004**, *23*, 109–115.

63. Torto, N.; Mwatseteza, J.; Sawula, G. A study of microdialysis sampling of metal ions. *Anal. Chim. Acta* **2002**, *456*, 253–261.

64. Rossell, S.; Gonzalez, L.E.; Hernandez, L. One-second time resolution brain microdialysis in fully awake rats. Protocol for the collection, separation and sorting of nanoliter dialyzate volumes. *J. Chromatogr. B Anal. Technol. Biomed. Life Sci.* **2003**, *784*, 385–393.

65. Tucci, S.; Rada, P.; Sepulveda, M.J.; Hernandez, L. Glutamate measured by 6-s resolution brain microdialysis: Capillary electrophoretic and laser-induced fluorescence detection application. *J. Chromatogr. B Biomed. Sci. Appl.* **1997**, *694*, 343–349.

66. Mecker, L.C.; Martin, R.S. Integration of microdialysis sampling and microchip electrophoresis with electrochemical detection. *Anal. Chem.* **2008**, *80*, 9257–9264.

67. Huff, J.K.; Davies, M.I. Microdialysis monitoring of methylphenidate in blood and brain correlated with changes in dopamine and rat activity. *J. Pharm. Biomed. Anal.* **2002**, *29*, 767–777.

68. Grabb, M.C.; Sciotti, V.M.; Gidday, J.M.; Cohen, S.A.; van Wylen, D.G. Neurochemical and morphological responses to acutely and chronically implanted brain microdialysis probes. J. Neurosci. *Methods* **1998**, *82*, 25–34.

69. Benveniste, H.; Drejer, J.; Schousboe, A.; Diemer, N.H. Regional cerebral glucose phosphorylation and blood flow after insertion of a microdialysis fiber through the dorsal hippocampus in the rat. *J. Neurochem.* **1987**, *49*, 729–734.

70. de Lange, E.C.; Danhof, M.; de Boer, A.G.; Breimer, D.D. Methodological considerations of intracerebral microdialysis in pharmacokinetic studies on drug transport across the blood-brain barrier. *Brain Res. Rev.* **1997**, *25*, 27–49.

71. Benveniste, H.; Drejer, J.; Schousboe, A.; Diemer, N.H. Elevation of the extracellular concentrations of glutamate and aspartate in rat hippocampus during transient cerebral ischemia monitored by intracerebral microdialysis. *J. Neurochem.* **1984**, *43*, 1369–1374.

72. de Lange, E.C.; Danhof, M.; de Boer, A.G.; Breimer, D.D. Critical factors of intracerebral microdialysis as a technique to determine the pharmacokinetics of drugs in rat brain. *Brain Res.* **1994**, *666*, 1–8.

73. Major, O.; Shdanova, T.; Duffek, L.; Nagy, Z. Continuous monitoring of blood-brain barrier opening to Cr51-EDTA by microdialysis following probe injury. *Acta Neurochir. Suppl.* **1990**, *51*, 46–48.

74. Mitala, C.M.; Wang, Y.; Borland, L.M.; Jung, M.; Shand, S.; Watkins, S.; Weber, S.G.; Michael, A.C. Impact of microdialysis probes on vasculature and dopamine in the rat striatum: A combined fluorescence and voltammetric study. *J. Neurosci. Methods* **2008**, *174*, 177–185.

75. Lu, Y.; Peters, J.L.; Michael, A.C. Direct comparison of the response of voltammetry and microdialysis to electrically evoked release of striatal dopamine. *J. Neurochem.* **1998**, *70*, 584–593.

76. Peters, J.L.; Michael, A.C. Modeling voltammetry and microdialysis of striatal extracellular dopamine: The impact of dopamine uptake on extraction and recovery ratios. *J. Neurochem.* **1998**, *70*, 594–603.

77. Borland, L.M.; Shi, G.; Yang, H.; Michael, A.C. Voltammetric study of extracellular dopamine near microdialysis probes acutely implanted in the striatum of the anesthetized rat. *J. Neurosci. Methods* **2005**, *146*, 149–158.

78. Yang, H.; Peters, J.L.; Allen, C.; Chern, S.-S.; Coalson, R.D.; Michael, A.C. A theoretical description of microdialysis with mass transport coupled to chemical events. *Anal. Chem.* **2000**, *72*, 2042–2049.

79. Zielke, H.R.; Zielke, C.L.; Baab, P.J. Direct measurement of oxidative metabolism in the living brain by microdialysis: A review. *J. Neurochem.* **2009**, *109*, 24–29.

80. Torregrossa, M.M.; Kalivas, P.W. Microdialysis and the neurochemistry of addiction. *Pharmacol. Biochem. Behav.* **2008**, *90*, 261–272.

81. McAdoo, D.J.; Wu, P. Microdialysis in central nervous system disorders and their treatment. *Pharmacol. Biochem. Behav.* **2008**, *90*, 282–296.

82. Linthorst, A.C.E.; Reul, J.M. Stress and the brain: Solving the puzzle using microdialysis. *Pharmacol. Biochem. Behav.* **2008**, *90*, 163–173.

83. Fillenz, M. In vivo neurochemical monitoring and the study of behaviour. *Neurosci. Biobehav. Rev.* **2005**, *29*, 949–962.

84. Maurer, M.H.; Haux, D.; Unterberg, A.W.; Sakowitz, O.W. Proteomics of human cerebral microdialysate: From detection of biomarkers to clinical application. *Proteomics Clin. Appl.* **2008**, *2*, 437–443.

85. Ungerstedt, U.; Rostami, E. Microdialysis in neurointensive care. *Curr. Pharm. Des.* **2004**, *10*, 2145–2152.

86. Lanckmans, K.; Sarre, S.; Smolders, I.; Michotte, Y. Quantitative liquid chromatography/mass spectrometry for the analysis of microdialysates. *Talanta* **2008**, *74*, 458–469.

87. Buck, K.; Voehringer, P.; Ferger, B. Rapid analysis of GABA and glutamate in microdialysis samples using high performance liquid chromatography and tandem mass spectrometry. *J. Neurosci. Methods* **2009**, *182*, 78–84.

88. Zhang, M.-Y.; Beyer, C.E. Measurement of neurotransmitters from extracellular fluid in brain by in vivo microdialysis and chromatography-mass spectrometry. *J. Pharm. Biomed. Anal.* **2006**, *40*, 492–499.

89. Lonnroth, P.; Jansson, P.A.; Fredholm, B.B.; Smith, U. Microdialysis of intercellular adenosine concentration in subcutaneous tissue in humans. *Am. J. Physiol.* **1989**, *256*, E250–E255.

90. Jansson, P.A.; Smith, U.; Lonnroth, P. Interstitial glycerol concentration measured by microdialysis in two subcutaneous regions in humans. *Am. J. Physiol.* **1990**, *258*, E918–E922.

91. Elmquist, W.F.; Sawchuk, R.J. Application of microdialysis in pharmacokinetic studies. *Pharm. Res.* **1997**, *14*, 267–288.

92. Scott, D.O.; Sorensen, L.R.; Lunte, C.E. In vivo microdialysis sampling coupled to liquid chromatography for the study of acetaminophen metabolism. *J. Chromatogr.* **1990**, *506*, 461–469.

93. Scott, D.O.; Sorenson, L.R.; Steele, K.L.; Puckett, D.L.; Lunte, C.E. In vivo microdialysis sampling for pharmacokinetic investigations. *Pharm. Res.* **1991**, *8*, 389–392.

94. Steele, K.L.; Scott, D.O.; Lunte, C.E. Pharmacokinetic studies of aspirin in rats using in vivo microdialysis sampling. *Anal. Chim. Acta* **1991**, *246*, 181–186.

95. Lunte, C.E.; Scott, D.O.; Kissinger, P.T. Sampling living systems using microdialysis probes. *Anal. Chem.* **1991**, *63*, 773A–774A, 776A–778A, 780A.

96. McDonald, S.; Lunte, C. Determination of the dermal penetration of esterom components using microdialysis sampling. *Pharm. Res.* **2003**, *20*, 1827–1834.

97. Davies, M.I.; Lunte, C.E. Microdialysis sampling for hepatic metabolism studies: Impact of microdialysis probe design and implantation technique on liver tissue. *Drug Metab. Dispos.* **1995**, *23*, 1072–1079.

98. Palsmeier, R.K.; Lunte, C.E. In the MD literature: Microdialysis. Microdialysis sampling in tumor and muscle: Study of the disposition of 3-amino-1,2,4-benzotriazine-1,4-di-N-oxide (SR 4233). *Curr. Sep.* **1994**, *13*, 93.

99. Hosoya, K.; Makihara, A.; Tsujikawa, Y.; Yoneyama, D.; Mori, S.; Terasaki, T.; Akanuma, S.; Tomi, M.; Tachikawa, M. Roles of inner blood-retinal barrier organic anion transporter 3 in the vitreous/retina-to-blood efflux transport of p-aminohippuric acid, benzylpenicillin, and 6-mercaptopurine. *J. Pharmacol. Exp. Ther.* **2009**, *329*, 87–93.

100. Dey, S.; Gunda, S.; Mitra, A.K. Pharmacokinetics of erythromycin in rabbit corneas after single-dose infusion: Role of P-glycoprotein as a barrier to in vivo ocular drug absorption. *J. Pharmacol. Exp. Ther.* **2004**, *311*, 246–255.

101. Anand, B.S.; Katragadda, S.; Gunda, S.; Mitra, A.K. In vivo ocular pharmacokinetics of acyclovir dipeptide ester prodrugs by microdialysis in rabbits. *Mol. Pharm.* **2006**, *3*, 431–440.

102. Macha, S.; Mitra, A.K. Ocular pharmacokinetics in rabbits using a novel dual probe microdialysis technique. *Exp. Eye Res.* **2001**, *72*, 289–299.

103. Waga, J.; Nilsson-Ehle, I.; Ljungberg, B.; Skarin, A.; Stahle, L.; Ehinger, B. Microdialysis for pharmacokinetic studies of ceftazidime in rabbit vitreous. *J. Ocul. Pharmacol. Ther.* **1999**, *15*, 455–463.

104. de la Pena, A.; Liu, P.; Derendorf, H. Microdialysis in peripheral tissues. *Adv. Drug Deliv. Rev.* **2000**, *45*, 189–216.

105. Lonnroth, P.; Carlsten, J.; Johnson, L.; Smith, U. Measurements by microdialysis of free tissue concentrations of propranolol. *J. Chromatogr. B Biomed. Appl.* **1991**, *568*, 419–425.

106. Lonnroth, P.; Smith, U. Microdialysis—A novel technique for clinical investigations. *J. Intern. Med.* **1990**, *227*, 295–300.

107. Flodgren, G.M.; Crenshaw, A.G.; Gref, M.; Fahlstrom, M. Changes in interstitial noradrenaline, trapezius muscle activity and oxygen saturation during low-load work and recovery. *Eur. J. Appl. Physiol.* **2009**, *107*, 31–42.

108. Barbour, A.; Schmidt, S.; Rout, W.R.; Ben-David, K.; Burkhardt, O.; Derendorf, H. Soft tissue penetration of cefuroxime determined by clinical microdialysis in morbidly obese patients undergoing abdominal surgery. *Int. J. Antimicrob. Agents* **2009**, *34*, 231–235.

109. Gerdle, B.; Hilgenfeldt, U.; Larsson, B.; Kristiansen, J.; Sogaard, K.; Rosendal, L. Bradykinin and kallidin levels in the trapezius muscle in patients with work-related trapezius myalgia, in patients with whiplash associated pain, and in healthy controls—A microdialysis study of women. *Pain* **2008**, *139*, 578–587.

110. Hamrin, K.; Henriksson, J. Interstitial glucose concentration in insulin-resistant human skeletal muscle: influence of one bout of exercise and of local perfusion with insulin or vanadate. *Eur. J. Appl. Physiol.* **2008**, *103*, 595–603.

111. Traunmuller, F.; Zeitlinger, M.; Zeleny, P.; Muller, M.; Joukhadar, C. Pharmacokinetics of single- and multiple-dose oral clarithromycin in soft tissues determined by microdialysis. *Antimicrob. Agents Chemother.* **2007**, *51*, 3185–3189.

112. Kurosaki, Y.; Tagawa, M.; Omoto, A.; Suito, H.; Komori, Y.; Kawasaki, H.; Aiba, T. Evaluation of intramuscular lateral distribution profile of topically administered acetaminophen in rats. *Int. J. Pharm.* **2007**, *343*, 190–195.

113. Marchand, S.; Dahyot, C.; Lamarche, I.; Plan, E.; Mimoz, O.; Couet, W. Lack of effect of experimental hypovolemia on imipenem muscle distribution in rats assessed by microdialysis. *Antimicrob. Agents Chemother.* **2005**, *49*, 4974–4979.

114. Tettey-Amlalo, R.N.O.; Kanfer, I.; Skinner, M.F.; Benfeldt, E.; Verbeeck, R.K. Application of dermal microdialysis for the evaluation of bioequivalence of a ketoprofen topical gel. *Eur. J. Pharm. Sci.* **2009**, *36*, 219–225.

115. Mathy, F.X.; Ntivunwa, D.; Verbeeck, R.K.; Preat, V. Fluconazole distribution in rat dermis following intravenous and topical application: A microdialysis study. *J. Pharm. Sci.* **2005**, *94*, 770–780.

116. Klimowicz, A.; Farfal, S.; Bielecka-Grzela, S. Evaluation of skin penetration of topically applied drugs in humans by cutaneous microdialysis: Acyclovir vs. salicylic acid. *J. Clin. Pharm. Ther.* **2007**, *32*, 143–148.

117. Obreja, O.; Rukwied, R.; Steinhoff, M.; Schmelz, M. Neurogenic components of trypsin- and thrombin-induced inflammation in rat skin, in vivo. *Exp. Dermatol.* **2006**, *15*, 58–65.

118. Delacher, S.; Derendorf, H.; Hollenstein, U.; Brunner, M.; Joukhadar, C.; Hofmann, S.; Georgopoulos, A.; Eichler, H.G.; Muller, M. A combined in vivo pharmacokinetic-in vitro pharmacodynamic approach to simulate target site pharmacodynamics of antibiotics in humans. *J. Antimicrob. Chemother.* **2000**, *46*, 733–739.

119. Joukhadar, C.; Derendorf, H.; Muller, M. Microdialysis. A novel tool for clinical studies of anti-infective agents. *Eur. J. Clin. Pharmacol.* **2001**, *57*, 211–219.

120. Joukhadar, C.; Thallinger, C.; Poppl, W.; Kovar, F.; Konz, K.H.; Joukhadar, S.M.; Traunmuller, F. Concentrations of voriconazole in healthy and inflamed lung in rats. *Antimicrob. Agents Chemother.* **2009**, *53*, 2684–2686.

121. Wesson, D.E.; Simoni, J. Increased tissue acid mediates a progressive decline in the glomerular filtration rate of animals with reduced nephron mass. *Kidney Int.* **2009**, *75*, 929–935.

122. Price, K.E.; Larive, C.K.; Lunte, C.E. Tissue-targeted metabonomics: Biological considerations and application to doxorubicin-induced hepatic oxidative stress. *Metabolomics* **2009**, *5*, 219–228.

123. Lefeuvre, S.; Marchand, S.; Pariat, C.; Lamarche, I.; Couet, W. Microdialysis study of imipenem distribution in the peritoneal fluid of rats with experimental acute pancreatitis. *Antimicrob. Agents Chemother.* **2008**, *52*, 1516–1518.

124. Wu, Y.T.; Lin, L.C.; Tsai, T.H. Measurement of free hydroxytyrosol in microdialysates from blood and brain of anesthetized rats by liquid chromatography with fluorescence detection. *J. Chromatogr. A* **2009**, *1216*, 3501–3507.

125. Sawchuk, R.J.; Cheung, B.W.; Ji, P.; Cartier, L.L. Microdialysis studies of the distribution of antibiotics into chinchilla middle ear fluid. *Pharmacotherapy* **2005**, *25*, 140S–145S.

126. Huang, Y.; Yang, Z.; Cartier, L.; Cheung, B.; Sawchuk, R.J. Estimating amoxicillin influx/efflux in chinchilla middle ear fluid and simultaneous measurement of antibacterial effect. *Antimicrob. Agents Chemother.* **2007**, *51*, 4336–4341.

127. Miro, M.; Hansen, E.H. On-line sample processing methods in flow analysis. *Adv. Flow Anal.* **2008**, 291–320.

128. Jen, J.F.; Chang, C.T.; Yang, T.C. On-line microdialysis-high-performance liquid chromatographic determination of aniline and 2-chloroaniline in polymer industrial wastewater. *J. Chromatogr. A* **2001**, *930*, 119–125.

129. Torto, N.; Peloewetse, E.; Lobelo, B.; Mwatseteza, J. Profiling of wastewater effluent from a beverage industry by microdialysis sampling and a combination of analytical techniques. *Environ. Sci. Indian J.* **2006**, *1*, 1–7.

130. Miro, M.; Frenzel, W. A novel flow-through microdialysis separation unit with integrated differential potentiometric detection for the determination of chloride in soil samples. *Analyst* **2003**, *128*, 1291–1297.

131. Mosetlha, K.; Torto, N.; Wibetoe, G. Determination of Cu and Ni in plants by microdialysis sampling: Comparison of dialyzable metal fractions with total metal content. *Talanta* **2007**, *71*, 766–770.

132. Miro, M.; Hansen, E.H. Miniaturization of environmental chemical assays in flowing systems: The lab-on-a-valve approach vis-a-vis lab-on-a-chip microfluidic devices. *Anal. Chim. Acta* **2007**, *600*, 46–57.

133. Fang, Z.-L.; Fang, Q.; Liu, X.-Z.; Chen, H.-W.; Liu, C.-L. Continuous monitoring in drug dissolution testing using flow injection systems. *Trends Anal. Chem.* **1999**, *18*, 261–271.

134. Shah, K.P.; Chang, M.; Riley, C.M. Automated analytical systems for drug development studies. 3. Multivessel dissolution testing system based on microdialysis sampling. *J. Pharm. Biomed. Anal.* **1995**, *13*, 1235–1241.

135. Shah, K.P.; Chang, M.; Riley, C.M. Automated analytical systems for drug development studies. II—A system for dissolution testing. *J. Pharm. Biomed. Anal.* **1994**, *12*, 1519–1527.

136. Dash, A.K.; Haney, P.W.; Garavalia, M.J. Development of an in vitro dissolution method using microdialysis sampling technique for implantable drug delivery systems. *J. Pharm. Sci.* **1999**, *88*, 1036–1040.

137. Fang, Q.; Liu, S.-S.; Wu, J.-F.; Sun, Y.-Q.; Fang, Z.-L. A stopped-flow microdialysis sampling-flow injection system for automated multivessel high resolution drug dissolution testing. *Talanta* **1999**, *49*, 403–414.

138. Torto, N.; Gorton, L.; Laurell, T.; Marko-Varga, G. Technical issues of in vitro microdialysis sampling in bioprocess monitoring. *Trends Anal. Chem.* **1999**, *18*, 252–260.

139. Buttler, T.; Nilsson, C.; Gorton, L.; Marko-Varga, G.; Laurell, T. Membrane characterization and performance of microdialysis probes

intended for use as bioprocess sampling units. *J. Chromatogr. A* **1996**, *725*, 41–56.

140. Torto, N.; Laurell, T.; Gorton, L.; Varga, G.M. A study of a polysulfone membrane for use in an in-situ tunable microdialysis probe during monitoring of starch enzymic hydrolyzates. *J. Memb. Sci.* **1997**, *130*, 239–248.

141. Modi, S.J.; LaCourse, W.R. Monitoring carbohydrate enzymatic reactions by quantitative in vitro microdialysis. *J. Chromatogr. A* **2006**, *1118*, 125–133.

142. Huynh, B.H.; Fogarty, B.A.; Martin, R.S.; Lunte, S.M. On-line coupling of microdialysis sampling with microchip-based capillary electrophoresis. *Anal. Chem.* **2004**, *76*, 6440–6447.

143. Xiang, F.; Lin, Y.; Wen, J.; Matson, D.W.; Smith, R.D. An integrated microfabricated device for dual microdialysis and online ESI-ion trap mass spectrometry for analysis of complex biological samples. *Anal. Chem.* **1999**, *71*, 1485–1490.

144. Yang, L.; Lee, C.S.; Hofstadler, S.A.; Smith, R.D. Characterization of microdialysis acidification for capillary isoelectric focusing-microelectrospray ionization mass spectrometry. *Anal. Chem.* **1998**, *70*, 4945–4950.

145. Xu, N.; Lin, Y.; Hofstadler, S.A.; Matson, D.; Call, C.J.; Smith, R.D. A microfabricated dialysis device for sample cleanup in electrospray ionization mass spectrometry. *Anal. Chem.* **1998**, *70*, 3553–3556.

146. Liu, C.; Wu, Q.; Harms, A.C.; Smith, R.D. On-line microdialysis sample cleanup for electrospray ionization mass spectrometry of nucleic acid samples. *Anal. Chem.* **1996**, *68*, 3295–3299.

147. Jiang, Y.; Wang, P.-C.; Locascio, L.E.; Lee, C.S. Integrated plastic microfluidic devices with ESI-MS for drug screening and residue analysis. *Anal. Chem.* **2001**, *73*, 2048–2053.

148. Herrera, A.M.; Scott, D.O.; Lunte, C.E. Microdialysis sampling for determination of plasma protein binding of drugs. *Pharm. Res.* **1990**, *7*, 1077–1081.

149. Lunte, C.E.; Scott, D.O.; Herrera, A.M. Determination of drug binding to proteins by microdialysis perfusion sampling. *Curr. Sep.* **1991**, *10*, 41–44.

150. Blakeley, J.O.; Olson, J.; Grossman, S.A.; He, X.; Weingart, J.; Supko, J.G. Effect of blood brain barrier permeability in recurrent high grade gliomas on the intratumoral pharmacokinetics of methotrexate: A microdialysis study. *J. Neurooncol.* **2009**, *91*, 51–58.

151. Sammeta, S.M.; Vaka, S.R.; Murthy, S.N. Dermal drug levels of antibiotic (cephalexin) determined by electroporation and transcutaneous sampling (ETS) technique. *J. Pharm. Sci.* **2009**, *98*, 2677–2685.

152. Goodman, J.C.; Robertson, C.S. Microdialysis: Is it ready for prime time? *Curr. Opin. Crit. Care* **2009**, *15*, 110–117.

153. Bengtsson, J.; Bostroem, E.; Hammarlund-Udenaes, M. The use of a deuterated calibrator for in vivo recovery estimations in microdialysis studies. *J. Pharm. Sci.* **2008**, *97*, 3433–3441.

154. Wang, Y.; Wong, S.L.; Sawchuk, R.J. Microdialysis calibration using retrodialysis and zero-net flux: Application to a study of the distribution of zidovudine to rabbit cerebrospinal fluid and thalamus. *Pharm. Res.* **1993**, *10*, 1411–1419.

155. Wong, S.L.; Wang, Y.; Sawchuk, R.J. Analysis of zidovudine distribution to specific regions in rabbit brain using microdialysis. *Pharm. Res.* **1992**, *9*, 332–338.

156. Stenken, J.A.; Lunte, C.E.; Southard, M.Z.; Staahle, L. Factors That influence microdialysis recovery. Comparison of experimental and theoretical microdialysis recoveries in rat liver. *J. Pharm. Sci.* **1997**, *86*, 958–966.

157. Ao, X.; Wang, X.; Lennartz, M.R.; Loegering, D.J.; Stenken, J.A. Multiplexed cytokine detection in microliter microdialysis samples obtained from activated cultured macrophages. *J. Pharm. Biomed. Anal.* **2006**, *40*, 915–921.

158. Mosetlha, K.; Torto, N.; Wibetoe, G. Enhancing the microdialysis recovery for sampling of Cu and Ni by incorporating humic acid in the perfusion liquid. *Anal. Chim. Acta* **2006**, *562*, 158–163.

159. Yao, T.; Okano, G. Simultaneous determination of L-glutamate, acetylcholine and dopamine in rat brain by a flow-injection biosensor system with microdialysis sampling. *Anal. Sci.* **2008**, *24*, 1469–1473.

160. Boutelle, M.G.; Fellows, L.K.; Cook, C. Enzyme packed bed system for the on-line measurement of glucose, glutamate, and lactate in brain microdialysate. *Anal. Chem.* **1992**, *64*, 1790–1794.

161. Stenken, J.A.; Topp, E.M.; Southard, M.Z.; Lunte, C.E. Examination of microdialysis sampling in a well-characterized hydrodynamic system. *Anal. Chem.* **1993**, *65*, 2324–2328.

162. Lunte, S.M.; Lunte, C.E. Microdialysis sampling for pharmacological studies: HPLC and CE analysis. *Adv. Chromatogr.* **1996**, *36*, 383–432.

163. Khandelwal, P.; Beyer, C.E.; Lin, Q.; McGonigle, P.; Schechter, L.E.; Bach, A.C. 2nd. Nanoprobe NMR spectroscopy and in vivo microdialysis: New analytical methods to study brain neurochemistry. *J. Neurosci. Methods* **2004**, *133*, 181–189.

164. Zhou, S.Y.; Zuo, H.; Stobaugh, J.F.; Lunte, C.E.; Lunte, S.M. Continuous in vivo monitoring of amino acid neurotransmitters by microdialysis sampling with on-line derivatization and capillary electrophoresis separation. *Anal. Chem.* **1995**, *67*, 594–599.

165. Cellar, N.A.; Burns, S.T.; Meiners, J.C.; Chen, H.; Kennedy, R.T. Microfluidic chip for low-flow push-pull perfusion sampling in vivo with on-line analysis of amino acids. *Anal. Chem.* **2005**, *77*, 7067–7073.

166. Kennedy, R.T.; Watson, C.J.; Haskins, W.E.; Powell, D.H.; Strecker, R.E. In vivo neurochemical monitoring by microdialysis and capillary separations. *Curr. Opin. Chem. Biol.* **2002**, *6*, 659–665.

167. Lada, M.W.; Kennedy, R.T. Quantitative in vivo measurements using microdialysis on-line with capillary zone electrophoresis. *J. Neurosci. Methods* **1995**, *63*, 147–152.

168. Hernandez, L.; Tucci, S.; Guzman, N.; Paez, X. In vivo monitoring of glutamate in the brain by microdialysis and capillary electrophoresis with laser-induced fluorescence detection. *J. Chromatogr. A* **1993**, *652*, 393–398.

169. Kennedy, R.T.; Thompson, J.E.; Vickroy, T.W. In vivo monitoring of amino acids by direct sampling of brain extracellular fluid at ultralow flow rates and capillary electrophoresis. *J. Neurosci. Methods* **2002**, *114*, 39–49.

170. O'Shea, T.J.; Weber, P.L.; Bammel, B.P.; Lunte, C.E.; Lunte, S.M.; Smyth, M.R. Monitoring excitatory amino acid release in vivo by microdialysis with capillary electrophoresis-electrochemistry. *J. Chromatogr.* **1992**, *608*, 189–195.

171. Venton, B.J.; Robinson, T.E.; Kennedy, R.T. Transient changes in nucleus accumbens amino acid concentrations correlate with individual responsivity to the predator fox odor 2,5-dihydro-2,4,5-trimethylthiazoline. *J. Neurochem.* **2006**, *96*, 236–246.

172. Haskins, W.E.; Watson, C.J.; Cellar, N.A.; Powell, D.H.; Kennedy, R.T. Discovery and neurochemical screening of peptides in brain extracellular fluid by chemical analysis of in vivo microdialysis samples. *Anal. Chem.* **2004**, *76*, 5523–5533.

173. Freed, A.L.; Cooper, J.D.; Davies, M.I.; Lunte, S.M. Investigation of the metabolism of substance P in rat striatum by microdialysis sampling and capillary electrophoresis with laser-induced fluorescence detection. *J. Neurosci. Methods* **2001**, *109*, 23–29.

174. Kostel, K.L.; Lunte, S.M. Evaluation of capillary electrophoresis with post-column derivatization and laser-induced fluorescence detection for the determination of substance P and its metabolites. *J. Chromatogr. B Biomed. Sci. Appl.* **1997**, *695*, 27–38.

175. Baseski, H.M.; Watson, C.J.; Cellar, N.A.; Shackman, J.G.; Kennedy, R.T. Capillary liquid chromatography with MS3 for the determination of enkephalins in microdialysis samples from the striatum of anesthetized and freely-moving rats. *J. Mass Spectrom.* **2005**, *40*, 146–153.

176. Arnett, S.D.; Osbourn, D.M.; Moore, K.D.; Vandaveer, S.S.; Lunte, C.E. Determination of 8-oxoguanine and 8-hydroxy-2′-deoxyguanosine in

the rat cerebral cortex using microdialysis sampling and capillary electrophoresis with electrochemical detection. *J. Chromatogr. B Analyt. Technol. Biomed Life Sci.* **2005**, *827*, 16–25.

177. Sood, P.; Cole, S.; Fraier, D.; Young, A.M. Evaluation of metaquant microdialysis for measurement of absolute concentrations of amphetamine and dopamine in brain: A viable method for assessing pharmacokinetic profile of drugs in the brain. *J. Neurosci. Methods* **2009**, *185*, 39–44.

178. Deng, Q.; Watson, C.J.; Kennedy, R.T. Aptamer affinity chromatography for rapid assay of adenosine in microdialysis samples collected in vivo. *J. Chromatogr. A* **2003**, *1005*, 123–130.

179. Presti, M.F.; Watson, C.J.; Kennedy, R.T.; Yang, M.; Lewis, M.H. Behavior-related alterations of striatal neurochemistry in a mouse model of stereotyped movement disorder. *Pharmacol. Biochem. Behav.* **2004**, *77*, 501–507.

180. Prokai, L.; Frycak, P.; Stevens, S.M.; Nguyen, V. Measurement of acetylcholine in rat brain microdialysates by LC-isotope dilution tandem MS. *Chromatographia* **2008**, *68*, s101–s105.

181. Shackman, H.M.; Shou, M.; Cellar, N.A.; Watson, C.J.; Kennedy, R.T. Microdialysis coupled on-line to capillary liquid chromatography with tandem mass spectrometry for monitoring acetylcholine in vivo. *J. Neurosci. Methods* **2007**, *159*, 86–92.

182. Dempsey, E.; Diamond, D.; Smyth, M.R.; Urban, G.; Jobst, G.; Moser, I.; Verpoorte, E.M.J.; Manz, A.; Michael Widmer, H.; Rabenstein, K.; Freaney, R. Design and development of a miniaturised total chemical analysis system for on-line lactate and glucose monitoring in biological samples. *Anal. Chim. Acta* **1997**, *346*, 341–349.

183. Yao, T.; Yano, T.; Nishino, H. Simultaneous in vivo monitoring of glucose, -lactate, and pyruvate concentrations in rat brain by a flow-injection biosensor system with an on-line microdialysis sampling. *Anal. Chim. Acta* **2004**, *510*, 53–59.

184. Sun, Y.; Dai, J.; Hu, Z.; Du, F.; Niu, W.; Wang, F.; Liu, F.; Jin, G.; Li, C. Oral bioavailability and brain penetration of (-)-stepholidine, a tetrahydroprotoberberine agonist at dopamine D(1) and antagonist at D(2) receptors, in rats. *Br. J. Pharmacol.* **2009**, *158*, 1302–1312.

185. Feng, J.; Li, F.; Zhao, Y.; Feng, Y.; Abe, Y. Brain pharmacokinetics of tetramethylpyrazine after intranasal and intravenous administration in awake rats. *Int. J. Pharm.* **2009**, *375*, 55–60.

186. Fu, J.; Fang, C.; Cui, Y.-Y.; Yang, L.-M.; Zhu, L.; Feng, X.-M.; Zheng, P.-L.; Lu, Y.; Chen, H.-Z. Quantitative determination of a novel enantiomeric tropane analog, (-)-satropane, in biological fluids using liquid chromatography/tandem mass spectrometry. *Biomed. Chromatogr.* **2009**, *23*, 1044–1050.

187. Barbour, A.; Schmidt, S.; Sabarinath, S.N.; Grant, M.; Seubert, C.; Skee, D.; Murthy, B.; Derendorf, H. Soft-tissue penetration of ceftobiprole in healthy volunteers determined by in vivo microdialysis. *Antimicrob. Agents Chemother.* **2009**, *53*, 2773–2776.

188. Traunmuller, F.; Fille, M.; Thallinger, C.; Joukhadar, C. Multiple-dose pharmacokinetics of telithromycin in peripheral soft tissues. *Int. J. Antimicrob. Agents* **2009**, *34*, 72–75.

189. Chang, J.-C.; Lee, W.-C.; Wu, Y.-T.; Tsai, T.-H. Distribution of blood-muscle for clenbuterol in rat using microdialysis. *Int. J. Pharm.* **2009**, *372*, 91–96.

190. Zheng, H.; Chen, G.; Shi, L.; Lou, Z.; Chen, F.; Hu, J. Determination of oxymatrine and its metabolite matrine in rat blood and dermal microdialysates by high throughput liquid chromatography/tandem mass spectrometry. *J. Pharm. Biomed. Anal.* **2009**, *49*, 427–433.

191. de Araujo, B.V.; da Silva, C.F.; Haas, S.E.; Dalla Costa, T. Free renal levels of voriconazole determined by microdialysis in healthy and Candida sp.-infected Wistar rats. *Int. J. Antimicrob. Agents* **2009**, *33*, 154–159.

Chapter 7

Liquid-Phase Microextraction (LPME) Utilizing Porous Hollow Fibers

Stig Pedersen-Bjergaard, Knut Einar Rasmussen, and Jan Åke Jönsson

7.1. OVERVIEW

In liquid-phase microextraction (LPME) utilizing porous hollow fibers, target analytes are extracted from aqueous samples, through a thin supported liquid membrane (SLM) sustained in the pores in the wall of a porous hollow fiber, and further into an acceptor solution placed inside the lumen of the hollow fiber. After extraction, the acceptor solution is directly subjected to a final chemical analysis. The acceptor solution can be an organic solvent providing a two-phase extraction system, and, in this case, the acceptor solution is directly compatible with capillary gas chromatography (GC). Alternatively, the acceptor solution can be aqueous providing a three-phase extraction system, where the acceptor solution is compatible with high-performance liquid chromatography (HPLC) or capillary electrophoresis (CE). The volume of sample in LPME may be highly flexible, from 50 µL to more than 1 L, whereas the volume of acceptor solution in most cases is in the range of 2–30 µL. Because of this high sample-to-acceptor volume ratio, very high analyte enrichments may be obtained by LPME. The volume of the SLM is normally only 5–30 µL, resulting in an extremely low consumption of organic solvent per extraction. LPME can therefore be utilized to reduce the consumption of hazardous organic solvents in the chemical laboratory. Also, LPME is a very efficient technique for sample cleanup, reducing or eliminating potential problems from matrix components. This is especially valuable for the analysis of complicated biological and environmental samples. Finally, LPME can easily be automated, providing a fully automated approach to sample preparation. In this chapter, focus will be directed to the historical development of LPME and to fundamental theory, equipment, solvents, method development, selected applications, performance, and future trends.

7.2. HISTORICAL DEVELOPMENT

The invention of solid-phase microextraction (SPME) by Arthur and Pawliszyn in 1990[1] initiated a considerable interest for analytical microextraction techniques, which has increased substantially during recent years. In SPME, target analytes of low or medium polarity are extracted from aqueous or gaseous samples and onto a solid polymeric fiber attached to a thin needle on a syringe. Extraction occurs by passive diffusion, and the extraction yield is essentially determined by the fiber to sample distribution coefficient. After extraction, the analytes are thermally desorbed in a gas chromatographic injection port and finally analyzed by GC. Alternatively, the target analytes can be desorbed from the SPME fiber by elution with an organic solvent, which is subsequently injected into HPLC. SPME rapidly gained widespread interest, especially because it eliminated the use of organic solvents for sample preparation.

In parallel to the development of SPME, research was also conducted to miniaturize liquid–liquid extractions (LLEs) into LPME. Again, a major philosophy was to reduce the use of organic solvents required for sample preparation. In 1996, Liu and Dasgupta[2] and Jeannot and Cantwell[3] presented the first papers on LPME, and later, He and Lee[4] and de Jager and Andrews[5] were also involved in this early development. In general, the LPME systems at that time were based on the extraction of analytes from aqueous samples and into a small drop of an organic solvent. Typically, the small drop of organic solvent was suspended from the tip of a GC syringe and placed in the aqueous solution for extraction. During extraction, target analytes were extracted from the aqueous sample and into this hanging drop based on passive diffusion, and extraction recoveries were essentially determined by the organic solvent to water

Handbook of Sample Preparation, Edited by Janusz Pawliszyn and Heather L. Lord

distribution coefficients. Although hanging drop LPME is very simple and efficient, and reduces the consumption of organic solvents per sample to a few microliters, it has currently not gained widespread interest and general acceptance. The principal reason for this is the low stability of the hanging drop, which is easily lost into the sample during extraction.

As an attempt to improve the stability and reliability of LPME, Pedersen-Bjergaard and Rasmussen introduced hollow fiber-based LPME in 1999,[6] where the extracting phase was placed inside the lumen of a porous hollow fiber. In this system, the extracting phase (acceptor solution) was mechanically protected inside the hollow fiber, and dissolution into the sample was eliminated. In this hollow fiber LPME system, analytes are extracted from aqueous samples, into an organic solvent immobilized as an SLM in the pores in the wall of the hollow fiber, and into an acceptor solution placed inside the lumen of the hollow fiber. Subsequently, the acceptor solution is removed by a microsyringe and transferred to a chromatographic or electrophoretic analysis.

The idea of hollow fiber LPME was essentially a combination of the early hanging drop LPME work of Dasgupta, Cantwell, and Lee reported above and the pioneering work of Jönsson and coworkers[7–13] directed to the development of SLM extraction. In the latter, extraction was accomplished from aqueous samples, through an SLM sustained in a flat sheet of porous polymeric membrane or in a hollow fiber, and into an acceptor phase on the other side of the SLM. The chemical principle was identical to that of hollow fiber LPME. However, the technical configuration differed considerably from hollow fiber LPME. In the SLM extraction work, the sample was pumped with a peristaltic or syringe pump to the SLM, providing a flow system. On the other side of the membrane, the acceptor solution was pumped with another pump. In total, the SLM extractions required substantial instrumentation. The first publication on hollow fiber extraction by Pedersen-Bjergaard and Rasmussen in 1999[6] was pioneering in the way that it was the first paper to show the use of SLM extractions with a small piece of hollow fiber in a stagnant system without the use of pumps to deliver the sample and the acceptor phase. In the period since 1999, the interest for hollow fiber LPME has increased significantly, and several research groups have been involved in the early development of the technique, including Lee and coworkers[14] and de Jager and Andrews.[15] During recent years, several other research groups have been involved,[16–37] and more that 90 publications have been published in the area from 27 different research groups as illustrated in Figure 7.1. In addition to this, a related technique named membrane-assisted solvent extraction (MASE) has been developed in the same period of time.[38] MASE is based on the use of a very thin nonporous membrane, and analytes are extracted from aqueous samples, through the membrane which is solvent modified, and into an acceptor phase consisting of an organic solvent. This concept is dedicated to two-phase extractions and will not be covered in the present text. This chapter will only focus on LPME based on porous hollow fibers.

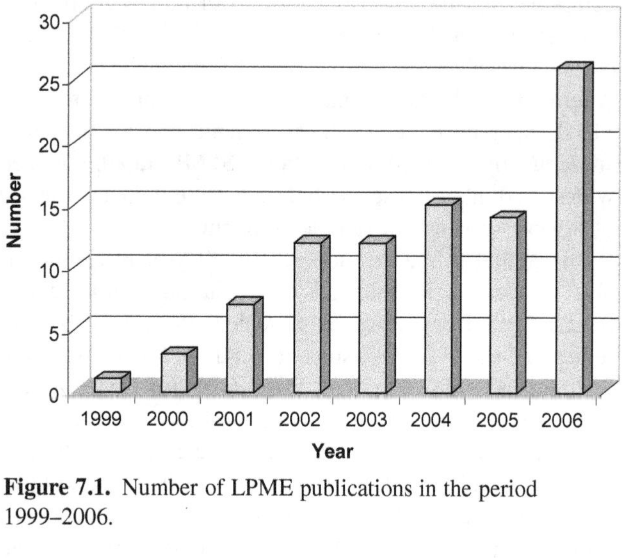

Figure 7.1. Number of LPME publications in the period 1999–2006.

7.3. FUNDAMENTAL THEORY

The basic principle of LPME is illustrated in Figure 7.2. LPME can be accomplished either in the two- or three-phase mode. Prior to extraction, the aqueous sample is filled into a sample vial, and an internal (surrogate) standard is added if required. In addition, for acidic or basic compounds, pH in the sample is adjusted in order to suppress the ionization of the target analytes. A short piece of a porous hollow fiber is used for LPME, and this may either be a rod configuration with a closed bottom or a U-configuration where both ends

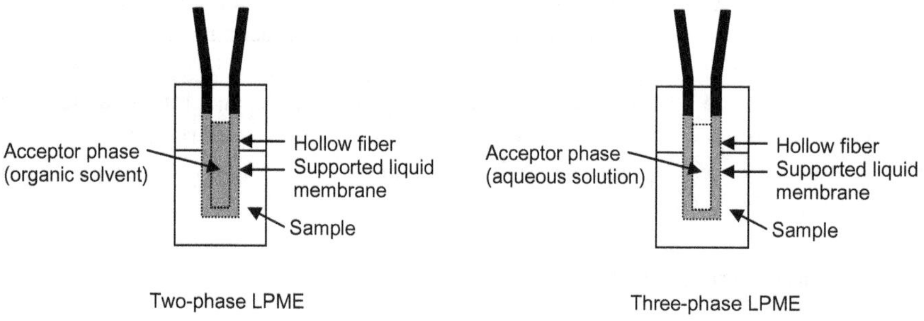

Figure 7.2. Principle of two- and three-phase LPME.

of the hollow fiber are connected to guiding tubes. Also, other configurations are used, for example, a loop made of hollow fiber or a short piece of hollow fiber closed at both ends (so-called solvent bar). First, a SLM is formed in the pores in the wall of the hollow fiber. This may be accomplished by dipping the hollow fiber into an organic solvent of low polarity (like *n*-octanol) for a few seconds. During this short period of time, the organic solvent is immobilized in the pores by capillary forces. Alternatively, a small quantity of the organic solvent may be injected into the lumen of the hollow fiber and immobilized from the inside of the hollow fiber. After the immobilization of the SLM, the acceptor solution is filled into the lumen of the hollow fiber. This acceptor solution may be an organic solvent (same as used for the SLM) resulting in a two-phase extraction system, or the acceptor solution may be aqueous providing a three-phase extraction system.

7.3.1. Two-Phase LPME Systems

In the two-phase LPME system, the target analytes are extracted from the aqueous sample and into the organic solvent present in the membrane pores and inside the lumen of the hollow fiber (acceptor phase). This process may be illustrated with the following equation:

$$A_{\text{sample}} \rightleftarrows A_{\text{acceptor}}, \tag{7.1}$$

where A represents the target analyte. The distribution coefficient for A is defined as

$$D_{\text{acceptor/sample}} = \frac{C_{\text{eq,acceptor}}}{C_{\text{eq,sample}}} = \alpha_{\text{D}} \cdot K_{\text{acceptor/sample}}, \tag{7.2}$$

where $C_{\text{eq,acceptor}}$ is the total concentration of A in the acceptor phase at equilibrium (organic phase) and $C_{\text{eq,sample}}$ is the total concentration of A in the sample at equilibrium (aqueous phase). $K_{\text{acceptor/sample}}$ is the corresponding partition coefficient between the phases, that is, the ratio of extractable species of the analyte in the two phases, and α_{D} signifies the extractable fraction of the total concentration of A in the sample. For example, if A is a base, the extractable fraction is the proportion of noncharged, nonprotonated species. Charged species, such as acid anions or protonated bases, are not extractable. Depending on the type of analytes, the extractable fraction can be calculated from relevant values for pH and pK_a according to Equation 7.3, which is valid for one-protonic acids or bases and which follows directly from the definition of pK_a:

$$\alpha = \frac{1}{1+10^{s(\text{pH}-\text{pKa})}}. \tag{7.3}$$

Here, s is 1 for acids and −1 for bases. In the case of amines, the pK_a value is that of the corresponding acid. In practice, experimental conditions in the sample are usually set so that all of the analyte is extractable (e.g., acid analytes are typically extracted from acidic solutions, and then α_{D} is close to 1). In the case of a two-phase extraction, only extractable species are present in the acceptor.

The extraction efficiency (E), sometimes called recovery, is defined as

$$E = \frac{n_{\text{extr}}}{n_{\text{total}}}, \tag{7.4}$$

where n_{extr} and n_{total} are the extracted and the total number of moles of A, respectively.

Based on Equation 7.2 and a mass balance of the two-phase LPME system, the extraction efficiency *at equilibrium* may be calculated by the following equation:[39]

$$E = \frac{D_{\text{acceptor/sample}} \cdot V_{\text{org}}}{D_{\text{acceptor/sample}} \cdot V_{\text{org}} + V_{\text{sample}}}, \tag{7.5}$$

where V_{org} is the total volume of organic phase in the system (sum of organic solvent present in the pores in the wall of the hollow fiber [V_{mem}] and in the lumen [V_{acceptor}] of the hollow fiber) and V_{sample} is the volume of the sample. In some cases, only the organic liquid in the lumen (the acceptor) is taken for analysis, leaving a portion of the extracted sample in the membrane pores. As the volume of the liquid in the pores can be of a similar magnitude as that of the lumen, it is in those cases necessary to modify Equation 7.5 as follows:

$$E = \frac{D_{\text{acceptor/sample}} \cdot V_{\text{acceptor}}}{D_{\text{acceptor/sample}} \cdot V_{\text{org}} + V_{\text{sample}}}. \tag{7.6}$$

From these equations, it may be predicted that the extraction efficiency (recovery) is dependent on the distribution coefficient and on the volumes of organic solvent and sample. High recoveries are obtained for compounds with high distribution coefficients, and small sample volumes are beneficial in order to obtain high recoveries for equilibrium extractions.

The analyte enrichment factor (Ee) in two-phase LPME can be calculated by the following equation at any time:[39]

$$Ee = \frac{C_{\text{acceptor,end}}}{C_{\text{initial}}} = \frac{V_{\text{sample}} \cdot E}{V_{\text{org}}}, \tag{7.7}$$

where $C_{\text{acceptor,end}}$ is the concentration of A in the acceptor phase at the end of the extraction ($C_{\text{acceptor,end}} = C_{\text{eq,acceptor}}$ for equilibrium extractions) and C_{initial} is the concentration of A in the sample at the start of the extraction. If E is calculated only on the acceptor volume as in Equation 7.6, the volume V_{org} in the denominator of Equation 7.7 must be exchanged for V_{acceptor}.

The extraction kinetics for a two-phase LPME system may be described by the following equation:[3]

$$C_{\text{acceptor}} = C_{\text{eq,acceptor}}\left(1 - e^{-kt}\right), \tag{7.8}$$

where k is the rate constant (s^{-1}) defined as

$$k = \frac{A_{\text{i}}}{V_{\text{org}}}\beta_0\left(D_{\text{acceptor/sample}}\frac{V_{\text{org}}}{V_{\text{sample}}} + 1\right). \tag{7.9}$$

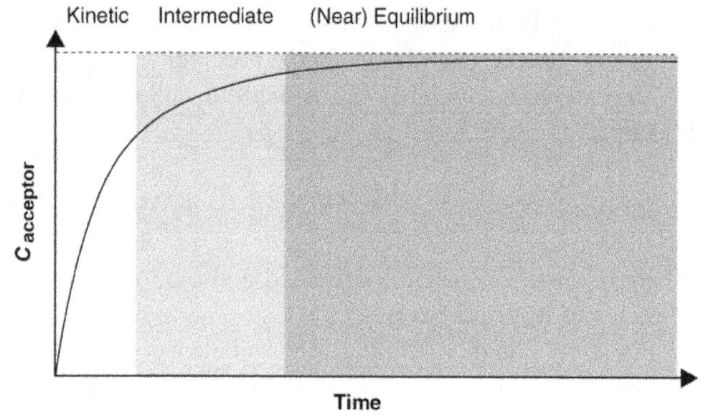

Figure 7.3. Generalized extraction profile.

$C_{acceptor}$ is the concentration of A in the acceptor phase (organic phase) at time t, A_i is the interfacial area, and β_0 is the overall mass transfer coefficient with respect to the organic phase (in centimeter per second). Equation 7.9 reveals that for rapid extractions, A_i and β_0 should be maximized and V_{sample} should be minimized. The effects of V_{org} and $D_{acceptor/sample}$ on equilibration time depend on the magnitude of the factor $D_{acceptor/sample}$ (V_{org}/V_{sample}), relative to 1. In Figure 7.3, a schematic graph of the extraction kinetics according to Equation 7.8 is shown.[40] It is seen that the acceptor concentration initially rises essentially linearly, and after enough time, it reaches a constant equilibrium level. Between these two regions, which are both utilized in practice, the acceptor concentration rises in a way that is less reproducible and therefore not practically useful.

7.3.2. Three-Phase LPME Systems

In three-phase LPME, the analytes are extracted from the aqueous sample, through the organic solvent immobilized in the pores of the hollow fiber, and further into the aqueous acceptor solution present inside the lumen of the hollow fiber. This process may be illustrated by the following equation:

$$A_{sample} \underset{\leftarrow k_2}{\overset{\rightarrow k_1}{}} A_{org} \underset{\leftarrow k_4}{\overset{\rightarrow k_3}{}} A_{acceptor}, \tag{7.10}$$

where k_1, k_2, k_3, and k_4 are the first-order extraction rate constants. To establish an equation for calculation of equilibrium recovery, distribution coefficients both between the organic phase and the sample as well as between the acceptor phase and the organic phase have to be considered:

$$D_{org/sample} = \frac{C_{eq,org}}{C_{eq,sample}} = \alpha_D \cdot K_{org/sample} \tag{7.11}$$

$$D_{org/acceptor} = \frac{C_{eq,org}}{C_{eq,acceptor}} = \alpha_A \cdot K_{org/acceptor}. \tag{7.12}$$

Here, α_D signifies the extractable fraction of the total concentration of A in the sample, and α_A that in the acceptor. Both of them can be calculated by Equation 7.3. It is further possible to define a combined distribution coefficient between the two aqueous phases by combining the Equations 7.11 and 7.12:

$$D_{acceptor/sample} = \frac{C_{eq,acceptor}}{C_{eq,sample}} = \frac{D_{org/sample}}{D_{org/acceptor}} = \frac{\alpha_D \cdot K_{org/sample}}{\alpha_A \cdot K_{org/acceptor}}. \tag{7.13}$$

It can be noted that in situations where the ion strength and other conditions (other than pH) are similar in the sample and in the acceptor, then $K_{org/sample} \approx K_{org/acceptor}$ and the distribution coefficient between the two aqueous phases takes a very simple form:

$$D_{acceptor/sample} = \frac{C_{eq,acceptor}}{C_{eq,sample}} \approx \frac{\alpha_D}{\alpha_A}. \tag{7.14}$$

This shows that in three-phase SLM extraction, the partition coefficients to the organic phase has little or no significance for the position of the equilibrium, which is instead mainly determined by the ionization situations in the aqueous phases.

Based on Equations 7.11 and 7.12 and considering the total mass balance for the three-phase LPME system, the following equation may be derived for calculation of extraction efficiency (E) in three-phase LPME *at equilibrium*:[39]

$$E = \frac{D_{acceptor/sample} V_{acceptor}}{D_{acceptor/sample} V_{acceptor} + D_{org/sample} V_{mem} + V_{sample}}, \tag{7.15}$$

where $V_{acceptor}$ is the volume of aqueous acceptor phase in the lumen and V_{mem} is the volume of organic phase immobilized in the pores of the hollow fiber. From Equation 7.15, it may be concluded that extraction efficiencies in three-phase LPME are controlled by the two individual distribution coefficients, and by the volumes of the sample, the organic phase and the acceptor phase, respectively. In general, high partition coefficients are beneficial, and this may be obtained by proper selection of organic solvent and pH conditions in the aqueous solutions. Also, for high efficiencies, the volumes of sample and organic phase should be as small as possible. The analyte enrichment factor (Ee) in three-phase LPME can be calculated by the following equation at any time:

$$Ee = \frac{C_{acceptor,end}}{C_{initial}} = \frac{V_{sample} E}{V_{acceptor}}, \tag{7.16}$$

where $C_{acceptor,end}$ is the concentration of A in the acceptor phase at the end of the extraction ($C_{acceptor,end} = C_{eq,acceptor}$ for equilibrium extractions) and $C_{initial}$ is the concentration of A in the sample at the start of the extraction.

For three-phase LPME, equations describing the kinetics are more complicated, but may be summarized to[41]

$$C_{acceptor} = C_{initial} \frac{V_{sample}}{V_{acceptor}} \left\{ \frac{k_1 k_3}{\lambda_2 \lambda_3} + \frac{k_1 k_3}{\lambda_2 (\lambda_2 - \lambda_3)} e^{-\lambda_2 t} + \frac{k_1 k_3}{\lambda_3 (\lambda_2 - \lambda_3)} e^{-\lambda_3 t} \right\}, \tag{7.17}$$

where

$$\lambda_2 = \frac{1}{2}(p+q), \tag{7.18}$$

$$\lambda_3 = \frac{1}{2}(p-q), \tag{7.19}$$

$$p = (k_1 + k_2 + k_3 + k_4), \tag{7.20}$$

$$q = [p_2 - 4(k_1k_3 + k_2k_4 + k_1k_4)]^{1/2}. \tag{7.21}$$

Equation 7.17 predicts that the extraction speed in three-phase LPME is determined by the individual rate constants and by the volume ratio of sample to acceptor phase. The rate constants, like in Equation 7.9 for two-phase LPME, are controlled by the interfacial area, by the overall mass transfer coefficient, and by the partition coefficient. Also, in this case, Figure 7.3 is generally applicable.

A closer examination of Equation 7.3 may give important information about LPME as compared with traditional LLE. In Table 7.1, this equation has been utilized to calculate recoveries in a typical two-phase LPME system as function of the distribution coefficient. In this case, the sample volume was 2 mL, the acceptor phase volume was 5 μL, and the volume of organic solvent immobilized in the pores of the hollow fiber was 5 μL. As seen from the table, recoveries were low in the case of low distribution coefficients, and in order to reach a 50% value for the recovery, the distribution coefficient should exceed 200. However, if enrichment factors were calculated, it is observed that LPME provides excellent enrichment values of compounds with high distribution coefficient. In Table 7.1, similar calculations were done for traditional LLE. In this case, the sample volume was 2 mL, the acceptor phase volume was 2 mL, and after extraction, the acceptor phase was evaporated and reconstituted in 200 μL. As seen from the results, LLE provided high recoveries even for analytes with low distribution coefficients due to the large volume of the extracting solvent. On the other side, no high enrichments were obtained in the case of LLE. In conclusion, LPME is superior to LLE for substances of low polarity, providing high enrichments and excellent discrimination of polar substances. The latter may explain the strong cleanup properties of LPME. On the other hand, LPME is relatively inefficient for the most polar substances, and for those, more efficient extractions may be accomplished by traditional LLE. This comparison is mainly valid for two-phase LPME.

Table 7.1. Calculated Recoveries in LPME and Traditional Liquid–Liquid Extraction (LLE) as Function of the Distribution Coefficient ($D_{acceptor/sample}$)

	LPME[a]		LLE[b]	
$D_{acceptor/sample}$	Recovery (%)	Enrichment	Recovery (%)	Enrichment
1	0.5	1.0	50	5.0
10	4.8	9.6	90.9	9.1
100	33.3	66.6	99.0	9.9
1,000	83.3	166.6	99.9	10.0
10,000	98.0	196.0	100.0	10.0
100,000	98.0	196.0	100.0	10.0

[a] V_{sample} = 2 mL, V_{org} = 10 μL (5 μL in pores and 5 μL in lumen).

[b] V_{sample} = 2 mL, V_{org} = 2 mL, which is reconstituted to 200 μL following evaporation.

7.3.3. Extraction End Points

Three different end points are commonly used for LPME extractions:

- exhaustive extraction,
- kinetic extraction, and
- equilibrium, nondepletive extraction.

For exhaustive extraction, the extraction is performed until the amount of analyte in the sample is exhausted. The conditions for this are that the sample volume is limited and that the distribution coefficient $D_{acceptor/sample}$ is large, for both two-phase and three-phase LPME. In the case of three-phase LPME, the latter condition is achieved by assuring that the pH conditions are such that $\alpha_A \approx 0$ and $\alpha_D \approx 1$ (c.f. Eq. 7.3). For the case of extraction of acidic compounds, this means an extraction from a sufficiently acidic sample to a sufficiently basic acceptor. These conditions are called "infinite sink" conditions. If the analyte takes part in secondary equilibria, if it is reversibly bound to other components in the sample (drug–protein binding is an important example), all such equilibria will be shifted during the extraction, so the total concentration of the analyte is extracted and taken to analysis. It is easily seen from Equations 7.14 to 7.16 that the above conditions lead to $D_{acceptor/sample} \rightarrow \infty$; $E \rightarrow 1$ and $Ee \rightarrow V_{sample}/V_{acceptor}$, that is, practically all of the analyte is extracted, and the enrichment factor approaches the volume ratio. The uncertainty of the enrichment factor is determined by the volume ratio, so factors like pH and extraction time will not appreciably influence the result, as long as the conditions for exhaustive extraction are met. In the case of two-phase LPME, the outcome is analogous, with the observation that if only the acceptor liquid is recovered for analysis (and not the organic content of the fiber pores) as was discussed above, the observed E and Ee are decreased correspondingly.

For kinetic extraction, the extraction is carried out in the linear range (c.f. Fig. 7.3) and stopped far before equilibrium, so $E \ll 100\%$. With high distribution coefficients, it is also in this case possible to obtain high enrichment factors in a reasonable time, but it is essential that the extraction conditions are kept constant in terms of pH, extraction time, and so on, in order to ensure reproducibility. Secondary equilibria can influence the results.

For nondepletive equilibrium extraction, the sample volume is largely compared with the acceptor volume, and with a finite value of $D_{acceptor/sample}$, equilibrium can be attained without appreciably depleting the sample. In this case, $E \approx 0$ and $Ee \approx D_{acceptor/sample}$. Thus, all secondary equilibria are undisturbed, and only the extractable, freely dissolved concentration of the analyte is sensed. The actual value of $D_{acceptor/sample}$ can either be calibrated or calculated from the relevant pK_a and pH values. This principle, related to the so-called equilibrium sampling devices,[40] has been called the equilibrium sampling through membrane (ESTM), and it has been used for both the environmental speciation and to study drug–protein binding in pharmacological applications.[42–44]

7.4. EQUIPMENT AND SOLVENTS

The equipment required for LPME is very simple, and for manual extractions, it consists of the following:

- hollow fiber,
- microsyringe,
- sample vial, and
- agitator.

The hollow fiber is of key importance since it works as support for the SLM and as container for the acceptor phase. Most LPME work in the literature has been accomplished with polypropylene hollow fibers. Polypropylene appears to be an excellent material for the purpose. It is resistant to organic solvents, it provides the required flexibility and mechanical strength, the organic phases used in LPME are strongly immobilized in pores of polypropylene, and no serious adsorptive problems are known between target analytes and the polymer itself. The most popular has been a 600-μm internal diameter (ID) hollow fiber with 200-μm walls, 0.2-μm pores, and 75% porosity. With this ID, narrow microsyringe needles may be inserted into the lumen, to inject and to remove the acceptor phase. The wall thickness of 200 μm is appropriate and is a compromise between mechanical stability and extraction performance. A thinner wall is advantageous for the extraction performance since this provides a very thin SLM and therefore faster mass transfer, but this sacrifices the mechanical stability. The pore size of 0.2 μm also appears to be appropriate; the organic solvents used as SLM are strongly immobilized in this type of porous structure, and the relatively open structures ensure efficient diffusion of target analytes.

A few other hollow fibers have been tested. In several reports, polypropylene hollow fibers with IDs in the range 1.0- to 1.2-mm ID have been used.[25,34,45] These fibers are not very different from the most common ones in terms of performance, but the larger ID simplifies the handling of the acceptor solution, and this type of hollow fiber has been used in combination with automation for the same reason.[29] Polypropylene hollow fibers with narrower ID (280–400 μm) have also been used.[24,46,47] Due to the larger contact area and thinner walls, these hollow fibers have provided improved extraction kinetics as compared with the most popular one. However, they are less convenient in practical terms because of the small dimensions. Also, a polypropylene hollow fiber with a reported pore size of 0.64 μm has been used,[14] but the effect of the increased pore size is unclear. On the other hand, it was observed that fibers with very small pore size (0.03 μm) can be unsuitable for the extraction of drugs from blood plasma, as the protein molecules tend to block the pores.[9] In addition to polypropylene hollow fibers, hollow fibers also made of polyvinyldene difluoride[15] and polysulfone[36] have been used for LPME, but in both cases, the results were inferior relative to polypropylene. Most probably, this was due to lower porosity or smaller pore size rather than the nature of the polymeric material. In conclusion, the most popular polypropylene hollow fiber appears to be close to optimum.

Different configurations of the hollow fiber have been reported in the literature. In the first LPME publication,[6] an 8-cm piece of hollow fiber was inserted in the sample in a U-shaped configuration as shown in Figure 7.4a. Each end of the hollow fiber was connected to a guiding needle. One needle was used for the introduction of the acceptor phase, and the second one was used for the removal of the acceptor phase. The major advantage of the U-shaped configuration is the large contact area between the sample and the hollow fiber, but this concept is probably less suited for the automation of LPME. For automation, a rodlike configuration with a 1.0- to 1.2-mm ID hollow fiber is more appropriate as illustrated in Figure 7.4b. In this case, a syringe in a robotic system may be inserted directly to the bottom of the hollow fiber to inject and remove the acceptor phase.[29] In the rodlike configuration, LPME may also be developed into the 96-well format as illustrated in Figure 7.4c. This enables simultaneous extraction of a large number of samples in parallel.

Several modifications of the U-shaped configuration have been reported. In one configuration, a 27-cm piece of a 330-μm polypropylene hollow fiber was coiled around a paper clip as illustrated in Figure 7.4d. The purpose of this was to increase the contact area between the hollow fiber and the sample, providing enhanced extraction kinetics.[46] In another configuration, the U-shaped fiber was connected to a funnel-shaped injection guide at the inlet end and to a small dent at the outlet end.[17] With this system, the microsyringe needle was guided into the hollow fiber, and the open outlet ensured that the acceptor phase was not pushed out by the intruding syringe needle. Also, with this configuration, no problems occurred with air bubbles when filling the autosampler syringe.

Another interesting hollow fiber configuration is the attachment of a small piece of a hollow fiber to a microsyringe needle as illustrated in Figure 7.4e.[48] In this configuration, the syringe served both to introduce the acceptor phase to support the hollow fiber, and as the injection syringe for the subsequent chromatographic analysis. The hollow fiber can be closed in the bottom to enable three-phase extractions,[48] or open at the bottom for two-phase extractions[49] and for three-phase extractions.[50] This configuration is very simple and well suited for manual extractions. To summarize, both the U-shaped and the rodlike configurations

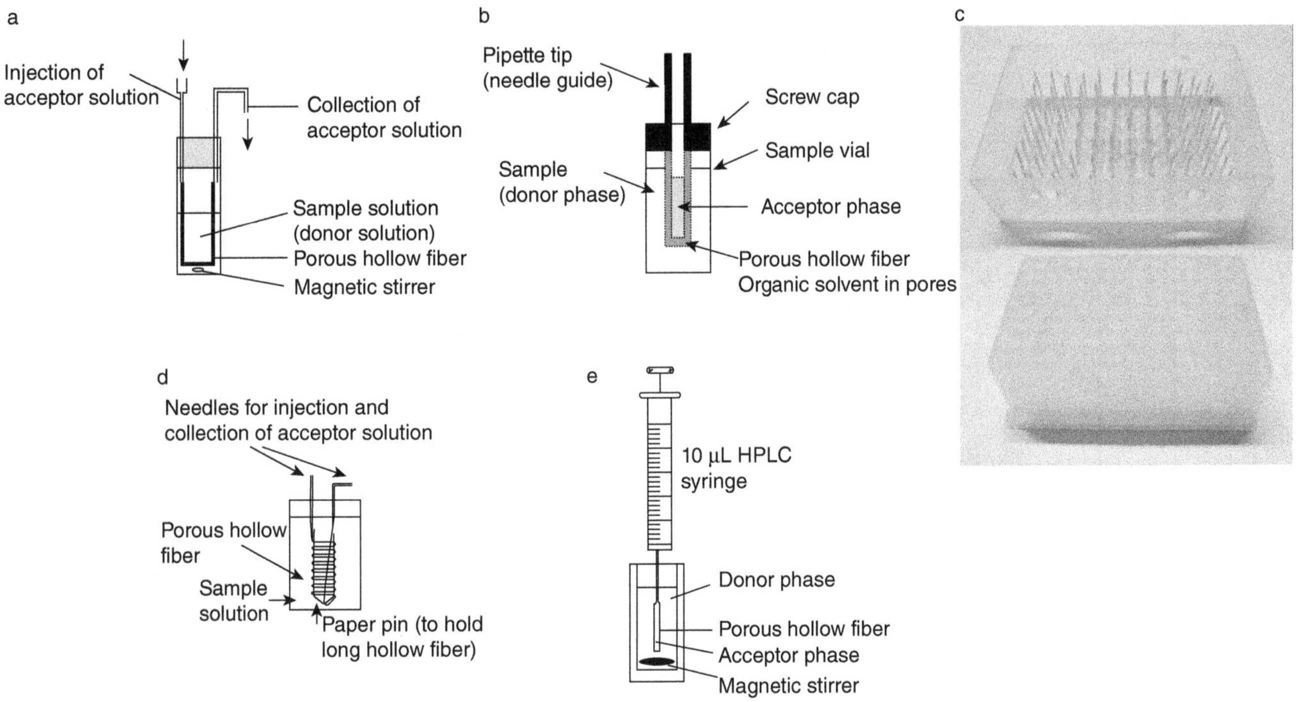

Figure 7.4. Different LPME configurations: (a) U-shaped configuration (reprinted from reference 6 with permission from the American Chemical Society. Copyright 1999); (b) rodlike configuration (reprinted from reference 45 with permission from Elsevier. Copyright 2003); (c) 96-well configuration (image courtesy of Varian, Lake Forest, CA); (d) long fiber configuration (reprinted from reference 46 with permission from Elsevier. Copyright 2001); (e) fiber connected to syringe (reprinted from reference 48 with permission from Elsevier. Copyright 2001).

are convenient for LPME providing similar results. The type of configuration used will normally be determined by the physical dimensions of the fiber and of the sample vial, and if extractions are to be performed manually or automatically.

In addition to a hollow fiber, a microsyringe is required in order to perform LPME. The purpose of the syringe is to inject the acceptor phase into the lumen of the hollow fiber and, in some cases, also to inject the organic solvent used as SLM from the inside of the hollow fiber. In addition, the syringe is used to collect the acceptor phase after extraction. The microsyringe may be one of a robotic system for automated LPME[17,29] or a conventional microsyringe for manual LPME. A broad range of different syringes have been used, but in general, the needle should be blunt in order not to perforate the hollow fiber. Furthermore, extraction vials and some type of agitator should be available. A broad range of sample vials have been used, and no special recommendations apply to LPME in that respect. Agitation on the other hand is very important to promote and to speed up LPME, and both vibration and magnetic stirring may be used.[6,51] In general, vibration is recommended as magnetic stir bars may contaminate or alter the composition of the sample. Vibration or stirring at rates in the range of 800–1200 rpm is typically used.

Organic solvents are used in small quantities in LPME as the SLM in the pores of the hollow fiber, and in the case of two-phase LPME, the acceptor phase is also an organic solvent. For the SLM, the organic solvent should first of all provide high partition coefficients for the target analytes to ensure efficient extraction. Furthermore, the solvents should have a relatively low polarity to avoid leakage to the aqueous sample and to ensure an efficient immobilization in the pores of the polypropylene material. Finally, the solvent should have low volatility to avoid evaporation during LPME, and the viscosity should be low to ensure high diffusion coefficients across the SLM. As seen from Table 7.2, 1-octanol, dihexyl ether, and toluene have been the most popular LPME solvents used. These solvents fulfill the criteria and are easily available in high purity at an acceptable price. However, they are not stable in the hollow fibers for more than a few hours and should be used immediately after immobilization. Several alternatives have also been reported. In some cases, mixtures of different solvents have been used to optimize recovery, like combinations of benzyl alcohol and ethyl acetate[52] or 1-nonanol and dihexyl ether.[27] Also, different additives have been used to adjust the polarity of the organic solvent, or to facilitate ion pairing within the SLM. An example of the former has been the addition of trioctylphosphine oxide (TOPO) to dihexyl ether for the extraction of pesticides.[35] Ion pairing within the SLM was reported with dihexyl ether as the organic phase where Aliquat 336 (methyltrioctylammonium chloride) was added to the SLM to facilitate ion pairing with the basic drugs salbutamol and terbutaline.[33] Ion pairing may also take place in the aqueous sample solution, where different ion-pair reagents have been studied for efficient extraction of polar basic drugs.[53]

Table 7.2. Overview of LPME Applications

Analytes	Two-Phase/ Three-Phase	Sample	SLM	Acceptor Phase	Analytical Technique	Reference
Drug analysis/pharmaceuticals						
Amino alcohols	Three-phase	Urine	1-Octanol	100 mM HCl	CE	89
Amphetamines	Three-phase	Blood, urine	Dihexyl ether	10 mM HCl	FIA-MS	81
Anabolic steroids	Two-phase	Urine	1-Octanol	1-Octanol	LC-MS	31
Antidepressants	Three-phase	Human breast milk	Polyphenyl-methylsiloxane	10 mM HCl	CE	45
Antidepressants	Three-phase	Blood, plasma	Dodecyl acetate	200 mM HCOOH	LC-MS	61
Anti-inflammatory drugs	Three-phase	Urine	Dihexyl ether	10 mM NaOH	CE	51
Basic drugs	Two-phase	Plasma	1-octanol	1-Octanol	GC	90
	Three-phase	Plasma, urine	1-octanol	10 mM phosphate (pH 3)	CE, HPLC	
Basic drugs	Three-phase	Blood, plasma, urine	Dihexyl ether	100 mM HCl	CE	46
Basic drugs	Three-phase	Plasma	Dihexyl ether	10 mM HCl	CE	63
Basic drugs	Three-phase	Plasma, urine	1-Octanol	50 mM HCl	CE	74
Basic drugs	Three-phase	Water, plasma	Dihexyl ether	10 mM HCl	CE	58
Basic drugs	Three-phase	Hair	1-Octanol	10 mM HCl	HPLC	25
Benzodiazepines	Two-phase	Plasma, urine	Dihexyl ether, 1-octanol, butyl acetate	Dihexyl ether, 1-octanol, butyl acetate	GC	77
Benzodiazepines	Three-phase	Blood	Nonanol	1 M HCl	HPLC	76
Cannabinol	Two-phase	Urine	BSTFA + 1-octane	BSTFA + 1-octane	GC-MS	56
Chiral drugs	Three-phase	Plasma	Dihexyl ether	10 mM HCl	CE	79
Citalopram	Three-phase	Plasma	Dihexyl ether	20 mM phosphate (pH 2.8)	CE	91
Citalopram	Three-phase	Plasma	Dodecyl acetate	20 mM phosphate (pH 2.8)	CE	80
Cocaine	Two-phase	Urine	Chloroform	Chloroform	GC	15
Cocaine	Two-phase	Saliva	Chloroform	Chloroform	GC	78
Dopamine	Two-phase	Urine	1-Octanol	1-Octanol	MALDI-MS	92
Doping agents	Three-phase	Urine	1-Octanol	50 mM NH_3	LC-MS	93
Drugs (various)	Three-phase (equil)	Plasma	Dihexyl ether + TOPO	Phosphate buffer	HPLC	43
Fatty acids	Three-phase	Serum	Dihexyl ether + TOPO	0.1 M NaOH	GC	94
Methamphetamine	Three-phase	Plasma, urine	1-Octanol	100 mM HCl	CE	6

(Continued)

Table 7.2. *(Continued)*

Analytes	Two-Phase/ Three-Phase	Sample	SLM	Acceptor Phase	Analytical Technique	Reference
Mirtazapine	Two-phase	Plasma	Toluene	Toluene	HPLC	26
PhIP	Three-phase	Plasma, urine	1-Octanol	0.1 M H_2SO_4	HPLC	95
Polar drugs	Three-phase	Plasma	1-Octanol	50 mM HCl	LC-MS	75
Polar drugs	Three-phase	Plasma	1-Octanol	50 mM HCl	CE	53
Salbutamol, terbutaline	Three-phase	Tablets, urine	Dihexyl ether + Aliquat 336	1 M NaBr	LC-MS	33
Triphenylphosphine oxide	Two-phase	Pharmaceuticals	1-Octanol	1-Octanol	HPLC	28
Environmental						
Acidic drugs	Three-phase	Wastewater	1-Octanol	10 mM $(NH_4)_2CO_3$	LC-MS	19
Acidic herbicides	Two-phase	Water	1-Octanol	1-Octanol	GC-MS	72
Anti-inflammatory drugs	Three-phase	Wastewater	1-Octanol	10 mM NaOH	HPLC	73
Anti-inflammatory drugs	Three-phase	Water	1-Octanol	100 mM NaOH	HPLC	96
Aromatic amines	Three-phase	Tap, surface water	Dihexyl ether	0.5 M HCl + 18-crown-6 ether	HPLC	57
Aromatic amines	Three-phase	Water	1-Octanol	500 mM HCl	CE	50
Aromatic amines	Three-phase	Water	Benzyl alcohol + ethyl acetate	10 mM HCl	HPLC	52
BTEX	Two-phase	Water	1-Octanol	1-Octanol	GC-MS	97
Carbamate pesticides	Two-phase	Water	1-Octanol	1-Octanol	GC-MS	98
Chlorophenols	Three-phase	Water	Ionic liquids	100 mM NaOH	HPLC	37
Chlorophenols	Three-phase (equil)	Water, leachate	Undecane	Carbonate buffers	HPLC	42
Copper	Three-phase (equil)	Water	Dihexyl ether + crown ether	PAR reagent	Spectrophotometry	44
Dichlorphenol	Two-phase	Water, urine	1-Octanol	1-Octanol	GC-MS	99
Dinitrophenols	Three-phase	Water	Undecane	600 mM $NaHCO_3$	HPLC	47
Dinitrophenols	Three-phase	Water, leachate	Undecane + toluene	600 mM $NaHCO_3$	HPLC	100
Dinitrophenols	Two-phase	Water	Dihexyl ether + TOPO	Dihexyl ether + TOPO	HPLC	101
Fatty acids	Two-phase	Water	1-Octanol	1-Octanol	GC-MS	71
Fluoxetine, norfluoxetine	Three-phase	Sewage	Dihexyl ether	Phosphate buffer	HPLC	102
Fungicides	Two-phase	Farm water	Toluene	Toluene	GC	21
Haloacetic acids	Three-phase	Drinking water	Dihexyl ether + TOPO	10 mM NaOH	HPLC	24
Hydroxyaromatic compounds	Three-phase	River water	Nonanol + dihexyl ether	1 M NaOH	HPLC	27

(Continued)

Table 7.2. (*Continued*)

Analytes	Two-Phase/ Three-Phase	Sample	SLM	Acceptor Phase	Analytical Technique	Reference
4-Isobutylacetophenone	Two-phase	River water, sewage	2-Heptanone	–	GC-MS	103
Inorganic selenium	Two-phase	Water	Tetrachloromethane	Tetrachloromethane	ETV-ICP-MS	30
Insecticides	Two-phase	Water	Toluene	Toluene	GC	104
Nitroaromatics	Two-phase	Water	Toluene	Toluene	GC-MS	105
Nitrophenols	Three-phase	Seawater	1-Octanol	10 mM NaOH	HPLC	48
Organic pollutants	Two-phase	Rainwater	Toluene	Toluene	GC-MS	67
Organic pollutants	Two-phase	Marine sediments	Toluene	Toluene	GC-MS	69
Organochlorine pesticides	Two-phase	Seawater	Toluene	Toluene	GC-MS	66
Organochlorine pesticides	Two-phase	Water	Toluene	Toluene	GC-MS	18
Organophosphorus pesticides	Two-phase	Lake water	Cyclohexane	Cyclohexane	GC-MS	106
Pesticides	Two-phase	Pond water	Toluene	Toluene	GC-MS	59
Pesticides	Two-phase	Soil	Toluene	Toluene	GC-MS	107
Phenols	Three-phase	Tap water	1-Octanol	8 mM NaOH	CE	14
Phenols	Two-phase	Seawater	Toluene	Toluene	GC-MS	108
Phenols	Three-phase	Water	1-Octanol	100 mM NaOH	HPLC	109
Phenols	Gaseous diffusion	Water	Air	500 mM NaOH	HPLC	70
Phenoxyacetic acid	Three-phase	River water	1-Octanol	500 mM NaOH	HPLC	110
Phenoxy acid herbicides	Three-phase	Water	1-Octanol	10 mM NaOH	HPLC	111
Phenoxy acid herbicides	Three-phase	Industrial wastewater	Dihexyl ether + TOPO	0.2 M NaOH	HPLC	112
Phthalate esters	Two-phase	Potable water	Toluene	Toluene	GC-MS	16
PAH metabolites	Two-phase	Water	1-Octanol	1-Octanol	Fluorescence, CE	23
Polyaromatic hydrocarbons	Two-phase	Water	1-Octanol	1-Octanol	GC-MS	49
Polyaromatic hydrocarbons	Two-phase	Soil slurry	1-Octane	1-Octane	GC	68
Polyaromatic hydrocarbons	Headspace	Water	1-Octanol	1-Octanol	GC-MS	60

(*Continued*)

Table 7.2. (*Continued*)

Analytes	Two-Phase/ Three-Phase	Sample	SLM	Acceptor Phase	Analytical Technique	Reference
Polyaromatic hydrocarbons	Two-phase	Wastewater	Toluene	Toluene	GC-MS	113
Polyaromatic hydrocarbons	Three-phase	Water	*n*-Decane	*n*-Decane	HPLC	114
Polybrominated biphenyls	Two-phase	Water	*n*-Undecane	*n*-Undecane	GC-MS	115
Polychlorinated biphenyls	Two-phase	Plasma	Toluene	Toluene	GC-MS	116
Primary amines	Two-phase	River water	Toluene	Toluene	GC-MS	117
Salbutamol, terbutaline	Three-phase	Water	Dihexyl ether + Aliquat 336	1 M NaBr	LC-MS	33
Triazine herbicides	Two-phase	Soil slurry	Toluene	Toluene	GC-MS	64
Trihalomethanes	Two-phase	Water	1-Octanol	1-Octanol	GC	36
Vinclozolin	Two-phase	Natural water	Toluene	Toluene	GC	20
Warfare agents	Two-phase	Water	Chloroform	Chloroform	GC-MS	118
Warfare agents	Two-phase	Water	Trichloroethylene	Trichloroethylene	GC-MS	32
Food and beverages						
Carbaryl	Two-phase	Red wine	1-Octanol	1-Octanol	GC-MS	29
Ochratoxin A	Two-phase	Wine	1-Octanol	1-Octanol	HPLC	22
Organochlorine pesticides	Two-phase	Tea	1-Octanol	1-Octanol	GC	119
Pesticides	Three-phase	Vegetables	Dihexyl ether + TOPO	100 mM HCl	LC-MS	35
Phenoxy herbicides	Three-phase	Bovine milk	1-Octanol	100 mM NaOH	HPLC	120
Peptides						
Peptides	Three-phase	Water	1-Octanol	HCl	HPLC	83
Peptides	Three-phase	Water	1-Octanol	100 mM HCl	HPLC	82
Miscellaneous						
Octanol–water partition	Two-phase	Water	1-Octanol	1-Octanol	HPLC	34

BTEX, benzene, toluene, ethylbenzene, and xylene; BSTFA, N,O-bis-(trimethylsilyl)trifluoroacetamide; MALDI, matrix-assisted laser desorption ionization; PAR, 4-(pyridyl-2-azo)resorcinol; TOPO, trioctylphosphine oxide; PhIP, 2-amino-1-methyl-6-phenylimidazo[4,5-b]pyridine.

Recently, the ionic liquid 1-octyl-3-methylimidazolium hexafluorophosphate was evaluated as organic phase in three-phase LPME.[37] This room temperature ionic liquid was easily immobilized in the pores of a porous polypropylene hollow fiber, and the polar character of this SLM provided very efficient extraction of chlorophenols. The use of room temperature ionic liquids, highly stable SLMs, and green solvents may be interesting future directions in LPME. Polyphenyl-methylsiloxane was used for the extraction of basic drugs from breast milk, and this solvent was found to be stable in the hollow fiber for several months.[45,54] Also, different plant oils (e.g., soybean oil) have been used as organic solvent for the SLM.[55] With the latter type of liquids, the use of hazardous organic solvents may be eliminated in the near future.

For two-phase LPME, an organic solvent serves as the acceptor phase. Normally, the same solvent is used as SLM and as acceptor phase. In addition to the SLM considerations, selection of acceptor phase solvent should include considerations for the GC characteristics. As illustrated in Table 7.2, most two-phase applications have utilized either 1-octanol or toluene as the acceptor phase. In addition to pure solvents, derivatization reagents may be added to the acceptor phase to simultaneously accomplish extraction and derivatization.[56]

For three-phase LPME, the acceptor solution is aqueous. For basic analytes, the acceptor phase should be acidic to efficiently ionize the target analytes and to trap them in this solution. As illustrated in Table 7.2, most reports have utilized diluted hydrochloric acid as the acceptor phase, with concentrations ranging from 10 mM to 1 M and with 10 mM as the most popular. In addition, acidic phosphate buffers or formic acid have been used. In a few cases, additions have been made to the acceptor phase to improve analyte solubility including crown ethers[57] and methanol.[51] For acidic analytes, an alkaline acceptor phase has to be used. As seen from Table 7.2, most work has involved the use of sodium hydroxide, with concentrations ranging from 8 mM to 1 M and with 10 mM as the most frequent one. In addition to NaOH, ammonia solutions and carbonate buffers have also been used.

7.5. METHOD DEVELOPMENT

Method development in LPME normally includes the following points. A comprehensive discussion of each point is provided below:

- selection of extraction mode,
- selection of sample volume,
- optimization of pH in the sample,
- optimization of agitation,
- optimization of extraction time,
- salt addition,
- optimization of organic phase,
- optimization of pH in the acceptor, and
- selection of acceptor volume.

7.5.1. Extraction Mode

A decision about the extraction mode should be taken first in the LPME method development. LPME can be performed in the two-phase mode, where the acceptor phase is organic, or in the three-phase mode, where the acceptor phase is an aqueous solution. In addition, LPME can be performed in a static mode, where both the sample and the acceptor are stagnant, or in a dynamic version where the acceptor solution is moving by movements of the plunger of a microsyringe, and also this point has to be considered during method development. Finally, the selection of extraction mode also includes a decision whether to immerse the SLM/hollow fiber in the sample solution (immersed LPME) or to extract from the headspace above the sample solution (headspace LPME).

Normally, two-phase LPME is performed in combination with GC. Thus, for the extraction of target analytes with relatively low polarity from aqueous samples and subsequent analysis by GC, two-phase LPME should be selected. The target analytes can be both uncharged compounds or acidic/basic components. The major benefit of this is that the acceptor phase can be injected directly into the GC. On the other hand, if HPLC or CE is the final analytical method, three-phase LPME should normally be selected. Here, the acceptor phase is aqueous, and this is normally directly injected into HPLC or CE. In the case of three-phase LPME, the analytes must be either acidic or basic to accomplish extraction. Uncharged compounds are not efficiently extracted in a three-phase system. Table 7.2 gives an overview of different LPME applications including the extraction mode.

For basic drugs, one recent publication demonstrated that analyte extractability may be predicted from computer-calculated solubilities in acidic and alkaline aqueous solution as illustrated in Figure 7.5.[58] In this case, 58 different drug substances were extracted in a three-phase LPME system from alkaline solution, through dodecyl acetate as the SLM and into 15 µL, 10 mM HCl as the acceptor solution. As seen from the figure, target analytes with calculated solubilities in the range 5–150 mg/mL at pH 2 were all efficiently extracted in the three-phase system. For these substances, the corresponding solubility in alkaline solution (sample) was low, and the compounds were readily extracted into the SLM by their organic phase/sample distribution coefficients. Because of their substantial solubility in acidic aqueous solution, they were efficiently extracted into the acceptor phase as well. On the other hand, for target analytes with solubilities below 1–5 mg/mL at pH 2, low recoveries were obtained in three-phase LPME because of low acceptor phase/organic phase distribution coefficients; these compounds were principally trapped within the SLM. For this

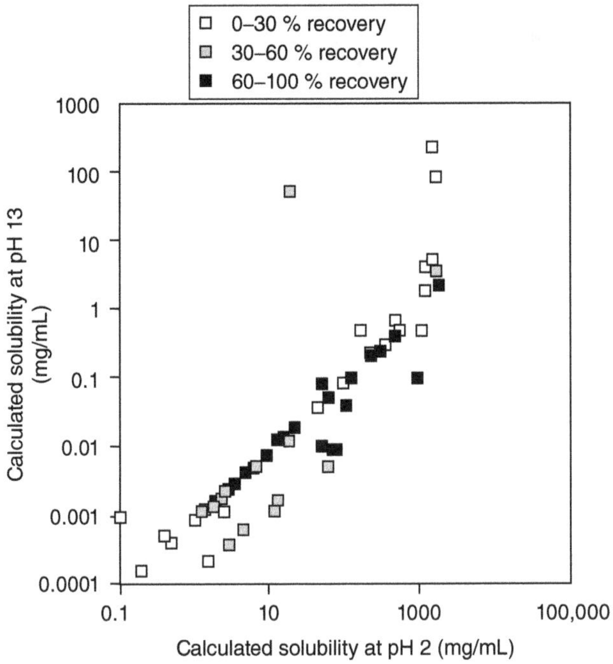

Figure 7.5. Extractability of drugs (low, medium, or high recovery) plotted against calculated solubility at pH 2 (x-axis) and pH 13 (y-axis).[58]

type of analytes, two-phase LPME is more appropriate. Finally, to the right of the diagram, we find the compounds with very high water solubility. For these compounds, three-phase LPME is inefficient because the extraction into the SLM fails. For the most polar analytes, this problem may be circumvented by addition of ion-pair reagents to the sample or to the SLM as discussed above. The selection of extraction mode in LPME may be aided by diagrams such as Figure 7.5.

To expand the discussion on extraction mode, it should also be mentioned that LPME can be accomplished in a static mode (static LPME), where both the sample and the acceptor are stagnant. The advantage of static LPME is that this version is very simple and requires only a simple vibrator or magnetic stirrer. Alternatively, LPME can be performed in a dynamic mode (dynamic LPME). The setup for dynamic LPME is illustrated in Figure 7.4e. The hollow fiber is attached to the needle of a microsyringe, and the acceptor phase situated inside the lumen of the hollow fiber can be moved up and down by the plunger. This movement has been demonstrated to improve extraction kinetics[59] but requires a programmable syringe pump to control the movements of the plunger.

To complete the discussion on extraction mode, it should be mentioned that LPME can be accomplished either as immersed LPME or as headspace LPME. To date, all LPME publications have described immersed LPME. However, a recent publication has discussed the possibility for headspace LPME,[60] where polycyclic aromatic hydrocarbons were extracted from the headspace of the sample by dynamic two-phase LPME. Most probably, the near future will see more research in this area.

7.5.2. Sample Volume

The selection of sample volume is to a large extent determined (1) by the amount of sample available, (2) by the concentration level of the target analytes, and (3) by the detection/quantification limit of the final analytical method. Basically, the sample volume is highly flexible in LPME, and sample volumes in the range 50 µL[61] to 1.1 mL[62] have been reported. In general, the sample volume should be kept as small as possible because the sample volume affects the extraction; increasing the sample volume results in reduced recovery and slower kinetics. This may be elucidated from Equations 7.5, 7.9, 7.15, and 7.17. A practical example of this is illustrated in Figure 7.6, where five basic drugs were extracted from a 1 and 4 mL sample volumes, respectively.[63] As seen from the figure, both extraction speed and equilibrium recoveries were superior in the case of the smallest sample volume.

7.5.3. Sample pH

For uncharged compounds, which are only extracted in two-phase systems, pH plays no significant role and pH adjustment is not required. For two-phase extractions of basic or acidic analytes, and for all types of three-phase LPME, pH in the sample plays a vital role. As seen from Equations 7.5 and 7.15, recovery is dependant on the distribution coefficient between the organic phase and the sample, and in order to maximize this, pH in the sample should be adjusted to deionize the target analytes. For basic analytes, this is accomplished by using alkaline conditions in the sample, whereas acidic conditions in the sample are used for acidic analytes. The practical effect of pH in the sample in three-phase LPME is illustrated in Figure 7.7 for methamphetamine.[6] This compound has a pK_a value of 10.1. As seen from the figure, the recovery increased from 40% at pH 10 to 80% at pH 14. Thus, a pH value that is several units above the pK_a value was required to maximize the recovery. In general, it is recommended to use a pH value at least 2–3 units above the pK_a value for basic analytes, and at least 2–3 pH units below the pK_a value for acidic compounds.

7.5.4. Agitation

During LPME, agitation/stirring of the whole system induces convection in the sample. Convection in the sample is important to promote mass transfer into the acceptor solution. However, a small part of the sample solution in close contact with the SLM is stagnant, and this boundary layer limits the mass transfer into the SLM and the acceptor solution. Agitation/stirring reduces this boundary layer and improves the kinetics of the system, and extraction

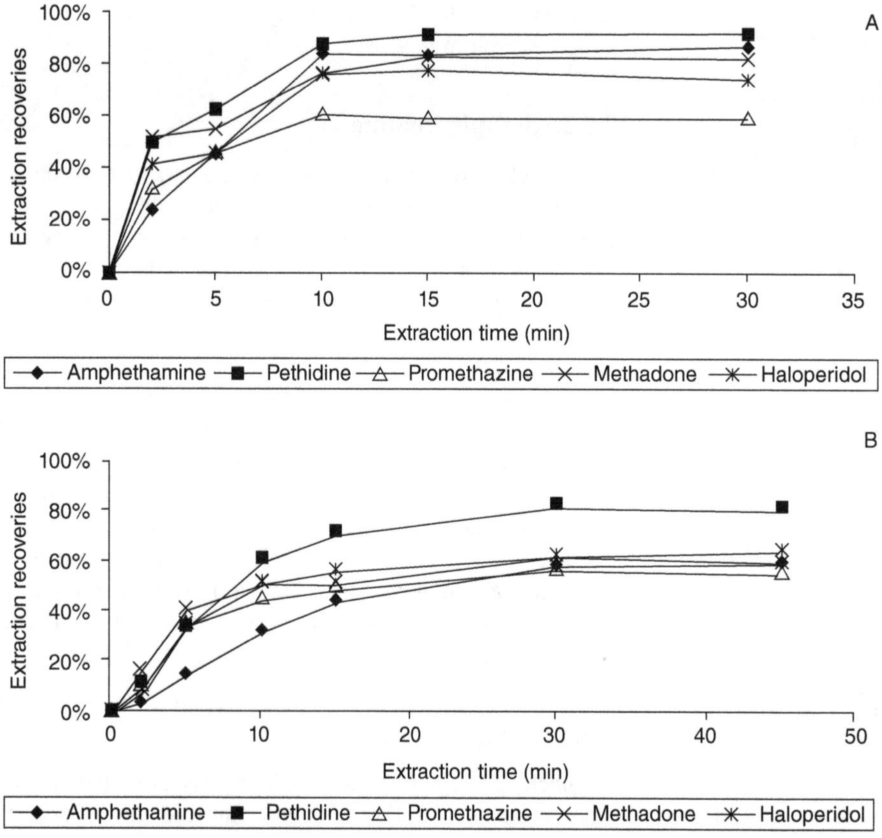

Figure 7.6. Extraction time profiles for five model drugs in 1 mL (A) and 4 mL (B) aqueous solution. Reprinted from reference 63 with permission from The Royal Society of Chemistry. Copyright 2002.

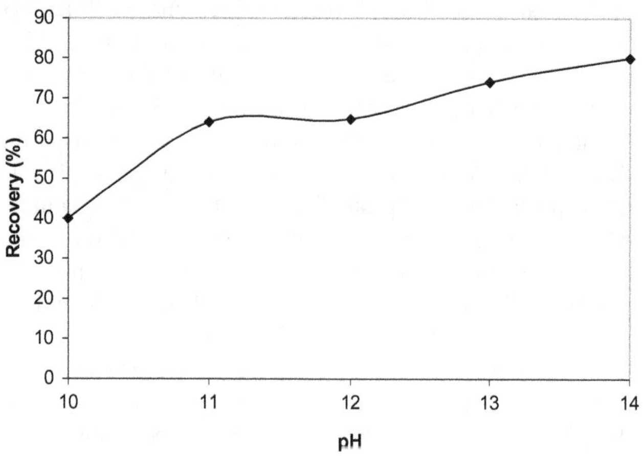

Figure 7.7. Effect of pH in the sample on recovery of methamphetamine in three-phase LPME. Reprinted from reference 6 with permission from the American Chemical Society. Copyright 1999.

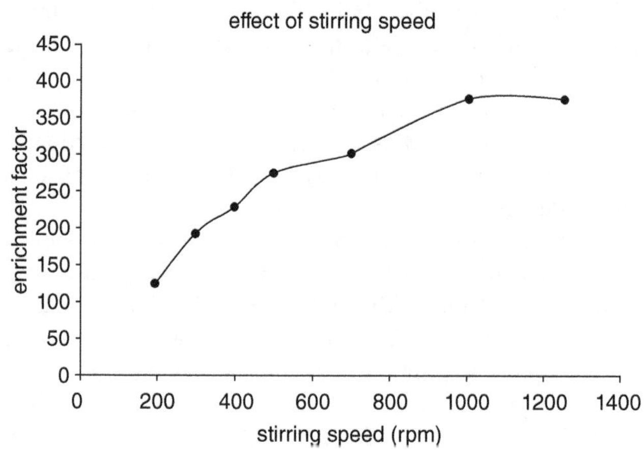

Figure 7.8. Effect of agitation/stirring speed on the extraction recovery of 3,4-dinitrophenol in three-phase LPME. Reprinted from reference 48 with permission from Elsevier. Copyright 2001.

recoveries generally increase with increasing agitation/stirring rate up to approximately 1000–1200 rpm as illustrated in Figure 7.8.[48] Above this point, bubble formation in the sample normally causes stability problems, and operation in this area is not recommended.

7.5.5. Extraction Time

As discussed earlier, LPME is usually used as an equilibrium extraction technique, where the recovery increases with time up to a certain level. After this, the system enters equilibrium and no further gain in extraction yield is observed versus time. This relationship is evidenced from Equations 7.8 and 7.17 as discussed earlier. Typical extraction curves are illustrated in Figures 7.3 and 7.6.[63] Most LPME extractions have been performed at equilibrium, and in order to do this, the equilibrium time has to be determined experimentally. Typical equilibrium times are in the range 15–45 min. The equilibrium time depends on several factors like (1) distribution coefficients, (2) sample volume, (3) agitation, and (4) the surface area of the hollow fiber. LPME may also be accomplished under nonequilibrium conditions to save time.[63] Thus, LPME may be accomplished for 5, 10, or 15 min to reduce the sample preparation time but with some loss of recovery. Nonequilibrium extractions have demonstrated validation data comparable with equilibrium extraction, but exact timing of the extraction time is highly important under nonequilibrium conditions.

7.5.6. Salt Addition

Addition of salt (sodium chloride) to the sample has frequently been used in LPME to improve recoveries. Generally, addition of salt can decrease the solubility of target analytes in the aqueous sample and enhance their partitioning into the organic phase. In some cases, significant improvements can be achieved by the addition of salt; in other cases, there are no effects, and in some cases, the addition of salt even suppresses extraction. This is difficult to predict, and experiments have to be conducted in most cases to verify the effect of salt. Figure 7.9 illustrates a practical example on the addition of salt. In this case, triazine herbicides were extracted by two-phase LPME.[64] For some of the compounds, a positive effect of salt was observed, whereas for others, the extractability decreased with increasing salt concentration.

7.5.7. Organic Phase Selection

Selection of an appropriate organic phase is crucial in both two- and three-phase LPME because both kinetics and recoveries are dependent on partition coefficients, and partition coefficients are dependent on the solvent used as SLM. As mentioned above, the organic solvent has to be water immiscible, and generally, 1-octanol, dihexyl ether, and toluene are the most commonly used. The final choice of the organic solvent is often based on practical experiments and should be based on several considerations. First, the water solubility should be low to prevent leakage into the aqueous sample. Second, the volatility should be low to avoid evaporation during extraction. In case of two-phase LPME combined with GC, the solubility of the target analytes within the organic solvent should be high to maximize partition coefficients, and the GC properties of the solvent should also be considered to ensure efficient chromatography. In three-phase LPME, the selection of organic solvent becomes even more critical as the organic phase only serves as an intermediate extraction medium. Thus, the selected solvent should provide a relatively high organic phase to sample distribution coefficient, and more importantly, it should also provide a high acceptor phase to organic phase distribution coefficient. This point is illustrated in Table 7.3, where three acidic drugs were extracted with either dihexyl ether, 1-octanol, or 2-octanone.[51] Finally, the organic solvent should have a polarity matching that of the polypropylene fiber to ensure efficient immobilization in the pores. The solvent may be immobilized in the pores of the hollow fiber by dipping it into the solvent for a few seconds, or it can be injected with a microsyringe from the inside lumen of the hollow fiber.[54] The former technique is very simple, but the exact amount of organic solvent loaded into the hollow fiber may vary from sample to sample. Also, when dipping the fiber in the solvent, excess solvent should be removed from the surface of the hollow fiber to reduce the volume of the SLM. Injection from the inside is easy to automate and provides a repeatable volume of the SLM from sample to sample. In the latter case, however, care should be undertaken to ensure that the whole length of the hollow fiber is covered by the organic solvent.

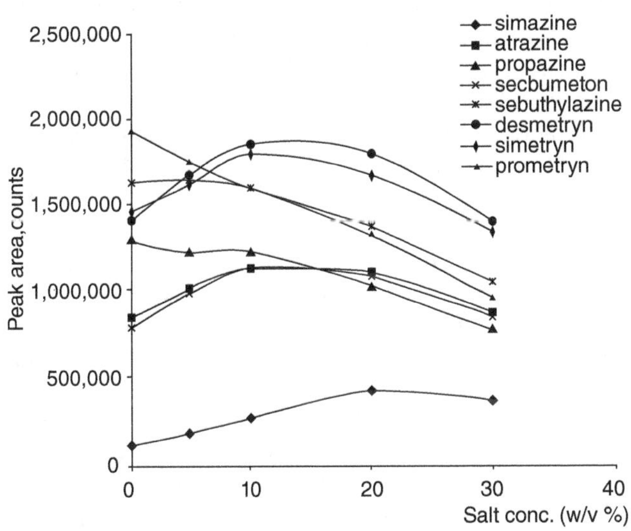

Figure 7.9. Effect of sodium chloride concentration on LPME for eight triazines. Reprinted from reference 64 with permission from the American Chemical Society. Copyright 2002.

7.5.8. Acceptor Phase pH

In three-phase LPME, the acceptor phase is a new aqueous solution, and again pH becomes critical for the extraction process. To efficiently extract the analytes from the SLM, pH conditions in the acceptor phase should be selected to ensure total ionization of the analytes in the acceptor phase. With this, the acceptor phase to organic phase distribution coefficient is maximized, and when extracted into the acceptor phase, the ionized target analytes are prevented from back-extraction into the SLM again (see Eqs. 7.13 and 7.14). Thus, for acidic analytes, an alkaline acceptor phase should

Table 7.3. Effect of Organic Solvent on the Extraction Recovery in Three-Phase LPME[51]

	Recovery (%)		
Organic Solvent	Ibuprofen	Naproxen	Ketoprofen
Dihexyl ether	101	84	77
1-Octanol	37	74	71
2-Octanone	14	45	68

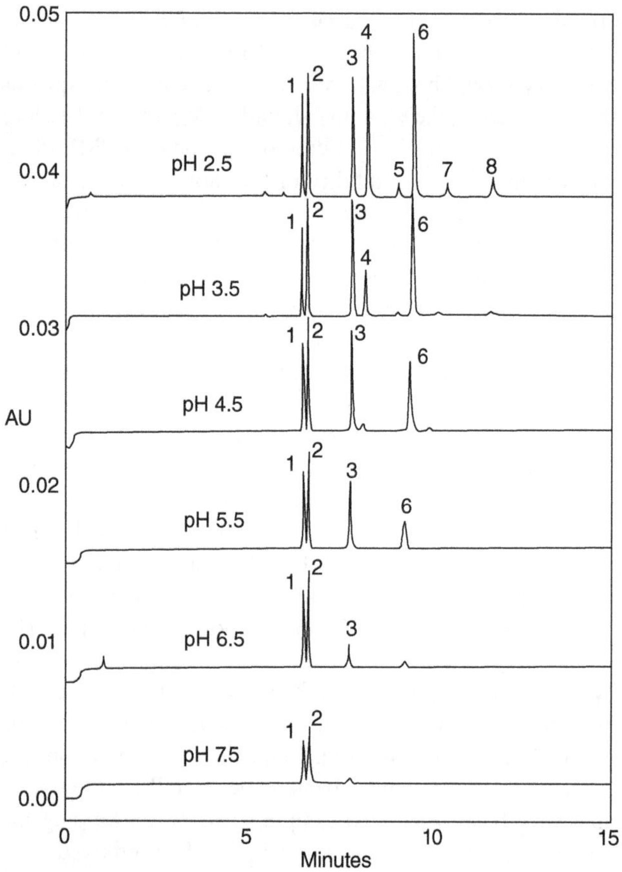

Figure 7.10. Effect of pH in the acceptor phase for LPME of (1) amphetamine ($pK_a = 9.9$), (2) methamphetamine ($pK_a = 10.1$), (3) pethidine ($pK_a = 8.7$), (4) chlorcyclizine ($pK_a = 7.8$), (5) noscapine ($pK_a = 6.2$), (6) haloperidol ($pK_a = 8.3$), (7) diazepam ($pK_a = 3.3$), and (8) reserpine ($pK_a = 6.6$). Reprinted from reference 65 with permission from Wiley-VCH. Copyright 2002.

be used, whereas an acidic acceptor phase is used for basic analytes. A practical example of the importance of appropriate pH conditions is illustrated in Figure 7.10 for eight different basic drugs.[65] With increasing pH value, the recovery rapidly dropped for the compounds with low pK_a values. At pH 2.5, all basic model compounds were extracted, whereas at pH 7.5, only amphetamine and methamphetamine were extracted. Because of the small volume of the acceptor phase, pH here is even more critical than in the sample. Thus, pH should be three to four units above the pK_a value for acids, and three to four units below the pK_a value for basic analytes if possible. It is also important that the buffer capacity in the acceptor phase is enough, so that the pH is not appreciably changed during the course of the extraction.

7.5.9. Acceptor Phase Volume

Normally, the volumes of acceptor phase used in LPME are in the range of 2–30 μL. With volumes in the lower end of this scale, evaporation may be a serious problem, which may limit the robustness of the LPME method. Therefore, it is generally recommended to use volumes above 10 μL. Recoveries at equilibrium are affected by the acceptor phase volume as discussed earlier (Eqs. 7.5 and 7.15), and in general, recoveries increase with increasing acceptor phase volume. On the other hand, the enrichment of analytes decreases with increasing acceptor phase volume (Eqs. 7.7 and 7.16), and both effects should be considered before selecting the final acceptor phase volume.

7.6. SELECTED APPLICATIONS

In Table 7.2, current applications of LPME reported in the scientific literature are summarized. As seen from the table, most applications have been reported within the field of environmental analysis. In addition, several applications within drug and pharmaceutical analysis have been published. Finally, a few papers have been focused on food, beverages, and peptides. The majority of applications will not be discussed in this chapter but can easily be found in the literature based on the list of references. However, the text below will focus on selected applications demonstrating some of the LPME frontiers, where difficult samples, alternative extraction principles, or specialized technical solutions have been addressed.

Substantial attention has been devoted to LPME of organochlorine pesticides (OCPs) from environmental waters. In one paper,[66] 12 different OCPs were extracted from seawater by two-phase LPME, utilizing toluene as the SLM. The proposed method required minimal sample preparation time and solvent consumption, and represented a significant advantage over conventional analytical methods. Combined with GC–mass spectrometry (MS) analysis, a high level of linearity was obtained ($R > 0.9995$) for OCPs over a range of analyte concentrations between 5 and 100 μg/L. Comparison between the proposed LPME method and United States Environmental Protection Agency Method 508 showed that the novel LPME method had comparable detection limits between 0.013 and 0.059 μg/L in seawater. Another paper demonstrated the simultaneous extraction of polycyclic aromatic hydrocarbons (PAHs) and OCPs from rainwater by two-phase LPME.[67] Again, toluene was utilized as the organic phase, and with GC-MS, several PAHs and OCPs were detected down to the 0.005–0.162 μg/L level.

In addition to different environmental waters (e.g., like tap water, river water, lake water, and seawater), solid samples have also been processed by LPME. In one paper, soil was extracted for PAHs by two-phase LPME utilizing *n*-octane as the organic phase.[68] In this procedure, 1 g of soil was placed in the LPME vial and soaked in a mixture of 7 mL acetone and 15 mL water to release the PAHs from the solid material. Thereafter, the hollow fiber containing *n*-octane was directly inserted in the suspension, and LPME was conducted without filtering the sample. The extraction time was only 8 min, and the total consumption of extraction solvent per sample was 8 μL. As another example of solid environmental samples, OCPs and polychlorinated biphenyls (PCBs) have been extracted from marine sediments with a combination of microwave-assisted solvent extraction with water (MWE) and two-phase LPME.[69] The purpose of the MWE step was to effectively release the analytes from the solid material, whereas the LPME step served to cleanup the sample and enrich the analytes prior to GC-MS.

While many LPME papers have utilized static extraction systems, some environmental applications have been conducted with dynamic systems, both in the two- and three-phase modes. Dynamic LPME has been claimed to provide better extraction efficiency and improved reproducibility as compared with the static mode. In one paper, dynamic LPME was conducted in the two-phase mode for the extraction of pesticides.[59] A conventional microsyringe with a 1.3 cm length of hollow fiber attached to its needle was connected to a syringe pump to perform the extraction. The microsyringe was used both as the microextraction device as well as the sample introduction device for GC-MS analysis. The attached hollow fiber served as the "holder" and "protector" of the 3 µL of toluene used for extraction. The solvent was repeatedly withdrawn from and discharged into the hollow fiber by the syringe pump. The pesticides were extracted from 4-mL water samples into the toluene impregnated in the hollow fiber, and subsequently, the organic solvent was injected into GC-MS. Slurry samples of soil and water were extracted with this setup, which provided excellent precision and linearity of data. A similar setup was utilized in the three-phase mode for extraction of aromatic amines prior to CE.[50]

The dynamic LPME approach has also been tested for headspace extractions,[60] where selected PAHs were extracted from soil samples. Water was added to the soil samples in order to release the target analytes to the headspace, the sample was heated to 90°C to support analyte volatilization, and extraction was accomplished with 1-octanol as the organic phase. This solvent provided excellent extractions, it was directly compatible with the GC-MS system, and very importantly, the volatility of this solvent was low in order to avoid partial evaporation during LPME. The system was simple and inexpensive, and provided good analyte enrichment factors, linear range, limits of detection, and repeatability.

An interesting variant of the headspace LPME system was recently published under the name liquid–gas–liquid microextraction (LGLME).[70] In this technique, a small amount (6 µL) of aqueous acceptor solution (0.5 M NaOH) was introduced into the lumen of a 2.7-cm polypropylene hollow fiber. The hollow fiber was then immersed in an aqueous sample solution (70°C). The aqueous acceptor solution in the channel of the hollow fiber was separated from the sample solution by the hydrophobic microporous hollow fiber wall with air inside its pores. Different phenols passed through the microporous hollow fiber membrane by gas diffusion and were trapped by the basic acceptor solution in the lumen of the hollow fiber. After extraction, the acceptor solution was withdrawn into a microsyringe and injected into a small sample vial for CE analysis. This system, which was totally solvent free, showed comparable results with traditional LPME with an extraction time of 20 min.

In LPME, ionic substances have typically been extracted under pH conditions where their ionization is suppressed. Thus, basic substances have been extracted from alkaline solutions, whereas acidic substances have been extracted from acidic solution. Recently, long-chain fatty acids were extracted by two-phase LPME from water samples as ion pairs with tetrabutylammonium hydrogen sulfate.[71] The ion-pair complexes were extracted into *n*-octanol as the solvent and were subsequently derivatized quantitatively to butyl esters in the injection port of a GC. This procedure was successfully applied to measure long-chain fatty acids in real water samples. A similar report with injection port derivatization of acidic herbicides has also been published,[72] and ion-pair LPME may be an interesting future option for difficult analytes.

One of the major advantages of LPME is the possibility of high analyte enrichments, as target analytes are extracted into a very small volume of acceptor phase. In two papers, LPME has been accomplished from large sample volumes in two individual steps to provide very high enrichment in relatively short time.[52,73] In the former report, nonsteroidal anti-inflammatory drugs (NSAIDs) were extracted from a 100-mL sample, through *n*-octanol as the SLM immobilized in 10 individual hollow fibers, and into 10 mM NaOH as acceptor solution present in the lumen of each of the 10 hollow fibers. The acceptor solutions were combined, acidified, and subsequently extracted with a single hollow fiber containing *n*-octanol as the SLM and 2 µL, 10 mM NaOH as the acceptor solution. In this system the analytes were transferred from a 100-mL sample to 2 µL of acceptor solution, and enrichments exceeding 15,000 were reported. This was further refined in a recent paper, where basic drugs in wastewater were extracted from 1100-mL samples, through dihexyl ether as SLM, and into 20 µL, 10 mM HCl as acceptor solution in a single step.[62] In the latter report, enrichments exceeding 27,000 were obtained after 2 h of extraction, and subsequently, the extracts were directly subjected to liquid chromatography (LC)–MS without further sample preparation. This enabled drug residues in municipal wastewater to be detected down to the low picogram per liter level.

Until now, nearly all LPME reports have been based on manual techniques. For the future, however, automation is required in order for LPME to be accepted as a potential routine sample preparation method. In a few papers, semi-automation[17] or full automation[29] has been reported. In the latter case, two-phase LPME was successfully automated with a CTC CombiPal autosampler. All steps of the LPME technique, including the filling of the extraction solvent, sample transfer and agitation, withdrawing the solvent to a syringe, and introducing the extraction phase into the GC injector, were automated by the CTC system. The same paper also explored the possibility of kinetic calibration in two-phase LPME, to correct for matrix effects in the carbaryl analysis of red wine samples. In kinetic calibration, desorption of standard preloaded in the acceptor phase was used to calibrate the extraction of analytes.

As seen from Table 7.2, most environmental applications have been carried out with GC, GC-MS, HPLC, or LC-MS as the final analytical method. In one paper, however, LPME was coupled directly to electrothermal vaporization inductively coupled plasma mass spectrometry (ETV-ICP-MS) for the determination of inorganic selenium in natural waters.[30] The organic chelating reagent ammonium pyrrolidine dithiocarbamate was used both as the extracting solvent in the hollow fiber and as a chemical modifier for the ETV-ICP-MS determination. Following extraction, the organic phase was directly loaded into the ETV-ICP-MS system for quantification of inorganic selenium.

In the field of drug analysis, the majority of LPME work has been focused on three-phase extractions of basic drugs. Drugs of low polarity (log $P > 1.5$), like antidepressants and antiepileptic drugs, are normally extracted with medium (>30%) or high (>60%) recoveries in three-phase LPME[58] based on their favorable distribution coefficients. On the other hand, polar substances provide low recoveries (0–30%) due to their low partition coefficient between the organic SLM and the aqueous sample. In order to enhance the extraction of hydrophilic drugs, carrier-mediated LPME has been reported.[74] In carrier-mediated LPME, sodium octanoate was added to the sample (pH 7.0) to form ion pairs with the hydrophilic basic drugs. The ion-pair complexes were easily extracted into the SLM, and at the interface between the SLM and the acidic acceptor solution, the hydrophilic drugs were released from the ion-pair reagent and extracted into the acceptor solution. From the acceptor solution, protons were back-extracted into the SLM to protonate the octanoate ions. Thus, the extraction was forced both by high partition coefficients of the drug–ion pair complexes, and by a high proton concentration in the acceptor solution. Several reports have investigated carrier-mediated LPME,[53,75] and several hydrophilic drugs with log P values below 1.0 were extracted from the human plasma with recoveries exceeding 45%.

Another challenge in LPME is the three-phase extraction of weakly basic drugs like benzodiazepines. Due to their low pK_a values, the acceptor solution has to be highly acidic in order to trap them efficiently from the SLM and to protonate the basic compounds. One paper on this[76] demonstrated that pH in the acceptor solution should be below 1.0 in order to effectively extract benzodiazepines. Under these conditions, the stability of the analytes was studied to confirm that they were not degraded during extraction. Alternatively, weakly basic drugs like benzodiazepines can be extracted by two-phase LPME[77] to avoid strong pH conditions in the acceptor solution.

In addition to basic drugs, acidic drugs have also been extracted in three-phase LPME. In one early paper, three acidic NSAIDs were extracted from human urine.[51] In this case, the pH gradient over the SLM was reversed; the sample was acidified in order to suppress the ionization of the acidic drugs and to promote their extraction into the SLM, whereas the acceptor solution was alkaline in order to ionize the target analytes and to promote their partitioning into this phase. The NSAIDs were extracted with recoveries in the range from 77% to 101%, and they were enriched by a factor between 77 and 101 times from 2.5-mL sample volumes. This high enrichment enabled the compounds to be detected and quantified down to the low nanogram per milliliter level by CE with UV detection.

Most LPME applications on drugs have been conducted from plasma and urine samples. For plasma samples, drugs are often highly bound to proteins, and in principle, this reduces the extraction recovery. However, prior to LPME, pH is adjusted in the sample, and this normally suppresses the drug–protein interactions and high recoveries can be obtained from plasma samples. However, in a few cases, pH adjustment is not sufficient, and in these cases, methanol can be added to the sample to release the drugs and to further increase the extraction recovery.[63] In addition to plasma and urine, three-phase LPME has also been conducted successfully from whole blood[46] and human breast milk.[45] For whole blood, LPME worked very efficiently after a simple dilution of the sample. In the case of breast milk, the high content of fat partially suppressed the extraction, and high recoveries were achieved after acidification and centrifugation of the sample prior to LPME. Finally, LPME has also been accomplished from saliva.[78]

In most cases, the drug applications of LPME have been accomplished using HPLC, LC-MS, or CE as the final method of analysis. Also, some two-phase applications have been reported for drugs where GC was used as the analytical method (see Table 7.2). In a few instances, chiral separations have been accomplished following three-phase LPME. Thus, three-phase LPME of both mianserin[79] and citalopram[80] was followed by chiral CE to monitor chiral metabolism in humans. Although the CE analysis was conducted with UV detection, therapeutically relevant concentrations were easily measured due to the high enrichment obtained by LPME. In addition to chiral CE, chiral HPLC has also been utilized in combination with LPME.[26] Finally, LPME has been combined with flow injection analysis–mass spectrometry (FIA-MS) for the identification of drug abuse in humans.[81] Due to the excellent sample cleanup capacity of three-phase LPME, no chromatography was required for reliable identification by tandem mass spectrometry (MS/MS), and the whole LPME extract was introduced directly into the mass spectrometer in a simple FIA system. With elimination of the chromatographic step, the system enabled a very high sample throughput, with analysis of one or two samples every minute.

A few applications of LPME have been reported for sample preparation of food and beverages (see Table 7.2). In one application pesticide residues were successfully extracted in a three-phase system from matrices such as cucumber, tomato, and pepper, and determined by LC coupled with electrospray MS.[35] LPME has also been used for the extraction of peptides[82,83] and for the determination of log P values.[34] The peptide applications were focused on relatively hydrophobic substances, and carrier-mediated LPME was accomplished to obtain acceptable recoveries. This area is highly interesting, and more efforts should be directed to LPME of peptides in the future.

7.7. PERFORMANCE

LPME extractions are typically characterized by recovery and enrichment. In general, LPME is an equilibrium extraction technique, where the concentration of analyte in the acceptor solution increases to a certain level, where after the system enters equilibrium and no further gains in recovery are obtained versus time. In other words, LPME is not an exhaustive extraction technique like LLE and SPE. The extraction recovery in LPME is determined by the actual distribution coefficients, and the volumes of sample, SLM, and acceptor phase, and, typically, recoveries range between 10% and 90%. The actual recovery for a certain application has to be measured and built into the calibration of the analytical method in a similar way as performed in SPME.

Because LPME is accomplished with a very small volume of acceptor solution, most LPME applications provide substantial analyte enrichment. In fact, this is one of the major advantages of LPME, and high enrichments are obtained directly without the need for solvent evaporation and reconstitution as required for high enrichment by LLE and SPE. The enrichment factor in LPME is basically determined by the analyte recovery and by the volume of the sample. As the volume of the sample increases, the enrichment factor also increases. From 1-mL sample volumes, typical enrichments range between 20 and 50, but values up to 27,000 have been reported from 1100-mL samples[62] supporting the high potential of LPME for analyte enrichment.

Several reports in the literature have focused on the excellent sample cleanup performance of LPME. Especially in the three-phase mode, LPME provides very clean extracts from a variety of samples. This is illustrated in Figure 7.11, where citalopram and methamphetamine were spiked to human urine, plasma, and whole blood.[46] In spite of the complex matrices of these samples, almost no other components than the two basic drugs were recovered in the acceptor solution. The high sample cleanup performance is a second major advantage of LPME.

In addition to high-analyte enrichment and excellent sample cleanup, a major advantage of LPME is the low consumption of organic solvent per sample. Typically, the volume of organic solvent immobilized in the pores of a hollow fiber segment range between 5 and 20 μL, and this is used for a single extraction. In the two-phase mode, also the acceptor solution inside the lumen of the hollow fiber is an organic solvent, with a volume typically ranging from 2 to 20 μL. In other words, the total consumption of organic solvent per analysis normally ranges between 10 and 40 μL. One recent paper even demonstrated that in the three-phase mode, where the organic solvent only serves as an intermediate extraction medium, soybean oil or olive oil can be used as the SLM providing a totally green chemistry approach to sample preparation.[55]

While enrichment, cleanup, and low-solvent consumption are the major advantages of LPME, the relatively long extraction time is perhaps the major disadvantage of the concept. Normally, extraction times to equilibrium range between 15 and 45 min for sample volumes below 2 mL,[63] whereas for 1-L samples even 2 h may be required to reach equilibrium.[62] On the other hand, when the LPME process is finished, the acceptor phase is directly subjected to the final analysis with no further sample preparation steps required. Also, many samples may be extracted in parallel to speed up the throughput, or extractions may be carried out under nonequilibrium conditions.[63] In the latter case, recoveries are somewhat sacrificed, but validation data are comparable with equilibrium extractions provided that the timing of the extractions are carried out with high accuracy. Recently, a paper demonstrated that LPME extraction can be performed with increased extraction speed utilizing an electrical potential applied across the SLM,[84] and this approach may eliminate the speed disadvantage of LPME in the future.

As demonstrated in this chapter, LPME enables a high degree of flexibility. With the same extraction device, either two-phase or three-phase extractions can be performed, providing compatibility with all of GC, HPLC, CE, and MS. Neutral substances can be extracted in a two-phase system, and acidic and basic compounds can be extracted either in a two- or three-phase system. This high degree of flexibility should be of high interest in the near future. In addition, validation data reported demonstrate acceptable repeatability, reproducibility, accuracy, linearity, and robustness, even though all reports until now have been accomplished with home-built equipment. With the development of commercial equipment, and with further automation of the technique, the figures of merit are expected to further improve.

7.8. FUTURE TRENDS

The current chapter has focused on hollow fiber LPME, including historical development, fundamental theory, method development, applications, and performance. From this, and from recent review articles on the technique,[85–88] we have a strong theoretical understanding of the technique, and we have a good overview of the applicability of LPME. In general, the technique is very well suited for extraction of nonpolar and medium-polar substances from different types of aqueous samples. Also, several reports have demonstrated acceptable validation data for the technique, in most cases with home-built equipment. With all this

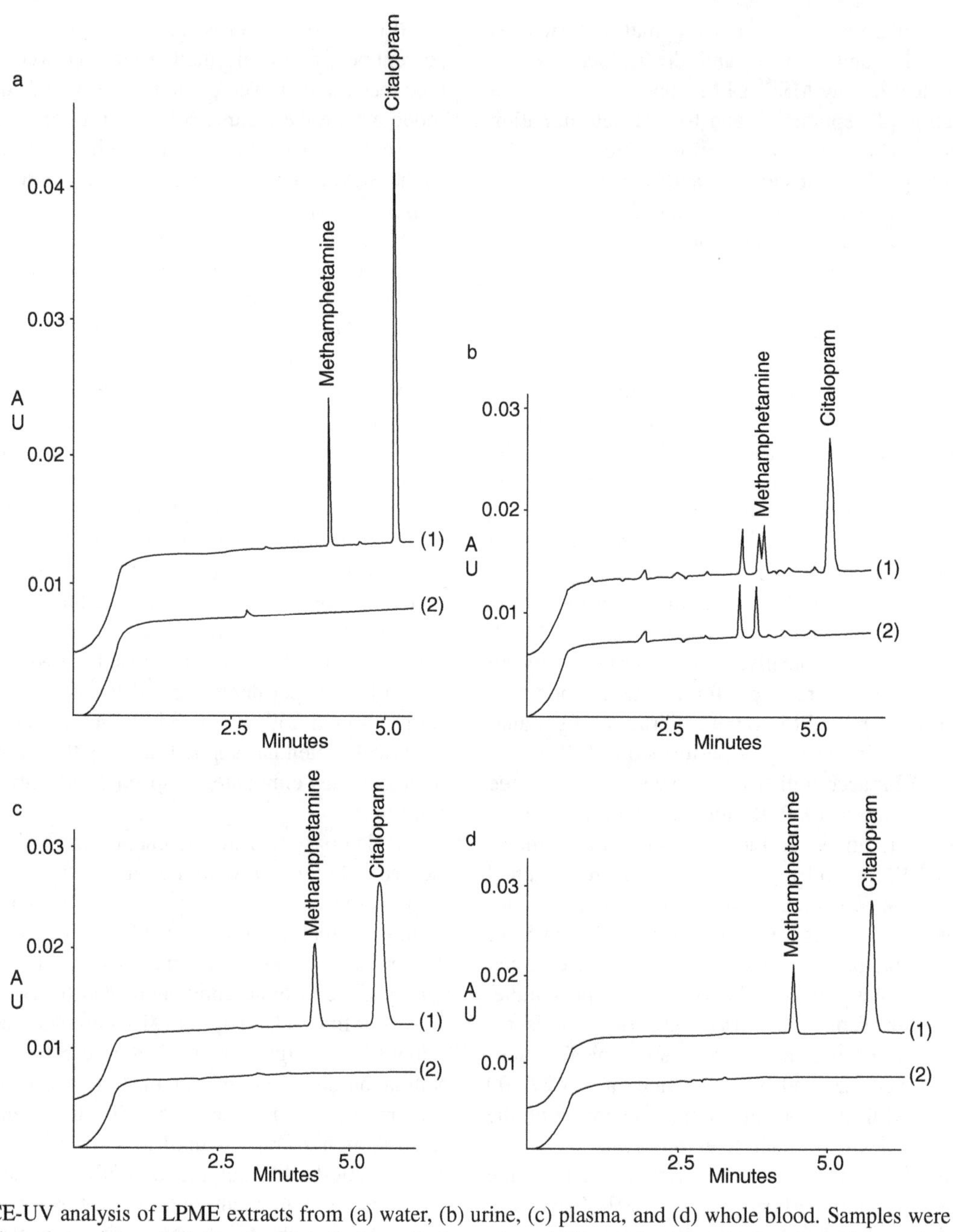

Figure 7.11. CE-UV analysis of LPME extracts from (a) water, (b) urine, (c) plasma, and (d) whole blood. Samples were spiked with 100 ng/mL methamphetamine and citalopram. (1) indicates spiked sample and (2) unspiked sample. Reprinted from reference 46 with permission from Elsevier. Copyright 2001.

information, LPME is, in principle, mature for implementation in routine analytical laboratories.

However, the implementation of LPME is currently limited by the unavailability of commercial equipment. Work is in progress in this area, and hopefully, the near future will show commercial equipment for LPME. This equipment should be fully automated and compatible with the most abundant laboratory robotics and autosamplers. Also, reports with this type of equipment, with special focus on robustness and validation, are mandatory for future general acceptance and implementation of the technique. Finally, development of generic methods for different types of analytes is important to reduce the time required for individual laboratories to develop LPME methods. For this purpose, the applications and theory reviewed in this chapter will be of high value.

Based on the strong advantages of LPME, it is expected to be an important future sample preparation technique complementing existing techniques like LLE, solid-phase extraction, and SPME. Most probably, the majority of applications will be within environmental chemistry and for pharmaceutical analysis. However, analysis of foods and beverages is an increasing application field and may also benefit from utilizing LPME. Finally, LPME may be used in more untraditional ways, for example, for the rapid determination of log P values, for estimating transport properties of chemical entities through biological and artificial membranes, and for passive samplers in the environment to mention only a few alternative applications.

REFERENCES

1. Arthur, C.L.; Pawliszyn, J. Solid phase microextraction with thermal desorption using fused silica optical fibers. *Anal. Chem.* **1990**, *62*, 2145–2148.

2. Liu, H.; Dasgupta, P.K. Analytical chemistry in a drop. Solvent extraction in a microdrop. *Anal. Chem.* **1996**, *68*, 1817–1821.

3. Jeannot, M.A.; Cantwell, F.F. Solvent microextraction into a single drop. *Anal. Chem.* **1996**, *68*, 2236–22240.

4. He, Y.; Lee, H.K. Liquid-phase microextraction in a single drop of organic solvent by using a conventional microsyringe. *Anal. Chem.* **1997**, *69*, 4634–4640.

5. de Jager, L.S.; Andrews, A.R.J. Solvent microextraction of chlorinated pesticides. *Chromatographia* **1999**, *50*, 733–738.

6. Pedersen-Bjergaard, S.; Rasmussen, K.E. Liquid-liquid-liquid microextraction for sample preparation of biological fluids prior to capillary electrophoresis. *Anal. Chem.* **1999**, *71*, 2650–2656.

7. Audunsson, G.A. Aqueous/aqueous extraction by means of a liquid membrane for sample cleanup and preconcentration of amines in a flow system. *Anal. Chem.* **1986**, *58*, 2714–2723.

8. Jönsson, J.Å.; Mathiasson, L. Supported liquid membrane techniques for sample preparation and enrichment in environmental and biological analysis. *Trends Anal. Chem.* **1992**, *11*, 106–114.

9. Thordarson, E.; Palmarsdottir, S.; Mathiasson, L.; Jönsson, J.Å. Sample preparation using a miniaturized supported liquid membrane device connected online to packed capillary liquid chromatography. *Anal. Chem.* **1996**, *68*, 2559–2563.

10. Jönsson, J.Å.; Mathiasson, L. Liquid membrane extraction in analytical sample preparation. I. Principles. *Trends Anal. Chem.* **1999**, *18*, 318–325.

11. Jönsson, J.Å.; Mathiasson, L. Liquid membrane extraction in analytical sample preparation. II. Applications. *Trends Anal. Chem.* **1999**, *18*, 325–334.

12. Jönsson, J.Å.; Mathiasson, L. Membrane-based techniques for sample enrichment. *J. Chromatogr. A* **2000**, *902*, 205–225.

13. Jönsson, J.Å.; Mathiasson, L. Membrane extraction in analytical chemistry. *J. Sep. Sci.* **2001**, *24*, 495–507.

14. Zhu, L.; Tu, C.; Lee, H.K. Liquid-phase microextraction of phenolic compounds combined with on-line preconcentration by field-amplified sample injection at low pH in micellar electrokinetic chromatography. *Anal. Chem.* **2001**, *73*, 5655–5660.

15. de Jager, L.S.; Andrews, A.R.J. Preliminary studies of a fast screening method for cocaine and cocaine metabolites in urine using hollow fiber membrane solvent microextraction (HFMSME). *Analyst* **2001**, *126*, 1298–1303.

16. Psillakis, E.; Kalogerakis, N. Hollow-fibre liquid-phase microextraction of phthalate esters from water. *J. Chromatogr. A* **2003**, *999*, 145–153.

17. Müller, S.; Möder, M.; Schrader, S.; Popp, P. Semi-automated hollow-fibre membrane extraction, a novel enrichment technique for the determination of biologically active compounds in water samples. *J. Chromatogr. A* **2003**, *985*, 99–106.

18. Yan, C.-H.; Wu, H.-F. A liquid-phase microextraction method, combining a dual gauge microsyringe with hollow fiber membrane, for the determination of organochlorine pesticides in aqueous solution by gas chromatography/ion trap mass spectrometry. *Rapid Commun. Mass Spectrom.* **2004**, *18*, 3015–3018.

19. Quintana, J.B.; Rodil, R.; Reemtsma, T. Suitability of hollow fibre liquid-phase microextraction for the determination of acidic pharmaceuticals in wastewater by liquid chromatography-electrospray tandem mass spectrometry without matrix effects. *J. Chromatogr. A* **2004**, *1061*, 19–26.

20. Lambropoulou, D.A.; Albanis, T.A. Sensitive trace enrichment of environmental andiandrogen vinclozolin from natural waters and sediment samples using hollow-fiber liquid-phase microextraction. *J. Chromatogr. A* **2004**, *1061*, 11–18.

21. Pan, H.-J.; Ho, W.-H. Determination of fungicides in water using liquid phase microextraction and gas chromatography with electron capture detection. *Anal. Chim. Acta* **2004**, *527*, 61–67.

22. González-Penas, E.; Leache, C.; Viscarret, M.; Perez de Obanos, A.; Araguás, C.; López de Cerain, A. Determination of ochratoxin A in wine using liquid-phase microextraction combined with liquid chromatography with fluorescence detection. *J. Chromatogr. A* **2004**, *1025*, 163–168.

23. Marlow, M.; Hurtubise, R.J. Liquid-liquid-liquid microextraction for the enrichment of polycyclic aromatic hydrocarbon metabolites investigated with fluorescence spectroscopy and capillary electrophoresis. *Anal. Chim. Acta* **2004**, *526*, 41–49.

24. Kou, D.; Wang, X.; Mitra, S. Supported liquid membrane microextraction with high-performance liquid chromatography-UV detection for monitoring trace haloacetic acids in water. *J. Chromatogr. A* **2004**, *1055*, 63–69.

25. Sarafraz Yazdi, A.; Es'haghi, Z. Surfactant enhanced liquid-phase microextraction of basic drugs of abuse in hair combined with high performance liquid chromatography. *J. Chromatogr. A* **2005**, *1094*, 1–8.

26. Malagueno de Santana, F.J.; de Oliveira, A.R.M.; Bunato, P.S. Chiral liquid chromatographic determination of mirtazapine in human plasma using two-phase liquid-phase microextraction for sample preparation. *Anal. Chim. Acta* **2005**, *549*, 96–103.

27. Melwanki, M.B.; Huang, S.-D. Extraction of hydroxyaromatic compounds in river water by liquid-liquid-liquid microextraction with automated movement of the acceptor and the donor phase. *J. Sep. Sci.* **2006**, *29*, 2078–2084.

28. Richoll, S.M.; Colón, I. Determination of triphenylphosphine oxide in active pharmaceutical ingredients by hollow-fiber liquid-phase microextraction followed by reversed-phase liquid chromatography. *J. Chromatogr. A* **2006**, *1127*, 147–153.

29. Ouyang, G.; Pawliszyn, J. Kinetic calibration for automated hollow fiber-protected liquid-phase microextraction. *Anal. Chem.* **2006**, *78*, 5783–5788.

30. Xia, L.; Hu, B.; Jiang, Z.; Wu, Y.; Chen, R.; Li, L. Hollow fiber liquid phase microextraction combined with electrothermal vaporization ICP-MS for the speciation of inorganic selenium in natural waters. *J. Anal. At. Spectrom.* **2006**, *21*, 362–365.

31. Leinonen, A.; Vuorensola, K.; Lepola, L.-M.; Kuuranne, T.; Kotiaho, T.; Ketola, R.A.; Kostiainen, R. Liquid-phase microextraction for sample preparation in analysis of unconjugated anabolic steroids in urine. *Anal. Chim. Acta* **2006**, *559*, 166 172.

32. Dubey, D.K.; Pardasani, D.; Gupta, A.K.; Palit, M.; Kanaujia, P.K.; Tak, V. Hollow fiber-mediated liquid-phase microextraction of chemical warfare agents from water. *J. Chromatogr. A* **2006**, *1107*, 29–35.

33. Yamini, Y.; Reimann, C.T.; Vatanara, A.; Jönsson, J.Å. Extraction and preconcentration of salbutamol and terbutaline from aqueous samples using hollow fiber supported liquid membrane containing anionic carrier. *J. Chromatogr. A* **2006**, *1124*, 57–67.

34. Guo, Y.G.; Zhang, J.; Liu, D.N.; Fu, H.F. Determination of n-octanol-water partition coefficients by hollow-fiber membrane solvent microextraction coupled with HPLC. *Anal. Bioanal. Chem.* **2006**, *386*, 2193–2198.

35. Romero-González, R.; Pastor-Montoro, E.; Martinez-Vidal, J.L.; Garrido-Frenich, A. Application of hollow fiber supported liquid membrane extraction to the simultaneous determination of pesticide residues in

vegetables by liquid chromatography/mass spectrometry. *Rapid Commun. Mass Spectrom.* **2006**, *20*, 2701–2708.

36. Vora-Adisak, N.; Varanusupakul, R. A simple supported liquid hollow fiber membrane microextraction for sample preparation of trihalomethanes in water samples. *J. Chromatogr. A* **2006**, *1121*, 236–241.

37. Peng, J.-F.; Liu, J.-F.; Hu, X.-L.; Jiang, G.-B. Direct determination of chlorophenols in environmental water samples by hollow fiber-supported ionic liquid membrane extraction coupled with high-performance liquid chromatography. *J. Chromatogr. A* **2007**, *1139*, 165–170.

38. Hauser, B.; Popp, P. Membrane-assisted solvent extraction of organochlorine compounds in combination with large-volume injection/gas chromatography-electron capture detection. *J. Sep. Sci.* **2001**, *24*, 551–560.

39. Ho, T.S.; Pedersen-Bjergaard, S.; Rasmussen, K.E. Recovery, enrichment and selectivity in liquid-phase microextraction comparison with conventional liquid-liquid extraction. *J. Chromatogr. A* **2002**, *963*, 3–17.

40. Mayer, P.; Tolls, J.; Hermens, J.L.M.; Mackay, D. Equilibrium sampling devices. *Environ. Sci. Technol.* **2003**, *37*, 270A.

41. Ma, M.; Cantwell, F.F. Solvent microextraction with simultaneous back-extraction for sample cleanup and preconcentration: Quantitative extraction. *Anal. Chem.* **1998**, *70*, 3912–3919.

42. Liu, J.-F.; Jönsson, J.Å.; Mayer, P. Equilibrium sampling through membranes of freely dissolved chlorophenols in water samples with hollow fiber supported liquid membrane. *Anal. Chem.* **2005**, *77*, 4800–4809.

43. Trtic-Petrovic, T.; Liu, J.-F.; Jönsson, J.Å. Equilibrium sampling through membrane based on a single hollow fibre for determination of drug-protein binding and free drug concentration in plasma. *J. Chromatogr. B* **2005**, *826*, 169–176.

44. Romero, R.; Liu, J.-F.; Jönsson, J.Å. Equilibrium sampling through membranes of freely dissolved copper concentrations with selective hollow fiber membranes and the spectrophotometric detection of a metal stripping agent. *Anal. Chem.* **2005**, *77*, 7605–7611.

45. Bjørhovde, A.; Grønhaug Halvorsen, T.; Rasmussen, K.E.; Pedersen-Bjergaard, S. Liquid-phase microextraction of drugs frum human breast milk. *Anal. Chim. Acta* **2003**, *491*, 155–161.

46. Grønhaug Halvorsen, T.; Pedersen-Bjergaard, S.; Rasmussen, K.E. Reduction of extraction times in liquid-phase microextraction. *J. Chromatogr. B* **2001**, *760*, 219–226.

47. Berhanu, T.; Liu, J.-F.; Romero, R.; Megersa, N.; Jönsson, J.Å. Determination of trace levels of dinitrophenolic compounds in environmental water samples using hollow fiber supported liquid membrane extraction and high performance liquid chromatography. *J. Chromatogr. A* **2006**, *1103*, 1–8.

48. Zhu, L.; Zhu, L.; Lee, H.K. Liquid-liquid-liquid microextraction of nitrophenols with a hollow fiber membrane prior to capillary liquid chromatography. *J. Chromatogr. A* **2001**, *924*, 407–414.

49. Zhao, L.; Lee, H.K. Liquid-phase microextraction combined with hollow fiber as a sample preparation technique prior to gas chromatography/mass spectrometry. *Anal. Chem.* **2002**, *74*, 2486–2492.

50. Hou, L.; Lee, H.K. Dynamic three-phase microextraction as a sample preparation technique prior to capillary electrophoresis. *Anal. Chem.* **2003**, *75*, 2784–2789.

51. Pedersen-Bjergaard, S.; Rasmussen, K.E. Liquid-phase microextraction and capillary electrophoresis of acidic drugs. *Electrophoresis* **2000**, *21*, 579–585.

52. Sarafraz Yazdi, A.; Es'haghi, Z. Two-step hollow fiber-based, liquid-phase microextraction combined with high-performance liquid chromatography: A new approach to determination of aromatic amines in water. *J. Chromatogr. A* **2005**, *1082*, 136–142.

53. Ho, T.S.; Pedersen-Bjergaard, S.; Rasmussen, K.E. Experiences with carrier-mediated transport in liquid-phase microextraction. *J. Chromatogr. Sci.* **2006**, *44*, 308–316.

54. Bårdstu, K.F.; Ho, T.S.; Rasmussen, K.E.; Pedersen-Bjergaard, S.; Jönsson, J.Å. Supported liquid membranes in hollow fiber liquid-phase microextraction (LPME). Practical considerations in the three-phase mode. *J. Sep. Sci.* **2007**, *30*, 1364–1370.

55. Pedersen-Bjergaard, S.; Rasmussen, K.E. Liquid-phase microextraction utilising plant oils as intermediate extraction medium—Towards elimination of synthetic organic solvents in sample preparation. *J. Sep. Sci.* **2004**, *27*, 1511–1516.

56. Kramer, K.E.; Andrews, A.R.J. Screening method for 11-nor-Δ^9-tetrahydrocannabinol-9-carboxylic acid in urine using hollow fiber membrane solvent microextraction with in-tube derivatization. *J. Chromatogr. B* **2001**, *760*, 27–36.

57. Zhao, L.; Zhu, L.; Lee, H.K. Analysis of amines in water samples by liquid-liquid-liquid microextraction with hollow fibers and high-performance liquid chromatography. *J. Chromatogr. A* **2002**, *963*, 239–248.

58. Pedersen-Bjergaard, S.; Rasmussen, K.E.; Brekke, A.; Ho, T.S.; Grønhaug Halvorsen, T. Liquid-phase microextraction of basic drugs—Selection of extraction mode based on computer calculated solubility data. *J. Sep. Sci.* **2005**, *28*, 1195–1203.

59. Hou, L.; Shen, G.; Lee, H.K. Automated hollow fiber-protected dynamic liquid-phase microextraction of pesticides for gas chromatography-mass spectrometric analysis. *J. Chromatogr. A* **2003**, *985*, 107–116.

60. Jiang, X.; Basheer, C.; Zhang, J.; Lee, H.K. Dynamic hollow fiber-supported headspace liquid-phase microextraction. *J. Chromatogr. A* **2005**, *1087*, 289–294.

61. Grønhaug Halvorsen, T.; Pedersen-Bjergaard, S.; Reubsaet, J.L.E.; Rasmussen, K.E. Liquid-phase microextraction combined with liquid chromatography-mass spectrometry. Extraction from small volumes of biological samples. *J. Sep. Sci.* **2003**, *26*, 1520–1526.

62. Ho, T.S.; Vasskog, T.; Andressen, T.; Jensen, E.; Rasmussen, K.E.; Pedersen-Bjergaard, S. 25,000-fold pre-concentration in a single step with liquid-phase microextraction. *Anal. Chim. Acta* **2007**, *592*, 1–8.

63. Ho, T.S.; Pedersen-Bjergaard, S.; Rasmussen, K.E. Liquid-phase microextraction of protein-bound drugs under non-equilibrium conditions. *Analyst* **2002**, *127*, 608.

64. Shen, G.; Lee, H.K. Hollow fiber-protected liquid-phase microextraction of triazine herbicides. *Anal. Chem.* **2002**, *74*, 648–654.

65. Pedersen-Bjergaard, S.; Ho, T.S.; Rasmussen, K.E. Fundamental studies on selectivity in 3-phase liquid-phase microextraction (LPME) of basic drugs. *J. Sep. Sci.* **2002**, *25*, 141–146.

66. Basheer, C.; Lee, H.K.; Obbard, J.P. Determination of organochlorine pesticides in seawater using liquid-phase hollow fiber membrane microextraction and gas chromatography-mass spectrometry. *J. Chromatogr. A* **2002**, *968*, 191–199.

67. Basheer, C.; Balasubramanian, R.; Lee, H.K. Determination of organic micropollutants in rainwater using hollow fiber membrane/liquid-phase microextraction combined with gas chromatography-mass spectrometry. *J. Chromatogr. A* **2003**, *1016*, 11–20.

68. King, S.; Meyer, J.S.; Andrews, A.R.J. Screening method for polycyclic aromatic hydrocarbons in soil using hollow fiber membrane solvent microextraction. *J. Chromatogr. A* **2002**, *982*, 201–208.

69. Basheer, C.; Obbard, J.P.; Lee, H.K. Analysis of persistent organic pollutants in marine sediments using a novel microwave assisted solvent extraction and liquid-phase microextraction technique. *J. Chromatogr. A* **2005**, *1068*, 221–228.

70. Zhang, J.; Su, T.; Lee, H.K. Development and application of microporous hollow fiber protected liquid-phase microextraction via gaseous diffusion to the determination of phenols in water. *J. Chromatogr. A* **2006**, *1121*, 10–15.

71. Wu, J.; Lee, H.K. Ion-pair dynamic liquid-phase microextraction combined with injection-port derivatization for the determination of long-chain fatty acids in water samples. *J. Chromatogr. A* **2006**, *1133*, 13–20.

72. Wu, J.; Lee, H.K. Injection port derivatization following ion-pair hollow fiber-protected liquid-phase microextraction for determining acidic herbicides by gas chromatography/mass spectrometry. *Anal. Chem.* **2006**, *78*, 7292–7301.

73. Wen, X.; Tu, C.; Lee, H.K. Two-step liquid-liquid-liquid microextraction of nonsteroidal anti-inflammatory drugs in waste water. *Anal. Chem.* **2004**, *76*, 228–232.

74. Ho, T.S.; Grønhaug Halvorsen, T.; Pedersen-Bjergaard, S.; Rasmussen, K.E. Liquid-phase microextraction of hydrophilic drugs by carrier-mediated transport. *J. Chromatogr. A* **2003**, *998*, 61–72.

75. Ho, T.S.; Reubsaet, J.L.E.; Anthonsen, H.S.; Pedersen-Bjergaard, S.; Rasmussen, K.E. Liquid-phase microextraction based on carrier mediated transport combined with liquid chromatography-mass spectrometry new concept for the determination of polar drugs in a single drop of human plasma. *J. Chromatogr. A* **2005**, *1072*, 29–36.

76. Ugland, H.G.; Krogh, M.; Reubsaet, J.L.E. Three-phase liquid-phase microextraction of weakly basic drugs from whole blood. *J. Chromatogr. B* **2003**, *798*, 127–135.

77. Ugland, H.G.; Krogh, M.; Rasmussen, K.E. Liquid-phase microextraction as a sample preparation technique prior to capillary gas chromatographic-determination of benzodiazepines in biological matrices. *J. Chromatogr. B* **2000**, *749*, 85–92.

78. de Jager, L.; Andrews, A.J.R. Development of a screening method for cocaine and cocaine metabolites in saliva using hollow fiber membrane solvent microextraction. *Anal. Chim. Acta* **2002**, *458*, 311–320.

79. Andersen, S.; Grønhaug Halvorsen, T.; Pedersen-Bjergaard, S.; Rasmussen, K.E. Liquid-phase microextraction combined with capillary electrophoresis, a promising tool for the determination of chiral drugs in biological matrices. *J. Chromatogr. A* **2002**, *963*, 303–312.

80. Andersen, S.; Grønhaug Halvorsen, T.; Pedersen-Bjergaard, S.; Rasmussen, K.E.; Tanum, L.; Refsum, H. Stereospecific determination of citalopram and desmethylcitalopram by capillary electrophoresis and liquid-phase microextraction. *J. Pharm. Biomed. Anal.* **2003**, *33*, 263–272.

81. Grønhaug Halvorsen, T.; Pedersen-Bjergaard, S.; Reubsaet, J.L.E.; Rasmussen, K.E. Liquid-phase microextraction combined with flow-injection tandem mass spectrometry rapid screening of amphetamines from biological matrices. *J. Sep. Sci.* **2001**, *24*, 615–622.

82. Reubsaet, J.L.E.; Loftheim, H.; Gjelstad, A. Ion-pair mediated transport of angiotensin, neurotensin, and their metabolites in liquid phase microextraction under acidic conditions. *J. Sep. Sci.* **2005**, *28*, 1204–1210.

83. Reubsaet, J.L.E.; Paulsen, J.V. Ion-pair mediated transport of small model peptides in liquid phase micro extraction under acidic conditions. *J. Sep. Sci.* **2005**, *28*, 295–300.

84. Pedersen-Bjergaard, S.; Rasmussen, K.E. Electrokinetic migration across artificial liquid membranes new concept for rapid sample preparation of biological fluids. *J. Chromatogr. A* **2006**, *1109*, 183–190.

85. Psillakis, E.; Kalogerakis, N. Developments in liquid-phase microextraction. *Trends Anal. Chem.* **2003**, *22*, 565–574.

86. Rasmussen, K.E.; Pedersen-Bjergaard, S. Developments in hollow fibre-based, liquid-phase microextraction. *Trends Anal. Chem.* **2004**, *23*, 1–10.

87. Pedersen-Bjergaard, S.; Rasmussen, K.E. Bioanalysis of drugs by liquid-phase microextraction coupled to separation techniques. *J. Chromatogr. B* **2005**, *817*, 3–12.

88. Flanagan, R.J.; Morgan, P.E.; Spencer, E.P.; Whelpton, R. Microextraction techniques in analytical toxicology: Short review. *Biomed. Chromatogr.* **2006**, *20*, 530–538.

89. Hou, L.; Wen, X.; Tu, C.; Lee, H.K. Combination of liquid-phase microextraction and on-column stacking fort race analysis of amino alcohols by capillary electrophoresis. *J. Chromatogr. A* **2002**, *979*, 163–169.

90. Rasmussen, K.E.; Pedersen-Bjergaard, S.; Krogh, M.; Ugland, H.G.; Grønhaug, T. Development of a simple in-vial liquid-phase microextraction device for drug analysis compatible with capillary gas chromatography, capillary electrophoresis and high-performance liquid chromatography. *J. Chromatogr. A* **2000**, *873*, 3–11.

91. Grønhaug Halvorsen, T.; Pedersen-Bjergaard, S.; Rasmussen, K.E. Liquid-phase microextraction and capillary electrophoresis of citalopram, an antidepressant drug. *J. Chromatogr. A* **2001**, *909*, 87–91.

92. Wu, H.-F.; Lin, C.-H. Direct combination of immersed single-drop microextraction with atmospheric pressure matrix-assisted laser desorption/ionization tandem mass spectrometry for rapid analysis of a hydrophilic drug via hydrogen-bonding interaction and comparison with liquid-liquid extraction and liquid-phase microextraction using a dual gauge microsyringe with a hollow fiber. *Rapid Commun. Mass Spectrom.* **2006**, *20*, 2511–2515.

93. Kuuranne, T.; Kotiaho, T.; Pedersen-Bjergaard, S.; Rasmussen, K.E.; Leinonen, A.; Westwood, S.; Kostiainen, R. Feasibility of a liquid-phase microextraction sample clean-up and liquid chromatographic/mass spectrometric screening method for selected anabolic steroid glucoronides in biological samples. *J. Mass Spectrom.* **2003**, *38*, 16–26.

94. Zhao, G.; Liu, J.-F.; Nyman, M.; Jönsson, J.Å. Determination of short-chain fatty acids in serum by hollow fiber supported liquid membrane extraction coupled with gas chromatography. *J. Chromatogr. B* **2007**, *846*, 202–208.

95. Lezamiz, J.; Barri, T.; Jönsson, J.Å.; Skog, K. A simplified hollow-fibre supported liquid membrane extraction method for quantification of 2-amino-1-methyl-6-phenylimidazo[4,5-b]pyridine (PhIP) in urine and plasma samples. *Anal. Bioanal. Chem.* **2008**, *390*, 689–696.

96. Wu, J.; Lee, H.K. Orthogonal array designs for the optimization of liquid-liquid-liquid microextraction of nonsteroidal anti-inflammatory drugs combined with high-performance liquid chromatography-ultraviolet detection. *J. Chromatogr. A* **2005**, *1092*, 182–190.

97. Ouyang, G.; Zhao, W.; Pawliszyn, J. Automation and optimization of liquid-phase microextraction by gas chromatography. *J. Chromatogr. A* **2007**, *1138*, 47–54.

98. Zhang, J.; Lee, H.K. Application of liquid-phase microextraction and on-column derivatization combined with gas chromatography-mass spectrometry to the determination of carbamate pesticides. *J. Chromatogr. A* **2006**, *1117*, 31–37.

99. Lai, B.-W.; Liu, B.-M.; Malik, P.K.; Wu, H.-F. Combination of liquid-phase hollow fiber membrane microextraction with gas chromatography-negative chemical ionization mass spectrometry for the determination of dichlorophenol isomers in water and urine. *Anal. Chim. Acta* **2006**, *576*, 61–66.

100. Lezamiz, J.; Jönsson, J.Å. Development of a simple hollow fibre supproted liquid membrane extraction method to extract and preconcentrate dinitrophenols in environmental samples at ng/mL level by liquid chromatography. *J. Chromatogr. A* **2007**, *1152*, 226–233.

101. Bartolomé, L.; Lezamiz, J.; Etxebarria, N.; Zuloaga, O.; Jönsson, J.Å. Determination of trace levels of dinitrophenolic compounds by microporous membrane liquid-liquid extraction in environmental water samples. *J. Sep. Sci.* **2007**, *30*, 2144–2152.

102. Zorita, S.; Mårtensson, L.; Mathiasson, L. Hollow-fibre supported liquid membrane extraction for determination of fluoxetine and norfluoxetine concentration at ultra trace level in sewage samples *J. Sep. Sci.* **2007**, *30*, 2513–2521.

103. Zorita, S.; Barri, T.; Mathiasson, L. A novel hollow-fibre microporous membrane liquid–liquid extraction for determination of free 4-isobutylacetophenone concentration at ultra traces level in environmental aqueous samples. *J. Chromatogr. A* **2007**, *1157*, 30–37.

104. Lambropoulou, D.A.; Albanis, T.A. Application of hollow fiber liquid phase microextraction for the determination of insecticides in water. *J. Chromatogr. A* **2005**, *1072*, 55–61.

105. Psillakis, E.; Mantzavinos, D.; Kalogerakis, N. Development of a hollow fibre liquid phase microextraction method to monitor the sonochemical degradation of explosives in water. *Anal. Chim. Acta* **2004**, *501*, 3–10.

106. Chen, P.-S.; Huang, S.-D. Determination of ethoprop, diazinon, disulfoton and fenthion using dynamic hollow fiber-protected liquid-phase microextraction coupled with gas chromatography-mass spectrometry. *Talanta* **2006**, *69*, 669–675.

107. Hou, L.; Lee, H.K. Determination of pesticides in soil by liquid-phase microextraction and gas chromatography-mass spectrometry. *J. Chromatogr. A* **2004**, *1038*, 37–42.

108. Basheer, C.; Lee, H.K. Analysis of endocrine disrupting alkylphenols, chlorophenols and bisphenol-A using hollow fiber-protected liquid-phase microextraction coupled with injection port-derivatization gas chromatography-mass spectrometry. *J. Chromatogr. A* **2004**, *1057*, 163–169.

109. Jiang, X.; Oh, S.Y.; Lee, H.K. Dynamic liquid-liquid-liquid microextraction with automated movement of the acceptor phase. *Anal. Chem.* **2005**, *77*, 1689–1695.

110. Chen, C.-C.; Melwanki, M.B.; Huang, S.-D. Liquid-liquid-liquid microextraction with automated movement of the acceptor and the donor phase for the extraction of phenoxyacetic acids prior to liquid chromatography detection. *J. Chromatogr. A* **2006**, *1104*, 33–39.

111. Wu, J.; Ee, K.H.; Lee, H.K. Automated dynamic liquid-liquid-liquid microextraction followed by high-performance liquid chromatography-ultraviolet detection for the determination of phenoxy acid herbicides in environmental waters. *J. Chromatogr. A* **2005**, *1082*, 121–127.

112. Liu, J.-F.; Toräng, L.; Mayer, P.; Jönsson, J.Å. Passive extraction and clean-up of phenoxy acid herbicides in samples from a groundwater plume using hollow fiber supported liquid membranes. *J. Chromatogr. A* **2007**, *1160*, 56–63.

113. Charalabaki, M.; Psillakis, E.; Mantzavinos, D.; Kalogerakis, N. Analysis of polycyclic aromatic hydrocarbons in wastewater treatment plant effluents using hollow fiber liquid-phase microextraction. *Chemosphere* **2005**, *60*, 690–698.

114. Wang, X.; Mitra, S. Enhancing micro-scale membrane extraction by implementing a barrier film. *J. Chromatogr. A* **2006**, *1122*, 1–6.

115. Fontanals, N.; Barri, T.; Bergström, S.; Jönsson, J.Å. Determination of polybrominated diphenyl ethers at trace levels in environmental waters using hollow-fiber microporous membrane liquid–liquid extraction and gas chromatography–mass spectrometry. *J. Chromatogr. A* **2006**, *1133*, 41–48.

116. Basheer, C.; Lee, H.K.; Obbard, J.P. Application of liquid-phase microextraction and gas chromatography-mass spectrometry for the determination of polychlorinated biphenyls in blood plasma. *J. Chromatogr. A* **2004**, *1022*, 161–169.

117. Chia, K.-J.; Huang, S.-D. Simultaneous derivatization and extraction of primary amines in river water with dynamic hollow fiber liquid-phase microextraction followed by gas chromatography-mass spectrometric detection. *J. Chromatogr. A* **2006**, *1103*, 158–161.

118. Lee, H.S.N.; Basheer, C.; Lee, H.K. Determination of trace level chemical warfare agents in water and slurry samples using hollow fiber-protected liquid-phase microextraction followed by gas chromatography-mass spectrometry. *J. Chromatogr. A* **2006**, *1124*, 91–96.

119. Huang, S.-P.; Huang, S.-D. Dynamic hollow fiber protected liquid phase microextraction and quantification using gas chromatography combined with electron capture detection of organochlorine pesticides in green tea leaves and ready-to-drink tea. *J. Chromatogr. A* **2006**, *1135*, 6–11.

120. Zhu, L.; Huey, K.; Zhao, L.; Lee, H.K. Analysis of phenoxy herbicides in bovine milk by means of liquid-liquid-liquid microextraction with a hollow-fiber membrane. *J. Chromatogr. A* **2002**, *963*, 335–343.

Chapter 8

Sample Preparation in Membrane Introduction Mass Spectrometry

Raimo A. Ketola, Tapio Kotiaho, and Frants R. Lauritsen

8.1. INTRODUCTION TO MEMBRANE INTRODUCTION MASS SPECTROMETRY (MIMS)

MIMS can be used for a direct sampling and analysis of atmospheric gases and organic compounds from soils, liquids, and gaseous samples. The direct sampling is achieved through an inert polymer membrane that is used as the only interface between a sample and a mass spectrometer. The analytes dissolve into the membrane, diffuse through it, and vaporize into the mass spectrometer, where they are ionized and analyzed according to their *m/z* ratio. Thus, most volatile organic compounds (VOCs) can be analyzed directly, and a continuous flow of the sample can be monitored online with MIMS. The membranes used in MIMS are usually very hydrophobic, and hydrophobic compounds dissolve very well into the membrane. They have low detection limits (nanogram per liter in water and microgram per cubic meter in air), whereas hydrophilic compounds have detection limits in the high microgram per liter range in water and milligram per cubic meter in air. At the moment, the detection of VOCs (bp < 200°C) has become a routine matter, and new methods in which less volatile organic compounds are preconcentrated inside a cold membrane before they are thermally desorbed from the membrane into the mass spectrometer have made the analysis of many less volatile organic compounds, such as pesticides (atrazine, $C_8H_{14}ClN_5$), polycyclic aromatic hydrocarbons (PAHs) (benzo[e]pyrene, $C_{20}H_{12}$), and some steroids (α-tocopherol, $C_{29}H_{50}O_2$), possible. The detection of hydrophilic compounds has also been enhanced using charge exchange ionization and different types of membranes.

8.1.1. Membrane Inlet Configurations

Various membrane inlets have been designed to fit specific applications. They can be divided into six major categories (Fig. 8.1): (a) a direct insertion membrane probe, with a tubular or sheet membrane, mounted inside the vacuum and preferentially inside the ion source itself; (b) a flow-over design using a flat sheet membrane, where one side of the membrane is exposed to the sample and the other to the vacuum. In the optimal design, the membrane directly constitutes a part of the wall in the ion source of the mass spectrometer; (c) a helium-purged inlet, where molecules permeate from the sample through a capillary membrane into the inner volume of the capillary and are then purged with helium to the mass spectrometric ion source; (d) a membrane probe, where the membrane is mounted at the end of a probe, which is inserted into the sample matrix; (e) a sample cell, where the membrane simultaneously constitutes a part of the wall of sample container, for example, a mini-reactor, and the interface to the mass spectrometer; (f) a trap-and-release design, where a capillary membrane is mounted inside the ion source of the mass spectrometer in a fashion that makes direct heating of the membrane via radiation from the filament possible.

The membrane probe (Fig. 8.1d) is probably the simplest inlet to use. It consists of a steel capillary perforated and covered by a polymer membrane in one end, whereas the other end is connected to the mass spectrometer via an evacuated tube. The probe can be inserted directly into almost any sample. For example, they have been inserted into plants[1] and sediments.[2] The drawback of the membrane probe is that it can only be used to analyze gases and highly volatile organic compounds (boiling point below 100°C). Compounds of lower volatility interact with the surfaces in the evacuated tube that connects the inlet to the mass spectrometer, and very long response times are often obtained. In order to circumvent the problem with condensation of sample molecules in the tube, the direct insertion membrane probe (Fig. 8.1a) was invented.[3] In these inlets, the membrane is mounted at the end of a probe that is inserted directly into the mass spectrometer. The liquid or gaseous

Handbook of Sample Preparation, Edited by Janusz Pawliszyn and Heather L. Lord

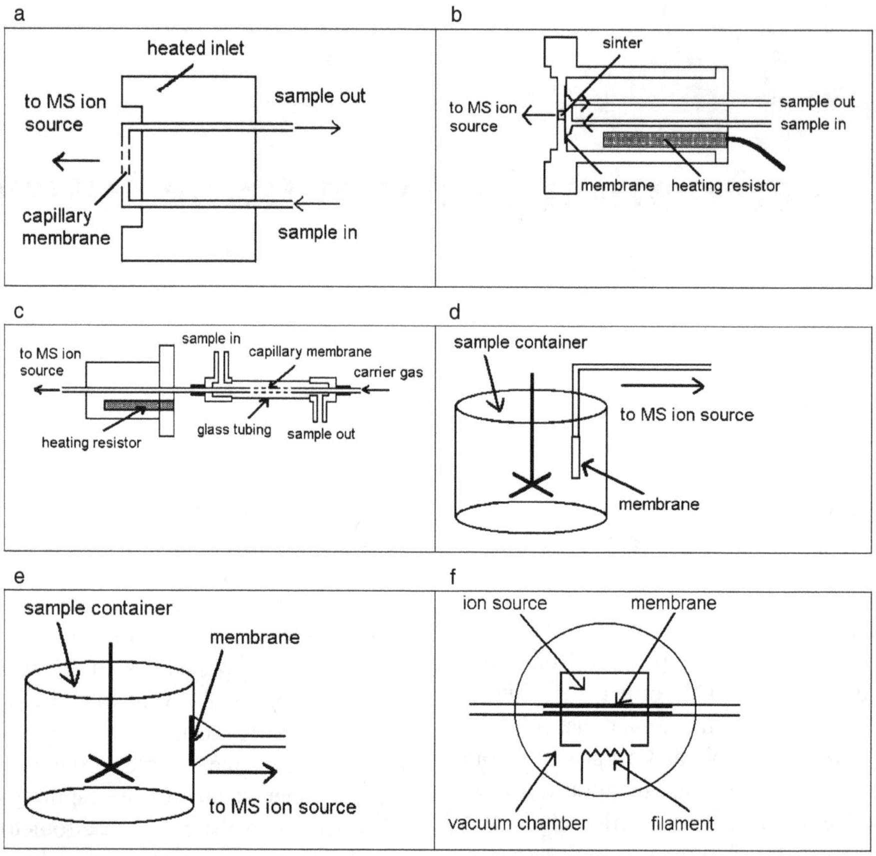

Figure 8.1. The most common membrane inlets used. (a) A direct insertion membrane probe, (b) a flow-over design using a flat sheet membrane, (c) a helium-purged inlet, (d) a membrane probe, (e) a sample cell, and (f) a trap-and-release design.

sample is flushed through the probe. The condensation of sample molecules in a connecting vacuum tube can be practically eliminated[4] and VOCs with a boiling point up to around 200°C can be measured. Furthermore, these probes have the advantage that the sample can be chemically modified online prior to the detection, and a calibration using an external standard is straightforward. An alternative to the close coupling of the membrane inlet to the ionization region of the mass spectrometer is the helium-purged membrane inlet (Fig. 8.1c).[5] In this inlet, a liquid or gaseous sample is flushed across one side of the capillary membrane, whereas the other side is continuously purged by a helium stream that carries the permeated molecules to the ion source of the mass spectrometer. The purging of molecules from the membrane to the ion source reduces the problem with condensation effects in connection tubes, and the inlet has almost the same applications as the direct insertion membrane probe. A microbial broth or a muddy aqueous sample might block the tubes in direct insertion membrane probes and the helium-purged inlet, and in these cases, the sample cell (Fig. 8.1e) can be used instead.[6] A small volume (a few milliliters) of untreated sample is transported to the sample cell for direct analysis. The sample cell can be easily heated, if necessary. Chemical and biological reactions are best studied with MIMS using this kind of a mini-reactor sample cell.

The efficiency of the membrane inlet system depends on the evaporation of analyte molecules from the membrane into the vacuum. As long as the analyte has a fairly high vapor pressure, this is not a problem, but it limits the use of traditional membrane inlets to the analysis of compounds with a boiling point below 250°C. Compounds with a boiling point above 250°C have to be analyzed using the trap-and-release technique.[7,8] In this technique, the aqueous sample is passed through a membrane inlet with a cold capillary membrane (<100°C), and less volatile compounds dissolve into the membrane and accumulate therein. After an accumulation period of 2–20 min, the less volatile compounds are released from the membrane into the mass spectrometer by rapid heating of the membrane. The trap-and-release system reminds of an inverted solid-phase microextraction (SPME).[9] Instead of inserting a silicone fiber into a sample for collection of hydrophobic organic compounds and then transferring the fiber with the analytes to an instrument for release and analysis, the sample is transferred to a silicone membrane mounted inside the mass spectrometer for collection and analysis of the analytes. Using this technique, high-molecular-weight and less volatile organic contaminants, such as PAHs, certain pesticides, and many estrogenic compounds, can be analyzed directly from water.[10]

Selectivity or sensitivity of MIMS instruments can be enhanced by combining them with another analytical technique for separation or concentration. The two-stage inlets include systems where another concentration or separation is made before or after the MIMS inlet, but two-step membrane separation systems have also been developed. One combination that is frequently used is the combination of a helium-purged membrane inlet and a jet separator.[11] In this combination, the jet separator is used for selective removal of helium from the purge stream before it reaches the mass spectrometer.

8.1.2. Limitations of MIMS

The most important limitation of traditional MIMS techniques is its inability to analyze polar compounds, which excludes the analysis of most biological metabolites and many pharmaceuticals. Polar compounds do not permeate very well through the traditionally used membranes, and they are generally not analyzed with MIMS. However, recent research using a new type of membrane inlet system, where the membrane has a donor liquid (the sample) on one side and an acceptor liquid on the other side, has been investigated. Such a liquid-membrane-liquid system has two major advantages: (1) polar membranes can be used for improved permeation of hydrophilic compounds and (2) advanced inlet and ionization principles, such as atmospheric pressure chemical ionization (APCI)[12] and electrospray ionization,[13] can be used. With the water-membrane-acceptor solution approach, MIMS might enter a new era, where practically any compound can be analyzed directly from an aqueous sample.

Another drawback of the membrane inlets is caused by memory effects; that is, analytes or matrix components from one run are carried over to the following analysis. The main reason for this memory effect is the sticking of sample molecules to the membrane or surfaces in the analytical system. In worst cases, where extremely hydrophobic compounds (e.g., PAHs) are analyzed, the carryover can be observed for minutes or even hours. The memory effect is the most severe with aqueous samples, where the inlet temperature has to be kept below 80°C and the least severe with air samples, where high temperatures above 150°C can be used. To prevent or decrease this memory effect, several procedures can be applied, for example, dilution of the sample, calibration and analysis using nonsteady state (see Section 8.2.1), thinner membranes, elevated membrane temperature, and analysis of blank samples between the real samples, for example, in trap-and-release MIMS.

8.2. APPLICATIONS OF MIMS TO WATER ANALYSIS

The capability for direct analysis of gases and VOCs in aqueous solution has been the major advantage of MIMS, and it is also here most applications are found. MIMS was first applied by Hoch and Kok[14] for the study of photosynthetic processes in algae and used to study the kinetics of oxygen evolution in photosynthesis. In the years to come, the analysis of contaminants in ground and drinking water and monitoring and characterization of fermentation processes should be the dominant applications.[15–17]

Fast and direct detection of contaminants in water samples is very useful for the characterization of drinking water supply systems, but many contaminants are dangerous at very low levels and high sensitivity is required. It took almost 30 years of development of the technique before several research groups in 1995[18–20] reached a goal of detection limits in the 1–10 ppt range for solvents such as chloroethylenes and gasoline residuals in drinking water. Today the analysis of solvents in aqueous samples is straightforward with response times between 30 and 60 s and detection limits in the upper nanogram per liter level are routine.

Whereas MIMS is often used in the field for gas analysis (see Section 8.3.3), there are only a few examples of water analysis using transportable MIMS systems. The most prominent example is the characterization of a petrochemical pollution in a river using an MIMS system mounted on a boat.[21] Recently, a renewed interest in *in situ* applications has arisen. For example, an underwater and remote-controlled MIMS system has been built that can be used to detect chemical plumes in harbors,[22] and recently a deep water compatible MIMS system was used to create depth profiles of gases in the ocean up to 500-m depth.[23] In some countries, MIMS instruments are used for online surveillance of drinking water supply systems for the content of common pollutants like chloroethylenes and methyl *tert*-butyl-ether (MTBE).

A major application of MIMS is the characterization and monitoring of fermentation processes.[15] Important gases, such as oxygen, carbon dioxide, and methane, can be monitored online and used for feedback control of the fermentation process. In some cases, VOCs of low molecular weight, such as ethanol, acetic acid, acetoin, and butanediol, have also been monitored.[24] The analysis of VOCs generally requires a short distance from the membrane to the ionization chamber of the mass spectrometer, and samples are therefore often drawn from the reactor and transported to a flow injection analysis (FIA) system. The FIA system offers the extra advantage that the sample can be online acidified/alkalized for improved detection of acids and bases. Membrane inlet mass spectrometers have also been incorporated directly into large fermentors,[25] and small bioreactors for online monitoring of VOCs have been fully integrated with the membrane inlet mass spectrometers.[26]

Larger organic compounds are not so easily monitored online with MIMS, due to slow response times and interactions at surfaces in vacuum.[27] However, such compounds are quite often analyzed from microbial media using MIMS. In these experiments, a sample from the fermentation broth containing organic compounds, salts, organisms, and so on, is simply transferred to a small measuring cell equipped with a membrane inlet,[6] and after a few minutes, a mass spectrum of the volatiles can be recorded. Using this technique, many hitherto unknown organic metabolites have been found and identified. In one example, high concentrations of pyrazines (including seven novel natural products) were found in a fermentation broth of an otherwise well-characterized industrial polymyxine-producing organism.[28] Also, the measurement of polar compounds can be enhanced by the analysis of derivatized products using MIMS and polydimethylsiloxane (PDMS) membrane. An example of this procedure is the acetylation of phenolic compounds in the aqueous phase and subsequent analysis with MIMS.[29] The results obtained showed that with acetylation phenolic compounds, such as phenol and 4-nitrophenol, can be detected at one to two orders of magnitude lower levels compared with the direct analysis.

8.2.1. Quantification

Aqueous samples are generally quantified using an external standard method. This is possible with most environmental samples, where concentrations are very low. Under these circumstances, the permeation through the membrane is unaffected by changes in the aqueous matrix; thus, the standards can be prepared in distilled/deionized water. This simple calibration cannot be done when analyzing microbial suspensions, where the concentration of sugars and salts can be quite high.[26,30] Figure 8.2 shows a series of linearity curves for 3-methylbutanal obtained from various matrices (distilled water, phosphate buffer, growth medium, and acidified broth before and after fermentation). The linearity curves are individually linear, but for a given concentration, the signal varies with up to a factor of three, depending on the matrix. Here, it is very important that the external standard is prepared in the growth medium and not in pure water. For quantification during online fermentations, it is

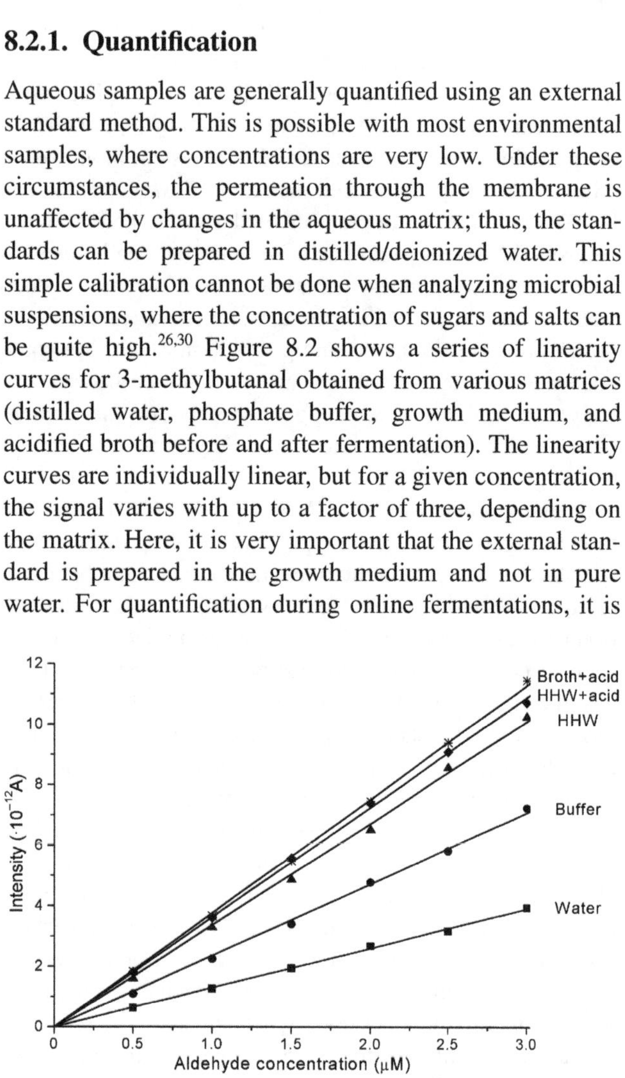

Figure 8.2. The matrix effects of buffer, growth medium, acid, and broth on the response of 3-methylbutanal. Online monitoring of important organoleptic methyl-branched aldehydes during batch fermentation of the starter culture *Staphylococcus xylosus* reveals new insight into their production in a model fermentation. HHW, Hussein, Hastings, and White medium. Reprinted from reference 26 with permission from John Wiley & Sons Limited. Copyright 2004.

very important that the signal is unaffected by matrix changes during the fermentation, and this seems generally to be the case. Off-site analysis of microbial suspensions is best done using the method of standard addition.

An important aspect of quantification with an MIMS system is the sample flow rate around the membrane. If the flow is sufficiently slow the sample solution will be completely depleted for analyte during its passage through the inlet and the signal becomes linearly dependent on the flow rate. With increasing flow rate, the signal dependency on the flow rate reduces until it disappears at a relatively high flow rate. If possible the membrane inlet should be operated under the latter condition, but quantification can be obtained as long as the flow rate is kept constant.

With a constant sample flow rate around the membrane, quantification can be performed using three different approaches, namely steady-state analysis, FIA, and dynamic FIA. The three different approaches are illustrated in Figure 8.3, which shows the flow of sample molecules through the membrane following a step change in concentration at the membrane surface. Initially, there will be a lag phase before the first sample molecules have diffused through the membrane and into the mass spectrometer. The signal then rises rapidly as an increasing number of analyte molecules enter the mass spectrometer. After some time, typically 30–90 s for VOCs, the signal reaches a steady state, where the net number of molecules entering the membrane equals the number of molecules entering the mass spectrometer. Under this steady-state condition, the mass spectrometric response is directly proportional to the concentration of the analyte in the aqueous sample.[31]

If the time it takes to reach the steady state is too long, the amount of sample is limited or if there is a drift associated with the establishment of the steady-state signal, then the FIA principle can be advantageously used. In FIA, the membrane is continuously exposed to a blank sample solution except for a short well-defined period, where it is exposed to the sample solution. The exposure period is normally shorter than the time it takes to reach steady state. Under such circumstances, the sample signal will initially follow the same trend as shown in Figure 8.3 for a step change in concentration, but it will reach a maximum signal, which is smaller than that observed at steady state and disappear again, resulting in a peak signal. When the membrane is exposed to the sample solution for exactly the same time in each FIA injection, then the peak height is directly proportional to the analyte concentration in the sample.[32] In some applications, such as a moving membrane inlet system passing through an uncharacterized pollution plume, the exposure time is not known and quantification cannot be performed using the peak height alone. In dynamic FIA,[33] both the peak height and the peak width are used for quantification in a relatively simple algorithm.

Temperature influences quantification of aqueous samples in two ways. First, the signal rises with increasing temperature until water starts to boil at around 100°C and the signal drops abruptly.[34] As a rule of thumb, the analyte signal will increase with a factor of two for every 10°C increase. Second, the gases dissolved in the water will start to form bubbles as the temperature is increased and the bubbles formed on the membrane surface create very erratic signals that make quantification impossible. Generally, the bubble formation prevents the analysis of environmental aqueous samples at temperatures above 70°C and biological media with high carbon dioxide content at temperatures above 50°C. The problem with bubble formation is reduced if the sample is pumped through the membrane inlet instead of drawn by suction.

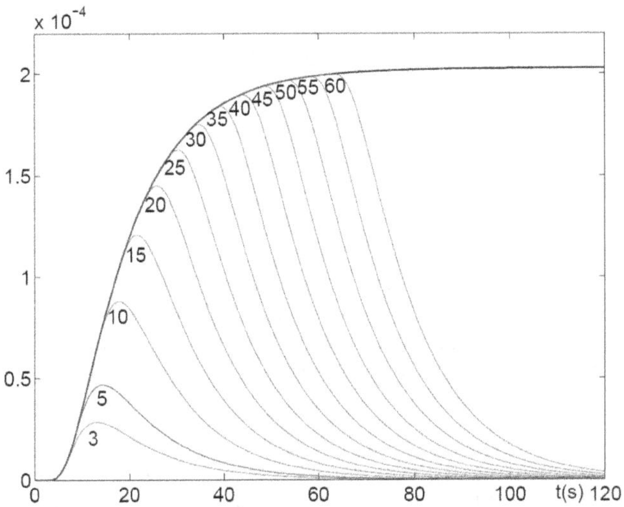

Figure 8.3. Theoretical MIMS response following a step change in concentration in front of the membrane (thick curve) and responses to flow injections performed with varying times of exposures to the membrane (thin curves). The simulations were kindly performed by C. Janfelt, Department of Chemistry, Copenhagen University.

8.3. ANALYSIS OF AIR SAMPLES

8.3.1. Analysis of VOCs in Air

One of the main applications of MIMS has been the measurement of VOCs from air. Table 8.1 shows the performance characteristics of selected VOCs from air using a flow-over design of MIMS inlet with a single quadrupole mass spectrometry (MS) instrument, demonstrating detection limits at low microgram per cubic meter levels or below. Even lower detection limits at nanogram per cubic meter levels have been reached with more complicated membrane inlets and MS instruments. Importantly, the response times that can be obtained using an MIMS system (a few seconds) are very short when measurements are done from air using a thin membrane (25 µm); still, wide linear dynamic ranges are obtainable, typically three to five orders of magnitude.[35,36]

Air is an easier sample matrix than water for MIMS analysis because the analytes are already in the gas phase and the air samples usually contain only small amounts of particles; thus, sampling is easy. By far, the most important application in MIMS analysis of air samples is the measurement of VOCs. Most of the different inlet types have been utilized in the analysis of VOCs from air samples, showing

Table 8.1. Detection Limits, Response Times, and Linear Ranges Measured by MIMS for a Selected Set of Environmentally Significant Compounds from Air Samples

Compound	Response Time (s)[a]	Detection Limit, µg/m^3 (pptrv)[b]	Linear Range (µg/m^3)
Benzene	1	1 (310)	1–10,000
Toluene	1	0.5 (130)	1–10,000
Xylenes	2	2 (450)	2–10,000
Chlorobenzene	2	0.5 (110)	2–10,000
Chloroform	1	5 (1000)	5–30,000
Carbon tetrachloride	1	3 (480)	10–20,000
1,2-Dichloroethene	1	1 (250)	5–10,000
1,1-Dichloroethane	1	1 (250)	1–5000
1,1,1-Trichloroethane	2	4 (730)	10–30,000
1,1,2,2-Tetrachloroethane	3	1 (150)	1–5000
Trichloroethene	1	1 (180)	1–10,000
Tetrachloroethene	1	0.5 (70)	1–10,000
Dimethyl sulfide	1	1 (390)	1–40,000
Dimethyl disulfide	2	2 (510)	2–40,000
Ethanethiol	2	5 (1900)	5–50,000

[a]Membrane (polydimethylsiloxane) thickness 25 µm. Response time is defined as the time it takes for the signal to rise from 10% to 90%.

[b]S/N = 3; pptrv = parts per trillion by volume.

that gaseous samples are easy and suitable for various MIMS configurations and instruments. The most common membrane used is PDMS since, as with water samples, it has one of the highest concentration factors and it is very durable, also at high temperature and high pressure. In addition to variation of concentrations of different VOCs, the parameters that mostly affect the signals and thus calibration with MIMS are temperature, humidity, and pressure of the air sample. Very high concentrations have previously been shown to cause memory effects with water samples, and this effect is similar with air samples. However, the magnitude of memory effect is much smaller, due to fast response times with air samples compared with those with water samples, being approximately one order of magnitude faster. With air samples, the memory effect can be further diminished using elevated membrane temperature, up to 200°C, which increases the desorption of VOCs from the membrane. Thus, the temperature range of the samples can vary from 0°C, or even below that, to over 200°C. If the temperature of the original sample is higher than that of the transfer lines and the membrane inlet, there is a risk of condensation of water and analytes with high boiling point inside the membrane inlet, thus contaminating the transfer lines and the inlet. Lower detection limits (20 pptrv [parts per trillion by volume] toluene in air) were detected with an ion trap mass spectrometer, using a membrane inlet design incorporating a jet separator for a second enrichment stage.[37] Still, no evidence of long-term memory effects after membrane exposure to high concentrations was observed when three sample matrices, namely, air, water, and soil, containing 1 ppb toluene were analyzed in sequence.

The calibration is normally performed either using gas bags, static dilution bottles, permeation tubes, or different gas calibration devices for VOC or gas analysis, for example, a gas calibrator.[35] The calibration is normally performed using steady-state conditions, as described in Section 8.2.1, because the response times are short in air analysis and the steady state is reached very rapidly.

In order to increase selectivity in the VOC analysis, an MIMS technique called temperature-programmed desorption (TPD-MIMS) was used to obtain analyte separation.[38,39] The gaseous sample was flushed through a solid adsorbent in a steel tube, which was then heated at a controlled rate releasing analytes into a helium stream entering a membrane inlet, resulting in good peak separation with 10-s desorption profiles at half height. Figure 8.4 shows desorption profiles of nine compounds obtained from analysis of a mixture containing a few micrograms of each contaminant.

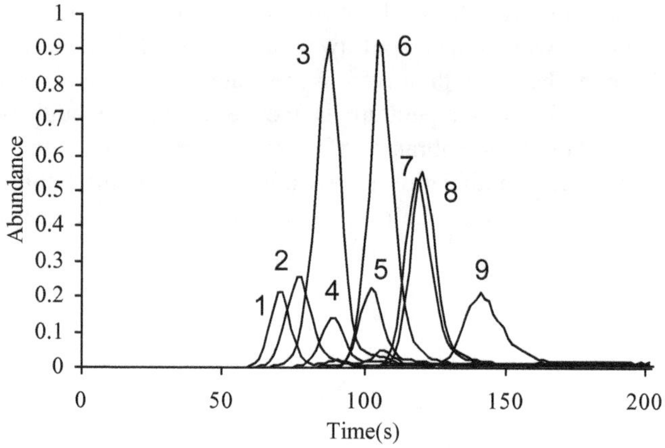

Figure 8.4. Desorption profiles of nine compounds obtained from the analysis of a mixture containing a few micrograms of each contaminant with TPD-MIMS. (1) *trans*-1,2-Dichloroethene, (2) chloroform, (3) carbon tetrachloride, (4) trichloroethene, (5) toluene, (6) tetrachloroethene, (7) xylenes, (8) styrene, and (9) 1,2-dibromo-1,2-dichloroethene in selected in monitoring (SIM) mode using HayeSepD adsorbent. Reprinted from reference 38 with permission from John Wiley & Sons Limited. Copyright 1998.

8.3.2. Analysis of Semivolatile and Polar Organic Compounds in Air

The analysis of polar or semivolatile organic compounds (SVOCs) with MIMS is not as straightforward as that of VOCs either because the membrane used is hydrophobic (polar organic compounds) or because their vapor pressures are too low. To improve sensitivity, especially for polar VOCs, chemical ionization (CI) using water or air as the reagent gas proved to be useful. Taking advantage of the presence of water vapor in the samples or using supplemental water, with CI-MIMS, it has been possible to obtain detection limits for airborne polar VOCs at 50% relative humidity from 31 ppbv for MTBE to 104 ppbv for methyl ethyl ketone.[40] Also, PAHs can be analyzed using a membrane inlet connected to a gas chromatography (GC)–MS.[41] After sampling into a HEPA glass fiber filter tape sampler, the compounds adsorbed were then desorbed into a membrane inlet through a GC column. In tests at an oil combustion plant, a production plant for coal tar-based electrodes, and a motor test site, a good correlation was found between the on-site method and literature-derived data.[41] To increase the permeation rates of airborne SVOCs through the membrane, ultrathin composite membranes have been employed.[42] The ultrathin fiber reduced the amount of membrane-dissolved analyte present, an advantage for compounds with slow diffusion rates and high solubilities. SVOCs from several compound classes were tested including nitrobenzene, methyl salicylate, 2-chlorophenol, and malathion, an organophosphate pesticide.

Several other SVOCs, including lindane (a pesticide), dimethylmethyl phosphonate (DMMP), naphthalene, and hexahydro-1,3,5-trinitro-1,3,4-triazine (RDX, an explosive), in air have been analyzed using a technique called single-sided membrane introduction mass spectrometry (SS-MIMS).[43] To overcome the slow diffusion through PDMS, the method utilized adsorption and desorption from the same side of the membrane. Rise times of 4–7 s and fall times of 12–36 s were achieved with cycle times from sampling to data acquisition of approximately 1 min; limits of detection were in the pptrv range with a linear response of over four orders of magnitude.

8.3.3. Online/On-Site/*In Situ* Applications

Due to the minimal need for sample treatment, MIMS is very suitable for online and on-site applications. This is also possible due to the recent development of low power consumption portable MS instruments. A transportable time-of-flight MS incorporating a membrane inlet system was tested for environmental analysis.[44] Converging annular geometry in the MS allowed for lower source pressure, smaller pumps, low power consumption, and a more efficient membrane inlet design in which two membranes operated in series, providing a secondary enrichment stage with enrichment factors as high as 15,600.

Another portable vapor detection quadrupole analyzer, incorporating a silicone membrane separator for pollution monitoring, solvent detection in degreasing operations, identification of pesticides, and monitoring of cholinesterase-inhibiting chemicals, was tested already in the 1970s.[45,46] The membrane separator was pumped, using evacuated canisters containing zeolite, thus eliminating mechanical pumping of this stage in the field. Detection limits for VOCs in ambient air ranged from 0.1 ppbv tetrachloroethene to 5 ppbv phosdrin, an organophosphate compound.

For the measurement of isoprene (2-methyl-1,3-butadiene), a highly reactive short-lived species emitted in significant amounts in the environment, a combination of a membrane inlet, cryogenic focusing, and CI in the ion trap MS was used.[47] An automated membrane sampler was used to collect permeate in a cryogenic sample loop, and ballistic heating under a helium stream delivered the sample plug to the MS. Vinyl methyl ether, a selective CI reagent for alkenes, was leaked into the MS and allowed to react with preconcentrated sample, providing linear isoprene calibrations over a range typical in deciduous forest environments[48] and a detection limit of 80 pptrv.

With an optimized MIMS method, paint shop exhaust air was analyzed, using a quadrupole MS and a silicone sheet membrane inlet.[35] Linear dynamic ranges for 14 VOCs extended over four orders of magnitude, and limits of detection ranged from 0.5 to 5 μg/m^3. The analytical results obtained with MIMS were compared with those obtained with an online flame ionization detector (FID) analyzer, and comparable results were achieved with differences in VOC totals ranging from −6.7% to 15.0%. Throughput in MIMS as high as 150 samples per hour was possible.

Air in five laboratory rooms was analyzed using a membrane inlet and a quadrupole MS.[49] Samples were collected twice a day in passive sampling tubes, which were then desorbed into a silicone sheet membrane interface. The results obtained with the continuous monitoring MIMS technique over 7 days were in good agreement with those obtained from passive sampling tubes. MIMS also provided sensitivity high enough for real-time monitoring of many volatile organic sulfur compounds (VOSCs) including carbon disulfide, ethanethiol, dimethyl sulfide, thiophene, and dimethyl sulfoxide.[36] Mixed VOSC samples containing three compounds were analyzed using a deconvolution program. The average of the absolute differences between calculated and measured concentrations was 6.1%.

Another interesting demonstration of the wide applicability of MIMS is online measurement of industrial solvents (e.g., styrene, 1,1,1-trichloroethane, and toluene) in breath[50] to noninvasively monitor solvent exposure in workers. Researchers combined a silicone membrane inlet and gas-handling system with a transportable quadrupole MS for direct analysis of VOCs in breath. The calibration must be performed carefully as it was noticed that water vapor in breath air affects the signals obtained with MIMS, that is, the amount of water vapor typically found in breath lowered the signals of solvents by 24–27%. This is due to the fact that water molecules hinder the transport of other molecules across the membrane, probably blocking the active sites. Thus, gaseous standards must contain approximately the same amount of water vapor than the real breath samples contain. This also applies to other air samples with high humidity.

8.4. ANALYSIS OF SOIL SAMPLES

MIMS applications in soil analysis can be divided into two areas, that is, the analysis of organic compounds, namely, VOCs[51–55] or SVOCs,[56] and the analysis of gases.[1,2,57–66] In the case of analysis of organic compounds, the soil sample is transferred to a vessel that is attached to the mass spectrometer in various ways, whereas the gas analysis is performed using a membrane probe that is inserted directly into the soil (inclusively peat and sediment) to be analyzed. The latter technique works very well with gas analysis, since the gases do not stick to the soil and the design prevents interferences from the atmosphere during the sampling step. Organic compounds on the other hand bind to the soil and need to be thermally released prior to analysis. In addition, very long response times are to be expected because of interactions between the organic compounds and the vacuum surfaces in the long, narrow connection tube to the mass spectrometer. The sample vessels (similar to that in Fig. 8.1e) used for the analysis of organics in soil samples are specially made for the different techniques,[51–54,56] and they normally operate with small samples (<10 g). This could cause large variations to the results because of the inhomogeneous nature of soil. In one case, a membrane probe[55] was inserted into the soil for direct analysis of VOCs. In this system, sample collection is avoided, but it is uncertain from how large a volume sample is collected and matrix effects are not well understood. Even so, good correlation between an United States Environmental Protection Agency (EPA) Method 8260 and the penetrometer MIMS method was observed for some samples ($r^2 = 0.9$).

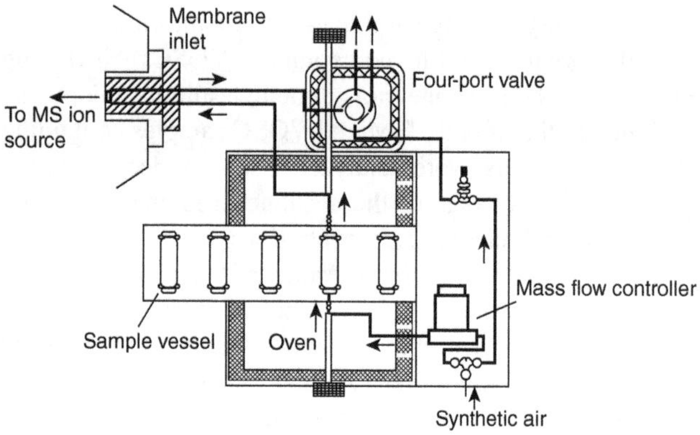

Figure 8.5. A schematic picture of the purge-and-membrane mass spectrometry device. Reprinted from reference 54 with permission from the American Chemical Society. Copyright 2001.

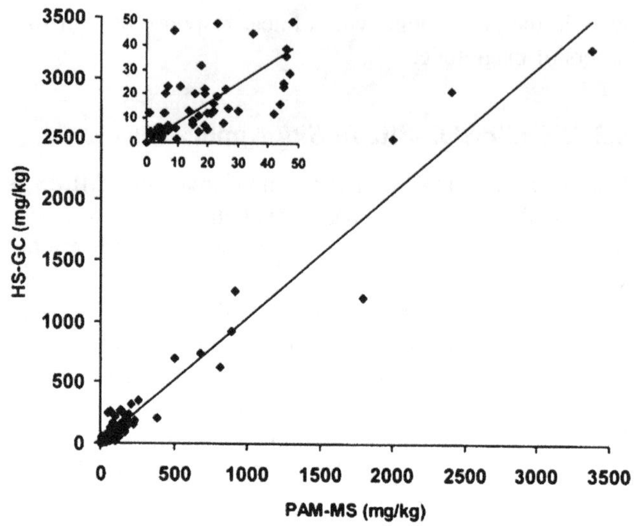

Figure 8.6. Correlation of the quantitative results obtained by the purge-and-membrane mass spectrometric method and a headspace gas chromatographic method. Reprinted from reference 54 with permission from the American Chemical Society. Copyright 2001.

As an example, the performance characteristics of the dynamic headspace membrane inlet method, and the purge-and-membrane mass spectrometry (PAM-MS) method, are presented in more detail since they have been studied the most thoroughly.[51,53,54] Figure 8.5 shows in detail the apparatus used in the PAM-MS technique; all the materials used in the device were selected so that the possible memory effects would be minimized.[54] In a typical PAM-MS analysis cycle, a soil sample is mounted to a sample vessel, and after a certain sample preheating time (e.g., 10 min at 80°C), the four-port valve is switched to the sampling position and simultaneous sampling line needles are punctured through the sample vessel septa, and finally, the purge gas with the desorbed VOCs are directed via the membrane inlet into the mass spectrometer for analysis.

One of the main points studied in the PAM-MS method development was the effect of soil type and the moisture content on the signal observed. These studies were made using different moisture contents (0%, 10%, and 20%) and sand (organics 0%), garden soil (organics 9%), and 1:1 mixture of sand and garden soil samples. Some general observations were (1) desorption times increased as the organic matter content was increased; (2) desorption times normally decreased as the water content was increased; however, for sand samples moisture content did not have much effect; (3) the highest peak areas were observed for dry sand and lowest for dry garden soil; and (4) peak areas for sand and garden soil at moisture content of 10% and 20% were about the same. As a conclusion, it was stated that the results are rather independent of soil type, especially if the moisture is higher than 10% as is normally for authentic samples. This means that quantification can be done using only one standard soil sample. Eberlin et al.[52] also reported that with their headspace (HS)-MIMS method, responses at the same level are obtained for different types of soil samples if the moisture content is adjusted to a 25% level.

For routine analysis, it is also of interest to know how long samples can be stored before the analysis is made. This was studied in the PAM-MS method development by preparing various types of samples (sand and moistened garden soil) and storing them up to 6 months at 5°C before analysis.[54] It was observed that peak areas decreased on average about 14% for dry sand within 2 weeks and about 50% after 6 months of storage, but for garden soil, there was no significant effect within 2 weeks and only about 14% decrease of peak areas after 6 months of storage. In conclusion, soil samples can be stored in closed bottles at 5°C for 2 weeks before analysis without a very significant effect on the results. Finally, the reliability of the PAM-MS method was demonstrated by comparing results obtained for 79 samples made by a vapor fortification method and 35 authentic samples to those measured with a headspace gas chromatographic method.[54] Figure 8.6 shows that the results obtained with the different methods are in relatively good agreement (coefficient of regression [*r*] is 0.953). Depending on the sample and analyte type, the PAM-MS method allows analysis of chlorinated and aromatic solvents with detection limits at the level of microgram per kilogram and with linearities of three to four orders of magnitude and with relative standard deviations of 5–10%.

Measurement of gases from soil samples[1,57] and especially measurement of dissolved gases from sediments[58,60,61,63–66] and peat samples[2,59,60,62] to produce information about the biological and microbiological processes (e.g., denitrification, characteristics of life in the soil and of the quality of the soil) occurring in these samples are very important application areas of MIMS. Gases that are measured include, for example, CH_4, N_2, N_2O, O_2, and CO_2. An important point to note while measuring the gases is that there are overlapping m/z values in the mass spectra at these low mass values; for example, methane is measured using a fragment ion at m/z 15 and not at m/z 16, due to overlapping oxygen signal. The possible interference caused by water vapor and carbon dioxide can be prevented by using a liquid nitrogen trap.[64,66] In addition, in many cases when measuring metabolically significant gases, it is better to estimate the amount of a particular gas in relation to a chemically and biologically inert standard such as argon.[57,64–66]

Most of the gas measurement studies are performed with membrane inlet probes, which can be directly inserted into the samples. For example, gas concentrations can be measured at 1 μM levels with spatial resolution of the order of 1 mm using a mini-probe, which has a 4-cm piece of a 0.3 mm (o.d.) hypodermic needle with a 0.07-mm hole covered with silicone rubber membrane at the end of a 1-m-long stainless steel tubing.[60] A very important advantage of the MIMS technique compared with microelectrodes used for gas measurements, even though MIMS does not provide the same spatial resolution, is the possibility for simultaneous measurement of multiple species and to measure gases for which no electrode method exists.[59,63] An additional advantage of the MIMS method is the robustness, which is needed since sediment or peat samples are often firm and contain many roots. A membrane probe does not break down easily when moved in the sample.[63]

Calibration of the gas measurements is typically done using aqueous solutions with known gas concentrations[2,58,63–65] and without stirring.[2,58,63] This is expected to work well for peat and sediment samples since they are typically saturated with water, but for soil samples, this does not necessarily hold.[57] Sheppard and Lloyd made a study by comparing various calibration methods in order to find out the best method for soil gas analysis.[57] The MIMS output was calibrated using gas-phase calibration, with a gas-equilibrated aqueous solution with stirring, with a gas-equilibrated aqueous solution without stirring, or with a water-glass bead slurry equilibrated with a gas. From these methods, the last one clearly deviated from the others. Additionally, the mass spectrometer output was calibrated using a sterile soil with known amount of water (reflecting the natural rainfall level) and by introducing a known gas concentration through the soil until a constant signal was obtained. A correlation between the sterile soil calibration method and the calibration with stirred water column was reported; that is, the most useful calibration method for soil analysis could be the use of a gas-equilibrated aqueous solution with stirring.

An example of the measured gas concentration in a peat sample is shown in Figure 8.7. A relatively good agreement between the oxygen signal measured by the mass spectrometric method and the oxygen electrode method is observed.

A dynamic calibration procedure has been developed for the analysis of dissolved gases with MIMS.[67] By this dynamic calibration, it is possible to measure the mass signal–concentration relationship, as well as the mass transfer coefficients and equilibrium of all dissolved species of carbon dioxide, and thus enabling the quantitation of oxygen

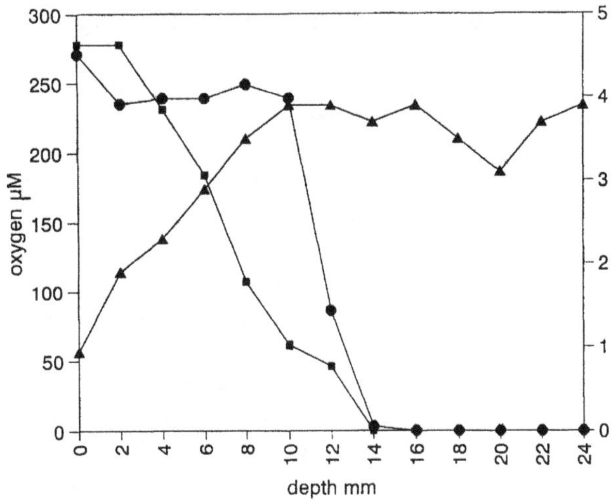

Figure 8.7. Dissolved gas profiles (CH_4 and O_2) for a peat sample measured using a mass spectrometer equipped with a mini-probe and with an oxygen electrode: CH_4 mass spectrometer (▲), O_2 mass spectrometer (■), and O_2 electrode (●). Reprinted from reference 59 with permission from Elsevier. Copyright 1995.

and carbon dioxide. The dynamic calibration was shown to give accurate results for cell respiration rates of a bacterial culture in a small-scale reactor.

8.5. ANALYSIS OF OTHER TYPES OF SAMPLES

8.5.1. Biological Samples

Measurement of dissolved gases and organic compounds from biological matrices, such as blood, bone, urine, saliva, stomach fluid, and plants, is a very important application of MIMS.[68–81] Also, in these measurements, one of the major advantages of the MIMS technique is that typically no sample treatment is needed or it is minimal. As a few examples, oxygen (O_2) and carbon dioxide (CO_2) were already measured *in vivo* from circulating blood of animals in 1966;[68] oxygen was measured *in vivo* and *in vitro* from rabbit blood;[69] partial pressures of O_2 and CO_2 have been measured from bone tissue;[70] haloethanes from dogs' blood using a stainless steel intra-arterial cannula;[71] VOCs have been measured from rat blood;[72] oxygen from tissue samples of rabbits, pigs, and humans;[73] NO has been measured with a special miniaturized membrane inlet from human plasma samples;[74] and an MIMS technique was developed to measure ventilation-to-perfusion ratios for several gases (SF_6, Kr, desfluorane, enflurane, diethyl ether, acetone).[78]

In most of the applications mentioned above, a small cannula (e.g., 0.4 mm [i.d.], 1.1 mm [o.d.], 1 m [length]) type of membrane inlet was used for the monitoring of gases and the membrane often used was a silicone membrane,[71–74,78] even though other polymers have also been tested and used (e.g., polyethylene and Teflon).[69,70,73] These studies revealed some important points to note in order to make sure that the measurement results obtained are relevant. The design of the membrane probe should be such that the signal is sample flow rate/stirring rate independent.[69,78] In order to minimize this effect, the membrane area should be smaller than the area available for diffusion in the liquid sample[78] or mixing should be used. In most biological MIMS applications, due to the small samples sizes or due to the fact that *in vivo* measurements are done, small membrane areas are used, but in some cases, a membrane with lower permeability has alternatively been selected (e.g., Teflon instead of silicon).[73] The small sampling area can lead to lower sensitivity, but sensitivity can be increased by using multiple separate diffusion areas.[69,78] The small sampling area also ensures that the MIMS probe's response is insensitive to changes in gas solubility in the sample.[78] This is important since blood gas solubilities are known to vary from subject to subject. Only a small dependence on gas solubility was observed by Baumgardner et al.[78] for their stainless steel probe (1.6 mm [o.d.], 1 mm [i.d.], 220 mm [length]) with a silicone membrane (membrane area estimated 28 μm) when comparing gas responses in water, rabbit blood, and 20% intralipid solution. Reliability of the oxygen measurements made by MIMS was demonstrated already quite early. In 1980, Lundsgaard et al.[69] showed very good correlation (correlation coefficient 0.997) between the *in vivo* MIMS measurement and *in vitro* oxygen electrode measurement. They also reported that the measurement cannula is reusable and stable at least for half a year and that argon signal can be used as an indicator of the reliability of the measurements.

MIMS has also been used to measure a human toluene metabolite, *p*-cresol, in urine,[75] using simple sample preparation steps, namely hydrolysis of the urine sample, extraction with an organic solvent and reconstitution into 1-propanol. Similarly, a trap-and-release MIMS method was used to quantify total cysteine and homocysteine in human plasma, after a sample preparation, that is, disulfide bond reduction with dithiothreitol, protein precipitation with trichloroacetic acid, and derivatization with ethyl chloroformate.[79] On the contrary, the time persistence of monochloramine in human saliva and stomach fluid was measured using a direct insertion membrane probe and continuously monitoring the decay of monochloramine signal when an aliquot of a saliva or stomach fluid sample was added to the monochloramine solution,[77] circulating through the membrane probe. Rumen fluid is another biological sample, which has been intensively studied with MIMS by *in situ* (fistulated sheep) or *in vitro* measurements of dissolved gases (CH_4, CO_2, H_2, and O_2).[76]

In vivo monitoring of gases in plants (e.g., wheat, corn, sugar beet, and reed) has also been performed with MIMS using the same type of small cannula or a capillary type of membrane probe as used for blood gas analysis.[1] If possible, the membrane probe should be inserted into a cavity in the plant since the probe is then mounted in the gas phase, and therefore, the stirring effect observable in the liquid phase measurements is lacking.

8.5.2. MIMS Analysis of Organic Matrices

In most MIMS applications, organic compounds are measured from aqueous samples or from air, but very good results have also been obtained with MIMS, when measuring water or organic compounds from an organic matrix.[12,81–90] The name reversed-phase membrane introduction mass spectrometry (RP-MIMS) has been adapted for this application.[81] RP-MIMS has been performed with many different membranes such as polyethylene terephtalate,[82–84] polyimide,[89] silicone,[85,86,90] hydrophilic Nafion,[87] microporous polypropylene,[81] microporous polyvinylidine fluoride,[91] microporous cellulose,[88] and microporous polyether sulfone.[88] In general, the performance characteristics of RP-MIMS are similar to those of normal-phase MIMS, and no special sample treatment methods before the MIMS analysis are needed. However, when microporous membranes are used, the mass spectrometer itself has to tolerate higher pressure, due to a much higher flow of molecules into the mass spectrometer.

In the first example of RP-MIMS, a hydrophilic polyethylene terephthalate membrane was used to measure water activity in organic solvents.[82–84] The terephthalate membrane was so restrictive in its permeation of molecules that only water passed through it and dissolved water in organic solvents could be directly detected at parts per million levels. In one example, the rate of hydrolysis of diphenyl carbonate by porcine liver esterase in diisopropyl ether was monitored online.[83] The restrictive nature of the polyethylene terephthalate membrane made it possible to replace the mass spectrometer with a simple vacuum gauge to create a very simple method for water analysis in organic liquids.[84]

One study using a hydrophilic Nafion membrane demonstrated a linear detection of small alcohols in chloroform at concentration levels down to 0.1% and with a linear range up to 2.5%.[87] The RP-MIMS system was used for online monitoring of the alcohols during distillation of chloroform, and the online RP-MIMS results were found to agree well with offline GC-MS data. In another study, a polyimide membrane was used to detect acetone, ethyl acetate, water, acetic acid, chloroform, and ethanol in hexane.[89] There have also been applications that can be considered as RP-MIMS, in which a silicone membrane was used. For example, styrene and tetrachloroethene residues were detected directly from olive oil,[85] and Kathon CG (an antimicrobial agent) was measured from cosmetic emulsions.[86] In addition, it has been demonstrated that thin (25 μm) silicone membrane with a supporting matrix can be used for RP-MIMS.[90]

As mentioned above, the requirements for the mass spectrometer are more demanding when microporous membranes are used in RP-MIMS, because of the 100- to 1000-fold increase in the number of molecules that enter the instrument. In the first example of RP-MIMS[81] using a microporous membrane, the problem was solved by using CI, where the pressure in the ion source is typically 1000 times higher than that in the ion source used for electron ionization. In this system, a missing enrichment of the sample is compensated by the higher sample flux into the mass spectrometer, and in its most advanced form, the solvent itself can be used as a reagent gas in the CI process. High selectivity in this type of RP-MIMS can be obtained if MS/MS is used, and even highly complex matrices, such as gasoline, can be analyzed directly as exemplified by the analysis of MTBE in gasoline.

An RP-MIMS method based on a microporous polyvinylidene fluoride membrane has also been developed for real-time monitoring of pharmaceutical processes, the Michael addition reaction of phenylethylamine being the model reaction in the study.[12] In this experiment, the microporous membrane was used to transfer analytes of interest from an ethanol solution to a water–acetonitrile solution, with subsequent analysis by APCI MS.

Recently, four membrane materials (cellulose, polyether sulfone, microporous polypropylene, and silicone membrane) were characterized for their potential use for measurement of organic compounds from air, water, and organic solvents using a miniaturized mass spectrometer.[88] With the standard silicone membrane, the mini-MIMS apparatus behaved like most standard MIMS systems, that is, VOCs could be easily measured from water. The microporous polypropylene membrane was not useful for any MIMS modes; in the RP-MIMS mode, the limiting factor was too high pressure caused by the solvents. Both cellulose and polyether sulfone membranes were good for RP-MIMS, but cellulose gave a clearly better performance and organic contaminants and water could be measured in organic solvents at 10–100 ppm levels by weight.

8.5.3. Other Types of Solid Samples

Solid samples mounted in various forms of vessels attached to an MIMS system have formed the basis for many interesting studies. For example, volatile pyrolysis products have been monitored from Kraton 1107 (an isoprene-styrene copolymer),[91] peptides,[92] pathogenic bacteria,[93] and even free radicals from aromatic diazoamino compounds.[94] The PAM-MS method was successfully used to detect residual solvents in pharmaceutical products.[95] The dry solid pharmaceutical material was simply mounted inside the PAM sample vessel for the analysis, and solvents such as benzene, toluene, and chloroform could be analyzed at around 0.1 mg/kg levels. When combined with a custom-made data interpretation program, the PAM-MS identified solvents almost as well as purge-and-trap GC-MS. Water slurries of birth control pills were analyzed for their content of steroid hormones at parts per billion levels using desorption CI-MIMS.[96]

Studies of membrane permeation properties are relatively common MIMS applications.[97–99] In these studies, MIMS provided fundamental knowledge of permeation parameters, such as diffusion coefficients and relative permeation coefficients for organic compounds diffusing through different types of membrane materials. In addition, MIMS has been used to study breakthrough curves for various sorbents.[100] MIMS allowed studies of VOC mixtures, and it was observed that the breakthrough curves can depend on the composition of mixtures.

8.6. CONCLUSIONS

The ability to analyze practically any type of sample for its content of volatile organic compounds without or with very little sample preparation makes MIMS a remarkable technique with applications as diverse as *in vivo* analysis of anesthetics in blood, online monitoring of chemical and biological processes, analysis of contaminants in environmental and food samples, surveillance of air quality, analysis of gas distributions in peat with 1-mm spatial resolution and studies of membrane permeation properties. The common element in all these applications is a membrane that separates the sample from the vacuum in the mass spectrometer. This simple feature encloses a large number of specially designed inlets that are optimized and dedicated to a particular application. Quite often, a difficult part of different applications is proper calibration to obtain reliable concentrations for the analytes. In many cases a simple external standard can be used, but it is important to ensure that the physical conditions regarding the exposure of the membrane to the sample and to the standard are identical and to keep in mind that matrix effects are common. Despite the obvious possibilities of MIMS for field applications due to simple sample preparation in many applications, most MIMS applications in the past have been performed inside a laboratory. A major reason for this is the size and complexity of the complete mass spectrometric system. The membrane inlet may be very simple to operate, but the rest of the mass spectrometric system is not field portable. This is about to change. Since the millennium shift, there has been an intense focus on the development of miniature mass spectrometers, and the first field portable instruments (10–20 kg, everything included) have become commercially available.

REFERENCES

1. Bohátka, S. Process monitoring in fermentors and living plants by membrane inlet mass spectrometry. *Rapid Commun. Mass Spectrom.* **1997**, *11*, 656–661.

2. Benstead, J.; Lloyd, D. Direct mass spectrometric measurement of gases in peat cores. *FEMS Microbiol. Ecol.* **1994**, *13*, 233–240.

3. Bier, M.E.; Cooks, R.G. Membrane interface for selective introduction of volatile compounds directly into the ionization chamber of a mass spectrometer. *Anal. Chem.* **1987**, *59*, 597–601.

4. Lauritsen, F.R. A new membrane inlet for on-line monitoring of dissolved, volatile organic compounds with mass spectrometry. *Int. J. Mass Spectrom. Ion Process.* **1990**, *95*, 259–268.

5. Slivon, L.E.; Bauer, M.R.; Ho, J.S.; Budde, W.L. Helium-purged hollow fiber membrane mass spectrometer interface for continuous measurement of organic compounds in water. *Anal. Chem.* **1991**, *63*, 1335–1340.

6. Lauritsen, F.R.; Lunding, A. A study of the bioconversion potential of the fungus *Bjerkandera adusta* with respect to a production of chlorinated aromatic compounds. *Enzyme Microb. Technol.* **1998**, *22*, 459–465.

7. Leth, M.; Lauritsen, F.R. A fully integrated trap-membrane inlet mass spectrometry system for the measurement of semivolatile organic compounds in aqueous solution. *Rapid Commun. Mass Spectrom.* **1995**, *9*, 591–596.

8. Matz, G.; Kesners, P. Thermal membrane desorption application (TMDA) method for on-line analysis of organics in water by GC-MS. *Analusis* **1995**, *23*, M12–M16.

9. Arthur, C.L.; Pawliszyn, J. Solid phase microextraction with thermal desorption using fused silica optical fibers. *Anal. Chem.* **1990**, *62*, 2145–2148.

10. Aggerholm, T.; Lauritsen, F.R. Direct detection of polyaromatic hydrocarbons, estrogenic compounds and pesticides in water using desorption chemical ionization MIMS. *Rapid Commun. Mass Spectrom.* **2001**, *15*, 1826–1831.

11. Dejarme, L.E.; Bauer, S.J.; Cooks, R.G.; Lauritsen, F.R.; Kotiaho, T.; Graf, T. Jet separator/membrane introduction mass spectrometry for on-line quantitation of volatile organic compounds in aqueous solutions. *Rapid Commun. Mass Spectrom.* **1993**, *7*, 935–942.

12. Clinton, R.; Creaser, C.S.; Bryan, D. Real-time monitoring of a pharmaceutical process reaction using a membrane interface combined with atmospheric pressure chemical ionization mass spectrometry. *Anal. Chim. Acta* **2005**, *539*, 133–140.

13. VanHassel, E.; Bier, M.E. An electrospray membrane probe for the analysis of volatile and semi-volatile organic compounds in water. *Rapid Commun. Mass Spectrom.* **2007**, *21*, 413–420.

14. Hoch, G.; Kok, B. A mass spectrometer inlet system for sampling gases dissolved in liquid phase. *Arch. Biochem. Biophys.* **1963**, *101*, 160–170.

15. Degn, H.; Cox, R.P.; Lloyd, D. Continuous measurement of dissolved gases in biochemical systems with the quadrupole mass spectrometer. *Methods Biochem. Anal.* **1985**, *31*, 165–195.

16. Johnson, R.C.; Cooks, R.G.; Allen, T.M.; Cisper, M.E; Hemberger, P.H. Membrane introduction mass spectrometry: Trends and applications. *Mass Spectrom. Rev.* **2000**, *19*, 1–37.

17. Ketola, R.A.; Kotiaho, T.; Cisper, M.E.; Allen, T.M. Environmental applications of membrane introduction mass spectrometry. *J. Mass Spectrom.* **2002**, *37*, 457–476.

18. Soni, M.; Bauer, S.; Amy, J.W.; Wong, P.; Cooks, R.G. Direct determination of organic compounds in water at parts-per-quadrillion levels by membrane introduction mass spectrometry. *Anal. Chem.* **1995**, *70*, 3103–3113.

19. Lauritsen, F.R.; Gylling, S. On-line monitoring of biological reactions at low part per trillion levels by membrane inlet mass spectrometry. *Anal. Chem.* **1995**, *67*, 1418–1420.

20. Virkki, V.T.; Ketola, R.A.; Ojala, M.; Kotiaho, T.; Komppa, V.; Grove, A.; Facchetti, S. On-site environmental analysis by membrane inlet mass spectrometry. *Anal. Chem.* **1995**, *67*, 1421–1425.

21. Harland, B.J.; Nicholson, P.J.D. Continuous measurement of volatile organic chemicals in natural waters. *Sci. Total Environ.* **1993**, *135*, 37–54.

22. Kibelka, G.P.G.; Short, R.T.; Toler, S.K.; Edkins, J.E.; Byrne, R.H. Field-deployed underwater mass spectrometers for investigations of transient chemical systems. *Talanta* **2004**, *64*, 961–969.

23. Short, R.T. SRI, St. Petersburg, FL. Personal Communication; 2007.

24. Hayward, M.J.; Kotiaho, T.; Lister, A.K.; Cooks, R.G.; Austin, G.D.; Narayan, R.; Tsao, G.T. On-line monitoring of bioreactions of *Bacillus polymyxa* and *Klebsiella oxytoca* by membrane introduction tandem mass spectrometry with flow injection analysis sampling. *Anal. Chem.* **1990**, *62*, 1798–1804.

25. Schou, M.; Graf, T.; Degn, H. Construction and performance of a plug-in membrane inlet mass spectrometer for fermentation monitoring. *Biotechnol. Bioeng.* **1997**, *54*, 535–542.

26. Petersen, C.; Beck, H.C.; Lauritsen, F.R. On-line monitoring of important organoleptic methyl-branched aldehydes during batch fermentation of the starter culture *Staphylococcus xylosus* reveal new insight into their production in a model fermentation. *Biotechnol. Bioeng.* **2004**, *85*, 299–305.

27. Hansen, K.F.; Gylling, S.; Lauritsen, F.R. Time and concentration dependent relative peak intensities observed in EI-membrane inlet mass spectra. *Int. J. Mass Spectrom. Ion Process.* **1996**, *152*, 143–155.

28. Beck, H.C.; Hansen, A.M.; Lauritsen, F.R. Novel pyrazine metabolites found in polymyxin biosynthesis by *Paenibacillus polymuxa*. *FEMS Microbiol. Lett.* **2003**, *220*, 67–73.

29. Ojala, M.; Ketola, R.A.; Virkki, V.; Sorsa, H.; Kotiaho, T. Determination of phenolic compounds in water using membrane inlet mass spectrometry. *Talanta* **1997**, *44*, 1253–1259.

30. Tarkiainen, V.; Kotiaho, T.; Mattila, I.; Virkajärvi, I.; Aristidou, A.; Ketola, R.A. On-line monitoring of continuous beer fermentation process using automatic membrane inlet mass spectrometric system. *Talanta* **2005**, *65*, 1254–1263.

31. Kotiaho, T.; Lauritsen, F.R.; Choudhury, T.K.; Cooks, R.G.; Tsao, G.T. Membrane introduction mass spectrometry. *Anal. Chem.* **1991**, *63*, 875A–883A.

32. Tsai, G.J.; Austin, G.D.; Tsui, M.J.; Tsao, G.T.; Hayward, M.J.; Kotiaho, T.; Cooks, R.G. Theoretical analysis of probe dynamics in flow injection/membrane introduction mass spectrometry. *Anal. Chem.* **1991**, *63*, 2460–2465.

33. Janfelt, C.; Lauritsen, F.R.; Toler, S.K.; Bell, R.J.; Short, R.T. A method for quantification of chemicals in a pollution plume using a moving membrane-based sensor exemplified by mass spectrometry. *Anal. Chem.* **2007**, *79*, 5336–5342.

34. Bier, M.E.; Kotiaho, T.; Cooks, R.G. Direct insertion membrane probe for selective introduction of organic compounds into a mass spectrometer. *Anal. Chim. Acta* **1990**, *231*, 175–190.

35. Ketola, R.A.; Ojala, M.; Sorsa, H.; Kotiaho, T.; Kostiainen, R.K. Development of a membrane inlet mass spectrometric method for analysis of air samples. *Anal. Chim. Acta* **1997**, *349*, 359–365.

36. Ketola, R.A.; Mansikka, T.; Ojala, M.; Kotiaho, T.; Kostiainen, R. Analysis of volatile organic sulfur compounds in air by membrane inlet mass spectrometry. *Anal. Chem.* **1997**, *69*, 4536–4539.

37. Cisper, M.E.; Gill, C.G.; Townsend, L.E.; Hemberger, P.H. On-line detection of volatile organic compounds in air at parts-per-trillion levels by membrane introduction mass spectrometry. *Anal. Chem.* **1995**, *67*, 1413–1417.

38. Ketola, R.A.; Grøn, C.; Lauritsen, F.R. Temperature-programmed desorption for membrane inlet mass spectrometry. *Rapid Comm. Mass Spectrom.* **1998**, *12*, 773–778.

39. Ketola, R.A.; Kiuru, J.T.; Tarkiainen, V.; Kokkonen, J.T.; Räsänen, J.; Kotiaho, T. Detection of volatile organic compounds by temperature-programmed desorption combined with mass spectrometry and Fourier transform infrared spectroscopy. *Anal. Chim. Acta* **2006**, *562*, 245–251.

40. Gordon, S.M.; Callahan, P.J.; Kenny, D.V.; Pleil, J.D. Direct sampling and analysis of volatile organic compounds in air by membrane introduction and glow discharge ion trap mass spectrometry with filtered noise fields. *Rapid Commun. Mass Spectrom.* **1996**, *10*, 1038–1046.

41. Munchmeyer, W.; Walte, A.; Matz, G. Online measurement of polycyclic aromatic hydrocarbons by fast GC/MS. *Polycycl. Aromat. Compd.* **1996**, *9*, 299–306.

42. Cisper, M.E.; Hemberger, P.H. The direct analysis of semi-volatile organic compounds by membrane introduction mass spectrometry. *Rapid Commun. Mass Spectrom.* **1997**, *11*, 1449–1453.

43. Riter, L.S.; Takáts, Z.; Cooks, R.G. Single-sided membrane introduction mass spectrometry for on-line determination of semi-volatile organic compounds in air. *Analyst* **2001**, *126*, 1980–1984.

44. White, A.J.; Blamire, M.G.; Corlett, C.A.; Grifftihs, B.W.; Martin, D.M.; Spencer, S.B.; Mullock, S.J. Development of a portable time-of-flight membrane inlet mass spectrometer for environmental analysis. *Rev. Sci. Instrum.* **1998**, *69*, 565–571.

45. Evans, J.E.; Arnold, J.T. Monitoring organic vapors. *Environ. Sci. Technol.* **1975**, *9*, 1134–1137.

46. Meier, R.W. A field portable mass spectrometer for monitoring organic vapors. *Am. Ind. Hyg. Assoc. J.* **1978**, *39*, 233–239.

47. Colorado, A.; Barket D.J. Jr.; Hurst, J.M.; Shepson, P.B. A fast-response method for determination of atmospheric isoprene using quadrupole ion trap mass spectrometry. *Anal. Chem.* **1998**, *70*, 5129–5135.

48. Martin, R.S.; Westberg, H.; Allwine, E.; Ashman, L.; Farmer, J.C.; Lamb, B. Measurement of isoprene and its atmospheric oxidation products in a central Pennsylvania deciduous forest. *J. Atmos. Chem.* **1991**, *13*, 1–32.

49. Ristoiu, D.; Cosma, C.; Voros, A.; Ristoiu, T. Evaluation of the volatile organic contents in aqueous samples by MI-QMS technique. *Vacuum* **1998**, *50*, 359–362.

50. Wilson, H.K.; Ottley, T.W. The use of a transportable mass spectrometer for the direct measurement of industrial solvents in breath. *Biomed. Mass Spectrom.* **1981**, *8*, 606–610.

51. Kostiainen, R.; Kotiaho, T.; Mattila, I.; Mansikka, T.; Ojala, M.; Ketola, R.A. Analysis of volatile organic compounds in water and soil samples by purge-and-membrane mass spectrometry. *Anal. Chem.* **1998**, *70*, 3028–3032.

52. Mendes, M.A.; Sparrapan, R.; Eberlin, M.N. Headspace membrane introduction mass spectrometry for trace level analysis of VOCs in soil and other solid matrixes. *Anal. Chem.* **2000**, *72*, 2166–2170.

53. Ojala, M.; Mattila, I.; Särme, T.; Ketola, R.A.; Kotiaho, T. A new purge-and-membrane mass spectrometric (PAM-MS) instrument for analysis of volatile organic compounds in soil samples. *Analyst* **1999**, *124*, 1421–1424.

54. Ojala, M.; Mattila, I.; Tarkiainen, V.; Särme, T.; Ketola, R.A.; Määttänen, A.; Kostiainen, R.; Kotiaho, T. Purge-and-membrane mass spectrometry, a rapid screening method for analysis of VOCs from soil samples. *Anal. Chem.* **2001**, *73*, 3624–3631.

55. Costanza, J.; Davis, W.M. Rapid detection of volatile organic compounds in the subsurface by membrane introduction into a direct sampling ion-trap mass spectrometer. *Field Anal. Chem. Technol.* **2000**, *4*, 246–254.

56. Frandsen, H.; Janfelt, C.; Lauritsen, F.R. Fast and direct screening of polyaromatic hydrocarbon (PAH)-contaminated sand using a miniaturized membrane inlet mass spectrometer (mini-MIMS). *Rapid Commun. Mass Spectrom.* **2007**, *21*, 1574–1578.

57. Sheppard, S.K.; Lloyd, D. Direct mass spectrometric measurement of gases in soil monoliths. *J. Microbiol. Methods* **2002**, *50*, 175–188.

58. Thomas, K.L.; Lloyd, D. Measurement of denitrification in estuarine sediment using membrane inlet mass spectrometry. *FEMS Microbiol. Ecol.* **1995**, *16*, 103–114.

59. Thomas, K.L.; Price, D.; Lloyd, D. A comparison of different methods for the measurement of dissolved gas gradients in waterlogged peat cores. *J. Microbiol. Methods* **1995**, *24*, 191–198.

60. Lloyd, D.; Thomas, K.; Price, D.; O'Neil, B.; Oliver, K.; Williams, T.N. A membrane-inlet mass spectrometer miniprobe for the direct simultaneous measurement of multiple gas species with spatial resolution of 1 mm. *J. Microbiol. Methods* **1996**, *25*, 145–151.

61. Cowie, G.; Lloyd, D. Membrane inlet ion trap mass spectrometry for the direct measurement of dissolved gases in ecological samples. *J. Microbiol. Methods* **1999**, *35*, 1–12.

62. Beckmann, M.; Lloyd, D. Mass spectrometric monitoring of gases (CO_2, CH_4, O_2) in a mesotrophic peat core from Kopparås Mire, *Sweden. Glob. Chang. Biol.* **2001**, *7*, 171–180.

63. Cartaxana, P.; Lloyd, D. N_2, N_2O and O_2 profiles in a tagus estuary salt marsh. *Estuar. Coast. Shelf Sci.* **1999**, *48*, 751–756.

64. Laursen, A.E.; Seitzinger, S.P. The role of denitrification in nitrogen removal and carbon mineralization in Mid-Atlantic Bight sediments. *Contin. Shelf Res.* **2002**, *22*, 1397–1416.

65. Kana, T.M.; Sullivan, M.B.; Cornwell, J.C.; Groszkowski, K.M. Denitrification in estuarine sediments determined by membrane inlet mass spectrometry. *Limnol. Oceanogr.* **1998**, *43*, 334–339.

66. McCarthy, M.J.; Gardner, W.S. An application of membrane inlet mass spectrometry to measure denitrification in a recirculating mariculture system. *Aquaculture* **2003**, *218*, 341–355.

67. Yang, T.H.; Wittmann, C.; Heinzle, E. Dynamic calibration and dissolved gas analysis using membrane inlet mass spectrometry for the quantification of cell respiration. *Rapid Commun. Mass Spectrom.* **2003**, *17*, 2721–2731.

68. Woldring, S.; Owens, G.; Woolford, D.C. Blood gases: Continuous in vivo recording of partial pressures by mass spectrography. *Science* **1966**, *153*, 885–887.

69. Lundsgaard, J.S.; Jensen, B.; Gronlund, J. Fast-responding flow-independent blood gas catheter for oxygen measurement. *J. Appl. Physiol.* **1980**, *48*, 376–381.

70. Kiær, T.; Dahl, B.; Lausten, G. Partial pressures of oxygen and carbon dioxide in bone and their correlation with bone-blood flow: Effect of decreased arterial supply and venous congestion on intraosseous oxygen and carbon dioxide in an animal model. *J. Orthop. Res.* **1992**, *10*, 807–812.

71. Roberts, M.; Colton, E.T. III; Owens, G.; Thomas, D.D.; Watkins, G.M. Continuous mass spectrometric measurement of halothane partial pressure in blood. *Med. Biol. Eng.* **1975**, *13*, 535–538.

72. Brodbelt, J.S.; Cooks, R.G.; Tou, J.C.; Kallos, G.J.; Dryzga, M.D. In vivo mass spectrometric determination of organic compounds in blood with a membrane probe. *Anal. Chem.* **1987**, *59*, 454–458.

73. Donavan, W.E.; Myers, B. Measurement of tissue gas levels with a mass spectrometer. In *Oxygen Transport to Tissue, Instrumentation, Methods and Physiology*, Bicher, H.I.; Bruley, D.F., Eds. New York: Plenum Press; 1973; pp. 67–72.

74. Trushina, E.V.; Clarke, N.J.; Benson, L.M.; Tomlinson, A.J.; McMurray, C.T.; Naylor, S. A miniaturized membrane inlet mass spectrometry interface for analysis of nitric oxide in human plasma. *Rapid Commun. Mass Spectrom.* **1998**, *12*, 985–987.

75. Sturaro, A.; Doretti, L.; Parvoli, G.; Traldi, P. Determination of human toluene metabolites by mass spectrometry using a silicone membrane source. *Anal. Chim. Acta* **1989**, *224*, 119–122.

76. Lloyd, D.; Ellis, J.E.; Hillman, K.; Williams, A.G. Membrane inlet mass spectrometry: Probing the rumen ecosystem. *J. Appl. Bacteriol. Symp. Suppl.* **1992**, *73*, 155S–163S.

77. Kotiaho, T.; Wood, J.M.; Wick, P.L. Jr.; Dejarme, L.E.; Ranasinghe, A.; Cooks, R.G.; Ringhand, H.P. Time persistence of monochloramine in human saliva and stomach fluid. *Environ. Sci. Technol.* **1992**, *26*, 302–306.

78. Baumgardner, J.E.; Choi, I.-C.; Vonk-Noordegraaf, A.; Frasch, H.F.; Neufeld, G.R.; Marshall, B.E. Sequential VA/Q distributions in the normal rabbit by micropore membrane inlet mass spectrometry. *J. Appl. Physiol.* **2000**, *89*, 1699–1708.

79. Vellasco, A.P.; Haddad, R.; Eberlin, M.N.; Höehr, N.F. Combined cysteine and homocysteine quantitation in plasma by trap and release membrane introduction mass spectrometry. *Analyst* **2002**, *127*, 1050–1053.

80. Riter, L.S.; Laughlin, B.C.; Nikolaev, E.; Cooks, R.G. Direct analysis of volatile organic compounds in human breath using a miniaturized

cylindrical ion trap mass spectrometer with a membrane inlet. *Rapid Commun. Mass Spectrom.* **2002**, *16*, 2370–2373.

81. Lauritsen, F.R.; Kotiaho, T.; Choudhury, T.K.; Cooks, R.G. Direct detection and identification of volatile organic compounds dissolved in organic solvents by reversed-phase membrane introduction tandem mass spectrometry. *Anal. Chem.* **1992**, *64*, 1205–1211.

82. Bohátka, S.; Degn, H. Continuous measurement of water in organic solution by membrane inlet mass spectrometry. *Rapid Commun. Mass Spectrom.* **1991**, *5*, 433–436.

83. Degn, H.; Bohátka, S.; Lloyd, D. Enzyme activity in organic solvent as a function of water activity determined by membrane inlet mass spectrometry. *Biotechnol. Tech.* **1992**, *6*, 161–164.

84. Degn, H.; Frigaard, N.-U. Membrane inlet manometry for the measurement of water activity in aqueous and organic solutions. *Anal. Chim. Acta* **1993**, *274*, 355–359.

85. Kotiaho, T.; Gylling, S.; Lunding, A.; Lauritsen, F.R. Direct determination of styrene and tetrachloroethylene in olive oil by membrane inlet mass spectrometry. *J. Agric. Food Chem.* **1995**, *43*, 928–930.

86. Favretto, D.; Traldi, P.; Benassi, C.A.; Bettero, A. Silicon membrane interface for the direct analysis of Kathon CG in aqueous solutions and cosmetic emulsions. *Biol. Mass Spectrom.* **1991**, *20*, 669–676.

87. Creaser, C.S.; Lamarca, D.G.; Brum, J.; Werner, C.; New, A.P.; Freitas dos Santos, L.M. Reversed-phase membrane inlet mass spectrometry applied to the real-time monitoring of low molecular weight alcohols in chloroform. *Anal. Chem.* **2002**, *74*, 300–304.

88. Janfelt, C.; Frandsen, H.; Lauritsen, F.R. Characterization of a mini membrane inlet mass spectrometer for on-site detection of contaminants in both aqueous and liquid organic samples. *Rapid Commun. Mass Spectrom.* **2006**, *20*, 1441–1446.

89. Maden, A.J.; Hayward, M.J. Sheet materials for use as membranes in membrane introduction mass spectrometry. *Anal. Chem.* **1996**, *68*, 1805–1811.

90. Kasthurikrishnan, N.; Cooks, R.G.; Bauer, S. Thin pervaporation membranes for improved performance in on-line flow injection analysis membrane introduction mass spectrometry. *Rapid Commun. Mass Spectrom.* **1996**, *10*, 751–756.

91. Vreeken, R.J.; Houriet, R. Membrane interface for the direct coupling of pyrolysis and mass spectrometry. *Anal. Chim. Acta* **1995**, *313*, 237–241.

92. Hendricker, A.D.; Basile, F.; Voorhees, K.J. A study of protein oxidative products using a pyrolysis-membrane inlet quadrupole ion trap mass spectrometer with air as the buffer gas. *J. Anal. Appl. Pyrolysis* **1998**, *46*, 65–82.

93. Hendricker, A.D.; Abbas-Hawks, C.; Basile, F.; Voorhees, K.J.; Hadfield, T.L. Rapid chemotaxonomy of pathogenic bacteria using in situ thermal hydrolysis and methylation as a sample preparation step coupled with a field-portable membrane-inlet quadrupole ion trap mass spectrometer. *Int. J. Mass Spectrom.* **1999**, *190/191*, 331–342.

94. Kasthurikrishnan, N.; Cooks, R.G.; Thompson, M.J. Sampling of aryldiazonium, anilino, and aryl radicals by membrane introduction mass spectrometry: Thermolysis of aromatic diazoamino compounds. *J. Am. Soc. Mass Spectrom.* **1998**, *9*, 234–241.

95. Ojala, M.; Poutanen, M.; Mattila, I.; Ketola, R.A.; Kotiaho, T.; Kostiainen, R. Analysis of residual solvents in pharmaceuticals with purge-and-membrane mass spectrometry. *Rapid Commun. Mass Spectrom.* **2000**, *14*, 994–998.

96. Lauritsen, F.R.; Rose, J. Determination of steroid hormones by membrane inlet mass spectrometry and desorption chemical ionization. *Analyst* **2000**, *125*, 1577–1581.

97. LaPack, M.A.; Tou, J.C.; McGuffin, V.L.; Enke, C.G. The correlation of membrane permselectivity with hildebrand solubility parameters. *J. Memb. Sci.* **1994**, *86*, 263–280.

98. Boscaini, E.; Alexander, M.L.; Prazeller, P.; Märk, T.D. Investigation of fundamental physical properties of a polydimethylsiloxane (PDMS) membrane using a proton transfer reaction mass spectrometer (PTRMS). *Int. J. Mass Spectrom.* **2004**, *239*, 179–186.

99. Schäfer, T.; Vital, J.; Crespo, J.G. Coupled pervaporation/mass spectrometry for investigating membrane mass transport phenomena. *J. Memb. Sci.* **2004**, *241*, 197–205.

100. Peters, R.J.B.; Bakkeren, H.A. Sorbents in sampling. Stability and breakthrough measurements. *Analyst* **1994**, *119*, 71–74.

Chapter 9

Pressurized Fluid Extraction

John R. Dean and Renli Ma

9.1. INTRODUCTION

The removal of persistent organic pollutants (POPs) from environmental matrices can be done by a variety of approaches that use combinations of solvent, heat, and pressure. The effectiveness of any approach is determined by its ability to remove the POPs from solid or semisolid environmental matrices that include soil and sludge. The chemical nature of POPs means that they are retained by soil and sludge by a variety of approaches that range from weak forces of attraction (e.g., van der Waals forces) in the case of hydrocarbons (e.g., polycyclic aromatic hydrocarbons) to strong interactions (e.g., electrostatic) in the case of pesticides (e.g., organochlorine pesticides [OCPs]). These interactions, between POPs and soil, can be overcome by the use of organic solvents that themselves range in different polarities from the nonpolar, for example, hexane, to the polar, for example, methanol. Assistance with the recovery of POPs is often done by the use of heat and pressure to speed up the processes.

As sample preparation is often the bottleneck of laboratory operations, particularly with the advent of fast gas chromatography (GC) with a flame ionization detector (FID) that allows extracts to be analyzed in a few seconds, the ability to perform faster extraction of samples is beneficial and required. The main traditional and often described as the "benchmark," technique for recovery of POPs from soils and sludges is Soxhlet extraction. However, this approach is slow (typically up to 24 h) and uses large quantities of organic solvent (e.g., up to 250 mL) particularly when compared with new instrumental approaches with the capability for automation or multiple sequential sample extraction, which use smaller quantities of organic solvent, for example, 10 mL, and are faster (e.g., minutes). This chapter focuses on one of the alternative approaches to Soxhlet extraction, namely, pressurized fluid extraction (PFE) and its use for the recovery of POPs from environmental matrices. Other approaches including hot water extraction, supercritical fluid extraction (SFE), and microwave-assisted extraction (MAE) are described in Chapters 10, 11, and 12, respectively.

PFE relies on the effect of heat and pressure, on a sample matrix, in the presence of solvent to liberate POPs. However, this concept is not new and was formerly exploited by the technique of SFE. Extensive developments in the use of analytical SFE were done in the late 1980s and early 1990s.[1] In SFE, however, the solvent used was carbon dioxide, which, in order to be supercritical, was required to be heated above 31.1°C and pressurized above 1070.4 psi (74.8 atm). The major limitation of the approach, however, was the lack of a dipole moment for CO_2, which highlighted its inability to extract moderately polar and polar compounds. This was, however, circumvented by the addition of an organic modifier, typically methanol, in the 1–10%, v/v range. Developments and commercialization of PFE can be viewed as an extension of the SFE technique in which the CO_2 with or without an organic modifier was replaced in favor of an organic solvent or mixture only.

It is important to identify at the start of this chapter that PFE is also known as accelerated solvent extraction (ASE), pressurized liquid extraction (PLE), or pressurized solvent extraction (PSE). This confusion in terms to describe this approach that uses organic solvents at elevated temperatures and pressures to extract compounds from matrices can be troublesome in identifying key literature in the field of study. The first commercial instrument was launched by Dionex Corporation in 1995 and it used the trade name "accelerated solvent extraction." However, as the United States Environmental Protection Agency (USEPA) adopted the name "pressurized fluid extraction" in its USEPA Method 3545,[2] it is reasonable to use this term as the generic phrase for the process. However, this has not been universally adopted as the phrase to describe the approach and the scientific literature includes other terms including pressurized

Handbook of Sample Preparation, Edited by Janusz Pawliszyn and Heather L. Lord

liquid extraction and pressurized solvent extraction. This means that when searching the scientific literature for work that has used this process, a range of terms need to be considered and used to source all relevant information in electronic databases. This chapter will use the generic term "pressurized fluid extraction" throughout.

The first commercial instrument to perform PFE was launched by Dionex Corporation in spring 1995 (Accelerated

Table 9.1. Specific Compounds Highlighted in USEPA Method 3545[2]

	(a) Base, Neutral, Acids (BNAs)	
Phenol	bis(2-Chloroisopropyl)ether	4-Nitrophenol
2-Chlorophenol	Isophorone	Dibenzofuran
1,4-Dichlorobenzene	2-Nitrophenol	*N*-nitrosodiphenylamine
2-Methylphenol	bis(Chlorethoxy)methane	Hexachlorobenzene
o-Toluidine	1,2,4-Trichlorobenzene	Phenanthrene
Hexachloroethane	4-Chloroaniline	Carbazole
2,4-Dimethylphenol	4-Chloro-3-methylphenol	Pyrene
bis(2-Chloroethyl)ether	Hexachlorocyclopentadiene	Benz[a]anthracene
1,3-Dichlorobenzene	2,4,5-Trichlorophenol	Benzo[b]fluoranthene
1,2-Dichlorobenzene	2-Nitroaniline	Benzo[a]pyrene
2,4-Dichlorophenol	2,4-Dinitrotoluene	Dibenz[a,h]anthracene
Naphthalene	4-Nitroaniline	Nitrobenzene
Hexachlorobutadiene	4-Bromophenyl-phenylether	3-Nitroaniline
2-Methylnaphthalene	Pentachlorophenol	Fluorene
2,4,6-Trichlorophenol	Anthracene	Chrysene
2-Chloronaphthalene	Fluoranthene	Benzo[k]fluoranthene
Acenaphthene	3,3′-Dichlorobenzidine	Indeno[1,2,3-cd]pyrene
Benzo[g,h,i]perylene	Acenaphthylene	4-Chlorophenyl-phenylether
	(b) Organochlorine Pesticides (OCPs)	
Alpha-BHC	Endosulfan II	Dieldrin
Beta-BHC	Endrin aldehyde	*p,p*′-DDD
Delta-BHC	Methoxychlor	*p,p*′-DDT
Heptachlor epoxide	Gamma-BHC-lindane	Endosulfan sulfate
Alpha chlordane	Heptachlor	Endrin ketone
p,p′-DDE	Gamma-chlordane	Aldrin
Endrin	Endosulfan I	
	(c) Organophosphorus Pesticides (OPPs)	
Dichlorvos	Fenthion	Disulfoton
Demeton O&S	Tetrachlorvinphos	Dimethoate
TEPP	Fensulfothion	Chlorpyrifos
Sulfotep	Azinfos methyl	Parathion ethyl
Diazinon	Mevinphos	Tokuthion
Monocrotophos	Ethoprop	Bolstar
Ronnel	Phorate	EPN
Parathion methyl	Naled	Coumaphos
	(d) Chlorinated Herbicides	
2,4-D	Dichloroprop	Dicamba
2,4,5-T	2,4-DB	Dinoseb
Dalapon	2,4,5-TP	
	(e) Polychlorinated Biphenyls (PCBs)	
PCB 28	PCB 101	PCB 153
PCB 52	PCB 138	PCB 180

BHC, benzenehexachloride; DDD, dichlorodiphenyldichloroethane; TEPP, tetraethylpyrophosphate; EPN, ethyl p-nitrophenylthionobenzene phosphonate.

Solvent Extraction, ASE®), with the first noncompany research publication appearing in early 1996.[3] In addition, other commercial systems are now available from Fluid Management Systems, Inc. (Pressurised Liquid Extraction, PLE™) and Applied Separations, Inc. (Pressurised Solvent Extraction, PSE). The basic principle of PFE is that organic solvents, at high temperature and pressure, are used to extract pollutants from environmental matrices. It was first proposed as a method (Method 3545) in Update III of the USEPA SW-846 Methods (1995).[2] This USEPA method (3545) was developed for the application of PFE to the following classes of compounds: base, neutral, acids (BNAs); OCPs; organophosphorus pesticides (OPPs); chlorinated herbicides; and polychlorinated biphenyls (PCBs) (Table 9.1). The choice of extraction solvent, as recommended in the USEPA Method 3545,[2] corresponds to the class of compound to be extracted. For example, for the extraction of BNAs and OPPs from solid matrices, use dichloromethane : acetone (1:1, v/v); for OCPs, use acetone : hexane (1:1, v/v); for PCBs, use hexane : acetone (1:1, v/v); and for chlorinated herbicides, use acetone : dichloromethane : phosphoric acid solution (250:125:15, v/v/v). Within the guidelines, the term solid matrices normally refers to samples of sewage sludge, soil, clays, and (marine/river) sediment.

9.2. THEORETICAL CONSIDERATIONS RELATING TO THE EXTRACTION PROCESS

The extraction process to recover POPs from solid matrices, using an organic solvent, is a complicated system in terms of the individual processes that occur to desorb, solvate, and diffuse POPs from a solid matrix into the bulk fluid (organic solvent) but can be illustrated in the form of a generalized diagram, which shows the typical two-stage extraction profile observed (Fig. 9.1). It is shown that the process (stage 2) of recovering the POP from the matrix is slow due to three processes:

1. desorption of POP from matrix active sites,
2. solvation of POP by organic solvent, and
3. diffusion of POP through static solvent layer,

whereas stage 1 is relatively fast in which the POP is transported away from its initial matrix site by the flowing (bulk) solvent. This situation is enhanced in PFE due to the use of elevated temperatures and pressures.

Organic (liquid) solvents at elevated temperatures and pressures provide enhanced extraction capabilities compared with their use at or near room temperature and atmospheric pressure for two main reasons:[4] (1) solubility and mass transfer effects and (2) disruption of surface equilibria.

9.2.1. Solubility and Mass Transfer Effects

The factors of importance when the temperature of the system is raised are the following:

- higher temperature, which increases the capacity of solvents to solubilize analytes. This is illustrated in Figure 9.2, which shows the influence of temperature on the solubility of glycine in water.[5] It is noted that as the temperature increases, so does the solubility of glycine;
- faster diffusion rates, which occur as a result of increased temperature.

In addition, the addition of fresh solvent during the PFE process improves the mass transfer of POPs from the matrix leading to enhanced extraction rates due to a greater concentration gradient between the fresh solvent and the surface of the sample matrix.

9.2.2. Disruption of Surface Equilibria

The combined use of temperature and pressure has concurrent and interrelated benefits, which lead to improved extraction efficiencies of POPs from sample matrices. As the temperature within the sample-containing cell increases, it can cause disruption of the strong solute–matrix interactions caused by van der Waals forces, hydrogen bonding, and dipole attractions of the POP molecules and active sites on the matrix. In addition, both solvent viscosity and surface tension decrease at higher temperature, thus allowing improved penetration within the sample matrix, which will lead to improved extraction efficiency. The influence of temperature (in the range of 0–100°C) on the viscosity and surface tension of water is illustrated in Figures 9.3 and 9.4, respectively.[6] It is noted that as the temperature increases, both the viscosity and surface tension decrease.

In addition, as the pressure within the sample-containing cell increases, this has the benefit of allowing the organic solvents used in the extraction process, to remain liquefied above their (atmospheric pressure) boiling points. Also, the use of a pressurized system allows the organic solvent to penetrate within the sample matrix, thereby enhancing extraction recovery.

9.3. INSTRUMENTATION

9.3.1. Dionex System

This commercial PFE system (ASE 350®) is a fully automated sequential extractor. A schematic diagram of a PFE system from Dionex Corporation is shown in Figure 9.5. By employing a carousel, the extractor can operate with up to 24 sample-containing extraction vessels and up to 26 collection vials plus an additional 4 vial positions for rinse/waste collection. The sample vessels range in volume between 1, 5, 10, 34, 66, and 100 mL, but all with the same internal diameter of 19 mm. The system is capable of heating samples in the temperature range of 50–200°C and pressures of up to 2000 psi. In addition, an alternative system is available, the ASE 150®, which is a single-cell extraction system.

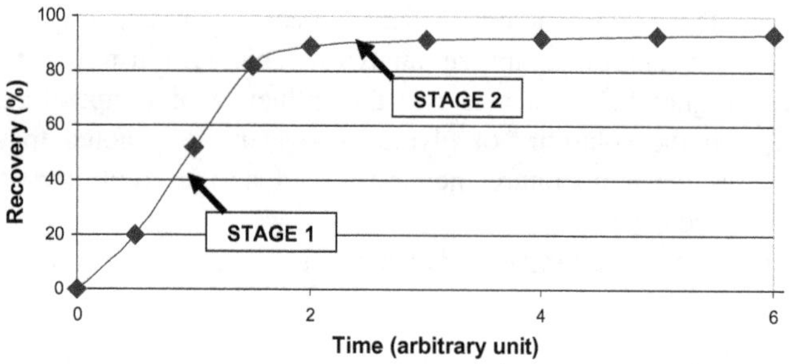

Figure 9.1. Extraction profile of POP from a solid matrix.

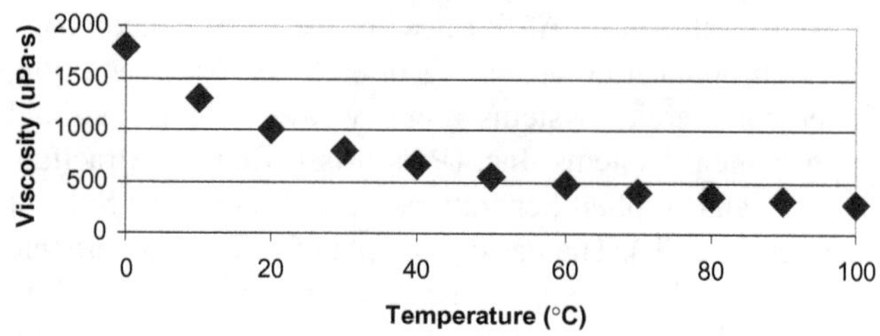

Figure 9.3. Influence of temperature on the viscosity of water.

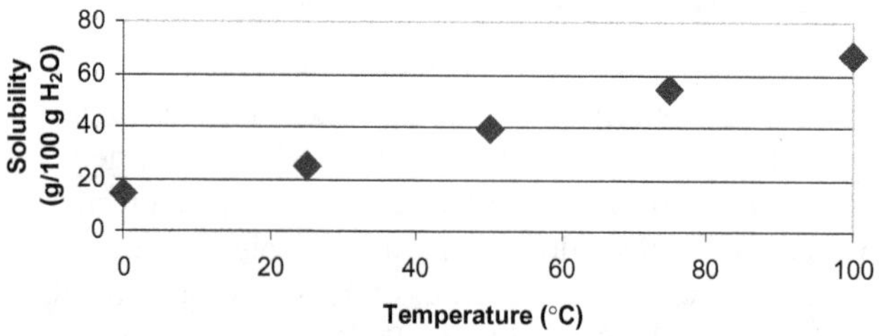

Figure 9.2. Influence of temperature on the solubility of glycine.

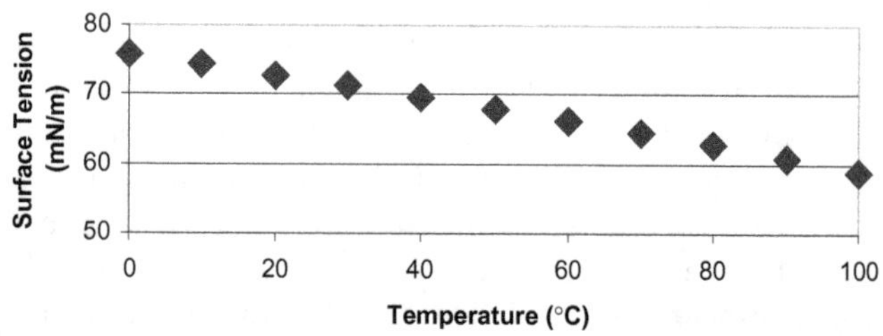

Figure 9.4. Influence of temperature on the surface tension of water.

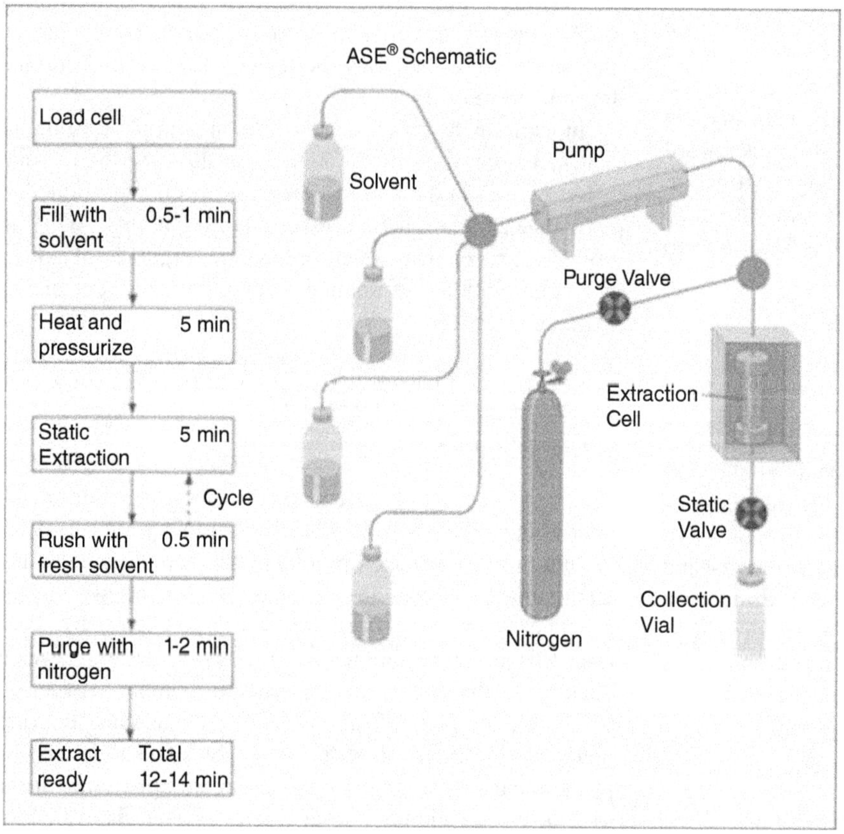

Figure 9.5. Schematic diagram of PFE (ASE system from Dionex Corporation). Image courtesy Dionex Corporation.

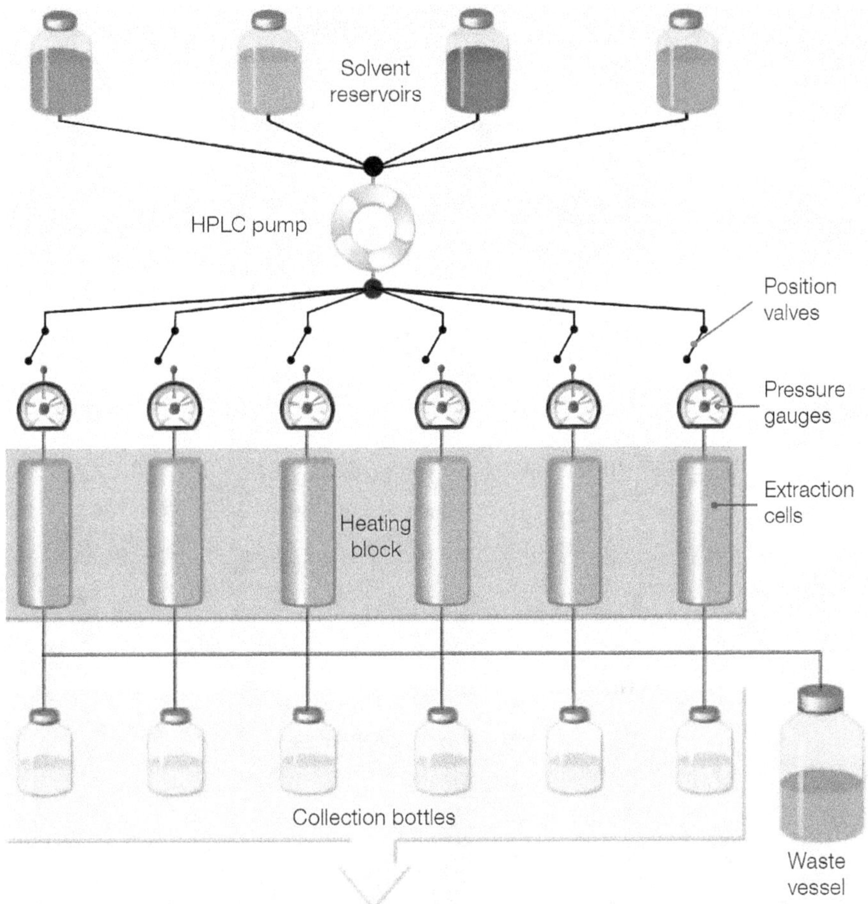

Figure 9.6. Schematic diagram of PFE demonstrating its use for multiple sample operation (SpeedExtractor system from Buchi Corp.). Image courtesy Buchi Corporation.

9.3.2. Buchi

This commercial system (Speed Extractor E-916) is a fully automated simultaneous extractor capable of processing six samples at the same time. A schematic diagram of a PFE system from Buchi is shown in Figure 9.6. The sample vessels range in volume between 10, 20, and 40 mL. The system is capable of heating samples in the temperature range of 50–200°C and pressures of up to 2175 psi. In addition, the SpeedExtractor E-914 is capable of extracting four samples simultaneously. In this system the sample vessels range in volume between 40, 80, and 120 mL. In each case the sample extracts are collected and can be simultaneously evaporated without sample transfer.

9.3.3. Fluid Management Systems System

This commercial system (PLE) is a fully automated simultaneous extractor capable of processing between one and six samples at the same time. A schematic diagram of a PFE system from Fluid Management Systems, Inc., indicating the with or without additional sample cleanup options, is shown in Figure 9.7. The system is capable of heating samples in the temperature range of 70–200°C and pressures of up to 3000 psi.

9.3.4. Common Features to All Three Systems

All commercial systems consist of a solvent supply system, extraction cell, oven, collection system, and purge system, all of which are under computer control. Figure 9.8 shows images of the Buchi Corporation, PSE System, and the Dionex Corporation, ASE system.

9.4. TYPICAL EXTRACTION PROCEDURE (BASED ON ASE 350)

The sample vessel has removable end caps that allow for ease of cleaning and sample filling. Each sample vessel is fitted with two finger-tight caps with compression seals for high-pressure closure. The step-by-step sample-filling procedure is as follows: screw one of the sample vessel's end caps into finger tightness; then, introduce a filter paper (Whatman, grade D28, 1.98-cm diameter) into the vessel

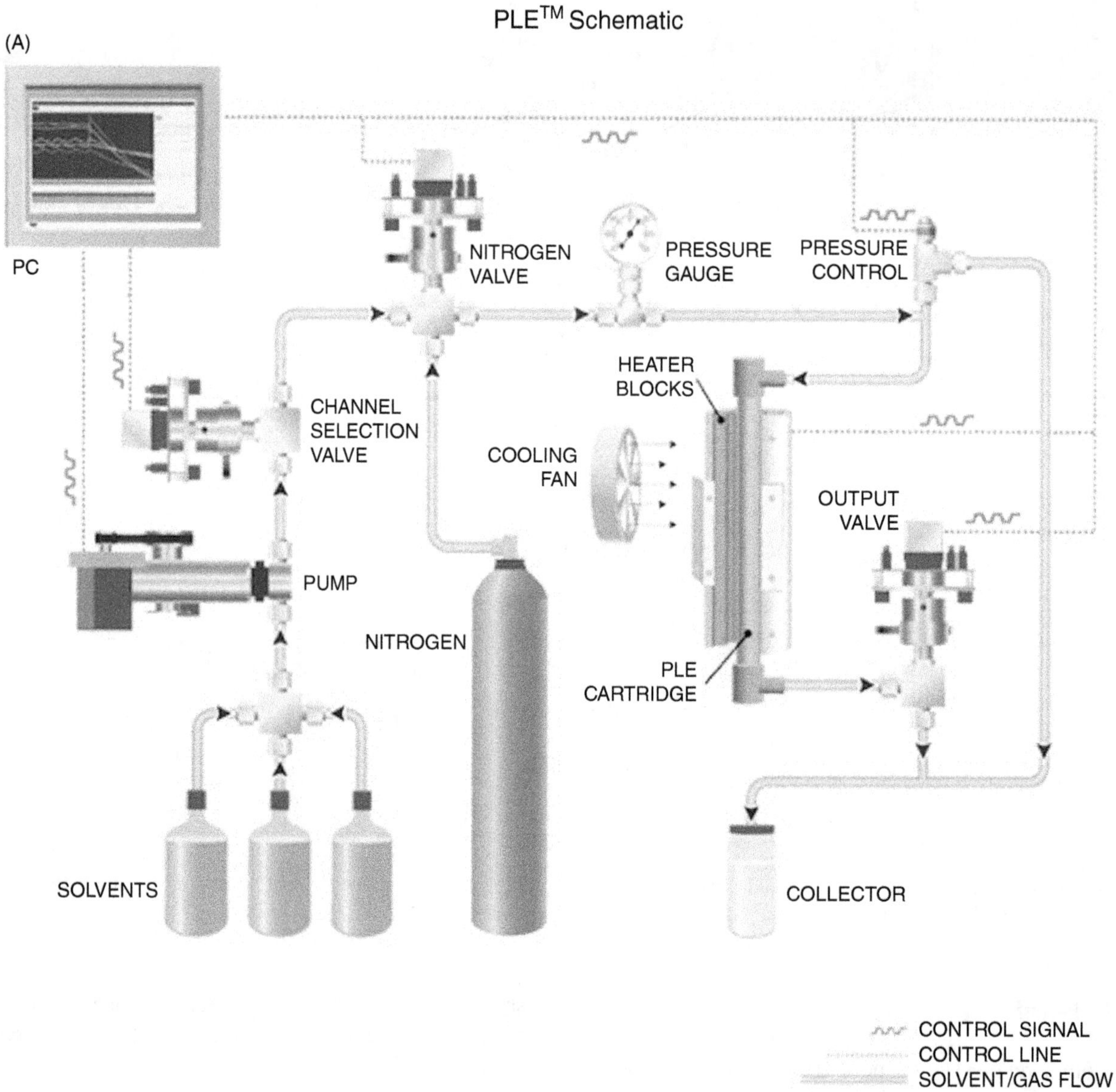

Figure 9.7. Schematic diagram of PFE (A) without and (B) with additional sample cleanup (PLE system from Fluid Management Systems, Inc.). Image courtesy FMS-Fluid Management Systems.

followed by the sample itself; lastly, screw the other end cap to finger tightness and then place the vessel in the carousel. Before performing an extraction, an autoseal actuator places the selected extraction vessel into the oven.

For extraction, the sample cell vessel is loaded into the oven and filled with the selected solvent (or solvent mixture) by the solvent supply system. Then, the cell is heated in the temperature range (up to 200°C) and pressurized for a few minutes (typically 5 min). After completion, the static valves are released. Then, a few milliliters of clean solvent pass through the extraction cell to exclude the existing solvent(s) and extracted analytes. Finally, a gas-purging process follows in which N_2 gas (45 s at 150 psi) evacuates both the sample vessel and the stainless steel transfer line. All extracted analytes pass through the stainless steel tubing that punctures a septum (solvent resistant) located on the top of the collection vial (40- or 60-mL capacity). The vessel will be automatically returned to the carousel after extraction. In-built safety features include an infrared (IR) sensor to monitor the arrival and level of solvent in the collection vial, as well as an automatic shutoff procedure that initiates in the case of system failure. An overview of the PFE method is shown in Figure 9.9.

9.5. METHOD DEVELOPMENT

A suggested extraction method development process for PFE in the recovery of POPs from soil samples is given below based on the scientific literature and our own experience in the use of a commercial PFE instrument since 1996:[3]

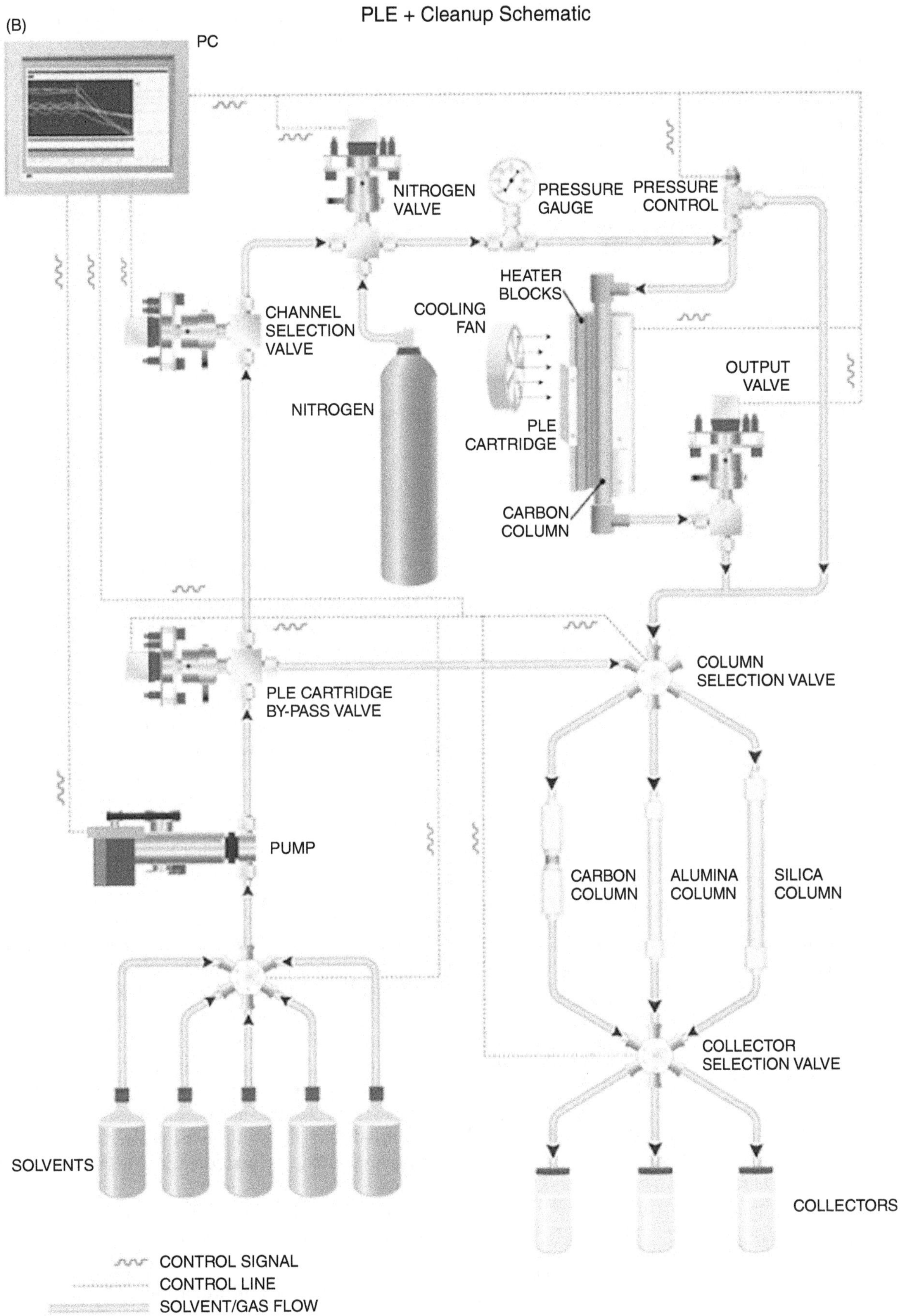

Figure 9.7. *Continued*

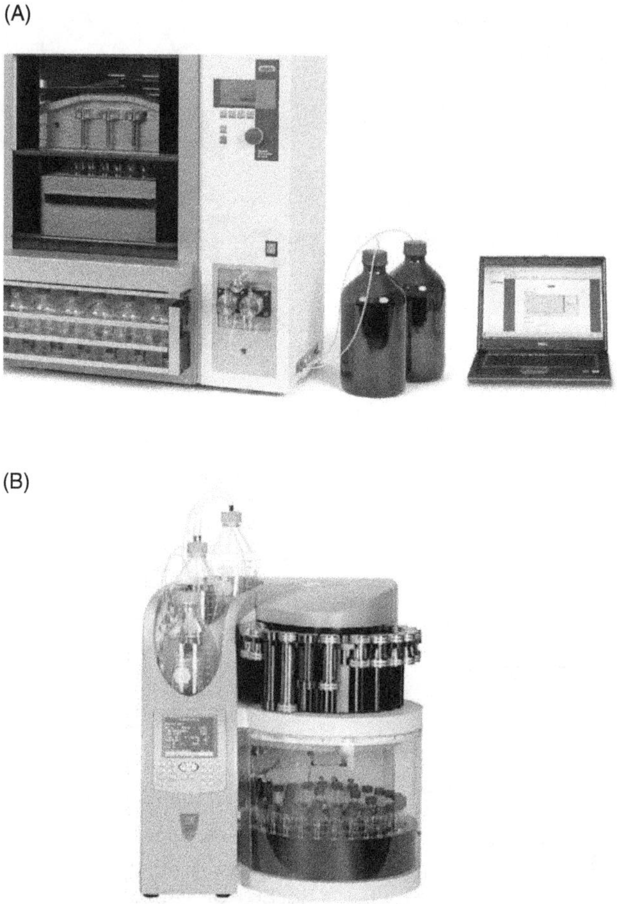

Figure 9.8. Images of the PSE and ASE system: (A) PSE system from Buchi Corporation and (B) ASE system from Dionex Corporation. Image courtesy Dionex Corporation.

1. assess and identify the POPs to be determined
2. pre-extraction considerations
   a. air-dry the sample at room temperature for 24 h
   b. sieve the sample (2 mm) to remove any debris, for example, stones
   c. grind the sample to a small particle size ($<15\,\mu m$), then mix well to ensure homogeneity
   d. mix the ground soil sample with either copper powder and/or a dispersing agent (e.g., hydromatrix or anhydrous sodium sulfate) prior to extraction
   e. fit a filter paper inside the bottom of the sample cell prior to introducing the sample
   f. ensure sample cell is comfortably full (i.e., remove the dead volume of the cell)
3. extraction considerations
   a. consider the influence of operating variables (time, temperature, and pressure) on recovery; be aware of potential POP degradation via use of elevated temperatures
   b. consider the choice of organic solvents; polar solvents will be more effective than nonpolar solvents; only use high-purity solvents to minimize chromatographic artifacts
4. overall criteria
   a. use certified reference materials (CRMs) to assess recoveries of POPs from matrices
   b. ensure that a CRM is regularly extracted/analyzed to assess long-term performance
5. maintenance
   a. ensure regular maintenance of cells and fittings occurs and replace, as necessary
   b. check alignment of collection vial carousel regularly
   c. replace tubing connection between extraction cell and collection vial on a periodic basis (can be easily blocked if sample contains a high sulfur content—problem can be alleviated if copper powder is mixed with sample pre-extraction)

9.6. APPLICATIONS OF PFE

There has been a sustained interest in PFE over the past decade for the determination of various organic and organometallic pollutants in a range of solid environmental samples such as soil, sediment, sewage sludge, fly ash, air particulate matter, plant materials, and animal tissues.[7–10] The technique is applicable to the extraction of phenols, polycyclic aromatic hydrocarbons (PAHs), OPPs/OCPs, chlorinated herbicides, PCBs, polychlorinated dibenzodioxins/dibenzofurans (PCDD/Fs), and diesel range organics (DROs) prior to chromatographic analysis. Applications to other classes of organic compounds such as ingredients of personal care products in environmental matrices[11] or to matrix components in food and biological samples[9] have also been described. In this section, recently published reports (since 2000), with a study focus on the extraction stage are discussed, to reflect on the progress and current status of PFE, particularly in contrast to other widely used extraction techniques. Operating parameters (solvents, temperature, pressure, times and numbers of static and dynamic cycles, and flush volume) have been optimized to improve extraction yield and selectivity for targeted analyte compounds and/or sample types. Some works were directed at miniaturization, *in situ* derivatization or cleanup to further reduce the solvent consumption and to perform faster extraction. Validation has been commonly achieved by analyzing CRMs or spiked and aged samples, or comparing against conventional/benchmark extraction techniques such as Soxhlet.

9.6.1. Parameter Optimization

High extraction efficiency of PFE for POPs (including chlorobenzenes, hexachlorocyclohexane (HCH), dichlorodi-

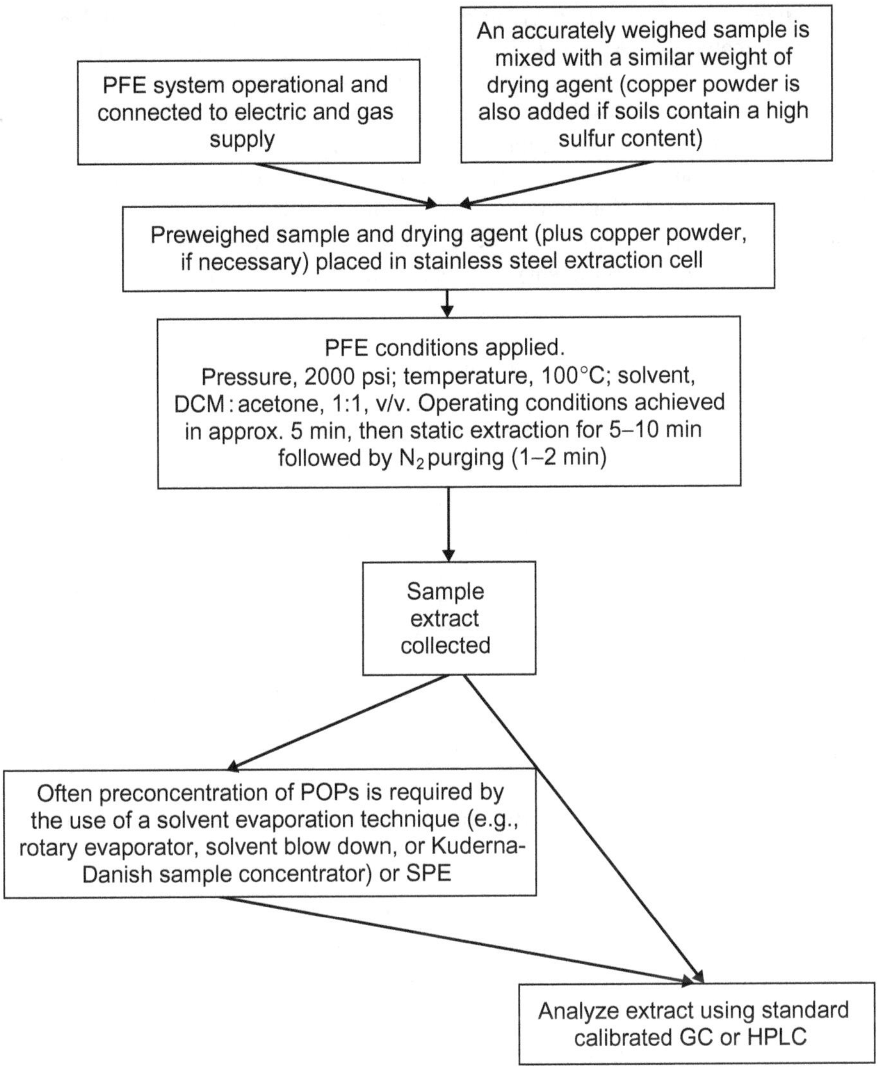

Figure 9.9. General analytical procedure for PFE.

phenyltrichloroethane (DDT) and its isomers, PCB congeners, and PAHs) in real contaminated soil samples was achieved at an early stage by systematic investigations of solvent and temperature.[12] Solvents with different polarity were tested between 20 and 180°C. The optimal conditions proved to be two extraction steps at 80 and 140°C (average relative standard deviation [RSD], 10.7%) with three static cycles (extraction time, 35 min) using toluene as solvent and at a pressure of 15 MPa. In many cases, analytical values were better by up to one order of magnitude or even more compared with Soxhlet results. PFE has been used to extract 4-nonylphenol in sediment.[13] The analytes were completely extracted by methanol at 100°C and 100 atm combined with 15-min static and then 10-min dynamic extraction steps prior to the determination by large-volume injection GC–mass spectrometry (MS). Recoveries in spiked sediment samples from a reservoir and a polluted river were 111% with 4% RSD and 106% with 5% RSD, respectively. The extraction efficiency was comparable with conventional Soxhlet and bath ultrasonication extraction methods for the spiked sediment samples. Recently, PFE was optimized in terms of solvent composition, temperature, pressure, number, and duration of static cycles, and flush volume for PAHs (hexane : acetone, 1:3 [v/v], 140°C, 1500 psi, 3 × 5 min, and 50%, respectively) and PCBs (toluene : acetone, 1:3 [v/v], 120°C, 2000 psi, 3 × 5 min, and 50%, respectively) in compost.[14] Optimal extraction conditions for PAHs were more difficult to determine due to their inhomogeneous distribution in samples. Detection limits and method precision were between 0.16 and 2.46 μg/kg dry weight and 6–17%, respectively, for PCBs; while for PAHs, they were between 1.1 and 37.2 μg/kg dry weight and 12–34%, respectively. Similarly, PFE has been optimized for DDT and dichlorodiphenyldichloroethylene (DDE) in Ethiopian soils.[15] *n*-Heptane with acetone was used to replace more hazardous organic solvents such as *n*-hexane, toluene, and dichloromethane. PFE for 3 × 10 min at 100°C was proven to be more effective than Soxhlet extraction for DDE and to be equally effective for DDT. The procedure was also adapted for the determination of alpha-endosulfan, beta-endosulfan, and endosulfan sulfate.[16] A cleanup step using a Florisil column and a relatively low volume of solvent (ethyl acetate : n-heptane) was required. The analytical method was validated by analyzing a certified reference soil material (CRM 811-050).

Harb and Aldstadt performed the optimization for PAHs in sediment based on a response function (recovery/precision) for three representative PAHs (naphthalene, pyrene, and benzo[g,h,i]perylene).[17] Small amounts of sample (0.5 g) and solvent (3 mL of n-hexane) were heated in a closed stainless steel extraction cell for 60 min at 200°C. The cell was then placed in an ice bath for 5 min, and the PAH content of the collected extract was determined by GC-MS. Detection limits ranged from 0.051 ± 0.011 to $0.96 \pm 0.08\,\mu g/g$ for the 16 PAHs classified as priority pollutants by the USEPA. The method was validated by analyzing a standard reference material (SRM 1944). Previously, Preud'homme and Potin-Gautier used experimental designs to evaluate the influence of five parameters (solvent nature, temperature, pressure, static time, and number of cycles) on the extraction of PCDD/Fs and mixed bromine/chlorine-dibenzodioxins/dibenzofurans (MXDD/Fs) in samples from municipal solid waste incinerators.[18] The extraction profiles and the optimal operating conditions were determined for each matrix from the modeling of the extraction performance. The two following effects, the relative peak area and the coextracted matrix (CEM), were screened. The temperature of extraction was found to be the most important parameter. Automated PFE was as efficient as Soxhlet or Soxtec extraction but four times faster. Quantitative results were obtained within 15 min for all congeners at 130°C with a mixed solvent (n-hexane : acetone [1:1, v/v]). Under these operating conditions, the CEM and the degradation of the highly brominated compounds were minimized.

Kiguchi et al. examined several solvents including toluene, n-hexane, acetone, acetone : toluene, and acetone : n-hexane (1:1, v/v), for extraction of PCDD/Fs and coplanar polychlorinated biphenyls (Co-PCBs) using a CRM (CRM 0422 forest soil).[19] Mixed solvents produced better recoveries than the single solvents; however, these results were lower than those by Soxhlet extraction. Additional extraction cycles using mixed solvents were necessary for satisfactory extraction of more tightly bound PCDD/Fs and Co-PCBs from soil. Six different solvents and temperatures from 125 to 200°C were investigated for the extraction of hydrocarbon contaminants from wet and dry soils.[20] Nonpolar solvents could not achieve complete recovery from wet soils at the temperatures studied. Quantitative recoveries for DROs and waste oil organics were obtained using the optimized conditions (dichloromethane : acetone [1:1, v/v], 175°C, 8-min heat-up time, and 5-min static time). Zhu et al. investigated the extraction of selected chloroacetanilide and nitrogen heterocyclic herbicides (alachlor, bromacil, hexazinone, metribuzin, and tebuthiuron) in Hawaiian clayey soil.[21] Pretreatment of the soil with 37.6% water was required prior to a two static cycle extraction with a total of 32 mL of acetone at 1500 psi and 100°C. Water was the most effective modifier for quantitative recoveries of the five herbicides, with average recoveries ranging from 93% to 103% compared with only 68–83% when no water was used. The extraction time and total organic solvent consumption were less than 22 min and 80 mL, respectively, relative to 18 h and 300 mL for Soxhlet extraction.

Alonso-Rodriguez et al. reviewed the application of PFE for simultaneous organometallic compounds speciation.[22] Two CRMs (GBW-08751 mussel tissue and PACS-2 marine sediment) were subjected to pressurized carboxylic acids. Acetic acid and formic acid appeared effective to extract metal species from the tested environmental materials. Marcic et al. evaluated a PFE procedure for organotin compounds (OTCs) in vegetable plants (French bean, lettuce leaves, and potato tubers).[23] The study focused on the four most toxic trisubstituted OTCs: tributyltin (TBT), triphenyltin (TPhT), tricyclohexyltin (TcHexT), and trioctyltin (TOcT). Methanol, ethyl acetate, and a mixture of methanol and ethyl acetate were tested, and the mixture was found to give the most quantitative results while preserving the species. The influence of pressure and temperature on the recovery of the process, and on species preservation, was evaluated using the experimental design methodology. The results from this procedure (an extraction duration of 15 min) were compared with those from the technique usually employed, that is, solid–liquid extraction (SLE) performed in an acidic solvent by mechanical shaking (an extraction duration of 2–12 h). The detection limits with optimized PFE were 1–2 ng/g (Sn), which were higher than those obtained by SLE (0.1–1 ng/g). The repeatability was similar for both PFE and SLE (2–12%) but appeared to be highly time dependent in the case of SLE.

9.6.2. Effect of Matrix Characteristics

Josefsson et al. evaluated the influence of sediment characteristics on PFE exhaustiveness for PCBs.[24] No correlation was found between recoveries and sediment PCB concentration, total organic carbon, soot carbon, or amorphous carbon. Statistically significant ($p < 0.05$) correlations were observed for water content and the carbon/nitrogen (C/N) ratio; the decrease in recovery with decreased water content was attributed to less access of the solvent to the analytes due to less matrix swelling, whereas the lowered recoveries with increased C/N ratio might be due to stronger binding between the analytes and the matrix organic matter. Overall, the modified USEPA method (100°C, 100 bar, n-heptane : acetone (1:1, v/v), 2×5 min) was found to be exhaustive, and the excellent recoveries confirmed that sediment characteristics had no marked influence on the extraction. The effectiveness of the PFE-based USEPA Method 3545 was also evaluated for the extraction of PCBs in Baltic Sea sediments with different characteristics such as organic carbon, soot carbon, sulfur, water content, and PCB concentration range.[25] Sediments containing relatively low amounts of organic and soot carbon could be quantitatively extracted in 5 min with n-hexane : acetone as proposed in the USEPA Method.

However, for sediments with larger amounts of carbon, the total amount of PCBs present in the sample might be underestimated when using *n*-hexane : acetone, for which toluene seemed to increase the extraction efficiency.

9.6.3. Ultrasound (US)-Assisted Extraction

US-assisted PFE was devised for PAHs and aliphatic DROs in soil.[26] The extraction cell was inserted into a US bath for US-PFE, whereas into an oven for usual dynamic PFE. For a spiked and aged test soil, analyte extraction by US-PFE was quantitative with dichloromethane for 10 min; in contrast, dynamic PFE with dichloromethane required an extraction time of 20 min. For a natural contaminated sample, the extraction time had to be increased to 20 min or the solvent (dichloromethane) had to be replaced by a mixture of dichloromethane : acetone (1:1, v/v) to reach comparability with Soxhlet extraction.

9.6.4. Sequential Extraction

A sequential PFE method was developed to extract PAHs and their oxygenated derivatives (oxy-PAHs) from contaminated soil and simultaneously separate them into two fractions.[27] The extraction cells were packed with 2% deactivated silica gel (a chromatographic adsorbent) and extraction solvents of increasing polarity. The PAHs were extracted with cyclohexane : dichloromethane (9:1, v/v) at 120°C, after which the oxy-PAHs were extracted with cyclohexane : dichloromethane (1:3, v/v) at 150°C. The recoveries were mostly above 70% for PAHs and above 90% for oxy-PAHs. The method was applied to seven authentic soil samples, and the results were in good agreement with those of a reference method, based on Soxhlet extraction, and silica gel cleanup and also in good agreement with the certified reference values available for one of the soils. The sequential PFE method was faster and consumed less solvents than a traditional method based on separate extraction and fractionation steps. Therefore, it was suitable for the concurrent analysis of PAHs and oxy-PAHs during large-scale soil contamination studies by providing more information than an ordinary PAH analysis.

9.6.5. Subsequent Cleanup

The high extracting power of PFE leads to the coextraction of numerous interfering organic substances; therefore, an efficient cleanup of the extract is often necessary. Burkhardt et al. coupled PFE with subsequent solid-phase extraction (SPE) cleanup to determine the distribution of PAH and alkylated homolog groups in sediment samples, with reduced background interferences for full-scan GC-MS analysis.[28] Method detection limits ranged from 20 to 302 μg/kg based on 25-g samples. Of 28 sediment samples, 35 of 41 compounds, excluding homologs, were detected in at least one sample with concentrations ranging from 20 to 100,000 μg/kg. The most frequently detected compound, 2,6-dimethylnaphthalene, was detected in 23 samples with a concentration ranging from 15 to 907 μg/kg. Results for the alkylated homolog groups in 96% of samples indicated concentrations ranging from 20 to 89,000 μg/kg. The most frequently detected homolog series, C-2-alkylated naphthalene, was found in 26 samples with a concentration ranging from 25 to 3900 μg/kg. Several sorbents (Florisil, silica, alumina, and carbon, as well as their combinations) have been tested to clean up PFE extracts for the determination of OCPs in soil.[29] The validation was achieved by analyzing a certified reference soil material (CRM 804-050) and comparing results with Soxhlet and MAE.

When PFE was used for the extraction of PCBs and OCPs in fish muscle tissue containing 3.2% (w/w) lipids, crude extracts had to be purified by gel permeation chromatography before analysis by GC–electron capture detection (ECD).[30] The recoveries obtained by PLE operated at 90–120°C were either comparable (for higher-chlorinated PCB congeners and the DDT group) or better (for lower-chlorinated analytes) compared with Soxhlet extraction using the same solvent mixtures hexane : dichloromethane (1:1, v/v) and hexane : acetone (4:1, v/v). The highest recoveries were obtained using three 5-min static cycles.

9.6.6. *In Situ* Cleanup

In situ simultaneous cleanup for biota samples to reduce sample preparation time and solvent consumption has been achieved by inclusion of sorbents in the extraction cell.[31] Initially, five types of fat retainers (Florisil, basic alumina, neutral alumina, acidic alumina, and sulfuric acid-impregnated silica) were investigated for lipid-free extraction of PCBs from fat-containing matrices (naturally contaminated fish meal and certified cod-liver oil CRM 349).[32] All retainers generated fat-free extracts when the fat/retainer ratio was 0.025. Sulfuric acid-impregnated silica and Florisil gave completely clear extracts, and the former was the only one not to show any reaction when treated with sulfuric acid after the extraction. A detailed study then established the best extraction conditions with sulfuric acid-impregnated silica to oxidize coextracted fat: a fat/retainer ratio of 0.025, 100°C, and a flush volume of 60% as suggested in the USEPA Method 3545.[33] Both *n*-pentane and *n*-heptane were capable of replacing *n*-hexane as the extraction solvent. Further work investigated the effects of lipid type and amount.[34] A fat/retainer ratio of 0.050 resulted in fat-free extracts for both fish triglycerides and oil matrices at 100°C using *n*-heptane. The method was tested on naturally contaminated cod-liver homogenate and naturally contaminated cod-liver oil spiked to animal feed for poultry, as well as CRMs (BCR 49 cod-liver oil and IRMM 445 spiked pork fat). Subsequently, the approach was adapted for the extraction of PCDD/Fs and dioxin-like PCBs at 100°C with *n*-heptane for 5 min in two cycles combined with high-resolution GC-MS analysis.[35] Results for spiked vegetable oil, naturally contaminated crude fish oil, and oil-containing feed samples were generally within ±20% of the spiked levels or values obtained by reference laboratories using conventional sample preparation. The precision, measured as the %RSD for 2,3,7,8-tetrachlorodibenzo-*p*-dioxin toxic equivalency values (TEQs), was below 10% in all cases. Satisfactory data were obtained for naturally contaminated herring tissue, chicken tissue, and pork tissue. However, in the case of sepiolitic clay, a sufficient amount of sodium sulfate was needed to be added to dehydrate the samples, as well as a loosely packed extraction cell in order to avoid suppressed extraction efficiency.

In a separate study,[36] freeze-dried samples were dispersed with Florisil and loaded in the extraction cell containing an extra amount of Florisil. The extraction was performed under mild conditions using 55 mL of a dichloromethane : pentane (15:85, v/v) mixture, a temperature of 40°C, a static extraction time of 10 min, and two extraction cycles. The Florisil retained coextracted lipids from the matrix, and the extract, after preconcentration, was clean enough for direct injection into GC-MS and GC-ECD. Quantitative recoveries (from 90% to 106%) were obtained for both native and spiked PCB congeners in samples with a high lipidic content (up to 42% dry mass, in spoonbill eggs). The reproducibility of replicate extractions was better than 11% RSD. Method detection limits were in the ranges of 0.001–0.004 and 0.002–0.07 ng/g dry mass for GC-ECD and GC-MS-MS, respectively. The method was validated using a standard reference material (SRM 2974 mussel tissue), compared with Soxhlet and matrix solid-phase dispersion (MSPD) extraction methods, and used to evaluate the PCB contamination in bivalves from the south of Spain.

Thirty-nine polybrominated diphenyl ether (PBDE) congeners in sediment were determined by GC-MS after PFE with alumina cleanup in the extraction cell.[37] Recoveries between 47% and 82% were obtained for spiked tri- to hepta-PBDE congeners in sediment samples. The repeatability of replicate extractions was better than 15% RSD. The detection limits were between 1 and 46 pg/g dry weight, while detected concentrations in sediment samples ranged between 0.86 and 2.49 ng/g dry mass. The sample preparation (extraction and cleanup) time was reduced from days to 30 min with efficiency similar to that afforded by conventional Soxhlet extraction plus SPE cleanup.

9.6.7. *In Situ* Derivatization

In situ derivatization with the addition of common derivatizing agents directly into the extraction cell was developed for polar herbicides in soil analyzed by GC-MS.[38,39] Six acid herbicides (2,4-D, 2,4,5-T, dicamba, silver, trichlopyr, and bentazone) were extracted at 100°C and 1500 psi and derivatized with pentafluorobenzyl bromide (PFBBr) in the presence of Na_4EDTA as a single step in the PFE cell. The use of Na_4EDTA significantly improved the analyte recoveries, with an overall average of 92 ± 16%. The analysis time was reduced from several hours (up to 24 h) for the conventional extraction to less than 1 h. The method was adapted for PAHs in sediment.[40] Furthermore, PFE with *in situ* derivatization was expanded for polar analytes such as phenols (including chlorophenols), sterols, and carboxylic acids.[41] The one-step protocol used acetic anhydride as an acetylation agent, *N*,*O*-bis(trimethylsilyl)trifluoroacetamide (BSTFA) as an silylation agent, and boron trifluoride-methanol, phenyltrimethyl ammoniumhydroxide, and trimethyl sulfoniumhydroxide as methylation agents. The addition of a silylation agent also facilitated the extraction kinetics of analytes not accessible to silylation (e.g., PAHs or alkylbenzenes), probably due to a dissociative action of the agent to weaken analyte–matrix interactions.

9.6.8. Miniaturization

A miniaturized PFE in a stainless steel cell was developed for PAHs in soil and sediment with subsequent large-volume injection GC-MS analysis.[42] Toluene was selected as solvent and a total volume of 100 µL was used for the 10-min static–dynamic extraction of 50-mg samples. Additional cleanup or filtration of the extracts was not required. Method detection limits were below 9 ng/g for the 13 PAHs more volatile than indeno[1,2,3-cd]pyrene in real soil samples, and the repeatability for the analysis of endogenous PAHs was better than 15%. The PFE efficiency for nonspiked samples was the same as or better than Soxhlet and liquid-partitioning extraction. Later, the miniaturized PFE procedure was optimized for the determination of chloroanilines in soil by GC-MS equipped with both large-volume on-column (LV-OC) and programmed temperature vaporization (PTV) injections.[43] Hexane was the only solvent suited for LV-OC, while hexane : acetone gave the best results when using PTV. Both approaches allowed the detection of chloroanilines down to the 5–50 ng/g range in complex soil samples. The latter was robust and showed overall better results in terms of analyte recovery (36–109% vs. 5–87%), repeatability (8–13% vs. 4–31%), and detection limits.

9.6.9. Comparative Studies

The performance of PFE has been repeatedly evaluated in comparison with other commonly applied extraction techniques. Recently, Ryno et al. compared PFE with Soxhlet and ultrasonic extractions based on extraction efficiency and recovery using three solvents for 16 USEPA PAHs in diesel exhaust particulate and urban ambient air particulate samples collected on filters followed by HPLC analysis with a multiple wavelength shift fluorescence detector.[44] The most effective extraction was achieved by PFE using dichloromethane as solvent. Another recent study on PAHs in contaminated urban soil revealed significant between-method and sometimes within-method variability in the stable carbon isotope signatures obtained for individual PAHs from a certified contaminated soil using GC-isotope ratio mass spectrometry (IRMS) despite accurate and highly reproducible concentration data.[45] PFE compared with Soxhlet, Soxtherm, and sonication gave more consistent and reproducible stable carbon isotope data after two cleanup procedures using silica/dichloromethane and alumina/hexane–toluene to remove coextracted material. In an earlier study also involving supercritical fluid and microwave extractions for PCBs in soil, when the extraction conditions were chosen wisely, accurate data were produced by all extraction techniques for native PCB-contaminated soils.[46] A comparative investigation on extraction recovery and selectivity for environmental solids reported in 2000 found that, although quantitative agreement between all tested methods was generally good (despite minor differences in recoveries) for a PAH-contaminated soil, the extract quality differed greatly.[47] The organic solvent extracts by PFE (and Soxhlet extraction) were much darker and also yielded more artifact peaks in the GC-MS and GC-FID chromatograms, especially compared with SFE with CO_2. PFE along with subcritical water and Soxhlet extraction presented poor selectivity for PAHs (one-fourth to one-third of the bulk soil organic matter also extracted), while SFE with pure CO_2 removed only 8% of the bulk organic matrix. For urban air particulate matter, organic solvent extractions yielded lower levels of PAHs but very high concentrations of *n*- and branched alkanes (C-18 to C-30) from diesel exhaust; no selectivity between PAHs and the bulk alkanes was obtained. PFE was also found to offer recoveries similar to Soxhlet extraction but slightly lower compared with SFE for 10 selected polycyclic aromatic nitrogen heterocycles (PANHs) in soil.[48]

9.7. COMPARISON OF PFE WITH OTHER EXTRACTION TECHNIQUES

It is possible to compare the advantages and disadvantages of PFE with other extraction techniques. Table 9.2 compares a range of extraction techniques that are used to recover POPs from environmental matrices. The extraction techniques compared are Soxhlet, shake flask, sonication, MSPD, SFE, and MAE with PFE. Eleven different parameters have been selected to compare each of the extraction techniques. While the list of parameters is not exhaustive, the 10 chosen ones are considered to be the most appropriate to allow a wide ranging comparison of the extraction techniques with PFE. Each of the parameters is now considered.

9.7.1. Sample Mass

This is often a balance between obtaining a representative and homogenous sample that can be extracted versus the total amount of sample available. In some cases, the amount of sample available may be large, whereas in other cases, only a limited quantity is available. In this case, Soxhlet extraction has the ability to extract the largest quantity of sample, whereas most of the other extraction techniques generally extract smaller quantities. PFE offers the ability to extract a range of samples masses up to 30 g.

9.7.2. Extraction Time

The ability to extract multiple samples simultaneously and quickly is of major benefit. In this case, shake flask, sonication, MSPD, and MAE offer short extraction times alongside PFE. In this category, PFE compares favorably with other approaches.

9.7.3. Solvent Type and Consumption

With the exception of SFE, most other extraction techniques utilize organic solvents that are generally polar and contain chlorine, which provides affective properties to solvate and recover POPs from environmental samples. In addition, with the exception of Soxhlet extraction, most approaches generally use small quantities of organic solvents, which make them cost-effective and potentially more environmentally friendly. PFE uses both minimal quantity and polar organic solvents.

9.7.4. Extraction Method

The use of elevated temperature is the most common single approach to facilitate the recovery of POPs from environmental matrices. This effect is also enhanced in PFE by the use of pressure (this is also the case for MAE and SFE).

9.7.5. Sequential or Simultaneous

Soxhlet, shake flask, sonication, and MSPD can extract more than one sample simultaneously simply by duplicating, triplicating, and so on, the apparatus used. The main simultaneous extraction technique is MAE. Modern instruments from, for example, CEM Corp. (www.cem.com) and Milestone (www.milestonesrl.com) allow approximately 40 samples to be extracted simultaneously. In the case of PFE, only the system from Applied Separations (PSE) can allow up to six samples to be extracted simultaneously (see also "extraction time").

9.7.6. Method Development Time

The higher the level of instrumentation, the more is required in terms of method development. However, the existence of standard methods does alleviate most of the development time for most sample types and POPs. PFE has a USEPA method.

9.7.7. Operator Skill

As above (see "method development time"), the more instrumental approaches, for example, MAE, SFE, and PFE,

Table 9.2. Comparison of Extraction Techniques for Recovery of POPs from Environmental Matrices

	Soxhlet	Shake Flask	Sonication	MSPD	SFE	MAE	PFE
Sample mass	Normally up to 10 g	0.5–5 g	2–30 g	0.5–5 g	1–10 g	2–5 g	Normally up to 30 g
Extraction time	6, 12, or 24 h per sample	3–5 min per sample	3–5 min per sample	5–20 min per sample	30 min–1 h per sample	20 min (plus 30 min cooling and pressure reduction) for up to 40 samples	12–15 min per sample
Solvent type	Acetone : hexane (1:1, v/v); acetone : DCM (1:1, v/v); DCM only; or toluene : methanol (10:1, v/v)	Typically, acetone : DCM (1:1, v/v)	Acetone : DCM (1:1, v/v) or acetone : hexane (1:1, v/v) for semivolatile organics and OCPs; acetone : DCM (1:1, v/v), acetone : hexane (1:1, v/v), or hexane for PCBs	Typically, DCM, hexane, ethyl acetate, acetonitrile, methanol, or acetone	CO_2 (plus organic modifier). Tetrachloroethene used as the collection solvent for TPHs for determination by FTIR, otherwise DCM	Typically, acetone : hexane (1:1, v/v). The solvent(s) is/are required to be able to absorb microwave energy.	Acetone : hexane (1:1, v/v) or acetone : DCM (1:1, v/v) for OCPs, semivolatile organics, PCBs, or OPPs; acetone : DCM : phosphoric acid (250:125:15, v/v/v) for chlorinated herbicides
Solvent consumption	150–300 mL	5–20 mL	5–20 mL	5–50 mL	10–20 mL	25–45 mL	25 mL
Extraction method	Heat	Agitation	Ultrasound	Solid-phase extraction	Heat + pressure	Heat + pressure	Heat + pressure
Sequential or simultaneous	Sequential (but multiple assemblies can operate simultaneously)	Sequential (but possible to shake several flasks at the same time)	Sequential (but possible to sonicate several flasks at the same time)	Sequential (but possible to use a manifold vacuum system for up to 12 samples at the same time)	Sequential	Simultaneous (up to 40 vessels can be extracted simultaneously)	Sequential + simultaneous (up to six vessels can be extracted simultaneously)
Method development time	Low	Low	Low	Medium	High	High	High
Operator skill	Low	Low	Low	Medium	High	Moderate	Moderate
Equipment cost	Low	Low	Low	Low	High	Moderate	High
Level of automation	Minimal	Minimal	Minimal	Minimal	Minimal to high	Minimal	Can be fully automated
USEPA method	3540	–	3550	–	3560 for TPHs, 3561 for PAHs, and 3562 for PCBs and OCPs	3546	3545

MSPD, matrix solid-phase dispersion; SFE, supercritical fluid extraction; MAE, microwave-assisted extraction; TPHs, total petroleum hydrocarbons; PAHs, polycyclic aromatic hydrocarbons; OCPs, organochlorine pesticides; DCM, dichloromethane (or methylene chloride); USEPA, United States Environmental Protection Agency; FTIR, Fourier transform infrared spectroscopy.

Table 9.3. SWOT Analysis of PFE

SWOT Analysis	Technique: PFE
Strengths	• range of compounds can be extracted from a range of matrices (environmental, pharmaceutical, industrial, clinical, food, and forensic) • range of suppliers of instruments • uses a small quantity of organic solvent • relatively rapid (12–15 min per sample) • range of sample masses can be extracted (0.5 to >30 g) • relatively small solvent consumption • no filtration required post extraction • full automation possible • efficient extraction recoveries
Weaknesses	• high relative capital cost • only one sample extracted per set time (ASE only) • acceptance as an alternative to Soxhlet extraction • often requires chlorinated solvent to be effective • different solvents required for different POPs • extraction of coextractive from matrices can lead to chromatographic artifacts • nonselectivity of extraction process • regular instrument maintenance program required
Opportunities	• *in situ* derivatization • *in situ* cleanup • miniaturization • selective extraction by altering solvents
Threats	• possible longer-term threat that it will be superseded • high relative capital cost

often require more operator skill. However, this is often alleviated by the use of specific technique software that is both simple to use and easy to understand.

9.7.8. Equipment Cost

The instrumental extraction approaches, for example, MAE, SFE, and PFE, have a higher capital purchase cost. However, the ability to recover this capital equipment cost can often only be justified by the high sample throughput of the extraction technique. This is a reasonable assumption in the case of PFE.

9.7.9. Level of Automation

The use of any degree of automation is beneficial to the operation of the laboratory as it can lead to enhanced productivity (provided the instrumentation is appropriately maintained and serviced). PFE offers a degree of automation that allows samples to be (normally) sequentially extracted and collected prior to analysis.

9.7.10. USEPA Method

The existence of specific, dedicated methods for most of the extraction techniques shown means that (1) method development time is reduced, (2) the extraction technique is acceptable, and (3) the extraction process should be transferable between laboratories.

In addition, another approach that can be used in considering the virtues of PFE is to perform a strengths, weaknesses, opportunities, and threats (SWOT) analysis (Table 9.3).

9.8. SUPPLIERS OF THE PFE EQUIPMENT

- Dionex Corporation, 1228 Titan Way, Sunnyvale, CA 94085 (www.dionex.com)
- Fluid Management Systems, Inc., 56 Felton Street, Waltham, MA 02453 (www.fmsenvironmental.com)
- Applied Separations, Inc., 930 Hamilton Street, Allentown, PA 18101 (www.appliedseparations.com)

REFERENCES

1. Dean, J.R., Ed. *Applications of Supercritical Fluids in Industrial Analysis*. Glasgow: Blackie Academic and Professional; 1993.
2. USEPA. *Test Methods for Evaluating Solid Waste*, Method 3545, USEPA SW-846, 3rd ed. Washington, DC: Update III, U.S. GPO; 1995.
3. Dean, J.R. Accelerated solvent extraction of polycyclic aromatic hydrocarbons from contaminated land. *Anal. Comm.* **1996**, *33*, 191–192.
4. Richter, B.E.; Jones, B.A.; Ezell, J.L.; Avdalovic, N.; Pohl, C. Accelerated solvent extraction: A technique for sample extraction. *Anal. Chem.* **1996**, *68*, 1033–1039.
5. Lide, D.R., Ed. *CRC Handbook of Chemistry and Physics*, 52nd ed. Boca Raton, FL: CRC Press; 1971–1972.
6. Lide, D.R., Ed. *CRC Handbook of Chemistry and Physics*, 73rd ed. Boca Raton, FL: CRC Press; 1992–1993; pp. 6–10.
7. Bjorklund, E.; Nilsson, T.; Bowadt, S. Pressurised liquid extraction of persistent organic pollutants in environmental analysis. *Trends Anal. Chem.* **2000**, *19*, 434–445.
8. Giergielewicz-Mozajska, H.; Dabrowski, L.; Namiesnik, J. Accelerated solvent extraction (ASE) in the analysis of environmental solid samples—Some aspects of theory and practice. *Crit. Rev. Anal. Chem.* **2001**, *31*, 149–165.
9. Carabias-Martinez, R.; Rodriguez-Gonzalo, E.; Revilla-Ruiz, P.; Hernandez-Mendez, J. Pressurized liquid extraction in the analysis of food and biological samples. *J. Chromatogr. A* **2005**, *1089*, 1–17.
10. Schantz, M.M. Pressurized liquid extraction in environmental analysis. *Anal. Bioanal. Chem.* **2006**, *386*, 1043–1047.
11. Peck, A.M. Analytical methods for the determination of persistent ingredients of personal care products in environmental matrices. *Anal. Bioanal. Chem.* **2006**, *386*, 907–939.
12. Hubert, A.; Wenzel, K.D.; Manz, M.; Weissflog, L.; Engewald, W.; Schürmann, G. High extraction efficiency for POPs in real contaminated soil samples using accelerated solvent extraction. *Anal. Chem.* **2000**, *72*, 1294–1300.

13. Ding, W.; Fann, J.C.H. Application of pressurized liquid extraction followed by gas chromatography-mass spectrometry to determine 4-nonylphenols in sediments. *J. Chromatogr. A* **2000**, *866*, 79–85.

14. Brandl, R.C.; Bucheli, T.D.; Kupper, T.; Stadelmann, F.X.; Tarradellas, J. Optimised accelerated solvent extraction of PCBs and PAHs from compost. *Int. J. Environ. Anal. Chem.* **2006**, *86*, 505–525.

15. Hussen, A.; Westbom, R.; Megersa, N.; Retta, N.; Mathiasson, L.; Bjorklund, E. Optimisation of pressurised liquid extraction for the determination of p,p′-DDT and p,p′-DDE in aged contaminated Ethiopian soils. *Anal. Bioanal. Chem.* **2006**, *386*, 1525–1533.

16. Hussen, A.; Westbom, R.; Megersa, N.; Mathiasson, L.; Bjorklund, E. Development of a pressurized liquid extraction and clean-up procedure for the determination of alpha-endosulfan, beta-endosulfan and endosulfan sulfate in aged contaminated Ethiopian soils. *J. Chromatogr. A* **2006**, *1103*, 202–210.

17. Harb, J.G.; Aldstadt, J.H. An improved pressurized liquid extraction method for the determination of polycyclic aromatic hydrocarbons in freshwater sediments by gas chromatography-mass spectrometry. *Anal. Lett.* **2004**, *37*, 2835–2850.

18. Preud'homme, H.; Potin-Gautier, M. Optimization of accelerated solvent extraction for polyhalogenated dibenzo-p-dioxins and benzo-p-furans in mineral and environmental matrixes using experimental designs. *Anal. Chem.* **2003**, *75*, 6109–6118.

19. Kiguchi, O.; Kobayashi, T.; Saitoh, K.; Ogawa, N. Pressurized liquid extraction of polychlorinated dibenzo-p-dioxins, polychlorinated dibenzofurans and coplanar polychlorinated biphenyls from contaminated soil. *J. Chromatogr. A* **2006**, *1108*, 176–182.

20. Richter, B.E. Extraction of hydrocarbon contamination from soils using accelerated solvent extraction. *J. Chromatogr. A* **2000**, *874*, 217–224.

21. Zhu, Y.; Yanagihara, K.; Guo, F.; Li, Q. Pressurized fluid extraction for quantitative recovery of chloroacetanilide and nitrogen heterocyclic herbicides in soil. *J. Agric. Food Chem.* **2000**, *48*, 4097–4102.

22. Alonso-Rodriguez, E.; Moreda-Pineiro, J.; Lopez-Mahia, P.; Prada-Rodriguez, D.; Fernandez-Fernandez, E.; Muniategui-Lorenzo, S.; Moreda-Pineiro, A.; Bermejo-Barrera, A.; Bermejo-Barrera, P. Pressurized liquid extraction of organometals and its feasibility for total metal extraction. *Trends Anal. Chem.* **2006**, *25*, 511–519.

23. Marcic, C.; Lespes, G.; Potin-Gautier, M. Pressurised solvent extraction for organotin speciation in vegetable matrices. *Anal. Bioanal. Chem.* **2005**, *382*, 1574–1583.

24. Josefsson, S.; Westbom, R.; Mathiasson, L.; Bjorklund, E. Evaluation of PLE exhaustiveness for the extraction of PCBs from sediments and the influence of sediment characteristics. *Anal. Chim. Acta* **2006**, *560*, 94–102.

25. Bandh, C.; Bjorklund, E.; Mathiasson, L.; Naf, C.; Zebuhr, Y. Comparison of accelerated solvent extraction and Soxhlet extraction for the determination of PCBs in Baltic Sea sediments. *Environ. Sci. Technol.* **2000**, *34*, 4995–5000.

26. Richter, P.; Jimenez, M.; Salazar, R.; Marican, A. Ultrasound-assisted pressurized solvent extraction for aliphatic and polycyclic aromatic hydrocarbons from soils. *J. Chromatogr. A* **2006**, *1132*, 15–20.

27. Lundstedt, S.; Haglund, P.; Oberg, L. Simultaneous extraction and fractionation of polycyclic aromatic hydrocarbons and their oxygenated derivatives in soil using selective pressurized liquid extraction. *Anal. Chem.* **2006**, *78*, 2993–3000.

28. Burkhardt, M.R.; Zaugg, S.D.; Burbank, T.L.; Olson, M.C.; Iverson, J.L. Pressurized liquid extraction using water/isopropanol coupled with solid-phase extraction cleanup for semivolatile organic compounds, polycyclic aromatic hydrocarbons (PAH), and alkylated PAH homolog groups in sediment. *Anal. Chim. Acta* **2005**, *549*, 104–116.

29. Concha-Grana, E.; Turnes-Carou, M.I.; Muniategui-Lorenzo, S.; Lopez-Mahia, P.; Fernandez-Fernandez, E.; Prada-Rodriguez, D. Development of pressurized liquid extraction and cleanup procedures for determination of organochlorine pesticides in soils. *J. Chromatogr. A* **2004**, *1047*, 147–155.

30. Suchan, P.; Pulkrabova, J.; Hajslova, J.; Kocourek, V. Pressurized liquid extraction in determination of polychlorinated biphenyls and organochlorine pesticides in fish samples. *Anal. Chim. Acta* **2004**, *520*, 193–200.

31. Bjorklund, E.; Sporring, S.; Wiberg, K.; Haglund, P.; von Hillist, C. New strategies for extraction and clean-up of persistent organic pollutants from food and feed samples using selective pressurized liquid extraction. *Trends Anal. Chem.* **2006**, *25*, 318–325.

32. Bjorklund, E.; Muller, A.; von Holst, C. Comparison of fat retainers in accelerated solvent extraction for the selective extraction of PCBs from fat-containing samples. *Anal. Chem.* **2001**, *73*, 4050–4053.

33. Sporring, S.; Bjorklund, E. Selective pressurized liquid extraction of polychlorinated biphenyls from fat-containing food and feed samples—Influence of cell dimensions, solvent type, temperature and flush volume. *J. Chromatogr. A* **2004**, *1040*, 155–161.

34. Sporring, S.; von Holst, C.; Bjorklund, E. Selective pressurized liquid extraction of PCBs from food and feed samples: Effects of high lipid amounts and lipid type on fat retention. *Chromatographia* **2006**, *64*, 553–557.

35. Wiberg, K.; Sporring, S.; Haglund, P.; Bjorklund, E. Selective pressurized liquid extraction of polychlorinated dibenzo-p-dioxins, dibenzofurans and dioxin-like polychlorinated biphenyls from food and feed samples. *J. Chromatogr. A* **2007**, *1138*, 55–64.

36. Gomez-Ariza, J.L.; Bujalance, M.; Giraldez, I.; Velasco, A.; Morales, E. Determination of polychlorinated biphenyls in biota samples using simultaneous pressurized liquid extraction and purification. *J. Chromatogr. A* **2002**, *946*, 209–219.

37. de la Cal, A.; Eljarrat, E.; Barcelo, D. Determination of 39 polybrominated diphenyl ether congeners in sediment samples using fast selective pressurized liquid extraction and purification. *J. Chromatogr. A* **2003**, *1021*, 165–173.

38. David, M.D.; Campbell, S.; Li, Q. Pressurized fluid extraction of nonpolar pesticides and polar herbicides using in situ derivatization. *Anal. Chem.* **2000**, *72*, 3665–3670.

39. Campbell, S.; Li, Q. Na_4EDTA-assisted in situ derivatization pressurized fluid extraction of polar herbicides in soil. *Anal. Chim. Acta* **2001**, *434*, 283–289.

40. Kim, J.; Moon, J.; Li, Q.; Cho, J. One-step pressurized liquid extraction method for the analysis of polycyclic aromatic hydrocarbons. *Anal. Chim. Acta* **2003**, *498*, 55–60.

41. Porschmann, J.; Plugge, J.; Toth, R. In situ derivatisation using pressurized liquid extraction to determine phenols, sterols and carboxylic acids in environmental samples and microbial biomasses. *J. Chromatogr. A* **2001**, *909*, 95–109.

42. Ramos, L.; Vreuls, J.J.; Brinkman, U.A.T. Miniaturised pressurised liquid extraction of polycyclic aromatic hydrocarbons from soil and sediment with subsequent large-volume injection-gas chromatography. *J. Chromatogr. A* **2000**, *891*, 275–286.

43. Kristenson, E.M.; Angioi, S.; Vreuls, R.J.J.; Gennaro, M.C.; Brinkman, U.A.T. Miniaturised pressurised liquid extraction of chloroanilines from soil with subsequent analysis by large-volume injection-gas chromatography-mass spectrometry. *J. Chromatogr. A* **2004**, *1058*, 243–249.

44. Ryno, M.; Rantanen, L.; Papaioannou, E.; Konstandopoulos, A.G.; Koskentalo, T.; Savela, K. Comparison of pressurized fluid extraction, Soxhlet extraction and sonication for the determination of polycyclic aromatic hydrocarbons in urban air and diesel exhaust particulate matter. *J. Environ. Monit.* **2006**, *8*, 488–493.

45. Graham, M.C.; Allan, R.; Fallick, A.E.; Farmer, J.G. Investigation of extraction and clean-up procedures used in the quantification and stable isotopic characterisation of PAHs in contaminated urban soils. *Sci. Total Environ.* **2006**, *360*, 81–89.

46. Sporring, S.; Bowadt, S.; Svensmark, B.; Bjorklund, E. Comprehensive comparison of classic Soxhlet extraction with Soxtec extraction, ultrasonication extraction, supercritical fluid extraction, microwave assisted extraction and accelerated solvent extraction for the determination of polychlorinated biphenyls in soil. *J. Chromatogr. A* **2005**, *1090*, 1–9.

47. Hawthorne, S.B.; Grabanski, C.B.; Martin, E.; Miller, D.J. Comparisons of Soxhlet extraction, pressurized liquid extraction, supercritical fluid extraction and subcritical water extraction for environmental solids: Recovery, selectivity and effects on sample matrix. *J. Chromatogr. A* **2000**, *892*, 421–433.

48. Koci, K.; Petrovska, H.; Simek, Z.; Varadova, E.; Syslova, A. Extraction of polycyclic aromatic nitrogen heterocycles from spiked soil samples. *Int. J. Environ. Anal. Chem.* **2007**, *87*, 111–123.

Chapter 10

Superheated Water Extraction

Roger M. Smith

10.1. INTRODUCTION

Water, as a river, oceanic, waste, or environmental sample, has frequently been the subject of an analysis, but in recent years, it is also being seen as a potential extraction solvent for solid or plant samples as an alternative to conventional organic solvents. Normally, the perception is that water is a poor solvent for organic compounds capable of only dissolving ionized or very polar analytes. However, on raising the temperature of water, there is a marked change in its properties, and the solubility of even quite nonpolar analytes increases markedly. Under these conditions, it has been used for the extraction of both polar and nonpolar analytes from a wide range of matrices.

The application of water at temperatures above 100 up to 250°C as a liquid under pressure has become known as subcritical water, pressurized water, or superheated water extraction (SHWE), depending on the authors, and the methods and applications have been reviewed in recent years.[1,2] As part of the suite of sample preparation techniques available to the analyst, it is a specialized version of pressurized fluid extraction (Chapter 9), and because the use of organic solvents can be reduced or eliminated, it has attracted interest as an environmentally friendly and green approach.[3,4] These changes in the properties of water at elevated temperatures have also lead to the use of superheated water as an organic solvent-free mobile phase for reversed-phase high-performance liquid chromatography (HPLC).[5–10]

10.2. HIGH-TEMPERATURE WATER AS AN EXTRACTION SOLVENT

Hawthorne and his coworkers first raised the awareness of the analytical potential of superheated water as an extraction solvent in the early 1990s[11] and, over the next few years, carried out subsequent extensive series of studies on environmental samples looking at polycyclic aromatic hydrocarbons (PAHs), polychlorinated biphenyls (PCBs), *n*-alkanes, and other constituents. They noted the marked changes that occur in the properties of water with increasing temperature and suggested that these held considerable analytical potential. For example, they demonstrated[12] that the solubility of chrysene increased by five orders of magnitude on raising the temperature from 25 to 225°C. Similar increases have been observed by other workers.

The properties of water at ambient temperature are unusual because of the extent of hydrogen bonding, which causes it to be much less volatile than its molecular mass would suggest, but makes liquid water a good solvent for polar hydrogen-bonding analytes, such as carbohydrates. These interactions also cause the viscosity to be high and diffusion rates to be restricted. However, when the temperature is raised, the hydrogen bonding is weakened and boiling takes place around 100°C at ambient pressure. In a pressurized system, the boiling point can be increased up to the critical point at more than 374°C and 218 atm. As the temperature is increased across this range, there is a marked drop in viscosity and a decrease in the relative permittivity (Fig. 10.1).[1] By 200°C, water has a similar relative permittivity to common extraction solvents, such a methanol ($\varepsilon = 32$) or acetonitrile ($\varepsilon = 37$) and thus should be a good solvent for many organic analytes.

Under these conditions, the vapor pressure of the water is relatively low and, even at 200°C, only reaches 15 bar so that it can be maintained in the liquid state with either a restriction in a flow system or a modest back-pressure controller. Across most of the temperature range of interest, the density of water is virtually unchanged with temperature or pressure. This makes the extraction process insensitive to the applied back pressure so this is not a critical parameter in extraction studies and, if altered, usually has no influence on the extraction efficiency. On raising the temperature further, water will reach its critical point, but by this

Handbook of Sample Preparation, Edited by Janusz Pawliszyn and Heather L. Lord

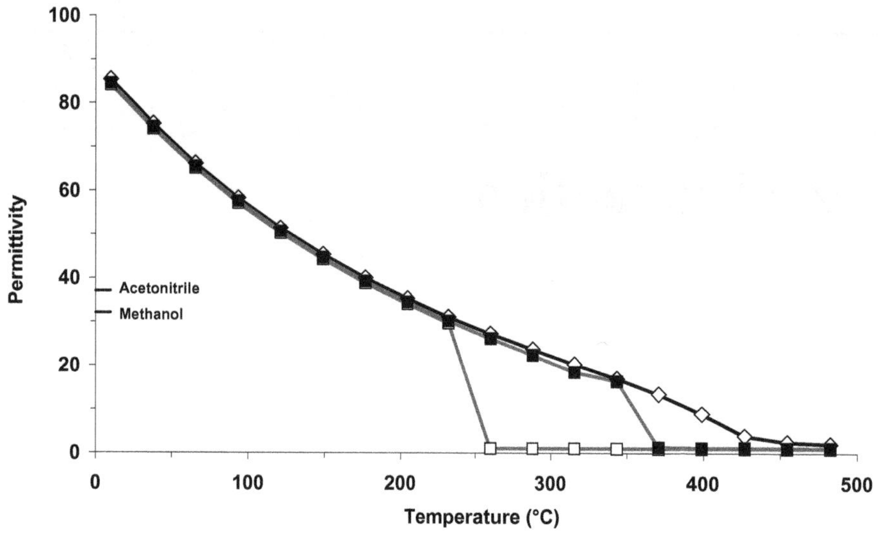

Figure 10.1. Changes in the permittivity (ε) of water with temperature at three pressures: □, 33 bar; ■, 129 bar; and ◇, 322 bar. The drop to low values corresponds to the formation of superheated steam. Reprinted from reference 1 with permission from Elsevier. Copyright 2002.

temperature, the conditions are more aggressive and the principal role of supercritical water has been to provide a condensed state for the oxidative destruction of noxious chemicals, such as chemical warfare agents or explosives[13–15] or to remove pollutants in bioremediation studies,[16,17] which have the advantage of leaving the soil structure undamaged[18] with zero earthworm toxicity.

10.2.1. Practical Aspects

Chemical processes utilizing water as either an extraction solvent or reaction medium are promising green chemistry alternatives to conventional organic solvent-based techniques but, as a solvent, water presents both practical problems and advantages. The principal advantage is that the raw material is readily available in high purity at low cost and presents no disposal problems, unlike most organic solvents, which can also be toxic and/or flammable and which need special methods of disposal. As an extraction solvent, water is also compatible with biological techniques for assays, such as immunoassays, and can be injected directly into reversed-phase HPLC systems. However, the product of an extraction is often a dilute aqueous solution, which is difficult to concentrate, and a subsequent concentration/extraction stage may be needed.

As yet, there is no equipment on the market dedicated to SHWE. Although it is similar in approach to pressurized liquid extraction, the current commercial systems have only a limited temperature range. As a result most SHWE has been carried out using homemade systems (Fig. 10.2), usually based around a gas chromatography (GC) oven, which normally has the capability of reaching 300–350°C, with a conventional HPLC pump for the water supply and either a back-pressure regulator or a length of narrow-bore tubing (3 m and 0.07 mm—typically red-line PEEK—at 1 mL/min flow) to maintain a sufficiently high pressure to keep the water in a liquid state.

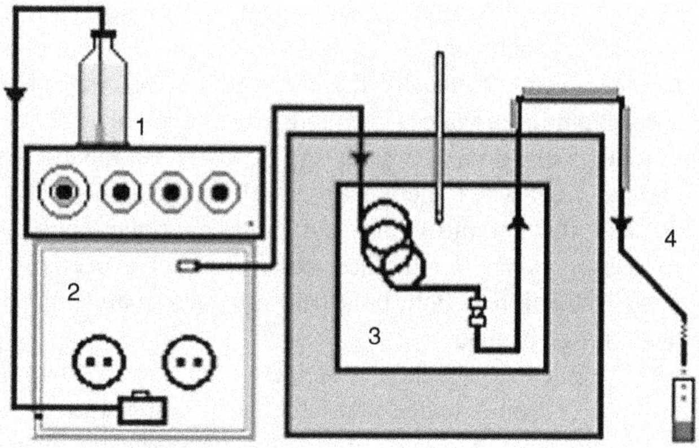

Figure 10.2. Dynamic superheated water extraction system. Components: 1, water supply; 2, pump; 3, oven with preheated coil and extraction vessel; 4, back-pressure restrictor and collection vessel.

10.2.2. Static and Dynamic Systems

Both static and dynamic or combined extraction systems have been used. In a static system, a simple extraction cell is filled with water and heated but has the disadvantage that the vessel has to be cooled back to room temperature before the water can be removed, which can result in readsorption of the analyte onto the matrix. However, the sample can be relatively concentrated as little water is used. Hawthorne and coworkers demonstrated how soil samples could be extracted in the field by heating a short pipe closed at both ends.[19]

More commonly, a dynamic system is used in which the water is pumped through a sample cell; this may be preceded by a static period while the sample under pressure is brought up to the extraction temperature. The exit tubing can also include a cooling system so that the eluted water is below 100°C. The advantage of the dynamic system is that that the sample is being continuously extracted with fresh water, and the extracted analyte is transported away from the matrix so readsorption cannot occur, but this generates a larger volume of extraction solvent.

Either pure water can be used as the extraction solvent or it can be modified with buffers[20,21] to increase selectivity as there is little change in the pH as temperature is increased. Some authors have included a small proportion of organic solvents, such as the addition of ethanol to enhance the extraction of atrazine from beef kidney.[22]

10.2.2.1. Sample recovery from the extraction solvent. For many applications, the principal disadvantage of SHWE is that the resulting extract is often a dilute aqueous solution. For subsequent assays, it is therefore desirable to concentrate the solution or extract the organic components from the water. However, direct evaporation of water is a slow process, and, although lyophilization (freeze-drying under vacuum) can be effective, it is also a slow process and may result in the loss of volatile components. The preferred method is therefore usually a trapping or extraction technique.

Traditionally, liquid–liquid extraction methods in a separating funnel with an immiscible organic solvent are used to obtain organic analytes from an aqueous matrix. However, this is a slow, labor intensive, and often inefficient method, usually requiring repeated extractions. A number of alternative methods have been employed, including solid-phase extraction (SPE), solid-phase microextraction (SPME),[23] and membrane extraction, and these have been reviewed for superheated water extracts by Lüthje and coworkers.[24] They concluded that trapping into a solvent gave the highest recovery, but that membrane extraction yielded the best selectivity. SPE requires further elution with an appropriate organic solvent. Solvent-free methods, such as SPME fibers, have the advantage that the extract can be directly transferred to a GC, but extraction is often incomplete. Other authors have examined the use of a stir-bar solvent extraction, which has a higher recovery because of its larger extraction volume.[25] An elegant method was employed by Hawthorne et al.[19] who placed a polystyrene-divinylbenzene (PS-DVB) extraction disk in a static system before it was heated. On cooling the extraction vessel, the disk trapped the sample preventing the analytes from returning to the soil matrix. This was convenient for field studies as only heating was required to carry out the extraction stages. They also used an anion exchange disk to trap chlorinated acid herbicides.[26]

In a model system, Tajuddin and Smith[27] showed that pharmaceuticals could be trapped from a dilute aqueous solution on a short PS-DVB trap. Then, by raising the temperature of the trap in stages, the analytes could be successively released according to their polarity onto a PS-DVB analytical column for superheated water chromatography (SHWC) (Fig. 10.3).

10.2.2.2.* In situ *derivatization. In order to enhance the extraction process, *in situ* derivatization in the aqueous extract has been employed by a number of groups, the usual aim being to convert the analytes to a more extractable form to enhance recovery. An adaption of this has been the use of a phase transfer derivatization reaction by employing a micellar mixture as a medium to methylate acids to the corresponding esters, and simultaneously extracting them into a small volume of organic solvent.[28]

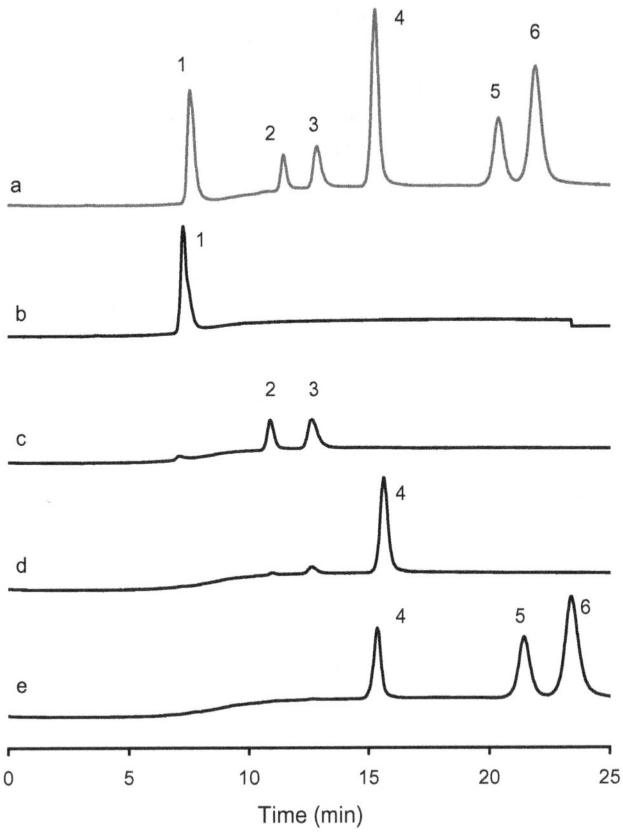

Figure 10.3. Sequential thermal release of trapped pharmaceuticals at increasing temperature and their chromatography using superheated water chromatography. Trap PS-DVB. Column: PS-DVB column at 75–185°C at 15°C/min. Analytes: 1, paracetamol; 2, salicylamide; 3, caffeine; 4, methyl paraben; 5, phenacetin; 6, ethyl paraben. Chromatograms: a, fraction untrapped at ambient temperature; b, fraction released from trap at 70°C; c, released at 90°C; d, released at 110°C.[27] Figure reprinted with permission.

10.2.3. Coupled Extraction Systems to Chromatography

Rather than collecting and extracting the analytes in a separate step before chromatography, a more efficient method is to couple the extraction stage directly to an analytical separation method so that all the extracted material is transferred to the assay method. Riekkola and coworkers demonstrated the coupling of SHWE and GC.[2] They have also coupled SHWE to HPLC and then GC for the analysis of polynuclear aromatic hydrocarbons in sediments.[29] Alternatively, an online microporous membrane extraction stage can be included to link SHWE and GC systems.[30,31]

An intermediate solid-phase trap (Fig. 10.4) was used by Li and coworkers[32] to couple SHWE and HPLC for the determination of a range of analytes, including nitrotoluenes, PCBs, and anilines. Crescenzi and coworkers[20] also used an intermediate online C-18 trap to link SHWE and a liquid chromatography (LC)–mass spectrometry (MS) system for the extraction and identification of pesticides from soils.

Lamm and Yang showed that an intermediate trap can be used between SHWE and SHWC[33] in a solvent-free system. In a totally integrated system, using a single source of water, a single pump, and three ovens, Tajuddin and Smith[34] extracted triazines from spiked compost by SHWE, followed by an online trap, clean up, and thermal gradient SHWC, with the trapping and release at each stage being temperature controlled (Fig. 10.5).

10.3. APPLICATIONS OF SHWE

Over the last 20 years, SHWE has been employed in a number of fields. Comprehensive listings of applications can be found in the recent reviews[1,2] covering environmental samples, pesticides, plant and food materials, and metal ions. A concern in both extraction and chromatography has been the thermal stability of analytes, but in most cases, little degradation has been observed within the time span of the extraction. In a detailed study, Andersson and coworkers[35] examined a number of PAHs and found some degradation to oxygenated compounds after some hours at 200°C. They suggested that the possibility for degradation should always be taken into account in extraction studies.

The extracts can be analyzed by GC, GC-MS, HPLC, HPLC-MS, SHWC, and supercritical fluid chromatography (SFC), and as indicated earlier, often the extraction can be integrated directly with the separation stage, reducing losses and improving sensitivity. As with other flowing solvent extraction methods, SHWE is most appropriate for the extraction of powdered materials, such as soils or dried and ground plant material, and is inappropriate for bulkier tissue samples or fresh plant materials as these limit the flow of the extraction fluid through the sample matrix.

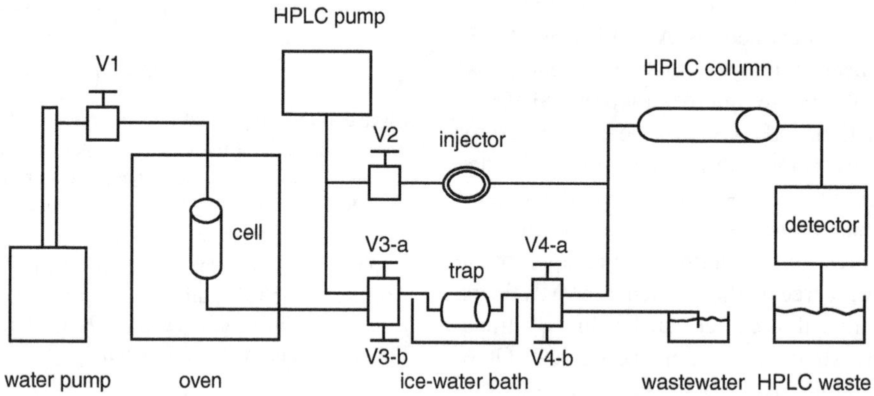

Figure 10.4. Online coupled extraction systems for SHWE coupled to HPLC using an intermediate trap.[32] Figure reprinted with permission.

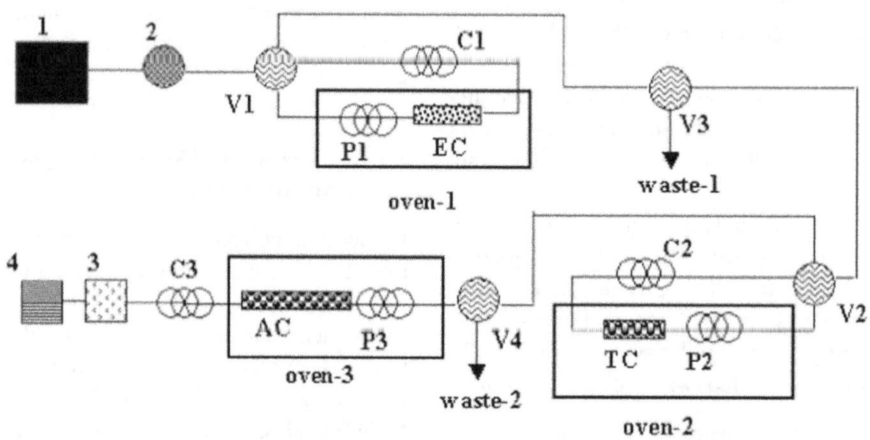

Figure 10.5. Online coupled SWE-SWC system: 1, water pump; 2, injector; 3, UV detector; 4, back-pressure regulator; EC, extraction cell; TC, X-Terra trap column; AC, PGC analytical column; V1, V2, V3, V4, switching valves; P1, P2, P3, preheating coils; C1, C2, C3, cooling coils used for the extraction of triazines from compost.[34] Figure reprinted with permission.

10.3.1. Environmental Samples

Most of the early work on SHWE was directed to environmental assays to determine hydrocarbons and PAHs[11] in soils and sediments and was usually followed by GC-MS analysis and sometimes by LC. Raising the temperature to 250 or 300°C enabled the extraction of PCBs from the soils.[36] However, a faster extraction was obtained by reducing the pressure and using steam at 300°C and 50 bar. By adjusting the temperature and pressure, Yang and coworkers[37] could selectively and successively remove phenols and benzene, toluene, ethylbenzene, and xylene (BTEX) compounds at up to 150°C. At higher temperatures of 250–300°C, the PAHs were extracted, but the *n*-alkanes > C_{20} were not extracted until superheated steam at 250–300°C and 5 bar was used. Hawthorne and coworkers compared extractions using superheated water with supercritical fluid extraction, and Soxhlet extraction, which is effectively the reference method, and found that the superheated water gave a cleaner extraction, mainly of the PAHs, and did not extract the *n*-alkanes (Fig. 10.6).[38]

Brominated flame retardants have been extracted[39] from sediment samples (Fig. 10.7) and were analyzed by online LC-GC to clean up and separate the components. In other studies, alkylbenzene sulfonates have been rapidly and effectively extracted from sediments samples[40] and enabled a fully automated extraction and detection system to be developed.

This work has been followed by a number of other groups, and a range of matrices (soils, sediments, sand, urban air particulates, and oil shales[1,2]) have been examined for hydrocarbons, PAHs, and agrochemicals, including pesticides, herbicides, pyrethroids, and chlorophenols.

10.3.2. Food and Plant Material

Because extraction from dried plant matrices by solvents or by water is traditionally often a messy and lengthy process, the relative speed of SHWE is an advantage, and the method has been employed in a wide range of studies. Early work examined the extraction of rosemary[41] and looked at the kinetics of the extraction process. A later report examined

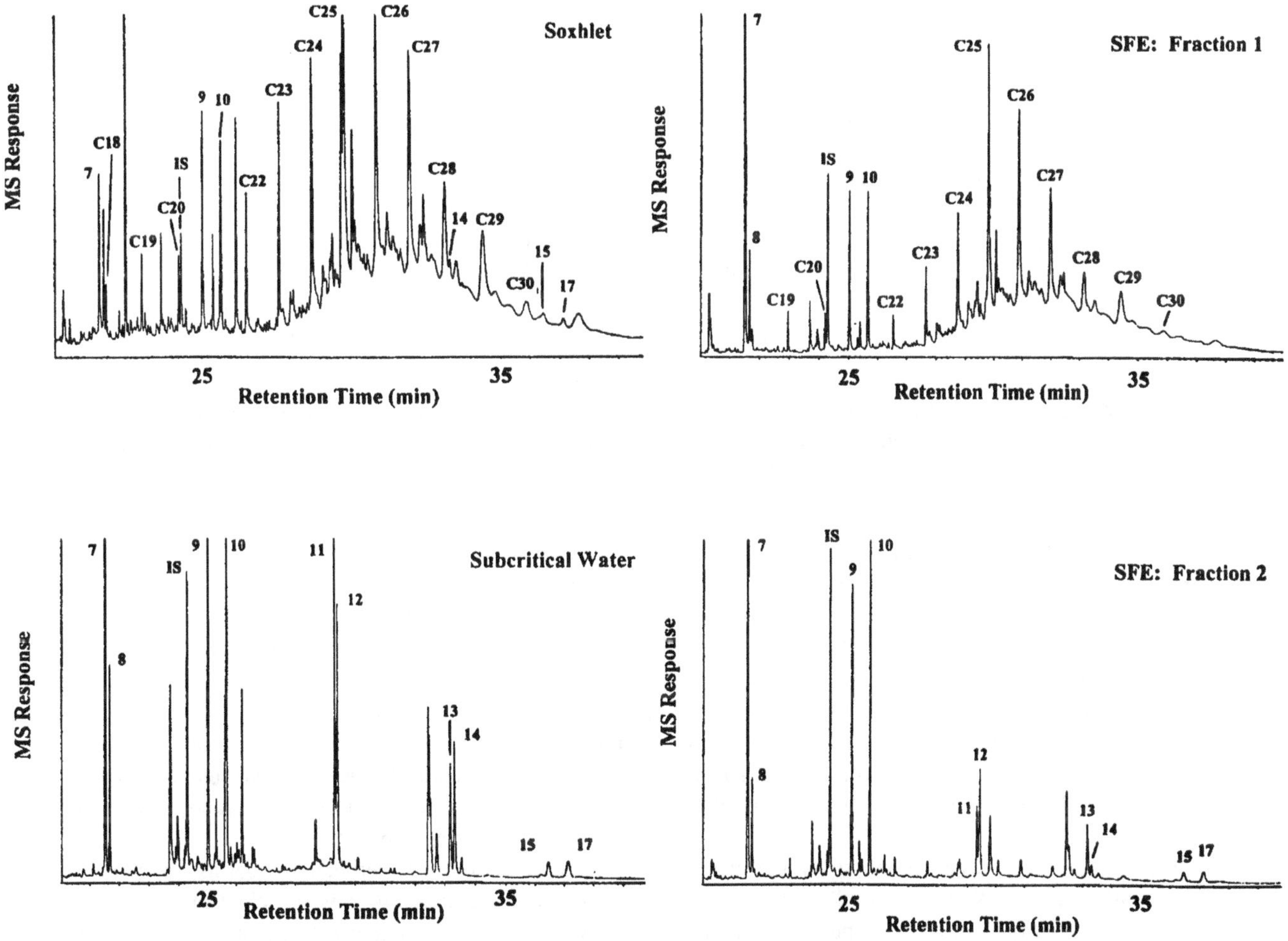

Figure 10.6. Comparison by GC-MS of the selectivity of SFE with carbon dioxide (upper right 200 bar, 50°C and lower right 400 bar, 150°C), Soxhlet extraction (dichloromethane-acetone 18 h), and SWE (250°C) for the extraction of urban air particulates (C_n, chain length of *n*-alkanes; number refers to PAHs).[38] Figure reprinted with permission.

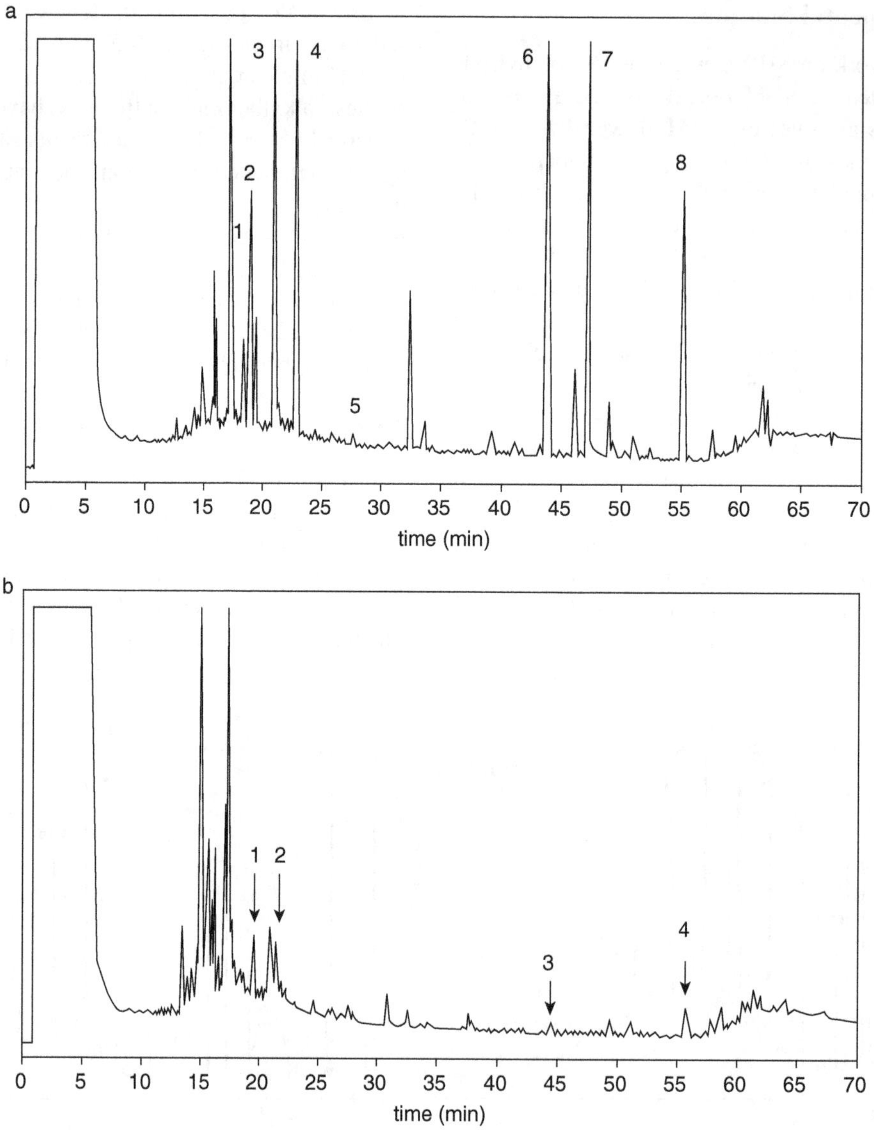

Figure 10.7. Superheated water LC-GC of brominated analytes extracted from (a) spiked sand and (b) sediment samples. Compounds: 1, tribromotrichlorocyclohexane; 2, tetrabromodichlorocyclohexane; 3, pentabromochlorocyclohexane; 4, pentabromotoluene; 5, PHT4; 6, BP6; 7, BP4A and T23P; 8, heptabromobiphenyl.[39] Figure reprinted with permission.

the use of SHWE to extract the antioxidant components.[42] Since then, numerous other herbs and spices have been examined for their essential oils or flavoring components, including basil and oregano,[43] marjoram,[44] and lime peel.[45] It has been suggested[44,45] that the oil provided by the SHWE contains a higher proportion of the oxygenated constituents, which provide the commercially valuable fragrance, than the oil obtained by steam distillation. Hence, the superheated water extract might be of higher quality and potentially of higher value.

The extraction of eucalyptus essential oil was compared[46] by hydrodistillation and SHWE (Fig. 10.8), and the latter method was found to be quicker and cheaper. An extraction of fennel found that SHWE gave a higher yield than hydrodistillation or solvent extraction, and altering the temperature enabled the composition of the oil to be manipulated.[47] The extraction at 50 min and 150°C was also faster than hydrodistillation or solvent extraction with dichloromethane. In an extraction of roses, Özel and coworkers compared direct thermal desorption (DTD) with water distillation and SHWE.[48] The three methods gave different compositions of the oils, where DTD and SHWE predominately yielded phenylethanol but the water distillation yielded benzyl alcohol and geraniol. Some work has also been carried out on the extraction of organochlorine and other pesticides from fruits and vegetables.[1,2]

The attraction of superheated water as a solvent has also lead to the examination of waste plant products, including winery waste[49] and quercetin glycosides from onion waste.[50] In these cases, the interest is often not just analytical but there is also the potential for a scale-up to a useful industrial process.

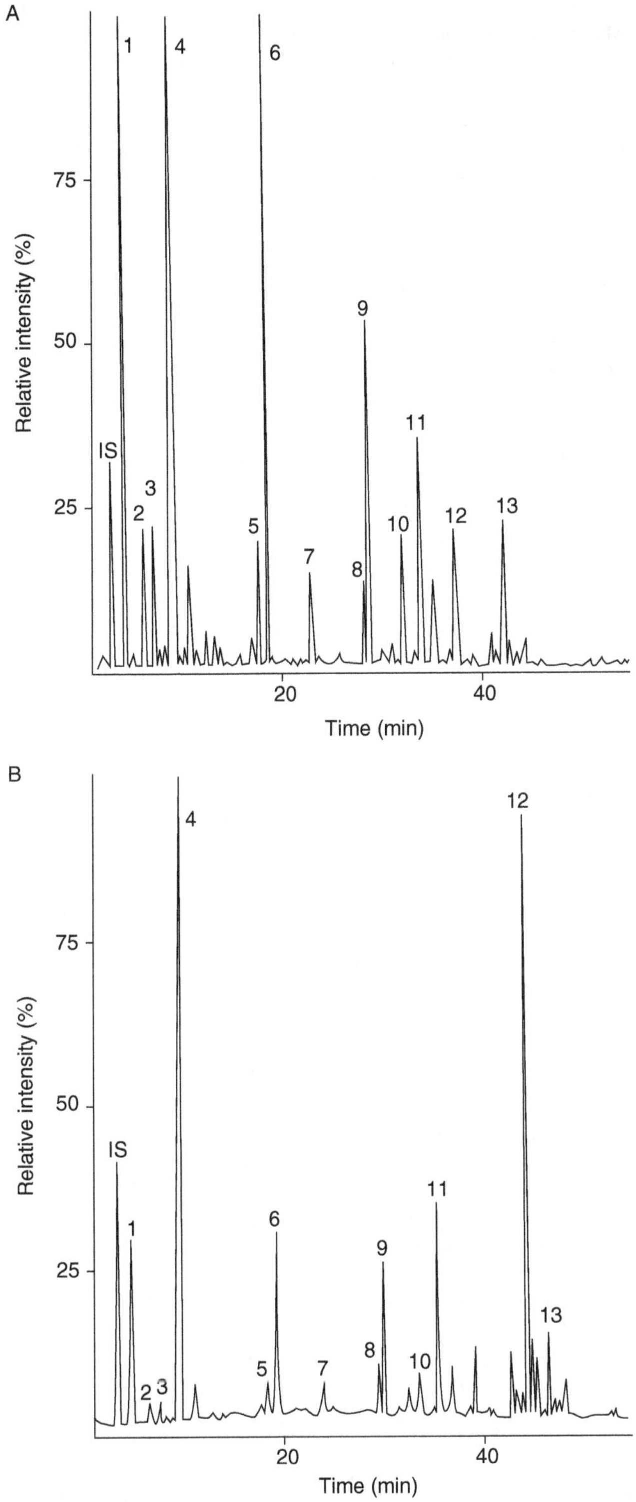

Figure 10.8. Comparison of (A) SHWE and (B) hydrodistillation of eucalyptus leaves. Compounds: 1, α-pinene; 2, β-pinene; 3, β-myrcene; 4, eucalyptol; 5, 1-isopropyl-4-methyl-3-cyclohexen-1-ol; 6, linalyl propanoate; 7, 3,7-dimethyl-2,6-octadien-1-ol; 8, 2-hydroxicineol acetate; 9, unidentified; 10, tricyclo[7.3.0^{6-8}.0^{1-9}]-2,7,7′,10-tetramethyl-9-dodecaene; 11, unidentified; 12, unidentified; 13, bicyclo[4.4.0] [2-2-hydroxypropyl]-6-methyl-1-decene.[46] Figure reprinted with permission.

10.3.3. Pharmaceuticals and Herbal Medicines

There has been some interest in the extraction of pharmaceuticals as superheated water is miscible with body fluids and tissue samples. Most interest has been in the extraction of herbal medicines especially traditional Chinese medicines.[51] Typical applications included the extraction of Danshen (*Salvia miltiorrhiza*) followed by LC and LC-ESI-MS.[52] Other studies have included the extraction of the lactone from kava roots (Fig. 10.9).[53]

10.3.4. Miscellaneous Applications

In more unusual applications, SHWE has been employed for the extraction of paper whitening.[54] Other studies have looked at the extraction of metal ions, for example, in the ash-forming elements from coal.[21]

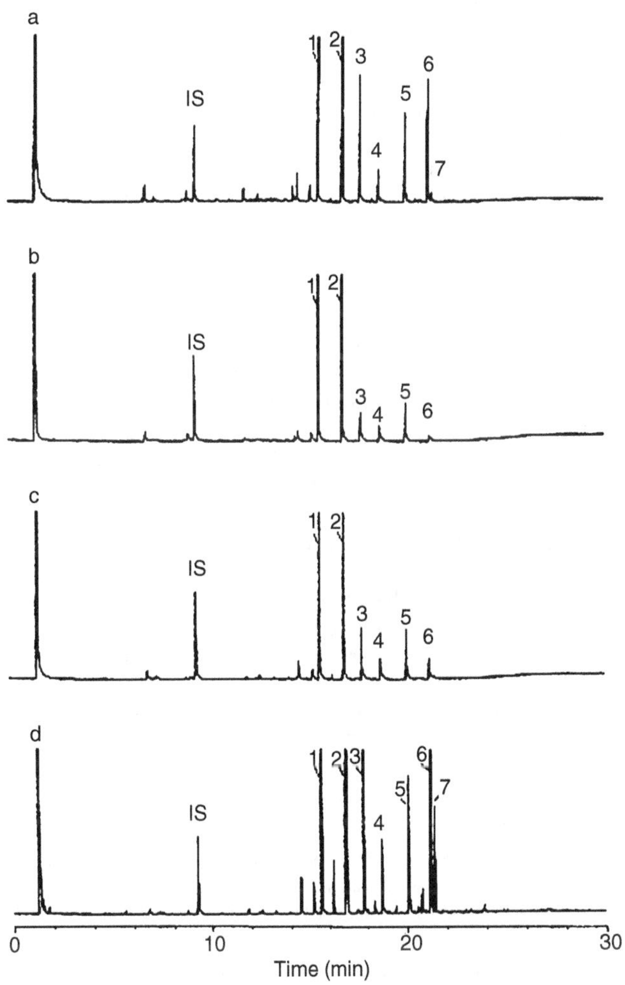

Figure 10.9. Comparison of kava lactone extracts with (a) superheated water at 175°C, (b) Soxhlet extraction with water, (c) boiling water, and (d) acetone extractions. Compounds: 1, dihydrokawain; 2, kawain; 3, desmethoxyyangonin; 4, tetrahydroyangonin; 5, dimethylmethyticin; 6, yangonin; 7, methysticin.[53] Figure reprinted with permission.

10.4. CONCLUSION

SHWE at up to 200°C is a relatively simple and solvent-free method for the extraction of many matrices including plant material, both as herbs and traditional medicines, for environmental samples. It causes little sample degradation and is compatible with subsequent analytical stages or can be directly coupled to HPLC and GC or SHWC.

REFERENCES

1. Smith, R.M. Extractions with superheated water. *J. Chromatogr. A* **2002**, *975*, 31–46.
2. Kronholm, J.; Hartonen, K.; Riekkola, M.L. Analytical extractions with water at elevated temperatures and pressures. *Trends Anal. Chem.* **2007**, *26*, 396–412.
3. Smith, R.M. Superheated water chromatography—A green technology for the future. *J. Chromatogr. A* **2008**, *1184*, 441–455.
4. Smith R.M. Superheated water: The ultimate green solvent for separation science. *Anal. Biochem. Anal.* **2007**, *385*, 419–425.
5. Smith, R.M.; Burgess, R.J. Superheated water as an eluent for reversed-phase high-performance liquid chromatography. *J. Chromatogr. A* **1997**, *785*, 49–55.
6. Yang, Y. Subcritical water chromatography: A green approach to high-temperature liquid chromatography. *J. Sep. Sci.* **2007**, *30*, 1131–1140.
7. Smith, R.M.; Burgess, R.J.; Chienthavorn, O.; Stuttard, J.R. Superheated water: A new look at chromatographic eluents for reversed-phase liquid chromatography. *LC-GC Int.* **1999**, *12*, 30–36.
8. Coym, J.W.; Dorsey, J.G. Superheated water chromatography: A brief review of an emerging technique. *Anal. Lett.* **2004**, *37*, 1013–1023.
9. Yang, Y. Practice and theory of high temperature liquid chromatography. *J. Sep. Sci.* **2007**, *30*, 1672–1685.
10. Hartonen, K.; Riekkola, M.L. Liquid chromatography at elevated temperatures with pure water as the mobile phase. *Trends Anal. Chem.* **2008**, *27*, 1–14.
11. Hawthorne, S.B.; Yang, Y.; Miller, D.J. Extraction of organic pollutants from environmental solids with sub- and supercritical water. *Anal. Chem.* **1994**, *66*, 2912–2920.
12. Miller, D.J.; Hawthorne, S.B.; Gizir, A.M.; Clifford, A.A. Solubility of polycyclic aromatic hydrocarbons in subcritical water from 298 K to 498 K. *J. Chem. Eng. Data* **1998**, *43*, 1043–1047.
13. Veriansyah, B.; Kim, J.D.; Lee, J.C. Destruction of chemical agent simulants in a supercritical water oxidation bench-scale reactor. *J. Hazard Mat.* **2007**, *147*, 8–14.
14. Chang, S.-J.; Liu, Y.-C. Degradation of TNT by supercritical water oxidation. *Chin. J. Expl. Prop.* **2007**, *30*, 34–36.
15. Zhao, B.-G.; Liu, Y.-C.; Chang, S.-J. Study on 2,4-dinitrotoluene waste water disposal by supercritical water oxidation (SCWO). *Chin. J. Expl. Prop.* **2007**, *30*, 71–74.
16. Kronholm, J.; Kalpala, J.; Hartonen, K.; Riekkola, M.-L. Pressurized hot water extraction coupled with supercritical water oxidation in remediation of sand and soil containing PAHs. *J. Supercrit. Fluids* **2002**, *23*, 123–134.
17. Lagadec, A.J.M.; Miller, D.J.; Lilke, A.V.; Hawthorne, S.B. Pilot-scale subcritical water remediation of polycyclic aromatic hydrocarbons- and pesticide-contaminated soil. *Environ. Sci. Technol.* **2000**, *34*, 1542–1548.
18. Kronholm, J.; Kuosmanen, T.; Hartonen, K.; Riekkola, M.L. Destruction of PAHs from soil by using pressurized hot water extraction coupled with supercritical water oxidation. *Waste Manag.* **2003**, *23*, 253–260.
19. Hawthorne, S.B.; Trembley, S.; Moniot, C.L.; Grabanski, C.B.; Miller, D.J. Static subcritical water extraction with simultaneous solid-phase extraction for determining polycyclic aromatic hydrocarbons on environmental solids. *J. Chromatogr. A* **2000**, *886*, 237–244.
20. Crescenzi, C.; Di Corcia, A.; Nazzari, M.; Samperi, R. Hot phosphate-buffered water extraction coupled on-line with liquid chromatography/mass spectrometry for analyzing contaminants in soil. *Anal. Chem.* **2000**, *72*, 3050–3055.
21. Jiménez-Carmona, M.M.; Fernández-Pérez, V.; Gualda-Bueno, M.J.; Cabanás-Espejo, J.M.; Luque de Castro, M.D. Acidified subcritical water extraction of major ash-forming elements from coal. *Anal. Chim. Acta* **1999**, *395*, 113–118.
22. Curren, M.S.S.; King, J.W. Ethanol-modified subcritical water extraction combined with solid-phase microextraction for determining atrazine in beef kidney. *J. Agric. Food Chem.* **2001**, *49*, 2175–2180.
23. Hageman, K.J.; Mazeas, L.; Grabanski, C.B.; Miller, D.J.; Hawthorne, S.B. Coupled subcritical water extraction with solid-phase microextraction for determining semivolatile organics in environmental soils. *Anal. Chem.* **1996**, *68*, 3892–3898.
24. Lüthje, K.; Hyötyläinen, T.; Riekkola, M.L. Comparison of different trapping methods for pressurised hot water extraction. *J. Chromatogr. A* **2004**, *1025*, 41–49.
25. Rodil, R.; Popp, P. Development of pressurized subcritical water extraction combined with stir bar sorptive extraction for the analysis of organochlorine pesticides and chlorobenzenes in soils. *J. Chromatogr. A* **2006**, *1124*, 82–90.
26. Lou, X.; Miller, D.J.; Hawthorne, S.B. Static subcritical water extraction combined with anion exchange disk sorption for determining chlorinated acid herbicides in soil. *Anal. Chem.* **2000**, *72*, 481–488.
27. Tajuddin, R.; Smith, R.M. On-line coupled superheated water extraction (SWE) and superheated water chromatography (SWC). *Analyst* **2002**, *127*, 883–885.
28. Chienthavorn, O.; Pengpumkiat, S.; Noomhorm, A.; Smith, R.M. Superheated water extraction and phase transfer methylation of phenoxy acid herbicides from solid matrices. *J. Chromatogr. A* **2007**, *1152*, 268–273.
29. Hyötyläinen, T.; Andersson, T.; Hartonen, K.; Kuosmanen, K.; Riekkola, M.-L. Pressurized hot water extraction coupled on-line with LC-GC: Determination of polyaromatic hydrocarbons in sediment. *Anal. Chem.* **2000**, *72*, 3070–3076.
30. Hyötyläinen, T.; Riekkola, M.-L. On-line combinations of pressurized hot-water extraction and microporous membrane liquid-liquid extraction with chromatography. *LC-GC Eur.* **2002**, *15*, 298–306.
31. Kuosmanen, K.; Hyotylainen, T.; Hartonen, K.; Jonsson, J.A.; Riekkola, M.-L. Analysis of PAH compounds in soil with on-line coupled pressurised hot water extraction-microporous membrane liquid-liquid extraction-gas chromatography. *Anal. Bioanal. Chem.* **2003**, *373*, 389–399.
32. Li, B.; Yang, Y.; Gan, Y.; Eaton, C.D.; He, P.; Jones, A.D. On-line coupling of subcritical water extraction with high-performance liquid chromatography via solid-phase trapping. *J. Chromatogr. A* **2000**, *873*, 175–184.
33. Lamm, L.J.; Yang, Y. Off-line coupling of subcritical water extraction with subcritical water chromatography via a sorbent trap and thermal desorption. *Anal. Chem.* **2003**, *75*, 2237–2242.
34. Tajuddin, R.; Smith, R.M. On-line coupled extraction and separation using superheated water for the analysis of triazine herbicides in spiked compost samples. *J. Chromatogr. A* **2005**, *1084*,194–200.
35. Andersson, T.; Hartonen, K.; Hyötyläinen, T.; Riekkola, M.-L. Stability of polycyclic aromatic hydrocarbons in pressurised hot water. *Analyst* **2003**, *128*, 150–155.
36. Yang, Y.; Bøwadt, S.; Hawthorne, S.B.; Miller, D.J. Subcritical water extraction of polychlorinated biphenyls from soil and sediment. *Anal. Chem.* **1995**, *67*, 4571–4576.

37. Yang, Y.; Hawthorne, S.B.; Miller, D.J. Class-selective extraction of polar, moderately polar, and non-polar organics from hydrocarbon wastes using subcritical water. *Environ. Sci. Technol.* **1997**, *31*, 430–437.

38. Hawthorne, S.B.; Grabanski, C.B.; Martin, E.; Miller, D.J. Comparisons of Soxhlet extraction, pressurized liquid extraction, supercritical fluid extraction and subcritical water extraction for environmental solids: Recovery, selectivity and effects on sample matrix. *J. Chromatogr. A* **2000**, *892*, 421–433.

39. Kuosmanen, K.; Hyötyläinen, T.; Hartonen, K.; Riekkola, M.-L. Pressurised hot water extraction coupled on-line with liquid chromatography-gas chromatography for the determination of brominated flame retardants in sediment samples. *J. Chromatogr. A* **2002**, *943*, 113–122.

40. Priego-López, E.; Luque de Castro, M.D. Superheated water extraction of linear alquilbenzene sulfonates from sediments with on-line preconcentration/derivatisation/detection. *Anal. Chim. Acta* **2004**, *511*, 249–254.

41. Basile, A.; Jiménez-Carmona, M.M.; Clifford, A.A. Extraction of rosemary by superheated water. *J. Agric. Food Chem.* **1998**, *46*, 5205–5209.

42. Ibañez, E.; Kubátová, A.; Señoráns, F.J.; Cavero, S.; Reglero, G.; Hawthorne, S.B. Subcritical water extraction of antioxidant compounds from rosemary plants. *J. Agric. Food. Chem.* **2003**, *51*, 375–382.

43. Yang, Y.; Kayan, B.; Bozer, N.; Pate, B.; Baker, C.; Gizir, A.M. Terpene degradation and extraction from basil and oregano leaves using subcritical water. *J. Chromatogr. A* **2007**, *1152*, 262–267.

44. Jiménez-Carmona, M.M.; Ubera, J.L.; Luque de Castro, M.D. Comparison of continuous subcritical water extraction and hydrodistillation of marjoram essential oil. *J. Chromatogr. A* **1999**, *855*, 625–632.

45. Chienthavorn, O.; Insuan, W. Superheated water extraction of lime peel: A comparison with conventional methods. *Anal. Lett.* **2004**, *37*, 2393–2409.

46. Jiménez-Carmona, M.M.; Luque de Castro, M.D. Isolation of eucalyptus essential oil for GC-MS analysis by extraction with subcritical water. *Chromatographia* **1999**, *50*, 578–582.

47. Gámiz-Gracia, L.; Luque de Castro, M.D. Continuous subcritical water extraction of medicinal plant essential oil: comparison with conventional techniques. *Talanta* **2000**, *51*, 1179–1185.

48. Özel, M.Z.; Göğüş, F.; Lewis, A.C. Comparison of direct thermal desorption with water distillation and superheated water extraction for the analysis of volatile components of *Rosa damascena* Mill. using GCxGC-TOF/MS. *Anal. Chim. Acta* **2006**, *566*, 172–177.

49. Garcia-Marino, M.; Rivas-Gonzalo, J.C.; Ibanez, E.; Garcia-Moreno, C. Recovery of catechins and proanthocyanidins from winery by-products using subcritical water extraction. *Anal. Chim. Acta* **2006**, *563*, 44–50.

50. Turner, C.; Turner, P.; Jacobson, G.; Almgren, K.; Waldeback, M.; Sjoberg, P.; Karlsson, E.N.; Markidas, K.E. Subcritical water extraction and beta-glucosidase-catalyzed hydrolysis of quercetin glycosides in onion waste. *Green Chem.* **2006**, *8*, 949–959.

51. Ong, E.S.; Cheong, J.S.H.; Goh, D. Pressurized hot water extraction of bioactive or marker compounds in botanicals and medicinal plant materials. *J. Chromatogr. A* **2006**, *1112*, 92–102.

52. Ong, E.S.; Len, S.M. Evaluation of pressurized liquid extraction and pressurised hot water extraction for tanshinone I and IIA in *Salvia miltiorrhiza* using LC and LC-ESI-MS. *J. Chromatogr. Sci.* **2004**, *42*, 211–216.

53. Kubátová, A.; Miller, D.J.; Hawthorne, S.B. Comparison of subcritical water and organic solvents for extracting kava lactones from kava root. *J. Chromatogr. A* **2001**, *923*, 187–194.

54. Chen, H.C.; Ding, W.H. Hot-water and solid-phase extraction of fluorescent whitening agents in paper materials and infant clothes followed by unequivocal determination with ion-pair chromatography-tandem mass spectrometry. *J. Chromatogr. A* **2006**, *1108*, 2002–2007.

Chapter 11

Supercritical Fluid Extraction

Jeremy J. Kroon and Douglas E. Raynie

11.1. INTRODUCTION

Supercritical fluid extraction (SFE) was the first popularly developed technique in a series, including superheated water extraction (Chapter 10), pressurized fluid extraction (PFE) (Chapter 9), and microwave-assisted extraction (MAE) (Chapter 12), which uses elevated temperatures and pressures to enhance the extraction process. From both a physical chemical and instrumental perspective, each of these methods shares several similarities. That is, applied pressure allows temperatures greater than the atmospheric boiling point to be used. In each case, the elevated temperature increases solvent diffusivity, reduces surface tension, and increases solute solubility. These extraction processes each possess a green aspect as well. Usually, the green claims are due to a minimization of (often toxic or hazardous) extraction solvents, though energy use, wastewater generation, life cycle, and other green attributes are typically ignored. PFE and MAE can use either aqueous or organic solvents; as such, they are somewhat familiar to most analytical chemists in that knowledge of traditional solvent properties guides the extraction. The robustness and applicability of PFE and MAE to acid hydrolysis and other sample preparation procedures have led to broad acceptability of these methods. Superheated water is the sole method of those listed where the chemical properties of the extracting solvent may deviate from those found under ambient conditions. As temperature increases to about 200°C and beyond, water takes on more of a nonpolar quality. However, the high temperatures, lack of commercial instrumentation, and oxidative potential have limited the growth of the technology.

SFE differs from the techniques described above in that the extracting solvent is a supercritical fluid. At temperatures and pressures near or above the critical point, the fluid possesses fast diffusion, minimal surface tension, and increased solubility favorable for extraction processes. Moreover, these properties may be varied by changes in fluid temperature or pressure. For analytical purposes, carbon dioxide, often modified with an organic cosolvent, is generally used. Since carbon dioxide is nonpolar, organic cosolvents (or modifiers) in concentrations of 5–20% alter the fluid polarity. Diffusion-limited extractions may be dramatically faster via SFE, while solubility-limited extractions use significantly less organic extracting solvent. These properties are the major advantages of SFE.

11.2. THEORY

At pressures and temperatures above the critical point, a single-phase region exists and the substance subjected to these conditions is called a supercritical fluid. The fluid possesses properties bounded by the extremes of the gaseous and liquid states, and these properties may be adjusted with alteration of the applied pressure and temperature. An example of the supercritical properties important for analytical separations is summarized in Table 11.1,[1] which also exhibits a comparison of these properties with the gaseous and liquid states. Thus, supercritical phases retain solvating power approximating liquids as well as the transport properties common to gases.

The behavior of a fluid in the supercritical state can be described as that of a very mobile liquid. The solubility behavior approaches that of the liquid phase, while penetration into a solid matrix is facilitated by the gas-like transport properties. As a consequence, the rates of extraction and phase separation can be significantly faster than for conventional extraction processes. The typical SFE takes 10–60 min to complete, while a similar liquid extraction, such as Soxhlet, can take anywhere from an hour to days. The SFE operating conditions can be controlled to affect extraction selectivity. SFE is known to be dependent on the density of the supercritical fluid, which in turn can be manipulated through the control of the system pressure and temperature. At a constant temperature, extraction at lower pressures will favor less polar analytes, and extraction at higher pressures

Handbook of Sample Preparation, Edited by Janusz Pawliszyn and Heather L. Lord

Table 11.1. Comparison of Properties of Gaseous, Liquid, and Supercritical Phases

	Gas (STP)	Supercritical Fluid	Liquid
Density (g/cm^3)	10^{-3}	0.2–0.5	0.6–2
Diffusion coefficient (cm^2/s)	10^{-1}	10^{-3}–10^{-4}	10^{-5}
Viscosity (g/cm/s)	10^{-4}	10^{-4}	10^{-2}

STP, standard temperature and pressure (0°C, 1 atm).

Table 11.2. Critical Points of Common Solvents[a]

Fluid	Critical Temperature (°C)	Critical Pressure (atm)
CO_2	31	73
N_2O	36	72
NH_3	132	112
MeOH	240	78
$CClF_3$	29	38
Ethane	32	48
Ethylene	10	51

[a]Modified from Reference 2.

will favor more polar and higher-molecular-weight analytes. This also allows for the collection of multiple distinct fractions from a single sample by using different pressures. However, unlike conventional extraction, once the conditions are returned to ambient, the quantity of residual solvent in the extracted material is negligible. This also removes the long and energy intensive concentration steps typically required in liquid extraction, making direct coupling to chromatography simpler. Supercritical fluids also have the advantage of usually being relatively inert, pure, nontoxic, and inexpensive. With fluids such as CO_2, SFE can be performed at low temperatures to extract thermally unstable compounds. In practice, SFE is generally performed with CO_2 at temperatures of 35–200°C (lower for liquid CO_2) and pressures of about 800 psi for liquid CO_2 or 1500–10,000 psi for the supercritical fluid.

How do you choose an appropriate solvent for the SFE? The critical points of several solvents that have been used for SFE are listed in Table 11.2.[2] Supercritical CO_2 is the most common choice for SFE, primarily because of its relatively low critical temperature and pressure, low toxicity, low reactivity, high purity, and low cost. The major drawback of CO_2 is that it is unable to quantitatively extract analytes that are more than moderately polar. CO_2 is an appropriate solvent for nonpolar species such as alkanes and terpenes. It is also reasonably good for moderately polar species, including polycyclic aromatic hydrocarbons (PAHs), polychlorinated biphenyls (PCBs), aldehydes, esters, alcohols, organochlorine pesticides, and fats.

When working with polar analytes, fluid selection becomes more difficult. One choice is NH_3, but this creates more problems than it solves. It is difficult to pump as it is highly corrosive, is chemically reactive, and is too dangerous for routine use. Methanol is another possible choice but is not practical because of the high critical temperature and because it is a liquid at room temperature, creating the necessity for postextraction analyte collection. Nitrous oxide is a powerful oxidizing agent that should be avoided at all costs.[3]

Due to the lack of a good polar solvent for SFE, the most common practice is to add a small percent of an organic solvent called a modifier. Modifiers can be added using a second pump or by simply injecting the organic solvent into the vessel before beginning the extraction. Methanol is the most common modifier, although a variety of other organic solvents have been used, ranging from alcohols to methylene chloride. The selection of the proper modifier is an educated guess due to the lack of analyte solubility data that exist for modified supercritical fluids. The best method for choosing a modifier would be to start with one that is a good solvent in its liquid state for the target analyte. Methanol is a good starting point, since the ultimate goal, in most cases, is to increase the fluid polarity.

As noted above, the solubility characteristics of supercritical fluids can be adjusted by pressure and temperature. This makes it possible to perform multiple extractions on a single sample with a single solvent by running the extractions under different conditions and collecting the appropriate fractions. It is also possible to extract a target analyte from a soluble matrix. For example, organochlorine pesticides can be extracted from fatty samples at CO_2 pressures low enough to avoid coextraction of lipids and allowing direct gas chromatography (GC) analysis.[4]

The SFE process is substantially similar on both the analytical and processing scales, so much can be learned from engineering studies. For example, a mathematical model of SFE, which accommodates phase equilibrium, flow, and diffusion in complex natural systems where intact and lysed cell walls limit extraction, has been developed,[5] and Langmuir kinetics appear to apply in SFE when the rate-limiting step is desorption from the solid surface.[6] Many researchers claim that SFE suffers from deleterious matrix effects or strong interactions between analyte molecules and the sample particle. However, in actuality, these matrix effects are present in all extractions; it is just that SFE uses a single solvent under different temperature and pressure conditions that can be considered weak (and exhibit these matrix effects) or stronger (and overcome the matrix effects).

11.3. INSTRUMENTATION

SFE instrumentation, displayed schematically in Figure 11.1, can be relatively simple. In addition to a fluid source, instrumentation usually contains a high-pressure pump, which preferably has a pressure rating of 10,000 psi, allowing a flow of the pressurized fluid of at least 2 mL/min. These pumps are commonly used in supercritical fluid chro-

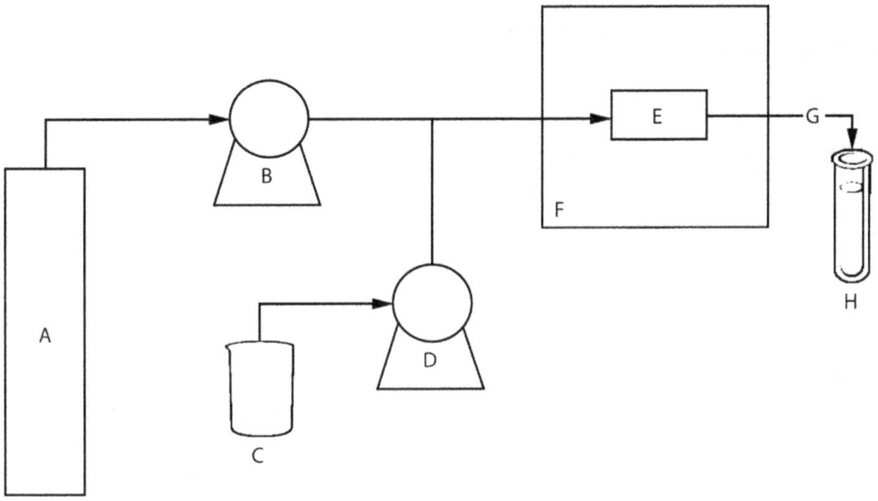

Figure 11.1. Generalized schematic of an SFE system: (A) carbon dioxide cylinder, (B) high-pressure pump, (C) modifier (cosolvent) supply, (D) modifier pump, (E) sample vessel, (F) thermostated oven, (G) flow restrictor, and (H) collection tube.

matography (SFC) and high-performance liquid chromatography (HPLC), although less expensive models can be used because SFE is usually done at a constant pressure, so there is no need for pressure/density ramp controllers. Two-pump systems are recommended to accommodate the addition of CO_2 modifiers. The SFE system also contains a valve to control the flow of the fluid into a heated extraction cell and an exit valve leading to a flow restrictor that depressurizes the fluid and transfers the extracted material into a collection device. Extraction cells can be purchased or made from common tubing fittings, but care must be taken to ensure that all of the components are rated for the appropriate pressure. The temperature of the extraction cell is normally controlled by placing the cell in a chromatographic oven or a simple tube heater. Flow restrictors can be made from micrometering valves, short lengths of fused-silica tubing, or a variety of automated controlled-orifice devices.

An SFE system can be operated in two modes. In the dynamic extraction mode, the valve between the extraction cell and the restrictor remains open. The sample is continually supplied with fresh fluid and the extracted material flows through the restrictor, which maintains the pressure in the cell, allowing the fluid to depressurize into the collection vessel. In the static extraction mode, the valve between the extraction cell and the restrictor is closed and the extraction cell is pressurized under static (nonflow) conditions. After a suitable period, the exit valve is opened and the cell contents are transferred through the restrictor by a dynamic flow to the collection vessel. One advantage of dynamic SFE is that the supercritical fluid is constantly renewed during the extraction, but this technique requires more fluid than static SFE, particularly when extracting large samples.

SFE has been used on sample sizes ranging from 1 mg to hundreds of grams, although samples of less than 10 g are the most common analytically. As sample size increases, more supercritical fluid is needed for extraction. This in turn makes trapping of the analyte more difficult. SFE is most commonly done using fluids that are gases under ambient conditions. The success of the trapping method is dependent on recovering the analytes from the flowing gas upon depressurization. For supercritical CO_2, a flow of 1 mL/min results in a gas flow of approximately 500 mL/min. Thus, it is easier to collect analytes from slower flow rates, especially for volatile components. This indicates that to reduce extraction time, SFE of small samples is preferable unless larger samples are needed to ensure sample homogeneity or to obtain sufficient mass of trace analytes. It is also usually necessary to pack extraction cells with inert material when they are not fully loaded with sample.

Methods that have been used to collect extracted analytes upon depressurization of the supercritical fluid fall into two general categories: "offline" SFE and "online" or "coupled" SFE. For offline SFE, the analytes are collected for subsequent analysis, whereas for online SFE, the analytes are directly transferred to a chromatographic system. The offline collection method is the simplest to perform because only the extraction step must be understood, and the extract can be analyzed by any appropriate method. Online SFE requires an understanding of both the SFE and the chromatographic conditions, and the sample extract is not available for analysis by a different method. The benefits of online SFE are the elimination of sample handling between extraction and chromatographic analysis and the potential to achieve maximum sensitivity by quantitatively transferring all of the extracted analyte into the chromatographic column.

Offline SFE collection is commonly performed by flowing the fluid into a few milliliters of liquid solvent, thus allowing the extract to be analyzed by any conventional method that would be used for a liquid sample solution. This can be done because the liquid is cooled by the expanding fluid, thereby decreasing the rapid evaporation that might be expected from the high gas flows. This method is capable of trapping volatile analytes at quantitative levels. The collection of decane was quantitatively retained in 3 mL of methylene chloride during a 15-min extraction using 1 mL/min of supercritical CO_2.[2] Alternatively, SFE analytes are collected onto bonded-phase packings similar to solid-phase extractions. They are then eluted with liquid solvents or supercritical fluids for subsequent analysis. When dealing with large quantities of nonvolatile analyte, such as fats from foods, direct depressurization into an empty receiving vessel can be used.

Online SFE has most often been coupled with capillary GC and both packed and capillary SFC. There are also reports of SFE/HPLC, but these are less common. GC and SFC are more common as the majority of analytes that are extracted with SFE naturally lend themselves to these types of chromatography. The proper choice of online coupling is based on the chromatographic technique best suited to the analyte.

SFE is coupled to capillary GC by directly depressurizing the fluid inside a conventional split/splitless injection port or by inserting the restrictor through an on-column injection port into the capillary column itself. The extracted analytes are cryogenically focused in the column stationary phase during the SFE step, and then analysis is performed in a normal manner. On-column SFE/GC yields the best sensitivity possible with small samples because the entire extract is transferred onto the GC column. Split SFE/GC, however, can use larger samples and is simpler to perform with wet samples.

SFE/SFC coupling can be achieved by several different methods. The extract can be passed through an injection loop or the analytes can be collected on an accumulator trap containing a sorbent material. Thermally trapping the analytes at the head of the SFC column, as in SFE/GC, is also an option. Once SFE is completed, the extracted analytes are swept from the accumulating device into the column using the SFC mobile phase, and the analysis is carried under appropriate chromatographic conditions. SFE/SFC has also been used as an injection method for samples that are difficult to dissolve in liquid solvents, as demonstrated by the analysis of a heavy hydrocarbon that was placed in the extraction cell as a slurry in chloroform.[2]

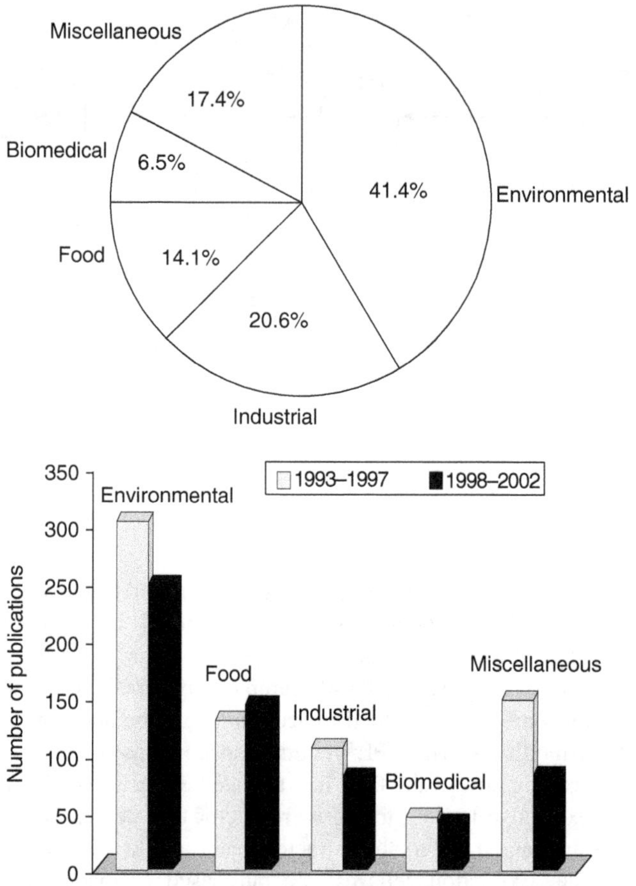

Figure 11.2. Summary of SFE application literature references. Reprinted from reference 6 with permission from Elsevier. Copyright 2004.

11.4. APPLICATIONS

Since the alternative technologies (i.e., PFE, MAE, and superheated water extraction) have become more fully developed, SFE applications have trended to those cases where neat CO_2 becomes advantageous. A 2004 review[7] uncovered these trends, as exemplified in Figure 11.2. While the United States Environmental Protection Agency methods for the SFE of total recoverable petroleum hydrocarbons (Method 3560), polynuclear aromatic hydrocarbons (Method 3561), and PCBs and organochlorine pesticides (Method 3562) exist and environmental samples represent the largest application area for SFE, the commercialization of PFE and MAE in the mid-1990s led to a drop-off in the number of publications of environmental SFE applications. On the other hand, the CO_2 extraction of lipophilic compounds has continued to be a major use of SFE in the food and fragrance industries. The American Oil Chemists' Society has issued Method AM 3-96 for the SFE of oils from oilseeds. The potential for hydrolysis or oxidation of extracted compounds from food samples is lessened with SFE, and the use of potentially objectionable organic solvents is also minimized. However, one must be careful during the extraction of bioactive samples; for example, a gradual decrease in the antioxidant activity of lycopenes extracted from tomato was noted as temperatures greater than 70°C were used, presumably due to a temperature-dependent isomerization.[8] While in most cases sample particle size and internal mass transfer have significant influence on extraction rates,[9] researchers have demonstrated that particle size may not influence the SFE of plant essential oils if the secretory structure is considered.[10]

Other applications of SFE seem to be less routine. A few recent, nonconventional SFE applications demonstrate the wide adaptability of the technique. For instance, liquid (especially aqueous) samples are generally not directly extracted via SFE, but researchers demonstrated the direct SFE of alkylphenols from water by placing a hydrophobic membrane in the sample cell.[11] Additives can also provide

Table 11.3. Comparison of Extraction Technologies

	Soxhlet	Modified Soxhlet	SFE	PFE	MAE	Sonication
Extraction time	High	Medium	Low	Low	Low	Medium
Organic solvent consumption	High	Medium	Low	Low	Low	Medium
Manpower (time/sample)	High	Medium	Low	Low	Medium	Medium
Operating cost per sample	High	Medium	Low	Low	Low	Medium
Equipment acquisition cost	Low	Medium	High	High	High	Medium
Extract filtration or cleanup	Cleanup	Cleanup	None	Cleanup	Both	Both
Automation	No	Partial	Yes	Yes	Yes	Partial

novel extraction selectivity. Chelating agents to isolate metal ions via SFE are typically fluorinated compounds, but sugar-derived ligands have also been reported.[12] Chiral selectors added to the SFE system, in addition to the direct extraction of the enantiomer of interest, may also be used in an inverse mode[13,14] or by precipitation of one of the enantiomers.[15,16]

As new SFE applications are developed, one must keep in mind the goal of the extraction. Typically, stated goals may be the reduction of organic solvent use or the minimization of extraction time. While the use of neat CO_2 can eliminate the need for organic solvents (except in the extract workup for analysis), if a modifier is used in SFE, about same amount of organic solvent is used as in PFE or MAE. In other general respects, SFE compares favorably to other newer extraction technologies. Table 11.3 displays this general comparison, though these characteristics may vary for an individual extraction method. Factors that must be considered in SFE method development include temperature, pressure, flow rate, time, modifier type and amount, sample matrix, vessel/geometry, and collection of the extracted analyte. Analytes that can be analyzed by GC can generally be extracted via SFE with neat CO_2, whereas other samples may require a modifier if SFE is to be considered applicable. When developing new methods, a plot of extraction yield as a function of time (or CO_2 volume) can be used to assess whether the extraction is diffusion limited or solubility limited. For diffusion-limited cases, increases in temperature and reduction of sample particle size will improve extraction yield; while in solubility-limited situations, increases in pressure, addition of a modifier, or increases in flow rate will improve extraction yield.[17] Then, temperature, pressure, modifier content, and CO_2 flow rate can be optimized followed by evaluation of the analyte collection conditions.[18] Finally, keep in mind that spiked samples do not necessarily behave like aged or native samples. A well-characterized sample is best for the development and continued evaluation of SFE methods. More comprehensive approaches to method development and optimization for SFE have been reported.[19] Evaluation of pressure stability, flow, and analyte recovery are vital to method development, as well as performance testing and preventive maintenance.[20]

11.5. CONCLUSION

SFE has matured in the 20 or so years of analytical practice, with routine applications in the environmental and food industry becoming common. Modern SFE is generally performed with neat CO_2 to provide high-throughput advantages coupled with an inexpensive and safe inorganic solvent.

REFERENCES

1. Westwood, S.A., Ed. *Supercritical Fluid Extraction and Its Use in Chromatographic Sample Preparation*. Chapman & Hall: New York; 1993.

2. Hawthorne, S.B. Analytical-scale supercritical fluid extraction. *Anal. Chem.* **1990**, *62*, 633A–642A.

3. Raynie, D.E. Warning concerning the use of nitrous oxide in supercritical fluid extractions. *Anal. Chem.* **1993**, *65*, 3127–3128.

4. King, J.W. Fundamentals and applications of supercritical fluid extraction in chromatographic science. *J. Chromatogr. Sci.* **1989**, *27*, 355–364.

5. Sovova, H. Mathematical model for supercritical fluid extraction of natural products and extraction curve evaluation. *J. Supercrit. Fluids* **2005**, *33*, 35–52.

6. Al-Jabari, M. Modeling analytical tests of supercritical fluid extraction from solids with Langmuir kinetics. *Chem. Eng. Commun.* **2003**, *190*, 1620–1640.

7. Zougagh, M.; Valcarcel, M.; Rios, A. Supercritical fluid extraction: A critical review of its analytical usefulness. *Trends Anal. Chem.* **2004**, *23*, 399–405.

8. Yi, C.; Shi, J.; Xue, S.J.; Jiang, Y.; Li, D. Effects of supercritical fluid extraction parameters on lycopene yield and antioxidant activity. *Food Chem.* **2009**, *113*, 1088–1094.

9. Fiori, L.; Calcagno, D.; Costa, P. Sensitivity analysis and operative conditions of a supercritical fluid extractor. *J. Supercrit. Fluids* **2007**, *41*, 31–42.

10. Zizovic, I.; Stamenic, M.; Orlovic, A.; Skala, D. Supercritical carbon dioxide extraction of essential oils from plants with secretory ducts: Mathematical modelling on the micro-scale. *J. Supercrit. Fluids* **2007**, *39*, 338–346.

11. Eskilsson, C.S.; Agren, A.; Mathiasson, L.; Bjorklund, E. Direct supercritical fluid extraction of alkylphenols from spiked and wastewater samples using extraction cells equipped with hydrophobic membrane-assemblies. *J. Liq. Chromatogr. Relat. Technol.* **2004**, *27*, 2871–2888.

12. Yang, H.J. Nonfluorous, highly CO_2-soluble chelating ligands for $scCO_2$ metal ion extraction. *Chem. Lett.* **2006**, *35*, 1000–1001.

13. Szekely, E.; Simandi, B.; Illes, R.; Molnar, P.; Gebefugi, I.; Kmecz, I.; Fogassy, E. Application of supercritical fluid extraction for fractionation of enantiomers. *J. Supercrit. Fluids* **2004**, *31*, 33–40.

14. Kmecz, D.; Simandi, M.; Szekely, E.; Fogassy, E. Resolution of N-methylamphetamine enantiomers with tartaric acid derivatives by supercritical fluid extraction. *Tetrahedron Asymmetry*, **2004**, *15*, 1841–1845.

15. Bauza, R.; Rios, A.; Gomez-Hens, A.; Valcarcel, M. Supercritical fluid immunoextraction: A new approach for immunoassay automation. *Anal. Chim. Acta* **2004**, *518*, 151–156.

16. Bauza, R.; Rios, A.; Valcarcel, M. Enantioselective supercritical fluid extraction from racemic mixtures by use of chiral selectors. *Sep. Sci. Technol.* **2004**, *39*, 459–478.

17. Lou, X.; Janssen, H.-G.; Cramers, C.A. Investigation of parameters affecting the on-line combination of supercritical fluid extraction with capillary gas chromatography. *J. Chromatogr. A* **1996**, *750*, 215–226.

18. Hawthorne, S.B.; Miller, D.J.; Burford, M.D.; Langenfeld, J.J.; Eckert-Tilotta, S; Louie, P.I. Factors controlling quantitative supercritical fluid extraction of environmental samples. *J. Chromatogr.* **1993**, *642*, 301–317.

19. Dean, J.R. Essential guides to method development in extraction. In *Encyclopedia of Separation Science*, Wilson, I.D., Ed. London, UK: Elsevier Science; 2000; pp. 4592–4608.

20. Palenzuela, B.; Manganiello, L.; Bauza, R.; Rios, A.; Valcarcel, M. Performance testing activities for analytical assessment of supercritical fluid extractors. *Accredit. Qual. Assur.* **2005**, *10*, 219–228.

Chapter 12

Microwave-Assisted Extraction

J.R. Jocelyn Paré and Jacqueline M.R. Bélanger

12.1. INTRODUCTION

Traditionally, more efforts have been devoted to the determinative steps of chemical analysis than to the sample preparation ones. This may appear surprising if one thinks of the need to have samples relatively clean to achieve optimal separation and determination of a complex mixture. Commercial issues are likely to explain, in a large part, this fact however, as the market value is grossly biased toward apparatus and instruments performing separation and analysis (determination) over those preparing matrices to render them amenable to such analytical procedures.

In this respect, the 1990s have been marked by an unusual level of development activities in the area of sample preparation. A number of techniques have been introduced, targeting a relatively large range of matrices to be treated. Solid-phase microextraction (SPME), pressurized fluid extraction (PFE), and microwave-assisted extraction (MAE) come to mind. No doubt others will be validated and further developed over time. The need for faster sample preparation has been the "raison d'être" for such a vivid interest into these new approaches to sample preparation.

MAE is the subject of this chapter. MAE is a technology within the microwave-assisted processes (MAP™) family of technologies that uses microwave energy to enhance or accelerate chemical, physical, and biological processes.[1–20] MAE is a technology that is amenable to the preparation of a broad and diversified range of matrices, targeting an equally broad range of analytes, often applicable to numerous analytes through a single operational procedure.

This chapter does not purport to be a complete review of the literature on MAE. Neither does it present an in-depth discussion of the fundamental principles underlying the technology. The main goal of this chapter is to review summarily the basic principles of the MAE and present a variety of examples of application. The authors have had the privilege of being intimately involved on the early development of the technology and to pioneer a number of applications as well as some apparatus and instruments currently available commercially or coming to market soon.

Consequently, this chapter will focus on applications based on selective heating. This is because no other technologies to date can lay claims on this parameter. We admit to a certain bias toward applications that are unique to the use of microwaves. A caveat is made however to the effect that applications where nonselective heating is used are widely available. Hence, for the convenience of the readers, and in order to make this information available within a single document, we also include two tables containing more thorough literature searches covering bulk as well as selective heating applications. The information is presented as two distinct tables because they target two distinct areas of applications, namely the agri-food sector in Table 12.1, and the environmental sector in Table 12.2. They are both formatted as follows: matrix, analyte, and reference. Although by no means exhaustive, these tables of references were prepared with the aim of being complete enough to cover most questions sample preparation workers may have on existing publications that can guide them on their choice of protocols. They were prepared by building on listings found in recent review articles from our laboratories[13–15] to which we have added some additional references from the literature that appeared since the original publications. The selection of references was based on range and variety rather than completeness. The references contained in the tables are an integral part of the reference list that can be found at the end of this chapter where they bear numbers 16–258 as noted in the tables themselves.

12.2. BASIC PRINCIPLES

There are basically two ways to apply microwave energy in sample preparation: bulk heating or selective heating. By bulk heating, one thinks of using microwave in a fashion that is similar to other heating techniques; for example,

Handbook of Sample Preparation, Edited by Janusz Pawliszyn and Heather L. Lord

Table 12.1. Literature Survey of MAE Protocols for Food Analysis

Matrices	Analysis/Preparation	Closed/Open	Reference
Meat, poultry, and fish			
Processed meat	Fat	O	11,20
Pork fat, codfish liver oil	Saponification of animal greases	a	21
Chicken nuggets, hake fingers	Oil quality	O	22
Sausage products (meat and fat from pork), fillets of mackerel, livers of cod	Lipids	O	23,30
Granulated and nongranulated meat extract, chicken extract, meat soup cubes	Heterocyclic amines	O	24
Catfish tissues	Geosmin, methylisoborneol	O	25–28
Fish tissue	Qualitative listing of 174 compounds, geosmin, methylisoborneol	O	29
Fillets of mackerel and livers of cod	Lipids		30
Dairy and eggs			
Milk powder, egg powder, cheddar cheese, powder cheese (semicured), cream cheese (fresh cheese), cured cheese	Fat	O	11,20,31,33
Milk (cow, goat, sheep)	Lipids	O	32,33
Greek cheeses, sheep milk yogurt	Organic acids	C	34
Dairy products	Furosine	C	45
Fruits and vegetables			
Grapes, grape juice	Glycosides	a	35
Orange peels	Pectin	a	36,37
Lime		C	38
Orange peels	Essential oils	O	39
Dried menthol mint and orange peels		a	40
Red raspberries	Anthocyanins	a	41
Apple pomace	Polyphenols	a	42
	Pectin	O	43
Plants	Nutraceuticals	C/O	44
Cereals and oilseeds			
Cereals products (pasta)	Furosine	C	45
Olives	Oil	O	46
Oleaginous	Oil	O	47
	Oil	*	48,50
Rice bran	Antioxidant components	a	49
Soybean	Soluble proteins	a	51
Cereals	Dimethyl sulfide	a	52
Herbs and spices			
Paprika (*Capsicum annuum* L.)	Pigments	O	11,14
		C	53
		a	124
Licorice root	Glycyrrhizic acid	a	54
Ginseng root	Ginsenosides	O	55–57
		C	58,59
Acanthopanax senticosus leaves	Ginsenosides, flavonoids	C	60
Chickpea (*Cicer arietinum* L.)	Saponins	C	61
Capsicum fruit	Capsaicinoids	C	62
Herbs	Essential oils	O	2–9,11 63–73 79,97
Black pepper (*Piper nigrum*)	Piperine	O	74,123
Vanilla planifolia	Vanillin	a, O	75,76
	p-Hydroxybenzaldehyde	a	76

Table 12.1. (*Continued*)

Matrices	Analysis/Preparation	Closed/Open	Reference
Curcuma longa (turmeric)	Curcuminoids [curcumin, demethoxy curcumin, bis-(demethoxy) curcumin]	a	77
Garlic juice with 2,4-decadienals	Volatile compounds	a	78
Neem seed kernel	Azadirachtin-related limonoids	O	80,81
Various	Essential oils	O	125
Food ingredients			
Chocolate and chocolate formulations,	Fat	C	82
Cocoa powder, *Theobroma* cocoa nibs		O	83
Coffee, beverages (coca cola), chewing gums, potato chips	Veltol® and Veltol-Plus®	C	84
Olive leaves	Oleuropein, verbacoside, apigenin-7-glucoside, luteolin-7-glucoside	O	85
Green tea leaves	Tea polyphenols, tea caffeine	a	86
Grape seeds	Phenolic compounds	O	87
Opuntia Milpa Aha, *Porphyra yezoensis*	Polysaccharides	a	88,89
Characterization, reaction, and bakery products			
Casein, gelatin, and BioProtein	Proteins	C	90
Salami, cheese, broccoli, cauliflower	Free amino acids	C	91
Maillard reactions: L-phenylalanine model	Benzene acetaldehyde 2-(5′-(hydroxyl-methyl)-2′-formylpyrrol-1′-yl)-3-phenyl-propionic acid lactone 3,5-diphenyl-pyridine	O	92,93
Maillard reactions: glycine model	1,6-Dimethyl-2(1H)-pyrazinone 1,5,6-trimethyl-2(1H)-[pyrazinone-5-hydroxy-1,3-dimethyl-2(1H)-quinoxa-linone		
Dipeptide aspartame, coke, and Maillard reaction	Racemization of amino acids	a	94
Cape jasmine (*Gardenia jasminoides* Ellis)	Pigments	O	95
Flour, bread	Free sugars	O	96
Ginger (*Zingiber officinale*)	Total ginger extracts	O	97
(*all-E*)-Astaxanthin	Stability study	a	98
Bakery products (cookies and snacks)	Total fat and *trans*-fatty acids	O	99–102
Food safety			
Egg albumen and yolk	Chloramphenicol	a	103
Chicken and egg	Salinomycin	a	104
Fruits and vegetables	Pesticides	C	105,107
		O	108,109
		a	110,111
Oleaginous seeds	Pesticides	O	106
		C	116
Frozen vegetables, fruit juice, jam	Atrazine	a	112
Corn and wheat	Zearalenone	C	113,114
Beans, white and black [*Vigna unguiculata* (L.)]	Fenitrothion (*O*,*O*-dimethyl *O*-4-nitro-*m*-tolyl phosphorothionate)	a	115
Cod liver, fish fillets	Polychlorinated biphenyls, DDT, toxaphene, chlordane, hexachloro-benzene, hexachlorocyclohexanes, dieldrin	O	117
Cod	Chloro and bromobenzene	C	118
Fish and seafood	Polybrominated diphenyl ethers	C	119
Turbot, salmon, pumpkin seed oils	PAHs	C	120,122
Sausage, lamb liver	Benzo[a]pyrene	a	121

[a]Indicates that an instrument not certified for laboratory purpose was used. In early development, commercial instruments were not available. There is no reason anymore to use such equipment.

DDT, dichlorodiphenyltrichloroethane.

Table 12.2. Literature Survey of MAE Protocols for Environmental Analysis

Analytes	Matrices	Closed/Open	Reference
Aliphatic hydrocarbons	Soil and sediment	a	257
Brominated flame retardants (BFR)	Maternal adipose tissue	C	187
	Semipermeable membrane	C	188
	House dust samples	C	189
	Sewage sludge	C	190
Carbamate pesticides (propoxur, proham, methiocarb, chlorpropham)	Soil	C	248
Chlorophenols	Soils samples	a	235
	Wood, leather, textiles	O	255
	Sewage sludge	C	216
Endocrine disrupting chemicals (EDCs)	River sediments	C	218
Fungicides (HCB)	Sand, soils, air filter	C	136
	Sediments	C	128
	Soils	C	130,131
Herbicide (triazine)	Soil samples	C	245
	Animal tissue	C	129
Herbicides (atrazine, alachlor), α-cypermethrin	Water	C	241
	Animal tissue	C	129
Herbicides (irgarol and M1)	Marine sediments	C	256
Herbicides (phenylurea)	Soils	C	246
Herbicides (triazine and chloroacetanilide)	Soils	C	243
	Animal tissue	C	129
Insecticides (aldrin)	Water and marine sediments	C	126
	Vegetables	C	127
	Sediments	C	128
	Soils	C	130,131
	Sand, soil, air filter	C	136
Insecticides (chlordane)	Cod-liver oil, mussel tissue	O	132
	Skua eggs, penguin eggs	O	133
	Blubber of fur seals	C	134
	Vegetables	C	127
	Leaves of nutty plants	C	135
	Sand, soil, air filter	C	136
	Soils	C	131
Insecticides (DDT)	Marine sediment	C	137
	Water and marine sediments	C	126
	Atmospheric samples	C	138
	Adipose and muscle tissues	C	139
	Cod-liver oil, mussel tissue	O	132
	Skua eggs, penguin eggs	O	133
	Blubber of fur seals	C	134
	Vegetables	C	127
	Leaves of nutty plants	C	135
	Sediments	C	128
	Sediment, mallow powder	C	140
	Soils	C	130,131,141
	Marine sediments, SRM 1944	C	142
	Soils	O	254
	Sand, soil, air filter	C	136

Table 12.2. (*Continued*)

Analytes	Matrices	Closed/Open	Reference
Insecticides (dieldrin)	Water and marine sediments	C	126
	Skua eggs, penguin eggs	O	133
	Blubber of fur seals	C	134
	Vegetables	C	127
	Leaves of nutty plants	C	135
	Sediments	C	128
	Soils	C	130,131,141
	Bark, cement, grass, vegetables	C	143
	Wool products	C	144
	Sand, soil, air filter	C	136
Insecticides (endrin)	Compost	C	145
	Water and marine sediments	C	126
	Vegetables	C	127
	Leaves of nutty plants	C	135
	Sand, soil, air filter	C	136
	Sediments	C	128
	Soils	C	130,131
Insecticides (heptachlor)	Water and marine sediments	C	126
	Vegetables	C	127
	Leaves of nutty plants	C	135
	Sand, soil, air filter	C	136
	Soils	C	130,131
Insecticides (hexachlorocyclohexanes)	Soil samples	C	202
Insecticides (mirex)	Compost	C	145
Insecticides (parathion, methyl parathion), *p,p′*-DDE, HCB, simazine, paraquat dichloride	Soils	a	244
Insecticides (pyrethroid)	Semipermeable membrane (SPMD)	C	208
	Soils	C	120,209
Insecticides (toxaphene)	Skua eggs, penguin eggs	O	133
	Blubber of fur seals	C	134
Organochlorine pesticides (OCPs)	Vegetation samples	C	229
	Medicinal plants	C	231
	Horticultural samples	C	227
	Soils	C	212
	Soils, sediments	C	220,233,240
	Vegetation samples	C	228
	Water	a	234
	Bark, cement, grass vegetables	C	230
Organophosphate esters	Glass fiber filters	O	238
OCPs	Soil samples	C	119,225
PAHs	Pure, tap, and sea water	C	159
	PM SRM 1648, SRM 1649a	C	160
	PM SRM 1649a, soils	O	161
	Sewage sludge	O	162
	Marine sediments	C	163,173,203, 223,224
	Soils, sediments	O	164
	Pollen, spruce needles	C	165
	Soil	C	166
	Air particulate, sediments	C	126,167
	Soils	O	168
	Soils	C	130,169
	Marine sediment	O	170

Table 12.2. (*Continued*)

Analytes	Matrices	Closed/Open	Reference
	Sediment	O	171
	Sewage sludge	C	172
	Sewage sludge, soil	C	174
	Particulate matter	C	175
	Marine biota	C	176
	Sediments	C	177
	Soils	O	178,179
	Industrial soil/organics	O	180
	Sediments and rocks	O	181
	Soil and sediment	C	182
	Air particulate	C	167
	Soil and sediment	C/O	183
	Marine sediment	C	184
	Sewage sludge	O	185
	Sediments, soils, air particulates	O	186
	Sand, soil, air filter	C	136
	Airborne particles	O	251
	Atmospheric particulate	C, O	201,253
	Diesel particulate	C	200
	Harbor sediments	C	204
	River sediments	C	222
	Sediments	O	207,236
	Soil and sediment	O	237
	Soils	O	252
	Wood samples	C	205
PAHs (simultaneous analysis of PAHs and metals)	Samples of atmospheric particulate matter	C	239
PAHs, PCBs	Sediments	O	250
PAHs, PCBs, chlorinated pesticides	Marine sediments and tissues	O	211
PAHs, PCBs, OCPs	Marine sediments	O	249
	Semipermeable membrane (SPMD)	a	207
PCBs	Ash samples	C	153,154
	Dry sediments	C	199
	Marine sediments	C	198,221,137,142,146,147
	Soils	C	131,197,213
	Soil	C/O	150
	Soils and sediments	C	206,232
	Marine and lake sediments	C	148
	Municipal sewage sludge	O	149
	Compost	C	145
	River water	O	151
	Pure, tap, and seawater	C	152
	Atmospheric samples	C	138
	Fish tissues, algae, invertebrates	C	155
	Marine mussels	C	156
	Adipose and muscle tissues	C	139
	Cod-liver oil, mussel tissue	O	132
	Skua eggs, penguin eggs	O	133
	Blubber of fur seals	C	134
	Sand, soil, air filter	C	136
	Sediment, mallow powder	C	140
	Fly ash	C	157

Table 12.2. (*Continued*)

Analytes	Matrices	Closed/Open	Reference
PCBs, OCPs	Marine sediments	C	219
PCBs, PAHs	Sediments	O	250
PCBs, PAHs, OCPs	Marine sediments and tissues	O	211
	Marine sediments	O	249
	Semipermeable membrane (SPMD)	a	207
PCBs, PCDD	Fly ash and sea sediments	C	210
Pentachlorophenols (PCP)	Organic solid matrices and soils	C	242
Phenolic compounds	Soils	C	226
Phenols	Soils	a	258
Polybrominated diphenyl ethers (PBDEs), polybrominated biphenyls (PBBs), polychlorinated naphthalenes (PCNs))	Sediment samples	C	214
Polychlorinated dibenzo-*p*-dioxins (dioxins) and polychlorinated dibenzofurans (furans)	Lake sediment	C	148
	EC DX-1 and NIST SRM-1944	C	136
	Sediment DX-2 and SRM-X, Urban Dust SRM-1649	C	158
Short-chain chlorinated alkanes (commonly named short-chain chlorinated paraffins [SCCPs])	River sediment samples	C	247
Total petroleum hydrocarbons	Soil	C	153
Tributyltin (TBT)	Sediment	O	191
	Marine biota and sediments	C	192
	Mussel tissues	C	193
	Mussel tissues and sediments	C	194
	Marine biomaterials	C	195
	Clays, sediments, CRMs	O	196
Xenobiotics (phthalate esters [Pes], nonylphenols [NPs], nonylphenol mono- and diethoxylates [NP1EOs and NP2EOs])	Sediment samples	C	206,217
Xenobiotics (nonylphenol [NP], nonylphenol polyethoxylates [NPnEOs])	Sewage sludge	C	215

[a]Indicates that an instrument not certified for laboratory purpose was used. In early development, commercial instruments were not available. There is no reason anymore to use such equipment.

HCB, hexachlorobenzene; DDT, dichlorodiphenyltrichloroethane; DDE, dichlorophenyldicholoroethylene; PCDD, polychlorinated dibenzodioxin.

resistive heating is widely used in sample preparation. The advantage of this approach is that it is basically transparent to the user; there is no need to change one's regular approach to "extraction" namely, provide randomized energy to a system. By system we refer to the container, the matrix of interest, and the solvent being used to recover the target analytes. Here microwaves offer a faster heating rate and sometimes higher temperatures can be achieved. The latter is especially true when closed vessels are used as the boiling point of the solvent is affected by the increase in pressure occurring in the container.

The main drawback of this approach is that the unique characteristics of microwaves, namely, inability to heat uniformly a mixture, or selective heating of certain components making up a mixture, are being removed. One is merely reproducing thermally controlled extraction. This is a viable approach for several applications where total extraction is desired and where selectivity is not an issue.

This approach, however, is clearly against natural behavior of microwaves. In fact, microwaves are inherently unable to heat the various components that make up a system to the same rate and temperature. Somewhat difficult to grasp fundamental parameters such as dielectric properties of materials play an important role in the ability microwaves have to heat mixtures. With the advent and increase in use of domestic microwave ovens, an analogy can be drawn to help substantiate this statement. Select a foodstuff that has clearly marked differences in phases, for example, a jam-filled pastry where the dough is relatively solid and the jam is a viscous liquid. Take that "matrix" from a refrigerator to amplify the phenomenon we want to observe. Submit the pastry while contained into a suitable microwave-safe glassware to, say, 30 s of microwave irradiation at full power. When the pastry is removed, one clearly notices that the inner jam is much hotter than the surrounding pastry. The gradient can be so high as to cause serious burns when the jam is contacted with the mouth tissues. This is because the dielectric properties of the pastry are significantly different from that of the water–sugar mixture that makes up the bulk of the jam. This occurs even though the jam is not physically accessible by normal resistive heat currents. The physical treatment of microwave behavior is beyond the scope of this sample preparation handbook, but the readers can refer to References 259–261 for more in-depth treatment and discussion of these fundamental aspects and their relevance when designing processes and apparatus making use of microwave energy.

To alleviate this undesirable result, commercial manufacturers call on a number of technical features. The description of these is clearly outside the scope of this chapter. The analogy, however, serves well to indicate why end users have favored the bulk heating approach. There is limited, if any, need to modify the protocol when changing matrices. This "universal" sample preparation is, then, driven almost exclusively by selecting a solvent pair that comprises one solvent that can heat when subjected to microwaves and one that offers suitable capacity to dissolve the target analytes. The main disadvantage then remains that this exhaustive extraction leads to relatively dirty extracts that often require significant subsequent cleanup and separation procedures.

This approach is by far the most widely spread in sample preparation laboratories and is associated with so-called closed-vessel technology. Time for sample processing though can be significant because there is a need to reach that thermal equilibrium, and if closed vessels are used, there is also a need to wait for the pressure to drop off. Time is not considered a drawback by most users however, because the sample processing time is still reduced to within 1 h as compared with say Soxhlet extraction where it may require as much as 48 h.

Selective heating on the other hand offers a more versatile tool to the sample preparation practitioner. The greatest challenge of this approach is to have an access to instrumentation that allows a good control on the application of microwave energy to ensure reproducible results. In as much as it is relatively easy to randomize the applied energy when using a closed-vessel, solvent-induced heating system, selective heating poses a completely different set of desirable parameters.

In bulk heating, the solvent represents the major mass and the major component that absorbs microwaves. Hence, heating the solvent and allowing thermal diffusion to the matrix is simple. No need to calculate energy requirements; simply ensure that enough solvent is present so that the influence of the matrix is less important than that of the solvent. In selective heating, one chooses a solvent that is transparent to microwaves relative to the matrix. Thus, the energy is applied directly and selectively to the matrix. As a general rule of thumb, nonpolar solvents and gases are poor microwave absorbers. Hence, if they possess adequate dissolving power, they are solvents of choice for true MAE applications.

For the purpose of further detailing the concepts and principles involved in MAE, it is preferable to introduce liquid-phase extraction and gas-phase extraction using separate sets of visualization examples and associated discussion. The following sections will provide such a basic introduction. It will be followed subsequently by complete description of specific applications for both approaches.

12.2.1. Liquid-Phase Extraction

The principles at play are simple to visualize. Let us take the case of a system consisting of fresh plant tissue as the matrix, hexane as the solvent, and terpenes as the target analytes. They are submitted to microwave irradiation while contained into a suitable microwave-transparent laboratory glassware. A 5-g fresh tissue sample contains on average 4 g of water. The water is well contained into the tissue, and water is a strong microwave absorber. Hexane is very microwave transparent. Hence, as microwaves are being applied, they can freely reach the matrix and start heating the water contents of the tissue. The rate of heating is dependent on the electrical field being applied. In laboratory instrumentation, where energy efficiency is not an issue from a cost standpoint, the field is relatively strong in relation to the small mass of absorbing materials (even though the equipment is not very efficient). This leads to quasi-instantaneous rise in temperature. The latter eventually reaches the boiling point of the water (somewhat over the normal one because of the fact that there is some internal pressure buildup). Once the water begins changing phase, the internal pressure exceeds the pressure that can be contained by the cellular walls of the tissue causing it to rupture, freeing the contents into the relatively cold surrounding hexane that can dissolve them. This overall procedure is in itself MAE.

The speed of extraction here is within seconds, hence the need for highly efficient instrumentation. A difference in irradiation time will bring about variations in final temperatures that may affect the overall success of the extraction. The energy applied is the key parameter. If absorption was complete (it is not!), the energy (E) could be calculated as the product of the applied power (P) and the time (t). If no boiling of the solvent takes place and if we ignore the losses due to surface evaporation of the solvent, this would equate to the product of the mass of the materials (m) and their specific heat (Cp) and the rise in temperature (T), as per Equation 12.1:

$$E = P \cdot t = m \cdot Cp \cdot \Delta T. \quad (12.1)$$

This relation explains the need for good control on the applied microwave power. Leaving aside the details of these theoretical calculations, one can still easily see that several unique parameters are available for the development of MAE protocols. One can control the temperature by selecting a solvent that has a boiling point in the range desired and simply install a reflux column. The heat of vaporization being much higher than the specific heat, on a per gram basis, and the refluxing behavior of the solvent, caused solely by the thermal diffusion of the microwave-absorbing component, water in our case, makes for an elegant temperature control mechanism without the need for expensive temperature measurement devices.

It can be said that the technology is deceivingly simple. Yet, despite that, the trend to date has been for manufacturers to offer equipment that does not make use of all these unique features of microwaves and rather to concentrate in bulk heating. Hence, to further use microwaves, the user has been provided with technical ancillary approaches such as that of adding microwave susceptors into relatively transparent media so as to offer the characteristic heating required even if the matrix is relatively nonabsorbing from a microwave standpoint. Hence, we had shown in the mid-1980s that one can rehydrate materials when needed. If water is not a desirable cosolvent, then one can use other susceptors such as carbon. The latter may provoke side reactions or act as a strong absorber toward some analytes; hence, some manufacturers embedded such materials into inert Teflon pieces that can be added to a mixture to cause them to heat and, by diffusion, heat the mixture.

The readers will have readily understood that these ancillary technical derivatives work almost exclusively with closed-vessel protocols characterized by relatively nonselective extraction and long extraction times. Unfortunately, there is no open-vessel, focused energy, microwave instrumentation on the market today dedicated exclusively to open-vessel extraction. This has greatly limited the development of so-called official or normalized or standardized protocols based on this approach. Nevertheless, in this chapter, we will spend some considerable time and space to this approach as it offers a much broader range of applications, and most importantly, allows for an analyte- or a matrix-tailored protocol. The latter may appear demanding from a routine sample preparation standpoint but offers unparalleled flexibility.

12.2.2. Gas-Phase Extraction

The concept of gases as solvents is not new in itself. One can think of headspace (HS) or purge-and-trap (P&T) as examples of relatively well-used gas-phase extraction technologies. MAE represents a very efficient alternative from time and sensitivity standpoints. Although no commercial instrumentation is available at press time, it is expected that some will be released over the years to come. Our laboratory has introduced the technology along with application-specific data in the past.[4,6–8,10,11,14,19,262–266]

The principles here are related to that of conventional HS sampling technology. One selects a matrix, let's say a fresh sample of Japanese rice wine known as sake. A precise quantity of an aliquot is transferred into a container that is hermetically sealed. The container is left with a significant HS above the sample. The HS can be air or an inert gas if needed. In conventional technology, the sample would be heated to say 90°C for a relatively long period of time (30–60 min are widely used incubation times according to the matrix). During that heating time, the volatiles are volatilized into gas phase (the HS).

Their concentration depends on the concentration of each component in the liquid phase and on the partial vapor pressure of each component. This relationship can be described by the so-called Raoult's law. Hence, quantification is not easy if one does not have a relatively well quantified set of standards. This phenomenon is further complicated by the fact that partial vapor pressures vary with temperature, and not necessarily at the same rate for each component. Hence, again calibration is not straightforward as it is also temperature dependent. This relative difficulty has greatly hindered the acceptance of HS as a standard reference method in several areas. Yet, it is on paper a very simple method from an operational standpoint, and as such, it gained popular acceptance as a screening method, especially in the food and fragrance industries where the volatile profile of materials is of prime importance.

It must also be stressed that the relatively long time needed in conventional HS for reaching good "equilibrium" state is not really an issue for the sample preparation practitioners as most instruments manufacturers offer an automated version that can preheat a number of samples simultaneously. This allows for the delivery of adequately incubated samples "just in time" for their injections into a suitable analytical determination means, usually a gas chromatograph equipped with a suitable injection interface system that can handle the relatively large volume of gas being subjected to the analysis.

The heating feature of the matrix is the only thing in common between HS and what we termed MAP-HS (for microwave-assisted process–headspace). In fact, in MAP-HS, one uses microwaves to generate the heat in the sample. If we refer to the sake example again, one can see that applying microwaves onto a microwave-transparent sealed container will cause the aqueous sample to heat. The difference in this approach is that the microwaves will *selectively* heat the aqueous phase over the gaseous phase. This is due, again, to the fact that the dielectric properties of the water—the main component of sake—are very different from those of the surrounding air (HS) component. This is true even after the boiling point of water is reached. Water has a dielectric constant of roughly 80 at room temperature. It decreases to 55 at 100°C in the still liquid state, whereas it falls to nearly 1 at 100°C in the gaseous state!

This phenomenon, specific to microwaves, results in a continuous supply of energy selectively to the aqueous or liquid phase. Hence, the process is a dynamic HS process, not a static HS process as in conventional HS where the energy being provided to the system is uniform and randomized. In MAP-HS, then, the temperature of the liquid can rapidly reach the boiling point of the water. This can lead to overheating situations. However, as will be shown later, this is easy to control when adequate instrumentation is used and does not constitute a safety issue any more significant than currently used technologies such as Soxhlet extraction or sample dissolution for inorganic analysis.

If one continues the analysis of the process, one must take into account some basic parameters such as specific heat, enthalpy of vaporization, and boiling points. Let us return to the sake example. Let us assume one brings the temperature of the liquid near to 100°C. The latter is easily achieved within a few seconds with for instance a 200-W microwave generator—remember Equation 12.1! At that point, the application of microwave energy is stopped. At that temperature, the system is in full effervescence in terms of distinct processes occurring simultaneously. Among those processes, we note the vaporization of the volatile component having a boiling point less than the actual temperature of the liquid phase. Remember that this vaporization is also affected by the relative increase in pressure present in the sealed container. The heat capacity of most of these organic components forming the aromatic profile of sake is less—actually nearly half—than that of the water. Hence at the surface, the dynamic process favors the vaporization of these components. Further, the energy that was applied affected only the liquid phase, and the condensation that occurred above the HS also favored the release of the organic components.

All in all, these processes are more complex than those in conventional HS. Even Raoult's law no longer applies under these operational conditions! In fact, we are facing a process whereby molecules are excited via an electromagnetic interaction and allowed to relax via thermal diffusion. The relationship between them is not evident. However, their importance is not truly applicable to the practice of MAP-HS. What we mean here is that provided one has a good control on the energy being applied to—and most importantly being absorbed by the system—one can reproduce very precisely the conditions that prevail in the container. As such, the technique has been shown to exceed conventional HS in terms of precision and accuracy. The same obviously holds true for sensitivity.

Maybe the most important feature and value of MAP-HS, however, resides in its relative nondependence on the matrix. Matrix effects are less numerous, and even when there are some, they are less important. This is easily explained in terms of energy being applied exceeding the hydrogen bonding between an alcohol and water for example, thus affecting the partial vapor pressure of the species, thus affecting the concentration in the gas phase. This powerful advantage of MAP-HS over conventional HS has also caused its delay in being introduced commercially. In fact, since data obtained from MAP-HS and HS cannot be compared directly, it has caused some misunderstanding with the end user and affected the timely release of commercial instrumentation. New parameters available for calibrating MAP-HS are simple; they consist primarily of the emitted power and the time period for which that power is applied. The same calibration will hold for a broad range of matrices and equally broad range of concentrations of the volatile components. This leads to the need for high-performance instrumentation.

The preceding statements dealing with calibration are dependent on the fact that one knows how much energy is applied to the matrix under study. While it may appear difficult to the noninitiated, this is not the case. It is relatively easy to exercise proper control over that parameter. We will discuss this issue in greater detail under the applications section of this chapter and we will introduce how to design and validate a single method or protocol that will apply directly to an unusually broad range of matrices.

12.3. EQUIPMENT FOR LIQUID-PHASE MAE

The availability of commercial instrumentation to implement MAE has been and remains a major issue for the sample preparation practitioner. Microwave apparatus may appear complicated to many users despite the fact that it is relatively straightforward. The real challenge is to have an extraction chamber that is optimized for the sample being treated. In principle, this is simple, but the reality is somewhat more demanding. Since the efficiency of the microwave is dependent on the dielectric properties of the materials being treated, and the latter vary a lot between matrices, it calls for the use of sophisticated, potentially difficult-to-use apparatus. Consequently, the approach has been to manufacture equipment that offers lower performance standards but are more universal and much simpler in their use.

The issue of specialized MAE equipment is further complicated by the lack of standard or reference methods. In the environmental sector, for example, the approval of an official method quite often indirectly will support the development of commercial instrumentation. The size of the market is also a factor in developing one or a series of instruments. Hence, in the case of MAE, a single approach, namely that promoted by the Method 3546 of SW-846 of the United States Environmental Protection Agency (USEPA) (see the details later in this chapter under the section dealing with examples), has proven so universal that it limited the need in the manufacturer's eyes to offer several types of instruments. It is always the philosophical question of market pull versus market push. The situation here may have also been further limited by the small number of manufacturers in that area and by the fact that only one manufacturer had legitimate access to patent rights to manufacture such equipment.

Hence, for all practical purposes, the only type of equipment commercially available today, we refer to current production, is what is known as a multimode apparatus. This equipment is characterized by the fact that, from a microwave design standpoint, it is a direct extension of microwave ovens sold for domestic use. Of course, they are provided with numerous additional safety and operational features that range from solvent sensors, fiber-optic temperature measurement devices, solvent-resistant materials, and so on.

This type of instrument does not warrant extensive discussion for the purpose of this book as their description can be found easily. It is mostly effective for closed-vessel MAE applications; in fact, this is the only recommended use. It has been successful in the market, and apparatus can be found worldwide in a variety of sample preparation laboratories dealing with food, cosmetics, consumer products, plastics, environmental samples, and numerous other applications.

The popularity and widespread acceptance of the closed-vessel instruments has hindered greatly the development of other types of instruments. While it would serve little purpose to discuss previously available open-vessel instruments that are not current production, it is worthwhile to spend some time to introduce what is required to design and build high-performance instruments that could process basically any matrix under open-vessel MAE conditions and would offer a range of operational parameters necessary for performing selective extraction for example. Let us say that most commercially available multimode oven-type MAE apparatus available today are designed for a relatively strong microwave coupler, something that varies between, say, acetone and water. By doing this, although not optimized for any specific matrix, it is able to heat almost all matrices to a certain degree. That is because most samples contain some residual water—a very strong microwave absorber when compared with most organic compounds—or the operator will use a solvent pair that will include acetone or another solvent that absorbs microwaves to a reasonable extent. In the laboratory, energy efficiency is not truly an issue from an operational cost standpoint. Nevertheless, this lack of control on the energy applied versus absorbed greatly limits any energy calculations and has contributed significantly to the never-ending existentialist question as to whether microwaves do anything else than simply heat, albeit heat in a very efficient and rapid way.

A microwave-assisted apparatus is basically made up of a microwave generator, a means to carry the wave (waveguides and coaxial cables being among the most popular) and an applicator, often called cavity, which in effect is, for MAE, the extraction chamber. There is little to be said about the generator and the waveguides. These are standard equipment that can be procured off the shelf from numerous manufacturers. The cavity is the key component from an efficiency standpoint. In many cases, it is the part that will make or break the applications. Numerous publications can be found in the open literature where authors report a failure to obtain successful extraction recoveries. A close scrutiny of the instrumentation used, however, shows that the apparatus in question was conceptually flawed from an MAE standpoint.

The difficulty stems from the fact that to be adapted from an electrical impedance standpoint, the cavity needs to be tuned or optimized for each sample type, potentially even each sample. This, as stated earlier, is due to the variations in dielectric properties between samples to be treated. This problem is further compounded by the fact that there exists very limited data dealing with dielectric properties of materials other than pure gaseous materials.[259,269–272] To perform such an adaptation, necessitates specialized instrumentation such as a network analyzer, which is not common in a chemistry laboratory.

That being said, it is possible to optimize a cavity without using such instrumentation. While that empirical approach will not allow for a precise determination of the electrical field position nor of its amplitude, it will maximize the interaction between the field and the matrix. A relatively simple approach to an efficient MAE cavity consists in using a generator equipped with emitted and reflected power instrumentation. It is then coupled to a resonant cavity through the use of circulators and waveguide components. These parts are all available off the shelf and are not specific to MAE. The portion of real interest is the cavity. The latter can be made by using a section of a standard waveguide. A hole that will serve as sample holder is made into the wide side of the waveguide, the center being at half the width of that side. Adequate leakage protection must be built-in by providing chimneys that will surround the sample container. A variable size and movable electrical iris is fitted between the microwave source and the sample holder. On the other side of the sample holder, the system is terminated using a short circuit made up of a movable piston. This is all that is needed to have a sample-specific MAE cavity. Figure 12.1 presents such an extraction cavity. When the sample is subjected to microwave irradiation, the operator will monitor emitted and reflected power. By moving the piston until minimal reflected power is measured, the operator will maximize energy deposition into the sample. Under these conditions, maximal interaction between the field and the sample will occur, thus allowing for greater thermal gradients to be produced if so desired.

The optional use of the iris is to further enhance that field interaction by creating so-called stationary waves. This subject matter will not be discussed here, but Figure 12.1 shows such a device for informational purpose.

12.4. APPLICATIONS

This section provides the description of the rationale behind the design of a number of MAE protocols. Although less challenging and more universal, closed-vessel applications are numerous and will be briefly introduced. They are very well spread and widely used for an ever-increasing number of analytes extracted from an equally broad range of matrix types. Although more limited in their variety, there are some variations on that theme. We will introduce one here, namely, the one-pot simultaneous extraction and derivatization of phenols in highly adsorbing samples such as high coke soils.

Open-vessel examples developed in this section will focus on the simplicity and the versatility of this much superior approach to MAE. Examples will range from

Figure 12.1. Photograph A shows a typical high-efficiency MAE cavity. The cavity itself is made by making a hole into a segment of a standard waveguide (WG, a WR340 in this case). The components include a sample holder (SH), a short circuit (SC) (electronically controlled in this case), and, optionally, an electrical iris (I). Photograph B shows further details about the nature of the SC; a manual SC is depicted to better see the relatively simple design.

simple direct extraction of animal tissues to a more sophisticated use of microwave susceptors to allow for a better control of the extracting conditions.

Finally, MAP-HS examples will be introduced to highlight the potential of this upcoming technology and to underline how universal its use may be if one uses microwaves for what they do best, namely, nonuniform heating!

12.4.1. Closed-Vessel Liquid-Phase MAE

The use of closed-vessel MAE is widely implemented in sample preparation laboratories around the world. Clearly favored by the availability of commercial instrumentation, it has developed into a routine workhorse for the determination of several target analytes in a variety of matrices. Yet, closed-vessel MAE has seen limited development in terms of variations, and a large portion of the literature to date deals with a single approach, namely, the use of a solvent pair that combines microwave absorption and ability to dissolve the target analyte. This may well be because MAE is a victim of its own success, since a single method, that we introduce hereafter, can be used for well over 100 contaminants from a broad range of solid or semisolid matrices.

Despite this, there is ample room for some elegant work to be done with closed-vessel MAE, although to date, very few reports have ventured into this area. Much work remains to be done for its widespread adoption. Nevertheless, we will introduce one such example hoping that it may serve as guide to users facing recalcitrant matrices for which they do not obtain good MAE results thus far.

12.4.1.1. MAE reference method for the extraction of contaminants. The USEPA publication SW-846, entitled *Test Methods for Evaluating Solid Waste, Physical/Chemical Methods*, is the Office of Solid Waste's official compendium of analytical and sampling methods that have been evaluated and approved for use in complying with the Resource Conservation and Recovery Act (RCRA) regulations. This manual has had a major impact on the acceptance of several technologies by sample preparation laboratories because of the need to meet certain performance standards for accreditation or recognition of the value of the data so obtained. Methods listed in SW-846 are deemed to meet de facto these performance standards; hence, the analysts do not have to demonstrate their level of performance. In 1999, the USEPA published Method 3546 entitled Microwave Extraction.[136] The latter has been incorporated into final Update IV of the SW-846 Manual in January 2008.[270] The method had been developed in our laboratory as part of a joint effort with our sister organization, the USEPA. Validation was made by drawing the services of several independent and contract laboratories based in Canada and in the United States. In fact, to the best of our knowledge, it was the first method ever to be promulgated in Canada and the United States simultaneously. Further, the collaborative team that pursued the development of this method has been formally recognized by both countries (the Green Technology Award of the CEM Corp. in 1999, the Award of Excellence of the Treasury Board of Canada in 2000, and the USEPA Office of Solid Waste and Emergency Response Science & Technical Award in 2001).

The popularity of this method calls for its inclusion here as an example, although it is the simplest application of MAE. The principle is simple: the method was mimicked after the standard Soxhlet technique in the sense that use is made of a solvent to dissolve the target analytes and one other solvent to induce heating if needed since microwaves may not be able to heat some solvents.

For example, in the simplest case of extracting alkanes from sand, one would think of hexane as a good solvent from a solubility standpoint. But hexane does not absorb microwaves, hence does not heat efficiently under normal microwave irradiation conditions. Sand, as in beach sand, is also devoid of significant microwave absorption. Hence, to overcome these limitations and to remove potential slight variations in the dielectric properties of different sample sizes, the method simply calls for the addition of a microwave-absorbing solvent that will interfere neither with the subsequent sample preparation steps, if any, nor with the final determinative analytical method. Acetone is a good prospect, especially when we take into account that it can also help in preventing the formation of phases or emulsions in cases where residual water may be present at various quantities in the sand samples.

This deceptively simple approach can be applied to a surprisingly broad range of matrices and analytes as it meets the basic requirements of several extraction applications. The official method lists well over 100 contaminants under a single sample preparation protocol applicable to an equally broad number of matrices. Of course, the analytical determinative step varies with the nature of the target analyte. This is because hexane offers good solubility characteristics for numerous pollutants (polycyclic aromatic hydrocarbons [PAHs], polychlorinated biphenyls [PCBs], etc.), whereas the acetone allows an even heating of the matrix for a broad range of matrices. Further, the acetone adds an additional valuable characteristic, namely that of providing a more uniform electrical load to the electrical field of the microwaves. This can be easily understood if one looks at matrices where the residual water content may vary slightly between samples. Water, being so important dielectrically, may account for significant dielectric behavior (thus, heating behavior) of the sample. The relatively large amount of acetone in relation to the residual water has the effect of making this residual water a much less significant contributor to the overall dielectric properties of the system consisting of all the matrix components plus the solvent pair.

In fact, the limitations of this approach stem from the fact that matrices that contain significant amounts of free metals may cause heating problems as the latter may become the most important microwave-absorbing species. Other obvious drawbacks are the relatively long time needed to heat the overall mixture—several minutes are usually required (15–20 in the case of the official MAE method)—and the time to allow for pressure decrease before being able to open the closed vessel and proceed with the subsequent sample preparation and analytical steps. This heating time is the result of the fact that since only acetone heats, it must transfer that thermal energy to the surrounding medium; hence, the relative concentration of acetone to hexane is an important parameter not only for solubility purposes. Further, with acetone having a lower boiling point than hexane, one must ensure that enough acetone is present to avoid reaching a point where all energy applied would barely serve the purpose of "distilling" the acetone. This is especially important because acetone in the gas phase is basically transparent to microwaves.

The drawback from this approach remains the relative insensitivity to microwave basics in the sense that under these conditions, no one can do any energy calculations. The equipment used for this purpose is not instrumented in such a way that one can calculate energy efficiency, field intensity, and true irradiation time (or power for that matter!). Hence, to overcome this method, one simply makes use of those long irradiation times of several minutes. They have the effect of randomizing the energy deposition, and over time, this randomization equates to uniform total energy absorbed—yet most of the energy applied is simply wasted.

12.4.1.2. Phenols in soils by simultaneous in situ derivatization and MAE. Despite the limitations associated with the use of closed-vessel MAE, it must be highlighted that it is still possible to perform a limited number of more refined applications when making use of closed-vessel MAE. For example, relatively simple applications making use of one-pot chemistry are amenable to closed-vessel MAE. Multiple-step work is also possible, but the significant periods required between each step to allow the system to return to safe operating temperature and pressure conditions make them less than practical, albeit possible.

An early example of such an approach came from our laboratories when we reported on the determination of phenol and methylphenol isomers in soils by a simultaneous *in situ* MAE and derivatization procedure.[271] To the best of our knowledge, this was the first literature reference reporting on the potential to develop closed-vessel MAE protocols involving one-pot extraction derivatization.

The nature of the problem to be addressed is derived from the previously mentioned inherent limitations of closed-vessel MAE. For example, let us consider the treatment of recalcitrant matrices offering greater adsorption characteristics than normal. This behavior, combined with the limited solubility of the target analyte and the relatively long period required for a closed-system vessel to return to room temperature and pressure, can lead to complete or partial readsorption of the target analytes. In our own example, the aim was the extraction of phenols from high-carbon-content soils. The very high adsorption affinity of phenols toward this type of matrix leads to their readsorption onto the surface of the soils, thus yielding poor and irreproducible recoveries. The use of Method 3546 of SW-846 is greatly hindered, perhaps even impossible, according to the level of readsorption taking place, thus the need for an alternative approach.

To begin the analysis of this problem, let us clarify that despite all the polemic and remaining uncertainty surrounding microwaves and their mode of action from an energy standpoint, microwaves do no perform magic. The chemistry available to the sample preparation practitioner remains basically the same, only the heating characteristics will change according to the dielectric properties of the materials being treated. This statement holds true for the matrix as well as the solvent or solvents being used. The efficiency of the heating step, its uniformity, and other details may vary, but the basic nature of the chemistry taking place remains the same.

In our example, then, the challenge is to find a means to prevent the phenols from having access to the surface of the matrix. Preliminary work showed that some limited recoveries could be obtained. Hence, the problem could be incomplete extraction, readsorption of the analytes, or a combination of both. A simple test whereby a standard phenol solution of known concentration is allowed to rest over the matrix for a time period similar in duration to that of the wait period of the MAE protocol showed that readsorption was the likely cause of the poor recoveries.

To circumvent this problem, one must change the nature of the analyte. This can be done by performing a derivatization of the material. The performance of a test similar to that above but using a solution of acetylated phenols of known concentration showed that readsorption of this species did not cause a problem. Hence, the design of an *in situ* one-pot MAE derivatization protocol was the obvious next step to overcome this problem.

The chemistry that can be used in this case is simple and can be taken straight out of a basic organic chemistry textbook. With phenols being acidic in nature, one can simply use this characteristic to perform a chemical reaction on them as they are released from the matrix and before the temperature and pressure conditions are such that readsorption can occur. Hence, the basic, quasi-universal closed-vessel MAE protocol can be modified by the addition of a catalytic amount of a base along with an adequate derivatization agent. Pyridine and acetic anhydride, respectively, were used in our original work. Under these conditions recoveries obtained were basically complete with excellent relative standard deviations. The high degree of success of this approach was further evidenced by using solutions of phenol, *o*-, *m*-, and *p*-cresol of known concentrations and by determining the recoveries for each concentration. It was shown that the protocol can be used over a linear range of more than four orders of magnitude while maintaining complete recoveries and precision better than current official methods even near the detectors limit of below 40 ng/g.

12.4.2. Open-Vessel Liquid-Phase MAE

Open-vessel liquid-phase MAE is the most powerful liquid-phase extraction approach within the MAE technology. It makes use of the unique heating characteristics of microwaves and is based on its ability to selectively heat some substances over others, or some phases over others. Originally developed with industrial applications in mind[2,3,5] where energy efficiency and low solvent consumption are primary drivers for success or failure in the implementation of new technologies, MAE has evolved into a most powerful extraction tool for the sample preparation practitioner.[10–15]

In its purest form, an MAE protocol makes use of solvents and containers that are transparent to microwaves relative to the matrix being treated. This approach provides the broadest range of control parameters to the user. In this laboratory, it is used whenever the sample lends itself to it. This applies to a broad range of matrices with special emphasis on plant and animal tissues, cellular organisms, or any other matrices where the residual water is in sufficient quantity to act as the component of the system that will generate the heat through selective interaction with the microwaves. Lack of such residual water is not necessarily a limiting step because in many cases, it is possible to rehydrate the matrix, as for dried tissues for example. This latter approach will be the subject of a specific application to be further discussed later in this chapter.

The process is, again as it is the case for most MAE applications, deceptively simple. Let us look at the example of the extraction of a high value-added component from a fresh plant tissue, or of a contaminant contained in that plant tissue. The sample preparation system would consist of the matrix, a solvent, and a container that are transparent to microwaves. The matrix is inserted into the container and immersed into the solvent. The system is placed into a suitable microwave apparatus and submitted to irradiation at a known power for a predetermined amount of time.

The microwaves travel freely across the container and the solvent and reach the matrix with minimal disruption or diminution of intensity. The residual water in the matrix interacts with the microwaves and begins heating selectively. The matrix starts to dissipate some of its heat to the surrounding system components. The rate at which the energy reaches the matrix is higher than that of the matrix dissipating its heat, thus temperature gradients are formed. Eventually, the residual water may get hot enough to reach boiling point at which time it will change phase, thus creating extreme localized pressures resulting in the alteration of the physical structure in which it is contained and the migration of the water outside the structure. This movement of water outbound has two effects: one is to entrain with it some of the contents of the matrix; the other is to facilitate the access to the inner part of the system by the immiscible solvent. The former is akin to an *in situ* steam distillation process, whereas the latter leads to enhanced solubility characteristics resulting from the expelling of the water that was creating a physical barrier to the solvent due to its immiscibility. The temperature of the overall system remains relatively low. In fact, that temperature can be controlled by the volume of solvent being used to perform the recovery of the desired analytes (remember Eq. 12.1!) since for all practical purposes, only the residual water is the source of heat for the whole system. This relatively low operational temperature coupled to the short treatment times that characterize MAE protocols is synonym of short exposure times to significant heat. This is another potentially important feature, especially when treating matrices or analytes that are thermally labile.

Of course, all systems may not be as clearly well defined from a physical structure standpoint as this one, and the number and complexity of the processes involved may prove higher. But no matter how many such processes are taking place simultaneously, they are all based on this fundamental behavior that can be seen and followed by electron microscopy.[2,3,5] Hence, the level of complexity increases with the nature and number of components that may interact with the microwaves, as well as the degree of intensity of that interaction. Also, whether the water can migrate prior to reaching the boiling point and changing phase will also have an impact on the efficiency of the process. All these issues can be addressed when using equipment capable to control electrical field density, but as mentioned previously these are not commercially available for MAE at this time. Hence, they will not be discussed in any detail here as we will focus on applications that can be implemented with currently marketed technology. Section 12.3, dealing with instrumentation, presented a short description on how to design an MAE-specific apparatus that can provide enhanced process efficiency and some control over the level of energy absorbed by the system, thus allowing the control of the level of heat being generated by the microwave treatment.

Clearly, it is not always easy to visualize the narrative description introduced above, especially when the matrix used may have been preprocessed; the tissue may have been cut and ground, hence leaving little if any evidence to the naked eye to observe the system characteristics described above. To look at this in greater detail and to help visualize it pictorially, we will choose an example where the water is well retained in the matrix and where the structure is well defined. Let us look at the behavior of a grape when treated by MAE to recover, say, essential oils. The protocol is simple: we insert the grape into a suitable container, we add the solvent, hexane in this case is appropriate, and we place the system into an MAE equipment.

For the purpose of this example, we have used a variable power monomode Soxwave-100 apparatus (formerly manufactured by Prolabo, Fontenay-sous-Bois, France) operating at 2450 MHz, the most common frequency. The power was set to 240 W, and irradiation was performed for a total of 30 s (yes seconds!). Figure 12.2 offers a dramatic visualization of the mechanism of action of MAE. The three photographs that make up Figure 12.2 show the intact system prior to irradiation, the appearance of water (and air) bubbles escaping from the grape when irradiation has been long enough to bring the water to its boiling point, and the final product where two phases are clearly visible now that the water has escaped and the system has been allowed to cool and rest.

Let us look in greater details at the second and third photographs as they provide even more valuable information indicative of the physical processes that take place and that make MAE such a powerful sample preparation tool.

In the second photograph, it can be seen that several bubbles are now leaving the grape. This example was chosen because the skin of the grape acts in effect as yet another membranous layer that further restricts and slows the movement of water outside the matrix (the grape). It can also be seen that turbidity is being created as the result of the water vapor being released into the hexane where it is cooled and condensed back to its liquid state. In fact, the effect of the temperature gradient can be seen in a most spectacular fashion if one watches the production and evolution of these bubbles. In a video—that cannot be inserted into a book—one could use the slow motion feature and would clearly see the water vapor bubbles reduce in size as they migrate toward the top of the solvent layer. Some will vanish as they will completely condense; some will reach the top and condense on the surface of the container. Some of these microemulsions can also be seen above the solvent layer on the surface of the container in photograph B. It must be noted that in this specific case, even when this same photograph was taken, the system bulk temperature—measured with a fiber-optic system shown in the first photograph (lost in the

A B C

Figure 12.2. Mechanism of action of MAE as visualized from the treatment of a grape in hexane. Photograph A shows a fresh grape immersed in hexane in a quartz vessel prior to treatment. The temperature is monitored using a fiber optics system highlighted by the arrow. Photograph B shows the grape after having been submitted to 30-s, 240-W microwave irradiation in a Prolabo Soxwave-100 apparatus. The arrow indicates the appearance of water vapor bubbles forming at the surface of the skin of the grape from which they will eventually escape and migrate out. At this point, the temperature raised only by about 15°C. Photograph C shows the grape and the solvent a few minutes after the end of irradiation. Hexane and the water expelled from the grape are immiscible, and the two phases separate as indicated by the arrow.

turbidity in photograph B)—only reached 35°C. The system was at 22°C when submitted to irradiation; yet clearly, the water is escaping at much higher temperatures. In fact, as can be observed in photograph B, one seed was even able to escape the skin of the grape under the influence of the high internal pressure that prevailed during irradiation!

In the third photograph, taken a few minutes after the end of the irradiation and after the system was allowed to cool and rest, it can be seen that the morphology of the grape has changed significantly. Its size is greatly reduced due to the loss of the water. The condensed water has now settled, and the two immiscible phase systems hexane–water are clearly evident.

Concluding this introduction to liquid-phase open-vessel MAE, one must draw attention to the most significant difference between MAE and any other extraction technology reported to date. All extraction technologies—except for MAE—are based on a similar set of physical phenomena and parameters. Energy is applied to a solvent, be it a liquid as in Soxhlet or a gas as in supercritical fluid extraction (SFE), that energized medium is then allowed to diffuse into the matrix, where it will dissolve the target analytes, then migrate out. Parameters of importance therefore are solubility (that can be affected by temperature or pressure or both depending on the extraction technology), diffusivity, and time! When using MAE, energy is applied to the matrix exclusively or at least in greater part, thus causing thermal gradients favoring the expelling of the target analytes. Thus, the importance of diffusivity is greatly reduced and time as well since energy is required for only a fraction of the overall system—and of course microwave heating is generally faster than resistive heating. Hence, when using CO_2, for example, the trend has been to increase pressure to overcome the poor diffusivity of the liquid CO_2, yet by doing this, one compromises the solubility. With MAE, one can use liquid CO_2 and not worry about diffusivity. This simplifies greatly the extraction system and brings about a significant cost reduction associated with high-pressure operation. This can have an important impact on the selection of an extraction technology especially at the industrial scale where operational costs associated with energy consumption and operational pressures are important.

12.4.2.1. MAE of fats from foodstuff under atmospheric pressure conditions. The previous discussion showed how deceptively simple MAE protocols can be. Of course, not all matrices lend themselves to simply being immersed into a suitable microwave-transparent solvent and subjected to microwave irradiation. While this approach may well prove successful for numerous applications, there remains a need to understand the nature of the matrix. In other words, to extract fats from foodstuff may not be a single-method affair for all foodstuffs. The following description deals with using atmospheric pressure, open-vessel MAE to extract fats from various types of foodstuffs, namely, meat, dairy products, and egg products. The key difference between these matrices, from a fat content perspective, is the form under which the fat is spread throughout the matrix. Hence, a difference in availability of the fat implies a difference on how to access it. This is by no means different than using conventional extraction techniques. In fact, users who perform routine fat extraction work know very well that different methods apply to different matrices.

In meat, the fat is essentially "free." It can be seen visually as it normally can be found on the edges surrounding the sample and in localized pockets of various shapes and sizes, the quantity and dispersion of which vary with the source of the sample (animal type, which part or parts of the animal, etc.). Hence, the problem appears to be strictly that of finding a suitable solvent that will dissolve the fat while in contact with the matrix. Conventional extraction techniques require that the sample be homogeneous and dry.[273–277] While the former is a requirement for any technique, the latter is not necessary for MAE. In fact, homogeneity is important since as stated, the fat can be observed as localized pockets of lipids dispersed more or less randomly throughout the meat, hence the sample must be homogenized so that a reproducible average result can be established and considered representative of the overall matrix.

In MAE however, diffusivity into the matrix is less critical since the analytes are expelled outbound toward the solvent. Thus, the need to dry the sample does not apply. In fact, on the contrary, it is desirable to have some residual water as we have seen before. Hence, a very simple protocol can be put forward by immersing a homogenous sample into a suitable solvent, again hexane works, although one could use petroleum ether as it is popular with conventional techniques and subjected to microwave irradiation for a short amount of time, in the order of a few seconds. If desired, one can provide a series of short microwave irradiations with a rest period in between. The latter can be useful if the time required to properly dissolve the fat is long. Subsequent steps are filtering, washing, and gravimetric determination or more thorough analysis through appropriate derivatization and analytical determination.

Of course, the residual water that was not removed in the preliminary step, although in relatively small quantity compared with the solvent, may cause some issues in terms of emulsions or yielding a higher-fat-content number due to the fact that although immiscible, there is still a certain degree of miscibility between hexane and water. As per conventional approaches therefore, it is necessary to wash the residue with acetone to circumvent this problem.

The problem with dairy products is more complicated due to the fact that in such materials, some fat is bound to the proteins. Further, these materials can give rise to the formation of microemulsions that make an exhaustive recovery much more difficult to achieve. However, again an approach similar to that of the conventional methods is possible. The chemistry used in traditional extraction techniques can be applied to MAE. As per the phenols example presented earlier under the closed-vessel section, *in situ* chemistry is readily available in open-vessel MAE. In fact, it is even broader in scope since the atmospheric nature of these applications lends itself to easy implementation of multiple steps if need be. In the case of dairy products, one can copy the basic method just discussed for meat products and amend it by adding steps required to bring the sample to be under the same "free" fat conditions. As per conventional techniques, this can be achieved by simply hydrolyzing the matrix *in situ* with a suitable catalyst. In the original work dealing with this issue,[20] ammonium hydroxide was used to effect the base-catalyzed hydrolysis of the protein-bound dairy products.

Again, the basic chemistry does not change; hence, this hydrolysis brings about the potential for protein components to enter the extract. Therefore, an additional step consisting of adding an ethanol aliquot after the microwave step and submitting this to a centrifugation step to precipitate and remove the protein material prior to final washing and fat content determination must be added to the overall protocol. One important point to stress, however, is that in order to be efficient, the hydrolysis requires that the matrix be in good contact with the catalyst. Hence, with ammonium hydroxide being relatively volatile (release of ammonia), it is necessary to apply relatively low power to the system. Failure to do so will lead to the vaporization of the ammonia, which will evolve from the solution and simply condense on the walls of the container above the solution, thus reducing the effective concentration to the point of limiting the efficiency of the hydrolysis or requiring much more time to be complete.

The approach to egg yolk is a further variation on that theme, where the hydrolysis can be effected via acid-catalyzed conditions. In the original work,[20] hydrochloric acid was used as the catalyst. As one might expect based on the same issue of good contact between the matrix and the catalyst, and because HCl is even more volatile, power application must be kept minimal.

12.4.2.2. Use of susceptors to enhance range and efficiency of extraction. All open-vessel examples discussed so far were based on matrices offering adequate microwave susceptibility to effect a selective heating of the matrix. In closed vessels, we saw that due to the nature of the approach, many reported applications do not make use of all the power of MAE and use solvents or solvent pairs that are in effect responsible for the heating, thus bringing the technique closer to conventional passive heating ones. What happens then when one wants to use selective heating but the matrix does not lend itself to such a behavior? For example, how does MAE deal with dried plant tissues or other matrices devoid of sufficient residual water to produce selective heating?

Closed-vessel MAE using solvents that absorb microwaves is certainly a viable approach. It is simple and relatively universal in its implementation, but all selectivity that characterizes MAE in terms of extraction is basically lost. A more sophisticated approach is to add materials to the matrix that will cause the desired selective heating to occur. These materials, generally termed microwave susceptors, are selected for their relative inertness to the matrix and the target analytes, their noninterference with the subsequent analytical procedures, and, of course, their high level of interaction with microwaves.

Water is by far the simplest and most common such material. It is considered a natural candidate because in many applications, it is simply a question of rehydrating matrices and bringing them back to a state nearing their original one.[1,3,5] While adding water can be seen as a nearly universal approach to enhance the selective heating characteristics of the matrix, it is not always the best one. There are additional advantages in making the selection of the susceptor based on its ability to modify the extract contents based on solubility for example. The degree to which dielectric properties can be modified will vary with the nature of the matrix itself.

Let us take the case of extracting a sample of dry spices. We presented such a detailed approach to modifying the dielectric characteristics of dry ginger by adding various susceptors into the matrix including water.[97] In this particular example, work was carried out to determine how refined of a modification can be made to obtain valuable extract results. MAE performed on intact matrices was compared with conventional Soxhlet extraction using solvents offering a range of dielectric properties, namely, hexane, petroleum ether, dichloromethane, and ethanol. MAE was also performed on various susceptor-modified samples that were also compared with the Soxhlet work. In this latter work, only the two extremes in terms of dielectrics were used as solvents, that is, hexane and ethanol. To ensure that comparable quality extracts were obtained, all extracts were monitored by gas chromatography–mass spectrometry (GC-MS). It is important to ensure that greater yield is not simply the result of extracting a broader range of potentially undesirable compounds. Clearly, due to polarity and solubility parameters, a hexane extract is different in nature than an ethanol extract performed on the same raw material.

When the matrix dielectric properties were modified by adding known amounts of ethanol and using hexane as the extraction solvent, a significant improvement in the extract yield was obtained. The yield went as high as that obtained by Soxhlet using ethanol as a solvent, and extract quality was similar. Of course, using ethanol as a modifier and using ethanol as solvent led to no benefits. This was expected since under these conditions, there was basically no difference between the dielectric properties of the modified matrix and those of the solvent, hence no additional benefits to be derived from selective heating.

Modifying the matrix by the simple addition of water also clearly led to significantly improved extract yields when both hexane and ethanol were used as solvents. In fact, higher yields were obtained and quality was maintained when compared with the respective conventional Soxhlet extracts irrespective whether hexane or ethanol was used as solvent. This was no surprise for hexane as a solvent given the relative strength of water as a microwave absorber compared with hexane. However, the results with ethanol as a solvent may appear surprising to some. If one takes into account that the specific heat of water is almost twice that of ethanol, and that its density is higher by nearly 25% (experiments we carried out using volume of solvent as the parameter, not mass of solvent, and heat capacity of course is mass dependent, see Eq. 12.1), one may have expected better heating characteristics when using ethanol as a modifier than water with hexane as a solvent. The main parameter at play was shown to be a dielectric constant and not a dielectric loss when the heating profiles were recorded using adequate *in situ* fiber-optic temperature measurement system. These data support the fact that the greater the thermal gradient—without exceeding the lability of the target analyte—the more efficient the MAE process is.[9]

The latter conclusion is not to be underestimated as it shows clearly the potential for tailoring MAE protocols according to the matrix. In fact, the nonavailability of open-vessel focused commercial instruments manufactured for MAE is the main limitation of the technology and is responsible for the relatively small number of reports dealing with the more matrix-specific applications. Of course, this situation is also the result of the general lack of understanding of the fundamentals underlying microwaves.

For those with an interest to further probe the limits of the MAE technology, it can be noted that there are numerous materials that offer heating characteristics by far superior to water such as some carbides, especially silicon and tungsten. They will not be discussed further here because of their limited use to date in sample preparation for analytical purposes, the subject of this book. Literature is readily available, however, as they are more and more widely used in catalyzed and noncatalyzed chemical synthesis applications. Their efficiency in terms of energy management lends them to today's interest toward the so-called green or benign chemistry and makes them candidates of choice for industrial MAE applications.

12.4.3. Gas-Phase Extraction

As mentioned earlier in this chapter, gas-phase MAE is a younger technology than its liquid-phase counterpart.[4,6–8,10,11] There are few publications on the subject in the literature. This is mainly due to the relative novelty of the method and the lack of commercial instrumentation. Only recently has such an instrument been made available by a commercial manufacturer (Shimadzu), and even so, it is produced in limited quantities for the time being and has not been offered at large yet. Also, while the technology can be used in a P&T mode or an HS mode, only the latter can be performed by the current model in an automated mode. Manual sampling in a P&T mode is however possible. Examples are presented below for both approaches to underline the fundamental nature of gas-phase MAE, but attention of the readers and sample preparation practitioners interested to test the technology is drawn to the need to strictly adhere to the following safety disclaimer and abide by all its directions.

Safety Disclaimer

Experiments must be carried out by highly skilled personnel well abreast of risks and hazards associated with subjecting volatilized materials to microwave energy while within a confined volume. Special care should be exercised if and when attempting to reproduce work presented and discussed herein. This hazard level will be reduced, if not totally removed, when commercial instrumentation is made available to the laboratory community. By the same token, when such instrumentation is available, even higher sensitivity will be achieved as the containers used therefore will have been designed specifically for that purpose.

Despite the seemingly simple nature of the procedure reported herein, workers must refrain from using domestic microwave ovens. The wide availability of such equipment that is not approved for use in a laboratory makes it tempting but at no time must the safety of operators be overlooked, and extreme care must be exercised as a result of the inherent danger of explosion. For example, using some of the procedures discussed herein, the surface temperature of the HS vial was measured to be above 110°C and the pressure up to above 12 psi. There should be no vigorous boiling or excessive distortion of the septum seal. The tab at the center of the safety cap is designed to peel back by the bulging septum if the pressure limit is exceeded. During the microwave exposure period, interrupt the power application immediately upon hearing any knocking or bumping noise, which may be indicative of imminent release of excessive pressure. In handling the vial after irradiation, use tongs with long handles and always wear proper face protection in the form of a clear face shield with an apron to protect the neck area.

Commercially available domestic ovens have so-called "duty cycles" (i.e., the magnetron is not always *on*), hence making it possible to give rise to rapid increases in pressure and enhance considerably the risk of explosion. This characteristic of domestic ovens also has the effect of reducing significantly the level of reproducibility.

12.4.3.1. MAP-P&T of fragrances from plant tissues. One of the simplest forms of gas-phase MAE is an alternative to conventional P&T technology. Again, as is usually the case with MAE, the approach is deceptively simple; one uses microwaves to induce thermal gradients between a matrix and the surrounding gaseous environment. These gradients lead to volatile and semivolatile materials to be released from the matrix into that environment. Let us take a simple case, that of determining the fragrance contents of fresh plant materials. The plant material is inserted into a container suitable for microwave radiation. The container and the materials are located into the microwave cavity and the power is applied. The fresh plant material has a relatively high moisture content. That water will interact with microwaves and heat to the point of vaporizing and escaping the matrix. Depending on the rate at which one applies the microwaves (a combination of power and time), it is possible to use that heat and let it dissipate to other components that in turn will also vaporize and escape to the surrounding gaseous environment. If one allows this process to continue, there will be a pressure buildup into that container. Hence, it is a relatively minor task to make a small orifice into the lid of the container and insert a trap in that orifice—internally to the container if the trap is transparent to microwaves, or externally if it is not transparent to microwaves—and allow these volatile compounds to exit the container via the orifice and through the trap. To be effective, the trap must be selected from those available that will retain the target compounds. In the present case where essentials oils are targeted, one can use commercial filtration or sample preparation cartridges. Then, it is a simple task to elute the contents of the trap and inject an aliquot into an appropriate separation and characterization instrument such as GC-MS for example.

Provided that the management of water is well controlled, the technique lends itself to the addition of water into the container to further enhance the efficiency of the microwave treatment and ensure that exhaustive extraction takes place. It is obvious from this brief description that one can also set up a cryogenic trap instead of a chemical trap. The latter has the disadvantage of collecting the water as well and necessitates an extra separation step once desorption of the extract is performed. All these steps, however, are common and routine as in other P&T technologies. This approach was used in our original work[4,6–8,19] that was carried out much before the advent of equipment specifically designed for this purpose.

12.4.3.2. MAP-HS for the analysis of volatile organic compounds (VOCs) in water. Most of the limited literature reporting on MAP-HS (microwave-assisted HS) has been for work dealing with the determination of VOCs in water. [4,6–8,10,11,262,263] The protocols used were all based on the selective heating of the water in its liquid phase over that of it gaseous phase. Basically, it is a simple question of subjecting the aqueous sample to microwave irradiation under a well-controlled application of energy. The phenomena involved were discussed at length earlier in Section 12.2. The protocol can be summarized as measuring an exact amount of water sample into a container suitable for use into a microwave apparatus. The sample is hermetically sealed and located into the apparatus. The key parameter here is to have good control over the applied energy (see Eq. 12.1 again). If one uses a microwave reactor, or cavity, that has been well tuned for water—which is relatively easy to do, given the good microwave absorption characteristics of water—then precise control of the time is essential to ensure reproducibility. Of course, time here is used generally to represent a precise control over the duration of the irradiation *and* the amount of time between the end of the irradiation and the actual sample injection into an analytical device such as GC. As explained earlier, for a set level of power, a given duration of microwave irradiation will lead to very reproducible vaporization of the VOCs from the water. Assuming a constant applied power, then calibration would be performed at a constant time. Sensitivity will be directly dependent on the time selected for irradiation. Provided that one performs the various treatments for the same duration and maintains the postmicrowave treatment time constant, then analytical data obtained for the various samples can be compared together. This control over time is required to ensure that the same amount of energy is applied to the sample (hence, same temperature will be reached) and that similar losses due to thermal diffusion are basically constant.

12.4.3.3. MAP-HS for the analysis of VOCs in soils. This example is selected because it is a simple extrapolation of the previous one and serves well the purpose of showing the quasi-universality of MAP-HS as a sample preparation for gas-phase extraction. We have discussed before the need to have a cavity, or reactor, well tuned for the sample to be treated. This stems from the fact that to be able to compare data obtained between different matrices, one must ensure that the absorbed energy is similar between samples. That is almost impossible to achieve with conventional multimode oven-type apparatus, especially when we have seen that the interaction between the matrix and the microwaves depends on the matrix dielectric properties. These vary greatly between, say, sand and pot soils.

However, we also saw that water is a very good microwave absorber, so good in fact that when placed in sufficient quantity compared with an organic substance, the latter has basically no effect on the dielectrics effectively seen by the microwave field. Hence, for the purpose of having a simple MAP-HS protocol that would apply not only to water but equally well to a variety of soil matrices, one can simply repeat the procedures described above for VOCs in water on soil samples to which a known quantity of water has been added.

Ideally, one would like to use as small an amount of water as possible to reduce the competitive vaporization that takes place between water itself and the VOCs. That however must be balanced against the desire to establish a protocol that will apply to both sand and pot soils, for example. If sensitivity is the main criterion guiding the method development needs, then one can optimize the protocol by adding various amounts of water to the matrix prior to subjecting the mixture to microwave treatment and by so doing establish the optimal quantity for the specific sample type. We have reported extensively on how to achieve this earlier.[4,6–8,264] In these papers, we have also shown that typically, the addition of a 10-mL aliquot of water to 1-g soil sample will prove effective for most soil matrices.

Again, sample irradiation here will be very short—a few seconds—as one can calculate when using Equation 12.1 and applying it to their samples, targeting a temperature of say, 130°C, to account for thermal losses and some degree of nonabsorption.

12.5. CONCLUSION

Microwaves offer a viable alternative means to heat samples. It offers some intrinsic advantages, namely, in terms of speed and universality that are undeniable.

Ironically, it is this universality that has limited their use in microwave-specific applications. Because of the ease to successfully heat practically any matrix via the selection of a microwave-absorbing substance (usually the solvent) in combination with overheated and overpressurized operating conditions, work aiming at using microwaves for what they do best, namely, selective heating, has been limited. This situation was compounded by the lack of instruments capable to offer the means to control such selective heating parameters.

Nevertheless, MAE, like several other MAP technologies, has made it into the analytical laboratory including its commercial component. If the interest and the current efforts deployed by the analytical instrumentation manufacturer can be taken as a sign of things to come, then the use of microwaves in the analytical laboratory can be seen as barely reaching its teenage years. The best of its usefulness to sample preparation remains in the developments to come!

REFERENCES

1. MAP is a trademark of Her Majesty The Queen in Right of Canada as represented by the Minister of the Environment.

2. Paré, J.R.J.; Sigouin, M.; Lapointe, J. U.S. Patent No. 5,002,784; 1991.

3. Paré, J.R.J. U.S. Patent No. 5,338,557; 1994.

4. Paré, J.R.J. U.S. Patent No. 5,377,426; 1995.

5. Paré, J.R.J. U.S. Patent No. 5,458,897; 1995.

6. Paré, J.R.J. U.S. Patent No. 5,519,947; 1996.

7. Paré, J.R.J. U.S. Patent No. 5,675,909; 1997.

8. Paré, J.R.J. U.S. Patent No. 5,732,476; 1998.

9. Paré, J.R.J.; Bélanger, J.M.R.; Punt, M.M. U.S. Patent No. 6,061,926; 2000.

10. Paré, J.R.J.; Bélanger, J.M.R.; Stafford, S.S. Microwave-assisted process (MAP™): A new tool for the analytical laboratory. *Trends Anal. Chem.* **1994**, *13*, 176–184.

11. Paré, J.R.J.; Bélanger, J.M.R. Microwave-assisted process (MAP™): Principles and applications. In *Instrumental Methods in Food Analysis*, Paré, J.R.J.; Bélanger, J.M.R., Eds. Amsterdam: Elsevier Science; **1997**; pp. 395–420.

12. Paré, J.R.J.; Bélanger, J.M.R. Microwave-assisted processes (MAP™) at Environment Canada: Innovative sustainable "made-in-Canada" tools for the processing industry. *AMPERE Newsl.* **2006**, *51*, 1–4.

13. Bélanger, J.M.R.; Paré, J.R.J. Applications of microwave-assisted processes (MAP™) to environmental analysis. *Anal. Bioanal. Chem.* **2006**, *386*, 1049–1058.

14. Bélanger, J.M.R.; Paré, J.R.J. Microwave-assisted processes (MAP™) in food analysis. In *Handbook of Food Analysis Instruments*, Otles, S., Ed. Boca Raton, FL: Taylor & Francis Publishers; **2008**; pp. 57–83.

15. Bélanger, J.M.R.; Siu, M.; Paré, J.R.J. Microwave-assisted extraction in the analysis of persistent organic pollutants (POPs) in environmental samples. In *Methods & New Trends in Environmental, Biological and Food Analysis of Persistent Organic Pollutants (POPs)*, Olivella Costa, M.A.; Salvadó Martín, V., Eds. Kerala: Research Signpost; **2009**.

16. Bélanger, A.; Landry, B.; Dextraze, L.; Bélanger, J.M.R.; Paré, J.R.J. Extraction et détermination de composés volatils de l'ail (*Allium sativum*). *Riv. Ital. EPPOS* **1991**, *2*, 455–461.

17. Paré, J.R.J.; Bélanger, J.M.R. Microwave-assisted process (MAP™): Applications to the extraction of natural products. *Proceedings of the 28th Microwave Power Symposium*. Manassas, VA: International Microwave Power Institute; **1993**; pp. 62–67.

18. Akhtar, M.H.; Croteau, L.; Bélanger, J.M.R.; Paré, J.R.J. High performance liquid chromatography determination following microwave-assisted extraction of 3-nitro-4-hydroxyphenylarsonic acid from swine liver, kidney and muscle. *J. Liq. Chromatogr.* **1994**, *17*, 2971–2981.

19. Paré, J.R.J.; Bélanger, J.M.R.; Bélanger, A.; Ramarathnam, N.A. Rapid method for monitoring food volatiles. In *Food Science and Nutrition*, Charalambous, G., Ed. Amsterdam: Elsevier; **1992**; pp. 141–144.

20. Paré, J.R.J.; Matni, G.; Bélanger, J.M.R.; Li, K.; Rule, C.; Thibert, B.; Yaylayan, V.; Liu, Z.; Mathé, D.; Jacquault, P. Use of the microwave-assisted process in extraction of fat from meat, dairy, and egg products under atmospheric pressure conditions. *J. AOAC Int.* **1997**, *80*, 928–933.

21. Piñeiro-Avila, G.; Salvador, A.; de la Guardia, M. Microwave-assisted saponification of animal greases for cholesterol determination. *Anal. Chim. Acta* **1998**, *371*, 297–303.

22. Luque-Garcìa, J.L.; Velasco, J.; Dobarganes, M.C.; Luque de Castro, M.D. Fast quality monitoring of oil from prefried and fried foods by focused microwave-assisted Soxhlet extraction. *Food Chem.* **2002**, *76*, 241–248.

23. Priego-Lòpez, E.; Velasco, J.; Dobarganes, M.C.; Ramis-Ramos, G.; Luque de Castro, M.D. Focused microwave-assisted Soxhlet extraction: An expeditive approach for the isolation of lipids from sausage products. *Food Chem.* **2003**, *83*, 143–149.

24. Martìn-Calero, A.; Pino, V.; Ayala, J.H.; Gonzalez, V.; Afonso, A.M. Focused microwave-assisted extraction and HPLC with electrochemical detection to determine heterocyclic amines in meat extracts. *J. Liq. Chromatogr. Relat. Technol.* **2007**, *30*, 27–42.

25. Conte, E.D.; Shen, C.Y.; Miller, D.W. Microwave distillation-solid phase adsorbent trapping device for the determination of off-flavors, geosmin and methylisoborneol, in catfish tissue below their rejection levels. *Anal. Chem.* **1996**, *68*, 2713–2716.

26. Conte, E.D.; Shen, C.Y.; Perschbacher, P.W.; Miller, D.W. Determination of geosmin and methylisoborneol in catfish tissue (*Ictalurus punctatus*) by microwave-assisted distillation-solid phase adsorbent trapping. *J. Agric. Food Chem.* **1996**, *44*, 829–835.

27. Zhu, M.; Aviles, F.J.; Conte, E.D.; Miller D.W.; Perschbacher, P.W. Microwave mediated distillation with solid-phase micro extraction: Determination of Off-flavors, geosmin and methylisoborneol, in catfish tissue. *J. Chromatogr. A* **1999**, *833*, 223–230.

28. Lloyd, S.W.; Grimm, C.C. Analysis of 2-methylisoborneol and geosmin in catfish by microwave distillation-solid-phase micro extraction. *J. Agric. Food Chem.* **1999**, *47*, 164–169.

29. Grimm, C.C.; Lloyd, S.W.; Batista, R.; Zimba, P.V. Using microwave distillation-solid-phase micro extraction-gas chromatography-mass

spectrometry for analyzing fish tissue. *J. Chromatogr. Sci.* **2000**, *38*, 289–296.

30. Batista, A.; Vetter, W.; Luckas, B. Use of focused open vessel microwave-assisted extraction as prelude for the determination of the fatty acid profile of fish—A comparison with results obtained after liquid-liquid extraction according to Bligh and Dyer. *Eur. Food Res. Technol.* **2001**, *212*, 377–384.

31. Garcia-Ayuso, L.E.; Velasco, J.; Dobarganes, M.C.; Luque de Castro, M.D. Accelerated extraction of the fat content in cheese using a focused microwave-assisted Soxhlet device. *J. Agric. Food Chem.* **1999**, *47*, 2308–2315.

32. Garcia-Ayuso, L.E.; Velasco, J.; Dobarganes, M.C.; Luque de Castro, M.D. Double use of focused microwave irradiation for accelerated matrix hydrolysis and lipid extraction in milk samples. *Int. Dairy J.* **1999**, *9*, 667–674.

33. Garcia-Ayuso, L.E.; Luque de Castro, M.D. Employing focused microwaves to counteract conventional Soxhlet extraction drawbacks. *Trends Anal. Chem.* **2001**, *20*, 28–34.

34. Papadakis, E.N.; Polychroniadou, A. Application of a microwave-assisted extraction method for the extraction or organic acids from Greek cheeses and sheep milk yoghurt and subsequent analysis by ion-exclusion liquid chromatography. *Int. Dairy J.* **2005**, *15*, 165–172.

35. Bureau, S.; Razungles, A.; Baumes, R.; Boyonove, C. Glycosylated flavor precursor extraction by microwaves from grape juice and grapes. *J. Food Sci.* **1996**, *61*, 557–559.

36. Kratchanova, M.; Pavlova, E.; Panchev, I. the effect of microwave heating of fresh orange peels on the fruit tissue and quality of extracted pectin. *Carbohydr. Polym.* **2004**, *56*, 181–185.

37. Zhongdong, L.; Guohua, W.; Yunchang, G.; Kennedy, J.F. Image study of pectin extraction from orange skin assisted by microwave. *Carbohydr. Polym.* **2006**, *64*, 548–552.

38. Fishman, M.L.; Chau, H.K.; Hoagland, P.D.; Hotchkiss, A.T. Microwave-assisted extraction of lime pectin. *Food Hydrocolloids* **2006**, *20*, 1170–1177.

39. Ferhat, M.A.; Meklati, B.Y.; Smadja, J.; Chemat, F. An improved microwave clevenger apparatus for distillation of essential oils from orange peel. *J. Chromatogr. A* **2006**, *1112*, 121–126.

40. Wang, Z.M.; Ding, L.; Wang, L.; Feng, J.; Li, T.C.; Zhou, X.; Zhang, H.Q. Fast determination of essential oil from dried menthol mint and orange peel by solvent free microwave extraction using carbonyl iron powder as the microwave absorption medium. *Chin. J. Chem.* **2006**, *24*, 649–652.

41. Sun, Y.; Liao, X.; Wang, Z.; Hu, X.; Chen, F. Optimization of microwave-assisted extraction of anthocyanins in red raspberries and identification of anthocyanin of extracts using high-performance liquid chromatography-mass spectrometry. *Eur. Food Res. Technol.* **2006**, *225*, 511–523.

42. Ai, Z.; Quo, J.; Wang, Y.; Liu, Y.; Zhao, Q. Microwave-assisted extraction technique of apple polyphenols in apple pomace. *Nongye Gongcheng Xuebao/Trans. Chin. Soc. Agric. Eng.* **2006**, *22*, 188–191.

43. Wang, S.; Chen, F.; Wu, J.; Wang, Z.; Liao, X.; Hu, X. Optimization of pectin extraction assisted by microwave from apple pomace using response surface methodology. *J. Food Eng.* **2007**, *78*, 693–700.

44. Wang, L.; Weller, C.L. Recent advances in extraction of nutraceuticals from plants. *Trends Food Sci. Technol.* **2006**, *17*, 300–312.

45. Acquistucci, R.; Panfili, G.; Marconi, E. Application of the microwave hydrolysis to furosine determination in cereal and dairy foods. *J. Agric. Food Chem.* **1996**, *44*, 3855–3857.

46. Garcia-Ayuso, L.E.; Luque de Castro, M.D. A multivariate study of the performance of a microwave-assisted Soxhlet extractor for olive seeds. *Anal. Chim. Acta* **1999**, *382*, 309–316.

47. Garcia-Ayuso, L.E.; Velasco, J.; Dobarganes, M.C.; Luque de Castro, M.D. Determination of the oil content of seeds by focused microwave-assisted Soxhlet extraction. *Chromatographia* **2000**, *52*, 103–108.

48. Li, H.; Pordesimo, L.O.; Weiss, J.; Wilhelm, L.R. Microwave and ultrasound assisted extraction of soybean oil. *Trans. ASAE* **2004**, *47*, 1187–1194.

49. Duvernay, W.H.; Assad, J.M.; Sabliov, C.M.; Lima, M.; Xu, Z. Microwave extraction of antioxidant components from rice bran. *Pharm. Eng.* **2005**, *25*, 126–130.

50. Matthäus, B.; Brühl, L. Comparison of different methods for the determination of the oil content of oilseeds. *J. Am. Oil Chem. Soc.* **2001**, *78*, 95–102.

51. Choi, I.; Choi, S.J.; Chun, J.K.; Moon, T.W. Extraction yield of soluble protein and microstructure of soybean affected by microwave heating. *J. Food Process. Preserv.* **2006**, *30*, 407–419.

52. Ren, Y.L. Comparison of solvent extraction and microwave extraction for release of dimethyl sulfide from cereals and canola. *J. Agric. Food Chem.* **2001**, *49*, 1737–1739.

53. Kiss, G.A.C.; Forgács, E.; Cserháti, T.; Mota, T.; Morais, H.; Ramos, A. Optimisation of the microwave-assisted extraction of pigments from paprika (*Capsicum annuum* L.) powders. *J. Chromatogr. A* **2000**, *889*, 41–49.

54. Pan, X.; Liu, H.; Jia, G.; Shu, Y.Y. Microwave-assisted extraction of glycyrrhizic acid from licorice root. *Biochem. Eng. J.* **2000**, *5*, 173–177.

55. Kwon, J.-H.; Bélanger, J.M.R.; Paré, J.R.J.; Yaylayan, V.A. Application of the microwave-assisted process (MAP™) to the fast extraction of ginseng saponins. *Food Res. Int.* **2003**, *36*, 491–498.

56. Kwon, J.-H.; Lee, G.-D.; Bélanger, J.M.R.; Paré, J.R.J. *Int. J. Food Sci. Technol.* **2003**, *38*, 615–622.

57. Shu, Y.Y.; Ko, M.Y.; Chang, Y.S. Microwave-assisted extraction of ginsenosides from ginseng root. *Microchem. J.* **2003**, *74*, 131–139.

58. Yang, Y.; Chen, L.; Zhang, X.-X.; Guo, Z. Microwave assisted extraction of major active ingredients in *Panax quinquefolium* L. *J. Liq. Chromatogr. Relat. Technol.* **2004**, *27*, 3203–3211.

59. Zhao, Y.; Chen, B.; Yao, S. Separation of 20(S)-protopanaxdiol type ginsenosides and 20(S)-protopanaxtriol type ginsenosides with the help of macroporous resin adsorption and microwave assisted desorption. *Sep. Purif. Technol.* **2007**, *52*, 533–538.

60. Liu, Z.; Yan, G.; Bu, F.; Sun, J.; Hu, X.; Zhang, H.; Liu, Z. Analysis of chemical composition of *Acanthopanax senticosus* leaves applying high-pressure microwave-assisted extraction. *Chem. Analityczna* **2005**, *50*, 851–861.

61. Kerem, Z.; German-Shashoua, H.; Yarden, O. Microwave-assisted extraction of bioactive saponins from chickpea (*Cicer arietinum* L.). *J. Sci. Food Agric.* **2005**, *85*, 406–412.

62. Williams, O.J.; Raghavan, G.S.V.; Orsat, V.; Dai, J. Microwave-assisted extraction of capsaicinoids from capsicum fruit. *J. Food Biochem.* **2004**, *28*, 113–122.

63. Lucchesi, M.E.; Chemat, F.; Smadja, J. An original solvent free microwave extraction of essential oils from spices. *Flavour Fragr. J.* **2004**, *19*, 134–138.

64. Lucchesi, M.E.; Chemat, F.; Smadja, J. Solvent-free microwave extraction of essential oil from aromatic herbs: Comparison with conventional hydro-distillation, *J. Chromatogr. A* **2004**, *1043*, 323–327.

65. Lucchesi, M.E.; Chemat, F.; Smadja, J. Solvent-free microwave extraction: An innovative tool for rapid extraction of essential oil from aromatic herbs and spices. *J. Microw. Power Electromagn. Energy* **2004**, *39*, 135–139.

66. Lucchesi, M.E.; Smadja, J.; Bradshaw, S.; Louw, W.; Chemat, F. Solvent free microwave extraction of *Elletaria cardamomum* L.: A multivariate study of a new technique for the extraction of essential oil. *J. Food Process. Eng.* **2007**, *79*, 1079–1086.

67. Chemat, F.; Lucchesi, M.E.; Smadja, J.; Favretto, L.; Colnaghi, G.; Visinoni, F. Microwave accelerated steam distillation of essential oil from

lavender: A rapid, clean and environmentally friendly approach. *Anal. Chim. Acta* **2006**, *555*, 157–160.

68. Iriti, M.; Colnaghi, G.; Chemat, F.; Smadja, J.; Faoro, F.; Visinoni, F. Histo-cytochemistry and scanning electron microscopy of lavender glandular trichomes following conventional and microwave-assisted hydrodistillation of essential oils: A comparative study. *Flavour Fragr. J.* **2006**, *21*, 704–712.

69. Tigrine-Kordjani, N.; Meklati, B.Y.; Chemat, F. Microwave "dry" distillation as a useful tool for extraction of edible essential oils. *Int. J. Aromatherapy* **2006**, *16*, 141–147.

70. Lo Presti, M.; Ragusa, S.; Trozzi, A.; Dugo, P.; Visinoni, F.; Fazio, A.; Dugo, G.; Mondello, L. A comparison between different techniques for the isolation of rosemary essential oil. *J. Sep. Sci.* **2005**, *28*, 273–280.

71. Chemat, F.; Aït-Amar, H.; Lagha, A.; Esveld, D.C. Microwave-assisted extraction kinetics of terpenes from caraway seeds. *Chem. Eng. Process* **2005**, *44*, 1320–1326.

72. Wang, Z.; Ding, L.; Li, T.; Zhou, X.; Wang, L.; Zhang, H.; Liu, L.; Li, Y.; Liu, Z.; Wang, H.; Zeng, H.; He, H. Improved solvent-free microwave extraction of essential oil from dried *Cuminum cyminum* L. and *Zanghoxylum bungeanum* Maxim. *J. Chromatogr. A* **2006**, *1102*, 11–17.

73. Wang, Z.; Wang, L.; Li, T.; Zhou, X.; Ding, L.; Yu, Y.; Yu, A.; Zhang, H. Rapid analysis of the essential oils from dried *Illicium verum* Hook. f. and *Zingiber officinale* Rosc. by improved solvent-free microwave extraction with three types of microwave-absorption medium. *Anal. Bioanal. Chem.* **2006**, *386*, 1863–1868.

74. Raman, G.; Gaikar, V.G. Microwave-assisted extraction of piperine from *Piper nigrum*. *Ind. Eng. Chem. Res.* **2002**, *41*, 2521–2528.

75. Sharma, A.; Verma, S.C.; Saxena, N.; Chadda, N.; Singh, N.P.; Sinha, A.K. Microwave- and ultrasound-assisted extraction of vanillin and its quantification by high-performance liquid chromatography in *Vanilla planifolia*. *J. Sep. Sci.* **2006**, *29*, 613–619.

76. Longares-Patrón, A.; Cañizares-Macías, M.P. Focused microwaves-assisted extraction and simultaneous spectrophotometric determination of vanillin and *p*-hydroxybenzaldehyde from *Vanilla fragans*. *Talanta* **2006**, *69*, 882–887.

77. Dandekar, D.V.; Gaikar, V.G. Microwave assisted extraction of curcuminoids from *Curcuma longa*. *Sep. Sci. Technol.* **2002**, *37*, 2669–2690.

78. Chyau, C.-C.; Mau, J.-L. Release of volatile compounds from microwave heating of garlic juice with 2,4-decadienals. *Food Chem.* **1999**, *64*, 531–535.

79. Stashenko, E.E.; Jaramillo, B.E.; Martínez, J.R. Analysis of volatile secondary metabolites from Colombian *Xylopia aromatica* (Lamarck) by different extraction and headspace methods and gas chromatography. *J. Chromatogr. A* **2004**, *1025*, 105–113.

80. Dai, J.; Yaylayan, V.A.; Raghavan, G.S.V.; Paré, J.R.J. Extraction and colorimetric determination of azadirachtin-related limonoids in neem seed kernel. *J. Agric. Food Chem.* **1999**, *47*, 3738–3742.

81. Dai, J.; Yaylayan, V.A.; Raghavan, G.S.V.; Paré, J.R.J.; Liu, Z.; Bélanger, J.M.R. Influence of operating parameters on the use of the microwave-assisted process (MAP) for the extraction of azadirachtin-related limonoids from neem under atmospheric pressure conditions. *J. Agric. Food Chem.* **2001**, *49*, 4584–4588.

82. Simoneau, C.; Naudin, C.; Hannaert, P.; Anklan, E. Comparison of classical and alternative extraction methods for the quantitative extraction of fat from plain chocolate and the subsequent application to the detection of added foreign fats to plain chocolate formulations. *Food Res. Int.* **2000**, *33*, 733–741.

83. El Khori, S.; Paré, J.R.J.; Bélanger, J.M.R.; Pérez, E. The microwave-assisted process (MAP™): Extraction and determination of fat from cocoa powder and cocoa nibs. *J. Food Eng.* **2007**, *79*, 1110–1114.

84. Wang, Y.; Bonilla, M.; McNair, H.M.; Khaled, M. Solid phase microextraction associated with microwave assisted extraction of food products. *J. High Resolut. Chromatogr.* **1997**, *20*, 213–216.

85. Japón-Luján, R.; Luque-Rodríguez, J.M.; Luque de Castro, M.D. Multivariate optimisation of the microwave-assisted extraction of oleuropein and related biophenols from olive leaves. *Anal. Bioanal. Chem.* **2006**, *385*, 753–759.

86. Pan, X.; Niu, G.; Liu, H. Microwave-assisted extraction of tea polyphenols and tea caffeine from green tea leaves. *Chem. Eng. Process.* **2003**, *42*, 129–133.

87. Hong, N.; Yaylayan, V.A.; Raghavan, V.G.S.; Paré, J.R.J.; Bélanger, J.M.R. Microwave-assisted extraction of phenolic compounds from grape seed. *Nat. Prod. Lett.* **2001**, *15*, 197–204.

88. Liang, Y.; Ying, M.; Lu, Y.; Lu, X.; Zhang, Y. Microwave-assisted technology for extracting opuntia polysaccharide. *Nongye Gongcheng Xuebao/Trans. Chin. Soc. Agric. Eng.* **2006**, *22*, 159–162.

89. Liu, Q.; Yang, X.; Deng, H.; Yang, L.; Wu, Y. Technology for extraction of polysaccharide by using microwave technology from *Porphyra yezoensis*. *Nongye Gongcheng Xueboa/Trans. Chin. Soc. Agric. Eng.* **2005**, *21*, 153–156.

90. Joergensen, L.; Thestrup, H.N. Determination of amino acids in biomass and protein samples by microwave hydrolysis and ion-exchange chromatography. *J. Chromatogr. A* **1995**, *706*, 421–428.

91. Kovács, Á.; Ganzler, K.; Simon-Sarkadi, L. Microwave-assisted extraction of free amino acids from foods. *Z. Lebensm. Unters. Forsch.* **1998**, *207*, 26–30.

92. Yaylayan, V.A.; Matni, G.; Paré, J.R.J.; Bélanger, J.M.R. Microwave-assisted synthesis and extraction of selected Maillard reaction products. *J. Agric. Food Chem.* **1997**, *45*, 149–152.

93. Yaylayan, V.A.; Keyhani, A. Application of microwave-assisted process and Py-GC-MS to the analysis of Maillard reaction products. *ACS Symp. Ser.* **1998**, *705*, 223–232.

94. Stenberg, M.; Marko-Varga, G.; Öste, R. Racemization of amino acids during classical and microwave oven hydrolysis—Application to aspartame and a Maillard reaction system. *Food Chem.* **2001**, *74*, 217–224.

95. Jun, S.J.; Chun, J.K. Design of U-columns microwave-assisted extraction system and its application to pigment extraction from food. *Trans. IChemE* **1998**, *76*, Part C, 231–236.

96. Caballo-López, A.; Luque de Castro, M.D. Fast microwave-assisted free sugars washing and hydrolysis pre-treatment for the flow injection determination of starch in food. *Talanta* **2003**, *59*, 837–843.

97. Alfaro, M.J.; Bélanger, J.M.R.; Padilla, F.C.; Paré, J.R.J. Influence of solvent, matrix dielectric properties, and applied power on the liquid-phase microwave-assisted processes (MAP™) extraction of ginger (*Zingiber officinale*). *Food Res. Int.* **2003**, *36*, 499–504.

98. Zhao, L.; Zhao, G.; Chen, F.; Wang, Z.; Wu, J.; Hu, X. Different effects of microwave and ultrasound on the stability of (*all-E*)-astaxanthin. *J. Agric. Food Chem.* **2006**, *54*, 8346–8351.

99. Priego-Capote, F.; Ruiz-Jiménez, J.; García-Olmo, J.; Luque de Castro, M.D. Fast method for the determination of total fat and *trans* fatty-acids content in bakery products based on microwave-assisted Soxhlet extraction and medium infrared spectroscopy detection. *Anal. Chim. Acta* **2004**, *517*, 13–20.

100. Priego-Capote, F.; Luque de Castro, M.D. Focused microwave-assisted Soxhlet extraction: A convincing alternative for total fat isolation from bakery products. *Talanta* **2005**, *65*, 81–86.

101. Ruiz-Jiménez, J.; Priego-Capote, F.; Luque de Castro, M.D. FT-midIR determination of fatty acid profiles, including *trans* fatty acids, in bakery products after focused microwave-assisted Soxhlet extraction. *Anal. Bioanal. Chem.* **2006**, *385*, 1532–1537.

102. Priego-Capote, F.; Ruiz-Jiménez, J.; Luque de Castro, M.D. Identification and quantification of trans fatty acids in bakery products by gas chromatography-mass spectrometry after focused microwave Soxhlet extraction. *Food Chem.* **2007**, *100*, 859–867.

103. Akhtar, M.H.; Croteau, L.G.; Dani, C.; Elsooud, K.A. Development and validation of microwave-assisted extraction of fortified and incurred

chloramphenicol residues in egg albumen and yolk. *Spectroscopy* **1997**, *13*, 33–40.

104. Akhtar, M.H. Comparison of microwave assisted extraction with conventional (homogenization, vortexing) for the determination of incurred salinomycin in chicken eggs and tissues. *J. Environ. Sci. Health B* **2004**, *B39*, 835–844.

105. Pylypiw, H.M. Jr.; Arsenault, T.L.; Thetford, C.M.; Incorvia Mattina, M.J. Suitability of microwave-assisted extraction for multiresidue pesticide analysis of produce. *J. Agric. Food Chem.* **1997**, *45*, 3522–3528.

106. Prados-Rosales, R.C.; Luque Garcia, J.L.; Luque de Castro, M.D. Rapid analytical method for the determination of pesticide residues in sunflower seeds based on focused microwave-assisted Soxhlet extraction prior to gas chromatography-tandem mass spectrometry. *J. Chromatogr. A* **2003**, *993*, 121–129.

107. Bouaid, A.; Martín-Esteban, A.; Fernández, P.; Cámara, C. Microwave-assisted extraction method for the determination of atrazine and four organophosphorus pesticides in oranges by gas chromatography (GC). *Fresenius J. Anal. Chem.* **2000**, *367*, 291–294.

108. Falqui-Cao, C.; Wang, Z.; Urruty, L.; Pommier, J.-J.; Montury, M. Focused microwave assistance for extracting some pesticide residues from strawberries into water before their determination by SPME/HPLC/DAD. *J. Agric. Food Chem.* **2001**, *49*, 5092–5097.

109. Sanusi, A.; Guillet, V.; Montury, M. Advanced method using microwaves and solid-phase microextraction coupled with gas chromatography-mass spectrometry for the determination of pyrethroid residues in strawberries. *J. Chromatogr. A* **2004**, *1046*, 35–40.

110. Chen, Y.-I.; Su, Y.-S.; Jen, J.-F. Determination of dichlorvos by on-line microwave-assisted extraction coupled to headspace solid-phase microextraction and gas chromatography-electron-capture detection. *J. Chromatogr. A* **2002**, *976*, 349–355.

111. Padrón-Sanz, C.; Sosa-Ferrera, Z.; Santana-Rodríguez, J.J. An approach to the application of microwave-assisted micellar extraction and liquid chromatography with ultraviolet detection to the extraction and determination of organophosphorus pesticides in tomato samples. *J. AOAC Int.* **2005**, *88*, 1485–1490.

112. El-Saeid, M.H.; Kanu, I.; Anyanwu, E.C.; Saleh, M.A. Impacts of extraction methods in the rapid determination of atrazine residues in foods using supercritical fluid chromatography and enzyme-linked immunosorbent assay: Microwave solvent *vs.* supercritical fluid extractions. *Sci. World J.* (electronic resource) **2005**, *5*, 11–19.

113. Pallaroni, L.; von Holst, C. Comparison of alternative and conventional extraction techniques for the determination of zearalenone in corn. *Anal. Bioanal. Chem.* **2003**, *376*, 908–912.

114. Pallaroni, L.; von Holst, C.; Eskilsson, C.S.; Björklund, E. Microwave-assisted extraction of zearalenone from wheat and corn. *Anal. Bioanal. Chem.* **2002**, *374*, 161–166.

115. Diagne, R.G.; Foster, G.D.; Khan, S.U. Comparison of Soxhlet and microwave-assisted extractions for the determination of fenitrothion residues in beans. *J. Agric. Food Chem.* **2002**, *50*, 3204–3207.

116. Papadakis, E.N.; Vryzas, Z.; Papadopoulou-Mourkidou, E. Rapid method for the determination of 16 organochlorine pesticides in sesame seeds by microwave-assisted extraction and analysis of extracts by gas chromatography-mass spectrometry. *J. Chromatogr. A* **2006**, *1127*, 6–11.

117. Weichbrodt, M.; Vetter, W.; Luckas, B. Microwave-assisted extraction and accelerated solvent extraction with ethyl acetate-cyclohexane before determination of organochlorines in fish tissue by gas chromatography with electron-capture detection. *J. AOAC Int.* **2000**, *83*, 1334–1343.

118. Wittmann, G.; Huybrechts, T.; Van Langenhove, H.; Dewulf, J.; Nollet, H. Trace analysis of trichlorobenzenes in fish by microwave-assisted extraction and gas chromatography-electron-capture detection. *J. Chromatogr. A* **2003**, *993*, 71–78.

119. Bayen, S.; Lee, H.K.; Obbard, J.P. Determination of polybrominated diphenyl ethers in marine biological tissues using microwave-assisted extraction. *J. Chromatogr. A* **2004**, *1035*, 291–294.

120. Pena, T.; Pensado, L.; Casais, C.; Mejuto, C.; Phan-Tan-Luu, R.; Cela, R. Optimization of a microwave-assisted extraction method for the analysis of polycyclic aromatic hydrocarbons from fish samples. *J. Chromatogr. A* **2006**, *1121*, 163–169.

121. García-Falcón, M.S.; Simal-Gándara, J.; Carril-González-Barros, S.T. Analysis of benzo[a]pyrene in spiked fatty foods by second derivative synchronous spectrofluorimetry after microwave-assisted treatment of samples. *Food Addit. Contam.* **2000**, *17*, 957–964.

122. Gfrerer, M.; Lankmayr, E. Microwave-assisted saponification for the determination of 16 polycyclic aromatic hydrocarbons from pumpkin seed oils. *J. Sep. Sci.* **2003**, *26*, 1230–1236.

123. Bélanger, J.M.R.; Bélanger, A.; Paré, J.R.J. Procédé Assisté par Micro-Ondes (MAP™): Oléorésines de graines de céleri et de poivre noir. *Riv. Ital. EPPOS* **1994**, *5*, 114–125.

124. Bélanger, J.M.R.; Paré, J.R.J. Procédé Assisté par Micro-Ondes (MAP™): Application au paprika. *Riv. Ital. EPPOS* **1994**, *5*, 126–135.

125. Bélanger, J.M.R.; Aubry, C.; Chagonda, L.; Paré, J.R.J. Procédé Assisté par Micro-Ondes (MAP™): Extraction d'huiles essentielles au Zimbabwe. *Riv. Ital. EPPOS* **1994**, *5*, 136–148.

126. Chee, K.K.; Wong, M.K.; Lee, H.K. Determination of organochlorine pesticides in water by membranous solid-phase extraction, and in sediment by microwave-assisted solvent extraction with gas chromatography and electron-capture and mass spectrometric detection. *J. Chromatogr. A* **1996**, *736*, 211–218.

127. Barriada-Pereira, M.; González-Castro, M.J.; Muniategui-Lorenzo, S.; López-Mahía, P.; Prada-Rodríguez, D.; Fernández-Fernández, E. Comparison of pressurized liquid extraction and microwave assisted extraction for the determination of organochlorine pesticides in vegetables. *Talanta* **2007**, *71*, 1345–1351.

128. Silgoner, I.; Krska, R.; Lombas, E.; Gans, O.; Rosenberg, E.; Grasserbauer, M. Microwave assisted extraction of organochlorine pesticides from sediments and its application to contaminated sediment samples. *Fresenius J. Anal. Chem.* **1998**, *362*, 120–124.

129. Papadakis, E.N.; Vryzas, Z.; Papadopoulou-Mourkidou, E. Rapid method for the determination of 16 organochlorine pesticides in sesame seeds by microwave-assisted extraction and analysis of extracts by gas chromatography–mass spectrometry. *J. Chromatogr. A* **2006**, *1127*, 6–11.

130. Lopez-Avila, V.; Young, R.; Benedicto, J.; Ho, P.; Kim, R. Extraction of organic pollutants from solid samples using microwave energy. *Anal. Chem.* **1995**, *67*, 2096–2102.

131. McMillan, R.; Miner, L.C.; Hurst, L. Abbreviated microwave extraction of pesticides and PCBs in soil. *Spectrosc. Int. J.* **1996**, *13*, 41–50.

132. Thompson, S.; Budzinski, H. Determination of polychlorinated biphenyls and chlorinated pesticides in environmental biological samples using focused microwave-assisted extraction. *Int. J. Environ. Anal. Chem.* **2000**, *76*, 49–60.

133. Weichbrodt, M.; Vetter, W.; Scholz, E.; Luckas, B.; Reinhard, K. Determination of organochlorine levels in antarctic skua and penguin eggs by application of combined focused open-vessel microwave extraction, gel-permeation chromatography, adsorption chromatography, and GC/ECD. *Int. J. Environ. Anal. Chem.* **1999**, *73*, 309–328.

134. Vetter, W.; Weichbrodt, M.; Scholz, E.; Luckas, B.; Oelschläger, H. Levels of organochlorines (DDT, PCBs, toxaphene, chlordane, dieldrin, and HCHs) in blubber of South African fur seals (*Arctocephalus pusillus pusillus*) from Cape Cross/Namibia. *Mar. Pollut. Bull.* **1999**, *38*, 830–836.

135. Barriada-Pereira, M.; González-Castro, M.J.; Muniategui-Lorenzo, S.; López-Mahía, P.; Prada-Rodríguez, D.; Fernández-Fernández, E. Determination of 21 organochlorine pesticides in tree leaves using solid-phase extraction clean-up cartridges. *J. Chromatogr. A* **2004**, *1061*, 133–139.

136. EPA Method 3546. *Microwave Extraction, Test Methods for Evaluating Solid Waste*. Prerelease version. Washington, DC: EPA; September 1999.

137. Numata, M.; Yarita, T.; Aoyagi, Y.; Tsuda, Y.; Yamazaki, M.; Takatsu, A.; Keiichiro, I.; Chiba, K.; Okamaoto, K. Sediment certified reference materials for the determination of polychlorinated biphenyls and organochlorine pesticides from the National Metrology Institute of Japan (NMIJ). *Anal. Bioanal. Chem.* **2007**, *387*, 2313–2323.

138. Wurl, O.; Obbard, J.P. Sediment certified reference materials for the determination of polychlorinated biphenyls and organochlorine pesticides from the National Metrology Institute of Japan (NMIJ). *Atmos. Environ.* **2005**, *39*, 7207–7216.

139. Li, Q.Q.; Loganath, A.; Chong, Y.S.; Tan, J.; Obbard, J.P. Levels of persistent organic pollutant residues in human adipose and muscle tissues in Singapore. *J. Toxicol. Environ. Health A* **2006**, *69*, 1927–1937.

140. Gfrerer, M.; Lankmayr, E. DDT degradation during enhanced solid–liquid extractions: A consideration. *J. Chromatogr. A* **2005**, *1072*, 117–125.

141. Moreno, D.V.; Ferrera, Z.S.; Santana Rodríguez, J.J. Microwave assisted micellar extraction coupled with solid phase microextraction for the determination of organochlorine pesticides in soil samples. *Anal. Chim. Acta* **2006**, *571*, 51–57.

142. Numata, M.; Yarita, T.; Aoyagi, Y.; Takatsu, A. Evaluation of a microwave-assisted extraction technique for the determination of polychlorinated biphenyls and organochlorine pesticides in sediments. *Anal. Sci.* **2004**, *20*, 793–798.

143. Sanghi, R.; Kannamkumarath, S.S. Comparison of extraction methods by Soxhlet, sonicator, and microwave in the screening of pesticide residues from solid matrices. *J. Anal. Chem.* **2004**, *59*, 1145–1149.

144. Ohto, M.; Kodama, S.; Saito, Y.; Yamamoto, A. Determination of dieldrin in wool products by gas chromatography with microwave-assisted extraction. *J. Sep. Sci.* **2006**, *29*, 2759–2764.

145. Losier, J.; Girard, L. Microwave-assisted extraction method for organochlorinated pesticides and polychlorinated biphenyls from compost. *Int. J. Environ. Anal. Chem.* **2001**, *80*, 1–11.

146. Eiguren Fernández, A.; Sosa Ferrera, Z.; Santana Rodriguez, J.J. Microwave-assisted extraction of organochlorine compounds in marine sediments with organized molecular systems. *Chromatographia* **2001**, *53*, 375–379.

147. Carro, N.; Saavedra, Y.; García, I.; Llompart, M. Optimization of microwave assisted solvent extraction of polychlorinated biphenyls from marine sediments. *J. Microcolumn Sep.* **1999**, *11*, 544–549.

148. Chiu, C.; Poole, G.; Tardif, M.; Miles, W.; Turle, R. Determination of PCDDs/PCDFs and PCBs in sediments using microwave-assisted solvent extraction. *Organohalogen Compd.* **1997**, *31*, 175–180.

149. Dupont, G.; Delteil, C.; Camel, V.; Bermond, A. The determination of polychlorinated biphenyls in municipal sewage sludges using microwave-assisted extraction and gas chromatography-mass spectrometry. *Analyst* **1999**, *124*, 453–458.

150. Lopez-Avila, V.; Benedicto, J.; Charan, C.; Young, R. Determination of PCBs in soil sediments by microwave-assisted extraction and GC/ECD or ELISA. *Environ. Sci. Technol.* **1995**, *29*, 2709–2712.

151. Shu, Y.Y.; Wang, S.S.; Tardif, M.; Huang, Y. Analysis of polychlorinated biphenyls in aqueous samples by microwave-assisted headspace solid-phase microextraction. *J. Chromatogr. A* **2003**, *1008*, 1–12.

152. Chee, K.K.; Wong, M.K.; Lee, H.K. Microwave-assisted solvent elution technique for the extraction of organic pollutants in water. *Anal. Chim. Acta* **1996**, *330*, 217–222.

153. Ramil Criado, M.; Rodríguez Pereiro, I.; Cela Torrijos, R. Optimization of a microwave-assisted extraction method for the analysis of polychlorinated biphenyls in ash samples. *J. Chromatogr. A* **2003**, *985*, 137–145.

154. Ramil Criado, M.; Rodríguez Pereiro, I.; Cela Torrijos, R. Determination of polychlorinated biphenyls in ash using dimethylsulfoxide microwave assisted extraction followed by solid-phase microextraction. *Talanta* **2004**, *63*, 533–540.

155. Ruddy, B.A.; Aldstadt, J.H. Challenges in the application of conventional PCB quantitation methods to studies of "invasive" freshwater organisms. *J. Environ. Monit.* **2007**, *9*, 260–265.

156. Carro, N.; García, I.; Llompart, M. Closed-vessel assisted microwave extraction of polychlorinated biphenyls in marine mussels. *Analusis* **2000**, *28*, 720–724.

157. Sun, Y.; Takaoka, M.; Takeda, N.; Matsumoto, T.; Oshita, K. Application of microwave-assisted extraction to the analysis of PCBs and CBzs in fly ash from municipal solid waste incinerators. *J. Hazard. Mater.* **2006**, *A137*, 106–112.

158. Chiu, C.; Poole, G.; Shu, Y.; Thomas, R.; Turle, R. Microwave *vs.* Soxhlet for the extraction of dioxins and furans from solid samples. *Organohalogen Compd.* **1996**, *27*, 333–338.

159. Chee, K.K.; Wong, M.K.; Lee, H.K. Optimization of microwave-assisted solvent extraction of polycyclic aromatic hydrocarbons in marine sediments using a microwave extraction system with high-performance liquid chromatography fluorescence detection and gas chromatography-mass spectrometry. *J. Chromatogr. A* **1996**, *723*, 259–271.

160. Karthikeyan, S.; Balasubramanian, R.; See, S.W. Optimization and validation of a low temperature microwave-assisted extraction method for analysis of polycyclic aromatic hydrocarbons in airborne particulate matter. *Talanta* **2006**, *69*, 79–86.

161. Shu, Y.Y.; Lai, T.L.; Lin, H.S.; Yang, T.C.; Chang, C.P. Study of factors affecting the extraction efficiency of polycyclic aromatic hydrocarbons from soils using open-vessel focused microwave-assisted extraction. *Chemosphere* **2003**, *52*, 1667–1676.

162. Flotron, V.; Houessou, J.; Bosio, A.; Delteil, C.; Bermond, A.; Camel, V. Rapid determination of polycyclic aromatic hydrocarbons in sewage sludges using microwave-assisted solvent extraction comparison with other extraction methods. *J. Chromatogr. A* **2003**, *999*, 175–184.

163. Prevot, A.B.; Gulmini, M.; Zelano, V.; Pramauro, E. Microwave-assisted extraction of polycyclic aromatic hydrocarbons from marine sediments using nonionic surfactant solutions. *Anal. Chem.* **2001**, *73*, 3790–3795.

164. Budzinski, H.; Letellier, M.; Garrigues, P.; Le Menach, K. Optimisation of the microwave-assisted extraction in open cell of polycyclic aromatic hydrocarbons from soils and sediments study of moisture effect. *J. Chromatogr. A* **1999**, *837*, 187–200.

165. Tomaniová, M.; Hajšlová, J.; Pavelka, J. Jr.; Kocourek, V.; Holadová, K.; Klímova, I. Microwave-assisted solvent extraction—A new method for isolation of polynuclear aromatic hydrocarbons from plants. *J. Chromatogr. A* **1998**, *827*, 21–29.

166. Saim, N.; Dean, J.R.; Abdullah, M.P.; Zakaria, Z. Extraction of polycyclic aromatic hydrocarbons from contaminated soil using Soxhlet extraction, pressurised and atmospheric microwave-assisted extraction, supercritical fluid extraction and accelerated solvent extraction. *J. Chromatogr. A* **1997**, *791*, 361–366.

167. Chee, K.K.; Wong, M.K.; Lee, H.K. Microwave-assisted solvent extraction of air particulates for the determination of PAHs. *Environ. Monit. Assess.* **1997**, *44*, 391–403.

168. Dupeyron, S.; Dudermel, P.M.; Couturier, D. Extraction sous champ micro-ondes focalisées d'hydrocarbures aromatiques polycycliques issus de sols contaminés: Rôle de l'acetone et impact de la teneur en eau sur l'éfficacité des micro-ondes. *Analusis* **1997**, *25*, 286–292.

169. Barnabas, I.J.; Dean, J.R.; Fowlis, I.A.; Owen, S.P. Extraction of polycyclic aromatic hydrocarbons from highly contaminated soils using microwave energy. *Analyst* **1995**, *120*, 897–904.

170. Letellier, M.; Budzinski, H.; Charrier, L.; Capes, S.; Dorthe, A.M. Optimization by factorial design of focused microwave assisted extraction of polycyclic aromatic hydrocarbons from marine sediment. *Fresenius J. Anal. Chem.* **1999**, *364*, 228–237.

171. Letellier, M.; Budzinski, H. Influence of sediment grain size on the efficiency of focused microwave extraction of polycyclic aromatic hydrocarbons. *Analyst* **1999**, *124*, 5–14.

172. Villar, P.; Callejón, M.; Alonso, E.; Jiménez, J.C.; Guiraúm, A. Optimization and validation of a new method of analysis for polycyclic aromatic hydrocarbons in sewage sludge by liquid chromatography after microwave assisted extraction. *Anal. Chim. Acta* **2004**, *524*, 295–304.

173. González-Piñuela, C.; Alonso-Salces, R.M.; Andrés, A.; Ortiz, I.; Vigri, J.R. Validated analytical strategy for the determination of polycyclic aromatic compounds in marine sediments by liquid chromatography coupled with diode-array detection and mass spectrometry. *J. Chromatogr. A* **2006**, *1129*, 189–200.

174. Pena, M.T.; Pensado, L.; Casais, C.; Mejuto, M.C.; Cela, R. Sample preparation of sewage sludge and soil samples for the determination of polycyclic aromatic hydrocarbons based on one-pot microwave-assisted saponification and extraction. *Anal. Bioanal. Chem.* **2007**, *387*, 2559–2567.

175. Bruno, P.; Caselli, M.; de Gennaro, G.; Tutino, M. Determination of polycyclic aromatic hydrocarbons (PAHs) in particulate matter collected with low volume samplers. *Talanta* **2007**, *72*, 1357–1361.

176. Hernández-Borges, J.; Rodríguez-Delgado, M.A.; García-Montelongo, F.J. Optimization of the microwave-assisted saponification and extraction of organic pollutants from marine biota using experimental design and artificial neural networks. *Chromatographia* **2006**, *63*, 155–160.

177. Bangkedphol, S.; Sakultantimetha, A.; Keenan, H.E.; Songsasen, A. *J. Environ. Sci. Health A* **2006**, *41*, 1105–1116.

178. Buco, S.; Moragues, M.; Doumenq, P.; Noor, A.; Mille, G. Analysis of polycyclic aromatic hydrocarbons in contaminated soil by Curie point pyrolysis coupled to gas chromatography-mass spectrometry, an alternative to conventional methods. *J. Chromatogr. A* **2004**, *1026*, 223–229.

179. Vallejo-Pecharromán, B.; García-Ayuso, L.E.; Luque de Castro, M.D. Efficient extraction of PAHs in aqueous organised media assisted by focused microwaves. *Chromatographia* **2001**, *53*, 5–10.

180. García-Ayuso, L.E.; Luque-García, J.L.; Luque de Castro, M.D. Approach for independent-matrix removal of polycyclic aromatic hydrocarbons from solid samples based on microwave-assisted Soxhlet extraction with on-line fluorescence monitoring. *Anal. Chem.* **2000**, *72*, 3627–3634.

181. Letellier, M.; Budzinki, H.; Bellocq, J.; Connan, J. Focused microwave-assisted extraction of polycyclic aromatic hydrocarbons and alkanes from sediments and source rocks. *Org. Geochem.* **1999**, *30*, 1353–1365.

182. Noordkamp, E.R.; Grotenhuis, J.T.C.; Rulkens, W.H. Selection of an efficient extraction method for the determination of polycyclic aromatic hydrocarbons in contaminated soil and sediment. *Chemosphere* **1997**, *35*, 1907–1917.

183. Lopez-Avila, V.; Benedicto, J. Microwave-assisted extraction combined with gas chromatography and enzyme-linked immunosorbent assay. *TrAC* **1996**, *15*, 334–341.

184. Pino, V.; Ayala, J.H.; Afonso, A.M.; González, V. Micellar microwave-assisted extraction combined with solid-phase microextraction for the determination of polycyclic aromatic hydrocarbons in a certified marine sediment. *Anal. Chim. Acta* **2003**, *477*, 81–91.

185. Miège, C.; Dugay, J.; Hennion, M.C. Optimization, validation and comparison of various extraction techniques for the trace determination of polycyclic aromatic hydrocarbons in sewage sludges by liquid chromatography coupled to diode-array and fluorescence detection. *J. Chromatogr. A* **2003**, *995*, 87–97.

186. Letellier, M.; Budzinski, H.; Garrigues, P.; Wise, S. Focused microwave-assisted extraction of polycyclic aromatic hydrocarbons in open cell from reference materials (sediment, soil, air particulates). *Spectrosc. Int. J.* **1996**, *13*, 71–80.

187. Li, Q.Q.; Loganath, A.; Chong, Y.S.; Obbard, J.P. Determination and occurrence of polybrominated diphenyl ethers in maternal adipose tissue from inhabitants of Singapore. *J. Chromatogr. B* **2005**, *819*, 253–257.

188. Yusà, V.; Pastor, A.; de la Guardia, M. Microwave-assisted extraction of polybrominated diphenyl ethers and polychlorinated naphthalenes concentrated on semipermeable membrane devices. *Anal. Chim. Acta* **2006**, *565*, 103–111.

189. Regueiro, J.; Llompart, M.; García-Jares, C.; Cela, R. Determination of polybrominated diphenyl ethers in domestic dust by microwave-assisted solvent extraction and gas chromatography-tandem mass spectrometry. *J. Chromatogr. A* **2006**, *1137*, 1–7.

190. Shin, M.; Svoboda, M.L.; Falletta, P. Microwave-assisted extraction (MAE) for the determination of polybrominated diphenylethers (PBDEs) in sewage sludge. *Anal. Bioanal. Chem.* **2007**, *387*, 2923–2929.

191. Rodríguez Pereiro, I.; Wasik, A.; Łobinski, R. Speciation of organotin in sediments by multicapillary gas chromatography with atomic emission detection after microwave-assisted leaching and solvent extraction-derivatization. *Fresenius J. Anal. Chem.* **1999**, *363*, 460–465.

192. Tutschku, S.; Schantz, M.M.; Wise, S.A. Determination of methylmercury and butyltin compounds in marine biota and sediments using microwave-assisted acid extraction, solid-phase microextraction, and gas chromatography with microwave-induced plasma atomic emission spectrometric detection. *Anal. Chem.* **2002**, *74*, 4694–4701.

193. Rodríguez-González, P.; Ruiz Encinar, J.; García Alonso, J.I.; Sanz-Medel, A. Development of a triple spike methodology for validation of butyltin compounds speciation analysis by isotope dilution mass spectrometry part I. Synthesis of the spike, characterization and development of the mathematical equations. *J. Anal. At. Spectrom.* **2004**, *19*, 685–691.

194. Jimenez Moreno, M.; Pacheco-Arjona, J.; Rodríguez-González, P.; Preud'Homme, H.; Amouroux, D.; Donard, O.F.X. Simultaneous determination of monomethylmercury, monobutyltin, and tributyltin in environmental samples by multi-elemental-species-specific isotope dilution analysis using electron ionization GC-MS. *J. Mass Spectrom.* **2006**, *41*, 1491–1497.

195. Rodríguez, I.; Santamarina, M.; Bollaín, M.H.; Mejuto, M.C.; Cela, R. Gas chromatography-atomic emission determination of organotin compounds in marine biomaterials after microwave extraction. *Spectrosc. Int. J.* **1996**, *13*, 51–59.

196. Schmitt, V.O.; de Diego, A.; Cosnier, A.; Tseng, C.M.; Moreau, J.; Donard, O.F.X. Open focused microwave-assisted sample preparation procedures: Fundamentals and application to the speciation of tin and mercury in environmental samples. *Spectrosc. Int. J.* **1996**, *13*, 99–111.

197. Sporring, S.; Bøwadt, S.; Svensmark, B.; Björklund, E. Comprehensive comparison of classic Soxhlet extraction with Soxtec extraction, ultrasonication extraction, supercritical fluid extraction, microwave assisted extraction and accelerated solvent extraction for the determination of polychlorinated biphenyls in soil. *J. Chromatogr. A* **2005**, *1090*, 1–9.

198. Cicero, A.M.; Pietrantonio, E.; Romanelli, G.; Di Muccio, A. Comparison of Soxhlet, shaking, and microwave-assisted extraction techniques for determination of PCB congeners in a marine sediment. *Bull. Environ. Contam. Toxicol.* **2000**, *65*, 307–313.

199. Martens, D.; Gfrerer, M.; Wenzl, T.; Zhang, A.; Gawlik, B.M.; Schramm, K.W.; Lankmayr, E.; Kettrup, A. Comparison of different extraction techniques for the determination of polychlorinated organic compounds in sediment. *Anal. Bioanal. Chem.* **2002**, *372*, 562–568.

200. Piñeiro-Iglesias, M.; López-Mahía, P.; Vázquez-Blanco, E.; Muniategui-Lorenzo, S.; Prada-Rodríguez, D. Problems in the extraction of polycyclic aromatic hydrocarbons from diesel particulate matter. *Polycycl. Aromat. Compd.* **2002**, *22*, 129–146.

201. Piñeiro-Iglesias, M.; López-Mahía, P.; Vázquez-Blanco, E.; Muniategui-Lorenzo, S.; Prada-Rodríguez, D.; Fernández-Fernández, E. Microwave assisted extraction of polycyclic aromatic hydrocarbons from atmospheric particulate samples. *Fresenius J. Anal. Chem.* **2000**, *367*, 29–34.

202. Zuloaga, O.; Etxebarria, N.; Fernández, L.A.; Madariaga, J.M. Optimization and comparison of MAE, ASE and Soxhlet extraction for the

determination of HCH isomers in soil samples. *Fresenius J. Anal. Chem.* **2000**, *367*, 733–737.

203. Vázquez-Blanco, E.; López-Mahía, P.; Muniategui-Lorenzo, S.; Prada-Rodríguez, D.; Fernández-Fernández, E. Optimization of microwave-assisted extraction of hydrocarbons in marine sediments: Comparison with the Soxhlet extraction method. *Fresenius J. Anal. Chem.* **2000**, *366*, 283–288.

204. Shu, Y.Y.; Lao, R.C.; Chiu, C.H.; Turle, R. Analysis of polycyclic hydrocarbons in sediment reference materials by microwave-assisted extraction. *Chemosphere* **2000**, *41*, 1709–1716.

205. Pensado, L.; Casais, C.; Mejuto, C.; Cela, R. Optimization of the extraction of polycyclic aromatic hydrocarbons from wood samples by the use of microwave energy. *J. Chromatogr. A* **2000**, *869*, 505–513.

206. Bartolomé, L.; Cortazar, R.; Raposo, J.C.; Usobiaga, A.; Zuloaga, O.; Etxebarria, N.; Fernández, L.A. Simultaneous microwave-assisted extraction of polycyclic aromatic hydrocarbons, polychlorinated biphenyls, phthalate esters and nonylphenols in sediments. *J. Chromatogr. A* **2005**, *1068*, 229–236.

207. Yusà, V.; Pastor, A.; de la Guardia, M. Microwave-assisted extraction of OCPs, PCBs and PAHs concentrated by semi-permeable membrane devices (SPMDs). *Anal. Chim. Acta* **2005**, *540*, 355–366.

208. Esteve-Turrillas, F.A.; Pastor, A.; de la Guardia, M. Microwave-assisted extraction of pyrethroid insecticides from semi-permeable membrane devices (SPMDs) used to indoor air monitoring. *Anal. Chim. Acta* **2006**, *560*, 118–127.

209. Esteve-Turrillas, F.A.; Aman, C.S.; Pastor, A.A.; de la Guardia, M. Microwave-assisted extraction of pyrethroid insecticides from soil. *Anal. Chim. Acta* **2004**, *522*, 73–78.

210. Yang, J.S.; Lee, D.W.; Lim, H. Microwave-assisted extraction (MAE) of polychlorinated biphenyls and polychlorinated dibenzo-*p*-dioxins from fly ash and sea sediments: Effects of water and removal of interferences. *J. Liq. Chromatogr. Relat. Technol.* **2003**, *26*, 803–818.

211. Jayaraman, S.; Pruell, R.J.; McKinney, R. Extraction of organic contaminants from marine sediments and tissues using microwave energy. *Chemosphere* **2001**, *44*, 181–191.

212. Concha-Graña, E.; Barriada-Pereira, M.; Turnes-Carou, M.I.; Muniategui-Lorenzo, S.; López-Mahía, P.; Prada Rodríguez, D. Microwave extraction of organochlorine pesticides from soils. *Anal. Bioanal. Chem.* **2003**, *375*, 1225–1228.

213. Düring, R.-A.; Gäth, S. Microwave assisted methodology for the determination of organic pollutants in organic municipal wastes and soils: Extraction of polychlorinated biphenyls using heat transformer disks. *Fresenius J. Anal. Chem.* **2000**, *368*, 684–688.

214. Yusà, V.; Pardo, O.; Pastor, A.; de la Guardia, M. Optimization of a microwave-assisted extraction large-volume injection and gas chromatography-ion trap mass spectrometry procedure for the determination of polybrominated diphenyl ethers, polybrominated biphenyls and polychlorinated naphthalenes in sediments. *Anal. Chim. Acta* **2006**, *557*, 304–313.

215. Fountoulakis, M.; Drillia, P.; Pakou, C.; Kampioti, A.; Stamatelatou, K.; Lyberatos, G. Analysis of nonylphenol and nonylphenol ethoxylates in sewage sludge by high performance liquid chromatography following microwave-assisted extraction. *J. Chromatogr. A* **2005**, *1089*, 45–51.

216. Morales, S.; Canosa, P.; Rodríguez, I.; Rubí, E.; Cela, R. Microwave assisted extraction followed by gas chromatography with tandem mass spectrometry for the determination of triclosan and two related chlorophenols in sludge and sediments. *J. Chromatogr. A* **2005**, *1082*, 128–135.

217. Cortazar, E.; Bartolomé, L.; Delgado, A.; Etxebarria, N.; Fernández, L.A.; Usobiaga, A.; Zuloaga, O. Optimisation of microwave-assisted extraction for the determination of nonylphenols and phthalate esters in sediments samples and comparison with pressurized solvent extraction. *Anal. Chim. Acta* **2005**, *534*, 247–254.

218. Liu, R.; Zhou, J.L.; Wilding, A. Microwave-assisted extraction followed by gas chromatography-mass spectrometry for the determination of endocrine disrupting chemicals in river sediments. *J. Chromatogr. A* **2004**, *1038*, 19–26.

219. Numata, M.; Yarita, T.; Aoyagi, Y.; Takatsu, A. Microwave-assisted steam distillation for simple determination of polychlorinated biphenyls and organochlorine pesticides in sediments. *Anal. Chem.* **2003**, *75*, 1450–1457.

220. Gfrerer, M.; Lankmayr, E. Screening, optimization and validation of microwave-assisted extraction for the determination of persistent organochlorine pesticides. *Anal. Chim. Acta* **2005**, *533*, 203–211.

221. Figuren Fernández, A.; Sosa Ferrera, Z.; Santana Rodríguez, J.J. Application of microwave-assisted extraction using micellar media to the determination of polychlorinated biphenyls in marine sediments. *Anal. Chim. Acta* **2001**, *433*, 237–244.

222. Gulmini, M.; Prevot, A.B.; Pramauro, E.; Zelano, V. Surfactant micellar solutions as alternative solvents for microwave-assisted extraction of polycyclic aromatic hydrocarbons from a spiked river sediment. *Polycycl. Aromat. Compd.* **2002**, *22*, 55–70.

223. Pino, V.; Ayala, J.H.; Afonso, A.M.; González, V. Determination of polycyclic aromatic hydrocarbons in marine sediments by high-performance liquid chromatography after microwave-assisted extraction with micellar media. *J. Chromatogr. A* **2000**, *869*, 515–522.

224. Pino, V.; Ayala, J.H.; Afonso, A.M.; González, V. Cloud-point preconcentration and HPLC determination of polycyclic hydrocarbons in sediments. *Fresenius J. Anal. Chem.* **2001**, *371*, 526–531.

225. Padrón-Sanz, C.; Halko, R.; Sosa-Ferrera, Z.; Santana-Rodríguez, J.J. Combination of microwave assisted micellar extraction and liquid chromatography for the determination of organophosphorous pesticides in soil samples. *J. Chromatogr. A* **2005**, *1078*, 13–21.

226. Mahugo Santana, C.; Sosa Ferrera, Z.; Santana Rodríguez, J.J. An environmentally friendly method for the extraction and determination of priority phenols in soils using microwave-assisted micellar extraction. *Anal. Bioanal. Chem.* **2005**, *382*, 125–133.

227. Barriada-Pereira, M.; González-Castro, M.J.; Muniategui-Lorenzo, S.; López-Mahía, P.; Prada-Rodríguez, D.; Fernández-Fernández, E. Determination of organochlorine pesticides in horticultural samples by microwave assisted extraction followed by GC-ECD. *Int. J. Environ. Anal. Chem.* **2005**, *85*, 325–333.

228. Barriada-Pereira, M.; González-Castro, M.J.; Muniategui-Lorenzo, S.; López-Mahía, P.; Prada-Rodríguez, D.; Fernández-Fernández, E. Organochlorine pesticides accumulation and degradation products in vegetation samples of a contaminated area in Galicia (NW Spain). *Chemosphere* **2005**, *58*, 1571–1578.

229. Barriada-Pereira, M.; Concha-Graña, E.; González-Castro, M.J.; Muniategui-Lorenzo, S.; López-Mahía, P.; Prada-Rodríguez, D.; Fernández-Fernández, E. Microwave-assisted extraction *versus* Soxhlet extraction in the analysis of 21 organochlorine pesticides in plants. *J. Chromatogr. A* **2003**, *1008*, 115–122.

230. Sanghi, R.; Kannamkumarath, S.S. Comparison of extraction methods by Soxhlet, sonicator, and microwave in the screening of pesticides residues from solid matrices. *J. Anal. Chem.* **2004**, *59*, 1032–1036.

231. Ho, W.H.; Hsieh, S.J. Solid microextraction associated with microwave assisted extraction of organochlorine pesticides in medicinal plants. *Anal. Chim. Acta* **2001**, *428*, 111–120.

232. Herbert, P.; Morais, S.; Paíga, P.; Alves, A.; Santos, L. Analysis of PCBs in soils and sediments by microwave-assisted extraction, headspace-SPME and high resolution gas chromatography with ion-trap tandem mass spectrometry. *Int. J. Environ. Anal. Chem.* **2006**, *86*, 391–400.

233. Herbert, P.; Morais, S.; Paíga, P.; Alves, A.; Santos, L. Development and validation of a novel method for the analysis of chlorinated pesticides in soils using microwave-assisted extraction-headspace solid phase microextraction and gas chromatography-tandem mass spectrometry. *Anal. Bioanal. Chem.* **2006**, *384*, 810–816.

234. Li, H.P.; Li, G.C.; Jen, J.F. Determination of organochlorine pesticides in water using microwave assisted headspace solid-phase microextraction and gas chromatography. *J. Chromatogr. A* **2003**, *1012*, 129–137.

235. Wei, M.C.; Jen, J.F. Determination of chlorophenols in soil samples by microwave-assisted extraction coupled to headspace solid-phase microextraction and gas chromatography-electron-capture detection. *J. Chromatogr. A* **2003**, *1012*, 111–118.

236. Ericsson, M.; Colmsjö, A. Dynamic microwave-assisted extraction. *J. Chromatogr. A* **2000**, *877*, 141–151.

237. Ericsson, M.; Colmsjö, A. Dynamic microwave-assisted extraction coupled on-line with solid-phase extraction: Determination of polycyclic aromatic hydrocarbons in sediment and soil. *J. Chromatogr. A* **2002**, *964*, 11–20.

238. Ericsson, M.; Colmsjö, A. Dynamic microwave-assisted extraction coupled on-line with solid-phase extraction and large-volume injection gas chromatography: Determination of organophoshate esters in air samples. *Anal. Chem.* **2003**, *75*, 1713–1719.

239. Piñeiro-Iglesias, M.; López-Mahía, P.; Muniategui-Lorenzo, S.; Prada-Rodríguez, D.; Querol, X.; Alastuey, A.A. New method for the simultaneous determination of PAH and metals in samples of atmospheric particulate matter. *Atmos. Environ.* **2003**, *37*, 4171–4175.

240. Basheer, C.; Obbard, J.P.; Lee, H.K. Analysis of persistent organic pollutants in marine sediments using a novel microwave assisted solvent extraction and liquid-phase microextraction technique. *J. Chromtogr. A* **2005**, *1068*, 221–228.

241. Mutavdžić, D.; Horvat, A.J.M.; Babić, S.; Kaštelan-Macan, M. SPE-microwave-assisted extraction coupled system for the extraction of pesticides from water samples. *J. Sep. Sci.* **2005**, *28*, 1485–1492.

242. Düring, R.A.; Zhang, X.; Hummel, H.E.; Czynski, J.; Gäth, S. Microwave-assisted steam distillation with simultaneous liquid/liquid extraction of pentachlorophenol form organic wastes and soils. *Anal. Bioanal. Chem.* **2003**, *375*, 584–588.

243. Vryzas, Z.; Papadopoulou-Mourkidou, E. Determination of triazine and chloroacetanilide herbicides in soils by microwave-assisted extraction (MAE) coupled to gas chromatographic analysis with either GC-NPD or GC-MS. *J. Agric. Food Chem.* **2002**, *50*, 5026–5033.

244. de Andréa, M.M.; Papini, S.; Nakagawa, L.E. Optimizing microwave-assisted solvent extraction (MASE) of pesticides from soil. *J. Environ. Sci. Health* **2001**, *B36*(1), 87–93.

245. Shen, G.; Lee, H.K. Determination of triazines in soil by microwave-assisted extraction followed by solid-phase microextraction and gas chromatography-mass spectrometry. *J. Chromatogr. A* **2003**, *985*, 167–174.

246. Molins, C.; Hogendoorn, E.A.; Dijkman, E.; Heusinkveld, H.A.G.; Baumann, R.A. Determination of linuron and related compounds in soil by microwave-assisted solvent extraction and reversed-phase liquid chromatography with UV detection. *J. Chromatogr. A* **2000**, *869*, 487–496.

247. Parera, J.; Santos, F.J.; Galceran, M.T. Microwave-assisted extraction *versus* Soxhlet extraction for the analysis of short-chain chlorinated alkanes in sediments. *J. Chromatogr. A* **2004**, *1046*, 19–26.

248. Sun, L.; Lee, H.K. Optimization of microwave-assisted extraction and supercritical fluid extraction of carbamates pesticides in soil by experimental design methodology. *J. Chromatogr. A* **2003**, *1014*, 165–177.

249. Thompson, S.; Budzinski, H.; LeMenach, K.; Letellier, M.; Garrigues, P. Multi-residue analysis of polycyclic aromatic hydrocarbons, polychlorobiphenyls, and organchlorine pesticides in marine sediments. *Anal. Bioanal. Chem.* **2002**, *372*, 196–204.

250. Budsinski, H.; Letellier, M.; Thompson, S.; LeMenach, K.; Garrigues, P. Combined protocol for the analysis of polycyclic aromatic hydrocarbons (PAHs) and polychlorobiphenyls (PCBs) from sediments using focused microwave assisted (FMW) extraction at atmospheric pressure. *Fresenius J. Anal. Chem.* **2000**, *367*, 165–171.

251. Shu, Y.Y.; Tey, S.Y.; Wu, D.K.S. Analysis of polycyclic aromatic hydrocarbons in airborne particles using open-vessel focused microwave-assisted extraction. *Anal. Chim. Acta* **2003**, *495*, 99–108.

252. Shu, Y.Y.; Lai, T.L. Effect of moisture on the extraction efficiency of polycyclic aromatic hydrocarbons from soils under atmospheric pressure by focused microwave-assisted extraction. *J. Chromatogr. A* **2001**, *927*, 131–141.

253. Bruno, P.; Caselli, M.; de Gennaro, G.; de Rienzo, M.; Traini, A. Analysis of polycyclic aromatic hydrocarbons in the atmospheric particulate: Focused microwaves for a faster extraction method. *J. Environ. Monit.* **2000**, *2*, 223–227.

254. Zuloaga, O.; Etxebarria, N.; Fernández, L.A.; Madariaga, J.M. Optimization of focused microwave-assisted extraction of DDT and derivatives from soil samples. *J. High Resolut. Chromatogr.* **2000**, *23*, 681–687.

255. Ganeshjeevan, R.; Chandrasekar, R.; Sugumar, P.; Kadigachalam, P.; Radhakrishnan, G. Focused microwave aqueous extraction of chlorophenols from solid matrices and their analysis by chromatographic techniques. *J. Chromatogr. A* **2005**, *1069*, 275–280.

256. Gatidou, G.; Zhou, J.L.; Thomaidis, N.S. Microwave-assisted extraction of Irgarol 1051 and its main degradation product from marine sediments using water as the extractant followed by gas chromatography-mass spectrometry determination. *J. Chromatogr. A* **2004**, *1046*, 41–48.

257. Serrano, A.; Gallego, M. Continuous microwave-assisted extraction coupled on-line with liquid-liquid extraction: Determination of aliphatic hydrocarbons in soil and sediments. *J. Chromatogr. A* **2006**, *1104*, 323–330.

258. Crespín, M.A.; Gallego, M.; Valcárcel, M. Continuous microwave-assisted extraction, solvent changeover and preconcentration of monophenols in agricultural soils. *J. Chromatogr. A* **2000**, *897*, 279–293.

259. Thuéry, J. *Microwaves: Industrial, Scientific, and Medical Applications*, Grant, E.H., Ed. London: Artech House Publishers; **1991**; p. 692.

260. Roussy G.; Pierce, J.A. *Foundations* and *Industrial Applications* of Microwaves *and Radio Frequency Fields: Physical and Chemical Processes*. West Sussex: Wiley; **1995**; p. 492.

261. Paré, J.R.J.; Rochas, J.-F.; Jacomino, J.-M.; Sánchez L., F.N.; Bélanger, J.M.R. The dielectmeter: A new powerful tool in microwave-assisted chemistry. *Proceedings of the International Symposium on Microwave Science and its Application to Related Fields*. Takamatsu, Japan; **2004**; pp. 25–28.

262. Paré, J.R.J.; Bélanger, J.M.R.; Li, K.; Stafford, S.S. Microwave-assisted process (MAP™): Application to the headspace analysis of VOCs in water. *J. Microcolumn Sep.* **1995**, *7*, 37–40.

263. Paré, J.R.J.; Bélanger, J.M.R.; Thornton, D.E.; Li, K.; Llompart, M.; Fingas, M.; Blenkinsopp, S.A. Gas-phase extraction using the microwave-assisted process (MAP™) for the analysis of dissolved volatile organic compounds (VOCs) in water. *Spectrosc. Int. J.* **1996**, *13*, 23–32.

264. Paré, J.R.J.; Li, K.; Bélanger, J.M.R.; Llompart, M. Gas-phase extraction method using the microwave-assisted process (MAP™) for the determination of aromatic contaminants in soil. *Spectrosc. Int. J.* **1996**, *13*, 89–98.

265. Bélanger, J.M.R.; Paré, J.R.J.; Turpin, R.D.; Schaefer, J.; Chuang, C.W. Evaluation of microwave-assisted process technology for Hapsite's headspace analysis of volatile organic compounds (VOC). *J. Hazard. Mater.* **2007**, *145*, 336–338.

266. Bélanger, J.M.R.; Paré, J.R.J.; Sánchez, L.F.N.; Komori, K.; Rochas, J.-F. New approach in gas-phase microwave-assisted extraction. *Anal. Bional. Chem.* **2008**, *391*, 929–932.

267. Debye, P. *Polar Molecules*. New York: The Chemical Catalog Co.; **1929**.

268. Debye, P. *Polar Molecules*. New York: Dover Publications; **1945**.

269. von Hippel, A.R., Ed. *Dielectric Materials and Applications*. New York: Technology Press MIT/Wiley; **1954**; p. 438.

270. U.S. Federal Register. January 3, **2008**, *73*(2), 486–489. Available from U.S. Federal Register Online via GPO Access [wais.access.gpo.gov] [DOCID: fr03ja08-36].

271. Llompart, M.P.; Lorenzo, R.A.; Cela, R.; Paré, J.R.J.; Bélanger, J.M.R.; Li, K. Phenol and methylphenol isomers determination in soils by

in-situ microwave-assisted extraction and derivatisation. *J. Chromatogr. A* **1997**, *757*, 153–164.

272. Buckley, F.; Maryott, A.A. *Tables of Dielectric Dispersion Data for Pure Liquids and Dilute Solutions*, Circular 589. Washington, DC: US National Bureau of Standards; 1958.

273. James, S.C. *Analytical Chemistry of Foods*. London: Blackie Academic and Professional; 1995.

274. Varnam, A.H.; Sutherland, J.P. *Meat and Meat Products, Technology, Chemistry and Microbiology*, Vol. 3—Food Products Series. London: Chapman and Hall; 1995.

275. American Dry Milk Institute, Inc. *Standard for Grades of Milk Including Methods of Analysis*, Bull 916. Chicago, IL: American Dry Milk Institute; 1971; pp. 24–25.

276. Richardson, G.H. *Standard Methods for the Examination of Dairy Products*, 15th Ed. New York: American Public Health Association; **1985**; pp. 358–362.

277. *Official Methods of Analysis*, Methods *17.012, 17.013, 920.15, 933.05, and 985.15, 16th Ed.* Arlington, VA: Association of Official Analytical Chemists; 1995.

Chapter 13

Chemical Derivatizations in Analytical Extractions

Jack Rosenfeld

13.1. INTRODUCTION

Developments in the science and technology of instrumentation have pushed detection limits to the zeptomole range and lower with superior detectors being responsible for the sensitivity. In high-performance liquid chromatography (HPLC) or capillary zone electrophoresis (CZE), fluorescence (FLU), electrochemical detectors (EChDs), ultraviolet (UV) detectors, and mass spectrometry (MS) are the most widely used group of detectors. In gas chromatography (GC), the flame ionization detector (FID) and its variant the nitrogen phosphorous detector (NPD) are stable, well-understood, and widely used detectors. The electron capture detector (ECD) is less widely used but is worthy of note because it first enabled the detection of halogenated hydrocarbons in the environment, which helped link their presence to the decline in bird populations, and was also important in understanding ozone depletion.[1] The FLU, EChD, and UV for liquid chromatography (LC), and the FID, NPD, and ECD for GC are small, relatively inexpensive instruments yet, when combined with chromatography, they are sufficient for a wide range of applications.

MS represents the highest-end instrumentation, combining high sensitivity with high specificity and selectivity as well as powerful software for data handling. In certain applications, the high resolution (HR) of some mass spectrometer techniques obviates the need for either chromatographic separation or sample preparation. The cost, however, is commensurate with high technology and its capabilities.

Despite the impressive scientific and technological achievements in laboratory instrumentation, there is no single universal technique applicable to all problems. The complexity of problems presented to the analytical chemist can present challenges to the best instrumentation. For instance, HR-MS cannot readily differentiate between isomeric *cis* and *trans* unsaturated fatty acids, which is important to the study of nutrition. Resolution of these isomers requires chromatographic separation[2] before submitting the analyte to the mass spectrometer. Similarly, the three glucuronides of estriol have identical tandem mass spectrometry (MS/MS) spectra, and differentiation of these three metabolites requires separation prior to analysis by MS/MS.[3] In addition, while normal estriol (16α-estradiol) is a normal part of the common pathway for metabolism, its isomer, 2-hydroxyestradiol, is a potential biomarker of oxidative stress leading to potential neurodegenerative disorders.[4]

Field analyses and, in particular, point-of-care analysis in medical practice raise additional issues. MS in its more sophisticated forms (MS/MS, HR-MS, etc.) is rarely an option. In such cases, sensitivity, selectivity, and specificity must come from chromatography and spectrophotometric or EChDs. The lab-on-a-chip technology while allowing an inexpensive highly mobile analytical system puts a very high premium on sensitivity due to limited sample size and correspondingly minimal mass of analyte.

In brief, despite the sophistication of modern laboratory instrumentation, innovations and development of other aspects in analysis remain important and active areas of investigation.

Arguably, the most controversial discipline in biomedical or environmental analysis is analytical derivatizations (ADs). The additional cost in time, manipulation, and reagent argue against the use of this technique. A closer examination of the literature, however, presents a valid case on the utility of these techniques. Derivatization of many classes of analyte for analysis by GC is essential to provide sufficient volatility. In HPLC techniques, derivatization adds chromophores and electrophores to analytes that are otherwise invisible to detectors. Additionally, derivatization stabilizes labile molecules.[5] It also proves effective for the

Handbook of Sample Preparation, Edited by Janusz Pawliszyn and Heather L. Lord

sample preparation of polar analytes or volatile analytes by converting them to lipophilic derivatives.[6–8] Where MS is involved, depending on the analyte derivatization, sensitivity can increase by 5- to 2000-fold.[9–18] Since derivatization is a form of functional group analysis, there is also an increase in specificity of analysis; the detector will "see" only those analytes with the tagged functional groups. There is sufficient published data to argue that there is reason to pursue AD both for the immediate positive impact on analysis and also for the development of techniques that overcome the problems inherent in using an additional step in sample preparation.

This report will address the problem by functional group analysis. It will deal with selected classes of analytes within each group. It will also discuss reagents targeted for various instrumental techniques. Exemplars of methods will illustrate both the nature of and solutions to the problems in derivatization and analysis.

13.2. ORGANIC ACIDS

13.2.1. Carboxylic Acids Overview

Reactions of carboxylic acids fall into five broad classes that produce either esters or hydrazones. The first derives from the pK of these weak acids, which ranges from 3.8 to 4.5. Thus, under relatively mild conditions, the acid converts to the carboxylate anion, which reacts with electrophilic alkylating agents. The second is *trans*-esterification with anhydrides. The third is reaction with alcohols and amines via a formal dehydration to form esters and amides. The fourth is reaction with diazo compounds. The fifth is reaction with silylating reagents to form silylesters. Although these reactions are simple, many analytes with the carboxylic acid functionality also contain substructures that reaction conditions, as well as the overall analytical procedure, must preserve.

13.2.2. Lipid-Bound and Free Fatty Acids: Carboxylic Acids in Biology

In biology, fatty acids exist either in the free form or as mono-, di-, and triacylglycerol esters, phospholipid esters, or as amides of sphingomyelin. Esterification is one of the most frequently used derivatizations in the studies of the free and lipid-bound fatty acids. A number of reactions find utility for different aspects of the problem. Reaction of fatty acids with alcohols under acidic conditions or with diazo compounds in diethyl ether will derivatize free fatty acids. Acidified alcohol will also catalyze transesterification of lipid-bound fatty acids and thus measure total—esterified and free—fatty acids. In contrast, sodium alkoxide in the corresponding anhydrous alcohol will react only with esterified fatty acids. Although the preceding summary was presented in general terms, formation of fatty acid methyl esters (FAMEs) is the predominant method used in biomedical research. In subsequent discussion, reactions producing FAME will allow for illustration of the issues in the analysis of fatty acids via derivatization.

Investigators have long recognized the complexities of the problem. This analyte class includes subclasses of saturated, mono- and polyunsaturated acids, *cis*/*trans* mono- and polyunsaturated, and cyclopropyl and hydroxyl fatty acids. In addition, fatty acids occur as carboxylic acids or esterified to biological alcohols or amines. The chemistry of the alkyl chain of these subclasses and not just that of the functional group determines the preferred analytical method.

More than a decade ago, Kramer and coworkers[19] reported a detailed study on methylation of polyunsaturated fatty acids (PUFAs). Under acidic conditions, they indentified *cis–trans* isomerizations, as well as methoxylation of the conjugated double bonds by Michael addition. From their detailed studies, they found that derivatization with sodium methoxide, followed by a brief reaction in acidified methanol, produced optimal yield of the FAME but with minimal isomerization and artifact creation.

Basconcillo and McCarry[20] addressed these problems and compared three methods for the determination of fatty acids from the soil bacterium *Sinorhizobium meliloti*, which is responsible for nitrogen fixation. The bacterium produces both saturated and unsaturated fatty acids, hydroxyl fatty acids, and cyclopropyl acids. These workers developed a miniaturized technique for the determination of the fatty acids and used it to evaluate three methods of preparing FAMEs. In the first method, the esterified fatty acids were first hydrolyzed with alkali and then methylated with 6 M HCl in methanol. The second method was acid-catalyzed transmethylation with 3% sulfuric acids in methanol. The third method was a base-catalyzed transesterification with sodium methoxide in methanol. All three procedures were followed by a reaction with *N*-methyl-*N*-(trimethylsilyl)-trifluoroacetamide (MSTFA) to form the trimethylsilyl esters of free fatty acids and to determine the degree of hydrolysis. With isopropyl tetradecanoate esters as models, the absolute recovery of the fatty acids was equivalent in the first two methods (88%) and 77% in the third method. There was a relatively small amount of hydrolysis (0.1–7%) in all three methods, but the difference was statistically significant. Methylation was lowest in the transesterification reaction. The type of lipid strongly affected the yield. Recoveries from cardiolipin were equivalent to those found with standards. Not surprisingly, since the fatty acids in sphingomyelin are bound via an amide, the recoveries were lowest in the analysis of this. Derivatization of 3 hydroxy was most effective with hydrolysis followed by methylation, while transesterification with sodium methoxide/methanol gave superior recoveries of the cyclopropyl acids. These results were a caution that there may be no universal method for determining absolute amounts of fatty acids. There are losses due to volatility and due to possible side reactions such as Claisen condensation to produce higher-molecular-weight compounds. Such condensations may be responsible for the observed decreased yield with increased temperature. With the proper standards that correct for losses, however, it is possible to determine relative amounts and therefore the profile of the fatty acids.

Ultrasonic mixing enables the base-catalyzed transmethylation in alkaline methanol coupled with simultaneous extraction into hexane.[21] With this technique, it is possible to determine bound fatty acids directly from the cell lysate, sodium dodecylsulphate (SDS)-resistant cells, or plant oils (rapeseed or cottonseed). Sample preparation is minimal. Addition of such lipid-rich samples to the hexane dissolves the materials. Contacting the hexane solution with 0.5 N methanolic NaOH and ultrasonic agitation completes the transesterification. Because this method of mixing increases rates of mass transfer between the two phases, the reaction is complete in minutes. By using perdeuterated CD_3OH, it is possible to label the FAMEs and to allow for accurate internal standardization as well as comparison of fatty acid profiles between two different samples of lipids.

Ishida and coworkers investigated the injector port thermally assisted hydrolysis and methylation (TAHM) of fatty acids in lipids (triglycerides, phospholipids, free fatty acids, and cholesteryl esters). The methylating agents were tetramethylammonium hydroxide (TMAH) and trimethylsulfonium hydroxide (TMSH).[22] These authors established a relationship between yield and methylating agent, lipid structure, and degree of unsaturation. They found that regardless of the lipid, TMAH affected quantitative recovery of saturated fatty acids as the FAME. This reagent, however, caused isomerization and degradation of the labile PUFA. TMSH also produced high yield of FAME but retained the unsaturated structures. The reagent, however, was unreactive toward cholesteryl esters, which the authors attributed to steric hindrance.

Injector port reactions with TMAH were also effective in the analysis of fatty acids in plant material (hemicellulose from larchwood, freshly dried pine needles [*Pinus strobes*], kraft lignin), foods derived from plants (green tea leaves, cacao powder), and herbal medications (St. John's wort). Both TMAH or TMSH methylated the fatty acids (as well as other organic acids including alcohols).[23] Depending on the matrix, leaching with either water or organic solvent followed by evaporation was sufficient for sample preparation. Addition of methanolic solutions of either reagent to a sample of extract completed sample preparation. Injection, however, required a specialized syringe for distributing the extract on a metal surface and modification of the injector port. Matrix effects mandate careful attention to the selection of reagent. While TMSH was effective in most matrices, its use in green tea extract produced no methylation. This particular matrix issue is unresolved. In contrast to the injector port reaction, thermolysis with TMAH in a microfurnace did not methylate the same extract.[24]

TAHM has an important clinical application in the rapid detection of tuberculostearic and hexacosanoic acid as biomarkers of tuberculosis (TB).[25] Rapid detection of these analytes derives from direct injection of sputum samples with TAHM. It is an important addition to the diagnosis and prevention of the spread of TB.

Development of an optimized method for the determination of fatty acids from fertilizer and the accompanying microbial cultures investigated the interactions between five highly complex matrices: soil, cattle manure, pig slurry, compost, and vermicompost.[26] They tested three extraction methods (Folch, Bligh, and microwave-assisted extraction [MAE] as well as two methylation techniques [KOH in methanol and TMSH]). The outcomes measured were total fatty acids and various subclasses viz. straight-chain saturated, straight-chain monounsaturated, straight-chain polyunsaturated, branched chain (iso and anteiso), and cyclopropyl. Within the class of polyunsaturated acids, the 18:2ω6c is of particular interest as it is a fungal biomarker. The optimal method for all matrices proved to be the combination of Folch extraction and methylation with TMSH. In this work, the KOH-catalyzed reaction occurred at 37°C in a reaction time of 15 min but required extraction and concentration. In contrast, the TMSH reaction occurs at ambient temperature in a reaction time of 30 min and the solution in *tert*-butylether can be injected directly. It is likely that the absence of postderivatization sample handling contributes to the higher quality of the results.

Eras and coworkers[27] used pentylation esters to determine fatty acids including the cyclopropyl and hydroxy fatty acids. Chlorotrimethylsilane (ClTMS) served as a catalyst and pentanol as the reagent for transesterification of acylglycerols *and* esterification of free fatty acids. Derivatization requires elevated temperatures (60–80°C) and a long reaction time (~1 h) but is simpler than other techniques as it just requires a single wash with bicarbonate to remove any acid formed during the reaction. The authors compared the methanol/sodium methoxide transesterification with the optimized ClTMS/pentanol procedure for analysis of fatty acids in the bacteria *Escherichia coli*, *Burkholderia cepacia*, and *Lactobacillus brevis*. The latter showed 50% higher yield for the C14–C16 fatty acids with moderate to no increase for the higher-carbon-numbered acids. This was probably due to the decreased volatility of the C14–C16 pentyl esters compared with the methyl esters. Moreover, derivatization with ClTMS/pentanol detected the 3-hydroxy C14:0 acid in *E. coli* and *B. cepacia*. Methylation did not find this fatty acid in the former bacterium, and in the latter, the yield was 1/10 that obtained with pentylation. In the case of the fungi *Aspergillus niger* and *Gibberella fujikuro*, the ClTMS/pentanol procedure gave 1.5- to 2-fold higher yield of almost all fatty acids.

Another comparative study provided data on four different methods for determining fatty acids in meat to determine suitability for analysis of different saturated and unsaturated C12–C20 fatty acids.[28] In the first method, esterification was performed *in situ* by treating the meat sample with 5% by weight methanolic HCl at 70°C for 90 min. The second method involved the hydrolysis of the sample with 5 M

KOH in methanol : water (50:50, v/v) at 60°C for 1 h under protection against oxidation. An extraction of the alkaline phase with pentane removed the nonsaponifiable fraction and following evaporation of the organic solvent, treatment of the residue with methanol : toluene (2:1, v/v) at 40°C for 10 min with trimethylsilyldiazomethane (TMS-DM) produced the trimethylsilylmethyl ester. The third, referred to as the classic method, involved an extraction with Folch's solvent, isolation of the residue, dissolution into hexane, and esterification with 50 µ of 2 M KOH in methanol. In the fourth method, the lipids were isolated via Folch extraction and derivatized with TMS-DM.

The *in situ* method had the lowest recoveries ranging from 49% to 100%, and an increase in yield may have needed repeat of the procedure. Use of elevated temperatures to increase yield was not an option as it increased isomerization of the conjugated linoleic acid (CLA). Although time-consuming and requiring elevated temperature, the method, in general, had good precision. In addition, it does not use chloroform or a diazo compound; thus, it is safer and more environmentally friendly. The classical method and modified classical method were by comparison imprecise with the relative standard deviations for most analytes above 10% at the three quality control concentrations tested. The saponification method followed by derivatization with TMS-DM provided the best compromise of high yield, precision, and lowest isomerization. TMS-DM has some advantages over diazomethane, which is volatile, toxic, and potentially explosive. Nevertheless, it must be used with recognition of potential problems such as artifact formation and silylation side reactions.[29,30]

13.2.3. Free Carboxylic Acids

Unlike the measurement of total fatty acids in biological samples, derivatization of free carboxylic acids focuses on the reaction of the carboxyl group. Reactions of carboxylic acids with alcohols under acid catalysis provide a simple esterification.[20] One approach is reaction between the nucleophilic carboxylate and the electrophilic alkylating group. Another approach is amidation of the carboxyl group by a formal dehydration.

13.2.4. Derivatization for GC

13.2.4.1. Silylation. This classical procedure continues to be the basis for innovations for application to the analysis of carboxylic acids. Use of liquid-phase microextraction (LPME) with *in situ* derivatizations is effective and has considerable potential in automation. This was as demonstrated by the device described by Sun and coworkers for silylation of volatile carboxylic acids from cigarette smoke.[31]

13.2.4.2. Alkylation. As noted, methylation is an important yet simple reaction. Methanol acidified with sulfuric acid affords rapid derivatization in several applications. The acid conditions did not alter the double bond structure of CLAs.[32] This methylation technique successfully prepared the methyl esters of CLA isolated by Folch extraction from rumen with preservation of stereochemistry. Determination of free tuberculostearic acid from sputum by solid-phase microextraction (SPME) also relied on methanol acidified with sulfuric acid.[33] Cha's group tested both *in situ* methylation and postextraction on-fiber methylation. In the latter procedure, they adsorbed the acids by direct insertion into a solution of fatty acid. The acids on the fiber were then methylated by exposure to the headspace (HS) above the acidified methanolic solution. In the analysis of the sputum, the sample was hydrolyzed and the derivatives were prepared using sorption by direct insertion and on-fiber methylation.[33]

LPME coupled with injector port derivatization formed the butyl esters of acidic herbicides. The acidic herbicides were extracted as ion pairs with tetrabutylammonium hydroxide (TBAH) into octanol as the extraction phase impregnated in a hollow fiber. Upon injection, pyrolytic alkylation with the TBAH butylated the carboxyl group of these analytes.[34]

An analogous LPME reaction used octanol as both the receiving solvent and the reagent for the derivatization of haloacetic acids. Following extraction, in-vial reaction catalyzed by trifluoroacetic anhydride converted these acids to their *n*-octyl esters.[35] Saraji and coworkers selected this procedure over silylation and pentafluorobenzylation (PFBation), based on fewer side reactions and cleaner chromatograms.

Alkylation with pentaflourobenzyl bromide (PFBBr) is a classical AD. This highly fluorinated alkylating agent still finds considerable utility as the pentaflourobenzyl (PFB) group produces an electrophoric ester. Traditionally developed for ECD,[1] this also proved useful for negative ion chemical ionization (NICI) detection at high sensitivity. In general, the reaction occurs post extraction and isolation. In the standard procedures, PFBBr reacts with isolated analyte in solution requiring only room temperature and a strong, hindered organic base to complete the reaction in 15–30 min.[36–38] The more current techniques combine PFBation with SPME, which is more time effective as extraction and derivatization occur without the need of an intermediate isolation. Acids can be derivatized in water followed by SPME[39,40] of the PFB ester, or the protonated acids can be sorbed onto an SPME coated with PFBBr, which effects derivatization. Clearly, this procedure is limited to analytes that are volatile and stable at moderate temperatures.[39,40] Because of the mild conditions, derivatization with PFBBr is suitable for more complex molecules such as eicosanoids.[36,37,41] Although the PFBation of the carboxyl group occurs at room temperature, the hindered amines used as a base can catalyze reactions at positions other than the carboxyl group. Jain and coworkers[37] reported the formation of an iso-oxazol from a nitro group alpha to a double bond. The initiating step leading to the cyclization was abstraction of a proton on carbon bearing the nitro group.

Other miniaturized techniques also offer simplification and automation of ADs of carboxylic acids. Saraji and Farajmand[38] extracted and derivatized acidic herbicides with a combination of LPME and derivatization in a microvial with 1.1-mm inner diameter. Triethylamine catalyzed the in-vial PFBation and a temperature of 100°C provided a reaction time of 5 min. While effective at reducing turn-around time, it is possible that such conditions would degrade more labile compounds.

*13.2.4.3. **Chloroformates.*** Derivatization of organic acids with chloroformates produces alkyl esters in water-based solutions or solutions in organic solvents. When the derivatization occurs in an aqueous matrix, the derivatives are then extracted or sorbed onto a solid-phase solution.[42–44] The combination of the mild, water-compatible reaction conditions and facile integration with SPME or other solid-phase extraction (SPE) methodologies makes this is an attractive derivatization for carboxylic acids. There are several likely reaction pathways from the reagents to the esters. One is the formation of the alkoxycarbonyl esters or mixed anhydrides, which then break down to alkyl esters. This is the most likely mechanism in the absence of alcohols. The other is an exchange reaction between the alcohol and the mixed anhydride to form the ester. The latter mechanism is preferred because it allows the formation of alkyl esters from chloroformates that have different alkyl substituents. For instance, Deng and coworkers[42] reported the formation of the ethyl ester of valproic acid in a reaction mixture of plasma, ethanol, and isobutyl chloroformate with pyridine as a catalyst. Similarly, derivatization of a carboxyl functionality with ethylchloroformate but in the presence of methanol and catalyzed by pyridine produced the methyl ester.[44]

An interesting application of derivatization of fatty acids and amino acids by chloroformate comes from the discipline of art history.[45,46] Binding media used by artists in the post-Byzantine period consist of materials such as egg yolk and egg/oil emulsion. Hydrolysis of samples from post-Byzantine panel paintings provided the fatty acids from the triglycerides and the amino acids from the proteins. Derivatization with ethyl chloroformate yielded the volatile derivatives suitable for the determination of fatty acids and amino acids by GC.

Although a facile and useful reaction, the composition of the aqueous phase and the ratio of moles reagent/mole base are determinants of the mechanism. Optimization of these variables is therefore a frequent[43] component of any method development.

13.2.5. Derivatization for HPLC

*13.2.5.1. **Flourescent/chromophoric derivatives for HPLC.*** Derivatizations preparatory to HPLC analysis are exemplars of reactions of carboxylic acids. These include reaction with diazo compounds, nucleophilic displacement of a sulfonate leaving group, and amidation via carbodiimide dehydration.

*13.2.5.2. **Diazo functionality.*** The diazo functionality is a mild, effective method for preparing esters with the ethylene moiety RCH_2 where R = H, trimethylsilyl or anthryl. For detection by analysis of carboxylic acids by HPLC, derivatization with 9-anthrlydiazomethane (ADAM) yields a highly fluorescent 9-anthrylmethyl ester.[47,48] The reaction conditions are mild requiring only an hour at room temperature but in the absence of light. Based on reports by Mueller[33] and Mueller and Binz,[49,50] the diazo functionality will not likely react with unsaturated double bonds that are not activated such as those found in unsaturated fatty acids. Although a cleaner reagent than trimethylsilyl diazomethane, ADAM is not volatile like diazomethane and can be subject to interference from impurities. These interferences are sufficiently polar that they do not interfere with the determination of acetic acid.[51]

*13.2.5.3. **Sulfonate displacement.*** The fluorophoric reagent 2-(naphthoxy)ethyl-2-(piperidino) ethanesulfonate (NOEPES)[52] has several interesting features. Derivatization occurs via nucleophilic displacement of the 2-piperinoethanesulfonate group, which produces a neutral derivative as well as the 2-piperinosulfonic acid, an amphoteric compound that is water soluble at all pH. The reagent itself is basic and is therefore extractable at acid pH. As a result, it is relatively easy to separate the derivatives from reagent and side products. In derivatization of very long-chain fatty acids (VLCFAs)—which are the C22, C24, C26 saturated fatty acids—and extraction from plasma, the reaction occurs in the toluene extract and is catalyzed by K_2CO_3 solubilized in the lipophilic solvent by an 18-crown-6 ether. The reaction requires 1 h at 95°C. These relatively harsh conditions are appropriate for VLCFA but may be detrimental to more labile analytes.

*13.2.5.4. **Dehydration.*** Du and coworkers recently reported dehydration with 6-oxy-(acetyl ethylenediamine [EDA]) fluorescein (AEF), and 6-oxy-(ethylpiperazine)-9-(2′-methoxycarbonyl)fluorescein (EPMF).[53,54] The reactive group in both these reagents is an amine and is condensed via a formal dehydration with 1-ethyl-3-(3-dimethylaminopropyl) carbodiimide (EDC). Conditions require elevated temperatures (50–55°C) and reaction times (40–50 min). Again, these conditions are appropriate for fatty acids, but they could degrade more labile compounds.

Derivatization with 2,4-dinitrophenylhydrazine (DNPH) provides chromophoric amides suitable for the determination by HPLC with UV detection. A recent report describes the use of DNPH and inorganic-acid-impregnated silica gel cartridges to derivatize C1–C4 acids in air.[55,56] The silica gel sorbed the volatile acids, which then reacted with the DNPH. Formation of the hydrazides required 80°C and reaction time of 4–8 h. Optimal recoveries of formic and acetic acid DNPH derivatives required substantially different flow rates and optimization balances between these different requirements. Although carried out in solution, Bravo and coworkers[57] found that microwave irradiation appears to afford approximately a 10-fold increase in reaction rate without excess increase in temperature.

*13.2.5.5. **Separation-oriented derivatization.*** A group at the Faculty of Pharmaceutical Sciences of Fukuoka University is developing analytical concepts based on fluorous chemistry.[58] They exploited the strong retention of fluorine-containing compounds on fluorous stationary phases in HPLC. Condensation of carboxylic acids with 4,4,5,5,6,6,7,7,8,8,9,9,10,10,11,11-heptadecafluoro-*n*-undecylamine (HFUA) and catalyzed by 4-(4,6-dimethoxy-1,3,5-triazin-2-yl)-4-methylmorpholininium chloride (DMT-MM) produced an amide, which was strongly retained on the fluorous stationary phase, while other underivatized compounds eluted in or close to the solvent front. The parent fluorescent carboxylic acids (homovanillic acid, vanillymandelic acid, and 5-hydroxyindoleacetic acid), upon elution, were readily detected by fluorimetry.

Although condensation of a carboxylic with an amine to form an amide is a well-known reaction, this work presented an additional challenge. HFUA is highly lipophilic, while DMT-MM chloride is hydrophilic as are the carboxylic acids. Dissolution of the amine, acid, and condensation catalyst was problematic but was achieved through the use of aqueous tetrahydrofuran. In addition, the authors reported that the amidation catalyzed by DMT-MM had a faster reaction rate and proceeded under milder conditions than those found in condensations catalyzed by EDC, hydroxybenzotriazole (HOBt), and diphenylphosphoryl azide (DPPA).

13.3. ORGANIC ACIDS: PHOSPHATES AND PHOSPHONIC ACIDS

These phosphorus derivatives are degradation products of insecticides to form organic phosphates or nerve gases, which degrade to phosponic acids. Phosphates are also moieties of the nucleotides and drugs based on nucleotide structure and biochemistry. Their determination provides risk assessment with respect to excess release of these highly toxic compounds into the environment but can also be an important tool in the treatment of AIDS. Although structurally different from carboxylic acids, their derivatization follows the same general class of nucleophilic displacement of leaving groups or condensation to esters or amides.

13.3.1. Derivatization for GC

*13.3.1.1. **PFBBr.*** Reactions between organic phosphates (e.g., dimethylphosphate, diethylthiophosphate) and PFBBr occurring under alkaline conditions either in solution[59,60] or by solid-phase analytical derivatization (SPAD)[61] provide facile post extraction. In solution, the reaction requires 30 min at 80°C followed by extraction of the derivative from the reaction mixture.[59] The SPAD reaction[61] proceeds at room temperature and is complete in 15 min. It is also a fully automated procedure. (*Note*: In the report by DeAlwis and coworkers, the experimental section states a reaction time of 15 min, while the scheme states a reaction time of 1 h. Personal communication confirmed the reaction time of 15 min.)

*13.3.1.2. **Alkylation.*** Nucleophilic displacement was also the reaction mechanism for preparing the pentyl derivatives of the phosphonic acids.[62,63] With surfactant-less microemulsion (SLME), reaction between phosponic acids (e.g., ethylphosphonic acid, isopropylethylphosphonic acid) and propyl bromide or pentyl iodide required 100°C and 2–4 h, depending on the analyte and alkylating agent. These analytes are stable to such conditions, and the resulting derivatives are suitable for GC.

13.3.2. Derivatization for HPLC

*13.3.2.1. **DANSYLation.*** Many phosphates, particularly those relating to nucleic acids, cannot readily be determined by GC, and their measurement requires HPLC. One such example is 2′,3′-dideoxyadenosine-5′-triphosphate (ddATP), which is the active metabolite of the prodrug[64] didanosine (2′3′-dideoxyinosine [ddI]). The chemistry of the acid functionality permits diimide-mediated condensation of the phosphate with DANSYL ethylenediamine (DANSYLeda). In this reaction, the phosphate group was activated by *N*-(3-dimethylaminopropyl)-*N*′-ethylcarbodiimide hydrochloride (EDAC). The water solubility of EDAC permits derivatization in aqueous solution. The reaction takes place in the dark at a temperature of 60°C over 90 min. Detection of the DANSYL derivative by fluorescence provided 160 times the sensitivity of detection by UV absorption. By virtue of ease of reaction, the authors considered DANSYLation as superior to EDAC derivatization with 4,4-difluoro-5,7-dimethyl-4-bora-3a,4a-diaza-*s*-indacene-3-propionyl ethylenediamine (BODIPY FL EDA), which is a frequently used reagent for the detection of 5′-monophosphates of nucleosides. Although the authors claim a faster reaction for DANSYLeda than for BODIPY FL EDA, this may be insufficiently supported since the latter took place at ambient temperature and the former at 60°C.

13.4. PHENOLS AND ALCOHOLS

Derivatization of phenols and alcohols shares many characteristics with those of carboxylic acids. Typically, these analytes are very weak acids. The pK_a for phenols is $10 < pK_a < 11$, and for alcohols, it is $pK_a > 12$ and so these require more alkaline conditions for derivatizations. Substitution of electron-withdrawing groups (F, Cl, Br, NO_2) on phenols, however, can substantially lower the pK_a of the phenols into the range of the carboxylic acids (e.g., for dinitrophenol $pK_a = 3.9$) and reaction conditions would change accordingly.

13.4.1. Derivatization for GC

13.4.1.1. Silylation. Trialkylsilyl ethers of phenols have greater volatility than their parent compounds and provide sharp, well-resolved peaks. It is a simple, very well-established procedure initially used for derivatization of both phenols and alcohols in anhydrous solution.[65] Under these conditions, the rapid reaction of silylation and relative ease of derivatization are attractive features in sample preparation of phenols[66,67] and alcohols.[68] Silylation converts more complex in-volatile molecules such as steroids[69–72] and sugars[68] into derivatives sufficiently volatile for gas chromatographic analysis. Typical reaction solutions are bis(trimethylsilyl)trifluoroacetamide (BSTFA), MSTFA, and *N*-(*t*-butyldimethylsilyl)-*N*-methyltrifluoracetamide (*t*BDMSMTFA).[70]

Although a long used classical reaction, new techniques and ideas continue to appear in the literature. Shareef et al. studied silylation of estrone (E1) and ethinylestradiol (EE2). They compared trimethylsilylation with MSTFA and BSTFA and *t*-butyldimethylsilylation with MTSTFA in several solvents. They found that solvents as well as silylating reagents affected the ratio of E1-TMS, EE2TMS, and EE2diTMS. Formation of the *t*-butyldimethylsilyl ethers (TBDMS) in pyridine or dimethylformamide (DMF) provided reproducible yields of the monoTBDMS of both analytes. Kovacs and coworkers[73] introduced a new reagent, trimethylsilyl-*N*,*N*-dimethylcarbamate (TMSDMC), for preparing trimethylsilyl ethers of phenols and alcohols. While standard derivatizations with reagents like *N*-(*t*-butyldimethylsilyl)-*N*-methyltrifluoracetamide requires 80°C and 1 h, derivatization of phenols with TMSDMC occurs at room temperature. The reaction appears to be complete within minutes, but there is no detailed study of the reaction kinetics. An additional advantage over and above the fast reaction at room temperature is that the reagent and by-products elute very early in the chromatogram. Like all trimethylsilylations, reactions using TMSDMC remain sensitive to water.

Silylations on solid supports are a current focus as the basis of automation.[71,74–76] In these techniques, the phenols in aqueous solutions are typically sorbed onto the fiber followed by exposure of the fiber to the vapors of the silylating reagent. An alternative was reported by Vaughan et al.[77] in which the phenols from cigarette smoke were trapped on a filter pad, which was then coated with BSTFA, and the analytes were silylated at 50°C. The silylated derivatives were then desorbed onto a fiber in the HS for injection onto the GC column.

In more complex molecules, it is important to consider that steric factors control the reaction rates of silylations at different positions, altering the mix of products. Although silylation is well established for steroids, silylation for hindered tertiary 17α-hydroxyl mestranol and norethindrone is incomplete. In a study on endocrine disrupting phenols in sediment, an alternative derivatization with pentafluoropropionic anhydride was proposed as a solution to this problem.[78] Reaction conditions required heating the extracted natural and xenoestrogenic phenols at 60°C for 2 h. There is, however, little evidence that these conditions were sufficient to derivatize the hindered hydroxyl at the 17 position that also has the 17α-ethinyl group.

13.4.1.2. Acylation. Another classical derivatization is acylation with acid anhydrides (Ac_2O) or an acyl chloride (AcCl). Acetic Ac_2O[79–82] is sufficiently stable in water to allow reaction *in situ* following which the now more lipophilic derivatives are extracted.

Stopforth and coworkers determined estrone (E1), estradiol (E2), and equilin (Eq), and, with the internal standard, equilenin (Eqn) demonstrated several principles of the process. The reaction occurs in buffered urine at room temperature to form the acetate at the three phenolic position found in all four estrogens. The reaction times, however, are critical and tightly controlled by stopwatch. Following this step, stir-bar sorptive extraction (SBSE) recovers the derivatives. Further acetylation of the sorbed compounds, but at the secondary alcohols, occurs in the sorbent in a HS vial containing acetic Ac_2O and pyridine and heated to 80°C for 30 min. These conditions are required to derivatize the secondary hydroxyl at the C17 position of the E2, Eq, and Eqn. The final step is thermal desorption onto the GC column thus avoiding any further workup. Another variant combining acetylation and solid-phase sorption relies on the purge–trap technology. Zhao et al.[83] reported a detailed study on the analysis of phenols and chlorophenols from water. *In situ* derivatization produced lipophilic, volatile acetates which following sparging and trapping on a solid phase of Tenax removed from these compounds from the aqueous layer. They described the reaction and the purge–trap conditions (including salting out) that provided the maximum recovery of the acetates.

LPME following *in situ* acetylation afforded a simplified sample preparation procedure[82] for the determination of bisphenol A in river water. A critical factor in optimizing reaction conditions was the use of NaOH rather than a carbonate/bicarbonate butter. The latter buffer reacted with the acetic acid formed as the anhydride reacted with water to form bubbles of carbon dioxide. These destabilized the drop. Another important point was the identification of toluene as the preferred solvent. Although LPME requires 150 min to complete the extraction/reaction, acetylation increases the sensitivity by more than two orders of magnitude over the analysis of underivatized bisphenol A.

13.4.1.3. Tosylation. Tosylation is a recent addition to the options in the AD of phenols. Fiamegos and coworkers[84] coupled this with LPME. In this extensive study using single-drop microextraction (SDME) coupled with phase transfer catalysis (PTC), the authors identified tetrabutyl ammonium bromide (TBAB) as the preferred catalyst and optimized concentrations of the catalyst, tosyl chloride, reaction times, and mixing rates.[84] Comparison of the optimal SDME conditions to those of classical PTC conditions found that the former reduced the reaction time and increased both the precision and the sensitivity. Detection limits with SDME were 6- to 16-fold lower than the classical method.

13.4.2. Derivatization for HPLC

13.4.2.1. DANSYLation. Derivatization of hydroxyl groups with DANSYL chloride (DANSYL-Cl) substantially increases sensitivity of analysis by HPLC–electrospray ionization (ESI)–MS/MS.[85–90] The reaction conditions, however, must be adapted to the different functional groups in a molecule. Determination of vanillin from plasma illustrated the importance of this point. DANSYL-Cl requires alkaline conditions for reaction with a hydroxyl, but the aldehyde is susceptible to Cannizzaro oxidation if the base is too strong. Use of triethylamine rather than NaOH as well a 10-min reaction time resulted in a highly reproducible yield of the DANSYL derivative at the phenolic position without Cannizzaro reaction forming the carboxylic acid.[87]

Determination of estrogens is enhanced and automated via DANSYLation at the phenolic position.[90–92] DANSYLation for the measurement of ethinyl estradiol from rat plasma is a manual, postextraction procedure and requires elevated temperatures to reduce reaction time. Nevertheless, it increases sensitivity by an order of magnitude compared with that of steroids that lack the phenolic group and are therefore *not* DANSYLated.[92] Ninety-six-well plate technology was a successful approach to the determination of estrogens, in this case, from human plasma.[90] This too used a liquid–liquid extraction, but since it was semiautomated, it greatly improved the sample throughput. SPAD was an enabling technique for the automated analysis of estradiol from sewage effluent and influent via derivatization with DANSYL-Cl.[91] This technique combined sorption, preliminary sample cleanup, derivatization, sample cleanup, and elution into the LC-MS.

A study compared the sensitivity of methods using DANSYL-Cl, 2-fluoro-1-methylpyridinium *p*-toluenesulfonate (FMPTS), PFBBr, and DANSYL-Cl from environmental water.[89] Of these three reagents, DANSYL-Cl gave the highest sensitivity of analysis by LC-MS.

13.4.2.2. Specific derivatizations. Complex compounds with aryl and alkyl hydroxyl moieties are metabolized via glucuronidation at either position. Derivatization with 1,2-dimethylimidazole-4-sulfonyl chloride (DMISC-Cl) in acetone and catalyzed with sodium carbonate occurred only with the phenolic hydroxyl, thus distinguishing between the different metabolic pathways.[3,93–95] For instance, glucuronidation of morphine occurs at the 3-phenol to give 3-morphine glucuronide (3-MG) or the 6-hydroxyl positions to form 6-morphine glucuronide (6-MG). 6-MG, but not 3-MG, shows the DMISC group and confirms the position of glucuronidation at 6.

The hydroxyl steroids include the phenolic estrogens and alcoholic hydroxyls that include the progestagens, androgens, and corticosteroids. Nishio et al.[14] developed 1-(2,4-dinitro-5-fluorophenyl)-4-methylpiperazine (PPZ) as a phenol-specific reagent and 4-(4-methyl-1-piperazyl)-3-nitrobenzoyl azide (APZ) as a reagent for the alkyl hydroxyls. The derivatives were further reacted with methyl iodide to produce quaternary amines that were suited for analysis by LC-ESI-MS. Relative to underivatized analyte, these derivatization techniques increased the sensitivity for detection of the estrogens by 2000-fold for derivatization of phenols with PPZ and 500-fold for the determination of steroids with a nonphenolic hydroxyl group derivatized with APZ.

Anhydride chemistry further provided derivatization at the alkyl hydroxyls and increased sensitivity. 2,3-Pyridinedicarboxylic anhydride (PCA) reacts with the hydroxyls of testosterone and dihydrotestosterone (DHT). This derivatization and ultra high performance liquid chromatography (UPLC)-MS enabled the detection of DHT in human serum. The combination of this reaction with the 96-well plate technology gave the high throughput necessary for basic and clinical studies.

Aldosterone controls the reabsorption of sodium and water as well as the release of potassium in the kidneys. With the retention of sodium in the circulation, there is a concomitant increase in blood volume and, therefore, an increase in blood pressure. Measurement of this steroid is important in developing an understanding of hypertension and aldosteronism but presents a significant problem in analysis. The molecule has three carbonyls (at the 3, 20, and 18 positions) and two hydroxyls (at the 11β and 21 positions). In addition, the 11β-hydroxyl and the 18-aldehyde are in equilibrium with the hemiacetal form. The AD of aldosterone preparatory to LC-MS is complex. Yamashita's group stabilized the hemiacetal by forming either ethyl or methyl ethers.[96] Subsequently, using the mixed anhydride method with picolinic acid (PA) and 2-methyl-6-nitrobenzoic anhydride (MNBA) to derivatize the 21-hydroxyl group with picolinoyl esters,[97–99] the pyridinyl function of the picolinoyl ester is readily ionized. This can increase sensitivity of HPLC-ESI-MS by two orders of magnitude.

13.5. AMINES

13.5.1. Amines Overview

The basic amine functionality has three variants: primary, secondary, and tertiary. Each will require different reagents with corresponding functionalities. Typical problems often require the analysis of, at least, the primary and secondary amines. One example is the biogenic amines, including for instance primary amines such as the amino acids, putriscene, and cadaverine, and secondary amines such as serpmine and spermidine. Analysis of such mixtures exemplifies the issues involved in the analysis of complex mixtures. Two general approaches are available from the literarure: mixed reagents and single reagents that react with both primary and secondary amines. As with the case of the carboxylic acids, there are well-established derivatization techniques that are used in innovative ways as well as those with newly developed reagents.

13.5.2. Ortho Dialdehydes

The *o*-phthalaldehyde (OPA)[100,101] and naphthalene-2,3-dicarboxaldehyde (NDA)[102,103] are classical reagents for derivatizing primary amines. These difunctional reagents react with the two hydrogens on the primary amine to form isoindoles, which are fluorescent but are also unstable. Reaction of the isoindoles with thiols forms derivatives that are sufficiently stable for analysis by HPLC with FLU. The reactions require basic conditions to permit the removal of hydrogen in the intermediate steps. The recent extensive review by Hanczko et al.[100] discussed the use of different thiols (ethanethiol, 2-mercaptoethanol, methanethiol, alkylcysteines) and identified ethanethiol as providing advantages in stability and low side reactions. Reaction rates are high even at room temperature, and because of the homogenous reaction solution, modern autosamplers have the capability to prepare these derivatives in the injection vial prior to injection. Successful use of these derivatizations requires control of conditions such as buffers, volumes of reagent, and volume of methanol. Under optimized conditions, it is possible to achieve reaction times of 1 or 2 min and minimize any unwanted side reactions.

Use of thiols to stabilize the isoindole offers an additional advantage in allowing facile integration of additional groups. The derivative of OPA derivatization with thiol, 5-((2-(and-3)-*S*-(acetylmercapto)succinoyl)amino)fluorescein (SAMSA-F),[104] was excited at the 488-nm laser line of an Ar(+) laser rather than at UV wavelengths. OPA/SAMSA-F has high epsilon values (78,000 M^{-1}) and quantum yield (0.98). It is suited for capillary electrophoresis (CE) with laser-induced fluorescence (LIF) and gives a sensitivity of 2 nM.

13.5.2.1. Fluorenylmethyl chloroformate (FMOC). This reagent is an exemplar of issues and solution in AD of amines. With primary and secondary amines the reactive hydrogen is replaced with the formyl group to yield carbamates. Reaction conditions are mild. In solution derivatization, the buffer is in the range of pH 9–10, with borate being the buffer of choice. Under these conditions, reaction times (depending on the structure of the analyte) of 1–15 min provide optimal yield for primary or secondary amines. It is not surprising that the tertiary amines react more slowly and require elevated temperature, and yields can be low. Applications of this reagent include analysis of primary, secondary,[105–113] and, with some caution, tertiary amines.[110,114] This caution stems from the quaternization of tertiary amines upon reaction with FMOC followed by subsequent dealkylation by the chloride produced in the reaction (or any other nucleophile). In such instance, what is measured is the FMOC-dialkylamine. If it is certain that the dialkyl amine is not present in the mixture of analytes then this derivatization is appropriate for determining tertiary amines.

SPME[106,110,111] and SPE[112,113] are frequently used techniques for the automation of sample preparation involving FMOC. For volatile analytes, the advantageous approach is to coat the adsorbent phase with FMOC and then contact the analyte with this loaded solid phase. In the analysis of low-molecular-weight alkyl amines, the preloading of FMOC on solid phase was advantageous for air and water analysis using SPE[112,113] and SPME in its fiber[110] and in-tube variants.[106]

Derivatization of complex and polar compounds with FMOC requires reaction in solution, but this is not a particular drawback due to ease of reaction. Selective hydrolysis of glycoproteins releases *N*-glycans for derivatization with FMOC.[105,115] This provides an alternative to reductive amination, a derivatization procedure widely used in glycomics. Reaction between *N*-glycans and FMOC is simple, rapid, and suitable for high-throughput analysis. When combined with CZE-ESI-MS/MS, the technique provides useful structural information for the glycans.[105] Determination of glucosamine sulfate is another example of the utility of derivatization of the amino group with FMOC. Again, the procedure is straightforward, with precipitation of plasma proteins with acetonitrile preceding derivatization with FMOC and subsequent analysis by HPLC with FLU detection. Lozanov and coworkers[108] also reported high-quality figures of merit for the determination of small, polar amines (polyamines, catecholeamines, metanephrines, and amino acids) with coefficients of correlations = 0.994 and limits of sensitivity in the picomole range. The calibration range and limit of quantitation (LOQ) were marginally higher for the amino acids (e.g., LOQ = 2.6–10 pmol) than for the other amines (e.g., LOQ = 2–4 pmol). This probably reflected the presence of the carboxyl functionality in the amino acids, which, because of its polarity, could cause loses due to sorption onto solid surfaces.

Combining FMOC with OPA-thiol offers advantages in the analysis of mixtures containing both primary and secondary amines. Derivatization with the latter combination did not provide the required high sensitivity for determining spermine. The necessary detection limit arose from the reaction sequence of OPA-thiol followed by FMOC, producing a mixed isoindole/FMOC derivative. A similar sequence proved useful for achieving high sensitivity of ornithine and lysine. Again, optimization of reagent ratios, volumes of solvent, and so on, is crucial for the successful use of the OPA-thiol/FMOC reaction, but once these conditions are known, both reactions require only 1 min each for a total reaction time of 2 min.[116,117] This combined derivatization technique is sufficiently rugged for application to the analysis of complex mixtures such as those involved in food chemistry.[118]

13.5.3. Succinimides

The *N*-hydroxysuccinimydil activating group permits derivatization of both primary *and* secondary amines with a range of fluorescent moieties[119–125] as well as with groups that offer other advantageous properties.[126] Since succinimides react with both primary and secondary amines, they have wider application than OPA/thiol reagents, which react only with primary amines.

Reaction of amines with *N*-hydroxysuccinimidyl fluorescein-*O*-acetate (SIFA) labeled amines with flourescein for CZE and detection with LIF at higher sensitivity than those found with other reagents.[119] The ferrocenyl group introduced by reaction of amines with succinimidyl-ferrocenyl propionate improved the detection of amines in foodstuffs by increasing the speed of separations by fivefold over the OPA methods.[126] Validation of this new technique was by comparison to the OPA method, which is the European standard for determining amines in fish.

A carbamate linkage for the incorporation of the 6-aminoquinolyl group produces a reaction rate with a half-life on the order of seconds and the derivatives have a high FLU[122,124] as well as being compatible with mass spectrometric determination. The flexibility of the activating group was evident in the work by Shimbo et al.[123] who prepared *p-N,N,N*-timethylammonioanilyl *N-O*-hydroxysuccinimidyl carbamate iodide (TAHS) as a precharged reagent for mass spectrometric determination of amines. The derivatives, however, did not have good chromatographic properties. To circumvent this problem, they prepared 3-aminopyridyl-*N*-hydroxysuccinimidyl carbamate (APDS), which had superior peak shape and width, which enabled the determination of a hundred amines in 10 min using HPLC-ESI-MS/MS.

A new fluorescent group, Pacific Blue (PB), provided the nanogram sensitivity believed necessary for the detection of life on Mars. Estimates of this sensitivity came from basal levels of amines found in Yungay Hills region in Atacama Desert, Chile, and from the Murchison meteorite. Under alkaline conditions (lithium carbonate), PB succinimidyl ester converts amines to a derivative that has almost six times higher extinction coefficient and seven times the quantum yield compared with fluorescamine, which is better known as fluorophore.[121] Although the fluorophore is superior, the reagent, however, has an ester linkage and therefore required reaction times of several hours rather than the minutes required for the carbamate reactive groups.

13.5.3.1. 4-Fluoro-7-Nitrobenzo-2,1,3-Oxadiazol (NBD-F). Preparation of fluorescent derivatives of amines with NBD-F is well established.[127–131] In this reagent, the nitro group and the fluorine activate the four positions by electron withdrawal, making it susceptible to nucleophilic attack. Although useful in the AD of amines, recent applications to sugars demonstrate the utility of ADs in manipulating analytes to suit the needs of the investigator. Mono-, di-, and oligosaccharides, which are aldehydes, underwent reductive amination in aqueous solution with ammonia and dimethylamine–borane complex at 70°C to convert the aldehydic reducing group to primary amine. The amines then reacted with NBD-F in acetonitrile also at 70°C.[128] The sensitivity of the method, as well as its relatively simple reaction conditions, makes this double derivatization an attractive approach.

13.5.3.2. Fluorescent reagents. Because there are numerous reagents available, it is important to evaluate, which provides the best results given the analyte and matrix. In studying the asparginyl oligosaccharides in glycoprotein, Kurihara and coworkers[132] evaluated eight different fluorescent reagents for reaction with the amine of asparginine and for optimizing ultrahigh-performance liquid chromatography and detection by FLU and structural identification by electrospray–time-of-flight MS. The set consisted of 4-(*N,N*-dimethylaminosulfonyl)-7-fluoro-2,1,3-benzoxadiazole (DBDF), NBD-F, FMOC-Cl, DANSYL-Cl, NDA, 1-pyrenesulfonyl chloride (PSC), fluorescein-5-isothiocyanate (FITC), and 3-chlorocarbonyl-6,7-dimethoxy-1-methyl-2(*1H*)-quinoxalinone (DMEQ-COCl). Although not widely used, the PSC derivative of disialo asparginine (the model compound) exhibited a detection limit at 3 fM by FLU. Detection by MS gave a sensitivity of 58 nM for the PSC derivative. Applying this derivatization to the study of ovalbumin gave 15 PSC-labeled oligosaccharides.

Consideration of sensitivity, however, did not show a clear rational for the selection of reagent particularly for mass spectrometric detection. For instance, the FLU detection limit for the PSC derivative was more than fivefold lower than that for DMEQ-COCl, but the latter reagent required a reaction time of only 1 min. In contrast, derivatization with PSC required 2 h at 50°C for reaction. Prolonged reaction times at equal or similar temperatures were also found for FMOC, FITC, DANSYL-Cl, and DBDF. Reaction times for NDA, NBD-F, and DMEQ-COCl were on the order of 1–10 min.

13.5.3.3. Removal of excess reagent. Exploitation of fluorous chemistry provided an elegant derivatization of amines to a fluorescent derivative while eliminating interferences from excess reagent.[133] The pyrene fluorophore, 1-pyrene-4-butyric acid, was esterified with 4-(*1H,1H,2H,2H*-perfluorodecylsulfanyl)phenol to form the F-trap pyrene. Primary amines including amino acids reacted with the ester group to form the fluorescent 1-pyrene-4 butyric alkyl amide or the corresponding amide of the amino acids. The excess reagent containing the perfluorodecyl and the reaction by-product (4-(*1H,1H,2H,2H*-perfluorodecylsulfanyl)phenol) were retained on a fluorous solid-phase extraction (F-SPE) cartridge, while the derivatized analytes were eluted in acetonitrile.

13.6. CARBONYLS

13.6.1. Carbonyls Overview

The carbonyl functionality occurs widely in nature as well as in anthropogenic products. Analytes containing this functionality cover an impressive range of compounds. The carbonyls contain one of the simplest organic molecules, but then also include complex molecules such as progesterone or the corticosteroids (both C21) and carbonylated proteins. Concentrations of these analytes have import in the clinical, environmental, and food production domains.

Carbonylated proteins and malondialdehyde are accepted markers of oxidative stress and are elevated in neurodegenerative diseases such as Parkinson's, Alzheimer's,[134–137] and multiple sclerosis.[138] They are also elevated in diabetes and during aging.[139]

Derivatization plays a major role in the determination of carbonyls because most do not have chromophores, fluorophores, or electrophores that permit detection at the required concentrations. In determining the protein carbonyls, formation of the derivatives also enhances specificity by reacting only with those proteins that were subject to oxidative stress or other forms of oxidation. For the measurement of low-molecular-weight carbonyls, it also provides a lipophilic and more readily extractable compound.

The fundamental process in the derivatization of carbonyls is the nucleophilic attack of a primary amine on the electrophilic carbon of the functionality to form a hydroxylamine followed by dehydration to an imine. The groups attached to the amines determine the instrumental technique.

13.6.2. GC

13.6.2.1. Pentafluorobenzylhydroxylamine (PFBONH2). PFBONH2 forms pentafluorobenzyloximes (PFBoximes) of aldehydes and ketones that are visible to ECDs[140,141] and mass spectrometers.[141–156] The mass spectrometer is the detection instrument of choice for most analyses and both GC[144–146,148,149,152,157] and HPLC[146,154,155,158] provide useful separations. Jakober and coworkers investigated the formation of airborne carbonyls derived from the combustion of diesel fuel.[154–156] They compared GC-ion trap MS (ITMS) and HPLC–atmospheric pressure chemical ionization (APCI)–ITMS in both the positive and negative ionization mode. For 20 of the 25 model carbonyls studied, derivatization produced a substantial increase in method sensitivity in both ionization modes. The pattern appeared to be that carbonyls detectable at high response factors without derivatizations usually did not exhibit an increase in sensitivity upon derivatization. Substantial increases in sensitivity upon derivatization appeared for those compounds that were detectable at low response factors or were not detectable at all. The mass spectra in these two detection modes produced pseudomolecular ions as well as fragmentation that gave structural information. Similarly, Long et al.[145] reported that both electron impact or positive-ion chemical ionization can provide more structural information. They studied the formation of toxic α,β-unsaturated aldehydes produced by oxidative stress and found that these two ionization modes provided specific fragments that permitted the differentiation of 4-hydroxy-2-alkenals and 2-alkenals.

Reactions occur in aqueous[143,154,156] and solid phases[6,142,144,159–161] under relatively mild conditions, although there are often substantial differences in reaction rates depending on the structure of the analytes.[154,155] Aldehydes react at a faster rate than ketones,[6] and derivatization of dicarbonyls require a higher concentration than monocarbonyls.[156] Reactions in aqueous phase can require as much as 24 h for optimal derivatizations under conditions that inhibit oxidation over that period of time.[154–156] Consequent to reactions occurring in two phases, there are a number of variants in the sample preparation techniques.

Following derivatization in aqueous phase, classical liquid–liquid extraction effectively isolates the PFBoximes both in the synthesis of standards and on the analytical scale.[142,162] LPME avoids the necessity of concentrating the derivatized compounds (often to residue followed by reconstitution).[152,163] Chen and Huang reported both derivatization of the carbonyl to the PFBoxime and subsequent silylation of the hydroxyl groups in a single microdrop. The technique required due attention to avoid drawing in any of the water-containing emulsion into the syringe prior to the second step, to protect the silylating agent.

SPME finds extensive use in the determination of carbonyls as their pentafluorobenzylhydroxylamine (PFBHA) derivatives.[142,143,151,157,159,160] In the most common instance, the analytes are derivatized directly in aqueous phase (e.g., plasma or urine) or are collected trapped from air and transferred to an aqueous phase. Beranek and Kubatova[143] identified three variations of the SPME technique in the analysis of such aqueous samples: HS aldehyde extraction with their on-fiber derivatization (HS-SPME-OFD); direct derivatization (D) in aqueous solutions followed by SPME in HS (D-HS-SPME); direct derivatization (D) liquid phase followed by SPME in liquid phase (D-L-SPME). In the on-fiber techniques, PFBHA is loaded onto the fiber and the carbonyls are transferred into the HS for reaction either because of inherent volatility or through heating. For carbonyls whose PFBoximes are sufficiently soluble in water (i.e., low-molecular-weight carbonyls, hydroxycarbonyls, etc.), these authors reported that D-L-SPME is the most effective technique providing the highest yields.

Another application of derivatization by on-fiber SPME is determination of carbonyls in air[164] including human breath.[148] The latter technique may be an important technique supporting studies in asthma and its pathophysiology.

Breckenridge et al.[6] reported a SPAD, which is an alternative to SPME. In SPAD, there is a simultaneous extraction derivatization of carbonyls in aqueous solution with PFBHA onto XAD2. Since the solid phase is always in direct contact with the aqueous phase, SPAD allows the isolation of highly water-soluble compounds by directly converting them to the more lipophilic PFBoximes.

13.6.3. HPLC

13.6.3.1. Hydrazine reagents. Derivatization of carbonyls with DNPH is a classical reaction in undergraduate organic chemistry laboratory. Nevertheless, formation of 2,4-dinitrophenylhydrazone has extensive utility in the determination of this class of analytes, and recent reviews are in the literature.[165–168] Despite its long use in AD,[65] this reagent remains a subject of investigation.[169–184]

As in the case derivatization with PFBHA, the reaction with DNPH can occur in liquid (usually),[169,172,174,178,179,182,183,185] or solid phases.[56,170,171,173,180,181,186] Biomedical investigations most often focus on the former sample, while environmental analyses use DNPH on solid supports to isolate carbonyls from air. There are, however, several reports of solid-phase ADs for the determination of these analytes from aqueous biomedical samples.[169]

Uchiyama's group addressed two important problems in the determination of carbonyls: the ozone problem and the *cis/trans* isomerism. In the determination of carbonyls in air, it is necessary to remove ozone from the sample prior to derivatization. These investigators reported sequential reactions with 1,2-bis-(4-pyridyl)ethylene (4-BPE) or 1,2-bis-(2-pyridyl)ethylene (2-BPE) impregnated on silica gel to trap ozone and DNPH on silica gel to trap the carbonyls.[180,186] 2-BPE proved superior performance showing a more rapid reaction rate with ozone, which converted it to pyridine-2-aldehyde (2PA) and showed greater tolerance to humidity. The 2-BPE and DNPH cartridges were placed in series with the 2-BPE cartridge in front, thus protecting reagents and derivatives from oxidation.

The second issue is the formation of the Z and E isomers at the imine position of the hydrazones, which produces two peaks for each aldehyde or ketone in the analyte. This produces a multiplicity of peaks, making the interpretation of chromatograms difficult. Splitting an analyte signal into two peaks also reduces the sensitivity. Addition of 2-picoline borane to either the eluate or the eluting solvent reduces the imine to an amine, thus eliminating the Z and E isomers, which collapse into a single stable compound, which retains a strong absorption in the 350-nm region.

Derivatization with DANSYL hydrazine imparts FLU into the corresponding hydrazone. This reagent has a dual role. Due to the DANSYL, fluorophore methods exhibit high sensitivity.[7,187–191] In addition, since both reagent and hydrazones are tertiary amines, they are readily protonated in an acidic mobile phase and enter the ion source as a precharged species and produce substantial increases in sensitivity.[188,190]

Reaction with quaternized ammonium reagents also produces precharged derivatives, which increases detection sensitivity. Girard P [1-(carboxymethyl)pyridinium chloride hydrazine)] and Girard T [(carboxymethyl)trimethylammonium chloride hydrazide] are the prototypical reagents in this class. The hydrazine group reacts with carbonyls under relatively mild conditions[9,10,192] with reaction times of 18 h at ambient temperature or 1 h at 60°C. Precharged derivatives are amenable to a variety of ionization techniques including electrospray,[193] APCI,[192,193] and matrix-assisted laser desorption/ionization (MALDI),[10] which provides considerable flexibility. Higashi and coworkers noted the poor chromatographic properties of the hydrazone derivatives.[192,194] They developed 2-hydrazino-1-methylpyridine (HMP), which also had the methylpyridinium functionality to provide the charged derivative but also had superior chromatographic properties. They used HPM to develop a method for the determination of testosterone and dehydroepiandrosterone (DHEA) in human saliva,[194] which had excellent figures of merit. This method demonstrated the expected diurnal and age-related variations of T and DHEA in humans. They also detected changes of T and DHEA following supplementation with DHEA, which is a potential treatment for late-onset hypogonadism (LOH).

13.7. SULFHYDRYL GROUPS

13.7.1. Sulfhydryls Overview

Thiols are critical compounds in biology as they are major determinants of redox status. As free radical traps, thiols counter the reactive oxygen species (ROS) that arise during oxidative stress. Thiols also act as nucleophiles in biosynthesis and in biotransformation reactions. Glutathione reacts with the epoxide of leukotriene A4 as the first step in the biosynthesis of leukotrienes C4, D4, and E4. The sulfhydryl group of glutathione (GSH) also reacts with inactivating reactive electrophilic metabolites formed during biotransformation of xenobiotics. Determination of the thiols is therefore an important part of biological studies. It is also a difficult analytical problem. Two recent reviews provide thorough and extensive discussions of the field.[195,196]

13.7.2. Thiol Reactivity

Reactivity of thiols with oxygen and subsequent dimerization are key issues in the development of methods for measuring this class of analyte. In biological systems, thiols are present either as free form, that is, RSH, or as dimers Ra-S-S-Rb. An example is the couple GSH and oxidized glutathione (GSSG). The ratio of GSH/GSSG is a measure of redox status or ROS activity. Determination of GSH gives an overview of the issues in thiol analysis. Since GSH reacts with atmospheric oxygen, sampling methodologies for plasma, tissue homogenate, urine, and so on, becomes crucial. Addition of metal chelators such as ethylenediamine tetraacetic acid (EDTA) and proper control of pH in the sampling procedures can reduce the oxidation reactions. AD stabilizes the SH group, and there are numerous reagents available for this purpose. Iwasaki and coworkers[197] used a classical reagent, *N*-ethylmaleimide (NEM), to stabilize GSH for determining this analyte from saliva. In developing a method for measuring GSH from cell lysate, Zhu et al.[198] studied a number of reagents including Ellman's reagent (dithionitrobenzoic acid), vinylpyridine, iodoacetamide, and 7-fluoro-2,1,3-benzoxadiazole (ABD-F) and reported that ABD-F provided the fast reaction rate necessary to stabilize the GSH. In more general applications, 7-fluoro-2,1,3-benzoxadiazole-4-sulfonate (SBD-F) as well as ABD-F provide a useful combination of water solubility and rapid reaction.[195,196]

There are a number of reports regarding new reagents for derivatization of thiols as well as interesting new uses of more classical reagents.

Guo's group[199] combined the iodoacetamide functionality with a fluorescent moiety to prepare 1,3,5,7-tetramethyl-8-phenyl-(4-iodoacetamido) difluoroboradiaza-*s*-indacene (TMPAB-I). This reagent utilizes the iodacetamide functionality to give rapid reaction with thiols and the indacene moiety then produces the fluorescent derivatives. The reagent exhibits the necessary characteristics of rapid reactivity and stability of reaction products.

Continuing work on thiol analysis by Kusmierek and Bald[200] demonstrated the utility of 1-benzyl-2-chloropyridinium bromide (BClPB) for rapid and quantitative derivatization of thiols. The derivatives were stable and had exhibited intense absorption in the UV. The resulting HPLC method detected thiols from plasma.

Owen described propiolic acid as a reagent for thiol derivatization.[201] This reagent minimized scrambling of thiol groups, which can occur if disulfides and/or derivatives are unstable under the reaction conditions. The resulting acrylate derivatives absorb in the region of 295 nm with molar absorption of 12,500.

Seiwert and Karst expanded their work[202–204] on ferrocenyl reagents and prepared ferrocene-based maleimides for derivatizing thiols. They tested *N*-(2-ferroceneethyl) maleimide (FEM), which completely reacted with thiols within 5 min and did not show any unspecific reaction of free amino functions. The FEM derivatives were detectable by both electrochemistry and MS.

PFBBr is more commonly associated with the derivatization of organic acids but was successfully exploited in the derivatization thiols. The analytes were first sorbed onto a column and derivatized with PFBBr *in situ* using a highly hindered and lipophilic base 1,8-diazabicyclo[5.4.0]undec-7-ene (DBU). This SPAD facilitated cleanup and minimized the use of organic solvent.

Determination of total GSH (i.e., GSH + GSSG) demonstrates another aspect of thiol analysis particularly—although not solely—in the biomedical domain. Because of the ready oxidation and the GSH–GSSG couple as well as the binding of GSH to the SH groups of proteins, the measurement of total GSH requires the cleavage of the disulfide. The nature of the optimal reagents for this reducing reaction is a topic of active discussion.[195,205] Reagents typically used include sodium borohydride ($NaBH_4$), dithiothreitol (DTT), tributylphosphine (TBP), and tris-(2-carboxyethyl) phosphine (TCEP). The phoshpines, which are powerful reducing agents, are the preferred class. The last reagent is more attractive because it is water soluble, relatively in-volatile, has minimal odor, and compatible with reduction in plasma. Dimercaptopropanesulfonate (DMPS) is a newer reducing reagent, which is also water soluble and produces cleaner extracts than other dithiols such as DTT.[205]

13.8. CONCLUSION

A number of paths forward recommend themselves to investigators in this field. Automation is one, if not the first, of the priorities. Methods, however sensitive or specific, will not find a widespread use if they are too expensive. Automation is the most effective approach to cost reduction. When developing ADs, it is therefore important to focus on the final need for application to automated systems. To this end, development of reagents that produce derivatives detectable at increased sensitivity is necessary.[119,121,126,206] Despite the acknowledged sensitivity and specificity of MS, derivatization still adds substantial sensitivity to methods utilizing this high technology instrumentation. Accordingly, the development of reagents suited to this instrumentation presents an important challenge and opportunity.[61,74,88,93,207–209] Investigation of such reagents includes metal-containing reagents that provide precharged derivatives for detection by MS and opens the scope of the periodic table to the discipline. Investigation of microtechniques such as SPME[74,210–212] and SDME[35,213] (or LPME[214–216]) continue to provide useful methods that easily incorporate ADs without adding complexity to the sample preparation procedures. These should see further development particularly in automated methods.

DERIVATIZATION AGENT ACRONYMS

Ac_2O	acid anhydride
AcCl	acyl chloride
ADAM	9-anthrlydiazomethane
AEF	6-oxy-(acetyl ethylenediamine) fluorescein
APDS	3-aminopyridyl-N-hydroxysuccinimidyl carbamate
APZ	4-(4-methyl-1-piperazyl)-3-nitrobenzoyl azide
BClPB	1-benzyl-2-chloropyridinium bromide
*t*BDMSMTFA	*N*-(*t*- butyldimethylsilyl)-*N*-methyltrifluoracetamide
BODIPY FL EDA	4,4-difluoro-5,7-dimethyl-4-bora-3a,4a-diaza-*s*-indacene-3-propionyl

2-BPE	1,2-bis-(2-pyridyl)ethylene
4-BPE	1,2-bis-(4-pyridyl)ethylene
BSTFA	bis(trimethylsilyl)trifluoroacetamide
ClTMS	chlorotrimethylsilane
DPPA	diphenylphosphoryl azide
DANSYL-Cl	DANSYL chloride
DANSYLeda	DANSYL ethylenediamine
DBDF	4-(*N,N*-dimethylaminosulfonyl)-7-fluoro-2,1,3-benzoxadiazole
DMEQ-COCl	3-chlorocarbonyl- 6,7-dimethoxy-1-methyl-2(*1H*)-quinoxalinone
DMISC-Cl	1,2-dimethylimidazole-4-sulfonyl chloride
DMT-MM	4-(4,6-dimethoxy-1,3,5-triazin-2-yl)-4-methylmorpholsininium chloride
DNPH	2,4 dinitrophenylhydrazine
DMPS	dimercaptopropanesulfonate
DTT	dithiothreitol
EDA	ethylenediamine
EDAC	*N*-(3-dimethylaminopropyl)-*N*'-ethylcarbodiimide hydrochloride
EDC	1-ethyl-3-(3-dimethylaminopropyl) carbodiimide
EPMF	6-oxy-(ethylpiperazine)-9-(2'-methoxycarbonyl) fluorescein
NBDF	4-fluoro-7-nitrobenzo-2,1,3-oxadiazol
FEM	N-(2-ferroceneethyl)maleimide
FITC	fluorescein-5-isothiocyanate
FMOC	fluorenylmethyl chloroformate
FMPTS	2-fluoro-1-methylpyridinium p-toluenesulfonate
Girard P	1-(carboxymethyl)pyridinium chloride hydrazine)
Girard T	(carboxymethyl)trimethylammonium chloride hydrazide
HMP	2-hydrazino-1-methylpyridine
HFUA	4,4,5,5,6,6,7,7,8,8,9,9,10,10,11,11-heptadecafluoro-n-undecylamine
HOBt	hydroxybenzotriazole.
MNBA	2-methyl-6-nitrobenzoic anhydride
MSTFA	*N*-methyl-*N*-(trimethylsilyl)-trifluoroacetamide
NDA	naphthalene-2,3-dicarboxaldehyde
NOEPES	2 - (naphthoxy)ethyl-2-(piperidino) ethanesulfonate
OPA	o-phthalaldehyde
PA	picolinic acid
PCA	2,3-pyridinedicarboxylic anhydride
PFBBr	pentafluorobenzylbromide
PFBONH2	pentafluorobenzylhydroxylamine
PPZ	1-(2,4-dinitro-5-fluorophenyl)-4-methylpiperazine
PSC	1-pyrenesulfonyl chloride
SIFA	N-hydroxysuccinimidyl fluorescein-O-acetate
$NaBH_4$	sodium borohydride
TAHS	p-N,N,N-timethylammonioanilyl N-O-hydroxysuccinimidyl carbamate iodide
TBAB	tetrabutyl ammonium bromide
TBP	tributylphosphine
TCEP	tris-(2-carboxyethyl) phosphine
TMAH	tetramethylammonium hydroxide
TMPAB-I	1,3,5,7-tetramethyl-8-phenyl-(4-iodoacetamido) difluoroboradiaza-s-indacene
TMS-DM	trimethylsilyldiazomethane
TMSDMC	trimethylsilyl-N,N-dimethylcarbamate
TMSH	trimethylsulfonium hydroxide

REFERENCES

1. Lovelock, J.E. Midwife to the greens: The electron capture detector. *Microbiologia* **1997**, *13*, 11–22.
2. Kaufman, M.; Wiesman, Z. Pomegranate oil analysis with emphasis on MALDI-TOF/MS triacylglycerol fingerprinting. *J. Agric. Food Chem.* **2007**, *55*, 10405–10413.
3. Lampinen-Salomonsson, M.; Bondesson, U.; Petersson, C.; Hedeland, M. Differentiation of estriol glucuronide isomers by chemical derivatization and electrospray tandem mass spectrometry. *Rapid Commun. Mass Spectrom.* **2006**, *20*, 1429–1440.
4. Nilsen, J. Estradiol and neurodegenerative oxidative stress. *Front Neuroendocrinol.* **2008**, *29*, 463–475.
5. Farid, N.A.; McIntosh, M.; Garofolo, F.; Wong, E.; Shwajch, A.; Kennedy, M.; Young, M.; Sarkar, P.; Kawabata, K.; Takahashi, M.; Pang, H. Determination of the active and inactive metabolites of prasugrel in human plasma by liquid chromatography/tandem mass spectrometry. *Rapid Commun. Mass Spectrom.* **2007**, *21*, 169–179.
6. Breckenridge, S.M.; Yin, X.; Rosenfeld, J.M.; Yu, Y.H. Analytical derivatizations of volatile and hydrophilic carbonyls from aqueous matrix onto a solid phase of a polystyrene-divinylbenzene macroreticular resin. *J. Chromatogr. B* **1997**, *694*, 289–296.
7. Lord, H.L.; Rosenfeld, J.; Raha, S.; Hamadeh, M.J. Automated derivatization and analysis of malondialdehyde using column switching sample preparation HPLC with fluorescence detection. *J. Sep. Sci.* **2008**, *31*, 387–401.
8. Rosenfeld, J.; Kim, M.; Rullo, A. Development of an impregnated reagent and automation of solid-phase analytical derivatization for carbonyls: Proof of principle. *J. Chromatogr. Sci.* **2006**, *44*, 333–339.
9. Hong, H.; Wang, Y. Derivatization with Girard reagent T combined with LC-MS/MS for the sensitive detection of 5-formyl-2′-deoxyuridine in cellular DNA. *Anal. Chem.* **2007**, *79*, 322–326.
10. Wang, Y.; Hornshaw, M.; Alvelius, G.; Bodin, K.; Liu, S.; Sjovall, J.; Griffiths, W.J. Matrix-assisted laser desorption/ionization high-energy collision-induced dissociation of steroids: Analysis of oxysterols in rat brain. *Anal. Chem.* **2006**, *78*, 164–173.
11. Lee, S.H.; Pettinella, C.; Blair, I.A. LC/ESI/MS analysis of saturated and unsaturated fatty acids in rat intestinal epithelial cells. *Curr. Drug Metab.* **2006**, *7*, 929–937.
12. Mirzaei, H.; Regnier, F. Enrichment of carbonylated peptides using Girard P reagent and strong cation exchange chromatography. *Anal. Chem.* **2006**, *78*, 770–778.
13. Griffiths, W.J.; Alvelius, G.; Liu, S.; Sjovall, J. High-energy collision-induced dissociation of oxosteroids derivatised to Girard hydrazones. *Eur. J. Mass Spectrom. (Chichester, Eng.)* **2004**, *10*, 63–88.

14. Nishio, T.; Higashi, T.; Funaishi, A.; Tanaka, J.; Shimada, K. Development and application of electrospray-active derivatization reagents for hydroxysteroids. *J. Pharm. Biomed. Anal.* **2007**, *44*(3), 786–795.

15. Wang, D.; Fang, S.; Wohlhueter, R.M. N-terminal derivatization of peptides with isothiocyanate analogues promoting Edman-type cleavage and enhancing sensitivity in electrospray ionization tandem mass spectrometry analysis. *Anal. Chem.* **2009**, *81*, 1893–1900.

16. Kushnir, M.M.; Rockwood, A.L.; Bergquist, J. Liquid chromatography-tandem mass spectrometry applications in endocrinology. *Mass Spectrom. Rev.* **2010**, *29*, 480–502.

17. Kushnir, M.M.; Naessen, T.; Kirilovas, D.; Chaika, A.; Nosenko, J.; Mogilevkina, I.; Rockwood, A.L.; Carlstrom, K.; Bergquist, J. Steroid profiles in ovarian follicular fluid from regularly menstruating women and women after ovarian stimulation. *Clin. Chem.* **2009**, *55*, 519–526.

18. Guo, K.; Li, L. Differential 12C-/13C-isotope dansylation labeling and fast liquid chromatography/mass spectrometry for absolute and relative quantification of the metabolome. *Anal. Chem.* **2009**, *81*, 3919–3932.

19. Kramer, J.K.; Fellner, V.; Dugan, M.E.; Sauer, F.D.; Mossoba, M.M.; Yurawecz, M.P. Evaluating acid and base catalysts in the methylation of milk and rumen fatty acids with special emphasis on conjugated dienes and total trans fatty acids. *Lipids* **1997**, *32*, 1219–1228.

20. Basconcillo, L.S.; McCarry, B.E. Comparison of three GC/MS methodologies for the analysis of fatty acids in *Sinorhizobium meliloti*: Development of a micro-scale, one-vial method. *J. Chromatogr. B* **2008**, *871*, 22–31.

21. Li, J.; Yue, Y.; Hu, X.; Zhong, H. Rapid transmethylation and stable isotope labeling for comparative analysis of fatty acids by mass spectrometry. *Anal. Chem.* **2009**, *81*, 5080–5087.

22. Ishida, Y.; Katagiri, M.; Ohtani, H. Reaction efficiency of organic alkalis with various classes of lipids during thermally assisted hydrolysis and methylation. *J. Chromatogr. A* **2009**, *1216*, 3296–3299.

23. Shadkami, F.; Helleur, R. Use of an injection port for thermochemolysis-gas chromatography/mass spectrometry: Rapid profiling of biomaterials. *J. Chromatogr. A* **2009**, *1216*, 5903–5910.

24. Shadkami, F.; Esteves, S.; Helleur, R. Methylation by micro furnace pyrolysis. *J. Anal. Appl. Pyrolysis* **2009**, *85*, 54–60.

25. Kaal, E.; Kolk, A.H.; Kuijper, S.; Janssen, H.G. A fast method for the identification of *Mycobacterium tuberculosis* in sputum and cultures based on thermally assisted hydrolysis and methylation followed by gas chromatography-mass spectrometry. *J. Chromatogr. A* **2009**, *1216*, 6319–6325.

26. Gomez-Brandon, M.; Lores, M.; Dominguez, J. Comparison of extraction and derivatization methods for fatty acid analysis in solid environmental matrixes. *Anal. Bioanal. Chem.* **2008**, *392*, 505–514.

27. Eras, J.; Oro, R.; Torres, M.; Canela, R. Direct quantitation of fatty acids present in bacteria and fungi: Stability of the cyclopropane ring to chlorotrimethylsilane. *J. Agric. Food Chem.* **2008**, *56*, 4923–4927.

28. Juarez, M.; Polvillo, O.; Conto, M.; Ficco, A.; Ballico, S.; Failla, S. Comparison of four extraction/methylation analytical methods to measure fatty acid composition by gas chromatography in meat. *J. Chromatogr. A* **2008**, *1190*, 327–332.

29. Park, Y.; Albright, K.J.; Cai, Z.Y.; Pariza, M.W. Comparison of methylation procedures for conjugated linoleic acid and artifact formation by commercial (trimethylsilyl) diazomethane. *J. Agric. Food Chem.* **2001**, *49*, 1158–1164.

30. Ostrowska, E.; Dunshea, F.R.; Muralitharan, M.; Cross, R.F. Comparison of silver-ion high-performance liquid chromatographic quantification of free and methylated conjugated linoleic acids. *Lipids* **2000**, *35*, 1147–1153.

31. Sun, S.H.; Xie, J.P.; Xie, F.W.; Zong, Y.L. Determination of volatile organic acids in oriental tobacco by needle-based derivatization headspace liquid-phase microextraction coupled to gas chromatography/mass spectrometry. *J. Chromatogr. A* **2008**, *1179*, 89–95.

32. Luna, P.; Juarez, M.; de la Fuente, M.A. Gas chromatography and silver-ion high-performance liquid chromatography analysis of conjugated linoleic acid isomers in free fatty acid form using sulphuric acid in methanol as catalyst. *J. Chromatogr. A* **2008**, *1204*, 110–113.

33. Cha, D.; Cheng, D.; Liu, M.; Zeng, Z.; Hu, X.; Guan, W. Analysis of fatty acids in sputum from patients with pulmonary tuberculosis using gas chromatography-mass spectrometry preceded by solid-phase microextraction and post-derivatization on the fiber. *J. Chromatogr. A* **2009**, *1216*, 1450–1457.

34. Wu, J.; Lee, H.K. Injection port derivatization following ion-pair hollow fiber-protected liquid-phase microextraction for determining acidic herbicides by gas chromatography/mass spectrometry. *Anal. Chem.* **2006**, *78*, 7292–7301.

35. Saraji, M.; Bidgoli, A.A. Single-drop microextraction with in-microvial derivatization for the determination of haloacetic acids in water sample by gas chromatography-mass spectrometry. *J. Chromatogr. A* **2009**, *1216*, 1059–1066.

36. Lee, C.Y.; Jenner, A.M.; Halliwell, B. Rapid preparation of human urine and plasma samples for analysis of F2-isoprostanes by gas chromatography-mass spectrometry. *Biochem. Biophys. Res. Commun.* **2004**, *320*, 696–702.

37. Jain, K.; Siddam, A.; Marathi, A.; Roy, U.; Falck, J.R.; Balazy, M. The mechanism of oleic acid nitration by *NO(2). *Free Radic. Biol. Med.* **2008**, *45*, 269–283.

38. Saraji, M.; Farajmand, B. Application of single-drop microextraction combined with in-microvial derivatization for determination of acidic herbicides in water samples by gas chromatography-mass spectrometry. *J. Chromatogr. A* **2008**, *1178*, 17–23.

39. Scheyer, A.; Briand, O.; Morville, S.; Mirabel, P.; Millet, M. Analysis of trace levels of pesticides in rainwater by SPME and GC-tandem mass spectrometry after derivatisation with PFBBr. *Anal. Bioanal. Chem.* **2007**, *387*, 359–368.

40. Stashenko, E.E.; Mora, A.L.; Cervantes, M.E.; Martinez, J.R. HS-SPME determination of volatile carbonyl and carboxylic compounds in different matrices. *J. Chromatogr. Sci.* **2006**, *44*, 347–353.

41. Chu, K.O.; Wang, C.C.; Rogers, M.S.; Pang, C.P. Quantifying F2-isoprostanes in umbilical cord blood of newborn by gas chromatography-mass spectrometry. *Anal. Biochem.* **2003**, *316*, 111–117.

42. Deng, C.; Li, N.; Ji, J.; Yang, B.; Duan, G.; Zhang, X. Development of water-phase derivatization followed by solid-phase microextraction and gas chromatography/mass spectrometry for fast determination of valproic acid in human plasma. *Rapid Commun. Mass Spectrom.* **2006**, *20*, 1281–1287.

43. Citova, I.; Sladkovsky, R.; Solich, P. Analysis of phenolic acids as chloroformate derivatives using solid phase microextraction-gas chromatography. *Anal. Chim. Acta* **2006**, *573–574*, 231–241.

44. Shin, H.S.; Ahn, H.S.; Lee, B.H. Determination of thiazolidine-4-carboxylates in urine by chloroformate derivatization and gas chromatography-electron impact mass spectrometry. *J. Mass Spectrom.* **2007**, *42*, 1225–1232.

45. Kouloumpi, E.; Vandenabeele, P.; Lawson, G.; Pavlidis, V.; Moens, L. Analysis of post-Byzantine icons from the Church of the Assumption in Cephalonia, Ionian Islands, Greece: A multi-method approach. *Anal. Chim. Acta* **2007**, *598*, 169–179.

46. Kouloumpi, E.; Lawson, G.; Pavlidis, V. The contribution of gas chromatography to the resynthesis of the post-Byzantine artist's technique. *Anal. Bioanal. Chem.* **2007**, *387*, 803–812.

47. Kristl, J.; Veber, M.; Krajnicic, B.; Oresnik, K.; Slekovec, M. Determination of jasmonic acid in *Lemna minor* (L.) by liquid chromatography with fluorescence detection. *Anal. Bioanal. Chem.* **2005**, *383*, 886–893.

48. Nishimura, K.; Suzuki, T.; Momchilova, S.; Miyashita, K.; Katsura, E.; Itabashi, Y. Analysis of conjugated linoleic acids as 9-anthrylmethyl

esters by reversed-phase high-performance liquid chromatography with fluorescence detection. *J. Chromatogr. Sci.* **2005**, *43*, 494–499.

49. Mueller, H.W.; Binz, K. Glass capillary gas chromatography of the serum fatty acid fraction via automatic injection of lipid extracts. *J. Chromatogr.* **1982**, *228*, 75–93.

50. Mueller, H.W. Diazomethane as a highly selective fatty acid methylating reagent for use in gas chromatographic analysis. *J. Chromatogr. B* **1996**, *679*, 208–209.

51. Ferreira Do Nascimento, R.; Rodrigues Cardoso, D.; De Keukeleire, D.; dos Santos Lima-Neto, B; Wagner Franco, D. Quantitative HPLC analysis of acids in Brazilian cachacas and various spirits using fluorescence detection of their 9-anthrylmethyl esters. *J. Agric. Food Chem.* **2000**, *48*, 6070–6073.

52. Chung, T.C.; Kou, H.S.; Chao, M.C.; Ou, Y.J.; Wu, H.L. A simple and sensitive liquid chromatographic method for the analysis of free docosanoic, tetracosanoic and hexacosanoic acids in human plasma as fluorescent derivatives. *Anal. Chim. Acta.* **2008**, *611*, 113–118.

53. Du, X.L.; Zhang, H.S.; Deng, Y.H.; Wang, H. Design and synthesis of a novel fluorescent reagent, 6-oxy-(ethylpiperazine)-9-(2′-methoxycarbonyl) fluorescein, for carboxylic acids and its application in food samples using high-performance liquid chromatography. *J. Chromatogr. A* **2008**, *1178*, 92–100.

54. Du, X.L.; Wang, H.; Ma, M.; Deng, Y.H.; Zhang, H.S. 6-Oxy-(acetyl ethylenediamine) fluorescein, a novel fluorescent derivatization reagent for carboxylic acids and its application in HPLC. *J. Sep. Sci.* **2008**, *31*, 999–1006.

55. Uchiyama, S.; Matsushima, E.; Aoyagi, S.; Ando, M. Simultaneous determination of C1-C4 carboxylic acids and aldehydes using 2,4-dinitrophenylhydrazine-impregnated silica gel and high-performance liquid chromatography. *Anal. Chem.* **2004**, *76*, 5849–5854.

56. Possanzini, M.; Tagliacozzo, G.; Cecinato, A. Simultaneous determination of formic acid and lower carbonyls in air samples by DNPH derivatization. *J. Sep. Sci.* **2007**, *30*, 2460–2465.

57. Bravo, B.; Chavez, G.; Pina, N.; Ysambertt, F.; Marquez, N.; Caceres, A. Developing an on-line derivatization of FAs by microwave irradiation coupled to HPLC separation with UV detection. *Talanta* **2004**, *64*, 1329–1334.

58. Sakaguchi, Y.; Yoshida, H.; Todoroki, K.; Nohta, H.; Yamaguchi, M. Separation-oriented derivatization of native fluorescent compounds through fluorous labeling followed by liquid chromatography with fluorous-phase. *Anal. Chem.* **2009**, *81*, 5039–5045.

59. Ueyama, J.; Saito, I.; Kamijima, M.; Nakajima, T.; Gotoh, M.; Suzuki, T.; Shibata, E.; Kondo, T.; Takagi, K.; Miyamoto, K.; Takamatsu, J.; Hasegawa, T.; Takagi, K. Simultaneous determination of urinary dialkylphosphate metabolites of organophosphorus pesticides using gas chromatography-mass spectrometry. *J. Chromatogr. B* **2006**, *832*, 58–66.

60. Schindler, B.K.; Forster, K.; Angerer, J. Determination of human urinary organophosphate flame retardant metabolites by solid-phase extraction and gas chromatography-tandem mass spectrometry. *J. Chromatogr. B* **2009**, *877*, 375–381.

61. De Alwis, G.K.; Needham, L.L.; Barr, D.B. Automated solid phase extraction, on-support derivatization and isotope dilution-GC/MS method for the detection of urinary dialkyl phosphates in humans. *Talanta* **2009**, *77*, 1063–1067.

62. Pardasani, D.; Kanaujia, P.K.; Gupta, A.K.; Tak, V.; Shrivastava, R.K.; Dubey, D.K. In situ derivatization hollow fiber mediated liquid phase microextraction of alkylphosphonic acids from water. *J. Chromatogr. A* **2007**, *1141*, 151–157.

63. Pardasani, D.; Palit, M.; Gupta, A.K.; Kanaujia, P.K.; Sekhar, K.; Dubey, D.K. Microemulsion mediated in situ derivatization-extraction and gas chromatography-mass spectrometric analysis of alkylphosphonic acids. *J. Chromatogr. A* **2006**, *1108*, 166–175.

64. Mallampati, S.; Van, A.A.; Hoogmartens, J.; Van, S.A. Analysis of dideoxyadenosine triphosphate by CE with fluorescence detection. I. Derivatization through the phosphate group. *Electrophoresis* **2007**, *28*, 3948–3956.

65. Knapp, D.K. *Handbook of Analytical Derivatizations*. New York: Wiley; 1979.

66. Arditsoglou, A.; Voutsa, D. Determination of phenolic and steroid endocrine disrupting compounds in environmental matrices. *Environ. Sci. Pollut. Res. Int.* **2008**, *15*, 228–236.

67. Ballesteros, O.; Zafra, A.; Navalon, A.; Vilchez, J.L. Sensitive gas chromatographic-mass spectrometric method for the determination of phthalate esters, alkylphenols, bisphenol A and their chlorinated derivatives in wastewater samples. *J. Chromatogr. A* **2006**, *1121*, 154–162.

68. Medeiros, P.M.; Simoneit, B.R. Analysis of sugars in environmental samples by gas chromatography-mass spectrometry. *J. Chromatogr. A* **2007**, *1141*, 271–278.

69. Chimchirian, R.F.; Suri, R.P.; Fu, H. Free synthetic and natural estrogen hormones in influent and effluent of three municipal wastewater treatment plants. *Water Environ. Res.* **2007**, *79*, 969–974.

70. Shareef, A.; Angove, M.J.; Wells, J.D. Optimization of silylation using N-methyl-N-(trimethylsilyl)-trifluoroacetamide, N,O-bis-(trimethylsilyl)-trifluoroacetamide and N-(tert-butyldimethylsilyl)-N-methyltrifluoroacetamide for the determination of the estrogens estrone and 17alpha-ethinylestradiol by gas chromatography-mass spectrometry. *J. Chromatogr. A* **2006**, *1108*, 121–128.

71. Yang, L.; Lan, C.; Liu, H.; Dong, J.; Luan, T. Full automation of solid-phase microextraction/on-fiber derivatization for simultaneous determination of endocrine-disrupting chemicals and steroid hormones by gas chromatography-mass spectrometry. *Anal. Bioanal. Chem.* **2006**, *386*, 391–397.

72. Zuo, Y.; Zhang, K.; Lin, Y. Microwave-accelerated derivatization for the simultaneous gas chromatographic-mass spectrometric analysis of natural and synthetic estrogenic steroids. *J. Chromatogr. A* **2007**, *1148*, 211–218.

73. Kovacs, A.; Kende, A.; Mortl, M.; Volk, G.; Rikker, T.; Torkos, K. Determination of phenols and chlorophenols as trimethylsilyl derivatives using gas chromatography-mass spectrometry. *J. Chromatogr. A* **2008**, *1194*, 139–142.

74. Cai, L.; Koziel, J.A.; Dharmadhikari, M.; Hans van Leeuwen, J. Rapid determination of trans-resveratrol in red wine by solid-phase microextraction with on-fiber derivatization and multidimensional gas chromatography-mass spectrometry. *J. Chromatogr. A* **2009**, *1216*, 281–287.

75. Vinas, P.; Campillo, N.; Martinez-Castillo, N.; Hernandez-Cordoba, M. Solid-phase microextraction on-fiber derivatization for the analysis of some polyphenols in wine and grapes using gas chromatography-mass spectrometry. *J. Chromatogr. A* **2009**, *1216*, 1279–1284.

76. Yang, L.; Luan, T.; Lan, C. Solid-phase microextraction with on-fiber silylation for simultaneous determinations of endocrine disrupting chemicals and steroid hormones by gas chromatography-mass spectrometry. *J. Chromatogr. A* **2006**, *1104*, 23–32.

77. Vaughan, C.; Stanfill, S.B.; Polzin, G.M.; Ashley, D.L.; Watson, C.H. Automated determination of seven phenolic compounds in mainstream tobacco smoke. *Nicotine Tob. Res.* **2008**, *10*, 1261–1268.

78. Peng, X.; Wang, Z.; Yang, C.; Chen, F.; Mai, B. Simultaneous determination of endocrine-disrupting phenols and steroid estrogens in sediment by gas chromatography-mass spectrometry. *J. Chromatogr. A* **2006**, *1116*, 51–56.

79. Stopforth, A.; Burger, B.V.; Crouch, A.M.; Sandra, P. The analysis of estrone and 17beta-estradiol by stir bar sorptive extraction-thermal desorption-gas chromatography/mass spectrometry: Application to urine samples after oral administration of conjugated equine estrogens. *J. Chromatogr. B* **2007**, *856*, 156–164.

80. Fattahi, N.; Assadi, Y.; Hosseini, M.R.; Jahromi, E.Z. Determination of chlorophenols in water samples using simultaneous dispersive liquid-liquid microextraction and derivatization followed by gas chromatography-electron-capture detection. *J. Chromatogr. A* **2007**, *1157*, 23–29.

81. Ganeshjeevan, R.; Chandrasekar, R.; Kadigachalam, P.; Radhakrishnan, G. Rapid, one-pot derivatization and distillation of chlorophenols from solid samples with their on-line enrichment. *J. Chromatogr. A* **2007**, *1140*, 168–173.

82. Kawaguchi, M.; Ito, R.; Endo, N.; Okanouchi, N.; Sakui, N.; Saito, K.; Nakazawa, H. Liquid phase microextraction with in situ derivatization for measurement of bisphenol A in river water sample by gas chromatography-mass spectrometry. *J. Chromatogr. A* **2006**, *1110*, 1–5.

83. Zhao, R.S.; Cheng, C.G.; Yuan, J.P.; Jiang, T.; Wang, X.; Lin, J.M. Sensitive measurement of ultratrace phenols in natural water by purge-and-trap with in situ acetylation coupled with gas chromatography-mass spectrometry. *Anal. Bioanal. Chem.* **2007**, *387*, 687–694.

84. Fiamegos, Y.C.; Kefala, A.P.; Stalikas, C.D. Ion-pair single-drop microextraction versus phase-transfer catalytic extraction for the gas chromatographic determination of phenols as tosylated derivatives. *J. Chromatogr. A* **2008**, *1190*, 44–51.

85. Beaudry, F.; Guenette, S.A.; Winterborn, A.; Marier, J.F.; Vachon, P. Development of a rapid and sensitive LC-ESI/MS/MS assay for the quantification of propofol using a simple off-line dansyl chloride derivatization reaction to enhance signal intensity. *J. Pharm. Biomed. Anal.* **2005**, *39*, 411–417.

86. Beaudry, F.; Guenette, S.A.; Vachon, P. Determination of eugenol in rat plasma by liquid chromatography-quadrupole ion trap mass spectrometry using a simple off-line dansyl chloride derivatization reaction to enhance signal intensity. *Biomed. Chromatogr.* **2006**, *20*, 1216–1222.

87. Beaudry, F.; Ross, A.; Vachon, P. Development of a LC-ESI/MS/MS assay for the quantification of vanillin using a simple off-line dansyl chloride derivatization reaction to enhance signal intensity. *Biomed. Chromatogr.* **2007**, *21*, 113–115.

88. Lien, G.W.; Chen, C.Y.; Wang, G.S. Comparison of electrospray ionization, atmospheric pressure chemical ionization and atmospheric pressure photoionization for determining estrogenic chemicals in water by liquid chromatography tandem mass spectrometry with chemical derivatizations. *J. Chromatogr. A* **2009**, *1216*, 956–966.

89. Lin, Y.H.; Chen, C.Y.; Wang, G.S. Analysis of steroid estrogens in water using liquid chromatography/tandem mass spectrometry with chemical derivatizations. *Rapid Commun. Mass Spectrom.* **2007**, *21*, 1973–1983.

90. Licea-Perez, H.; Wang, S.; Bowen, C.L.; Yang, E. A semi-automated 96-well plate method for the simultaneous determination of oral contraceptives concentrations in human plasma using ultra performance liquid chromatography coupled with tandem mass spectrometry. *J. Chromatogr. B* **2007**, *852*, 69–76.

91. Salvador, A.; Moretton, C.; Piram, A.; Faure, R. On-line solid-phase extraction with on support derivatization for high-sensitivity liquid chromatography tandem mass spectrometry of estrogens in influent/effluent of wastewater treatment plants. *J. Chromatogr. A* **2007**, *1145*, 102–109.

92. Xiong, Z.; Sun, X.; Huo, T.; Li, N.; Zheng, Y.; Sun, Y. Development and validation of UPLC-MS/MS method for simultaneous determination of gestodene and ethinyl estradiol in rat plasma. *Biomed. Chromatogr.* **2010**, *24*(2), 160–168.

93. Lampinen, S.M.; Bondesson, U.; Hedeland, M. In vitro formation of phase I and II metabolites of propranolol and determination of their structures using chemical derivatization and liquid chromatography-tandem mass spectrometry. *J. Mass Spectrom.* **2009**, *44*, 742–754.

94. Salomonsson, M.L.; Bondesson, U.; Hedeland, M. Structural evaluation of the glucuronides of morphine and formoterol using chemical derivatization with 1,2-dimethylimidazole-4-sulfonyl chloride and liquid chromatography/ion trap mass spectrometry. *Rapid Commun. Mass Spectrom.* **2008**, *22*, 2685–2697.

95. Xu, L.; Spink, D.C. 1,2-Dimethylimidazole-4-sulfonyl chloride, a novel derivatization reagent for the analysis of phenolic compounds by liquid chromatography electrospray tandem mass spectrometry: Application to 1-hydroxypyrene in human urine. *J. Chromatogr. B* **2007**, *855*, 159–165.

96. Yamashita, K.; Okuyama, M.; Nakagawa, R.; Honma, S.; Satoh, F.; Morimoto, R.; Ito, S.; Takahashi, M.; Numazawa, M. Development of sensitive derivatization method for aldosterone in liquid chromatography-electrospray ionization tandem mass spectrometry of corticosteroids. *J. Chromatogr. A* **2008**, *1200*, 114–121.

97. Yamashita, K.; Okuyama, M.; Watanabe, Y.; Honma, S.; Kobayashi, S.; Numazawa, M. Highly sensitive determination of estrone and estradiol in human serum by liquid chromatography-electrospray ionization tandem mass spectrometry. *Steroids* **2007**, *72*, 819–827.

98. Yamashita, K.; Takahashi, M.; Tsukamoto, S.; Numazawa, M.; Okuyama, M.; Honma, S. Use of novel picolinoyl derivatization for simultaneous quantification of six corticosteroids by liquid chromatography-electrospray ionization tandem mass spectrometry. *J. Chromatogr. A* **2007**, *1173*, 120–128.

99. Honda, A.; Yamashita, K.; Miyazaki, H.; Shirai, M.; Ikegami, T.; Xu, G.; Numazawa, M.; Hara, T.; Matsuzaki, Y. Highly sensitive analysis of sterol profiles in human serum by LC-ESI-MS/MS. *J. Lipid Res.* **2008**, *49*, 2063–2073.

100. Hanczko, R.; Jambor, A.; Perl, A.; Molnar-Perl, I. Advances in the o-phthalaldehyde derivatizations. Comeback to the o-phthalaldehyde-ethanethiol reagent. *J. Chromatogr. A* **2007**, *1163*, 25–42.

101. Markowski, P.; Baranowska, I.; Baranowski, J. Simultaneous determination of L-arginine and 12 molecules participating in its metabolic cycle by gradient RP-HPLC method: Application to human urine samples. *Anal. Chim. Acta* **2007**, *605*, 205–217.

102. Rammouz, G.; Lacroix, M.; Garrigues, J.C.; Poinsot, V.; Couderc, F. The use of naphthalene-2,3-dicarboxaldehyde for the analysis of primary amines using high-performance liquid chromatography and capillary electrophoresis. *Biomed. Chromatogr.* **2007**, *21*, 1223–1239.

103. Lamba, S.; Pandit, A.; Sanghi, S.K.; Gowri, V.S.; Tiwari, A.; Baderia, V.K.; Singh, D.K.; Nigam, P. Determination of aliphatic amines by high-performance liquid chromatography-amperometric detection after derivatization with naphthalene-2,3-dicarboxaldehyde. *Anal. Chim. Acta* **2008**, *614*, 190–195.

104. Hapuarachchi, S.; Aspinwall, C.A. Design, characterization, and utilization of a fast fluorescence derivatization reaction utilizing o-phthaldialdehyde coupled with fluorescent thiols. *Electrophoresis* **2007**, *28*, 1100–1106.

105. Nakano, M.; Higo, D.; Arai, E.; Nakagawa, T.; Kakehi, K.; Taniguchi, N.; Kondo, A. Capillary electrophoresis-electrospray ionization mass spectrometry for rapid and sensitive N-glycan analysis of glycoproteins as 9-fluorenylmethyl derivatives. *Glycobiology* **2009**, *19*, 135–143.

106. Prieto-Blanco, M.C.; Chafer-Pericas, C.; Lopez-Mahia, P.; Campins-Falco, P. Automated on-line in-tube solid-phase microextraction-assisted derivatization coupled to liquid chromatography for quantifying residual dimethylamine in cationic polymers. *J. Chromatogr. A* **2008**, *1188*, 118–123.

107. You, J.; Liu, L.; Zhao, W.; Zhao, X.; Suo, Y.; Wang, H.; Li, Y. Study of a new derivatizing reagent that improves the analysis of amino acids by HPLC with fluorescence detection: Application to hydrolyzed rape bee pollen. *Anal. Bioanal. Chem.* **2007**, *387*, 2705–2718.

108. Lozanov, V.; Benkova, B.; Mateva, L.; Petrov, S.; Popov, E.; Slavov, C.; Mitev, V. Liquid chromatography method for simultaneous analysis of amino acids and biogenic amines in biological fluids with simultaneous gradient of pH and acetonitrile. *J. Chromatogr. B* **2007**, *860*, 92–97.

109. Huang, T.M.; Deng, C.H.; Chen, N.Z.; Liu, Z.; Duan, G.L. High performance liquid chromatography for the determination of glucosamine

sulfate in human plasma after derivatization with 9-fluorenylmethyl chloroformate. *J. Sep. Sci.* **2006**, *29*, 2296–2302.

110. Herraez-Hernandez, R.; Chafer-Pericas, C.; Verdu-Andres, J.; Campins-Falco, P. An evaluation of solid phase microextraction for aliphatic amines using derivatization with 9-fluorenylmethyl chloroformate and liquid chromatography. *J. Chromatogr. A* **2006**, *1104*, 40–46.

111. Chafer-Pericas, C.; Herraez-Hernandez, R.; Campins-Falco, P. A new selective method for dimethylamine in water analysis by liquid chromatography using solid-phase microextraction and two-stage derivatization with o-phthalaldialdehyde and 9-fluorenylmethyl chloroformate. *Talanta* **2005**, *66*, 1139–1145.

112. Chafer-Pericas, C.; Herraez-Hernandez, R.; Campins-Falco, P. Selective determination of trimethylamine in air by liquid chromatography using solid phase extraction cartridges for sampling. *J. Chromatogr. A* **2004**, *1042*, 219–223.

113. Chafer-Pericas, C.; Herraez-Hernandez, R.; Campins-Falco, P. Liquid chromatographic determination of trimethylamine in water. *J. Chromatogr. A* **2004**, *1023*, 27–31.

114. Herraez-Hernandez, R.; Campins-Falco, P. Derivatization of tertiary amphetamines with 9-fluorenylmethyl chloroformate for liquid chromatography: Determination of N-methylephedrine. *Analyst* **2000**, *125*, 1071–1076.

115. Kamoda, S.; Nakano, M.; Ishikawa, R.; Suzuki, S.; Kakehi, K. Rapid and sensitive screening of N-glycans as 9-fluorenylmethyl derivatives by high-performance liquid chromatography: A method which can recover free oligosaccharides after analysis. *J. Proteome Res.* **2005**, *4*, 146–152.

116. Koros, A.; Hanczko, R.; Jambor, A.; Qian, Y.; Perl, A.; Molnar-Perl, I. Analysis of amino acids and biogenic amines in biological tissues as their o-phthalaldehyde/ethanethiol/fluorenylmethyl chloroformate derivatives by high-performance liquid chromatography. A deproteinization study. *J. Chromatogr. A* **2007**, *1149*, 46–55.

117. Hanczko, R.; Koros, A.; Toth, F.; Molnar-Perl, I. Behavior and characteristics of biogenic amines, ornithine and lysine derivatized with the o-phthalaldehyde-ethanethiol-fluorenylmethyl chloroformate reagent. *J. Chromatogr. A* **2005**, *1087*, 210–222.

118. Koros, A.; Varga, Z.; Molnar-Perl, I. Simultaneous analysis of amino acids and amines as their o-phthalaldehyde-ethanethiol-9-fluorenylmethyl chloroformate derivatives in cheese by high-performance liquid chromatography. *J. Chromatogr. A* **2008**, *1203*, 146–152.

119. Deng, Y.H.; Wang, H.; Zhong, L.; Zhang, H.S. Trace determination of short-chain aliphatic amines in biological samples by micellar electrokinetic capillary chromatography with laser-induced fluorescence detection. *Talanta* **2009**, *77*, 1337–1342.

120. Shimbo, K.; Oonuki, T.; Yahashi, A.; Hirayama, K.; Miyano, H. Precolumn derivatization reagents for high-speed analysis of amines and amino acids in biological fluid using liquid chromatography/electrospray ionization tandem mass spectrometry. *Rapid Commun. Mass Spectrom.* **2009**, *23*, 1483–1492.

121. Chiesl, T.N.; Chu, W.K.; Stockton, A.M.; Amashukeli, X.; Grunthaner, F.; Mathies, R.A. Enhanced amine and amino acid analysis using Pacific Blue and the Mars Organic Analyzer microchip capillary electrophoresis system. *Anal. Chem.* **2009**, *81*, 2537–2544.

122. Pena-Gallego, A.; Hernandez-Orte, P.; Cacho, J.; Ferreira, V. Biogenic amine determination in wines using solid-phase extraction: A comparative study. *J. Chromatogr. A* **2009**, *1216*, 3398–3401.

123. Shimbo, K.; Yahashi, A.; Hirayama, K.; Nakazawa, M.; Miyano, H. Multifunctional and highly sensitive precolumn reagents for amino acids in liquid chromatography/tandem mass spectrometry. *Anal. Chem.* **2009**, *81*, 5172–5179.

124. Mayer, H.K.; Fiechter, G.; Fischer, E. A new ultra-pressure liquid chromatography method for the determination of biogenic amines in cheese. *J. Chromatogr. A* **2009**, *1217*, 3398–3401.

125. Tsumoto, H.; Murata, C.; Miyata, N.; Kohda, K.; Taguchi, R. Efficient identification and quantification of proteins using isotope-coded 1-(6-methylnicotinoyloxy)succinimides by matrix-assisted laser desorption/ionization time-of-flight mass spectrometry. *Rapid Commun. Mass Spectrom.* **2007**, *21*, 3815–3824.

126. Bomke, S.; Seiwert, B.; Dudek, L.; Effkemann, S.; Karst, U. Determination of biogenic amines in food samples using derivatization followed by liquid chromatography/mass spectrometry. *Anal. Bioanal. Chem.* **2009**, *393*, 247–256.

127. Aoyama, C.; Tsunoda, M.; Funatsu, T. Determination of selenomethionine by high-performance liquid chromatography-fluorescence detection coupled with on-line oxidation. *Anal. Sci.* **2009**, *25*, 63–65.

128. Tseng, H.M.; Gattolin, S.; Pritchard, J.; Newbury, H.J.; Barrett, D.A. Analysis of mono-, di- and oligosaccharides by CE using a two-stage derivatization method and LIF detection. *Electrophoresis* **2009**, *30*, 1399–1405.

129. Zhou, L.; Zhou, X.; Luo, Z.; Wang, W.; Yan, N.; Hu, Z. In-capillary derivatization and analysis of ephedrine and pseudoephedrine by micellar electrokinetic chromatography with laser-induced fluorescence detection. *J. Chromatogr. A* **2008**, *1190*, 383–389.

130. Zhou, L.; Wang, S.; Tian, K.; Dong, Y.; Hu, Z. Simultaneous determination of catecholamines and amino acids by micellar electrokinetic chromatography with laser-induced fluorescence detection. *J. Sep. Sci.* **2007**, *30*, 110–117.

131. Zhang, W.Z.; Lang, C.; Kaye, D.M. Determination of plasma free 3-nitrotyrosine and tyrosine by reversed-phase liquid chromatography with 4-fluoro-7-nitrobenzofurazan derivatization. *Biomed. Chromatogr.* **2007**, *21*, 273–278.

132. Kurihara, T.; Min, J.Z.; Toyo'oka, T.; Fukushima, T.; Inagaki, S. Determination of fluorescence-labeled asparaginyl-oligosaccharide in glycoprotein by reversed-phase ultraperformance liquid chromatography with electrospray ionization time-of-flight mass spectrometry. *Anal. Chem.* **2007**, *79*, 8694–8698.

133. Todoroki, K.; Etoh, H.; Yoshida, H.; Nohta, H.; Yamaguchi, M. A fluorous tag-bound fluorescence derivatization reagent, F-trap pyrene, for reagent peak-free HPLC analysis of aliphatic amines. *Anal. Bioanal. Chem.* **2009**, *394*, 321–327.

134. Barnham, K.J.; Bush, A.I. Metals in Alzheimer's and Parkinson's diseases. *Curr. Opin. Chem. Biol.* **2008**, *12*, 222–228.

135. Casado, A.; Encarnacion Lopez-Fernandez, M.; Concepcion Casado, M.; de La Torre, R. Lipid peroxidation and antioxidant enzyme activities in vascular and Alzheimer dementias. *Neurochem. Res.* **2008**, *33*, 450–458.

136. Greilberger, J.; Koidl, C.; Greilberger, M.; Lamprecht, M.; Schroecksnadel, K.; Leblhuber, F.; Fuchs, D.; Oettl, K. Malondialdehyde, carbonyl proteins and albumin-disulphide as useful oxidative markers in mild cognitive impairment and Alzheimer's disease. *Free Radic. Res.* **2008**, *42*, 633–638.

137. Pamplona, R.; Dalfo, E.; Ayala, V.; Bellmunt, M.J.; Prat, J.; Ferrer, I.; Portero-Otin, M. Proteins in human brain cortex are modified by oxidation, glycoxidation, and lipoxidation. Effects of Alzheimer disease and identification of lipoxidation targets. *J. Biol. Chem.* **2005**, *280*, 21522–21530.

138. Smerjac, S.M.; Bizzozero, O.A. Cytoskeletal protein carbonylation and degradation in experimental autoimmune encephalomyelitis. *J. Neurochem.* **2008**, *105*, 763–772.

139. Alexandru, N.; Constantin, A.; Popov, D. Carbonylation of platelet proteins occurs as consequence of oxidative stress and thrombin activation, and is stimulated by ageing and type 2 diabetes. *Clin. Chem. Lab. Med.* **2008**, *46*, 528–536.

140. Lee, I.S.; Tsai, S.W. Passive sampling of ambient ozone by solid phase microextraction with on-fiber derivatization. *Anal. Chim. Acta* **2008**, *610*, 149–155.

141. Ferreira, V.; Cullere, L.; Loscos, N.; Cacho, J. Critical aspects of the determination of pentafluorobenzyl derivatives of aldehydes by gas chromatography with electron-capture or mass spectrometric detection:

Validation of an optimized strategy for the determination of oxygen-related odor-active aldehydes in wine. *J. Chromatogr. A* **2006**, *1122*, 255–265.

142. Schmarr, H.G.; Potouridis, T.; Ganss, S.; Sang, W.; Kopp, B.; Bokuz, U.; Fischer, U. Analysis of carbonyl compounds via headspace solid-phase microextraction with on-fiber derivatization and gas chromatographic-ion trap tandem mass spectrometric determination of their O-(2,3,4,5,6-pentafluorobenzyl)oxime derivatives. *Anal. Chim. Acta* **2008**, *617*, 119–131.

143. Beranek, J.; Kubatova, A. Evaluation of solid-phase microextraction methods for determination of trace concentration aldehydes in aqueous solution. *J. Chromatogr. A* **2008**, *1209*, 44–54.

144. Schmarr, H.G.; Sang, W.; Ganss, S.; Fischer, U.; Kopp, B.; Schulz, C.; Potouridis, T. Analysis of aldehydes via headspace SPME with on-fiber derivatization to their O-(2,3,4,5,6-pentafluorobenzyl)oxime derivatives and comprehensive 2D-GC-MS. *J. Sep. Sci.* **2008**, *31*, 3458–3465.

145. Long, E.K.; Smoliakova, I.; Honzatko, A.; Picklo, M.J. Sr. Structural characterization of alpha,beta-unsaturated aldehydes by GC/MS is dependent upon ionization method. *Lipids* **2008**, *43*, 765–774.

146. Beilin, E.; Baker, L.J.; Culbert, P.; Toltl, N.P. Quantitation of acetol in common pharmaceutical excipients using LC-MS. *J. Pharm. Biomed. Anal.* **2008**, *46*, 316–321.

147. Liou, S.W.; Chen, C.Y.; Yang, T.T.; Lin, J.M. Determination of particulate-bound formaldehyde from burning incense by solid phase microextraction. *Bull. Environ. Contam. Toxicol.* **2008**, *80*, 324–328.

148. Svensson, S.; Larstad, M.; Broo, K.; Olin, A.C. Determination of aldehydes in human breath by on-fibre derivatization, solid-phase microextraction and GC-MS. *J. Chromatogr. B* **2007**, *860*, 86–91.

149. Li, Z.; Kozlowski, B.M.; Chang, E.P. Analysis of aldehydes in excipients used in liquid/semi-solid formulations by gas chromatography-negative chemical ionization mass spectrometry. *J. Chromatogr. A* **2007**, *1160*, 299–305.

150. Kawai, Y.; Takeda, S.; Terao, J. Lipidomic analysis for lipid peroxidation-derived aldehydes using gas chromatography-mass spectrometry. *Chem. Res. Toxicol.* **2007**, *20*, 99–107.

151. Pacolay, B.D.; Ham, J.E.; Wells, J.R. Use of solid-phase microextraction to detect and quantify gas-phase dicarbonyls in indoor environments. *J. Chromatogr. A* **2006**, *1131*, 275–280.

152. Chen, P.S.; Huang, S.D. Coupled two-step microextraction devices with derivatizations to identify hydroxycarbonyls in rain samples by gas chromatography-mass spectrometry. *J. Chromatogr. A* **2006**, *1118*, 161–167.

153. Stopforth, A.; Burger, B.V.; Crouch, A.M.; Sandra, P. Urinalysis of 4-hydroxynonenal, a marker of oxidative stress, using stir bar sorptive extraction-thermal desorption-gas chromatography/mass spectrometry. *J. Chromatogr. B* **2006**, *834*, 134–140.

154. Jakober, C.A.; Charles, M.J.; Kleeman, M.J.; Green, P.G. LC-MS analysis of carbonyl compounds and their occurrence in diesel emissions. *Anal. Chem.* **2006**, *78*, 5086–5093.

155. Jakober, C.A.; Riddle, S.G.; Robert, M.A.; Destaillats, H.; Charles, M.J.; Green, P.G.; Kleeman, M.J. Quinone emissions from gasoline and diesel motor vehicles. *Environ. Sci. Technol.* **2007**, *41*, 4548–4554.

156. Jakober, C.A.; Robert, M.A.; Riddle, S.G.; Destaillats, H.; Charles, M.J.; Green, P.G.; Kleeman, M.J. Carbonyl emissions from gasoline and diesel motor vehicles. *Environ. Sci. Technol.* **2008**, *42*, 4697–4703.

157. Pacolay, B.D.; Ham, J.E.; Slaven, J.E.; Wells, J.R. Feasibility of detection and quantification of gas-phase carbonyls in indoor environments using PFBHA derivatization and solid-phase microextraction (SPME). *J. Environ. Monit.* **2008**, *10*, 853–860.

158. Nemet, I.; Varga-Defterdarovic, L.; Turk, Z. Methylglyoxal in food and living organisms. *Mol. Nutr. Food Res.* **2006**, *50*, 1105–1117.

159. Saison, D.; De Schutter, D.P.; Delvaux, F.; Delvaux, F.R. Determination of carbonyl compounds in beer by derivatisation and headspace solid-phase microextraction in combination with gas chromatography and mass spectrometry. *J. Chromatogr. A* **2009**, *1216*, 5061–5068.

160. Saison, D.; De Schutter, D.P.; Delvaux, F.; Delvaux, F.R. Optimisation of a complete method for the analysis of volatiles involved in the flavour stability of beer by solid-phase microextraction in combination with gas chromatography and mass spectrometry. *J. Chromatogr. A* **2008**, *1190*, 342–349.

161. Rodriguez, M.C.; Rosenfeld, J.; Tarnopolsky, M.A. Plasma malondialdehyde increases transiently after ischemic forearm exercise. *Med. Sci. Sports Exerc.* **2003**, *35*, 1859–1865.

162. Cancilla, D.A.; Que Hee, S.S. O-(2,3,4,5,6-pentafluorophenyl)methylhydroxylamine hydrochloride: A versatile reagent for the determination of carbonyl-containing compounds. *J. Chromatogr.* **1992**, *627*, 1–16.

163. Deng, C.; Li, N.; Wang, X.; Zhang, X.; Zeng, J. Rapid determination of acetone in human blood by derivatization with pentafluorobenzyl hydroxylamine followed by headspace liquid-phase microextraction and gas chromatography/mass spectrometry. *Rapid Commun. Mass Spectrom.* **2005**, *19*, 647–653.

164. Tsai, S.W.; Kao, K.Y. Diffusive sampling of airborne furfural by solid-phase microextraction device with on-fiber derivatization. *J. Chromatogr. A* **2006**, *1129*, 29–33.

165. Gao, S.; Zhang, Z.P.; Karnes, H.T. *J Chromatogr B Analyt Technol Biomed Life Sci.* **2005**, *825*, 98–110.

166. Rosenfeld, J.M. Derivatization in the current practice of analytical chemistry. *Trends Anal. Chem.* **2003**, *22*, 785–798.

167. Rosenfeld, J. In *Sampling and Sample Preparation for Field and Laboratory: Fundamentals and New Directions in Sample Preparation*, J. Pawliszyn, J., Ed., Amsterdam: Elsevier; 2002; pp. 609–668.

168. Vogel, M.; Buldt, A.; Karst, U. Hydrazine reagents as derivatizing agents in environmental analysis—A critical review. *Fresenius J. Anal. Chem.* **2000**, *366*, 781–791.

169. Zhang, H.J.; Huang, J.F.; Lin, B.; Feng, Y.Q. Polymer monolith microextraction with in situ derivatization and its application to high-performance liquid chromatography determination of hexanal and heptanal in plasma. *J. Chromatogr. A* **2007**, *1160*, 114–119.

170. Herrington, J.S.; Fan, Z.H.; Lioy, P.J.; Zhang, J.J. Low acetaldehyde collection efficiencies for 24-hour sampling with 2,4-dinitrophenylhydrazine (DNPH)-coated solid sorbents. *Environ. Sci. Technol.* **2007**, *41*, 580–585.

171. Pal, R.; Kim, K.H. Experimental choices for the determination of carbonyl compounds in air. *J. Sep. Sci.* **2007**, *30*, 2708–2718.

172. Knighton, W.B.; Herndon, S.C.; Shorter, J.H.; Miake-Lye, R.C.; Zahniser, M.S.; Akiyama, K.; Shimono, A.; Kitasaka, K.; Shimajiri, H.; Sugihara, K. Laboratory evaluation of an aldehyde scrubber system specifically for the detection of acrolein. *J. Air Waste Manag. Assoc.* **2007**, *57*, 1370–1378.

173. Chi, Y.; Feng, Y.; Wen, S.; Lu, H.; Yu, Z.; Zhang, W.; Sheng, G.; Fu, J. Determination of carbonyl compounds in the atmosphere by DNPH derivatization and LC-ESI-MS/MS detection. *Talanta* **2007**, *72*, 539–545.

174. Duan, X.; Zhong, D.; Chen, X. Derivatization of beta-dicarbonyl compound with 2,4-dinitrophenylhydrazine to enhance mass spectrometric detection: Application in quantitative analysis of houttuynin in human plasma. *J. Mass Spectrom.* **2008**, *43*, 814–824.

175. Hong, S.B.; Kim, G.S.; Jung, Y.G.; Lee, J.H. The determination of ambient formaldehyde using a dual coil system and an assessment of dominant factors that influence its abundance in Korea. *Environ. Monit. Assess.* **2008**, *138*, 1–15.

176. Soman, A.; Qiu, Y.; Chan, L.Q. HPLC-UV method development and validation for the determination of low level formaldehyde in a drug substance. *J. Chromatogr. Sci.* **2008**, *46*, 461–465.

177. Sun, H.; Chan, K.Y.; Fung, Y.S. Determination of gaseous and particulate carbonyls in air by gradient-elution micellar electrokinetic capillary chromatography. *Electrophoresis* **2008**, *29*, 3971–3979.

178. Chen, L.; Jin, H.; Wang, L.; Sun, L.; Xu, H.; Ding, L.; Yu, A.; Zhang, H. Dynamic ultrasound-assisted extraction coupled on-line with solid support derivatization and high-performance liquid chromatography for the determination of formaldehyde in textiles. *J. Chromatogr. A* **2008**, *1192*, 89–94.

179. Elias, R.J.; Laurie, V.F.; Ebeler, S.E.; Wong, J.W.; Waterhouse, A.L. Analysis of selected carbonyl oxidation products in wine by liquid chromatography with diode array detection. *Anal. Chim. Acta* **2008**, *626*, 104–110.

180. Uchiyama, S.; Otsubo, Y. Simultaneous determination of ozone and carbonyls using trans-1,2-bis(4-pyridyl)ethylene as an ozone scrubber for 2,4-dinitrophenylhydrazine-impregnated silica cartridge. *Anal. Chem.* **2008**, *80*, 3285–3290.

181. Pal, R.; Kim, K.H. Influences of sampling volume and sample concentration on the analysis of atmospheric carbonyls by 2,4-dinitrophenylhydrazine cartridge. *Anal. Chim. Acta* **2008**, *610*, 289–296.

182. Chen, L.; Jin, H.; Xu, H.; Sun, L.; Yu, A.; Zhang, H.; Ding, L. Microwave-assisted extraction coupled online with derivatization, restricted access material cleanup, and high-performance liquid chromatography for determination of formaldehyde in aquatic products. *J. Agric. Food Chem.* **2009**, epub ahead of print.

183. Wu, J.Y.; Shi, Z.G.; Feng, Y.Q. Determination of 5-hydroxymethylfurfural using derivatization combined with polymer monolith microextraction by high-performance liquid chromatography. *J. Agric. Food Chem.* **2009**, epub ahead of print.

184. Uchiyama, S.; Inaba, Y.; Matsumoto, M.; Suzuki, G. Reductive amination of aldehyde 2,4-dinitorophenylhydrazones using 2-picoline borane and high-performance liquid chromatographic analysis. *Anal. Chem.* **2009**, *81*, 485–489.

185. Mateos, R.; Lecumberri, E.; Ramos, S.; Goya, L.; Bravo, L. Determination of malondialdehyde (MDA) by high-performance liquid chromatography in serum and liver as a biomarker for oxidative stress. Application to a rat model for hypercholesterolemia and evaluation of the effect of diets rich in phenolic antioxidants from fruits. *J. Chromatogr. B* **2005**, *827*, 76–82.

186. Uchiyama, S.; Naito, S.; Matsumoto, M.; Inaba, Y.; Kunugita, N. Improved measurement of ozone and carbonyls using a dual-bed sampling cartridge containing trans-1,2-bis(2-pyridyl)ethylene and 2,4-dinitrophenylhydrazine-impregnated silica. *Anal. Chem.* **2009**, *81*, 6552–6557.

187. Visser, S.A.; Smulders, C.J.; Gladdines, W.W.; Irth, H.; van der Graaf, P.H.; Danhof, M. High-performance liquid chromatography of the neuroactive steroids alphaxalone and pregnanolone in plasma using dansyl hydrazine as fluorescent label: Application to a pharmacokinetic-pharmacodynamic study in rats. *J. Chromatogr. B* **2000**, *745*, 357–363.

188. Herrington, J.; Zhang, L.; Whitaker, D.; Sheldon, L.; Zhang, J.J. Optimizing a dansylhydrazine (DNSH) based method for measuring airborne acrolein and other unsaturated carbonyls. *J. Environ. Monit.* **2005**, *7*, 969–976.

189. Peng, X.D.; Xu, D.H.; Jin, J.; Mei, X.T.; Lv, J.Y.; Xu, S.B. Determination of a new active steroid by high performance liquid chromatography with laser-induced fluorescence detection following the pre-column derivatization. *Int. J. Pharm.* **2007**, *337*, 25–30.

190. Al-Dirbashi, O.Y.; Rashed, M.S.; Brink, H.J.; Jakobs, C.; Filimban, N.; Al-Ahaidib, L.Y.; Jacob, M.; Al-Sayed, M.M.; Al-Hassnan, Z.; Faqeih, E. Determination of succinylacetone in dried blood spots and liquid urine as a dansylhydrazone by liquid chromatography tandem mass spectrometry. *J. Chromatogr. B* **2006**, *831*, 274–280.

191. Lord, H.L.; Rosenfeld, J.; Volovich, V.; Kumbhare, D.; Parkinson, B. Determination of malondialdehyde in human plasma by fully automated solid phase analytical derivatization. *J. Chromatogr. B* **2009**, *877*, 1292–1298.

192. Higashi, T.; Yamauchi, A.; Shimada, K. 2-Hydrazino-1-methylpyridine: A highly sensitive derivatization reagent for oxosteroids in liquid chromatography-electrospray ionization-mass spectrometry. *J. Chromatogr. B* **2005**, *825*, 214–222.

193. Griffiths, W.J.; Wang, Y.; Alvelius, G.; Liu, S.; Bodin, K.; Sjovall, J. Analysis of oxysterols by electrospray tandem mass spectrometry. *J. Am. Soc. Mass Spectrom.* **2006**, *17*, 341–362.

194. Shibayama, Y.; Higashi, T.; Shimada, K.; Odani, A.; Mizokami, A.; Konaka, H.; Koh, E.; Namiki, M. Simultaneous determination of salivary testosterone and dehydroepiandrosterone using LC-MS/MS: Method development and evaluation of applicability for diagnosis and medication for late-onset hypogonadism. *J. Chromatogr. B* **2009**, *877*, 2615–2623.

195. McMenamina, E.; Himmelfarb, J.; Nolin, T.D. Simultaneous analysis of multiple aminothiols in human plasma by high performance liquid chromatography with fluorescence detection. *J. Chromatogr. B* **2009**, *877*, 3274–3281.

196. Toyo'oka, T. Recent advances in separation and detection methods for thiol compounds in biological samples. *J. Chromatogr. B* **2009**, *877*, 3318–3330.

197. Iwasaki, Y.; Hoshi, M.; Ito, R.; Saito, K.; Nakazawa, H. Analysis of glutathione and glutathione disulfide in human saliva using hydrophilic interaction chromatography with mass spectrometry. *J. Chromatogr. B* **2006**, *839*, 74–79.

198. Zhu, P.; Oe, T.; Blair, I.A. Determination of cellular redox status by stable isotope dilution liquid chromatography/mass spectrometry analysis of glutathione and glutathione disulfide. *Rapid Commun. Mass Spectrom.* **2008**, *22*, 432–440.

199. Guo, X.F.; Wang, H.; Guo, Y.H.; Zhang, Z.X.; Zhang, H.S. Simultaneous analysis of plasma thiols by high-performance liquid chromatography with fluorescence detection using a new probe, 1,3,5,7-tetramethyl-8-phenyl-(4-iodoacetamido)difluoroboradiaza-s-indacene. *J. Chromatogr. A* **2009**, *1216*, 3874–3880.

200. Kusmierek, K.; Bald, E. Reversed-phase liquid chromatography method for the determination of total plasma thiols after derivatization with 1-benzyl-2-chloropyridinium bromide. *Biomed. Chromatogr.* **2009**, *23*, 770–775.

201. Owen, T.C. Thiol detection, derivatization and tagging at micromole to nanomole levels using propiolates. *Bioorg.Chem.* **2008**, *36*, 156–160.

202. Seiwert, B.; Karst, U. Analysis of cysteine-containing proteins using precolumn derivatization with N-(2-ferroceneethyl)maleimide and liquid chromatography/electrochemistry/mass spectrometry. *Anal. Bioanal. Chem.* **2007**, *388*, 1633–1642.

203. Seiwert, B.; Karst, U. Ferrocene-based derivatization in analytical chemistry. *Anal. Bioanal. Chem.* **2008**, *390*, 181–200.

204. Seiwert, B.; Karst, U. Simultaneous LC/MS/MS determination of thiols and disulfides in urine samples based on differential labeling with ferrocene-based maleimides. *Anal. Chem.* **2007**, *79*, 7131–7138.

205. Simek, P.; Husek, P.; Zahradnickova, H. Gas chromatographic-mass spectrometric analysis of biomarkers related to folate and cobalamin status in human serum after dimercaptopropanesulfonate reduction and heptafluorobutyl chloroformate derivatization. *Anal. Chem.* **2008**, *80*, 5776–5782.

206. Deng, Y.H.; Zhang, H.S.; Du, X.L.; Wang, H. Quantification of biogenic amines in human plasma based on the derivatization with N-hydroxy-succinimidyl fluorescein-O-acetate by high-performance liquid chromatography. *J. Sep. Sci.* **2008**, *31*, 990–998.

207. Ito, R.; Kawaguchi, M.; Sakui, N.; Okanouchi, N.; Saito, K.; Seto, Y.; Nakazawa, H. Stir bar sorptive extraction with in situ derivatization and thermal desorption-gas chromatography-mass spectrometry for trace analysis of methylmercury and mercury(II) in water sample. *Talanta* **2009**, *77*, 1295–1298.

208. Jiang, Z.; Liang, Q.; Luo, G.; Hu, P.; Li, P.; Wang, Y. HPLC-electrospray tandem mass spectrometry for simultaneous quantitation of eight plasma aminothiols: Application to studies of diabetic nephropathy. *Talanta* **2009**, *77*, 1279–1284.

209. Masood, M.A.; Xu, X.; Acharya, J.K.; Veenstra, T.D.; Blonder, J. Enhanced detection of sphingoid bases via divalent ruthenium bipyridine

complex derivatization and electrospray ionization tandem mass spectrometry. *Anal. Chem.* **2009**, *81*, 495–502.

210. Liu, W.; Zhang, L.; Chen, S.; Duan, H.; Chen, X.; Wei, Z.; Chen, G. A method by homemade OH/TSO-PMHS fibre solid-phase microextraction coupling with gas chromatography-mass spectrometry for analysis of antiestrogens in biological matrices. *Anal. Chim. Acta* **2009**, *631*, 47–53.

211. Zhang, Z.; Duan, H.; Zhang, L.; Chen, X.; Liu, W.; Chen, G. Direct determination of anabolic steroids in pig urine by a new SPME-GC-MS method. *Talanta* **2009**, *78*, 1083–1089.

212. Liu, Y.; Muralidhara, S.; Bruckner, J.V.; Bartlett, M.G. In situ derivatization/solid-phase microextraction coupled with gas chromatography/negative chemical ionization mass spectrometry for the determination of trichloroethylene metabolites in rat blood. *Rapid Commun. Mass Spectrom.* **2008**, *22*, 1023–1031.

213. Saraji, M.; Mirmahdieh, S. Single-drop microextraction followed by in-syringe derivatization and GC-MS detection for the determination of parabens in water and cosmetic products. *J. Sep. Sci.* **2009**, *32*, 988–995.

214. Saaid, M.; Saad, B.; Ali, A.S.; Saleh, M.I.; Basheer, C.; Lee, H.K. In situ derivatization hollow fibre liquid-phase microextraction for the determination of biogenic amines in food samples. *J. Chromatogr. A* **2009**, *1216*, 5165–5170.

215. Wang, X.; Luo, L.; Ouyang, G.; Lin, L.; Tam, N.F.; Lan, C.; Luan, T. One-step extraction and derivatization liquid-phase microextraction for the determination of chlorophenols by gas chromatography-mass spectrometry. *J. Chromatogr. A* **2009**, *1216*, 6267–6273.

216. Xu, L.; Basheer, C.; Lee, H.K. Chemical reactions in liquid-phase microextraction. *J. Chromatogr. A* **2009**, *1216*, 701–707.

Part II

Application Considerations

Chapter 14

Sample Preparation Techniques for Environmental Organic Pollutant Analysis

Ray E. Clement and Chunyan Hao

14.1. INTRODUCTION: SPECIAL CONSIDERATIONS FOR ENVIRONMENTAL ANALYSES AND SAMPLE PREPARATION OVERVIEW

There are two overriding considerations for the analysis of organic pollutants in environmental samples: the complexity of sample extracts, which could contain hundreds of coextracted organic compounds, and the ultralow detection limits (DLs) at which some of the more toxic analytes must be determined. In addition, target compounds must generally be quantitatively measured in the widest variety of sample types including air, industrial stack emissions, surface water, drinking water, industrial effluents, fish, vegetation and other biota, soil, sediment, industrial products and formulations, sewage sludge, combustion products, and just about everything else. Because possible human exposure to toxic chemicals and other regulatory considerations often drive the necessity for performing environmental analyses, it is important to fully understand the intended use of the analytical data before selecting the specific methods used to perform the analyses. The data quality objectives (DQOs) selected for a specific application include considerations of cost and speed.

In light of the above considerations, sample preparation for environmental analyses requires extreme care in contamination control and in all sample manipulation steps. Typical trace analyses require detection at concentrations ranging from parts per billion (ppb: 10^{-9} g analyte/g sample) to parts per trillion (ppt: 10^{-12} g analyte/g sample) or parts per quadrillion (ppq: 10^{-15} g analyte/g sample). About 20 years ago, parts per million (ppm: 10^{-6} g analyte/g sample) would have been considered a low concentration, but modern instrumentation with ultrasensitive detectors means that sample extracts containing analytes at such relatively high concentrations may need to be diluted before instrumental analysis. In fact, the very best laboratories today are able to achieve quantitative analysis of selected analytes in environmental samples at concentrations in the parts-per-quintillion (ppqt: 10^{-18} g analyte/g sample) range. Such low concentrations are generally achieved by a combination of high-volume sampling methods and/or ultrasensitive detectors, but cannot be considered "routine." In fact—to achieve widespread capability for a large number of laboratories to achieve such detection capability—the most difficult barrier to overcome may be the ability to consistently obtain a clean "blank."

Although tremendous advances have been made over the past 20 years for the chromatography of complex mixtures and in the various configurations of mass spectrometers used for analyte detection and quantitation, the basic sample preparation techniques and methods have changed to a much lesser extent. The introduction of various solid-phase extraction (SPE) and solid-phase microextraction (SPME) methods represent significant advances in the determination of organics in aqueous and air sample types, but adaptation of these methods is still ongoing. Basic liquid–liquid extraction (LLE) methods are still widely employed. For solid samples, Soxhlet extraction still represents a significant proportion of the extractions performed, although the allure of the more efficient pressurized fluid extraction (PFE, or accelerated solvent extraction [ASE]) is rapidly growing. Advances in the cleanup of sample extracts to eliminate analytical interferences have not kept pace, although in recent years, some success has been made in the construction of fully automated cleanup systems. Current research is moving to develop a fully automated sample extraction/cleanup system, but the variety and complexity of environmental sample types is such that a universal automated sample preparation system for environmental analysis is not yet available.

Handbook of Sample Preparation, Edited by Janusz Pawliszyn and Heather L. Lord

In this chapter, the basic sample preparation tools and most widely used methods will be described. References to esoteric or seldom-used methods are largely omitted. Although our coverage of sample preparation methods was not intended to be comprehensive, the methods described herein are those used by a large majority of modern environmental laboratories to prepare the majority of their samples for instrumental analysis.

14.2. THE ISOTOPE DILUTION (ID) METHOD

Tremendous advances over the past 30 years have been made in the determination of organic compounds such as the chlorinated dibenzo-*p*-dioxins (dioxins) and related compounds at trace concentrations in environmental samples. The top laboratories can detect less than 100 femtograms (fg: 10^{-15} g) of individual dioxins on a regular basis, and a few laboratories can detect 10–100 times lower than this. Most nonspecialists attribute these advances to developments in modern ultrasensitive high-resolution mass spectrometers (HRMS) and related improvements in gas chromatography (GC). However, the use of the ID technique is just as important as other analytical developments in the improved performance of current ultratrace methods.

14.2.1. What Is ID?

Some elements are found in nature in two or more isotopic forms, which are present in known, fixed ratios. For example, stable carbon isotopes are found with mass 12 and mass 13—the carbon 13 (C^{13}) isotope is only present at about 1.1% of the abundance of the C^{12} isotope. A molecule such as one of the dioxins contains 12 carbon atoms. Because the relative abundance of C^{13} is so low, the odds that one of the dioxin carbons is C^{13} is only about 1:100. By specialized synthetic procedures, a dioxin molecule can be prepared so all of the carbons are C^{13}. If the fully C^{13}-labeled dioxins are added to a sample containing native dioxins, the synthetic dioxins will all be 12 mass units (daltons) greater than the masses of corresponding dioxin molecules originally present in the sample. By the use of a mass spectrometer detector, we can precisely separate the signals from the native and isotopically labeled dioxins. We know exactly how much of the labeled dioxins were added to samples before processing, so we can determine their percent recoveries. If we make the generally safe assumption that the chemistry and physical properties of native and corresponding isotopically enriched compounds are the same through the entire analytical process, then factors affecting analyte detection such as low extraction recovery, losses during sample cleanup, and sample injection to the GC-HRMS detection system, can be adjusted according to the recovery of the corresponding isotopically enriched compound. In a similar manner, compounds can also be prepared that are enriched with the hydrogen isotope—deuterium. The natural abundance of deuterium is much less than that of hydrogen, so it is also a good candidate for use in analytical ID applications. The precision and accuracy of analytical results can be significantly improved by employing ID methods. There are several other advantages. The isotopically labeled standards can act as markers for the identification of native analytes in samples. As the ID standards are often added at concentrations 10–100 times more than expected for the analytes, the standards can act as "carriers" to improve the recoveries of analytes at ultratrace concentrations. In some cases, the presence of the ID internal standards can correct for gross blunders during preparation, such as spilling a portion of a sample (if the sample was homogeneous!). The isotopically enriched standards are great for use in method development, as they can be added at various stages of the analytical process to track down trouble spots. Mass spectrometry (MS) methods based on the ID method are only going to grow in importance. The continuing requirements of improved precision and accuracy of analytical results, lower DLs, and a greater need to defend the overall quality of analytical results reported suggest that this trend will continue for the foreseeable future.

For applications that require the determination of organic compounds in environmental samples, the ID approach in combination with the use of mass spectrometer-based instrumentation is preferred. This approach cannot always be employed as only a few isotopically enriched, pure analytical standards are commercially available, compared with the total number of organic analytes of interest. It is possible to use an existing isotopically enriched standard as surrogate for a different compound of similar chemical structure, although in such cases the achievable analytical precision and accuracy are not expected to be as good compared with the results when the enriched standard is an exact analog of the nonenriched analyte. Other key considerations of the ID method of analysis include the following:

- *Matrix considerations:* Extraction recoveries of spiked isotopically enriched surrogates on a sample surface may not exactly represent the recoveries of target analytes initially present; for example, spherically shaped particulates formed during combustion processes may consist of hollow "cores," which may contain analytes. If such particles are not physically or chemically broken down before extraction, then high recoveries of surrogates spiked onto the surfaces of particles may not reflect poor overall analyte recoveries.
- *Spiking procedures used:* Care must be taken to document and validate spiking procedures as part of the overall method to ensure consistency; for example, care should be taken to match as closely as possible the solvent used for the surrogate standard solution and the extraction solvent. Ideally, these solvents would be the same, but this is not always practical.
- *When spiking should be performed:* Most current methods recommend that isotopically enriched analytes used to correct for sample preparation recovery losses

should be spiked onto samples just before the extraction step. In some cases, spiking is performed before sampling to estimate possible analyte breakthrough during sampling—for example, a polyurethane foam plug (PUF) used for air sampling may be spiked by enriched standards before drawing air through the PUF for ambient air sampling applications. Very poor recoveries of the spiked surrogates may indicate analyte breakthrough of the PUF during sampling. Spiking of isotopically enriched analytes can be performed at any stage of a method as part of the quality control procedures of the method, for analyte loss correction, to troubleshoot problems, or during the initial method development and validation processes.

- *Validity of correction factors:* Methods that rely on the ID-MS approach for organic compound determinations in environmental samples have been shown to be very effective over decades of use. In principle—if the only sources of errors are random losses during the sample extraction and recovery steps, and random errors related to sample injection to the instrumental system and random variations in the detection system response—then analyte correction by a factor based directly on the recovery of a spiked surrogate that is an isotopically enriched analog of the analyte should be valid. In practice, limits are generally set for the extent of correction deemed to be acceptable. For most reported methods, recovery corrections ranging from 60% to 120% of the spiked quantities are used. Recoveries of surrogates much greater than 100% may be indicative of unresolved interferences from coextracted compounds, while recoveries much less than 50% may reflect gross errors during sample preparation, poor spiking techniques, or problems in recovery resulting from analysis of a highly absorptive sample matrix. Thus, to set upper and lower limits on the use of recovery correction factors also depends on the experience of analysts with specific sample types in specific applications.

14.3. COMPOUND CLASSES OF ENVIRONMENTAL INTEREST

Organic contaminants in the environment may be classified as volatile organic compounds (VOCs), semivolatile organic compounds (SVOCs), and persistent organic pollutants (POPs). These include compounds with a wide range of chemical properties that determine whether they are most likely associated with air, aqueous, or solid sample types. Many of the compounds of greatest environmental concern are aromatic substances containing several halogens, or aromatic or aliphatic hydrocarbons. These substances include the following compound classes: organochlorine pesticides (OCPs), polychlorinated biphenyls (PCBs), polycyclic aromatic hydrocarbons (PAHs), polychlorinated dibenzo-*p*-dioxins (PCDDs)/polychlorinated dibenzofurans (PCDFs) and other endocrine-disrupting compounds (EDCs), halogenated (brominated and chlorinated) flame retardants (HFRs), pharmaceuticals and personal care products (PPCPs), water disinfection by-products (DBPs), and perfluorinated chemicals (PFCs). Table 14.1 is a summary of environmental compound classes of greatest current interest and gives some of the properties of each with respect to the sample preparation approaches required.

Quantitative analysis of these organic contaminants in environmental samples consists of several general processes: representative sampling, sample preparation (extraction and cleanup), gas or liquid chromatographic separation, instrument detection, and data processing. Sample preparation is almost always the most time-consuming step in environmental sample analysis and is a key factor determining the speed, cost, and quality of the final results. Thus, despite advanced improvements in instrumentation, target analytes still need to be isolated from sample matrices and enriched to a suitable concentration level prior to instrument detection, which are the two principal goals of sample preparation procedures.

An effective sample preparation procedure must meet the following requirements:

- good recovery of target analytes (minimal loss),
- efficient removal of coextracted matrix components (potential interferences),
- compatible with subsequent instrumental analysis, and
- robust and rugged so results are repeatable.

To accomplish the goals of sample preparation—to provide the target analyte at a concentration appropriate for positive detection and free from interfering matrix elements—a representative portion of the sample has to be processed by several procedures that may include filtration, drying, dissolution, homogenization, extraction, evaporation, chemical cleanup (i.e., through column chromatography), and derivatization. The fewer steps used in a sample preparation protocol, the more convenient, cost- and time-effective, and rugged it is likely to be. Simpler sample preparation methods are more likely to be subject to automation, which may lead to even more accurate, reliable, and reproducible methods. Different sample preparation techniques have been applied to various sample types that generally include air, water, soil, and biota. These will be discussed here. The United States Environment Protection Agency (USEPA) methods will be principally used in these discussions as representative examples to illustrate practical sample preparation methods used for real-world sample analyses because of the extensive recognition and acceptance of these methods.

14.4. SAMPLING AND SAMPLE PREPARATION FOR AIR SAMPLES

Air samples can be complicated to handle because of the reactivity of some components, the interaction between these components, and the sampling equipment and sampling media employed. There are two common ways to

Table 14.1. Description and Basic Properties of Organic Compounds of Environmental Interest

Pollutant Class	Basic Description and Structure	Properties Related to Sample Preparation
Water disinfection by-products (DBPs); most common DBPs are trihalomethanes (THMs) and haloacetic acids (HAAs)	THMs are compounds in which three hydrogen atoms of methane (CH_4) are replaced by halogen atoms. THMs with the same halogen atoms are called haloforms. HAAs are carboxylic acids in which hydrogen atom(s) of acetic acid (CH_3COOH) are replaced by halogen atom(s).	THMs are highly volatile with low water solubility, thus the best candidates for the purge-and-trap technique; HAAs are strong organic acids, readily extracted to organic phase under acidic pH (<1) conditions. Common solvents are methyl *tert*-butyl ether (MTBE) or *tert*-amyl methyl ether (TAME).
Chlorinated benzenes (CBs), organochlorines (OCs), and polychlorinated biphenyls (PCBs)	OCs are halogen derivatives of alicyclic (BHCs, lindane, dieldrin, and aldrin) or aromatic (DDTs, endosulfans, and methoxychlor) hydrocarbons; PCBs (chemical formula $C_{12}H_{10-x}Cl_x$) consist of 209 congeners with 1–10 chlorines attached to biphenyl (two benzene rings connected by a carbon–carbon bond).	OCs, CBs, and PCBs are neutral chlorinated organic compounds with low water solubility and high lipid solubility; these compounds are easily extracted/eluted by organic solvents such as hexane, toluene, and dichloromethane.
Polycyclic aromatic hydrocarbons (PAHs)	PAHs are formed during the incomplete burning of organic substances and those of environmental interest consist of fused four-, five-, six- or seven-member rings; PAH molecules consist of only C, H atoms.	PAHs are highly lipophilic and highly aromatic organic compounds; they can be extracted/eluted by aromatic solvents like toluene or strong organic solvents like dichloromethane (DCM).
Polychlorinated dibenzo-*p*-dioxins (PCDDs) and polychlorinated dibenzofurans (PCDFs)	PCDDs are a group of compounds with chlorine atoms attached to a dibenzo-*p*-dioxin skeletal core structure in which two benzene rings are joined by two oxygen bridges; PCDFs have the similar core structure and properties as PCDDs with two benzene rings fused to one furan ring in the middle; PCDDs and PCDFs with chlorines in the 2, 3, 7, and 8 positions are most toxic.	PCDDs/PCDFs are aromatic diethers/aromatic ethers; hexane, toluene, acetone, ethanol, or the combination of these solvents is a good choice for extraction/elution; extensive chromatographic cleanup procedures are followed to remove other chemical interferences from extracts before analysis to achieve ultralow detection limits (parts per trillion or lower).
Halogenated flame retardants (HFRs); major ones are brominated flame retardands (BFRs): polybrominated diphenyl ethers (PBDEs)	HFRs are halogenated organic compounds that can inhibit or resist the spread of fire; PBDEs have similar structures as PCDDs but with two brominated benzene rings linked by an ether group.	Due to the similar structure and properties, the sample preparation procedure for dioxins can also be used for PBDEs.
Perfluorinated (or polyfluoroalkyl) compounds (PFCs)	PFCs are a family of fluorocarbon-containing organic chemicals; PFCs are named by the number of carbon atoms in the linear chain and by the terminal group, for example, sulfonate or carboxyl groups; studies have been focused mainly on the C8 molecules, perfluorooctanoic acid (PFOA), and perfuorooctane sulfonate (PFOS) as PFOA exhibits the highest degree of bioaccumulation and toxicity in PFCs.	Fluorocarbons are more hydrophobic and lipophobic than hydrocarbons; the acid/sulfonate part of PFOA/PFOS is hydrophilic, so they can be extracted/eluted by polar organic solvents like methanol.
Chlorophenols (CPs) and phenoxyacids (PAs)	CPs are a group of chemicals in which 1–5 chlorines are added to phenol (C_6H_5OH); 19 different CPs in total; PAs are phenoxy derivatives of fatty acids; common members of the group are 2,4-D; 2,4,5-T; MCPA; 2,4-DB; MCPB; and fenoprop.	CPs and PAs are acidic chlorinated organics. These are usually extracted/eluted under acidic pH (≤2) with ethers, alcohol, and petroleum solvents. Tetrahydrofuran/ethyl acetate/hexane combination gives better efficiency.

Table 14.1. *(Continued)*

Pollutant Class	Basic Description and Structure	Properties Related to Sample Preparation
Carbamate (CAR) pesticides	CAR pesticides are derivatives of carbamic acid (NH_2COOH), so they are a group of organic compounds sharing a common group of –NH(CO)O– with substituted groups on both acid and amine sides; commonly used CARs include aldicarb, bendiocarb, carbaryl, carbofuron, and triallate.	CARs are not persistent in the environment; solubility in organic solvents depends on different substitution groups in the molecules; commonly used solvents for extraction/elution are DCM, acetone, methanol, and acetonitrile.
Organophosphate (OP) pesticides	OP pesticides are derivatives of phosphoric acid (H_3PO_4), so they are a group of organic compounds containing P(V), a terminal oxygen or sulfur double-bonded to phosphorus, and organic groups are connected to phosphorus through three –O– or –S– bridges; commonly used OPs include parathion, malathion, chlorpyrifos, diazinon, and azinphos methyl.	OPs solubility in organic solvents depends on substitution groups; commonly used solvents for extraction/elution are similar to carbamate pesticides.
Phenylurea (PHU) herbicides	PHU herbicides are derivatives of urea (H_2NCONH_2) with a phenyl group (C_6H_5) and are thermally unstable; commonly used PHUs include linuron, diuron, and metobromuron.	PHUs have both lipophilic and hydrophilic groups, so they are soluble in a wide range of solvents; commonly used ones are methanol and acetonitrile.
Glyphosate (GLY) herbicide	GLY, *N*-(phosphonomethyl) glycine, is an aminophosphonic analog of the natural amino acid glycine; usually used as salt forms: ammonium salt, potassium salt, and isopropylamine salt; GLY can also be attributed to OPs group, but it does not inhibit cholinesterase activity as do most OPs.	Due to the high ionic strength, GLY has high solubility in water (~900 g/L), and insoluble in common organic solvents; derivatizaon is needed most of the time and silanized or plastic labwares should be used.
Quaternary ammonium cation (Quat) herbicides: diquat (DQ) and paraquat (PQ)	Quats are positively charged polyatomic ions of the structure NR_4^+. DQ (6,7-dihydrodipyrido-[1,2-a:2′,1′-c] pyrazidinumion) and PQ (*N,N′*-dimethyl-4,4′-bipyridiniumion) are common Quats.	As ionic compounds, DQ and PQ are highly soluble in water, but have little to no solubility in organic solvents; ion-pair reagents in methanol needed for elution and silanized or plastic labwares should be used.
Triazine (TRI) herbicides	TRI pesticides are a group of compounds consisting of an *s*-triazine-ring ($C_3H_3N_3$), which is analogous to the benzene ring but with three carbons replaced by nitrogens. Commonly used TRIs include atriazine, simazine, and cyanazine.	TRI molecules contain a heterocyclic ring and exhibit basic characteristics due to the multiple nitrogen atoms; usually extracted/eluted under basic pH (>12) conditions with DCM or chloroform.
Endocrine-disrupting compounds (EDCs) and pharmaceuticals and personal care products (PPCPs)	These compounds are grouped together because of their biological effects or mechanisms. EDCs and PPCPs cover a wide range of chemicals with different compositions and structures like hormones, antibiotics, and fragrances.	Due to the diversity of chemical types, various sample preparation methods have been chosen for studies on EDCs and PPCPs depending on target compound types, detection instruments, and sample matrices.

gather chemicals in air: collection of air into an evacuated container, or drawing air through a sampling media such as a filter, sorbent, or solution. An air pump is usually used to draw air through different sample media. Due to the sample nature (i.e., most sample analytes are VOCs or SVOCs), air samples are normally analyzed by GC-based methods.

14.4.1. Direct Air Analysis

If air samples are collected in a container, either passively by evacuating the container prior to sampling or actively by using a pump, the sample is referred to as a "whole air" sample and the compounds remain in the gaseous phase inside the container. Commonly used sampling vessels are stainless steel containers called Summa canisters (passive sampling) and inexpensive Tedlar® bags (requires a pump for sampling, active sampling). Once collected, whole air samples can be transported to the laboratory for analysis without further treatment. As whole air samples do not undergo any preconcentration, these sampling methods suffer from low sensitivity and are applicable to applications where target analytes are stable and have vapor pressures greater than about 0.1 Torr. The recoveries of target compounds in whole air samples depend on the humidity of the sample, the inertness of the container, and the chemical stability of the target analytes.

Tedlar bags are made from a polyvinyl fluoride (PVF) film that is chemically inert to a wide range of compounds. Bags used in environmental applications are typically tough, durable, and applicable for use over a wide temperature range (about −35–105°C). Bags have heat-sealed, leak-proof seams and can be used to collect both liquid and air samples for applications such as stack sampling, soil gas sampling, ambient and indoor air sampling, and groundwater testing by active pumping of samples into the bags. They are not recommended for use with reactive compounds and compounds that might adhere to the bags' surfaces. Due to the chemical structure of Tedlar, highly polar compounds may adhere to the inner surfaces, and low-molecular-weight compounds may permeate the bags. Therefore, holding times are relatively short (about 3 days for most VOCs). Black bags can be used to protect the contents from light that may cause sample degradation, but these bags contain carbon black, which can adsorb some compounds. Layered bags can be used with a layer of black Tedlar on the outside and clear Tedlar on the inside of the bag to eliminate the possibility of sample component adsorption. Bags can be reused, but extreme care must be taken to evacuate, clean, and thoroughly flush bags after each use with purified air or nitrogen. An analysis of the final flush should be performed before additional sampling is attempted.

Summa canisters are initially evacuated and are filled in the field by simply opening a valve. They are inert, rugged, and suitable for longer holding times (up to 30 days) compared with Tedlar bags, but are considerably more expensive and are more difficult to transport. Standard sampling sizes are 1 or 6 L. The name "Summa" refers to the process by which the internal surfaces of the canister were deactivated by means of an electropolishing step combined with chemical deactivation. In practice, it is recommended to incorporate a particulate filter with the inlet of a Summa canister. For DLs in the parts-per-billion range, Summa canisters must be employed. Tedlar bags may be used for applications where VOCs are expected to be at parts-per-million concentrations.

14.4.2. Air Preconcentration Sampling

To analyze air for concentrations of organic chemicals at parts-per-billion or lower levels, sample preconcentration is necessary in order to acquire sufficient material for identification and quantification. The two major preconcentration techniques for air samples are cryogenic collection and the use of filters or sorbents. Cryogenic collection can be used together with GC. In practice, a collection trap is submerged in nitrogen; an air sample is admitted into the trap by opening a valve; and an in-line GC analysis column oven is cooled to subambient temperatures (−50°C). Once sample collection is completed, a valve is switched so that the GC carrier gas sweeps the contents of the trap onto the head of the cooled GC column. Meanwhile, the liquid cryogen is removed, and the trap is heated to enable rapid sample transfer. The GC column is temperature programmed, and the component peaks eluting from the columns are identified and quantified by (usually) a mass spectrometer detector.

When using a sorbent for sampling, air is drawn through the sorbent by a pump while target compounds are retained. By this method, target compounds need to be extracted from the sorbent after collection. Target compound levels are calculated based on the air volume that passed through the collection media; the airflow and collection data must be accurately recorded to precisely measure the air volume. A variety of sorbent materials have been employed including Tenax, Carbotrap, and Florisil.

Headspace is a traditional sampling technique for VOCs and SVOCs, which can be used for liquid and solid sample as well as gaseous samples by allowing target analytes to partition into a gas phase (headspace) from the sample matrix, followed by GC or GC-MS analysis. There are two types of headspace methods: static and dynamic. Dynamic headspace (DHS) is also known as purge and trap, where the outgassed analytes are purged with ultrapure nitrogen and collected onto an adsorbent material in a Teflon tube, then flushed onto a cold trap. Following collection, the cold trap is heated quickly, and the analytes are swept by gas flow into the detection system. Typical DLs for compounds analyzed by DHS are in the nanogram per gram range. Static headspace (SHS) refers to the sealed environment in which a liquid or solid sample is placed into a vial and the outgassed products are collected. All of the components that are volatile at or below the preset temperature escape from the sample to form a gaseous "headspace" above the sample.

SHS can also be used to "sniff" air directly from an area of concern for investigation. The DLs for SHS applications depend on the volatility of the compounds and are typically in the submicrogram range. Complex sample types, which would otherwise require sample extraction or preparation or which are difficult to analyze directly, are ideal candidates for headspace as they can be placed directly in a vial with little or no sample preparation. Headspace is an ideal choice for VOCs and SVOCs, such as residual solvents or low-molecular-weight additives.

Many method examples, like those presented in the *Compendium of Methods for the Determination of Toxic Organic Compounds in Ambient Air*, have been developed by the USEPA to deal with organic pollutants in air samples.[1] Method TO-13A "Determination of Polycyclic Aromatic Hydrocarbons (PAHs) in Ambient Air Using Gas Chromatography/Mass Spectrometry (GC/MS)" and Method TO-10A "Determination of Pesticides and Polychlorinated Biphenyls in Ambient Air Using Low Volume (PUF) Sampling Followed by Gas Chromatographic/Multi-Detector Detection (GC/MD)" use *solid sorbents* to sample organic pollutants including PAHs, organochlorines, carbamates, ureas, pyrethroids, phenolics, organophosphorus pesticides, and PCBs. Air is drawn through a filter and sorbent cartridge by using a high-volume flow rate, and then the filters and sorbent cartridges are extracted by appropriate solvents. The solvent extract is further concentrated then analyzed. Similar approaches have been used in sampling OCPs, PCBs, PCDDs, and naphthalene. For some compounds, like ketene, sampling and derivatization processes can be combined together. This is the case in Method TO-5 "Method for the Determination of Aldehydes and Ketones in Ambient Air Using High-Performance Liquid Chromatography (HPLC)."

14.5. SAMPLE PREPARATION FOR AQUEOUS SAMPLES

Aqueous sample types are likely analyzed more frequently for environmental analysis than any other, because of the regular monitoring performed worldwide for the protection of drinking water supplies. Other aqueous sample types include groundwater, surface water, wastewater, industrial effluents, atmospheric precipitation, stack sampling condensate, and the aqueous fraction of industrial waste. Depending on the specific sample matrix, some aqueous samples need to be filtered to remove suspended matter content and particulates (e.g., influent and effluent from wastewater treatment plants or surface water). For cleaner sample types such as drinking water, filtration is unnecessary and may in fact cause biased results. Some target analytes in aqueous samples are actually hydrophobic (e.g., chlorinated dibenzo-*p*-dioxins and other highly chlorinated species) and may adsorb to glassware surfaces, so that the recoveries of these compounds may be low unless special precautions are followed during all sample preparation steps. In such cases, silanization of glassware before use may be performed to reduce surface activity, and/or plastic labware may be used.[2] Hydrophobic analytes may be associated with fine particulates in aqueous sample types, so in some cases, the particulates may be filtered from the aqueous fraction, extracted separately by a procedure used for solid sample types, and the extracts combined to produce the total analyte quantity. Unfortunately, in such cases, it is difficult to determine precisely which proportion of the analyte was initially associated with the particulate matter, and which was dissolved in water. Some target analytes, like the tetracycline antibiotics, can form chelating complexes with metal ions, and the addition of ethylenediaminetetraacetic acid (EDTA) disodium salt to the sample is highly recommended to maximize analyte recovery.[3,4]

14.5.1. LLE

Traditionally, organic solvents were used to extract organic compounds from water samples based on their solubility difference in two immiscible liquid phases—an aqueous sample and an organic solvent. In LLE, the organic solvent in which target compounds have greater affinity is chosen so that efficient mass transfer of analytes from the water matrix to the organic solvent occurs. The two immiscible liquid phases having different densities are thoroughly mixed for efficient transfer of target compounds between the two liquid phases, then the two immiscible phases are separated by gravity (lower layer allowed to drain through a stopcock) or by suction of the top layer. This procedure is also known as solvent extraction or partitioning. The ideal solvent is able to selectively extract target analytes from the aqueous samples. However, it is not an easy job to find a solvent to avoid coextraction of possible interfering sample components. As a practical rule, polar solvents are chosen to extract polar targets, and nonpolar solvents are chosen to extract less polar targets. Separatory funnels are usually employed for LLE to easily separate the two phases after the extraction, and then the organic phase is collected and concentrated by reducing volume through solvent evaporation. One or a combination of rotary evaporator, nitrogen blowdown apparatus, or Kuderna–Danish (K-D) concentrator are commonly employed in the laboratory to concentrate sample extracts. Suitable organic solvents for LLE should meet the following requirements: (1) low boiling point; (2) good dissolving ability for target compounds; (3) low miscibility with water; (4) higher or lower density than water; and (5) safe, inexpensive, and nonreactive with target compounds. Depending on the distribution ratio of analytes and the desired recovery, several stages of extraction may be required. To obtain better recoveries in practice, several smaller portions of the solvent are used instead of one or two large volumes. To avoid excessive use of toxic organic solvents—when the distribution ratios of target analytes are high for the solvent chosen—liquid–liquid microextraction (LLME) can been employed. Significantly smaller volumes of solvents are required in LLME to obtain the required extraction efficiency. LLME techniques have been demonstrated to work well for hydrophobic classes of compounds such as the chlorinated dibenzo-*p*-dioxins, PCBs, OCPs, and other highly chlorinated compound classes. Typical LLE extraction schemes for USEPA methods are summarized in Table 14.2.

14.5.2. SPE

Since its advent in 1978, SPE has represented a revolution in sample preparation techniques. SPE, where a solid sorbent replaces one of the immiscible liquid phases, is very similar in practice to LLE, but provides advantages such as higher analyte recoveries, less organic solvent usage, shorter preparation times, easier operation, and more effective analyte enrichment. Now, SPE has replaced the traditional LLE to become the most commonly used sample preparation technique for aqueous sample types in environmental analysis. A typical SPE setup is illustrated in Figure 14.1. To perform SPE, extraction cartridges are first put onto a manifold and preconditioned with solvent. As the aqueous samples are then passed through cartridges assisted by positive pressure or vacuum, the components of interest are adsorbed onto the solid sorbents—separating them from the aqueous matrix. After this, cartridges are washed with one or more solvents

Table 14.2. LLE (LLME) Schemes Used in USEPA Drinking Water and Wastewater Analysis Methods (500, 600, and 1600 Series)

Method	Pollutants	LLE Scheme (~1 L of Sample Unless Stated Otherwise)
504.1	1,2-Dibromoethane (EDB); 1,2-dibromo-3-chloropropane (DBCP); and 1,2,3-trichloropropane (123TCP)	Microextraction with 2 mL hexane, 35-mL sample
505	Organohalide pesticides and commercial PCBs	
506	Phthalate and adipate esters	Extract with 60 mL dichloromethane (DCM) for three times, and then 40 mL hexane
507	Nitrogen-, phosphorus-containing pesticides (pH 7)	Extract three times with 60 mL DCM
550	PAHs	
553	Benzidines and nitrogen-containing pesticides (pH 7)	
604	Phenol and substituted phenols (pH 1–2)	
606	Phthalate esters	
607	Nitrosamines (pH 5–9)	
609	Nitroaromatics and isophorone (pH 5–9)	
611	Haloethers	
612	Chlorinated hydrocarbons	
613, 1613	Dioxins	
616	Cycloprate, kinoprene, methoprene, resmethrin (pH 6.8)	
619, 629	Triazine pesticides	
622.1	Organophosphorus pesticides (pH 6–8)	
627	Dinitroaniline pesticides	
632	Carbamate and urea pesticides	
633	Organonitrogen pesticides	
635, 636 638, 639	Rotenone, bensulide, oryzalin, bendiocarb (pH 7)	
637, 640	Mercaptobenzothiazole disulphide (MBTS) and 2-(thiocyanomethylthio) benzothiazole (TCMTB), mercaptobenzothiazole (pH 6–8)	
642	Biphenyl and *o*-phenylphenol	
644	Picloram (pH 1.5–2.5)	
645	Amine pesticides and lethane (pH 5–9)	
1614	Polybrominated diphenyl ether (PBDE)	
1668	Chlorinated biphenyl congeners (CBs)	

Table 14.2. (*Continued*)

Method	Pollutants	LLE Scheme (~1 L of Sample Unless Stated Otherwise)
508	Chlorinated pesticides	Extract with 60 mL DCM twice
515.1	Chlorinated acids	pH ≤ 2, extract with 120, 60, and 60 mL of ethyl ether
515.4		pH ~1, microextraction with 4 mL methyl *tert*-butyl ether (MTBE), 40-mL sample
551.1	Chlorination DBPs, chlorinated solvents, and halogenated pesticides/herbicides	Microextraction with 3 mL MTBE or 5 mL pentane, 50-mL sample
552	Haloacetic acids	pH ≤ 0.5, extract with 15 mL MTBE twice, 100-mL sample
552.3	Haloacetic acids and dalapon	pH < 0.5, microextraction with 4 mL MTBE or *tert*-amyl methyl ether (TAME), 40-mL sample
556.1	Derivatized carbonyl compounds	Microextraction with 4 mL hexane, 20-mL sample
600.1	Hexachlorophene and dichlorophen	pH 4.0–4.5, extract with 200 mL DCM for three times
605	Benzidines	pH 6.5–7.5, extract with 100, 50, and 50 mL chloroform; back-extract with 25 mL, 1 M H_2SO_4 for three times; neutralize pH; extract again with 30, 20, and 20 mL chloroform
614.1	Organophosphorus pesticides (pH 5–9)	Extract with 60 mL of 15% (v/v) DCM in hexane for three times
617	Organohalide pesticides and PCBs	
646	Amine pesticides and lethane (pH 5–9)	
615	Chlorinated herbicides	pH ≤ 2, extract with 150, 50, and 50 mL ethyl ether
618	Volatile pesticides: chloropicrin and ethylene dibromide	pH 6–8, microextraction with 4 mL cyclohexane, 20-mL sample
620	Diphenylamine	pH 6–8, extract with 400 mL DCM for 18–24 h using continuous extractor
634	Thiocarbamate pesticides	
643	Bentazon	pH 2.5–3.5, extract with 60 mL DCM for three times, back-extract with 2 mL of 0.1 M NaOH twice
1614	Polybrominated diphenyl ether (PBDE)	Extract with 300–450 mL DCM for 16–24 h using continuous extractor
1625	Semivolatile toxic organics (pH 12–13, then at pH < 2)	
1656	Organohalide pesticides and PCBs (pH 5–9)	
1657	Organophosphorus pesticides (pH 5–9)	
1658	Phenoxy-acid herbicides (pH < 2)	
1668	Chlorinated biphenyl congeners (CBs)	
1698	Steroids and hormones	
1699	Organochlorine, organophosphorus, triazine, and pyrethroid pesticides	
1659	Dazomet (base hydrolysis to methyl isothiocyanate)	After hydrolysis, saturated with salt and extracted with 2.5 mL ethyl acetate, 50-mL sample
1660	Pyrethrins and pyrethroids	Saturated with salt and extracted by stirring with 160 mL acetonitrile, 750-mL sample
1664	Oil and grease and nonpolar material	pH < 2, extract with 30 mL *n*-hexane for three times
1698	Steroids and hormones	Extract with 100 mL DCM for three times
1699	Organochlorine, organophosphorus, triazine, and pyrethroid pesticides	

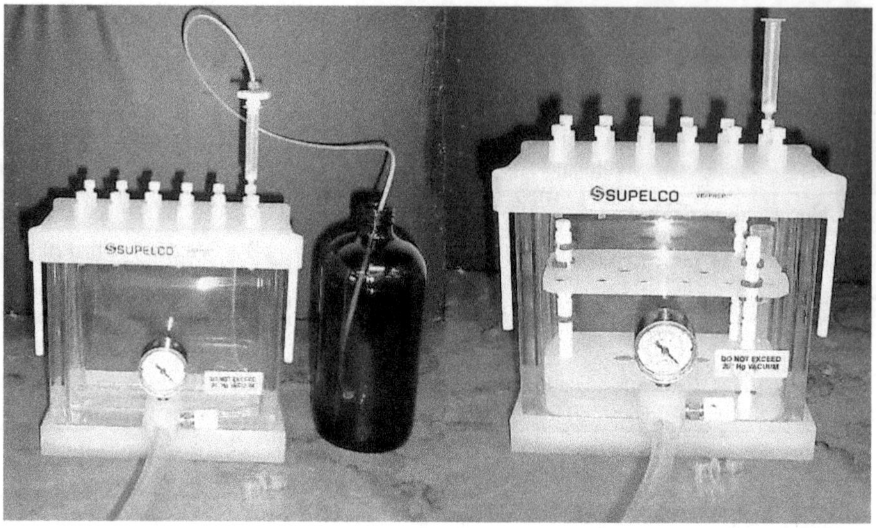

Figure 14.1. A typical solid-phase extraction (SPE) setup. Left side: aqueous sample is passing through an SPE cartridge; target compounds are retained by sorbent. Right side: retained target compounds are eluted from the cartridge into a collection tube for further analysis.

Table 14.3. SPE Sorbents Used in USEPA Drinking Water and Wastewater Analysis Methods (500, 600, and 1600 Series)

Method	Pollutants	SPE Sorbent(s) Used
508.1	Chlorinated pesticides/herbicides and organohalides	Octadecyl (C18) groups
525.2	Organic compounds	
532	Phenylurea pesticides	
550.1	PAHs	
554	Derivatized carbonyl compounds	
1614	Polybrominated diphenyl ether (PBDE)	
1668	Chlorinated biphenyl congeners (CBs)	
1699	Organochlorine, organophosphorus, triazine, and pyrethroid pesticides	
509	Ethylene thiourea (ETU)	Modified large pore Kieselguhr with a granular structure
515.2	Chlorinated acids	Polystyrenedivinylbenzene (SDVB)
526	Semivolatile organic compounds	
527	Selected pesticides and flame retardants	
537	Selected perfluorinated alkyl acids (PFAAs)	
528	Phenol	Cross-linked and chemically modified styrene divinyl benzene copolymer
521	Nitrosamines	Activated carbon
522	1,4-Dioxane	
535	Chloroacetanilide and other acetamide herbicide degradates	
529	Explosives and related compounds	Divinylbenzene/vinylpyrrolidone copolymer or sulfonated SDVB
548.1	Endothall	Intermediate strength, primarily tertiary amine anion exchanger
549.2	Diquat and paraquat	C8 sorbent conditioned and eluted with ion-pairing reagents
552.1	Haloacetic acids and dalapon	AG-1-X8 anion exchange resin
553	Benzidines and nitrogen-containing pesticides	C18 or neutral SDVB
1694	Pharmaceuticals and personal care products (PPCPs)	Hydrophilic-lipophilic balance (HLB) polymer

For method details: http://www.epa.gov/osa/fem/methcollectns.htm.

to get rid of matrix interferences as much as possible. Finally, components of interest are eluted with an appropriate elution solvent and are collected into collection vessels. In some cases, both sample extraction and partial cleanup are achieved at the same time. By carefully tailoring the specific materials used to construct SPE cartridges, they can be designed to preferentially adsorb either the target analytes, or the potential interfering compounds. Analytes are then either recovered by using a small volume of solvent to wash them from the SPE cartridge, or by LLME from the aqueous sample passed through the cartridge.

SPE cartridges are packed with different types of sorbents such as nonpolar phases (C18, C8), polar phases (silica, alumina, Florisil), ion exchange phases (Accell Plus QMA and Accell Plus CM), and polymeric phases (hydrophilic–lipophilic balance [HLB]). Different SPE cartridges have been utilized to selectively retain various groups of organic pollutants by a variety of adsorptive forces from van der Waals, hydrogen bonding, and dipole–dipole interactions to hydrophilic–lipophilic and ion exchange processes. Table 14.3 gives a summary of SPE adsorbents used in USEPA methods.

Effective extraction by SPE depends mainly on the appropriate choice of cartridges and eluting solvents. In practice, an appropriate pH can be chosen to achieve specific retention of the analytes of interest on the SPE cartridge depending on the pK_a values of target analytes, and a suitable elution solvent should be employed to realize the selective elution from SPE cartridges. For the compounds with amphoteric characteristics, extraction on SPE cartridges can be performed under either acidic or basic conditions for the different types of cartridges in use. Waters oasis HLB cartridge has been the cartridge of choice for the preconcentration of both polar and nonpolar compounds using the same extraction conditions—a prerequisite for multiresidue analysis of different targets. Recently, the use of online SPE coupled directly to GC- or liquid chromatography (LC)–MS is another big achievement and has been successfully applied in the analysis of organic targets at low nanogram per liter levels. This automation technique enables high-precision, high-throughput, and cost-effective sample preparation for routine analysis.

A rather unique method for extracting organics from water using an adsorbent (Ambersorb 572®) for analysis of nitrosamines in water and wastewater has been reported,[5] in which 500 mL of sample is spiked with isotopically labeled surrogates, then extracted via adsorption by using 200 mg of Ambersorb, a carbonaceous resin. The sorbent is recovered by filtering the sample, and the analytes recovered from the Ambersorb by using only 400 μL of dichloromethane, followed by determination by GC–tandem mass spectrometry (GC-MS/MS), or GC–high-resolution MS.[6] The initial Ambersorb extraction can be automated by simply adding the resin to each sample bottle, then gently rolling the sample bottles via a roller apparatus (see Fig. 14.2). This method is not in widespread use, but illustrates the advantages of sorbent resins for effective extractions using very small quantities of extraction solvents.

Figure 14.2. A simple roller device to extract organics from water samples by using Ambersorb solid-phase extraction. Only 200 mg of Ambersorb are required for effective extraction, because gentle rolling of the sample containers for several hours ensures close contact of the Ambersorb surface with all of the samples. The gentle rolling also guards against the formation of emulsions, which can occur if strong physical shaking of samples is performed.

14.5.3. SPME

The SPME technology is simple, fast, economical, versatile, and solvent free. It integrates sampling, extraction, concentration, and sample introduction into a single step and is a good prechromatographic sample preparation technique allowing relatively simple automation and field analysis. In SPME, a fiber coated with extraction phase, either polymeric liquid or solid sorbent or a combination of both, stored inside a steel syringe needle and an assembly holder is used to adsorb target compounds directly from the sample by depressing the plunger to expose the fiber to the sample. When the SPME needle is inserted into a sample, the target analytes partition from the sample matrix into the extraction phase until equilibrium is reached. Two types of SPME techniques can be used in practice: headspace (HS)-SPME and direct immersion (DI)-SPME. In HS-SPME, the extraction phase is exposed in the headspace of gaseous, liquid, or solid samples. In DI-SPME, the extraction phase is directly immersed in liquid samples. After extraction, the SPME needle with concentrated analytes can be directly coupled with GC or LC for desorption and analysis. Coatings used in SPME typically have strong affinities for certain organic compounds to give good selectivity and concentration effect. A suitable polarity and thickness of the coating need to be chosen according to the target analytes. Depending on the target analytes, derivatization agents or extraction-enhancing additives can also be put into the extraction phase for optimized operation. SPME has gained wide acceptance for many applications of environmental organic analysis from volatile to nonvolatile compounds in different sample types.

14.6. SAMPLE PREPARATION FOR SOLID SAMPLES

Environmental solid samples encompass a much wider variety of variations in matrix composition than any other type of environmental sample. Major types include sediments, a variety of soils, fly ash and other combustion products, industrial powders and related products, solid sewage, and industrial wastes. Solid samples are also frequently derived from the particulates separated from aqueous or air samples by various types of filter media including cellulite filters, glass-fiber filters, and PUF cartridges. Many solid samples contain water that will lessen the extraction ability of nonpolar solvents. Polar solvents (e.g., acetone, methanol) or solvent mixtures containing polar solvents (e.g., hexane/acetone) can aid for wetting samples for extraction, but in general, drying samples prior to extraction is the most common approach prior to extraction in sample preparation methods. If the desired compounds can be easily dissolved in a selective solvent, then a simple sample dissolution and filtration procedure can be used to separate the target analytes from other insoluble matrix substances. However, in most cases, target analytes are strongly sorbed onto solid surfaces of the sample matrix, so it is important to use more aggressive extraction approaches.

14.6.1. Soxhlet Extraction Method

Franz von Soxhlet described a procedure to extract lipids from foods in 1879, and it has been used for solid sample extraction ever since.[7] A typical Soxhlet extractor is shown in Figure 14.3. In Soxhlet extraction, the sample is usually dried, ground into small particles, and placed in a porous cellulose thimble made from thick filter paper or a glass thimble with a porous frit. The extraction thimble is placed in a glass extraction chamber suspended above a flask containing the extracting solvent and below a condenser. The solvent is heated, and solvent vapors evaporate through a distillation path onto the condenser (which is generally water cooled). The cooled solvent vapors condense into a liquid that trickles onto the sample surface in the extraction chamber. When the solvent—which thoroughly percolates through the sample—exceeds a certain level, it overflows and trickles back down into the boiling flask by a siphon sidearm. The solvent continuously recirculates until the extraction is stopped. During each cycle, a portion of the target analytes dissolves in the solvent and is carried to the solvent flask. After many cycles, target compounds are quantitatively removed from the sample and are concentrated in the distillation flask. After extraction, the solvent is carefully evaporated to enrich analytes. The nonsoluble portion of the solid sample remains in the thimble, and is usually discarded. In this manner, Soxhlet extraction can separate analytes from the insoluble solid matrix components and as the sample is always extracted by refluxed pure solvent, extraction efficiency can be great even when target compounds have limited solubility in the solvent employed. The Soxhlet extraction method suffers from the disadvantages of requiring use of large amounts of glassware (difficult to clean) and relatively large solvent volumes (ca. 100–200 mL). However, once set up, samples can be left unattended for many hours. It is common to leave samples for Soxhlet extraction overnight, although quantitative extraction of analytes in some applications may only require a few hours. If samples are not dried properly it is possible that the Soxhlet extraction can be rendered completely ineffective, as solvents not miscible with water may flow around the extraction thimble rather than percolate through the sample.

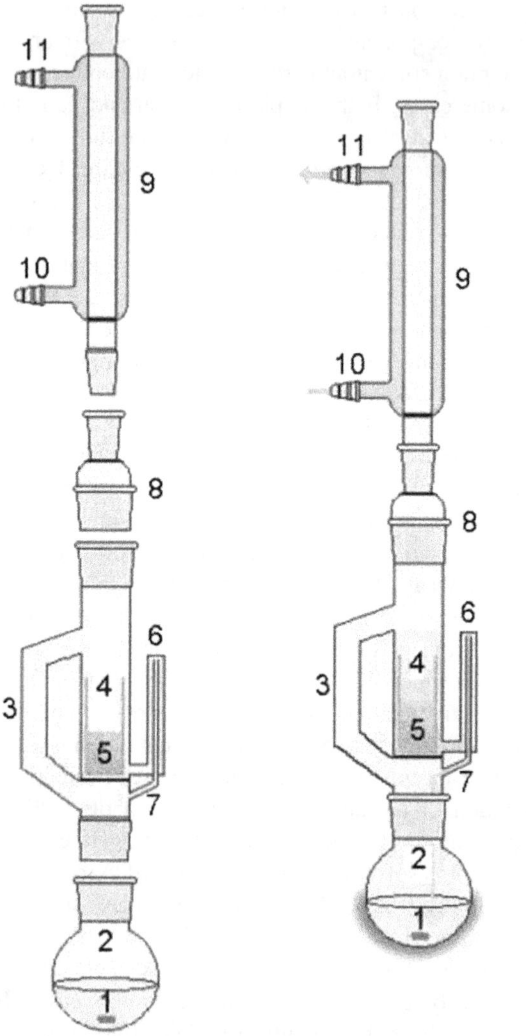

Figure 14.3. A typical Soxhlet extractor: 1, stirrer bar; 2, flask; 3, distillation path; 4, Soxhlet thimble; 5, sample; 6, siphon arm inlet; 7, siphon arm outlet; 8, expansion adapter; 9, condenser; 10, cooling water in; 11, cooling water out (diagram reprinted from Wikimedia Commons open source).

14.6.2. Use of Microwave for Organics Extraction

Microwave energy has been used to improve the extraction efficiency of target compounds in solid environmental samples.[8] The microwave energy affects molecules by ionic conduction and dipole rotation, so that heating appears right in the core of the sample and spreads from the inside to the outside in contrast to a conventional heating process where the heat penetrates from outside to inside. When microwave-assisted extraction (MAE) is conducted in closed vessels made of microwave transparent materials (e.g., Teflon liners), the temperatures reached during the extraction will be much greater than the boiling points of the solvents, and the pressures can be as high as 1500–1600 psi. To select MAE solvents, analyte solubility, and microwave-absorbing properties should both be considered. The larger the dipole moment of the solvent, the faster the solvent will heat under microwave irradiation. Nonpolar solvents like hexane will not heat, and polar solvents like acetone will heat in seconds. A mixture of hexane and acetone is an ideal solvent for most environmental MAE applications.

14.6.3. PLE

ASE, also called PLE[9] or PFE, is an innovative sample preparation technique that combines elevated temperatures and pressures to achieve fast and efficient extraction of organic analytes from environmental samples. The USEPA recently adopted PLE as an accepted method for the extraction of test compounds from solid sample types.[10] High temperatures (50–200°C) and high pressures (1000–2000 psi) in PLE can significantly improve the ability of extraction solvents to penetrate into the sample matrix, so small volumes of solvent and short extraction times can still provide good extraction efficiencies. When developing a new PLE method, it is recommended to start at about 100°C, or 20°C below the thermal degradation point of the target analytes. Typical PLE procedures include the following steps:

- Load the sample into the cell.
- Fill the cell with solvent.
- Heat and pressurize the cell.
- Hold the pressure and temperature for extraction.
- Pump clean solvent into the sample cell.
- Purge solvent from the cell to the collection vessel.

For the majority of reported methods where PLE is employed, all steps can be completed within about 20 min. Figure 14.4 shows a PLE device in a working environmental laboratory. Although the initial cost of a PLE device is great, significant improvements in extraction efficiencies for solid samples make the PLE technique a worthy addition to the environmental analysis laboratory. The high temperatures and pressures engaged in both MAE and PLE can significantly improve the extraction efficiencies of target analytes from environmental solid samples. However, this also results in the major disadvantage of the two extraction

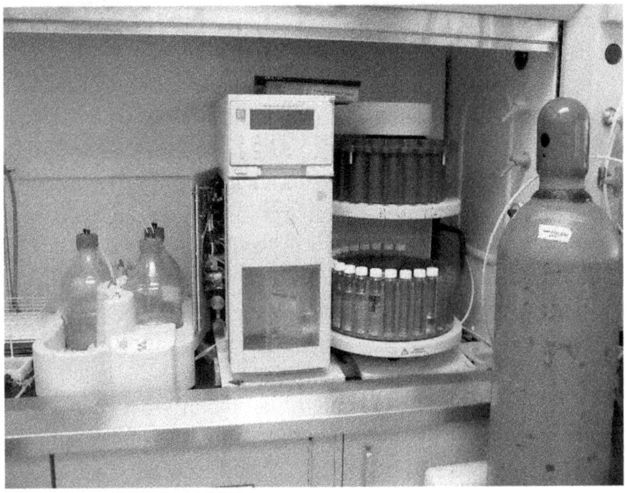

Figure 14.4. Dionex device for pressurized fluid extraction (or accelerated solvent extraction). Several samples may be set up for automated sequential extraction.

techniques: they are so efficient that they extract almost all of the organics present—in addition to the target analytes. A cleanup step after MAE or ASE extraction is required in almost all cases. In early USEPA methods, Soxhlet extraction was used to deal with solid sample types, but the efficient ASE or MAE techniques are more common now. In USEPA Method 3545A "Pressurized Fluid Extraction (PFE),"[10] which was updated in 1998, ASE is employed to extract organic analytes, like organophosphorus (OP) pesticides, OCPs, chlorinated herbicides, PCBs, and PCDDs/PCDFs from soils, clays, sediments, sludges, and waste solids to achieve analyte recoveries equivalent to those from Soxhlet extraction with less solvent and significantly less extraction time.

14.7. SAMPLE PREPARATION FOR BIOTA SAMPLES

Environmental biota samples include fish, animal tissues, plants, and vegetables. Prior to processing tissue samples, a determination of the exact nature of the tissue sample to be analyzed should be made. For example, common requests for analysis of fish tissue can include whole fish–skin on, whole fish–skin removed, edible fish fillets (filleted in the field or by the laboratory), specific organs, and other portions. All biota samples are complex, and sample extracts generally contain a wide variety of organic compounds in addition to the target analytes.

Environmental biota samples generally require initial homogenization by a grinder, blender, or mixer. After homogenization, another important step is to remove water. After that, samples can often be treated as described for solid samples for further treatment. Drying of biota samples can be simply accomplished by direct addition of a drying agent. The choice of drying agent depends on the sample type: while sodium sulfate works well for soil and sediment samples, hydromatrix (a special grade of flux-calcined, high-purity, inert diatomaceous earth) is a good choice for animal tissues, and cellulose may be used for fruits and vegetables. Magnesium sulfate should not be used together with PLE due to its potential for melting at higher temperatures, while sodium sulfate should not be used with polar solvents, due to its solubility in the solvents and the possibility of recrystallization during the experiment. Oven-drying and freeze-drying are other viable alternatives for sample drying prior to extraction; however, the recovery of more volatile compounds may be decreased by employing these procedures.

The natural lipid content of animal tissues can cause problems during extraction and analysis. Levels of lipids in different species and portions of tissues can vary widely. As lipids are soluble in organic solvents, they may be present in sufficient quantity in sample extracts to damage the chromatographic column used for analysis. Acid digestion (normally with 6 N HCl) and acid/base back-extraction are used commonly for the removal of lipids from tissue samples. An anthropogenic isolation column or acidified silica gels are used to remove lipid from sample extracts. Centrifugation can also be used to separate high mass biomolecules (lipid, protein, etc.) from target small organic molecules in sample extracts. Gel permeation chromatography (GPC) can be effective for lipid removal, but this method is not widely used. The HCl digestion/extraction, acid/base back-extraction procedure used in USEPA Method 1613, Tetra-through Octa-Chlorinated Dioxins and Furans by Isotope Dilution HRGC/HRMS, is an example for animal tissue sample preparation.[11]

14.8. CLEANUP AND CONTAMINATION CONTROL

After the extraction of samples, additional cleanup may not be necessary for relatively clean sample types (e.g., groundwater and drinking water). However, it can be beneficial when dealing with complex sample types, especially for the detection of highly toxic analytes at ultratrace concentrations, like the chlorinated dioxins. Due to the high toxicity of dioxins, regulations usually require low DLs in the range of picogram per liter in water and picogram per gram in soils. To meet this requirement, in USEPA Method 1613,[11] GPC was used to remove high-molecular-weight interferences that caused GC column performance to degrade. Acid, neutral, and basic silica gel, alumina, and Florisil were used to remove nonpolar and polar interferences; alumina and Florisil were used to remove chlorodiphenyl ethers; Carbopak/Celite was used to remove nonpolar interferences, and HPLC was used to provide specificity for the 2,3,7,8-substituted and other PCDD and PCDF isomers. Cleanup is generally required to achieve better results when the extraction procedure is efficient but not selective. As discussed earlier, SPE cartridges can be used for cleanup after MAE or PLE.

14.8.1. Contamination Control

To achieve better analytical results, it is essential to avoid cross-contamination during analysis. Reuse of labwares should be minimized, and all reusable labwares must be scrupulously cleaned as soon as possible after each use. Sonication of glassware containing a detergent solution may aid in the removal of substances adsorbed by glass surfaces. During glassware cleaning, baking reusable glassware in an oven may be warranted after highly contaminated samples are encountered, but it is not recommended as a routine part of cleaning as repeated baking of glassware may cause active sites on the glass surface that will irreversibly adsorb analytes. An essential element of contamination control in modern ultratrace organics laboratories is to employ separate sets of glassware for low-, medium-, and high-concentration sample types. For example, for the determination of the chlorinated dibenzo-*p*-dioxins in various environmental samples, the glassware used to prepare incinerator fly-ash samples for analysis (high levels of analytes expected) should not be used for the analysis of sediment samples (medium levels of analytes expected), and the glassware used for any solid sample type should never be used for the determination of dioxins in surface or drinking water samples (very low analyte concentrations expected).

14.8.2. Quality Control during Sample Processing

For contamination control, several blanks should be employed during environmental analysis applications. A field blank consists of a clean sample matrix (e.g., pure water) brought to the sampling location, which was processed through the complete sampling and sample analysis steps to monitor possible contamination introduced during sampling and sample transport. A sample preparation blank is the clean sample matrix that goes through the identical sample preparation procedure in parallel to samples to monitor possible contamination during sample preparation. An instrument blank usually consists of injection of the pure solvent used for the final analytical sample solution being analyzed at the same time as the samples—usually one at the beginning and one at the end of the sequence of the analytical runs to determine whether there is any contamination in the instrument detection system. Before being used to process samples, all labwares should be rinsed by appropriate solvents. The final rinsing solution should usually be collected as a labware rinse blank, which is analyzed to check for potential contamination.

Contamination control is a critical procedure to determine the quality of environmental analysis data. Analytical interferences can be introduced during the sampling and transportation process, or by the solvents, reagents, and glassware used, and from any other sample processing step during the analysis. The use of high-purity water, reagents, and solvents is essential to minimize interference problems when determinations of analytes in environmental samples at parts-per-billion or lower concentrations are performed. The various types of blanks described here should be used routinely to monitor possible contamination at different stages of sample processing. If the results of any blanks are positive, this must be considered during the interpretation of analytical data. If results of blanks are above certain predetermined levels, the validity of any analytical data generated can be suspect for certain applications. If results from blank determinations are significantly lower than the analytical data generated (say, 1%), then they may not indicate significant contamination. However, the interpretation of blank results depends somewhat on the DQOs of the work performed. For example, significant blank results for data to be used for legal proceedings or for the determination of health risk are far more serious than for general survey data or for the determination of general environmental trends.

14.9. SAMPLE PREPARATION METHOD DEVELOPMENT AND VALIDATION

Method development encompasses the process of setting up an analytical procedure that will be appropriate for the analysis of certain types of samples for selected analytes, within stated DQOs. Some practical "rules of thumb" for sample preparation method development could include the following: develop a method that is as simple as possible by eliminating unnecessary sample transfers and combining several steps into a multifunction purpose wherever possible; ensure the final method is rugged and rapid by reducing the sample size and using advanced technologies to save time, labor, and cost; and, employ the ID-MS approach wherever possible.

To develop a sample preparation method, an analyst first needs to think about the purpose of the method. The intended sample matrix will determine the possible sample preparation techniques and instrumentation being used. Then, the analyst needs to obtain knowledge about the chemical and physical properties of the target analytes and possible matrix interferences, then choose the most suitable sample preparation technique(s) for the best performance, including appropriate extraction solvent/sorbent/method, the most appropriate chromatographic–MS instrumentation to be used for the detection of analytes, and the degree of quality control blanks and procedures that must be introduced to assure that the DQOs for the method have been attained. For convenience, the commonly used solvents are listed here in order of increasing polarity and the ones miscible in water are highlighted with ***bold and italic font***: hexane < isooctane < carbon tetrachloride < chloroform < methylene chloride (dichloromethane) < ***tetrahydrofuran*** < diethyl ether < ethyl acetate < ***acetone*** < ***acetonitrile*** < ***isopropanol*** < ***methanol*** < ***water*** < ***acetic acid***. With the wide range of available sorbents, the selection of SPE cartridges is complex. The interaction between the sorbent and the target analytes not only depends on their own chemical nature, but also the properties of sample matrices. Other important factors to consider should include the compatibility between the extraction solvent and the analytical instrument used in the measurement of the target analytes. To achieve optimum method performance, sometimes it is necessary for solvent exchange after extraction. The most common way to change solvent is to evaporate the original solvent just to dryness and reconstitute with the new solvent. However, labile or volatile analytes may suffer from low recoveries.

Method validation is the process to prove an analytical method is acceptable for its intended purpose. After a method is developed, important method attributes need to be established to demonstrate the method is fit for purpose. Basic method validation for an environmental analytical method usually includes the determination of selectivity, accuracy, precision, spike recovery, instrument linearity and repeatability, and *DL*. Except for instrument linearity and repeatability, most of these parameters depend on the sample preparation procedure employed. Selectivity is the ability of an analytical method to differentiate and quantify the analyte in the presence of other components in the sample. Sample preparation blanks mentioned earlier should be tested for possible interferences during sample preparation. The accuracy of an analytical measurement can be determined by replicate analysis of samples containing known amounts of the analyte to demonstrate the closeness of mean test results obtained by the method to the true value (concentration) of the analyte. Precision represents the closeness of individual measurements to each other among replicate analyses. The spike recovery is a measure of how much of the original amount of the standard put into the spiked sample was detected at the end. The DL or limit of detection (LOD) is the smallest amount of an analyte that can be reliably distinguished from background noise. The way a DL is calculated is not universally accepted, but in practice, most definitions of DL specify an analyte signal that is three to five times greater than the average signal due to noise. This should not be confused with sensitivity, which is the minimum difference in concentration that can be reliably distinguished between two samples having slightly different concentrations. In practice, these terms are often used interchangeably, although sensitivity and LD refer to fundamentally different characteristics of a method. A related method characteristic is the limit of quantitation (LOQ), or quantitation limit. The LOQ is important because although we may be able to reliably detect the presence of an analyte at a very low DL, that does not mean we can necessarily know how much of the analyte is actually present. The LOQ is the limit at which we can reliably determine how much of the analyte is present. In practice, LOQ is often chosen to be five times the DL.

The most important method attributes for a sample preparation protocol are method accuracy and precision. The accuracy and precision of a method are usually determined by repeatedly measuring some traceable reference standard—certified reference material (CRM) or standard reference material (SRM). The use of reference materials is required by agencies and programs such as the American Society for Testing Materials (ASTM), the International Organization for Standardization (ISO), the International Union of Pure and Applied Chemistry (IUPAC), the European Economic Community (EEC), and others. Reference materials are usually supplied with a certificate of analysis and a materials safety data sheet (MSDS). These give the well-characterized composition or properties of the standards used to perform instrument calibrations for quality assurance, to verify the accuracy of specific measurements, and to support the development of new measurement methods. The need for reference materials for the validation and control of the analysis of environmental samples is steadily increasing as part of the effort to improve the quality of analyses. The natural-matrix reference materials provide the metrology community with means by which they can (1) validate analytical methods, (2) control the quality of their measurement process; (3) compare measurement results within projects, programs, within laboratory, and between laboratories over an extended period of time; and (4) support the traceability and credibility of measurement results.

14.10. TRENDS AND FUTURE DEVELOPMENTS

The information provided in this chapter represents the most basic and important knowledge for the preparation of environmental samples for trace organic analysis in the modern laboratory. Coverage has in no means been comprehensive, but the methods described herein cover the majority of common methods used today.

In the environmental field—in general—the number and variety of organic pollutants has grown tremendously over the past decade, and this trend should continue for the foreseeable future. A short list of compounds that are now monitored by environmental agencies worldwide, or that are being investigated, include PCBs, PCDDs/PCDFs, PPCPs, pesticides, herbicides, HFR chemicals, PFCs, PAHs, and many others. To get an idea of how complex the environment can be, one only needs to consider the Chemical Abstracts Service Chemical Registry Database (http://www.cas.org/expertise/cascontent/registry/index.html), where as of June 25, 2009, there were 48,079,635 organic and inorganic substances listed. The list is updated daily, so it will be interesting for the reader to see how many new substances have been added from the time of the completion of this chapter (June 25, 2009) to the date when these words are read. Although some of these substances appear only in research laboratories, many of them, if produced in any significant quantity for commercial purposes, can now be found worldwide in environmental samples—if analytical DLs are low enough.

In response to the pressing need for analyzing environmental samples for more and more chemical pollutants at lower and lower concentrations, automated chemical cleanup procedures are now being developed. One example of this effort is shown in Figure 14.5. In this apparatus are six parallel modules each controlled by a series of solvent lines, valves, pumps, and solvents that can be programmed to push solvent through a series of cartridges containing solid sorbents for chemical cleanup. Very complex cleanup procedures can be adapted for this apparatus. Once the details have been developed and the programming completed, a set of six samples can be processed without operator intervention until completed. Work is under way to add an extraction apparatus to the cleanup modules so the entire sample extraction/cleanup process can be automated. Although initial capital cost is great, the advantages in sample throughput, consistency of performance, optimized processing speed, and savings of technicians' time make this an attractive approach for routine environmental laboratories. A summary of early work in the automation of sample preparation methods for PCDDs and related compounds in biological sample types has been presented elsewhere.[12]

Figure 14.5. Device for automated cleanup of samples for chlorinated dioxin analysis manufactured by Fluid Management Systems. The device consists of a system of cleanup modules, connecting lines, and solvent pumps that can be customized to process samples through a variety of cleanup columns.

With so many analytes to monitor at DLs in the parts-per-trillion range or lower, issues of cost, speed, and automation are of increasing importance. Contamination control at all stages of the analysis, including sampling and sample preparation, is crucial to generating valid data for regulatory purposes. Although much attention is given to the new separation and detection tools such as multidimensional chromatography, fast GC, and various chromatography–MS technologies (time-of-flight, LC-MS, Fourier transform mass spectrometry), it is ultimately the sample preparation techniques that will determine the overall precision and accuracy of the results, and the speed and cost of the analyses.

REFERENCES

1. *Compendium of Methods for the Determination of Toxic Organic Compounds in Ambient Air*, 2nd ed., EPA/625/R-96/010b. Cincinnati, OH: USEPA Office of Research and Development; 1999.

2. Bruno, F.; Curini, R.; Di Corcia, A.; Nazzari, M.; Samperi, R. Method development for measuring trace levels of penicillins in aqueous environmental samples. *Rapid Commun. Mass Spectrom.* **2001**, *15*, 1391–1400.

3. Lindsey, M.E.; Meyer, M.; Thurman, E.M. Analysis of trace levels of sulfonamide and tetracycline antimicrobials in groundwater and surface water using solid-phase extraction and liquid chromatography/mass spectrometry. *Anal. Chem.* **2001**, *73*, 4640–4646.

4. Yang, S.; Cha, J.; Carlson, K. Quantitative determination of trace concentrations of tetracycline and sulfonamide antibiotics in surface water using solid-phase extraction and liquid chromatography/ion trap tandem mass spectrometry. *Rapid Commun. Mass Spectrom.* **2004**, *18*, 2131–2145.

5. Taguchi, V.Y.; Jenkins, S.W.D.; Wang, D.T.; Palmentier, J.F.P.; Reiner, E.J. Determination of N-nitrosodimethylamine by isotope dilution, high-resolution mass spectrometry. *Can. J. Appl. Spectrosc.* **1994**, *39*, 87–93.

6. Jenkins, S.W.D.; Koester, C.J.; Taguchi, V.Y.; Wang, D.T.; Palmentier, J.F.P.; Hong, K.P. N-nitrosodimethylamine in drinking water using a rapid, solid-phase extraction method. *Environ. Sci. Pollut. Res.* **1995**, *2*, 207–210.

7. Soxhlet, F. Die gewichtsanalytische Bestimmung des Milchfettes. *Polytechnisches J. (Dingler's)* **1879**, *232*, 461–465.

8. U.S. Environmental Protection Agency Method 3546. *Microwave Extraction, Test Methods for Evaluating Solid Waste*, EPA SW-846. Washington, DC: USEPA; 2007.

9. Ternes, T.A.; Bonerz, M.; Herrmann, N.; Löffler, D.; Keller, E.; Lacida, B.B.; Alder, A.C. Determination of pharmaceuticals, iodinated contrast media and musk fragrances in sludge by LC tandem MS and GC/MS. *J. Chromatogr. A* **2005**, *1067*, 213–223.

10. U.S. Environmental Protection Agency Method 3545. *Pressurized Fluid Extraction, Test Methods for Evaluating Solid Waste*, Update III, EPA SW-846. Washington, DC: USEPA; 1998.

11. U.S. Environmental Protection Agency Method 1613. Tetra- through Octa-Chlorinated Dioxins and Furans by Isotope Dilution HRGC/HRMS. Washington, DC: USEPA; 1994.

12. Focant, J.-F.; Pirard, C.; De Pauw, E. Automated sample preparation-fractionation for the measurement of dioxins and related compounds in biological matrices: A review. *Talanta* **2004**, *63*, 1101–1113.

Chapter 15

Sample Preparation for the Study of Flavor Compounds in Food

Henryk H. Jeleń

15.1. SPECIFICITY OF FLAVOR ANALYSIS—AN INTRODUCTION

Flavor is considered one of the most important attributes determining the acceptance of food. Aroma together with taste, texture, and color gives food its palatability. The flavor of food depends on the presence of various numbers and amounts of volatile compounds. Some foods have an aroma that depends almost solely on the presence of one or a few main compounds, whereas there are many products in which tens or hundreds of compounds form their characteristic aroma.

Flavor research nowadays focuses on several main subjects: (1) identification of key odorants and their formation pathways; (2) investigation of the release mechanisms of aroma compounds from food, which is of special importance in the development of functional foods, such as low-fat foods; and (3) changes in aroma compounds during food processing and storage. To meet all these goals, there is a need for reliable techniques for flavor compounds analysis. Several important issues make the analysis of flavor compounds a challenging task:

- *Odor:* Aroma compounds have to be thought of as molecules that are perceived by the human olfactory system. The human nose is still, for many compounds, the most sensitive detector. Aroma compounds are characterized by odor notes and odor thresholds that are highly dependent on compound concentration, molecule chirality, and position of functional groups. Finally, odor thresholds depend on the matrix from which the compound is released.
- *Concentration:* Aroma volatiles normally constitute only a very minor proportion of the weight of foodstuff, ranging in total from a few hundred parts per million down to a few parts per million. However, particular aroma compounds are usually present in foods in extremely broad concentration ranges, from many parts per million down to parts per trillion levels. Therefore, methods for their analysis besides isolation usually involve preconcentration to facilitate chromatographic analysis. Concentrations of aroma compounds also influence the subsequent chromatographic methods.
- *Complexity:* Food aromas can be very complex: coffee contains as many as 1000 volatile compounds belonging to different chemical classes, characterized by different volatility, broad ranges of boiling points, solubility, and polarity. Some food aroma compounds have boiling points below room temperature, whereas others, such as vanillin, are solid at room temperature.
- *Instability:* Aroma compounds are often very unstable. They are prone to oxidation by air and are often thermally unstable, influenced by pH, and susceptible to light-induced changes. To complicate the issue even further, the food aroma itself is very "volatile": The profile of aroma compounds can vary within a short time period. Many flavors form in heat-induced reactions. Many compounds formed during these processes are not the final reaction products. In addition, enzymatic reactions, which take place in certain foods, especially fruits and vegetables, cause a rapid change of the profile of volatiles that can be extracted. There are also seasonal variations in the content of many volatile compounds in fruits and vegetables.
- *Matrix:* Food aroma compounds are constituents that are usually present in a very low concentration in a matrix, which is usually rich in either sugars, fats, or proteins, and far more often in a matrix that comprises all these food macroconstituents in different proportions. The phenomenon of flavor binding and flavor release should be understood when dealing with different food matrices.

Handbook of Sample Preparation, Edited by Janusz Pawliszyn and Heather L. Lord

The main obstacles in the isolation and analysis of food aroma compounds are matrix constituents coextracted and interfering in the analysis. This makes some of the isolation methods such as liquid–liquid extraction of limited use for some matrices, and subsequent fractionation is often required. An additional problem caused by matrix is foaming. Due to the matrix rich in flavor precursors, often artifacts are formed, when isolation involves high temperatures or prolonged time.

Aroma compounds can be regarded as food trace constituents depending on the concentrations in which they appear. Although there are numerous methods that apply to trace compounds isolation,[1,2] not all of them apply to food aroma compounds extraction mainly due to the volatility and the thermal instability of these compounds. Various methods of food odoriferous compounds analysis have been summarized in several works.[3–5]

The approach to the analysis of volatile compounds can take two paths:

- The first is a total volatiles analysis approach where all volatile compounds present in a food are isolated and analyzed.
- The second approach focuses on the volatiles present in headspace, as the headspace phase reflects the profile of compounds perceived by our olfactory system when eating food.

This chapter aims at helping researchers who intend to undertake food flavor analysis to familiarize them with currently available methods for aroma compounds isolation, and to make them aware of the difficulties in flavor compounds extraction caused by the nature of flavor compounds. It also aims to present the complexity of food as a matrix and the interactions of food macroconstituents with aroma compounds, factors of great importance in aroma compounds analysis.

15.2. ODOR: A KEY FACTOR IN AROMA COMPOUNDS ANALYSIS

Crucial in flavor analysis is the compound's odor threshold, the attribute of aroma compounds present in food. This term refers to the lowest concentration of a compound required to recognize its odor, also known as recognition threshold. The detection threshold, the concentration at which a compound can be detected but its flavor note cannot be determined, is lower than the odor threshold.[6] Some of the food aroma compounds are presented in Table 15.1. Cited odor thresholds indicate the required performance of potential methods of their analysis. Odor threshold concentrations depend on the compound vapor pressure. It is influenced by temperature and the type of matrix from which the odorant is released. In addition, the presence of other odorants in a matrix influences the perception of particular compounds.

Table 15.1. Sensory Properties of Selected Food Aroma Compounds[6]

Compound	Odor Note	Odor Threshold (μg/L)
Vanillin	Vanilla	20
γ-Decalactone	Peach-like	7
Hexanal	Green	4.5
2-Methylbutanal	Malty	4
2-Isobutyl-thiazole	Green, tomato	3
3-Methylbutanal	Malty	0.2
2-Acetyl-1-pyroline	Popcorn	0.1
Filbertone	Hazelnut-like	0.05
(all-E)-α-sinensal	Orange-like	0.05
2-Methyl-3-furanthiol	Boiled meat	0.007
2-Isobutyl-3-methoxypyrazine	Hot paprika	0.002
2-Furfurythiol	Roasted coffee	0.0012
bis(2-Methyl-3-furyl)disulfide	Meat-like	0.00002
1-*p*-Menthene-8-thiol	Grapefruit	0.00002

Odor thresholds provided in the literature are usually determined for water, but they may be very different in other matrices. In 1963, Rothe and Thomas[7] developed the concept of aroma values and related in a simple form the odor threshold of a compound and its concentration to be a useful indicator of a compound's flavor importance. Aroma value, defined as the ratio of concentration of compound X in the food to its odor threshold ($AV = C_x/OT_x$), enabled an estimate of the role of a particular compound in the overall aroma of a product.

Some of the odoriferous compounds are responsible for the flavor or off-flavor of a product when present in low but still detectable concentrations by the human olfactory system. In beer, 3-mercapto-3-methylbutylformate can cause an off-odor in a concentration of 5 ng/L, so ideally, the analytical method for this compound should provide isolation and quantification in this concentration range.[6] Trace concentration of key food odorants having very low odor thresholds requires the development of sophisticated methods for the odorants' isolation, separation, detection, and description. To address the specificity of aroma compounds analysis, gas chromatography (GC) was combined with the human perception of odorant. GC–olfactometry (GC-O) was introduced as a routine tool in flavor analysis in the early 1960s, especially after the development in the 1980s of methods that enabled the quantification of odorants using GC-O: CharmAnalysis™[8] and Aroma Extract Dilution Analysis (AEDA)[9] became valuable aids in the analysis of key odorants.

An additional challenge for the analysis of odoriferous compounds is their chiral character. Enantiomers can differ either in odor quality, eliciting different odor notes, or in odor intensity, influencing their odor thresholds. The examples of carvone ((+) isomer having caraway aroma, whereas

(–) isomer has a spearmint odor note) and limonene ((+) isomer of limonene smells like orange, whereas (–) isomer smells like turpentine) are often cited. Sometimes, as in the case of methyl jasmonate (1R,2R)-(–) isomer and its diastereoisomer (+)-(1R,2S), they occur in nature and contribute to the flavor of jasmine flower, whereas remaining enantiomers are virtually odorless. Odor properties of *cis* rose oxide differ slightly: (2S,4R)-isomer has floral, green notes, whereas (2R,4S)-isomer has somewhat spicy notes; however, their odor thresholds are 0.5 and 160 ppb, respectively.[10] Therefore, two isomers, detectable within the same limits by gas chromatographic detectors can be perceived extremely differently when sniffed in a food product.

15.3. FOOD AS A MATRIX FOR VOLATILE COMPOUNDS

15.3.1. Release and Binding of Aroma Molecules

The release of volatile flavor molecules from the food into the air gives the characteristic volatile profile, which we sense as aroma. The compounds have to be released into the headspace. Phase partitioning and mass transport are the factors determining flavor release from food. According to Henry's law, the mass of vapor dissolved in a certain volume of solvent is directly proportional to the partial pressure of the vapor that is in equilibrium with the solution. Aroma compounds present in foods in very low concentrations obey Henry's law and are almost infinitely diluted. Henry's law applies to pure solvents, but food cannot be treated as a simple two-phase system—usually it is a complicated, often nonhomogeneous mixture of different solutes or phases. Exceptions to Henry's law result from excessive concentration of volatiles, lack of homogeneity, poor solubility in water leading to the accumulation of volatiles on the water–air border, or phase partitioning between aqueous and lipid (hydrophobic) phase. Vapor pressure in the medium is also influenced by such factors as temperature, which affects gas/product partition coefficient; presence of some macroconstituents such as sugars or alcohol in an aqueous phase; acid–base equilibria; and binding of flavor molecules to some biopolymers that can dramatically influence liquid phase viscosity. Foaming of phase can also influence the partition of volatiles into headspace.[11,12]

Thermodynamics determine the retention and release of aroma compounds from the matrix in equilibrium conditions. In the equilibrium state, the concentration of the compound in two phases can be expressed as

$$P_{ap} = C_a / C_p, \quad (15.1)$$

where P_{ap} is an air–product partition coefficient and C_a and C_p are the concentrations of the flavor compounds in air and product, respectively. On the other hand, under dynamic conditions, flavor release is directed by thermodynamics as well as kinetics. The process of releasing flavor from the food product will take place only if the phase equilibria are disturbed, so nonequilibrium is the crucial factor for mass transport. The expression of the velocity at which a compound diffuses through the phase–mass transfer coefficient determines the rate at which equilibrium is achieved.[11] Compounds characterized by high air–product partition coefficients are rapidly depleted from the phase interface.

For complicated food matrix systems containing lipids, emulsions, proteins, and other bio-macromolecules, the behavior of aroma compounds and their release is hard to predict, and the isolation of flavor compounds and their analysis, especially by headspace techniques, has to take into consideration the specificity of the matrix.

Flavor binding and release are subject to underestimation or overestimation of some volatiles in analytical procedures. Interactions between flavor compounds and food ingredients have been reviewed in detail.[13,14] The main food macroconstituents—lipids, proteins, and carbohydrates—are the most important in terms of flavor molecules–matrix interactions.

15.3.1.1. Lipids. Lipids help food products to maintain their flavors. Aroma compounds are often lipophilic, bound to the lipid molecules by hydrophobic interactions and van der Waals forces. Hydrophobicity is one of the main factors influencing the reactions of volatile compounds in their liquid phase. Octanol–water partition coefficient, or log *P*, may serve as a relative indicator of the hydrophobicity of a compound. Lipids act as solvents for lipophilic compounds; therefore, fat-containing solutions would have a greater effect on the hydrophobic compounds and their concentration in the headspace than on the hydrophilic ones. The presence of only 1% of vegetable oil in water causes a decrease in odor threshold of aldehydes.[15] Because of the partition coefficients, the odor thresholds of aliphatic aldehydes are usually significantly higher in oil than in water.[6] Even the type of vegetable oil influences the binding of flavor compounds and their release from oil/water emulsions.[16] In emulsion systems such as milk, the binding of odor compounds to fat has a much larger aroma-retaining effect than to milk proteins or carbohydrates. Milk with a fat content of only 0.7% showed substantial (up to 91%) absorption of volatiles.[17] The ability of fats to mask certain off-flavors is also attributed to the hydrophobicity of most flavors; so, being fat soluble, their volatility decreases.

15.3.1.2. Proteins. Emulsifying and stabilizing properties of proteins define them as functional constituents in foods. The interaction of proteins with aroma compounds usually has two modes: noncovalent interactions, which are reversible, and covalent bonding, which is irreversible. The flavor-binding capacity of flavor molecules with proteins depends on the aroma compound class, chain length, and the position of the functional group in a chain.[18,19] In proteins, the most probable binding site for aldehydes, ketones, esters, and fatty acids is the hydrophobic cavity of the protein.[20] As proteins are sensitive to heat treatment and pH changes, their aroma-binding capabilities rely on these factors and, as a consequence, depend on protein-type structure. Dramatic differences in binding of hexyl acetate by native legumin (11S globulin), both heat treated and acid denaturated, were noticed.[18] Upon heat treatment, the binding affinity of β-lactoglobulin for 2-nonanone was reduced, which was related to aggregation of β-lactoglobulin.[21] An increase in retention of volatiles—methylketones, esters, and terpenes—with the increase of pH in a range of 3–9 was observed for the same protein.[22] It is evident that binding constants are related to the type of protein.[14] The recovery of disulfides from nonheated albumin systems at pH 6.7 is similar to aqueous control; however, losses in extracted disulfides were noted at higher pH and in heat-denaturated albumin.[23]

Changes induced in labile protein matrix during sample preparation can influence the recoveries of analyzed aroma compounds. It has to be emphasized that binding to proteins can have different mechanisms: (1) reversible due to hydrophobic interactions and hydrogen bonding (ketones, hydrocarbons, and alcohols) and (2) irreversible such as in the case of sulfur compound reactions with proteins.

15.3.1.3. Carbohydrates. Carbohydrates provide several flavor-associated functions, being used as sweeteners, Maillard reaction substrates, viscosity modifiers, and fat replacements. Compared with water solutions of flavor compounds, the addition of mono- or disaccharides to water increases the volatility of flavor compounds. Polysaccharides act in an opposite way, usually altering flavor partition in solutes due to the modification of viscosity.[24] In water solutions, hydrogen bonds are formed between xanthan and alcohols,[25] which can cause a decrease in their aroma perception (headspace concentration). Retention of flavor by carbohydrates increases with molecular weight and decreases with increased polarity and volatility of the flavor compounds within the same class.[26]

Large carbohydrate molecules have structures favoring hydrophobic attraction of odorants. Starch is probably the most investigated polysaccharide, and it is assumed that the binding of flavor with amylose is based on inclusion complexes. In amylose, flavor compounds can be located in the hydrophobic cavity or in free spaces between the helices. In diluted starch solutions, competitive binding is observed; the addition of a second flavor compound can disturb the equilibrium and replace the first one in the cavity of amylose helix. Amylopectin plays a minor role in flavor retention, as the potential binding and complex stability are rather weak.[27] Native starch granules show sufficient surface porosity to adsorb flavor compounds; this type of flavor retention is experienced in, for example, flours. When starch is gelatinized by means of thermal treatment, not only are flavor compounds sorbed but also retention is achieved by minimizing diffusion rates.[27] In starch with higher moisture content, solubility of flavor compounds in starch water systems also has to be taken into account.

15.4. METHODS FOR FLAVOR COMPOUNDS ISOLATION FROM FOOD

The main analytical tool for the separation, identification, and quantitation of flavor compounds is GC. GC–mass spectrometry (MS) offers an especially powerful device for compound separation and identification. Together with GC-O, these techniques are indispensable for aroma compounds analysis. For complicated flavor mixtures, when the separation power of single-dimensional chromatography is insufficient, more sophisticated GC separation techniques are used: two-dimensional GC (2D-GC) or comprehensive GC (GC × GC). For the unequivocal identification of aroma molecules, flash chromatography, preparative liquid chromatography (LC) and GC, and nuclear magnetic resonance (NMR) techniques are often required.

Though separation and identification of odoriferous molecules can be challenging, the prerequisite for successful analysis is the proper isolation of volatiles from food. Identification and proper quantitation of key odorants depends on the choice of techniques used for isolation of the volatile compounds from foods.

Rarely can aroma compounds be analyzed by direct injection into a gas chromatograph. Direct injection is used in case of essential oils that are otherwise obtained from plant sources by one of numerous industrial- or laboratory-scale isolation methods. They have to be diluted or divided due to the high concentration of analytes. Aqueous aroma compound samples, if concentrated, can be injected into a gas chromatograph, but it has to be kept in mind that water expands to the greatest extent among solvents used in the injection port of the gas chromatograph, which can cause problems in quantitation. Villen and coworkers[28] used programmed temperature vaporization (PTV) injector in a solvent elimination mode to inject a large volume (50 μL) of wine samples directly into the gas chromatograph. However, volatiles usually need to be isolated from the matrix and prepared for subsequent transfer to GC.

In a review paper on the isolation of aroma compounds from food, Weurmann summarized methods used in 300 papers published between 1960 and 1967. Distillation methods were dominant and appeared in 78% of papers. They were followed by extraction, chemical reactions to isolate volatiles, adsorption, and freeze concentration.[29] Nowadays, among many techniques in use for isolation of aroma chemicals, there is a tendency to shift toward solventless, rapid methods, mainly headspace analysis. All methods used for the isolation of food aroma compounds can be grouped broadly into solvent extraction methods, steam distillation (SD) methods, headspace techniques, and methods based on sorption mechanisms.

15.4.1. Solvent Extraction Methods

Solvent extraction techniques are the oldest in the field of flavor analysis. The extract is obtained by mixing a liquid or solid sample with organic solvent, separation of phases, and concentration of organic phase for the subsequent chromatographic analysis. The usefulness of solvent extraction (liquid–liquid extraction or extraction of solids) for isolation of aroma compounds is sometimes limited due to matrix complexity. Organic solvents usually coextract other matrix constituents (especially lipids) together with flavor compounds, so further sample fractionation is required. The choice of solvents for extraction is governed by their polarity, density, and boiling point. Usually, dichloromethane (bp 45°C), pentane (bp 35°C), isopentane (bp 28°C), or diethyl ether (bp 35°C) are used. To extract aroma compounds of wide polarity, range mixtures such as pentane/ethyl ether are often used. Low boiling point solvents are used to minimize losses of the most volatile aroma compounds in sample concentration. Sample concentration is performed using rotary evaporators, and more often using Kuderna–Danish concentrators, Vigreaux columns, or concentration under a stream of nitrogen. Emulsion formation can be an obstacle in phase separation; therefore, saturated salt solutions are used to separate layers. To improve the efficiency of liquid–liquid extraction and shorten its time, ultrasound can be applied.[30] Vila and coworkers tested three solvents for extraction of volatiles from wine with the use of ultrasound: dichloromethane, *n*-pentane : diethyl ether (2:1), and diethyl ether : pentane (2:1), and obtained the highest extraction efficiency for the last extractant.[31] Of four tested salts, $MgSO_4$ provided the highest extraction yields.

Supercritical fluid extraction (SFE) has a special position in solvent extraction. A detailed description of the technique is provided in Chapter 11. It has received increasing attention, especially in the extraction of essential oils, but also in many other food products such as cheese and strawberries.[32–34] Supercritical fluids can provide improved mass transfer rates. The operation can be influenced by changing basic parameters of the system: temperature and pressure. These parameters are easy to control and influence the density of the fluid. As there is a direct relationship between a solvent's strength and density, by manipulating temperature and pressure of CO_2, solvent strengths of pentane or carbon tetrachloride or methylene chloride can be mimicked. In the development of SFE methods, pressure, temperature of the fluid, percentage of the added modifier, and dynamic and static extraction times are considered crucial in obtaining the highest extraction yield.[35] Increasing pressure of supercritical CO_2 enhances extraction yield, as its density increases at higher pressures and compounds of higher molecular weight become soluble. However, at higher pressures, diffusion rates of the solutes are reduced. Temperature can also affect the extractability of compounds; density of supercritical CO_2 is reduced at higher temperature. The profile of compounds extracted using supercritical CO_2 is dependent on the solubility of aromas in CO_2 and the influence of the modifier, so the composition of the extract can differ from that obtained by other methods. SFE extraction of essential oils from thyme (*Thymus zygis* L.) and mentha (pennyroyal; *Mentha pulegium* L.) yielded lower recoveries for thymol in thyme and pulegone and menthone in mentha compared with traditional distillation in the Clevenger apparatus.[36]

Liquid-phase microextraction (LPME), in particular the automated version, can be a potentially attractive microextraction method, still not fully explored in food flavor analysis. It utilizes very low amounts of solvent to extract analyte. This technique is based on the passive distribution of analytes in the aqueous matrix.[37] Different variations of this method exist: static and dynamic LPME, solvent bar microextraction and hollow fiber membrane LPME, or drop to drop solvent microextraction.

15.4.2. Distillation Methods

These methods are used for isolation of steam distillable components from foods. Steam can be produced in the same vessel (direct distillation) or in a separate one (indirect steam distillation). Steam-assisted distillation/hydrodistillation are procedures used for the isolation of thermally stable flavor compounds from plants and foods. This method is especially suitable for the isolation of thermally stable essential oils from various plant parts. A broader application range is attained by a combination of steam distillation and extraction (SDE) in one apparatus.[38] SDE was described by Nickerson and Likens, and the apparatus designed by them provides simultaneous condensation of steam distillate and an immiscible organic solvent.[39] There is a relatively small amount of organic solvent used, which recirculates in the system, and the recoveries of aroma compounds are usually high. The aroma compounds are transferred from the matrix by water vapor into organic phase. Both phases are in a constant reflux. Low boiling point solvents allow for a subsequent evaporation and condensation of extract prior to GC analysis. The main disadvantage of distillation–extraction performed in the Likens–Nickerson apparatus is the duration of extraction, which usually takes 1–2 h. An even more important problem results when water, which accompanies the extraction process, reaches the boiling point. This can easily lead to oxidation of more thermally labile compounds, especially unsaturated terpene hydrocarbons or aldehydes. Therefore, the Likens–Nickerson extraction, though initially used for hop volatiles, should preferably be used for the extraction of flavor compounds that are thermally stable, such as some essential oils,[40] or compounds responsible for the aroma of cooked/boiled foods. To overcome the problems associated with the formation of artifacts, the distillation can be performed at lower pressure. High-vacuum distillation at pressures <1.5 Pa allows the isolation of flavor compounds at 35–45°C. Two types of micro Likens–Nickerson apparatuses—for extracting with solvents denser than water (dichloromethane, chloroform) or for solvents lighter than water (usually pentane or ethyl ether/pentane mixtures)—are commercially available. There are also variations for use at reduced pressure or with an aeration mode. Such instruments were used for wine analysis where all SDE and GC procedures were done within 3 h.[41] The recoveries of volatile compounds in SDE methods are dependent on the organic solvent used, and normally should be above 90%; however, there are compounds for which a given solvent will provide much lower recoveries.[40] Many food products such as fresh vegetables, fruits, and juices emit flavor compounds that are thermally unstable; therefore, isolation of these compounds requires mild conditions. For many food products, such as vegetable oils and cheese, SDE provides satisfactory results,[42,43] especially for the analysis of free fatty acids in the latter example.

An interesting tool was introduced by Schieberle and coworkers called solvent-assisted flavor evaporation (SAFE) for the isolation of volatiles under reduced pressure from solvent extracts, aqueous foods, suspensions, and fat-containing matrices.[44] As the procedure is performed under high vacuum, this method provides minimized artifact formation. It was used by the authors to isolate polar and unstable 4-hydroxy-2,5-dimethyl-3(2H)-furanone in strawberries and tomatoes, and is commonly used for the isolation of aroma compounds under reduced pressure. Krings and coworkers also introduced a modified vacuum distillation apparatus (thin-layer high-vacuum distillation [TLHVD]) for the isolation of volatiles from lipid matrices and obtained high recoveries for the majority of analyzed flavor compounds.[45]

15.4.3. Headspace Analysis (Static Headspace [SHS])

In SHS analysis, the sample is placed in a sealed vessel and allowed to stand until equilibrium is attained between the concentration of analyte in the liquid (or solid) phase and the headspace above it.[46] The sample is assayed by injection of an aliquot of the headspace into a gas chromatograph. As SHS does not include preconcentration of the analyte, its use is restricted to the analysis of very volatile compounds or compounds present in food in relatively high concentrations (routinely high parts per billion or low parts per million, but applications with lower limits of detection (LODs) can be encountered). The performance of the SHS method is related to the amount of analyte present in a headspace, which depends on the volatility of compound, its relation with the matrix, temperature, and volumes of the phases. It is a method of choice if the partition coefficient of an analyte favors it for the migration from the matrix to the headspace. As equilibrium state is required for SHS, and the partition coefficient depends on the temperature, samples placed in sealed vials are usually thermostated for a time required to attain equilibrium and then transferred into the GC. However, the temperature used for sample incubation should not induce changes in analyzed compounds or in matrix. In an automated version, the sampling is reproducible and done in different ways, depending on the autosampler construction, either with gastight syringe or by using a loop and valve allowing the transfer of headspace into the loop, then flushing it into the injection port of GC. The main advantages of SHS are the absence of a solvent peak, easy automation, and minimal sample preparation.

A simple procedure was proposed by Schreyen et al. to analyze leek volatiles.[47] The finely chopped product was placed on a glass plate covered with a glass jar fitted with a cold finger maintained slightly above 0°C. In a collecting flask, aromas were collected in a headspace condensation process, which is in fact an SD at ambient temperature. The extraction time, however, was 1 day long.

15.4.4. Methods Based on Sorption Mechanisms

Methods based on sorption mechanisms are probably the most rapidly evolving in the analysis of volatile flavor compounds in foods. They involve methods based on adsorption as well as absorption mechanisms and include such techniques as purge and trap (P&T), solid-phase extraction (SPE), solid-phase microextraction (SPME), stir-bar sorptive extraction (SBSE), and headspace sorptive extraction (HSSE), to name those most popular in aroma applications. A detailed review on the mechanisms of extraction of food aroma compounds has been recently published.[48] Adsorbents provide a preconcentration step, nonexistent in SHS methods; therefore, sorption methods gained enormous popularity within the last 35 years since their introduction.

Volatile compounds can be bound to a phase according to two mechanisms: adsorption and absorption. In the adsorption process, the analyte remains on the surface of porous material used as an adsorbent, whereas the absorption process involves analyte dissolving in a bulk liquid phase. Adsorbents have porous structures with a high surface area of often more than $100\,m^2/g$ and small pores varying in size from <2 nm (micropores) to 50 nm (mesopores). Molecules interact with sorbent either physically (weak reversible interaction with sorbent such as van der Waals and electrostatic forces) or chemically, in which analyte reacts irreversibly with surface of adsorbent; therefore, for analytical separation purposes, this type of reaction is of limited use. Absorption takes place when the analyte dissolves into liquid phase (polymer). It has an advantage over adsorption on polymeric sorbents, as there is no phenomenon of analyte displacement in liquid phases governed by absorption mechanisms. Moreover, for liquid polymers (polydimethylsiloxane [PDMS] in SPME fiber) distribution coefficients can be calculated relatively easily using retention indices on a GC column of the same phase coating.[49]

All sorbents used for extraction can be divided into silica gels, activated aluminas, zeolites, activated carbons, and

Table 15.2. Sorbents Used for the Extraction of Aroma Compounds of Different Polarity

Compounds Extracted	Adsorbent Preferred
Hydrophobic, nonpolar	Porapak P, Chromosorb 101, Carboxen 1006, Carbotrap, Carbopack X, Carbosieve III, *polydimethylsiloxane (PDMS)*, Tenax TA, Tenax GR
Low polarity	Amberlite XAD-2, Amberlite XAD-4
Slightly polar	Porapak Q, Chromosorb 102,
Moderately polar	Chromosorb 105, Amberlite XAD-7, Porapak R
Hydrophilic, polar	Porapak T, *Carbowax*, Divinylbenzene (DVB), *polyacrylate (PA)* Carboxen 1006, Carbotrap, Carbopack X, Carbosieve III, Tenax TA, Tenax GR

Liquid phases in italics.

polymers. As volatile flavor compounds are usually hydrophobic, hydrophobic sorbents play a more important role in their extraction than polar ones, such as silica gels and activated aluminas. The most frequently used sorbents are activated carbon and polymers that can be used for the extraction of volatiles of a broad spectrum of polarity. Activated carbon has the largest surface of adsorption and smallest pore diameters. Carbons are very effective for the extraction of especially low-molecular-mass compounds; however, desorption requires the highest temperatures of all sorption materials and sometimes carryover effects are encountered. Tenax polymers (2,6-diphenylene oxide polymers) have low affinity for water and are universal. Desorption temperature can be as high as 350°C. Degradation temperatures of Porapak and Chromosorb polymers are lower than that of Tenax and are in a range of 180–275°C. Table 15.2 lists sorbents commonly used in the extraction of aroma compounds. Adsorbents are used in the form of granules that can be mixed with a liquid matrix to extract volatiles (rarely), or placed in tubular traps for P&T applications (usually 50–250 mg) or for closed-loop stripping analysis (CLSA) (from 1 to 50 mg of absorbent). Compared with other extraction techniques, SPE cartridges are not often used in aroma compounds research. Their main application is fractionation of free and bound volatiles in various foods, mainly fruit and wines. Mangas et al. used SPE for the extraction of trace aroma compounds in cider using Extrelut-20 columns; this method is quantitative for alcohols, esters, phenols, and for medium- and long-chain fatty acids, not for the short-chain fatty acids.[50] Together with the development of SPME and then SBSE, the adsorbent began being used as coatings in devices enabling convenient and fast extraction and subsequent desorption of aroma compounds.

15.4.4.1. Dynamic headspace. In this technique, volatile compounds are swept from the product by a stream of inert gas and trapped in various types of sorbents, solvents, or trapping devices. When an inert gas bubbles through a liquid, the technique is often called "purge and trap." The purging can be performed in an open- or closed-loop system. The trapped compounds are usually thermally desorbed, less frequently eluted from the trap by solvent. In the most popular version, compounds are trapped into a polymer material, which is usually placed in a short tube. As adsorbents, Tenax TA, Tenax GR, Carboxen, Carbosieve, Carbotrap, or Chromosorb are usually used.[51] Tenax TA is probably the most popular bead for flavor isolation due to its good adsorbing properties for many compound classes, low water retention, low bleed, and resistance to relatively high temperatures. As the trap can be treated as a short column, which has certain retention parameters, an important issue in using P&T methods is a breakthrough volume for analyzed compounds (volumes of gas that exceed the loading capacity of the trap). Highly volatile compounds usually have small breakthrough volumes. To overcome problems related to selectivity of adsorbents and problems with breakthrough volume, multibed adsorbents are often used.

Dynamic headspace provides the preconcentration required for aroma compounds analysis, contrary to SHS analysis. It proved to be a suitable tool for the measurement of volatile compounds released during the disintegration of fruit as in the case of strawberries, where enzymatic processes led to dynamic formation of flavors, or for the monitoring of formation of flavor compounds in intact whole apples.[52] The vessels used for dynamic trapping of specific products can be suited to its size and form, as for baguettes, and thus facilitate the sampling from whole or chopped fruit (Fig. 15.1). CLSA developed by Grob and Zürcher in 1976 initially for air–water stripping is an excellent tool for investigation of food and living systems. In the system, air volume circulates through a closed system by means of a membrane pump, and volatiles are adsorbed usually on a charcoal trap, from which usually dichloromethane extraction is performed (Fig. 15.1).

15.4.4.2. SPME and related methods. Among several solventless sampling techniques used for food flavor analysis, the method used most often and which is the foundation for further developments is SPME, developed in the early 1990s by Pawliszyn and coworkers.[49] SPME, due to its simplicity, robustness, and low cost, gained popularity as a fast screening method for qualitative analysis. As shown in many applications, it is also a reliable method in quantitative analyses. It is characterized by high sensitivity and is often compared with dynamic headspace in this respect. There are numerous review papers on SPME, and several cover its applications in food volatiles analysis.[5,54,55]

In food flavor analysis, SPME is used mainly as headspace–solid-phase microextraction (HS-SPME), not a

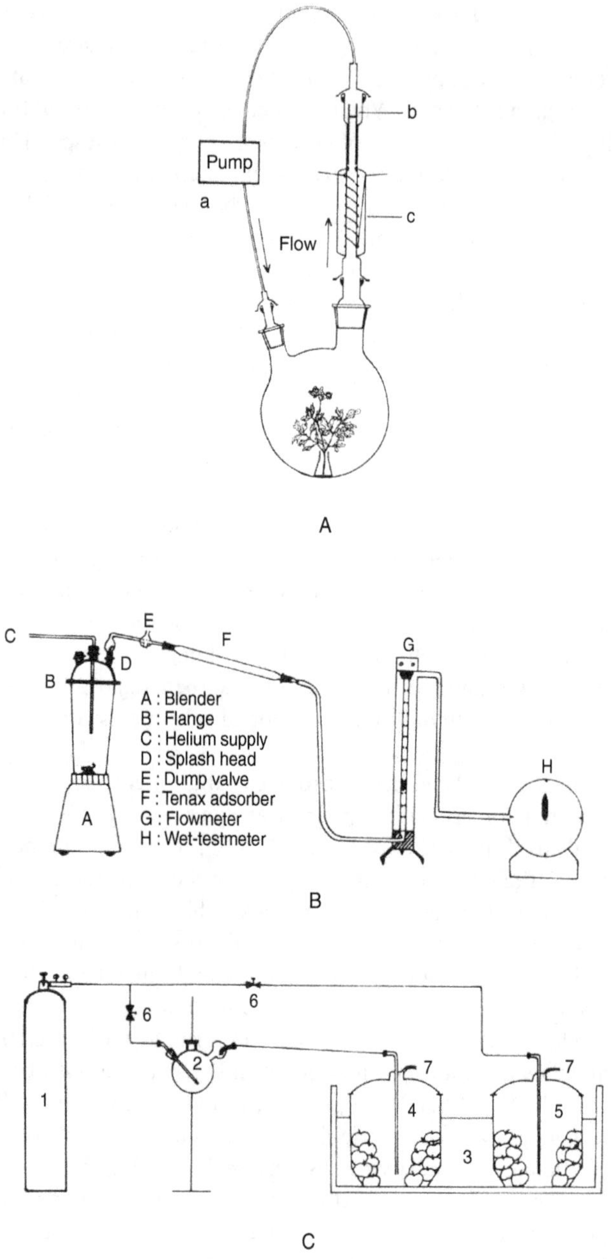

Figure 15.1. Setups for dynamic headspace sampling of plants and food products. From top to bottom: (A) closed-loop stripping apparatus,[53] (B) apparatus for isolation of volatiles during fruit disintegration, and (C) apparatus for headspace collection from intact apples. Reprinted from references 52 and 53 with permission from Walter de Gruyter. Copyright 1984.

direct extraction, as liquid foods are rich in compounds that may interfere with adsorbed compounds or undergo thermal degradation during desorption. For solid matrices, usually a piece of material, ground or chopped, is placed in a vial, then sampled using SPME fiber. The other approach is to prepare a slurry with water and analyze its headspace.[56] Andrés and coworkers used a direct-extraction device (DED), which enables the insertion of SPME fiber into the solid material (dry cured ham) without damaging the fiber.[57] In the majority of research, SPME is used as an equilibrium technique, and the extraction time is usually optimized. However, for many compounds and adsorbing polymers, equilibrium for all analytes is difficult to obtain in a reasonable time, so extraction time is a compromise between analysis time and extraction efficiency. SPME fiber has been used as a trapping device for dynamic sampling where PDMS fiber was placed in a funnel stem over volatiles escaping from a blender where cucumber samples were prepared for analysis.[58] Sometimes SPME is used as a secondary step in the extraction, as shown by Lloyd and Grimm for the analysis of 2-methylisoborneol and geosmin in catfish with muddy and musty flavor.[59] In their experiment setup, compounds were distilled off using microwave from the lipophilic matrix into water, from which they were subsequently isolated using SPME. Microwave distillation (MD)-SPME allowed the isolation of these off-odorants from fish tissues at the concentrations of 0.1 µg/kg—better than P&T and MD-SE in terms of sensitivity and analysis time. Rengarajan and Seitz[60] used a combination of supercritical fluid (CO_2) extraction (SFE) with subsequent dynamic headspace purge and with SPME for the isolation of flavor compounds from microwave popcorn and compared it with these methods used independently. A combination of SFE and subsequent extraction of the SFE extract using SPME proved to be a very effective tool in the isolation of pyrazines.

The crucial factor influencing selectivity of extraction using SPME is fiber choice. The universal fiber used for the absorption of aroma compounds is PDMS; however, in an increasing number of papers, a DVB/CAR/PDMS fiber is recommended due to the high sensitivity obtained using this fiber. It has been used in the analysis of aldehydes in oil and water emulsions,[61] potato crisps,[62] and many compounds causing off-flavor and appearing in trace quantities.[5] Song et al.[63] ranked the polarity of SPME fiber coatings as Carbowax/DVB > PDMS/DVB > PDMS and showed that SPME used with fast gas chromatography–time-of-flight (GC-TOF) MS can be a very rapid method (GC analysis completed within 140 s). As the profile of extracted volatile compounds can be highly influenced by fiber type, it has to be considered in elucidation of aroma profile of foods by SPME. It has to be remembered that abundances of extracted compounds are related to their affinities to fiber, so for the complete aroma profile, additional methods of headspace analysis and exhaustive extraction are recommended. It has also been suggested to perform simultaneous extraction with two different fibers to obtain a broader aroma profile. In the analysis of cooked pork volatiles, low boiling point aroma volatiles dominated on Carboxen/PDMS fiber, whereas DVB/Carboxen/PDMS provided high levels of high boiling point compounds.[64] SPME is also considered as a method for extraction of high-molecular-weight compounds not isolated by other headspace methods, as in the case of dry cured ham,[65] although for this purpose, SDE often performs more efficiently.

The question of how representative SPME is in terms of total aroma extraction was explored by Rega and coworkers.[66] They used direct GC-O (with no compound separation) to evaluate the extracted flavor of orange juice and found out that SHS and aqueous distillates produced odor closely resembling that of the original juice, while SPME produced odor with lower similarity to the reference. Different fibers induced different odors perceived by GC-O, the best being DVB/CAR/PDMS.

SPME is frequently used for quantitative analysis. The most difficulties arise from food matrix complexity, mainly the flavor/matrix interactions described earlier, and the specificity of adsorption process. In the HS-SPME analysis using adsorbent-type fibers, competition effects between analytes[67] or linearity deviations dependent on the matrix composition[68] were observed. The adsorption of volatiles from olive oil using PDMS-DVB fiber was linear for 27 out of 44 compounds in a range of 0.1–20 ppm, while the remaining ranged from 7.1 to 17.7 ppm. The dilution of olive oil with deodorized oil was suggested to overcome displacements that resulted from fiber saturation.[69] Also in the case of essential oils analysis, where the contents of odorants is very high (milligram per gram of seeds), small samples and short extraction times are recommended to maintain linearity of the method.[70] Very often in preparation of calibration mixtures for external standard calibration, a matrix mimicking the real one but devoid of compounds of interest can be used. Examples of such matrices include deodorized orange juice;[71] artificial model wines[72] or wine, from which sulfur compounds were eliminated (in the analysis of sulfur compounds);[73] cucumbers in which enzymes mediating formation of key C9 aldehydes were thermally inactivated;[58] or spices from which essential oils were removed using distillation.[70] The alternative approach is quantitation using standard addition method or stable isotope dilution analysis.

The volume of fiber coating phase influences SPME limits of detection; therefore, developments in this direction resulted in the invention of SBSE, which was then used in trace aroma analysis.[74,75] The magnetic stirrer is covered with PDMS film, the volume of which is 100 times the volume of typical SPME fiber. SBSE offers a benefit over SPME in terms of sensitivity, as the volume of the accessible phase is much larger than in SPME. SBSE was used for the analysis of wine aroma compounds[76] and low levels of off-flavor compounds such as 2-methylisoborneol, geosmin, and 2,4,6-trichloroanisole in drinking water.[77] The drawback of this technique is a relatively large phase thickness, so it takes a long time for volatiles to penetrate the thickness of the absorbent, and as a result, equilibrium times are longer. This drawback is overcome by designing the adsorbing material in the form of thin film, where the contact surface with aroma compounds is larger.[78] Solid-phase aroma concentrate extraction (SPACE) was proposed by Ishikawa et al., similar to SPME but using a larger adsorbent surface, which is mounted on a stainless steel rod. The authors tested it for coffee powder analysis and compared it with SPME, dynamic headspace (DHS), and solvent extraction; the highest performance was with SPACE compared with the other methods.[79] Recently, two phase twisters offering higher selectivity and sensitivity due to PDMS and activated carbon were also proposed.[80] Grob and Habich were probably the first to develop open tubular columns coated with liquid stationary phase (PDMS).[81] However, due to the limited loading capacity and low sampling rate, this method did not gain popularity.

15.4.5. Short Path Thermal Desorption

Thermal desorption analysis can be carried out directly in the desorption tube: a minute amount (a few microliters) of compound is placed on glass wool in the desorption tube, which is subsequently heated from the ambient temperature to a temperature allowing aroma compounds' migration to headspace.[82,83] This technique is used mainly in the analysis of lipids, especially oils, where it provides information in addition to that provided by other techniques on volatiles as well as semivolatile compounds. It has also been used for cheese samples, where samples as large as 50 mg were used.[84] This method offers total transfer of volatiles into the GC system and exists in various setups.

15.4.6. Further Sample Cleanup and Fractionation

The mixture of volatile compounds obtained after extraction often contains numerous compounds that would interfere in compound separation by chromatographic techniques. Coeluting compounds of different character, present in much higher concentrations, can hide minor flavor compounds. For complicated flavor mixtures such as coffee, a procedure was proposed by Parliament.[85] The aqueous phase obtained after distillation (either direct, indirect steam, or vacuum steam) is prefractionated in the following way. Extraction is performed in a Mixxor extraction device, and the total sample is analyzed by injecting the diethyl ether layer. Then acids are removed by the addition of sodium hydroxide, with the sample being extracted with the same ether portion. In the next step, bases are removed by the addition of sulfuric acid, and in the last step, 2,4-DNPH is added to remove carbonyl compounds. After each step, the organic layer is analyzed. A similar approach was proposed for the unified analysis of Maillard reaction products.[86] The sample fractionation scheme is shown in Figure 15.2. The aroma compounds in alcoholic beverages, important from the sensory point of view, are often masked by the presence of ethanol and other major fermentation products, such as fusel alcohols and ethyl esters of medium-chain-length fatty acids. Semipreparative high-performance liquid chromatography (HPLC) was proposed for the fractionation of volatiles in water/ethanol system as a mobile phase.[87]

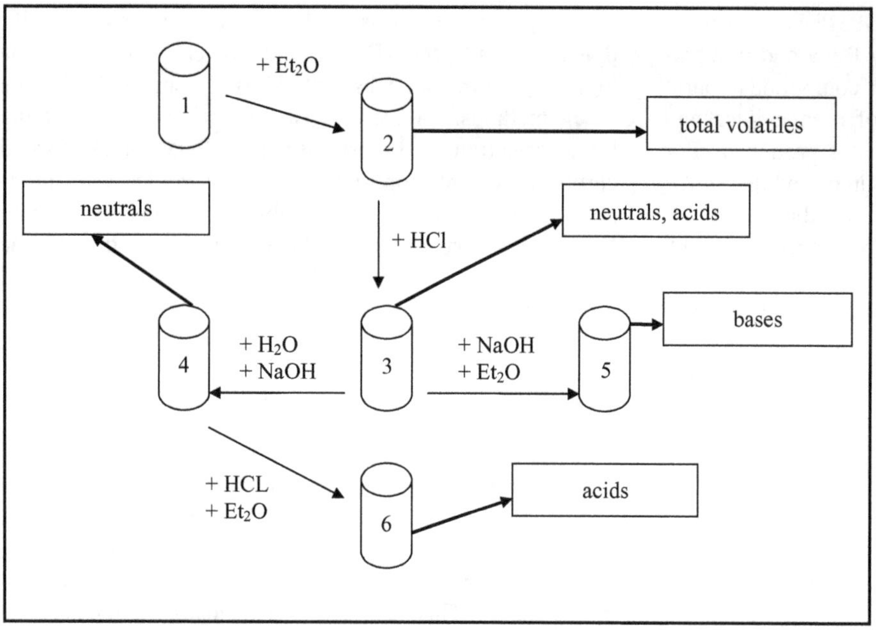

Figure 15.2. Sample fractionation scheme for complex aroma extracts analysis:[86] 1, reaction mixture; 2, diethyl ether (Et_2O) extract (neutral); 3, acidified extract; 4, extract treated with NaOH (neutral); 5, basic extract; 6, acidic extract.

15.5. COMPARISON OF METHODS USED FOR FOOD AROMA COMPOUNDS ISOLATION

The profile of aroma compounds extracted from food is influenced by the extraction method. There are several studies on aqueous model systems that show the differences in profiles of volatile compounds obtained using various isolation techniques. Jennings and Filsoof found that for model mixtures, distillation/extraction provides a profile of aroma compounds that most resembles the neat solution injection.[88] Leahy and Reineccius conducted a comparison of various methods most commonly used at that time—direct headspace (split, on-column, splitless), headspace concentration (adsorbent trap, freon reflux), Likens–Nickerson (atmospheric and reduced pressure), and solvent extraction (pentane, dichloromethane, diethyl ether)—for the isolation of a volatiles mixture ranging from ethanol (bp 78.5°C) to isoeugenol (bp 266°C) and ethyl methyl phenyl glycidate (bp 272–275°C).[89] Results presenting recoveries or peak areas (for headspace) in Figure 15.3 show the dependence of the profile of volatiles on the extraction method. The most effective for flavor isolation was solvent extraction, where even at 50 ppb, excellent recoveries were obtained. For more complicated matrices, SDE was an efficient extraction alternative.

The goal of many investigations into aroma compounds was a comparison of various methods for their isolation from different matrices. Wine volatiles are a mixture of over 800 compounds, some of which are present in parts per million concentrations; some, often the important odorants, are in parts per billion or parts per trillion range. Various methods have been compared for the isolation of aroma compounds. In such complicated matrix, there is a need to combine different methods to obtain a complete extract of volatiles without altering them or to use a specific one aimed at a particular compound or group of compounds. For wine volatiles analysis, liquid–liquid extraction is often used. Ortega-Heras et al.[90] compared liquid–liquid extraction using dichloromethane with SHS, stripping with nitrogen in a specially designed apparatus with a cold trapping (liquid nitrogen) of volatiles. The highest number of volatile compounds was obtained by liquid–liquid extraction: the coefficients of variation (CVs) were the lowest; SHS yielded fewer compounds, mainly the most volatile; and the stripping method gave the poorest reproducibility. Liquid–liquid extraction was compared with SPME in the analysis of guaiacol in spiked wine and in oak extracts. Using diethyl ether as a solvent and high injector temperature, artifacts were formed, leading to higher results. Artifact formation can be eliminated using pentane or pentane ethyl ether (2:1) for extraction and lower temperatures of injection, or alternatively SPME.[91] Extraction with dichloromethane can be combined with ultrasound to provide higher yields.[30] Comparison of SPME (CAR/PDMS/DVB) and SPE (ENV+, 1-g cross-linked styrene-DVB) for the analysis of 3-mercaptohexan-1-ol and 3-mercaptohexyl acetate in wine provided similar results in terms of LODs and reproducibility, slightly lower for SPME. To quantify these compounds, wine treated with charcoal to remove any sulfur compounds was used as a matrix, with a subsequent reset of SO_2 level.[73] Both SPME (PDMS) and SPE (Oasis HLB sorbents) provided comparable results in isolation of volatile phenols in wine and cork macerate (LODs for SPME in a range of 0.8–1.0 ng/L for tri-, tetra-, and pentachlorophenols [TCP, TeCP, PCP]) when investigated by Insa et al.[92] Elution of compounds adsorbed on SPE cartridges/columns using solvents of different polarity allows fractionation of aroma compounds.[93] Caven-Quantrill and Buglass compared microscale SDE with SBSE for the determination of volatile organic constituents in grape juice. Optimization was performed using synthetic grape juice. SBSE sensitivity was

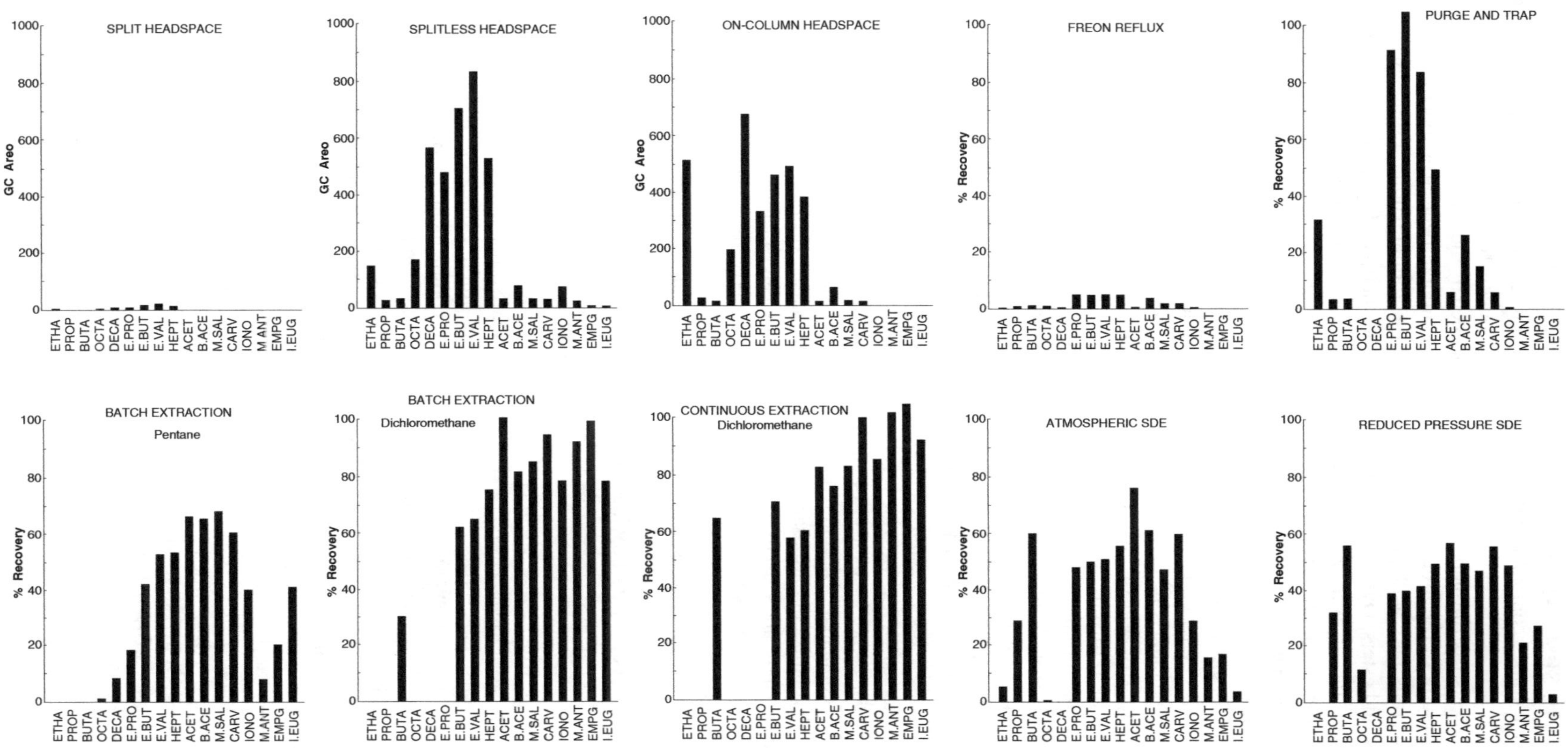

Figure 15.3. Comparison of methods for the isolation of volatile compounds from the aqueous model system. Reprinted from reference 89 with permission from Walter de Gruyter. Copyright 1984.

higher than SDE. For example, SBSE extracted 126 compounds from Muscat-derived grapes compared with 98 in SDE.[74]

Comparative studies were also performed for isolation methods of dairy product volatiles. The aroma fraction of milk products contains numerous fatty acids, ketones, diketones, alcohols, aldehydes, and esters. Dynamic headspace and SDE were compared for the analysis of ewes' milk cheeses (Roncal, Pecorino Sardo, and Fiore Sardo) at various stages of ripening. SDE was more efficient in extraction of less volatile compounds such as phenols, free fatty acids, long-chain carbonyl compounds, alcohols, and esters. DHS, complementary to SDE, provided better efficiency for more volatile compounds.[94] Complementary results were also obtained for SPME and P&T extraction of volatiles from cheeses produced from raw milk (Gruyere, Manchego, and Ragusano).[95] Different volatile profiles were obtained: SPME (Carboxene/DVB/PDMS) was more effective in extracting medium and high boiling point compounds, whereas P&T (Tenax) was better for extracting highly volatile compounds. Similar results were obtained for butter.[96] Tenax had a higher affinity for aldehydes, alcohols, and low boiling point compounds, whereas SPME provided good recovery for ketones and acids, as well as those of high molecular mass. Careri et al.[97] compared SDE and dynamic headspace for the isolation of volatiles in parmesan cheese, and out of more than 100 compounds, about 50 were isolated using both procedures. DHS performed better for the more volatile compounds, whereas SDE performed better for the long-chain volatiles. When distillation, dialysis, and solvent extraction were compared for the analysis of cheddar cheese flavors, solvent extraction with acetonitrile provided most characteristic and typical odor of the three examined methods.[98]

Vegetable oils are foods in which, due to fatty acid autoxidation, mainly aldehydes are formed, as well as alcohols, ketones, and hydrocarbons. Cold-pressed oils, not subjected to a refining process, can contain hundreds of aroma compounds. Cavalli et al.[82] compared the suitability of SHS, HS-SPME, HSSE, and direct thermal desorption (DTD) for the isolation of volatiles from olive oil. They found a similar number of compounds isolated by SPME and HSSE, though sensitivity of the latter was better (due to the volume of the adsorbing/absorbing phase). The DTD method provided the highest number of flavor compounds extracted, mainly semivolatiles. Also, when HS-SPME, SDE, and CLSA were compared, SPME showed the highest affinity for alcohols and ketones, CLSA yielded the highest percentages of esters and hydrocarbons, and SDE and CLSA yielded the highest total terpenoids. SDE caused thermal degradation of samples and an increase in aldehydes.[99] More compounds were extracted from olive oil by DHS-thermal desorption (TD) than by HS-SPME. PDMS-DVB fiber extracted and retained a broader class of compounds from olive oil than PDMS, in particular more ketones and higher-molecular-weight aldehydes.[100]

Essential oils are extracted from plants by distillation methods if they are thermally stable. Comparing SDE and SFE in general, similar profiles of volatile compounds were obtained by these methods for fennel and thyme;[101] however, more monoterpenes were obtained by SDE. Also for oregano, basil, and mint, similar profiles were obtained: SFE did not create thermal degeneration and provided better sensitivity.[101] Essential oils of clove buds obtained by SFE, hydrodistillation, SD, and Soxhlet extraction showed similar composition; however, relative concentrations were different and SFE was assumed to be an optimal method.[102] Comparing concentrations of volatiles extracted from rosemary leaves using distillation and SPME, similar results were obtained for borneol, camphor, and 1,8-cineole, and significant differences were noted for α-terpineol and verbenone.[103]

Various isolation methods were compared for the analysis of beverages. Similar reproducibility for SPME and DHS using Tenax TA was obtained with a higher sensitivity for the latter method observed for cola flavor.[64] When SDE under reduced pressure and adsorptive column extraction was used for the analysis of 219 Sencha tea volatile compounds, 10 times lower amounts of terpenes were detected using adsorption column compared with SDE. However, the odor of the column concentrate resembled green tea, whereas SDE caused decomposition of some volatiles (coumarins, vanillin, and lactones), and others, such as aldehydes and ionone derivatives, were formed in SDE.[104] The method that resulted in extract most resembling the original for coffee was vacuum stripping with water. (SFE, SDE, pressurized oil recovery, steam stripping under vacuum with water, and methylene chloride were compared.)[105] In alcoholic beverages of high ethanol content, this compound can influence SPME adsorption of other aroma compounds. Due to the affinity to ethanol, 3-phase polymers were not used, in favor of PDMS fiber in brandy aroma extractions by SPME and liquid–liquid extraction with freon. Both methods were found complementary: SPME was more selective for esters and acids, and LLE was more efficient in alcohol extraction.[106] For higher alcohols and esters in beer, a high correlation was observed for all compounds analyzed by the SHS method and SPME, the latter being more sensitive.[107] When analyzing pyrazines and thiazoles in distillates, lipophilicity was found to be a determining parameter for vacuum distillation, whereas basicity was a determining parameter for liquid–liquid extraction.[108]

15.6. SPECIFICITY OF SELECTED GROUPS OF FOOD ODORANTS

15.6.1. Aldehydes

Aldehydes in food are formed in three main pathways: lipid peroxidation, degradation of monosaccharides, and amino acid degradation by Strecker mechanism.[6] The most common and abundant in fat-containing foods are products of fatty acid oxidation. Aldehydes formed from hydroperoxide degradation are mainly alkanals, alkenals, and alkadienals, as well as some ketones.[109] Depending on the fatty acids' composition, products of their autoxidation vary. The reaction is of a free radical mechanism, and when the sample is prepared for analysis, this has to be taken into consideration. A review of volatile lipid oxidation products analysis was recently published.[110] Heating a sample for SHS, dynamic headspace, or SPME, especially for a prolonged time, can lead to the formation of additional volatile lipid oxidation products. Therefore, addition of antioxidant to the sample prior to analysis is recommended.[111]

Two unsaturated fatty acids that occur in plant tissues, linoleic and linolenic, are precursors of numerous volatile products of their breakdown in plants: soya beans, cucumber, tomato, and olives. Their formation is catalyzed by the appropriate lipoxygenase, and then lyases provide hydroperoxide decomposition to characteristic volatile compounds. Disruption of cucumber tissues leads to the production of volatile compounds responsible for the characteristic taste and aroma of this vegetable: hexanal, *trans*-2-nonenal, and *trans*-2, *cis*-6-nonadienal. This is a dynamic process and must be taken into consideration in the choice of isolation methods.

Aldehydes that are formed in lipid autoxidation are unstable in the product and may undergo further decomposition or oxidation reactions. They are also prone to react with the matrix, and in products such as yogurt vast alkanal losses (95% for octanal) were noted due to their conversion by microbial nonspecific hydrogenases to corresponding alcohols. The conversion took place in the first days of storage.[112] The most abundant aldehyde resulting from lipid oxidation, and a good marker for this process, is hexanal; the SHS method is used for its measurement in various foods. However, for the determination of other volatile aldehydes, more sensitive methods have to be used. In fat-containing foods, SPME is a popular tool guaranteeing high sensitivity and the isolation of a broad spectrum of compounds.[5] For the analysis of carbonyl compounds, their derivatization, usually with PFBOA (O-(2,3,4,5,6-pentafluorobenzyl) hydroxylamine), has to be mentioned. Formation of aldoxymes as a pair of syn- and anti-isomer peaks provides a more reliable detection by electron capture detector (ECD) and fluorine atoms present in the derivatized molecule increase sensitivity of their determination. Derivatization of carbonyl compounds is often used with SPME.[113] Derivatization has been used for detection of beer off-odorant (*E*)-2-nonenal and other aged beer aldehydes.[114] An alternative to derivatization and SPME may be a single-drop extraction of derivatized carbonyls as shown for spirits and vodkas, for which good sensitivity (<1 μg/L) was achieved.[115]

15.6.2. Sulfur Compounds

It is estimated that about 10% of food and beverage flavor compounds contain sulfur. Main sources of volatile sulfur compounds in foods are sulfur-containing amino acids—cysteine and methionine—as well as cystine, glutathione, and some vitamins. Their degradation may be enzymatic or nonenzymatic, with heat-mediated processes being some of the most important pathways for sulfur-containing volatiles' formation. Many sulfur-containing compounds in vegetables are formed from the breakdown of nonvolatile precursors. They include alkyl and *S*-alkenyl sulfoxides, which are precursors of *Allium* genus plants (garlic, onion, leek), and glucosinolates, which lead to the characteristic flavor of Cruciferae crops due to the release of thiocyanates, isothiocyanates, and cyanides. Odor thresholds of sulfur compounds are usually very low. Sulfur compounds are among the characteristic substances in meat,[116] *Allium* genus vegetables,[117] wine,[118] beer,[119,120] and many fruits such as grapefruit, blackcurrant, grape, and pineapple. The concentration of sulfur compounds in foods is usually very low. This, combined with usually low odor thresholds for these compounds, poses a challenge for both isolation and detection methods. Therefore, selective detectors are often required to attain desirable limits of detection.[120,121]

Sulfur compounds have different polarities, from polar thiols to moderately polar thiazoles to nonpolar methyl sulfides, which would influence the extraction efficiency when liquid–liquid extraction is applied. The combination of extraction solvents can increase recoveries. Liquid–liquid extraction can be used for the extraction of most sulfur compounds from wine, but not for low boiling point compounds. Preferred solvents are dichloromethane, pentane-dichloromethane, or ethyl acetate. Concentration factors of up to 150,000 were reported.[118] Many sulfur compounds are unstable. As presented by Mottram et al. for meat or ovalbumin, some disulfides are broken down significantly to corresponding thiols or are completely lost.[122] Very often dimethyl sulfide (DMS) oxidizes to dimethyl sulfoxide or MeSH can be oxidized to dimethyl disulfide (DMDS). SPME sampling minimizes oxidation of thiols in wine.[123] When simultaneous distillation extraction is used, thermal liability and reactivity of sulfur compounds must be considered; therefore, high vacuum is recommended to avoid artifact formation. For allylisothiocyanate extraction from mustard paste, SDE provides a similar proportion to SPME.[124] Garlic flavor components were analyzed by SD, simultaneous distillation/extraction (SDE), solid-phase trapping solvent extraction (STPE), and HS-SPME. When SD and SDE methods were applied, thermal degradation of some of the compounds—allyl methyl sulfide, DMDS, and thiirane—took place, which shows the detrimental influence of high temperature on the analyzed compounds.[125] Therefore, for both SHS and dynamic headspace special care must be taken when heating the sample to improve partitioning between a solid phase and headspace. In dynamic headspace methods, carbon traps can release adsorbed sulfur compounds incompletely.

Due to the relatively low sensitivity of SHS, this technique is used for the extraction of the most volatile sulfur

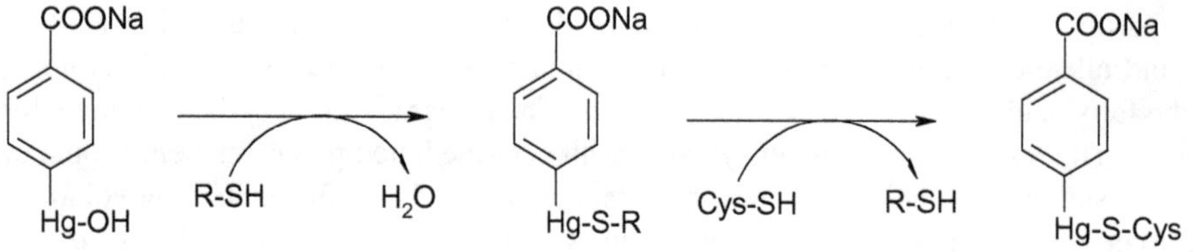

Figure 15.4. Reaction scheme for thiol trapping using *p*-hydroxymercuribenzoate (pHMB).

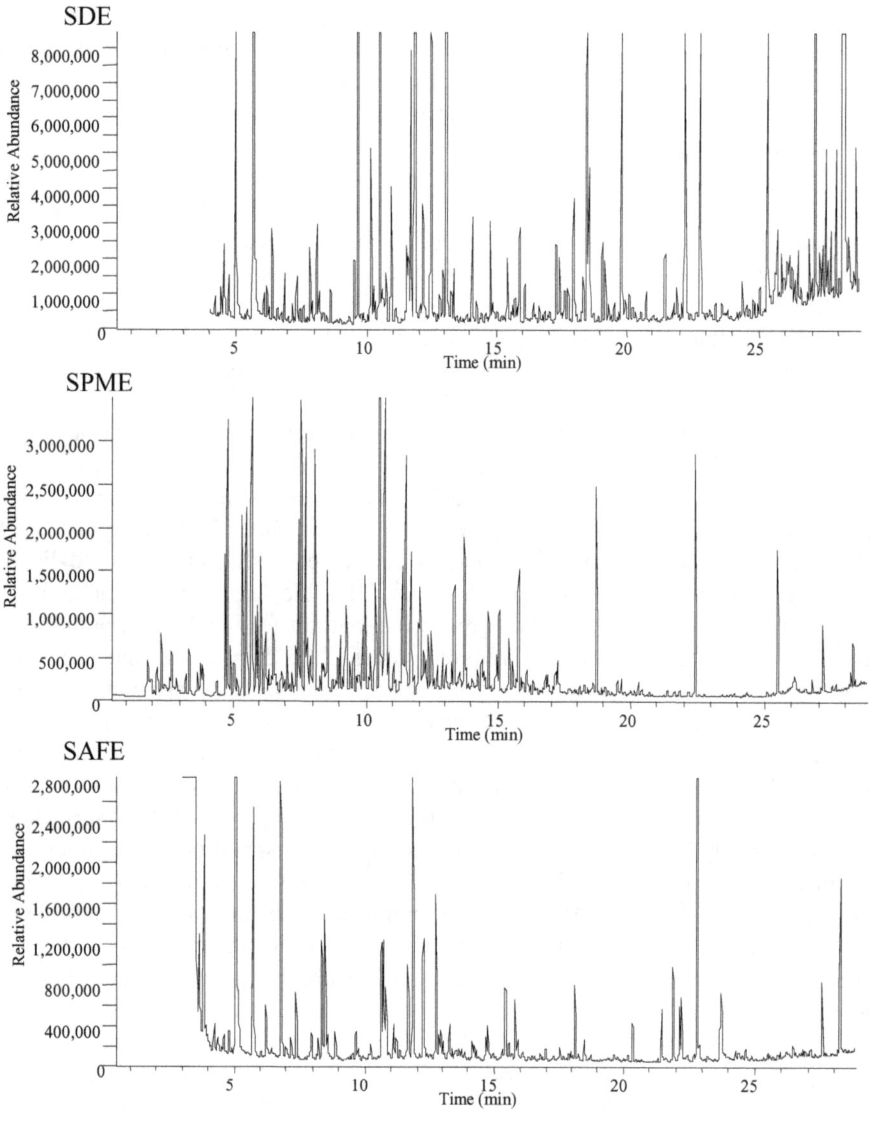

Figure 15.5. Chromatogram of volatile compounds isolated from extruded potato snacks using (from top to bottom) SDE, SPME, and SAFE methods. Reprinted from reference 127 with permission from Elsevier. Copyright 2009.

compounds, occurring in relatively high concentrations in food, such as DMS in malt and beer. A unique trap for thiols is a solution of pHMB (*p*-hydroxymercuribenzoate) added directly to wine, which forms thiol conjugates (Fig. 15.4). Trapped thiols are then released by adding glutathione or cysteamine. This method was used for the analysis of 2-furanmethanethiol.[126]

15.7. AROMA COMPOUNDS ISOLATION FROM FOOD—A CONCLUSION

In the analysis of food aroma compounds, two approaches dominate: (1) estimation of total volatiles, a profile of aroma compounds, and (2) detection and quantitation of specific/target compounds, usually responsible for characteristic notes or the key odorants.

Aroma profiles can be used for the comparison of various fruit cultivars, fruit in different ripening stages, and comparison of foods processed or stored. Aroma profiles can also be used for traceability or authenticity purposes. In such comparisons, chromatographic data are often subjected to multivariate analysis, and often with the chemometric approach, a detailed knowledge of the identity of each compound is not required. Therefore, changes in the material induced during sample preparation are of lesser importance, assuming the comparison is performed in the same manner for all samples. Profiles of aroma compounds can be obtained by various methods. Exhaustive extraction may be used, as in SDE or liquid–liquid extraction. For aroma profiling, rapid extraction methods are in demand; therefore, automated headspace analysis methods, preferably SPME or dynamic headspace, can be used. SPME is especially promising where equilibrium is attained rapidly; a nonequilibrium method can also be used, particularly in combination with fast chromatography. For subsequent chemometric data treatment of aroma profiles, sample preparation should be as rapid as possible, guaranteeing high sample throughput, and should be sensitive enough to facilitate the comparison of compounds present in minute concentrations.

On the other hand, for the determination of key odorants in food, the most important issue is the extraction of volatiles from the matrix in such a way that compounds remain unchanged during the isolation process. Obviously, the extraction method should not alter the aroma composition of the analyzed food by inducing chemical changes in some compound or causing the formation of artifacts nonexistent in the analyzed sample. Therefore, vacuum distillation methods should be used for thermally unstable compounds. To prevent extracted compounds from oxidation, an antioxidant can be added. To obtain complete information on the aroma compounds present in a sample, several extraction methods should be applied. As seen in Figure 15.5, depending on the isolation method used, different compounds are often isolated. Based on the literature data for the majority of products, such methods as SPME, SAFE, SDE, or SHS are complementary, and the more methods applied for the isolation of volatile compounds, the more information can be obtained about the odorants.

REFERENCES

1. Ridgway, K.; Lalljie, S.P.D.; Smith, R.M. Sample preparation techniques for the determination of trace residues and contaminants in foods. *J. Chromatogr. A* **2007**, *1153*, 36–53.
2. Buldini, P.L.; Ricci, L.; Sharma, J.L. Recent applications of sample preparation techniques in food analysis. *J. Chromatogr. A* **2002**, *975*, 47–70.
3. Wilkes, J.G.; Conte, E.D.; Kim, Y.; Holcomb, M.; Sutherland, J.B.; Miller, D.W. Sample preparation for the analysis of flavors and off-flavors in foods. *J. Chromatogr. A* **2000**, *880*, 3–33.
4. Pillonel, L.; Bosset, O.; Tabacchi, R. Rapid preconcentration and enrichment techniques for the analysis of food volatile. A review. *Lebensm.-Wiss. Technol.* **2002**, *35*, 1–14.
5. Jeleń, H.H. Solid phase microextraction in the analysis of food taints and off-flavors. *J. Chromatogr. Sci.* **2006**, *44*, 399–415.
6. Belitz, H.D.; Grosch, W.; Schieberle, P. *Food Chemistry*. Berlin Heidelberg: Springer-Verlag; 2004.
7. Rothe, M.; Thomas, B. Aromastoffe des brotes. Versuch einer auswertung chemischer geschmacksanalysen mit hilfe des schwellenwertes. *Z. Lebensm.-Unters. Forsch.* **1963**, *119*, 302–310.
8. Acree, T.; Barnard, J.; Cunningham, D.G. A procedure for the sensory analysis of gas chromatographic effluents. *Food Chem.* **1984**, *14*, 273–286.
9. Grosch, W. Detection of potent odorants in food by aroma extract dilution analysis. *Trends Food Sci. Technol.* **1993**, *4*, 68–73.
10. Brenna, E.; Fuganti, C.; Serra, S. Enantioselective perception of chiral odorants. *Tetrahedron Asymmetry* **2003**, *14*, 1–42.
11. de Roos, K.B. Physiochemical models of flavor release from foods. In *Flavor Release, ACS Symposium Series 763*, Roberts, D.D.; Taylor, A.J., Eds. Washington, DC: ACS; 2000; pp. 126–141.
12. Taylor, A. Physical chemistry of flavour. *Int. J. Food Sci. Technol.* **1998**, *33*, 53–62.
13. Druaux, C.; Voilley, A. Effect of food composition and microstructure on volatile flavour release. *Trends Food Sci. Technol.* **1997**, *8*, 364–368.
14. Guichard, E. Interactions between flavor compounds and food ingredients and their influence on flavor perception. *Food Rev. Int.* **2002**, *18*, 49–70.
15. Buttery, R.G.; Guadani, D.G.; Ling, L.C. Flavor compounds: Volatilities in vegetable oils and oil-water mixture. Estimation of odor thresholds. *J. Agric. Food Chem.* **1973**, *21*, 198–201.
16. Welsh, F.W.; Williams, R.E. The use of vegetable oils to recover compounds from aqueous solutions. *J. Chem. Technol. Biotechnol.* **1989**, *46*, 169–178.
17. Roberts, D.D.; Pollien, P. Relative influence of milk components on flavor compound volatility. In *Flavor Release, ACS Symposium Series 763*, Roberts, D.D.; Taylor, A.J., Eds. Washington, DC: ACS; 2000; pp. 321–332.
18. Semenova, M.G.; Antipova, A.S.; Misharina, T.A.; Golovnya, R.V. Factors determining binding of aroma esters with legumin in aqueous medium. In *Flavor Release, ACS Symposium Series 763*, Roberts, D.D.; Taylor, A.J., Eds. Washington, DC: ACS; 2000; pp. 293–307.
19. Lubbers, S.; Landy, P.; Voilley, A. Retention and release of aroma compounds in foods containing proteins. *Food Technol.* **1998**, *52*, 68–74.
20. Wu, S.Y.; Perez, M.D.; Puyol, P.; Saweyr, L. Beta-lactoglobulin binds palmitate within its central cavity. *J. Biol. Chem.* **1999**, *274*, 170–174.
21. O'Neill, T.E.; Kinsella, J.E. Effect of heat treatment and modification on conformation and flavor binding by beta-lactoglobulin. *J. Food Sci.* **1988**, *53*, 906–909.
22. Jouenne, E.; Crouzet, J. Effect of pH on retention of aroma compounds by beta-lactoglobulin. *J. Agric. Food Chem.* **2000**, *48*, 1273–1277.
23. Adams, R.L.; Mottram, D.S.; Parker, J.K.; Brown, H.M. Flavor–protein binding: Disulfide interchange reactions between ovoalbumin and volatile disulfides. *J. Agric. Food Chem.* **2001**, *49*, 4333–4336.
24. Godshall, M.A. How carbohydrates influence food flavor. *Food Technol.* **1997**, *51*, 63–67.
25. Yven, C.; Guichard, E.; Giboreau, A.; Roberts, A.D. Assessment of interactions between hydrocolloids and flavor compounds by sensory, headspace and binding methodologies. *J. Agric. Food Chem.* **1998**, *46*, 1510–1514.
26. Goubet, I.; Le Quere, J.L.; Voilley, A.J. Retention of aroma compounds by carbohydrates: Influence of their physicochemical characteristics and of their physical state. *J. Agric. Food Chem.* **1998**, *46*, 1981–1990.
27. Escher, F.E.; Nuessli, J.; Conde-Petiti, B. Interaction of flavor compounds with starch in food processing. In *Flavor Release, ACS Symposium Series 763*, Roberts, D.D.; Taylor, A.J., Eds. Washington, DC: ACS; 2000; pp. 230–245.
28. Villen, J.; Senorans, F.J.; Reglero, G.; Herraiz, M. Analysis of wine aroma by direct injection in gas chromatography without previous extraction. *J. Agric. Food Chem.* **1995**, *43*, 717–722.
29. Weurman, C. Isolation and concentration of volatiles in food odor research. *J. Agric. Food Chem.* **1969**, *17*, 370–384.
30. Cocito, C.; Gaetano, G.; Delfini, C. Rapid extraction of aroma compounds in must and wine by ultrasound. *Food Chem.* **1995**, *52*, 311–320.
31. Vila, D.H.; Mira, F.J.H.; Lucena, R.B.; Recamales, M.A.F. Optimization of an extraction method of aroma compounds in white wine using ultrasound. *Talanta* **1999**, *50*, 413–421.

32. Kerrola, K. Literature review: Isolation of essential oils and flavor compounds by dense carbon dioxide. *Food Rev. Int.* **1995**, *11*, 547–573.

33. Larráyoz, P.; Carbonell, M.; Ibáñez, F.; Torre, P.; Barcina, Y. Optimization of indirect parameters which affect the extractability of volatile aroma compounds from Idiazábal cheese using analytical supercritical fluid extractions (SFE). *Food Chem.* **1999**, *64*, 123–127.

34. Polesello, S.; Lovati, F.; Rizzolo, A.; Rovida, C. Supercritical fluid extraction as a preparative tool for strawberry aroma analysis. *J. High Resolut. Chromatogr.* **1993**, *16*, 555–559.

35. Kazazi, H.; Rezaei, K.; Ghotb-Sharif, S.J.; Emam-Djomeh, Z.; Yamini, Y. Supercritical fluid extraction of flavors and fragrances from *Hyssopus officinalis* L. cultivated in Iran. *Food Chem.* **2007**, *105*, 805–811.

36. Moldão-Martins, M.; Trigo, R.; Nolasco, M.A.; Bernardo Gil, M.G.; Beirão da Costa, M.L. Influence of extraction procedure on the aroma composition of *Thymus zygis* L. and *Mentha pulegium* L. In *Food Flavors: Formation, Analysis and Packaging Influences. Developments in Food Science Series Vol. 40*, Contis, E.T.; Ho, C.-T.; Mussinan, C.J.; Parliament, T.H.; Shahidi, F.; Spanier, A.M., Eds. Amsterdam: Elsevier; 1998; pp. 133–143.

37. Ouang, G.; Zhao, W.; Pawliszyn, J. Automation and optimization of liquid phase microextraction by gas chromatography. *J. Chromatogr. A* **2007**, *1138*, 47–54.

38. Chaintreau, A. Simultaneous distillation-extraction: From birth to maturity—Review. *Flavour Fragr. J.* **2001**, *16*, 136–148.

39. Nickerson, G.B.; Likens, S.T. Gas chromatographic evidence for the occurrence of hop oil components in beer. *J. Chromatogr. A* **1966**, *21*, 1–3.

40. Jayatilaka, A.; Poole, S.K.; Poole, C.F.; Chichila, T.M.P. Simultaneous microsteam distillation/solvent extraction for the isolation of semivolatile flavor compounds from cinnamon and their separation by series coupled-column gas chromatography. *Anal. Chim. Acta* **1995**, *302*, 147–162.

41. Blanch, G.P.; Reglero, G.; Herraiz, M. Rapid extraction of wine aroma compounds using a new simultaneous distillation-solvent extraction device. *Food Chem.* **1996**, *56*, 439–444.

42. Caja, M.M.; Ruiz del Castillo, M.L.; Martinez-Alvarez, R.; Herraiz, M.; Blanch, G.P. Analysis of volatile compounds in edible oils using simultaneous distillation-solvent extraction and direct coupling of liquid chromatography with gas chromatography. *Eur. Food Res. Technol.* **2000**, *211*, 45–51.

43. Escriche, I.; Serra, J.; Guardiola, V.; Mulet, A. Composition of medium volatility (simultaneous distillation extraction—SDE) aromatic fraction of pressed uncooked paste cheese (Mahon cheese). *J. Food Compost. Anal.* **1999**, *12*, 63–69.

44. Engel, W.; Bahr, W.; Schieberle, P. Solvent assisted flavour evaporation—A new and versatile technique for the careful and direct isolation of aroma compounds from complex food matrices. *Eur. Food Res. Technol.* **1999**, *209*, 237–241.

45. Krings, U.; Banavara, D.S.; Berger, R. Thin layer high vacuum distillation to isolate the flavor of high-fat food. *Eur. Food Res. Technol.* **2003**, *217*, 70–73.

46. Kolb, B. Headspace sampling with capillary columns. *J. Chromatogr. A* **1999**, *842*, 163–205.

47. Schreyen, L.; Dirinck, P.; Van Wassenhove, F.; Schamp, N. Analysis of leek volatiles by headspace condensation. *J. Agric. Food Chem.* **1976**, *24*, 1147–1152.

48. Nongonierma, A.; Cayot, P.; Le Quéré, J.-L.; Springett, M.; Voilley, A. Mechanisms of extraction of aroma compounds from foods, using adsorbents. Effect of various parameters. *Food Rev. Int.* **2006**, *22*, 51–94.

49. Pawliszyn, J. *Solid Phase Microextraction. Theory and Practice.* New York: Wiley-VCH; 1997.

50. Mangas, J.J.; Gonzales, M.P.; Rodriguea, R.; Blanco, D. Solid phase extraction and determination of trace aroma and flavor components in cider by GC-MS. *Chromatographia* **1996**, *42*, 101–105.

51. Nunez, A.; Gonzalez, L.F.; Janak, J. Preconcentration of headspace volatiles for trace organic analysis by gas chromatography. *J. Chromatogr. A* **1984**, *300*, 127–162.

52. Dirinck, P.; De Pooter, H.; Willaert, G.; Schamp, N. Application of dynamic headspace procedure in fruit flavour analysis. In *Analysis of Volatiles. Methods and Applications*, Schreier, P., Ed. Berlin; New York: Walter de Gruyter; 1984; pp. 381–400.

53. Boland, W.; Ney, P.; Jaenicke, L. A "closed-loop-stripping" technique as versatile tool for metabolic studies of volatiles. In *Analysis of Volatiles. Methods and Applications*, Schreier, P., Ed. Berlin; New York: Walter de Gruyter; 1984; pp. 371–380.

54. Kataoka, H.; Lord, H.L.; Pawliszyn, J. Application of solid phase microextraction in food analysis. *J. Chromatogr. A* **2000**, *880*, 35–62.

55. Wardencki, W.; Michulec, M.; Curyło, J. A review of theoretical and practical aspects of solid phase microextraction in food analysis. *Int. J. Food Sci. Technol.* **2004**, *39*, 703–717.

56. Brunton, N.P.; Cronin, D.A.; Monahan, F.J.; Durcan, R. A comparison of solid phase microextraction (SPME) fibers for measurement of hexanal and pentanal in cooked turkey. *Food Chem.* **2000**, *68*, 339–345.

57. Andrés, A.I.; Cava, R.; Ruiz, J. Monitoring volatile compounds during dry-cured ham ripening by solid-phase microextraction coupled to a new direct extraction device. *J. Chromatogr. A* **2002**, *963*, 83–88.

58. Palma-Harris, C.; McFeeters, R.F.; Fleming, H.P. Solid phase microextraction (SPME) technique for measurement of generation of fresh cucumber flavor compounds. *J. Agric. Food Chem.* **2001**, *49*, 4203–4207.

59. Lloyd, S.W.; Grimm, C.C. Analysis of 2-methylisoborneol and geosmin in catfish by microwave distillation-solid phase microextraction. *J. Agric. Food Chem.* **1999**, *47*, 164–169.

60. Rengarajan, R.; Seitz, L.M. Analysis of flavor compounds from microwave popcorn using supercritical fluid CO_2 followed by dynamic/static headspace techniques. In *Handbook of Flavor Characterization*, Deibler, K.D.; Delwiche, J., Eds. New York; Basel: Marcel Dekker; 2004; pp. 311–334.

61. Beltran, G.; Aguilera, M.P.; Gordon, M. Solid phase microextraction of volatile oxidation compounds in oil-in-water emulsions. *Food Chem.* **2005**, *92*, 401–406.

62. Sanchez-Silva, A.; Lopez-Hernández, J.; Paseiro-Losada, P. Profiling flavor compounds of potato crisps during storage using solid-phase microextraction. *J. Chromatogr. A* **2005**, *1064*, 239–245.

63. Song, J.; Fan, L.; Beaudry, R.M. Application of solid phase microextraction and gas chromatography/time-of-flight mass spectrometry for rapid analysis of flavor volatiles in tomato and strawberry fruits. *J. Agric. Food Chem.* **1998**, *46*, 3721–3726.

64. Elmore, J.S.; Erbahadir, M.A.; Mottram, D.S. Comparison of dynamic headspace concentration on Tenax with solid phase microextraction for the analysis of aroma volatiles. *J. Agric. Food Chem.* **1997**, *45*, 2638–2641.

65. Ruiz, J.; Cava, R.; Ventanas, J.; Jensen, M.T. Headspace solid phase microextraction for the analysis of volatiles in meat products: Dry cured Iberian ham. *J. Agric. Food Chem.* **1998**, *46*, 4688–4694.

66. Rega, B.; Fournier, N.; Guichard, E. Solid phase microextraction (SPME) of orange juice flavor: Odor representativeness by direct chromatography olfactometry (D-GC-O). *J. Agric. Food Chem.* **2003**, *51*, 7092–7099.

67. Roberts, D.D.; Pollien, P.; Milo, C. Solid phase microextraction method development for headspace analysis of volatile flavor compounds. *J. Agric. Food Chem.* **2000**, *48*, 2430–2437.

68. Vichi, S.; Castellote, A.I.; Pizzale, L.; Conte, L.S.; Buxaderas, S.; Lopez-Tamames, E. Analysis of virgin olive oil volatile compounds by headspace solid phase microextraction coupled to gas chromatography with

mass spectrometric and flame ionization detection. *J. Chromatogr. A* **2003**, *983*, 19–33.

69. Contini, M.; Esti, M. Effect of the matrix volatile composition in the headspace solid-phase microextraction analysis of extra virgin olive oil. *Food Chem.* **2006**, *94*, 143–150.

70. Zawirska-Wojtasiak, R.; Wąsowicz, E. Estimation of the main dill seeds odorant carvone by solid phase microextraction and gas chromatography. *Nahrung/Food* **2002**, *46*, 357–359.

71. Jia, M.; Zhang, Q.H.; Min, D.B. Optimization of solid phase microextraction analysis for headspace flavor compounds of orange juice. *J. Agric. Food Chem.* **1998**, *46*, 2744–2747.

72. Segurel, M.A.; Razungles, A.J.; Riou, C.; Salles, M.; Baumes, R.L. Contribution of dimethylsulfide to the aroma of Syrah and Grenache Noir wines and estimation of its potential in grapes of these varieties. *J. Agric. Food Chem.* **2004**, *52*, 7084–7093.

73. Fedrizzi, B.; Versini, G.; Lavangini, I.; Nicolini, G.; Magno, F. Gas chromatography–mass spectrometry determination of 3-mercaptohexan-1-ol and 3-mercaptohexyl acetate in wine. A comparison of headspace solid phase microextraction and solid phase extraction methods. *Anal. Chim. Acta* **2007**, *596*, 291–297.

74. Caven-Quantrill, D.J.; Buglass, A.J. Comparison of microscale simultaneous distillation extraction and stir bar sorptive extraction for the determination of volatile organic constituents of grape juice. *J. Chromatogr. A* **2006**, *1117*, 121–131.

75. Hayasaka, Y.; MacNamara, K.; Baldock, G.A.; Taylor, R.L.; Pollnitz, A.P. Application of stir bar sorptive extraction for wine analysis. *Anal. Bioanal. Chem.* **2003**, *375*, 948–955.

76. Zalacain, A.; Marin, J.; Alonso, G.L.; Salinas, M.R. Analysis of primary aroma compounds by stir bar sorptive extraction. *Talanta* **2007**, *71*, 1610–1615.

77. Ochiai, N.; Sasamoto, K.; Takino, M.; Yamashita, S.; Daishima, S.; Heiden, A.; Hoffman, A. Determination of trace amounts of off-flavor compounds in drinking water by stir bar sorptive extraction and thermal desorption GC-MS. *Analyst* **2001**, *126*, 1652–1657.

78. Bicchi, C.; Cordero, C.; Liberto, E.; Rubiolo, P.; Sgorbini, B.; Sandra, P. Sorptive tape extraction in the analysis of the volatile fraction emitted from biological solid matrices. *J. Chromatogr. A* **2007**, *1148*, 137–144.

79. Ishikawa, M.; Ito, O.; Ishizaki, S.; Kurobayashi, Y.; Fujita, A. Solid phase aroma concentrate extraction (SPACE): A new headspace technique for more sensitive analysis of volatiles. *Flavour Fragr. J.* **2004**, *19*, 183–187.

80. Bicchi, C.; Cordero, C.; Liberto, E.; Rubiolo, P.; Sgorbini, B.; Dawid, F.; Sandra, P. Dual-phase twisters: A new approach to headspace sorptive extraction and stir bar sorptive extraction. *J. Chromatogr. A* **2005**, *1094*, 9–16.

81. Grob, K.; Habich, A. Headspace gas analysis: The role and design of concentration traps specifically suitable for capillary gas chromatography. *J. Chromatogr.* **1985**, *321*, 45–58.

82. Cavalli, J.F.; Fernandez, X.; Lizzani-Cuvelier, L.; Loiseau, A.-M. Comparison of static headspace, headspace solid phase microextraction, headspace sorptive extraction, and direct thermal desorption techniques on chemical composition of French olive oils. *J. Agric. Food Chem.* **2003**, *51*, 7709–7716.

83. Hartman, T.G.; Overton, S.V.; Manura, J.J.; Baker, C.W.; Manos, J.N. Short path thermal desorption: Food sciences applications. *Food Technol.* **1991**, *45*, 104–105.

84. Valero, E.; Sanz, J.; Martinez-Castro, I. Direct thermal desorption in the analysis of cheese volatiles by gas chromatography and gas chromatography-mass spectrometry: Comparison with simultaneous distillation-extraction and dynamic headspace. *J. Chromatogr. Sci.* **2001**, *39*, 222–228.

85. Parliament, T. Solvent extraction and distillation techniques. In *Flavor, Fragrance and Odor Analysis*, Marsili, R., Ed. New York; Basel: Marcel Dekker; 2004; pp. 1–24.

86. Parliament, T.H. A concerted procedure for the generation, concentration fractionation and sensory evaluation of Maillard reaction products. In *Flavor Chemistry. Thirty Years of Progress*, Teranishi, R.; Wick, E.L.; Hornstein, I., Eds. New York: Kluwer Academic/Plenum Publishers; 1999; pp. 43–54.

87. Ferreira, V.; Hernández-Orte, P.; Escudero, A.; López, R.; Cacho, J. Semipreparative reversed-phase liquid chromatographic fractionation of aroma extracts from wine and other alcoholic beverages. *J. Chromatogr. A* **1999**, *864*, 77–88.

88. Jennings, W.; Filsoof, M. Comparison of sample preparation techniques for gas chromatographic analysis. *J. Agric. Food Chem.* **1977**, *25*, 440–446.

89. Leahy, M.M.; Reineccius, G.A. Comparison of methods for the isolation of volatile compounds from aqueous model systems. In *Analysis of Volatiles. Methods and Applications*, Schreier, P., Ed. Berlin; New York: Walter de Gruyter; 1984; pp. 19–47.

90. Ortega-Heras, M.; González-San Jose, M.L.; Beltrán, S. Aroma composition of wine studied by different extraction methods. *Anal. Chim. Acta* **2002**, *458*, 85–93.

91. Pollnitz, A.P.; Pardon, K.H.; Sykes, M.; Sefton, M.A. The effect of sample preparation and gas chromatograph injection techniques on the accuracy of measuring guaiacol, 4-methylguaiacol and other volatile oak compounds in oak extracts by stable isotope dilution analyses. *J. Agric. Food Chem.* **2004**, *52*, 3244–3252.

92. Insa, S.; Salvadó, V.; Anticó, E. Development of solid-phase extraction and solid-phase microextraction methods for the determination of chlorophenols in cork macerate and wine samples. *J. Chromatogr. A* **2004**, *1047*, 15–20.

93. Cullere, L.; Aznar, M.; Cacho, J.; Ferreira, V. Fast fractionation of complex organic extracts by normal phase chromatography on a solid phase extraction polymeric sorbent. Optimization of a method to fractionate wine flavor extracts. *J. Chromatogr. A* **2003**, *1017*, 17–26.

94. Larráyoz, P.; Addis, M.; Gauch, R.; Bosset, J.O. Comparison of dynamic headspace and simultaneous distillation extraction techniques used for the analysis of the volatile components in three European PDO ewe's milk cheeses. *Int. Dairy J.* **2001**, *11*, 911–926.

95. Mallia, S.; Fernandez-Garcia, E.; Bosset, J.O. Comparison of purge and trap and solid phase microextraction techniques for studying the volatile aroma compounds of three European PDO hard cheeses. *Int. Dairy J.* **2005**, *15*, 741–758.

96. Povolo, M.; Contarini, G. Comparison of solid phase microextraction and purge-and-trap methods for the analysis of the volatile fraction of butter. *J. Chromatogr. A* **2003**, *985*, 117–125.

97. Careri, M.; Manini, P.; Spagnoli, S.; Barbieri, G.; Bolzoni, L. Simultaneous distillation-extraction and dynamic headspace methods in the gas chromatographic analysis of parmesan cheese volatiles. *Chromatographia* **1994**, *38*, 386–394.

98. Vandeweghe, P.; Reineccius, G.A. Comparison of flavor isolation techniques applied to cheddar cheese. *J. Agric. Food Chem.* **1990**, *38*, 1549–1552.

99. Vichi, S.; Guadayol, J.M.; Caixach, J.; Lopez-Tamames, E.; Buxaderas, S. Comparative study of different extraction techniques for the analysis of virgin olive oil aroma. *Food Chem.* **2007**, *105*, 1171–1178.

100. Kanavouras, A.; Hernandez, R.J. The analysis of volatiles from thermally oxidized virgin olive oil using dynamic sorption-thermal desorption and solid phase micro-extraction technique. *Int. J. Food Sci. Technol.* **2006**, *41*, 743–750.

101. Diaz-Maroto, M.C.; Perez-Coello, M.S.; Cabezudo, M.D. Supercritical carbon dioxide extraction of volatiles from spices. Comparison with simultaneous distillation-extraction. *J. Chromatogr. A* **2002**, *947*, 23–29.

102. Guan, W.; Li, S.; Yan, ZR.; Tang, S.; Quan, C. Comparison of essential oils of clove buds extracted with supercritical carbon dioxide and other three traditional extraction methods. *Food Chem.* **2007**, *101*, 1575-1581.

103. Zawirska-Wojtasiak, R.; Wąsowicz, E. GC analysis of rosemary aroma isolated traditionally by distillation and by SPME. *J. Essent. Oil Res.* **2008**, *21*, 1–8.

104. Shimoda, M.; Shigematsu, H.; Shiratsuchi, H.; Osajima, Y. Comparison of the odor concentrates by SDE and adsorptive column method from green tea infusion. *J. Agric. Food Chem.* **1995**, *43*, 1616–1620.

105. Sarrazin, C.; Quere, J.L.; Gretsch, C.; Liardon, R. Representativeness of coffee aroma extracts: A comparison of different extraction methods. *Food Chem.* **2000**, *70*, 99–106.

106. Ebeler, S.E.; Terrien, M.B.; Butzke, C.E. Analysis of brandy aroma by solid phase microextraction and liquid-liquid extraction. *J. Sci. Food Agric.* **2000**, *80*, 625–630.

107. Jeleń, H.H.; Wlazły, K.; Wąsowicz, E.; Kamiński, E. Solid-phase microextraction for the analysis of some alcohols and esters in beer: Comparison with static headspace method. *J. Agric. Food Chem.* **1998**, *46*, 1469–1473.

108. Herent, M.F.; Collin, S. Pyrazine and thiazole structural properties and their influence on the recovery of such derivatives in aroma extraction procedures. *J. Agric. Food Chem.* **1998**, *46*, 1975–1980.

109. Frankel, E.N. Chemistry of autoxidation: Mechanism, products and flavor significance. In *Flavor Chemistry of Fats and Oils*, Min, D.B.; Smouse, T.H., Eds. Urbana-Champaign, IL: AOCS; 1985; p. 33.

110. Ross, C.F.; Smith, D.M. Use of volatiles as indicators of lipid oxidation in muscle foods. *Compr. Rev. Food Sci. Food Saf.* **2006**, *5*, 18–25.

111. Wu, Y.; Hamouz, F.; Schnepf, M. Factors affecting static HS-gas chromatographic analysis of lipid oxidation in precooked meat. *J. Agric. Food Chem.* **1998**, *46*, 3677–3682.

112. Gassenmeier, K. Interactions of selected flavor compounds with selected dairy products. In *Handbook of Flavor Characterization*, Deibler, K.D.; Delwiche, J., Eds. New York; Basel: Marcel Dekker; 2004; pp. 259–266.

113. Stashenko, E.E.; Martínez, J.R. Derivatization and solid-phase microextraction. *Trends Anal. Chem.* **2004**, *23*, 553–561.

114. Vesely, P.; Lusk, L.; Basarova, G.; Seabrooks, J.; Ryder, D. Analysis of aldehydes in beer using solid phase microextraction with on-fiber derivatization and gas chromatography/mass spectrometry. *J. Agric. Food Chem.* **2003**, *51*, 6941–6944.

115. Curylo, J.; Wardencki, W. Application of single drop extraction (SDE) gas chromatography method for the determination of carbonyl compounds in spirits and vodkas. *Anal. Lett.* **2006**, *39*, 2629–2642.

116. Mottram, D.S. Flavour formation in meat and meat products: A review. *Food Chem.* **1998**, *62*, 415–424.

117. Yu, T.-H.; Ho, C.-T. Chemistry and stability of sulfur containing compounds in the genus *Allium*. In *Shelf Life Studies of Foods and Beverages. Developments in Food Science 33*, Charalambous, G., Ed. Amsterdam: Elsevier; 1993; pp. 501–547.

118. Mestres, M.; Busto, O.; Guasch, J. Analysis of organic sulfur compounds in wine aroma. *J. Chromatogr. A* **2000**, *881*, 569–581.

119. Scarlata, C.J.; Ebeler, S.E. Headspace solid phase microextraction for the analysis of dimethyl sulfide in beer. *J. Agric. Food Chem.* **1999**, *47*, 2505–2508.

120. Hill, P.G.; Smith, R.M. Determination of sulfur compounds in beer using headspace solid phase microextraction and gas chromatographic analysis with pulsed flame photometric detection. *J. Chromatogr. A* **2000**, *872*, 203–213.

121. Mistry, B.S.; Reineccius, G.A.; Jasper, B.L. Comparison of chromatographic detectors for the analysis of volatile sulfur compounds in foods. In *Sulfur Compounds in Foods, ACS Symposium Series 564*, Mussinan, C.J.; Keelan, M.E., Eds. Washington, DC: ACS; 1994; pp. 8–21.

122. Mottram, D.S.; Szauman-Szumski, C.; Dodson, A.T. Interaction of thiol and disulfide flavor compounds with food components. *J. Agric. Food Chem.* **1996**, *44*, 2349–2351.

123. Majcenovic, A.B.; Schneider, R.; Lepoutre, J.-P.; Lempereur, V.; Baumes, R. Synthesis and stable isotope dilution assay of ethanethiol and diethyldisulfide in wine using solid phase microextraction. Effect of aging on their levels in wine. *J. Agric. Food Chem.* **2002**, *50*, 6653–6658.

124. Cai, J.; Liu, B.; Su, Q. Comparison of simultaneous distillation extraction and solid-phase microextraction for the determination of volatile flavor components. *J. Chromatogr. A* **2003**, *930*, 1–7.

125. Lee, S.-N.; Kim, N.-S.; Lee, D.-S. Comparative study of extraction techniques for determination of garlic flavor components by gas chromatography-mass spectrometry. *Anal. Bioanal. Chem.* **2003**, *377*, 749–756.

126. Tominaga, T.; Dubourdieu, D. A novel method for quantification of 2-methyl-3-furanthiol and 2-furanmethanethiol in wines made from *Vitis vinifera* grape varieties. *J. Agric. Food Chem.* **2006**, *54*, 29–33.

127. Majcher, M.; Jeleń, H.H. Comparison of suitability of SPME, SAFE and SDE methods for isolation of flavor compounds from extruded potato snacks. *J. Food Compost. Anal.* **2009**, *22*, 606–612.

Chapter 16

Sampling and Sample Preparation for Clinical and Pharmaceutical Analysis

Hiroyuki Kataoka, Keita Saito, and Atsushi Yokoyama

16.1. INTRODUCTION

Analysis of drugs in biological samples and pharmaceutical products is becoming more important owing to the need to understand the therapeutic and toxic effects of drugs and the continuing efforts to develop more selective and effective drugs.[1–3] Knowledge of drug levels in body fluids, such as serum and urine, can optimize pharmacotherapy and provide the basis for studies on patient compliance, bioavailability, pharmacokinetics and genetics, organ function, and the influences of comedication. The quantitative and qualitative analysis of drugs and metabolites is extensively applied to pharmacokinetic studies; in addition, variables such as time to reach maximal concentration in plasma, clearance, and bioavailability have to be known for the approval of a new drug.[3–5] Furthermore, therapeutic drug monitoring (TDM), based on the measurements of drug concentration in blood, can be used to improve patient care.[3,6,7] On the other hand, the screening and confirmation of drugs of abuse in body fluids can be used to identify potential users of illegal drugs and to monitor drug addicts following withdrawal therapy. Reliable qualitative and quantitative toxicological analysis is the basis of competent toxicological judgment and consultation in doping control, and clinical and forensic toxicology.[8–11]

Although drug analyses have been performed using various analytical instruments, most of these instruments cannot handle the sample matrices directly, because pharmaceutical products and biological materials may contain acids, bases, salts, proteins, and other organic compounds with similar properties to the analytes. In addition, the analytes in these complex matrices are often present at low concentrations. Despite the development of highly efficient analytical instrumentation for the endpoint determination of analytes in pharmaceutical products and biological samples, sample pretreatment is usually necessary to extract, isolate, and concentrate the analytes of interest from complex matrices. In general, sampling and sample preparation steps constitute over 80% of the total analysis time, and these steps are important in determining the success of analyzing compounds in complex matrices.

The main objective in any sampling strategy is to obtain a representative portion of the sample. The sampling step includes deciding where to get samples that properly define the object being analyzed and choosing methods that obtain appropriate amounts of samples. Sampling for subsequent trace analysis is the most crucial step in an analytical procedure. Objectives of sample preparation for clinical and pharmaceutical analysis are summarized in Table 16.1. In developing an analytical method, it is important to determine how much time and effort is necessary for sample preparation. For example, TDM usually requires specificity to distinguish the drug being monitored from similar compounds, metabolites, or coadministered drugs. In contrast, pharmacokinetic study of a potential drug candidate requires a specific and sensitive analytical method. Therefore, in carrying out efficient sampling and sample preparation, it is important that sample loss be kept to a minimum, so that the analyte of interest can be recovered in good yield; coexisting components be efficiently removed; problems do not occur in the chromatography and electrophoresis systems; the procedure can be performed conveniently and quickly; and the cost of analysis be kept to a minimum.

Sampling and sample preparation for clinical and pharmaceutical analysis can be classified into five operation groups: sample collection, release of drugs from the matrix, liquid handling, removal of endogenous compounds, and enhancement of sensitivity and selectivity (Table 16.2). For sample collection, procedures are adopted to ensure the integrity of the determinant during collection, transport, and

Handbook of Sample Preparation, Edited by Janusz Pawliszyn and Heather L. Lord

Table16.1. Objectives of Sample Preparation for Clinical and Pharmaceutical Analysis

1. Removal of unwanted macromolecular contaminants (mainly proteins) that may would interfere with analyte determination.
2. Removal of material if the resolving power of the chromatographic or electrophoretic column is insufficient to separate all the components in the sample completely or in a reasonably practical time.
3. Removal of material that would affect chromatographic and electrophoretic resolution or reproducibility.
4. Removal of material that could block the chromatograph and electrophoresis tubing, valve, column, or frits.
5. Quantitative extraction of analytes from complex matrices.
6. Solubilization of analytes to enable injection under the initial chromatographic and electrophoretic conditions.
7. Dilution to reduce solvent strength or avoid solvent incompatibility.
8. Concentration of the analyte within the detection limits of the analytical instrument.
9. Stabilization of the analyte to avoid hydrolytic or enzymatic degradation.
10. Derivatization of analytes for enhancement of sensitivity and selectivity.

storage. Various sampling techniques have been used for different samples, such as blood, urine, milk, hair, and tissues. These samples must be collected without contamination from the sampling equipment (e.g., syringes, pipettes, cannulas, tubes, and vessels) and without using the wrong sampling technique. Biological matrices are complex mixtures of compounds and usually have high protein content. The drugs may be bound to protein, glucuronide, and/or sulfate, and may be present in low concentrations. Therefore, the analyst must initially release drugs from the matrix by hydrolysis. Hydrolysis can be performed either by enzymes or by acids or bases. Sonication may also improve the recovery of drugs from biological samples. Liquid handling procedures mainly provide the links between each operation group. Procedures for liquid handling include the addition, mixing, separation, and/or removal of liquids. They can often be the rate-limiting steps in a sample preparation process because they are both labor intensive and tedious.

A biological matrix may be solid, particulate, or a mixed composition of organic compounds in an aqueous solution, and may consist of many components, such as macromolecules and small molecules. Ultrafiltration and protein precipitation can be used to remove protein from biological samples. The membrane filters used for ultrafiltration are efficient, although care must be taken to avoid binding of the drug to the membrane. Precipitation using organic solvents and inorganic acids or salts is effective for the removal of proteins. Dialysis can separate an analyte from the matrix by diffusion through a semipermeable membrane. In contrast, liquid–liquid extraction (LLE), solid-phase extraction (SPE), and solid-phase microextraction (SPME) are useful

Table16.2. Classification of Sampling and Sample Preparation in Clinical and Pharmaceutical Analysis

1. Sample collection	• Suction • Drawing • Cutting • Pipetting • Headspace sampling
2. Release of drugs from matrix	• Hydrolysis Acid Base Enzyme Proteases Lipase β-Glucosidase Arylsulfatase • Sonication
3. Liquid handling	• Aspiration • Centrifugation • Dilution • Evaporation • Filtering • Mixing • Salting out • Separation
4. Removal of endogenous compounds	• Precipitation Organic solvents Inorganic acids and salts Ammonium sulfate • Ultrafiltration • Dialysis • Membrane extraction • Liquid–liquid extraction • Supercritical fluid extraction • Solid-phase extraction • Solid-phase microextraction • Immunoaffinity extraction • Liquid chromatography
5. Procedures for enhancement of sensitivity and selectivity	• Precolumn derivatization • Postcolumn derivatization • Use of selective detectors

sample preparation techniques, and can efficiently produce clean extracts for analysis. Supercritical fluid extraction (SFE), immunoaffinity extraction (IAE), molecularly imprinted polymer (MIP)-based extraction, and membrane-based extraction are also specific and efficient sample preparation techniques. These techniques, which are based on the adsorption or partition of analytes, are responsible for removing the majority of the biological material from the sample matrix prior to analysis. For other sample preparation techniques, derivatization of the analyte can enhance its sensitivity and/or selectivity. The use of selective detectors for gas chromatography (GC), high-performance liquid chromatography (HPLC), and capillary electrophoresis (CE) are also effective for this purpose. As shown in Table 16.2, the final aim of sample preparation must be to isolate and purify the analyte and to introduce it to the GC, HPLC, or CE in a manner compatible with each instrument. Thus, sample preparation from biological samples for drug analysis is usually carried out using a combination of the techniques described above. Furthermore, these techniques may be used for automated sample preparation systems.

In this chapter, we review advances after 2000 in sampling and sample preparation techniques for clinical and pharmaceutical analysis. We first describe the general aspects of biological matrices and pharmaceutical products for sampling and sample preparation. We next consider sample preparation techniques for clinical and pharmaceutical analysis according to method of extraction. This review focuses primarily on newly developed sample preparation techniques, including LLE, SPE, and SPME. The details of sampling and sample preparation for clinical and pharmaceutical analysis have also been described in books[1–3,8] and well-documented reviews.[5–7,9–43]

16.2. SAMPLING AND STORAGE OF BIOLOGICAL SPECIMENS

Various biological matrices, such as plasma, serum, blood, urine, saliva, milk, sweat, breath, hair, skin, tissues, feces, and stomach contents, are used for drug analysis in clinical and forensic applications (Table 16.3). In sampling from these matrices, factors that must be considered include collection method, collection time, and the characteristics of the individual (e.g., age, sex, circadian rhythm, before or after a meal or exercise, season of the year). Furthermore, in storing biological specimens, it is necessary to consider the effects of temperature, pH, ionic strength, light, oxygen, humidity, or enzymes. Factors and measures used to avoid contamination and alterations during sampling, storage, and sample preparation are shown in Table 16.4.

Blood is perhaps the most useful sample for the identification and quantification of drugs and for the interpretation of data of toxicological significance. Blood is a complex fluid containing solubilized proteins, dissolved fats and salts, and suspended cells. The major constituents, the red

Table16.3. Biological Matrices Used for Drug Analysis in Clinical and Forensic Applications

1. Liquids and gases	• Blood	• Whole blood, plasma, serum
	• Urine	
	• Secretion and effusion	• Saliva, tears, sweat, milk, cerebrospinal fluid, gastric juice, bile, semen, liquor amnii
	• Breath	
2. Solids and semisolids	• Feces	
	• Keratinaceous tissues	• Hair, skin, nails
	• Organs and other tissues	• Brain, liver, lung, kidney, muscle, fat, bone, stomach contents

Table16.4. Factors and Measures for Contaminations and Alterations during Sampling, Storage, and Sample Preparation in Clinical and Pharmaceutical Analysis

	Factors	Measures
Contamination and adsorption	1. Contamination	
	• Experiment apparatus	• Washing of apparatus
	• Reagents (impurities, additives, decomposition compounds)	• Conservation in airtight container
		• Environmental clarification
	• Environment (volatile materials, dust, microbial products, experiment person origin thing)	• Choice of appropriate materials
	2. Adsorption to experiment apparatus	• Correction by addition of analog
	3. Drying and moisture absorption	• Operation in closed system
Chemical alteration	1. Oxygen	• Addition of antioxidants chelating agents
	2. Temperature	• Storage in liquid nitrogen, freezer and refrigerator
	3. Moisture	• Dry disposal of sample, desiccant addition to a container
	4. Light	• Shading, use of brown pot
Biological alteration	1. Vital reaction, metabolism	• Addition of metabolic inhibitor, storage at low temperature
	2. Enzyme reaction	• Addition of enzyme inhibitor, storage at low temperature
	3. Breeding of a microbe	• Addition of fungicides, storage at low temperature

blood cells, can be separated from the clear fluid (plasma) by centrifugation. If blood is allowed to stand without the addition of anticoagulating agents, the red cells will eventually clot, and the resulting fluid, serum, can be decanted. As matrices for drug analysis, the significant difference between plasma and serum is that serum does not contain fibrinogen and some clotting factors (2.5–5% of proteins). A method developed for plasma can normally be applied without modification to serum. However, whole blood may be the sample of choice when trying to detect the presence of a drug, or when quantitative information is needed on a deteriorated blood sample where complete separation of plasma or serum from red cells is not possible. A common error in blood collection is sample hemolysis, causing analyte dilution and interference in a serum assay. Because serum and plasma contain high concentrations of protein, drugs may bind to these proteins, with the degree of binding dependent on the individual physicochemical properties of each drug. In general, acidic and neutral drugs bind primarily to albumin, whereas basic drugs bind primarily to acid glycoprotein. Only free drug is available for extravascular distribution and elimination, as well as being able to cross cellular membranes and interact with drug receptors. This direct analysis of drugs in serum or plasma from which protein has been removed may not reflect total drug concentration. Therefore, prior to extraction, proteins are often denatured with an organic solvent. Alternatively, drugs strongly bound to protein may be freed by the appropriate adjustment of pH.

Urine is also one of the most commonly used biological matrices for drug analysis, particularly because of its relative ease of collection. Usually, urine is sampled over a 24-h period into precleaned bottles according to detailed instructions. Preservatives may be required to stabilize the analyte, particularly if it is susceptible to bacterial degradation. To guarantee the stability of the urine, the time between sampling and analysis should be kept as short as possible. Results are typically normalized by expressing drug concentration per gram creatinine, in order to circumvent the highly variable dilution that results from the body's attempts to maintain water balance. Compared with blood, plasma, or serum, urine is relatively free of protein, thus making possible the direct extraction of an analyte with an organic solvent. However, the interpretation of results obtained from urine samples is complicated by several factors, such as the amount of urine excreted, variations in pH and ionic strength, and the time lapse after intake of the drug. Variations in urine pH can affect the elimination of many drugs, as well as their extraction efficiency. Variations in ionic strength can also have a significant effect on drug extraction. This may be controlled for by using an appropriate internal standard, or adding a standard compound to normalize quantification. The internal standard should be structurally related to the other compounds. If urine samples must be significantly diluted prior to extraction, because of very high drug concentrations, variability in the matrix becomes of less significance.

Saliva is a colorless fluid excreted into the oral cavity from three principle glands: the parotid, sublingual, and submandibular glands. Unstimulated saliva pH is in the range of 5.6–7 and increases with stimulation to a maximum of 8.0. Therefore, drug concentrations in saliva partially depend on the pH of the saliva and the degree of stimulation. Compared with urine collection, saliva sample collection is less invasive and causes fewer concerns about violation of privacy and adulteration of samples. Compared with blood, saliva samples can be directly analyzed without a prior extraction step. Limited amounts of mixed saliva may be collected by spitting. Larger amounts of saliva may be collected by stimulating saliva flow by mastication with rubber bands, wax, Teflon tape, or gum; sucking on pebbles, marbles or candy; placing citric acid (sapid stimulus) on the tongue; and adsorption on cotton rolls or administration of pilocarpine. Material for stimulation of saliva must be carefully selected because lipophilic drugs may be adsorbed into the material. For identification purposes, adsorption (similar to SPE or SPME) could be employed to directly extract the drug from the saliva while inside the mouth. Although selected gland secretions can reduce saliva/plasma ratios and minimize oral contamination, most studies utilize mixed saliva specimens because collection of the latter is less invasive. An ultrafiltration device has been developed to reduce the viscosity of saliva for easier analysis. The dialysis membrane is also chosen to exclude most higher-molecular-mass substances, including mucopolysaccharides, food particles, and bacteria.

Hair is a popular target for drug analysis as its collection is relatively noninvasive, and it provides a historical record of exposure. However, drug concentration may vary by hair type, due to variations in drug affinity. At present, hair is recognized as a third fundamental biological specimen for drug testing, after urine and blood.[41,42] The type of hair determines its length and texture, which are dependent on sex and age. Hair patterns, colors, textures, diameter, and growth rates vary considerably between the genders and among races. The most important considerations in hair sampling are collection from a preferable anatomical location (posterior vertex) where hairs are relatively uniform, collection at a uniform distance from scalp, collection of a sufficient sample for the number of tests to be performed, prevention of contamination, and accurate identification of the samples. The problem of external contamination is very important when assaying for drugs of abuse as people may be exposed to a smoky or dusty environment. Another important concern is the change in drug concentration induced by cosmetic hair treatments. Various pretreatment procedures have been proposed including washing with water, buffer, sodium dodecylsulfate, methanol or acetone; extraction with 1 M sodium hydroxide, 25% ammonium hydroxide in methanol, or 5 M hydrochloric acid in methanol; and exzymatic treatment with proteinase K or β-glucuronidase-arylsulfatase. With the exception of methanolic sonication, the resulting mixture from the pretreatment step must be further purified prior to analytical examination. To date, hair analysis has been applied primarily to forensic toxicology and drug abuse studies, followed by clinical toxicology and clinical chemistry. The largest number of papers on hair analysis has dealt with assays for cocaine, opiates, and amphetamines. However, recent hair analysis studies have assessed other drug types, such as doping agents.

16.3. CONSIDERATIONS FOR PHARMACEUTICAL PRODUCTS AND ILLICIT DRUG PREPARATIONS

Quality control of pharmaceutical products includes assays for the confirmation, purity, and content of the active ingredient(s). Pharmaceutical products can be divided into various types based on formulation, such as tablets, elixirs, creams, ointments, gels, lotions, emulsions, suspensions, injectables, and inhalants, or their excipient action such as bulk, capsules, coatings, flavorings, colorings, buffer, and preservatives. The nature of the active ingredient is important in the design of pharmaceutical products. Optimal drug formulation is crucial in the design of products that produce optimal therapeutic effect. In addition to meeting regulatory standards for active ingredient quality, identity, purity, and stability, products are typically optimized for utility and bioavailability to assure the best competitive advantage. The nature of the formulation procedures and/or auxiliary (non-medical) components of a product can, for example, make a product easier to handle, limit degradation, maintain homogeneity, enhance bioavailability and targeting of the active ingredient to receptor sites, provide for additional routes of administration, and target the product for additional therapeutic indications. Auxiliary inactive ingredients, known collectively as excipients, can solubilize the active ingredient, suppress the growth of microorganisms, provide bulk or coating for a tablet or capsule, provide color, or mask an unpleasant taste or odor. It is therefore essential that, in analyzing pharmaceutical products, the nature of these excipients be taken into account, due to their potential impact on the analysis of the active ingredient. For sample preparation of liquid formulations, LLE or SPE are generally used. For sample preparation of solid formulations, the sample is usually crushed, dissolved, or suspended in water or buffer, and subsequently extracted by LLE; or it may be directly dissolved in a polar organic solvent such as methanol or acetonitrile. For sample preparation of ointments and suppositories, the active ingredient in the matrix is directly extracted into a polar organic solvent with or without water, or dispersed into a less polar organic solvent such as hexane or chloroform, followed by extraction with methanol–water, acetonitrile–water, or water.

Drugs of abuse may be prepared by various synthetic routes, for each of which incomplete or side reactions may also occur. The active ingredients are usually not extensively purified following synthesis. As a result, several impurities are typically present in these preparations, and impurity profiling may be used to track the source of these products. LLE or, at times SPE, has been used to prepare samples of confiscated materials for profiling by GC or HPLC. SPME has recently been evaluated for impurity profiling of confiscated methamphetamine.[44] Headspace (HS) extraction was preferred to avoid contamination of the SPME fiber by tablet components. Compounds related to methamphetamine such as benzaldehyde, benzyl alcohol, amphetamine, benzyl methyl ketone, *cis-* and *trans-*1,2-dimethyl-3-phenylaziridine, dimethylamphetamine, *N*-acetylamphetamine, *N*-acetylmethamphetamine, and *N*-formylmethamphetamine were detected in the chromatograms. Trace amounts of ethanol, diethyl ether, and acetic acid were also detected in some of the chromatograms. The numbers and intensities of the peaks detected were different, depending on the samples from nine different origins. SPME has also been used to profile manufacturing by-products and impurities from an illicit drug seizure of 4-methoxyamphetamine. The results indicated the synthetic route likely used in the preparation of the compound. In addition, the SPME method was found to be rapid and non-destructive and to give results complementary to those from conventional LLE.

16.4. SAMPLE PREPARATION TECHNIQUES IN CLINICAL AND PHARMACEUTICAL ANALYSIS

Biological materials and pharmaceutical products are very complex, often containing proteins, salts, acids, bases, and numerous organic compounds with similar chemistry to the analytes. In addition, these analytes are often present in the samples at low concentrations. Thus, sample preparation is usually necessary in order to extract and isolate the analytes in complex matrices. Many sample preparation methods, including LLE,[12,13,29,31,39,40] SPE,[12–14,29,39,40,42,43] SPME,[15–30,32–43] affinity sorbent extraction,[45,46] and membrane-based extraction,[14,47,48] are used for drug analysis. Each step may be used alone or in combination with other methods. In this section, we review the most recently developed sample preparation techniques used in clinical and pharmaceutical analysis, especially LLE, SPE, and SPME. Clinical and pharmaceutical applications of these techniques reported during the past 7 or 8 years are summarized in Tables 16.5–16.7.

16.4.1. Protein Removal

For drug analysis in biological fluids or tissues, protein must often be removed from the sample matrix before a suitable extraction process. Deproteinization approaches include the use of (1) extreme temperatures, osmotic or ionic strengths, or pH; (2) organic precipitation agents; or (3) proteolytic enzymes. Protein precipitation, ultrafiltration, and dialysis

Table 16.5. Selected Applications for Drug Analysis Using Liquid–Liquid Extraction

Analyte	Matrix	Extraction	Analytical Method	Reference
Methenolone	Hair	Pentane	GC-MS/MS	49
Amphetamine	Plasma	Hexane	GC-MS	50
Tramadol	Milk	Hexane	HPLC-UV	51
Catecholamines	Urine	Heptane	LC-MS-MS	52
Atenolol	Plasma, urine	Hexane/ethanol	HPLC-UV	53
Atovaquone	Plasma	Hexane/isoamyl alcohol	HPLC-UV	54
Clozapine and metabolite	Plasma	Hexane/isoamyl alcohol	HPLC-UV	55
Pseudoephedrine	Plasma	Hexane/isoamyl alcohol	HPLC-UV	56
Mianserin and metabolite	Plasma	Hexane/isoamyl alcohol	LC-MS	57
Amlodipine	Plasma	Hexane/ethyl acetate	LC-MS/MS	58
Paroxetine	Plasma	Hexane/ethyl acetate	LC-MS-MS	59
Tofisopam	Plasma	Hexane/dichloromethane	GC-NPD	60
Clemastine	Plasma	Toluene	GC-NPD	61
Disopyramide and metabolite	Plasma	Toluene	CE-UV	62
Quinine and 3-hydroxyquinine	Blood	Toluene/butanol	HPLC-UV	63
Dextrophane, guaifenesin	Plasma	Chloroform	HPLC-FL	64
Disopyramide and metabolite	Plasma, urine	Dichloromethane	HPLC-UV	65
Anabolic steroids	Urine	Dichloromethane	HPLC-DAD	66
Carvedilol	Plasma	Dichloromethane	HPLC-FL	67
Levodropropizine	Plasma	Dichloromethane/diethyl ether	LC-MS-MS	68
Armillarisin	Plasma	Dichloromethane/diethyl ether	LC-MS-MS	69
Iptakalim	Plasma	Dichloromethane/diethyl ether	LC-MS-MS	70
Sildenafil citrate (Viagra)	Saliva	Chloroform/2-propanol/heptane	LC-MS	71
Phenmetrazine	Urine	1-Chlorobutane	GC-MS	72
Narcotics	Urine	Diethyl ether	LC-MS-MS	73
Desloratadine and metabolite	Plasma	Diethyl ether	LC-MS-MS	74
Rapamycin	Whole blood	Diethyl ether/butyl chloride	HPLC-UV	75
HIV protease inhibitors	Plasma	*tert*-Butyl methyl ether	HPLC-FL	76
Arbidol	Plasma	*tert*-Butyl methyl ether	HPLC-UV	77
Itraconazole and metabolite	Plasma	Acetonitrile/methyl *t*-butyl ether	LC-MS-MS	78
Cimetidine	Plasma	Ethyl acetate	HPLC-UV	79
Seratodas and metabolite	Serum, urine	Ethyl acetate	HPLC-UV	80
Clindamycin	Plasma	Ethyl acetate	LC-MS-MS	81
Clindamycin	Plasma	Ethyl acetate	HPLC-UV	82
Budesonide	Urine	Ethyl acetate	LC-MS	83
Tamsulosin	Serum	Ethyl acetate	LC-MS-MS	84
Tamsulosin	Plasma	Butyl acetate	HPLC-FL	85
Mycophenolate mofetil	Serum	Ethyl acetate/2-propanol	HPLC-UV	86
Fluoxetine	Plasma	Ethyl acetate/96-well plate	LC-MS-MS	87
Donepezil	Plasma	Hexane/96-well plate	LC-MS-MS	88
Amphetamine-type stimulants	Urine	1-Chlorobutane/microchip	GC-FID	89
Clenbuterol	Urine	Three-phase LLLME	HPLC-UV	90
Methamphetamine, amphetamine	Urine	Three-phase LLLME	HPLC-UV	91
Phenylacetic acid, phenylpropionic acid	Plasma, serum, urine	Three-phase LLLME	HPLC-UV	92
Antifouling agents	Water	SDME	GC-ECD	93
Fentanyl	Water	SDME	GC-MS	94
Traditional Chinese medicines	Medicine	HS-SDME	LC-MS	95
Z-ligustilide	Plasma	HS-SDME	GC-MS	96
Hypericin, hyperforin	Plasma, urine	SD-LPME	HPLC-UV-FL	97
Local anesthetics	Urine	SD-LPME	HPLC-UV	98
Methamphetamine, amphetamine	Urine	HS-LPME	HPLC-UV	99

Table 16.5. (*Continued*)

Analyte	Matrix	Extraction	Analytical Method	Reference
Promethazine, methadone, haloperidol	Plasma, urine	Two-phase LPME	GC-FID, CE-UV	100
Chiral drugs: mianserine	Plasma	Two-phase LPME	CE-UV	101
Abused basic drugs	Hair	Surfactant-enhanced LPME	HPLC-UV	102
Angiotensin, neurotensin, and their metabolites	Water	Ion-pair-mediated LPME	HPLC-UV	103
Amphetamine, morphine, practorol	Plasma, urine	Carrier-mediated LPME	CE-UV	104
Polar drugs	Plasma	Carrier-mediated LPME	LC-MS	105
Basic drugs	Plasma	Carrier-mediated LPME	CE-UV	106

LPME, liquid-phase microextraction; LLLME, liquid–liquid–liquid microextraction; HS-SDME, headspace–single-drop microextraction; GC, gas chromatography; GC-MS, gas chromatography–mass spectrometry; HPLC, high-performance liquid chromatography; LC-MS, liquid chromatography–mass spectrometry; CE, capillary electrophoresis; NPD, nitrogen–phosphorus detector; FID, hydrogen flame ionization detector; DAD, diode array detector; FL, fluorescence detector; ECD, electron capture detector.

Table 16.6. Selected Applications for Drug Analysis Using Solid-Phase Extraction

Analyte	Matrix	SPE Sorbent	Hyphenated Analysis	Reference
Haloperidol, perphenazine	Serum	C2	HPLC-UV	107
Citalopram	Plasma	C4	HPLC-FL	108
Atovaquone	Plasma, whole blood	C8	HPLC-UV	109
Olanzapine and metabolite	Plasma	C8	HPLC-ED	110
Cilazapril and metabolite	Pharmaceuticals, urine	C8/C18	HPLC-ED, DAD	111
Anabolic steroids	Urine	C18	HPLC-UV	112
HIV protease inhibitors	Plasma	C18	HPLC-UV	113
Vitamins	Pharmaceuticals	C18	HPLC-UV	114
Carbamazepine and metabolite	Plasma, milk	C18	HPLC-UV	115
Corticosteroids	Hair	C18	LC-MS	116
Retinoids	Pharmaceuticals	C18	HPLC-FL	117
Tacrolimus, sirolimus	Blood	C18	LC-MS	118
Tramadol and metabolite	Plasma	C18	HPLC-UV	119
Ranitidine and cimetidine	Plasma	C18	HPLC-UV	120
Phenytoin and dextromethorphan	Urine	C18	HPLC-DAD	121
Antiretroviral drugs	Plasma	C18	LC-MS-MS	122
Abused drugs	Urine	C18	LC-MS-MS	123
Antihyperglycemic biguanides	Serum, urine	C18/Oasis HLB	HPLC-UV	124
Xylazine	Blood	Oasis HLB	GC-MS	125
Antiepileptic drugs	Serum	Oasis HLB	HPLC-DAD	126
Rivanol and mifepristone	Plasma	Oasis HLB	HPLC-UV	127
Propofol	Plasma	Oasis HLB	LC-MS-MS	128
Hydroxytyrosol	Plasma	Oasis HLB	HPLC-DAD	129
bis-Thiazolium	Plasma, whole blood	Oasis HLB	LC-MS	130
Cocaine and heroin	Meconium	Oasis HLB	GC-MS	131
Drugs and toxic compounds	Serum	Oasis MCX	LC-MS	132
Mefenamic acid, flufenamic acid, telfenamic acid	Pharmaceuticals, urine	SAX	HPLC-UV	133
Abused drugs	Urine	Bakerbond NARC-2	GC-MS	134
Neuroleptics	Hair	Chromabond drug	LC-MS-MS	135
Domperidone	Plasma	Nitrile SPE	HPLC-FL	136
Ribavirin	Serum	PBA	HPLC-UV	137

Table 16.6. (*Continued*)

Analyte	Matrix	SPE Sorbent	Hyphenated Analysis	Reference
Ketorolac	Plasma	PH	HPLC-UV	138
Quinfamide	Plasma, urine, feces	Sep-Pack	HPLC-UV	139
Flunitrazepam and metabolite	Urine	Bond-Elute Certify	GC-MS	140
Raclopride	Plasma	SPE disk	LC-MS	141
Methotrexate	Urine, plasma	384-well SPE disk	LC-MS-MS	142
Glybenclamide	Serum	SPE disk	LC-MS	143
Indolocarbazole	Plasma	96-well SPE disk	LC-MS-MS	144
Oxazepam	Plasma	SPE disk	LC-MS-MS	145
GABA receptor modulator	Biosamples	C18 disk plates	LC-MS-MS	146
Amphetamine and methamphetamine	Blood	Pipette tip SPE	GC-MS	147
Cyproterone acetate	Plasma	LiChrospher RP-4 ADS	LC-MS-MS	148
Voriconazole	Serum	LiChrospher RP-8 ADS	LC-MS-MS	149
Furosemide	Serum	LiChrospher RP-18 ADS	HPLC-UV	150
Rofecoxib	Plasma	LiChrospher RP-18 ADS	LC-MS	151
Fluoxetine	Plasma	RAM BSA-C18	HPLC-UV	152
Naproxen, ibuprofen, diclofenac	Plasma	LiChrospher 60 XDS	HPLC-UV	153
Cloxacillin	Plasma	LiChrospher 60 XDS	HPLC-UV	154
Atropine	Plasma	LiChrospher 60 XDS	HPLC-UV	155
Warfarin	Serum	IAA	HPLC-FL	156
Fluoroquinolones	Serum	IAA	HPLC-FL	157
Estradiol, estrone, estriol	Urine	IAA	ELISA	158
Morphine and glucuronides	Blood	IAA	HPLC-LIFL	159
Bupivacaine	Plasma	MIP	GC	160
Phenytoin	Plasma	MIP	GC	161
Verapamil, gallopamil	Urine, plasma	MIP/RAM	LC-MS	162
β-Agonists	Urine	MIP	LC-MS-MS	163
Ciprofloxacin	Urine	MIP	LC-MS	164
Naproxen	Urine	MIP/RAM	LC-MS	165
Caffeine	Urine	MIP	HPLC-UV	166
Alprenolol	Plasma	MIP/RAM	HPLC-UV	167

PH, phenyl; SAX, trimethyl aminopropyl; PBA, phenylboronic acid; ADS, alkyl-diol silica; XDS, exchange diol silica; IAA, immunoaffinity adsorbent; MIP, molecularly imprinted polymer; RAM, restricted access material; GC, gas chromatography; GC-MS, gas chromatography–mass spectrometry; HPLC, high-performance liquid chromatography; LC-MS, liquid chromatography–mass spectrometry; DAD, diode array detector; ED, electrochemical detector; FL, fluorescence detector; LIFL, laser-induced native fluorescence detection; HLB, hydrophilic-lipophilic balance sorbent [poly(divinylbenzene-co-*N*-vinylpyrrolidone)]; ELISA, enzyme-linked immunosorbent assay; MCX, strong cation-exchange sulfonic groups on the surface of the HLB polymeric resin.

Table 16.7. Selected Applications for Drug Analysis Using SPME and Related Microextraction Techniques

Analyte	Matrix	Extraction Device	Extraction Mode	Hyphenated Analysis	Reference
Cannabinoids	Hair	PDMS fiber	HS + OFD	GC-MS	168
Cannabinoids	Hair	PDMS fiber	HS + OFD	GC-MS	169,170
Cannabinoids	Saliva	PDMS fiber	DI	GC-MS	171
Amphetamines	Hair	PDMS fiber	D + HS	GC-MS	172
Amphetamines, designer drugs	Hair	PDMS fiber	HS + OFD	GC-MS	173
Amphetamine-like drugs	Hair, saliva	PDMS fiber	HS	GC-MS	174
MDA, MDMA, MDEA	Saliva	PDMS fiber	DI	GC-MS	175
Amphetamine, fenfluramines	Blood	PDMS fiber	D + DI	GC-MS	176
Amphetamines	Blood	PDMS fiber	HS	GC-MS	177
Amphetamines, MDA, MDMA, MDEA	Serum	PDMS fiber	D + HS	GC-MS	178

Table 16.7. *(Continued)*

Analyte	Matrix	Extraction Device	Extraction Mode	Hyphenated Analysis	Reference
Amphetamines, MDA, MDMA, MDEA	Urine	PDMS fiber	HS	GC-MS	179
Amphetamine-related drugs	Urine	PDMS fiber	D + HS	GC-MS	180
Amphetamines, MDA, MDMA, MDEA	Urine	PDMS fiber	D + DI	FAI-MS-MS	181
Amphetamines	Urine	PDMS fiber	D + HS	GC-MS	182
Amphetamines	Urine	PDMS fiber	HS	GC-FID	183
Amphetamine, MDMA	Urine	CW/TPR fiber	DI + OFD D + DI	HPLC-FLD	184
Amphetamines	Drugs	PDMS fiber	DI + OFD	GC-MS	185
Recreational drugs	Hair	PDMS fiber	HS	GC-MS	186
Club drugs	Urine	PDMS fiber	HS + OFD	GC-MS	187
Selegiline, norselegiline	Blood, urine	DVB/CRB/PDMS fiber	HS	GC-MS	188
Cocaine, benzoylecgonine	Hair	PDMS fiber	DI	GC-MS	189
Cocaine, cocaethylene	Sweat	PDMS fiber	DI	GC-MS	190
Cocaine, cocaethylene	Plasma	PDMS fiber	DI	GC-MS	191
Cocaine, cocaethylene	Urine	PDMS fiber	DI	GC-MS	192
Illicit drugs, impurities	Tablets	DVB/CRB/PDMS fiber	HS	GC-MS	44
Illicit drugs	Serum, urine	PA fiber	HS	GC-MS	193
Methadone	Hair	PDMS fiber	DI	GC-MS	194
Methadone, metabolite	Hair	PDMS/DVB fiber	HS	GC-MS	195
Methadone	Saliva	PDMS fiber	DI	GC-MS	196
Methadone, metabolite	Plasma	PDMS fiber	DI	GC-MS	197
Basic drugs	Hair	PA fiber	HS	GC-MS	15
Tramadol	Plasma	PDMS/DVB fiber	HS	GC-MS	198
Fentanyl	Plasma	PDMS fiber	HS	GC-MS	199
Sufentanil	Plasma	PDMS/DVB fiber	DI	GC-MS	200
Halothane	Blood	PDMS fiber	HS	GC-MS	201
Lidocaine	Hair	CW/DVB fiber	HS	GC-MS	202
Lidocaine	Plasma	PDMS fiber	DI	GC-FID	203
Lidocaine	Urine	PDMS fiber	DI	MS-MS	204
Thiopental, pentobarbital	Hair	CW/TPR fiber	DI	GC-MS-MS	205
Barbiturates	Blood, urine	PA fiber	DI	GC-MS	206
Ketamine	Hair	PDMS fiber	HS	GC-MS	185
Propofol	Blood, tissues	PDMS fiber	HS	GC-MS	207
Antiepileptics	Plasma	CW/DVB fiber	DI	GC-TSD	208
Anticonvulsants	Plasma	CW/TPR fiber	DI	HPLC-UV	209
Valproic acid	Plasma	PDMS fiber	HS + OFD	GC-MS	210
Midazolam	Plasma	PA fiber	DI	GC-MS	211
Delorazepam	Urine	CW/TPR, PDMS/DVB fiber	DI	HPLC-UV	212
7-Aminoflunitrazepam	Urine	Immunoaffinity fiber	DI	LC-MS-MS	213
Benzodiazepines	Blood	ADS fiber	DI	HPLC-UV	214
Benzodiazepines	Urine	ADS fiber	DI	HPLC-UV	215
Phenothiazines	Blood, urine	PA fiber	DI	LC-MS-MS	216
Levomepromazine	Plasma	PDMS fiber	DI	GC-NPD	217
Amitriptyline	Urine	PDMS fiber	DI	HPLC-UV	218
Propranolol	Urine	Sol–gel fiber	HS and DI	GC-FID	219
Propranolol enantiomers	Urine	Sol–gel fiber	HS	CZE-DAD	220
Tricyclic antidepressants	Hair	PA fiber	HS	GC-MS	221
Tricyclic antidepressants	Plasma	CW/TPR fiber	DI	HPLC-UV	209

Table 16.7. (*Continued*)

Analyte	Matrix	Extraction Device	Extraction Mode	Hyphenated Analysis	Reference
Tricyclic antidepressants	Plasma	PDMS/DVB fiber	DI	LC-MS	222
Selective serotonin reuptake inhibitor antidepressants	Urine	PDMS/DVB fiber	DI	GC-MS	223
Fluoxetine, norfluoxetine	Plasma	PDMS/DVB fiber	DI	GPLC-UV	224
Amitraz	Plasma	PDMS fiber	DI	GC-TSD	225
Naproxen and metabolites	Urine	CW/TPR fiber	DI	HPLC-UV	226
Ibuprofen and glucuronide	Urine	PDMS/DVB fiber	DI	HPLC-UV	227
Rivastigmine	Plasma	PDMS/DVB fiber	HS	GC-MS	228
Thymol	Plasma	PDMS/DVB fiber	HS	GC-FID	229
Menthol	Plasma, urine	PDMS/DVB fiber	HS	GC-MS	230
Busulphan	Plasma	CW/DVB fiber	DI	GC-MS	231
Mycophenolic acid	Serum	CW/TPR fiber	DI	HPLC-UV	232
Anti-inflammatory drugs	Drugs	PA fiber	HS + OFD	GC-MS	233
Sterols	Serum	PA fiber	DI + OFD	GC-FID	234
Pharmaceuticals	Urine	PDMS, PA fiber	DI	HPLC-FLD	235
Ethanol	Blood, urine	PDMS/DVB fiber	HS	GC-FID	236
Fatty acid ethyl esters	Hair	PDMS/DVB fiber	HS	GC-MS	237,238
Acrolein	Urine	CW/DVB fiber	HS	GC-MS	239
Acetone	Breath, plasma	PDMS/DVB fiber	HS-OFD	GC-MS	240,241
Volatiles	Blood	CRB/PDMS fiber	HS	GC-MS	242,243
Chloroform, toluene, xylenes	Blood	PDMS fiber	HS	GC-MS	244
Dichloromethane, perchloroethylene	Urine	CRB/PDMS fiber	HS	GC-MS	245
Residual solvents	Pharmaceuticals	PDMS fiber	HS	GC-FID	246
Residual solvents	Pharmaceuticals	DVB/PDMS, CRB/PDMS/ DVB fiber	HS	GC-FID	247
Residual solvents	Pharmaceuticals	CRB/PDMS fiber	HS	GC-FID	248
Residual solvents	Pharmaceuticals	CRB/PDMS/DVB fiber	HS	GC-FID	249
Tetramine	Urine	PA fiber	DI	GC-FTD	250
Organophosphorous pesticides	Blood	PDMS fiber	HS	GC-MS	251
Organophosphorous pesticides	Blood	PA fiber	HS	GC-NPD	252
Parathion, paraxon	Blood	CW/DVB fiber	DI	GC-MS	253
Parathion-methyl	Blood	PA fiber	HS	GC-MS	254
Quinalphos	Blood, urine	CW/DVB fiber	DI	GC-MS	255
Fluoroacetic acid	Serum, urine	DVB/CRB/PDMS fiber	HS + OFD	GC-NPD	256
Fluoroacetamide	Blood, urine	PDMS fiber	DI	GC-MS	257
Chlorovinylarsonous acid	Urine	PDMS fiber	HS + OFD	GC-MS	258
Cyanide	Blood	CRB/PDMS fiber	HS	GC-MS	259
Strychnine	Blood	CW/DVB fiber	DI	GC-MS	260
Amphetamines, MDA, MDMA, MDEA	Hair, urine	PPY coated	IT	LC-MS	261
Amphetamines, MDA, MDMA, MDEA	Urine	Omegawax 250 capillary	IT	LC-MS	262
Benzodiazepines	Serum	Supel-Q plot capillary	IT	LC-MS	263
Benzodiazepines	Serum	ADS-packed capillary	IT	HPLC-UV	264
Ketamine	Urine	Monolithic capillary	IT	HPLC-UV	265
Tricyclic antidepressants	Urine	Wire-packed DB-1 capillary	IT	HPLC-UV	266
Tricyclic antidepressants	Urine	Zylon DB-1 capillary	IT	CZE-UV	267
β-Blockers	Serum, urine	PPY-coated capillary	IT	LC-MS	268
Propranolol	Serum	MIP-packed capillary	IT	HPLC-UV	269
Verapamil	Cell culture	PPY-coated capillary	IT	LC-MS	270

Table 16.7. (*Continued*)

Analyte	Matrix	Extraction Device	Extraction Mode	Hyphenated Analysis	Reference
Xanthines	Serum	Monolithic capillary	IT	HPLC-UV	271
Angiotensin II receptor agonists	Plasma, urine	Monolithic capillary	IT	HPLC-DAD and FLD	272
Angiotensin II receptor agonists	Plasma, urine	Monolithic capillary	IT	CZE-UV	273
Nonsteroidal anti-inflammatory drugs	Urine	β-Cyclodextrin-coated capillary	IT	HPLC-UV	274
Fluoxetine	Serum	Immunoaffinity capillary	IT	LC-MS	275
Drugs	Serum, urine	Omegawax 250 capillary	IT	LC-MS	276
Cortisol	Saliva	Supel-Q plot capillary	IT	LC-MS	277
Phthalates	Liquid medicines	Supel-Q plot capillary	IT	HPLC-UV	278,279
Barbiturates, benzodiazepines	Urine	PDMS	SBSE	GC-MS	280
Caffeine and metabolites	Plasma	PDMS	SBSE	HPLC-UV	281
Antidepressants	Plasma	PDMS	SBSE	GC-MS	282
4-Hydroxynonenal	Urine	PDMS	SBSE	GC-MS	283
Basic drugs	Blood	PDMS	SBSE	GC-MS	284
Steroid sex hormones	Urine	PDMS	SBSE	HPLC-DAD	285
Testosterone, epitestosterone	Urine	PDMS	SBSE	GC-MS	286
Phenolic xenoestrogens	Urine	PDMS	SBSE	GC-MS	287
Alkyl phenols and glucuronides	Urine	PDMS	SBSE	GC-MS	288,289
Cannabinoids	Hair	PDMS-coated syringe	HS-SPDE	GC-MS	290
Amphetamines, designer drugs	Hair	PDMS-coated syringe	HS-SPDE	GC-MS	291
Illicit drugs	Hair	PDMS-coated syringe	HS-SPDE	GC-MS	292
Local anesthetics	Plasma	Silica C2-packed syringe	MEPS	GC-MS	293,294
Ropivacaine and metabolites	Plasma	ISOLUTE ENV$^+$	MEPS	LC-MS-MS	295
Roscovitine	Plasma, urine	ISOLUTE ENV$^+$	MEPS	LC-MS-MS	296

PDMS, polydimethylsiloxane; CRB, carboxen; PA, polyacrylate; CW, carbowax; DVB, divinylbenzene; TPR, templated resin; PPY, polypyrrole; MIP, molecularly imprinted polymer; ADS, alkyl-diol silica; HS, headspace fiber SPME; DI, direct immersion fiber SPME; IT, in-tube SPME; SBSE, stir-bar sorptive extraction; SPDE, solid-phase dynamic extraction; MEPS, microextraction in a packed syringe; D, derivatization; OFD, on-fiber derivatization; GC-MS, gas chromatography–mass spectrometry; LC-MS, liquid chromatography–mass spectrometry; HPLC: high-performance liquid chromatography; CZE, capillary zone electrophoresis; FID, flame ionization detection; NPD, nitrogen–phosphorus detection; FTD, flame thermionic detector; TSD, thermionic specific detection; FLD, fluorescence detection; FAI, field asymmetric waveform ion mobility spectrometry; DAD, diode array detector.

can be used to remove protein from biological samples. Common deproteinization procedures involve protein denaturation by acids such as trichloroacetic acid, perchloric acid, and sulfosalicylic acid; the use of extreme temperature of 90–100°C or freezing–thawing cycles; and coprecipitation with bases such as zinc hydroxide. However, such precipitants rarely contain >80% of the protein and frequently contain less. Water-miscible organic solvents, such as acetone, methanol, and acetonitrile, have several advantages. The mild conditions used for precipitation minimize the possibility of decomposition of labile drugs. Methanol is often preferred because it produces a flocculent precipitate, resulting in a clearer supernatant suitable for direct injection into the HPLC. Although acetonitrile is the most popular solvent, its use is associated with several problems, including poor recovery, late-eluting peaks, incomplete protein precipitation, and the presence of some globulins in solution, which can react in the analytical system. Furthermore, such procedures by their nature dilute the sample, suggesting they are best utilized for high-concentration determinants. The precipitants can react in the analytical system, eluting as a peak on HPLC or interfering in a spectrophotometric assay. The membrane filters used for ultrafiltration are also efficient, although care must be taken to avoid binding of the drug to the membrane. Ultrafiltration may be used to measure total drug in serum, but only if it is bound marginally to proteins. Ultrafiltration works best for acidic and neutral drugs; basic drugs tend to bind to the membrane. Dialysis can separate an analyte from the matrix by diffusion through a semipermeable membrane rather than the centrifugal force used in ultrafiltration. Dialysis is the classic procedure used to separate proteins from small molecular weight analytes; the automation of this technique in continuous flow analyzers revolutionized clinical chemistry, enabling the explosion in sample processing and analysis. The combination of this technique with a trace enrichment column is available commercially for linkage to an HPLC. As described in Section 16.4.3.2, drugs in blood can be separated from proteins by a column switching technique using a restricted access material (RAM) sorbent, enabling continuous and automatic sample preparation and HPLC analysis of blood samples.

16.4.2. LLE

16.4.2.1. Conventional LLE. LLE is a traditional sample preparation technique that depends on the partitioning of the determinant between two immiscible liquids. Many methods have been developed for the extraction of drugs from biological fluids, involving extraction under acidic (pH 3) or basic (pH 8) conditions, followed by a sequential back-extraction with bicarbonate and sodium hydroxide to remove the strong and weak acids, respectively. Neutral drugs will remain in the organic solvent. Important parameters that must be considered when selecting an appropriate solvent for the intended extraction system include drug density, volatility, polarity, selectivity, and solubility. Obviously, solvents that are harmful to the health or environment should be avoided. The selected solvent must also be compatible with the subsequent analytical step. The concentration process will be facilitated by using a solvent of high volatility. Solvent polarity determines the solubility of the drug. In using LLE, many factors must be considered, including safety (e.g., the risk of explosion, flammability, toxicity, and carcinogenicity) and the association of each technique with imprecision and recovery. LLE is also labor intensive and difficult to automate and to connect online with analytical instruments, and there are problems of analyte adsorption on glassware and solvent-mediated decomposition. Furthermore, impurities in the solvent can cause interference, emulsion formation increases complexity and decreases recovery, and there are delays caused by the often lengthy evaporation stage. Despite several drawbacks, however, LLE is widely used because many of its applications efficiently achieve both cleanup and enrichment. Furthermore, LLE for drug analysis has recently become semiautomated, and multiwell plates can now be used. Details of the LLE techniques are described in Chapter 3.

Selected applications of LLE techniques reported recently are summarized in Table 16.5. Among the solvents used for LLE are hexane, chloroform, dichloromethane, diethyl ether, ethyl acetate, and mixtures of these solvents. LLE of various drugs is used in combination with GC, GC–mass spectrometry (MS), HPLC, and liquid chromatography (LC)–MS. These solvents are difficult to distinguish in terms of their extraction activity, although use of polar solvents often results in higher backgrounds. LC-MS-MS has been utilized increasingly for drug analysis because of its selectivity and sensitivity, although a serious drawback is the large consumption of pure solvents. In an analytical method for the doping agent methenolone in hair by GC-MS-MS, the sample was decontaminated with methylene chloride, solubilized with 1 M NaOH, and the drug was extracted with pentane.[49] Hexane : isoamyl alcohol (98:2 v/v) was used for the LLE of the antidepressant mianserin and its metabolite in human plasma, which were analyzed by LC-MS after back-extraction with 5 mM formic acid.[57] Dichloromethane : diethyl ether (2:3 v/v) was used for LLE of the antihypertensive drug iptakalim in human plasma, and the drug was analyzed by LC-MS-MS using sildenafil as an internal standard.[70] A rapid, sensitive, and specific HPLC method for the simultaneous TDM of HIV protease inhibitors (indinavir, neldinavir, ritonavir, and saquinavir) in human plasma utilized LLE with *tert*-butyl methyl ether and sequential washing of the reconstituted sample with hexane.[76] A number of flow-system LLE approaches have been based on mixing the aqueous and organic phases in a tube coil and their subsequent separation. These approaches are not widely used, especially not in sample preparation for chromatographic analysis. In contrast, semiautomated LLE methods using 96-deep-well plates have been developed for drug analysis.[86–88] All liquid transfers during sample preparation were automated using an autosampler, thus shortening the extraction time. An automated high-throughput LLE method, using a 96-well LLE plate and a 96-channel robotic liquid handling workstation, was developed to prepare plasma samples.[88] Each well has a filter composed of inert diatomaceous earth particles, allowing continuous and efficient extraction of analytes between the aqueous biological sample and the organic extraction solvent. Recently, a microchip LLE for the analysis of amphetamine-type stimulants in urine was developed, based on the pressure-driven manipulation of liquids in fluoloalkylated microchannels in a glass microchip.[89] The organic solvent and the fortified urine sample were pumped through the microchip, resulting in the extraction of the compounds into the organic solvent, which was collected at the end of the microchannels and analyzed by GC– flame ionization detector (FID).

16.4.2.2. Liquid-phase microextraction (LPME). LPME is a newly developed sample preparation technique, using minimal amounts of solvent. It is quick and inexpensive, with minimal exposure to toxic organic solvents. This technique is based on the suspension of a single droplet of organic solvent from the end of a microsyringe needle in an aqueous solution. The droplet containing analytes extracted by passive diffusion is directly injected into the GC, CE, or HPLC. LPME provides extracts highly enriched for analytes and excellent cleanup of endogenous compounds. Compared with traditional LLE, solvent consumption is reduced by 99%, and there is no need for evaporation of solvent and reconstitution of analytes prior to injection into the instrument. Recently, three-phase liquid–liquid–liquid microextraction and hollow fiber-based LPME were developed for the extraction of drugs in biological fluids, with ionic liquids also used as extraction solvents for LPME.[104] In the hollow fiber-based LPME, target drugs extracted from aqueous biological samples pass through a thin layer of organic solvent immobilized within the pores of the walls of a porous hollow fiber and into a microliter volume of acceptor solution inside the lumen of the hollow fiber. Details of these LPME techniques are also described in Chapter 7.

Applications of LPME to drug analysis have been reviewed.[31] LPME can be used to prepare biological samples for analysis of basic drugs,[91,95,99,102,105,106] antidepressants,[97,100,101] narcotics,[93,94,98] and other drugs.[90,92,96,103] Using hollow fiber-based LPME of polar drugs, selected analytes were extracted as ion pairs from small volumes of biological samples into microliter volumes of an acidic aqueous acceptor solution placed inside the lumen of the hollow fiber (liquid membrane) and then analyzed by LC-MS.[106] Sodium octanoate was added to the sample as a carrier, and extraction was performed from 50 μL of plasma sample. This carrier-mediated LPME was found to provide acceptable extraction recoveries and excellent cleanup from small volumes of plasma. An HS-LPME method, using a single drop in combination with HPLC-UV detection, has been used to analyze amphetamines in urine samples.[95] In addition, a three phase liquid–liquid–liquid microextraction of these compounds has also been developed.[99] In this method, the target compounds were extracted from a sodium hydroxide modified sample solution into a thin layer of organic solvent membrane, and back-extracted into an acidic acceptor drop suspended on the tip of a 50-μL HPLC syringe in the aforementioned organic layer; and the syringe was directly inserted into the HPLC.

16.4.3. SPE

SPE is a method for the extraction and concentration of analytes from a liquid matrix. SPE is based on the partitioning of compounds between a solid (extraction) phase and a liquid (sample) phase, whereby the intermolecular forces between the phases influences retention and elution. SPE offers the following advantages over conventional liquid–liquid procedures: (1) higher recovery, (2) more effective concentration, (3) less organic solvent usage, (4) no foaming or emulsion problems, (5) shorter sample preparation time, (6) easier operation, (7) easier incorporation into an automated process, and (8) simultaneous processing of many samples. SPE also has several disadvantages, including (1) poor reproducibility, (2) difficulty of standardizing the use of a vacuum, (3) variable nature of drying steps, and (4) expensive SPE materials. Details of the SPE technique are described in Chapter 4.

The fundamental concept of SPE is similar to that of column chromatography. The SPE procedure consists of four consecutive steps: (1) conditioning, (2) sample loading, (3) rinsing, and (4) eluting of analytes. For method development, a suitable adsorbent material and suitable rinsing and eluting solvents must be selected, in accordance with the characteristics of the analytes and the matrix and the purpose of the analysis. The final extraction solvent should be compatible with the analytical system (GC or HPLC); if not, solvent evaporation and further redissolving the residue in a suitable solvent are necessary. Various SPE products are now available, including column cartridges, disks, membranes, and well plates. The SPE packed into syringe barrels is usually used in combination with vacuum manifolds. The volume of the syringe barrel depends on sample volume, and the amount of adsorbent determines sample capacity. Disk and membrane type SPEs were developed to avoid the drawbacks of classical barrels (such as large void volume). These techniques require smaller volumes of conditioning and eluting solvents. This trend of minimizing the volume of the adsorbent bed, resulting in smaller elution volumes and highly concentrated extracts, is also apparent from the microscale formats of the SPE pipette tips and the SPME fibers. SPE can be performed on- or offline. Online techniques, with direct connection to the chromatographic system, do not require further sample handling between the trace-enrichment and separation steps and are, therefore, highly suitable for fully automated techniques. In online fully automatic SPE equipment, the column switching system is adopted, and the precolumn is fixed in the sample loop part in the injection value; preconcentration is performed by loading a comparatively large sample and by the retention of the target solute into the solid phase. The details of automated SPE techniques for clinical and pharmaceutical analysis has also been described by Wells.[297]

16.4.3.1. Commercial SPE sorbents. The fundamental mechanisms of retention are based on nonpolar, polar, and ionic interactions, leading to a wide selection in commercial SPE sorbents. The sorbent should be selected to obtain the optimal extraction. Various SPE sorbents are now available, including diatomaceous earth Extrelut, Chem Elut, Bond Elut Certify, and Chromabond mixed-mode columns. The mixed-phase extraction columns (e.g., Bond-Elut Certify and Chromabond) show good recoveries and allow retention of all functional groups and differing polarities. Bond-Elut Certify is a mixture of C8 and SCX (strong cation exchanger, benzenesulfonyl propyl), which has both hydrophobicity and ion exchange, and is therefore suitable for the extraction of basic drugs dissociated in blood and urine. SPE disks (Empore disk cartridge) can also be used to rapidly extract drugs from liquid specimens, with elution by the mobile phases of HPLC following direct injection into the HPLC system. The use of a 96- or 384-well format for SPE automated with a robotic liquid handling system facilitates high-throughput analysis of biological samples. Most of the laborious steps in traditional manual extraction procedures can be automated.

SPE has been widely adopted for clinical and pharmaceutical analyses, and its position is now ensured as a preparation method. SPE has been applied to the isolation, concentration, and cleanup of objective substances from biological specimens. Recent selected applications of SPE are summarized in Table 16.6. Apolar sorbents such as C18 are widely used for drug analysis. C8 and C18 SPE cartridges were used for the extraction of the antihypertensive drug cilazapril and its active metabolite cilazaprilat in pharmaceuticals and urine samples, and the extracts analyzed by HPLC–diode array detector (DAD) or electrochemical detector (ED).[111] Flunitrazepam (Rohypnol) and its metabolites have been extracted with RP-18[112] and Bond-Elut Certify[140] from plasma and urine samples, respectively. The LC-MS method has been used to analyze for doping agents and corticosteroids in hairs by methanolic extraction followed by SPE on a C18 cartridge.[116] In addition, SPE was applied to the cleanup of milk samples.[115]

SPE disks[298] are also a useful and rapid method to extract drugs from liquid specimens. Various applications of the well-plate technique have been used for drug analysis in urine, serum, and plasma.[140–146] In addition, a dual column switching system has been used for the online concentration of a GABA (γ-aminobutyric acid) receptor modulator and its metabolites in plasma samples.[146] Since extraction and elution can be performed simultaneously using dual trapping columns, toggling between these two stages of operation provides a run cycle time of 3 min. In contrast, pipette tip SPE with MonoTip C18 tips fixed with C18-bonded monolithic silica gel has been used for the extraction of methamphetamine and amphetamine from human whole blood samples.[147]

16.4.3.2. Specific SPE sorbents. A highly biocompatible SPE packed with alkyl-diol silica (ADS) particles was developed as a RAM sorbent, with the bifunctionality of the ADS extraction phase preventing fouling of the column by protein adsorption while simultaneously trapping the analytes in the hydrophobic porous interior. In this type of column, the internal surface is covered with a bonded reversed-phase material and the external surface is covered with a nonadsorptive but hydrophilic material. This dual-phase column can effectively separate the analyte of interest from macromolecules in the sample matrix; drugs and other small molecules enter the pores of the hydrophobic reversed phase to become retained by partitioning, while proteins and larger matrix components are excluded by the outer, hydrophilic phase and pass through as waste. RAM adsorbents containing hydrophilic diol and diethylaminoethyl groups, with weak anion exchange capacity, have been used to extract acidic compounds such as naproxen, ibuprofen, and diclofenac from plasma.[153]

Immunoaffinity adsorbents (IAAs) consist of biological antibodies covalently linked to silica, controlled-size glass particles, or agarose or other soft gels, and result in high affinity and selective antigen–antibody interactions. The antibodies can be compound specific but usually have some cross-reactivity with structural analogs, and are denatured by contact with an organic solvent, or by bacteria if stored incorrectly. Therefore, the adsorbent should be preserved in buffer solution containing sodium azide and stored in a cool environment. These adsorbents should also be used under mild SPE conditions, with regard to pH and organic content. These materials, however, are expensive, and only a limited number are commercially available. Details of the IAE have been described.[299–302] Use of IAAs exploits molecular-recognition mechanisms for selective extraction of trace amounts of the analytes of interest from matrices containing large amounts of interfering compounds. Ultrafast IAE of warfarin was accomplished using a 2.1-mm i.d. sandwich microcolumn containing a 1.1-mm layer of an anti-warfarin antibody support.[156] In addition, the IAE technique was used for the specific cleanup of fluoroquinolines[157] from biological samples.

MIPs are "tailor-made" adsorbents with high selectivity for a target molecule or structural analogs due to their molecular-recognition mechanisms, and are therefore useful for sensitive and specific analysis of drugs. MIPs are sufficiently stable in organic solvents, at high temperature, and over a wide pH range, and they are easily prepared and inexpensive. However, they may have difficulty removing target molecules completely and residual template material may leak into sample extracts, leading to incorrect quantification, especially at trace levels.[303] Different modes of MIP-based SPE have been demonstrated, including various modes of offline and online SPE for preconcentration or pretreatment of analytes, as well as for conventional SPE where the MIP is packed into columns or cartridges. Several applications of MIP use for biological samples have now appeared. Preconcentration of bupivacaine[160] from plasma samples prior to GC analysis has been performed with an MIP, and the specificity of the MIP-SPE was high compared with a C18 SPE method. RAM-MIP was used for propranolol, which was applied to the direct injection analysis of β-blockers in biological fluids.[167] All these examples demonstrate the high potential of MIP-SPE to become a broadly applicable sample preparation tool. The details of these techniques have been described in many reviews.[12–14,29,39,40–43,45,46,297,299–302,304,305]

16.4.4. SPME and Related Microextraction Techniques

SPME is a new and effective sample preparation technique. Fibers and capillary tubes coated with an appropriate stationary phase are usually used for SPME. In addition, alternative microextraction techniques have been developed, including solid-phase dynamic extraction (SPDE) using an internal coated needle, microextraction in a packed syringe (MEPS), and stir-bar sorptive extraction (SBSE) using a coated magnetic stir bar. These microextraction techniques are considered advantageous for the pretreatment of complex sample matrices prior to chromatographic and capillary electrophoretic processes because they enable rapid analysis at low operating costs and with no environmental pollution. Recent trends in sample preparation processes focus on methods to miniaturize the process and on the medium to use for the extraction and preconcentration of sample components. In this section, we review fiber SPME, in-tube SPME (or capillary microextraction), SPDE, MEPS, and SBSE. Of these, fiber SPME is the most widely used technique. Recently described clinical and pharmaceutical applications of these techniques are summarized in Table 16.7.

Details of the SPME techniques are also described in Chapters 5 and 21.

16.4.4.1. Fiber SPME. Fiber SPME is a sample preparation technique using a fused-silica fiber coated on the outside with an appropriate stationary phase. The fiber SPME device consists of a fiber assembly with a built-in extraction fiber protected inside a needle and an assembly holder. When the fiber is inserted into the sample and exposed, the target analytes partition from the sample matrix into the polymeric stationary phase coated onto the surface of the fiber until equilibrium is reached. In contrast to conventional SPE with packed-bed columns and micro- or nonmicrocolumns, this arrangement combines all the steps of sample preparation into one step. Fiber SPME has been used routinely in combination with GC and GC-MS, and successfully applied to a wide variety of compounds in gaseous, liquid, and solid samples. It is used especially to extract volatile and semivolatile organic compounds from pharmaceutical, biological, environmental, and food samples. Fiber SPME has also been coupled directly with HPLC and LC-MS to analyze weakly volatile or thermally labile compounds not amenable to GC or GC-MS. In place of thermal desorption in the injection port of a GC, an SPME/HPLC interface equipped with a special desorption chamber is utilized for solvent desorption prior to HPLC analysis or offline desorption is performed. Fiber SPME, in combination with HPLC and LC-MS, has been applied to the analysis of various polar compounds such as drugs and pesticides. The main advantages of fiber SPME are simplicity, rapidity, solvent elimination, high sensitivity, small sample volume, relatively low cost, and simple automation.

Two types of fiber SPME techniques can be used to extract analytes: HS and direct immersion (DI) SPME. In HS-SPME, the fiber is exposed in the HS of gaseous, liquid, or solid samples. In DI-SPME, the fiber is directly immersed in liquid samples. Minor variations in these methods depend on whether or not derivatization is applied and in which phase, the type of sample agitation, and whether or not additives are required to optimize extraction. The fiber with concentrated analytes is transferred to an instrument for desorption, followed by separation and quantitation. HS- and DI-SPME techniques can be used in combination with any GC, GC-MS, HPLC, or LC-MS system.

Stationary phases in the fiber are immobilized as nonbonded, bonded, partial cross-linked, or highly cross-linked films. Nonbonded phases are stable when used with some water-miscible organic solvents, but slight swelling may occur when used with nonpolar solvents. Bonded phases are stable with all organic solvents except for some nonpolar solvents. Partially cross-linked phases are stable in most water-miscible organic solvents and some nonpolar solvents. Highly cross-linked phases are equivalent to partially cross-linked phases, except that some bonding to the core may occur. The advantages of these phases for SPME applications are similar to their advantages as GC stationary phases. The most important factor in fiber SPME is the affinity of the fiber coating for an analyte. Depending on the substance being analyzed, a suitable polarity and thickness of the fiber coating can be selected. Most nonpolar drugs in biological samples are extracted with a polydimethylsiloxane (PDMS), polyacrylate (PA), porous polymer divinylbenzene (DVB), or PDMS/DVB fiber, and analyzed in combination with GC-MS. Besides commercial fibers, unique coating fibers, such as polypyrrole (PPY), immunosorbent, ADS, and sol-gel porous silica, have been developed.[213–215,219,220]

A number of fiber SPME methods have been developed to extract drugs from various biological samples (Table 16.7). Illicit drugs, amphetamine, methamphetamine, methylenedioxy amphetamine, methylenedioxy methamphetamine, and methylenedioxy ethamphetamine, are extracted from hair, serum, and urine samples by HS- or DI-SPME techniques using PDMS fibers and analyzed by GC-MS.[172–182] A home-assembled desorption chamber (inner volume ca. 3 μL), designed for direct coupling of SPME with electrospray ionization (ESI)–high field asymmetric waveform ion mobility spectrometry (FAI-MS)-MS, has been used to analyze amphetamine, methamphetamine, and their analogs in urine.[181] This method was found to be sensitive and selective without requiring chromatographic separation, and results can be obtained in less than 20 min after a urine sample arrives at the laboratory. Hair analysis has proven reliable in clinical and forensic toxicology for the retrospective detection of chronic drugs of abuse. Drugs in hair samples are analyzed using several microextraction techniques.[168–170,172–174,186,189] The SPME method was used for the toxicological analysis of the local anesthetic lidocaine in biological fluids obtained from a suspected victim of lidocaine poisoning.[203,204] The SPME was coupled directly to an atmospheric pressure chemical ionization (APCI)–MS/ion trap for rapid analysis of the drug in urine samples, thereby applying MS-MS [fragmentation of $[M + H]^+$ (m/z 235) to a fragment with m/z 86]. Throughput of samples was increased using nonequilibrium SPME with PDMS fibers, and desorption was performed with a homemade chamber allowing thermostating. Benzodiazepines are frequently used in clinical practice as tranquilizers, sleep inducers, antiepileptic hypnotics, anticonvulsants, and muscle relaxants. These drugs, however, can cause intoxication due to accidental overdosage or intentional abuse. Benzodiazepines in blood and urine can be analyzed in combination with DI-SPME/HPLC-UV using PDMS/DVB and carbowax (CW)/templated resin (TPR) fibers.[212] A new approach consists of using biocompatible SPME fibers coated with immunosorbent and ADS-RAM for direct sampling in biological matrixes.[213–215]

Volatile compounds in biological samples are important markers of disease. For example, a new diagnostic tool for diabetes consists of the determination of breath acetone using HS-SPME with on-fiber derivatization and GC-MS.[240,241] The HS-SPME method was also used to analyze volatile compounds in the blood as biomarkers of lung cancer.[242] In addition, the HS-SPME technique has been used to assay residual solvents in pharmaceutical products.[246–249] PDMS, PDMS/DVB, and carboxen (CRB)/PDMS/DVB fibers were used to extract impurities from solvents such as ethanol, cyclohexane, toluene, benzyl chloride, and triethylamine, and these impurities were analyzed in combination with GC-FID. Fiber SPME techniques have also been widely used for the analysis of various contaminants in biological samples, including pesticides[251–257] and some toxic compounds.[258–260]

16.4.4.2. In-tube SPME. In-tube SPME is a newer sample preparation technique using a wall-coated open tubular capillary as an SPME device. In addition, in-tube SPME can be coupled on-line with HPLC or LC-MS and is suitable for automation. Extraction, desorption, and injection can be performed continuously using a standard autosampler. Automated sample handling procedures not only shorten the total analysis time, but they are more accurate and precise than manual techniques. Using in-tube SPME, organic compounds in aqueous samples are directly extracted from the sample into the internally coated stationary phase of a capillary. These compounds are then desorbed by introducing a stream of mobile phase, or by using a static desorption solvent when the analytes are more strongly adsorbed to the capillary coating. The desorbed compounds are subsequently injected into the LC column for analysis. It is therefore necessary to prevent plugging of the capillary column and flow lines by filtering the sample solution before extraction. Although the extraction yields are generally low, these compounds may be extracted reproducibly using an autosampler, and all of the extracts may be introduced into the LC column after in-tube SPME, typically resulting in good precision and sensitivity.

In-tube SPME is an extraction method whereby the analyte is transferred; thus, extraction depends on the distribution coefficient of each analyte as well as its affinity for the fiber SPME. It is therefore important to raise the distribution factor in the stationary phase to optimize the rapidity and efficiency of extraction. The amount of analyte extracted into the stationary phase of the capillary column depends on the polarity of the capillary coating, the number and volume of draw/eject cycles, and the pH of the sample. Several commercially available capillary columns, which differ according to the selectivity of the stationary phase, internal diameter, length, and film thickness, are now available. Some unique phases and technical solutions have been developed to improve extraction efficiency and selectivity. Extraction phases better suited to the extraction of relatively polar compounds from aqueous samples have been found to enhance the sensitivity and overall utility of this method. For example, an MIP was synthesized for use as an in-tube SPME adsorbent, and a capillary packed with MIP particles in a PEEK tube was used for the selective analysis of propranolol.[269] In addition, a highly biocompatible SPME capillary packed with ADS particles was developed as a RAM.[264] Furthermore, a simple SPME device has been fabricated for use in online immunoaffinity capillary.[275] An alternative approach consists of in-tube SPME using a monolithic capillary column comprised of one piece of organic polymer or silica with a unique flow-through double-pore structure.[265,271–273] Other techniques include wire-in-tube SPME using a modified capillary column with inserted stainless-steel wires or fiber-in-tube SPME using PEEK tubes packed with fibrous rigid-rod heterocyclic polymers, which improve extraction efficiency while extending the method to microscale applications.[266,267]

The in-tube SPME method can also be applied to polar and nonpolar drugs in liquid samples and can be easily coupled with various analytical methods such as HPLC and LC-MS (Table 16.7). An automated in-tube SPME coupled with LC-ESI-MS using an Omegawax capillary has been developed for the determination of amphetamines and β-blockers in serum and urine samples.[262,276] In addition, a Supel-Q PLOT capillary coated with the porous DVB polymer has been used for the automated in-tube SPME of seven benzodiazepines in serum.[263] A biocompatible in-tube SPME method using a PEEK capillary packed with ADS particles has also been used for direct extraction of several benzodiazepines from serum samples.[264] The bifunctionality of the ADS extraction phase prevented fouling of the capillary by protein adsorption while simultaneously trapping the analytes in the hydrophobic porous interior. This approach required a more simplified apparatus compared with existing RAM column switching procedures, as well as overcoming the need for ultrafiltration or another deproteinization step prior to handling biological samples, thus further minimizing sample preparation requirements. A fiber-in-tube SPME technique has been developed using a short capillary packed longitudinally with several hundred polymer filaments or wires as extraction medium.[266,267] Because of the parallel arrangement of the filaments to the outer tubing, a large number of narrow coaxial channels can be formed in the capillary. This method, in combination with HPLC and CE, was used to analyze tricyclic antidepressants in urine samples.

A new in-tube SPME technique was recently developed using a PEEK capillary packed with propranolol MIP particles.[269] In addition to its inherent selectivity and chemical and physical robustness, the MIP material was demonstrated to be an effective stationary phase material for in-tube SPME. Using a propranolol MIP-packed capillary, an automated online in-tube SPME/HPLC system was developed for the selective analysis of propranolol in serum samples. In addition, an immunoaffinity capillary containing immobilized antibody was used for in-tube SPME of the selective serotonin inhibitor fluoxetine in serum samples.[275] An in-tube SPME method was developed using a monolithic capillary column as extraction device. Hydrophobic main chains and acidic pendant groups of poly (methacrylic acid-ethylene glycol dimethacrylate) make it superior for extracting basic analytes from an aqueous matrix. Online monolithic capillary in-tube SPME methods have been developed for the determination of ketamine in urine samples,[265] basic drugs in serum samples,[271] and angiotensin II receptor agonists in plasma and urine.[272,273] These methods showed high extraction efficiencies for all analytes.

16.4.4.3. SBSE. SBSE is a new sample preparation technique that overcomes the limited capacity of SPME fibers. This technique uses a magnetic stir bar coated with PDMS phase, similar to SPME but in a thicker layer (0.3–1.0 mm), which means that the phase is 50–250 times greater in volume than in SPME.[33] PDMS-coated stir bars are now commercially available as Twister® stir bars (Gerstel, Mülheim a/d Ruhr, Germany) of lengths 10 and 40 mm, coated with 55 and 219 μL of PDMS liquid phase, respectively. The 10-mm stir bars are best suited for stirring samples of 10–50 mL, whereas 40-mm stir bars are better for sample volumes up to 250 mL. After sampling, the stir bar is removed with tweezers, briefly applied to clean paper tissue to remove residual water droplets, and placed in an empty glass tube 187 mm long, 6-mm o.d., and 4-mm i.d. for thermal desorption. In SBSE, sample volume and stirring speed influence extraction efficiency, and typical stirring times to attain equilibration are between 30 and 60 min.

SBSE in combination with thermal desorption online coupled to capillary GC-MS was recently developed for the analysis of pharmaceutical drugs and metabolites in urine samples.[280] Urine samples can be extracted directly or after enzymatic hydrolysis. This technique is very versatile and sensitive for analyzing a wide range of substances. Moreover, the relatively high enrichment efficiencies of SBSE permitted the use of mass spectrometric detection in the full scan mode. More than 200 compounds could be enriched and detected at the same time, resulting in a complete profile of GC-amenable compounds in the urine sample. An SBSE/GC-MS method has been developed for the detection of basic drugs in blood, urine and tissue samples for routine drug screening in the forensic toxicology laboratory.[284] Furthermore, SBSE in combination with HPLC-DAD could simultaneously measure nine steroid sex hormones in urine samples.[285]

16.4.4.4. SPDE and MEPS. SPDE is an inside-needle technique for vapor and liquid sampling. Stainless-steel needles (8 cm) coated with a 50-μm film of PDMS and 10% activated carbon are used. Dynamic sampling is performed by passing the HS through the tube using a syringe. The volume of the stationary phase of the SPDE needle is approximately 6.0 mm^3, whereas a 100-μm PDMS SPME fiber has a volume of 0.94 mm^3. In SPDE, the analytes are concentrated onto PDMS and activated carbon coated onto the inside wall of the stainless-steel needle of a 2.5-mL gastight syringe. When used for HS-SPDE, an amount of analyte sufficient for a reliable GC or GC-MS analysis is accumulated in the polymer coating of the inside-needle wall by pulling in and pushing out a fixed volume of the sample HS for an appropriate number of times. Thus, SPDE sampling permits operation under dynamic conditions while keeping the HS volume constant. The trapped analytes are recovered by heat desorption directly into a GC injector body. A great advantage of SPDE over SPME is the robustness of the capillary. In contrast to the fragile SPME fibers, the SPDE device is nearly impossible to damage mechanically. SPDE has been successfully applied to the analysis of volatile compounds, pesticides, and some drugs.[290–292] However, SPDE may have carryover, because the analytes tend to remain in the inside-needle wall after heat desorption in the GC injection port.

MEPS is a newly developed technique for sample preparation online with LC and GC-MS. This form of miniaturized SPE uses a procedure similar to in-tube SPME and SPDE. In MEPS, approximately 1 mg of solid packing material is inserted into a syringe (100–250 μL) as a plug with a filter on both sides, and sample preparation takes place on the packed bed. After conditioning with methanol and water, sample solution is drawn through the syringe three times (50 μL each) by the autosampler, which pumps the sample up and down three times. After washing with water to remove any interfering materials such as proteins, the extracted analytes are eluted with an organic solvent or LC mobile phase directly into the GC or LC injector. The process is fully automated. Any absorption materials, including silica based (C2, C8, C18), RAM, and MIPs, can be used.

SPDE and MEPS are new techniques using a microsyringe as an extraction device and have not yet been extensively applied. Similar to in-tube SPME, an automated SPDE technique using a hollow needle with an internal coating of PDMS instead of a capillary column may be suitable for HS extraction coupled with GC-MS. This method can be used for the determination of illicit drugs, such as amphetamines, cannabinoids, and methadone, in hair samples.[290–292] A new online MEPS sample preparation technique has been applied to the analysis of the local anesthetics lidocaine, prilocaine, ropivacaine, and mepivacaine in human plasma.[293,294] Compared with traditional LLE and SPE, MEPS reduced sample preparation time and organic solvent consumption. In addition, MEPS can treat small samples, requiring only 1 min for each sample. MEPS is more easily automated than SPE and more rugged than SPME. Compared with SPME, MEPS is more stable and has a higher recovery. Although the sampling fibers used in SPME are very sensitive to sample matrix, the new technique can be used without difficulty for complex matrices, including plasma, urine, and organic solvents. In addition, this method has a much higher extraction recovery rate (60–90%) compared with SPME (1–10%).

16.5. CONCLUSIONS

Drug analysis in biological samples and pharmaceutical products is very important for TDM, pharmacokinetic studies, screening of illicit drugs, and development of new medicines. These substances can be assayed in various biological specimens, particularly plasma, serum, and urine.

Although blood (plasma, serum) is most useful for simultaneous screening and quantification, it usually requires deproteinization and, if necessary, cleavage of conjugates of the drugs and their metabolites, prior to sample pretreatment and chromatography or electrophoresis. Due to the relatively high concentrations of drugs in urine, this substance is the sample of choice for the comprehensive screening and identification of unknown drugs and their metabolites. Urine collection is easy, but there are risks of contamination and drug degradation during storage. Saliva and hair analyses are considered useful adjuncts to conventional drug testing. Particularly, toxicological analysis in hair samples allows the detection of past, chronic drug use. However, the interpretation of results may be unclear, particularly concerning external contamination, cosmetic treatment, and ethnic bias. Pharmaceutical products are manufactured in various formulation types, including solids, liquids, topicals, injectables, and inhalants, and have various excipient activities, including tablets, capsules, coatings, flavorings, colorings, buffers, suspendings, and preservatives. Cleanup of these samples for the drug analysis is commonly performed by LLE, SPE, SPME, and their combination after precipitation, centrifugation, or dialysis.

The choice of analytical method depends on the concentration of the target analytes and the variety and complexity of the sample. Since most analytical instruments cannot handle complex matrices directly, sample preparation methods are necessary to isolate the components of interest from sample matrices. This greatly influences the reliability and accuracy of the analytical method, but it has always been a somewhat neglected problem in analytical chemistry. Suitable sample preparation is an important prerequisite for the analysis of biological samples. Many of the techniques currently used for the preparation of complex samples prior to chromatographic or electrophoretic analysis, such as filtration, precipitation, and extraction with organic solvents, have been utilized for several decades with essentially no modifications. Universal LLE procedures are preferable for general screening, because substances with very different physicochemical properties must be isolated from heterogeneous matrices. In general, sample preparation is still regarded as "low tech" and in many laboratories is assigned to the least trained staff, individuals who may often be reluctant to accept new technologies. This is one of the main reasons for the slow implementation of new sample preparation techniques, making sample preparation frequently the most time-consuming and often rate-limiting step of the entire analytical procedure.

As described in this chapter, LPME, SPE, SPME, SBSE, SPDE, and MEPS are newer sample preparation techniques. Among these, SPE is the most popular for drug analysis and has largely replaced older techniques in laboratories throughout the world. Sample pretreatment for SPE depends on the sample type: whole blood and tissue need deproteinization and filtration/centrifugation steps before application to the SPE columns, whereas for urine, usually a simple dilution step and/or centrifugation is satisfactory. However, clotting, channeling and percolation are typically encountered problems of SPE. Some promising approaches in SPE are based on special packings, such as RAMs, and MIPs.

SPME is becoming a modern alternative to SPE and LLE. Fiber SPME is a solvent-free, concentrating extraction technique that has been applied widely to the analysis of drugs and metabolites in biological fluids. In-tube SPME is a useful technique for the construction of online automated systems. Additional customized coatings may become available in the future, including chirally active phases, various derivatized cyclodextrins, ion exchangers, HPLC stationary phase particles, monolithic capillary, and sol–gel porous silicas. As evidenced with SPE, immunoaffinity sorbents and MIPs can be applied as new coating materials for highly efficient extraction of drugs from various biological samples. The details of developments in new coating materials have also been described in well-documented reviews.[306,307] As the market for SPME increases, disposable low-cost extraction fibers (e.g., in the form of a carousel) or tubes similar to those used in other areas of sample preparation (e.g., SPE multiwell plates) may become available. SBSE using coated magnetic stirring bars, with similar types of phase but a thicker layer, is compatible with both GC and HPLC and is as or more sensitive than fiber SPME. SPDE and MEPS are new microextraction techniques using a needle internally coated with PDMS or packed with adsorption materials. These miniaturized techniques are rapid and easy to automate, and can be easily coupled to various analytical methods such as GC-MS and LC-MS.

The use of more selective extraction procedures has shown clear trends toward the simplification of sampling and sample preparation methods, an increase in their reliability and precision, and the elimination of the cleanup step. With the development of more sensitive and selective phases, it may be possible to further miniaturize these techniques. Furthermore, there is an increasing interest in automating sample preparation, thus speeding these procedures and improving precision and cost-effectiveness. The key attractive features of automated sample preparation techniques include miniaturization, high throughput, reproducibility, and traceability. In the last decade, sample preparation devices have been coupled online to separation and detection systems especially designed for automation. In the future, better integration of sampling/sample preparation and instrumental analysis will allow wider use of automated online analysis in forensic, clinical, and pharmaceutical analysis. Finally, we hope that this review will serve as a guide to choosing the most effective sample preparation techniques for pharmaceutical and biomedical analysis.

REFERENCES

1. Valko, K., Ed. Separation methods in drug synthesis and purification. In *Handbook of Analytical Separation*, Vol. 1. Amsterdam: Elsevier; 2000.

2. Pawliszyn, J., Ed. Sampling and sample preparation for field and laboratory: Fundamentals and new directions in sample preparation. In *Comprehensive Analytical Chemistry*, Vol. XXXVII. Amsterdam: Elsevier, 2002.

3. Hempel, G., Ed. Drug monitoring and clinical chemistry. In *Handbook of Analytical Separations*, Vol. 5. Amsterdam: Elsevier; **2004**.

4. Watt, A.P.; Mortishire-Smith, R.J.; Gerhard, U.; Thomas, S.R. Metabolite identification in drug discovery. *Curr. Opin. Drug Discov. Dev.* **2003**, *6*, 57–65.

5. Staack, R.F.; Hopfgartner, G. New analytical strategies in studying drug metabolism. *Anal. Bioanal. Chem.* **2007**, *388*, 1365–1380.

6. Baumann, P.; Hiemke, C.; Ulrich, S.; Eckermann, G.; Gaertner, I.; Gerlach, M.; Kuss, H.-J.; Laux, G.; Muller-Oerlinghausen, B.; Rao, M.L.; Riederer, P.; Zernig, G. The AGNP-TDM expert group consensus guidelines: Therapeutic drug monitoring in psychiatry. *Pharmacopsychiatry* **2004**, *37*, 243–265.

7. Saint-Marcoux, F.; Sauvage, F.L.; Marquet, P. Current role of LC-MS in therapeutic drug monitoring. *Anal. Bioanal. Chem.* **2007**, *388*, 1327–1349.

8. Bogusz, M.J., Ed. Forensic science. In *Handbook of Analytical Separation*, Vol. 2. Amsterdam: Elsevier; 2000.

9. Wood, M.; Laloup, M.; Samyn, N.; Del Mar Ramirez Fernandez, M.; de Bruijn, E.A.; Maes, R.A.A.; De Boeck, G. Recent applications of liquid chromatography-mass spectrometry in forensic science. *J. Chromatogr. A* **2006**, *1130*, 3–15.

10. Maurer, H.H. Current role of liquid chromatography-mass spectrometry in clinical and forensic toxicology. *Anal. Bioanal. Chem.* **2007**, *388*, 1315–1325.

11. Thevis, M.; Schanzer, W. Current role of LC-MS(/MS) in doping control. *Anal. Bioanal. Chem.* **2007**, *388*, 1351–1358.

12. Pedersen-Bjergaard, S.; Rasmussen, K.E.; Halvorsen, T.G. Liquid-liquid extraction procedures for sample enrichment in capillary zone electrophoresis. *J. Chromatogr. A* **2000**, *902*, 91–105.

13. Kumazawa, T.; Suzuki, O. Separation methods for amino group-possessing pesticides in biological samples. *J. Chromatogr. B* **2000**, *747*, 241–254.

14. Gilar, M.; Bouvier, E.S.P.; Compton, B.J. Advances in sample preparation in electromigration, chromatographic and mass spectrometric separation methods. *J. Chromatogr. A* **2001**, *909*, 111–135.

15. Sporkert, F.; Pragst, F. Use of headspace solid-phase microextraction (HS-SPME) in hair analysis for organic compounds. *Forensic Sci. Int.* **2000**, *107*, 129–148.

16. Theodoridis, G.; Koster, E.H.M.; de Jong, G.J. Solid-phase microextraction for the analysis of biological samples. *J. Chromatogr. B* **2000**, *745*, 49–82.

17. Snow, N.H. Solid-phase micro-extraction of drugs from biological matrices. *J. Chromatogr. A* **2000**, *885*, 445–455.

18. Lord, H.; Pawliszyn, J. Microextraction of drugs. *J. Chromatogr. A* **2000**, *902*, 17–63.

19. Ulrich, S. Solid-phase microextraction in biomedical analysis. *J. Chromatogr. A* **2000**, *902*, 167–194.

20. Mills, G.A.; Walker, V. Headspace solid-phase microextraction procedures for gas chromatographic analysis of biological fluids and materials. *J. Chromatogr. A* **2000**, *902*, 267–287.

21. Furton, K.G.; Wang, J.; Hsu, Y.-L.; Walton, J.; Almirall, J.R. The use of solid-phase microextraction—Gas chromatography in forensic analysis. *J. Chromatogr. Sci.* **2000**, *38*, 297–306.

22. Eskilsson, C.S.; Bjorklund, E. Analytical-scale microwave-assisted extraction. *J. Chromatogr. A* **2000**, *902*, 227–250.

23. Dehon, B.; Humbert, L.; Devisme, L.; Stievenart, M.; Mathieu, D.; Houdret, N.; Lhermitte, M. Tetrachloroethylene and trichloroethylene fatality: Case report and simple headspace SPME-capillary gas chromatographic determination in tissues. *J. Anal. Toxicol.* **2000**, *24*, 22–26.

24. Pawliszyn, J. Solid phase microextraction. *Adv. Exp. Med. Biol.* **2001**, *488*, 73–87.

25. Kataoka, H. Automated sample preparation using in-tube solid-phase microextraction and its application—A review. *Anal. Bioanal. Chem.* **2002**, *373*, 31–45.

26. Baltussen, E.; Cramers, C.A.; Sandra, P.J.F. Sorptive sample preparation—A review. *Anal. Bioanal. Chem.* **2002**, *373*, 3–22.

27. Saito, Y.; Jinno, K. Miniaturized sample preparation combined with liquid phase separations. *J. Chromatogr. A* **2003**, *1000*, 53–67.

28. Zambonin, C.G. Coupling solid-phase microextraction to liquid chromatography. A review. *Anal. Bioanal. Chem.* **2003**, *375*, 73–80.

29. Kataoka, H. New trends in sample preparation for clinical and pharmaceutical analysis. *Trends Anal. Chem.* **2003**, *22*, 232–254.

30. Vas, G.; Vekey, K. Solid-phase microextraction: A powerful sample preparation tool prior to mass spectrometric analysis. *J. Mass Spectrom.* **2004**, *39*, 233–254.

31. Pedersen-Bjergaad, S.; Rasmussen, K.E. Bioanalysis of drugs by liquid-phase microextraction coupled to separation techniques. *J. Chromatogr. B* **2005**, *817*, 3–12.

32. Kataoka, H. Recent advances in solid-phase microextraction and related techniques for pharmaceutical and biomedical analysis. *Curr. Pharm. Anal.* **2005**, *1*, 65–84.

33. Kawaguchi, M.; Ito, R.; Saito, K.; Nakazawa, H. Novel stir bar sorptive extraction methods for environmental and biomedical analysis. *J. Pharm. Biomed. Anal.* **2006**, *40*, 500–508.

34. David, F.; Sandra, P. Stir bar sorptive extraction for trace analysis. *J. Chromatogr. A* **2007**, *1152*, 54–69.

35. Lord, H.L. Strategies for interfacing solid-phase microextraction with liquid chromatography. *J. Chromatogr. A* **2007**, *1152*, 2–13.

36. Pragst, F. *Anal. Bioanal. Chem.* **2007**, *388*, 1393–1414.

37. Musteata, F.M.; Pawliszyn, J. In vivo sampling with solid phase microextraction. *J. Biochem. Biophys. Methods* **2007**, *70*, 181–193.

38. Musteata, F.M.; Pawliszyn, J. Bioanalytical applications of solid-phase microextraction. *Trends Anal. Chem.* **2007**, *26*, 36–45.

39. Deng, C.; Liu, N.; Gao, M.; Zhang, X. Recent developments in sample preparation techniques for chromatography analysis of traditional Chinese medicines. *J. Chromatogr. A* **2007**, *1153*, 90–96.

40. Wille, S.M.R.; Lambert, W.E.E. Recent developments in extraction procedures relevant to analytical toxicology. *Anal. Bioanal. Chem.* **2007**, *388*, 1381–1391.

41. Musshoff, F.; Madea, B. Analytical pitfalls in hair testing. *Anal. Bioanal. Chem.* **2007**, *388*, 1475–1494.

42. Musshoff, F.; Madea, B. New trends in hair analysis and scientific demands on validation and technical notes. *Forensic Sci. Int.* **2007**, *165*, 204–215.

43. Mullett, W.M. Determination of drugs in biological fluids by direct injection of samples for liquid-chromatographic analysis. *J. Biochem. Biophys. Methods* **2007**, *70*, 263–273.

44. Kuwayama, K.; Tsujikawa, K.; Miyaguchi, H.; Kanamori, T.; Iwata, Y.; Inoue, H.; Saitoh, S.; Kishi, T.M. Identification of impurities and the statistical classification of methamphetamine using headspace solid phase microextraction and gas chromatography-mass spectrometry. *Forensic Sci. Int.* **2006**, *160*, 44–52.

45. Delaunay, N.; Pichon, V.; Hennion, M.-C. Immunoaffinity solid-phase extraction for the trace-analysis of low-molecular-mass analytes in complex sample matrices. *J. Chromatogr. B* **2000**, *745*, 15–37.

46. Stevenson, D. Immuno-affinity solid-phase extraction. *J. Chromatogr. B* **2000**, *745*, 39–48.

47. Moreno Cordero, B.; Perez Pavon, J.L.; Pinto, C.G.; Fernandez Laespada, M.E.; Carabias Martinez, R.; Rodriguez Gonzalo, E. Analytical applications of membrane extraction in chromatography and electrophoresis. *J. Chromatogr. A* **2000**, *902*, 195–204.

48. Jonsson, J.A.; Mathiasson, L. Membrane-based techniques for sample enrichment. *J. Chromatogr. A* **2000**, *902*, 205–225.

49. Kintz, P.; Cirimele, V.; Dumestre-Toulet, V.; Villain, M.; Ludes, B. Doping control for methenolone using hair analysis by gas chromatography-tandem mass spectrometry. *J. Chromatogr. B* **2002**, *766*, 161–167.

50. Leis, H.J.; Rechberger, G.N.; Fauler, G.; Windischhofer, W. Enantioselective trace analysis of amphetamine in human plasma by gas chromatography/negative ion chemical ionization mass spectrometry. *Rapid Commun. Mass Spectrom.* **2003**, *17*, 569–575.

51. Kmetec, V.; Roskar, R. HPLC determination of tramadol in human breast milk. *J. Pharm. Biomed. Anal.* **2003**, *32*, 1061–1066.

52. Kushnir, M.M.; Urry, F.M.; Frank, E.L.; Roberts, W.L.; Shushan, B. Analysis of catecholamines in urine by positive-ion electrospray tandem mass spectrometry. *Clin. Chem.* **2002**, *48*, 323–331.

53. Iha, M.H.; Martinez, A.S.; Bonato, P.S. Enantioselective analysis of atenolol in biologic fluids: Comparison of liquid-liquid and solid-phase extraction methods. *J. Chromatogr. B* **2002**, *767*, 1–9.

54. Tolokan, A.; Horvath, V.; Horvai, G.; Egresi, A.; Nemes, K.B.; Klebovich, I. Determination of deramciclane and N-desmethylderamciclane in human plasma by liquid chromatography-tandem mass spectrometry using off-line robotic sample pretreatment. *J. Chromatogr. A* **2000**, *896*, 279–290.

55. Macek, J.; Ptacek, P.; Klima, J. Rapid determination of citalopram in human plasma by high-performance liquid chromatography. *J. Chromatogr. B* **2001**, *755*, 279–285.

56. Macek, J.; Ptacek, P.; Klima, J. Rapid determination of pseudoephedrine in human plasma by high-performance liquid chromatography. *J. Chromatogr. B* **2002**, *766*, 289–294.

57. Chauhan, B.; Rani, S.; Guttikar, S.; Zope, A.; Jadon, N.; Padh, H. Analytical method development and validation of mianserin hydrochloride and its metabolite in human plasma by LC-MS. *J. Chromatogr. B* **2005**, *823*, 69–74.

58. Massaroti, P.; Moraes, L.A.; Marchioretto, M.A.; Cassiano, N.M.; Bernasconi, G.; Calafatti, S.A.; Barros, F.A.; Meurer, E.C.; Pedrazzoli, J. Development and validation of a selective and robust LC-MS/MS method for quantifying amlodipine in human plasma. *Anal. Bioanal. Chem.* **2005**, *382*, 1049–1054.

59. Massaroti, P.; Cassiano, N.M.; Duarte, L.F.; Campos, D.R.; Marchioretto, M.A.; Bernasconi, G.; Calafatti, S.; Barros, F.A.; Meurer, E.C.; Pedrazzoli, J. Validation of a selective method for determination of paroxetine in human plasma by LC-MS/MS. *J. Pharm. Pharm. Sci.* **2005**, *8*, 340–347.

60. Tóth, M.; Bereczki, A.; Drabant, S.; Nemes, K.B.; Varga, B.; Grézal, G.; Tömlo, J.; Lakner, G.; Klebovich, I. Gas chromatography nitrogen phosphorous detection (GC-NPD) assay of tofisopam in human plasma for pharmacokinetic evaluation. *J. Pharm. Biomed. Anal.* **2006**, *41*, 1354–1359.

61. Llerena, A.; Berecz, R.; Norberto, M.J.; de la Rubia, A. Determination of clozapine and its N-desmethyl metabolite by high-performance liquid chromatography with ultraviolet detection. *J. Chromatogr. B* **2001**, *755*, 349–354.

62. Davydova, N.N.; Yasuda, S.U.; Woosley, R.L.; Wainer, I.W. Determination of clemastine in human plasma by gas chromatography with nitrogen-phosphorus detection. *J. Chromatogr. B* **2000**, *744*, 177–181.

63. Jansson, A.; Gustafsson, L.L.; Mirghani, R.A. High-performance liquid chromatographic method for the determination of quinine and 3 hydroxyquinine in blood samples dried on filter paper. *J. Chromatogr. B* **2003**, *795*, 151–156.

64. Jabor, V.A.; Lanchote, V.L.; Bonato, P.L. Simultaneous determination of disopyramide and mono-N-dealkyldisopyramide enantiomers in human plasma by capillary electrophoresis. *Electrophoresis* **2001**, *22*, 1406–1412.

65. Prochazkova, A.; Chouki, M.; Theurillat, R.; Thormann, W. Therapeutic drug monitoring of albendazole: Determination of albendazole, albendazole sulfoxide, and albendazole sulfone in human plasma using nonaqueous capillary electrophoresis. *Electrophoresis* **2000**, *21*, 729–736.

66. Bortocan, R.; Lanchote, V.L.; Cesarino, E.J.; Bonato, P.S. Enantioselective analysis of disopyramide and mono-N-dealkyldisopyramide in plasma and urine by high-performance liquid chromatography on an amylose-derived chiral stationary phase. *J. Chromatogr. B* **2000**, *744*, 299–306.

67. Rathod, R.; Prasad, L.P.; Rani, S.; Nivsarkar, M.; Padh, H. Estimation of carvedilol in human plasma by using HPLC-fluorescence detector and its application to pharmacokinetic study. *J. Chromatogr. B* **2007**, *857*, 219–223.

68. Tang, Y.; Zhao, L.; Wang, Y.; Fawcett, J.P.; Gu, J. Rapid and sensitive liquid chromatography-tandem mass spectrometry method for the quantitation of levodropropizine in human plasma. *J. Chromatogr. B* **2005**, *819*, 185–189.

69. Wang, Y.; Wang, Y.; Li, P.; Tang, Y.; Fawcett, J.P.; Gu, J. Quantitation of Armillarisin A in human plasma by liquid chromatography-electrospray tandem mass spectrometry. *J. Pharm. Biomed. Anal.* **2007**, *43*, 1860–1863.

70. Teng, G.; Wang, Y.; Tang, Y.; Wang, R.; Fang, Y.; Fawcett, J.P.; Gu, J. Quantitation of iptakalim in human plasma by high-performance liquid chromatography-tandem mass spectrometry. *J. Chromatogr. B* **2007**, *859*, 256–260.

71. Tracqui, A.; Ludes, B. HPLC-MS for the determination of sildenafil citrate (Viagra) in biological fluids. Application to the salivary excretion of sildenafil after oral intake. *J. Anal. Toxicol.* **2003**, *27*, 88–94.

72. Gonzalo-Lumbreras, R.; Pimentel-Trapero, D.; Izqierdo-Hornillos, R. Solvent and solid-phase extraction of natural and synthetic anabolic steroids in human urine. *J. Chromatogr. B* **2001**, *754*, 419–425.

73. Deventer, K.; Pozo, O.J.; Van Eenoo, P.; Delbeke, F.T. Development of a qualitative liquid chromatography/tandem mass spectrometric method for the detection of narcotics in urine relevant to doping analysis. *Rapid Commun. Mass Spectrom.* **2007**, *21*, 3015–3025.

74. Xu, H.R.; Li, X.N.; Chen, W.L.; Chu, N.N. Simultaneous determination of desloratadine and its active metabolite 3-hydroxydesloratadine in human plasma by LC/MS/MS and its application to pharmacokinetics and bioequivalence. *J. Pharm. Biomed. Anal.* **2007**, *45*, 659–666.

75. Ciccolini, J.; Catalin, J.; Blachon, M.F.; Durand, A. Rapid high-performance liquid chromatographic determination of docetaxel (Taxotere) in plasma using liquid-liquid extraction. *J. Chromatogr. B* **2001**, *759*, 299–306.

76. Bouley, M.; Briere, C.; Padoin, C.; Petitjean, O.; Tod, M. Sensitive and rapid method for the simultaneous quantification of the HIV-protease inhibitors indinavir, nelfinavir, ritonavir, and saquinavir in human plasma by reversed-phase liquid chromatography. *Ther. Drug Monit.* **2001**, *23*, 56–60.

77. El Mahjoub, A.; Staub, C. Stability of benzodiazepines in whole blood samples stored at varying temperatures. *J. Pharm. Biomed. Anal.* **2000**, *23*, 1057–1063.

78. Kousoulos, C.; Tsatsou, G.; Apostolou, C.; Dotsikas, Y.; Loukas, Y.L. Development of a high-throughput method for the determination of itraconazole and its hydroxy metabolite in human plasma, employing automated liquid-liquid extraction based on 96-well format plates and LC/MS/MS. *Anal. Bioanal. Chem.* **2006**, *384*, 199–207.

79. Gauvin, A.; Pinguet, F.; Poujol, S.; Astre, C.; Bressolle, F. High-performance liquid chromatographic determination of vinorelbine in human plasma and blood: Application to a pharmacokinetic study. *J. Chromatogr. B* **2000**, *748*, 389–399.

80. Kerbusch, T.; Jeuken, M.J.; Derraz, J.; van Putten, J.W.; Huitema, A.D.; Beijnen, J.H. Determination of ifosfamide, 2- and 3-dechloroethyifosfamide using gas chromatography with nitrogen-phosphorus or mass spectrometry detection. *Ther. Drug Monit.* **2000**, *22*, 613–620.

81. Rechberger, G.N.; Fauler, G.; Windischhofer, W.; Köfeler, H.; Erwa, W.; Leis, H.J. Quantitative analysis of clindamycin in human plasma by liquid chromatography/electrospray ionisation tandem mass spectrometry using d1-N-ethylclindamycin as internal standard. *Rapid Commun. Mass Spectrom.* **2003**, *17*, 135–139.

82. Na-Bangchang, K.; Banmairuroi, V.; Kamanikom, B.; Kiod, D. An alternative high-performance liquid chromatographic method for determination of clindamycin in plasma. *Southeast Asian J. Trop. Med. Public Health* **2006**, *37*, 177–184.

83. Deventer, K.; Mikulcíková, P.; Van Hoecke, H.; Van Eenoo, P.; Delbeke, F.T. Detection of budesonide in human urine after inhalation by liquid chromatography-mass spectrometry. *J. Pharm. Biomed. Anal.* **2006**, *42*, 474–479.

84. Keski-Rahkonen, P.; Pärssinen, O.; Leppänen, E.; Mauriala, T.; Lehtonen, M.; Auriola, S. Determination of tamsulosin in human aqueous humor and serum by liquid chromatography-electrospray ionization tandem mass spectrometry. *J. Pharm. Biomed. Anal.* **2007**, *43*, 606–612.

85. Macek, J.; Klíma, J.; Ptácek, P. Rapid determination of tamsulosin in human plasma by high-performance liquid chromatography using extraction with butyl acetate. *J. Chromatogr. B* **2004**, *809*, 307–311.

86. Bahrami, G.; Mohammadi, B. An isocratic high performance liquid chromatographic method for quantification of mycophenolic acid and its glucuronide metabolite in human serum using liquid-liquid extraction: Application to human pharmacokinetic studies. *Clin. Chim. Acta* **2006**, *370*, 185–190.

87. Shen, Z.; Wang, S.; Bakhtiar, R. Enantiomeric separation and quantification of fluoxetine (Prozac) in human plasma by liquid chromatography/tandem mass spectrometry using liquid-liquid extraction in 96-well plate format. *Rapid Commun. Mass Spectrom.* **2002**, *16*, 332–338.

88. Apostolou, C.; Dotsikas, Y.; Kousoulos, C.; Loukas, Y.L. Quantitative determination of donepezil in human plasma by liquid chromatography/tandem mass spectrometry employing an automated liquid-liquid extraction based on 96-well format plates. Application to a bioequivalence study. *J. Chromatogr. B* **2007**, *848*, 239–244.

89. Miyaguchi, H.; Tokeshi, M.; Kikutani, Y.; Hibara, A.; Inoue, H.; Kitamori, T. Microchip-based liquid-liquid extraction for gas-chromatography analysis of amphetamine-type stimulants in urine. *J. Chromatogr. A* **2006**, *1129*, 105–110.

90. Melwanki, M.B.; Hsu, W.-H.; Huang, S.-D. Determination of clenbuterol in urine using headspace solid phase microextraction or liquid-liquid-liquid microextraction. *Anal. Chim. Acta* **2005**, *552*, 67–75.

91. He, Y.; Kang, Y.-J. Single drop liquid-liquid-liquid microextraction of methamphetamine and amphetamine in urine. *J. Chromatogr. A* **2006**, *1133*, 35–40.

92. Shariati, S.; Yamini, Y.; Darabi, M.; Amini, M. Three phase liquid phase microextraction of phenylacetic acid and phenylpropionic acid from biological fluids. *J. Chromatogr. B* **2007**, *855*, 228–235.

93. Lambropoulou, D.A.; Albanis, T.A. Application of solvent microextraction in a single drop for the determination of new antifouling agents in waters. *J. Chromatogr. A* **2004**, *1049*, 17–23.

94. Gupta, P.K.; Manral, L.; Ganesan, K.; Dubey, D.K. Use of single-drop microextraction for determination of fentanyl in water samples. *Anal. Bioanal. Chem.* **2007**, *388*, 579–583.

95. Deng, C.; Yao, N.; Wang, B.; Zhang, X. Development of microwave-assisted extraction followed by headspace single-drop microextraction for fast determination of paeonol in traditional Chinese medicines. *J. Chromatogr. A* **2006**, *1103*, 15–21.

96. Dong, L.; Deng, C.; Wang, B.; Shen, X. Fast determination of Z-ligustilide in plasma by gas chromatography/mass spectrometry following headspace single-drop microextraction. *J. Sep. Sci.* **2007**, *30*, 1318–1325.

97. Gioti, E.M.; Skalkos, D.C.; Fiamegos, Y.C.; Stalikas, C.D. Single-drop liquid-phase microextraction for the determination of hypericin, pseudohypericin and hyperforin in biological fluids by high performance liquid chromatography. *J. Chromatogr. A* **2005**, *1093*, 1–10.

98. Ma, M.; Kang, S.; Zhao, Q.; Chen, B.; Yao, S. Liquid-phase microextraction combined with high-performance liquid chromatography for the determination of local anaesthetics in human urine. *J. Pharm. Biomed. Anal.* **2006**, *40*, 128–135.

99. He, Y.; Vargas, A.; Kang, Y.-J. Headspace liquid-phase microextraction of methamphetamine and amphetamine in urine by an aqueous drop. *Anal. Chim. Acta* **2007**, *589*, 225–230.

100. Ho, T.S.; Pedersen-Bjergaard, S.; Rasmussen, K.E. Recovery, enrichment and selectivity in liquid-phase microextraction comparison with conventional liquid-liquid extraction. *J. Chromatogr. A* **2002**, *963*, 3–17.

101. Andersen, S.; Halvorsen, T.G.; Pedersen-Bjergaard, S.; Rasmussen, K.E. Liquid-phase microextraction combined with capillary electrophoresis, a promising tool for the determination of chiral drugs in biological matrices. *J. Chromatogr. A* **2002**, *963*, 303–312.

102. Yazdi, A.S.; Es'haghi, Z. Two-step hollow fiber-based, liquid-phase microextraction combined with high-performance liquid chromatography: A new approach to determination of aromatic amines in water. *J. Chromatogr. A* **2005**, *1094*, 136–142.

103. Reubsaet, J.L.E.; Loftheim, H.; Gjelstad, A. Ion-pair mediated transport of angiotensin, neurotensin, and their metabolites in liquid phase microextraction under acidic conditions. *J. Sep. Sci.* **2005**, *28*, 1204–1210.

104. Ho, T.S.; Halvorsen, T.G.; Pedersen-Bjergaard, S.; Rasmussen, K.E. Liquid-phase microextraction of hydrophilic drugs by carrier-mediated transport. *J. Chromatogr. A* **2003**, *998*, 61–72.

105. Ho, T.S.; Reubsaet, J.L.E.; Anthonsen, H.S.; Pedersen-Bjergaard, S.; Rasmussen, K.E. Liquid-phase microextraction based on carrier mediated transport combined with liquid chromatography-mass spectrometry. New concept for the determination of polar drugs in a single drop of human plasma. *J. Chromatogr. A* **2005**, *1072*, 29–36.

106. Pedersen-Bjergaard, S.; Rasmussen, K.E.; Brekke, A.; Ho, T.S.; Halvorsen, T.G. Liquid-phase microextraction of basic drugs—Selection of extraction mode based on computer calculated solubility data. *J. Sep. Sci.* **2005**, *28*, 1195–1203.

107. Angelo, H.R.; Petersen, A. Therapeutic drug monitoring of haloperidol, perphenazine, and zuclopenthixol in serum by a fully automated sequential solid phase extraction followed by high-performance liquid chromatography. *Ther. Drug Monit.* **2001**, *23*, 157–162.

108. Ohman, D.; Carlsson, B.; Norlander, B. On-line extraction using an alkyl-diol silica precolumn for racemic citalopram and its metabolites in plasma. Results compared with solid-phase extraction methodology. *J. Chromatogr. B* **2001**, *753*, 365–373.

109. Lindegardh, N.; Bergqvist, Y. Automated solid-phase extraction method for the determination of atovaquone in plasma and whole blood by rapid high-performance liquid chromatography. *J. Chromatogr. B* **2000**, *744*, 9–17.

110. Raggi, M.A.; Casamenti, G.; Mandrioli, R.; Volterra, V. A sensitive high-performance liquid chromatographic method using electrochemical detection for the analysis of olanzapine and desmethylolanzapine in plasma of schizophrenic patients using a new solid-phase extraction procedure. *J. Chromatogr. B* **2001**, *750*, 137–146.

111. Prieto, J.A.; Jimenez, R.M.; Alonso, R.M.; Ortiz, E. Determination of the antihypertensive drug cilazapril and its active metabolite cilazaprilat in pharmaceuticals and urine by solid-phase extraction and high-performance liquid chromatography with photometric detection. *J. Chromatogr. B* **2001**, *754*, 23–34.

112. El Mahjoub, A.; Staub, C. High-performance liquid chromatography determination of flunitrazepam and its metabolites in plasma by use of column-switching technique: Comparison of two extraction columns. *J. Chromatogr. B* **2001**, *754*, 271–283.

113. Marzolini, C.; Telenti, A.; Buclin, T.; Biollaz, J.; Decosterd, L.A. Simultaneous determination of the HIV protease inhibitors indinavir, amprenavir, saquinavir, ritonavir, nelfinavir and the non-nucleoside reverse

transcriptase inhibitor efavirenz by high-performance liquid chromatography after solid-phase extraction. *J. Chromatogr. B* **2000**, *740*, 43–58.

114. Moreno, P.; Salvado, V. Determination of eight water- and fat-soluble vitamins in multi-vitamin pharmaceutical formulations by high-performance liquid chromatography. *J. Chromatogr. A* **2000**, *870*, 207–215.

115. Shimoyama, R.; Ohkubo, T.; Sugawara, K. Monitoring of carbamazepine and carbamazepine 10,11-epoxide in breast milk and plasma by high-performance liquid chromatography. *Ann. Clin. Biochem.* **2000**, *37*, 210–215.

116. Bevalot, F.; Gaillard, Y.; Lhermitte, M.A.; Pepin, G. Analysis of corticosteroids in hair by liquid chromatography-electrospray ionization mass spectrometry. *J. Chromatogr. B* **2000**, *740*, 227–236.

117. Gatti, R.; Gioia, M.G.; Cavrini, V. Analysis and stability study of retinoids in pharmaceuticals by LC with fluorescence detection. *J. Pharm. Biomed. Anal.* **2000**, *23*, 147–159.

118. Taylor, P.J.; Salm, P.; Lynch, S.V.; Pillans, P.I. Simultaneous quantification of tacrolimus and sirolimus, in human blood, by high-performance liquid chromatography-tandem mass spectrometry. *Ther. Drug Monit.* **2000**, *22*, 608–612.

119. Gan, S.H.; Ismail, R. Validation of a high-performance liquid chromatography method for tramadol and o-desmethyltramadol in human plasma using solid-phase extraction. *J. Chromatogr. B* **2001**, *759*, 325–335.

120. Zendelovska, D.; Stafilov, T. Development of an HPLC method for the determination of ranitidine and cimetidine in human plasma following SPE. *J. Pharm. Biomed. Anal.* **2003**, *33*, 165–173.

121. Santagati, N.A.; Gotti, R.; Ronsisvalle, G. Simultaneous determination of phenytoin and dextromethorphan in urine by solid-phase extraction and HPLC-DAD. *J. Sep. Sci.* **2005**, *28*, 1157–1162.

122. Koal, T.; Sibum, M.; Koster, E.; Resch, K.; Kaever, V. Direct and fast determination of antiretroviral drugs by automated online solid-phase extraction-liquid chromatography-tandem mass spectrometry in human plasma. *Clin. Chem. Lab. Med.* **2006**, *44*, 299–305.

123. Cheng, W.C.; Yau, T.S.; Wong, M.K.; Chan, L.P. Mok, V.K. A high-throughput urinalysis of abused drugs based on a SPE-LC-MS/MS method coupled with an in-house developed post-analysis data treatment system. *Forensic Sci. Int.* **2006**, *162*, 95–107.

124. Tahara, K.; Yonemoto, A.; Yoshiyama, Y.; Nakamura, T.; Aizawa, M.; Fujita, Y.; Nishikawa, T. Determination of antihyperglycemic biguanides in serum and urine using an ion-pair solid-phase extraction technique followed by HPLC-UV on a pentafluorophenylpropyl column and on an octadecyl column. *Biomed. Chromatogr.* **2006**, *20*, 1200–1205.

125. Barroso, M.; Gallardo, E.; Margalho, C.; Devesa, N.; Pimentel, J.; Vieira, D.N. Solid-phase extraction and gas chromatographic-mass spectrometric determination of the veterinary drug xylazine in human blood. *J. Anal. Toxicol.* **2007**, *31*, 165–169.

126. Vermeij, T.A.; Edelbroek, P.M. Robust isocratic high performance liquid chromatographic method for simultaneous determination of seven antiepileptic drugs including lamotrigine, oxcarbazepine and zonisamide in serum after solid-phase extraction. *J. Chromatogr. B* **2007**, *857*, 40–46.

127. Guo, Z.; Wei, D.; Yin, G.; Wang, S.; Zhao, S.; Chu, Y.; Zhai, J. Simultaneous determination of rivanol and mifepristone in human plasma by a HPLC-UV method with solid-phase extraction. *J. Chromatogr. B* **2007**, *856*, 312–317.

128. Bajpai, L.; Varshney, M.; Seubert, C.N.; Dennis, D.M. A new method for the quantitation of propofol in human plasma. Efficient solid-phase extraction and liquid chromatography/APCI-triple quadrupole mass spectrometry detection. *J. Chromatogr. B* **2004**, *810*, 291–296.

129. Ruiz-Gutierrez, V.; Juan, M.E.; Cert, A.; Planas, J.M. Determination of hydroxytyrosol in plasma by HPLC. *Anal. Chem.* **2000**, *72*, 4458–4461.

130. Nicolas, O.; Margout, D.; Taudon, N.; Calas, M.; Vial, H.J.; Bressolle, F. Liquid chromatography-electrospray mass spectrometry determination of a bis-thiazolium compound with potent antimalarial activity and its neutral bioprecursor in human plasma, whole blood and red blood cells. *J. Chromatogr. B* **2005**, *820*, 83–93.

131. López, P.; Bermejo, A.M.; Tabernero, M.J.; Fernández, P.; Alvarez, I. Determination of cocaine and heroin with their respective metabolites in meconium by gas chromatography-mass spectrometry. *J. Appl. Toxicol.* **2007**, *27*, 464–471.

132. Venisse, N.; Marquet, P.; Duchoslav, E.; Dupuy, J.L.; Lachâtre, G. A general unknown screening procedure for drugs and toxic compounds in serum using liquid chromatography-electrospray-single quadrupole mass spectrometry. *J. Anal. Toxicol.* **2003**, *27*, 7–14.

133. Mikami, E.; Goto, T.; Ohno, T.; Matsumoto, H.; Inagaki, K.; Ishihara, H.; Nishida, M. Simultaneous analysis of anthranilic acid derivatives in pharmaceuticals and human urine by high-performance liquid chromatography with isocratic elution. *J. Chromatogr. B* **2000**, *744*, 81–89.

134. Paterson, S.; Cordero, R.; McCulloch, S.; Houldsworth, P. Analysis of urine for drugs of abuse using mixed-mode solid-phase extraction and gas chromatography-mass spectrometry. *Ann. Clin. Biochem.* **2000**, *37*, 690–700.

135. Weinmann, W.; Müller, C.; Vogt, S.; Frei, A. LC-MS-MS analysis of the neuroleptics clozapine, flupentixol, haloperidol, penfluridol, thioridazine, and zuclopenthixol in hair obtained from psychiatric patients. *J. Anal. Toxicol.* **2002**, *26*, 303–307.

136. Kobylinska, M.; Kobylinska, K. High-performance liquid chromatographic analysis for the determination of domperidone in human plasma. *J. Chromatogr. B* **2000**, *744*, 207–212.

137. Svensson, J.O.; Bruchfeld, A.; Schvarcz, R.; Stahle, L. Determination of ribavirin in serum using highly selective solid-phase extraction and high-performance liquid chromatography. *Ther. Drug Monit.* **2000**, *22*, 215–218.

138. Wang, Z.; Dsida, R.M.; Avram, M.J. Determination of ketorolac in human plasma by reversed-phase high-performance liquid chromatography using solid-phase extraction and ultraviolet detection. *J. Chromatogr. B* **2001**, *755*, 383–386.

139. Morales, J.M.; Jung, C.H.; Alarcon, A.; Barreda, A. Solid-phase extraction and liquid chromatographic quantitation of quinfamide in biological samples. *J. Chromatogr. B* **2000**, *746*, 133–139.

140. Nguyen, H.; Nau, D.R. Rapid method for the solid-phase extraction and GC-MS analysis of flunitrazepam and its major metabolites in urine. *J. Anal. Toxicol.* **2000**, *24*, 37–45.

141. Forngren, B.H.; Tyrefors, N.; Markides, K.E.; Langstrom, B. Determination of raclopride in human plasma by on-column focusing packed capillary liquid chromatography-electrospray ionisation mass spectrometry. *J. Chromatogr. B* **2000**, *748*, 189–195.

142. Rule, G.; Chapple, M.; Henion, J. A 384-well solid-phase extraction for LC/MS/MS determination of methotrexate and its 7-hydroxy metabolite in human urine and plasma. *Anal. Chem.* **2001**, *73*, 439–443.

143. Hsieh, S.; Selinger, K. High-throughput bioanalytical method using automated sample preparation and liquid chromatography-atmospheric pressure ionspray mass spectrometry for quantitative determination of glybenclamide in human serum. *J. Chromatogr. B* **2002**, *772*, 347–356.

144. Wang, A.Q.; Zeng, W.; Musson, D.G.; Rogers, J.D.; Fisher, A.L. A rapid and sensitive liquid chromatography/negative ion tandem mass spectrometry method for the determination of an indolocarbazole in human plasma using 96-well diatomaceous earth plates for solid-liquid extraction [correction of using internal standard (IS) 96-well diatomaceous earth plates for solid-liquid extraction]. *Rapid Commun. Mass Spectrom.* **2002**, *16*, 975–981.

145. Deng, Y.; Wu, J.-T.; Lloyd, T.L.; Chi, C.L.; Olah, T.V.; Unger, S.E. High-speed gradient parallel liquid chromatography/tandem mass spectrometry with fully automated sample preparation for bioanalysis: 30 seconds per sample from plasma. *Rapid Commun. Mass Spectrom.* **2002**, *16*, 1116–1123.

146. Hopfgartner, G.; Husser, C.; Zell, M. High-throughput quantification of drugs and their metabolites in biosamples by LC-MS/MS and CE-MS/MS: Possibilities and limitations. *Ther. Drug Monit.* **2002**, *24*, 134–143.

147. Hasegawa, C.; Kumazawa, T.; Lee, X.-P.; Marumo, A.; Shinmen, N.; Seno, H.; Sato, K. Pipette tip solid-phase extraction and gas chromatography-mass spectrometry for the determination of methamphetamine and amphetamine in human whole blood. *Anal. Bioanal. Chem.* **2007**, *389*, 563–570.

148. Christiaens, B.; Fillet, M.; Chiap, P.; Rbeida, O.; Ceccato, A.; Streel, B.; De Graeve, J.; Crommen, J.; Hubert, P. Fully automated method for the liquid chromatographic-tandem mass spectrometric determination of cyproterone acetate in human plasma using restricted access material for on-line sample clean-up. *J. Chromatogr. A* **2004**, *1056*, 105–110.

149. Egle, H.; Trittler, R.; Konig, A.; Kummerer, K. Fast, fully automated analysis of voriconazole from serum by LC-LC-ESI-MS-MS with parallel column-switching technique. *J. Chromatogr. B* **2005**, *814*, 361–367.

150. Huclova, J.; Satinsky, D.; Maia, T.; Karlicek, R.; Solich, P.; Araujo, A.N. Sequential injection extraction based on restricted access material for determination of furosemide in serum. *J. Chromatogr. A* **2005**, *1087*, 245–251.

151. Vintiloiu, A.; Mullett, W.M.; Papp, R.; Lubda, D.; Kwong, E. Combining restricted access material (RAM) and turbulent flow for the rapid on-line extraction of the cyclooxygenase-2 inhibitor rofecoxib in plasma samples. *J. Chromatogr. A* **2005**, *1082*, 150–157.

152. Yang, N.C.; Hwang, K.L.; Shen, C.H.; Wang, H.F.; Ho, W.M. Simultaneous determination of fluorinated inhalation anesthetics in blood by gas chromatography-mass spectrometry combined with a headspace autosampler. *J. Chromatogr. B* **2001**, *759*, 307–318.

153. Rbeida, O.; Christiaens, B.; Hubert, P.; Lubda, D.; Boos, K.S.; Crommen, J.; Chiap, P. Evaluation of a novel anion-exchange restricted-access sorbent for on-line sample clean-up prior to the determination of acidic compounds in plasma by liquid chromatography. *J. Chromatogr. A* **2004**, *1030*, 95–102.

154. Rbeida, O.; Chiap, P.; Lubda, D.; Boos, K.S.; Crommen, J.; Hubert, P. Development and validation of a fully automated LC method for the determination of cloxacillin in human plasma using anion exchange restricted access material for sample clean-up. *J. Pharm. Biomed. Anal.* **2005**, *36*, 961–968.

155. Rbeida, O.; Christiaens, B.; Hubert, P.; Lubda, D.; Boos, K.S.; Crommen, J.; Chiap, P. Integrated on-line sample clean-up using cation exchange restricted access sorbent for the LC determination of atropine in human plasma coupled to UV detection. *J. Pharm. Biomed. Anal.* **2005**, *36*, 947–954.

156. Clarke, W.; Chowdhuri, A.R.; Hage, D.S. Analysis of free drug fractions by ultrafast immunoaffinity chromatography. *Anal. Chem.* **2001**, *73*, 2157–2164.

157. Holtzapple, C.K.; Buckley, S.A.; Stanker, L.H. Determination of fluoroquinolones in serum using an on-line clean-up column coupled to high-performance immunoaffinity-reversed-phase liquid chromatography. *J. Chromatogr. B* **2001**, *754*, 1–9.

158. Su, P.; Zhang, X.X.; Chang, W.B. Development and application of a multi-target immunoaffinity column for the selective extraction of natural estrogens from pregnant women's urine samples by capillary electrophoresis. *J. Chromatogr. B* **2005**, *816*, 7–14.

159. Hupka, Y.; Beike, J.; Roegener, J.; Brinkmann, B.; Blaschke, G.; Köhler, H. HPLC with laser-induced native fluorescence detection for morphine and morphine glucuronides from blood after immunoaffinity extraction. *Int. J. Legal Med.* **2005**, *119*, 121–128.

160. Andersson, L.I. Efficient sample pre-concentration of bupivacaine from human plasma by solid-phase extraction on molecularly imprinted polymers. *Analyst* **2000**, *125*, 1515–1517.

161. Bereczki, A.; Tolokan, A.; Horvaia, G.; Horvath, V.; Lanza, F.; Hall, A.J.; Sellergren, B. Determination of phenytoin in plasma by molecularly imprinted solid-phase extraction. *J. Chromatogr. A* **2001**, *930*, 31–38.

162. Mullett, W.M.; Walles, M.; Levsen, K.; Borlak, J.; Pawliszyn, J. Multidimensional on-line sample preparation of verapamil and its metabolites by a molecularly imprinted polymer coupled to liquid chromatography-mass spectrometry. *J. Chromatogr. B* **2004**, *801*, 297–306.

163. Widstrand, C.; Larsson, F.; Fiori, M.; Civitareale, C.; Mirante, S.; Brambilla, G. Evaluation of MISPE for the multi-residue extraction of beta-agonists from calves urine. *J. Chromatogr. B* **2004**, *804*, 85–91.

164. Caro, E.; Marce, R.M.; Cormack, P.A.G.; Sherrington, D.C.; Borrull, F. Direct determination of ciprofloxacin by mass spectrometry after a two-step solid-phase extraction using a molecularly imprinted polymer. *J. Sep. Sci.* **2006**, *29*, 1230–1236.

165. Caro, E.; Marce, R.M.; Cormack, P.A.G.; Sherrington, D.C.; Borrull, F. A new molecularly imprinted polymer for the selective extraction of naproxen from urine samples by solid-phase extraction. *J. Chromatogr. B* **2004**, *813*, 137–143.

166. Theodoridis, G.; Zacharis, C.K.; Tzanavaras, P.D.; Themelis, D.G.; Economou, A. Automated sample preparation based on the sequential injection principle. Solid-phase extraction on a molecularly imprinted polymer coupled on-line to high-performance liquid chromatography. *J. Chromatogr. A* **2004**, *1030*, 69–76.

167. Sanbe, H.; Haginaka, J. Restricted access media-molecularly imprinted polymer for propranolol and its application to direct injection analysis of beta-blockers in biological fluids. *Analyst* **2003**, *128*, 593–597.

168. Musshoff, F.; Junker, H.P.; Lachenmeier, D.W.; Kroener, L.; Madea, B. Fully automated determination of cannabinoids in hair samples using headspace solid-phase microextraction and gas chromatography-mass spectrometry. *J. Anal. Toxicol.* **2002**, *26*, 554–560.

169. Nadulski, T.; Pragst, F. Simple and sensitive determination of delta(9)-tetrahydrocannabinol, cannabidiol and cannabinol in hair by combined silylation, headspace solid phase microextraction and gas chromatography-mass spectrometry. *J. Chromatogr. B* **2007**, *846*, 78–85.

170. Dizioli Rodrigues de Oliveira, C.; Yonamine, M.; Lucia de Moraea Moreau, R. Headspace solid-phase microextraction of cannabinoids in human head hair samples. *J. Sep. Sci.* **2007**, *30*, 128–134.

171. Fucci, N.; De Giovanni, N.; Chiarotti, M.; Scarlata, S. SPME-GC analysis of THC in saliva samples collected with "EPITOPE" device. *Forensic Sci. Int.* **2001**, *119*, 318–321.

172. Liu, J.; Hara, K.; Kashimura, S.; Kashiwagi, M.; Kageura, M. New method of derivatization and headspace solid-phase microextraction for gas chromatographic-mass spectrometric analysis of amphetamines in hair. *J. Chromatogr. B* **2001**, *758*, 95–101.

173. Musshoff, F.; Junker, H.P.; Lachenmeier, D.W.; Kroener, L.; Madea, B. Fully automated determination of amphetamines and synthetic designer drugs in hair samples using headspace solid-phase microextraction and gas chromatography-mass spectrometry. *J. Chromatogr. Sci.* **2002**, *40*, 359–364.

174. Gentili, S.; Torresi, A.; Marsili, R.; Chiarotti, M.; Macchia, T. Simultaneous detection of amphetamine-like drugs with headspace solid-phase microextraction and gas chromatography-mass spectrometry. *J. Chromatogr. B* **2002**, *780*, 183–192.

175. Fucci, N.; De Giovanni, N.; Chiarotti, M. Simultaneous detection of some drugs of abuse in saliva samples by SPME technique. *Forensic Sci. Int.* **2003**, *134*, 40–45.

176. Namera, A.; Yashiki, M.; Liu, J.; Okajima, K.; Hara, K.; Imamura, T.; Kojima, T. Simple and simultaneous analysis of fenfluramine, amphetamine and methamphetamine in whole blood by gas chromatography-mass spectrometry after headspace-solid phase microextraction and derivatization. *Forensic Sci. Int.* **2000**, *109*, 215–223.

177. Okajima, K.; Namera, A.; Yashiki, M.; Tsukue, I.; Kojima, T. Highly sensitive analysis of methamphetamine and amphetamine in human whole blood using headspace solid-phase microextraction and gas chromatography-mass spectrometry. *Forensic Sci. Int.* **2001**, *116*, 15–22.

178. Lee, M.R.; Song, Y.S.; Hwang, B.H.; Chou, C.C. Determination of amphetamine and methamphetamine in serum via headspace derivatization solid-phase microextraction-gas chromatography-mass spectrometry. *J. Chromatogr. A* **2000**, *896*, 265–273.

179. Jurado, C.; Gimenez, M.P.; Soriano, T.; Menendez, M.; Repetto, M. Rapid analysis of amphetamine, methamphetamine, MDA, and MDMA in urine using solid-phase microextraction, direct on-fiber derivatization, and analysis by GC-MS. *J. Anal. Toxicol.* **2000**, *24*, 11–16.

180. Namera, A.; Yashiki, M.; Kojima, T.; Ueki, M. Automated headspace solid-phase microextraction and in-matrix derivatization for the determination of amphetamine-related drugs in human urine by gas chromatography-mass spectrometry. *J. Chromatogr. Sci.* **2002**, *40*, 19–25.

181. McCooeye, M.A.; Mester, Z.; Ells, B.; Barnett, D.A.; Purves, R.W.; Guevremont, R. Quantitation of amphetamine, methamphetamine, and their methylenedioxy derivatives in urine by solid-phase microextraction coupled with electrospray ionization-high-field asymmetric waveform ion mobility spectrometry-mass spectrometry. *Anal. Chem.* **2002**, *74*, 3071–3075.

182. Huang, M.K.; Liu, C.; Huang, S.D. One step and highly sensitive headspace solid-phase microextraction sample preparation approach for the analysis of methamphetamine and amphetamine in human urine. *Analyst* **2002**, *127*, 1203–1206.

183. Raikos, N.; Christopoulou, K.; Theodoridis, G.; Tsoukali, H.; Psaroulis, D. Determination of amphetamines in human urine by headspace solid-phase microextraction and gas chromatography. *J. Chromatogr. B* **2003**, *789*, 59–63.

184. Chafer-Pericas, C.; Campins-Falco, P.; Herraez-Hernandez, R. Application of solid-phase microextraction combined with derivatization to the determination of amphetamines by liquid chromatography. *Anal. Biochem.* **2004**, *333*, 328–335.

185. Brown, H.; Kirkbride, K.P.; Pigou, P.E.; Walker, G.S. New developments in SPME, part 1: The use of vapor-phase deprotonation and on-fiber derivatization with alkylchloroformates in the analysis of preparations containing amphetamines. *J. Forensic Sci.* **2003**, *48*, 1231–1238.

186. Gentili, S.; Cornetta, M.; Macchia, T. Rapid screening procedure based on headspace solid-phase microextraction and gas chromatography-mass spectrometry for the detection of many recreational drugs in hair. *J. Chromatogr. B* **2004**, *801*, 289–296.

187. Brown, S.D.; Rhodes, D.J.; Pritchard, B.J. A validated SPME-GC-MS method for simultaneous quantification of club drugs in human urine. *Forensic Sci. Int.* **2007**, *171*, 142–150.

188. Kuriki, A.; Kumazawa, T.; Lee, X.P.; Hasegawa, C.; Kawamura, M.; Suzuki, O.; Sato, K. Simultaneous determination of selegiline and desmethylselegiline in human body fluids by headspace solid-phase microextraction and gas chromatography-mass spectrometry. *J. Chromatogr. B* **2006**, *844*, 283–291.

189. de Toledo, F.C.; Yonamine, M.; de Moraes Moreau, R.L.; Silva, O.A. Determination of cocaine, benzoylecgonine and cocaethylene in human hair by solid-phase microextraction and gas chromatography-mass spectrometry. *J. Chromatogr. B* **2003**, *798*, 361–365.

190. Follador, M.J.D.; Yonamine, M.; de Moraes Moreau, R.L.; Silva, O.A. Detection of cocaine and cocaethylene in sweat by solid-phase microextraction and gas chromatography/mass spectrometry. *J. Chromatogr. B* **2004**, *811*, 37–40.

191. Alvarez, I.; Bermejo, A.M.; Tabernero, M.J.; Fernandez, P.; Lopez, P. Determination of cocaine and cocaethylene in plasma by solid-phase microextraction and gas chromatography-mass spectrometry. *J. Chromatogr. B* **2007**, *845*, 90–94.

192. Yonamine, M.; Saviano, A.M. Determination of cocaine and cocaethylene in urine by solid-phase microextraction and gas chromatography-mass spectrometry. *Biomed. Chromatogr.* **2006**, *20*, 1071–1075.

193. Staerk, U.; Kulpmann, W.R. High-temperature solid-phase microextraction procedure for the detection of drugs by gas chromatography-mass spectrometry. *J. Chromatogr. B* **2000**, *745*, 399–411.

194. Lucas, A.C.; Bermejo, A.M.; Tabernero, M.J.; Fernandez, P.; Strano-Rossi, S. Use of solid-phase microextraction (SPME) for the determination of methadone and EDDP in human hair by GC-MS. *Forensic Sci. Int.* **2000**, *107*, 225–232.

195. Sporkert, F.; Pragst, F. Determination of methadone and its metabolites EDDP and EMDP in human hair by headspace solid-phase microextraction and gas chromatography-mass spectrometry. *J. Chromatogr. B* **2000**, *746*, 255–264.

196. Lucas, A.C.; Bermejo, A.; Fernandez, P.; Tabernero, M.J. Solid-phase microextraction in the determination of methadone in human saliva by gas chromatography-mass spectrometry. *J. Anal. Toxicol.* **2000**, *24*, 93–96.

197. Bermejo, A.M.; Seara, R.; dos Santos Lucas, A.C.; Tabernero, M.J.; Fernandez, P; Marsili, R. Use of solid-phase microextraction (SPME) for the determination of methadone and its main metabolite, EDDP, in plasma by gas chromatography-mass spectrometry. *J. Anal. Toxicol.* **2000**, *24*, 66–69.

198. Sha, Y.F.; Shen, S.; Duan, G.L. Rapid determination of tramadol in human plasma by headspace solid-phase microextraction and capillary gas chromatography-mass spectrometry. *J. Pharm. Biomed. Anal.* **2005**, *37*, 143–147.

199. Bagheri, H.; Es-Haghi, A.; Khalilian, F.; Rouini, M.R. Determination of fentanyl in human plasma by head-space solid-phase microextraction and gas chromatography-mass spectrometry. *J. Pharm. Biomed. Anal.* **2007**, *43*, 1763–1768.

200. Paradis, C.; Dufresne, C.; Bolon, M.; Boulieu, R. Solid-phase microextraction of human plasma samples for determination of sufentanil by gas chromatography-mass spectrometry. *Ther. Drug Monit.* **2002**, *24*, 768–774.

201. Musshoff, F.; Junker, H.; Madea, B. Rapid analysis of halothane in biological samples using headspace solid-phase microextraction and gas chromatography-mass spectrometry—A case of a double homicide. *J. Anal. Toxicol.* **2000**, *24*, 372–376.

202. Sporkert, F.; Pragst, F. Determination of lidocaine in hair of drug fatalities by headspace solid-phase microextraction. *J. Anal. Toxicol.* **2000**, *24*, 316–322.

203. Koster, E.H.M.; Wemes, C.; Morsink, J.B.; de Jong, G.J. Determination of lidocaine in plasma by direct solid-phase microextraction combined with gas chromatography. *J. Chromatogr. B* **2000**, *739*, 175–182.

204. van Hout, M.W.J.; Jas, V.; Niederlander, H.A.G.; de Zeeuw, R.A.; de Jong, G.J. Ultra-rapid non-equilibrium solid-phase microextraction at elevated temperatures and direct coupling to mass spectrometry for the analysis of lidocaine in urine. *J. Sep. Sci.* **2003**, *26*, 1563–1568.

205. Frison, G.; Favretto, D.; Tedeschi, L.; Ferrara, S.D. Detection of thiopental and pentobarbital in head and pubic hair in a case of drug-facilitated sexual assault. *Forensic Sci. Int.* **2003**, *133*, 171–174.

206. Iwai, M.; Hattori, H.; Arinobu, T.; Ishii, A.; Kumazawa, T.; Noguchi, H.; Noguchi, H.; Suzuki, O.; Seno, H. Simultaneous determination of barbiturates in human biological fluids by direct immersion solid-phase microextraction and gas chromatography-mass spectrometry. *J. Chromatogr. B* **2004**, *806*, 65–73.

207. Strehler, M.; Preuss, J.; Wollersen, H.; Madea, B. Lethal mixed intoxication with propofol in a medical layman. *Arch. Kriminol.* **2006**, *217*, 153–160.

208. Queiroz, M.E.; Silva, S.M.; Carvalho, D.; Lancas, F.M. Determination of lamotrigine simultaneously with carbamazepine, carbamazepine epoxide, phenytoin, phenobarbital, and primidone in human plasma by SPME-GC-TSD. *J. Chromatogr. Sci.* **2002**, *40*, 219–223.

209. Cantu, M.D.; Toso, D.R.; Lacerda, C.A.; Lancas, F.M.; Carrilho, E.; Queiroz, M.E. Optimization of solid-phase microextraction procedures for the determination of tricyclic antidepressants and anticonvulsants in plasma samples by liquid chromatography. *Anal. Bioanal. Chem.* **2006**, *386*, 256–263.

210. Deng, C.; Li, N.; Ji, J.; Yang, B.; Duan, G.; Zhang, X. Development of water-phase derivatization followed by solid-phase microextraction and gas chromatography/mass spectrometry for fast determination of valproic acid in human plasma. *Rapid Commun. Mass Spectrom.* **2006**, *20*, 1281–1287.

211. Frison, G.; Tedeschi, L.; Maietti, S.; Ferrara, S.D. Determination of midazolam in human plasma by solid-phase microextraction and gas chromatography/mass spectrometry. *Rapid Commun. Mass Spectrom.* **2001**, *15*, 2497–2501.

212. Aresta, A.; Monaci, L.; Zambonin, C.G. Determination of delorazepam in urine by solid-phase microextraction coupled to high performance liquid chromatography. *J. Pharm. Biomed. Anal.* **2002**, *28*, 965–972.

213. Lord, H.L.; Rajabi, M.; Safari, S.; Pawliszyn, J. Development of immunoaffinity solid phase microextraction probes for analysis of sub ng/mL concentrations of 7-aminoflunitrazepam in urine. *J. Pharm. Biomed. Anal.* **2006**, *40*, 769–780.

214. Walles, M.; Mullett, W.M.; Pawliszyn, J. Monitoring of drugs and metabolites in whole blood by restricted-access solid-phase microextraction coupled to liquid chromatography-mass spectrometry. *J. Chromatogr. A* **2004**, *1025*, 85–92.

215. Mullett, W.M.; Pawliszyn, J. Direct determination of benzodiazepines in biological fluids by restricted-access solid-phase microextraction. *J. Anal. Chem.* **2002**, *74*, 1081–1087.

216. Kumazawa, T.; Seno, H.; Watanabe-Suzuki, K.; Hattori, H.; Ishii, A.; Sato, K.; Suzuki, O. Determination of phenothiazines in human body fluids by solid-phase microextraction and liquid chromatography/tandem mass spectrometry. *J. Mass Spectrom.* **2000**, *35*, 1091–1099.

217. Kruggel, S.; Ulrich, S. Solid-phase microextraction for the assay of levomepromazine in human plasma. *Ther. Drug Monit.* **2000**, *22*, 723–728.

218. Jinno, K.; Kawazoe, M.; Hayashida, M. Solid-phase microextraction coupled with microcolumn liquid chromatography for the analysis of amitriptyline in human urine. *Chromatographia* **2000**, *52*, 309–313.

219. Li, X.; Zeng, Z.; Hu, M.; Mao, M. High operationally stable sol-gel diglycidyloxycalix[4]arene fiber for solid-phase microextraction of propranolol in human urine. *J. Sep. Sci.* **2005**, *28*, 2489–2500.

220. Zhou, X.; Li, X.; Zeng, Z. Solid-phase microextraction coupled with capillary electrophoresis for the determination of propranolol enantiomers in urine using a sol-gel derived calix[4]arene fiber. *J. Chromatogr. A* **2006**, *1104*, 359–365.

221. Sporkert, F. Analysis of hair by headspace SPME and GC-MS. *Chromatogr. Sep. Technol.* **2001**, *17*, 4–8.

222. Alves, C.; Santos-Neto, A.J.; Fernandes, C.; Rodrigues, J.C.; Lancas, F.M. Analysis of tricyclic antidepressant drugs in plasma by means of solid-phase microextraction-liquid chromatography-mass spectrometry. *J. Mass Spectrom.* **2007**, *42*, 1342–1347.

223. Salgado-Petinal, C.; Lamas, J.P.; Garcia-Jares, C.; Llompart, M.; Cela, R. Rapid screening of selective serotonin re-uptake inhibitors in urine samples using solid-phase microextraction gas chromatography-mass spectrometry. *Anal. Bioanal. Chem.* **2005**, *382*, 1351–1359.

224. Fernandes, C.; Neto, A.J.; Rodrigues, J.C.; Alves, C.; Lancas, F.M. Solid-phase microextraction-liquid chromatography (SPME-LC) determination of fluoxetine and norfluoxetine in plasma using a heated liquid flow through interface. *J. Chromatogr. B* **2007**, *847*, 217–223.

225. Queiroz, M.E.; Valadao, C.A.; Farias, A.; Carvalho, D.; Lancas, F.M. Determination of amitraz in canine plasma by solid-phase microextraction-gas chromatography with thermionic specific detection. *J. Chromatogr. B* **2003**, *794*, 337–342.

226. Aresta, A.; Palmisano, F.; Zambonin, C.G. Determination of naproxen in human urine by solid-phase microextraction coupled to liquid chromatography. *J. Pharm. Biomed. Anal.* **2005**, *39*, 643–647.

227. de Oliveira, A.R.; Cesarino, E.J.; Bonato, P.S. Solid-phase microextraction and chiral HPLC analysis of ibuprofen in urine. *J. Chromatogr. B* **2005**, *818*, 285–291.

228. Sha, Y.; Deng, C.; Liu, Z.; Huang, T.; Yang, B.; Duan, G. Headspace solid-phase microextraction and capillary gas chromatographic-mass spectrometric determination of rivastigmine in canine plasma samples. *J. Chromatogr. B* **2004**, *806*, 271–276.

229. Kohlert, C.; Abel, G.; Schmid, E.; Veit, M. Determination of thymol in human plasma by automated headspace solid-phase microextraction-gas chromatographic analysis. *J. Chromatogr. B* **2002**, *767*, 11–18.

230. Spichiger, M.; Muhlbauer, R.C.; Brenneisen, R. Determination of menthol in plasma and urine of rats and humans by headspace solid phase microextraction and gas chromatography-mass spectrometry. *J. Chromatogr. B* **2004**, *799*, 111–117.

231. Abdel-Rehim, M.; Hassan, Z.; Blomberg, L.; Hassan, M. On-line derivatization utilizing solid-phase microextraction (SPME) for determination of busulphan in plasma using gas chromatography-mass spectrometry (GC-MS). *Ther. Drug Monit.* **2003**, *25*, 400–406.

232. Zambonin, C.G.; Aresta, A.; Palmisano, F. Determination of the immunosuppressant mycophenolic acid in human serum by solid-phase microextraction coupled to liquid chromatography. *J. Chromatogr. B* **2004**, *806*, 89–93.

233. Rodriguez, I.; Carpinteiro, J.; Quintana, J.B.; Carro, A.M.; Lorenzo, R.A.; Cela, R. Solid-phase microextraction with on-fiber derivatization for the analysis of anti-inflammatory drugs in water samples. *J. Chromatogr. A* **2004**, *1024*, 1–8.

234. Domeno, C.; Ruiz, B.; Nerin, C. Determination of sterols in biological samples by SPME with on-fiber derivatization and GC/FID. *Anal. Bioanal. Chem.* **2005**, *381*, 1576–1583.

235. Theodoridis, G.; Lontou, M.A.; Michopoulos, F.; Sucha, M.; Gondova, T. Study of multiple solid-phase microextraction combined off-line with high performance liquid chromatography. Application in the analysis of pharmaceuticals in urine. *Anal. Chim. Acta* **2004**, *516*, 197–204.

236. De Martinis, B.S.; Martin, C.C. Automated headspace solid-phase microextraction and capillary gas chromatography analysis of ethanol in postmortem specimens. *Forensic Sci. Int.* **2002**, *128*, 115–119.

237. Pragst, F.; Auwaerter, V.; Sporkert, F.; Spiegel, K. Analysis of fatty acid ethyl esters in hair as possible markers of chronically elevated alcohol consumption by headspace solid-phase microextraction (HS-SPME) and gas chromatography-mass spectrometry (GC-MS). *Forensic Sci. Int.* **2001**, *121*, 76–88.

238. Auwarter, V.; Sporkert, F.; Hartwig, S.; Pragst, F.; Vater, H.; Diefenbacher, A. Fatty acid ethyl esters in hair as markers of alcohol consumption. Segmental hair analysis of alcoholics, social drinkers, and teetotalers. *Clin. Chem.* **2001**, *47*, 2114–2123.

239. Takamoto, S.; Sakura, N.; Namera, A.; Yashiki, M. Monitoring of urinary acrolein concentration in patients receiving cyclophosphamide and ifosphamide. *J. Chromatogr. B* **2004**, *806*, 59–63.

240. Deng, C.; Zhang, J.; Zhang, W.; Zhang, X. Rapid determination of acetone in human plasma by gas chromatography-mass spectrometry and solid-phase microextraction with on-fiber derivatization. *J. Chromatogr. B* **2004**, *805*, 235–240.

241. Deng, C.; Zhang, J.; Yu, X.; Zhang, W.; Zhang, X. Determination of acetone in human breath by gas chromatography-mass spectrometry and solid-phase microextraction with on-fiber derivatization. *J. Chromatogr. B* **2004**, *810*, 269–275.

242. Deng, C.; Zhang, X.; Li, N. Investigation of volatile biomarkers in lung cancer blood using solid-phase microextraction and capillary gas chromatography-mass spectrometry. *J. Chromatogr. B* **2004**, *808*, 269–277.

243. Blount, B.C.; Kobelski, R.J.; McElprang, D.O.; Ashley, D.L.; Morrow, J.C.; Chambers, D.M.; Cardinali, F.L. Quantification of 31 volatile organic compounds in whole blood using solid-phase microextraction and gas chromatography-mass spectrometry. *J. Chromatogr. B* **2006**, *832*, 292–301.

244. Musshoff, F.; Padosch, S.A.; Kroener, L.A.; Madea, B. Accidental autoerotic death by volatile substance abuse or nonsexually motivated accidents? *Am. J. Forensic Med. Pathol.* **2006**, *27*, 188–192.

245. Poli, D.; Manini, P.; Andreoli, R.; Franchini, I.; Mutti, A. Determination of dichloromethane, trichloroethylene and perchloroethylene in urine samples by headspace solid phase microextraction gas chromatography-mass spectrometry. *J. Chromatogr. B* **2005**, *820*, 95–102.

246. Coran, S.A.; Giannellini, V.; Furlanetto, S.; Bambagiotti-Alberti, M.; Pinzauti, S. Improving gas chromatographic determination of residual solvents in pharmaceuticals by the combined use of headspace solid-phase microextraction and isotopic dilution. *J. Chromatogr. A* **2001**, *915*, 209–216.

247. Raghani, A.R. High-speed gas chromatographic analysis of solvents in pharmaceuticals using solid phase microextraction. *J. Pharm. Biomed. Anal.* **2002**, *29*, 507–518.

248. Frost, R.P.; Hussain, M.S.; Raghani, A.R. Determination of pharmaceutical process impurities by solid phase microextraction gas chromatography. *J. Sep. Sci.* **2003**, *26*, 1097–1103.

249. Legrand, S.; Dugay, J.; Vial, J. Use of solid-phase microextraction coupled with gas chromatography for the determination of residual solvents in pharmaceutical products. *J. Chromatogr. A* **2003**, *999*, 195–201.

250. Yang, R.; Xie, W. Determination of cannabinoids in biological samples using a new solid phase micro-extraction membrane and liquid chromatography-mass spectrometry. *Forensic Sci. Int.* **2006**, *162*, 135–139.

251. Musshoff, F.; Junker, H.; Madea, B. Simple determination of 22 organophosphorous pesticides in human blood using headspace solid-phase microextraction and gas chromatography with mass spectrometric detection. *J. Chromatogr. Sci.* **2002**, *40*, 29–34.

252. Tsoukali, H.; Theodoridis, G.; Raikos, N.; Grigoratou, I. Solid phase microextraction gas chromatographic analysis of organophosphorus pesticides in biological samples. *J. Chromatogr. B* **2005**, *822*, 194–200.

253. Gallardo, E.; Barroso, M.; Margalho, C.; Cruz, A.; Vieira, D.N.; Lopez-Rivadulla, M. Determination of parathion in biological fluids by means of direct solid-phase microextraction. *Anal. Bioanal. Chem.* **2006**, *386*, 1717–1726.

254. Tsoukali, H.; Raikos, N.; Theodoridis, G.; Psaroulis, D. Headspace solid phase microextraction for the gas chromatographic analysis of methyl-parathion in post-mortem human samples. Application in a suicide case by intravenous injection. *Forensic Sci. Int.* **2004**, *143*, 127–132.

255. Gallardo, E.; Barroso, M.; Margalho, C.; Cruz, A.; Vieira, D.N.; Lopez-Rivadulla, M. Determination of quinalphos in blood and urine by direct solid-phase microextraction combined with gas chromatography-mass spectrometry. *J. Chromatogr. B* **2006**, *832*, 162–168.

256. Sporkert, F.; Pragst, F.; Hübner, S.; Mills, G. Headspace solid-phase microextraction with 1-pyrenyldiazomethane on-fibre derivatisation for analysis of fluoroacetic acid in biological samples. *J. Chromatogr. B* **2002**, *772*, 45–51.

257. Cai, X.; Zhang, D.; Ju, H.; Wu, G.; Liu, X. Fast detection of fluoroacetamide in body fluid using gas chromatography-mass spectrometry after solid-phase microextraction. *J. Chromatogr. B* **2004**, *802*, 239–245.

258. Wooten, J.V.; Ashley, D.L.; Calafat, A.M. Quantitation of 2-chlorovinylarsonous acid in human urine by automated solid-phase microextraction-gas chromatography-mass spectrometry. *J. Chromatogr. B* **2002**, *772*, 147–153.

259. Frison, G.; Zancanaro, F.; Favretto, D.; Ferrara, S.D. An improved method for cyanide determination in blood using solid-phase microextraction and gas chromatography/mass spectrometry. *Rapid Commun. Mass Spectrom.* **2006**, *20*, 2932–2938.

260. Barroso, M.; Gallardo, E.; Margalho, C.; Avila, S.; Marques, E.P.; Vieira, D.N.; Lopez-Rivadulla, M. Application of solid phase microextraction to the determination of strychnine in blood. *J. Chromatogr. B* **2005**, *816*, 29–34.

261. Wu, J.; Lord, H.; Pawliszyn, J. Determination of stimulants in human urine and hair samples by polypyrrole coated capillary in-tube solid phase microextraction coupled with liquid chromatography-electrospray mass spectrometry. *Talanta* **2001** *54*, 655–672.

262. Kataoka, H.; Lord, H.L.; Pawliszyn, J. Simple and rapid determination of amphetamine, methamphetamine, and their methylenedioxy derivatives in urine by automated in-tube solid-phase microextraction coupled with liquid chromatography-electrospray ionization mass spectrometry. *J. Anal. Toxicol.* **2000**, *24*, 257–265.

263. Yuan, H.; Mester, Z.; Lord, H.; Pawliszyn, J. Automated in-tube solid-phase microextraction coupled with liquid chromatography-electrospray ionization mass spectrometry for the determination of selected benzodiazepines. *J. Anal. Toxicol.* **2000**, *24*, 718–725.

264. Mullett, W.M.; Levsen, K.; Lubda, D.; Pawliszyn, J. Bio-compatible in-tube solid-phase microextraction capillary for the direct extraction and high-performance liquid chromatographic determination of drugs in human serum. *J. Chromatogr. A* **2002**, *963*, 325–334.

265. Fan, Y.; Feng, Y.Q.; Da, S.L.; Gao, X.P. In-tube solid-phase microextraction with poly(methacrylic acid-ethylene glycol dimethacrylate) monolithic capillary for direct high-performance liquid chromatographic determination of ketamine in urine samples. *Analyst* **2004**, *129*, 1065–1069.

266. Saito, Y.; Kawazoe, M.; Hayashida, M.; Jinno, K. Direct coupling of microcolumn liquid chromatography with in-tube solid-phase microextraction for the analysis of antidepressant drugs. *Analyst* **2000**, *125*, 807–809.

267. Jinno, K.; Kawazoe, M.; Saito, Y.; Takeichi, T.; Hayashida, M. Sample preparation with fiber-in-tube solid-phase microextraction for capillary electrophoretic separation of tricyclic antidepressant drugs in human urine. *Electrophoresis* **2001**, *22*, 3785–3790.

268. Wu, J.; Lord, H.L.; Pawliszyn, J.; Kataoka, H. Polypyrrole-coated capillary in-tube solid phase microextraction coupled with liquid chromatography-electrospray ionization mass spectrometry for the determination of β-blockers in urine and serum samples. *J. Microcol. Sep.* **2000**, *12*, 255–266.

269. Mullett, W.M.; Martin, P.; Pawliszyn, J. In-tube moleculary imprinted polymer solid-phase microextraction for the selective determination of propranolol. *Anal. Chem.* **2001**, *73*, 2383–2389.

270. Walles, M.; Mullett, W.M.; Levsen, K.; Borlak, J.; Wunsch, G.; Pawliszyn, J. Verapamil drug metabolism studies by automated in-tube solid phase microextraction. *J. Pharm. Biomed. Anal.* **2002**, *30*, 307–319.

271. Fan, Y.; Feng, Y.-Q.; Da, S.-L.; Shi, Z.-G. Poly (methacrylic acid-ethylene glycol dimethacrylate) monolithic capillary for in-tube solid phase microextraction coupled to high performance liquid chromatography and its application to determination of basic drugs in human serum. *Anal. Chim. Acta* **2004**, *523*, 251–258.

272. Nie, J.; Zhang, M.; Fan, Y.; Wen, Y.; Xiang, B.; Feng, Y.-Q. Biocompatible in-tube solid-phase microextraction coupled to HPLC for the determination of angiotensin II receptor antagonists in human plasma and urine. *J. Chromatogr. B* **2005**, *828*, 62–69.

273. Zhang, M.; Wei, F.; Zhang, Y.-F.; Nie, J.; Feng, Y.-Q. Novel polymer monolith microextraction using a poly(methacrylic acid-ethylene glycol dimethacrylate) monolith and its application to simultaneous analysis of several angiotensin II receptor antagonists in human urine by capillary zone electrophoresis. *J. Chromatogr. A* **2006**, *1102*, 294–301.

274. Fan, Y.; Feng, Y.-Q.; Da, S.-L.; Wang, Z.-H. In-tube solid phase microextraction using a beta-cyclodextrin coated capillary coupled to high performance liquid chromatography for determination of non-steroidal anti-inflammatory drugs in urine samples. *Talanta* **2005**, *65*, 111–117.

275. Queiroz, M.E.C.; Oliveira, E.B.; Breton, F.; Pawliszyn, J. Immunoaffinity in-tube solid phase microextraction coupled with liquid chromatography-mass spectrometry for analysis of fluoxetine in serum samples. *J. Chromatogr. A* **2007**, *1174*, 72–77.

276. Kataoka, H.; Lord, H.L.; Yamamoto, S.; Narimatsu, S.; Pawliszyn, J. Development of automated in-tube SPME/LC/MS method for drug analysis. *J. Microcol. Sep.* **2000**, *12*, 493–500.

277. Kataoka, H.; Matsuura, E.; Mitani, K. Determination of cortisol in human saliva by automated in-tube solid-phase microextraction coupled with liquid chromatography-mass spectrometry. *J. Pharm. Biomed. Anal.* **2007**, *44*, 160–165.

278. Mitani, K.; Narimatsu, S.; Izushi, F.; Kataoka, H. Simple and rapid analysis of endocrine disruptors in liquid medicines and intravenous injection solutions by automated in-tube solid-phase microextraction/high performance liquid chromatography. *J. Pharm. Biomed. Anal.* **2003**, *32*, 469–478.

279. Mitani, K.; Izushi, F.; Kataoka, H. Analysis of phthalate contamination in infusion solutions by automated on-line in-tube solid-phase microextraction coupled with high-performance liquid chromatography. *J. Anal. Toxicol.* **2004**, *28*, 575–580.

280. Tienpont, B.; David, F.; Benijts, T.; Sandra, P. Stir bar sorptive extraction-thermal desorption-capillary GC-MS for profiling and target component analysis of pharmaceutical drugs in urine. *J. Pharm. Biomed. Anal.* **2003**, *32*, 569–579.

281. Lambert, J.-P.; Mullett, W.M.; Kwong, E.; Lubda, D. Stir bar sorptive extraction based on restricted access material for the direct extraction of caffeine and metabolites in biological fluids. *J. Chromatogr. A* **2005**, *1075*, 43–49.

282. Chaves, A.R.; Silva, S.M.; Queiroz, R.H.; Lanças, F.M.; Queiroz, M.E. Stir bar sorptive extraction and liquid chromatography with UV detection for determination of antidepressants in plasma samples. *J. Chromatogr. B* **2007**, *850*, 295–302.

283. Stopforth, A.; Burger, B.V.; Crouch, A.M.; Sandra, P. Urinalysis of 4-hydroxynonenal, a marker of oxidative stress, using stir bar sorptive extraction-thermal desorption-gas chromatography/mass spectrometry. *J. Chromatogr. B* **2006**, *834*, 134–140.

284. Crifasi, J.A.; Bruder, M.F.; Long, C.W.; Janssen, K. Performance evaluation of thermal desorption system (TDS) for detection of basic drugs in forensic samples by GC-MS. *J. Anal. Toxicol.* **2006**, *30*, 581–592.

285. Almeida, C.; Nogueira, J.M. Determination of steroid sex hormones in water and urine matrices by stir bar sorptive extraction and liquid chromatography with diode array detection. *J. Pharm. Biomed. Anal.* **2006**, *41*, 1303–1311.

286. Stopforth, A.; Grobbelaar, C.J.; Crouch, A.M.; Sandra, P. Quantification of testosterone and epitestosterone in human urine samples by stir bar sorptive extraction-thermal desorption-gas chromatography/mass spectrometry: Application to HIV-positive urine samples. *J. Sep. Sci.* **2007**, *30*, 257–265.

287. Kawaguchi, M.; Sakui, N.; Okanouchi, N.; Ito, R.; Saito, K.; Izumi, S.; Makino, T.; Nakazawa, H. Stir bar sorptive extraction with in situ derivatization and thermal desorption-gas chromatography-mass spectrometry for measurement of phenolic xenoestrogens in human urine samples. *J. Chromatogr. B* **2005**, *820*, 49–57.

288. Kawaguchi, M.; To, R.; Hayatsu, Y.; Nakata, H.; Sakui, N.; Okanouchi, N.; Saito, K.; Yokota, H.; Izumi, S.; Makino, T.; Nakazawa, H. Stir bar sorptive extraction with in situ de-conjugation and thermal desorption gas chromatography-mass spectrometry for measurement of 4-nonylphenol glucuronide in human urine sample. *J. Pharm. Biomed. Anal.* **2006**, *40*, 82–87.

289. Kawaguchi, M.; Ito, R.; Sakui, N.; Okanouchi, N.; Saito, K.; Seto, Y.; Nakazawa, H. Stir-bar-sorptive extraction, with in-situ deconjugation, and thermal desorption with in-tube silylation, followed by gas chromatography-mass spectrometry for measurement of urinary 4-nonylphenol and 4-tert-octylphenol glucuronides. *Anal. Bioanal. Chem.* **2007**, *388*, 391–398.

290. Musshoff, F.; Lachenmeier, D.W.; Kroener, L.; Madea, B. Automated headspace solid-phase dynamic extraction for the determination of cannabinoids in hair samples. *Forensic Sci. Int.* **2003**, *133*, 32–38.

291. Musshoff, F.; Lachenmeier, D.W.; Kroener, L.; Madea, B. Automated headspace solid-phase dynamic extraction for the determination of amphetamines and synthetic designer drugs in hair samples. *J. Chromatogr. A* **2002**, *958*, 231–238.

292. Lachenmeier, D.W.; Kroener, L.; Musshoff, F.; Madea, B. Application of tandem mass spectrometry combined with gas chromatography and headspace solid-phase dynamic extraction for the determination of drugs of abuse in hair samples. *Rapid Commun. Mass Spectrom.* **2003**, *17*, 472–478.

293. Abdel-Rehim, M. New trend in sample preparation: On-line microextraction in packed syringe for liquid and gas chromatography applications. I. Determination of local anaesthetics in human plasma samples using gas chromatography-mass spectrometry. *J. Chromatogr. B* **2004**, *801*, 317–321.

294. Altun, Z.; Abdel-Rehim, M.; Blomberg, L.G. New trends in sample preparation: on-line microextraction in packed syringe (MEPS) for LC and GC applications part III: Determination and validation of local anaesthetics in human plasma samples using a cation-exchange sorbent, and MEPS-LC-MS-MS. *J. Chromatogr. B* **2004**, *813*, 129–135.

295. Abdel-Rehim, M.; Dahlgren, M.; Blomberg, L. Quantification of ropivacaine and its major metabolites in human urine samples utilizing microextraction in a packed syringe automated with liquid chromatography-tandem mass spectrometry (MEPS-LC-MS/MS). *J. Sep. Sci.* **2006**, *29*, 1658–1661.

296. Vita, M.; Skansen, P.; Hassan, M.; Abdel-Rehim, M. Development and validation of a liquid chromatography and tandem mass spectrometry method for determination of roscovitine in plasma and urine samples utilizing on-line sample preparation. *J. Chromatogr. B* **2005**, *817*, 303–307.

297. Wells, D.A., Ed. High throughput bioanalytical sample preparation: Methods and automation strategies. In *Progress in Pharmaceutical and Biomedical Analysis*, Vol. 5. Amsterdam: Elsevier; 2003.

298. Fritz, J.S.; Masso, J.J. Miniaturized solid-phase extraction with resin disks. *J. Chromatogr. A* **2001**, *909*, 79–85.

299. Weller, M.G. Immunochromatographic techniques—A critical review. *Fresenius J. Anal. Chem.* **2000**, *366*, 635–645.

300. Hennion, M.C.; Pichon, V. Immuno-based sample preparation for trace analysis. *J. Chromatogr. A* **2003**, *1000*, 29–52.

301. Walker, V.; Mills, G.A. Solid-phase extraction in clinical biochemistry. *Ann. Clin. Biochem.* **2002**, *39*, 464–477.

302. Pyrzynska, K. Novel selective sorbents for solid-phase extraction. *Chem. Anal. Warsaw* **2003**, *48*, 781–795.

303. Poole, C.F. New trends in solid-phase extraction. *Trends Anal. Chem.* **2003**, *22*, 362–373.

304. Souverain, S.; Rudaz, S.; Veuthey, J.L. Restricted access materials and large particle supports for on-line sample preparation: an attractive approach for biological fluids analysis. *J. Chromatogr. B* **2004**, *801*, 141–156.

305. Wei, S.; Mizaikoff, B. Recent advances on noncovalent molecular imprints for affinity separations. *J. Sep. Sci.* **2007**, *30*, 1794–1805.

306. Dietz, C.; Sanz, J.; Camara, C. Recent developments in solid-phase microextraction coatings and related techniques. *J. Chromatogr. A* **2006**, *1103*, 183–192.

307. Fontanals, N.; Marce, R.M.; Borrull, F. New materials in sorptive extraction techniques for polar compounds. *J. Chromatogr. A* **2007**, *1152*, 14–31.

Chapter 17

Statistics of Sampling and Sample Preparation

Byron Kratochvil

The results of chemical analyses are relied upon worldwide to monitor product purity, evaluate human health, and check environmental quality. These analyses involve countless daily sampling and measurement operations, all subject to a variety of errors that can create serious and costly problems if not identified and controlled. That the usefulness of an analytical result depends on the quality and integrity of the sample taken for measurement is self-evident, but too often analysts and users of analytical information fail to ensure that samples are sufficiently representative of the population under study to provide valid results. Blanks, standards, and reference materials can help identify and correct potential sources of error during sample preparation and measurement, but cannot correct for an unrepresentative sample produced by a poorly designed sampling plan, or by improper execution of a good one.

This chapter presents basic elements of representative sampling for chemical analysis from a statistical point of view, with examples of statistical sampling designs for selected applications. Also discussed are statistical considerations related to the preparation of representative test portions of solids from field or laboratory samples. It concludes with a general discussion of practical considerations in sample collection and handling. Although the focus is on sampling of solids because of greater problems with heterogeneity, the principles are also applicable to liquids and gases as well as phase mixtures. Further reading and definitions of terms commonly used in sampling are also provided.

17.1. SAMPLING ERRORS

Sampling errors, like those of chemical analysis, may be classified as indeterminate (random) or determinate (systematic). Indeterminate errors vary nonreproducibly around the true value and may be treated statistically by the laws of probability. For measurements on a set of samples collected from a population, the overall estimated standard deviation s_o of the average $\bar{x}$ includes random errors generated during both sampling and analysis. This uncertainty may be expressed as the square root of the sum of the variances of the sampling and analytical operations:

$$s_o = \left(s_s^2 + s_a^2\right)^{1/2}, \quad (17.1)$$

where the values s_o, s_s, and s_a are experimentally determined estimates of the true standard deviations σ_o, σ_s, and σ_a. The value of the indeterminate sampling error s_s can only be obtained indirectly through the analysis of homogeneous reference materials or standards to determine s_a, followed by substitution into Equation 17.1.

From Equation 17.1, it can be seen that if either s_a or s_s is much smaller than the other, further reduction of the smaller one is of little help in reducing s_o. Accordingly, when s_s is large relative to s_a, the most cost-effective way of determining the average of a sought-for component in a population may be to analyze a larger number of samples using a rapid, simple analytical method of lower precision. Ways of optimizing this approach are given in a later section on cost considerations. Usually optimum efficiency is achieved when sampling and analytical precisions are about equal.

Determinate error, or *bias*, in an analytical result is a systematic, nonrandom error introduced during sampling and measurement. Bias may be the result of (1) failure to include portions of the population in the samples due to discriminatory factors in the sampling plan or sampling operations; (2) contamination or alteration of samples during collection, transport, storage, or preparation for analysis; or (3) systematic errors in the analytical method. Determinate errors tend to be reproducible from sample to sample and cause the results to be biased in one direction or the other from the true value. They cannot be treated using statistical

Handbook of Sample Preparation, Edited by Janusz Pawliszyn and Heather L. Lord

probabilities. The overall bias B_o is the algebraic sum of the sampling and analysis bias:

$$B_o = B_s + B_a. \qquad (17.2)$$

17.2. OVERVIEW OF THE TYPES OF SAMPLING

17.2.1. Probability Sampling

Probability sampling involves the hypothetical division of a target population into parts or units from which a specified number are chosen for analysis. The sample is selected in such a way that every part of the population has an equal chance of being selected. It is this known probability that permits the statistical projection of the results of analyses done on these units to the population as a whole. In unsegregated populations, a normal (Gaussian) distribution, for which the average and standard deviation provide the best estimates of the true value and its uncertainty, is usually assumed. Other population distributions, such as Poisson, binomial, and lognormal, are occasionally encountered but are sufficiently uncommon that they will not be considered here; information on their occurrence and statistical treatment is available in the literature (see, e.g., *Biometry*, by Sokal and Rohlf[1]).

17.2.2. Simple Random Sampling

The most basic form of probability sampling is simple *random* sampling. In this method, the population, which may be one-, two-, or three-dimensional, is divided into real or imaginary segments (or units for discrete objects). Each segment is assigned a number, and those to be sampled are picked by using a random number table or generator.* Figure 17.1a illustrates a typical random sampling pattern. Variants on this approach exist, but all have in common the opportunity for every segment or unit of the population to have an equal chance of being selected. When part of a well-designed and executed sampling plan, random sampling avoids the selection of inappropriate sampling sites through shortcuts or personal bias.

From a practical perspective simple random sampling has some disadvantages. One is it does not take into account past experience on similar populations, thereby requiring more samples than might be otherwise selected to attain a desired precision. This means additional work and expense. Another

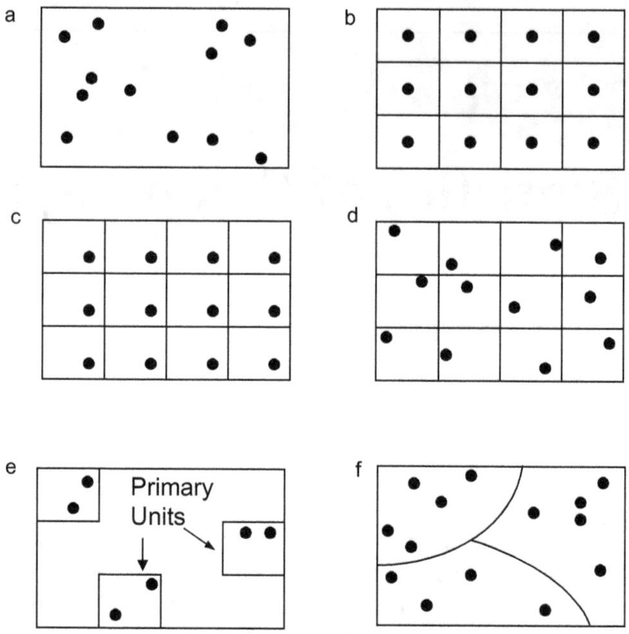

Figure 17.1. Some types of sampling designs.

is that it precludes the possibility of collecting samples from locations that good judgment indicates may be significant. Also, by chance, some parts of the population may be sampled little while others are heavily represented. The only remedy for these concerns with a strictly random sampling plan is to collect enough samples so that all parts are covered. The methods described below, when they include opportunity for all parts of the population to be selected, can reduce or eliminate most of these problems.

17.2.3. Systematic Sampling

In *systematic* sampling the samples are collected in a regular pattern throughout the population. To sample a surface, for example, a grid plan may be drawn up and samples are collected at the same position within each area (Fig. 17.1b). The spacing of the grid determines the number of sampling sites. The method is simple and straightforward, and avoids the possibility of missing significant sections of the population. In the systematic grid sampling of volumes, the population is divided by a three-dimensional grid into volume units, and samples are collected from each grid volume. Systematic sampling introduces little inherent bias and increases the probability of detecting possible hot spots in a population, but has the major disadvantage that statistical probability tests cannot be applied to the results because the sampling sites are not determined by chance.

Another use of systematic sampling is to test a hypothesis, such as changes in population composition with time or spatial location. Under these circumstances, each set of samples collected for a given set of conditions may be considered a separate discrete population, yet the results can still be tested statistically for the significance of apparent differences.

*Random numbers may be generated by a mathematical formula or extracted from a natural process such as radioactive decay or atmospheric noise. It has been shown that the number sequences produced by mathematical calculation are not completely random, so numbers based on a natural process are recommended. Two websites that provide random numbers involving natural sources are www.fourmilab.ch/hotbits/, which measures the times between successive pairs of beta decay events of cesium-137, and www.random.org, which measures differences in the amplitude of atmospheric white noise (radio static).

17.2.4. Systematic Sampling with Random Start Point

To circumvent the drawbacks of simple random and systematic sampling, a combination of the two is advantageous. By choosing the first sampling point within a systematic sampling plan by means of a random process, then collecting subsequent samples at the same relative position on the grid as shown in Figure 17.1c, all parts of the population have an equal chance of being selected. In this way, the simplicity and efficiency of the systematic design is retained, yet classical statistical tests can be used to evaluate the results.

Other ways of combining systematic and random sampling include random sampling within grid blocks, as illustrated in Figure 17.1d, and two-stage sampling (Fig. 17.1e), in which primary blocks or units are randomly selected within a population and two or more sample increments are taken from locations within each selected unit.

17.2.5. Stratified Sampling

Stratified random sampling involves the division of the population into sections, or strata (Fig. 17.1f). The number, size, and shape of the strata are selected on the basis of identifiable areas of known or expected uniformity in the substance of interest. Simple random sampling is performed within each stratum, and the overall estimate is the size-weighted average of the stratum estimates. The more uniform the analyte concentration within a stratum, the fewer samples are needed and the more efficient the sampling operation.

17.3. QUALITATIVE SAMPLING

When it is necessary to sample a population without prior information on which to base a sampling plan, or when poor precision in the result can be tolerated, judgment sampling may be used. *Judgment* samples are samples collected from a population on the basis of experience, intuition, and knowledge of the history or properties of the population. Sometimes, the goal is to obtain a single sample that may be termed "representative" to connote that it is expected to exhibit the average properties of the population. Other times, the purpose is to collect preliminary information on which to base a more rigorous sampling plan. While it is possible for a judgment sample to represent a population that is homogeneous, homogeneity should never be assumed. Where time or cost restricts a sampling operation, the shortcomings and the resulting limitations in data assessment should be clearly stated.

Judgment sampling requires assumptions about the degree to which the samples may be considered representative. For example, a series of judgment samples might be collected to assess the variability of analyte concentration in the population, or to identify local sites of high concentration. Advantages of this method include speed, simplicity, and lower cost. But since the quality of samples taken in this way depends on the knowledge and experience of the one making the judgment, the validity of the sample is necessarily problematic. Where circumstances do not permit implementation of a statistically valid sampling plan, a combination of judgment and random sampling may be considered as a compromise.

17.3.1. Protocol Sampling

Sometimes regulations or standard methods specify detailed sampling procedures, called *protocols*, that if not followed render the sample (and the resulting analysis) invalid for the intended purpose. Some protocols may not meet accepted general guidelines for sample quality, but have been shown adequate for a specific purpose. An example is sampling of highly complex environmental or biological populations for health or safety testing that, for reasons of time and expense, require an element of judgment in sample collection. Other protocols meet all established criteria for obtaining a truly representative sample that when followed produce samples of high quality. Regulatory agencies or groups such as the American Society for Testing and Materials (ASTM), the International Organization for Standardization (ISO), the American Water Works Association (AWWA), and the United States Environmental Protection Agency (USEPA) publish a wide variety of protocols that are widely used by manufacturers and commodity producers as a basis for quality control of products and for environmental and commercial decision making.

17.4. SAMPLING PLANS

Before sampling is initiated, it is necessary to develop a plan that defines clearly the population to be studied, the substance(s) to be measured, the precision required in the result, and the extent to which information on distribution within the population is needed. Any assumptions about the population should be clearly identified. The planning group should ideally include the client, those responsible for the sampling and analyses, and someone knowledgeable about statistics. The decisions made at this point establish the goals of the work and how to achieve them (Fig. 17.2); with this information in hand, a viable sampling plan can be drafted.

The variety of populations encountered in practical work mandates a customized sampling plan for most kinds of populations. If the analytical data involves compliance with environmental regulations, workplace safety, or commercial contracts, protocols provided by recognized associations or agencies such as those mentioned in the previous section should be used whenever available.

All valid plans for sampling and sample preparation have the following elements in common:

- clear statement of the objectives of the work, including the population to be sampled, the data to be collected, and the degree of accuracy and precision required;

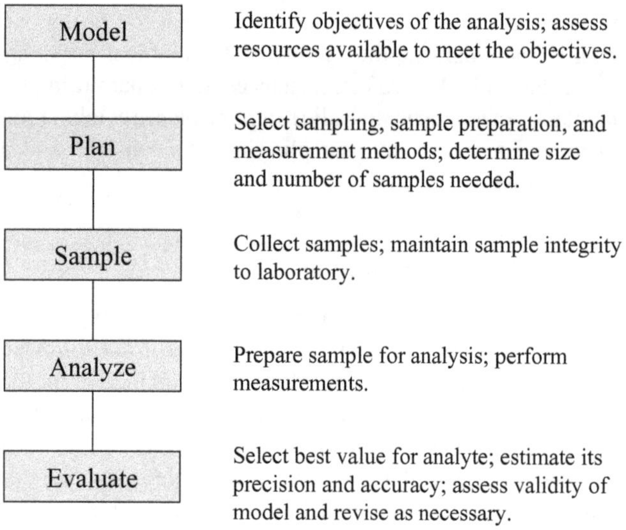

Figure 17.2. Sampling plan.

- selection of sampling design, including number and location of sampling sites, sample increment size, sampling apparatus, and extent of sample splitting or compositing to be performed;
- if needed, a procedure for presampling to assess the level of population heterogeneity;
- procedure for collection, labeling, storage, and transport of increments to the laboratory for analysis; and
- procedure for reduction of field increments to laboratory samples, and then to representative test portions for analysis.

The sampling plan should be in the form of a written protocol that includes detailed procedures for all sample collection. It should give criteria for rejection of collected material that is not to be treated as part of the population, specify the qualifications/training of those doing the sampling, and include details not only on the sampling operations but also on sample logging, labeling, and chain of custody procedures, the type and size of containers to be used, cleaning procedures for equipment and containers, any preservatives needed to stabilize the analyte, conditions of sample storage, and any necessary auxiliary information that might affect the results. Barcelona[2] provided an example of a checklist for this purpose. As mentioned previously, for new populations whose characteristics are not well known, it is advisable to collect and analyze a small preliminary set of samples, relying on experience and intuition to make them as representative as possible. On the basis of these data, a more efficient and cost-effective sampling plan can be prepared.

Ideally, the analyst should be involved not only in the development of the sampling protocol but also in the sampling itself. Though this is often impractical, the individual responsible for the analytical data should ensure that the sample collectors are properly trained and understand the importance of following the sampling plan and procedures carefully. Where appropriate, samplers should also be made aware of possible safety issues when working in potentially hazardous areas.

17.5. STATISTICAL SAMPLING OF SOLIDS

A large fraction of sampling uncertainties arise when sampling bulk solid materials, so it is not surprising to see a significant amount of effort spent on this topic. This section discusses several ways in which statistical principles can be applied to the sampling of particulate solids.

17.5.1. Local Heterogeneity in Particulate Materials

For a bulk solid consisting of particles of uniform composition, a single increment from any location in the population will be representative of the whole. If the particles vary in composition, a single increment may still be sufficient provided the particles are well mixed and the increment contains enough of each kind of particle to represent the average composition adequately. Well-mixed populations, though rarely encountered in nature, are often needed in commercial products such as drug formulations, where homogeneity may be critical for uniform dosage. Attainment of adequate homogeneity by operations such as grinding and sieving may also be necessary to assure representativeness before laboratory samples are taken from bulk samples, or test portions from laboratory samples.

17.5.2. Increment Size

Sometimes, the size of a bulk sample increment or test portion taken for analysis is based on convenience or on the analytical procedure, and not on ensuring that it is large enough to represent the population from which it was drawn. Significant sampling error may occur, even with well-mixed particulate solids, if too few particles are taken. The error may become especially significant if the component of interest is present only in a small fraction of the particles. The number of particles needed, and therefore the sample size, thus depends on both particle homogeneity and the precision desired in the result.

This problem may be handled in the following way. Consider a pharmaceutical mixture consisting of two kinds of particles, one is the active ingredient A and the other is an inactive matrix material I. Suppose a laboratory sample contains n_A particles of A and n_I particles of I. The probability p of drawing a particle of A from the sample is $n_A/(n_A + n_I)$, and of drawing a particle of I is $1 - p$, or $n_I/(n_A + n_I)$. Letting $q = 1 - p$, if n particles are drawn randomly from the population, the expected number of particles of n_A is np, with a standard deviation of

$$\sigma_s = (npq)^{1/2}. \tag{17.3}$$

A more general expression based on Equation 17.3 that includes variation in particle densities and composition is

$$n = pq\left[d_1 d_2 / d_{av}^2\right]^2 \left[100(P_1 - P_2)/(\sigma_s P_{av})\right]^2. \quad (17.4)$$

Here, d_1 and d_2 are the densities and P_1 and P_2 are the percentages of the substance of interest in the two kinds of particles, P_{av} is the overall average composition of the material, and σ_s is the percent relative standard deviation (sampling error) of the sampling operation. By incorporating the average densities and sizes of the two particle types, and assuming all particles to be spherical, the weight of n particles is given by

$$w = (4/3)\pi r_{av}^3 d_{av} n. \quad (17.5)$$

The curves in Figure 17.3, calculated from Equations 17.4 and 17.5, show the relation between the minimum weight of sample to be taken and the composition of a mixture of two types of spherical particles of density 3 and diameter 0.1 mm for sampling standard deviations of 0.1% and 1%. It can be seen that the *relative* difference in particle composition is the most important factor. This means that the minimum sample size needed to achieve a sampling standard deviation of, say, 0.1% for a given particle size and density is the same whether the analyte percentage is 100 and 0 or 0.1 and 0 in the two types of particles.

To illustrate the use of Figure 17.3, if 20% of the particles in a sample each contain 10% of the substance of interest, and the remaining 80% each contain 0% of that substance, from Figure 17.3A a sample weight of about 6 g would be required to hold the sampling error to 0.1%. If the remaining 80% each contained 1%, 5%, or 9% of the substance, the corresponding sample weights required would be 3, 0.2, and 0.003 g. Figure 17.3B shows that if the acceptable sampling standard deviation is increased 10-fold, the sample size required decreases a hundredfold. Note that the curves in Figure 17.3 can be used for any relative composition by substituting x% for 10% and $0.1x$, $0.5x$, and $0.9x$% for the curves corresponding to 1%, 5%, and 9%.

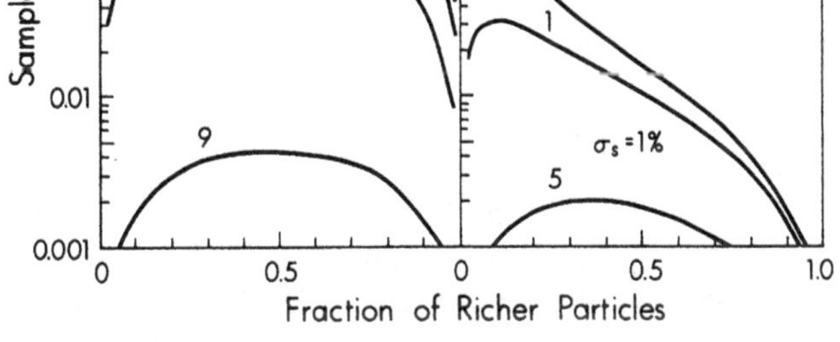

Figure 17.3. Relation between minimum sample size and fraction of richer particles in a mixture of two types of spherical particles (diameter 0.1 mm and density 3) for sampling standard deviation of (A) 0.1% and (B) 1%. Richer particles contain 10% of substance of interest, and leaner ones contain 0%, 1%, 5%, or 9%.[3,4] Adapted from reference 4 with permission from the American Chemical Society. Copyright 2001.

Figure 17.3 also shows that for mixtures of particles of similar composition (lower curves), the sample weight required is greatest when an approximately 1:1 mixture of the two types of particles is taken, but for dissimilar compositions (upper curves), the sample weight required increases as the fraction of particles richer in the substance of interest decreases. Note that the sample weight required is greatest in the region of a 1 to 1 mixture of the two types of particles, but that it increases greatly as the fraction of particles richer in the analyte decreases. This is why assays of gold and diamond ores require very large samples.

If the particles are not uniform in size, a mean effective particle mass can be calculated by sieving the material into several size classes and using the relation[5]

$$w = (P_1 - P_2)^2 (pq/\sigma_s^2)\left[y(\Sigma f_i m_i)_1 + x(\Sigma f_i m_i)_2\right], \quad (17.6)$$

where P_1, P_2, p, q, and w are as in Equation 17.4, f_i is the fraction by mass of the analyte in particle size class i (note that Σf_i = unity), and m_i is the mean mass of particles consisting of analyte in a given particle size class. Thus, $\Sigma f_i m_i$ is the mean effective particle mass of the analyte over all i particle size classes for a given particle type.

When the population consists of more than two types of particles, a more general equation that relates the relative sampling variance in analyte mass to the number of particles in the sample, the fractions of the different types of particles, and the mass and analyte concentrations of the individual particles is

$$\frac{\sigma_s}{w} = \frac{1}{2n}\left[\sum_{i=1}^{r}\sum_{j=1}^{r}\left(M_i c_i - M_j c_j\right)^2 p_i p_j\right]^{1/2} \Big/ \left[\sum_{i=1}^{r} p_i M_i c_i\right], \quad (17.7)$$

where n is the number of particles in the sample, M_i is the individual particle mass for type i, c_i is the concentration in terms of mass fraction of analyte in type i, n_i is the number of particles of type i, and p_i and p_j are the fractions of number of particles for types i and j. This equation, which is applicable to samples containing any number of particles, even if very small, was verified by sampling and analysis of mixtures of cereal grains for manganese, potassium, chlorine, and magnesium, as well as by Monte Carlo computer simulation.[6]

The impracticality of measuring relative fractions of analyte in different particle fractions has led to the development of indirect methods of relating sampling uncertainty to test portion size. A useful one is that of Ingamells and Switzer,[7,8] who showed that for a well-mixed population of particles of differing composition, such as a blended

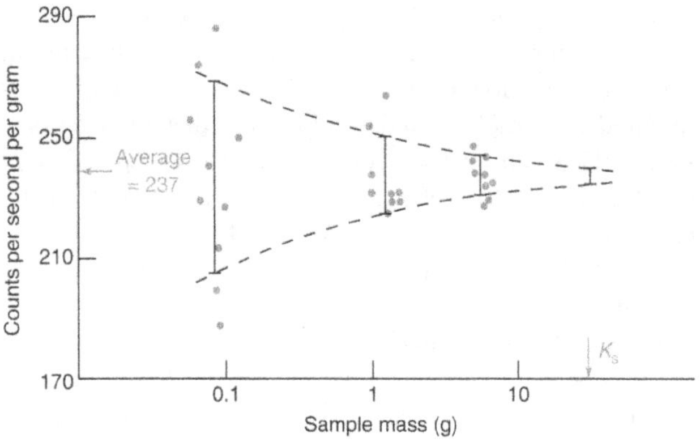

Figure 17.4. Sampling diagram of experimental results for [24]Na in liver homogenate.[9] Dots indicate experimental points, and error bars ±1 standard deviation about the mean. The value of the sampling constant K_s is about 36 g. Reprinted with permission from National Institute of Standards and Technology Internal Report NBSIR 80-2164.

laboratory sample, the sampling uncertainty is related to the size of the test portion taken for analysis by

$$wR^2 = K_s, \tag{17.8}$$

where w is the weight of the test portion, R is the relative standard deviation of the sampling operation in percent, and K_s is a constant equal to the weight of sample required to limit the sampling uncertainty to 1% relative at the 68% confidence level. K_s is best obtained by determining s_s for two or three sets of increments, each set consisting of test portions of different weight. An illustration of this approach is shown in Figure 17.4. With K_s known, the sample weight required for any desired relative standard deviation can be calculated. For poorly mixed or stratified populations, the value of K_s does not remain constant, but increases as w is increased. This provides a way of testing for segregation, which if present must be taken into account when sampling.

17.5.2.1. Number of increments to be collected. For a Gaussian distribution, the confidence interval within which the true value μ lies is given by

$$\mu - X \pm \left(ts/\sqrt{n_s}\right), \tag{17.9}$$

where X is the average of n measurements, s is the standard deviation, and t is a tabulated value for $N - 1$ samples at a specified probability level. By defining the maximum acceptable sampling uncertainty E_s as the difference between the true value μ and the average X of n measurements,

$$E_s = |\mu - X| = ts/\sqrt{n_s}, \tag{17.10}$$

which on rearrangement gives the number of increments to be collected

$$n_s = (ts_s/E_s)^2. \tag{17.11}$$

Initially, t may be set at 1.96 for 95% confidence limits and a first estimate of n calculated, using a value for s_s from either past work on the population or from a set of preliminary samples. The value of t for this n, obtained from a statistical t-table* is then substituted into Equation 17.11 and the system iterated to constant n. A confidence level of 95% is often used because it provides a balance between an acceptable degree of uncertainty in the result and the cost in time and labor of sample collection.

17.5.3. Sample Increment Size in Segregated Populations

Visman[10] developed an empirical expression relating sampling variance σ_s to the weight w and number of sample increments n collected from mined coal:

$$\sigma_s^2 = (A/w_{\text{sum}}n) + (B_{\text{seg}}/n). \tag{17.12}$$

Here, w_{sum} is the total weight of n increments of the sample, A is a constant related to local heterogeneity, and B_{seg} is constant related to the degree of segregation of the population. Values of A and B_{seg} may be obtained experimentally from the bulk population in two ways. In the first, two sets of sample increments, one small and the other large, are collected (a minimum 10-fold difference in weight between the two is recommended to give sufficient precision). Each set is analyzed, and its sampling variance is calculated. The variances are then substituted into Equation 17.12 to give two equations that can be solved simultaneously to give values for A and B_{seg}. In the second method, proposed by Duncan and Visman,[11,12] pairs of increments are collected at random locations in the population, each pair being of the same size and taken from adjacent sites in the population. From the analytical data on the increment pairs, an intraclass correlation coefficient, r, is calculated, either directly or by analysis of variance (ANOVA). Values for A and B_{seg} are then calculated from Equation 17.12 and the relation $r = B_{\text{seg}}m/A$, where m is the average particle mass. Note that increasing either w or n reduces uncertainty due to random variability, but only increasing the number of increments reduces uncertainty due to segregation. An evaluation of the Visman equation to the sampling of Athabasca oil sand for the determination of bitumen, water, and solids[13] showed it to be applicable to this heterogeneous mixture of particulate and nonparticulate materials. The two-increment-size approach gave better results than the use of the intraclass correlation coefficient relation for the calculation of the constants A and B_{seg}.

The random component in all the sampling equations discussed in this section was derived assuming a normally distributed population. As mentioned earlier, not all populations follow a Gaussian distribution. Procedures for testing the normality of data, and for dealing with non-normality by data transformation or use of distribution functions, are available in the statistical literature. Almost all populations in nature have elements of segregation or stratification that

*t-tables may be found in most introductory books on statistics and analytical chemistry as well as in books of statistical tables and on the Internet.

must be taken into account in sampling. Thus, Visman developed Equation 17.12 to obtain a better way of estimating the variance in coal. Sometimes, segregation of the substance of interest may be severe. Examples include sampling ores for gold or diamonds, contaminated industrial sites for toxic deposits, and extremely small surface areas for trace contamination with high-resolution analytical instrumentation. Accordingly, all sampling plans should consider the possibility of hot spots and the potential effect on results if they are missed. There is the danger, too, of identifying a high analytical result due to a hot spot as an outlier in the measurement operations and rejecting it.

17.5.4. Sampling Sites

As mentioned earlier, all parts of a population must have an equal chance of being selected for analysis if the resulting data are to be considered representative of the population. To provide this equal chance of selection requires a random element in the sampling design. Strategies to meet this requirement include, in addition to simple random sampling, systematic grid sampling utilizing a random initial start point, and random sampling within specified grid units. To improve sampling efficiency, other schemes, including stratified and two-stage sampling, have been developed (Fig. 17.1). The target population may be one-dimensional (sections along a drill core, objects on a production line), two-dimensional (surfaces of solids or liquids), or three-dimensional (geological deposits, water bodies, smokestack plumes, etc.), so any of a variety of sampling plans may need to be devised to address the type of population encountered and to provide samples that can answer the questions under study with the needed precision in a timely and cost-effective way.

17.5.4.1. Model-based sampling. The sampling equations discussed in previous sections are all based on classical sampling theory.[14] This approach, sometimes called design-based sampling, makes no assumptions about the population other than that it is unchanged during sampling. Many sampling methodologies and statistical tools have been developed to handle various population distributions within this framework.

A second approach, termed model-based sampling, employs one of several models to describe variability within a population. This methodology is most developed in geostatistics. Borgman et al.[15] proposed that, since the model-based approach treats randomness as an inherent property of a population, pure random sampling is no longer required and, in fact, may not be desirable because regularly spaced observations usually provide the best information about the degree of randomness present. A drawback is that the model must include information on expected patterns of variability within the population, though these patterns need not be completely understood to achieve reliable results.

The biggest applications of model-based sampling have been for geostatistical estimations of underground ore reserves, but the method has also been applied to environmental studies.[16] A widely used form, called *kriging*, assumes a linear trend in concentration of the sought-for element.

Gy developed a sampling approach that includes elements of model design.[17] Although Gy employs classical random sampling statistics, he systematically considers all possible errors that might be encountered in the collection of a valid sample, including population variability, prior to sampling. In effect, all uncertainties that may affect the representativeness of samples are incorporated into the sampling design rather than assuming randomness to be the only source of variability.

17.5.4.2. Balancing economic factors against sample quality. Sampling may be expensive, especially in terms of the time commitment required for proper collection by qualified personnel. As a result, sampling plans often end with compromises in data quality. In some circumstances, the cost of determining a population mean within a desired variance may be minimized by using the approach of Marcuse.[18]

Suppose a stratified sampling design has been formulated consisting of n_1 strata, with n_2 samples from each stratum and n_3 analytical measurements on each sample. For strata equal in size and in variance, for the analyte of interest, the total cost of the operation, c, is the sum of the costs of selecting the strata c_1, sampling within the strata c_2, and performing the analyses c_3:

$$c = n_1c_1 + n_1n_2c_2 + n_1n_2n_3c_3. \tag{17.13}$$

The overall variance for the population may be expressed as the sum of the variance contributions from the two stages of sampling and the analyses by

$$\sigma_o^2 = (\sigma_1^2/n_1) + (\sigma_2^2/n_1n_2) + (\sigma_3^2/n_1n_2n_3). \tag{17.14}$$

The values of n_1, n_2, and n_3 to be used to minimize the total cost of a preselected overall variance are given by

$$n_1 = \left[(\sigma_1^2/c_1)^{1/2}/\sigma_o^2\right]\left[(\sigma_1^2c_1)^{1/2} + (\sigma_2^2c_2)^{1/2} + (\sigma_3^2c_3)^{1/2}\right], \tag{17.15}$$

$$n_2 = (\sigma_2^2c_1/\sigma_1^2c_2)^{1/2}, \tag{17.16}$$

$$n_3 = (\sigma_3^2c_2/\sigma_2^2c_3)^{1/2}. \tag{17.17}$$

Note that the optimum allocation of sampling effort after the first stage is independent of the desired overall variance. This means that when the goal is reduction in overall variance at minimum cost, the number of strata sampled should be increased and the other steps kept unchanged.

Similarly, for a fixed total cost, Marcuse showed that the optimum value for n_1 is given by

$$n_1 = c(\sigma_1^2/c_1)^{1/2}/\left[(\sigma_1^2/c_1)^{1/2} + (\sigma_2^2/c_2)^{1/2} + (\sigma_3^2/c_3)^{1/2}\right], \tag{17.18}$$

while the optimum values for n_2 and n_3 remain those given by Equations 17.16 and 17.17. Thus, the optimum allocation beyond the first stage is the same for fixed total cost as for fixed total variance. The same principles can be applied to any number of stages in a nested sampling design.

If strata vary in size or analyte distribution pattern, appropriate weighting factors can be incorporated into the above expressions.

17.5.5. Subsampling of Solid Samples

Solid samples are usually subsampled to provide laboratory samples, and laboratory samples in turn are typically subsampled to obtain test portions for analysis. In both instances, the subsampling operation must be done so as to produce fractions that accurately represent the composition of the material under study. This means particle size reduction by crushing or milling may be necessary. The time and effort depends on the heterogeneity and properties of the material, but must not introduce contamination or otherwise change the composition of the sample. If a sampling plan calls for bulk samples to be subsampled in the field, trained, committed personnel should do the operations. A thorough mixing should follow all particle reduction operations prior to sample withdrawal. For liquids or suspensions, a thorough mixing followed by immediate sample withdrawal is usually sufficient.

17.5.6. Composite Samples

Sometimes, sample increments taken from a population are combined to produce a *composite* sample that is defined as representative of the population. Advantages of compositing include reduced sample handling and analytical effort. A composite sample provides an estimate of the average concentration of the analyte in the population, but not of its distribution. A variety of sampling systems and mixing procedures have been developed to produce composites from both solid and liquid materials. Compositing of increments becomes more attractive when the costs of the analytical measurements exceed those of sampling to a significant extent, or when the population is known to be relatively homogeneous.

The main disadvantage of compositing is the loss of information on analyte distribution within the population. This can be especially significant if sites of high concentration ("hot spots") may be present. Garner and coauthors discussed the advantages and limitations of composite sampling for environmental monitoring.[19]

17.6. QUALITY ASSURANCE IN SAMPLING[20,21]

17.6.1. Overall Objectives

The goal of a quality assurance program in sampling is to identify and assess all sources of error in the sampling operations and to set up ways to reduce them to acceptable levels. Quality assurance is achieved through quality control and quality assessment.

17.6.2. Quality Control

Quality control involves a system of testing and corrective actions that allows, through inspection, an estimate of the quality of the results. The goal is to attain a level of data quality that is adequate for the purpose, dependable, and economical. The control system should specify whether changes are needed and, if they are, what measures should be taken to maintain a predetermined level of quality.

Factors important to establishing and maintaining control of sample quality include the following:

- a clear, complete, and detailed sampling protocol that includes well-defined criteria for on-site decisions such as rejection of foreign material;
- clean, well-maintained sampling tools and sampling containers appropriate for the purpose;
- a sample management system designed to ensure sample quality and integrity from collection through analysis; and
- trained, knowledgeable, and motivated samplers who understand and appreciate the need for producing samples according to the specified protocol.

17.6.3. Sampling Equipment

An important factor in the success of any sampling operation is the design and quality of the equipment used to collect and process samples. Devices used for sampling, for division of samples into smaller fractions, and for mixing should be as simple as possible for ease of sample collection, removal, and cleaning. The design should protect the integrity of the sample by minimizing bias or loss of components during use. The material of construction should be noncontaminating, durable, and easily cleaned. The wide range of materials that need to be sampled has led to the development of many specialized kinds of sampling apparatus. For example, Koerner discussed the selection and composition of equipment for environmental sampling.[22]

17.6.4. Handling and Storage of Samples

Samples, if not handled properly, may undergo a variety of chemical or physical changes during collection, transport, storage, and preparation for analysis. Possible changes that may affect analytical results include loss of sample components through volatilization, chemical reactions between sample components, and reactions with sampling tools, containers, and transfer lines. Other changes may include reactions of sample components with atmospheric oxygen, carbon dioxide, or water. Decomposition may be accelerated through microbial action or high temperatures during transport or storage.

Sample stability is enhanced by collection with clean sampling tools made of inert materials, immediate transfer to inert containers with tight caps, and transport and storage under controlled conditions. Where sample decomposition is a problem, preservatives that will not interfere with subsequent analyses may be needed. The best sample preservation method is storage at low temperatures in Teflon or Tedlar containers.

Samples likely to undergo change should be analyzed as soon as feasible after collection. If digestion or extraction is required prior to measurement, consideration should be given to performing these operations promptly after collection. More stable processed samples can be stored until convenient for measurement.

Procedures for sample collection, preservation, and storage are available from sources such as the USEPA for sampling of the environment, and ASTM and ISO for industrial and commercial materials.

17.6.5. Quality Assessment

Quality assessment involves continuous monitoring and evaluation of the quality control program to ensure that its goals are being met. For analytical measurements, this may be done through test samples, interlaboratory comparisons, control charts, and so on, but for sampling, it is not so simple. Monitoring of a sampling program may require that multiple increments be collected from both adjacent and widely spaced sites across the population to ensure representativeness and provide backup in the event of sample loss. It may also require the use of field blanks and spiking of samples in the field to detect bias from sample contamination, loss, or alteration. Maintaining of sample integrity through appropriate use of preservatives, containers, storage conditions, labeling, and logging should be regularly checked.

A typical quality assessment program might incorporate the following activities on a scheduled basis:

- collection and comparison of data on replicate samples;
- external audit of sampling procedures and their execution in the field, including appropriate safety precautions;
- review of sampling protocols, sample documentation procedures, and record keeping; and
- thorough and objective feedback to those involved in sampling operations.

Overall, the aims of quality assurance in sampling are to provide a mechanism to reduce sampling errors to acceptable limits, the means to assure the mechanism is operative, and the means to assure that the samples have a high probability of being of acceptable quality. Achieving these aims requires constant attention and maintenance, but with regular monitoring and review, a well-designed and implemented quality assurance program will ensure the quality of sampling operations indefinitely.

Glossary of Selected Terms Used in Sampling

Batch. A quantity of material that is known or assumed to be produced under uniform conditions. See Lot.

Composite sample. A sample composed of two or more increments collected from different locations within a population or from the same location more than one time.

Consignment. A quantity of material transferred on one occasion and covered by a single set of shipping documents. It may consist of one or more lots or portions of lots.

Determinate error. Sampling errors that tend to bias a result in one direction. They are usually reproducible from sample to sample in a population.

Field sample. A sample collected from a population. It may consist of a single increment, or a composite of several increments.

Grab sample. Also called convenience sample, it is a sample collected from a population on the basis of accessibility, expediency, cost, efficiency, or other reason not directly concerned with sampling parameters.

Homogeneity. The degree to which a property or substance is uniformly distributed throughout a material. Homogeneity depends on the size and composition of the units under consideration for a given analyte or property. The degree of heterogeneity (the opposite of homogeneity) is the determining factor of sampling error.

Hot spot. A localized part of the population in which the analyte is present in significantly higher concentration than elsewhere.

Increment. A portion of material, collected by a single operation of a sampling device, from parts of a population separated in time or space. Increments may be analyzed either individually or combined and tested as a composite sample, depending on whether it is the variability of a property throughout the population, or only the average, that is of interest.

Laboratory sample. A sample, intended for submission to a laboratory for testing or analysis. For inhomogeneous materials, its preparation from a primary sample often requires subdivision, mixing, drying, particle size reduction, or other operations.

Lot. A quantity of individual units or bulk material of similar composition that is assumed to be a single population for sampling purposes. A lot may consist of one or more batches.

Lot sample. A primary sample collected from a lot for analysis, testing, or archival purposes.

Population. A general term used for the quantity of parent material being sampled; it may consist of bulk material, individual items, or events in the broadest sense. It is used as a general term for the quantity of parent material being sampled when it is immaterial if the parent material is a consignment, lot, field, and so on.

Primary sample. A collection of one or more increments or units initially taken from a population. The primary sample may become a laboratory sample directly, or may undergo pretreatment and/or subsampling prior to being sent to the laboratory. Use of the terms bulk, gross, or lot sample to denote a primary sample are no longer recommended.

Random sample. A sample so selected that any portion of the population has an equal (or known) chance of being chosen.

Reduction. The process of preparing one or more representative subsamples from a sample.

Referee sample. Also called umpire sample or reserve sample, it is a sample collected, prepared, and stored in an agreed upon manner for the purpose of settling a dispute.

Replicate samples. Two or more samples collected from a population in an identical manner at the same time and place. A replicate sample consisting of two portions is called a duplicate sample. Replicates are usually used to estimate sample variability at the local level.

Representative sample. A sample collected from a population in a manner that ensures, to the extent possible, accurate representation of the population, or subset of the population, from which it was taken. A representative sample may be a random sample or, for example, a stratified sample, depending on the objective of the sampling and the properties of the population.

Riffling. The separation of a free-flowing sample into parts (usually equal) by means of a mechanical device composed of diverter chutes.

Sample. A portion of material collected from a larger quantity of material, usually to obtain information on its properties or for archival purposes.

Sampling error. The error associated with using only a fraction of the population and extrapolating to the whole, as distinct from analytical or test error. Sampling error arises from a lack of homogeneity in the parent population.

Sampling plan. A predetermined procedure for the selection, withdrawal, preservation, transport, and preparation of portions to be removed from a population as samples.

Sampling protocol. A detailed written description of the steps and procedures to be followed for the collection of valid samples.

Specimen. In analytical sampling, a portion of material taken from a dynamic system that is assumed to be representative of the parent material at the time taken. A specimen is taken primarily in time rather than in space, and therefore may not be reproducible if taken from a flowing stream or as a portion of blood.

Spiked sample. A sample to which has been added a known quantity of the analyte to test the extent of interference by the matrix with the analytical measurement.

Split sample. A sample divided into two or more representative parts for independent analysis or testing.

Strata. Specifically demarked subparts of a population, either actual or hypothetical, that may vary with respect to the property under study.

Stratified sample. A sample taken randomly from a stratum, or a sample consisting of portions obtained from identified strata of the parent population.

Subsample. A representative portion taken from a sample. A laboratory sample may be a subsample of a bulk sample; similarly, a test portion may be a subsample of a laboratory sample.

Test portion. Also called analytical portion, test unit, or aliquot, it is the quantity of a material of appropriate size for measurement of the property of interest. Test portions may be taken from a bulk sample directly, but preliminary operations such as mixing or further reduction in particle size are often necessary to ensure representativeness. See also Specimen.

Unit. Also called item or individual, it is a discrete, identifiable portion of material suitable for removal from a population as a sample or as a portion of a sample, and which can be individually examined or tested, or combined. When sampling bulk materials, the units are defined as the increments created by the sampling device.

Symbols

- A — a constant in the Visman equation related to the local heterogeneity of an analyte in a population
- B — determinate error, or bias; systematic, nonrandom error in a set of data
- B_o — overall determinate error
- B_s — determinate error in sampling
- B_a — determinate error in analytical measurement
- B_{seg} — a constant in the Visman equation related to the degree of segregation present in a population
- c — total cost of an analytical set of data
- c_i — concentration of analyte of type i expressed as mass fraction
- c_1 — cost of selecting strata in a stratified sampling plan
- c_2 — cost of sampling within strata in a stratified sampling plan
- c_3 — cost of performing analyses in a stratified sampling plan
- E_s — sampling uncertainty, defined as the difference between the true value μ and the average X of a set of measurements
- f_i — fraction by mass of an analyte in particle size class i
- g — sample mass
- K_s — sampling constant in the Ingamells subsampling equation. K_s is equal to the weight of sample required to limit sampling uncertainty to 1% relative at the 68% confidence level
- M_i — individual mass of a particle of type i in a sample
- m_i — mean mass of analyte particles of a given size class in a sample

m average mass of particles in a sample
n number of particles in a sample, or number of sample increments collected
n_i number of particles in a sample of type i
n_1 number of strata in a stratified sampling plan
n_2 number of samples collected within each stratum in a stratified sampling plan
n_3 number of analyses performed per sample in a stratified sampling plan
p_i fraction of particles in a sample of type i
p_j fraction of particles in a sample of type j
P_1 percentage of the first of two types of particles in a population
P_2 percentage of the second of two types of particles in a population
R the relative standard deviation of a sampling operation in percent
r intraclass correlation coefficient
s_o experimentally determined estimate of sigma, the true overall standard deviation in an analytical result
s_s experimentally determined estimate of standard deviation of sampling operations
s_a experimentally determined estimate of the standard deviation of analytical measurements on a sample
W the weight of a test portion taken for analysis
w the total weight of sample collected from a population, equal to the sum of the weights of all sample increments
X (also x) average concentration of an analyte in a population or sample thereof; usually considered the best estimate of the true concentration
y the fraction by mass of the second of two types of particles in a population
μ true concentration of an analyte in a population
σ_o true standard deviation of a population
σ_s true standard deviation of a sampling operation
σ_a true standard deviation of an analytical measurement
s_P true standard deviation of the analyte concentration in a sample in percent
s_1 true standard deviation of the between-strata sampling in a stratified sampling plan
s_2 true standard deviation of the within-strata sampling in a stratified sampling plan
s_3 true standard deviation of the analytical measurement in a stratified sampling plan

REFERENCES

1. Sokal, R.R.; Rohlf, F.J. *Biometry*, 3rd ed. New York: Freeman; 1995.
2. Barcelona, M.J. Overview of the sampling process. In *Principles of Environmental Sampling*, Keith, L.H., Ed., 2nd ed. Washington, DC: American Chemical Society; 1996; pp. 41–61.
3. Harris, W.E.; Kratochvil, B. *An Introduction to Chemical Analysis*. New York: Saunders; 1981.
4. Harris, W.E.; Kratochvil, B. Sampling variance in analysis for trace components in solids. Preparation of reference samples. *Anal. Chem.* **1974**, *46*, 313–315.
5. Zheng, L.; Kratochvil, B. Combined Benedetti-Pichler/Stange-Poole sampling equation for two-component particulate mixtures. *Analyst* **1996**, *121*, 163–168.
6. Gao, Z.; Kratochvil, B.; Duke, M.J.M. Uncertainty in analyte mass for samples containing small numbers of particles. *Analyst* **2001**, *126*, 947–952.
7. Ingamells, C.O.; Switzer, P. A proposed sampling constant for use in geochemical analysis. *Talanta* **1973**, *20*, 547–568.
8. Ingamells, C.O.; Switzer, P. Derivation of the sampling constant equation. *Talanta* **1976**, *23*, 263–264.
9. Kratochvil, B.; Taylor, J.K. Sampling for chemical analysis. *Anal. Chem.* **1981**, *53*, 924A–938A.
10. Visman, J. A general sampling theory. *Mat. Res. Stds.* **1969**, *9*, 9–13, 51–56, 62–64.
11. Visman, J.; Duncan, A.J.; Lerner, M. Further discussion: A general theory of sampling. *Mat. Res. Stds.* **1971**, *11*, 32.
12. Visman, J. Discussion 3: A general theory of sampling. *J. Mat.* **1972**, *7*, 345–350.
13. Wallace, D.; Kratochvil, B. Visman equations in the design of sampling plans for chemical analysis of segregated bulk materials. *Anal. Chem.* **1987**, *59*, 226–232.
14. Cochran, W.G. *Sampling Techniques*. New York: Wiley; 1977.
15. Borgman, L.E.; Kern, J.W.; Anderson-Sprecher, R.; Flatman, G.T. The sampling theory of Pierre Gy: Comparisons, implementation, and applications for environmental sampling. In *Principles of Environmental Sampling*, Keith, L.H., Ed., 2nd ed. Washington, DC: American Chemical Society; 1996; pp. 203–221.
16. Flatman, G.T.; Yfantis, A.A. Geostatistical sampling designs for hazardous waste sites. In *Principles of Environmental Sampling*, Keith, L.H., Ed., 2nd ed. Washington, DC: American Chemical Society; 1996; pp. 779–801.
17. Pitard, F.F. *Pierre Gy's Sampling Theory and Sampling Practice. Vol. 1, Heterogeneity and Sampling*. Boca Raton FL: CRC Press; 1989.
18. Marcuse, S. Optimum allocation and variance components in nested sampling with an application to chemical analysis. *Biometrics* **1949**, *5*, 189–206.
19. Garner, F.C.; Stapanian, M.A.; Williams, L.R. Composite sampling for environmental monitoring. In *Principles of Environmental Sampling*, Keith, L.H., Ed., 2nd ed. Washington, DC: American Chemical Society; 1996; pp. 679–690.
20. Taylor, J.K. *Quality Assurance of Chemical Measurements*. Chelsea, MI: Lewis Publishers; 1987.
21. Kulkarni, S.V.; Bertoni, M.J. Environmental sampling quality assurance. In *Principles of Environmental Sampling*, Keith, L.H., Ed., 2nd ed. Washington, DC: American Chemical Society; 1996; pp. 111–137.
22. Koerner, C.E. Effect of equipment on sample representativeness. In *Principles of Environmental Sampling*, Keith, L.H., Ed., 2nd ed. Washington, DC: American Chemical Society; 1996; pp. 155–168.

FURTHER READING

ASTM International Technical Standards. ASTM International, known as the American Society for Testing and Materials until 2005, provides a limited number of technical standards for sampling materials and products such as coal, iron, steel, and soils, as well as guidance on preservation and transport of samples and on sampling equipment. Information is available at www.astm.org.

ISO Technical Standards. The International Organization for Standardization (ISO), a network of the national standards institutes of 157 countries coordinated by a central secretariat in Geneva, Switzerland, produces consensus technical standards for business, trade, and regulatory use. Over 300 ISO standards involve a sampling component. Available standards and associated information can be found at www.iso.org.

Schweitzer, G.E.; Santolucito, J.A., Eds. *Environmental Sampling for Hazardous Wastes*, ACS Symposium Series 267. Washington, DC: American Chemical Society; 1984. Coverage of general aspects of sampling statistics and quality assurance as well as specific sampling problems in industry and in the environment.

Keith, L.H., Ed. *Principles of Environmental Sampling*, 2nd ed. Washington, DC: American Chemical Society; 1996. An excellent source of information on planning and design of sampling programs, statistical sampling, and quality assurance. Includes discussion of sampling specialized components of the environment such as air, water, biota, solids, and hazardous wastes.

Keith, L.H. *Environmental Sampling and Analysis, A Practical Guide*. Chelsea, MI: Lewis Publishers; 1991. A guidance manual on how to collect and analyze samples to produce quality data.

Pitard, F.F. *Pierre Gy's Sampling Theory and Sampling Practice, Vol. 1, Heterogeneity and Sampling, Vol. 2, Sampling Correctness and Sampling Practice*. Boca Raton, FL: CRC Press; 1989. A detailed presentation of the inclusive approach to sampling developed by Pierre Gy over his lifetime.

Gy, P. *Sampling for Analytical Purposes*. New York: Wiley; 1998. A brief, readable book summarizing the philosophy behind the systematic sampling methodology developed by the author.

Taylor, J.K. *Quality Assurance of Chemical Measurements*. Chelsea, MI: Lewis Publishers; 1987. A highly readable treatment of the basic concepts of quality assurance and how to plan and implement a quality assurance program for chemical analysis, including discussion of the importance of sampling in the overall process.

Chapter 18

SPME Devices Integrating Sampling with Sample Preparation for On-Site Analysis

Gangfeng Ouyang

18.1. INTRODUCTION

The sample preparation step in an analytical process typically consists of an extraction procedure that results in the isolation and enrichment of target analytes from a sample matrix. Proper design of on-site extraction devices and procedures reduces the use of solvent, and combines sampling, isolation, and enrichment into one step. Quantification is based on the amount of analyte extracted under appropriate conditions by using suitable calibration method.

On-site sampling can be classified as either spot (grab) sampling or continuous sampling. On-site analysis requires portable and solvent-free sample preparation approaches, which work well with micro-instruments.[1] Solid-phase microextraction (SPME) was developed to address the need for rapid sampling and sample preparation, both in the laboratory and on-site.[2] It is a solvent-free sample preparation technique and combines sampling, isolation, and enrichment into one step. In this approach, microquantities of the solid sorbent or liquid polymer in appropriate format are exposed to the sample. Quantification is based on the amount of analyte extracted at appropriate conditions.

It should be noted that SPME was originally named after the first experiment using an SPME device, which involved extraction on solid fused-silica fibers and, later as such, as a reference to the appearance of the extracting phase, relative to a liquid or gaseous phase, even though it is recognized that the extraction phase is not always technically a solid. The implementations of SPME include mainly open-bed extraction concepts, such as coated fibers, vessels, stirrers, and agitation mechanism disks, but in-tube approaches are also considered.[3]

The understanding of the fundamental principles governing mass transfer of analytes in multiphase is the key to rational choice, optimization, and design.[1] For on-site sampling, the extraction process generally follows the profile shown in Figure 18.1.[4] Calibration can be performed in the linear regime, equilibrium regime, or the kinetic regime.

The calibration based on equilibrium extraction is very simple and has been widely used for many on-site samplers. Using this calibration method, the sampler should be deployed long enough to ensure that the thermodynamic equilibrium is established between the environmental media and the receiving phase. The equilibrium times of different samplers range from seconds to months. The results obtained by equilibrium extraction are comparable to those obtained by grab sampling, and therefore, this type of device is unsuitable for the determination of time-weighted average (TWA) concentrations of pollutants in the environment.

Conventional SPME is performed by exposing a fiber coated with a liquid polymeric coating to a sample matrix, or its headspace, until an equilibrium is reached between the analyte partitioned in the fiber coating and the analyte dissolved in the sample matrix. The amount of the analyte extracted onto the fiber is linearly proportional to its initial concentration in the sample matrix.[5] The equilibrium extraction method is a widely used method of SPME for on-site sampling.[6–15] The on-site SPME passive sampling devices based on equilibrium extraction require a sampling time that is long enough to reach equilibrium. To noticeably shorten the equilibrium time for on-site sampling, SPME active sampling devices were developed. For on-site air sampling, it is easily accomplished with the use of an air pump. For field water sampling, stirring the fiber can markedly shorten the equilibration time.[15]

For the samplers with long equilibrium times, the calibration can be performed in the linear range in Figure 18.1. When this device is used for field sampling, the sampling

Handbook of Sample Preparation, Edited by Janusz Pawliszyn and Heather L. Lord

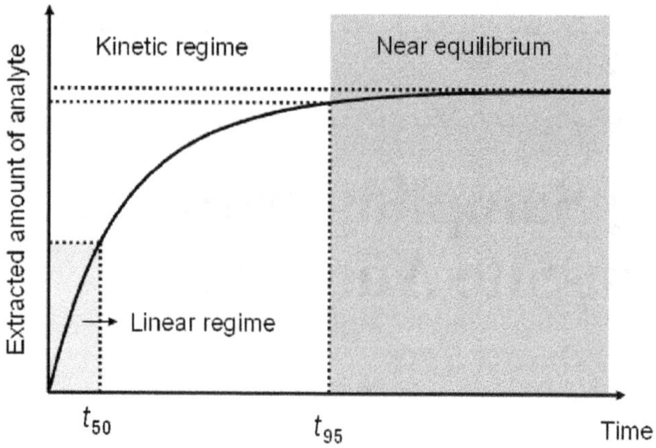

Figure 18.1. Typical extraction profile for on-site sampling. Reprinted from reference 4 with permission from Elsevier. Copyright 2007.

rate should be determined in the laboratory previously or predicted with empirical equations. This type of device provides the TWA concentration of target analytes in the sample.

Fick's first law of diffusion can be used for the calibration of the fiber-retracted SPME device when the inner diameter of the diffusion path is very small and the fiber coating acts as a "zero sink" for the target analytes.[16–19] Two other diffusion-based calibration methods have been proposed based on the interface model[20] and the cross-flow model.[21] These calibration methods are suitable for the quantification of fast air and water sampling.

Two kinetic calibration methods, in-fiber standardization technique[22,23] and standard-free kinetic calibration approach,[24] have been developed for the quantification of passive samplers for the entire sampling period. Both approaches are pre-equilibrium methods and can be used for TWA and fast passive on-site sampling, respectively.

In this chapter, the available SPME calibration methods and devices for the on-site monitoring of environmental pollutants are introduced. The characteristics of these calibration methods are also discussed. Calibration of SPME by liquid injection is also discussed.

18.2. STRATEGIES EMPLOYING EQUILIBRIUM EXTRACTION

The equilibrium extraction method is a widely used quantification method for SPME and involves exposing a fiber, which is coated with a liquid polymeric film, to a sample matrix until an equilibrium is reached. The extracted amount of the analyte, n, can be calculated with Equation 18.1,[5]

$$n = \frac{K_{\mathrm{fs}} V_{\mathrm{f}} V_{\mathrm{s}}}{K_{\mathrm{fs}} V_{\mathrm{f}} + V_{\mathrm{s}}} C_0, \tag{18.1}$$

where C_0 is the initial concentration of the target analyte in the sample, V_{s} is the sample volume, V_{f} is the fiber coating volume, and K_{fs} is the distribution coefficient of the analytes between the fiber coating and the sample. Equation 18.1 indicates that the amount of analyte extracted onto the coating (n) is linearly proportional to the analyte concentration in the sample (C_0), which is the analytical basis for quantification using SPME.

When the sample volume is very large, such as in-field sampling, then $V_s >> K_{\mathrm{fs}} V_{\mathrm{f}}$ and Equation 18.1 can be simplified to Equation 18.2,

$$n = K_{\mathrm{fs}} V_{\mathrm{f}} C_0, \tag{18.2}$$

which illustrates the advantage of the equilibrium extraction method for field applications. In this equation, the amount of the extracted analyte is independent of the sample volume. In practice, there is no need to collect a defined sample prior to analysis, since the fiber can be exposed directly to the ambient air, water, production stream, and so on. The amount of extracted analyte will correspond directly to its concentration in the matrix, without being dependent on the sample volume. When the sampling step is eliminated, the whole analytical process can be accelerated, and errors associated with analyte losses through the decomposition or adsorption on the sampling container walls will be prevented.

Equation 18.1 also illustrates another characteristic for field sampling with SPME: The concentration of the target analytes can be determined by the amount of the analytes on the fiber under extraction equilibrium, by knowing the distribution coefficients of the analytes between the fiber coating and the sample matrix.

The distribution coefficients between the fiber coating and air can be determined experimentally or estimated with retention indexes from a linear temperature-programmed capillary gas chromatography (GC).[25,26] Koziel et al. described a system for the generation of standard gas mixtures of volatile and semivolatile organic compounds, which can be used for determining the distribution coefficients between the fiber coating and air and the calibrations of SPME and other sampling devices.[27] Shurmer and Pawliszyn introduced a standard polycyclic aromatic hydrocarbon (PAH) aqueous solution-generating system based on dilution,[28] which has been used in determining the distribution coefficients between fiber coating and water. Recently, the system has been modified, and a new flow-through system, based on permeation,[29] for the generation of standard aqueous solution, has been developed.

18.2.1. Distribution Coefficients between SPME Fiber Coating and Air/Water

Numerous distribution coefficients for different compounds between SPME fiber coating and air/water have been reported. Tables 18.1–18.3 present the distribution coefficients of some compounds between different SPME fiber coating and air at 298 K. Table 18.4 presents the distribution coefficients of several compounds between different SPME fiber coating and water.

Table 18.1. Distribution Coefficients (K_{fa}) between 100-μm PDMS Coating and Air at 298 K

Compound	K_{fa}	Compound	K_{fa}
3-Methylpentane	159[a]	2-Methyloctane	2600[a]
2,4-Dimethylpentane	262[a]	3-Methyloctane	2890[a]
2,2,3-Trimethylbutane	280[a]	3,3-Dimethylpentane	2610[a]
2-Methylhexane	387[a]	2,2-Dimethyloctane	4320[a]
2,3-Dimethylpentane	412[a]	3,3-Dimethyloctane	5050[a]
2,2-Dimethylhexane	673[a]	2,3-Dimethyloctane	6100[a]
2,5-Dimethylhexane	587[a]	2-Methylnonane	6690[a]
2,2,3-Trimethylpentane	569[a]	3-Ethyloctane	6970[a]
2,3-Dimethylhexane	968[a]	3-Methylnonane	7100[a]
2-Methylheptane	993[a]	*n*-Pentane	95[b]
4-Methylheptane	1060[a]	*n*-Hexane	150[b]
3-Methylheptane	1090[a]	α-Pinene	4500[b]
3-Ethylhexane	990[a]	*d*-Limonene	10,300[b]
2,5-Dimethylheptane	1970[a]	*n*-Undecane	25,000[b]
3,5-Dimethylheptane (D)	1960[a]	Benzene	300[b]
3,3-Dimethylheptane	2090[a]	Toluene	880[b]
3,5-Dimethylheptane (L)	2100[a]	Ethylbenzene	2100[b]
2,3-Dimethylheptane	2390[a]	*p*-Xylene	2400[b]
3,4-Dimethylheptane (D)	2420[a]	*o*-Xylene	3100[b]
3,4-Dimethylheptane (L)	2620[a]	1,3,5-Trimethylbenzene	5800[b]
Compound	**$\log K_{fa}$[c]**	**Compound**	**$\log K_{fa}$[d]**
Isoprene	1.76	Naphthalene	5.5
α-Pinene	3.66	Biphenyl	6.4
Myrcene	3.93	Acenaphtylene	7.2
3-Carene	3.94	Acenaphtene	6.8
Limonene	4.04	Fluorene	7.0
γ-Terpinene	4.14	Phenanthrene	7.8
Terpinolene	4.25	Anthracene	8.2
Acetone	2.06	Fluoranthene	8.4
2-Butanone	2.4	Pyrene	8.2
2-Pentanone	2.99		
3-Hexanone	3.25		
2-Hexanone	3.27		
2-Heptanone	3.60		
3-Octanone	3.91		

[a]Reference 25.
[b]Reference 26.
[c]Reference 32.
[d]Reference 33.

18.2.2. Application of SPME Equilibrium Extraction for Field Sampling

Table 18.5 lists applications where SPME has been used for on-site sampling by equilibrium extraction in recent years.

Muller et al. described several SPME field sampling devices and presented the advantages and limitations of the samplers.[7] Mayer et al. used a glass fiber with a 15-μm coating of polydimethylsiloxane (PDMS) to measure dissolved concentrations of persistent and bioaccumulative pollutants (PBPs) in sediment pore water.[8] The equilibrium extraction time for the field sediment experiments was up to 30 days. Conder et al. explored the use of the SPME technique to nondestructively measure the explosive compound 2,4,6-trinitrotoluene (TNT) and its nitroaromatic degradation products in both laboratory and field sediments.[9] The sampling times that were required to achieve steady-state equilibrium were 48 h for room temperatures (23–25°C) and up to 7 days for cold (5°C) temperatures. Zeng et al. employed an SPME-based method for sampling persistent chlorinated hydrocarbons at three coastal locations off southern California.[11] The equilibrium for the target analytes

Table 18.2. K_{fa} Values between 30-μm PDMS Coating and Air at 298 K[25,26]

Compound	K_{fa}	Compound	K_{fa}
Benzene	301; 260	1-Methyl-2-*n*-propylbenzene	14,900
Toluene	818; 710	1,4-Dimethyl-2-ethylbenzene	15,900
Ethylbenzene	2070; 2000	1,2-Dimethyl-4-ethylbenzene	17,400
m-Xylene	2090	1,3-Dimethyl-2-ethylbenzene	18,100
p-Xylene	2500; 2300	1,2-Dimethyl-3-ethylbenzene	20,000
o-Xylene	2900; 3100	1,2,4,5-Tetramethylbenzene	24,700
Isopropylbenzene	3880	2-Methylbutylbenzene	24,100
n-Propylbenzene	5040	1-*tert*-Butyl-2-methylbenzene	26,200
1-Methyl-3-ethylbenzene	4750	*n*-Pentylbenzene	34,500
1-Methyl-4-ethylbenzene	6230	1-*tert*-butyl-3,5-dimethylbenzene	45,600
1,3,5-Trimethylbenzene	6480; 5900	1-*tert*-4-ethylbenzene	43,700
1-Methyl-2-ethylbenzene	6580	1,3,5-Triethylbenzene	67,300
Isobutylbenzene	8360	1,2,4-Triethylbenzene	75,600
sec-Butylbenzene	8590	*n*-Hexylbenzene	90,100
1-Methyl-3-isopropylbenzene	10,100	*n*-Pentane	100
1-Methyl-4-isopropylbenzene	10,200	*n*-Hexane	170
1-Methyl-2-isopropylbenzene	12,000	α-Pinene	4300
1-Methyl-3-n-propylbenzene	13,200	*d*-Limonene	10,300
1,3-Dimethyl-5-ethylbenzene	15,000	*n*-Undecane	25,000

Table 18.3. *log*K_{fa} Values between Carboxen/PDMS Coating and Air at 298 K[32]

Compound	logK_{fa}	Compound	logK_{fa}
Acetone	2.54	Ethyl acetate	2.90
2-Butanone	2.94	Butyl acetate	3.78
2-Pentanone	3.37	Methyl butyrate	3.38
2-Heptanone	4.33	Pentyl butyrate	4.37
Heptanal	4.42	Hexyl butyrate	4.87
Acrolein	2.72	Ethanol	2.57
Methylacrolein	2.91	1-Propanol	2.82
3-Buten-2-one	3.09	1-Butanol	3.42
3-Penten-2one	4.03	3-Methyl-1-butanol	3.54
Benzaldehyde	5.09		

Table 18.4. Distribution Coefficients between Different SPME Fiber Coating and Water

Compound	logK_{fw}[33] (100-μm PDMS, 25°C)	logK_{fw}[33] (85-μm PA, 25°C)
Naphthalene	3.02	3.37
Acenaphthylene	3.40	4.01
Acenaphthene	3.63	4.09
Fluorene	3.71	4.32
Phenanthrene	3.96	4.39
Anthracene	3.98	4.66
Fluoranthene	4.71	4.87
Pyrene	4.86	4.84
Benz[*a*]anthracene	5.26	5.34
Chrysene	5.69	4.95
Benzo[*b*]fluoranthene	5.17	4.34
Benzo[*k*]fluoranthene	5.33	4.39
Benzo[*a*]pyrene	5.39	5.62

Table 18.4. *(Continued)*

Compound	$\log K_{fw}$[33] (100-μm PDMS, 25°C)	$\log K_{fw}$[33] (85-μm PA, 25°C)
Dibenzo[*a,h*]anthracene	4.86	4.91
Benzo[*ghi*]perylene	4.28	4.03
Indeno[1,2,3-*cd*]pyrene	4.43	4.43
Compound	**$\log K_{fw}$[34] (7-μm PDMS, 25°C)**	**$\log K_{fw}$[34] (100-μm PDMS, 25°C)**
PCB 1	4.44	4.09
PCB 15	5.11	4.83
PCB 28	5.47	5.18
PCB 47	5.86	5.64
PCB 101	6.21	6.08
PCB153	6.68	6.45
PCB 202	6.77	6.20
PCB 180	6.76	6.54
PCB 206	7.04	6.16
PCB 209	6.84	5.59
Compound	**K_{fw}[35] (30-μm PDMS, 22°C)**	**K_{fw}[35] (85-μm PA, 22°C)**
Dimethyl phthalate	20	33
Diethyl phthalate	59	218
Di-*n*-propyl phthalate	320	1030
Diisobutyl phthalate	2090	3340
Di-*n*-butyl phthalate	2230	4945
Di-2-ethylhexyl phthalate	128,080	9535
Compound	**K_{fw}[36] (65-μm PDMS-DVB, 24°C)**	**K_{fw}[36] (65-μm CW-DVB, 25°C)**
1,2,3,4-Tetrachlorobenzene (TeCB)	304,000	189,000
Pentachlorobenzene (PeCB)	468,000	281,000
Hexachlorobenzene (HCB)	407,000	281,000
α-Hexachlorocyclohexane (α-HCH)	120,000	253,100
β-Hexachlorocyclohexane (β-HCH)	74,100	266,300
γ-Hexachlorocyclohexane (γ-HCH)	158,000	24,000
δ-Hexachlorocyclohexane (δ-HCH)	52,500	11,000
1,1-Dichloro-2,2-bis(*p*-chlorophenyl)ethylene (*p,p*′-DDE)	275,000	457,000
1,1-Dichloro-2,2-bis(*p*-chlorophenyl)ethane (*p,p*′-DDD)	256,000	226,000
PCB 28	516,000	620,000
PCB 52	477,000	674,000
PCB 101	494,000	860,000
PCB 138	427,000	769,000
PCB 153	504,000	942,000
PCB 180	320,000	367,000
Phenanthrene	275,000	238,000
Anthracene	227,000	206,000
Fluoranthene	240,000	240,000
Pyrene	206,000	212,000
Benzo[*a*]pyrene	20,000	27,500

PA, polyacrylate; DVB, divinylbenzene; CW, carbowax.

Table 18.5. Quantitative Analysis of SPME for On-Site Sampling by Equilibrium Extraction

Analytes	Matrix and Location	Experimental Details (Fiber, Sampling Time, Analysis Instrument, etc.)	Reference
VOCs	Air (laboratory)	SPME field sampler with a two-leaf closure, 100-µm PDMS, 75-µm carboxen (CAR)/PDMS, 65-µm PDMS/DVB, static, 3 min, GC/FID	7
PBPs	Sediment	15-µm PDMS, static, 30 days, GC/ECD	8
TNT and its degradation products	Sediment	85-µm PA, static, 48 h–7 days, HPLC	9
Organic pollutants	Soil	30-µm PDMS, static, within 20 days, GC/MS	10
Chlorinated hydrocarbons	Water (coastal locations off south California)	100-µm PDMS, static, within 23 days, GC/MS	11
Organophosphate triesters	Air (lecture room)	100- and 30-µm PDMS, dynamic, >18 h, GC/NPD	12,13
VOCs	Air (classrooms)	75-µm PDMS/CAR, static, 4 h, GC/MS	14
PAHs	Water	Rotated-fiber and rotated-membrane sampler, dynamic, 30–60 min, GC/MS	15

DVB, divinylbenzene; FID, flame ionization detector; ECD, electron capture detector; HPLC, high-performance liquid chromatography; NPD, nitrogen phosphorus detector.

could be reached within 18–23 days. The efficacy of the SPME sampler was also compared with an Infiltrex 100 water pump.

The SPME sampling method has also been applied as a tool to estimate bioavailability in soil.[10] For example, organic micropollutants were measured with 30-µm PDMS SPME fibers after exposure to two field-contaminated soils. The equilibrium was achieved within 20 days. In addition, freely dissolved concentrations of contaminants in pore water, derived from SPME measurements, were used to calculate concentrations of the target analytes in earthworms. Isetum et al. evaluated a dynamic SPME field air sampling device for organophosphate triester vapor under equilibrium.[12,13] The SPME fibers were placed in series and were exposed to a laminar flow of air through the sample. A constant linear airflow of 10 cm/s over the fibers was applied to increase the extraction rate. Field measurements were performed in a lecture room on two occasions, and the results were compared with those from active sampling on filters. Larroque et al. studied the measurement of volatile organic compounds (VOCs) in indoor air by SPME.[14] The air sample was enclosed in a 250-mL glass bulb where the SPME PDMS/carboxen (CAR) fiber was exposed until adsorption equilibrium was reached. The methodology was applied to compare the VOC content in the classrooms of two different schools.

Newly developed rotated-fiber and rotated-membrane samplers are also equilibrium extraction samplers designed for field water sampling.[15] The commercially available fiber or PDMS membrane is coupled with a battery-operated drill. This design is considered an active model and can markedly shorten the equilibrium time. Figure 18.2 is the schematic diagram of field water sampling with rotated-fiber and rotated-membrane samplers.

Agitation of the sample during fiber exposure has been shown to enhance the amount of analyte extracted, while

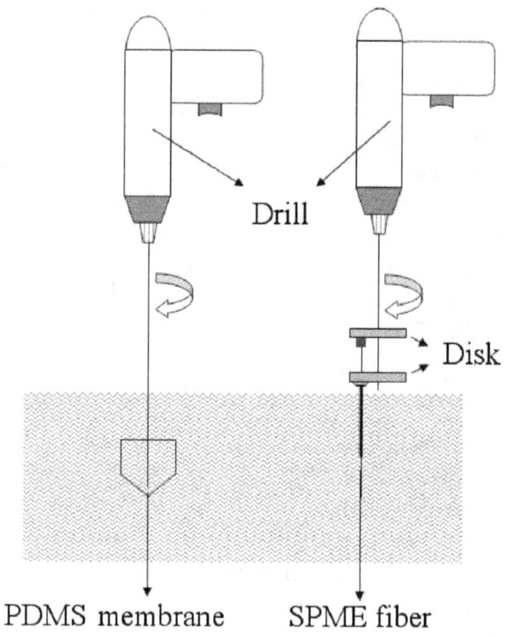

Figure 18.2. Schematic diagram of field water sampling with a rotated fiber and a rotated membrane.[2] Figure reprinted with permission.

reducing the extraction time.[38,39] For field water sampling, however, it is inconvenient to agitate the sample. Using a drill to circumrotate the fiber can obtain the same efficiency as agitation. When the 100-µm PDMS fiber and 125-µm-thick PDMS membrane were used, with a rotational speed of 570 rpm, the equilibrium times for the six target PAHs were between 20 and 40 min.[15] This was much shorter than the equilibrium time without agitation.[39] The difference between the rotated-fiber and the rotated-membrane samplers relates to the sensitivity of the membrane sampler, which is much higher than the fiber sampler, due to the higher capacity of the membrane. The volume of the membrane (62.5 μL) is 100 times the volume of the 100-μm PDMS fiber (0.61 μL). But the fiber sampler also provides a number of advantages. For example, it can be conveniently injected into a gas chromatograph with a commercial SPME holder. Conversely, the membrane sampler requires a special injection system. The ATAS (A Total Analytical Solution, Leap Technologies, Carrboro, NC) system can be used for the automated injection of the PDMS-membrane sampler.[15]

18.3. STRATEGIES EMPLOYING LINEAR MASS UPTAKE

18.3.1. Interface Model

Koziel et al. developed a rapid air sampling methodology using adsorptive SPME coatings and controlled air convection conditions. A theoretical model for rapid extraction was formulated based on the diffusion through the interface surrounding the fiber:[20]

$$C_0 = \frac{n \ln((b+\delta)/b)}{2\pi D_g L t}, \tag{18.3}$$

$$\delta = 9.52\left(b/Re^{0.62} Sc^{0.38}\right), \tag{18.4}$$

where C_0 is the analyte concentration in the bulk air, n is the amount of analyte extracted by the fiber coating during time t, L is the length of the coated rod, D_g is the gas-phase molecular diffusion coefficient (it can be found in the literature or can be reliably estimated from empirical equations), b is the outside radius of the fiber coating, δ is the thickness of the boundary layer surrounding the fiber coating, Re is the Reynolds number ($Re = 2ub/v$), u is the linear air velocity, v is the kinetic viscosity for air, and Sc is the Schmidt number ($Sc = v/D_g$).

Sukola et al. extended this interface model to the calibration of aqueous samples.[40] Portable SPME devices for rapid field air sampling with diffusion-based calibration were developed by Augusto et al. and used for the quantification of aromatic VOCs in indoor air.[41] Measured VOC concentrations were identified in low parts per billion by volume range using only a 30-s SPME fiber exposure, and the results were comparable to those obtained with the standard NIOSH-1501 method.

Tuduri et al. tested an approach based on this model for the sampling of VOCs in air with the PDMS/CAR fiber.[42] Isetun et al. evaluated a dynamic SPME field air sampling device for organophosphate triester vapor under nonequilibrium conditions.[13,43] A constant linear airflow of 10 cm/s over the fibers was applied to the TWA air sampling with interface model calibration, and the mass uptake rate was found to be constant for all tested air velocities over the fiber surface greater than 7 cm/s. Field measurements were performed in an office, and the results were compared with those of active sampling on filters. Table 18.6 presents some applications of SPME using the calibration method based on the interface model.

18.3.2. Cross-Flow Model

Chen et al. proposed another diffusion-based calibration method through a cross-flow model.[21] With this model, the target analyte concentration can be calculated with Equation 18.5,

$$C_0 = \frac{n}{\bar{h}At} = \frac{nd}{ERe^m Sc^{1/3} ADt}, \tag{18.5}$$

where A is the surface area of the fiber, n is the mass uptake onto the fiber during sampling time t, D is the diffusion coefficient of the analyte molecule, $\bar{h}$ is the average mass transfer coefficient, d is the outer diameter of the fiber, Re is the Reynolds number, and Sc is the Schmidt number. Constants E and m are dependent on the Reynolds number and are available in the literature.[21]

This diffusion-based calibration method was validated by on-site analysis of benzene, toluene, ethylbenzene, and xylene (BTEX) in both aqueous samples and indoor air. The results were confirmed by the NIOSH-1501 method.

18.3.3. Fick's First Law of Diffusion

These types of SPME on-site sampling devices are unlike conventional sampling with SPME, as the fiber is retracted

Table 18.6. Applications of SPME for the On-Site Monitoring of Environmental Pollutants Using Interface Model

Analytes	Matrix and Location	Experimental Details (Fiber, Sampling Time, Analysis Instrument, etc.)	Reference
BTEX	Air (standard gas generating system)	75-μm CAR/PDMS, 65-μm PDMS/DVB, dynamic, <1 min, GC/FID	20
VOCs	Indoor air, living plant aroma compounds	65-μm PDMS/DVB, dynamic, 30 s, GC/MS, FID	41
Organophosphate triesters	Air (standard gaseous sample, office)	100-μm PDMS, dynamic, 40–90 min, GC/NPD	13,43
VOCs	Air (standard gas-generating system)	75-μm PDMS/DVB, dynamic, 1 min, GC/FID	42

DVB, divinylbenzene; FID, flame ionization detector; NPD, nitrogen phosphorus detector.

to a known distance into its needle housing during the sampling period. Analyte molecules access the fiber coating only by means of diffusion through the static air/water gap between the opening and fiber coating. Therefore, Fick's first law of diffusion can be used for the calibration (Fig. 18.3).

If the sorbent is a "zero sink" for the target analytes, the concentrations of analytes in the sample can be calculated with Equation 18.6,

$$C = \frac{nZ}{ADt}, \quad (18.6)$$

where C is the TWA concentration of the target analyte in air or water during the sampling time t, Z is the diffusion path length, A is the cross-sectional area of the needle, D is the diffusion coefficient of the target analyte in air or water, and n is the amount of analyte that is extracted by the fiber during time t.

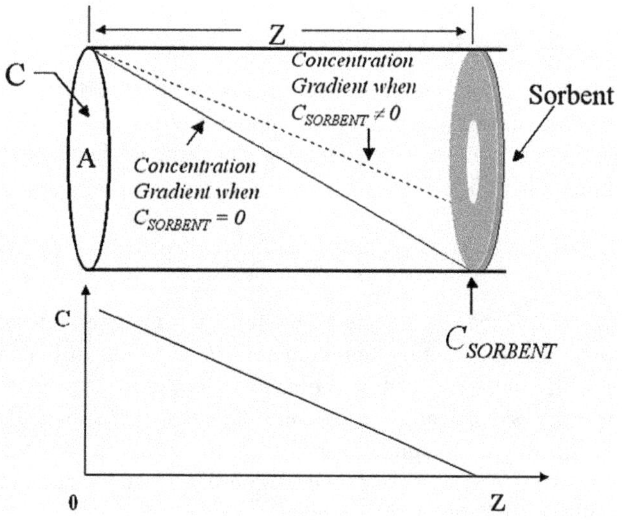

Figure 18.3. Schematic diagram of fiber retracted SPME device and concentration gradient. Reprinted from reference 44 with permission from Elsevier. Copyright 2007.

The commercially available SPME device was used as a TWA diffusive sampler by Martos and Pawliszyn.[16] The investigations demonstrated that the fiber-retracted SPME device can be used as a TWA passive sampler. The device was used for TWA sampling of hydrocarbons. Airborne (industrial) formaldehyde concentrations were measured with the device with on-fiber derivatization, and the results were compared with those produced with active air sampling using the NIOSH-2541 method. The device was also used for TWA air sampling of normal alkanes from C5 to C15.[17] Field sampling trials were performed at a house, an apartment, and a school. The results were compared with those of active sampling through charcoal tubes using the NIOSH-1550 method. Koziel et al. used the fiber-retracted SPME devices for TWA sampling of dodecane in indoor air.[45] Field dodecane sampling was conducted in a computer laboratory. An SPME 100-μm PDMS fiber was used for the indoor air survey, and the passive sampling time was 75 min. The sampling and analysis of formaldehyde in indoor air with the fiber-retracted SPME device was completed at a chemical store, a residential house, an office building, an engineering shop, an elementary school, and a rental apartment, using the SPME on-fiber derivatization with *o*-(2,3,4,5,6-pentafluorobenzyl) hydroxylamine hydrochloride (PFBHA). The TWA passive sampling time was 8 h.[46]

Later, Chen and Pawliszyn used the fiber-retracted SPME device to determine the TWA concentrations of VOCs in air,[18] and demonstrated that the face velocity of air across the needle opening does not affect the sampling, due to the extremely small inner diameter of the fiber needle.

Ouyang et al. extended the applications of this type of SPME device to TWA passive water sampling.[19] A Hamilton

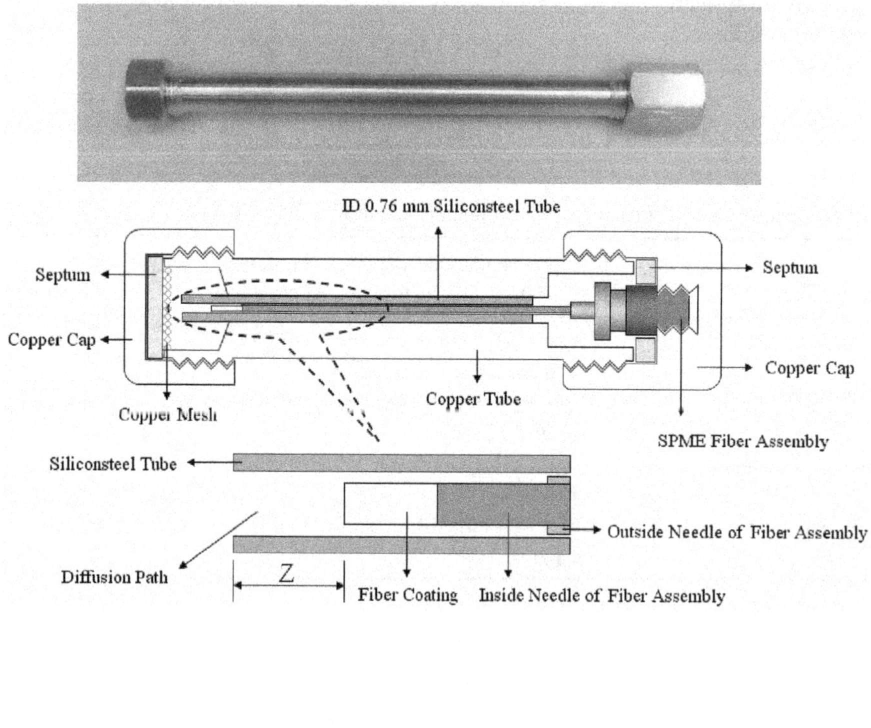

Figure 18.4. Schematic diagram and graphic of the SPME TWA field water sampler. Reprinted from reference 44 with permission from Elsevier. Copyright 2007.

500-μL gastight syringe was modified as a TWA water sampling device, to ensure all of the air in the SPME needle could be replaced with water. In addition, a removable needle was designed to avoid the effect of the adsorption of the target analytes on the outside wall of the needle.

A field TWA water sampling device have been designed with a commercial SPME fiber assembly (Fig. 18.4) and has been used to monitor PAHs in Hamilton Harbour (Hamilton, Canada) and Laurel Creek (Waterloo, Canada).[44,51]

18.4. STRATEGIES EMPLOYING KINETIC CALIBRATION

18.4.1. Kinetic Calibration with Standard in the Extraction Phase/In-Fiber Standardization Technique

In 1997, Ai Jiu proposed a dynamic model of SPME in nonequilibrium situations,[52,53] based on a diffusion-controlled mass transfer process. From this model, Chen et al. demonstrated the symmetry of absorption and desorption in the SPME liquid coating fiber and a new calibration method, kinetic calibration, was proposed.[22,23] This kinetic calibration method, also referred to as the in-fiber standardization technique,[54] uses the desorption of the standards, which are preloaded in the extraction phase, to calibrate the extraction of the analytes.

When an SPME liquid coating fiber is exposed to a sample matrix, mass transfer from the sample matrix to the surface of the SPME fiber coating and from the surface of the fiber coating to its inner layers is the rate-determining step (Fig. 18.5). This process follows Fick's first law of diffusion:

$$J \equiv \frac{dn}{dt} = -D_s A \frac{dC_s}{dx} = -D_f A \frac{dC_f}{dx}, \tag{18.7}$$

where J is the mass flux of the analyte from the sample matrix to the SPME fiber, A is the surface area of the fiber, dn is the amount of the analyte extracted during sampling time dt, and D_s and D_f are diffusion coefficients of the analyte in the sample matrix and the fiber coating, respectively. C_s and C_f are the concentrations of the analyte in the sample matrix and the surface of the fiber coating, respectively.

A linear concentration gradient in the fiber coating and boundary layer is assumed, then

$$\frac{dn}{dt} = \frac{D_s A}{\delta_s}(C_s - C_s') = \frac{D_f A}{\delta_f}(C_f - C_f'), \tag{18.8}$$

where δ_f is the thickness of the fiber coating and δ_s is the diffusion layer thickness in the sample matrix. It is assumed that a partition equilibrium will be quickly reached between the sample matrix and the fiber coating, then we have

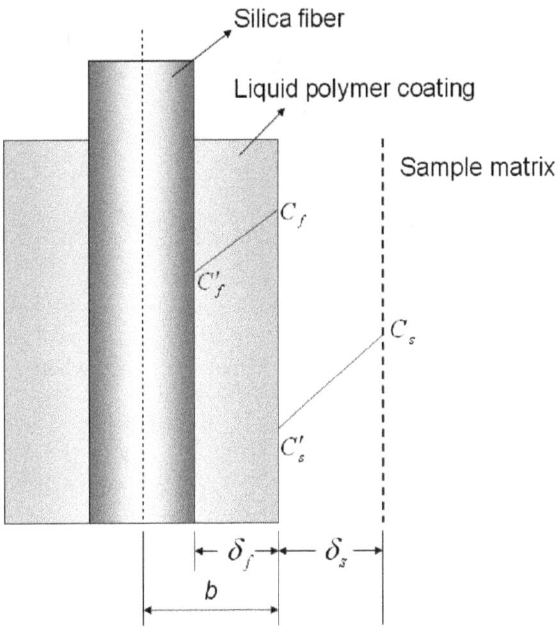

Figure 18.5. Schematic of the interface for SPME liquid coating fiber exposed in sample matrix. A linear concentration gradient is assumed in both the liquid coating and the boundary layer. Reprinted from reference 47 with permission from the American Chemical Society. Copyright 2009.

$$K = \frac{C_f}{C_s'} \Rightarrow C_s' = \frac{C_f}{K}, \tag{18.9}$$

where K is the distribution coefficient of the analyte between the fiber coating and water. Let

$$R_s = \frac{D_s A}{\delta_s}, \quad R_f = \frac{D_f A}{\delta_f},$$

Equation 18.8 can then be rewritten as

$$\frac{dn}{dt} = R_s\left(C_s - \frac{C_f}{K}\right) = R_f(C_f - C_f'). \tag{18.10}$$

The average concentration of the analyte in the polymer coating is approximately equal to $(C_f + C_f')/2$. Then

$$n = V_f(C_f + C_f')/2. \tag{18.11}$$

Combining Equations 18.10 and 18.11,

$$C_f = \frac{K(2nR_f + R_s V_f C_s)}{V_f(2KR_f + R_s)}. \tag{18.12}$$

By substituting Equation 18.12 into Equation 18.10, we have

$$C_f - C_f' = \frac{2KR_s C_s}{2KR_f + R_s} - \frac{2R_s n}{V_f(2KR_f + R_s)}. \tag{18.13}$$

Then

$$\frac{dn}{dt} = R_f(C_f - C_f') = R_f\left[\frac{2KR_s C_s}{2KR_f + R_s} - \frac{2R_s n}{V_f(2KR_f + R_s)}\right]. \tag{18.14}$$

Let

$$a = \frac{2R_f R_s}{V_f(2R_f K + R_s)}, \quad (18.15)$$

$$b = \frac{2KR_f R_s}{2KR_f + R_s}. \quad (18.16)$$

Equation 18.14 is converted to

$$\frac{dn}{dt} = bC_s - an. \quad (18.17)$$

The integrated result for Equation 18.17 when $t = 0$, $n = 0$ is

$$\int_0^t \frac{dn}{dt} = bC_s - an \Rightarrow n = [1 - \exp(-at)]\frac{b}{a}C_s. \quad (18.18)$$

Based on Equations 18.15 and 18.16, we have

$$\frac{b}{a} = KV_f. \quad (18.19)$$

Then, Equation 18.18 becomes

$$C_s = \frac{n}{KV_f[1 - \exp(-at)]}. \quad (18.20)$$

For field sampling, the dynamic process for the desorption of preloaded calibrant from an SPME fiber can be described by Reference 23:

$$\frac{Q}{q_0} = \exp(-a't), \quad (18.21)$$

where a' is the rate constant of the desorption of the preloaded calibrant, q_0 is the amount of preloaded calibrant in the fiber coating, and Q is the amount of the calibrant remaining in the fiber after the exposure of the fiber to the sample matrix for the sampling time t. When the rate constants a and a' have the same value for the absorption of target analytes and the desorption of preloaded calibrants, then Equations 18.20 and 18.21 can be combined to

$$C_s = \frac{n}{KV_f(1 - Q/q_0)}. \quad (18.22)$$

Equation 18.22 means that we can directly deploy the samplers, which have been preloaded with a certain amount of calibrant, into the sample matrix and retrieve them after different sampling times. Then, the TWA concentration of the target analyte during the sampling time can be calculated, since n and Q can be determined, and K, V_f, and q_0 are already known.

Equation 18.10 is the theoretical basis of Equation 18.22. It is worth to emphasize that C_s' is not assumed to be zero in Equation 18.10; therefore, there is no "zero sink" requirement for this calibration method. If we assume $C_s' \approx 0$, it is easy to derive from Equation 18.8:

$$C_s = \frac{AD_s n}{\delta_s t} = R_s \frac{n}{t}, \quad (18.23)$$

where R_s is the sampling rate of the sampler for target analyte. Equation 18.23 is the theoretical basis of the linear mass uptake passive sampler. The assumption of $C_s' \approx 0$ means this type of sampler should be a "zero sink" for the target analyte.

Although there is no "zero sink" requirement for this calibration method, the limitation for the application of Equation 18.22 is that C_s should be larger than C_f/K, otherwise, the extracted analytes will be desorbed, according to Equation 18.10, which will cause an unreliable result.

This technique is a pre-equilibrium method and can be used for the entire sampling period. The determined concentration before the sampling reaches equilibrium is a TWA concentration, because the desorption of the preloaded standard calibrated the extraction of the analytes, and the extraction is an integrated process. If the sampling reaches equilibrium, the determined data are the concentrations of the analytes in the sample at the time that the samplers were retrieved. The feasibility of this technique for TWA water sampling was demonstrated by both theoretical derivations and field trials.[48,50,51]

The in-fiber standardization technique makes it possible to use a simple PDMS rod or PDMS membrane as a passive sampler to obtain the TWA concentrations of target analytes in a sampling environment.[48,50,51]

A pure PDMS rod, which has been used as a passive sampler, is 1 cm long with a diameter of 1 mm. The SPME PDMS-membrane passive sampler has been made using a 127-μm-thick PDMS thin film for the extraction phase. The thin film was cut into a specific house-like shape, and the dimension of the thin films was 2 × 2 cm with a 1-cm-high triangle on top of the square. The volumes of the SPME PDMS-rod (7.85 μL) and the SPME PDMS-membrane (62.5 μL) samplers are much larger than commercial SPME PDMS fibers (0.61 μL), thus increasing the sensitivity of the passive samplers.

Using the kinetic calibration method, the samplers require the preloading of a certain amount of standard and are exposed in the sample matrix for a defined period of time. The TWA concentrations of the target analytes in the sample can then be calculated with Equation 18.22.[50]

Standard loading approaches include (1) headspace extraction of the standard dissolved in a solvent or vacuum pump oil, (2) headspace extraction of pure standard in a vial, (3) direct extraction in a standard solution, and (4) direct transfer of the standard solution from the syringe to the fiber.[49] The deployed samplers should be retrieved before all the preloaded standards are lost. Normally, the sampler should be retrieved in 1 month or less, and this period is determined by the flow velocity of the sample matrix.

The usefulness of these samplers has been demonstrated experimentally, both under laboratory and field conditions, in Hamilton Harbour, Hamilton, Canada. The effects of environmental factors, such as temperature and turbulence,

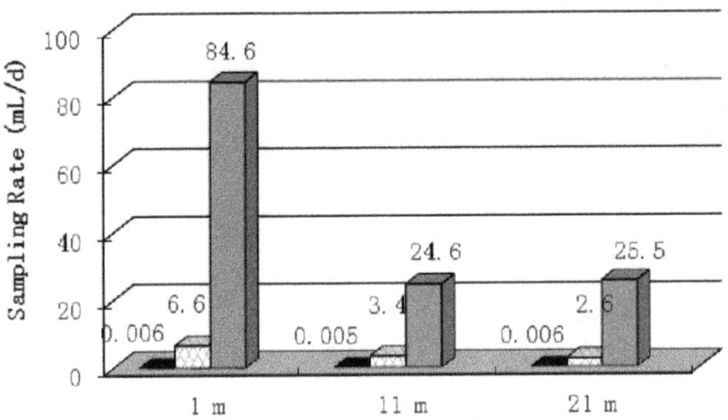

Figure 18.6. Comparison of the sampling rates of three types of SPME passive samplers at different sampling depths. Black, fiber-retracted device; gridding, PDMS rod; gray, PDMS membrane. Reprinted from reference 51 with permission from the American Chemical Society. Copyright 2007.

were successfully calibrated in the field studies, and the results are comparable with those achieved by traditional methods.[48,50,51]

The SPME PDMS-rod and PDMS-membrane passive samplers possess all of the advantages of SPME: They are solvent free; combine sampling, isolation, and enrichment into one step; and can be directly injected into a gas chromatograph for analysis without further treatment. Both samplers are simple and easy to deploy and retrieve. They have large sampling rates, and the sensitivity is much higher than the fiber-retracted SPME device, since the samplers are in direct contact with the sample matrix. The SPME PDMS membrane has a high surface-to-volume ratio compared with the PDMS rod, which results in higher sensitivity and mass uptake, leading to the detection of lower levels of analytes than many other techniques. Figure 18.6 illustrates the sampling rates of three types of SPME passive samplers at different sampling depths in Hamilton Harbour, Hamilton, Ontario.[51]

The SPME PDMS-membrane sampling rate is about 10 times higher than that achieved by the PDMS-rod sampler and more than 1000 times higher than the rate achieved by the fiber-retracted SPME device. An SPME PDMS-membrane passive sampler requires a large volume injector for analysis, but the PDMS-rod sampler can be analyzed by a GC–mass spectrometry (MS) with a normal injector.

Recently, the concept of calibrants in the extraction phase was extended to determinate concentrations of target analytes directly in the vein of animals, indicating that this approach is useful for *in vivo* studies as well.[55] It has also been successfully used for liquid-phase microextraction (LPME).[56,57]

18.4.2. Standard-Free Kinetic Calibration

Recently, Ouyang et al. proposed a standard-free kinetic calibration method for fast on-site passive sampling, based on Ai Jiu's dynamic model.[24] With this calibration method, all analytes can be directly calibrated with only two samplings. For sampling times t_1 and t_2, the kinetic process for the absorption can be described by

$$\frac{n_1}{n_e} = 1 - \exp(-at_1), \tag{18.24}$$

$$\frac{n_2}{n_e} = 1 - \exp(-at_2). \tag{18.25}$$

Equations 18.24 and 18.25 can be combined to produce

$$\frac{t_2}{t_1}\ln\left(1 - \frac{n_1}{n_e}\right) = \ln\left(1 - \frac{n_2}{n_e}\right). \tag{18.26}$$

Two methods, illustration and iteration, can be used to obtain the value of n_e with a spreadsheet.[24] Then, the concentration of target analytes in the sample can be calculated with Equation 18.1.

The feasibility of this calibration method was validated in a standard aqueous solution flow-through system and a standard gas flow-through system.[24] Using this standard-free kinetic calibration method, the sampling time for equilibrium extraction can be markedly shortened. In the reported study, sampling times for the equilibrium extraction of PAHs in a water environment, which typically range from 2 to 24 h, can be shortened to 2–5 min, and sampling times for BTEX in air can be shortened from 5–10 min to 5–10 s.[24] This calibration method can be used for the entire sampling period, without considering whether the system reaches equilibrium. This aspect of the technique is desirable for systems when the equilibrium time is not known, and particularly useful for instances when a number of compounds are measured simultaneously. The method is unsuitable for long-term monitoring of pollutants in the environment, since the method requires that the sampling rate remains constant and the determined concentration is therefore representative of a spot sampling.

18.5. STANDARD AQUEOUS SOLUTION-GENERATING SYSTEM

Using a flow-through system to evaluate SPME integrative sampling device in laboratory can significantly shorten the test time. A flow-through standard water system was introduced by Shurmer and Pawliszyn.[28] The system was constructed using a liquid chromatography pump to deliver water to a stirred SPME sampling cell. The SPME sampling cell consists of a modified 40-mL vial with 1/4-in. glass inlet and outlet tubes added to it. A standard mixture of naphthalene, anthracene, fluoranthene, pyrene, and benzo[a]-pyrene in acetone was injected into the water upstream from the SPME sampling cell using a syringe pump and 1-mL Hamilton gastight syringe. To ensure adequate mixing of the water and PAH, a 15-mL in-line glass impinger filled with 4-mm glass beads was inserted immediately after the PAH injection tee.

The flow-through standard water generator eliminates errors associated with both losses of analytes to the system

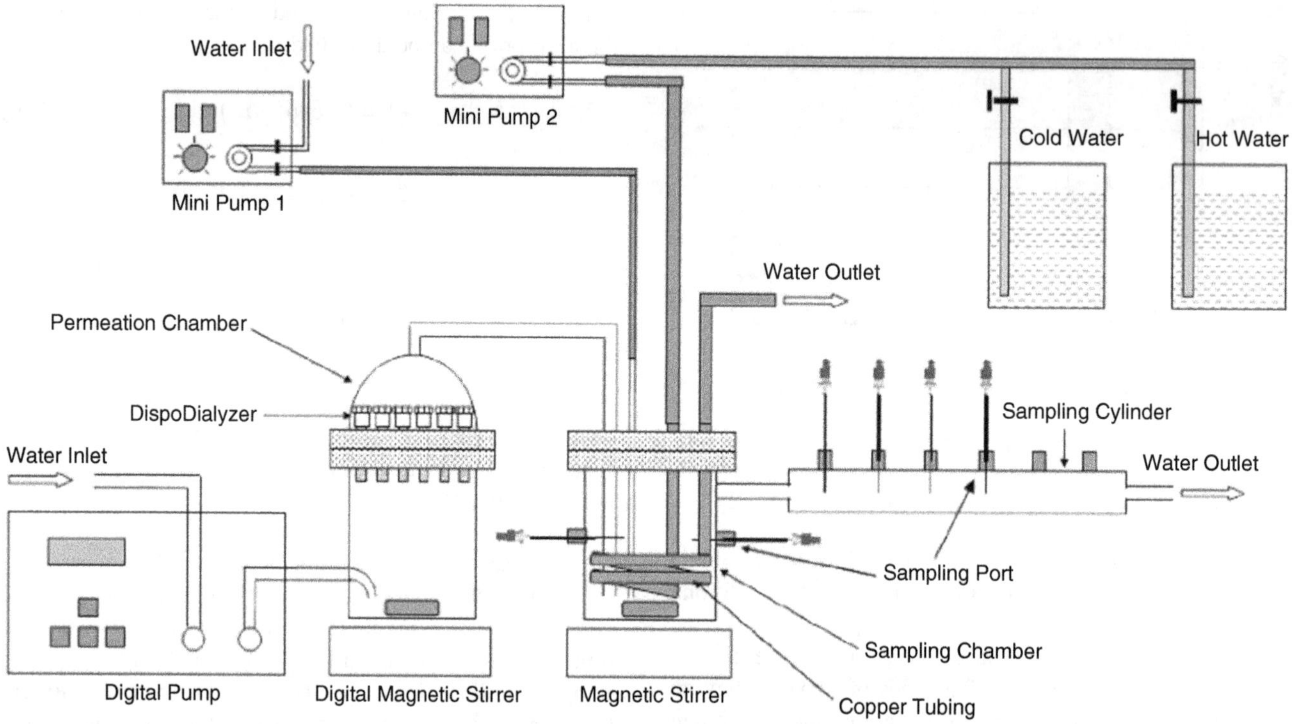

Figure 18.7. Schematic diagram of the new flow-through system for the generation of standard aqueous solution by permeation. Reprinted from reference 47 with permission from the American Chemical Society. Copyright 2009.

surfaces and limited sample volume. The system was successfully used to determine the distribution constant of PAHs between SPME fiber coating and water, a good agreement between K_{fw}, the distribution constant between SPME PDMS fiber coating water, and K_{ow}, the distribution coefficient between octanol and water, was observed.

A new flow-through standard water system has been developed[29] and has been modified recently.[47] Figure 18.7 presents the schematic diagram of the new flow-through standard water system based on permeation. It consists of a permeation chamber, a mixing chamber (this was also used as a sampling chamber), and a sampling cylinder and chamber (sampling cylinder is used for determining the effect of different linear velocities of the water). Water was filled in an approximately 12-L glass reservoir and delivered by a digital pump.

The key part of the generator is the DispoDialyzers (Spectrum Laboratories, Rancho Dominguez, CA) in the permeation chamber. Each DispoDialyzer was partially filled with pure standards (naphthalene, acenaphthene, fluorene, anthrancene, fluoranthene, pyrene) and pure water. The DispoDialyzers were then tightly sealed and placed into a 400-mL beaker filled with 300 mL of water and then sonicated for about 30 min. The prepared DispoDialyzers were deployed in the permeation chamber with an inlet close to its bottom and an outlet near its top. A little air was left inside the DispoDialyzers to suspend them at the top of the chamber.

A magnetic stir bar was placed at the bottom of the permeation chamber, and a digital magnetic stirrer was used for agitation. Analytes were dissolved in the water inside the DispoDialyzer. The dissolved analyte molecules further diffused through the membrane of the DispoDialyzer and were carried out by the water to the sampling cylinder and the sampling chamber. As the solids (PAHs) and liquid (water) coexist inside of the DispoDialyzer, the concentrations of analytes inside the DispoDialyzer will remain constant (saturated concentration) if the temperature remains constant. If the flow rate of water is also constant, the diffusion of analyte molecules from the inside of the DispoDialyzer to the outside water will reach steady state, creating a constant concentration of analytes. The observed variability in the concentration of the six PAH compounds was less than 20% over 3 months. Each DispoDialyzer was filled with 20–50 mg of each compound and the lifetime for the generator was more than 1 year.

The main difference between the new permeation system and a typical dilution system is that the syringe pump was exchanged with a permeation generator. The generator-based permeation approach offers many advantages, including convenience, low expense (DispoDialyzer is refillable), is solvent-free, has a long lifetime, and is capable of producing high concentrations of the target analytes. The ability to achieve high concentrations of analytes in the flow-through system is important, as it shortens the testing time for passive samplers. It is difficult to achieve high concentrations of analytes when a peristaltic pump or syringe pump was used for dilution, due to the presence of a high concentration of solvent and the frequent need to refill.

Table 18.7. Average Sample Transfer Efficiencies for Liquid Injection and SPME Injection with i.d. 0.8-mm SPME Liner

	Liquid Injection		SPME Injection	
Injection method	With wool, PTV, column o.d. 0.34 mm	With wool, PTV, column o.d. 0.67 mm	Z = 0.8 cm, column flow 1 mL/min, column o.d. 0.67 mm	Z = 0.8 cm, column flow 2 mL/min, column o.d. 0.34 mm
Average sample transfer efficiency	0.93	0.94	0.95	0.95

18.6. CALIBRATION OF SPME BY LIQUID INJECTION

Most quantitative approaches for SPME, especially for on-site sampling and analysis, require that the absolute amounts of the analytes are known, which are most often calibrated by the injection of liquid standards.

For both liquid injection and SPME fiber injection to a gas chromatograph, usually more than 95% of the sample is transferred into the capillary column with splitless injection. The feasibility of this calibration method relies on the assumption that sample transfer efficiencies are the same for both the liquid injection and the SPME injection. However, the rate of sample transfer into the column for liquid injection is affected by many factors, such as the dimensions of the liner, the presence of the wool, and the temperature of the injector. The sample transfer efficiency for SPME fiber injection is also affected by the carrier gas flow rate, the cross-sectional area of the space between the column and the liner, and the length of the column inside the liner.

An equation has been proposed to estimate the sample loss rate for SPME injection:[58]

$$\frac{n_L}{n_T} = \frac{DA}{Zv}, \tag{18.27}$$

where n_T and n_L are the total amount of the analyte in the fiber and the amount of any analyte lost in the process, respectively, D is the diffusion coefficient of the analyte, A is the cross-sectional area of the space between the liner and the column, Z is the length of the column inside the liner, and v is the flow rate of the carrier gas.

The correctness of Equation 18.27 has been validated experimentally through changing the values of A, Z, and v.[58] Equation 18.27 indicates that the sample transfer efficiency of SPME injection is affected by the cross-sectional area of the space between the column and the liner, the carrier gas flow rate, and the length of the column inside the liner.

The sample transfer efficiencies of liquid injection are affected by the dimensions of the liner and the presence of wool. Programmed temperature vaporizer (PTV) injection using a small inner diameter (i.d.). liner with wool can result in good sample transfer efficiency. It was found that less than 70% of the analytes were transferred into the column when high-temperature liquid injection method was used,[58] which suggests that high-temperature liquid injection may not be suitable for SPME calibration. No obvious difference in the response factors was found when methanol, acetone, and toluene were used as the solvent.

For both liquid injection and SPME injection, the cross-sectional area between the column inserted into the liner and the inside of the liner is the most important factor in sample transfer. The best way to avoid the deviation for SPME quantitative analysis caused by the calibration with liquid injection is if the analytes are completely transferred into the GC column, and PTV injection with a direct injection (DI) liner (SPI, Uniliner®, etc.) is the best choice to achieve this objective. Sample introduced into the DI liner can obtain high sample transfer efficiency, for both liquid injection and SPME fiber injection, and can be performed with a common autosampler and a regular syringe for liquid injection.

When DI liners are not available for the injectors, Table 18.7 summarizes the average sample transfer efficiencies for liquid injection and SPME injection when an SPME liner was used. The average sample transfer efficiencies were close to 95%, for both liquid injection and SPME injection. The close sample transfer efficiencies suggest that it is feasible to calibrate SPME injection by liquid injection with common SPME liners.

18.7. SUMMARY

Devices based on SPME are able to integrate sampling with sample preparation and sample introduction. They are simple but require careful calibration to deliver quantitative data. There are several strategies that can be used to obtain accurate and precise analytical data.

Equilibrium extraction of SPME is widely used for on-site sampling. The advantage of this method is that the sensitivity is higher than the nonequilibrium method. However, the disadvantage of this method is the length of time that is required to reach equilibrium, especially for compounds with lower volatility or for field water sampling. An air pump or fan is normally used for dynamic air sampling to shorten the sampling time. For field water sampling, because agitation of the sample is inconvenient, rotated-extraction phase samplers were developed to address the need.

The advantage of SPME diffusion-based calibration methods, including the calibration methods based on Fick's first law of diffusion, interface model, and cross-flow model, is that the concentration of target analytes in the sample can be directly calculated using the amount of analytes extracted by the extraction phase and no calibration curve or internal standard is required. This characteristic makes these calibration methods more suitable for on-site quantitative analysis, compared with traditional calibration methods.

The fiber-retracted (fiber-in-needle) SPME devices can be easily calibrated based on Fick's first law of diffusion. This type of SPME device is independent of the face velocity, affected only slightly by temperature, and the position of the SPME fiber can be easily adjusted to different diffusion path lengths in order to adapt to a large range of analyte concentrations. This approach is more desirable for long-term field sampling, especially where the convection conditions of air/water are difficult to measure and calibrate. But when the concentration of the target analyte is very low, such as in water for example, the sampling time for this type of SPME device will be very long to obtain an adequate amount of extracted analyte, because the diffusion coefficients of molecules are much lower in water than in air.

Diffusion-based calibration through an interface model and a cross-flow model are more suitable for fast sampling. These calibration methods require samples with a flowing medium. The sample velocity must be known and controlled for, necessitating additional equipment. The other approach is to use the rotated extraction phase devices to fix the convection conditions.

Recently proposed calibration methods, including the in-fiber standardization technique and standard-free kinetic calibration method, can be used for the quantification of passive sampling for the entire regime. Both approaches are pre-equilibrium techniques, and the analyte concentrations are calculated with an equilibrium equation, which requires that the distribution coefficients of analytes between the extraction phase and sample matrix are known. When the in-fiber standardization technique is used for quantification, the physicochemical properties of the preloaded standard and the target analyte should be similar. The results obtained before the sampling reaches equilibrium are representative of the TWA concentrations of the target analytes. Using the standard-free kinetic calibration method, the sampling time for equilibrium extraction can be markedly shortened. This calibration method can be used for the entire sampling period, without considering whether the system reaches equilibrium. This aspect of the technique makes it desirable for situations where the equilibration time profile is unknown or when a number of compounds are measured simultaneously.

REFERENCES

1. Pawliszyn, J. Sample preparation: Quo Vadis? *Anal. Chem.* **2003**, *75*, 2543–2558.
2. Ouyang, G.; Pawliszyn, J. Recent developments in SPME for on-site analysis and monitoring. *Trends Anal. Chem.* **2006**, *25*, 692–703.
3. Lord, H.; Pawliszyn, J. Evolution of solid-phase microextraction technology. *J. Chromatogr. A* **2000**, *885*, 153–193.
4. Ouyang, G.; Pawliszyn, J. Passive sampling techniques: Configurations and calibration methods. *J. Chromatogr. A* **2007**, *1168*, 226–235.
5. Pawliszyn, J. *Solid Phase Microextraction—Theory and Practice.* New York: Wiley-VCH; 1997.
6. Chai, M.; Pawliszyn, J. Analysis of environmental air samples by solid-phase microextraction and gas chromatography/ion trap mass spectrometry. *Environ. Sci. Technol.* **1995**, *29*, 693–701.
7. Muller, L.; Gorecki, T.; Pawliszyn, J. Optimization of the SPME device design for field applications. *Fresenius J. Anal. Chem.* **1999**, *364*, 610–616.
8. Mayer, P.; Vaes, W.; Wijnker, F.; Legierse, K.; Kraaij, R.; Tolls, J.; Hermens, J. Sensing dissolved sediment porewater concentrations of persistent and bioaccumulative pollutants using disposable solid-phase microextraction fibers. *Environ. Sci. Technol.* **2000**, *34*, 5177–5183.
9. Conder, J.; La Point, T.; Lotufo, G.; Steevens, J. Nondestructive, minimal-disturbance, direct-burial solid-phase microextraction fiber technique for measuring TNT in sediment. *Environ. Sci. Technol.* **2003**, *37*, 1625–1632.
10. van der Wal, V.; Jager, T.; Fleuren, R.; Barendregt, A.; Sinnige, T.; Gestel, C.; Hermens, J. Solid-phase microextraction to predict bioavailability and accumulation of organic micropollutants in terrestrial organisms after exposure to a field-contaminated soil. *Environ. Sci. Technol.* **2004**, *38*, 4842–4848.
11. Zeng, E.Y.; Tsukada, D.; Diehl, D.W. Development of a solid-phase microextraction-based method for sampling of persistent chlorinated hydrocarbons in an urbanized coastal environment. *Environ. Sci. Technol.* **2004**, *38*, 5737–5743.
12. Isetun, S.; Nilsson, U.; Colmsjo, A. Evaluation of solid-phase microextraction with PDMS for air sampling of gaseous organophosphate flame-retardants and plasticizers. *Anal. Bioanal. Chem.* **2004**, *380*, 319–324.
13. Isetun, S.; Nilsson, U. Dynamic field sampling of airborne organophosphate triesters using solid-phase microextraction under equilibrium and non-equilibrium conditions. *Analyst* **2005**, *130*, 94–98.
14. Larroque, V.; Desauziers, V.; Mocho, P. Development of a solid phase microextraction (SPME) method for the sampling of VOC traces in indoor air. *J. Environ. Monit.* **2006**, *8*, 106–111.
15. Qin, Z.; Bragg, L.; Ouyang, G.; Zhao, W.; Pawliszyn, J. Comparison of thin-film microextraction and stir bar sorptive extraction for the analysis of polycyclic aromatic hydrocarbons in aqueous samples with controlled agitation conditions. *J. Chromatogr. A* **2008**, *1196–1197*, 89–95.
16. Martos, P.A.; Pawliszyn, J. Time-weighted average sampling with solid-phase microextraction device: Implications for enhanced personal exposure monitoring to airborne pollutants. *Anal. Chem.* **1999**, *71*, 1513–1520.
17. Khaled, A.; Pawliszyn, J. Time-weighted average sampling of volatile and semi-volatile airborne organic compounds by the solid-phase microextraction device. *J. Chromatogr. A* **2000**, *892*, 455–467.
18. Chen, Y.; Pawliszyn, J. Time-weighted average passive sampling with a solid-phase microextraction device. *Anal. Chem.* **2003**, *75*, 2004–2010.
19. Ouyang, G.; Chen, Y.; Pawliszyn, J. Time-weighted average water sampling with a solid-phase microextraction device. *Anal. Chem.* **2005**, *77*, 7319–7325.
20. Koziel, J.; Jia, M.; Pawliszyn, J. Air sampling with porous solid-phase microextraction fibers. *Anal. Chem.* **2000**, *72*, 5178–5186.
21. Chen, Y.; Koziel, J.A.; Pawliszyn, J. Calibration for on-site analysis of hydrocarbons in aqueous and gaseous samples using solid-phase microextraction. *Anal. Chem.* **2003**, *75*, 6485–6493.
22. Chen, Y.; O'Reilly, J.; Wang, Y.; Pawliszyn, J. Standards in the extraction phase, a new approach to calibration of microextraction processes. *Analyst* **2004**, *129*, 702–703.

23. Chen, Y.; Pawliszyn, J. Kinetics and the on-site application of standards in a solid-phase microextraction fiber. *Anal. Chem.* **2004**, *76*, 5807–5815.

24. Ouyang, G.; Cai, J.; Zhang, X.; Li, H.; Pawliszyn, J. Standard-free kinetic calibration for rapid on-site analysis by solid-phase microextraction. *J. Sep. Sci.* **2008**, *31*, 1167–1172.

25. Martos, P.; Pawliszyn, J. Calibration of solid phase microextraction for air analyses based on physical chemical properties of the coating. *Anal. Chem.* **1997**, *69*, 206–215.

26. Martos, P.; Saraullo, A.; Pawliszyn, J. Estimation of air/coating distribution coefficients for solid phase microextraction using retention indexes from linear temperature-programmed capillary gas chromatography. Application to the sampling and analysis of total petroleum hydrocarbons in air. *Anal. Chem.* **1997**, *69*, 402–408.

27. Koziel, J.A.; Martos, P.A.; Pawliszyn, J. System for the generation of standard gas mixtures of volatile and semi-volatile organic compounds for calibrations of solid-phase microextraction and other sampling devices. *J. Chromatogr. A* **2004**, *1025*, 3–9.

28. Shurmer, B.; Pawliszyn, J. Determination of distribution constants between a liquid polymeric coating and water by a solid-phase microextraction technique with a flow-through standard water system. *Anal. Chem.* **2000**, *72*, 3660–3664.

29. Ouyang, G.; Chen, Y.; Pawliszyn, J. Flow-through system for the generation of standard aqueous solution of polycyclic aromatic hydrocarbons. *J. Chromatogr. A* **2006**, *1105*, 176–179.

30. Kolar, K.; Ciganek, M.; Malecha, J. Air/polymer distribution coefficients for polycyclic aromatic hydrocarbons by solid-phase microextraction sampling. *J. Chromatogr. A* **2004**, *1029*, 263–266.

31. Isidorov, V.A.; Vinogorova, V.T. Experimental determination and calculation of distribution coefficients between air and fiber with polydimethylsiloxane coating for some groups of organic compounds. *J. Chromatogr. A* **2005**, *1077*, 195–201.

32. Isidorov, V.; Purzynska, A.; Modzelewska, A.; Serowiecka, M. Distribution coefficients of aliphatic alcohols, carbonyl compounds and esters between air and Carboxen/polydimethylsiloxane fiber coating. *Anal. Chim. Acta* **2006**, *560*, 103–109.

33. Doong, R.; Chang, S. Determination of distribution coefficients of priority polycyclic aromatic hydrocarbons using solid-phase microextraction. *Anal. Chem.* **2000**, *72*, 3647–3652.

34. Yang, Z.Y.; Zeng, E.Y.; Xia, H.; Wang, J.Z.; Mai, B.X.; Maruya, K.A. Application of a static solid-phase microextraction procedure combined with liquid–liquid extraction to determine poly(dimethyl)siloxane–water partition coefficients for selected polychlorinated biphenyls. *J. Chromatogr. A* **2006**, *1116*, 240–247.

35. Kotowska, U.; Garbowska, K.; Isidorov, V.A. Distribution coefficients of phthalates between absorption fiber and water and its using in quantitative analysis. *Anal. Chim. Acta* **2006**, *560*, 110–117.

36. Paschkea, A.; Popp, P. Solid-phase microextraction fibre–water distribution constants of more hydrophobic organic compounds and their correlations with octanol–water partition coefficients. *J. Chromatogr. A* **2003**, *999*, 35–42.

37. Ouyang, G.; Pawliszyn, J. SPME in environmental analysis. *Anal. Bioanal. Chem.* **2006**, *386*, 1059–1073.

38. Arthur, C.L.; Killam, L.M.; Buchholz, K.D.; Pawliszyn, J. Automation and optimization of solid-phase microextraction. *Anal. Chem.* **1992**, *64*, 1960–1966.

39. King, J.; Readman, J.W.; Zhou, J.L. Determination of polycyclic aromatic hydrocarbons in water by solid-phase microextraction–gas chromatography–mass spectrometry. *Anal. Chim. Acta* **2004**, *523*, 259–267.

40. Sukola, K.; Koziel, J.; Pawliszyn, J. Diffusion-based calibration for SPME analysis of aqueous samples. *Anal. Chem.* **2001**, *73*, 13–18.

41. Augusto, F.; Koziel, J.; Pawliszyn, J. Design and validation of portable SPME devices for rapid field air sampling and diffusion-based calibration. *Anal. Chem.* **2001**, *73*, 481–486.

42. Tuduri, L.; Desauziers, V.; Fanlo, J.L. A simple calibration procedure for volatile organic compounds sampling in air with adsorptive solid-phase microextraction fibres. *Analyst* **2003**, *128*, 1028–1032.

43. Isetun, S.; Nilsson, U.; Colmsjo, A.; Johansson, R. Air sampling of organophosphate triesters using SPME under non-equilibrium conditions. *Anal. Bioanal. Chem.* **2004**, *378*, 1847–1853.

44. Ouyang, G.; Zhao, W.; Alaee, M.; Pawliszyn, J. Time-weighted average water sampling with a diffusion-base SPME device. *J. Chromatogr. A* **2007**, *1138*, 42–46.

45. Koziel, J.; Jia, M.; Khaled, A.; Noah, J.; Pawliszyn, J. Field air analysis with SPME device. *Anal. Chim. Acta* **1999**, *400*, 153–162.

46. Koziel, J.; Noah, J.; Pawliszyn, J. Field sampling and determination of formaldehyde in indoor air with solid-phase microextraction and on-fiber derivatization. *Environ. Sci. Technol.* **2001**, *35*, 1481–1486.

47. Ouyang, G.; Cui, S.; Qin, Z.; Pawliszyn, J. One-calibrant kinetic calibration for on-site water sampling with solid-phase microextraction. *Anal. Chem.* **2009**, *81*, 5629–5636.

48. Bragg, L.; Qin, Z.; Alaee, M.; Pawliszyn, J. Field sampling with a polydimethylsiloxane thin-film. *J. Chromatogr. Sci.* **2006**, *44*, 317–323.

49. Zhao, W.; Ouyang, G.; Pawliszyn, J. Preparation and application of in-fiber internal standardization solid-phase microextraction. *Analyst* **2007**, *132*, 256–261.

50. Zhao, W.; Ouyang, G.; Alaee, M.; Pawliszyn, J. On-rod standardization technique for time-weighted average water sampling with a polydimethylsiloxane rod. *J. Chromatogr. A* **2006**, *1124*, 112–120.

51. Ouyang, G.; Zhao, W.; Bragg, L.; Qin, Z.; Alaee, M.; Pawliszyn, J. Time-weighted average water sampling in Lake Ontario with solid-phase microextraction passive samplers. *Environ. Sci. Technol.* **2007**, *41*, 4026–4031.

52. Ai, J. Solid phase microextraction for quantitative analysis in nonequilibrium situations. *Anal. Chem.* **1997**, *69*, 1230–1236.

53. Ai, J. Headspace solid phase microextraction. Dynamics and quantitative analysis before reaching a partition equilibrium. *Anal. Chem.* **1997**, *69*, 3260–3266.

54. Wang, Y.; O'Reilly, J.; Chen, Y.; Pawliszyn, J. Equilibrium in-fibre standardization technique for solid-phase microextraction. *J. Chromatogr. A* **2005**, *1072*, 13–17.

55. Musteata, F.M.; Musteata, M.L.; Pawliszyn, J. Fast in vivo microextraction: A new tool for clinical analysis. *Clin. Chem.* **2006**, *52*, 708–715.

56. Ouyang, G.; Zhao, W.; Pawliszyn, J. Kinetic calibration for automated headspace liquid-phase microextraction. *Anal. Chem.* **2005**, *77*, 8122–8128.

57. Ouyang, G.; Pawliszyn, J. Kinetic calibration for automated hollow fiber-protected liquid-phase microextraction. *Anal. Chem.* **2006**, *78*, 5783–5788.

58. Ouyang, G.; Chen, Y.; Setkova, L.; Pawliszyn, J. Calibration of SPME for quantitative analysis by gas chromatography. *J. Chromatogr. A* **2005**, *1097*, 9–16.

Part III

Recent Developments

Chapter 19

Developments in the Use of Passive Sampling Devices for Monitoring Pollutants in Water

Graham A. Mills, Rocio Aguilar-Martínez, Richard Greenwood, Ian J. Allan, Janine Brümmer, Jesper Knutsson, and Branislav Vrana

19.1. INTRODUCTION

The development of sampling and associated preparative and analytical methods that are able to provide a representative picture of water quality presents an ongoing challenge to environmental chemists.[1–4] Most aquatic monitoring programs rely on the collection of discrete grab, spot, or bottle samples of water at a given time. Often, large volumes of water are required in situations where pollutants are present at only trace levels. The subsequent laboratory analysis of the sample provides merely a snapshot of the levels of pollutants at the time of sampling. There are, however, drawbacks to this approach in environments where contaminant concentrations vary over time, and where episodic pollution events can easily be missed. One solution to this problem is to increase the frequency of sampling or to install automatic sampling systems that can collect numerous water samples over a given time period. This is costly and, in many cases, impractical, since a secure site and a significant pretreatment of water is required. Such systems are rarely used in widespread sampling campaigns. In addition, spot sampling methods yield different apparent concentrations of pollutants depending on the pretreatment (e.g., filtering at 0.45 μm) applied to the sample, and they do not provide information on the truly dissolved, bioavailable, fractions of the contaminants. Interlaboratory exercises of analysis of priority pollutants found in spot water samples show that for many pollutants, there are recurring problems of reproducibility between laboratories, especially at the low concentrations that normally occur in the environment.[5] These problems are usually linked to the standard procedures applied to the samples.

Another approach that yields information on biologically relevant concentrations of pollutants in water is the use of biota. A number of test species (e.g., mussels, fish) can be used depending on the water body being investigated. These organisms can be deployed for extended periods of time during which they passively bioaccumulate pollutants from the surrounding water. Analysis of tissues or lipid extracts can give an indication of the equilibrium level of waterborne contamination. A number of factors can influence the results: metabolism, excretion, stress, viability, and the condition of test organisms. Furthermore, the extraction of analytes from the tissue of animals prior to instrumental analysis is complex.

Estimates of pollutant concentrations in water can also be made by measuring concentrations in benthic sediments and then using equilibrium distribution coefficients to derive levels for the dissolved, water-phase analytes. This approach is limited by the assumption of the attainment of equilibrium of the chemical between the sediment and water column, and the potential effects of differences in organic carbon quality among sediments or the formation of nonextractable sediment-bound residues that are not accounted for in the currently used equilibrium partition models.

Over the last two decades, alternative methods for monitoring water quality have been sought to overcome some of these difficulties. Of these, passive sampling techniques have shown much promise as tools for the measurement of aqueous concentrations of a wide range of priority pollutants. Passive samplers avoid many of the problems outlined above since they collect the target analyte *in situ* and without affecting the bulk solution. Depending on sampler design, the mass of pollutant accumulated by a sampler reflects either the concentration with which the device is at equilibrium or the time-weighted average (TWA) concentration to which the sampler was exposed. These devices can be deployed in the field for days to months. The technology is simple to use, and in contrast to active samplers, passive

Handbook of Sample Preparation, Edited by Janusz Pawliszyn and Heather L. Lord

samplers have no moving parts and require no power source and are therefore relatively inexpensive to employ. Furthermore, these devices can be deployed in most environmental conditions, thus making them ideal for ecological monitoring even in remote areas. As an analytical chemistry sampling tool, they also have the ability to preconcentrate analytes and thereby improve detection limits, and once sequestered can often preserve the integrity of the analyte until analysis and also help to reduce or, in some cases, eliminate the use of extraction solvents.

The use of passive samplers for environmental monitoring can be traced back to the early 1970s when they were used to measure concentrations of gaseous pollutants in air.[6] Since this time, their use in monitoring ambient air quality and workplace exposures to potentially harmful compounds has matured and become well recognized and accepted. A number of sampler designs are commercially available, and there are now established standards and official methods (e.g., American Society for Testing Materials [ASTM], United States Environment Protection Agency [USEPA], National Institute for Occupational Safety and Health (NIOSH), Comité Européen de Normalisation (CEN), and International Organization for Standardization [ISO] protocols) for the use of these devices, which form part of legal frameworks. In addition, worldwide monitoring networks have been set up using passive air monitors to follow the movement of persistent anthropogenic organic pollutants across the globe. In contrast, the application of passive samplers in monitoring water quality is some way behind the situation for air, and the technologies available for monitoring soils and sediments are even further from recognition. That said, the first patent for a passive sampler for monitoring waterborne pollutants was granted to Byrne and Aylott[7] in the United Kingdom in 1980, and since this time, many other devices have become commercially available or exist as advanced laboratory prototypes. A number of recent reviews have been published describing the design, calibration procedures, figures of merit, and applications of the different devices for monitoring the aquatic environment.[8–13] In addition, a book describing the semipermeable membrane device (SPMD),[14] and a general text on all passive sampling techniques for environmental monitoring[15] have recently become available.

Due to the wealth of existing literature, this chapter concentrates on the developments in passive sampling technology over the period 2000–2008. Its focus is limited to the application of passive sampling for monitoring priority and emerging pollutants in water and environmental issues that are driving current research in this area. It is not meant to be exhaustive but will highlight representative developments and papers in the area of focus.

19.2. PRINCIPLES OF OPERATION

Passive sampling can be defined in its broadest sense as any sampling technique based on the free flow of analyte molecules from the sampled medium to a receiving phase in a sampling device, as a result of a difference between chemical potentials of the analyte in the two media (Fig. 19.1). The net flow of analyte molecules from one medium to the other continues until equilibrium is established in the system, or until the sampling period is stopped.[16] Analytes are trapped or retained in a suitable medium within the passive sampler, known as a receiving or reference phase. This can be a solvent, chemical reagent, absorbent, or a porous adsorbent material. Unlike some extraction methods (e.g., solid-phase extraction), the receiving phase is exposed to the water phase, but *without* the aim of quantitatively extracting the dissolved contaminants. In most passive sampling systems, the absorption or adsorption of a pollutant from water generally follows the pattern shown in Figure 19.2. The exchange kinetics between a passive sampler and water phase can be described by a first-order, one-compartment mathematical model:

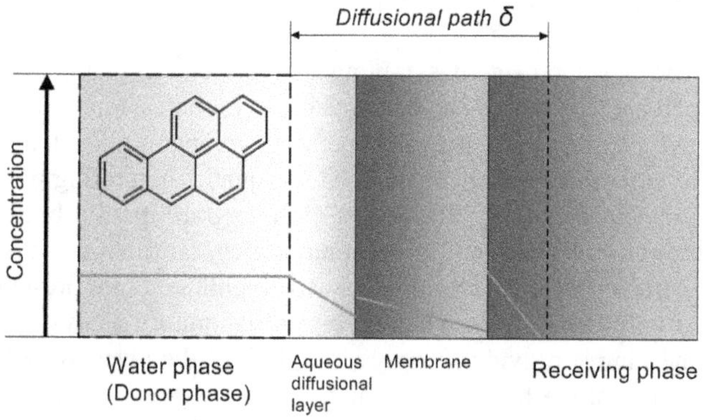

Figure 19.1. Schematic representation of concentration profiles in a passive sampling device. The driving force of the analyte accumulation is determined by the difference of chemical potentials of the analyte between the donor and receiving phase. The mass transfer of an analyte is governed by the overall resistance along the diffusional path, including contributions from particular resistances of individual barriers (boundary layers, potential biofilm, and membrane).

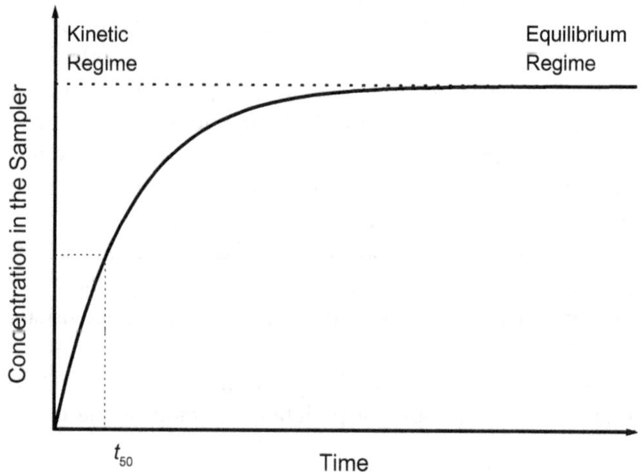

Figure 19.2. Passive sampling devices operate in two main regimes: kinetic and equilibrium.

$$C_s(t) = C_W \frac{k_1}{k_2}\left(1 - e^{-k_2 t}\right), \quad (19.1)$$

where $C_s(t)$ is the concentration of the analyte in the sampler at exposure time t, C_W is the concentration of the analyte in water, and k_1 and k_2 are the uptake and offload rate constants, respectively. Two main accumulation regimes, either equilibrium or kinetic, can be distinguished in the operation of a sampler during field deployment.

19.2.1. Equilibrium Sampling

In equilibrium sampling, the exposure time is sufficiently long to permit the establishment of thermodynamic equilibrium between the water and receiving phases. Knowledge of the receiving phase-water partition coefficient (K_{sw}) allows the estimation of the dissolved analyte concentration. An overview of equilibrium passive sampling devices has been published by Mayer et al.[17] The basic requirements of the equilibrium sampling approach are that equilibrium concentrations are reached after a known response time, and the device response time needs to be shorter than any fluctuations in the environmental medium. In order not to disturb the equilibrium, analyte depletion from the sampled medium during extraction must be avoided. Passive diffusion bag samplers (PDBS) based on equilibrium sampling have been extensively used for monitoring volatile organic compounds in groundwater.[18,19]

19.2.2. Kinetic Sampling

With kinetic sampling, it is assumed that the rate of mass transfer to the receiving phase is linearly proportional to the difference between the chemical activity of the contaminant in the water phase and that in the receiving phase. In the initial phase of sampler exposure, the rate of desorption of analyte from the receiving phase to water is negligible, and the sampler works in the linear uptake regime. Under these conditions, Equation 19.1 reduces to

$$C_s(t) = C_W k_1 t. \quad (19.2)$$

Equation 19.2 can be rearranged to an equivalent relationship:

$$m_S(t) = C_W R_S t, \quad (19.3)$$

where $m_S(t)$ is the mass of analyte accumulated in the receiving phase after an exposure time t, and R_S is the proportionality constant (sampling rate), which is the product of the first-order rate constant for uptake of pollutant k_1 and the volume of water that gives the same chemical activity as the volume of receiving phase. R_S may be interpreted as the volume of water cleared of analyte per unit of exposure time (e.g., milliliter per hour or liter per day) by the device.

For situations where aqueous concentrations may vary with time, the concentration is linearly proportional to TWA concentration, and sampling is called time integrative. When R_S is known, C_W (the TWA concentration of a pollutant in the water phase) may be calculated from the sampling rate (R_S), exposure time (t), and the amount (m_S) of the analyte collected by the receiving phase. For devices operating in the kinetic mode, R_S does not vary with the aqueous concentration C_W, but is often affected by water flow (turbulence), water temperature, and biofouling of the diffusive surface.

The advantage of kinetic or integrative sampling is that contaminants from episodic events commonly not detected with spot sampling are measured, and it can be used in situations with variable water concentrations. Kinetic samplers permit measurement of ultratrace, yet toxicologically relevant, contaminant concentrations over extended time periods. A sampling method that yields a TWA concentration over a comparatively extended period, using a single sample (extract from the passive sampling device), would give a more accurate picture of prevailing contaminant levels than that obtained with the use of infrequent spot samples.

19.3. MODELING AND CALIBRATION OF PASSIVE SAMPLERS

A number of models have been used to improve our understanding of the kinetics of contaminant transfer to passive samplers.[20–25] These models are essential for understanding how the amounts of contaminants sequestered relate to ambient concentrations, as well as for the design and evaluation of calibration experiments. Models differ in the number of phases and simplifying assumptions that are taken into consideration, such as the existence of (pseudo-) steady-state conditions, the presence or absence of linear concentration gradients within the membrane phase, the way in which transport within the aqueous boundary layer is modeled, and whether or not the aqueous concentration is constant during the sampler exposure.

The substance-specific kinetic constants k_1 and k_2 and distribution coefficient K can be determined in two ways. In theory, kinetic parameters characterizing the uptake of analytes can be estimated using semiempirical correlations between mass transfer coefficients, physicochemical properties (mainly diffusivities and permeabilities in various media) and hydrodynamic parameters.[22,25] However, because of the complexity of the water flow around passive sampling devices during exposure (usually nonstreamlined objects), it is difficult to estimate uptake parameters from the first principles. More substance-specific information is often available from the literature for the parameter K that characterizes the affinity of the receiving phase relative to that of water for a pollutant.

19.3.1. Calibration Techniques

In a practical approach, calibration of passive sampling exchange kinetics is performed in the laboratory at known exposure concentrations. In the static or batch exposure design, passive samplers are exposed in a single volume of contaminated water. This method has been used in the past for determining bioaccumulation factors and uptake rates of contaminants by fish and mussels and is suitable especially for samplers with low sampling rates. Static exposures have been used in the calibration of samplers[26–32] and is the calibration method often used in solid-phase microextraction (SPME)[33–35] (see Chapter 5). Calibrations using short equilibration times obtained with static exposures, performed in a small volume of water phase where a rapid depletion of analyte occurs, have sometimes been erroneously applied to field exposures, where depletion of water phase is usually insignificant.[29,30]

In the static renewal experimental design, the exposure water is refreshed batchwise.[36–38] This design can be used when static or continuous flow exposure designs are not practicable, that is, in case when a static exposure would result in depletion of the water phase below detection limits, or when it is difficult to maintain stable water concentrations in a flow-through exposure system. Analyte concentrations in the exposure water are determined at the beginning and at the end of each renewal cycle, to determine the average exposure concentration, assuming that the loss of analyte from water has been due to sorption by the sampler.

In the continuous flow design, the depletion of analyte in the water phase is prevented by a constant supply of freshly contaminated water to the exposure chamber. Stable aqueous concentrations can be maintained during the entire exposure if the daily exchanged volume of water in the exposure chamber is much larger (~20 times) than the total sampling rate (expressed as the analyte specific volume of water cleared by the device per unit time, e.g., liter per day) of all samplers in the tank. Components of the flow-through system are allowed to equilibrate (~48 h) with the exposure water prior to the calibration exposure in order to prevent the depletion of analytes due to sorption to surfaces. A further precaution is the prevention (e.g., by addition of biocides and/or filtration of water entering the tank) of the growth of microorganisms in the exposure chamber during long-term exposures. Mixing of stock solutions of test compounds in water-miscible solvents is the most widely used method for preparing contaminated water needed in the exposure experiments,[20,23,26,39,40,41] but generator column techniques,[42,43] partitioning controlled delivery from solid materials,[44–48] or permeation through a dialysis membrane[49] can also be used depending on the analytes of interest.

In order to predict TWA water concentrations of contaminants from levels accumulated in passive samplers, extensive calibration studies are necessary to characterize the uptake of chemicals under various exposure conditions. Uptake kinetics of chemicals depend not only on the physicochemical properties of the diffusand, but also on the sampler design and environmental variables such as water temperature,[43] water turbulence,[28,40] and biofouling of the diffusive surface of the sampler.[38]

19.3.2. Environmental Factors Affecting Passive Sampling

It is important to consider the mechanisms of the exchange process between the aqueous phase and the sampler components. The uptake to the receiving phase (in the absence of fouling) may be controlled by diffusion across the diffusion-limiting membrane or across the aqueous diffusive boundary layer at the membrane-water interface, forming the rate-limiting step.[22] Water turbulence affects the thickness of the unstirred layer of water that forms part of the diffusion-limiting barrier at the sampler surface and, as a result, affects the mass transfer of the analytes. The rate-limiting step depends on the type and properties of the membrane, the environmental conditions prevailing during sampling, and the properties of the compound being sampled.

A number of methods have been developed to compensate the effect of environmental variables on sampler performance. Huckins et al.[27] described a method to estimate the uptake kinetics in both laboratory and field situations by spiking sampling devices prior to exposure with a number of performance reference compounds (PRCs) that do not occur in the environment. Where factors influencing uptake kinetics affect the offloading kinetics of PRCs in an identical manner, the release rate of these compounds is a measure of the exchange kinetics between the sampler and water, and can be used in field exposures to compensate for variations in environmental conditions.[40] An implicit assumption in the use of PRCs in passive samplers is that the overall uptake and release of targeted chemicals and PRCs are governed by first-order kinetics and that the sum of resistances to mass transfer across sampler associated barriers is equal in both directions, that is, that isotropic exchange kinetics apply. These characteristics are observed in the partitioning of residues between immiscible liquid phases and between certain nonpolar polymeric films and water or air.[27,50] This assumption, however, may not be valid for samplers fitted with solid-phase sorbent receiving phases (e.g., Polar Organic Chemical Integrative Sampler [POCIS] or polar pollutant variant of Chemcatcher® sampler) because of the fundamental differences between solute partitioning and adsorption phenomena.[51] Also, because the solid-phase sorbents often act as an infinite sink with a very high sorption capacity, the selection of a PRC with a sufficient fugacity from such samplers is problematic. Similar problems exist with samplers used to measure metals, as often the receiving phases used are chelating agents and therefore do not permit the use of the PRC approach.

19.3.3. Biofouling

Unprotected surfaces submersed in water eventually become colonized by bacteria and various flora and fauna that may ultimately form a biofilm. The thickness and density of this biofilm vary not only from exposure to exposure but also from zone to zone on the same membrane. The composition of biofilms varies significantly depending on the properties of the aquatic system being monitored. Biofouling can affect the overall resistance to mass transfer by increasing the thickness of the barrier and by blocking any water-filled pores in the diffusion limiting membranes. Colonizing organisms may damage the surface of membranes that are made of a biodegradable material (e.g., cellulose acetate). Huckins et al.[14] reported a 20–70% impedance in uptake of polycyclic aromatic hydrocarbons (PAHs) in severe cases, but also showed for fouled SPMDs that PRCs can be used to correct for biofouling during deployment. Their model describing the mass transfer in a biofilm indicates that it behaves like an immobilized water layer, with a resistance that is independent of the biofilm/water partition coefficient, which would mean a similar mobility of compounds in biofilm, independent of their hydrophobicity.[12] Richardson et al.[38] observed that biofouling caused a reduction, by as much as 50%, in the uptake of PAHs and organochlorine pesticides by SPMDs. The problem of sampler fouling may be reduced by the selection of suitable construction materials. For example, the polyethersulfone diffusion membranes used in POCIS and one design of the Chemcatcher® are less prone to fouling than polyethylene used in SPMDs. This observation may be due to the low surface energy of polyethersulfone discouraging the initial onset of the biofouling process by creating unfavorable conditions for the settlement of colonizing microorganisms.[52] Alternatively, coating the membrane with a low surface energy material, for example, Nafion®, can be used to inhibit biofouling.[53] In addition, certain solvent-filled membrane devices are protected from fouling by a slow seeping of a fouling-inhibiting solvent (e.g., *n*-hexane) from the sampler during the exposure. Protective screens made of copper or bronze mesh have also been shown to inhibit biofouling; however, they cannot be used when monitoring for heavy metals. Recently, a novel approach using antibiotics to prevent the development of biofilms has been attempted with diffusive gradients in thin films (DGT) device.[54]

19.4. FIELD VALIDATION

For validation and acceptance of passive sampling methodology, demonstration of its field performance alongside conventional spot or grab sampling is essential. The performance of passive samplers may be affected by fluctuations in water temperature, flow velocity, acidity, and salinity at the deployment site. Knowledge of the effects of these environmental variables on the kinetic and equilibrium parameters of the sampler is necessary to evaluate sampler uptake data from field deployments. Robustness or at least a measurable responsiveness to these factors has to be demonstrated in order for passive sampling technology to be accepted for use in routine water quality monitoring programs.

It is difficult to compare repeated spot sampling data with those obtained using passive sampling. One possible reason for differences sometimes observed between spot sampling and passive sampling data is that the discontinuous spot samples may be taken at time points when concentrations of a pollutant were higher or lower than average concentrations, and large fluctuations may occur in the intervals between samples. Any change in ambient concentration during the intervals between samples is undetected when the conventional sampling approach is used, but passive samplers may reflect such events.

When comparing TWA concentrations calculated from spot samples and passive samplers, it is also important to consider the differences in contaminant fractions that are measured by the two methods. For example, TWA concentrations estimated using passive samplers usually do not include the fractions of pollutants bound to particles and colloids in the water. In contrast, filtered spot water samples can still contain a contaminant fraction that is bound to dissolved organic material (DOM) present in water. The truly dissolved fraction of hydrophobic analytes in water will depend on the level and quality of DOM, and these may fluctuate during the sampling period.[55] Booij[22] estimated that a twofold reduction in the freely dissolved concentration can be expected for compounds with log $K_{OW} = 7$ when the DOC level is as low as 1 mg/L. There is, however, some uncertainty associated with the calculations of the truly dissolved fraction using available models, and at this DOC level, a reduction such as that described above in the freely dissolved concentration could easily occur for compounds with log $K_{OW} > 6$.[22]

Some studies have tested the accuracy of field pollutant concentrations derived using passive samplers by comparing the data with those obtained by active sampling methods. Ellis et al.[56] found for a number of organochlorine pesticides that the measurements of concentrations derived from SPMDs were in close agreement with those found in spot samples processed through a tangential-flow ultrafilter. However, many compounds found in SPMD extracts remained undetected in filtered spot samples. Other field validation studies measured the concentrations of hydrophobic organic pollutants present in water at sub-parts per billion levels using SPMDs and high-volume solid-phase *in situ* extraction techniques such as Infiltrex® and compared the data from the two methods.[57,58] Rantalainen et al.[57] deployed SPMDs in the water column of the Lower Fraser River (BC, Canada) to compare the levels and congener profiles of polychlorinated dibenzo-*p*-dioxins, dibenzofurans, and non-*ortho*-chlorinated biphenyls against those sampled by an Infiltrex resin column water sampler. The calculated water concentrations obtained from the SPMDs were similar to those measured using the Infiltrex sampler. Axelman et al.[58] compared the concentrations of dissolved PAHs calculated from SPMDs and those obtained using an online filtration system at a site contaminated by discharges from an aluminum reduction plant. The concentrations obtained from SPMD data showed a systematic deviation from the active sampling method that increased with PAH hydrophobicity. Although the reason for this deviation was not explained, the method used for SPMD data evaluation was indicated as a potential source of a systematic error. Zeng et al.[59] compared the performance of an equilibrium sampling device based on SPME with that of an Infiltrex 100 *in situ* large-volume extractor in the monitoring of a range of polychlorinated biphenyls (PCBs) and organochlorine pesticides and found a very good agreement of results obtained using the two sampling techniques. Hyne et al.[60] compared the field performance of the integrative trimethylpentane-containing passive samplers (TRIMPS) with daily river water samples for several pesticides. The pesticide concentrations determined using the TRIMPS were within twofold of the cumulated mean concentrations of endosulfan and ethyl-chlorpyrifos in daily spot samples for deployment periods of 7–22 days. Other field validation trials for organic pollutants have assessed the performance of passive samplers: Chemcatcher®,[39,61–63] ceramic dosimeters,[64,65] POCIS,[36,37] SPMDs,[41,66] SPME,[67] and membrane-enclosed sorptive coating (MESCO)[50] devices alongside conventional frequent, discrete volume, spot sampling.

For inorganic pollutants, several field validation studies comparing the Chemcatcher and DGT with other sampling methods and theoretical models have been reported.[68–75]

Figure 19.3 shows this approach using an example from a field validation study performed in April 2005 at a sampling site on the River Meuse at Eijsden, The Netherlands. Here, TWA concentrations of cadmium calculated at various time intervals from amounts accumulated in passive samplers were compared with those determined from concentrations measured by standard spot water sampling. The TWA concentrations obtained from passive samplers over 14- and 28-day exposures were in reasonable agreement with those obtained from the filtered spot samples, but lower than those found in the unfiltered samples.

In order to validate and demonstrate the reliability of data obtained with passive sampling methodology, there is a need for reliable reference methods that can provide measures of concentrations of target contaminants in water that are equivalent to those obtained with passive samplers, and at a reasonable cost.

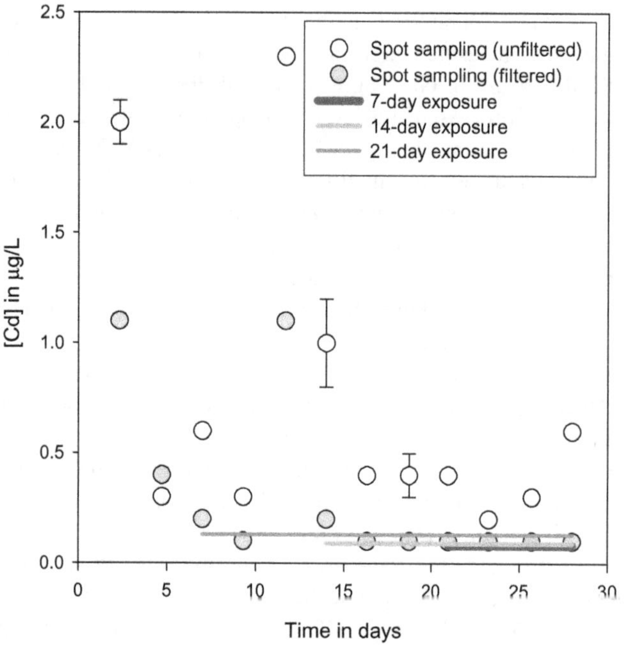

Figure 19.3. Comparison of concentration of cadmium in the River Meuse (April 2005) measured by standard spot sampling (nonfiltered and filtered through a 0.45-µm filter) and by the Chemcatcher sampler after exposures of 7, 14, and 21 days, respectively.

19.5. QUALITY CONTROL (QC)

The level of QC applied to passive sampling varies with project goals and analytical procedures involved. The application of appropriate QC procedures/parameters is an essential consideration in both sampler deployment and the subsequent sampler analysis. QC samples should address issues of purity of materials used to construct a device, and potential contamination during preparation, transport, deployment, retrieval, and subsequent storage. Furthermore, QC protocols are required for analyte recovery, further processing (enrichment and fractionation operations), and analysis. Control charts are recommended to monitor analyte recoveries throughout a project. The QC samples related to the performance of passive sampler studies include fabrication blanks, process blanks, reagent blanks, field blanks, sampler spikes, and procedural spikes. DeVita and Crunkilton[76] examined the QC issues associated with the use of SPMDs for monitoring PAHs in water. The results of their study showed that QC measures applied to SPMDs met or surpassed conventional guidelines (USEPA Method 610 for PAHs in water) for precision and accuracy. However, assessing the accuracy of determinations made by passive samplers may prove difficult, as the results may not be directly comparable with total concentrations found in spot samples or by other sampling techniques. This is because only very few methods other than passive samplers are capable of measuring the dissolved contaminant fractions. When environmental conditions at an exposure site differ from laboratory calibration conditions or calibration data are not available, samplers spiked with PRCs serve as a special type of QC sample. These provide information about *in situ* uptake kinetics.[27] Stuer-Lauridsen[8] reviewed the quality assurance that would be required for acceptance of passive samplers in water quality monitoring programs. He concluded that, at the moment, there is a lack of data from interlaboratory proficiency tests that are required for full method validation. Field studies demonstrating a good comparability of results obtained with passive samplers and those obtained with conventional sampling techniques also increase the body of evidence required for acceptance of passive sampling technology. The recently published British Standards Institute Publicly Available Specification on the use of passive samplers for monitoring pollutants in water gives helpful advice on this issue.[77]

19.6. RECENT DEVELOPMENTS IN FIELD USE OF PASSIVE SAMPLING DEVICES

Passive samplers usually combine sampling, selective analyte isolation, preconcentration, and preservation of speciation in one single step and simplify the operations performed at the sampling site. They eliminate the need for an energy/power supply and allow the entire sampling setup to be simplified and miniaturized. Once the sample is collected, further steps in its processing are usually the same as for other sampling/sample preconcentration analytical methods. They include extraction/desorption of the analytes, instrumental analysis, and processing of the data.

Passive sampling technology is widely applicable in monitoring studies and the results obtained can be interpreted at different levels of complexity. Passive samplers have been employed in field studies aimed at screening for the presence of pollutants, investigations of temporal and spatial trends in levels of aquatic pollutants, speciation of contaminants, assessment of pollutant fate, measurement of TWA concentrations of pollutants, comparison of contaminant patterns in biota and passive samplers (biomimetic approach), and toxicity assessment of pollutants in extracts from passive samplers.[8,9,11–15] As stated earlier, several reviews of passive samplers used for monitoring environmental pollutants have been published recently, and the reader should refer to these for a full description of the design and construction of these devices.[8–12,78] Tables 19.1 and 19.2 summarize some of the properties of the devices

Table 19.1. Overview of Passive Sampling Devices for Organic Contaminants

Sampler	Analytes	Sampling Purpose	Typical Deployment Period	Sample Preparation for Chemical Analysis	Reference
Ceramic dosimeter	PAHs, BTEX, chlorinated hydrocarbons	Integrative sampling in groundwater	Up to 1 year	Solvent extraction or thermal desorption	64
Chemcatcher	Polar and nonpolar organics	Integrative	14 days to 1 month	Solvent extraction	39
Dosimeter according DiGiano et al.	BTEX and atrazine	Integrative	Up to 2 months	Solvent extraction	151
Ecoscope	Hydrophobic organic compounds	Qualitative screening		Direct injection of solvent or after concentration	152
Gaiasafe sampler	Metals, anions, organic compounds	Screening	2 days to 2 months	Solvent extraction	153
Gore-Sorber	BTEX, MTBE, PAHs, VOCs, semi-VOCs	Equilibrium	14 days	Thermal desorption	154
Low density polyethylene (LDPE) and silicone strips	Hydrophobic organic compounds	Integrative	1 month	Solvent extraction	43,155
Membrane-enclosed sorptive coating (MESCO)	PAHs, PCBs, organochlorine pesticides	Integrative	2 weeks	Thermal desorption	66
Negligible depletion solid-phase microextraction (SPME)	Hydrophobic chemicals, including PAHs, PCBs, petroleum hydrocarbons, organochlorine pesticides, aniline, phenols	Equilibrium	Hours	Thermal desorption in GC inlet	156
Passive sampler according to Lee and Hardy	Chlorobenzenes, nitrobenzenes, and nitrotoluenes	Integrative	Up to 1 day	Solvent extraction	157
Passive diffusion bag (PDB)	Polar organic compounds, VOCs, metals, trace elements	Equilibrium sampling in groundwater	2 weeks	Conventional analyis of the receiving water phase	19
Passive in situ concentration-extraction sampler (PISCES)	PCBs	Integrative	2 weeks	Volume reduction of the receiving phase	158
Polar organic chemical integrative sampler (POCIS)	Herbicides and pharmaceuticals with log $K_{OW} < 3$	Integrative	Up to 2 months	Solvent extraction	36

Table 19.1. *(Continued)*

Sampler	Analytes	Sampling Purpose	Typical Deployment Period	Sample Preparation for Chemical Analysis	Reference
Sampler according to De Jonge and Rothenberg	Wide range of contaminants	Flux-proportional sampling in soil and groundwater	1 month	Solvent extraction	159
Sampler according to Kot-Wasik et al.	Phenols, acid herbicides, triazines	Integrative	1 month	Analysis of a subsample of solvent is taken and analyzed without further cleanup steps	160
Solvent-filled dialysis membranes	Hydrophobic organic compounds	Integrative	1 month	Volume reduction of the receiving phase	161
Solid-phase adsorption toxin tracking (SPATT)	Polar phytotoxins	Integrative	1 week	Solvent extraction	101
Semipermeable membrane device (SPMD)	Hydrophobic organic compounds	Integrative	1 month	Dialysis in organic solvents, size exclusion chromatography	14
Thin-layer chromatographic plate	Organophosphates	Screening	1 month	Solvent extraction	90
TRIMPS	Pesticides	Integrative	1 month	Direct analysis of the receiving phase solvent	92
Time-weighted average–solid-phase microextraction (TWA-SPME)	BTEX, PAHs	Integrative	Up to several days	Thermal desorption in GC inlet	145,162
Mobile passive sampling device	PCBs	Integrative	Several weeks	Solvent extraction	163

BTEX, benzene, toluene, ethylbenzene, and xylene.

Table 19.2. Overview of Passive Sampling Devices for Inorganic Pollutants

Sampler	Analytes	Sampling Purpose	Typical Deployment Period	Reference
Chemcatcher (inorganics version)	Range of heavy metals included in priority pollutant lists, including organotin and mercury compounds	*In situ* sampling, integrative, speciation	2–4 weeks	61,115
Diffusion gradient in thin films (DGT)	Most heavy metal pollutants, phosphorus, sulfide, and radioactive metal isotopes	Integrative, speciation, screening, mimicking biological uptake	1 day to 1 week	116
Ecoscope	Range of heavy metals included in priority pollutant lists	Qualitative screening for presence and absence of metal	1–4 weeks	152
Passive integrative mercury sampler (PIMS)	Neutral mercury species	Preconcentration, screening	Weeks to months	164
Permeation liquid membrane (PLM)	Heavy metals	Bioavailable metal species	Hours	165
Supported liquid membrane (SLM)	Divalent metal ions	Integrative field sampling, preconcentration of trace elements, mimicking biological membranes	Days	166
Solid-phase microextraction (SPME)	Organometallic compounds	Emerging technique for *in situ* time integrative sampling of organometallic compounds	Hours	145,146
Stabilized liquid membrane device (SLMD)	Divalent metal ions	Preconcentration, *in situ* sampling, determination of labile metal ions in grab samples	Days to several weeks	167

Table 19.3. List of Organic Priority and Emerging Pollutants and Selected References Describing Applications of Passive Sampling for Monitoring These Compounds

Pollutant Group	On the USEPA List of Priority Pollutants	On the EU WFD List of Priority Pollutants	Selected Reference to Passive Sampling Technique[a]
Algal toxins	—	—	101
Alkyl phenols	—	Yes	52
Aromatic hydrocarbons (BTEX)	Yes	Yes	64,162,168
Brominated diphenyl ethers	—	Yes	98,99
Chlorophenols	—	Yes	169
Dithiocarbamates and carbamates	—	—	170
Endocrine disruptors	—	—	94,95
Fluorinated surfactants	—	—	102
Methyl *tert*-butyl ether	—	—	113
Organochlorine pesticides	Yes	Yes	37,50,163,171,172
Organophosphate pesticides	—	Yes	37,90
Pharmaceuticals and illicit drugs	—	—	86
Phenoxy acid herbicides	—	—	173
Phenylurea pesticides	—	Yes	32,39,88
Phthalates	—	Yes	52
Polychlorinated biphenyls	Yes	—	169,174,175
Polychlorinated dioxins and furans	—	—	48,81
Polycyclic aromatic hydrocarbons	Yes	Yes	81
Pyrethroids	—	—	176
Sunscreen/UV filters	—	—	96,97
Synthetic musks	—	—	171,177
Triazines	—	Yes	32,88
Volatile organic chemicals	—	—	18

[a]Examples of papers published preferably after 2000 are listed.

USEPA, United States Environmental Protection Agency; EU WFD, European Union's Water Framework Directive; BTEX, benzene, toluene, ethylbenzene, and xylene.

used to measure organic and inorganic contaminants in water.

Passive samplers have been available for some time to measure a wide range of recognized pollutants of concern in water. An increasing number of compounds are being added to a list of emerging pollutants whose potential deleterious effects are being evaluated, and passive samplers are being developed to monitor those chemicals (Tables 19.3 and 19.4).

19.7. ORGANIC POLLUTANTS

19.7.1. Hydrophobic Organic Compounds

Persistent hydrophobic pollutants, such as organochlorine pesticides, PCBs, and PAHs are a concern, and have been classified as priority pollutants (e.g., by the USEPA [Clean Water Act] and European Union) for some time. Due to their low aqueous solubilities, the concentrations of nonpolar pollutants dissolved in water are very low; usually less than 1 ppb. These compounds bind strongly to particulate matter and are finally deposited in the sediment. Nevertheless, because organisms often bioconcentrate these low levels of contaminants in water to relatively high concentrations in their tissues, the determination of the dissolved portion of pollutants is critical for assessing the potential for detrimental biological impacts and for associated risk assessments.

Among other methods (e.g., online continuous monitoring, biomonitoring) attempting to provide representative long-term information on concentrations of these compounds, passive sampling technology has the potential to become a reliable, robust, and cost-effective tool that could be used in routine monitoring programs. A range of passive sampling devices has been developed and applied in the monitoring of hydrophobic organic pollutants in water.[8,9,11,12,78] Among the passive sampling devices for hydrophobic organic contaminants, SPMD is the most mature sampling technique.[20] The design of the SPMD was first published in 1990, and since then, over 200 studies have been reported.[79] Several reviews and one monograph have been recently published on this technology.[13,14,80,81]

19.7.2. Polar Organic Compounds

The focus of environmental research has recently been extended from persistent bioconcentratable organic chemicals to more water-soluble polar or hydrophilic organic

Table 19.4. List of Inorganic and Organometallic Priority Pollutants and selected References Describing Applications of Passive Sampling for Monitoring These Compounds

Pollutant	On the USEPA List of Priority Pollutants	On the EU WFD List of Priority Pollutants	Selected Reference to Passive Sampling Technique
Antimony	Yes	—	123
Arsenic	Yes	—	178
Beryllium	Yes	—	123
Cadmium	Yes	Yes	68,118,119,132,167,179–181
Chromium (III)	Yes	—	120,182
Chromium (VI)	Yes	—	118,120,182
Copper	Yes	—	53,69,118,119,167,179–181,183–188
Lead	Yes	Yes	118,119,165,167,179,184,186
Mercury	Yes	Yes	117,164
Nickel	Yes	Yes	70,119,167,179,184,186,188
Selenium	Yes	—	123
Silver	Yes	—	123
Thallium	Yes	—	123
Zinc	Yes	—	118,167,179,181,186,188
Organotin compounds	—	Yes	141–143,146

USEPA, United States Environmental Protection Agency; EU WFD, European Union's Water Framework Directive.

compounds including some pesticides, pharmaceuticals, personal care products, and endocrine disrupting compounds.[4] Polar organic compounds entering aquatic systems are often present at low concentrations, posing problems with most conventional sampling and analytical procedures, although the increasing use of liquid chromatography with mass spectrometry (LC-MS and LC-MS/MS) has led to a "revolution" in environmental analysis, providing a tool that enables the identification of these compounds and their determination down to low nanogram per liter levels. There has been a considerable effort directed toward the development of active sampling methods for polar organics in water. Much research has focused on the use of solid-phase extraction techniques[82] (see Chapter 4) and alternative sample preparation methods such as hollow fiber liquid-phase microextraction[83] (see Chapter 7). Although these novel approaches are advantageous over earlier liquid–liquid extraction methods, they often require transport, preservation, and processing of large volumes of water to satisfy the detection limit requirements of commonly used analytical methods. Alternative sampling methods such as on-site automatic systems can be expensive to operate and maintain.

The residence times of polar organic compounds in aquatic environments are generally lower than those of hydrophobic organic compounds. Fluctuations in concentrations, however, are more likely to occur for polar compounds as a result of seasonal, weekly, or diurnal patterns in pesticide applications or sewage effluent releases. Episodic events such as spills or storm water runoff may not be detected when conventional spot sampling methods are employed. Thus, there is a critical need for sampling methods capable of continuous *in situ* preconcentration of polar organic compounds and of providing a representative measure of levels of these chemicals during a given time period.

Passive samplers offer an attractive alternative to traditional sampling methods for polar pollutants. An advance in the development of a passive sampling technique for hydrophilic compounds has been achieved at the U.S. Geological Survey. A device called POCIS has been used to monitor hydrophilic contaminants such as pesticides, prescription and over-the-counter drugs, steroids, hormones, antibiotics, and personal care products.[36] The POCIS consists of a solid sorbent material contained between two microporous polyethersulfone membranes. These are held in position by two "O" rings that are clamped together by a series of bolts. The active sampling surface area is 41 cm^2. Two versions (i.e., generic or pharmaceutical configurations) of the device are available, and this depends on the composition of the receiving phase used. POCIS samples hydrophilic chemicals (with log $K_{OW} < 3$) from the dissolved phase and permits the determination of TWA concentration in water over extended (several weeks) periods. To date, more than 120 hydrophilic organic contaminants have been identified in POCIS extracts.[36,84,85]

Another design suitable for monitoring of a broad range of polar organic contaminants is one variant of the Chemcatcher sampler that consists of a C_{18} 3M Empore™ disk as the receiving phase combined with a polyethersulfone diffusion-limiting membrane.[39] This sampler is suitable for sampling of chemicals with log $K_{OW} < 3$, and its sampling capacity can be modulated by choosing the appropriate receiving phase disk material within the Empore range (e.g., one of the modified poly(styrene divinylbenzene) copolymer disks: SDB-XC, SDB-RPS, Cation-SR, or Anion-SR).

19.7.3. Pharmaceuticals and Drugs

Jones-Lepp et al.[86] assessed the ability of the POCIS sampler to accumulate four pharmaceuticals (azithromycin, fluoxetine, omeprazole, levothyroxine) and two illicit drugs (methamphetamine and methylenedioxymethamphetamine [MDMA or ecstasy]) in a wastewater effluent. The antibiotic azithromycin was detected in the treated effluent from three municipal wastewater treatment plants. Two illicit drugs, methamphetamine and MDMA, were detected and confirmed in the effluent from two wastewater treatment plants. The TWA aqueous concentrations estimated from the amounts accumulated in POCIS were at low nanogram per liter levels. In another study aimed at screening for 96 hydrophilic organic pollutants in wastewater, several pharmaceuticals were detected in POCIS extracts.[52] In that study, some of the pharmaceuticals detected in POCIS extracts remained undetected in extracts from standard spot samples taken from the water column. Mills et al.[85] recently reviewed the use of passive sampling devices for the measurement of pharmaceuticals and personal care products in different aquatic environments. A number of potentially valuable forensic niche applications of the technology were identified such as estimating illicit drug usage from the populous within a given wastewater catchment area. Accurate calibration data (in terms of uptake rates) for most pharmaceutical compounds and related products are, however, lacking for both the Chemcatcher and POCIS devices.[87]

19.7.4. Pesticides and Biocides

Kingston et al.[39] used the polar variant of the Chemcatcher sampler for monitoring diuron and Irgarol 1051, two herbicides that are used in antifouling paints, at two marine harbor sites. Shaw and Müller[88] used Empore disks to collect polar herbicides and SPMDs to sequester more hydrophobic compounds to measure their levels in the Wet Tropics region of the Great Barrier Reef, Australia. Moore et al.[89] deployed SPMDs to monitor and determine the fate of methyl-parathion in runoff from a simulated wetland outflow. Leblanc et al.[90] developed an *in situ* solid-phase extraction method based on thin-layer chromatography for screening pesticides in water. They tested the method in laboratory and field studies for several substances, including the organophosphorus pesticides, diazinon, and chlorpyrifos. Leonard et al.[91,92] employed the TRIMPS sampler to monitor concentrations of hydrophobic pesticides, including endosulfan, in the riverine environment of cotton-growing regions of Australia. Lindström et al.[93] studied the occurrence of triclosan, an important bactericide used in various personal care and consumer products, in surface waters in Switzerland. Passive sampling with SPMDs indicated the presence of methyl-triclosan, an environmental transformation product of triclosan in lakes with inputs from anthropogenic sources. However, SPMDs did not accumulate any parent triclosan.

Several pesticides (e.g., atrazine, metolachlor, prometon, diazinon, and pentachlorophenol) were detected during a screening for 96 targeted analytes in extracts from POCIS deployed in a stream that receives agricultural, municipal, and industrial wastewaters.[52] Some of the pesticides (e.g., diazinon and pentachlorophenol) detected in POCIS extracts remained undetected in extracts from spot water samples. Passive sampling techniques offer an efficient and effective alternative for detecting contaminants, and have a number of advantages compared with individual water-column samples: a greater number of pollutants detected, a greater mass of chemical residues sequestered, and the ability to detect chemicals, which dissipate rapidly. In addition, the ease of performing a single deployment instead of collecting and processing multiple water samples is a further benefit.

19.7.5. Endocrine Disrupting Compounds and Hormones

Estrogenicity of river water is highly variable, and it is difficult to obtain a representative measure of this. Passive samplers offer an option for continuous sampling of these compounds in situations of variable aqueous concentrations over an extended time period, enabling the detection of episodic contamination events.

Rastall et al.[94] coupled biomimetic sampling using SPMDs to a bioassay-directed chemical analysis scheme for the detection and identification of bioconcentratable hydrophobic estrogen receptor agonists (ERAs) in surface waters. SPMDs were deployed at a number of riverine sampling sites in Germany and the United Kingdom. SPMD extracts were fractionated using a reverse-phase high-performance liquid chromatography (HPLC) method calibrated to provide an estimation of target analyte hydrophobicity. A portion of each HPLC fraction was then subjected to the yeast estrogen screen (YES) assay to determine estrogenic potential. Results were plotted in the form of "estrograms," which displayed profiles of estrogenic potential as a function of HPLC retention time (i.e., hydrophobicity) for each of the samples. Where significant activity was elicited in the YES, the remaining portion of the respective active fraction was subjected to gas chromatography–mass spectrometry (GC-MS) analysis in an attempt to identify the compounds with estrogenic activity. The study indicated that improvements to the analytical methodology used to identify ERAs or other target analytes in active fractions could greatly enhance the applicability of the methodology to risk assessment and monitoring programs.

Vermeirssen et al.[95] used POCIS for integrative sampling of polar estrogens. Masses of estrogens accumulated in POCIS and concentrations of these compounds in spot samples were determined using YES assay (expressed as

17-β-estradiol equivalents [EEQ]) and chemical analysis of selected compounds using LC-MS.[3] They found that results from spot sampling, passive sampling, and bioaccumulation in caged fish were correlated and provided comparable values. POCIS provided an integrated and biologically meaningful measure of estrogenicity in that it accumulated estrogens in a pattern similar to that of brown trout.

19.7.6. Sunscreen and Ultraviolet (UV) Filters

The analysis of sunscreens/organic UV filters in water has increased substantially in the last 5 years.[4] Due to their use in a variety of personal care products, these compounds can enter the aquatic environment indirectly from showering, washing clothes, via wastewater treatment plants, and also directly from recreational activities.

Poiger et al.[96] detected four organic UV filters (80–950 ng/SPMD) in SPMDs deployed at Lakes Zurich and Greifensee, Switzerland. SPMD-derived water concentrations were in the range of 1–10 ng/L and corresponded well with those determined in spot samples of water. No UV filters were detected above blank levels in SPMDs deployed at a remote mountain lake used for background measurements. In a later study, Balmer et al.[97] investigated the occurrence of four important organic UV filter compounds in water, wastewater, and fish from various Swiss lakes. Data from passive sampling using SPMDs supported the presence of these UV filters in the lakes and the river and suggested some potential for accumulation of these compounds in biota.

19.7.7. Brominated Flame Retardants

Polybrominated diphenyl ethers (PBDEs) are widely used as flame retardants in products such as furniture, textiles, plastics, paints, and electronic appliances. Due to their extreme hydrophobicity (log K_{OW} values 4–10), these compounds are present in the aqueous phase at extremely low (sub-parts per billion) concentrations. Nevertheless, because of their possible environmental risks due to their persistence and bioaccumulation, the inclusion of certain PBDE congeners in monitoring programs is justified. Most studies have focused on the measurement of PBDEs in biomonitoring organisms and sediments.

Booij et al.[98] used SPMDs for sampling and *in situ* preconcentration of PBDEs from water at several sampling stations in the Scheldt estuary and the North Sea along the Dutch coast. The application of integrative sampling enabled the back-calculation of extremely low concentrations (in range 0.1–5 pg/L) of PBDE congeners in water from SPMD-accumulated amounts. Rayne and Ikonomou[99] employed SPMDs for sampling PBDEs in water in the Fraser River near Vancouver, Canada. The PBDE concentrations found in SPMDs in combination with their physicochemical properties and their SPMD uptake parameters were fed into an aquatic transport model to reconstruct the PBDE patterns in pollution sources. The reconstructed SPMD patterns closely approximated the composition of known technical mixtures.

19.7.8. Algal Toxins

There is evidence that the increase in frequency and intensity of harmful algal blooms has led to shellfish poisoning, large fish kills, and deaths of livestock and wildlife, as well as illness and death in humans. Toxins produced by these algae have been implicated in these adverse effects.[100]

Mackenzie et al.[101] reported the development of a simple and sensitive passive sampling technique for monitoring the occurrence of toxic algal blooms and shellfish contamination. The technique involves the passive adsorption of biotoxins onto porous synthetic resin-filled sachets (solid-phase adsorption toxin tracking [SPATT] bags) and their subsequent extraction and analysis. The success of the method is founded on the observation that during algal blooms, significant amounts of toxin, including lipophilic compounds such as pectenotoxins and okadaic acid complex toxins, are dissolved in the seawater. These data demonstrate that the technique provides a means of forecasting shellfish contamination events and predicting the net accumulation of polyether toxins by mussels. The method has several advantages over current monitoring techniques such as shellfish-flesh testing and phytoplankton monitoring. In contrast to the evidence provided by genetic probe technologies and conventional phytoplankton monitoring methods, it directly targets the toxic compounds of interest. The extracts from passive sampling devices eliminate the matrix problems associated with chemical and biological analysis shellfish tissue extracts. This method reduces the analytical and toxicological uncertainties associated with the multiplicity of toxin analogs produced by transformation in shellfish tissues, and allows the target parent compounds to be detected with confidence. Time-integrated sampling provides an estimate of biotoxin accumulation in filter feeders, and the high sensitivity provides an early warning and conservative estimates of contamination potential. The technique may reduce monitoring costs and provide improved spatial and temporal sampling opportunities. When coupled with appropriate analytical techniques (e.g., LC-MS multitoxin screens, enzyme-linked immunosorbent assays [ELISA], receptor binding assays), the technique has the potential to offer a powerful early warning method for marine and freshwater micro-algae toxins.

19.7.9. Fluorinated Surfactants

Fluorinated surfactants (also referred as fluorotelomeric acids, alcohols, and sulfonates) have been used to make stain repellents that are widely applied to fabrics, carpets, and paper. They are also used in the manufacture of paints, adhesives, waxes, polishes, metals, electronics, and caulks. Due to the concern of their persistence and global occurrence in human and wildlife, two of these fluorinated surfactants, perfluorooctanoic acid (PFOA) and perfluorooctanesulfonate (PFOS), are considered as emerging pollutants.[4] Although no quantitative studies aimed at monitoring of these substances have been reported, Casey et al.[102] reported the identification of these compounds in POCIS extracts at levels above associated controls.

19.7.10. Volatile Organic Chemicals (VOCs)

Passive samplers have found a broad application in monitoring VOCs in groundwater. They can be advantageous primarily because they have the potential to substantially reduce costs of monitoring from the high levels associated with pumping approaches to well sampling. The depth-specific characteristic of the samples can also be advantageous in certain investigations.

Equilibrium-type passive samplers used in groundwater monitoring typically consist of a closed receptacle composed of a semipermeable or permeable membrane containing vapor or water free of the target analytes. When these types of sampling devices are deployed in VOC-contaminated water, equilibrium develops between VOC concentrations in the ambient water and in the water or air of the sampling device. Once the VOC concentrations attain equilibrium, the VOC concentrations within the sampling device maintain equilibrium with the concentrations in the ambient water and can be used to track changes in the ambient water.[103] The equilibrated sampling device can be recovered, and the sample can either be sealed in the sampling device or transferred to sealable sample vials, depending on specific sampler requirements. An equilibration time of 1–7 days is typical for equilibrium-based membranes.[104–107] However, most equilibrium-based samplers are usually left in place for at least 2 weeks prior to recovery in order to allow the deployment disturbance to dissipate. Longer times may be required in poorly permeable sediments.[108] A useful database of equilibrium passive diffusion samplers and their application in monitoring of VOCs in groundwater is maintained at a website.[18]

Among the kinetic sampling devices, ceramic dosimeters have been successfully applied for long-term surveillance of many organic pollutants, including VOCs.[65,109] They use a ceramic tube as diffusion-limiting barrier enclosing a receiving phase consisting of solid sorbent beads. Recently, the utility of the ceramic dosimeter as a robust groundwater sampling device was demonstrated for benzene, toluene, ethylbenzenes, xylenes, and naphthalenes, using Dowex Optipore L-493 resin as the receiving phase.[64] In up to 90 days of sampling in a contaminated aquifer, the ceramic dosimeter provided TWA contaminant concentrations that closely matched those derived from frequent spot sampling.

Rogers et al.[110] reported a kinetic sampling device interfaced with two toxicity-based assays for screening for 17 volatile toxic industrial chemicals. The sampling device is based on SPMD; however, in this application, the normal triolein receiving phase is replaced with dimethylsulfoxide as this is more directly compatible with toxicity assays. The intention of the study was to demonstrate the feasibility of a passive sampling system that can be linked directly (i.e., without solvent extraction or solvent exchange before analysis) with the rapid and potentially field portable, prokaryotic (Microtox®) and eukaryotic (IQ-Tox) toxicity screening assays. Vorkamp et al.[111] developed a sorbent-based passive sampling device for qualitative detection and tracing of VOCs in the atmosphere of sewer systems.[112] Among the adsorption materials tested, Tenax® TA resin showed the best performance characteristics for sampling of various volatile chlorinated hydrocarbons including chloroform, trichloroethylene, and perchlorethylene.[111]

19.7.11. Methyl *tert*-Butyl Ether (MTBE)

MTBE is widely used as a gasoline oxygenate, and because of its high aqueous solubility and persistence, contamination of groundwater and surface water with MTBE and its major degradation product *tert*-butyl alcohol (TBA) has received attention.[4]

Divine and McCray[106] reported that the diffusivity of MTBE in low-density polyethylene (LDPE) is very low, and that this prevented MTBE from being accumulated by passive sampling devices such as SPMD and polyethylene diffusion bag samplers where analytes must permeate through an LDPE membrane before accumulation in the receiving phase. Bi et al.[113] studied sorption affinity and sorption mechanisms of MTBE and TBA to synthetic resins, which can be potentially used as a receiving phase in passive sampling devices. They found that the sorbent material Optipore L-493 has favorable sorption and desorption characteristics, and is a suitable sorbent for passive sampling of MTBE. For highly polar compounds such as TBA, the sorbent material Ambersorb 563 was recommended. There is therefore the potential for developing a passive sampler for this group of hydrophilic pollutants.

19.8. METALS AND INORGANIC POLLUTANTS

Most applications of passive sampling for inorganic pollutants have focused on the measurement of metal ions. There is also a significant interest in the use of passive sampling devices to characterize the speciation of metals in water, for example, differentiating between free, inorganic, and organic bound metal species and organometallic compounds (see Table 19.2).

19.8.1. Heavy Metals and Emerging Inorganic Pollutants

A number of lists of priority inorganic pollutants have been published both in Europe and the United States. The compounds on these lists have been monitored in the aquatic environment and included in national monitoring programs. Many of the pollutants can be measured using passive sampling techniques (Table 19.4).

Among the passive sampling techniques for inorganic pollutants, the most widely used is the DGT device with a large number of publications since the mid-1990s, although others such as the inorganic version of the Chemcatcher are finding an increasing range of applications. Usually, metals are accumulated on a Chelex-based binding phase. After the retrieval of the samplers, analytes are extracted using an acid, and concentrations are measured using a range of spectroscopic techniques.[114–116] Using the DGT, quantitative measurements of mercury have been made using a device fitted with a Chelex resin and a Spheron-Thiol resin.[117] Developments have also been made in speciation of chromium. Here, chromium (III) is selectively accumulated on the Chelex-resin binding phase, while chromium (VI) establishes a concentration equilibrium with the hydrogel. This makes it possible, after extraction, to determine the concentrations of both species.[118–120] Recently, a number of other techniques based on the permeation of analytes through a variety of membrane materials have been developed for the measurement of heavy metals (Table 19.2).

Some heavy metals (other than those on priority pollutant lists) and ionic substances are now thought of as emerging pollutants and warrant additional environmental monitoring and further ecotoxicological risk evaluation. Among these are the platinum group elements ([PGEs] platinum, palladium, rhodium, and osmium) whose environmental occurrence has increased with the introduction of catalytic converters on vehicles.[121] Environmental concentrations of PGEs are still low, but there is a concern over their ecotoxicity. There is presently no reported use of passive samplers for the monitoring of PGEs in water. Another emerging pollutant with increasing importance is thallium, which, in its trivalent state, has a comparable level of toxicity to mercury.[122] It has been observed to accumulate on a Chelex-resin binding phase of a DGT, but uptake was not quantitative.[123]

Perchlorate is used in many applications, such as solid rocket fuel, pyrotechnics, and in additives in the chemical industry. It has a known endocrine disrupting activity. Its emergence as a pollutant of concern coincided with improvements in ion chromatography technology in the late 1990s, allowing the detection of perchlorate in the environment at very low concentrations (<1 μg/L). The research on measuring perchlorate levels using passive sampling is limited. Tan et al.[124] used a sediment peeper (an equilibrium passive sampling device) to measure perchlorate in pore water with a detection limit of 1 μg/L.

19.8.2. Speciation of Metals

The importance of speciation for the determination of the fate and toxicity of metals in natural waters is recognized.[125] Recent developments in passive sampling have allowed a better understanding of the speciation of metals in the environment. It has been demonstrated that kinetic passive samplers accumulate mainly free hydrated metal ions and metal complexes with sufficiently high dissociation rates.[123,126–130] Most research has been undertaken using the DGT, in which by varying the pore size of the hydrogel diffusion layer and using a cross-linker modifier, it is possible to exclude or restrict (on size basis) diffusion of large organic complexes into the sampler.[126,127]

Another way to modify selectivity is to vary the thickness of the diffusion layer. Increasing the thickness allows the sampling of stable complexes since labile species will dissociate during the time taken to diffuse across the diffusion layer.[128,130] A numerical modeling of the ligand–metal complex lability based on a relatively simple copper citrate and copper–ethylenediaminetetraacetic acid (EDTA) system has been presented.[131] Using this model, information of metal speciation from DGT measurements was possible given a priori knowledge of the precise diffusion coefficients of the metal complexes involved.[131] Unsworth et al.[132] examined a similar system of cadmium complexes with fulvic and humic acids, and compared empirical results with existing speciation models. A good agreement was found between results obtained from DGTs and the speciation model output.

Selectivity of passive samplers can also be modified by employing different binding phases. Recently, an assessment of the impact of the binding strength (stability constant) of the receiving phase on the accumulation of metals has been undertaken.[133] Several binding phases were studied, including conventional Chelex 100 resin embedded polyacrylamide hydrogel, poly(acrylamide-co-acrylic acid) gel, poly(acrylamidoglycolic acid-co-acrylamide) gel, Whatman P-81 cellulose phosphate ion exchange membrane, and poly(4-styrenesulfonate) aqueous solution. Laboratory testing in metal solutions spiked with EDTA or humic acid suggested that all the DGT devices measured only free metal ions and inorganic metal complexes. Upon field testing at both freshwater and seawater sites, it was found that the DGT labile metal concentrations measured by different binding phases can be significantly different, suggesting that the DGT labile metal fractions were dependent on binding strength of the binding phase. By designing binding phases that can compete with different natural water complexing ligands to varying extents, it is possible to use these different DGT devices to measure metal speciation in natural waters.

19.8.3. Radionuclides

Radionuclides are released into the environment via a range of nuclear sources.[134] The DGT using receiving phases with specific binding agents has been used for the accumulation of radionuclides. French et al.[134] used a DGT with TEVA (specifically design to sequester tetravalent actinides) resin-gel receiving phase to measure technetium (^{99}Tc) in seawater. In seawater, the uptake of ^{99}Tc was independent of ionic strength and pH. Murdock et al.[135] used ammonium molybdophosphate as a binding agent in a DGT for measuring cesium radionuclides (^{137}Cs) in freshwater. Over long-term exposures (2 months), the gradual degradation of ammonium molybdophosphate influenced the performance of the device. The DGT configured with the standard Chelex binding phase has also been used to measure ^{60}Co and ^{65}Zn.[135]

19.8.4. Inorganic Anions

Most of the inorganic anions present in water originate from natural sources, but can also be of anthropogenic origin, for example, agricultural and urban runoff, industrial wastewater, and storm water. Anions together with cations contribute to bulk parameters such as conductivity and salinity in aquatic environment, vital for the functioning of aquatic organisms. At the same time, some anions (e.g., phosphate and sulfate) represent a sink for metal ions, producing insoluble solid phase which under certain conditions (e.g., acidification) can be remobilized back into solution.

Sulfides occur in the environment as a result of the microbially mediated reduction of metal sulfates in anoxic conditions or by decomposition of organic matter.[136,137] Motelica-Heino et al.[138] and Naylor et al.[136] measured sulfides (together with trace metals) in sediments using a DGT consisting of a layer of gel impregnated with silver iodide overlain with a layer of gel containing Chelex. The silver iodide immobilized the sulfide by the formation of insoluble silver sulfide inside the device.

Phosphorus anions occur in a number of forms that are in dynamic equilibrium in the environment. In spot water samples, prolonged storage can change the equilibrium, leading to changes in concentration of the individual phosphorus species. To overcome this problem of sampling, a DGT with ferrihydrite as a binding gel has been developed for measuring reactive phosphorus species in pore water and freshwater sediments.[139,140]

19.8.5. Organometallics

Due to their bioaccumulation potential and toxicity, organometallic substances are considered as emerging pollutants of concern. In some cases, organometallic compounds (e.g., some organic forms of tin) are more toxic than inorganic complexes or free forms of the parent metal. Passive sampling devices have been used to measure a number of organometallic species, including those of lead, mercury, and tin.

Følsvik et al.[141,142] reported the use of SPMDs for monitoring organotin compounds. Both dibutyl- and tributyltin were accumulated by the devices, but no accumulation of monobutyltin was observed during several weeks of SPMD exposure in a Norwegian fjord. Using this method, it was possible to identify concentration gradients of organotin compounds at the sampling site. A variant of the Chemcatcher sampler has been developed and calibrated for the TWA concentration measurement of organotin compounds.[143,144] Pawliszyn and coworkers demonstrated the feasibility of using SPME as an *in situ* passive sampler for the determination of TWA concentrations of a range of organic chemicals in water.[145] The SPME technique has been used for measuring organometallic species of lead, mercury, and tin in spot samples of water.[146] It may therefore be possible to use this technique for passive sampling of organometallic compounds.

19.9. FUTURE TRENDS

Several major trends are emerging in the development of passive sampling technology. The first is toward the miniaturization of devices. Smaller devices offer the advantages of inexpensive transportation to and from the sampling site, the need for only minimal deployment apparatus, and a low consumption of solvents and reagents during their processing. Many miniaturized samplers can be seen as disposable and/or recyclable items, and this can help in avoiding both cross-contamination and the need for rigorous cleaning procedures as is the case for reusable devices. Moreover, miniaturized devices can be used in situations where space and volume of water are limited, for example, in groundwater boreholes.[4] However, as the size of the device is reduced, so is the effective sampling surface area and hence the uptake rate of the analytes being investigated. This may become problematic for compounds present at trace concentrations in the aquatic environment. The miniaturization trend goes hand-in-hand with the tendency to develop solventless sample preparation techniques. Passive samplers based on *in situ* analyte preconcentration using SPME or similar techniques allow sample processing (following exposure) using thermal desorption GC[145] or solvent microextraction followed by HPLC.[147] However, the practical application of SPME-based techniques in *in situ* passive sampling of trace aqueous contaminants will require further enhancement of their robustness and sensitivity.

A second trend is the widening of the range of pollutants that can be monitored using passive sampling technology. Recently, attention has been focused on compounds with medium-to-high polarity, for example, polar pesticides, some pharmaceuticals and personal protection products, and other emerging pollutants of concern.[36,52,85] This will require the use of novel materials as receiving phases, together with membrane materials that permit the selective accumulation of these chemicals, together with the development of appropriate quantitative analytical methods. In addition, laboratory calibration experiments need to be performed to measure sampler uptake rates for these substances. These tests will need to be undertaken at a range of typical field conditions to enable accurate estimates of the aqueous concentrations of these pollutants to be determined.

As the number of organic compounds identified as concern increases, it becomes more important to be able to predict uptake parameters for passive samplers on the basis of physicochemical properties of the pollutants in order to avoid the need for time-consuming calibration studies. Further work is necessary to model the performance of passive samplers when concentrations of pollutants fluctuate. This is important where a lag phase in response of the sampler can affect the efficiency of the detection of short pollution event. An understanding of the factors determining the responsiveness to changing concentrations will form the basis of improvements to the design of samplers.

A further challenge is to improve robustness by reducing or controlling the impacts of environmental conditions and biofouling on the sampler performance and uptake rates. PRCs are being tested for improving the accuracy of the measurement of TWA concentrations of contaminants in the field. However, the application of the PRC approach for passive sampling devices equipped with sorbent-based receiving phases and the ability to fully account for the variable effects of the biofouling on diffusion surfaces remain two important problems to be solved.

Likewise, precise calibration data for passive sampling devices used in monitoring trace metals is essential for the quantification of the various metal species and complexes found in different aquatic environments. This requires knowledge of the uptake kinetics of different metal moieties. Configuration of specific devices for monitoring well-defined fractions and species of metals will increase their potential as regulatory tools and help to overcome some of the current limitations of spot water sampling.

Another trend is the linking of chemical and biological analyses of sample extracts collected using passive devices, with the aim of detecting/identifying toxicologically relevant compounds. The marriage of passive samplers and biomarker/bio-indicator tests offers new ways to provide information concerning the relative toxicological significance of waterborne contaminants and to improve risk assessments for different water bodies.

An essential requirement is the development of quality assurance (QA)/quality control (QC) and method validation schemes for passive sampling devices, and standards for the application of this technology[77] are needed. The use of interlaboratory trials with reference materials is problematic since large volumes of standard solutions are needed in tank trials of passive samplers. One way forward would be to use a dispersible tablet formulation of reference analytes. An alternative approach based on the use of a reference site may be more helpful. Here, a range of sampling technologies and the performance of participating laboratories can be evaluated simultaneously. A compromise between using artificial laboratory conditions with pure water, and *in situ* exposure to natural surface waters is to use a tank trial with circulating environmental water. This provides a test involving realistic matrix effects but allows some control over conditions (selection and concentrations of test analytes, temperature, and flow of water). These approaches together with well-designed field trials performed alongside existing sampling procedures will help in establishing the acceptance and increase the confidence in the use of passive sampling technology in routine regulatory monitoring programs in the future.[148–150]

ACKNOWLEDGMENT

We acknowledge the financial support of the European Commission (Contracts EVK1-CT-2002-00119, http://www.port.ac.uk/stamps/, and SSPI-CT-2003-502492, http://www.swift-wfd.com/).

REFERENCES

1. Koester, C.J.; Simonich, S.L.; Esser, B.K. Environmental analysis. *Anal. Chem.* **2003**, *75*, 2813–2829.

2. Koester, C.J.; Moulik, A. Trends in environmental analysis. *Anal. Chem.* **2005**, *77*, 3737–3754.

3. Pawliszyn, J. Sample preparation: Quo Vadis? *Anal. Chem.* **2003**, *75*, 2543–2558.

4. Richardson, S.D.; Ternes, T.A. Water analysis: Emerging contaminants and current issues. *Anal. Chem.* **2005**, *77*, 3807–3838.

5. Coquery, M.; Morin, A.; Becue, A.; Lepot, B. Priority substances of the European Water Framework Directive: Analytical challenges in monitoring water quality. *Trends Anal. Chem.* **2005**, *24*, 117–127.

6. Palmes, E.D.; Gunnison, A.F. Personal monitoring device for gaseous contaminants. *Am. Ind. Hyg. Assoc. J.* **1973**, *34*, 78–81.

7. Byrne, B.A.; Aylott, R.I. Concentrator for removing organic materials from aqueous systems. British Patent No. 1566253, London, UK; 1980.

8. Stuer-Lauridsen, F. Review of passive accumulation devices for monitoring organic micropollutants in the aquatic environment. *Environ. Pollut.* **2005**, *136*, 503–524.

9. Vrana, B.; Allan, I.J.; Greenwood, R.; Mills, G.A.; Dominiak, E.; Svensson, K.; Knutsson, J.; Morrison, G. Passive sampling techniques for monitoring pollutants in water. *Trends Anal. Chem.* **2005**, *24*, 845–868.

10. Ouyang, G.; Pawliszyn, J. Configurations and calibration methods for passive sampling techniques. *J. Chromatogr. A* **2007**, *1168*, 226–235.

11. Seethapathy, S.; Górecki, T.; Li, X. Passive sampling in environmental analysis. *J. Chromatogr. A* **2008**, *1184*, 234–253.

12. Kot-Wasik, A.; Zabiegała, B.; Urbanowicz, M.; Dominiak, E.; Wasik, A.; Namieśnik, J. Advances in passive sampling in environmental studies. *Anal. Chim. Acta* **2007**, *602*, 141–163.

13. Esteve-Turrillas, F.A.; Pastor, A.; Yusa, V.; De La Guardia, M. Using semi-permeable membrane devices as passive samplers. *Trends Anal. Chem.* **2007**, *26*, 703–712.

14. Huckins, J.N.; Petty, J.D.; Booij, K., Eds. *Monitors of Organic Chemicals in the Environment: Semipermeable Membrane Devices*. New York: Springer; 2006.

15. Greenwood, R.; Mills, G.A.; Vrana, B. Passive sampling techniques in environmental monitoring. In *Comprehensive Analytical Chemistry Series*, Barcelo, D., Ed. Amsterdam: Elsevier; 2007.

16. Gorecki, T.; Namieśnik, J. Passive sampling. *Trends Anal. Chem.* **2002**, *21*, 276–291.

17. Mayer, P.; Tolls, J.; Hermens, L.; Mackay, D. Equilibrium sampling devices. *Environ. Sci. Technol.* **2003**, *37*, 184A–191A.

18. http://diffusionsampler.itrcweb.org/.

19. Vroblesky, D.A.; Hyde, W.T. Diffusion samplers as an inexpensive approach to monitoring VOCs in ground water. *Ground Water Monit. Remediat.* **1997**, *17*, 177–184.

20. Huckins, J.N.; Manuweera, G.K.; Petty, J.D.; Mackay, D.; Lebo, J.A. Lipid-containing semipermeable membrane devices for monitoring organic contaminants in water. *Environ. Sci. Technol.* **1993**, *27*, 2489–2496.

21. Gale, R.W. Three-compartment model for contaminant accumulation by semipermeable membrane devices. *Environ. Sci. Technol.* **1998**, *32*, 2292–2300.

22. Booij, K. Theory and modelling. In *Monitors of Organic Chemicals in the Environment: Semipermeable Membrane Devices*, Huckins, J.N.; Petty, J.D.; Booij, K., Eds. New York: Springer; **2006**; pp. 45–85.

23. Vrana, B.; Popp, P.; Paschke, A.; Schüürmann, G. Membrane-enclosed sorptive coating. An integrative passive sampler for monitoring organic contaminants in water. *Anal. Chem.* **2001**, *73*, 5191–5200.

24. Chen, Y.; Pawliszyn, J. Kinetics and the on-site application of standards in a solid-phase microextraction fiber. *Anal. Chem.* **2004**, *76*, 5807–5815.

25. Vaes, H.J.; Hamwijk, C.; Ramos, E.U.; Verhaar, H.J.M.; Hermens, J.L.M. Partitioning of organic chemicals to polyacrylate-coated solid phase microextraction fibers: Kinetic behavior and quantitative structure-property relationships. *Anal. Chem.* **1996**, *68*, 4458–4462.

26. Huckins, J.N.; Petty, J.D.; Orazio, C.E.; Lebo, J.A.; Clark, R.C.; Gibson, V.L.; Gala, W.R.; Echols, K.R. Determination of uptake kinetics (sampling rates) by lipid-containing semipermeable membrane devices (SPMDs) for polycyclic aromatic hydrocarbons (PAHs) in water. *Environ. Sci. Technol.* **1999**, *33*, 3918–3923.

27. Huckins, J.N.; Petty, J.D.; Lebo, J.A.; Almeida, F.V.; Booij, K.; Alvarez, D.A.; Cranor, W.L.; Clark, R.C.; Mogensen, B.B. Development of the permeability/performance reference compound (PRC) approach for in situ calibration of semipermeable membrane devices (SPMDs). *Environ. Sci. Technol.* **2002**, *36*, 85–91.

28. Vrana, B.; Schüürmann, G. Calibrating the uptake kinetics of semipermeable membrane devices in water: Impact of hydrodynamics. *Environ. Sci. Technol.* **2002**, *36*, 290–296.

29. Xu, Y.P.; Wang, Z.J.; Ke, R.H.; Khan, S.U. Accumulation of organochlorine pesticides from water using triolein embedded cellulose acetate membranes. *Environ. Sci. Technol.* **2005**, *39*, 1152–1157.

30. Gustafson, K.E.; Dickhut, R.M. Distribution of polycyclic aromatic hydrocarbons in Southern Chesapeake Bay Region. *Environ. Toxicol. Chem.* **1997**, *16*, 452–461.

31. Green, C.E.; Abraham, M.H. Investigation into the effects of temperature and stirring rate on the solid-phase extraction of diuron from water using a C-18 extraction disk. *J. Chromatogr. A* **2000**, *885*, 41–49.

32. Stephens, B.S.; Kapernick, A.; Eaglesham, G.; Mueller, J. Aquatic passive sampling of herbicides on naked particle loaded membranes: Accelerated measurement and empirical estimation of kinetic parameters. *Environ. Sci. Technol.* **2005**, *39*, 8891–8897.

33. Pawliszyn, J. *Solid-Phase Microextraction: Theory and Practice*. New York: Wiley; 1997.

34. Baltussen, E.; Sandra, P.; David, F.; Cramers, C. Stir bar sorptive extraction (SBSE), a novel extraction technique for aqueous samples: Theory and principles. *J. Microcolumn Sep.* **1999**, *11*, 737–747.

35. Mayer, P.; Vaes, W.H.J.; Hermens, J.L.M. Absorption of hydrophobic compounds into the poly(dimethylsiloxane) coating of solid-phase microextraction fibers: High partition coefficients and fluorescence microscopy images. *Anal. Chem.* **2000**, *72*, 459–464.

36. Alvarez, D.A.; Petty, J.D.; Huckins, J.N.; Jones-Lepp, T.L.; Getting, D.T.; Goddard, J.P.; Manahan, S.E. Development of a passive, in situ, integrative sampler for hydrophilic organic contaminants in aquatic environments. *Environ. Toxicol. Chem.* **2004**, *23*, 1640–1648.

37. Petty, J.D.; Huckins, J.N.; Alvarez, D.A.; Brumbaugh, W.G.; Cranor, W.L.; Gale, R.W.; Rastall, A.C.; Jones-Lepp, T.L.; Leiker, T.J.; Rostad, C.E.; Furlong, E.T. A holistic passive integrative sampling approach for assessing the presence and potential impacts of waterborne environmental contaminants. *Chemosphere* **2004**, *54*, 695–705.

38. Richardson, B.J.; Lam, P.K.S.; Zheng, G.J.; McCellan, K.E.; De Luca-Abbott, S.B. Biofouling confounds the uptake of trace organic contaminants by semi-permeable membrane devices. *Mar. Pollut. Bull.* **2002**, *44*, 1372–1379.

39. Kingston, J.K.; Greenwood, R.; Mills, G.A.; Morrison, G.M.; Persson, B.L. Development of novel passive sampling system for the time-averaged measurement of a range of organic pollutants in aquatic environments. *J. Environ. Monitor.* **2000**, *2*, 487–495.

40. Vrana, B.; Mills, G.A.; Dominiak, E.; Greenwood, R. Calibration of the Chemcatcher passive sampler for the monitoring of priority organic pollutants in water. *Environ. Pollut.* **2006**, *142*, 333–343.

41. Luellen, D.R.; Shea, D. Calibration and field verification of semipermeable membrane devices for measuring polycyclic aromatic hydrocarbons in water. *Environ. Sci. Technol.* **2002**, *36*, 1791–1797.

42. Booij, K.; Sleiderink, H.M.; Smedes, F. Calibrating the uptake kinetics of semipermeable membrane devices using exposure standards. *Environ. Toxicol. Chem.* **1998**, *17*, 1236–1245.

43. Booij, K.; Hofmans, H.E.; Fischer, C.V.; van Weerlee, E.M. Temperature-dependent uptake rates of nonpolar organic compounds by semipermeable membrane devices and low-density polyethylene membranes. *Environ. Sci. Technol.* **2003**, *37*, 361–366.

44. Mayer, P.; Wernsing, J.; Tolls, J.; De Maagd, P.G.J.; Sijm, D.T.H.M. Establishing and controlling dissolved concentrations of hydrophobic organics by partitioning from a solid phase. *Environ. Sci. Technol.* **1999**, *33*, 2284–2290.

45. Kiparissis, Y.; Akhtar, P.; Hodson, P.V.; Brown, R.S. Partition-controlled delivery of toxicants: A novel in vivo approach for embryo toxicity testing. *Environ. Sci. Technol.* **2003**, *37*, 2262–2266.

46. Gerofke, A.; Komp, P.; McLachlan, M.S. Stir bar contamination: A method to establish and maintain constant water concentrations of poorly water-soluble chemicals in bioconcentration experiments. *Water Res.* **2004**, *38*, 3411–3419.

47. Brown, R.S.; Akhtar, P.; Akerman, J.; Hampel, L.; Kozin, I.S.; Villerius, L.A.; Klamer, H.J.C. Partition controlled delivery of hydrophobic substances in toxicity tests using poly(dimethylsiloxane) (PDMS) films. *Environ. Sci. Technol.* **2001**, *35*, 4097–4102.

48. Rantalainen, A.-L.; Cretney, W.J.; Ikonomou, M.G. Uptake rates of semipermeable membrane devices (SPMDs) for PCDDs, PCDFs. *Chemosphere* **2000**, *40*, 147–158.

49. Ouyang, G.; Chen, Y.; Pawliszyn, J. Flow-through system for the generation of standard aqueous solution of polycyclic aromatic hydrocarbons. *J. Chromatogr. A* **2006**, *1105*, 176–179.

50. Vrana, B.; Paschke, A.; Popp, P. Calibration and field performance of membrane-enclosed sorptive coating for integrative passive sampling of persistent organic pollutants in water. *Environ. Pollut.* **2006**, *144*, 296–307.

51. Alvarez, D.A. Personal communication; September 2005.

52. Alvarez, D.A.; Stackelberg, P.E.; Petty, J.D.; Huckins, J.N.; Furlong, E.T.; Zaugg, S.D.; Meyer, M.T. Comparison of a novel passive sampler to standard water-column sampling for organic contaminants associated with wastewater effluents entering a New Jersey stream. *Chemosphere* **2005**, *61*, 610–622.

53. Björklund Blom, L.; Morrison, G.M.; Rauch, M.S.; Mills, G.; Greenwood, R. Metal diffusion properties of a Nafion-coated porous membrane in an aquatic passive sampler system. *J. Environ. Monit.* **2003**, *5*, 404–409.

54. Pichette, C.; Zhang, H.; Davison, W.; Sauve, S. Preventing biofilm development on devices using metals and antibiotics. *Talanta* **2007**, *72*, 716–722.

55. Burkhard, L.P. Estimating dissolved organic carbon partition coefficients for nonionic organic chemicals. *Environ. Sci. Technol.* **2000**, *34*, 4663–4667.

56. Ellis, G.S.; Huckins, J.N.; Rostad, C.E.; Schmitt, C.J.; Petty, J.D.; MacCarthy, P. Evaluation of lipid-containing semipermeable membrane devices (SPMDs) for monitoring organochlorine contaminants in the upper Mississippi river. *Environ. Toxicol. Chem.* **1995**, *14*, 1875–1884.

57. Rantalainen, A.L.; Ikonomou, M.G.; Rogers, I.H. Lipid-containing semipermeable-membrane devices (SPMDs) as concentrators of toxic

chemicals in the lower Fraser River, British Columbia. *Chemosphere* **1998**, *37*, 1119–1138.

58. Axelman, J.; Naes, K.; Näf, C.; Broman, D. Accumulation of polycyclic aromatic hydrocarbons in semipermeable membrane devices and caged mussels (*Mytilus edulis* L.) in relation to water column phase distribution. *Environ. Toxicol. Chem.* **1999**, *18*, 2454–2461.

59. Zeng, E.Y.; Tsukada, D.; Diehl, D.W. Development of solid-phase microextraction-based method for sampling of persistent chlorinated hydrocarbons in an urbanized coastal environment. *Environ. Sci. Technol.* **2004**, *38*, 5737–5743.

60. Hyne, R.V.; Pablo, F.; Aistrope, M.; Leonard, A.W.; Ahmad, N. Comparison of time-integrated pesticide concentrations determined from field-deployed passive samplers with daily river-water extractions. *Environ. Toxicol. Chem.* **2004**, *23*, 2090–2098.

61. Blom, L.B.; Morrison, G.M.; Kingston, J.; Mills, G.; Greenwood, R.; Petersson, T.J.R.; Rauch, S. Performance of an in situ passive sampling system for metals in stormwater. *J. Environ. Monit.* **2002**, *4*, 258–262.

62. Vrana, B.; Mills, G.A.; Kotterman, M.; Leonards, P.; Booij, K.; Greenwood, R. Modelling and field application of the Chemcatcher passive sampler calibration data for the monitoring of hydrophobic organic pollutants in water. *Environ. Pollut.* **2007**, *145*, 895–904.

63. www.port.ac.uk/stamps

64. Martin, H.; Patterson, B.M.; Davis, G.B. Field trial of contaminant groundwater monitoring: Comparing time-integrating ceramic dosimeters and conventional water sampling. *Environ. Sci. Technol.* **2003**, *37*, 1360–1364.

65. Bopp, S.; Weiss, H.; Schirmer, K. Time-integrated monitoring of polycyclic aromatic hydrocarbons (PAHs) in groundwater using the ceramic dosimeter passive sampling device. *J. Chromatogr. A* **2005**, *1072*, 137–147.

66. Vrana, B.; Paschke, A.; Popp, P.; Schüürmann, G. Use of semipermeable membrane devices (SPMDs): Determination of bioavailable organic contaminants in the industrial region of Bitterfeld, Saxony-Anhalt, Germany. *Environ. Sci. Pollut. Res.* **2001**, *8*, 27–34.

67. Ouyang, G.F.; Zhao, W.N.; Bragg, L.; Qin, Z.P.; Alaee, M.; Pawliszyn, J. Time-weighted average water sampling in Lake Ontario with solid-phase microextraction passive samplers. *Environ. Sci. Technol.* **2007**, *41*, 4026–4031.

68. Alfaro-De la Torre, M.C.; Beaulieu, P.Y.; Tessier, A. In situ measurement of trace metals in lakewater using the dialysis and DGT techniques. *Anal. Chim. Acta* **2000**, *418*, 53–68.

69. Gimpel, J.; Zhang, H.; Davison, W.; Edwards, A.C. In situ trace metal speciation in lake surface waters using DGT, dialysis and filtration. *Environ. Sci. Technol.* **2003**, *37*, 138–146.

70. Zhang, H. In-situ speciation of Ni and Zn in freshwaters: Comparison between DGT measurements and speciation models. *Environ. Sci. Technol.* *38*, **2004**, 1421–1427.

71. Meylan, S.; Odzak, N.; Behra, R.; Sigg, L. Speciation of copper and zinc in natural freshwater: Comparison of voltammetric measurements, diffusive gradients in thin films (DGT) and chemical equilibrium models. *Anal. Chim. Acta* **2004**, *510*, 91–100.

72. Allan, I.J.; Knutsson, J.; Guigues, N.; Mills, G.A.; Fouillac, A.-M.; Greenwood, R. Evaluation of the Chemcatcher and DGT passive samplers for monitoring metals with highly fluctuating water concentrations. *J. Environ. Monit.* **2007**, *9*, 672–681.

73. Dunn, R.J.K.; Teasdale, P.R.; Warnken, J.; Arthur, J.M. Evaluation of the in situ, time-integrated DGT technique by monitoring changes in heavy metal concentrations in estuarine waters. *Environ. Pollut.* **2007**, *148*, 213–220.

74. Sigg, L.; Black, F.; Buffle, J.; Cao, J.; Cleven, R.; Davison, W.; Galceran, J.; Gunkel, P.; Kalis, E.; Kistler, D.; Martin, M.; Noel, S.; Nur, Y.; Odzak, N.; Puy, J.; Van Riemsdijk, W.; Temminghoff, E.; Tercier-Waeber, M.L.; Toepperwien, S.; Town, R.M.; Unsworth, E.; Warnken, K.W.; Weng, L.P.; Xue, H.B.; Zhang, H. Comparison of analytical techniques for dynamic trace metal speciation in natural freshwaters. *Environ. Sci. Technol.* **2006**, *40*, 1934–1941.

75. Unsworth, E.R.; Warnken, K.W.; Zhang, H.; Davison, W.; Black, F.; Buffle, J.; Cao, J.; Cleven, R.; Galceran, J.; Gunkel, P.; Kalis, E.; Kistler, D.; Van Leeuwen, H.P.; Martin, M.; Noel, S.; Nur, Y.; Odzak, N.; Puy, J.; Van Riemsdijk, W.; Sigg, L.; Temminghoff, E.; Tercier-Waeber, M.L.; Toepperwien, S.; Town, R.M.; Weng, L.P.; Xue, H.B. Model predictions of metal speciation in freshwaters compared to measurements by in situ techniques. *Environ. Sci. Technol.* **2006**, *40*, 1942–1949.

76. DeVita, W.M.; Crunkilton, R.L. Quality control associated with use of semipermeable membrane devices. *Environ. Toxicol. Risk Assess.* **1998**, *1333*, 237–245.

77. British Standards Institute (BSI). Publicly Available Specification: Determination of priority pollutants in surface water using passive sampling (PAS-61); May 2006.

78. Namieśnik, J.; Zabiegala, B.; Kot-Wasik, A.; Partyka, M.; Wasik, A. Passive sampling and/or extraction techniques in environmental analysis: A review. *Anal. Bioanal. Chem.* **2005**, *381*, 279–301.

79. http://wwwaux.cerc.cr.usgs.gov/spmd/SPMD_references.htm

80. Lu, Y.B.; Wang, Z.J.; Huckins, J.N. Review of the background and application of triolein-containing semipermeable membrane devices in aquatic environmental study. *Aquat. Toxicol.* **2002**, *60*, 139–153.

81. Petty, J.D.; Orazio, C.E.; Huckins, J.N.; Gale, R.W.; Lebo, J.A.; Meadows, J.C.; Echols, K.R.; Cranor, W.L. Considerations involved with the use of semipermeable membrane devices for monitoring environmental contaminants. *J. Chromatogr. A* **2000**, *879*, 83–95.

82. Zwiener, C.; Frimmel, F.H. Application of headspace GC/MS screening and general parameters for the analysis of polycyclic aromatic hydrocarbons. *Anal. Bioanal. Chem.* **2004**, *378*, 851–861.

83. Pedersen-Bjergaard, S.; Rasmussen, K.E. Bioanalysis of drugs by liquid-phase microextraction coupled to separation techniques. *J. Chromatogr. B* **2005**, *817*, 3–12.

84. Mazzella, N.; Dubernet, J.F.; Delmas, F. Determination of kinetic and equilibrium regimes in the operation of polar organic chemical integrative samplers—Application to the passive sampling of the polar herbicides in aquatic environments. *J. Chromatogr. A* **2007**, *1154*, 42–51.

85. Mills, G.A.; Vrana, B.; Allan, I.; Alvarez, D.A.; Huckins, J.N.; Greenwood, R. Trends in monitoring pharmaceuticals and personal-care products in the aquatic environment by use of passive sampling devices. *Anal. Bioanal. Chem.* **2007**, *387*, 1153–1157.

86. Jones-Lepp, T.L.; Alvarez, D.A.; Petty, J.D.; Huckins, J.N. Polar organic chemical integrative sampling and liquid chromatography-electrospray/ion-trap mass spectrometry for assessing selected prescription and illicit drugs in treated sewage effluents. *Arch. Environ. Contam. Toxicol.* **2004**, *47*, 427–439.

87. Togola, A.; Budzinski, H. Development of polar organic integrative samplers for analysis of pharmaceuticals in aquatic systems. *Anal. Chem.* **2007**, *79*, 6734–6741.

88. Shaw, M.; Müller, J.F. Preliminary evaluation of the occurrence of herbicides and PAHs in the wet tropics region of the Great Barrier Reef, Australia, using passive samplers: 4th International Conference on Marine Pollution and Ecotoxicology. *Mar. Pollut. Bull.* **2005**, *51*, 876–881.

89. Moore, M.T.; Bennett, E.R.; Cooper, C.M.; Smith, S.; Farris, J.L. Jr.; Drouillard, K.G.; Schulz, R. Influence of vegetation in mitigation of methyl parathion runoff. *Environ. Pollut.* **2006**, *142*, 288–294.

90. Leblanc, C.J.; Stallard, W.M.; Green, P.G.; Schroeder, E.D. Passive sampling screening method using thin-layer chromatography plates. *Environ. Sci. Technol.* **2003**, *37*, 3966–3971.

91. Leonard, A.W.; Hyne, R.V.; Lim, R.P.; Pablo, F.; Van Den Brink, P.J. Riverine endosulfan concentrations in the Mamoi River, Australia: Link to cotton field runoff and macroinvertebrate population densities. *Environ. Toxicol. Chem.* **2000**, *19*, 1540–1551.

92. Leonard, A.W.; Hyne, R.V.; Pablo, F. Trimethylpentane-containing passive samplers for predicting time-integrated concentrations of pesticides in water *Environ. Toxicol. Chem.* **2002**, *21*, 2591–2599.

93. Lindström, A.; Buerge, I.J.; Poiger, T.; Bergqvist, P.-A.; Müller, M.D.; Buser, H.-R. Occurrence and environmental behavior of the bactericide triclosan and its methyl derivative in surface waters and in wastewater. *Environ. Sci. Technol.* **2002**, *36*, 2322–2329.

94. Rastall, A.C.; Getting, D.; Goddard, J.; Roberts, D.R.; Erdinger, L. A biomimetic approach to the detection and identification of estrogen receptor agonists in surface waters using semipermeable membrane devices (SPMDs) and bioassay-directed chemical analysis. *Environ. Sci. Pollut. Res.* **2006**, *13*, 256–267.

95. Vermeirssen, E.L.M.; Korner, O.; Schonenberger, R.; Sutter, M.J.F.; Burkhardt-Holm, P. Characterization of environmental estrogens in river water using a three pronged approach: Active and passive water sampling and the analysis of accumulated estrogens in the bile of caged fish. *Environ. Sci. Technol.* **2005**, *39*, 8191–8198.

96. Poiger, T.; Buser, H.R.; Balmer, M.E.; Bergqvist, P.-A.; Muller, M.D. Occurrence of UV filter compounds from sunscreens in surface waters: Regional mass balance in two Swiss lakes. *Chemosphere* **2004**, *55*, 951–963.

97. Balmer, M.E.; Buser, H.R.; Muller, M.D.; Poiger, T. Occurrence of some organic UV filters in wastewater, in surface waters, and in fish from Swiss lakes. *Environ. Sci. Technol.* **2005**, *39*, 953–962.

98. Booij, K.; Zegers, B.N.; Boon, J.P. Levels of some polybrominated diphenyl ether (PBDE) flame retardants along the Dutch coast as derived from their accumulation in SPMDs and blue mussels. *Chemosphere* **2002**, *46*, 683–688.

99. Rayne, S.; Ikonomou, M.G. Reconstructing source polybrominated diphenyl ether congener patterns from semipermeable membrane devices in the Fraser River, British Columbia, Canada: Comparison to commercial mixtures. *Environ. Toxicol. Chem.* **2002**, *21*, 2292–2300.

100. http://www.chbr.noaa.gov

101. Mackenzie, L.; Beuzenberg, V.; Holland, P.; McNabb, P.; Selwood, A. Solid phase adsorption toxin tracking (SPATT): A new monitoring tool that simulates the biotoxin contamination of filter feeding bivalves. *Toxicon* **2004**, *44*, 901–918.

102. Casey, C.R.; Strattan, L.; Jones-Lepp, T.L.; Alvarez, D. EPA Science Forum 2004: Healthy Communities and Ecosystems. Washington, DC; 2004.

103. Vroblesky, D.A.; Robertson, J.F. Temporal changes in VOC discharge to surface water from a fractured rock aquifer during well installation and operation, Greenville, South Carolina. *Ground Water Monit. Remediat.* **1996**, *16*, 196–201.

104. Hare, P.W.; Cohen, R.; Forsell, J.; Gefell, M. Discussion of "active monitoring and thoughtful remediation," by Dr. Michael Barcelona, Ground Water Monitoring & Remediation, v. 20, no. 2, page 4. *Ground Water Monit. Remediat.* **2000**, *20*, 127–130.

105. Vroblesky, D.A.; Campbell, T.R. Equilibration times, compound selectivity, and stability of diffusion samplers for collection of groundwater VOC concentrations. *Adv. Environ. Res.* **2001**, *5*, 1–12.

106. Divine, C.E.; McCray, J.E. Estimation of membrane diffusion coefficients and equilibration times for low-density polyethylene passive diffusion samplers. *Environ. Sci. Technol.* **2004**, *38*, 1849–1857.

107. Ehlke, T.A.; Imbrigiotta, T.E.; Dale, J.M. Laboratory comparison of polyethylene and dialysis membrane diffusion samplers. *Ground Water Monit. Remediat.* **2004**, *24*, 53–59.

108. Harrington, G.A., Cook, P.G.; Robinson, N.I. Equilibration times of gas-filled diffusion samplers in slow-moving ground water systems. *Ground Water Monit. Remediat.* **2000**, *20*, 60–65.

109. Martin, H.; Piepenbrink, M; Grathwohl, P. Ceramic dosimeters for time-integrated contaminant monitoring. *Process Anal. Chem.* **2001**, *6*, 68–73.

110. Rogers, K.R.; Harper, S.L.; Robertson, G. Screening for toxic industrial chemicals using semipermeable membrane devices with rapid toxicity assays. *Anal. Chim. Acta* **2005**, *543*, 229–235.

111. Vorkamp, K.; Haas, B.S.; Herrmann, R. Development of passive samplers for the detection of chlorinated hydrocarbons in sewers. *Water Sci. Technol.* **1998**, *37*, 311–318.

112. Haas, B.S.; Hermann, R. Tracing volatile organic compounds in sewers. *Water Sci. Technol.* **1998**, *37*, 295–301.

113. Bi, E.; Haderlein, S.B.; Schmidt, T.C. Sorption of methyl tert-butyl ether (MTBE) and tert-butyl alcohol (TBA) to synthetic resins. *Water Res.* **2005**, *39*, 4164–4176.

114. Zhang, H.; Davison, W. Performance-characteristics of diffusion gradients in thin-films for the in-situ measurement of trace-metals in aqueous-solution. *Anal. Chem.* **1995**, *67*, 3391–3400.

115. Persson, L.B.; Morrison, G.M.; Friemann, J.U.; Kingston, J.; Mills, G.; Greenwood, R. Diffusional behaviour of metals in a passive sampling system for monitoring aquatic pollution. *J. Environ. Monit.* **2001**, *3*, 639–645.

116. Davison, W.; Zhang, H. In situ measurement of labile species in water and sediments using DGT. *Ocean Sci. Technol. (Chemical Sensors in Oceanography)* **2000**, *1*, 283–302.

117. Docekalova, H; Divis, P. Application of diffusive gradient in thin films technique (DGT) to measurement of mercury in aquatic systems. *Talanta* **2005**, *65*, 1174–1178.

118. Guéguen, C.; Gilbin, R.; Pardos, M.; Dominik, J. Water toxicity and metal contamination assessment of a polluted river: The Upper Vistula River (Poland). *Appl. Geochem.* **2004**, *19*, 153–162.

119. Jiann, K.-T.; Wen, L.-S.; Santschi, P.H. Trace metal (Cd, Cu, Ni and Pb) partitioning, affinities and removal in the Danshuei River estuary, a macro-tidal, temporally anoxic estuary in Taiwan. *Mar. Chem.* **2005**, *96*, 293–313.

120. Ernstberger, H.; Zhang, H.; Davison, W. Determination of chromium speciation in natural systems using DGT. *Anal. Bioanal. Chem.* **2002**, *373*, 873–879.

121. Rauch, S.; Hemond, H.F.; Peucker-Ehrenbrink, B. Recent changes in platinum group element concentrations and osmium isotopic composition in sediments from an urban lake. *Environ. Sci. Technol.* **2004**, *38*, 396–402.

122. Nriagu, J.O. *Thallium in the Environment*. New York: Wiley; 1998.

123. Garmo, O.A.; Royset, O.; Steinnes, E.; Flaten, T.P. Performance study of diffusive gradients in thin films for 55 elements. *Anal. Chem.* **2003**, *75*, 3573–3580.

124. Tan, K.; Anderson, T.A.; Jackson, W.A. Temporal and spatial variation of perchlorate in streambed sediments: Results from in-situ dialysis samplers. *Environ. Pollut.* **2005**, *136*, 283–291.

125. Buffle, J.; Tercier-Waeber, M.-L. Voltammetric environmental trace-metal analysis and speciation: From laboratory to in situ measurements. *Trends Anal. Chem.* **2005**, *24*, 172–191.

126. Zhang, H.; Davison, W. Direct in situ measurements of labile inorganic and organically bound metal species in synthetic solutions and natural waters using diffusive gradients in thin films. *Anal. Chem.* **2000**, *72*, 4447–4457.

127. Zhang, H; Davison, W. In situ speciation measurements. Using diffusive gradients in thin films (DGT) to determine inorganically and organically complexed metals. *Pure Appl. Chem.* **2001**, *73*, 9–15.

128. Scally, S.; Davison, W.; Zhang, H. In situ measurements of dissociation kinetics and labilities of metal complexes in solution using DGT. *Environ. Sci. Technol.* **2003**, *37*, 1379–1384.

129. Scally, S.; Zhang, H.; Davison, W. Measurements of lead complexation with organic ligands using DGT. *Aust. J. Chem.* **2004**, *57*, 925–930.

130. Li, W.; Zhao, H.; Teasdale, P.R.; Wang, F. Trace metal speciation measurements in waters by the liquid binding phase DGT device. *Talanta* **2005**, *67*, 571–578.

131. Tusseau-Vuillemin, M.-H.; Gilbin, R.; Taillefert, M.A. Dynamic numerical model to characterize labile metal complexes collected with diffusion gradient in thin films devices. *Environ. Sci. Technol.* **2003**, *37*, 1645–1652.

132. Unsworth, E.R.; Zhang, H.; Davison, W. Use of diffusive gradients in thin films to measure cadmium speciation in solutions with synthetic and natural ligands: Comparison with model predictions. *Environ. Sci. Technol.* **2005**, *39*, 624–630.

133. Li, W.; Zhao, H.; Teasdale, P.R.; John, R.; Wang, F. Metal speciation measurement by diffusive gradients in thin films technique with different binding phases. *Anal. Chim. Acta* **2005**, *533*, 193–202.

134. French, M.A.; Zhang, H.; Pates, J.M.; Bryan, S.E., Wilson, R.C. Development and performance of the diffusive gradients in thin-films technique for the measurement of technetium-99 in seawater. *Anal. Chem.* **2005**, *77*, 135–139.

135. Murdock, C.; Kelly, M.; Chang, L.-Y.; Davison, W.; Zhang, H. DGT as an in situ tool for measuring radiocesium in natural waters. *Environ. Sci. Technol.* **2001**, *35*, 4530–4535.

136. Naylor, C.; Davison, W.; Motelica-Heino, M.; Van Den Berg, G.A.; Van Der Heijdt, L.M. Simultaneous release of sulfide with Fe, Mn, Ni and Zn in marine harbour sediment measured using a combined metal/sulfide DGT probe. *Sci. Total Environ.* **2004**, *328*, 275–286.

137. Neal, A.L.; Techkarnjanaruk, S.; Dohnalkova, A.; McCready, D.; Peyton, B.M.; Geesey, G.G. Iron sulfides and sulfur species produced at hematite surfaces in the presence of sulfate-reducing bacteria. *Geochim. Cosmochim. Acta* **2001**, *65*, 223–235.

138. Motelica-Heino, M; Naylor, C; Zhang, H; Davison, W. Simultaneous release of metals and sulfide in lacustrine sediment. *Environ. Sci. Technol.* **2003**, *37*, 4374–4381.

139. Mentzer, H.M.; Peaslee G.F. Analysis of soluble reactive phosphorus by diffusive gradient in thin-films (DGT) and application to the resuspension of sedimentary phosphorus. Abstracts of Papers, *227th ACS National Meeting*, Anaheim, CA, March 28–April 1; 2004; p. CHED-359.

140. Zhang, H; Davison, W; Gadi, R; Kobayashi, T. In situ measurement of dissolved phosphorus in natural waters using DGT. *Anal. Chim. Acta* **1998**, *370*, 29–38.

141. Følsvik, N.; Brevik, E.M.; Berge, J.A. Monitoring of organotin compounds in seawater using semipermeable membrane devices. *J. Environ. Monit.* **2000**, *2*, 281–284.

142. Følsvik, N.; Brevik, E.M.; Berge, J.A. Organotin compounds in a Norwegian fjord. A comparison of concentration levels in semipermeable membrane devices (SPMDs), blue mussels (*Mytilus edulis*) and water samples. *J. Environ. Monit.* **2002**, *4*, 280–283.

143. Aguilar-Martínez, R.; Greenwood, R.; Mills, G.A.; Vrana, B.; Palacios-Corvillo, M.A.; Gómez-Gómez, M.M. Assessment of Chemcatcher passive sampler for the monitoring of inorganic mercury and organotin compounds in water. *Int. J. Environ. Anal. Chem.* **2008**, *88*, 75–90.

144. Aguilar-Martínez, R.; Palacios-Corvillo, M.A.; Greenwood, R.; Mills, G.A.; Vrana, B.; Gómez-Gómez, M.M. Calibration and use of the Chemcatcher® passive sampler for monitoring organotin compounds in water. *Anal. Chim. Acta* **2008**, *618*, 157–167.

145. Ouyang, G.; Chen, Y.; Pawliszyn, J. Time-weighted average water sampling with a solid-phase microextraction device. *Anal. Chem.* **2005**, *77*, 7319–7325.

146. De Smaele, T.; Moens, L.; Sandra, P.; Dams, R. Determination of organometallic compounds in surface water and sediment samples with SPME-CGC-ICPMS. *Mikrochim. Acta* **1999**, *130*, 241–251.

147. Popp, P.; Bauer, C.; Moder, M.; Paschke, A. Determination of polycyclic aromatic hydrocarbons in waste water by off-line coupling of solid-phase microextraction with column liquid chromatography. *J. Chromatogr. A* **2000**, *897*, 153–159.

148. Allan, I.J.; Booij, K.; Paschke, A.; Vrana, B.; Mills, G.A.; Greenwood, R. Field performance of seven passive sampling devices for monitoring of hydrophobic substances. *Environ. Sci. Technol.* **2009**, *43*, 5383–5390.

149. Allan, I.J.; Knutsson, J.; Guigues, N.; Mills, G.A.; Fouillac, A.-M.; Greenwood, R. Evaluation of the performance of the Chemcatcher and DGT passive sampling devices for monitoring heavy metal pollutants in water. *J. Environ. Monit.* **2008**, *10*, 821–829.

150. Allan, I.J.; Knutsson, J.; Guigues, N.; Mills, G.A.; Fouillac, A.-M.; Greenwood, R. Evaluation of the Chemcatcher and DGT passive samplers for monitoring metals with highly fluctuating water concentrations. *J. Environ. Monit.* **2007**, *9*, 672–681.

151. DiGiano, F.A.; Elliot, D.; Leith, D. Application of passive dosimetry to the detection of trace organics. *Environ. Sci. Technol.* **1988**, *22*, 1365–1367.

152. www.alcontrol.se

153. www.gaiasafe.de

154. Blind, T.; Gerschwitz, M.; Gessler, H.; Slama, H.; Sorge, H.; Göttzelmann, P.; Nallinger, M. Vergleichsuntersuchungen bei der grundwasserprobenahme: passives adsorptionsverfahren zur erkundung organischer kontaminationen. *Terra Tech.* **1994**, *4*, 26–28.

155. Carls, M.G.; Holland, L.G.; Short, J.W.; Heintz, R.A.; Rice, S.D. Monitoring polynuclear aromatic hydrocarbons in aqueous environments with passive low-density polyethylene membrane devices. *Environ. Toxicol. Chem.* **2004**, *23*, 1416–1424.

156. Heringa, M.B.; Hermens, J.L.M. Measurement of free concentrations using negligible depletion-solid phase microextraction (nd-SPME). *Trends Anal. Chem.* **2003**, *22*, 575–587.

157. Lee, H.L.; Hardy, J.K. Passive sampling of monocyclic aromatic priority pollutants in water. *Int. J. Environ. Anal. Chem.* **1998**, *72*, 83–97.

158. Litten, S.; Mead, B.; Hassett, J. Application of passive samplers (PISCES) to locating a source of PCBs on the Black River, New York. *Environ. Toxicol. Chem.* **1993**, *12*, 639–647.

159. De Jonge, H.; Rothenberg, G. New device and method for flux-proportional sampling of mobile solutes in soil and groundwater. *Environ. Sci. Technol.* **2005**, *39*, 274–282.

160. Kot-Wasik, A. A new passive sampler as an alternative tool for monitoring of water pollutants. *Chem. Anal.* **2004**, *49*, 691–705.

161. Södergren, A. Solvent-filled dialysis membranes simulate uptake of pollutants by aquatic organisms. *Environ. Sci. Technol.* **1987**, *21*, 855–859.

162. Sukola, K.; Koziel, J.; Augusto, F.; Pawliszyn, J. Diffusion-based calibration for SPME analysis of aqueous samples. *Anal. Chem.* **2001**, *73*, 13–18.

163. Heltsley, R.M.; Cope, W.G.; Shea, D.; Bringolf, R.B.; Kwak, T.J.; Malindzak, E.G. Assessing organic contaminants in fish: Comparison of a non-lethal tissue sampling technique to mobile and stationary passive sampling devices. *Environ. Sci. Technol.* **2005**, *39*, 7601–7608.

164. Brumbaugh, W.G.; Petty, J.D.; May, T.W.; Huckins, J.N. A passive integrative sampler for mercury vapour in air and neutral mercury species in water. *Chemosphere* **2000**, *2*, 1–9.

165. Slaveykova, V.I.; Parthasarathy, N.; Buffle, J.; Wilkinson, K.J. Permeation liquid membrane as a tool for monitoring bioavailable Pb in natural waters. *Sci. Total Environ.* **2004**, *328*, 55–68.

166. Parthasarathy, N.; Buffle, J.; Gassama, N.; Cuenod, F. Speciation of trace metals in waters: Direct selective separation and preconcentration of free metal ion by supported liquid membrane. *Chem. Anal.* **1999**, *44*, 455–470.

167. Brumbaugh, W.G.; Petty, J.D.; Huckins, J.N.; Manahan, S.E. Stabilized liquid membrane device (SLMD) for the passive, integrative sampling of labile metals in water. *Water Air Soil Pollut.* **2002**, *133*, 109–119.

168. Paschke, A.; Popp, P. Diffusion-based calibration for solid-phase microextraction of benzene, toluene, ethylbenzene, p-xylene and chlorobenzenes from aqueous samples. *J. Chromatogr. A* **2004**, *1025*, 11–16.

169. Granmo, A.; Ekelund, R.; Bergren, M.; Brorström-Lunden, E.; Bergqvist, P.-A. Temporal trend of organochlorine marine pollution indicated by concentrations in mussels, semipermeable membrane devices, and sediment. *Environ. Sci. Technol.* **2000**, *34*, 3323–3329.

170. Aulakh, J.S.; Malik, A.K.; Kaur, V.; Schmitt-Kopplin, P. A review on solid phase micro extraction-high performance liquid chromatography (SPME-HPLC) analysis of pesticides. *Crit. Rev. Anal. Chem.* **2005**, *35*, 71–85.

171. Scott, G.I.; Fulton, M.H.; Wirth, E.F.; Chandler, G.T.; Key, P.B.; Daugomah, J.W.; Bearden, D.; Chung, K.W.; Strozier, E.D.; DeLorenzo, M.; Sivertsen, S.; Dias, A.; Sanders, M.; MaCauley, J.M.; Goodman, L.R.; LaCroix, M.W.; Thayer, G.W.; Kucklick, J. Toxicological studies in tropical ecosystems: An ecotoxicological risk. *J. Agric. Food Chem.* **2002**, *50*, 4400–4408.

172. Setkova, L.; Hajslova, J.; Bergqvist, P.-A.; Kocourek, V.; Kazda, R.; Suchan, P. Fast isolation of hydrophobic organic environmental contaminants from exposed semipermeable membrane devices (SPMDs) prior to GC analysis. *J. Chromatogr. A* **2005**, *1092*, 170–181.

173. Kingston, J. The development of a passive sampling system for the determination of time-averaged concentrations of organic pollutants in aqueous environments. PhD Thesis, University of Portsmouth, Portsmouth, UK; 2002.

174. Echols, K.R.; Gale, R.W.; Schwartz, T.R.; Huckins, J.N.; Williams, L.L.; Meadows, J.C.; Morse, D.; Petty, J.D.; Orazio, C.E.; Tillitt, D.E. Comparing polychlorinated biphenyl concentrations and patterns in the Saginaw river using sediment, caged fish, and semipermeable membrane devices. *Environ. Sci. Technol.* **2000**, *34*, 4095–4102.

175. McCarthy, J.F.; Southworth, G.R.; Ham, K.D.; Palmer, J.A. Time-integrated, flux-based monitoring using semipermeable membrane devices to estimate the contribution of industrial facilities to regional polychlorinated biphenyls budgets. *Environ. Toxicol. Chem.* **2000**, *19*, 352–359.

176. Sabaliunas, D.; Lazutka, J.; Sabaliuniene, I.; Södergren, A. Use of semipermeable membrane devices for studying effects of organic pollutants: Comparison of pesticide uptake by semipermeable membrane devices and mussels. *Environ. Toxicol. Chem.* **1998**, *17*, 1815–1824.

177. Gatermann, R.; Biselli, S.; Hühnerfuss, H.; Rimkus, G.G.; Hecker, M.; Karbe, L. Synthetic musks in the environment. Part 1: Species-dependent bioaccumulation of polycyclic and nitro musk fragrances in freshwater fish and mussels *Arch. Environ. Contam. Toxicol.* **2002**, *42*, 437–446.

178. Zhang, H.; Davison, W.; Mortimer, R.J.G.; Krom, M.D.; Hayes, P.J.; Davies, I.M. Localized remobilization of metals in a marine sediment. *Sci. Total Environ.* **2002**, *296*, 175–187.

179. Larner, B.L.; Seen, A.J. Evaluation of paper-based diffusive gradients in thin film samplers for trace metal sampling. *Anal. Chim. Acta* **2005**, *539*, 349–355.

180. Buzier, R.; Tusseau-Vuillemin, M.-H., Mouchel, J.-M. Evaluation of DGT as a metal speciation tool in wastewater. *Sci. Total Environ.* **2006**, *358*, 277–285.

181. Gimpel, J.; Zhang, H.; Hutchinson, W.; Davison, W. Effect of solution composition, flow and deployment time on the measurement of trace metals by the diffusive gradient in thin films technique. *Anal. Chim. Acta* **2001**, *448*, 93–103.

182. Giusti, L.; Barakat, S. The monitoring of Cr (III) and Cr (VI) in natural water and synthetic solutions: An assessment of the performance of the DGT and DPC methods. *Water Air Soil Pollut.* **2005**, *161*, 313–334.

183. Apte, S.C.; Batley, G.E.; Bowles, K.C.; Brown, P.L.; Creighton, N.; Hales, L.T.; Hyne, R.V.; Julli, M.; Markich, S.J.; Pablo, F.; Rogers, N.J., Stauber, J.L.; Wilde, K.A. Comparison of copper speciation measurements with the toxic responses of three sensitive freshwater organisms. *Environ. Chem.* **2005**, *2*, 320–330.

184. Cleven, R.; Nur, Y.; Krystek, P.; Van Den Berg, G. Monitoring metal speciation in the rivers Meuse and Rhine using DGT. *Water Air Soil Pollut.* **2005**, *165*, 249–263.

185. Downard, A.J.; Panther, J.; Kim, Y.-C.; Powell, K.J. Lability of metal ion-fulvic acid complexes as probed by FIA and DGT: A comparative study. *Anal. Chim. Acta* **2003**, *499*, 17–18.

186. Dunn, R.J.K.; Teasdale, P.R.; Warnken, J.; Schleich, R.R. Evaluation of the diffusive gradient in a thin film technique for monitoring trace metal concentrations in estuarine waters. *Environ. Sci. Technol.* **2003**, *37*, 2794–2800.

187. Fones, G.R.; Moffett, J.W. In situ Cu speciation measurements in Boston Harbor using DGT. Abstracts of Papers, *224th ACS National Meeting*, Boston, MA, August 18–22; 2002; p. GEOC-018.

188. Garofalo, F.; Ceradini, S.; Winter, M. The use of diffusive gradients in thin-film (DGT) passive samplers for the measurement of bioavailable metals in river water. *Ann. Chim. (Rome, Italy)* **2004**, *94*, 515–520.

Chapter 20

Solid-Phase Microextraction for Drug Analysis

Heather L. Lord

20.1. INTRODUCTION

Studies on the application of solid-phase microextraction (SPME) technology for the analysis of drugs has been pursued for over a decade. In this time, significant advances have been made in the theory and application of the technology for both therapeutic pharmaceutical compounds and drugs of abuse. As has been observed for other applications of the technology, the simplified nature of SPME has been the primary attractant for its use, resulting in reduced sample handling, solvent use, time, and cost. Initial applications in drug analysis were primarily for forensic drugs (see Section 20.6.1). As these compounds are typically less polar and more volatile relative to therapeutic drugs, they were easily analyzed by polydimethylsiloxane (PDMS)-coated fibers interfaced to gas chromatography (GC). Numerous applications appeared for the analysis of compound classes such as the amphetamines, opiates, barbiturates, and cannabinoids. Initial reports also appeared on the analysis of therapeutic drugs such as anesthetics, antibiotics, and antidepressants. With the introduction of the solid sorbent coatings, primarily the divinylbenzene (DVB)-related products, in the mid-1990s, the analysis of a broader range of drugs became feasible, opening up the field of application to the semivolatiles such as cocaine, the benzodiazepines, and their metabolites. Application to nonvolatile drugs was somewhat later to appear, partially due to the limited availability of extraction phases for these compounds—only the carbowax–templated resin (CW-TPR) fiber was widely applicable—and due to the more cumbersome nature of interfacing SPME with condensed phase separations (e.g., high-performance liquid chromatography [HPLC]). In recent years, these challenges have spurred much research interest in the development of improved phases for nonvolatile extraction and strategies for interfacing SPME with liquid chromatography (LC) and capillary electrophoresis (CE) separations. These developments are now also enabling application of the technology to analysis of other compounds of biomedical and biological interest, such as proteins, peptides, endogenous biological compounds, and biomarkers of health status.

This chapter describes fundamental considerations specific for method development for drug analysis, the use of derivatization, a review of recent efforts in novel coating development, and a review of the application of SPME technology to various aspects of drug and biomedical analysis, with practical descriptions of its implementation and use provided throughout.

20.2. FUNDAMENTAL CONSIDERATIONS IN METHOD DEVELOPMENT

20.2.1. Selection of Sample Volume

Often for biological sample analysis, the volume of sample available for analysis is limited. Also, as analytical equipment is reduced in scale and with the developments of lab-on-a-chip devices, additional pressures are seen for efficient handling of very small sample volumes. Most conventional sample preparation strategies cannot be used with sample volumes on the microliter scale. The small dimension of SPME devices can be an advantage in this regard.[1] So long as a sampling strategy can be devised where the entire surface of the extraction phase is in contact with the sample, reproducible extraction can be achieved. For instance, Li and Weber demonstrated the extraction of barbiturates from just 50 µL of sample contained in a 1.5-mm internal diameter (i.d.) Teflon tube.[2] The further application of SPME to *in vivo* analysis has taken advantage of the devices' abilities to sample from very small spaces.[3,4] In such cases, sample volume may not be known.

While SPME devices are capable of extraction from very small sample volumes, analysts must carefully consider the consequences on calibration. The equilibrium concentration of analyte in the extraction phase is proportional to the equilibrium or *final* concentration in the sample phase according to the partition or distribution coefficient.

Handbook of Sample Preparation, Edited by Janusz Pawliszyn and Heather L. Lord

Analysts, however, typically calibrate an extraction by plotting the amount extracted versus the *initial* sample concentration. If a significant proportion of the analyte is extracted, the initial and final sample concentrations can be significantly different. This is particularly true of very small volume samples. In this case of course, calibration is still feasible, but the sample volume must be tightly controlled. Variation in the sample volume will directly impact analytical precision. Because liquid handling devices intended for microliter volumes are inherently less precise than those for milliliter volumes, significant effort must be put into volume measurement in these cases. It is often not possible to achieve acceptable precision with manual devices. Automated sample handling is more appropriate in these cases. These considerations point to the importance of an estimation of the conditions of *negligible depletion* and an understanding of whether sample volume must be controlled for a particular application.

20.2.2. Calculation of Negligible Depletion Sample Volume

Analyte depletion is considered negligible when the sample concentration changes by 10% or less during extraction, particularly where this is an amount in line with experimental error. This may be evaluated experimentally by performing extractions at different sample volumes and observing the range of volumes for which the extracted amount varies by ≤10%. Where analyte depletion is not negligible, sample volume must be carefully controlled during extraction. If the goal is to estimate drug binding affinity to one or more matrix components, the free analyte concentration in the sample at equilibrium must be calculated and employed in the binding affinity calculations. Depletion will obviously complicate the analysis, and so it is beneficial in method development to determine the minimal sample volume that will permit negligible depletion during extraction. The expression describing the amount of analyte extracted by the fiber at equilibrium (Equation 20.1) may be used for this purpose (for absorptive extraction phases):

$$n_f^{\infty} = \frac{K_{fs} V_f C_0 V_s}{K_{fs} V_f + V_s}. \tag{20.1}$$

Here, K_{fs} is the analyte partition coefficient between the fiber and sample, V_f is the extraction phase volume, C_0 is the initial sample concentration, and V_s is the sample volume. Sample depletion may be regarded as negligible where the capacity of the sample phase ($\sim V_s$) is 10× larger than the capacity of the extraction phase ($\sim K_{fs} V_f$). In this case, the $K_{fs} V_f$ term in the denominator is considered insignificant, and V_s cancels out of the equation. Note that $K_{fs} V_f$ may be estimated from the slope of the negligible depletion equilibrium calibration curve (n_f^{∞} vs. C_0) determined experimentally as described above. The minimal sample volume permitting negligible depletion may be calculated from the relationship $V_s = 10 K_{fs} V_f$. The calculation for solid sorbent fibers is analogous, except that the expression for equilibrium amount extracted for these fibers is slightly different.[5] The relationship used to calculate minimal sample volume permitting negligible depletion is $V_s = 10 K V_f (C_{f\max} - C_f^{\infty})$, where K here is the solid sorbent/sample distribution coefficient, $C_{f\max}$ is the maximum surface concentration of active sites, and C_f^{∞} is the surface concentration of occupied active sites.

20.2.3. Recovery: Comparison of Definitions

It is important to understand the degree of recovery of an SPME method for drug analysis prior to implementing it for a new application. Authors reporting the recovery of a method typically report either the proportion of total analyte extracted by a method or the proportion of analyte extracted from the target sample matrix relative to extraction from a matrix-free sample. For SPME, the former is useful in determining the degree of exhaustive extraction, or whether negligible depletion has been achieved. The latter indicates whether the sample matrix has an effect on the extraction process. For practitioners of solid-phase extraction (SPE) sample preparation, the former descriptor is important as the extraction by definition is assumed to be exhaustive. If it is not, a correction factor is applied for calibration. For some early SPME methods in the literature, the authors described the low observed "recovery" as a deficit in this regard. Care must be taken when reading literature methods to determine which descriptor of recovery is being described.

The second descriptor of recovery (degree of matrix influence) is important for decisions on appropriate calibration for SPME and is particularly important in bioanalysis where matrices are often complicated. Where matrix has little or no impact on amount extracted, external calibration is feasible, and it may not be necessary to ensure a close match between the matrix used for the standards and that of the samples. Where matrix does have a significant impact on the amount extracted, it is important to either take care in ensuring the matrix used for the standards is closely matched to that of the samples, or else employ an internal standard or standard addition calibration.

20.2.4. Equilibration Time

Several factors affect the time required for sufficient mass of analyte to reach the surface of the extraction phase including time to reach the boundary layer from the bulk sample and time to diffuse within the boundary layer surrounding the extraction phase to the surface of this phase. For analysis by direct immersion, diffusion both to and within the boundary layer can be relatively slow. Because many drugs of analytical interest are nonvolatile, this extraction mode is often employed, with the result that equilibrium extraction for these compounds takes much longer than for equilibrium extraction of volatile compounds from a gaseous sample where diffusion coefficients are much faster.

In a well-stirred liquid sample with direct immersion extraction, it is assumed that the analyte concentration at the interface between the boundary layer and the bulk of the sample is always equal to the concentration in the bulk of the sample. This is true of extraction from both liquid and gaseous samples. In this case, time required for a sufficient mass of analyte to reach the boundary layer is inconsequential, and diffusion within the boundary layer controls the rate of this step. Where extraction is from the headspace above a liquid or solid sample, however, mass transfer from the liquid or solid to the headspace may be slow. Again, because of their relatively low volatilities and slow diffusion in condensed phase, many drugs of analytical interest will suffer from this slow mass transfer to the boundary layer and have relatively long equilibrium times.

Finally, diffusion or establishment of an equilibrium within the extraction phase can be much slower for these compounds than for compounds more amenable to GC. In a partitioning sorbent, either the diffusion coefficient itself may be very small or the partition coefficient low. In the latter case, a larger volume of extraction phase (negligible depletion conditions met) may be required to provide the necessary sensitivity. Because diffusion equilibrium time within a partitioning sorbent is dependent on the square of the thickness of the extraction phase (Equation 20.2),[6] even a small increase in the sorbent thickness will cause a significant lengthening of extraction time:

$$t_e = t_{95\%} = \frac{(b-a)^2}{2D_f}. \tag{20.2}$$

In this equation, $(b - a)$ refers to the thickness of the sorbent material, and D_f refers to analyte diffusion in the sorbent.

Solid sorbents are widely used for SPME drug analysis, primarily because of the larger amounts of drug that can be extracted by these phases relative to the partitioning (liquid) phases; the fiber capacity terms for solid sorbents are generally larger than those for liquid sorbents. In the case of a nonporous solid sorbent, equilibration by diffusion within the sorbent does not occur within typical extraction times. As soon as an analyte molecule diffuses through the boundary layer and reaches the active site, it binds with a rate constant that is usually sufficiently fast that this step is not limiting for the overall extraction rate. Diffusion through the boundary layer is rate limiting for the extraction, but the path length is small so extraction rates can be fast. In most cases, however, such sorbents would not have sufficient capacity to be of much value. Generally, solid sorbents used for SPME have a high degree of porosity, in which case, analyte molecules must diffuse through the sorbent pores prior to binding with an active site. Thus, the diffusion path length can be quite long, resulting in a relatively long extraction equilibration time.

Finally, although in most cases the kinetics associated with an analyte binding to an active site are negligible compared with the time required for other steps in an extraction, they can be appreciable and limiting where the active site is a biological recognition unit. This has been demonstrated for immunoaffinity SPME sorbents.[7] A slow rate of derivatization may also limit the overall extraction time.

Several strategies have been used to address the long equilibration times encountered in many SPME drug analyses. First, effective stirring is essential in any direct immersion extraction. Also, strategies to efficiently force a sufficient analyte concentration to the headspace are often employed for headspace extractions, including limiting headspace volume, heating of the sample, and the addition of salt to the sample. For porous solid sorbents, limiting of the sorbent volume to just that required for the sensitivity needed can make a substantial improvement in equilibration time. Finally, significant effort has been put into developing methods that take advantage of pre-equilibrium extraction, as is described in Section 20.2.8.

20.2.5. Effect of Additional Phases in the Sample

When additional phases are present in a sample, analyte can also partition into those phases in addition to the sample matrix and the extraction phase. Any discontinuous phase in the sample having a significant affinity for the analyte of interest competes with the extraction phase for analyte and, therefore, causes a reduction in the amount extracted. Given the complexity in most biological matrices, consideration should be given to the potential for competing phases. Typically, this is evaluated by assessing recovery by the technique of degree of matrix influence as discussed above (Section 20.2.3). Where matrix interferences are present but predictable and reproducible from one sample to the next, sensitivity may be compromised but good accuracy can be obtained. On the other hand, method precision suffers significantly if the amount of such a competing phase varies from sample to sample. An understanding of the cause of this variability is useful in assessing viable means to reduce it.

At equilibrium, analyte will have distributed to all phases present according to their partition (affinity) coefficients and volumes. Thus, phases with either very low volume or low affinity can be disregarded. SPME will only extract free analyte, so where a competing phase with high volume or affinity is present, less analyte will be extracted. Where this phase is variable from one sample to the next, accurate analysis is compromised. A detailed discussion of these equations has been published.[6] In biological drug extractions, significant competing phases could include dissolved protein, tissue fragments, cells or cell fragments, or even receptors or antibodies to the analyte of interest. If the sample is a pharmaceutical compound, several competing phases may be present originating as excipients in the drug formulation. The analysis of timed- or controlled-release formulations must be considered carefully as has been described for the analysis of paclitaxel free concentration in solution from liposome formulations.[8]

20.2.6. Ionic Strength

The addition of a salt can often improve recovery when conventional extraction methods are used. Sodium chloride (NaCl) is commonly used for this purpose.[9–11] Occasionally, one sees an initial increase in the extraction yield with an increase in salt concentration, with a maximum being reached followed by a decrease in amount extracted with further increase in salt concentration. Hall and Brodbelt[12] noted that the effect varied through a series of barbiturate drugs, apparently related to the pK_a of the drug. Abdel-Rehim et al. noted that for ropivicaine and a series of its metabolites, 20% NaCl provided optimal extraction. Higher levels resulted in reduced extraction efficiency.[13] This behavior can be explained by considering two simultaneously occurring processes. Initially, analyte recovery increases due to a phenomenon termed salting out, whereby water molecules form hydration spheres around the ionic salt molecules. These hydration spheres reduce the concentration of water available to dissolve analyte molecules; thus, it is expected that this phenomenon will drive additional analytes into the fiber coating.[14] In competition with this process, however, is the fact that molecules may themselves participate in electrostatic interactions with the salt ions in solution,[15] thereby reducing their ability to move into the fiber coating. Initially, it is the interaction of the salt molecules with water that is the predominant process. As salt concentration increases further, salt molecules will begin to interact with analyte molecules. Thus, it is reasonable that analyte extraction initially increases with increasing salt concentration, followed by a decrease as the salt interaction with the analytes in solution predominates.

20.2.7. Extraction pH

The pH of the extraction mixture is important for drugs possessing a pH-dependent dissociable group. It is only the undissociated form of the drug that will be extracted by an absorptive fiber coating—PDMS, polyacrylate (PA), CW, or octadecylsilane ([ODS] C18)—or extraction processes governed by nonpolar associations in solid sorbent phases. This is important for drug extraction because drug that has partitioned into the fiber coating cannot participate in the Henderson–Hasselbalch equilibrium between the acid and base forms of the drug in the sample. In the case where the extraction mixture pH is controlled with a buffer, as the undissociated form of the drug is extracted by the fiber, more dissociated drug reassociates and is therefore available for extraction. Thus, in a buffered extraction mixture, more drug can be extracted than in an extraction mixture where the pH is not buffered. In an unbuffered extraction mixture, the ratio of undissociated to dissociated forms of the drug can vary. Therefore, one does not achieve the continual transfer of drug from the dissociated to the undissociated form, and then to the fiber coating. This effect is seen in Figure 20.1, which compares the amount of methamphetamine extracted

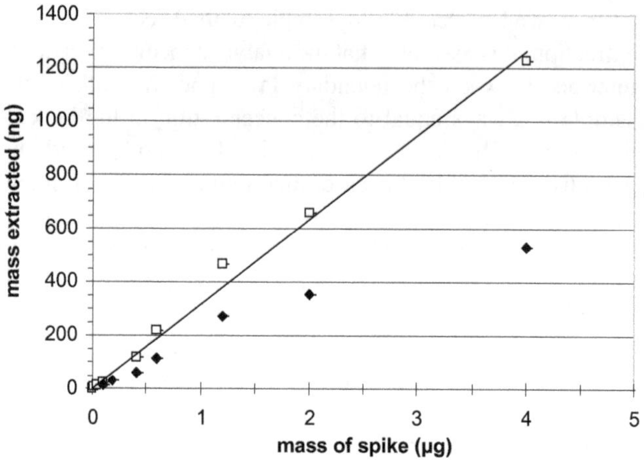

Figure 20.1. Effect of sample buffering on methamphetamine extraction linearity. Key: open squares, pH 12 buffered samples; closed diamonds, sample adjusted to pH 12 with potassium hydroxide. Adapted from Reference 11.

from a solution where the pH is base adjusted with potassium hydroxide, versus one where the pH is controlled at pH 12 with phosphate buffer. Note there is a significant deviation from linearity in amount extracted as drug concentration increases in the nonbuffered system. Queiroz et al. provide a good description and discussion of the effect of pH variation on a series of drugs used therapeutically in the control of seizures.[16]

Many drug formulations designed to be taken orally utilize the salt of the active ingredient to aid dissolution in the stomach. Control of extraction pH is important in such cases to ensure that the undissociated form of the drug is present for extraction. Additionally, the formulation matrix of a pharmaceutical preparation may itself act to buffer the extraction matrix to favor the dissociated form, so final extraction pH should be checked before extraction.

20.2.8. Pre-Equilibrium Extraction for Quantification with High Sample Concentrations

When the capacity of a solid sorbent is approached, that is, a high proportion of the active sites are occupied, the calibration of the extracted amount related to sample concentration deviates from linearity. Displacement of analytes with lower affinities may also occur.[17,18] Thus, the amount extracted can vary with concentrations of both the target and other analytes. This situation is of course more problematic with higher sample concentrations, significantly complicating calibration by equilibrium extraction in this case. For this reason, solid sorbents are best used only for trace analysis. Another solution to this dilemma is to employ pre-equilibrium extraction times. In this case, the extraction process is governed by extraction kinetics rather than equilibrium states.[19] Pre-equilibrium extraction is also helpful to improve sample throughput in the case of long equilibration times. In such cases, careful control of extraction time and agitation conditions may be required. Where this is not, practical alternative calibration strategies should be considered. A significant volume of literature is available describing these options,[20–23] and the issues are also described in more detail in Chapter 5.

20.3. METHOD DEVELOPMENT CONSIDERATIONS BASED ON DRUG PROPERTIES

20.3.1. Volatile Drugs

Drugs that have inherent volatility are typically analyzed by the same techniques as fragrance and environmental compounds. As for the terpenes in fragrance analysis or the benzene-like compounds in environmental samples, these analytes are easily extracted by either headspace or direct immersion extraction, using conventional extraction phases, primarily PDMS. Henry's constants are typically high and polarity is low, providing for high partition coefficients and relatively fast headspace extraction. An example is the analysis of halothane described by Musshoff et al.[24] The method employed was headspace–SPME (HS-SPME) with GC–mass spectrometry (MS) to determine halothane in the headspace of blood and other forensic samples. Total analysis time per sample including extraction, separation, and detection was 40 min.

20.3.2. Non- or Semivolatile Drugs

Drugs with low volatility have low headspace concentrations. They may also be polar and have strong interactions with aqueous sample matrices. Sorbents with better efficiency for polar compounds should be evaluated (i.e., PA, CW, C18, or one of the phases incorporating DVB). Because headspace diffusion coefficients are much higher than liquid diffusion coefficients, it may still be preferable to extract these with a sorbent located in the headspace. However, fiber extraction may be faster than the rate at which the headspace concentration is replenished by mass transfer from the sample. In this case, either direct immersion extraction must be employed or the compound's affinity for the headspace must be increased. Headspace affinity may be improved by the addition of salt to the sample as was discussed previously (Section 20.2.6) or by heating of the sample to increase Henry's constant and move more analyte to the headspace. If these options do not provide sufficient extraction, the analyte may also be derivatized to a form with a higher volatility or lower polarity. Alternatively, one of the novel extraction phases under development for polar analytes may be considered.

20.3.3. Presence of Ionized or Ionizable Groups

For drug compounds with ionized or ionizable groups, the ionized species have a very high affinity for aqueous medium and so are poorly extracted by most sorbents. As described above, pH adjustment to favor a nonionized form may be employed to achieve sufficient sensitivity. Amine-containing basic drugs such as the amphetamines have a neutral charge state only at very high pH, whereas acidic drugs such as ibuprofen are neutral at acidic pH. It should be noted that PDMS will dissolve from the fiber if immersed in a very basic solution (>pH 10). If adjustment to a highly basic condition is required, headspace extraction may be necessary.[11] Difficulty arises where pH adjustment alone is not sufficient to improve extraction.

It is also possible to improve the extraction efficiency of ionized analytes by incorporating an ion-pairing reagent in the sample matrix. Ion pairing allows for the extraction of ionized compounds by partitioning, by combining the analyte ions with counterions of opposite charge.[25] The inclusion of an ion-pairing reagent in the mobile phase is commonly used to improve HPLC separations. An example is the use of this strategy for the LC separation of antiretroviral drugs extracted from biological fluids by SPE.[26] Several articles have appeared in recent years on the use of ion pairing to improve extraction of polar drugs by SPE,[27,28] liquid–liquid extraction,[29] liquid-phase microextraction,[30] and liquid-phase membrane extraction,[31,32] with or without subsequent derivatization. A few articles have appeared describing this technique's application to SPME. Most of these do not describe the analysis of drugs, but the methods developed could be instructive in the development of drug analyses. Pan and Pawliszyn proposed the use of ion-pair SPME to convert long-chain fatty acids into their methyl esters using in-port derivatization.[33] This same strategy was employed for the analysis of alkylbenzene sulfonates from water samples using the 100-μm PDMS fiber.[34] The ion-pairing reagent employed was tetrabutylammonium hydrogen sulfate (TBA-HSO_4). In this case, the counterion association both promoted the extraction of these ionic compounds using the nonpolar PDMS sorbent, and facilitated the conversion of the analytes to their butyl esters in the hot injector port of the gas chromatograph.

In an example more closely related to drug analysis, the shellfish toxin saxatoxin was analyzed in water samples by SPME-HPLC using postcolumn derivatization fluorescence detection. In the method development, desorption of the toxin from the CW-TPR fiber was observed to have poor efficiency. The problem was resolved by employing static ion-pair desorption with heptane sulfonic acid.[35] Li et al. have employed *in situ* derivatization with ion pairing for the analysis of methylmalonic and glutaric acid in urine as a clinical tool for the diagnosis of organic acidemia.[36] The organic acids were converted to their ethyl esters using diethyl sulfate. The esters were then easily extracted from headspace using PDMS fibers, and GC-MS was employed for separation and detection. In this work, the ion-pairing reagent TBA-HSO_4 was included in the derivatization solution to activate the analytes during derivatization and to increase the derivatization yield.[37,38] A similar strategy was employed for the analysis of haloacetic acids in water.[39]

Additional discussion on the use of derivatization is provided below (Section 20.5).

20.3.4. Drug Metabolites Analysis

For the analysis of drugs in biological samples, it is often of interest to evaluate both the parent compound and a variety of drug metabolites. For phase I metabolites, the chemical changes to the molecule are minor (e.g., addition of oxygen, removal of hydrogen), although the resulting metabolites are typically more polar to aid in excretion. Thus, the same method developed for the analysis of the parent compound may be employed for the analysis of these metabolites as well, although lower sensitivity may be encountered for the more polar metabolites. Some examples where both the parent and phase I metabolites were analyzed are described for diazepam,[4] methamphetamine,[11,40] methadone,[41] and cocaine.[42–44] For phase II metabolism, compounds are conjugated with a polar group such as glucuronic acid, glutathione, a sulfonate, or an amino acid. These relatively high-molecular-weight polar metabolites are typically not well extracted by SPME sorbents, and so enzymatic deconjugation is typically employed prior to analysis.[45,46] By performing extraction both before and after deconjugation, a determination of both conjugated and unconjugated metabolites may be made.

20.4. NOVEL COATINGS FOR LC

Much of the effort around the development of novel extraction phases for SPME has been spurred by the need to more efficiently extract polar and semipolar molecules from aqueous (biological) matrices. The early SPME investigations of forensic drugs successfully employed the PDMS-coated fibers with GC separation, due to the relatively low polarity of most of those compounds. Many drugs cannot be analyzed by GC, however. When efforts turned to optimizing SPME-LC, researchers quickly realized that the somewhat more polar DVB-based fibers were far superior to any of the other commercially available fibers and that the CW-TPR fiber typically performed the best. Because only the one commercially available phase consistently performed well for these analyses, the need for more appropriate phases for these analyses became evident. Supelco is currently introducing a new line of coatings suitable for SPME-LC as discussed in Section 20.4.8.

Several experimental fibers have been reported for specific applications of analysis of polar molecules. For example, ion exchange coatings were used to remove metal ions and proteins from aqueous solutions,[47,48] CW for polar organic molecules,[49] metal rods to electrodeposit analytes,[50,51] pencil "leads" to extract pesticides,[52] and Nafion coatings to extract polar compounds from nonpolar matrices.[53] Polypyrrole (PPY) used initially for biosensors[54] and restricted access materials (RAMs) originally designed as SPE packing have specifically been investigated for the analysis of drugs from biological matrices. Additionally, the development of immunoaffinity fibers has been described, which should allow the application of SPME technology for the selective extraction of large biomolecules from complex matrices,[55,56] and interesting developments have been seen in the use of ionic liquids for SPME.[57–59] Some of the findings from these efforts are summarized below.

20.4.1. Restricted Access Materials

RAMs are particulate sorbents produced by Merck primarily for use in SPE cartridges. They feature an outer surface that is effective at repelling biomolecules, and pores sufficiently small enough to exclude biomolecules from the inside of the particle. The extraction phase, however, is located exclusively inside the pores. In this way, biomolecules are excluded from the extraction surface. The extraction phase, typically either C18 or an ion exchange phase, is ideally suited for extracting analytes typically present in biological matrices such as urine or blood. The particle size is similar to the DVB phases already used for SPME. Hence, these phases were evaluated for use in SPME analysis of therapeutic drugs from biological matrices.[60–62] While the fibers performed well, equilibration times were very long. This was likely due both to the time for diffusion into the interior of the particles and within the C18 liquid extraction phase. Regardless, the sorbent holds promise for high sensitivity analysis of therapeutic drugs where extraction time is not the primary concern or when pre-equilibrium extraction is used.

20.4.2. Octadecyl Silane: C18

Because C18 phases extract absorptively, they are of particular interest for the extraction of more polar analytes, which has, to date, been dominated by the solid sorbents. Given the good success seen with the C18 extraction phase of the RAM particles, it was of interest to determine if a coated C18 fiber would give promising results, without the limitations of long diffusion times. Although technical details for the preparation of such a sorbent directly on a fiber support have not yet been addressed, SPME fibers have been prepared by coating C18 modified silica particles onto a stainless steel wire by using polyethylene glycol (PEG) glue.[63] The properties of these fibers were investigated, with the observation that they have good stability, reproducibility, extraction times, and linear range. Because both PEG and immobilized C18 are expected to have good biocompatibility, these fibers should be useful for *in vivo* analysis of drugs by intravenous sampling. More recently, other biocompatible polymers have also been evaluated for gluing C18-silica particles to fibers.[64]

20.4.3. Sol–Gels

Sol–gels have been of interest in chemical separations for at least the past decade, due to their relative ease of preparation, high stability, and reproducibility of performance. Early reports described their use in GC capillaries with PDMS sol-gel,[65] CE capillaries with polyalkylene glycol sol–gel,[66] and SPME fibers with PDMS sol-gel.[67] The sol-gel SPME fibers were used for the extraction of polycyclic aromatic hydrocarbons (PAHs), alkanes, anilines, and alcohols from water. In 2000, Wang et al. demonstrated the use of SPME fibers coated with sol-gel incorporating PEG for the extraction of a range of compounds, including benzene, toluene, ethylbenzene, and xylene (BTEX), phenols, phthalate esters, and aromatic amines.[68] The porosity of the resulting phase was studied, and the fibers were demonstrated as useful for the extraction of compounds with a range of polarities. In 2005, Bagheri et al. described the preparation of SPME fibers incorporating all three strategies (PDMS, polyalkylene glycol, and PEG) and their use for the extraction of dextromethorphan and dextrorphan from plasma.[69] In this work, headspace extraction was employed, and the polar dextrorphan was derivatized prior to extraction to improve its volatility. Interestingly, the PDMS phase was preferred for these relatively polar compounds. The PEG coating demonstrated a low level of extraction, likely due to its low porosity. The polyalkylene glycol coating extracted more analyte than the PDMS coating, but at a longer (>30 min) extraction time than PDMS (30 min at 60°C). The good performance of PDMS was explained in that although conventional PDMS is quite nonpolar, sol–gel PDMS contains a significant number of residual hydroxyl groups after polymerization, imparting a more polar nature to the sorbent.

Recently, two articles have appeared describing the use of a calixarene sol-gel SPME fiber for the extraction of propranolol from urine with separation by GC[70] and CE.[71] The phase was shown to be highly base and temperature resistant and was used successfully for extraction by both direct and HS-SPME from highly basic matrix with heating up to 100°C during extraction and 280°C during desorption in the GC injector port. For coupling to CE, back-extraction of the absorbed analytes into 20% acetonitrile, 80% water at slightly acidic pH (6.2) was used to promote protonation of the analyte. Both articles describe excellent reproducibility and durability of the fibers.

The same group has also investigated the use of a butyl methacrylate sol-gel extraction phase for the analysis of ephedrine derivatives from water and urine with CE separation.[72] For extraction, 5 mL of sample was placed into a 10-mL vial. Optimal headspace extraction conditions included 2.0-g sodium hydroxide and 0.5-g sodium chloride added to the vial, with stirring and heating at 90°C during the 30-min extraction. Back-extraction was optimized at 20°C, 20% acetonitrile, 20 min. The methacrylate fiber performed the best compared with commercial PDMS and PA fibers and a sol-gel fiber prepared without methacrylate. The sol-gel had good solvent stability, extraction efficiency, and durability. Limits of detection were in the low nanogram per milliliter range with linearity between 20 and 5000 ng/mL.

While the experience with sol–gel fibers for drug analysis is currently in its early stages, the ruggedness, reproducibility, and customized properties appear to make it a most interesting phase for SPME drug analyses of the future. A detailed description of the use of sol–gels in sample preparation is presented in Chapter 22.

20.4.4. Monolithic Extraction Phases

Several reports have appeared describing the use of polymer monoliths prepared in fused-silica capillaries for sample preparation. Although the technique of in-tube SPME is quoted, the authors employ high sorbent capacity and high extraction efficiency. In some cases, near exhaustive extraction is accomplished, and the amount extracted is proportional to the total analyte mass in the sample volume (as per SPE), not initial sample concentration (as per SPME). In some cases, it is not possible to determine with certainty if exhaustive or equilibrium extraction is being used as the authors do not always report extraction efficiency or whether analyte breakthrough is occurring with the sorbent. However, given the general strategy presented of maximizing extraction efficiency, it is likely valid to consider these techniques as miniaturized SPE rather than SPME.

Regardless of whether a particular extraction is considered as exhaustive or nonexhaustive, the sorbents described could potentially be employed in either strategy. The sorbents have proved beneficial for a variety of bioanalyses including amphetamines in urine;[73] ketamine in urine;[74] camptothecin in plasma;[75] telmisartin in tissues;[76] antiotensin II receptor agonists in plasma and urine;[77] and theobromine, theophylline, and caffeine in serum.[78] Like the sol–gel phases, the high ruggedness and degree of tunable selectivity make these phases of significant interest in new device development.

20.4.5. Polypyrrole

The PPY-based conducting polymers have been used extensively in biosensor applications, and so good data are available both on their biocompatibility and on the abilities of these polymers for extracting analytes with polarities similar to those of therapeutic drugs. In addition to its use in sensors for neuroscience applications,[79–82] PPY has been investigated for its utility as an extraction phase for SPME applications.[83,84] It was shown to extract as an adsorptive phase.[85] In addition to its low toxicity,[86] it is insoluble in all solvents, tolerates elevated temperatures, and can be used in many chemical environments.[87]

The known biocompatibility of PPY made it an interesting candidate for initial efforts at *in vivo* SPME.[4] The SPME fibers commercially available at the time were unsuitable for this application, due to several factors including their large size, incompatibility with sterilization processes, and large volume of extraction phase, which caused long extraction and desorption times. PPY was electrochemically polymerized onto the surface of a fine (100 μm) stainless-steel wire in a thin (~10 μm) coating, making for a probe with an overall diameter of <130 μm. The reduced capacity allowed for fast equilibrium extraction and exhaustive desorption times (<5 min and <20 s, respectively, under optimized conditions) but still allowed for quantitative analysis of many drugs in their therapeutic ranges in circulating intravenous blood.

Several challenges also attended these efforts. First was the relatively fragile nature of the polymer and difficulty producing a suitably rugged probe. This appeared to be related to the poor quality of commercially available pyrrole monomer and the difficulty in distilling this to high purity in quantity. Other challenges included the difficulty in obtaining highly reproducible performance between probes and somewhat cumbersome nature of their preparation. Further development of these probes will likely be hampered until the challenges of coating preparation and quality can be addressed. Even so, a significant body of work has been amassed on the subject.[54,88–91]

20.4.6. Immunoaffinity Sorbents

SPME devices with immobilized antibodies have been investigated for the highly specific analysis of small-molecule drugs from complex biological matrices. The fibers were found to be useful for very high-sensitivity drug analysis, having some of the lowest limits of detection of any SPME device, but were limited in linear range and the effort required to prepare them.

An immunoaffinity SPME coating was initially prepared via the covalent immobilization of a theophylline antibody onto a fused-silica fiber surface. This surface was modified with 3-aminopropyl triethoxysilane (APTES) and glutaraldehyde, for the selective extraction of theophylline from serum samples.[55] No significant nonspecific binding (nsb) was observed, and the reproducibility of the fiber preparation and the immunoaffinity SPME analysis were good with a percent relative standard deviation (% RSD) of 6.1% for five analyses of the same fiber. Validation of the coating was completed with the successful analysis of theophylline-spiked serum samples (10 and 50 ng/mL). A limitation of this work was that all analyses had to be done with radioactive analytes as the probe did not have sufficient sensitivity (capacity) for analysis by LC-MS.

Subsequently, a technique for purifying the active fraction of a polyclonal antibody was developed. When probes with these antibodies were prepared using an optimized immobilization strategy, capacity and sensitivity were sufficient for the analysis of diazepam and related benzodiazepines from biological fluids by LC-MS.[7,56] Monoclonal antibodies performed similarly to purified polyclonals for extraction from aqueous samples. Another important advance was the development of a routine for drug desorption that maintained antibody activity. The probes could be reused numerous times (>25) and were sufficiently stable that performance was acceptable for at least 6 months with storage under refrigeration.

Nonspecific drug binding (nsb) was found to be insignificant for the probes when used in their dynamic range. At higher concentrations, a high proportion of drug binding was due to nsb and so selectivity was lost. Antibody affinity was evaluated both before and after immobilization. Immobilized antibodies had specific affinities in the range of 10^9–10^{10} M^{-1} with no observed reduction in affinity on immobilization. Cross-reactivity was evaluated both for a range of benzodiazepines as well as an unrelated molecule (erythromycin). Calibration was successful for single analytes at sorbent saturations up to 50%. Because of the nonlinear nature of the adsorption isotherm, nonlinear regression analysis was required for calibration. Accurate calibration of one benzodiazepine in the presence of another, particularly for evaluation of a lower affinity drug in the presence of a higher affinity drug, was problematic. This issue was partially resolved by the use of an antibody with higher specificity for one member of the benzodiazepines class over the others. In general, it was found that where a mixture of cross-reacting analytes is present, sorbent saturation should be limited to 10%, resulting in an extremely narrow dynamic range. For the analysis of benzodiazepines individually or in the presence of erythromycin, limits of detection were ca. 0.001 ng/mL and the dynamic range was extended to 0.5–1 ng/mL, depending on the antibody.

20.4.7. Molecularly Imprinted Polymers (MIPs)

MIPs are highly cross-linked polymer materials containing synthetic cavities or recognition sites with a predetermined selectivity, which have seen extensive application in chemical anslysis.[92] The method of polymer imprinting involves including a template molecule with sufficient structural similarity to a desired analytes in the polymerization matrix.[93] The resulting so-called antibody mimics can be generated with specificities to virtually any analyte of choice and a very high degree of selectivity toward the target analyte or analyte class. The MIP material is resistant to mechanical stress, heat, acids, bases, and organic solvents, while maintaining a storage life of years with no apparent reduction in its molecular recognition performance. Briefly, the desired affinity can be introduced by adding an amount of the compound of interest or a structural analog to the polymerization reaction. This "pattern" chemical may be removed after polymerization, leaving vacant sites of a specific size and shape, suitable for binding the same chemical again from an unknown sample. Since the cavities are based on the size, shape, and chemical complementary of the imprint molecule, chiral selectivity is also possible.

These sorbents also first saw application as an SPE packing material prior to application as an SPME sorbent. Several articles and reviews discussing the application of MIP for sample preparation for drug analysis are available.[94–97] For SPME application, the polymer is prepared as normal with a template related to the analyte of interest, and either coated on the inside surface of an in-tube SPME capillary or the outside of a fused-silica fiber, or crushed and sieved to obtain appropriately sized particulate material for gluing to an SPME fiber.

For fiber SPME, applications have appeared for the analysis is triazines,[98,99] clenbuterol,[100] and diacetyl morphine.[101] In the triazine application, the MIP was formed in a 20- to 53-μm-thick coating on the outer surface of a fused-silica fiber. A similar strategy was used in the preparation of the earlier clenbuterol MIP fibers, with a resulting 75-μm sorbent thickness. Finally, the diacetyl morphine fibers were prepared using methacrylic acid/dimethacrylate monolith imprinted with diacetyl morphine. In this work, the authors also performed extensive mathematical modeling of the adsorption isotherms for the final polymers.

The in-tube SPME option has been investigated for analysis of propranolol,[102] verapamil,[103] and related metabolites. As for SPE packings, the main interest in these phases was for a highly selective extraction of the desired analyte from a complex biological matrix. The main limitations have been the amount of work required to prepare the phase and the difficulty in obtaining efficient desorption. Also, if the target analyte is used as the template, it is sometimes difficult to completely eliminate the target analyte from blank analyses due to bleeding from the polymer. In the case of the automated online MIP in-tube SPME extraction method for propranolol determination in biological fluids,[102] the method showed improved selectivity in comparison to alternative in-tube SPME coatings. Preconcentration of the sample by the MIP adsorbent increased the sensitivity, yielding a limit of detection suitable for clinical analysis. Excellent method reproducibility (% RSD < 5.0%) and column reusability (>500 injections) were observed over a fairly wide linear dynamic range (0.5–100 μg/mL) in serum samples. The method was inexpensive (aside from the effort required to prepare the MIP), simple to set up, and simplified the choice of SPME adsorbent for in-tube extraction. More importantly, the approach can potentially be extended to other MIPs for the determination of a wide range of chemically significant analytes. A more complete description of the use of MIPs for sample preparation is presented in Chapter 23.

20.4.8. New Supelco Commercially Available Phases

Supelco is in the process of introducing new phases that are biocompatible, better suited for the analysis of polar analytes, and/or more amenable to solvent desorption and use with LC. A CW-PEG phase has been introduced that should allow better extraction of polar compounds in an absorptive phase, although to date no literature references are available on the application of this phase to drug analysis. In addition, a variety of phases are in the late stages of development and intended for commercial release in the near future. These include phases based on various SPE sorbent particles embedded in a proprietary biocompatible binder that resists swelling in solvent. The performance of three of these (octadecyl, cyano, and polar embedded sorbents) was recently compared with that of the conventional CW-TPR for the analysis of a variety of drugs (carbamazepine, propranolol, pseudoephedrine, ranitidine, and diazepam) from the plasma and urine.[104] The first of these was introduced in 2010 for both *in vivo* and *in vitro* applications.

20.5. DERIVATIZATION

Analyte derivatization is used in general chemical analysis for several purposes. Derivatization changes the physicochemical properties of an analyte to a form that is more easily analyzed. This may involve improvements in sample preparation efficiency, chromatographic properties, or enhancement of detection sensitivity. In particular, derivatization can produce a less polar product from a polar analyte, significantly improving extraction efficiency. Table 20.1 summarizes various derivatization techniques that can be implemented in combination with SPME. Some of the techniques, such as direct derivatization in the sample matrix, are analogous to well-established approaches used in solvent extraction. In the direct technique, the derivatizing agent is first added to the vial containing the sample. The derivatives are then extracted by SPME and introduced into the analytical instrument. For example, this approach has been applied to extract and separate phenols from aqueous samples by first converting the target analytes to their acetate derivatives.[105]

Where extraction efficiency for underivatized compounds is sufficient to reach the sensitivity required, there may still be problems associated with the separation or detection of these analytes. Good chromatographic performance and detection can be facilitated by in-coating derivatization following extraction. This has been shown with high-molecular-mass carboxylic acids.[33] After exposing an SPME coating containing extracted analytes to diazomethane, the resulting ester derivatives can be separated as narrow bands on a GC column. For improved detection, selective derivatization to analogs containing high detector response groups will result in enhancement in sensitivity and selectivity of detection. Derivatization in the GC injector is an analogous approach, but it is performed at high injection port temperatures. For example, long-chain carboxylic acids can be extracted onto the coating as ion pairs when tetramethylammonium hydrogen sulfite is added to the sample. During volatilization, analytes are converted to methyl esters.[33]

Another interesting and potentially very useful technique is simultaneous derivatization and extraction, performed directly in the coating. This approach allows high efficiencies and can be used in remote field applications. The simplest way to execute the process is to dope the fiber with a derivatization reagent and subsequently expose it to the sample. Then, the analytes are extracted and simultaneously

Table 20.1. Summary of SPME/Derivatization Schemes

Derivatization Strategy	Comments	Reference
Derivatization before extraction	Reagent added to sample matrix prior to extraction, allow sufficient time for reaction to complete before introducing SPME fiber, extraction proceeds according to the polarity/affinity of the product; beneficial where analyte has insufficient affinity for the extraction phase.	109
Derivatization during extraction	Reagent preloaded onto SPME sorbent, simultaneous extraction and derivatization; extraction proceeds according to polarity/affinity of the analyte, but sorbent acts as a zero sink until reagent is consumed; beneficial where analyte has some affinity for the extraction phase.	112
Derivatization on fiber after extraction	Fiber with extracted analyte exposed to reagent after being removed from the extraction vial, reaction normally proceeds on the fiber prior to desorption; beneficial to improve desorption, separation, and/or detection.	113
Derivatization after desorption	Analytes desorbed from fiber as normal, derivatization occurs after desorption, either in a desorption solvent, the injector port, or on column; beneficial where separation/detection characteristics of the analyte are suboptimal.	114

converted to analogs having high affinity for the coating. This is no longer an equilibrium process as derivatized analytes are collected in the coating as long as extraction continues. So long as the derivatization reaction kinetics are faster than the extraction process, the sorbent acts as a zero sink, as the concentration of the underivatized analyte in the sorbent is essentially zero at all times. This approach, which is used for low-molecular-mass carboxylic acids, results in exhaustive extraction of gaseous samples.[106] When 1-pyrenyldiazomethane is used as the derivatization reagent, it is introduced into the coating by first dissolving the reagent in a volatile solvent. The fiber is then immersed in the solution. The fiber coating swells and is doped with the reagent. After the evaporation of the solvent, the fiber is ready to perform extraction. The reagent, having low vapor pressure and high affinity toward the coating, remains on the fiber during its exposure to the sample. Volatilities of the pyrenylmethyl esters formed during the reaction are also low, resulting in the accumulation of the product onto the fiber until analyte or reagent is exhausted or decomposed. At a high injector temperature, the derivatized analytes are removed from the coating and the fiber can be reused. A similar approach is used for the analysis of formaldehyde from gaseous samples.[107]

This simple but powerful procedure as described above for this analysis is limited to low volatility reagents. The approach can be made more general by chemically attaching the reagent directly to the coating. The chemically bound product can then be released from the coating either by high temperature in the injector, light illumination, or change of an applied potential. The feasibility of this approach was recently demonstrated by synthesizing standards bonded to silica gel, which were released during heating.[108] This approach allowed solvent-free calibration of the instrument.

Numerous applications involving derivatization SPME including all of the typical strategies have been reported in the literature. These include derivatization in the sample prior to extraction,[109] simultaneous derivatization and extraction in sample headspace,[110–112] and derivatization on the fiber between extraction and desorption,[113] as well as injector port derivatization subsequent to desorption.[114] Examples of drugs and biologically relevant compounds for which derivatizaiton has proven useful include amphetamines,[110,115–117] cocaine,[44] methadone and cannabinoids,[118] barbiturates[119] clenbuterol,[120] ephedrine,[72] steroids,[113] busulfan,[121] and methylmalonic acid.[36]

20.6. APPLICATIONS

20.6.1. Forensics/Toxicology

The first applications of SPME technology to drug analysis were directed at monitoring drugs of abuse or toxic substances in clinical and forensic samples. The primary advantage noted by many authors was the very fast method development possible. This permitted fast turnaround of samples when needed for urgent patient care. A significant volume of review literature is available on this subject.[122–128] Helpful literature examples of methods for many of the main classes of forensic drugs are available including amphetamines,[11,129–131] benzodiazepines,[132,133] cannabinoids,[133,134] cocaine,[40,135] γ-hydroxybutyric acid (GHB),[136] opiates,[41,137] and phencyclidine.[138]

SPME has also seen significant use in profiling illicit drug preparations, to determine the synthetic route used or distribution pathways for a particular preparation. Illicit preparations of drugs of abuse may be prepared by various different synthetic routes, and for each of these, incomplete and side reactions may also occur. Extensive purification of the active ingredients is not normally performed following synthesis. As a result, several impurities are typically present in these preparations, and impurity profiling is convenient to "fingerprint" these products when they are confiscated. Liquid–liquid extraction or occasionally SPE have been used to prepare samples of confiscated materials for profiling by either GC or HPLC. SPME is often more convenient because of the faster turnaround and typically more detailed fingerprint obtained.

The use of SPME has been evaluated for impurity profiling in confiscated amphetamine powders and ecstasy tablets.[139] For ecstasy tablets, 10 mg of granulated tablet was mixed with 5 mL of 0.1 M acetate buffer (pH 5.0), sonicated for 10 min, and heated to 90°C. SPME headspace extraction was performed for 30 min. For amphetamine powders, 10 mg was mixed with 5 mL of 0.1 M acetate buffer (pH 5.0), and the SPME fiber was immersed in the sample for 30 min at 20°C. Acidic extraction pH was employed to reduce the amount of the basic active ingredient extracted in order that impurities could be more easily observed.

For ecstasy tablets, it was observed that the PDMS-DVB fiber extracted more impurities than the 100-μm PDMS fiber, substantially increasing the information content of the profile. Headspace extraction was preferred to avoid contamination of the fiber by tablet components. For amphetamine powders, direct immersion extraction was feasible and the two fibers provided equivalent profiles. The repeatability of profiles was consistent to that attained from liquid–liquid extraction, indicating similar abilities in differentiating between closely related but different drug seizures. Interestingly, for ecstasy, isosafrole was identified as a precursor for synthesis. With liquid–liquid extraction, this compound was difficult to analyze due to its high volatility. With SPME extraction, however, it could be quantified to 2 ng/g tablet. Koester et al. demonstrated the use of SPME for impurities profiling for illicit methamphetamine.[140] In an interesting additional application, SPME has been used to determine the odor signature of street cocaine samples to guide the rational training of certified drug dogs.[141]

To date, SPME has not seen significant application to screening of antidoping samples, likely due to the lack of automation. A first publication on this potential field has appeared recently for the screening of stimulants, narcotics, and local anesthetics.[142] The authors identified several advantages to SPME screening over conventional practices, including improved sensitivity, better use of human resources, and reduced solvent/consumables and sample volume requirements. Drawbacks identified were the high cost of the fibers and the requirement for a dedicated autosampler, although these were seen as acceptable given the identified benefits.

To date, SPME methods have been published on several classes of prohibited substances typically monitored in antidoping samples including anabolic steroids,[143,144] clenbuterol,[121,145] tamoxifen as a hormone antagonist,[146] a range of stimulants,[144] narcotics,[144,147] *and* β-blockers.[148]

20.6.2. Therapeutic Drug Monitoring

Therapeutic drug monitoring, or the monitoring of circulating drug levels in a patient, is important in some clinical situations. Although it can be prohibitive due to cost and inconvenience, it is advantageous in several situations such as monitoring for patient compliance, ascertaining appropriate drug levels where the therapeutic window is very narrow, or monitoring in cases when patient physiology is variable or outside of normal ranges, which may occur in critically ill, very young, or elderly patients. In these cases, it allows for optimized pharmacotherapy, or monitoring for genetic variability, organ function, or the influence of comedication. Efforts are also under way to perform the analyses *in vivo*.[149] A review of the clinical applications of SPME is available.[129]

Abdel-Rehim et al. demonstrated monitoring of busulfan from plasma using the CW-DVB fiber and GC-MS analysis.[121] This alkylating agent is administered prior to stem cell transplantation for the treatment of leukemia. The drug is characterized by a narrow therapeutic window, where either underdosing or overdosing may result in fatality. The benefits of the SPME method were reduced analysis time and automation of the analysis, with accuracy and precision comparable to conventional analysis. The fast analysis is particularly useful in this case where frequent high dosing is typically required.

In another example of therapeutic drug monitoring, Takamoto et al. demonstrated monitoring of urinary acrolein during administration of cyclophosphamide and ifosphamide.[150] Acrolein is a metabolite of these two chemotherapeutic agents and is responsible for mucosal irritation in the urinary tract when this compound is excreted. Left unchecked severe hemorrhagic cystitis can result. The authors demonstrated that the SPME GC-MS method developed allows simple frequent monitoring for acrolein, which could improve patient care by quickly signaling when to take heightened preventative measures against hemorrhagic cystitis.

The free form of the immunosuppressant mycophenolic acid has also been monitored in patient serum samples.[151] This example is of particular interest as the drug is extensively glucuronidated and plasma bound, so the free concentration is particularly hard to evaluate but highly relevant for patient care. SPME is of course ideal for monitoring free concentration.

20.6.3. *In Vivo* Monitoring

In vivo SPME has the potential to complement the range of technologies currently being employed for *in vivo* analysis of living systems. SPME may provide the selectivity and sensitivity of microdialysis analysis with improved time resolution of sampling. In the case of intravenous monitoring of drug concentrations, because no blood and only very small amounts of drugs are removed, minimal disruption to the chemical balance in blood occurs. Ultimately, this should result in lower stress levels on animals for pharmacokinetic studies, reduced exposure to blood for analytical personnel, and simplified, less disruptive sampling.

To date, commercial SPME devices have been used in some applications of direct *in vivo* analysis of living systems. For example, they have been applied for the *in vivo* analysis of pheromones and semiochemicals used in chemical communications by insects[152,153] and frogs,[154] respectively. In these cases, the living animals were noninvasively monitored over time, providing a convenient means to study complicated dynamic processes without interference. The commercially available CW-TPR phase has been used for monitoring pesticides levels in living plant tissues over time.[3] Recently, the commercial PDMS fibers were used for direct extraction of toluene from the brain tissue in a project to monitor its effect on neurotransmitters levels.[155,156] Although commercial SPME fibers were employed for the above-mentioned applications, typically some redesign of the SPME device is required for optimal *in vivo* analysis. In particular, for invasive sampling of living animals/humans, it is appropriate for the devices to be constructed of surgical-grade stainless steel rather than fused silica. Extraction polymers must have demonstrated biocompatibility, commonly defined as causing no toxic, immunological, or inflammatory response in the study system. The devices need to be sterilizable, and extraction coatings have to demonstrate high mechanical strength and resistance to both polymer bleed and decomposition.

Improvements in the preparation of SPME coatings for *in vivo* analysis are ongoing. Investigations into the use of novel chromatographic sorbent materials with proven biocompatibility such as RAMs have been conducted, as well as the application of a biocompatible coating to conventional or nonbiocompatible extraction phases.[60–62] Polymers not used before as SPME sorbent materials have been investigated.[54] Initial success with *in vivo* SPME has been achieved with a PPY solid sorbent coated on stainless-steel wires.[4] The devices have been shown to provide consistent response over a range of anticipated blood flow rates.[91] Absorptive fibers coated with PEG/C18 have also been used successfully, employing calibration by preloading standards on the fiber.[63] Polyacrylonitrile is of current interest as a biocompatible coating, which has been shown as beneficial for analysis of a variety of drugs from plasma with HPLC analysis.[64]

A membrane made from appropriate material and coated or placed around a sorbent can add a certain degree of selectivity to the extraction process, resulting in membrane-protected SPME.[62] Large unwanted biomolecules may be excluded from a sorbent that would otherwise extract them, or a non-biocompatible sorbent may be rendered biocompatible. However, it is a limitation that the kinetics of membrane extraction are substantially slower than direct extraction, because the analytes must diffuse through the membrane before they can reach the coating. This may be overcome by the use of thin membranes and increased extraction temperatures, which would result in faster extraction times.[157,158] Such experimental parameters as ruggedness, binding capacity, and biocompatibility of the resultant extraction material need to be evaluated and optimized for each case. A dedicated sampler for *in vivo* SPME is under development by Supelco and commercial release is expected in the near future. The performance of the device has been evaluated for studying pharmacokinetic profiles in both rats[159] and mice.[160]

Several advantages for such approaches are envisioned. The technique would ensure that the drug and metabolite concentrations determined are representative of those to which body tissues are exposed. Because sampling, sample preparation, and preconcentration occur *in vivo*, there is no concern that the sampling step would itself promote a change in the drug concentration or composition, due to biochemical reactions occurring during sampling such as the activation of clotting factors. In the case of monitoring of drugs in experimental animals, one would expect that such a technique would shorten and simplify drug dosing studies and significantly reduce the number of animals required per study. It would also reduce the impact of inconsistencies often observed among individual animals.

The field of *in vivo* SPME is still quite new, and doubtless, much still needs to be done to evaluate all the potentials it holds. For more information, the interested reader is referred to two recent reviews of the status of *in vivo* SPME.[149,161]

20.6.4. Binding Affinity

As discussed previously, the presence of competing phases in a sample matrix can compromise the analysis of analyte concentration (Section 20.2.5). Because of the nature of equilibrium extraction, it also provides an important advantage in assessing binding affinities in biological systems. So long as depletion is negligible, the extraction will not disturb the equilibrium in a system. The method can therefore be used relatively easily to assess the affinities of various components of a complex system. To briefly summarize, a microextraction of a multicomponent system will provide information on the concentration of free analyte present in the system, so long as the amount extracted is insignificant relative to the total free concentration. By adding a standard addition spike and allowing the system to re-equilibrate, one can then assess the degree of binding to system components. By performing this analysis in systems containing the sample phase and one discontinuous phase at a time, one can assess the binding affinities of all system components individually.

This technique has been described for analyzing protein and membrane binding and determination of the free amount of a drug or other chemical in a system. It is the free concentration that is most toxicologically and pharmacologically relevant, as opposed to the nominal or total concentration most commonly reported. The method is faster and more convenient than alternatives, which include equilibrium dialysis or headspace equilibration analysis. The method is particularly well suited to situations where the discontinuous phase of interest is dissolved (such as proteins and antibodies) or otherwise difficult to separate from the liquid phase. The basic theory and application of the method for fiber SPME analysis of several polar compounds in biological matrices has been reported.[162,163] The authors report criteria that must be met to successfully implement the method:

- The extraction itself must not influence the equilibrium between the aqueous phase of the sample and the discontinuous phase.
- The binding phase must not interfere with the extraction by binding to the extraction phase.
- The absorption profile of the compound should be the same in a calibration sample as in the sample containing the discontinuous phase.

The method was validated by comparing the amount of one of the compounds extracted by microextraction, both inside and outside of the membrane in an equilibrium dialysis experiment. There was no difference in the free concentrations determined in the samples from the inside and outside of the dialysis tube. Albumin concentrations were varied from 2.0 to 200 μM. They therefore concluded that the protein itself is not adsorbed to the fiber and that the protein does not influence the amount of chemical absorbed by the fiber in any other way.

In practice, the free fraction (f_X) of analyte X in solution is calculated as follows:

$$f_X = [X]_a/[X]_t = 1/(1 + K_M[M]), \tag{20.3}$$

where $[X]_a$ is the concentration of the freely dissolved chemical, $[X]_t$ is the total concentration (free and bound to M), K_M is the protein–water partition coefficient of X, and $[M]$ is the concentration of protein M in the aqueous phase.

Experimentally, f_X is determined by calculating the slope of the linear plot of $[X]_a/[X]_t$. Compound X is first added to an aqueous solution without protein, and the total concentration was determined. Protein is added to the system, and the system is allowed to re-equilibrate. Once equilibrated, the freely dissolved concentration is determined by microextraction. By varying $[X]_t$, the required linear correlation can be determined. The analyte/protein partition coefficient (K_M) can then be calculated. Equation 20.3 can be rearranged to give

$$K_M = \frac{([X]_0 - [X]_a) \times \frac{V_a}{V_M}}{[X]_a}, \tag{20.4}$$

where $[X]_0$ is the total (nominal) concentration, $[X]_a$ is the free concentration in the solution, and V_a and V_M are the volumes of the aqueous and protein phases, respectively.

A similar strategy has been employed for the determination of the equilibrium constant of protein binding between diazepam and human serum albumin (HSA).[164] In this work, a Scatchard plot was used to evaluate the binding data. An equilibrium constant and an indication of one independent binding site were determined, which was consistent with literature results obtained by gel filtration analysis. For diazepam analysis, direct extraction with a 100-μm PDMS fiber was performed from a 0.5 μg/mL solution of diazepam in 1 mg/mL HSA dissolved in 0.066 M phosphate buffer pH 7.4. Trace amounts of methanol were observed to affect the precision of analysis. Because methanol could not be eliminated entirely due to its use as a solvent for diazepam standards, methanol content was kept consistent for all analyses.

Recently, a method has been published for the automated determination of ligand-receptor binding for the determination of binding of diazepam to HSA.[165] A new multifiber SPME system was used to allow high-throughput analysis of binding from samples in a 96-well plate. Fiber constant calibration was used to remove interfiber variability from the binding data. The method is amenable to the high-throughput analysis of drug–protein binding in either plasma or whole blood.

For the interested reader, reviews of the use of SPME for binding affinity determinations and more advanced treatments of the subject are available.[62,166,167]

20.6.5. Pharmaceutical Product Analysis

A key aspect in the regulatory control of pharmaceutical products is monitoring for impurities in these products. These may arise either from the synthetic route employed or from degradation on storage. Because of the potential for adverse toxicity arising from the presence of such impurities careful monitoring is essential. SPME has proved useful for such monitoring, as the impurities are often volatile small molecules such as solvents.[168] Colón and Richoll recently demonstrated the use of a similar strategy for the monitoring of sulfonic acid ester impurities.[169]

In the design of pharmaceutical products, the nature of the active ingredient is only one of several important considerations. Optimal drug formulation is a crucial step in the design of products that produce optimal therapeutic effects. The nature of the formulation procedures and/or auxiliary (nonmedical) components of a product can, for example, make a product easier to handle, limit degradation, enhance bioavailability and targeting of the active ingredient to receptor sites, provide for additional routes of administration, and target the product for additional therapeutic indications for use.[170] These auxiliary inactive ingredients, known collectively as excipients, can also solubilize the active ingredient, suppress the growth of microorganisms, provide bulk or a coating for a tablet or capsule, provide color, or mask an unpleasant taste or odor.

With so many tasks to perform, it is obvious that the nature and numbers of excipients in various pharmaceutical products can vary widely. It is essential therefore, that for the analysis of pharmaceutical products, the nature of these excipients is taken into account, for their potential impact on the process of analysis of the active ingredient. In microextraction, it is the effects on the partitioning to an extraction phase that are of utmost concern.

Because excipients can be present in concentrations far in excess of those of the active ingredient, use of solid sorbents for microextraction is often precluded for direct extractions. For other extraction phases, care should be taken to ensure that the interface between the sample and the extraction phase does not become fouled by an excipient. It is recommended that headspace extraction be used where possible.

20.6.6. Enantiomeric Differentiation

Selectivity in analysis based on the differentiation of drug enatiomers is important for several reasons. First, different enantiomers may have significantly different therapeutic potency. Quantification on undifferentiated enantiomeric mixtures may give erroneous indications of therapeutic efficacy. Second, clandestine preparations of drugs can have significantly different enantiomeric ratios than legitimate drug preparations. Enantiomeric differentiation can be used both to determine if drug intake arose from legitimate or illicit sources or to determine the most likely synthetic pathways used to prepare illicit drug preparations. This can help investigators determine sources and distribution routes for these products.

SPME methods employing enantiomeric selectivity are commonly employed for fragrance analysis, with just a few reports having appeared for drug analysis. Enantiomeric selectivity may be introduced into an SPME analysis by first employing derivatization with a chiral reagent, followed by separation of the diasteriomers. Two reports are available on the use of this strategy for the analysis of amphetamines.[117,171] Alternatively, where enantiomers are sufficiently extracted on their own, selectivity can be introduced by the use of a chiral column, which has been demonstrated for hydroxychloroquin and its metabolites, with extraction by PDMS-DVB fiber and HPLC-UV separation and detection.

20.7. CONCLUSIONS

There has been significant advance over the past dozen years in the application of SPME technology to drug and bioanalysis. Much of this advance has come from innovation in sorbent and instrument design and configuration. At the time of writing, it appears that numerous applications exploiting these innovations have been described or are in limited application, although none have yet achieved wide acceptance or broad application. However, many of the innovations seen are highlighting new possibilities for chemical analysis, which had not been envisioned even just 5 years ago. As an enabling tool allowing analysts to ask questions that had not been possible previously, it appears the field will enjoy continued attention for new discoveries in chemical analysis, for both novel tools of investigation and new fields of application.

REFERENCES

1. Pawliszyn, J. Sample preparation: Quo Vadis? *Anal. Chem.* **2003**, *75*, 2543–2558.

2. Li, S.; Weber, S. Determination of barbiturates by solid-phase microextraction and capillary electrophoresis. *Anal. Chem.* **1997**, *69*, 1217–1222.

3. Lord, H.L.; Moder, M.; Popp, P.; Pawliszyn, J.B. In-vivo study of triazine herbicides in plants by SPME. *Analyst* **2004**, *129*, 107–108.

4. Lord, H.L.; Grant, R.; Walles, M.; Incledon, B.; Fahie, B.; Pawliszyn, J. Development and evaluation of a solid phase microextraction probe for in-vivo pharmacokinetic studies. *Anal. Chem.* **2003**, *75*, 5103–5115.

5. Gorecki, T.; Yu, X.; Pawliszyn, J. Theory of analyte extraction by selected porous polymer SPME fibres. *Analyst* **1999**, *124*, 643–649.

6. Pawliszyn, J. *Solid Phase Microextraction: Theory and Practice*. New York: Wiley; 1997.

7. Lord, H.L.; Rajabi, M.; Safari, S.; Pawliszyn, J. A study of the performance characteristics of immunoaffinity SPME probes for extraction of a range of benzodiazepines. *J. Pharm. Biomed. Anal.* **2007**, *44*, 506–519.

8. Musteata, F.M.; Pawliszyn, J. Determination of free concentration of paclitaxel in liposome formulations. *J. Pharm. Pharm. Sci.* **2006**, *9*, 231–237.

9. Yang, K.; Eisert, R.; Lord, H.; Pawliszyn, J. Determination of pesticides in foods by automated SPME-GC-MS. In *Applications of Solid-Phase Microextraction*, Pawliszyn, J., Ed. Letchworth, Hertfordshire, UK: Royal Society of Chemistry; 1999.

10. Boyd-Boland, A.; Gorecki, T.; Chai, M.; Luo, Y.; Zhang, Z.; Yang, M.; Pawliszyn, J. New solvent-free sample preparation techniques based on fibre and polymer technologies. *Environ. Sci. Technol.* **1994**, *28*, 569A–574A.

11. Lord, H.; Pawliszyn, J. Method optimization for the analysis of amphetamines in urine by SPME. *Anal. Chem.* **1997**, *69*, 3899–3906.

12. Hall, B.J.; Brodbelt, J.S. Determination of barbiturates by solid-phase microextraction (SPME) and ion trap gas chromatography-mass spectrometry. *J. Chromatogr. A* **1997**, *777*, 275–282.

13. Abdel-Rehim, M.; Andersson, M.; Portelius, E.; Norsten-Höög, C.; Blomberg, L.G. Determination of ropivacaine and its metabolites in human plasma using solid phase microextraction and GC-NPD/GC-MS. *J. Microcolumn Sep.* **2001**, *13*, 313–321.

14. Boyd-Boland, A.A.; Pawliszyn, J. New solvent-free sample preparation techniques based on fibre and polymer technologies. *J. Chromatogr. A* **1995**, *704*, 163–172.

15. Perrin, D.D.; Dempsey, B. *Buffers for pH and Metal Ion Control*. London: Chapman and Hall; 1994.

16. Queiroz, M.E.C.; Silva, S.M.; Carvalho, D.; Lanças, F.M. Determination of lamotrigine simultaneously with carbamazepine, carbamazepine epoxide, phenytoin, phenobarbital, and primidone in human plasma by SPME–GC–TSD. *J. Chromatogr. Sci.* **2002**, *40*, 219–223.

17. Liao, J.-L.; Zeng, C.-M.; Hjerten, S.; Pawliszyn, J. SPME of biopolymers: Adsorption of basic proteins onto a fibre coated with polyacrylic acid. *J. Microcolumn Sep.* **1996**, *8*, 1–4.

18. Gorecki, T. Solid versus liquid coatings. In *Applications of Solid-Phase Microextraction*, Pawliszyn, J., Ed. Letchworth, Hertfordshire, UK: Royal Society of Chemistry; 1999; pp. 92–108.

19. Ai, J. Quantitation by SPME before reaching a partition equilibrium. In *Applications of Solid-Phase Microextraction*, Pawliszyn, J., Ed. Letchworth, Hertfordshire, UK: Royal Society of Chemistry; 1999; pp. 22–37.

20. Zhou, S.; Zhang, X.; Ouyang G.; Pawliszyn J. On-fibre standardization technique for solid-coated SPME. *Anal. Chem.* **2007**, *79*, 1221–1230.

21. Ouyang, G.; Cai, J.; Zhang, X.; Li, H.; Pawliszyn, J. Standard-free kinetic calibration for rapid on-site analysis by solid-phase microextraction. *J. Sep. Sci.* **2008**, *31*, 1167–1172.

22. Ouyang, G.; Pawliszyn, J. A critical review in calibration methods for solid-phase microextraction. *Anal. Chim. Acta* **2008**, *627*, 184–197.

23. Ouyang, G.; Cui, S.; Qin, Z.; Pawliszyn, J. One-calibrant kinetic calibration for on-site water sampling with solid-phase microextraction. *Anal. Chem.* **2009**, *81*, 5629–5636.

24. Musshoff, F.; Junker, H.; Madea, B. Rapid analysis of halothane in biological samples using headspace solid-phase microextraction and gas chromatography-mass spectrometry—A case of a double homicide. *J. Anal. Toxicol.* **2000**, *24*, 372–376.

25. Blau, K.; Halket, J.M. *Handbook of Derivatives for Chromatography*, 2nd ed. New York: Wiley; 1993.

26. Alnouti, Y.; White, C.A.; Bartlett, M.G. Simultaneous determination of zidovudine and lamivudine from rat plasma, amniotic fluid and tissues by HPLC. *Biomed. Chromatogr.* **2004**, *18*, 641–647.

27. AbuRuz, S.; Millership, J.; McElnay, J. Determination of metformin in plasma using a new ion pair solid phase extraction technique and ion pair liquid chromatography. *J. Chromatogr. B* **2003**, *798*, 203–209.

28. Cawley, A.T.; Kazlauskas, R.; Trout, G.J.; George, A.V. Determination of urinary steroid sulfate metabolites using ion paired extraction. *J. Chromatogr. B* **2005**, *825*, 1–10.

29. Lan, K.; Jiang, X.; Wang, L.; Zhou, J.; Jiang, Q.; Ye, L. Quantitative determination of rosuvastatin in human plasma by ion pair liquid-liquid extraction using liquid chromatography with electrospray ionization tandem mass spectrometry. *J. Pharm. Biomed. Anal.* **2007**, *44*, 540–546.

30. Fiamegos, Y.C.; Stalikas, C.D. In-drop derivatization liquid-phase microextraction assisted by ion-pairing transfer for the gas chromatographic determination of phenolic endocrine disruptors. *Anal. Chim. Acta* **2007**, *597*, 32–40.

31. Ho, T.S.; Halvorsen, T.G.; Pedersen-Bjergaard, S.; Rasmussen, K.E. Liquid-phase microextraction of hydrophilic drugs by carrier-mediated transport. *J. Chromatogr. A* **2003**, *998*, 61–72.

32. Ho, T.S.; Reubsaet, J.L.E.; Anthonsen, H.S.; Pedersen-Bjergaard, S.; Rasmussen, K.E. Liquid-phase microextraction based on carrier mediated transport combined with liquid chromatography-mass spectrometry. New concept for the determination of polar drugs in a single drop of human plasma. *J. Chromatogr. A* **2005**, *1072*, 29–36.

33. Pan, L.; Pawliszyn, J. Derivatization/solid phase microextraction—New approach to polar analytes. *Anal. Chem.* **1997**, *69*, 196–205.

34. Alzaga, R.; Peña, A.; Ortiz, L.; Bayona, J.M. Determination of linear alkylbenzensulfonates in aqueous matrices by ion-pair solid-phase microextraction–in-port derivatization–gas chromatography–mass spectrometry. *J. Chromatogr. A* **2003**, *999*, 51–60.

35. Chan, I.O.M.; Lam, P.K.S.; Cheung, R.H.Y.; Lam, M.H.W.; Wu, R.S.S. Application of solid phase microextraction in the determination of paralytic shellfish poisoning toxins. *Analyst* **2005**, *130*, 1524–1529.

36. Li, N.; Deng, C.; Zhang, X. Determination of methylmalonic acid and glutaric acid in urine by aqueous-phase derivatization followed by headspace solid-phase microextraction and gas chromatography-mass spectrometry. *J. Sep. Sci.* **2007**, *30*, 266–271.

37. Neitzel, P.L.; Walther, W.; Nestler, W. In situ methylation of strongly polar organic acids in natural waters supported by ion-pairing agents for headspace GC-MSD analysis. *Fresenius J. Anal. Chem.* **1998**, *361*, 318–323.

38. Yu, X.; Pawliszyn, J. Speciation of alkyllead and inorganic lead by derivatization with deuterium-labeled sodium tetraethylborate. *Anal. Chem.* **2000**, *72*, 1788–1792.

39. Sarrión, M.N.; Santos, F.J.; Galceran, M.T. In situ derivatization/solid-phase microextraction for the determination of haloacetic acids in water. *Anal. Chem.* **2000**, *72*, 4865–4873.

40. Makino, Y.; Takagi, T.; Ohta, S.; Hirobe, M. Rapid analysis of methamphetamine and cocaine in the urine of drug abusers with solid-phase microextraction. *Chromatography* **1997**, *18*, 185–188.

41. Lucas, A.C.S.; Bermejo, A.M.; Tabernero, M.J.; Fernández, P.; Strano-Rossi, S. Use of solid-phase microextraction (SPME) for the determination of methadone and EDDP in human hair by GC-MS. *Forensic Sci. Int.* **2000**, *107*, 225–232.

42. Álvarez, I.; Bermejo, A.M.; Tabernero, M.J.; Fernández, P.; López, P. Detrmination of cocaine and cocaethylene in plasma by solid-phase microextraction and gas chromatography-mass spectrometry. *J. Chromatogr. B* **2007**, *845*, 90–94.

43. Yonamine, M.; Saviano, A.M. Determination of cocaine and cocaethylene in urine by solid-phase microextraction and gas chromatography-mass spectrometry. *Biomed. Chromatogr.* **2006**, *20*, 1071–1075.

44. de Toledo, F.C.P.; Yonamine, M.; de Moraes, M.; Regina, L.; Silva, O.A. Determination of cocaine, benzoylecgonine and cocaethylene in human hair by solid-phase microextraction and gas chromatography-mass spectrometry. *J. Chromatogr. B* **2003**, *798*, 361–365.

45. Spichiger, M.; Muhlbauer, R.C.; Brenneisen, R. Determination of menthol in plasma and urine of rats and humans by headspace solid phase microextraction and gas chromatography–mass spectrometry. *J. Chromatogr. B* **2004**, *799*, 111–117.

46. Aresta, A.; Palmisano, F.; Zambonin, C.G. Determination of naproxen in human urine by solid-phase microextraction coupled to liquid chromatography. *J. Pharm. Biomed. Anal.* **2005**, *39*, 643–647.

47. de Jong, G.J.; Koster, E.H.M. Solid-phase micro-extraction in bioanalysis, exemplified by lidocaine determination. *Chromatographia* **2000**, *52*(Suppl.), S12–S13.

48. Otu, E.; Pawliszyn, J. Solid phase micro-extraction of metal ions. *J. Mikrochim. Acta* **1993**, *112*, 41–46.

49. Belardi, R.G.; Pawliszyn, J. The application of chemically modified fused silica fibres in extraction of organics from water matrix samples, and their rapid transfer to capillary column. *Water Pollut. Res. J. Can.* **1989**, *24*, 179–191.

50. Zhang, Z.; Pawliszyn, J. Headspace solid phase microextraction. *Anal. Chem.* **1993**, *65*, 1843–1852.

51. Guo, F.; Gorecki, T.; Irish, D.; Pawliszyn, J. Solid phase microextraction combined with electrochemistry. *Anal. Commun.* **1996**, *33*, 361–364.

52. Wan, H.B.; Chi, H.; Wong, M.K.; Mok, C.Y. Solid-phase microextraction using pencil lead as sorbent for analysis of organic pollutants in water. *Anal. Chim. Acta* **1994**, *298*, 219–223.

53. Gorecki, T.; Martos, P.; Pawliszyn, J. Strategies for the analysis of polar solvents in liquid matrices. *Anal. Chem.* **1998**, *70*, 19–27.

54. Wu, J.; Pawliszyn, J. Preparation and applications of polypyrrole films in SPME. *J. Chromatogr. A* **2001**, *909*, 37–52.

55. Yuan, H.; Mullett, W.M.; Pawliszyn, J. Biological sample analysis with immunoaffinity SPME. *Analyst* **2001**, *126*, 1456–1461.

56. Lord, H.L.; Rajabi, M.; Safari, S.; Pawliszyn, J. Development of immunoaffinity SPME probes for analysis of sub ng/ml concentrations of 7-aminoflunitrazepam in urine. *J. Pharm. Biomed. Anal.* **2005**, *40*, 769–780.

57. Meng, Y.; Pino, V.; Anderson, J.L. Exploiting the versatility of ionic liquids in separation science: Determination of low-volatility aliphatic hydrocarbons and fatty acid methyl esters using headspace solid-phase microextraction coupled to gas chromatography. *Anal. Chem.* **2009**, *81*, 7107–7112.

58. He, Y.; Pohl, J.; Engel, R.; Rothman, L.; Thomas, M. Preparation of ionic liquid based solid-phase microextraction fiber and its application to forensic determination of methamphetamine and amphetamine in human urine. *J. Chromatogr. A* **2009**, *1216*, 4824–4830.

59. Wanigasekara, E.; Perera, S.; Crank, J.A.; Sidisky, L.; Shirey, R.; Berthod, A.; Armstrong, D.W. Bonded ionic liquid polymeric material for solid-phase microextraction GC analysis. *Anal. Bioanal. Chem.* **2010**, *396*, 511–524.

60. Mullett, W.; Pawliszyn, J. Direct determination of benzodiazepines in biological fluids by restricted-access solid-phase microextraction. *Anal. Chem.* **2002**, *74*, 1081–1087.

61. Walles, M.; Mullett, W.; Pawliszyn, J. Monitoring of drugs and metabolites in whole blood by restricted-access SPME coupled to LC/MS. *J. Chromatogr. A* **2004**, *1025*, 85–92.

62. Musteata, F.M.; Walles, M.; Pawliszyn, J. Fast assay of angiotensin 1 from whole blood by cation-exchange restricted-access solid-phase microextraction. *Anal. Chim. Acta* **2005**, *537*, 231–237.

63. Es-haghi, A.; Zhang, X.; Musteata, F.M.; Bagheri, H.; Pawliszyn, J. Evaluation of bio-compatible PEG-based SPME fibre for in-vivo pharmacokinetic studies of diazepam in dogs. *Analyst*, **2007**, *132*, 672–678.

64. Musteata, M.M.; Musteata, F.M.; Pawliszyn, J. Biocompatible solid-phase microextraction coatings based on polyacrylonitrile and solid-phase extraction phases. *Anal. Chem.* **2007**, *79*, 6903–6911.

65. Wang, D.; Chong, S.L.; Malik, A. Sol-gel column technology for single-step deactivation, coating, and stationary-phase immobilization in high-resolution capillary gas chromatography. *Anal. Chem.* **1997**, *69*, 4566–4576.

66. Hayes, J.D.; Malik, A. Sol-gel chemistry-based Ucon-coated columns for capillary electrophoresis. *J. Chromatogr. B* **1997**, *695*, 3–13.

67. Chong, S.L.; Wang, D.; Hayes, J.D.; Wilhite, B.W.; Malik, A. Sol-gel coating technology for the preparation of solid-phase microextraction fibers of enhanced thermal stability. *Anal. Chem.* **1997**, *69*, 3889–3898.

68. Wang, Z.; Xiao, C.; Wu, C.; Han, H. High-performance polyethylene glycol-coated solid-phase microextraction fibers using sol-gel technology. *J. Chromatogr. A* **2000**, *893*, 157–168.

69. Bagheri, H.; Es-haghi, A.; Rouini, M.R. High-performance polyethylene glycol-coated solid-phase microextraction fibers using sol-gel technology. *J. Chromatogr. B* **2005**, *818*, 147–157.

70. Li, X.; Zheng, Z.; Hu, M.; Mao, M. High operationally stable sol–gel diglycidyloxycalix[4]arene fiber for solid-phase microextraction of propranolol in human urine. *J. Sep. Sci.* **2005**, *28*, 2489–2500.

71. Zhou, X.; Li, X.; Zeng, Z. Solid-phase microextraction coupled with capillary electrophoresis for the determination of propranolol enantiomers in urine using a sol–gel derived calix[4]arene fiber. *J. Chromatogr. A* **2006**, *1104*, 359–365.

72. Fang, H.; Liu, M.; Zeng, Z. Solid-phase microextraction coupled with capillary electrophoresis to determine ephedrine derivatives in water and urine using a sol–gel derived butyl methacrylate/silicone fiber. *Talanta* **2006**, *68*, 979–986.

73. Fan, Y.; Feng, Y.-Q.; Zhang, J.-T.; Da, S.-L.; Zhang, M. Poly(methacrylic acid-ethylene glycol dimethacrylate) monolith in-tube solid phase microextraction coupled to high performance liquid chromatography and analysis of amphetamines in urine samples. *J. Chromatogr. A* **2005**, *1074*, 9–16.

74. Fan, Y.; Feng, Y.-Q.; Da, S.-L.; Gao, X.-P. In-tube solid-phase microextraction with poly(methacrylic acidethylene glycol dimethacrylate) monolithic capillary for direct high-performance liquid chromatographic determination of ketamine in urine samples. *Analyst* **2004**, *129*, 1065–1069.

75. Wen, Y.; Fan, Y.; Zhang, M.; Feng, Y.-Q. Determination of camptothecin and 10-hydroxycamptothecin in human plasma using polymer monolithic in-tube solid phase microextraction combined with high-performance liquid chromatography. *Anal. Bioanal. Chem.* **2005**, *382*, 204–210.

76. Nie, J.; Zhao, Q.; Huang, J.; Xiang, B.; Feng, Y.-Q. Determination of telmisartan in rat tissues by in-tube solid-phase microextraction coupled to high performance liquid chromatography. *J. Sep. Sci.* **2006**, *29*, 650–655.

77. Nie, J.; Zhang, M.; Fan, Y.; Wen, Y.; Xiang, B.; Feng, Y.-Q. Biocompatible in-tube solid-phase microextraction coupled to HPLC for the determination of angiotensin II receptor antagonists in human plasma and urine. *J. Chromatogr. B* **2005**, *828*, 62–69.

78. Fan, Y.; Feng, Y.-Q.; Da, S.-L.; Shi, Z.-G. Poly (methacrylic acid–ethylene glycol dimethacrylate) monolithic capillary for in-tube solid phase microextraction coupled to high performance liquid chromatography and its application to determination of basic drugs in human serum. *Anal. Chim. Acta* **2004**, *523*, 251–258.

79. Witkowski, A.; Brajter-Toth, A. Overoxidized polypyrrole films: A model for the design of permselective electrodes. *Anal. Chem.* **1992**, *64*, 635–641.

80. Pihel, K.; Walker, Q.D.; Wightman, R.M. Overoxidized polypyrrole-coated carbon fiber microelectrodes for dopamine measurements with fast-scan cyclic voltammetry. *Anal. Chem.* **1996**, *68*, 2084–2089.

81. Hsueh, C.; Brajter-Toth, A. Electrochemical preparation and analytical applications of ultrathin overoxidized polypyrrole films. *Anal. Chem.* **1994**, *66*, 2458–2464.

82. Gao, Z.; Chen, B.; Zi, M. Voltammetric determination of dopamine in the presence of ascorbic acid at over-oxidized polypyrrole-indigo carmine film-coated electrodes. *Analyst* **1994**, *119*, 459–464.

83. Wu, J.; Lord, H.; Kataoka, H.; Pawliszyn, J. Polypyrrole coated capillary in-tube SPME/LC/EIMS for the determination of β-blockers in urine and serum samples. *J. Microcolumn Sep.* **2000**, *12*, 255–266.

84. Wu, J.; Yu, X.; Lord, H.; Pawliszyn, J. Solid phase microextraction of inorganic anions based on polypyrrole film. *Analyst* **2000**, *125*, 391–394.

85. Wu, J. Solid phase microextraction based on polypyrrole films, PhD Thesis, University of Waterloo, Waterloo, ON; 2001; Section 4.3.3.

86. Haase, V.; Beck, F. Electrodeposition of N-substituted polypyrroles on iron and the CIPL strategy. *Electrochim. Acta* **1994**, *39*, 1195–1205.

87. Su, W.; Iroh, J.O. Electrodeposition mechanism of polypyrrole coatings on steel substrates from aqueous oxalate solutions. *Electrochim. Acta* **2000**, *46*, 1–8.

88. Wang, Y.; Walles, M.; Thomson B.; Nacson, S.; Pawliszyn, J. SPME combined with surface-enhanced laser desorption/ionization introduction for ion mobility spectrometry and mass spectrometry using polypyrrole coatings. *Rapid. Commun. Mass Spectrom.* **2004**, *18*, 157–162.

89. Wang, Y.; Schneider, B.B.; Covey, T.R.; Pawliszyn, J. High-performance SPME/AP MALDI system for high throughput sampling and determination of peptides. *Anal. Chem.* **2005**, *77*, 8095–8101.

90. Wang, Y.; Nacson, S.; Pawliszyn, J. The coupling of SPME/SELDI to ion mobility spectrometry for drug analysis. *Anal. Chim. Acta* **2007**, *582*, 50–54.

91. Schubert, J.K.; Miekisch, W.; Fuchs, P.; Scherzer, N.; Lord, H.; Pawliszyn, J.; Mundkowski, R.G. Determination of antibiotic drug concentration in circulating human blood by means of SPME. *Clin. Chim. Acta* **2007**, *386*, 57–62.

92. Ensing, K.; de Boer, T. Tailor-made materials for tailor-made applications: Application of molecular imprints in chemical analysis. *Trends Anal. Chem.* **1999**, *18*, 138–145.

93. Vlatakis, G.; Andersson, L.I.; Müller, R.; Mosbach, K. Drug assay using antibody mimics made by molecular imprinting. *Nature* **1993**, *361*, 645–647.

94. Chassaing, C.; Stokes, J.; Venn, R.F.; Lanza, F.; Sellergren, B.; Holmberg, A.; Berggren, C. Molecularly imprinted polymers for the determination of a pharmaceutical development compound in plasma using 96-well MISPE technology. *J. Chromatogr. B* **2004**, *804*, 71–81.

95. Dzygiel, P.; O'Donnell, E.; Fraier, D.; Chassaing, C.; Cormack, P.A.G. Evaluation of water-compatible molecularly imprinted polymers as solid-phase extraction sorbents for the selective extraction of sildenafil and its desmethyl metabolite from plasma samples. *J. Chromatogr. B* **2007**, *853*, 346–353.

96. Courtois, J.; Fischer, G.; Sellergren, B.; Irgum, K. Molecularly imprinted polymers grafted to flow through poly(trimethylolpropane trimethacrylate) monoliths for capillary-based solid-phase extraction. *J. Chromatogr. A* **2006**, *1109*, 92–99.

97. Dirion, B.; Cobb, Z.; Schillinger, E.; Andersson, L.I.; Sellergren, B. Water-compatible molecularly imprinted polymers obtained via high-throughput synthesis and experimental design. *J. Am. Chem. Soc.* **2003**, *125*, 15101–15109.

98. Hu, X.; Hu, Y.; Li, G. Development of novel molecularly imprinted solid-phase microextraction fiber and its application for the determination of triazines in complicated samples coupled with high-performance liquid chromatography. *J. Chromatogr. A* **2007**, *1147*, 1–9.

99. Turiel, E.; Tadeo, J.L., Martin-Esteban, A. Molecularly imprinted polymeric fibers for solid-phase microextraction. *Anal. Chem.* **2007**, *79*, 3099–3104.

100. Koster, M.E.H.; Crescenzi, C.; Hoedt, W.D.; Ensing, K.; de Jong, G.J. Fibers coated with molecularly imprinted polymers for solid-phase microextraction. *Anal. Chem.* **2001**, *73*, 3140–3145.

101. Djozan, D.; Baheri, T. Preparation and evaluation of solid-phase microextraction fibers based on monolithic molecularly imprinted polymers for selective extraction of diacetylmorphine and analogous compounds. *J. Chromatogr. A* **2007**, *1166*, 16–23.

102. Mullett, W.; Martin, P.; Pawliszyn, J. In-tube molecularly imprinted polymer SPME for selective determination of propranolol. *Anal. Chem.* **2001**, *73*, 2383–2389.

103. Mullett, W.; Walles, M.; Levsen, K.; Pawliszyn, J. Multidimensional on-line sample preparation of verpamil and its metabolites by a molecularly imprinted polymer coupled to LC/MS. *J. Chromatogr. B* **2004**, *801*, 297–306.

104. Vuckovic, D.; Shirey, R.; Chen, Y.; Sidisky, L.; Aurand, C.; Stenerson, K.; Pawliszyn, J. In vitro evaluation of new biocompatible coatings for solid phase microextraction: Implications for drug analysis and in vivo sampling applications. *Anal. Chim. Acta* **2009**, *638*, 175–185.

105. Buchholz, K.; Pawliszyn, J. Optimization of solid phase microextraction conditions for determination of phenols. *Anal. Chem.* **1994**, *66*, 160–167.

106. Pan, L.; Pawliszyn, J. Determination of fatty acids using solid phase microextraction. *Anal. Chem.* **1995**, *67*, 4396–4403.

107. Martos, P.; Pawliszyn, J. Sampling and determination of formaldehyde using solid phase microextraction with on-fiber derivatization. *Anal. Chem.* **1998**, *70*, 2311–2320.

108. Konieczka, P.; Wolska, L.; Luboch, E.; Namiesnik, J.; Przyjazny, A.; Biernat, J. Calibration of the thermal desorption-gas chromatography-mass spectrometry system using standards generated in the process of thermal decomposition of chemically modified silica gel. *J. Chromatogr. A* **1996**, *742*, 175–179.

109. Ugland, H.G.; Krogh, M.; Rasmussen, K.E. Automated determination of 'Ecstasy' and amphetamines in urine by SPME and capillary gas chromatography after propylchloroformate derivatisation. *J. Pharm. Biomed. Anal.* **1999**, *19*, 463–475.

110. Chia, K.-J.; Huang, S.-D. Simultaneous derivatization and extraction of amphetamine-like drugs in urine with headspace solid-phase microextraction followed by gas chromatography–mass spectrometry. *Anal. Chim. Acta* **2005**, *539*, 49–54.

111. Wang, Q.; Chong, J.M.; Pawliszyn, J. Determination of thiol compounds by automated headspace SPME with in-fiber derivatization. *Flavour Fragr. J.* **2006**, *21*, 385–394.

112. Wang, Q.; O'Reilly, J.; Pawliszyn, J. Determination of low-molecular mass aldehydes by automated headspace solid-phase microextraction with in-fibre derivatization. *J. Chromatogr. A* **2005**, *1071*, 147–154.

113. Okeyo, P.; Rentz, S.M.; Snow, N.H. Analysis of steroids from human serum by SPME with headspace derivatization and GCMS. *J. High Resolut. Chromatogr.* **1997**, *20*, 171–173.

114. del Olmo, M.; Suárez, A.Z.B.; Gonzalez-Casado, A.; Taoufiki, J.; Vilchez, J.L. Use of solid-phase microextraction followed by on-column silylation for determining chlorinated bisphenol A in human plasma by gas chromatography–mass spectrometry. *J. Chromatogr. B* **2005**, *817*, 167–172.

115. Namera, A.; Yashiki, M.; Kojima, T.; Ueki, M. Automated headspace solid-phase microextraction and in-matrix derivatization for the determination of amphetamine-related drugs in human urine by gas chromatography–mass spectrometry. *J. Chromatogr. Sci.* **2002**, *40*, 19–25.

116. Watanabe, K.; Okamoto, N.; Yamagishi, I.; Nozawa, H.; Ishii, A.; Suzuki, O. Simple analysis of amphetamines in human whole blood and urine by headspace capillary gas chromatography with large volume injection. *Chromatographia* **2003**, *58*, 455–458.

117. Cháfer-Pericás, C.; Campíns-Falcó, P.; Herráez-Hernández, R. Application of solid-phase microextraction combined with derivatization to the enantiomeric determination of amphetamines. *J. Pharm. Biomed. Anal.* **2006**, *40*, 1209–1217.

118. Lachenmeier, D.W.; Kroener, L.; Musshoff, F.; Madea, B. Application of tandem mass spectrometry combined with gas chromatography and headspace solid-phase dynamic extraction for the determination of drugs of abuse in hair samples. *Rapid Commun. Mass Spectrom.* **2003**, *17*, 472–478.

119. Staerk, U.; Külpmann, W.R. High-temperature solid-phase microextraction procedure for the detection of drugs by gas chromatography–mass spectrometry. *J. Chromatogr. B* **2000**, *745*, 399–411.

120. Engelmann, M.D.; Hinz, D.; Wenclawlak, B.W. Solid-phase microextraction (SPME) and headspace derivatization of clenbuterol followed by GC–FID and GC–SIMMS quantification. *Anal. Bioanal. Chem.* **2003**, *375*, 460–464.

121. Abdel-Rehim, M.; Hassan, Z.; Blomberg, L.; Hassan, M. On-line derivatization utilizing solid-phase microextraction (SPME) for determination of busulphan in plasma using gas chromatography–mass spectrometry (GC-MS). *Ther. Drug Monit.* **2003**, *25*, 400–406.

122. Degel, F. Comparison of new solid-phase microextraction methods for chromatographic identification of drugs in clinical toxicological analysis. *Clin. Biochem.* **1996**, *29*, 529–540.

123. Theodoridis, G.; Koster, E.H.M.; de Jong, G.J. Solid-phase microextraction for the analysis of biological samples. *J. Chromatogr. B* **2000**, *745*, 49–82.

124. Ulrich, S. Solid-phase microextraction in biomedical analysis. *J. Chromatogr. A* **2000**, *902*, 167–194.

125. Snow, N.H. Solid-phase micro-extraction of drugs from biological matrices. *J. Chromatogr. A* **2000**, *85*, 445–455.

126. Lord, H.L.; Pawliszyn, J. Microextraction of drugs. *J. Chromatogr. A* **2000**, *902*, 17–63.

127. Kataoka, H. New trends in sample preparation for clinical and pharmaceutical analysis. *Trends Anal. Chem.* **2003**, *22*, 232–244.

128. Queiroz, M.E.C.; Lanças, F.M. Practical tips on preparing plasma samples for drug analysis using SPME. *LC-GC Eur.* **2005**, *18*, 145–154.

129. Nagasawa, N.; Yashiki, M.; Iwasaki, Y.; Hara, K.; Kojima, T. Rapid analysis of amphetamines in blood using headspace solid phase microextraction and selected ion monitoring. *Forensic Sci. Int.* **1996**, *78*, 95–102.

130. Battu, C.; Marquet, P.; Fauconnet, A.L.; Lacassie, E.; Lachâtre, G. Screening procedure for 21 amphetamine-related compounds in urine using solid-phase microextraction and gas chromatography-mass spectrometry. *J. Chromatogr. Sci.* **1998**, *36*, 1–7.

131. Koide, I.; Noguchi, O.; Okada, K.; Yokoyama, A.; Oda, H.; Yamamoto, S.; Kataoka, H. Determination of amphetamine and methamphetamine in human hair by headspace solid-phase microextraction and gas chromatography with nitrogen-phosphorus detection. *J. Chromatogr. B* **1998**, *707*, 99–104.

132. Guan, F.; Seno, H.; Ishii, A.; Watanabe, K.; Kumazawa, T.; Hattori, H.; Suzuki, O. Solid-phase microextraction and GC-ECD of benzophenones for detection of benzodiazepines in urine. *J. Anal. Toxicol.* **1999**, *23*, 54–61.

133. Reubsaet, K.J.; Norli, H.R.; Hemmersbach, P.; Rasmussen, K.E. Determination of benzodiazepines in human urine and plasma with solvent modified solid phase microextraction and gas chromatography; rationalisation of method development using experimental design strategies. *J. Pharm. Biomed. Anal.* **1998**, *18*, 667–680.

134. Hall, B.J.; Satterfield-Doerr, M.; Parikh, A.R.; Brodbelt, J.S. Determination of cannabinoids in water and human saliva by solid phase microextraction and quadrupole ion trap gas chromatography/mass spectrometry. *Anal. Chem.* **1998**, *70*, 1788–1796.

135. Kumazawa, T.; Watanabe, K.; Sato, K.; Seno, H.; Ishii, A.; Suzuki, O. Detection of cocaine in human urine by solid-phase microextraction and capillary gas chromatography with nitrogen-phosphorus detection. *Jpn. J. Forensic Toxicol.* **1995**, *13*, 207–210.

136. Frison, G.; Tedeschi, L.; Maiette, S.; Ferrara, S.D. Determination of γ-hydroxybutyric acid (GHB) in plasma and urine by headspace solid-phase microextraction and gas chromatography/positive ion chemical ionization mass spectrometry. *Rapid. Commun. Mass Spectrom.* **2000**, *14*, 2401–2407.

137. Myung, S.-W.; Kim, S.; Park, J.-H.; Kim, M.; Lee, J.-C.; Kim, T.-J. Solid-phase microextraction for the determination of pethidine and methadone in human urine using gas chromatography with nitrogen-phosphorus detection. *Analyst* **1999**, *124*, 1283–1286.

138. Ishii, A.; Seno, H.; Kumazawa, T.; Watanabe, K.; Hattori, H.; Suzuki, O. Simple extraction of phencyclidine from human body fluids by headspace solid-phase microextraction (SPME). *Chromatographia* **1996**, *43*, 331–333.

139. Kongshaug, K.E.; Pedersen-Bjergaard, S.; Rasmussen, K.E.; Krogh, M. Solid-phase microextraction/capillary gas chromatography for the profiling of confiscated ecstasy and amphetamine. *Chromatographia* **1999**, *50*, 247–252.

140. Koester, C.J.; Andresen, B.D.; Grant, P.M. Optimum methamphetamine profiling with sample preparation by solid-phase microextraction. *J. Forensic Sci.* **2002**, *47*, 1002–1007.

141. Furton, K.G.; Hong, Y.-C.; Hsu, Y.L.; Luo, T.; Rose, S.; Walton, J. Identification of odor signature chemicals in cocaine using solid-phase microextraction–gas chromatography and detector-dog response to isolated compounds spiked on U.S. paper currency. *J. Chromatogr. Sci.* **2002**, *40*, 147–155.

142. Strano-Rossi, S.; Molaioni, F.; Botrè, F. Application of solid-phase microextraction to antidoping analysis: Determination of stimulants, narcotics, and other classes of substances excreted free in urine. *J. Anal. Toxicol.* **2005**, *29*, 217–222.

143. Zhang, Z.M.; Duan, H.B.; Zhang, L.; Chen, X.; Liu, W.; Chen, G.N. Direct determination of anabolic steroids in pig urine by a new SPME-GC-MS method. *Talanta* **2009**, *78*, 1083–1089.

144. Huang, G.; Chen, H.; Zhang, X.; Cooks, R.G.; Ouyang, Z. Rapid screening of anabolic steroids in urine by reactive desorption electrospray ionization. *Anal. Chem.* **2007**, *79*, 8327–8332.

145. Aresta, A.; Calvano, C.D.; Palmisano, F.; Zambonin, C.G. Determination of clenbuterol in human urine and serum by solid-phase microextraction coupled to liquid chromatography. *J. Pharm. Biomed. Anal.* **2008**, *47*, 641–645.

146. Liu, W.; Zhang, L.; Chen, S.; Duan, H.; Chen, X.; Wei, Z.; Chen, G. A method by homemade OH/TSO-PMHS fibre solid-phase microextraction coupling with gas chromatography–mass spectrometry for analysis of antiestrogens in biological matrices. *Anal. Chim. Acta* **2009**, *631*, 47–53.

147. Pragst, F. Application of solid-phase microextraction in analytical toxicology. *Anal. Bioanal. Chem.* **2007**, *388*, 1393–1414.

148. Wu, J.; Kataoka, H.; Lord, H.; Pawliszyn, J. Polypyrrole coated capillary in-tube solid phase microextraction coupled with liquid chromatography/electrospray ionization mass spectrometry for the determination of β-blockers in urine and serum samples. *J. Microcolumn Sep.* **2000**, *12*, 255–266.

149. Musteata, F.M.; Musteata, M.L.; Pawliszyn, J. Fast in-vivo microextraction: A new tool for clinical analysis. *Clin. Chem.* **2006**, *52*, 708–715.

150. Takamoto, S.; Sakura, N.; Namera, A.; Yashiki, M. Monitoring of urinary acrolein concentration in patients receiving cyclophosphamide and ifosphamide. *J. Chromatogr. B* **2004**, *806*, 59–63.

151. Zambonin, C.G.; Aresta, A.; Palmisano, F. Determination of the immunosuppressant mycophenolic acid in serum by solid-phase microextraction coupled to liquid chromatography. *J. Chromatogr. B* **2004**, *806*, 89–93.

152. Moneti, G.; Dani, F.R.; Pieraccini, G.T.S. Solid-phase microextraction of insect epicuticular hydrocarbons for gas chromatographic/mass spectrometric analysis. *Rapid Commun. Mass Spectrom.* **1997**, *11*, 857–862.

153. Frerot, B.; Malosse, C.; Cain, A.H. Solid-phase microextraction (SPME). A new tool in pheromone identification in Lepidoptera. *J. High Resolut. Chromatogr.* **1997**, *20*, 340–342.

154. Smith, B.P.; Zini, C.A.; Pawliszyn, J.; Tyler, M.J.; Hayasaka, Y.; Williams, B.; Caramao, E.B. The presence of limonene and eucalyptol in the skin secretion of the brown tree frog, *Litoria ewingi*. *Chem. Ecol.* **2000**, *17*, 215–225.

155. Nakajima, D.; Win-Shwe, T.-T.; Kakeyama, M.; Fujimaki, H.; Goto, S. Determination of toluene in brain of freely moving mice using solid-phase microextraction technique. *NeuroToxicology* **2006**, *27*, 615–618.

156. Win-Shwe, T.-T.; Mitsushima, D.; Nakajima, D.; Ahmed, S.; Yamamoto, S.; Tsukahara, S.; Kakeyama, M.; Goto, S.; Fujimaki, H. Toluene induces rapid and reversible rise of hippocampal glutamate and taurine neurotransmitter levels in mice. *Toxicol. Lett.* **2007**, *168*, 75–82.

157. Lord, H.; Pawliszyn, J. Evolution of SPME technology. *J. Chromatogr. A* **2000**, *885*, 153–193.

158. Zhang, Z.; Poerschmann, J.; Pawliszyn, J. Direct solid phase microextraction of complex aqueous samples with hollow fibre membrane protection. *Anal. Commun.* **1996**, *33*, 129–131.

159. Musteata, F.; de Lannoy, I.; Pawliszyn, J. Blood sampling without blood draws for in vivo pharmacokinetic studies in rats. *J. Pharm. Biomed. Anal.* **2008**, *47*, 907–912.

160. Vuckovic, D.; de Lannoy, I.; Gien, B.; Yang, Y.; Musteata, F.M.; Shirey, B.; Sidisky, L.; Pawliszyn, J. In vivo solid-phase microextraction for single rodent pharmacokinetics studies of carbamazepine and carbamazepine-10,11-epoxide in mice. *J. Chromatogr. A*, 2010 doi:10.1016/j.chroma.2010.07.060.

161. Musteata, F.M.; Pawliszyn, J. In-vivo sampling with solid phase microextraction. *J. Biochem. Biophys. Methods* **2006**, *70*, 181–193.

162. Vaes, W.H.J.; Ramos, E.U.; Verhaar, H.J.M.; Seinen, W.; Hermens, J.L.M. Measurement of the free concentration using solid phase microextraction: Binding to protein. *Anal. Chem.* **1996**, *68*, 4463–4467.

163. Vaes, W.H.J.; Ramos, E.U.; Hamwijk, C.; van Holsteijn, I.; Blaauboer, B.J.; Seinen, W.; Verhaar, H.J.M.; Hermens, J.L.M. Solid phase microextraction as a tool to determine membrane/water partition coefficients and bioavailable concentrations in vitro systems. *Chem. Res. Toxicol.* **1997**, *10*, 1067–1072.

164. Yuan, H.; Ranatunga, R.; Carr, P.; Pawliszyn, J. Determination of equilibrium constant of alkylbenzenes binding to bovine serum albumin by SPME. *Analyst* **1999**, *124*, 1443–1448.

165. Vuckovic, O.; Cudjoe, E.; Musteata, F.; Pawliszyn, J. Automated solid-phase microextraction and thin-film microextraction for high-throughput analysis of biological fluids and ligand-receptor binding studies. *Nat. Protocol.* **2010**, *5*, 140–161.

166. Musteata, F.M.; Pawliszyn, J. Study of ligand-receptor binding using SPME: Investigation of receptor, free and total ligand concentrations. *J. Proteome Res.* **2005**, *4*, 789–800.

167. Musteata, F.; Pawliszyn, J.; Qian, M.; Wu, J.; Niwa G. Determination of drug plasma protein binding by SPME. *J. Pharm. Sci.* **2006**, *95*, 1712–1722.

168. Raghani, A.R. High-speed gas chromatographic analysis of solvents in pharmaceuticals using solid phase microextraction. *J. Pharm. Biomed. Anal.* **2002**, *29*, 507–518.

169. Colón, I.; Richoll, S.M. Determination of methyl and ethyl esters of methanesulfonic, benzenesulfonic and p-toluenesulfonic acids in active pharmaceutical ingredients by solid-phase microextraction (SPME) coupled to GC/SIM-MS. *J. Pharm. Biomed. Anal.* **2005**, *39*, 477–485.

170. Racz, I. *Drug Formulation*. Budapest: John Wiley & Sons; 1989.

171. Wang, S.M. Enantiomeric determination of amphetamines: Exploring a novel one-step solid-phase microextraction-based approach. *J. Chromatogr. B* **2005**, *825*, 79–87.

Chapter 21

Sample Handling Protocols for Biosensor Applications

Andrew Chan, Teresa Artuso, and Ulrich J. Krull

21.1. INTRODUCTION

Pretreatment of a sample is often necessary in order to provide the target (analyte) in a form amenable to the detection strategy as well as to remove components that may interfere with the analytical process. This is the reality of matrix effects that are encountered in areas such as the detection of biological targets from environmental, clinical, agricultural, and food samples or products. Biosensors offer the potential to provide fast, sensitive, and selective detection methods for targets and have been touted as being suitable for hazard analysis critical control point (HACCP) issues, and for the development of field and point of care devices. Extensive pretreatment of samples would encumber fast analysis times, offsetting a primary advantage offered by the implementation of biosensor technology. However, many of the biosensor technologies reported in the literature currently require some form of sample pretreatment and purification to extract the target as well as to preconcentrate targets when attempting analysis of trace quantities.

The following chapter will review some of the pretreatment and purification technologies that can be used in conjunction with biosensors, and with implementation of microfluidics as an adjunct to biosensor operation. The focus will be on methods that deal with rapid extraction and purification of whole cells and intracellular targets such as proteins and DNA.

21.1.1. Introduction to Biosensors

A biosensor is a self-contained device that is designed to provide quantitative information about the presence of a specific target. It typically consists of a biological recognition element integrated onto a transducer.[1,2] Binding between a biological element that is located on a surface with its target will result in a physical or chemical change at this interface, and this change is then converted by the transducer into a quantifiable electrical signal.[3,4]

Ideally, biosensors are designed to be highly sensitive and selective devices. The intention is to achieve detection of low quantities of targets and to achieve excellent discrimination within mixtures of closely related compounds.[3,5] The terminology "biosensor" is meant to signal that the device is reusable, and typically, it is desirable that the sensor can quickly be regenerated to be ready for analysis of a series of samples.[1,3,5] Such operational capabilities have made biosensors attractive for applications such as rapid detection of pathogens and detection of contaminants in consumable products such as food and water supplies. One goal for using biosensors is in application to problems requiring real-time monitoring of chemical processes and the presence of pathogens.[1,2,5,6]

A wide variety of biorecognition molecules and transduction elements have been used to develop biosensors. Table 21.1 provides some representative examples of selective reagents and transduction technologies that are used to develop biosensors.

It is important to note that many of the examples presented in Table 21.1 require pretreatment protocols for purification or amplification, which can take longer than the actual process of measurement. Purification removes the majority of the contaminating species from the sample so that the target is in a relatively clean matrix. Amplification can be in the form of enrichment by culturing, where the target cells are allowed to multiply in growth media. Amplification of DNA targets can also be done by polymerase chain reaction (PCR) for DNA targets. However, such amplification steps can significantly add to the total analysis time. Therefore, there is a desire for the detection

Handbook of Sample Preparation, Edited by Janusz Pawliszyn and Heather L. Lord

Table 21.1. Examples of Biosensors Used for the Detection of Targets from Various Different Sample Matrices

Transduction Mode	Probe	Target	Sample	Pretreatment Protocol	LOD	Total Time	Reference
Optical	DNA	*E. coli*	In buffer	Culture enrichment, centrifugation, lysis, chloroform–phenol extraction, ultrasonication	1 pg/mL	>12 h (20-s detection)	7
Optical	Antibody	*E. coli* O157:H7	Water	Culture enrichment	10 cells per 0.075 mL	>20 h (3-h detection)	8
Optical	DNA	Telomerase	Human leukemia cells	Cell lysis, centrifugation	N/R	N/R	9,10
Optical	Antibody	*Salmonella typhimurium*	Pork (spiked)	Homogenization	10^5 CFU/g	5 min	11
SPR	DNA	Plant 35S promoter	GM maize	DNA extraction, fragmentation by enzyme digest, thermal denaturation	N/R	N/R	12
		Apo E polymorphism	Human blood				
SPR	DNA	Enrofloxacin	Milk (spiked)	Centrifuged, filtered (0.22-μm membrane), 100× dilution with buffer	3 μg/mL	N/R	13
SPR	Antibody	Beta2-glycoprotein I	Human blood	Centrifugation, 90× dilution, removal of IgG and IgM	N/R	1 h (4-min detection)	14
SPR	DNA	Tumor necrosis factor (TNF-alpha)	In buffer	PCR, denaturation by heat	0.677 pM	~120 min	15
Electrochemical	DNA	Apolipoprotein E polymorphism	Human blood	Extraction and purification, PCR, denaturation by heat	N/R	N/R	16
	Antibody	Sulfonamide	Milk (spiked)	4× dilution, immunomagnetic capture and preconcentration	5.92 nmol/L	N/R	17
	Antibody	*Vibrio cholerae*	Environmental water	Filtration (0.22 μm), enrichment	80 CFU/mL	7 h (55-min detection)	18
	Antibody	*E. coli* O157:H7	Lettuce wash (spiked)	Stomacher	10^6 CFU/mL	N/R	19
Piezoelectric	DNA	HPV	Clinical samples	Centrifugation, DNA extraction (Commercial Kit), PCR	N/R	~10 h (20-min detection)	20
	DNA	*Aeromonas hydrophila*	Water, vegetable, human	DNA extraction, PCR, denatured by heat	N/R	N/R	21
	Antibody	*E. coli* O157:H7	In buffer	Culture enrichment, centrifugation	N/R	>12 h	22
	DNA	*Viral orchid RNA*	Leaf sample	Homogenization, centrifugation, 100× dilution	10 ng	1 h	23
	DNA	*E. coli*	Water	Culture enrichment, lysis, phenol–chloroform extraction, centrifugation, PCR, denaturation	1 μg/mL	>12 h (10-min detection)	24
	DNA	*E. coli* O157:H7	In buffer	Culture enrichment, centrifugation, extraction, asymmetric PCR	2.67×10^2 CFU/mL	22 h	25

Steps for pretreatment protocols prior to detection are also outlined as well as the total time needed for pretreatment and detection. When available, detection time is also provided in parenthesis.
N/R, not reported; HPV, human papillomavirus; LOD, limit of detection.

of a biological target directly from its complex matrix with minimal sample preparation.

Mascini's group demonstrated a piezoelectric DNA biosensor for the detection of nonamplified bovine genomic DNA extracted from animal muscle. Large concentrations of genomic DNA (5–20 µg/mL) were needed for detection, and the extraction, fragmentation, and denaturation of the target DNA was still necessary.[26]

Grant's group developed an antibody-based optical biosensor for the detection of *Salmonella typhimurium*. A limit of detection of 10^3 cells/mL was obtained for standard solutions made in buffer. However, for samples of *S. typhimurium* in homogenized pork, the limit of detection for the sensor was 10^5 CFU/g. The high limit of detection was thought to be a result of interference from the complex nature of the food matrix with the antibody-based recognition.[11]

Fauchet's group presented a protein-based surface plasmon resonance (SPR) biosensor which required minimal sample pretreatment by implementation of macroporous silicon (pore size > 100 nm). Such macroporous silicon structures are very sensitive to refractive index changes. The biosensors used the translocated Intimin receptor-Intimin binding domain (Tir-IBD) as the probe, and Intimin-ECD (extracellular domain of Intimin) as the target. These are two key proteins responsible for the pathogenicity of *Escherichia coli*. The limit of detection was determined to be 130 fmol of Intimin-ECD. While the target protein was only about 23% of the total protein, the results showed minimal background interference from other proteins. However, this required a lengthy wash step (1–2 h), which could result in the loss of the captured target proteins, leading to a higher limit of detection.[27]

Mixed self-assembled monolayers of DNA and short oligoethylene glycol (OEG) on gold surfaces has also been used to improve the specific capture of DNA targets from complex media. OEG has been demonstrated to reduce nonspecific adsorption of proteins to the target DNA as well as the immobilized probes. Hybridization was detected by SPR from test mixtures containing 1 µM of target DNA in either bovine serum or salmon genomic DNA, demonstrating detection of DNA within a relatively complex matrix.[28] Detection of targets directly within complex matrices have, in most cases, been reported to provide inferior limits of detection when compared with targets that were prepared in clean buffers.[29] At present, some form of pretreatment and purification of "real" samples is still needed for low detection levels or trace analysis monitoring when using biosensors.[12]

21.1.2. Biosensors for Trace Analysis of Biological Targets

One of the potential applications of biosensors is in the real-time and in-field detection of trace amounts of analytes from clinical samples, forensics, environmental monitoring, and contamination of consumables.[30–32] This requires the ability of the biosensor to detect the analyte that may be present in extremely low number density as well as in complex matrices.[12,24,30]

Current strategies for the detection of trace amounts of biological targets include improving the sensitivity of the biosensor by increasing the amount of signal generated for each recognition event as described below, as well as examining methods for sample pretreatment and sample enrichment prior to detection.

Even though the focus of this chapter is on methods for sample pretreatment and enrichment, it is still useful to consider methods for increasing biosensor sensitivity as these represent adjuncts that offer advantages in accelerating the time required to measure the analytical signal.

21.1.3. Increasing Sensitivity by Signal Amplification

A strategy for increasing the sensitivity of biosensors has been the development of technologies which amplify the signals that are generated from one capture event.[33] Many of these amplification techniques are based on enzyme-linked immunosorbent assay (ELISA). In such a strategy, one type of antibody serves as a probe and is immobilized onto a solid surface. This probe first captures the target to localize it onto the surface. A second different antibody, which can bind to the immobilized target, is then introduced, creating a sandwich structure. This second probe is typically labeled with an enzyme. Extensive washing leaves only the antibody–target–antibody (enzyme) sandwich structure on the surface. Addition of substrate for the enzyme then takes place, and the enzyme-catalyzed reaction leads to product formation that can quantitatively be measured as an analytical signal.

This sandwich structure strategy has been used for the detection of *E. coli* cells. Antibodies targeting *E. coli* were immobilized onto indium tin oxide chips. Following the capture of *E. coli* cells onto the sensor surface, a second antibody modified with alkaline phosphatase was used to form the sandwich structure. The alkaline phosphatase catalyzed the precipitation of 5-boro-4-chloro-3-indolyl phosphate, which was then detected by impedance. A limit of detection of 6×10^3 cells/mL was reported.[34]

A modification of this immunoassay method has been reported where the second antibody used in the assay is coupled to a liposome which contains a large number of reporter molecules rather than an enzyme (Fig. 21.1). Following the formation of the sandwich structure, the liposome is lysed to release the reporter molecules, which are detected based on the transduction strategy used (typically, an electrochemical or optical signal). Since multiple reporter molecules are released per capture event, the signal that is generated is greatly amplified, with the amplification factor being dependent on the number of reporter molecules that can be encapsulated by the liposome.[5] A 500- to 1000-fold signal enhancement has been demonstrated using these dye

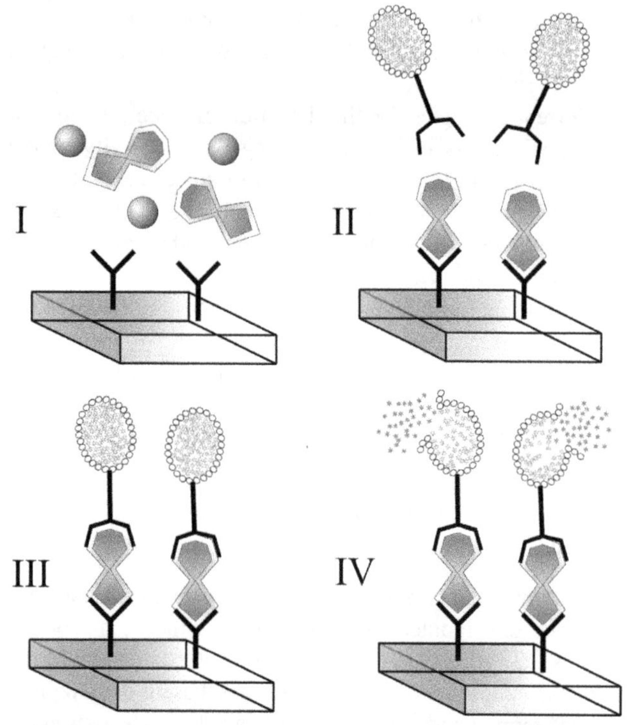

Figure 21.1. Schematic diagram of signal enhancement by liposome amplification. (I) Target analyte is selectively captured by the probe. (II) A second probe conjugated with a liposome filled with a large number of reporter molecules is added. (III) The second probe binds to the immobilized target, forming a sandwich structure. (IV) Lysis of the liposome releases the reporter molecules, which are then detected.

molecule-filled liposomes versus a single fluorophore attached to the second antibody.[35]

A variety of similar schemes have been devised to achieve amplification. For example, an electrochemical DNA-based biosensor that used carbon nanotubes (CNTs) and DNA to provide selective chemistry has been reported to have detection limits of 1.3 zmol from 25- to 50-μL samples. Rather than using CNTs to modify an electrode surface directly, the CNTs were covered with approximately 9600 alkaline phosphatases and modified with a DNA probe.

The DNA targets were first incubated with a solution containing magnetic beads immobilized with a different DNA probe than those on the CNTs. The probe on the magnetic bead was designed to be selective toward one end of the target DNA. Following hybridization of the target with the probe-modified magnetic beads, any unbound targets were washed off the beads. Next, a solution containing the modified CNTs was introduced to the magnetic beads. The DNA probe on the CNTs was designed to be selective to the remaining free end of the target DNA. The target was therefore sandwiched between the magnetic bead and a CNT. The sandwiched complexes were moved to an electrode surface using an external magnetic field, where a signal was generated based on detecting the enzymatic conversion product α-naphthol.[36]

DNA dendrimers have been used to obtain a larger number of fluorescent targets for signal amplification. The target DNA was first captured by beads that were coated with probe molecules. The beads were then sandwiched by three DNA dendrimers, which were multiply labeled with organic fluorescent dyes. Using this amplification protocol, 0.031 fmol of target DNA could be detected.[37]

The liposome-based signal amplification system has also been used to produce a double signal enhancement. In this immunoassay, target protein from *Bacillus anthracis* was first captured by immobilized antibodies. A second antibody conjugated with liposomes that were filled with fluorescein-labeled oligonucleotide was added to sandwich with the immobilized protein. Lysis of the liposome released the labeled oligonucleotides, and these hybridized with complementary sequences that were immobilized onto a second plate. A second set of liposomes modified with anti-fluorescein antibodies were used to sandwich with the hybridized oligonucleotides. The detection of the signal was based on the fluorescence from the dye contained inside the second set of liposomes, with the resulting signal being proportional to the original analyte concentration. The use of this double liposome signal amplification resulted in limits of detection, which were 400 times lower than a one-step amplification using the fluorescein-labeled probes alone. A limit of detection of 4.1 ng/mL of the *B. anthracis* antigen was reported.[35]

An immuno-PCR method has been used as an amplification technique for the detection of noroviruses from stool and food samples. Selective capture of the virus is achieved by a primary antibody, followed by the addition of a DNA reporter molecule that is coupled to a second antibody to form a sandwich. PCR is done using the DNA sequence as the template, and the DNA amplicons are subsequently detected. This detection strategy was tested with viruses mixed in blended food matrices. PCR required dilution of the samples to suppress inhibitor effects. Sensitivity was shown to be 10 times better than that obtained using the real-time polymerase chain reaction (RT-PCR) method.[38]

Nanoparticles have been used to provide an enhancement in sensitivity for a biosensor based on SPR. Oligonucleotide probes conjugated with gold nanoparticles formed a sandwich complex with DNA targets that had already been captured onto a gold film. The use of gold nanoparticles on an SPR-based sensor has been shown to improve the sensitivity to hybridization by 1000-fold in comparison to an unamplified detection strategy. The enhancement of sensitivity can be attributed to properties such as an increase in surface mass, the high dielectric constant of gold particles, and electromagnetic coupling between gold nanoparticles and the gold film. A limit of detection of 10 pM was achieved.[39]

A DNA-based piezoelectric sensor was reported with a limit of detection of 1×10^{-12} M of oligonucleotides. This was achieved by amplifying the frequency shift associated with hybridization using Fe_3O_4 nanoparticles as "mass enhancers."[25] The detection limit was four orders of magnitude lower than obtained using a piezoelectric biosensor without the nanoparticle signal amplification. PCR-amplified *E. coli* O157:H7 DNA targets were examined in conjunction with the "mass enhancers," and the system provided a limit of detection of 2.67×10^2 CFU/mL. Implementation of this method ameliorates the need for enrichment of the cells by culturing.[25]

The use of gold nanoshells has also been demonstrated for rapid immunoassays of immunoglobulins in saline, serum, and whole blood. Nanoshells consist of a dielectric core made of silica surrounded by a thin metal shell of either gold or silver. The plasmon resonance response of the nanoshells was modified to operate in the near-infrared (IR) region (700–1300 nm). This range allowed for optical transmission through tissue and was advantageous for direct analysis in samples of whole blood. Different antibodies that were selective toward only one particular target were immobilized onto different particles. This allowed for the binding of multiple particles to one target, leading to the formation of dimers and higher-order aggregates. The result was in a red shift of the plasmon resonance maximum, and a decrease in the amplitude of the original single particle plasmon resonance was observed. A linear range of response from 0.8 to 88 ng/mL was obtained in a time frame of 10–30 min.[40]

21.1.4. Goals of Pretreatment

There are three major goals in the pretreatment of a sample for detection by a biosensor: extraction of the target material from the sample matrix, purification of the target, and volume reduction.

21.1.4.1. Extraction. Extraction may be required to isolate the analyte of interest from the background matrix. Depending on the target of interest, extraction can be relatively straightforward or may be a highly involved series of steps. Typically, the extraction of target cells from solution only requires a method that can selectively target some marker on the surface of the cell that can be distinguished from the background matrix. However, for biological targets that are inside the cells, such as proteins and DNA, the intracellular contents must first be released and the target must be separated by an appropriate method.

21.1.4.2. Purification. Aside from releasing the target analyte from its cellular matrix, removal of background components other than the target analyte by purification may also be necessary. Components present in the matrix may inhibit or compete with the biorecognition between the probe and its target,[41,42] or may negatively influence quantitative analysis if contaminants can provide signals and are not removed prior to detection.[43] Since biological materials are generally used as the selective reagents to develop biosensors, issues related to the stability of the selective components in the background matrix may arise.[43] Additionally, for trace analysis of target within a complex matrix, nonselective adsorption of nontarget materials onto the biosensor surface can easily overwhelm the analytical signal. For transduction methods such as piezoelectric devices and SPR systems, which detect the change in mass on the surface, such nonselective adsorption can result in false positives.[44]

Purification may be required in order to differentiate between viable and dead cells for detection of intracellular targets such as DNA. This is important where only viable cells are of interest, for example, in the detection of a pathogen in consumable products where only live cells will adversely affect human health. One challenge is that DNA extraction methods will extract DNA from both viable and dead cells,[41] making it very difficult to differentiate whether there is any actual danger associated with biologicals in a consumable product.[41,45,46]

Purification of samples can also be important when considering the quantitative determination of a target. Variations in the components of the matrix between samples may pose a problem for quantitative analysis.[1] Calibration curves are often constructed using solutions of known concentrations of the target in water or buffer. Unless steps are taken to ensure that the background matrix from a "real" sample would not interfere with the quantitative determination, purification is necessary to ensure that the background matrix of the sample was compatible with that used with the calibration curve.[1] Alternatively, a standard addition approach might be considered, but this is of little interest if the matrix interferes with signal development.

Of concern in the purification of antigens and proteins is the elimination of cross-reactivity by components, which may be similar to the target analyte.[6] Additionally, enzymes such as proteinases may cause the degradation of certain targets and probes. Enzymes would need to be removed or their activity would need to be suppressed.[47] Purification can also help minimize the loss of target proteins resulting from adsorption onto particulate matter in the sample matrix. This is especially important for the detection of proteins that are present in low abundance.[48]

Interactions between antibodies and targets can be influenced by the pH, ionic strength, and competitive binding by matrix components. Dissolved organic carbon can interact weakly with antibodies, inducing an overestimation of the immunoassay response. Purification ensures that these interferences will be removed before detection.[6]

The majority of purification protocols to extract DNA are intended to remove compounds that might inhibit amplification of DNA targets by PCR. PCR is a common step for amplification of DNA prior to detection by a biosensor to improve sensitivity, and can also serve in itself as a detection strategy, confirming the presence or absence of the analyte.

Environmental samples such as those from water, soil, and food materials are known to contain compounds that can inhibit PCR or that can interfere with detection toward the end of the process at the detection stage.[49–51] These compounds include humic acids, polyphenolic compounds, polysaccharides, urea, soot, dust and pollen, silt, clay, metal ions, chelators, milk products, fat in foods, hemoglobin, iron, heparin, and acidic polysaccharides, as well as DNA from nontarget microorganisms.[42,49,50,52–56] Humic acids and polysaccharides can bind to DNA polymerase and to chelating agents, which may be cofactors for the enzyme. It has been shown that as little as 1 ng of humic acid can inhibit PCR.[42] Nucleic acids can also complex with proteins, polysaccharides, or polyphenolic compounds that are released during cell lysis.[57]

Some phenolic compounds will also absorb light at 260 nm, interfering with quantification of DNA by UV-VIS spectroscopy.[55] Inhibitor compounds can also affect quantification by spectrofluorimetry.[58] Fluorescent dyes such as ethidium bromide or Hoechst 33258 have been shown to be susceptible to quenching by impurities such as proteins and phenols. Newer dyes such as PicoGreen are less susceptible to quenching, but quantum yields can still be affected by concentrations of humic acids that are 100 ng/μL or greater.[58,59]

21.1.4.3. Volume reduction. Biosensors typically require small sample volumes for detection. Volumes usually range from nano- to microliters. However, for trace detection of analytes in "real" samples, collection of microliters may not be statistically representative of the sample.

Bacteria or viruses may not be homogenously distributed within samples, necessitating that large volumes be collected for representative sampling.[60] Additionally, the target pathogens may be present in low numbers, again requiring large volumes to be collected so as to have confidence in quantification.[45,61,62] For example, in *Mycobacterium ulcerans*, which is a human pathogen that causes chronic necrotic skin disease, the typical environmental load of concern is estimated to be approximately 0.5 cells/100 mL of water.[63] The infectious dose for many foodborne pathogenic bacteria is often a few cells.[64,65] For *E. coli* O157:H7 in ground beef, contamination levels are usually less than 100 CFU/g of beef.[52,66]

In the detection of cancerous colon cells from stool samples, the amount of DNA obtained from colon cells only represents 0.01–0.1% of the total DNA that is recovered from the samples. Of that small percentage, only about 1% of those cells may be cancerous.[67] The balance is primarily DNA from bacterial cells.

The problem of low abundance of the target for these cases results in the need for large sampling volumes. The challenge is to ensure that a sufficient quantity of pathogen has been collected to meet the limit of detection for the sensor, as well as to provide replicates for statistical confidence measures of results. For example, assume that an RNA-based biosensor had a limit of detection of 16 ng/μL of RNA for a 2-μL sample. Assuming a homogeneous sample and that the extraction and purification protocol yields an average of 0.02 ng of RNA per cell, then the sampling volume would be about 6.4 L if the solution contained 250 cells/L (i.e., 800 cells in total) to get a detectable amount of rRNA.[68] Volumes taken for environmental samples are often several orders of magnitude larger than the volumes actually required by a biosensor in the measurement step. This requires some form of volume reduction so that the volume is reduced without significant loss of the target. Volume reduction is usually done by selective capture or sequestering of the target, purification, followed by elution of the target into a smaller volume. This selective concentrating of the target results in enhancement in signal-to-noise (S/N) ratio by preconcentration.

In the previous example for the detection of RNA, the limit of detection was 16 ng/μL for a 2-μL sample. If the biosensor was designed to measure only 1 μL of processed sample, then the amount of RNA obtained from the 6.4-L sample, concentrated into a 1-μL sample would be 32 ng/μL, an enrichment factor of 2. Preconcentration not only improves detection sensitivity but also improves the reliability of analysis by significantly increasing S/N ratio and increasing the detection limit of the sensor system.[31,32]

21.1.5. Time Required for Total System Analysis

Purification protocols must be able to remove contaminants as well as interfering species. However, such protocols often require multiple steps.[42,69] The selection of the protocol depends on the type of sample, the efficiency of recovery, the purity of the extracted analyte, and also any constraints in time.[45,69]

Ultimately, the resources spent in preprocessing a sample must be balanced with the total time that the analytical procedure will take. With any detection technique, the desire is to have a sample processing and a purification protocol, which is fast and uses a minimal number of steps.[69] One of the advantages of a biosensor is that it can generate a selective response following the introduction of a sample in a relatively short time frame. Any extensive purification process would greatly reduce this advantage.

21.1.6. Purification and Quantification

Consideration must be given to the reality that with each purification step taken, there is the possibility of loss of the analyte by partial transfer and nonselective adsorption. An extensive multistep protocol may lead to significant loss of target, which can be disproportionately detrimental in cases where the quantity of target is low.[70]

Preprocessing and purification processes will inevitably result in some loss of target. If the loss induced by the purification protocol is not accounted for, then the quantitative information provided by the biosensor would become statistically meaningless. However, attempts at quantifying the efficiency of a purification protocol is not always a simple matter. For example, differences in cell wall structures may also result in differences of the adhesion behavior of cells with particulate matter in the sample, resulting in variation of loss due to adsorption.[69,71] Differences between bacterial cells will also alter the effectiveness of typical lysis protocols when attempting to release intracellular materials for assay.[69,71,72] While gram-negative organisms were successfully lysed using an alkaline agent, gram-positive organisms, having thicker cell walls, required the action of a surfactant (TritonX-100) and an enzyme (lysozyme) in order to efficiently lyse the cells.[63,73] It was reported by Honoré-Bouakline et al. and Kotlowski et al. that some commercially available kits do not allow for the complete lysis of mycobacterial cells such as *Mycobacterium tuberculosis*.[63,74]

The yield (quantity recovered) and purity (relative amount of target within sample) of extracted DNA often varies depending on the protocol that is being used, as well as the composition and type of sample.[41,55,69,75,76] The yield of DNA from standard phenol–chloroform extraction was demonstrated to be lower when using a sample such as canned tuna in comparison to DNA extracted from raw, fresh tuna.[77] Purification methods such as gradient centrifugation, glass bead extraction, chromatography, and spin columns may also result in a significant loss of extracted DNA.[50,55,78] As a general rule, the more extensive the purification protocol, the lower will be the yield of DNA.[55,79] The situation is the same for other intracellular biological targets. Incomplete lysis of the cells will result in loss of intracellular target. Additionally, since many biological molecules are charged, loss due to adsorption by electrostatic interaction with the materials used for purification is common.

Many purification protocols can only provide semiquantitative assays at best, unless steps have been taken to account for the loss of material during purification.[80] In order to obtain valid quantitative information, one must either ensure that the sample pretreatment protocol passes all the target forward, or must evaluate the loss to demonstrate that a quantitative correction factor can be applied.[41] Quantification at each stage of a multistep extraction procedure is considered important to account for loss during purification.[58]

For the case of handling of samples containing a DNA target, one solution is to track the purity and yield of the extracted DNA spectroscopically by measuring absorbance at 260 and 280 nm.[57,81] The absorbance at 260 nm can be related to the quantity of DNA present, while ratios of A260/280 from 1.5 to 1.8 are indicative of DNA samples that are relatively free of protein contamination. The change in absorbance at 260 nm can be used to track the loss of DNA over the course of a proposed purification protocol and can be used to develop a quantitative correct factor.[58,82] Targets can also be labeled in order to track the yield of a purification process, with radioisotope labeling being commonly used for this purpose.[83] Radioisotope labeling has been used to track the yield of a purification protocol that focused on the recovery of proteins from seawater samples.[83]

Each sample pretreatment protocol is optimized to suit each application and sample type, resulting in variations in the pretreatment protocols.[41,72,84] Given the myriad of pretreatment protocols and targets, it is challenging to create a compendium of detailed systematic information about the efficiency of extraction that is largely inclusive of all validated procedures.

21.2. SAMPLE PRETREATMENT FOR DETECTION OF BIOLOGICAL TARGETS

Conventional methods used for the purification and extraction of different types of biological targets associated with pathogenic cells, including whole cells, proteins, and DNA, will be reviewed. It should not be surprising that the protocols outlined in this section are not exclusively used to prepare only one type of target. It is also important to note that multiple pretreatment steps are often necessary for the isolation and purification of the target. When using highly selective biosensors, it is often the case that the total number of steps may be reduced.

In Figure 21.2, the flowchart outlines a sequential series of steps required for the extraction and purification of various biological targets of interest that are contained within cells. The methods listed are not intended to represent an exhaustive summary, but rather are chosen on the basis that they are better suited for use with biosensors, and contemplate the possibility of applications focused for on-field detection and point-of-care devices.

It is also important to note that many of the methods provided are conventional techniques used for sample pretreatment using a wide array of detection strategies besides biosensors. Therefore, these methods are not exclusively used with biosensor technology. However, examples for the use of some of the pretreatment protocols outlined in this section for use with biosensors can be found in Table 21.1.

Extraction and purification methods can be distinguished on the basis of the types of targets. For example, while the extraction of whole cells may only require the selective isolation of cells from the background matrix, the purification of DNA requires a more extensive protocol involving lysis, purification, and denaturation following selective isolation of cells.[69,76,85]

The methods outlined in Figure 21.2 will be briefly reviewed, and examples will be provided where appropriate. Many of the pretreatment protocols operate on the basis of capturing the target of interest, either through affinity interactions or by electrostatic or hydrophobic interactions, while all other materials are washed away and then the target is eluted. By eluting the target into a smaller volume than the original, enrichment of the sample is achieved.

21.2.1. Methods for Purification of Biological Targets from Real Samples for Detection Using Biosensors

21.2.1.1. Homogenization. Since most analytical techniques are solution based, heterogeneous or solid samples such as cellular tissues or food must be liquefied prior to analysis. The most convenient and fastest method is by

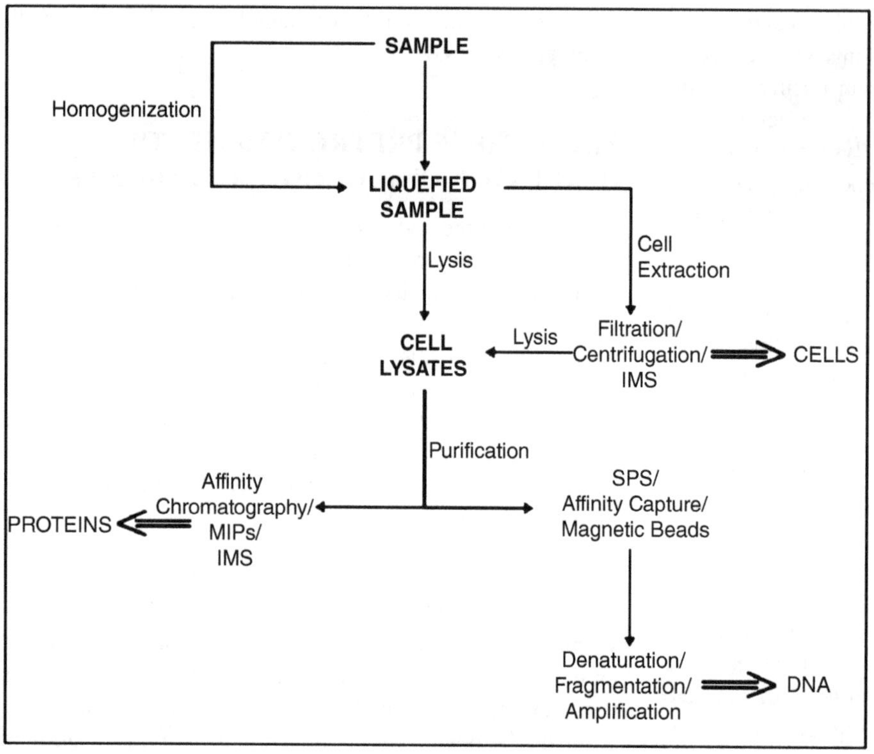

Figure 21.2. Flow diagram of pretreatment steps required for extraction and purification of various biological targets from environmental, consumable, and clinical samples.

adding a volume of buffer (usually several hundreds of milliliters) to the sample and grinding with a stomacher or blender.[52,54,86–88]

Blenders operate on a combination of washing, cutting, and cavitation motions directly on the sample, while stomachers work through a combination of washing and hammering actions where the action is applied externally to a sterile polyethylene bag containing the sample. This results in less labor to clean the equipment between runs with the use of stomachers. However, for certain types of sample, such as chicken skin, stomaching may not necessarily generate a homogeneous slurry as with blending.[87]

A study examining the efficiency between the two methods showed similar recovery of bacteria from various food samples, except for those containing more than 20% fat, where samples subjected to stomaching showed reduced recovery of bacteria. This was alleviated by the addition of a surfactant (1% Tween 80).[87]

Homogenization may also be accomplished through chemical means such as acid digestion or enzymatic digest. Although this chemically breaks down all of the solid material in the sample, it can also take a long time, especially when using an enzyme, and is incompatible with recovery of whole cells.[88]

21.2.1.2. Direct versus indirect extraction. Beyond the purification of cells, two approaches exist for the extraction of intracellular biological targets: direct or indirect extraction.[89]

In direct extraction, the microbial cells are lysed while still in the original sample, releasing their intracellular contents into the sample matrix.[89,90] The target analyte is then isolated and purified.[89] Methods that use direct extraction are generally faster, but problems arise when the target analyte becomes adsorbed onto any particulate matter that may be present in the sample matrix, reducing the recovery yield.[89] Many commercially available DNA extraction kits feature a direct cell lysis approach.[71]

21.2.1.3. Indirect extraction. For indirect extraction protocols, the microorganisms of interest are first separated from the matrix before being lysed to release the intracellular contents.[89–91] This requires additional steps to extract the cells from the matrix, which are usually more time-consuming. However, one must ensure that all of the target cells are removed from the sample matrix in order for the approach to be quantitative.[69,89]

The distinction between direct and indirect extraction methods is important when dealing with the extraction and purification of DNA. Differences in the purity and yield of the extracted DNA have been reported between the two methods. In general, indirect extraction has been shown to produce target DNA of higher purity than direct extraction, but better yields are observed using direct methods.[69,75,92] Typically, DNA recovery yields with cell fractionation are 40–60% versus 90–99% for direct lysis.[90]

There is little information in the literature as to whether direct or indirect extraction methods impact the purity or yield when the goal is to extract proteins. Presumably, loss of proteins by adsorption onto particulate matter will also occur in a manner similar to that observed for loss of DNA. As well, direct lysis will release intracellular proteins into a complex matrix that may already contain a wide array of other proteins. This may result in problems for purification downstream. Furthermore, if the target protein was also present in the sample matrix, it would be extremely difficult to distinguish between the protein from the cell and the sample matrix, creating false positive results.[48,93]

21.2.1.4. Filtration. Isolation of the cells following homogenization can be done by filtration. In filtration, the filter pore size is selected to be smaller than the pathogenic cells of interest so that they are entrapped. Typically, filters with a pore size of 0.2 μm are used. Microorganisms of similar size may also be captured along with the target pathogens on the filter. The captured cells can then be washed off the filter and used for biosensor detection or processed further by indirect extraction.

As can be imagined, any debris or particulate matter larger than the pore size of the filter will also be captured. This can be alleviated by the addition of a prefiltration step using a larger pore size filter, which is typically a 5-μm-pore filter. Prefiltration will remove the majority of the larger debris or particulate matter from the sample solution.[56,94–96] However, any target pathogens trapped within the particulate matter during prefiltration will be removed from the filtrate, causing an underestimate of the total pathogen count.[95,97,98] Additionally, any cells adsorbed onto the surface of the filter will also cause a negative deviation in the results.[56]

Smaller molecules such as proteins, nucleic acids, and small viruses can also be captured and concentrated by ultrafiltration, where filters with pore sizes from 1000 to 1,000,000 molecular weight cutoff (MWCO) are used.[99–103] Ultrafiltration has been demonstrated for the separation of lysozyme from chicken egg white using polysulfone membranes with MWCO of 50 kDa. Lysozyme has a molecular weight of 14 kDa and is approximately 3.4% of the total protein content in chicken eggs, while the other major proteins, ovalbumin and conalbumin, have molecular weights of 45 and 80 kDa, respectively. The lysozyme is allowed to pass though the filter membrane and collected in the permeate, while larger proteins are kept in the feed solution. However, other proteins with similar molecular weights will also pass through the membrane along with lysozyme. It was shown that optimal purification of the lysozyme can be obtained by a combination of surface passivation with myoglobin as well as adjusting the transmembrane pressure (TMP). An increase in TMP was shown to increase the purity of lysozyme. This was thought to be due to the increased pressure causing adsorption and fouling against the membrane, reducing the transmission of the larger proteins across the membrane.[100]

For dead-end filtration techniques such as gravity filtration or ultrafiltration, clogging of the filters by particulate matter can become a major concern as this reduces the efficiency of the filtration process. Rather than applying the filtration feed solution orthogonally on the filter, tangential flow filtration (TFF) can also be used as a method to reduce problems associated with clogging of the filters.[104] TFF prevents the accumulation of particles on the filter by moving the feedstock tangentially across the surface of the filter.[83,99,104] As the feed flows across the filter, particles that are smaller than the pore size passes through the membrane, while the remaining particles in the flow help to clean the surface, preventing the buildup of a "filter cake."[99] Physical damage of the cells caused by shearing effects is minimized, and high recovery yields are usually obtained.[83]

Since the flow is applied tangentially to the filter, this means that materials in the solution may not be given enough time to pass through the membrane pores. This situation necessitates cycling of the feed solution to allow for multiple passes through the filter and results in an increase of the time required for filtration. Increasing the size of the membranes increases the available surface area for interaction with the feed solution, but will also increase the number of sites available for nonselective adsorption, leading to loss of analyte.[83]

TFF has been used to recover biomolecules such as proteins, viruses, cells, and other particulates from large volumes of dilute water samples and has the capability of processing large volumes while being effective with low analyte concentrations.[83,105] TFF has been used for the concentration and purification of human influenza A virus. Particulate matter, such as host cell fragments and cell debris, as well as proteins and DNA from the host cells were successfully removed. To improve the system further, it was suggested that a stepwise TFF system could be implemented, where the particulate matter is first removed using a large pore size membrane (0.45 μm) followed by a 300-kDa membrane for the concentration of virus particles and removal of proteins and DNA.[103] TFF and ultrafiltration have also been demonstrated for purification and preconcentration of DNA samples. Plasmid DNA flowing through a 300-kDa MWCO membrane was retained, while RNA was removed due to size differences between the two materials.[101,102]

21.2.1.5. Centrifugation. Low-speed centrifugation, usually at less than 1000 × g is commonly used to sediment any large particulate matter while leaving bacterial cells suspended in solution. Application of high-speed centrifugation will cause the bacterial cells to sediment out of suspension. Through serial centrifugation, successively smaller-sized particles of debris are systematically removed from the sample at each step of the centrifugation.[97,106]

For indirect cell lysis, centrifugation can be used to remove solid particulate matter, while cells of interest are retained and lysed.[89] The most common method for this is by density gradient centrifugation.[69,104]

In density gradient centrifugation, the centrifugation cell is filled with a solution where the density increases going to the bottom of the cell. Common materials used to generate the density gradient include cesium chloride, sucrose, and Ficoll. Centrifugation forces the cells and particulate matter to migrate to a region where the density of the solution is at equilibrium with the density of the cell or particulate, resulting in the formation of bands. These bands can be physically separated and recovered for further processing or analysis.[97] Cells may become denatured if the osmotic strength of the density gradient used in the separation is very different than the osmotic pressure of the cells. This can impact detection methods that require living cells.[97]

Aside from cells, intracellular components such as proteins, polysaccharides, chromosomal and plasmid DNA, and RNA can also be separated by density gradient centrifugation. However, this method can be labor intensive and time-consuming, taking upwards of 48 h at centrifugation speeds of 150,000 × g and often requires further purification of the target.[69]

21.2.1.6. Immunomagnetic separation (IMS). IMS is a selective enrichment technique in which the pathogens are selectively targeted and captured while any debris or other microbial cells are removed.[107] This is in contrast to filtration or centrifugation where microbial cells with similar properties, such as size, are isolated concurrently with the target pathogen.[106]

In IMS, monoclonal antibodies selective toward the target pathogen are attached onto the surface of paramagnetic beads. Magnetic particles are mostly iron-oxide-based nanoparticles including maghemite (gamma-Fe_2O_3) and magnetite (Fe_3O_4). They can range from 10 nm to 10 μm in diameter.[108,109] When added to a sample, the target pathogens will be selectively captured by the antibodies on the magnetic beads. The beads are then isolated from the rest of the sample by the application of an external magnetic field, and are subsequently released into a different container.[97,106,107] The isolated beads with the captured pathogens can then be eluted into a smaller volume, increasing the concentration of the pathogens.[110]

The efficiency of IMS depends on the specific monoclonal antibody, the size and surface area of the magnetic particles, the recovery procedure, interference in the sample matrix, and nonselective binding blocking capture sites.[111] As well, due to the cost of monoclonal antibodies, IMS is usually performed on small volumes of samples.[49,97] The pathogen–antibody interaction utilized in IMS may also be disrupted by the presence of interfering components or debris in the sample.[112]

IMS has been demonstrated to capture cells such as *Salmonella* directly from spiked and real food samples such as milk powder and dried eggs as well as from stool samples.[110,113] IMS has also been used for the selective capture of *Campylobacter jejuni* from milk and chicken skin washes. Following the selective capture of the cells, detection was done by PCR amplification of the DNA following cell lysis.[114] This method was able to detect one *C. jejuni* cell present in 1 mL of sample, which also contained 10^6 cells of *E. coli*, *Salmonella serovar Enteritidis*, or *Vibrio parahaemolyticus*. The total procedure took 8 h to complete.[114,115]

21.2.2. Cell Lysis

For the detection of intracellular targets such as DNA and proteins, the cell must first be broken in order to release its components into solution.[75,116] Cell lysis protocols act on either plasma membranes or the cell walls present in plant cells. As a result, different lysis methods are available depending on the type of cells that are to be lysed.[116] Cell lysis can be performed through chemical or mechanical means or a combination of both.[89]

21.2.2.1. Chemical lysis. Chemical lysis often uses harsh chemicals in order to disrupt the cell wall and/or membrane. Agents such as detergents (sodium dodecyl sulfate [SDS]), high salt concentration solutions, or enzymes such as lysozyme have been used. Lysozyme hydrolyzes the polysaccharide component of a bacterial cell wall. The use of enzymes generally requires long incubation times.[69,89]

The use of chemical agents for lysis requires the use of consumable products and may also interfere with detection if not removed.[117] The chemical agents may also show selectively toward the lysis of a particular species of cell. The agents may also not be able to completely penetrate soil or sediment where the cells may be adsorbed when this method is used for direct extraction.[69,89]

21.2.2.2. Mechanical lysis. Mechanical methods for cell lysis include freeze/thawing, ultrasonication, and bead beating, as well as heating.[53,69,88,117] Mechanical treatments are generally more effective but less selective when compared with chemical lysis methods. Mechanical methods are also more effective at dislodging cells that may have been adsorbed onto solid particulate matter.[69]

Repeated cycles of freezing and thawing rupture cells due to the repeated formation of sharp ice crystals inside the cell.[99] This type of lysis is usually done by submerging the sample into liquid nitrogen or ice followed by heating in a water bath.[69] Such cycles of repeated freezing and thawing can become time-consuming.[99]

Application of ultrasonication energy to disrupt cells in a sample can be done with the use of an ultrasonic horn; beads may also be added during ultrasonication for cell lysis.[117] Ultrasonication creates cavitation in the solution, causing cell walls to disintegrate and releasing their cellular contents into the solution.[99] Ultrasonication can also help to release bacterial species adsorbed onto solid matter or to break up aggregates of cells.[69,76] Heating as a result of ultrasonication may denature proteins.[99,117]

Bead beating involves the addition of beads to the sample followed by vigorous mixing to disrupt cells. Bead beating can be done using glass beads, ceramic beads, or garnet powder.[69,118] Bead beating is especially effective for bacteria with thick cell walls and may also result in lower amounts of phenolic compounds, such as tannin and humic acid, being coextracted.[55] However, bead beating has resulted in shorter DNA fragments being generated.[69,89] Noteworthy is that small, round cells have been known to survive the bead-beating procedure.[58]

Heating of a sample can also be used to disrupt cell walls and membranes, but will denature proteins[117] and may release phenolic compounds from the sample, which can inhibit PCR if analyzing for nucleic acid.[89] Alternatively, microwave heating has been shown to be effective for lysing gram-positive bacteria, but extreme heating using a microwave may damage nucleic acids.[69]

21.2.3. Chromatography

21.2.3.1. DNA. Chromatographic separation can be used for the purification of biological targets. Separation of DNA can be done by size exclusion, ion exchange, or affinity chromatography.[69,119] Size exclusion separates molecules based on molecular weights. Molecules larger than the largest pores of the gels are eluted first since they cannot enter the gel. Gel filtration is a common method for size exclusion that can purify DNA with minimal loss of the target.[69,120]

Ion exchange columns can selectively bind or elute DNA based on the pH and ionic strength of the buffers used. Anion exchange has been the most prominent technique. Positively charged diethylaminoethyl (DEAE) tertiary or quaternary amine anion exchange resins will bind to negatively charged DNA.[119] Elution of the DNA from the resin can be done by adjusting the pH of the mobile phase.[121] One of the problems with the use of resins is that large molecules such as DNA cannot penetrate the pores of the resin, limiting the binding capacity of DNA.[119] The loss of DNA on ion exchange resin columns has been measured at 20–30%.[69]

Alternatively, polymeric monoliths can be used. Monoliths are single pieces of porous material where the pores are interconnected, forming channels with diameters ranging from 13 to 4000 nm and 60–80% porosity within the monolith material. The increased pore sizes increases the surface area available for binding with DNA. Monoliths tend to offer good mass transfer between mobile and stationary phases within channels and low back pressures at high flow rates.[119,122,123] Methacrylate-based monoliths have been demonstrated for the separation and purification of varying sizes of DNA molecules. For example, DNA binding was calculated to be 9 mg/mL of column for all types of DNA, as compared with <200 µg/mL on anionic Sepharose resins. Recovery of the DNA was achieved by adjusting pH and ionic strength of the elution buffer. Under optimal conditions, recoveries of up to 80% were obtained within 10 min.[119]

A DEAE monolithic anion exchange column has been used to purify genomic DNA.[123] Samples of bacterial and eukaryotic genomic DNA purified by a commercially available kit were tested on the monolithic column. Results showed that the genomic DNA could be bound onto the column. Elution of the DNA was performed by increasing the concentration of salt in the elution buffer. The 3 × 12 mm monolithic column could also be used to process large volumes of DNA samples. Binding of ~4 mg of DNA from a 30-mL sample was demonstrated with no DNA being detected in the wash fractions. It was also shown that RNA present in the purified samples could be removed by the column.[123]

Oligonucleotide probes selective toward a particular target can be incorporated into chromatography support materials for the selective capture and purification of DNA from a sample. Following the selective capture of the DNA targets, the column can be washed, and the captured DNA can be eluted as a relatively pure sample.[121,124]

Various research groups, including Mathies, Olsen, and ours, have utilized a commercially available oligonucleotide modification sold under the name of Acrydite™. Acrydite is a modification that adds an acrylamide monomer unit to the 5′ end of an oligonucleotide sequence. When mixed with a solution of acrylamide monomer and polymerized, the resulting polyacrylamide gel has the oligonucleotide probes incorporated into the gel matrix.[125–127] DNA targets are then driven through the modified gels electrophoretically.[67] Any debris or unwanted salts can be washed away, and the complementary targets are subsequently released by denaturation.[125–127]

Whitney et al. manufactured probe-modified polyacrylamide gels, which were cut into 1-cm-diameter disks. The affinity gel disks were used to capture DNA from a sample volume of 2.4 mL by electrophoresis, and the captured target was eluted into a volume of 50 µL by the addition of NaOH. A fivefold increase in the quantity of DNA was recovered from stool samples when compared with the use of immobilized probes on magnetic beads.[67]

Polyacrylamide gels modified with capture probe have also been used inside microfluidic channels by Olsen and Mathies.[126,127] After the running conditions were optimized, this method was capable of desalting and concentrating a sample of PCR amplification product in 120 s with a 100-fold increase in concentration based on volume reduction.[126]

Our group has reported the quantitative determination of the efficiency of capture by probe-modified gels loaded inside fused-silica capillaries (100 µm ID × 7.5 cm length). By measuring the fluorescence intensity of fluorescently labeled oligonucleotide targets, the capture efficiency was determined to be approximately 90% of the amount of probe molecules incorporated into the gel. Elution by heating and chemical denaturation released approximately 95% of the captured target. A constant loss of 8 fmol of the target DNA by adsorption onto the gels was observed after elution and was independent of probe and target concentration. This loss of DNA was still observed even after efforts to block active sites by preconditioning the capture gels with a solution of noncomplementary oligonucleotides.[125]

Besides using oligonucleotide probes as affinity ligands, peptide nucleic acids (PNAs) have also been demonstrated for affinity capture of DNA targets. PNAs are synthetic mimics of oligonucleotides, which are stable against the action of peptidases and nucleases.[128] Duplexes of PNA–DNA and PNA–RNA show higher thermal stability and faster hybridization kinetics versus their DNA–DNA and DNA–RNA counterparts.[129,130] The improvement in hybridization kinetics is believed to be due to the neutrality of the PNA backbone, eliminating the electrostatic repulsion observed for the hybridization of two DNA strands.[129] Unfortunately, PNAs are not highly soluble in water and have a tendency to self-aggregate and sediment in low salt concentrations.[128]

PNA immobilized onto beads has also been reported to offer enhanced recovery and detection of rRNA. rRNA was captured and detected on PNA-coated Lumavidin beads from 0.1 ng of RNA within 15 min. By comparison, the use of DNA-immobilized beads required an overnight hybridization before a detectable signal was obtained from the same amount of RNA.[129]

21.2.3.2. Proteins. Purification of proteins can be done by reversed-phase, ion exchange, size exclusion, and affinity chromatography.[116] Mills et al. demonstrated a one-step purification of TM1-99 from cell lysate by reversed-phase high-performance liquid chromatography (RP-HPLC). TM1-99 is a recombinant protein that can be expressed in *E. coli*. Intracellular contents were first extracted in 0.1% aqueous trifluoroacetic acid, followed by separation on a narrow bore C8 column that was eluted with a shallow acetonitrile : water gradient. The process resulted in 94% pure protein from cell lysate with 94% recovery efficiency. Purification of trace amounts (<0.1%) of the recombinant protein in the cell lysate was also demonstrated using this method.[70] Reversed-phase chromatography has been used for the purification of bovine serum albumin (BSA), ribonuclease A, and human tumor necrosis factor.[131]

Prostate secretory protein 94 (PSP94) from human seminal plasma has been purified in a three-step procedure by ammonium sulfate precipitation, cation flow through, and anion exchange chromatography.[132] The solubility of proteins decreases with increasing ionic strength. Ammonium sulfate was first used to salt out proteins as well as keep them in a stable state.[116] Cation flow through provided a quick method for the removal of any impurities. The pH of the elution buffer was adjusted so that impurities were adsorbed onto the cation column while PSP94 flowed through. The final anion exchange column was used for the adsorption of the proteins while impurities were washed away using a high-ionic-strength buffer. Washing with the high-ionic-strength buffer resulted in the loss of some of the target protein. The yield of the PSP94 protein following elution was approximately 53% with a 17-fold enrichment.[132]

Ion exchangers have been demonstrated to purify milk proteins and BSA, as well as plasma proteins.[131] Cytotoxin-like proteins from the venom of *Ophiophagus hannah* were isolated by a combination of ion exchange chromatography and RP-HPLC.[133]

Affinity ligands for the purification of proteins can include antibodies, oligonucleotide probes, metal ions, and synthetic textile dyes.[131] Antibodies are useful for affinity-based purification since they can selectively recognize an antigen from heterogeneous solutions such as cell lysates, urine, or serum.[44] However, they may be unstable under harsh elution conditions of extreme pH, detergents, organic solvents, or chaotropic salts. Such conditions often lead to denaturation of the capture molecules, limiting this purification method to single use only.[134]

Immunoaffinity ligands have been demonstrated to bind proteins such as BSA, lipoproteins, and human serum amyloid P.[131] Polyol-responsive monoclonal antibodies (PR-MAb) have been demonstrated to elute a captured antigen under gentle conditions, such that even multi-subunit proteins can be released in their active form. Following capture, the antigens can be eluted by using buffers containing polyol-(ethylene glycol, 1,2-propanediol, or 2,4-butanediol) and salt.[135] The purification of RNA polymerase from *E. coli*, *Bacillus subtilis*, *Pseudomonas aeruginosa*, and *Streptomyces coelicolor* was demonstrated by 8RB13 antibodies conjugated with Sepharose and packed inside a column.[135]

DNA molecules can also be used for the purification of DNA-binding proteins. Depending on the protein that is to be purified, different types of DNA can be used for affinity separation. Proteins such as histones and DNA polymerases are nonsequence selective and will bind to nonspecific DNA such as calf thymus DNA and salmon or herring sperm DNA. Certain proteins, such as DNA/RNA polymerase have higher affinity to single-stranded DNA (ssDNA) than double-stranded DNA (dsDNA). Specific DNA columns have also been used for proteins that have high affinity for a specific dsDNA sequence such as different transcription factors and restriction enzymes. Such a specific sequence is the footprint region for transcription factors, and the restriction site in the case of restriction endonucleases.[136]

Immobilized metal ion affinity chromatography can also be applied for isolation of phosphopeptides. Many cellular functions are regulated by phosphorylation and dephosphorylation on serine, threonine, and tyrosine residues. Phosphorylation has an influence on various cell functions such as signal transduction, metabolic maintenance, and cell division. However, only 1–2% of total protein is phosphorylated, and phosphorylation may occur at different sites and some proteins are transiently phosphorylated.[137,138] These issues can make isolation of phosphorylated proteins difficult.

The use of metal ions such as Cu^{2+} and Fe^{3+} has been used to retain histidine-containing peptides as well as phosphorylated proteins by interactions between the phosphate group and the nonbonding ion-pair electrons of the metal ions. The metal ions are immobilized by chelation with multidentate ligands, which have been coupled onto a support material. Common solid supports include silica, agarose, Sepharose, and cross-linked poly(styrene-divinylbenzene).[122,137,138]

Dye-ligand chromatography has also been used in protein purification. These dyes usually consist of a chromophore (azo, anthraquinon, or phathalocyananine) linked to a reactive group such as a mono- or dichloro-triazine ring. Interaction between the dye and protein is based on either electrostatic, hydrophobic, or hydrogen bonding. Cibacron Blue F3GA is an anthraquinone textile dye that interacts specifically with albumin. Başar et al. used magnetic beads modified with Cibacron Blue F3GA dye and demonstrated the purification of lysozyme from egg white. The purity of the lysozyme eluted from the dye-affinity beads as determined by sodium dodecyl sulphate–polyacrylamide gel electrophoresis (SDS-PAGE) was 87.4% with about 80% recovery.[139]

Magnetic beads modified by Reactive Red 120 have also been used for the selective capture of β-casein.[140] Ninety-five percent of the captured proteins were eluted following the addition of potassium thiocyanate. Some nonselective adsorption onto the beads was observed and was attributed to diffusion of the protein into the swollen polymeric matrix and interactions with the hydroxyl groups on the surface of the beads.[140]

21.2.4. Solid-Phase Extraction (SPE)

21.2.4.1. DNA. SPE is a commonly used technique for the purification of DNA from a complex matrix following cell lysis.[85,141] DNA will bind to silica or glass particles by hydrogen bonding or electrostatic interactions when in the presence of a chaotropic agents such as NaI, $NaClO_4$, guanidium hydrochloride, and GuSCN.[63,142–144] The captured DNA can then be eluted by the application of a low-salt buffer.[142]

The use of SPE eliminates the need for organic solvents associated with conventional extraction and purification techniques such as phenol–chloroform extraction.[141] Many commercially available kits for extraction and purification of DNA also use an SPE step.[85,141] SPE has been used for the purification of DNA from a wide variety of different types of samples including plants, fungi, and microorganisms, as well as human hair, teeth, bone, and blood.[141] Enrichment of the sample can be accomplished by eluting the DNA into a smaller volume than the original sample.[142,145]

Common SPE materials include hydrophobic surfaces, such as alkyl-bonded silicas (C18, C20), and copolymers, such as poly(styrene-divinylbenzene).[2,146] Normal-phase materials such as those containing diol, aminopropyl, or cyanopropyl functional groups have also been used.[146]

Proteins that are present in a biological sample can also bind to the SPE material, limiting the binding capacity of DNA onto the columns.[147] As well, since the capture is nonselective, any compounds with similar properties may also be captured by the resin along with the DNA, decreasing the purity of the extracted material.[121]

Landers' group demonstrated an SPE purification system for DNA by using monolithic extraction columns made from 3-(trimethoxysilyl)propyl methacrylate (TMSPM) modified with 85% v/v tetramethyl orthosilicate (TMOS). The addition of TMOS has been shown to increase DNA binding by the monolith. Elution of the captured targets into a final volume of 1 μL was done by using a low-ionic-strength Tris buffer at pH 8. Solid-phase capture of PCR-amplified 380 bp human genomic beta-globin fragments as well as human genomic DNA purified from blood was demonstrated. Both samples still required the addition of guanidine hydrochloride prior to binding. Extraction efficiency as quantified by a PicoGreen assay was 86% for the PCR-amplified genomic DNA and 60% for DNA from whole blood. These efficiencies were greater than DNA purified using a commercially available SPE kit.[148]

21.2.4.2. Proteins. Proteins have also been purified by SPE through the use of affinity ligands immobilized onto the solid support. Examples include immobilized antibodies and molecularly imprinted polymers (MIPs).[2] SPE cartridges filled with antibodies can be used to selectively bind a specific target.[149] MIPs have also been developed for the extraction of small organic molecules such as drugs,[150–153] antibodies from water samples,[154] and amino acids and proteins.[150]

RNA aptamers immobilized onto beads in a microchip device have been shown to provide for the selective capture of hepatitis C viral (HCV) RNA polymerase protein. The RNA aptamers were immobilized onto the beads via a photocleavable linker. The captured polymerase proteins were eluted by irradiation of the beads with UV radiation, which cleaved the linkage between the aptamer and the microbeads, releasing the polymerase–RNA aptamer complex. The released polymerase did not require further purification, and a detection limit of 9.6 fmol of HCV RNA polymerase was achieved based on detection by ELISA.[134]

21.2.4.3. Magnetic particle technology. The extraction of cells by IMS can also be extended toward the selective extraction and purification of other biological materials such as DNA and proteins.[155]

The use of magnetic beads can reduce processing times for sample pretreatment, can eliminate the need for equipment such as centrifuges, can eliminate organic solvents,[156] and has been used to collect target DNA from the blood, bone marrow, cultured cells, and other crude samples for processes such as further target amplification or direct detection.[108,156] The captured targets on the magnetic beads can be eluted into a smaller volume, allowing for preconcentration and purification of the samples prior to detection.[157]

As can be expected, the protocol for the purification of DNA and proteins is similar to the IMS of bacterial cells. The surface of the magnetic beads can be modified with silica to allow for capture by adsorption, or can be coupled with affinity ligands for the selective capture of targets. Surface modifications for the selective capture of DNA are identical to those used for SPE as well as in affinity chromatography.[121,144]

Capture of DNA targets onto magnetic beads by adsorption occurs in a similar fashion to SPE and requires modification of the beads. This could involve the use of a silica surface,[158] polymer surfaces,[159] polyvinyl alcohol (PVA), or hyperbranched polyamidoamine dendrimers.[85] The latter two provide an additional electrostatic component for the capture of charged material.[85] Carboxy-coated magnetic beads have also been used for bioaffinity adsorption of DNA.[160]

Anion exchange resins which contain a paramagnetic core are commercially available (Whatman DEAE-Magarose). These beads have been used to isolate plasmid DNA from bacterial cell lysate and genomic DNA from bacterial cells or blood samples by means of electrostatic attraction.[121] Magnetic beads can be functionalized with a dense coating of amino groups on the particle surface by an aminosilane reagent, 3-[2-2-aminoethyl(amino)-ethylamino-propyetrimethoxysilane]. DNA can then be captured by electrostatic interaction between the positively charged amino-modified beads and the negatively charged DNA.[108]

Magnetic beads can also be modified with oligonucleotide probes to capture a specific DNA target.[85] The captured DNA on the magnetic beads can also be used for direct detection without eluting the DNA.[157,161,162] Fuentes et al. demonstrated the detection of 10^{-18} g/mL (two molecules) of HCV cDNA from a 1-mL sample by PCR amplification of DNA targets labeled with dUTP-digoxigenin that were selectively captured on magnetic beads. This was followed by incubation of a solution containing anti-digoxigenin Fab fragments conjugated with horseradish peroxidase. Detection was based on measuring the absorbance of a green-colored product from the enzymatic reaction of 2,2′-azido-di(3-ethylbenzothiazoline)-6-sulfonic-acid. The detection was not impaired by the addition to the sample solution of 2.5 million fold excess of noncomplementary DNA. Nonselective adsorption of DNA was decreased by blocking active sites with aldehyde–aspartic–dextran. The use of the negatively charged polymer reduced the positive charge of the support surface and also allowed for a strong secondary amine bond to be formed between the oligonucleotide probe and the polymer-coated magnetic beads. However, some nonspecific adsorption was still observed.[157]

Even though purification using magnetic beads may offer a much simpler and faster extraction method than conventional phenol–chloroform extraction, the yield and purity of the extracted DNA may not be as high as conventional methods.[156] A recent study by Faggi et al. comparing purification of DNA from yeast using magnetic beads and conventional phenol–chloroform extraction found that phenol–chloroform resulted in higher yields of extracted DNA with better purity than magnetic beads. The amount of DNA extracted from three different types of fungi by phenol–chloroform was 6684, 2022, and 945 μg/mL. Purities as indicated by the ratio of optical densities of UV absorbance at 260 and 280 nm, were reported to be 1.8, 1.4, and 1.4, respectively. By comparison, the amounts of DNA extracted from the same three types fungi by magnetic beads were 443, 223, and 329 μg/mL with purities of 1.3, 1.2, and 1.4, respectively. Additionally, the cost per test using magnetic beads was higher than phenol–chloroform extraction.[156]

Figure 21.3 shows the schematic for an automated sample pretreatment using magnetic beads has been introduced for

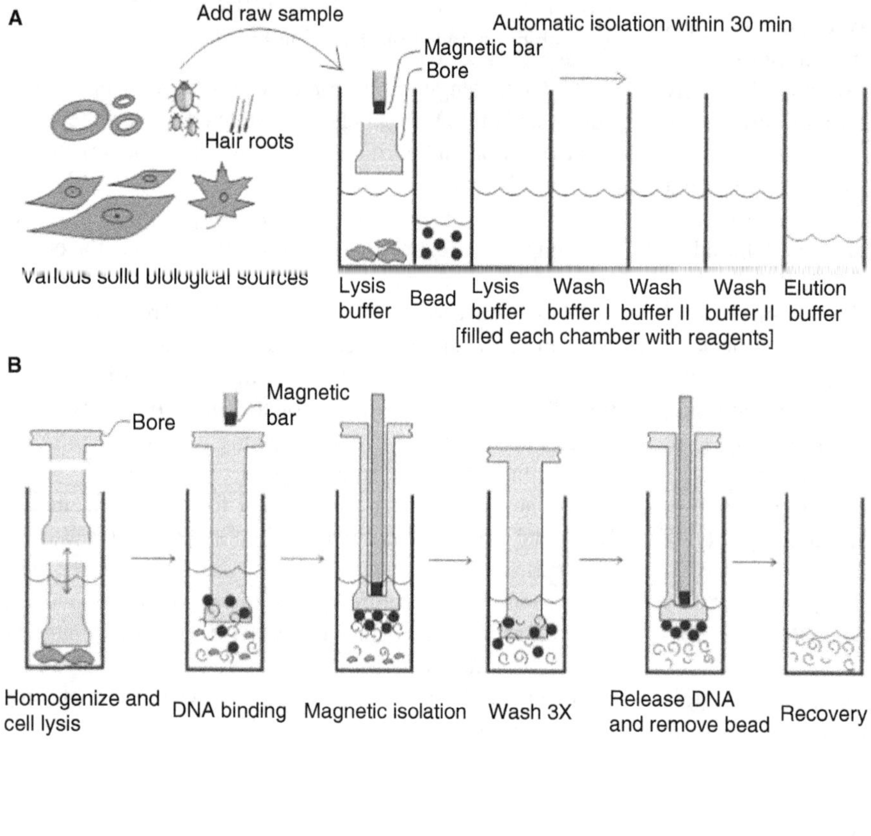

Figure 21.3. Schematic of an automated DNA extraction and purification system using magnetic beads. Reprinted from reference 144 with permission of the Society for Biotechnology, Japan. Copyright 2007.

the extraction of DNA from solid biomaterials developed by Ki et al. Since most extraction methods work on solution-based samples, the automated pretreatment combines the manipulation of magnetic beads with bore-mediated grinding for homogenization of solid samples. This allows for an automated method for mechanical grinding, extraction, and purification of DNA from solid samples. The magnetic bore also collected and moved the magnetic beads with DNA into different chambers for washing and finally elution within 30 min. Extraction and purification of genomic DNA from different materials such as plants, animal tissues, and human forensic samples have been demonstrated.[144]

21.2.5. Denaturation and Fragmentation of DNA

Denaturation and fragmentation of target DNA is a necessary step for detection by a DNA-based biosensor. In nature, DNA exists in a double-stranded form. Therefore, it must be denatured to generate single-stranded DNA targets that can hybridize with the immobilized single-stranded oligonucleotide probes that are associated with the selectivity of a DNA biosensor.

Additionally, the length of the DNA targets will also impact the hybridization kinetics. DNA fragments need to be sufficiently short to allow for fast hybridization with the immobilized probes of the biosensor. Lengths of DNA target commonly used are on the order of 25–300 base pairs. By comparison, the lengths of DNA from cells are orders of magnitude longer. For example, the genomic DNA for most *E. coli* strains is on the order of 4 million base pairs in length.[163] If the fragments are too long, it may form intramolecular structures, which prevent hybridization. Long fragments of DNA may also be hindered sterically from reaching the immobilized probes, preventing hybridization.[163,164]

21.2.5.1. Fragmentation. Various mechanical cell lysis methods have been shown to fragment the DNA to differing degrees.[53,69] Bead beating often results in the most fragmented DNA,[53,89,165] while ultrasonication can be used to shear as well as denature DNA.[89,163]

Fragmentation of DNA by ultrasonication occurs by mechanical damage through heating of the solution and by chemical damage by free radicals attack on the DNA. Our group has previously demonstrated that ultrasonication (85 W power, 20 kHz) can fragment genomic DNA extracted from *E. coli* cells to produce products in the 100–400 bp range within 30 s. Application of ultrasound will also heat the sample solution, denaturing the DNA fragments.[163]

Enzyme catalyzed scission of DNA involves use of restriction enzymes, and this approach has also been used to generate shorter DNA fragments than those obtained from ultrasonication.[12,163]

21.2.5.2. Denaturation. The most common method for denaturation of DNA targets is by heating the DNA solution at 95°C for 5 min followed immediately by cooling in ice.[12,166,167] A modification to thermal denaturation has been to introduce blocking oligonucleotides and involves the addition of blocking oligonucleotides directly to the denaturation mixture. These blocking oligonucleotides are 10–30 base pair length oligonucleotides that are complementary to a region on the DNA target away from the actual target region. Following thermal denaturation at 95°C, the solution is cooled to allow for annealing of the blocking oligonucleotides to the DNA targets. This prevents the annealing between the denatured DNA targets while still leaving the target region available for binding to the oligonucleotide probes on the biosensor.[12,166,167]

An SPR-based DNA biosensor was used to examine samples of bovine DNA processed by thermal denaturation with blocking oligonucleotides, and showed a 43% increase in signal in comparison to the same samples that were only thermally denatured.[12] Similar results were also observed for samples of apolipoprotein E gene in humans and the 35S promoter sequence, which is present in most genetically modified organisms.[12]

21.2.6. DNA Amplification by PCR

PCR allows for a small number of copies of DNA to be amplified exponentially. This step is often used for the amplification of DNA targets for applications in trace analysis. The methodology requires the target DNA to serve as a template that will be copied. DNA primers are used to flank the sequence to mark the initiation and termination locations for copying. Nucleotide triphosphates are used to build the new DNA copies and are stitched together by DNA polymerase. Heat-stable DNA polymerase enzyme is used, and amplification occurs by first annealing the primers to the target DNA, followed by an increase in temperature to 72°C for polymerase activity and 95°C to denature the newly formed dsDNA. The temperature of the reaction vial is lowered again to anneal the primers to the target DNA and the cycle is repeated and the quantity of DNA is amplified exponentially.[168] The length of the PCR amplicons can be controlled by designing the DNA primers to bind to different areas of the DNA target. However, this requires knowledge of the base sequence of the DNA target beforehand. Labeled-DNA targets, such as incorporation of fluorescent dyes, can also be generated by adding the specific modification to the nucleotides prior to PCR.

Theoretically, more than 10^9 copies of target DNA can be produced after 25–30 cycles.[169] On average, a typical 30- to 40-cycle PCR amplification can take upwards of 2–3 h to complete, but is wholly dependent on the reaction times for each step.[15,20,24] Amplification yields obtained are rarely truly exponential and will plateau over the course of the PCR amplification.[170] This is thought to be the result of the consumption of reagent at higher cycle numbers, poor template to primer hybridization, contamination in the sample, inefficient thermal cycling, and poor temperature control.[168] Presumably, the predominant factor in amplification efficiency during late cycles is the consumption of reagents. However, no significant improvement in the amplification efficiencies was observed after these reagents were replenished during late PCR cycles.[170]

Efficient amplification will also depend on rapid heat cycling to allow for quick temperature transitions. Temperature cycling in PCR is most commonly controlled using a block heater. For a typical reaction volume of 50 µL, there is a delay to equilibrate the temperature inside the PCR mixture following a temperature change in the block heater. This delay can result in temperature gradients within the reaction tube; 10°C temperature differentials between the top and bottom of the reaction tube have been observed. This temperature differential can result in loss of specificity, nonideal strand extension, and reduced annealing efficiencies.[168] More rapid and efficient heat cycling can be achieved by reducing the volume of the PCR mixture to less than 1 µL.[168]

Another major contributor to amplification efficiency is the purity of the template DNA. Inhibiting compounds can deactivate the thermostable DNA polymerase or degrade the target nucleic acids.[171] For any sample pretreatment protocols that include amplification by PCR, stringent purification protocols must be done in order to remove most, if not all, of the inhibitor compounds.

Short, single-stranded DNA targets can be deliberately designed and generated by asymmetric PCR. In conventional symmetric PCR, equal concentrations of forward and reverse primers are used in the reaction mixture. In asymmetric PCR, a higher concentration of one primer is added over the other.[172] Over the course of the amplification, the amplicons will be predominantly in a single-stranded form, extended from the excess primer.[173]

Use of asymmetric PCR provides an amplification yield that is significantly less than that from conventional PCR. For asymmetric PCR, amplification occurs exponentially until the lower abundance primer is exhausted. After that, the available number of templates remains the same for the remaining cycles, and amplification occurs in a linear fashion. Additionally, asymmetric PCR often requires extensive optimization to determine the ratio between the primers, amounts of starting materials, and number of cycles to provide sufficient amplification for a particular target.[173]

Generally, the forward and reverse primers are selected so that they have similar melt temperatures. However, the concentration of the primers will also affect the melt temperature. This is not an issue in conventional symmetric PCR, where the concentration of both primers is equal. However, in asymmetric PCR, the melt temperature of the limiting primer is usually a few degrees lower than the excess primer. This discrepancy often results in an inefficient asymmetric PCR amplification. Wangh's group introduced a modification to asymmetric PCR called linear-after-the-exponential PCR (LATE-PCR). In LATE-PCR, the length and nucleotide base composition of the two primers are adjusted so that the difference in melt temperatures is zero or the limiting primer has a higher melt temperature than the excess primer.

The amplification efficiency of LATE-PCR over asymmetric PCR was demonstrated for an application involving the cystic fibrosis CDΔ508 allele and the Tay-Sachs disease TSD 1278 allele, where more products were detected by fluorescence from LATE-PCR after 80 cycles.[173]

It is also important to mention briefly that PCR itself has been used as a detection strategy for DNA targets. RT-PCR allows for the measurement of the accumulation of PCR amplicons during each cycle of PCR in real time, eliminating the need for post-PCR detection. Quantitative information on the original number of target DNA template in the sample can also be obtained by RT-PCR.[169]

Intercalating dyes such as ethidium bromide, SYBR Green I, or LC will generate a large fluorescence signal when they intercalate into dsDNA.[174,175] When added to the PCR mixture, the fluorescent dyes intercalate into newly made copies of dsDNA following each PCR cycle, increasing the fluorescence signal.[174] Specific amplicon targets can also be identified using fluorescently labeled DNA probes that are complementary to the target of interest.[171,174]

RT-PCR has been used to detect *Salmonella enterica*. A 1 CFU/mL solution was detected following 8.4 h of enrichment.[175] Cultural enrichment followed by IMS allowed for the detection of *Listeria monocytogenes* and *Salmonella*[176] and *E. coli*[117] from food samples using RT-PCR.

21.3. MICROFLUIDIC DEVICES

Microfluidic devices are designed to manipulate volumes on the order of nanoliters by patterning structures in the micrometer scale. The reduction in volume reduces the need for reagent consumption while increasing assay speed and improving sample throughput.[1,97,177–182] Movement of solution through microfluidic channels is controlled either by application or pressure, or by electroosmotic flow generated within the channels upon the application of an electrical potential.[180]

Different processes such as sample pretreatment, transportation, separation, and detection can also be integrated into a single microfluidic device, providing a micrototal analysis system (µTAS) and can provide a high level of automation.[177,178,183] Fully integrated lab-on-a-chip systems are typically constructed from polydimethylsiloxane (PDMS) where systems of channels on the order of 10–40 µm deep and 60–200 µm wide are patterned into the PDMS material by methods such as injection molding, hot embossing, laser ablation, and soft lithography.[66,67,178] The small scale of microfluidic devices as well as the ability to integrate multiple functions on a single chip makes them good candidates as portable biosensors for on-field analysis or point-of-care systems.[178]

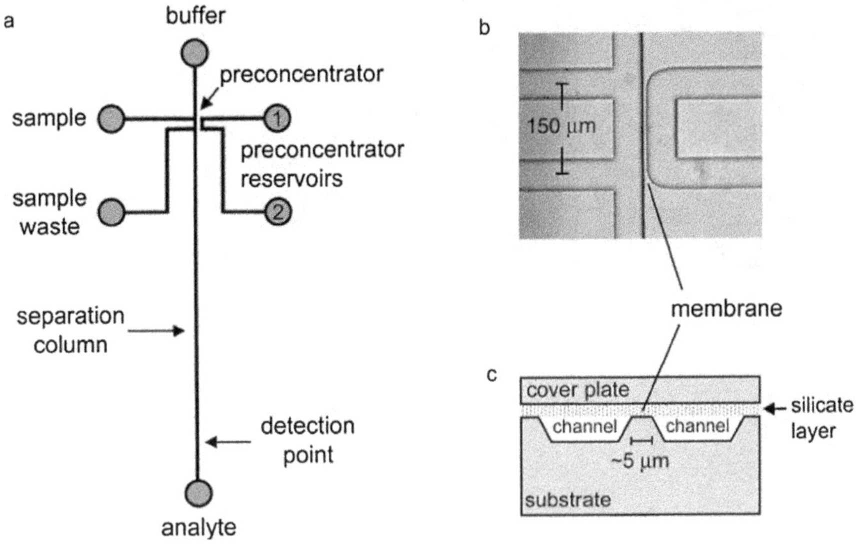

Figure 21.4. (a) Schematic diagram of the microchip layout for preconcentration, (b) image of the preconcentrator channel, and (c) schematic of how the filtration membrane is placed in between the microchip and the cover plate. Reprinted from reference 186 with permission from the American Chemical Society. Copyright 2005.

21.3.1. Methods for Pretreatment and Purification in Microfluidic Devices

For the development of microfluidics-based biosensors, there is a need to develop sample pretreatment and purification methods that can be used within a microfluidic device. Many of the pretreatment and purification technologies have aimed to miniaturize macroscale pretreatment techniques for use inside a microfluidic system. However, some of the conventional techniques such as centrifugation or precipitation, used in direct lysis procedures, are not currently readily adaptable into a microfluidic device.[85,160,183,184] Other techniques such as filtration, magnetic particle-based separation, and SPE have been miniaturized to operate on chips.[179,184] The following is a review of the purification methods as well as preconcentration methods that are available for microfluidic devices.

21.3.1.1. Filtration. Filtration in a microfluidic device can be accomplished either by incorporating a piece of membrane filter material, most commonly by sandwiching it in between two pieces of the microfluidic device, or by construction of microstructures such as micropillar arrays or microweirs.[183,185] Sample filtration and preconcentration in a microchip was demonstrated with porous membranes. Long et al. sandwiched a 10-nm-pore-diameter membrane between two pieces of PDMS, which had microchannels etched inside. Molecules small enough to pass through the nanopores were moved across the membrane by the application of a voltage. This allowed for the collection of larger molecules at the membrane by size exclusion, resulting in the concentration of the sample.[185] The filter was used to preconcentrate a sample of PCR-amplified DNA target. An 80-fold enhancement in the signal was observed after 40 s of filtration. This enhancement factor could be further increased by using longer sampling times as well as an increased applied voltage.[185]

Foote et al. demonstrated a nanoporous silica membrane, which is capable of a 100-fold enrichment of DNA after 5 min. As shown in Figure 21.4c, the silica filter membrane was placed between the silicon-based microchip and a glass cover plate, which trapped the DNA molecules while still allowing current to flow. When coupled with enrichment by field amplified stacking (FAS), a 600-fold preconcentration of fluorescently labeled ovalbumin in the signal was observed.[186]

Kirby's group also demonstrated the use of a nanoporous membrane for the concentration of proteins. The nanoporous membrane (MWCO 5700 Da) was fabricated inside the microchannel by photoinitiated polymerization of a monomer/solvent solution in a region of the channel. Following polymerization, the channel was washed, removing the unreacted monomer solution, leaving the filter behind. The concentration of a sample of fluorescently labeled BSA and phosphorylase b against the nanoporous membrane was observed to increase by four orders of magnitude within 100 s.[190] However, these membranes proved to be difficult to manufacture in a reproducible manner and were shown to contain many defects.[31,187]

Aside from using nanoporous membranes, micro- and nanostructures can also be used for filtration and concentration of targets.[188,189] Irimia et al. demonstrated a microfluidic device that could isolate cells from fluids by the fabrication of microweirs, which are dam-like structures, within the chip. Essentially, when a solution containing the cells is introduced into the chip, the microweirs trap the cells while allowing the remaining solution to flow through.[188]

Microposts set across a fluid channel can be used to trap cells in between the gaps of the microposts (Fig. 21.5A–C). The coarseness of the filter can be adjusted by decreasing the size of the posts. This allows for a series of microposts to be set up, going from coarse filtration to fine filtration, helping to remove large aggregates that can clog the finer

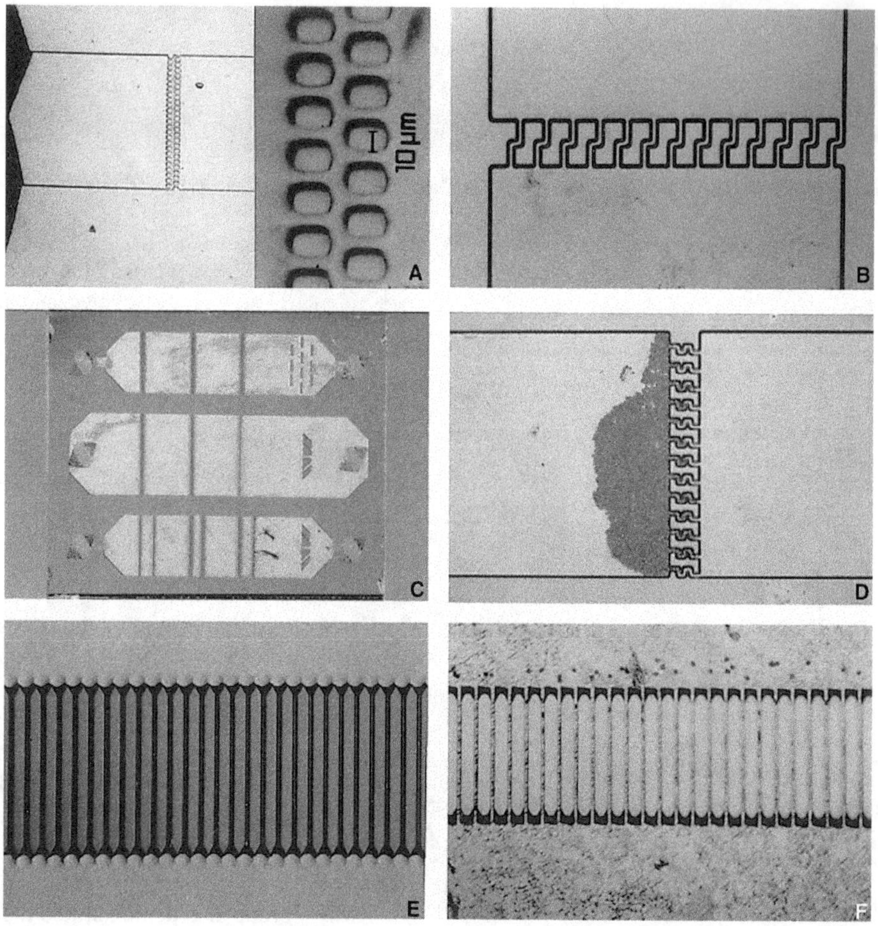

Figure 21.5. Examples of some type of microstructures used for filtration: (A) an array of microposts (13 × 20 μm) and spaced 7 μm apart; (B) an array of complex microposts; (C) filter chip containing three channels testing various serial filter and flow deflection designs; (D) filtration of a sample of 5.78-μm-diameter latex microspheres using the complex microposts (73-μm-wide post with 5-μ channels in between); (E) white blood cells isolated by a comb-type filter formed from an array of 120 posts (175 × 18 μm) with 6-μm channels between; and (F) release of the isolated white blood cells (stained with methylene blue) by reversing the flow through the filter. Reprinted from reference 190 with permission from Elsevier B.V. Copyright 1998.

filters along the channel. While these filters work well with solid particles such as latex microspheres, they are less effective for cells, since cells can deform and squeeze through the gaps.[190]

Comb-type filters have also been used for the filtration of cells (Fig. 21.5D–F). Wilding et al. constructed a series of 120 closely spaced long narrow pillars and aligned to the direction of flow in the channel. Using this structure, red blood cells can flow through the channels while white blood cells are captured at the front of the filter or within the gaps between the pillars as they try to squeeze through. The "pore size" of the filters can be changed by changing the distance between the gaps. These micropillar structures can also be difficult to fabricate inside the microfluidic channel.[190]

21.3.1.2. Lysis. Conventional methods of cell lysis, such as the use of chemical agents,[191–193] heating,[194] and ultrasonication,[192,195] have also been demonstrated for use in microfluidic devices. In addition to these methods, cell lysis can also be performed in the microchip by electroporation.[192,193,196] The application of an electric field creates pores in cell membranes, making such cells permeable to foreign DNA or RNA. At high enough electric fields (>1 kV/cm), the poration effect can become permanent, leading to lysis of the cell.[192,196] The use of AC pulses for lysis is preferred over DC pulses, as DC can cause the electrolysis of water, generating hydrogen and oxygen gas as well as large pH changes near the electrodes.[194]

Lee and Tai demonstrated a microchip for the lysis of cells by electroporation. Cells were first collected at sawtooth-shaped electrodes by dielectrophoresis (DEP) and were then lysed by the application of a pulsed electric field. The shape of the electrodes was designed such that the electric field strength was highest at the tip.[196] The magnitude of the electric field at the tip of the electrode from a 10 V power source was in the range of kilovolt per centimeter.[196] The lysis of yeast, *E. coli* cells, and Chinese cabbage and radish protoplasts were tested. Prior to electroporation, the cell walls present in the protoplasts were removed by enzyme digestion. Lysis of the protoplasts was demonstrated

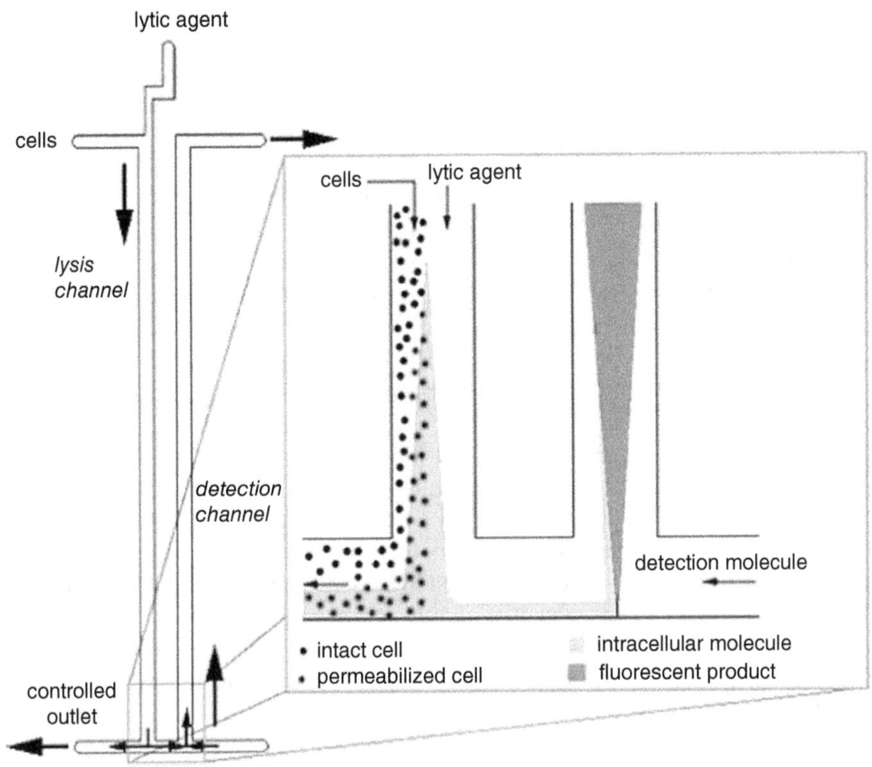

Figure 21.6. Schematic of a microfluidic cell lysis and fractionation device. Please refer to text for details. Reprinted from reference 191 with permission from the American Chemical Society. Copyright 2002.

by the application of either AC or DC pulses. Lysis of yeast and *E. coli* cells was accomplished with the application of high field strength DC square pulses.[196] The effectiveness for AC pulses to lyse protoplasts but not yeast or *E. coli* cells was attributed to the size difference between the types of cells tested. The diameters of the protoplasts were approximately 30 μm, while the yeast and *E. coli* were approximately 5 and 3 μm, respectively. The AC pulses generate a dielectrophoretic force, which stretches the cells; if the field strength is high enough, the cell is stretched until it breaks. Since the dielectrophoretic force is proportional to the volume inside the cell, it was more difficult to break the smaller yeast and *E. coli* cells using an AC pulse than the larger protoplasts.[196]

On-chip cell lysis has also been demonstrated by generating hydroxide ions in the aqueous buffer using the electrodes in the microchip. Hydroxide ions have been known to lyse cells by poration of membranes. Palladium electrodes have been integrated into devices, which were used to hydrolyze water at a potential of 2.6 V. Hydroxide ions were produced at the cathode and hydronium at the anode. Following lysis, the excess hydroxide ions were neutralized by the hydronium produced at the anode. This helped in maintaining a stable pH, minimizing denaturation of the secondary and tertiary structures of proteins in the sample. Lysis of HeLa, Chinese hamster ovary (CHO), and blood cells were demonstrated by hydroxide poration.[197]

Yager's group demonstrated a method for cell lysis followed by separation and detection of β-galactosidase based on differences in the diffusion coefficients of the various cellular components. The device consists of two main channels, a lysis and a detection channel, as shown in Figure 21.6.

Two streams of solutions, one containing cells and the other a chemical lysis agent, are introduced into the "lysis" channel. Due to laminar flow, the two streams do not mix. However, the lysis agents can diffuse into the stream containing the cells, resulting in lysis. The released intracellular components can diffuse into the stream containing the lysis agent, which then flows into the detection channel. Smaller molecules can diffuse further into the stream due to their smaller size and faster diffusion coefficients. By adjusting the flow rate, larger molecules such as DNA do not have sufficient time to diffuse out of the cell stream and are carried to the waste reservoir. Detection of β-galactosidase in the cells was done by introducing a stream of fluorogenic substrate into the detection channel. A fluorescence signal was produced as the substrate diffused into the stream containing the separated proteins, where it was converted into a fluorescent product. The experimental data indicated that 10% of the β-galactosidase present in the cell stream was fractionated and delivered into the detection channel.[191]

A novel, automated-microsystem integrating cell lysis and DNA isolation in one step using magnetic beads has been reported. Here, the magnetic beads are used to lyse the cells. Carboxylic-acid-terminated polystyrene magnetic beads were first added to a sample of cells that were located in a vibrating microchip. A high-power 808-nm laser was used to illuminate the microchip. Water exhibits a low molar absorptivity at this wavelength, and so the energy from the laser was absorbed by the magnetic beads in the sample, causing them to heat. The combination of mechanical motion of the vibrating microchip and the heated magnetic beads lysed the cells as well as denatured any proteins present. Instead of capturing the DNA targets onto the magnetic beads themselves, they are used to adsorb the denatured proteins and other cell lysates, following which the magnetic beads are moved out of the chamber, leaving the purified DNA targets behind for further analysis. The use of the negatively charged carboxylic acid polystyrene magnetic beads showed the highest amount of DNA released following lysis due to charge repulsion between the DNA and the magnetic beads when compared with amine-terminated polystyrene, unmodified polystyrene, and silica magnetic beads. The system was demonstrated to be effective for both gram-positive and gram-negative bacterial cells and could lyse and isolate DNA within 40 s. The limit of detection by PCR on DNA extracted from *E. coli* using this system was 20 cells/μL (~100 fg DNA).[117]

21.3.1.3. SPE. The incorporation of SPE in a microfluidic device has been demonstrated.[143,198] Like its conventional-scale counterpart, SPE can be used in a microscale environment for purification and preconcentration of biological targets such as DNA.[142] Capture efficiency can also be improved by repetitively cycling the sample through the SPE matrix in the microchannel.[160] Following capture, the target can be released for further downstream processing and detection.[199,200]

Methods of incorporating SPE into microfluidic devices include filling the microfluidic channels with silica resins or beads[142,160,199–201] as well as constructing monoliths or microstructures inside the channels.[199] Wolfe et al. demonstrated a silica material that was able to extract 500 base pair length DNA from a 25-μL sample of *Hind*III digested λ-phage DNA. The extraction efficiency was reported at 80%, and the material was eluted into a final volume of 5 μL. However, their results were not reproducible when considering repeated use of the same silica material.[202] Other reports using SPE for the purification of DNA samples in microfluidic devices have also been reported by Yu et al.[203] and Oleschuk et al.[204]

Binding capacity of DNA can be improved by going to smaller diameter beads to increase the surface area of capture. However, this results in an increase in back pressure and requires the application of higher pressure to achieve reasonable flow rates.[143,145,147] In addition, the use of microbeads in the channel requires the consideration of a mechanism to retain microbeads inside the microfluidic channel. This is most commonly done by placing a frit in the channels.[20,205]

Landers' group introduced an alternative approach based on the immobilization of the silica beads with sol–gels. The sol–gel is thought to act as an "interparticle glue" to hold the particles in place. SPE was tested using a 25-μL solution of λ-phage DNA (12.5 ng DNA) in guanidine stock solution. Following capture, isopropanol was used to wash away excess guanidine and contaminating substances, and the DNA was eluted by the addition of the Tris–ethylenediaminetetraacetic acid (EDTA) buffer. Extraction and purification of genomic DNA from samples of human whole blood in 25 min was demonstrated using an SPE material that was based on silica gel immobilized using a sol–gel within a microchannel.[202]

Alternatively, monoliths can be used for DNA isolation and preconcentration. Monoliths offer larger surface areas, controllable pore size, and higher mass transfer from porous structures than silica beads[145,147,206] and can be cast *in situ* by photopolymerization. Pore size and porosity can be adjusted by altering the concentration and type of porogenic solvent that is selected.[142,206] The use of porous monoliths for the purification of DNA has been reported by Bhattacharyya et al.[142] and Satterfield et al.[206] A sol-gel-based monolith has been demonstrated for the isolation of DNA from clinical samples. The monolith was a tetramethyl orthosilicate-based sol-gel loaded inside a microchannel. The addition of ethylene glycol as a porogen resulted in a monolith with pores in the micrometer scale, providing a large surface area for adsorption of DNA with little back pressure. Extraction efficiencies of 85% for λ-phage DNA and 70% for the extraction of human genomic DNA from human blood were observed. Using this system, a 200-μL solution containing DNA can be concentrated by eluting into a final volume of 12–18 μL. Extraction of viral DNA from the human spinal fluid using this monolithic material has also been demonstrated. Repeated extraction using a single device showed blockage of the pores by components in the lysed cells.[207]

Proteins present in the sample can affect the capture efficiency of DNA when using SPE materials. It was previously observed that the binding efficiency of DNA from whole blood was 35–40%, while the binding efficiency of prepurified genomic DNA was 88%. This discrepancy was attributed to the binding of proteins present in the blood to the monolith. Typically, 1 μL of lysed human blood contains 35 ng of DNA and 150 μg of protein. Many of these proteins will bind to the solid-phase material, reducing the binding capacity of DNA. Therefore, Landers' group added a C18 reversed-phase column for protein capture in series with a monolithic column for DNA extraction. From a 10-μL blood sample containing 350 ng DNA, 99% of the DNA moved through the C18 column while 70% of the proteins were retained. The overall extraction efficiency of the DNA from the two-stage extraction procedure was improved to 69%, which was significantly higher than previous reports. The two-stage column also eliminated a wash step using isopropanol that effectively removed proteins from the monolith.[147]

Landers' group also demonstrated a fully integrated microfluidic device that could generate a genetic profile from whole blood based on purification by SPE. Upon loading the sample into a glass-based microfluidic device, a chemical lysis agent was mixed with the blood followed by loading into a silica bed for purification of genomic DNA.

The eluted DNA was amplified by PCR. The entire process took less than 30 min, and 1500–2000 CFU of *Bacillus anthrax* was detected in 750 nL of whole blood.[208]

The use of photoinitiated polymerization of porous monolith material has been used for the preconcentration of very dilute solutions of Coumarin 519, a hydrophobic tetrapeptide Phe-Gly-Phe-Gly, and green fluorescent proteins (GFPs). A hydrophobic monolith was prepared by copolymerization of butyl methacrylate with ethylene dimethacrylate. Depending on the hydrophobicity of the target compound, the composition of the buffer associated with the initial sample and the eluting agent could be adjusted to change the degree of capture by the SPE. For example, the tetrapeptide is highly hydrophobic and did not require any additives to the buffer to observe complete adsorption to the column. However, GFP is less hydrophobic and required the addition of ammonium sulfate prior to adsorption onto the monoliths. Following elution, an increase in concentration by a factor of 1000 was observed for all three proteins.[203]

DNA can be captured onto silica-based microstructures such as pillars inside the microchannels.[142,160] The use of these pillars increases the surface area but requires complex fabrication procedures, increasing the cost of each chip.[143] Cady et al. demonstrated an increase in surface area available for the capture of DNA by using silica-coated pillars (Fig. 21.7). Depending on the etch depth (20–50 µm), an increase in surface area from 300% to 600% was observed. Selective binding of genomic DNA was observed in the presence of a chaotropic salt (guanidinium isothiocyanate), followed by washing with ethanol and elution in a low-ionic-strength buffer. Samples were moved through the extraction device by the application of positive pressure, eliminating the need for centrifugation. The binding capacity for DNA in the device was calculated to be approximately 82 ng/cm^2. Approximately 87% of the proteins present in the cell lysate were removed following the washing step.[201]

Polycarbonate-based microchips have also been used for the capture of DNA. Exposing the polycarbonate to UV light at 254–300 nm resulted in the formation of a range of carbonyl/carboxyl-containing groups, which could capture, preconcentrate, and purify DNA from complex sample matrices. The surface area for capture could also be increased by fabricating microposts inside the polycarbonate channels. Samples of lysed *E. coli* cells mixed with an immobilization buffer were introduced into a polycarbonate chip for DNA capture. A capture efficiency of 85% for the genomic DNA was observed, with 80% of the protein removed. The entire process took less than 25 min.[160]

21.3.1.4. Magnetic beads. Magnetic bead technology has been popular in terms of technology that can be incorporated into microfluidic devices. Functional magnetic nanoparticles are advantageous in microfluidic devices since they combine the requirement of adsorbent materials with a method to manipulate such materials.[184]

Cells can be captured by using antibody-conjugated magnetic beads. This allows for purification and enrichment prior to subsequent analysis. Dengue viruses were isolated by antibody-conjugated magnetic beads in a microfluidic device. Detection of the concentrated virus sample showed a limit of detection of 10^2 pfu (plaque forming unit)/mL.[32]

An impedance biosensor based on an interdigitated array microelectrode (IDAM) coupled with magnetic nanoparticle-antibody conjugates (MNACs) was developed for the rapid and specific detection of *E. coli* O157:H7 in ground beef. Detection of the *E. coli* cells by impedance showed a 35% improvement in sensitivity due to the preconcentration of the bacterial cells near the electrode. The lowest detection limit of *E. coli* in buffer was 7.4 × 10^4 CFU/mL and 8 × 10^5 CFU/mL for *E. coli* in ground beef. The total detection time was 35 min.[209]

The use of electromagnetic traps has also been demonstrated for the confinement of magnetic particles bound with DNA.[155] Target DNA was first captured by the probe-functionalized magnetic microbeads and confined into a small volume by the electromagnetic trap (Fig. 21.8). A

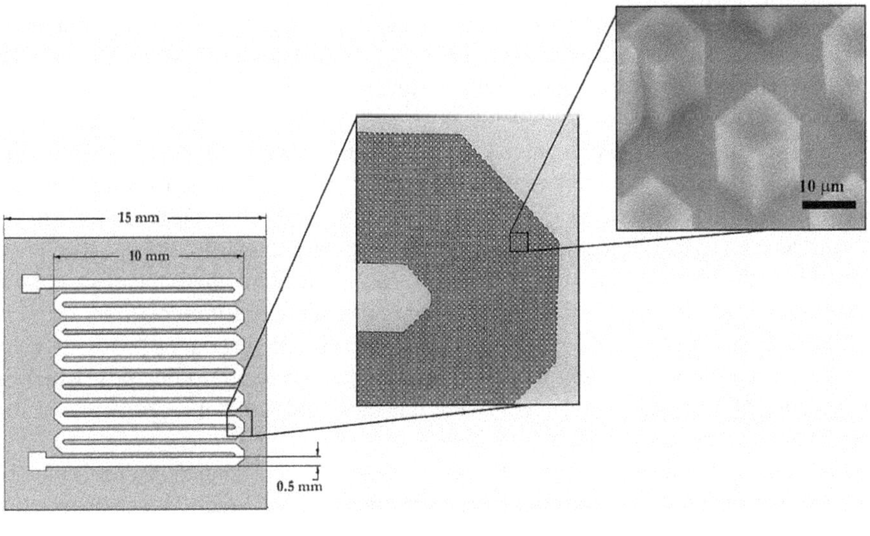

Figure 21.7. Schematic and scanning electron microscope (SEM) images of the microfabricated silica pillars for the SPE of DNA. Reprinted from reference 201 with permission from Elsevier B.V. Copyright 2003.

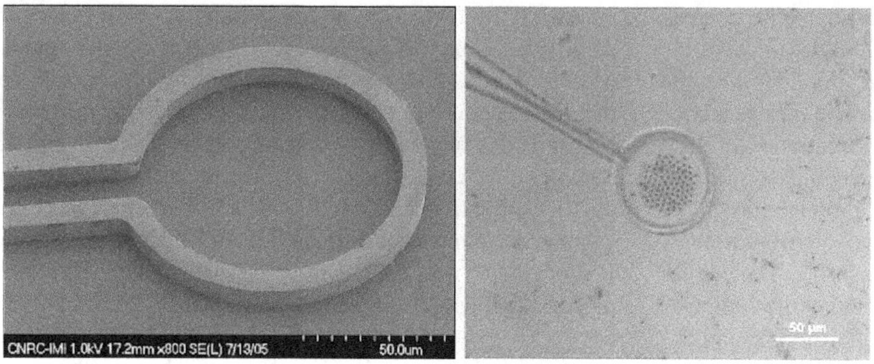

Figure 21.8. SEM image of the micro-electromagnetic trap (left) and image of the magnetic microbeads trapped on the micro-electromagnetic trap. Reprinted from reference 155 with permission from the American Chemical Society. Copyright 2006.

concentrating factor of 100 was demonstrated. Detection limits of 75 and 150 DNA target copies from yeast and *B. anthracis* were achieved, respectively, without the need for amplification of the DNA by PCR.[155]

A fully integrated microfluidic device was recently developed by Liu et al., where cell capture by immunomagnetic beads, preconcentration, purification, and thermal lysis were all done within the same device. The capture efficiency using the magnetic beads was determined to be at 40%. Discrimination of single-nucleotide polymorphism targets was demonstrated following on-chip PCR amplification. The detection of 10^3–10^6 *E. coli* K12 cells from a 1-mL whole blood sample was shown and required 2.7 h. It was observed that the presence of the magnetic beads in the PCR mixture reduced the amplification efficiency by 50%.[210]

Due to the small-sized scale of microfluidic devices, the use of large conventional-sized permanent or electromagnets may be a problem due to difficulties in obtaining a well-defined force on magnetic particles. Accurate manipulation of the magnetic particles inside microfluidic channels can also be difficult with large magnets. Additionally, high magnetic field gradients are often desired for efficient capture of magnetic nanoparticles flowing through a microchannel.[211]

It is possible to generate extremely large, local magnetic fields by constructing various structures inside the chip. Ferromagnetic pillar structures constructed inside a microchannel can be magnetized by using a low external magnetic field.[212,213] The magnetic field gradients induced at these micropillar structures are strong enough to trap and concentrate nanoparticles from a flowing stream.[212,213] Guo et al. was able to demonstrate the capture of magnetic nanoparticles using an array of nickel squares with dimensions of $50 \times 50 \times 1\,\mu m$ using an external magnetic field of 20 mT.[212]

Beyond microchannel technology, ferromagnetic wire (24 µm diameter) loosely packed into a region near the poles of a large magnet can also be used to generate high-magnetic-field gradients in order to trap cells, which are naturally paramagnetic. This setup has been used to retain red blood cells without the need to magnetically tag the cells.[109] Certain cells contain paramagnetic material. The most paramagnetic cell in the body is the deoxygenated red blood cell. Each red blood cell contains around 250 million hemoglobin proteins, which each contain four iron atoms. Each of the atoms can have up to four unpaired electrons, cumulatively giving the cell paramagnetic properties.[109]

Other biological cells can also be manipulated using a complementary metal-oxide semiconductor (CMOS)/microfluidic hybrid system. A microfluidic chip was placed on top of a CMOS chip, which consisted of an array of microelectromagnets. Movement of biological cells tagged with magnetic beads in the microfluidic chip was controlled by the CMOS chip through the generation of local magnetic fields using the array of micro-electromagnets. This approach for the manipulation of biological cells has only been demonstrated with bovine capillary endothelial cells that were "magnetized" by coculturing the cells with peptide-coated beads.[214]

21.3.1.5. Dielectrophoresis. DEP can be used for the manipulation and separation of a variety of particles, including dielectric materials and metallic particles, as well as biological cells and viruses inside microfluidic devices.[56,193,215–217] DEP is a phenomenon in which uncharged particles are polarized under a nonuniform external electric field and exhibit motion either toward or away from the electrode.[193,215,216] The magnitude and direction of the polarization are dependent on the applied electric field as well as the dielectric properties of the particle and the medium.[215,218] If the electric field is uniform, the attractive and repulsive forces on the polarized particle are equal and opposite, resulting in no net motion. In a nonuniform field, the attractive and repulsive forces no longer cancel, and there is net motion of the polarized particle.[218] If the particle is more polarizable than its medium, particles move toward a region where the electric field strength is highest (toward the electrode), which is referred to as positive DEP. Conversely, if the medium is more polarizable than the particles, then there is a net motion toward the location where the electric field strength is at a minimum. This is referred to as negative DEP.[215,217,219,220] DEP allows for sorting of types of cells based on differences in their dielectric properties, which is influenced by the chemical and structural properties of the cell.[56,160,215,216]

Many DEP devices use an interdigitated electrode pattern on the microchip substrate.[56,215–217] Figure 21.9 shows a sche-

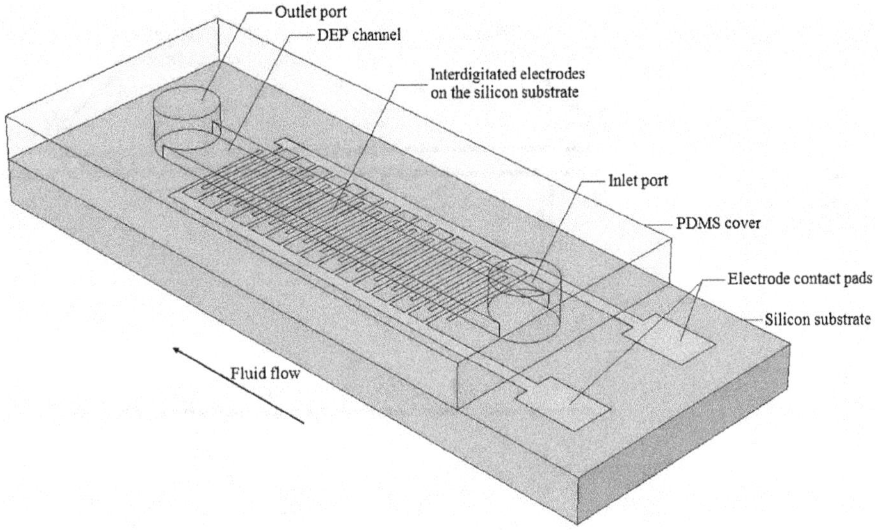

Figure 21.9. Schematic of a microchip with an interdigitated electrode pattern for the separation of particles by dielectophoresis. Reprinted from reference 216 with permission from The Royal Society of Chemistry. Copyright 2006.

matic of one such DEP device. DEP devices are constructed by patterning conducting materials such as gold onto an insulating substrate such as silicon dioxide.[217] The electrode structures are typically between fractions of a micrometer and a few micrometers thick, and are considered to be planar on the surface of the substrate.[215,218]

Particles experiencing positive DEP are trapped at the electrode while the other particles can be washed away.[218] By trapping the target cells at the electrode using positive DEP, the target can be concentrated from a dilute solution. This has been demonstrated for yeast, bacteria, and mammalian and other cell types.[221] DEP devices can also be used for the selective capture and purification of various biological targets.[216] However, if the dielectric properties of the different particle types are not significantly different, then mixtures of particles become entrapped. An example of this was reported for mixed mammalian cell systems.[221] In addition, cells may become permanently attached to the electrodes, meaning that they cannot be released simply by terminating the application of the electric field.[56]

The ability for DEP to collect cells is dependent on the geometry of the electrodes, the conductivity of the sample, and flow rate, as well as the applied field strength.[56,218] A larger electrode area allows for larger quantities of cells to be captured, but may also increase loss due to adsorption. Inhibitor compounds present in the sample can increase the conductivity of the solution until a point is reached such that the positive DEP ceases and the electrodes can no longer attract the cells. Maintaining a low conductivity in the solution allows for a larger DEP force and higher flow velocities to be applied, reducing the time for DEP.[218]

Trapping of cells from larger sample volumes (microliters) can be difficult since the magnitude of the DEP force decays exponentially from the planar electrodes.[218] Increasing the field strength to increase the magnitude of the DEP force can lead to joule heating of the solution, causing cell lysis.[56,183,218] Since the electrodes are deposited inside the microchannel, application of a voltage may cause side electrochemical reactions as well as gas generation at the surface of the electrodes due to electrolysis of water. These effects can be minimized through the use of AC rather than DC voltages.[222]

An alternative is to use electrodeless or insulating DEP (EDEP) traps. EDEP devices use structures made of an insulating material inside the microfluidic channel. These structures then constrict the electric field generated from a DC voltage by using the electrodes at the inlet and outlet reservoirs on the chip.[215,223] The structures used in EDEP are chemically inert and physically robust, and high electric fields can be used without issues of electrolysis of water.[215] The electric field gradient can also be fine-tuned by changing the physical parameters of the insulating structures.[222] Figure 21.10 illustrates how the electric field can be constricted by an insulating structure.

Separation of cells and enrichment has been demonstrated by EDEP from mixtures of *E. coli* with 2% blood. Upon application of voltage, the *E. coli* cells experienced positive DEP and were trapped at the field constrictions. Concurrently, blood cells experienced negative DEP and were repelled from the constrictions. It was shown that the purity of the trapped *E. coli* cells was approximately 100% with an enrichment factor of 10^3 10^5 being achieved within minutes.[215]

A microfluidics-based system has incorporated the use of DEP in conjunction with immunocapture for the concentration and capture of bacterial cells. The DEP device consisted of an array of interdigitated electrodes on a flat silicon substrate. Positive DEP was used to collect and concentrate *L. monocytogenes* from 5- to 20-μL samples into a volume of 12.5 nL, with concentration factors between 10^2 and 10^3. The concentration factor was a function of the sample volume and solution flow times. Antibodies immobilized on the insulator over the DEP electrode surface provided selective capture of the targets collected by DEP, while nontarget

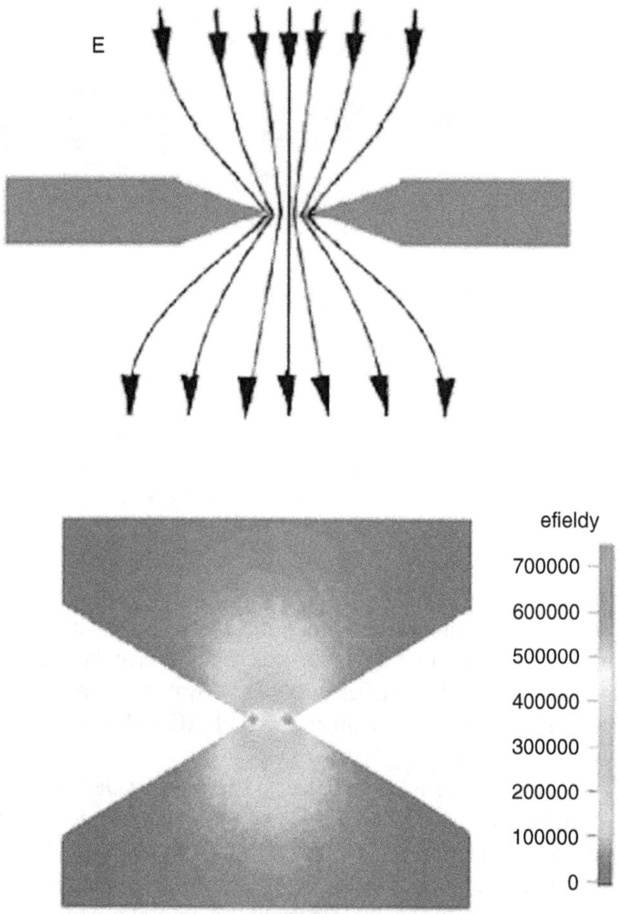

Figure 21.10. Constriction of the electric field by an insulating microstructure in a microfluidic chip (top) and a simulation of the electric field strength at the constriction (bottom). Reprinted from reference 215 with permission from the IEEE. Copyright 2003.

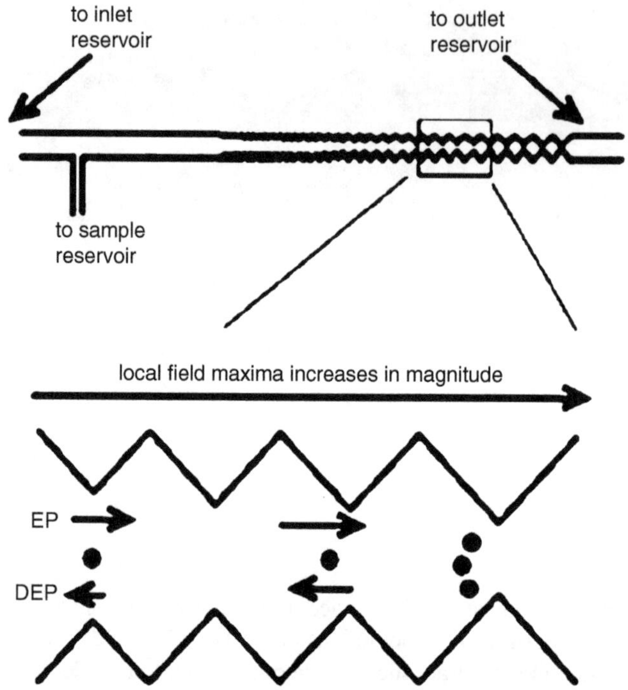

Figure 21.11. Schematic of the channel design for the separation of viable and dead cells (top) and the mechanism of trapping cells by balancing the electrophoretic (EP) and dielectrophoretic (DEP) forces. Reprinted from reference 222 with permission from the American Chemical Society. Copyright 2007.

cells would be flushed out by solution flow when DEP was turned off. The average capture efficiency of the cells by the immobilized antibodies was 20.7%.[216]

Hayes' group demonstrated the ability to separate viable and dead cells by using electrophoresis and DEP. DEP of cells in DC mode was governed mainly by the conductivity of their cell membranes. For dead cells, the permeability of the cell membrane changed, and the conductivity increased by a factor of 10^4.[219,223] This difference in conductivity allowed for the separation of viable and dead cells by DEP. The wall of the separation channel of the device that was reported was designed with a sawtooth pattern, which acted to constrict the electric field (Fig. 21.11). The shape of each sawtooth was adjusted such that the local field maximum at each tooth increased further along the channel. The cells in the channel were subjected to two forces: an electrophoretic force moving them through the channel and a negative DEP force acting in the opposite direction. If the electrophoretic force was greater than the negative DEP when a cell approached a field gradient trap (a sawtooth), then the cell passed through the trap. If the electrophoretic force was less than the negative DEP, the cell switched the direction of the movement. The cells become immobilized when the sum of the electrophoretic, the negative dielectrophoretic, and the electroosmotic flows all cancelled. Because the local field strength increased along the channel, cells with different polarizabilities were separated at different points along the channel. Separation between live and dead cells of *E. coli*, *B. subtilis*, and *Staphylococcus epidermidis* was demonstrated.[222]

Lapizco-Encinas et al. developed a method for the separation of live and dead *E. coli* cells by EDEP. The constriction of the electric field was done by fashioning an array of insulative microposts. Dead *E. coli* cells exhibited a lower negative DEP force than live cells. A sample of live and dead *E. coli* cells under the micropost array showed separation of the two populations at different areas around the micropost.[223]

Sorting of live and dead yeast cells has also been demonstrated by DEP.[220] Cells were streamed through a channel with an electrode array. Figure 21.12 shows the layout of the microfluidic cell sorter. As the cells stream through the electrode array, DEP results in positive DEP in the viable cells while the dead cells experience negative DEP. The applied DEP force on a cell is orthogonal to the fluid flow, which changes its trajectory. At the end of the DEP channel, the trajectories of the live cells were changed so that they were directed into one outlet port, while dead cells were directed to a different outlet port.

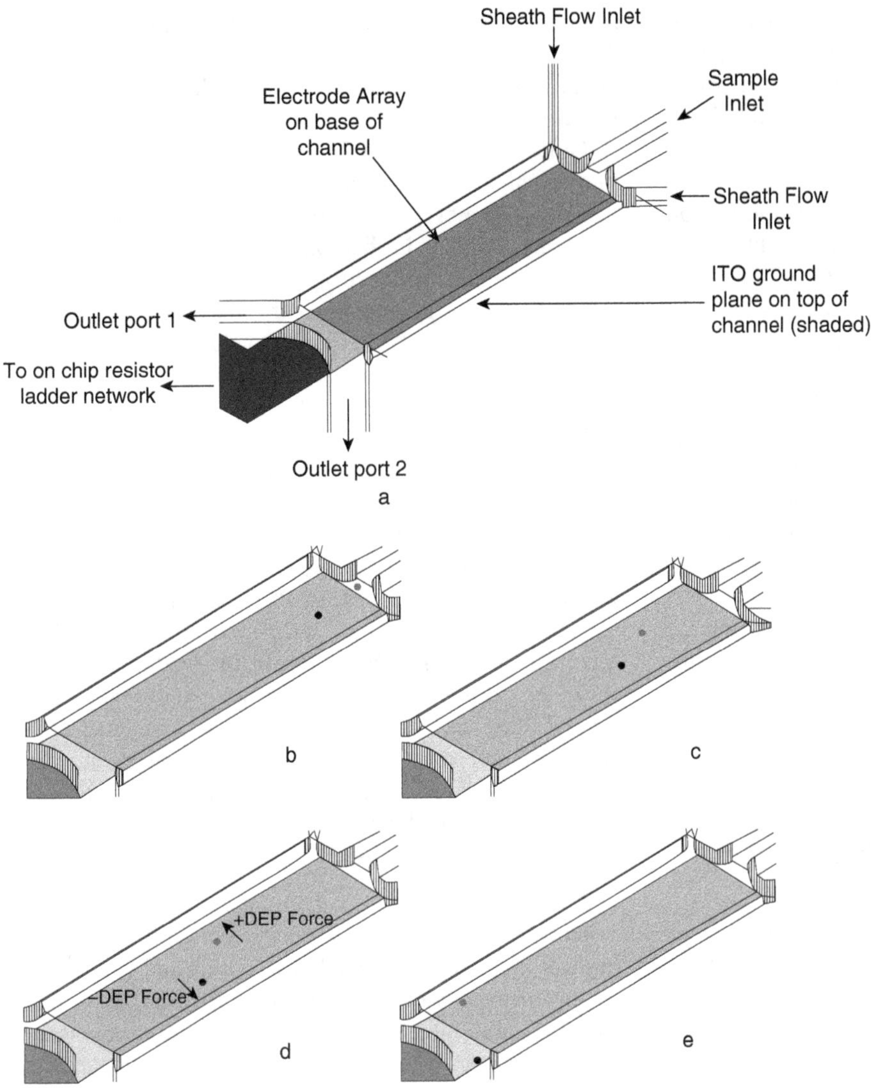

Figure 21.12. (a) Schematic of the dielectrophoresis cell sorter. (b,c) Viable and dead cells are introduced into the channel and experienced either an orthogonal positive or negative DEP force, which deflects their trajectory. (d,e) At the end of the channel, viable cells are deflected into one outlet, while dead cells are deflected into the other outlet. Reprinted from reference 219 with permission from The Royal Society of Chemistry. Copyright 2006.

It was noted that debris present in the cell stream may become adsorbed onto the DEP channel, which could in turn capture cells from the flow stream, reducing the separation yield. Additionally, cells could settle to the bottom of the channel and could become attached to the electrodes, reducing the yield. These problems can be reduced by the use of higher flow rates.[219]

21.3.2. PCR in Microfluidic Devices

On-chip PCR can be performed in two ways: time domain and space domain. In time domain, the PCR mixture is kept in one chamber and the temperature in the chamber is cycled through the denaturation, annealing, and extension steps.[192,224] This method requires careful optimization of thermal mass to optimize reaction times and power consumption.[192]

Landers' group demonstrated PCR in small volumes by IR-mediated temperature control. The microchip was constructed from polyimide, which is transparent from 600 to 3000 nm. Consequently, IR radiation could be transmitted through the chip, heating the solution inside. Since only the solution was heated, the thermal mass was reduced, allowing for faster heating rates. Faster cooling rates were afforded since cooling of the solution was independent of the heating source. One PCR cycle could be completed in 12 s. By comparison, this same cycle would take 40 s to complete with a conventional thermocycler. Amplification of a 500-bp fragment from λ-phage DNA in a 1.7-μL chamber was demonstrated. Detectable amounts of the PCR amplicons were observed after 15 cycles, which took a total of 240 s.[225]

Yang et al. demonstrated a high sensitivity on-chip PCR assay for the detection of *E. coli* K12 from the DNA of 10 cells (~50-fg genomic DNA) in 2% blood following thermal lysis. This was done without purification in a period of 30 min. Predetermined concentrations of *E. coli* were mixed with solutions of 2% blood and a commercially available PCR buffer, which contained components to block inhibiting factors. A solution of polyethylene glycol (average molecular weight 8000 g/mol) was also added to the PCR mixture to improve amplification efficiency.[226]

In space domain PCR, the reaction mixture is flowed continuously through different temperature regions inside a channel. PCR using space domain can be relatively fast, since one does not need to wait for cooling and heating. Such an approach can, however, take up a large space on the microfluidic chip as well as require relatively large sample volumes when considering microfluidic applications.[192]

Hashimoto et al. demonstrated a 20-cycle amplification of a 500-bp target from λ-DNA in 1.7 min using a continuous-flow PCR device. A 997-bp fragment required 3.2 min for the same number of reaction cycles. A longer time was required in the latter case because the kinetic limit of the Taq polymerase for deoxynucleotide incorporation was reached.[227] Amplification time can be reduced by increasing flow velocity through each temperature zone. However, it is possible to increase flow rates to a point where there is reduction of amplification efficiency. The amplification yield was reported to be 25% lower than that obtained from a conventional block thermal cycler with similar cycle times.[227]

Concerns with integrating PCR into a microfluidic device include the need for large and rapid changes in temperature, avoidance of bubble formation as a result of those temperature changes, and the denaturation of enzymes on the surfaces of the microfluidic devices.[228]

21.3.3. Sample Enrichment by Manipulation of Electrophoretic Mobility

The manipulation of solutions and targets by an electric field in microfluidic devices has provided potential for enrichment of low concentrations of targets by manipulation of their electrophoretic mobility inside a channel so that there is confinement in a smaller volume. Such techniques include isotachophoresis (ITP), isoelectric focusing (IEF), and FAS.[30,229]

21.3.3.1. ITP. In ITP, the target is concentrated and separated from other components in a discontinuous buffer system. The target plug is positioned between two different buffers. A leading buffer contains a large concentration of higher mobility ions (higher conductivity) than those of the targets, and a terminating buffer contains ions with lower mobility at the injection side (lower conductivity).[9,230] The velocity that ions will travel under an electric field is the product of their electrophoretic mobilities and the strength of the applied electric field. The local field strength can be modified by changing the ionic strength and conductivity, or by adding salts.[231] In general, an ion traveling through a low conductivity buffer will experience a higher field strength than through a high conductivity buffer. Therefore, the velocity of the ions will decrease on moving from a low conductivity buffer to a high conductivity buffer, given the same external electric field strength.[231]

During ITP, the charged targets in the sample plug having faster electrophoretic mobilities move ahead into the leading buffer. As the targets begin to separate from the other ions in the sample plug, they create a local region of high conductivity, reducing the local field strength. The velocity of the fast-moving targets decreases until they reach a constant velocity.[230,231]

Other targets with different mobilities present in the sample will also separate into different zones until a constant velocity is achieved.[231] At equilibrium, the sample components are contained in sharp zones with the same concentration as the leading electrolyte. Therefore, the degree of preconcentration is determined by the composition of the leading buffer.[205] ITP requires loading and maintaining zones of the different buffers, and knowledge of the conductivity of the targets is needed for an efficient separation.[31]

The preconcentration of digested *Hae*III dsDNA fragments has been demonstrated by ITP. The ITP device was constructed from poly(methyl methacrylate) (PMMA). The leading edge buffer consisted of 15 mM HCl and 36 mM imidazole (pH 7), and the terminating edge buffer was a solution of 20 mM HEPES (4-(2-hydroxyethyl)-1-piperazineethane sulfonic acid) and 36 mM imidazole (pH 7.2). The high mobility ion in the leading buffer was chloride, and the low mobility ion in the terminating buffer was HEPES. Injection and stacking was completed in 70 s. By measurement of fluorescence from an intercalating dye, stacking was shown to increase the fluorescence intensity by 40-fold. The high salt content used in ITP may interfere with PCR of the stacked DNA.[230]

21.3.3.2. IEF. IEF can be used to separate proteins in an electric field based on isoelectric point in a pH gradient formed by carrier ampholytes. The sample-containing targets is mixed with a solution of ampholytes and is loaded into a separation channel. After a voltage is applied, the ampholytes rearrange according to their pH while the targets focus into narrow zones within the channel where the net charge of the protein is zero.[231]

An affinity-based method of IEF has also been demonstrated. A fluorescent tag was used to selectively label a target through an affinity interaction. The free ligand could then be separated from the ligand–protein complex by IEF. G proteins were first incubated with a fluorescently labeled 5′-O-(3-thiotriphosphate). A pH 3–10 ampholyte solution was used for focusing. Following focusing for 8 min, a detection limit of the affinity ligand–protein complex of 30 pM was observed as compared with a limit of detection of 360 nM by conventional separation and detection by capillary electrophoresis.[234]

IEF in short columns has also been demonstrated. In IEF, the length of the column does not affect the resolution of separation for a given voltage. Therefore, IEF in shorter channels at higher voltages would give rise to faster separations and more concentrated peaks. IEF of enhanced green-fluorescent protein, R-phycoerythrin and carbonic anhydrase II inside a 4-mm-length channel was achieved in 26 s.[235]

21.3.3.3. FAS. FAS involves a concentrating effect that is achieved when the target is introduced electrokinetically into the separation channel. The targets are first dissolved in a buffer that has conductivity lower than that of the buffer in the channel. Upon electrokinetic injection, the targets move at a higher velocity in the sample buffer until they hit the interface between the sample buffer and the buffer in the separation channel. At this point, the targets experience a sharp decrease in velocity. Stacking into a smaller volume occurs, increasing the concentration of targets. Preconcentration enhancements of 3- to 100-fold have been observed, depending on the charge of the targets and conductivities of the buffer.[30,31,205,232]

The difference in conductivity between the two buffers should be as large as possible to achieve the largest stacking effect. A large difference in conductivity would allow the targets to stack much more quickly against the boundary while limiting the velocity of the stacked band in the high conductivity buffer. Ideally, the use of water as the low conductivity buffer would yield the greatest enhancement. However, for the processing of biological samples, this may not be possible[232] as real sample matrices often contain substances that could increase the conductivity of the buffer.[233]

A solution to this is to dilute the sample with water prior to injection to lower the conductivity of the sample.[231,233] Alternatively, a short plug of water can be introduced into the injection end of a channel for sample stacking. The target is injected by electrokinetic injection into the water and is stacked against the separation buffer. Preconcentration is improved since stacking occurs between the water and the separation buffer rather than with the sample's original buffer. Preconcentration factors of 1000-fold have been reported.[205]

Shim's group introduced a method for preconcentration of DNA to improve the sensitivity of downstream electrochemical detection.[30] The system first loads a short plug of water into the channel where DNA is loaded by electrokinetic injection. Stacking occurs against a buffer, which contains hydroxypropyl cellulose modified with gold nanoparticles (HPC/AuNP) and sodium citrate. Enhanced stacking occurs due to an increase in conductivity of the stacking buffer, as well as a decrease of the electrophoretic mobility of the DNA by adsorption onto the large gold nanoparticles. A preconcentration factor of 25,000 was observed using this method, which is 2–3 orders of magnitude higher than other stacking methods. When combined with electrochemical detection by oxidation of DNA on conducting polymer-modified electrodes that were also modified with AuNPs, a detection limit of 5.7 attomole from a 50-μL sample was observed. The stacking, separation, and detection of a sample of 100-bp DNA ladder was completed within 435 s.[30]

21.4. CONCLUSIONS

Methods for the pretreatment and purification of various biological targets such as whole cell, DNA, and proteins that can be used to process samples for detection by a biosensor have been presented. Methods that can be adapted into microfluidic devices have also been reviewed. Common to these techniques is the fact that the target material is captured, either through electrostatic or hydrophobic interactions or by selective capture using an affinity ligand. Following washing of the contaminants, elution of the captured targets into a volume smaller than the original sample provides a sample enrichment by preconcentration. This technique can help to improve sensitivity when trace analysis is desired.

Other methods exist for increasing the total quantity of the target either through cultural enrichment or by PCR. However, these methods often require long enrichment times, especially for cultural enrichments, which can take days. With the advent of microfluidics-based PCR, the time required for amplification is greatly reduced. However, applications are limited to DNA-based detection methods and require proper selection of primers and temperatures for efficient amplification. A challenge in applications to real samples is that PCR can be highly sensitive to inhibitor compounds present in the matrix. One goal for many of the DNA purification protocols is to remove these inhibitor compounds to allow for PCR.

Aside from sample enrichment by preconcentration or amplification by PCR or cultural enrichment, the sensitivity of the biosensor has also been improved through the introduction of various technologies to improve the signal obtained from one biorecognition event. Examples of biosensors have been reported in the literature where the sensitivity was such that it was possible to detect targets directly from the sample matrix, eliminating the need for PCR or enrichment.

The ideal biosensor would be a device with the ability for the quantitative determination of the target from a biological sample with no sample pretreatment or purification. This is not currently possible since biological targets such as DNA or proteins necessitate pretreatment to release the targets from within cells, and for genomic DNA to fragment and convert into single-stranded form.

The purification and sample pretreatment protocols presented in this chapter are intended to demonstrate some of the techniques that are available for purification or preparation of targets. It is not necessarily the case that each of the steps that have been outlined need to be followed in all cases. The choice of pretreatment and purification protocols will ultimately depend on the details of a particular application.

REFERENCES

1. Rodriguez-Mozaz, S.; Lopez de Alda, M.J.; Barceló, D. Biosensors as useful tools for environmental analysis and monitoring. *Anal. Bioanal. Chem.* **2006**, *386*, 1025–1041.

2. Rodriguez-Mozaz, S.; Lopez de Alda, M.J.; Barceló, D. Advantages and limitations of on-line solid phase extraction coupled to liquid chromatography–mass spectrometry technologies versus biosensors for

monitoring of emerging contaminants in water. *J. Chromatogr. A* **2007**, *1152*, 97–115.

3. Leonard, P.; Hearty, S.; Brennan, J.; Dunne, L.; Quinn, J.; Chakraborty, T.; O'Kennedy, R. Advances in biosensors for detection of pathogens in food and water. *Enzyme Microb. Technol.* **2003**, *32*, 3–13.

4. Castillo, J.; Gáspár, S.; Leth, S.; Niculescu, M.; Mortari, A.; Bontidean, I.; Soukharev, V.; Dorneanu, S.A.; Ryabov, A.D.; Csoregi, E. Biosensors for life quality design, development and applications. *Sens. Actuators B* **2004**, *102*, 179–194.

5. Hartley, H.A.; Baeumner, A.J. Biosensor for the specific detection of a single viable *B. anthracis* spore. *Anal. Bioanal. Chem.* **2003**, *376*, 319–327.

6. Rodriguez-Mozaz, S.; Lopez de Alda, M.J.; Barceló, D. Fast and simultaneous monitoring of organic pollutants in a drinking water treatment plant by a multi-analyte biosensor followed by LC–MS validation. *Talanta* **2006**, *69*, 377–384.

7. Almadidy, A.; Watterson, J.; Piuuno, P.A.E.; Raha, S.; Foulds, I.V.; Horgen, P.A.; Castle, A.; Krull, U.J. Direct selective detection of genomic DNA from coliform using a fiber optic biosensor. *Anal. Chim. Acta* **2002**, *461*, 37–47.

8. Zhu, P.; Shelton, D.R.; Karns, J.S.; Sundaram, A.; Li, S.; Amstutz, P.; Tang, C.-M. Detection of water-borne *E. coli* O157 using the integrating waveguide biosensor. *Biosens. Bioelectron.* **2005**, *21*, 678–683.

9. Schmidt, P.M.; Matthes, E.; Scheller, F.W.; Bienert, M.; Lehmann, C.; Ehrlick, A.; Bier, F.F. Real-time determination of telomerase activity in cell extracts using an optical biosensor. *Biol. Chem.* **2002**, *383*, 1659–1666.

10. Schmidt, P.M.; Lehmann, C.; Matthes, E.; Bier, F.F. Detection of activity of telomerase in tumor cells using fiber optical biosensors. *Biosens. Bioelectron.* **2002**, *17*, 1081–1087.

11. Ko, S.; Grant, S.A. A novel FRET-based optical fiber biosensor for rapid detection of *Salmonella typhimurium*. *Biosens. Bioelectron.* **2006**, *21*, 1283–1290.

12. Minunni, M.; Tombelli, S.; Mascini, M. A biosensor approach for DNA sequences detection in non-amplified genomic DNA. *Anal. Lett.* **2007**, *40*, 1360–1370.

13. Cao, L.; Lin, H.; Mirsky, V.M. Surface plasmon resonance biosensor for enrofloxacin based on deoxyribonucleic acid. *Anal. Chim. Acta* **2007**, *589*, 1–5.

14. Metzger, J.; von Landenberg, P.; Kehrel, M.; Buhl, A.; Lackner, K.J.; Lupper, P.B. Biosensor analysis of beta2-glycoprotein I–reactive autoantibodies: Evidence for isotype-specific binding and differentiation of pathogenic from infection-induced antibodies. *Clin. Chem.* **2007**, *52*, 1137–1143.

15. Endo, T.; Kerman, K.; Nagatani, N.; Takamura, Y.; Tamiya, E. Label-free detection of peptide nucleic acid-DNA hybridization using localized surface plasmon resonance based optical biosensor. *Anal. Chem.* **2005**, *77*, 6976–6984.

16. Mascini, M.; Palchetti, I.; Marrazza, G. DNA electrochemical biosensors. *Fresenius J. Anal. Chem.* **2001**, *369*, 15–22.

17. Zacco, E.; Adrian, J.; Galve, R.; Marco, M.-P.; Alegret, S.; Pividori, M.I. Electrochemical magneto immunosensing of antibiotic residues in milk. *Biosens. Bioelectron.* **2007**, *22*, 2184–2191.

18. Sharma, M.K.; Goel, A.K.; Singh, L.; Rao, V.K. Immunological biosensor for detection of *Vibrio cholerae* O1 in environmental water samples. *World J. Microbiol. Biotechnol.* **2006**, *22*, 1155–1159.

19. Radke, S.M.; Alocilja, E.C. A high density microelectrode array biosensor for detection of *E. coli* O157:H7. *Biosens. Bioelectron.* **2005**, *20*, 1662–1667.

20. Dell'Atti, D.; Zavaglia, M.; Tombelli, S.; Bertacca, G.; Cavazzana, A.O.; Bevilacqua, G.; Minunni, M.; Mascini, M. Development of combined DNA-based piezoelectric biosensors for the simultaneous detection and genotyping of high risk human papilloma virus strains. *Clin. Chim. Acta* **2007**, *383*, 140–146.

21. Tombelli, S.; Mascini, M.; Sacco, C.; Turner, A.P.F. A DNA piezoelectric biosensor assay coupled with a polymerase chain reaction for bacterial toxicity determination in environmental samples. *Anal. Chim. Acta* **2000**, *418*, 1–9.

22. Olsen, E.V.; Pathirana, S.T.; Samoylov, A.M.; Barbaree, J.M.; Chin, B.A.; Neely, W.C.; Vodyanoy, V. Specific and selective biosensor for *Salmonella* and its detection in the environment. *J. Microbiol. Methods* **2003**, *53*, 273–285.

23. Eun, A.J.-C.; Huang, L.; Chew, F.-T.; Li, S.F.-Y.; Wong, S.-M. Detection of two orchid viruses using quartz crystal microbalance-based DNA biosensors. *Virology* **2002**, *92*, 654–658.

24. Mo, X.-T.; Zhou, Y.-P.; Lei, H.; Deng, L. Microbalance-DNA probe method for the detection of specific bacteria in water. *Enzyme Microb. Technol.* **2002**, *30*, 583–589.

25. Mao, X.; Yang, L.; Su, X.-L.; Li, Y. A nanoparticle amplification based quartz crystal microbalance DNA sensor for detection of *Escherichia coli* O157:H7. *Biosens. Bioelectron.* **2006**, *21*, 1178–1185.

26. Minunni, M.; Mannelli, I.; Spiriti, M.M.; Tombelli, S.; Mascini, M. Detection of highly repeated sequences in non-amplified genomic DNA by bulk acoustic wave (BAW) affinity biosensor. *Anal. Chim. Acta* **2004**, *526*, 19–25.

27. Ouyang, H.; DeLouise, L.A.; Miller, B.L.; Fauchet, P.M. Label-free quantitative detection of protein using macroporous silicon photonic bandgap biosensors. *Anal. Chem.* **2007**, *79*, 1502–1506.

28. Lee, C.-Y.; Gamble, L.J.; Grainger, D.W.; Castner, D.G. Mixed DNA/oligo(ethylene glycol) functionalized gold surfaces improve DNA hybridization in complex media. *Biointerphases* **2006**, *1*, 82–92.

29. Sezgintürk, M.K.; Dinçkaya, E. Direct determination of sulfite in food samples by a biosensor based on plant tissue homogenate. *Talanta* **2005**, *65*, 998–1002.

30. Shiddiky, M.J.A.; Shim, Y.-B. Trace analysis of DNA: Preconcentration, separation, and electrochemical detection in microchip electrophoresis using Au nanoparticles. *Anal. Chem.* **2007**, *79*, 3724–3733.

31. Hatch, A.V.; Herr, A.E.; Throckmorton, D.J.; Brennan, J.S.; Singh, A.K. Integrated preconcentration SDS-PAGE of proteins in microchips using photopatterned cross-linked polyacrylamide gels. *Anal. Chem.* **2007**, *78*, 4976–4984.

32. Lien, K.-Y.; Lin, J.-L.; Liu, C.-Y.; Lei, H.-Y.; Lee, G.-B. Purification and enrichment of virus samples utilizing magnetic beads on a microfluidic system. *Lab Chip* **2007**, *7*, 868–875.

33. Wang, J.; Profitt, J.A.; Pugia, M.J.; Suni, I.I. Au nanoparticle conjugation for impedance and capacitance signal amplification in biosensors. *Anal. Chem.* **2006**, *78*, 1769–1773.

34. Ruan, C.; Yang, L.; Li, Y. Immunobiosensor chips for detection of *Escherichia coli* O157:H7 using electrochemical impedance spectroscopy. *Anal. Chem.* **2002**, *74*, 4814–4820.

35. Edwards, K.A.; Baeumner, A.J. DNA-oligonucleotide encapsulating liposomes as a secondary signal amplification means. *Anal. Chem.* **2007**, *79*, 1806–1815.

36. Jackson, G.W.; Wilson, R. Preparative electrophoresis with on-column optical fiber monitoring and direct elution into a minimized volume. *Biotechnol. Lett.* **2005**, *27*, 1739–1743.

37. Lowe, M.; Spiro, A.; Zhang, Y.-Z.; Getts, R. Multiplexed, particle-based detection of DNA using flow cytometry With 3DNA dendrimers for signal amplification. *Cytometry A* **2004**, *60A*, 135–144.

38. Tian, P.; Mandrell, R. Detection of norovirus capsid proteins in faecal and food samples by a real time immuno-PCR method. *J. Appl. Microbiol.* **2006**, *100*, 564–574.

39. He, L.; Musick, M.D.; Nicewarner, S.R.; Salinas, F.G.; Benkovic, S.J.; Natan, M.J.; Keating, C.D. Colloidal Au-enhanced surface plasmon resonance for ultrasensitive detection of DNA hybridization. *J. Am. Chem. Soc.* **2000**, *122*, 9071–9077.

40. Hirsch, L.R.; Jackson, J.B.; Lee, A.; Halas, N.J.; West, J.L. A whole blood immunoassay using gold nanoshells. *Anal. Chem.* **2003**, *75*, 2377.

41. Wolffs, P.; Knutsson, R.; Norling, B.; Radstrom, P. Rapid quantification of *Yersinia enterocolitica* in pork samples by a novel sample preparation method, flotation, prior to real-time PCR. *J. Clin. Microbiol.* **2004**, *42*, 1042–1047.

42. Menking, D.E.; Emanuel, P.A.; Valdes, J.J.; Kracke, S.K. Rapid cleanup of bacterial DNA from field samples. *Resour. Conserv. Recycl.* **1999**, *27*, 179–186.

43. Tekel, J.; Hudecová, T.; Pecníková, K. Isolation and purification techniques for pesticide residue analyses in samples of plant or animal origin. *Eur. Food Res. Technol.* **2001**, *213*, 250–258.

44. Kelly, L.S.; Kozak, M.; Walker, T.; Pierce, M.; Puett, D. Lectin immunoassays using antibody fragments to detect glycoforms of human chorionic gonadotropin secreted by choriocarcinoma cells. *Anal. Biochem.* **2005**, *338*, 253–262.

45. Rudi, K.; Nogva, H.K.; Moen, B.; Nissen, H.; Bredholt, S.; Moretro, T.; Naterstad, K.; Holck, A. Development and application of new nucleic acid-based technologies for microbial community analyses in foods. *Int. J. Food Microbiol.* **2002**, *78*, 171–180.

46. de Boer, E.; Beumer, R.R. Methodology for detection and typing of foodborne microorganisms. *Int. J. Food Microbiol.* **1999**, *50*, 119–130.

47. Okumura, S.; Akao, T.; Yamashita, S.; Ichimatsu, T.; Inouye, K. Determination of biotinylated proteins as an index for purification of plasma membrane using surface plasmon resonance-based optical biosensor. *Cytotechnology* **2005**, *47*, 59–67.

48. Josic, D.; Brown, M.K.; Huang, F.; Callahan, H.; Rucevic, M.; Nicoletti, A.; Clifton, J.; Hixson, D.C. Use of selective extraction and fast chromatographic separation combined with electrophoretic methods for mapping of membrane proteins. *Electrophoresis* **2005**, *26*, 2809–2822.

49. Jiang, J.; Alderisio, K.A.; Singh, A.; Xiao, L. Development of procedures for direct extraction of *Cryptosporidium* DNA from water concentrates and for relief of PCR inhibitors. *Appl. Environ. Microbiol.* **2005**, *71*, 1135–1141.

50. Roh, C.; Villatte, F.; Kim, B.-G.; Schmid, R.D. "In-gel patch electrophoresis:" A new method for environmental DNA purification. *Electrophoresis* **2005**, *26*, 3055–3061.

51. Schwab, K.J.; Neill, F.H.; Estes, M.K.; Atmar, R.L. Improvements in the RT-PCR detection of enteric viruses in environmental samples. *Water Sci. Technol.* **1998**, *38*, 83–86.

52. Cui, S.; Schroeder, C.M.; Zhang, D.Y.; Meng, J. Rapid sample preparation method for PCR-based detection of *Escherichia coli* O157:H7 in ground beef. *J. Appl. Microbiol.* **2003**, *95*, 129–134.

53. Lakay, F.M.; Botha, A.; Prior, B.A. Comparative analysis of environmental DNA extraction and purification methods from different humic acid-rich soils. *J. Appl. Microbiol.* **2007**, *102*, 265–273.

54. Liang, S.; Redlinger, T. A protocol for isolating putative *Helicobacter pylori* from fecal specimens and genotyping using *vacA* alleles. *Helicobacter* **2003**, *8*, 561–567.

55. Koide, R.T. Nucleic acid isolation from ecological samples—Fungal associations, Mycorrhizae. *Enzymology* **2005**, *395*, 58–72.

56. Perch-Nielsen, I.R.; Bang, D.D.; Poulsen, C.R.; El-Ali, J.; Wolff, A. Removal of PCR inhibitors using dielectrophoresis as a selective filter in a microsystem. *Lab Chip* **2003**, *3*, 212–216.

57. Remya, R.; Syamkumar, S.; Sasikumar, B. Isolation and amplification of DNA from turmeric powder. *Br. Food J.* **2004**, *106*, 673–678.

58. Howelera, M.; Ghiorseb, W.C.; Walker, L.P. A quantitative analysis of DNA extraction and purification from compost. *J. Microbiol. Methods* **2003**, *54*, 37–45.

59. Sandaa, R.-A.; Enger, O.; Torsvik, V. Rapid method for fluorometric quantification of DNA in soil. *Soil Biol. Biochem.* **1998**, *30*, 265–268.

60. López, M.M.; Bertolini, E.; Olmos, A.; Caruso, P.; Gorris, M.T.; Llop, P.; Penyalver, R.; Cambra, M. Innovative tools for detection of plant pathogenic viruses and bacteria. *Int. Microbiol.* **2003**, *6*, 233–243.

61. Estrada, C.S.M.L.; del Carmen Velazquez, L.; Di Genaro, S.; de Guzman, A.M.S. Comparison of DNA extraction methods for pathogenic *Yersinia enterocolitica* detection from meat food by nested PCR. *Food Res. Int.* **2007**, *40*, 637–642.

62. Kahler, A.M.; Thurston-Enriquez, J.A. Human pathogenic microsporidia detection in agricultural samples: Method development and assessment. *Parasitol. Res.* **2007**, *100*, 529–538.

63. Kotlowski, R.; Martin, A.; Anthony, A.; Chemlal, K.; Fonteyene, P.-A.; Portaels, F. One-tube cell lysis and DNA extraction procedure for PCR-based detection of *Mycobacterium ulcerans* in aquatic insects, molluscs and fish. *J. Med. Microbiol.* **2004**, *53*, 927–933.

64. Miyachi, H.; Yano, K.; Ikebukuro, K.; Kono, M.; Hoshina, S.; Karube, I. Application of chimeric RNA-DNA oligonucleotides to the detection of pathogenic microorganisms using surface plasmon resonance. *Anal. Chim. Acta* **2000**, *407*, 1–10.

65. Waller, D.F.; Ogata, S.A. Development and evaluation of a micro- and nanoscale proteomic sample preparation method. *Appl. Environ. Microbiol.* **2000**, *66*, 4115–4118.

66. Zhao, X.; Hilliard, L.R.; Mechery, S.J.; Wang, Y.; Bagwe, R.P.; Jin, S. A rapid bioassay for single bacterial cell quantitation using bioconjugated nanoparticles. *Proc. Natl. Acad. Sci. USA* **2004**, *101*, 15027–15032.

67. Whitney, D.; Skoletsky, J.; Moore, K.; Boynton, K.; Kann, L.; Brand, R.; Syngal, S.; Lawson, M.; Shuber, A. Enhanced retrieval of DNA from human fecal samples results in improved performance of colorectal cancer screening test. *J. Mol. Diagn.* **2004**, *6*, 386–395.

68. Metfies, K.; Huljic, S.; Lange, M.; Medlin, L.K. Electrochemical detection of the toxic dinoflagellate *Alexandrium ostenfeldii* with a DNA-biosensor. *Biosens. Bioelectron.* **2005**, *20*, 1349–1357.

69. Roose-Amsaleg, C.L.; Garnier-Sillam, E.; Harry, M. Extraction and purification of microbial DNA from soil and sediment samples. *Appl. Soil Ecol.* **2001**, *18*, 47–60.

70. Mills, J.B.; Mant, C.T.; Hodges, R.S. One-step purification of a recombinant protein from a whole cell extract by reversed-phase high-performance liquid chromatography. *J. Chromatogr. A* **2006**, *1133*, 248–253.

71. Mumy, K.L.; Findlay, R.H. Convenient determination of DNA extraction efficiency using an external DNA recovery standard and quantitative-competitive PCR. *J. Microbiol. Methods* **2004**, *57*, 259–268.

72. Niemi, R.M.; Heiskanen, I.; Wallenius, K.; Lindstrom, K. Extraction and purification of DNA in rhizosphere soil samples for PCR-DGGE analysis of bacterial consortia. *J. Microbiol. Methods* **2001**, *45*, 155–165.

73. Liao, J.C.; Mastali, M.; Li, Y.; Gau, V.; Suchard, M.A.; Babbitt, J.; Gornbein, J.; Landlaw, E.M.; McCabe, E.R.B.; Churchill, B.M.; Haake, D.A. Development of an advanced electrochemical DNA biosensor for bacterial pathogen detection. *J. Mol. Diagn.* **2007**, *9*, 158–168.

74. Honoré-Bouakline, S.; Vincensini, J.P.; Giacuzzo, V.; Herrmann, J.L. Rapid diagnosis of extrapulmonary tuberculosis by PCR: Impact of sample preparation and DNA extraction. *J. Clin. Microbiol.* **2003**, *41*, 2323–2329.

75. D'Souza, G.; Sugar, E.; Ruby, W.; Gravitt, P.; Gillison, M. Analysis of the effect of DNA purification on detection of human papillomavirus in oral rinse samples by PCR. *J. Clin. Microbiol.* **2005**, *43*, 5526–5535.

76. Lemarchand, K.; Berthiaume, F.; Maynard, C.; Harel, J.; Payment, P.; Bayardelle, P.; Masson, L.; Brousseau, R. Optimization of microbial DNA extraction and purification from raw wastewater samples for downstream pathogen detection by microarrays. *J. Microbiol. Methods* **2005**, *63*, 115–126.

77. Lin, W.-F.; Hwang, D.-F. Application of PCR-RFLP analysis on species identification of canned tuna. *Food Control* **2007**, *18*, 1050–1057.

78. Josefsen, M.H.; Krause, M.; Hansen, F.; Hoorfar, J. Optimization of a 12-hour TaqMan PCR-based method for detection of *Salmonella* bacteria in meat. *Appl. Environ. Microbiol.* **2007**, *73*, 3040–3048.

79. Long, J.-R.; Zhang, Y.-Y.; Hofer, E.; Seeman, C.; Schatz, M.; O'Brien, D.; Deng, H.-W. Rapid purification of high quality DNA from large whole blood samples using a new rapid protocol. *Eng. Life Sci.* **2004**, *4*, 372–374.

80. Scanga, L.; Chaing, S.; Powell, C.; Aylsworth, A.S.; Harrell, L.J.; Henshaw, N.G.; Civalier, C.J.; Thorne, L.B.; Weck, K.; Booker, J.; Gulley, M.L. Diagnosis of human congenital cytomegalovirus infection by amplification of viral DNA from dried blood spots on perinatal cards. *J. Mol. Diagn.* **2006**, *8*, 240–245.

81. Cler, L.; Bu, D.; Lewis, C.; Euhus, D. A comparison of five methods for extracting DNA from paucicellular clinical samples. *Mol. Cell. Probes* **2006**, *20*, 191–196.

82. Vanni, A.; Anfossi, L.; Giovannoli, C.; Oddenino, L.; Giraudi, G. Evaluation of purification procedures of DNA from maize-meal samples by exploiting different analytical techniques for the assessment of DNA quality. *Ann. Chim.* **2004**, *94*, 1–12.

83. Powell, M.J.; Timperman, A.T. Quantitative analysis of protein recovery from dilute, large volume samples by tangential flow ultrafiltration. *J. Memb. Sci.* **2005**, *252*, 227–236.

84. Haugland, R.A.; Brinkman, N.; Vesper, S.J. Evaluation of rapid DNA extraction methods for the quantitative detection of fungi using real-time PCR analysis. *J. Microbiol. Methods* **2002**, *50*, 319–323.

85. Yeung, S.W.; Hsing, I.-M. Manipulation and extraction of genomic DNA from cell lysate by functionalized magnetic particles for lab on a chip applications. *Biosens. Bioelecton.* **2006**, *21*, 989–997.

86. Sharpe, A.N.; Kilsby, D.C. Ultrasound and vortex stirring as bacteriological sampling methods for foods. *J. Appl. Bacteriol.* **1970**, *33*, 351–357.

87. Sharpe, A.N.; Harshman, G.C. Recovery of *Clostridium perfringens*, *Staphylococcus aureus* and molds from foods by the stomacher: Effect of fat content, surfactant concentration and blending time. *J. Can. Inst. Food Sci. Technol.* **1976**, *9*, 30–34.

88. Yu, C.; Cohen, L. Tissue sample preparation—Not the same old grind. *LC-GC* **2003**, *21*, 1038–1048.

89. Tien, C.C.; Chao, C.C.; Chao, W.L. Methods for DNA extraction from various soils: A comparison. *J. Appl. Microbiol.* **1999**, *86*, 937–943.

90. Bruns, M.A.; Buckley, D.H. Isolation and purification of microbial community nucleic acids from environmental samples. In *Manual of Environmental Microbiology*, Hurst, C.J.; Crawford, R.L.; Knudsen, G.R., Eds., 2nd ed. Washington, DC: ASM Press; 2002; pp. 564–572.

91. Tilsala-Timisjarvi, A.; Alatossava, T. Rapid DNA preparation from milk and dairy process samples for the detection of bacteria by PCR. *Food Microbiol.* **2004**, *21*, 365–368.

92. He, J.; Xu, Z.; Hughes, J. Pre-lysis washing improves DNA extraction from a forest soil. *Soil Biol. Biochem.* **2005**, *37*, 2337–2341.

93. Mayr, B.M.; Kohlbacher, O.; Reinert, K.; Sturm, M.; Gropl, C.; Lange, E.; Klein, C.; Huber, C.G. Absolute myoglobin quantitation in serum by combining two-dimensional liquid chromatography-electrospray ionization mass spectrometry and novel data analysis algorithms. *J. Proteome Res.* **2006**, *5*, 414–421.

94. Straub, T.M.; Chandler, D.P. Towards a unified system for detecting waterborne pathogens. *J. Microbiol. Methods* **2003**, *52*, 185–187.

95. Tortorello, M.L.; Stewart, D.S. Antibody-direct epifluorescent filter technique for rapid, direct enumeration of *Escherichia coli* O157:H7 in beef. *Appl. Environ. Microbiol.* **1994**, *60*, 3553–3559.

96. Pettipher, G.L.; Rodrigues, U.M. Rapid enumeration of microorganisms in foods by the direct epifluorescent filter technique. *Appl. Environ. Microbiol.* **1982**, *44*, 809–813.

97. Stevens, K.A.; Jaykus, L.-A. Bacterial separation and concentration from complex sample matrices: A review. *Crit. Rev. Microbiol.* **2004**, *30*, 7–24.

98. Entis, P.; Brodsky, M.H.; Sharpe, A.N. Effect of pre-filtration and enzyme treatment of membrane filtration of foods. *J. Food Prot.* **1982**, *45*, 8–11.

99. Walker, P.A. III; Morris, M.D. Isotachophoretic separations on a microchip. Normal raman spectroscopy detection. *Anal. Chem.* **1998**, *70*, 3766–3769.

100. Ghosh, R.; Cui, Z.F. Protein purification by ultrafiltration with pre-treated membrane. *J. Memb. Sci.* **2000**, *167*, 47–53.

101. Eon-Duval, A.; MacDuff, R.H.; Fisher, C.A.; Harris, M.J.; Brook, C. Removal of RNA impurities by tangential flow filtration in an RNase-free plasmid DNA purification process. *Anal. Biochem.* **2003**, *316*, 66–74.

102. Kendall, D.; Lye, G.J.; Levy, M.S. Purification of Plasmid DNA by an integrated operation comprising tangential flow filtration and nitrocellulose adsorption. *Biotechnol. Bioeng.* **2002**, *79*, 816–822.

103. Wickramasinghe, S.R.; Kalbfuß, B.; Zimmermann, A.; Thom, V.; Reichl, U. Tangential flow microfiltration and ultrafiltration for human influenza A virus concentration and purification. *Biotechnol. Bioeng.* **2005**, *92*, 199–208.

104. Casas, V.; Rohwer, F. Phage metagenomics. *Methods Enzymol.* **2007**, *421*, 259–268.

105. Jiang, S.; Noble, R.; Chu, W. Human adenoviruses and coliphages in urban runoff-impacted coastal waters of southern California. *Appl. Environ. Microbiol.* **2001**, *67*, 179–184.

106. Payne, M.J.; Kroll, R.G. Methods for the separation and concentration of bacteria from foods. *Trends Food Sci. Technol.* **1991**, *2*, 315–319.

107. Deisingh, A.K.; Thompson, M. Strategies for the detection of *Escherichia coli* O157:H7 in foods. *J. Appl. Microbiol.* **2004**, *96*, 419–429.

108. Hsing, I.-M.; Xu, Y.; Zhao, W. Micro- and nano-magnetic particles for applications in biosensing. *Electroanalysis* **2007**, *19*, 755–768.

109. Inglis, D.W.; Riehn, R.; Sturm, J.C.; Austin, R.H. Microfluidic high gradient magnetic cell separation. *J. Appl. Physics* **2006**, *99*, 08K101-1–08K101-3.

110. Steingroewer, J.; Knaus, H.; Bley, T.; Boschke, E. A rapid method for the pre-enrichment and detection of *Salmonella typhimurium* by immunomagnetic separation and subsequent fluorescence microscopical techniques. *Eng. Life Sci.* **2005**, *5*, 267–272.

111. Skjerve, E.; Rorvik, L.M.; Olsvik, O. High-volume extraction of nucleic acids by magnetic bead technology for ultrasensitive detection of bacteria in blood components. *Appl. Environ. Microbiol.* **1990**, *56*, 3478–3481.

112. Fitzmaurice, J.; Duffy, G.; Kilbride, B.; Sheridan, J.J.; Carroll, C.; Maher, M. Comparison of a membrane surface adhesion recovery method with an IMS method for use in a polymerase chain reaction method to detect *Escherichia coli* O157:H7 in minced beef. *J. Microbiol. Methods* **2004**, *59*, 243–252.

113. Španová, A.; Rittich, B.; Karpíšková, R.; Cechova, L.; Skapova, D. PCR identification of *Salmonella* cells in food and stool samples after immunomagnetic separation. *Bioseparation* **2001**, *9*, 379–384.

114. Waller, D.F.; Ogata, S.A. Quantitative immunocapture PCR assay for detection of *Campylobacter jejuni* in foods. *Appl. Environ. Microbiol.* **2000**, *66*, 4115–4118.

115. Kang, D.-H.; Barkocy-Gallagher, G.A.; Koohmaraie, M.; Siragusa, G.R. Screening bovine carcass sponge samples for *Escherichia coli* O157 using a short enrichment coupled with immunomagnetic separation and a polymerase chain reaction-based (BAX) detection step. *J. Food Prot.* **2001**, *64*, 1610–1612.

116. Majors, R.E. Sample preparation for large-scale protein purification. *LC-GC* **2005**, *23*, 730–737.

117. Lee, J.-G.; Cheong, K.H.; Huh, N.; Kim, S.; Choi, J.-W.; Ko, C. Microchip-based one step DNA extraction and real-time PCR in one chamber for rapid pathogen identification. *Lab Chip* **2006**, *6*, 886–895.

118. Yeates, C.; Gillings, M.R. Microchip-based one step DNA extraction and real-time PCR in one chamber for rapid pathogen identification. *Lett. Appl. Microbiol.* **1998**, *27*, 49–53.

119. Benčina, M.; Podgornik, A.; Štrancar, A. Characterization of methacrylate monoliths for purification of DNA molecules. *J. Sep. Sci.* **2004**, *27*, 801–810.

120. Trochimchuka, T.; Fotheringhama, J.; Topp, E.; Schraft, H.; Leung, K.T. A comparison of DNA extraction and purification methods to detect *Escherichia coli* O157:H7 in cattle manure. *J. Microbiol. Methods* **2003**, *54*, 165–175.

121. Levisona, P.R.; Badger, S.E.; Hathi, P.; Davies, M.J.; Bruce, I.J.; Grimm, V. New approaches to the isolation of DNA by ion-exchange chromatography. *J. Chromatogr. A* **1998**, *827*, 337–344.

122. Josic, D.; Clifton, J.G. Use of monolithic supports in proteomics technology. *J. Chromatogr. A* **2007**, *1144*, 2–13.

123. Forcica, D.; Branovic-Cakanic, K.; Ivancic, J.; Jug, R.; Barut, M.; Strancar, A. Purification of genomic DNA by short monolithic columns. *J. Chromatogr. A* **2005**, *1065*, 115–120.

124. Zou, H.; Luo, Q.; Zhou, D. Affinity membrane chromatography for the analysis and purification of proteins. *J. Biochem. Biophys. Methods* **2001**, *49*, 199–240.

125. Chan, A.; Krull, U.J. Capillary electrophoresis for capture and concentrating of target nucleic acids by affinity gels modified to contain single-stranded nucleic acid probes. *Anal. Chim. Acta* **2006**, *578*, 31–42.

126. Paegel, B.M.; Yeung, S.H.I.; Mathies, R.A. Microchip bioprocessor for integrated nanovolume sample purification and DNA sequencing. *Anal. Chem.* **2002**, *74*, 5092–5098.

127. Olsen, K.G.; Ross, D.J.; Tarlov, M.J. Immobilization of DNA hydrogel plugs in microfluidic channels. *Anal. Chem.* **2002**, *74*, 1436–1441.

128. Efimov, V.A.; Buryakova, A.A.; Chakhmakhcheva, O.G. Synthesis of polyacrylamides N-substituted with PNA-like oligonucleotide mimics for molecular diagnostic applications. *Nucleic Acids Res.* **1999**, *27*, 4416–4426.

129. Chandler, D.P.; Jarrell, A.E. Enhanced nucleic acid capture and flow cytometry detection with peptide nucleic acid probes and tunable-surface microparticles. *Anal. Biochem.* **2003**, *312*, 182–190.

130. Chandler, D.P.; Stults, J.R.; Cebula, S.; Schuck, B.L.; Weaver, D.W.; Anderson, K.K.; Egholm, M.; Brockman, F.J. Affinity purification of DNA and RNA from environmental samples with peptide nucleic acid clamps. *Appl. Environ. Microbiol.* **2000**, *66*, 3438–3445.

131. Ghosh, R. Protein separation using membrane chromatography: Opportunities and challenges. *J. Chromatogr. A* **2002**, *952*, 13–27.

132. Baijal-Gupta, M.; Fraser, J.E.; Clarke, M.W.; Xuan, J.W.; Finkelman, M.A. A new scalable purification procedure for prostatic secretory protein (PSP94) from human seminal plasma. *Protein Expr. Purif.* **1996**, *8*, 483–488.

133. Long, L.-S.; Chen, K.-C.; Lin, S.-R.; Huang, H.-B. Purification and characterization of *Ophiophagus hannah* cytotoxin-like proteins. *Toxicon* **2006**, *48*, 429–436.

134. Cho, S.; Lee, S.-H.; Chung, W.-J.; Kim, Y.-K.; Lee, Y.-S.; Kim, B.-G. Microbead-based affinity chromatography chip using RNA aptamer modified with photocleavable linker. *Electrophoresis* **2004**, *25*, 3730–3739.

135. Bergendahl, V.; Thompson, N.E.; Foley, K.M.; Olson, B.M.; Burgess, R.R. A cross-reactive polyol-responsive monoclonal antibody useful for isolation of core RNA polymerase from many bacterial species. *Protein Expr. Purif.* **2002**, *31*, 155–160.

136. Gadgil, H.; Oak, S.A.; Jarrett, H.W. Affinity purification of DNA-binding proteins. *J. Biochem. Biophys. Methods* **2001**, *49*, 607–624.

137. Azarkan, M.; Huet, J.; Baeyens-Volant, D.; Looze, Y.; Vandenbussche, G. Affinity chromatography: A useful tool in proteomics studies. *J. Chromatogr. B* **2007**, *849*, 81–90.

138. Aprilita, N.H.; Huck, C.W.; Bakry, R.; Feuerstein, I.; Stecher, G.; Morandell, S.; Huang, H.-L.; Stasyk, T.; Huber, L.A.; Bonn, G.K. Poly(glycidyl methacrylate/divinylbenzene)-IDA-FeIII in phosphoproteomics. *J. Proteome Res.* **2005**, *4*, 2312–2319.

139. Başar, N.; Uzun, L.; Güner, A.; Denizli, A. Lysozyme purification with dye-affinity beads under magnetic field. *Int. J. Bio. Macromol.* **2007**, *41*, 234–242.

140. Akgöl, S.; Bereli, N.; Denizli, A. Magnetic dye affinity beads for the adsorption of β-casein. *Macromol. Biosci.* **2005**, *5*, 786–794.

141. Elphinstone, M.S.; Hinten, G.N.; Anderson, M.J.; Nock, C.J. An inexpensive and high-throughput procedure to extract and purify total genomic DNA for population studies. *Mol. Ecol. Notes* **2003**, *3*, 317–320.

142. Bhattacharyya, A.; Klapperich, C.M. Removal of PCR inhibitors using dielectrophoresis as a selective filter in a microsystem. *Anal. Chem.* **2003**, *78*, 788–792.

143. Breadmore, M.C.; Wolfe, K.A.; Arcibal, I.G.; Leung, W.K.; Dickson, D.; Giordano, B.C.; Power, M.E.; Ferrance, J.P.; Feldman, S.H.; Norris, P.M.; Landers, J.P. Microchip-based purification of DNA from biological samples. *Anal. Chem.* **2003**, *75*, 1880–1886.

144. Ki, J.-S.; Chang, K.B.; Roh, H.J.; Lee, B.Y.; Yoon, J.Y.; Jang, G.Y. Direct DNA isolation from solid biological sources without pretreatments with proteinase-K and/or homogenization through automated DNA extraction. *J. Biosci. Bioeng.* **2007**, *103*, 242–246.

145. Svec, F. Less common applications of monoliths: Preconcentration and solid-phase extraction. *J. Chromatogr. B* **2006**, *841*, 52–64.

146. Fontanals, N.; Marcé, R.M.; Borrull, F. New hydrophilic materials for solid-phase extraction. *Trends Anal. Chem.* **2005**, *24*, 394–406.

147. Wen, J.; Guillo, C.; Ferrance, J.P.; Landers, J.P. Microfluidic-based DNA purification in a two-stage, dual-phase microchip containing a reversed-phase and a photopolymerized monolith. *Anal. Chem.* **2007**, *79*, 6135–6142.

148. Wen, J.; Guillo, C.; Ferrance, J.P.; Landers, J.P. DNA extraction using a tetramethyl orthosilicate-grafted photopolymerized monolithic solid phase. *Anal. Chem.* **2006**, *78*, 1673–1681.

149. Pichon, V. Selective sample treatment using molecularly imprinted polymers. *J. Chromatogr. A* **2007**, *1152*, 41–53.

150. Kandimalla, V.B.; Ju, H. Molecular imprinting: A dynamic technique for diverse applications in analytical chemistry. *Anal. Bioanal. Chem.* **2004**, *380*, 587–605.

151. Xu, X.; Zhu, L.; Chen, L. Separation and screening of compounds of biological origin using molecularly imprinted polymers. *J. Chromatogr. B* **2004**, *804*, 61–69.

152. Chapuis, F.; Mullot, J.-U.; Pichon, V.; Tuffal, G.; Hennion, M.-C. Molecularly imprinted polymers for the clean-up of a basic drug from environmental and biological samples. *J. Chromatogr. A* **2006**, *1135*, 127–134.

153. Li, H.; Liu, Y.; Zhang, Z.; Liao, H.; Nie, L.; Yao, S. Separation and purification of chlorogenic acid by molecularly imprinted polymer monolithic stationary phase. *J. Chromatogr. A* **2005**, *1098*, 66–74.

154. Urraca, J.L.; Moreno-Bondi, M.C.; Hall, A.J.; Sellergren, B. Direct extraction of penicillin G and derivatives from aqueous samples using a stoichiometrically imprinted polymer. *Anal. Chem.* **2007**, *79*, 695–701.

155. Dubus, S.; Gravel, J.-F.; Le Drogoff, B.; Nobert, P.; Veres, T.; Boudreau, D. PCR-free DNA detection using a magnetic bead-supported polymeric transducer and microelectromagnetic traps. *Anal. Chem.* **2006**, *78*, 4457–4464.

156. Faggi, E.; Pini, G.; Campisi, E. Use of magnetic beads to extract fungal DNA. *Mycoses* **2005**, *48*, 3–7.

157. Fuentes, M.; Mateo, C.; Rodriguez, A.; Casqueiro, M.; Tercero, J.C.; Riese, H.H.; Fernandez-Lafuente, R.; Guisan, J.M. Detecting minimal traces of DNA using DNA covalently attached to superparamagnetic nanoparticles and direct PCR-ELISA. *Biosens. Bioelectron.* **2006**, *21*, 1574–1580.

158. Bitner, R.; Koller, S.; Sankbeil, J.; Denhart, M.; Shenoi, H. Purifying genomic DNA from whole blood on automated, high-throughput, and moderate-throughput platforms. *JALA*, **2004**, *9(2)*, 64–71.

159. Fontaine, M.; Guillot, E. An immunomagnetic separation–real-time PCR method for quantification of *Cryptosporidium parvum* in water samples. *J. Microbiol. Methods* **2003**, *54*, 29–36.

160. Witek, M.A.; Llopis, S.D.; Wheatley, A.; McCarley, R.L.; Soper, S.A. Purification and preconcentration of genomic DNA from whole cell lysates using photoactivated polycarbonate (PPC) microfluidic chips. *Nucleic Acids Res.* **2006**, *34*, e74.

161. Erdem, A.; Pividori, M.I.; Lermo, A.; Bonanni, A.; del Valle, M.; Alegret, S. Genomagnetic assay based on label-free electrochemical detection using magneto-composite electrodes. *Sens. Actuators B* **2006**, *114*, 591–598.

162. Wang, J.; Xu, D.; Kawde, A.-N.; Polsky, R. Metal nanoparticle-based electrochemical stripping potentiometric detection of DNA hybridization. *Anal. Chem.* **2001**, *73*, 5576–5581.

163. Mann, T.L.; Krull, U.J. The application of ultrasound as a rapid method to provide DNA fragments suitable for detection by DNA biosensors. *Biosens. Bioelectron.* **2004**, *20*, 945–955.

164. Vinogradova, O.A.; Pyshnaya, I.A.; Zarytova, V.F.; Ivanova, E.M.; Pyshnyi, D.V. Enhancement of a hybridization analysis efficiency by the controlled DNA fragmentation. *Mol. Biol.* **2007**, *41*, 148–156.

165. LaMontagne, M.G.; Michel, F.C. Jr.; Holden, P.A.; Reddy, C.A. Evaluation of extraction and purification methods for obtaining PCR-amplifiable DNA from compost for microbial community analysis. *J. Microbiol. Methods* **2002**, *49*, 255–264.

166. Mannelli, I.; Minunni, M.; Tombelli, S.; Mascini, M. Quartz crystal microbalance (QCM) affinity biosensor for genetically modified organisms (GMOs) detection. *Biosens. Bioelectron.* **2003**, *18*, 129–140.

167. Tombelli, S.; Minunni, M.; Santucci, A.; Spiriti, M.M.; Mascini, M. A DNA-based piezoelectric biosensor: Strategies for coupling nucleic acids to piezoelectric devices. *Talanta* **2006**, *68*, 806–812.

168. de Mello, A.J. DNA amplification: Does "small" really mean "efficient"? *Lab Chip* **2001**, *1*, 24N–29N.

169. Toze, S. PCR and the detection of microbial pathogens in water and wastewater. *Water Res.* **1999**, *33*, 3545–3556.

170. Sachse, K. Specificity and performance of diagnostic PCR assays. In *Methods in Molecular Biology, Vol. 216: PCR Detection of Microbial Pathogens: Methods and Protocols*, Sachse, K.; Frey, J., Eds. Totowa: Humana Press; 2003; pp. 3–29.

171. Blackstone, G.M.; Nordstrom, J.L.; Vickery, M.C.L.; Bowen, M.D.; Meyer, R.F.; DePaolo, A. Detection of pathogenic *Vibrio parahaemolyticus* in oyster enrichments by real time PCR. *J. Microbiol. Methods* **2003**, *53*, 149–155.

172. Salk, J.J.; Sanchez, J.A.; Pierce, K.E.; Rice, J.E.; Soares, K.C.; Wangh, L.J. Direct amplification of single-stranded DNA for pyrosequencing using linear-after-the-exponential (LATE)–PCR. *Anal. Biochem.* **2006**, *353*, 124–132.

173. Sanchez, J.A.; Pierce, K.E.; Rice, J.E.; Wangh, L.J. Linear-after-the-exponential (LATE)–PCR: An advanced method of asymmetric PCR and its uses in quantitative real-time analysis. *Proc. Natl. Acad. Sci. USA* **2004**, *101*, 1933–1938.

174. Monis, P.T.; Giglio, S. Nucleic acid amplification-based techniques for pathogen detection and identification. *Infect. Genet. Evol.* **2006**, *6*, 2–12.

175. Levin, R.E. The application of real-time PCR to food and agricultural systems. A review. *Food Biotechnol.* **2004**, *18*, 97–133.

176. Hsih, H.-Y.; Tsen, H.-Y. Combination of immunomagnetic separation and polymerase chain reaction for the simultaneous detection of *Listeria monocytogenes* and *Salmonella* spp. in food samples. *J. Food Prot.* **2001**, *64*, 1744–1750.

177. Ache, H.J. Chemical microanalytical systems: Objectives and latest developments. *Fresenius J. Anal. Chem.* **1996**, *355*, 467–474.

178. Situma, C.; Hashimoto, M.; Soper, S.A. Chemical microanalytical systems: Objectives and latest developments. *Biomol. Eng.* **2006**, *23*, 213–231.

179. de Mello, A.J.; Beard, N. Focus. Dealing with "real" samples: Sample pre-treatment in microfluidic systems. *Lab Chip* **2003**, *3*, 11N–19N.

180. Bruin, G.J.M. Recent developments in electrokinetically driven analysis on microfabricated devices. *Electrophoresis* **2000**, *21*, 3931–3951.

181. Chiesl, T.N.; Shi, W.; Barron, A.E. Poly(acrylamide-co-alkylacrylamides) for electrophoretic DNA purification in microchannels. *Anal. Chem.* **2005**, *77*, 772–779.

182. Cygan, Z.T.; Cabral, J.T.; Beers, K.L.; Amis, E.J. Microfluidic platform for the generation of organic-phase microreactors. *Langmuir* **2005**, *21*, 3629–3634.

183. Chen, X.; Cui, D.; Liu, C.; Li, H.; Chen, J. Continuous flow microfluidic device for cell separation, cell lysis and DNA purification. *Anal. Chim. Acta* **2007**, *584*, 237–243.

184. Paegel, B.M.; Blazej, R.G.; Mathies, R.A. Microfluidic devices for DNA sequencing: Sample preparation and electrophoretic analysis. *Curr. Opin. Biotechnol.* **2003**, *14*, 42–50.

185. Long, Z.; Liu, D.; Ye, N.; Qin, J.; Lin, B. Integration of nanoporous membranes for sample filtration/preconcentration in microchip electrophoresis. *Electrophoresis* **2006**, *27*, 4927–4934.

186. Foote, R.S.; Khandurina, J.; Jacobson, S.C.; Ramsey, J.M. Preconcentration of proteins on microfluidic devices using porous silica membranes. *Anal. Chem.* **2005**, *77*, 57–63.

187. Song, S.; Singh, A.K.; Kirby, B.J. Electrophoretic concentration of proteins at laser-patterned nanoporous membranes in microchips. *Anal. Chem.* **2004**, *76*, 4589–4592.

188. Irimia, D.; Tompkins, R.G.; Toner, M. Single-cell chemical lysis in picoliter-scale closed volumes using a microfabricated device. *Anal. Chem.* **2004**, *76*, 6137–6143.

189. Hui, W.C.; Yobas, L.; Samper, V.D.; Heng, C.-K.; Liw, S.; Ji, H.; Chen, Y.; Cong, L.; Li, J.; Lim, T.M. Single-cell chemical lysis in picoliter-scale closed volumes using a microfabricated device. *Sens. Actuators A* **2007**, *133*, 335–339.

190. Wilding, P.; Kricka, L.J.; Cheng, J.; Hvichia, G.; Shoffner, M.A.; Fortina, P. Integrated cell isolation and polymerase chain reaction analysis using silicon microfilter chambers. *Anal. Biochem.* **1998**, *257*, 95–100.

191. Schilling, E.A.; Kamholz, A.E.; Yager, P. Cell lysis and protein extraction in a microfluidic device with detection by a fluorogenic enzyme assay. *Anal. Chem.* **2002**, *74*, 1798–1804.

192. Sun, Y.; Kwok, Y.C. Polymeric microfluidic system for DNA analysis. *Anal. Chim. Acta* **2006**, *556*, 80–96.

193. Lee, N.Y.; Yamada, M.; Seki, M. Development of a passive micromixer based on repeated fluid twisting and flattening, and its application to DNA purification. *Anal. Bioanal. Chem.* **2005**, *383*, 776–782.

194. Lu, H.; Schmidt, M.A.; Jensen, K.F. A microfluidic electroporation device for cell lysis. *Lab Chip* **2005**, *5*, 23–29.

195. Taylor, M.T.; Belgrader, P.; Furman, B.J.; Pourahmadi, F.; Kovacs, G.T.A.; Northrup, M.A. Lysing bacterial spores by sonication through a flexible interface in a microfluidic system. *Anal. Chem.* **2001**, *73*, 492–496.

196. Lee, S.-W.; Tai, Y.-C. A micro cell lysis device. *Sens. Actuators A* **1999**, *73*, 74–79.

197. Di Carlo, D.; Ionescu-Zanetti, C.; Zhang, Y.; Hung, P.; Lee, L.P. On-chip cell lysis by local hydroxide generation. *Lab Chip* **2005**, *5*, 171–178.

198. Bienvenue, J.M.; Duncalf, N.; Marchiarullo, D.; Ferrance, J.P.; Landers, J.P. Microchip-based cell lysis and DNA extraction from sperm cells for application to forensic analysis. *J. Forensic Sci.* **2006**, *51*, 266–273.

199. Bruckner-Lea, C.J.; Tsukuda, T.; Dockendorff, B.; Follansbee, J.C.; Kingsley, M.T.; Ocampo, C.; Stults, J.R.; Chandler, D.P. Renewable microcolumns for automated DNA purification and flow-through amplification: From sediment samples through polymerase chain reaction. *Anal. Chim. Acta* **2002**, *469*, 129–140.

200. Chandler, D.P.; Schuck, B.L.; Brockman, F.J.; Bruckner-Lea, C.J. Automated nucleic acid isolation and purification from soil extracts using renewable affinity microcolumns in a sequential injection system. *Talanta* **1999**, *49*, 969–983.

201. Cady, N.C.; Stelick, S.; Batt, C.A. Nucleic acid purification using microfabricated silicon structures. *Biosens. Bioelecton.* **2003**, *19*, 59–66.

202. Wolfe, K.A.; Breadmore, M.C.; Ferrance, J.P.; Power, M.E.; Conroy, J.F.; Norris, P.M.; Landers, J.P. Toward a microchip-based solid-phase extraction method for isolation of nucleic acids. *Electrophoresis* **2002**, *23*, 727–733.

203. Yu, C.; Davey, M.H.; Svec, F.; Frechet, J.M.J. Monolithic porous polymer for on-chip solid-phase extraction and preconcentration prepared by photoinitiated in situ polymerization within a microfluidic device. *Anal. Chem.* **2001**, *73*, 5088–5096.

204. Oleschuk, R.D.; Shultz-Lockyear, L.L.; Ning, Y.; Harrison, D.J. Trapping of bead-based reagents within microfluidic systems: On-chip solid-phase extraction and electrochromatography. *Anal. Chem.* **2000**, *72*, 585–590.

205. Lichtenberg, J.; Verpoorte, E.; de Rooij, N.F. Sample preconcentration by field amplification stacking for microchip-based capillary electrophoresis. *Electrophoresis* **2001**, *22*, 258–271.

206. Satterfield, B.C.; Stern, S.; Caplan, M.R.; Hukari, K.W.; West, J.A.A. Microfluidic purification and preconcentration of mRNA by flow-through polymeric monolith. *Anal. Chem.* **2007**, *79*, 6230–6235.

207. Wu, Q.; Bienvenue, J.M.; Hassan, B.J.; Kwok, Y.C.; Giordano, B.C.; Norris, P.M.; Landers, J.P.; Ferrance, J.P. Microchip-based macroporous silica sol-gel monolith for efficient isolation of DNA from clinical samples. *Anal. Chem.* **2006**, *78*, 5704–5710.

208. Easley, C.J.; Karlinsey, J.M.; Bienvenue, J.M.; Legendre, L.A.; Roper, M.G.; Feldman, S.H.; Hughes, M.A.; Hewitt, E.L.; Merkel, T.J.; Ferrance, J.P.; Landers, J.P. A fully integrated microfluidic genetic analysis system with sample-in–answer-out capability. *Proc. Natl. Acad. Sci. USA* **2006**, *103*, 19272–19277.

209. Varshney, M.; Li, Y. Interdigitated array microelectrode based impedance biosensor coupled with magnetic nanoparticle–antibody conjugates for detection of *Escherichia coli* O157:H7 in food samples. *Biosens. Bioelectron.* **2007**, *22*, 2408–2414.

210. Liu, Y.; Rauch, C.B.; Stevens, R.L.; Lenigk, R.; Yang, J.; Rhine, D.B.; Grodzinski, P. DNA amplification and hybridization assays in integrated plastic monolithic devices. *Anal. Chem.* **2002**, *74*, 3063–3070.

211. Gijs, M.A.M. Magnetic bead handling on-chip: New opportunities for analytical applications. *Microfluid. Nanofluidics* **2004**, *1*, 22–40.

212. Guo, S.S.; Zuo, C.C.; Huang, W.H.; Peroz, C.; Chen, Y. Response of super-paramagnetic beads in microfluidic devices with integrated magnetic micro-columns. *Microelectron. Eng.* **2006**, *83*, 1655–1659.

213. Ramadan, Q.; Samper, V.; Poenar, D.; Chen, Y. Magnetic-based microfluidic platform for biomolecular separation. *Biomed. Microdevices* **2006**, *8*, 151–158.

214. Lee, H.; Liu, Y.; Ham, D.; Westervelt, R.M. Integrated cell manipulation system-CMOS/microfluidic hybrid. *Lab Chip* **2007**, *7*, 331–337.

215. Chou, C.-F.; Zenhausern, F. Electrodeless dielectrophoresis for micro total analysis systems. *IEEE Eng. Med. Biol. Mag.* **2003**, November/December, 62–67.

216. Yang, L.; Banada, P.P.; Chatni, M.R.; Lim, K.S.; Bhunia, A.K.; Ladish, M.; Bashir, R. A multifunctional micro-fluidic system for dielectrophoretic concentration coupled with immuno-capture of low numbers of *Listeria monocytogenes*. *Lab Chip* **2006**, *6*, 896–906.

217. Jung, J.-Y.; Kwak, H.-Y. Separation of microparticles and biological cells inside an evaporating droplet using dielectrophoresis. *Anal. Chem.* **2007**, *79*, 5087–5092.

218. Hughes, M.P. Strategies for dielectrophoretic separation in laboratory-on-a-chip systems. *Electrophoresis* **2002**, *23*, 2569–2582.

219. Li, Y.; Dalton, C.; Crabtree, H.J.; Nilsson, G.; Kaler, K.V.I.S. Continuous dielectrophoretic cell separation microfluidic device. *Lab Chip* **2007**, *7*, 239–248.

220. Kralj, J.G.; Lis, M.T.W.; Schmidt, M.A.; Jensen, K.F. Continuous dielectrophoretic size-based particle sorting. *Anal. Chem.* **2006**, *78*, 5019–5025.

221. Gascoyne, P.R.C.; Vykoukal, J. Particle separation by dielectrophoresis. *Electrophoresis* **2002**, *23*, 1973–1983.

222. Pysher, M.D.; Hayes, M.A. Electrophoretic and dielectrophoretic field gradient technique for separating bioparticles. *Anal. Chem.* **2007**, *79*, 4552–4557.

223. Lapizco-Encinas, B.H.; Simmons, B.A.; Cummings, E.B.; Fintschenko, Y. Dielectrophoretic concentration and separation of live and dead bacteria in an array of insulators. *Anal. Chem.* **2004**, *76*, 1571–1579.

224. Lee, T.M.H.; Hsing, I.-M. Sequence-specific electrochemical detection of asymmetric PCR amplicons of traditional Chinese medicinal plant DNA. *Anal. Chem.* **2002**, *74*, 5057–5062.

225. Giordano, B.C.; Ferrance, J.; Swedberg, S.; Hunter, A.F.R.; Landers, J.P. Polymerase chain reaction in polymeric microchips: DNA amplification in less than 240 seconds. *Anal. Biochem.* **2001**, *291*, 124–132.

226. Yang, J.; Liu, Y.; Rauch, C.B.; Stevens, R.L.; Liu, R.H.; Lenigk, R.; Grodzinski, P. High sensitivity PCR assay in plastic micro reactors. *Lab Chip* **2002**, *2*, 179–187.

227. Hashimoto, M.; Chen, P.-C.; Mitchell, M.W.; Nikitopoulos, D.E.; Soper, S.A.; Murphy, M.C. Rapid PCR in a continuous flow device. *Lab Chip* **2004**, *4*, 638–645.

228. Tang, T.; Badal, M.Y.; Ocvirk, G.; Lee, W.E.; Bader, D.E.; Bekkaoui, F.; Harrison, D.J. Integrated microfluidic electrophoresis system for analysis of genetic materials using signal amplification methods. *Anal. Chem.* **2002**, *74*, 725–733.

229. Lee, J.H.; Chung, S.; Kim, S.J.; Han, J. Poly(dimethylsiloxane)-based protein preconcentration using a nanogap generated by junction gap breakdown. *Anal. Chem.* **2007**, *79*, 6868–6873.

230. Wainright, A.; Nguyen, U.T.; Bjornson, T.; Boone, T.D. Preconcentration and separation of double-stranded DNA fragments by electrophoresis in plastic microfluidic devices. *Electrophoresis* **2003**, *24*, 3784–3792.

231. Shihabi, Z.K. Stacking in capillary zone electrophoresis. *J. Chromatogr. A* **2000**, *902*, 107–117.

232. Cabrera, C.R.; Yager, P. Continuous separation of bacteria in a microfluidic flow cell using electrokinetic techniques. *Electrophoresis* **2001**, *22*, 355–362.

233. Quirino, J.P.; Terabe, S. Sample stacking of cationic and anionic analytes in capillary electrophoresis. *J. Chromatogr. A* **2000**, *902*, 119–135.

234. Cunliffe, J.M.; Liu, Z.; Pawliszyn, J.; Kennedy, R.T. Use of a native affinity ligand for the detection of G proteins by capillary isoelectric focusing with laser-induced fluorescence detection. *Electrophoresis* **2004**, *25*, 2319–2325.

235. Han, J.; Singh, A.K. Rapid protein separations in ultra-short microchannels: Microchip sodium dodecyl sulfate–polyacrylamide gel electrophoresis and isoelectric focusing. *J. Chromatogr. A* **2004**, *1049*, 205–209.

Chapter 22

Sol-Gel Coatings and Monoliths in Analytical Sample Preparation

Scott Segro and Abdul Malik

22.1. INTRODUCTION

Sol-gel materials are attracting pronounced attention in analytical sample preparation. However, there have been few review articles written about sol-gel materials in analytical sample preparation. Sol-gel materials[1] have been advantageously used as extraction media in various formats, including (1) surface coatings[2–103] and (2) monolithic beds.[104–117] As surface coatings, sol-gel materials have been used on fibers for solid-phase microextraction (SPME),[2–75] on capillary inner walls for in-tube SPME,[76–96] and on stir-bar sorptive extraction.[98–103] Sol-gel monoliths have been used for sample preconcentration and extraction.[104–117]

In this chapter, we will provide a comprehensive state-of-the-art review of sol-gel extraction materials in analytical microextraction. It will include an introduction to sol-gel chemistry, classification of sol-gel materials, specific information on the preparation and characterization of sol-gel materials, diverse applications of sol-gel materials in analytical extraction or sample preparation, instrumentation for the sol-gel microextraction techniques, types of compounds extracted and analyzed using the sol-gel materials, and analytical characteristics, including extraction capabilities, detection limits, relative standard deviation (RSD) values, percent recoveries, and desirable properties. Real-world applications of the sol-gel materials to sample preparation will also be discussed for various fields of analysis, including environmental, food, and biomedical areas.

22.2. SOL-GEL CHEMISTRY

Sol-gel precursors typically consist of silicon or transition metal alkoxides.[1] Transition metals whose alkoxides have been used as sol-gel precursors for analytical sample preparation include zirconium-,[75,94,95] aluminum-,[14,15] titanium-,[16,73,74,91–93] and germanium.[96] The sol-gel process involves mainly hydrolysis and condensation reactions.[1] The first step in the formation of a sol-gel material is the complete or partial hydrolysis of the sol-gel precursor. This is usually accomplished under acid or base catalysis.[1] Next, the hydrolyzed precursor undergoes condensation reactions, forming a three-dimensional polymeric network. During this time, any sol-gel active component in the sol-gel mixture can also condense with the evolving polymeric network. Also, particularly in the formation of sol-gel fiber SPME coatings, sol-gel capillary coatings, and sol-gel monoliths, the evolving sol-gel network can condense with hydroxyl groups on the surface of the substrate, which is typically fused silica. This covalent bonding to the surface gives sol-gel materials their high thermal and solvent stability and makes them very successful as extracting phases.[1,2]

22.3. PHYSICOCHEMICAL CHARACTERIZATION OF SOL-GEL MATERIALS

This chapter is focused primarily on sol-gel materials and their extraction properties and sorbent characteristics. However, to have a better insight into the sorbent behavior at the molecular level, many of these sol-gel materials have also been characterized by various physicochemical methods. Such methods include Fourier transform infrared spectroscopy,[17] Raman spectroscopy,[89] elemental analysis,[75] and X-ray studies.[75] These studies are typically used to identify specific chemical bonds and functional groups in the sol-gel materials.

Also, microscope images have been taken of the sol-gel materials to provide vivid illustration of the morphological aspects of these materials. Most often, a scanning electron microscope (SEM) is used to generate

Handbook of Sample Preparation, Edited by Janusz Pawliszyn and Heather L. Lord

these images. A fluorescence micrograph can also be used. These images show the physical appearance of the sol-gel materials and are extremely useful for examining properties such as porosity, cracking, and coating thicknesses.

Many of the studies on sol-gel extraction materials developed to date involved extensive studies on the physical properties of the coatings that served as the extracting phase and the performance of these coatings under different conditions. An important aspect of these studies was the elucidation of extraction profiles for the sol-gel sorbents. In the extraction profile, samples representative of particular chemical types are extracted multiple times for different extraction times. The peak areas for each extraction time are then averaged and plotted against their respective extraction times. The point on the graph in which the average peak area stops increasing with extraction time corresponds to the time required for equilibrium to be established between the sample solution and the sol-gel extracting phase. An example of an extraction profile is indicated in Figure 22.1.[84]

Other studies that have been done include extraction temperature profiles, pH studies, solvent tolerances, and lifetime studies. These results are typically presented in graphs and are conducted in order to determine the optimal extraction conditions for the sol-gel materials.

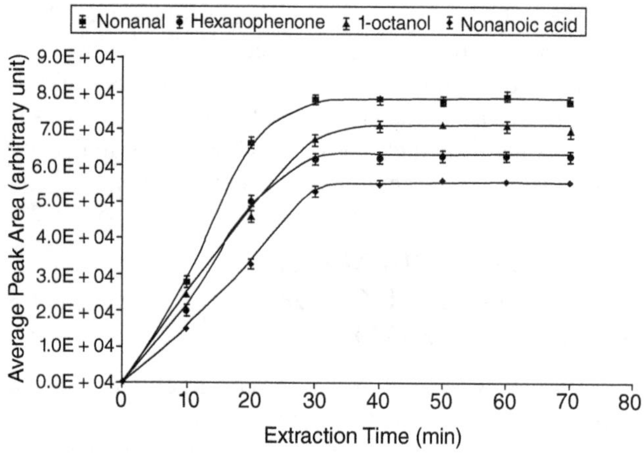

Figure 22.1. Example of an extraction profile. The point on the graph where the average peak area stops increasing with extraction time corresponds to the time required for an equilibrium to be established between the analyte in the sample and the sol-gel sorbent. This example illustrates the extraction profiles of nonanal (120 μg/L), hexanophenone (50 μg/L), 1-nonanol (100 μg/L), and nonanoic acid (100 μg/L) analytes extracted on a 12 cm × 250 μm i.d. sol-gel PEG-coated capillary from aqueous samples. Extraction conditions: triplicate extraction for 10, 20, 30, 40, 50, 60, and 70 min. GC analysis conditions: 5 m × 250 μm i.d. sol-gel PDMS column; splitless injection; injector temperature: initial 30°C, final 340°C, programmed at a rate of 60°C/min; GC oven temperature programmed from 35°C (5 min) to 320°C at a rate of 20°C/min; helium carrier gas; FID temperature 350°C. Reprinted from reference 84 with permission from Elsevier. Copyright 2007.

22.4. CLASSIFICATION OF SOL-GEL COATINGS IN SPME

Sol-gel technology allowed for the creation of many different types of sol-gel coatings for analytical microextraction. Such coatings were developed both for fiber SPME[2–65] and in-tube SPME (capillary microextraction [CME]),[76–96] with different extraction capabilities and applications. The reported sol-gel SPME coatings can be classified into (1) polysiloxane-based monofunctional sol-gel fiber coatings, (2) polysiloxane-based multifunctional sol-gel fiber coatings, (3) sol-gel coatings with cavity ligands, and (4) non-polysiloxane sol-gel coatings.

22.4.1. Fiber Preparation and Pretreatment Procedures

Most sol-gel fiber SPME coatings are made by coating the surface of a fused-silica fiber. Some important pretreatment steps must first be performed. Fused-silica fibers typically have a protective polyimide coating on them, which must be removed in order to expose the fused-silica surface on the portion of the fiber at one of its ends that is to be coated. One of the simplest ways to accomplish this is to burn the polyimide coating off of the fused-silica fiber, using either a match or a cigarette lighter. After burning off the polyimide layer, the fused-silica surface is rinsed in methanol.[2,6] Other ways to remove the protective coating are to dip the end of the fiber in acetone,[21,27,31,35,37–39,42,43,53,54,57,59,61,70] dichloromethane,[69] or concentrated sulfuric acid[9,17] for several hours.

Fused-silica fiber contains a small number of silanol groups on its surface. The chemical bonding of a sol-gel coating to a fiber is accomplished through condensation reactions between hydroxyl groups in the sol-gel network and silanol groups on the surface of the fused-silica fiber. Therefore, it would be advantageous to have more silanol groups on the fused-silica surface to allow for more chemical bonding between the sol-gel coating and the surface of the fiber. The formation of more silanol groups on the fused-silica surface can be accomplished by soaking the fiber in a strong base, such as 1 M NaOH.[8–11,21,27,31,35,37–39,42,43,53,54,57,59,61,70] After soaking in the base, the excess is subsequentially neutralized by soaking the fiber in 0.1 M HCl.[6,8,11,17,31,39,42,43,53,58,59,61,70] After cleaning, the fiber is ready to be coated with the sol-gel solution.

22.4.2. Sol-Gel Coating Procedures for SPME

The sol-gel fiber coating procedure was developed by Malık and coworkers.[2–4] The fiber is coated by dipping it vertically into the sol solution for a specific amount of time, then removing it from the sol solution and allowing it to gel. Sometimes, the dipping process is repeated several times until the desired coating thickness is obtained. In between dippings, the sol-gel fiber may be allowed to stand, heated,

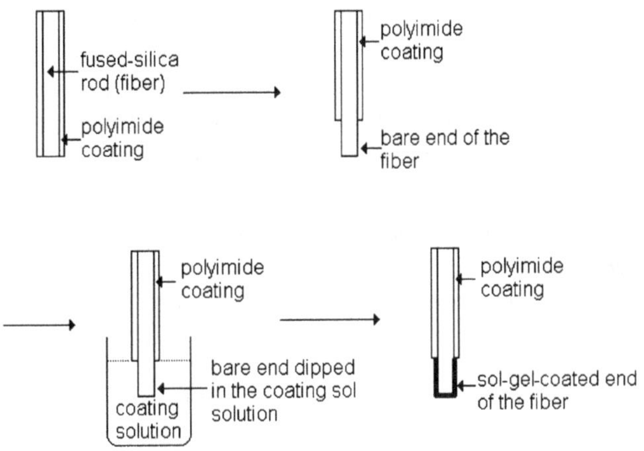

Figure 22.2. Sol-gel fiber pretreatment and coating procedure. The polyimide coating is removed from the fused-silica capillary in the pretreatment step. The sol-gel coating is made by vertically dipping the bare end of the fused-silica fiber into the sol solution for a period of time (typically 20–30 min). During this time, the sol-gel material chemically bonds to the silanol groups on the fused-silica surface of the fiber.

or dried with helium. Figure 22.2 illustrates the fiber pretreatment and coating procedure.

After the coating is made, it is usually conditioned in a gas chromatography (GC) injection port by heating to a specific temperature under helium or nitrogen purge for a certain amount of time to facilitate the sol-gel reactions to completion. The fibers may be cleaned or conditioned again prior to each use. Unless otherwise stated, all of the sol-gel SPME coatings discussed in this chapter were created following this general procedure.

22.4.3. Sol-Gel Polysiloxane-Based Monofunctional Coatings for SPME

22.4.3.1. Silica-based sol-gel polydimethylsiloxane (PDMS) coatings for SPME. The first sol-gel SPME coating used for analytical microextraction was developed by Malik and coworkers[2–4] in 1997. A fused-silica fiber was coated with sol-gel PDMS. The sol-gel PDMS coating solution was prepared by mixing 300 µL of methyltrimethoxysilane (MTMS), which served as the sol-gel precursor; 180 µL of hydroxy-terminated PDMS, which served as the coating polymer; 30 mg of polymethylhydrosiloxane (PMHS), which served as a deactivation reagent; and 200 µL of 95% trifluoroacetic acid, which served as a sol-gel catalyst. This mixture was vortexed for 2 min and centrifuged at 13,000 rpm for 5 min. The clear top sol solution was removed and used to make the coating.[2–4]

The coating process was followed by an end-capping process, in which the fiber was dipped into a (4:1 v/v) trimethylmethoxysilane/methanol solution for 1 min and then placed in a desiccator at room temperature. The sol-gel PDMS fiber was attached to a homemade SPME device and conditioned at 320°C for 1–2 hours under helium purge in a GC injection port. The conditioning process was then repeated, using 10-min conditioning cycles, until a stable GC baseline was established. The sol-gel PDMS coating was found to be thermally stable to over 320°C.[2–4]

GC–flame ionization detection (FID) was used in the analysis of compounds extracted using the sol-gel PDMS coating. For extraction, the coated fiber was dipped in aqueous samples at room temperature. Magnetic stirring was used to decrease the equilibration time. Equilibrium was found to take from about 10 min for naphthalene to about 1 hour for *N*-methylaniline. After the extraction, the fiber was withdrawn into the needle of a homemade SPME syringe and immediately inserted into the GC injection port held at 320°C for 5 min. The carrier gas flow transported the desorbed analytes into the GC column for separation and analysis. The desorption step was conducted in splitless mode, holding the GC column at a relatively low temperature (40°C), which allowed for analyte focusing on the front end of the GC column.[2–4]

Unlike traditional SPME PDMS coatings, the sol-gel PDMS fiber exhibited practically no bleeding or decomposition at 320°C. The sol-gel PDMS coating was successfully employed in extracting both polar and nonpolar compounds. Polar components (e.g., silanol groups) in the composition of the sol-gel network might have played a positive role in the extraction of polar compounds, such as dimethylphenol isomers, aliphatic alcohols, and aniline derivatives.[2]

Caruso and coworkers[5] applied a sol-gel PDMS fiber coating for the analysis of seleno amino acids using GC with inductively coupled plasma mass spectrometry (ICP-MS), with sub-parts per billion detection limits.[5]

The method described by Malik and coworkers[2] was used by Bagheri and coworkers[6] to prepare a sol-gel SPME fiber for the determination of dextromethorphan (DM) and dextrorphan (DP) in human plasma using headspace (HS)-SPME with GC-MS.[6] This method was capable of detecting concentrations of 10, 50, and 500 ng/mL of DM and DP with accuracies between 99.86% and 104.5%. The observed intraday run-to-run RSD values were between 3.29% and 4.81%. The observed interday run-to-run RSD values were between 3.38% and 5.04%. The authors advocated that this method can be used to determine DM and DP in clinical plasma samples with sufficient sensitivity and reproducibility.[6]

Bagheri and coworkers[7] also successfully used the PDMS sol-gel coating in the determination of fentanyl in human plasma by HS-SPME and GC-MS. The observed detection limit was 0.03 ng/mL. The interday and intraday run-to-run RSD values were both less than 5%. This method was also applied successfully to the determination of fentanyl in human plasma after a volunteer applied a 50 µg/h Duragesic fentanyl patch.[7]

de Oliveira and coworkers[8] developed a sol-gel coating using a thin glass-ceramic rod as a surface for SPME. These thin glass-ceramic rod coatings were compared with fiber coatings using the same sol-gel mixture. The thin, glass-ceramic rods were prepared by melting appropriate amounts of Li_2CO_3, $ZrOCl_2{\cdot}8H_2O$, $Ba(CH_3CO_2)_2$, and SiO_2 for 3h at 1100°C. Fused-silica fibers were also used for comparison.[8]

GC-FID was used with HS-SPME in the analysis of a benzene, toluene, ethylbenzene, and xylene (BTEX) sample. The sol-gel PDMS glass-ceramic-based fiber demonstrated detection limits of 0.3 µg/L for benzene, 0.7 µg/L for toluene, 0.2 µg/L for ethylbenzene, and 0.3 µg/L for xylene. The run-to-run RSD values were between 4.2% and 5.3%, and the fiber-to-fiber RSD values were between 6.1% and 7.2%. This glass-ceramic-based sol-gel fiber coating produced a thicker coating and better extraction ability than the fused-silica coating.[8]

Guan and coworkers[9,10] developed a sol-gel SPME fiber using PDMS containing 3% vinyl group in the coating solution.[9] GC-FID was used for the analysis in this study. HS-SPME was used in the extraction of BTEX compounds. Direct SPME was used for the extraction of organophosphorus pesticides in water, orange juice, and wine.[9]

This coating was capable of extracting the BTEX compounds with detection limits between 0.26 and 2.03 ng/mL with run-to-run RSD values between 2.8% and 5.7%. For organophosphorus pesticides, the detection limits ranged from 0.4 to 19.9 ng/L in water, from 0.7 to 32.9 ng/L in orange juice, and from 0.5 to 38.2 ng/L in wine. The run-to-run RSD values were between 1.0% and 27.2% in water, 5.1% and 27.6% in orange juice, and 1.1% and 19.2% in wine. In this coating, the PDMS is entangled in the sol-gel network and also cross-linked within itself. It is thermally stable at 290°C and has been used over 100 times without a significant decrease in coating thickness or performance.[9]

Azenha and coworkers[11] developed a sol-gel coating on an unbreakable titanium wire. A titanium wire was dipped in 1 M NaOH to promote the formation of titanol groups on the surface. The wire was then washed with 0.1 M HCl, water, and methanol. The sol-gel PDMS coating was prepared as described by Malik et al.[2] The titanium wire was attached to an SPME device. The wire was dipped into the sol solution for 20 min. The sol-gel-coated titanium wire was conditioned for 30 min at 300°C under nitrogen in a GC injection port. HS-SPME was used to perform the extractions. The titanium sol-gel wire was exposed to the HS at 40°C for 30 min, the time required for equilibrium to be established. The extracted analytes were then thermally desorbed into a GC injection port at 300°C for 1 min. This coating was capable of extracting benzene, toluene, ethylbenzene, benzaldehyde, 2-octanone, acetophenone, and 2,5-dimethylphenol. It was resistant to organic solvents and thermally stable to 350°C. This fiber combined the advantages of sol-gel stability with the robustness of titanium wire as a new substrate. This could be a solution to the problems associated with fiber breakage.[11] Recently, the performance of the sol-gel PDMS-coated titanium wire was improved using a sol-gel/silica particle blend.[12]

Recently, a novel SPME mode was developed.[13] For this, solid glass microspheres were coated with sol-gel PDMS and used to extract polycyclic aromatic hydrocarbons (PAHs). Using a homemade thermal desorption unit, the extraction was coupled with GC-FID analysis. For PAHs, the observed detection limits ranged from 0.01 to 0.045 ng/mL. The recoveries of PAHs from a river water sample ranged from 78% to 127%.[13]

22.4.3.2. Nonsilica-based sol-gel polysiloxane-based monofunctional coatings for SPME. Zeng and coworkers[14] developed a sol-gel alumina-based PDMS coating for the SPME of polar compounds. HS-SPME was used to perform the extractions for subsequent analysis by GC-FID. This coating demonstrated run-to-run RSD values less than 5.4% and fiber-to-fiber RSD values less than 6.0%. This coating was also applied to the analysis of alcohols and fatty acids in beer samples, where extraction recoveries were between 85.7% and 104% with run-to-run RSD values below 9%. The sol-gel coating was found to be thermally stable to 400°C and resistant to high pH liquids. This coating survived 12 h of soaking in pH 14 NaOH without significant loss in performance. Highly basic compounds and compounds in highly basic matrices can be successfully extracted using this sol-gel fiber coating. The coating was used successfully in over 160 extractions and desorptions.[14]

Zeng and Liu[15] developed a sol-gel alumina and hydroxyl silicone oil fiber coating, which has high extraction efficiency of polar compounds. Zeng and coworkers[16] also developed a titania–hydroxy-terminated silicone oil (OH-TSO) sol-gel coating for the SPME of polar compounds.[16]

Aromatic amines, phenols, and PAHs were extracted by HS-SPME followed by GC-FID analysis using nitrogen as the carrier gas. This titania-based sol-gel coating was found to be more efficient than an analogous in-house prepared sol-gel silica OH-TSO fiber in the extraction of PAHs, amines, and phenols. It was resistant to extreme pH conditions and was thermally stable to 320°C. This coating was applied to the analysis of aromatic amines in dye process wastewater. The detection limits ranged from 0.22 to 0.84 µg/L with run-to-run RSD values between 5.9% and 7% and extraction recoveries between 83.6% and 101.4%.[16]

22.4.4. Polysiloxane-Based Multifunctional Sol-Gel Fiber Coatings

22.4.4.1. PDMS/poly(vinyl alcohol) sol-gel fiber coatings. In 2004, Augusto and Lopes[17] developed a sol-gel PDMS/poly(vinyl alcohol)-coated fiber for HS-SPME of polychlorinated biphenyls (PCBs). GC–electron capture detection (ECD) using helium carrier gas was used for the analysis of the extracted PCBs. This coating demonstrated run-to-run peak area RSD values between 3.4% and 17%. This coating was thermally stable to over 350°C with less degradation at high temperatures than pure PDMS coatings. It was used successfully over 150 times without a significant loss in performance.[17]

This coating was applied to the SPME of pesticide residues in herbal infusions.[18] Chlorothlonil, methyl parathion, malathion, α-endosulfan, and β-endosulfan were analyzed. HS-SPME with GC-ECD was used in this study. The sol-gel coating demonstrated detection limits between 0.01 and 1.00 ng/mL for *Passiflora alata* infusions, 0.03 and 1.50 ng/mL for *Passiflora edulis* infusions, and between 0.03 and 1.10 ng/mL for *Passiflora incarnata* infusions. The run-to-run peak area RSD values were all between 1.2% and 12.1%. The sol-gel-coated fiber was used in more than 300 extractions over 5 months with no significant change in performance.[18] Most recently, Augusto and coworkers[19] applied this sol-gel coating to the determination of organochlorine and organophosphorus pesticides in herbal infusions using fiber introduction MS with detection limits between 0.3 and 3.9 ng/mL.[19] This coating was also used for the determination of pesticides in *P. alata* infusions using HS-SPME with GC-ECD.[20]

22.4.4.2. Poly(methylphenylvinylsiloxane) (PMPVS) sol-gel fiber coatings. Zeng and coworkers[21] developed a sol-gel PMPVS fiber coating for SPME with GC-FID analysis. In HS-SPME, the benzene, toluene, and xylene (BTX) analytes reached equilibrium quite fast, taking only 60 s for benzene and toluene and 120 s for xylene at 15°C. The analytes required a desorption time of 40 s at 150°C. The equilibrium time of extraction for PAHs ranged from 20 to 240 min. This coating had detection limits ranging from 0.01 to 0.1 μg/L for the BTX and from 0.05 to 0.29 μg/L for PAHs. This sol-gel fiber was thermally stable to 350°C and used successfully in over 150 extractions/desorption cycles.[21] This coating was applied to the analysis of toluene and xylenes in industrial wastewater samples from a paint mill.[22] Detection limits were between 0.01 and 0.1 μg/L.[22] Recently, Zeng and coworkers[23] applied this sol-gel coating to the determination of methylphosphonates and phosphates in air samples, with detection limits ranging from 27.2 to 28.3 μg/L and run-to-run RSD values between 4.7% and 6.8%.[23]

Dong and coworkers[24] used a sol-gel PMPVS-coated fiber for the determination of organochlorine pesticides and their derivations in water using HS-SPME with GC-ECD analysis. The observed detection limits for the organochlorine pesticides studied were between 0.835 and 13.0 ng/L. In lake water samples, recovery values between 71.5% and 115.5% were observed. The observed run-to-run RSD values were all less than 11.8%. This sol-gel SPME method was adequate for the determination of organochlorine pesticides in water at ultratrace levels.[24] In 2006, Dong and coworkers[25] applied this sol-gel coating to the GC analysis of organophosphorus pesticides in water, with detection limits between 0.14 and 1.89 μg/L[25] and to the GC analysis of organochlorine pesticides in water using HS-SPME with detection limits below 35.3 ng/L.[26]

Wu and coworkers[27] developed a sol-gel polyphenylmethylsiloxane-coated fiber for the determination of organochlorine pesticides in Chinese teas using microwave-assisted SPME-GC-ECD. The observed detection limits were below 0.081 ng/L, the extraction recovery values were between 39.05% and 94.35%, and the run-to-run RSD values were less than 16%.[27]

22.4.4.3. C11 PDMS sol-gel fiber coating. In 1998, Jinno and coworkers[28] used micro-liquid chromatography (LC) coupled with SPME for the analysis of benzodiazepines in human urine. Two traditional SPME fibers and a sol-gel C11 PDMS fiber were compared in performance. The technique required a short analysis time and very low solvent consumption.[28]

22.4.4.4. PMPVS/OH-TSO sol-gel fiber coating. Zeng and coworkers[29] developed a sol-gel PMPVS/OH-TSO fiber for SPME-GC-FID. Organophosphorus pesticides were analyzed with detection limits between 1.5 and 2.1 μg/L and RSD values below 7.4%.[29]

22.4.4.5. Sol-gel anilinemethyltriethoxysilane/PDMS coating for SPME. Li and coworkers[30] developed a sol-gel anilinemethyltriethoxysilane/PDMS coating for SPME. For monocyclic aromatic hydrocarbons (MAHs), the observed detection limits ranged from 0.6 to 3.8 μg/L, and the observed run-to-run RSD values were between 6.9% and 13.6%. For PAHs, the observed detection limits ranged from 0.2 to 1.5 μg/L, and the observed run-to-run RSD values were between 2.8% and 9.5%. This sol-gel coating was also applied to the extraction of PAHs in river water, and the extraction recoveries ranged from 95.2% to 113% with run-to-run RSD values between 4.1% and 10.5%. The phenyl group present in this coating makes it suitable for environmental analysis of real water samples for the determination of aromatic pollutants. This coating was also thermally stable to 300°C and demonstrated a lifetime of over 150 uses without significant change in performance.[30]

22.4.4.6. Sol-gel silicone-divinylbenzene (DVB) copolymer coating for SPME. Zeng and coworkers[31] developed a sol-gel-derived silicone-DVB copolymer fiber for SPME. It was capable of extracting dimethyl methylphosphonate, trimethyl phosphate, and tributyl phosphate from water using HS-SPME or directly from air samples. For water samples, this coating provided detection limits and run-to-run RSD values of 0.34 μg/mL and 5.04% for dimethyl methylphosphonate, 2.20 μg/mL and 3.67% for trimethyl phosphate, and 0.01 μg/mL and 6.44% for tributyl phosphate, respectively. For air samples, this coating gave detection limits and run-to-run RSD values of 1.11 μg/mL and 4.29% for dimethyl methylphosphonate, 1.46 μg/mL and 4.79% for trimethyl phosphate, and 1.64 μg/mL and 5.88% for tributyl phosphate, respectively. This coating might be useful in the analysis of chemical warfare agents using SPME.[31]

22.4.4.7. Sol-gel aminopropylsilica/PDMS coating for SPME. Recently, Biajoli and Augusto[32] developed sol-gel aminopropylsilica/PDMS-coated SPME fibers and applied them to the analysis of beer HS. These fibers offered superior extraction efficiency for both polar and semipolar compounds when compared with conventional fibers and were capable of extracting a broad range of analytes, including organic acids.[32]

22.4.4.8. Sol-gel 3-aminopropyltriethoxysilane/PDMS coating for SPME. Sol-gel 3-aminopropyltriethoxysilane/PDMS-coated fibers were recently developed for the analysis of PAHs at trace concentrations.[33] This coating offered enhanced thermal stability, good fiber-to-fiber reproducibility (RSD values lower than 6%), and low nanogram per liter detection limits.[33]

22.4.4.9. Sol-gel amino-functionalized PDMS coating for SPME. Recently, Bagheri and coworkers[34] developed an amino-functionalized PDMS coating for SPME using 3-(trimethoxysilyl) Pr amine and PDMS. Chlorophenols were extracted using these coated fibers, followed by GC-MS analysis. The phenols were derivatized using acetic anhydride under alkaline conditions prior to extraction. The observed detection limits ranged from 0.02 to 0.05 ng/mL and the RSDs were between 6.8% and 10%. This coating was also capable of extracting chlorophenols from spiked tap water samples, with recovery values of over 90%.[34]

22.4.5. Sol-Gel Coatings with Cavity Ligands

22.4.5.1. Crown-ether sol-gel fiber coatings. Zeng and coworkers[35] developed a sol-gel-derived hydroxyl-crown-ether fiber coating for SPME with GC-FID analysis. This coating demonstrated good limits of detection for phenols, ranging from 0.15 to 0.86 ng/mL and good run-to-run peak area RSD values between 2.9% and 4.6%.[35] The sol-gel crown-ether coating was thermally stable to 350°C and provided good detection limits and reproducibility. Using this coating, it was determined that a wastewater sample from a paper mill contained 1.3 µg/mL of phenol and 0.8 µg/mL of 2,4-dimethylphenol, with extraction recoveries of 87.4% and 95.3%, respectively.[35]

Zeng and coworkers[36] developed three more sol-gel fibers based on crown ethers and applied them to the SPME of monocyclic aromatic amines. The three fibers were made using dihydroxydibenzo-14-crown-4 (OH-DB14C4), dihydroxy-substituted urushiol crown ether (DHSU14C4), and 3,5-dibutyl-unsymmetry-dibenzo-14-crown-4 dihydroxy crown ether (DBUD14C4). In their study, it was determined that the sol-gel OH-DB14C4-coated fiber had the best affinity for aniline derivatives. This fiber demonstrated detection limits of 0.96 ng/mL for aniline, 0.98 ng/mL for *m*-toluidine, 0.23 ng/mL for *N,N* diethylaniline, 0.17 ng/mL for *N*-ethyl-*m*-toluidine, and 0.27 ng/mL for 3,4-dimethylaniline. The observed run-to-run RSD values were between 3.23% and 6.20% for all of the compounds tested. This coating was applied to the analysis of a wastewater sample obtained from a pharmaceutical factory, and it was determined that the wastewater sample contained 0.7 µg/mL of aniline and 0.3 µg/mL of 3,4-dimethylaniline. All three of the sol-gel crown-ether coatings exhibited excellent solvent resistance and thermal stability up to 340°C. The OH-DB14C4 demonstrated excellent detection limits for aromatic amine derivatives with good run-to-run RSD values. No significant decline in extraction performance was observed even after more than 150 uses.[36]

Yun[37] developed a sol-gel SPME coating using the open crown ether α, ω-diallyltriethylene glycol/OH-TSO, and vinyltriethoxysilane in the coating sol solution. Phenols reached equilibrium within 40 min of extraction time and demonstrated detection limits between 0.32 and 2.6 ng/mL. Run-to-run RSD values for phenols were between 2.5% and 3.5%. BTX compounds reached equilibrium within 15 min of extraction time, with detection limits between 0.03 and 0.3 ng/mL and run-to-run RSD values between 2.1% and 5.2%. This coating was also capable of extracting high boiling point compounds, such as phthalate esters. Direct SPME was used for the phthalate esters, which required between 60 and 90 min of extraction to reach equilibrium. A 10-min desorption time was used for these compounds. Detection limits ranged from 0.078 to 0.41 ng/mL for phthalate esters. Run-to-run RSD values were below 10% for phthalate esters. The high temperature resistance of this coating allowed it to extract phthalate esters successfully. Overall, this coating provided extractions of both polar and nonpolar analytes, using both HS and direct SPME.[37]

Wu and coworkers[38] developed a sol-gel dibenzo-18-crown-6 fiber for SPME with a new derivatizing reagent for the determination of aliphatic amines in lake water and human urine.[38] HS-SPME was used to extract derivatized aliphatic amine samples followed by GC-FID analysis. The observed detection limits for the amines were between 0.05 and 0.005 µg/L. All run-to-run RSD values were below 6%. This sol-gel coating could be useful in environmental and biomedical analysis. This coating was also solvent resistant and thermally stable at over 350°C.[38]

Wu and coworkers[39] developed a benzo-15-crown-5 sol-gel coating for SPME with GC-FID analysis. The benzo-crown-ether sol-gel coating exhibited thermal stability to 350°C and experienced less cracking when viewed under a microscope than other sol-gel coatings that did not contain benzo-crown ether. This coating extracted BTEX compounds with detection limits between 0.01 and 0.05 µg/L, and run-to-run RSD values below 4%. This coating was also capable of extracting phenols with detection limits between 0.05 and 1 µg/L. This coating was used to extract 6 chlorobenzenes and 18 carcinogenic arylamines. Overall, this coating demonstrated good selectivity toward both polar and nonpolar compounds.[39] Wu and coworkers[40] applied this sol-gel coating to the analysis of trace organochlorine pesticides in water, with detection limits between 0.01 and 0.5 ng/L.[40]

Wu and coworkers[41] also developed new SPME fibers using sol-gel technology for the determination of organophosphorus pesticide multiresidues in food. Allyloxy bis-benzo 16-crown-5 trimethoxysilane was synthesized and used as the sol-gel precursor. Honey, orange, juice, and vegetable samples were used in the SPME extraction. In honey, the detection limits for the organophosphorus pesticides ranged from 0.004 to 0.70 ng/g with run-to-run RSD values between 2.1% and 15%. In juice, the detection limits for the organophosphorus pesticides ranged from 0.003 to 0.20 ng/g with run-to-run RSD values between 2.0% and 9.2%. In Pakchoi, the detection limits for the organopesticides ranged from 0.10 to 1.0 ng/g with run-to-run RSD values between 2.3% and 9.1%. This fiber demonstrated solvent resistance and was used over 200 times without a significant decrease in performance.[41]

Wu and coworkers[42] developed a sol-gel vinyl crown ether cross-linked SPME fiber for the determination of organophosporus pesticides in food samples using GC-flame photometric detection (FPD) analysis. This coating demonstrated detection limits between 0.0015 and 0.081 ng/g in water samples, between 0.003 and 0.075 ng/g in apple juice, between 0.0032 and 0.09 ng/g in apple, and between 0.0042 and 0.076 ng/g in tomato. The observed run-to-run RSD values were between 1.8% and 5.3% in water samples. Extraction recoveries of organophosphorus pesticides ranged from 71.5% to 104.6% in apple juice, from 70.5% to 104.9% in apple, and from 55.3% to 106.4% in tomato. The detection limits for this sol-gel coating using the described method were below the maximum residue limits recommended by the European Union.[42]

22.4.5.2. Sol-gel calixarene coatings for SPME. Zeng and coworkers[43] developed a sol-gel calix[4]arene coating for SPME with GC-FID analysis. The sol-gel calix[4]arene coating had a long lifetime, showing no significant decrease in performance after over 170 uses. It was thermally stable to 380°C and resistant to organic solvents. The coating demonstrated detection limits from 4.7 to 35.2 ng/L for BTEX compounds, from 1.2 to 51.4 ng/L for PAHs, and from 15.6 to 72.4 ng/L for aromatic amines. This coating was also capable of extracting PAEs through direct SPME.[43]

The calix[4]arene sol-gel fiber was applied to the SPME of chlorophenols in deionized water, river water, and soil samples taken from lakes.[44] For the chlorophenols, detection limits were between 0.005 and 0.276 µg/L. Run-to-run RSD values were under 6.8%. Recoveries of 2-chlorophenol, 2,4-dichlorophenol, and 2,4,6-trichlorophenol were 94.7%, 97.4%, and 97% from river water, respectively, and 98.6%, 97%, and 90.2% from soil, respectively. Pentachlorophenol was recovered at 86.3% in river water and 81% in soil.[44]

Zeng and coworkers[45] used a sol-gel fiber coating made with amide bridged-calix[4]arene for the HS-SPME of aliphatic amines with GC-FID analysis. Six aliphatic amines in distilled water were analyzed, their detection limits ranged from 0.19 to 39.51 µg/L, and their run-to-run RSD values ranged from 1.4% to 5.1%. This coating was used to determine the amount of trimethylamine in fish samples, which accumulates as fish spoils, and can cause cancer in humans.[45]

The sol-gel calix[4]arene coating was also used for the determination of organochlorine pesticides and their metabolites in radish samples using HS-SPME.[46] It was capable of extracting organochlorine pesticides from distilled water and radish matrix with detection limits ranging from 0.159 to 21.7 ng/L. Run-to-run RSD values ranged between 6.83% and 13.1%. The extraction recoveries of organochlorine pesticides from radish matrix observed were between 78.39% and 112.5%. This coating demonstrated better extraction efficiency than commercially available PDMS fibers and was considered adequate in the determination of organochlorine pesticides at ultratrace levels in complex matrices.[46]

The same research group developed a sol-gel calix[4] open-chain crown-ether fiber for the SPME-GC-FID of polar aromatic and aliphatic compounds. Detection limits were in the parts per billion and parts per trillion range, with RSD values less than 7%.[47,48]

Zeng and coworkers[49] synthesized diglycidyloxy-C[4] then used it to prepare a sol-gel diglycidyloxycalix[4]arene coating for SPME-GC-FID. It demonstrated detection limits between 0.07 and 1.72 ng/L for PAHs, 1.53 and 4.50 ng/L for aromatic amines, and 0.01 and 0.48 µg/L for phenols. The coating had a lifetime of over 300 uses and was thermally stable to 350°C. Run-to-run RSD values were between 1.8% and 6.7% for phenols. It showed the potential for the analysis of trace contaminants in the environment.[49] This coating was also applied to the SPME of propranolol in human urine using GC-FID, with detection limits of 0.275 g/L in HS-SPME and 0.019 g/L in direct SPME.[50]

The sol-gel diglycidyloxycalix[4]arene coating was also used to determine propranolol enantiomers in urine using HS-SPME coupled with capillary electrophoresis (CE).[51] To perform the extractions, 10-mL silanized amber vials were filled with 5 mL of spiked water, 1.5 g of NaCl, and a magnetic spin bar. Depending on the sample, the pH was adjusted to the appropriate value using NaOH. The vial was sealed and the fiber was exposed to the HS at 90°C for 30 min, although the propranolol enantiomers required 50 min for equilibrium to be established. Quantification was possible under nonequilibrium conditions since a linear relationship was found to exist between the amount of analyte adsorbed by the SPME fiber and its initial concentration in the sample matrix in nonequilibrium conditions. The fiber was then removed and placed in a back-extraction device (Fig. 22.3), which was in-house designed.[51] The chamber of a 100-µL syringe was used to hold the back-extraction solution. The fiber was exposed to the back-extraction solution for 15 min. After heating to 40°C and using ultrasonic agitation to accelerate the back-extraction process, the fiber was removed from the syringe and the back-extraction solution was transferred by injection to a vial for CE analysis. After the back-extraction, the fiber was conditioned in a GC injection port at 320°C for 2 min to eliminate any carryover problems. CE with field-amplified sample injection (FASI) was performed, with the capillary temperature set at 25°C. Ultraviolet (UV)

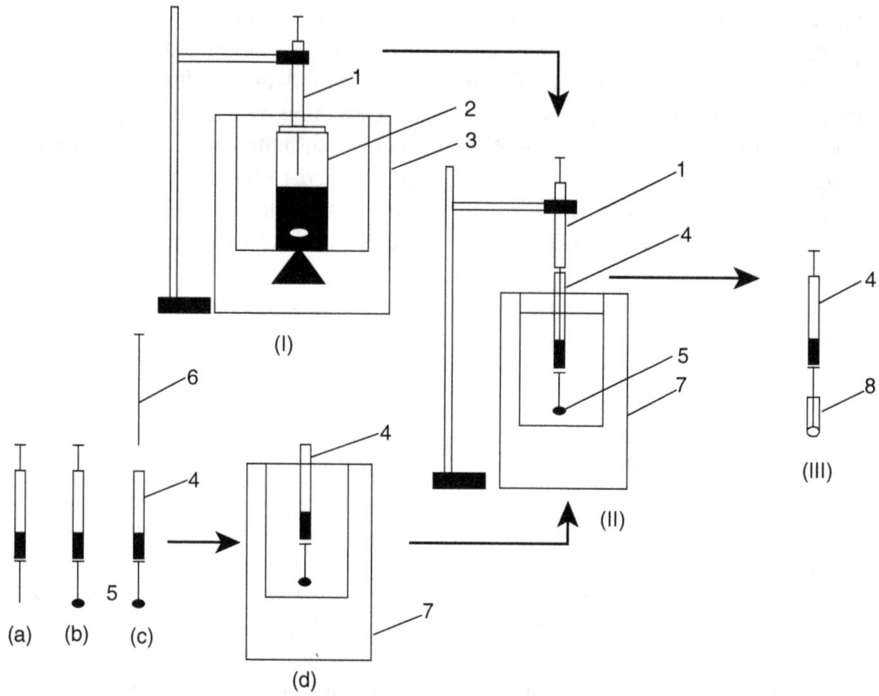

Figure 22.3. Schematic diagram of SPME and back-extraction procedures: (I) SPME facility; (II) back-extraction facility; and (III) back-extraction solution is transferred to an injection vial for CE analysis. (a), (b), (c), and (d) show the preparation procedures for a back-extraction chamber: (1) laboratory-made SPME syringe; (2) SPME vial; (3) facility for heating and stirring; (4) back-extraction chamber; (5) sealing septum; (6) plunger; (7) ultrasonator; and (8) 200-μL vial for CE sampling. Reprinted from reference 51 with permission from Elsevier. Copyright 2006.

absorption detection was used at 214 nm for the propranolol samples.[51]

This sol-gel fiber was capable of successfully extracting the propranolol enantiomers from both water and urine samples. The observed detection limits in urine samples were 8 ng/mL for *R*-(+)-propranol and 10 ng/mL for *S*-(−)-propranol. The fiber demonstrated extraction recovery values between 94% and 104% for *R*-(+)-propranol and between 86% and 107% for *S*-(−)-propranol. The fiber was thermally stable at 350°C, resistant to high alkali conditions, resistant to solvents, and was used over 100 times without any cracking or loss of performance. This fiber might find potential applications in drug testing. When combined with FASI–capillary zone electrophoresis (CZE), it could be used for the routine analysis of β-blockers in doping control laboratories.[51]

Zeng and coworkers[52] also applied the sol-gel diglycidyloxycalix[4]arene SPME coating to the analysis of chlorobenzenes in soil using GC-ECD.[52] It was capable of successfully extracting nine chlorobenzenes from soil samples with detection limits between 0.11 and 3.85 ng/g in kaleyard soil samples and between 0.011 and 0.200 ng/g in red clay soil samples. The observed run-to-run RSD values were between 4.4% and 7.6%. The recovery of chlorobenzenes from real kaleyard soil ranged from 64% to 109.6%.[52]

22.4.5.3. Sol-gel β-cyclodextrin coating for SPME. Zhou and Zeng[53] used a sol-gel coating with β-cyclodextrin for HS-SPME of ephedrine and methamphetamine in human urine. The detection limits were 0.33 ng/mL for ephedrine and 0.60 ng/mL for methamphetamine in GC-FID. The observed run-to-run RSD values were 3.9% and 5.0% for ephedrine and methamphetamine, respectively. The recoveries from human urine were 98% for ephedrine and 98.2% for methamphetamine. This sol-gel coating demonstrated thermal stability to 340°C without loss of extraction efficiency and was still stable and usable after 150 extractions. This coating showed potential in the extraction of ephedrine and methamphetamine in more difficult matrices, such as blood.[53]

22.4.5.4. Sol-gel hydroxyfullerene coating for SPME. Wu and coworkers[54] developed a sol-gel hydroxyfullerene fiber for SPME. PCBs were analyzed using GC-ECD. Detection limits were between 0.013 and 0.051 ng/L for all of the PCBs analyzed. The observed run-to-run peak area RSD values were between 1.8% and 4.6%. This coating was used to determine the concentration of PCBs in a contaminated sediment sample. PAHs were analyzed by GC-FID. The observed detection limits for PAHs were between 0.0049 and 0.125 ng/mL with run-to-run peak area RSD values between 1.9% and 8.9%. The sol-gel fullerene coating was applied to the analysis of a wastewater sample containing PAHs and was even capable of extracting aromatic amines.[54]

The sol-gel hydroxyfullerene fiber was thermally stable at 360°C. This was significant because it extended the use of SPME to the analysis of compounds with higher boiling points. The fiber was also resistant to polar and nonpolar solvents and was used over 190 times without a significant loss in performance.[54] This fiber was used in the determination of the leaching of phthalic diesters from polyvinyl chloride toys dipped in simulated saliva using SPME-GC. The detection limits ranged from 0.079 to 1.36 µg/L, the extraction recovery values were from 88.4% to 107.7%, and the run-to-run peak area RSD values were all less than 8%.[55]

22.4.5.5. Sol-gel quinoxaline-bridged cavitand coating for SPME. Recently, a quinoxaline-bridged cavitand-coated fiber was developed for the SPME of benzene and chlorobenzenes.[56] These fibers demonstrated excellent thermal stability and good batch-to-batch reproducibility with RSD values less than 6%. These coated fibers were capable of sub-nanogram per liter detection limits. They were also applied to the extraction of chlorobenzene in river water samples with recovery values between 87.4% and 94.7%.[56]

22.4.6. Non-Polysiloxane Sol-Gel Coatings

22.4.6.1. Sol-gel polyethylene glycol (PEG) coatings for SPME. Wu and coworkers[57] developed high-performance PEG-coated SPME fibers using sol-gel technology. First, BTEX was extracted and analyzed using the sol-gel PEG coating. The equilibrium time of extraction was found to be very short, approximately 30 s for benzene and toluene, 40 s for ethylbenzene and *p*-xylene, and 90 s for *o*-xylene. Commercial PDMS fibers normally take much longer to reach equilibration. Analytes can be thermally desorbed within 20 s at 280°C. The short extraction and desorption times were attributed to the porous structure of the sol-gel PEG fiber. Sample carryover and bleeding at high temperatures were not observed. This fiber coating remained stable and reusable after more than 150 desorptions at 280°C. This coating demonstrated detection limits of 50 pg/mL for benzene, 20 pg/mL for toluene, 20 pg/mL for ethylbenzene, 10 pg/mL for *p*-xylene, and 10 pg/mL for *o*-xylene in GC-FID.[57] Phenols were also extracted on the sol-gel PEG coating. Detection limits were 10 ng/mL for phenol, 0.5 ng/mL for 2-chlorophenol, 5 ng/mL for 2-nitrophenol, 1 ng/mL for 2,4-dimethylphenol, and 0.1 ng/mL for 2,4-dichlorophenol and 2,4,6-trichlorophenol. The sol-gel PEG coating was also effective in the extraction of phthalic diesters, naphthalene congeners, and pesticides.[57]

Gomes da Costa Silva and Augusto[58] developed a highly porous sol-gel SPME fiber based on PEG-modified ormosils. For the more volatile compounds tested, such as benzene and toluene, the sol-gel carbowax (CW) 20M fiber was comparable to commercial PDMS and CW-DVB counterparts, but for heavier compounds, such as *o*-xylene, the sol-gel fiber was able to extract 230–540% of the mass of the commercial fibers.[58]

This sol-gel coating is notably more porous than most other coatings, which provides faster equilibrium and desorption times. CW 20M and related polymers are often used as additives to improve porosity in other sol-gel SPME coatings. This coating was also applied to the screening of contaminants released by plastic containers in microwave ovens and in the monitoring of contamination of groundwater by fuel leakage from gas stations.[58]

22.4.6.2. Sol-gel acrylate coatings for SPME. Zeng and coworkers[59] developed a sol-gel-derived 3-(trimethoxysilyl)propyl methacrylate (TMSPMA) hydroxyl-terminated silicone oil SPME fiber for the HS-SPME of aroma compounds in beer. This sol-gel coating was capable of simultaneously extracting analytes of diverse polarity (alcohols, esters, and fatty acids) from beer samples by HS-SPME. The detection limits were calculated based on the analysis of these compounds in volatile-free beer samples. In volatile-free beer, the sol-gel coating had detection limits between 0.01 and 10.4 µg/L for alcohols, 0.01 and 35.2 µg/L for fatty acids, and 0.01 and 16.4 µg/L for esters. The observed run-to-run peak area RSD values were between 1.68% and 3.66% for alcohols, 3.00% and 6.18% for fatty acids, and 1.98% and 5.02% for esters. This method was used to determine volatile compounds in four beer types. The major aroma-contributing substances in the beer were identified based on their odor description and odor thresholds.[59]

Zeng and coworkers[60] also developed a sol-gel hydroxyl-terminated silicone oil–butyl methacrylate–divinylbenzene (OH-TSO-BMA-DVB) copolymer γ-methacryloxypropyltri methoxysilane (KH-570) coating for the SPME of volatile compounds in red wine. This coating was capable of extracting alcohols, esters, and fatty acids from wine samples. In volatile-free wine samples, the sol-gel coating provided detection limits between 0.0494 and 49.1 µg/L for alcohols, 0.0504 and 0.523 µg/L for fatty acids, and 0.0108 and 66.2 µg/L for esters. The observed run-to-run RSD values were between 0.22% and 3.59% for alcohols, 6.81% and 7.30% for fatty acids, and 1.73% and 5.16% for esters. This method was used to determine volatile compounds in store-bought wine samples without significant matrix effects.[60]

Zeng and coworkers[61] developed a sol-gel acrylate/silicone copolymer coating for the HS-SPME of 2-chloroethyl ethyl sulfide in soil. This sol-gel fiber coating was capable of extracting 2-chloroethyl ethyl sulfide with a detection limit of 2.7 ng/g. The run-to-run RSD value for the extraction was 2.19%. This coating was also used to recover 2-chloroethyl ethyl sulfide from red clay, sandy soil, and agricultural soil. The observed extraction recoveries were 88.06% in agricultural soil, 92.61% in red clay, and 101.95% in sandy soil. This coating demonstrated thermal stability to 350°C and excellent solvent stability since no appreciable change in performance was observed after 2 h of soaking in methylene chloride, acetonitrile, and acetone, and 12 h of soaking in distilled water. This coating was still stable after 150 extraction cycles. Since it is capable of successfully extracting 2-chloroethyl ethyl sulfide, this coating might be suitable for the extraction of sulfur mustard agent, which is used in chemical weapons, due to its similarity in structure.[61]

Zeng and coworkers[62] used a sol-gel BMA/silicone coating for SPME coupled with CE to determine ephedrine derivatives in water and urine. The analytes were desorbed by immersing the fiber into a back-extraction device (Fig. 22.3).[51] The back-extraction solution was then transferred to an injection vial for CE analysis. The sol-gel fiber was prepared for the next extraction by conditioning at 260°C for 5 min. Field amplification was performed to enhance the detection limits in CE. The observed detection limits were between 0.003 and 0.005 μg/mL for the ephedrine derivatives. The observed run-to-run RSD values were between 4.96% and 7.57% and extraction recoveries between 88.7% and 98.6%. This sol-gel fiber and method can detect ephedrine derivatives in human urine samples, despite natural variations in the urine matrix, with run-to-run RSD values between 4.38% and 7.76%. The fiber demonstrated high solvent stability, high extraction efficiency, and a long lifetime. It is suitable for SPME with CE or high-performance liquid chromatography (HPLC) analysis.[62]

This coating was applied to the extraction of medium- and long-chain fatty acids and to the analysis of fatty acids in lung tissues using GC-MS.[63] Detection limits, run-to-run RSD values, and extraction recoveries for the fatty acids ranged from 0.51 to 2000 μg/L, 3.33% to 13.33%, and 76.35% to 107%, respectively. It was determined that lung tissue mostly contained $C_{16:0}$, $C_{18:0}$, and $C_{24:0}$ fatty acids, with smaller amounts of other saturated and unsaturated fatty acids. Higher levels of saturated fatty acids (0.01–66.60 μg/mg) and lower levels of unsaturated fatty acids (<0.005–3.95 μg/mg) were found in cancerous lung tissue compared with normal lung tissue (<0.005–63.74 μg/mg for saturated fatty acids and <0.005–9.22 μg/mg for unsaturated fatty acids). This study demonstrates that lung cancer could be potentially diagnosed by the SPME analysis of fatty acids in lung tissue. The major objective of this work was to provide an approach to investigate whether fatty acids have any relationship with lung cancer, which could provide potentially useful information for preventing and curing lung cancer.[63]

22.4.6.3. Other silica-based non-polysiloxane-based sol-gel fiber coatings. Caruso and coworkers[64] prepared several different sol-gel fiber coatings consisting of different molar ratios of *n*-octyltriethoxysilane (C_8-TEOS) and MTMS. It was found that the 1:1 molar ratio of C_8-TEOS : MTMS demonstrated both the best extraction capability and run-to-run repeatability.[64]

HPLC was used for the analysis of organotin, organoarsenic, and organomercury compounds extracted using the sol-gel octyl-methyl coating. After extraction, the fiber was rinsed in distilled water and placed in the 200-μL desorption chamber containing 1.0 μg/mL benzoic acid dissolved in the mobile phase (80% acetonitrile and 20% water). Static desorption was performed for 5 min. The desorption solvent was then directly injected into the HPLC.[64] Such a desorption process results in the loss of analytes and decreases the extraction sensitivity. Sol-gel CME, which will be discussed later, can overcome these problems with online hyphenation with HPLC.

Diphenylmercury, trimethylphenyltin, and triphenylarsine extractions were tested with these sol-gel fibers. The fiber-to-fiber ($n = 5$) RSDs observed for these compounds were 29% for triphenylarsine, 22% for trimethylphenyltin, and 21% for diphenylmercury. These large RSD values are due to problems with the uniformity of the coatings. This sol-gel octyl-methyl coating demonstrated detection limits of 80 μg/L for triphenylarsine, 412 μg/L for diphenylmercury, and 647 μg/L for trimethylphenyltin. The sol-gel octyl-methyl coating was resistant to solvents and solutions of 1% trifluoroacetic acid (TFA) and 1% NaOH for periods of up to 16 h.[64]

Giardina and Olesik[65–67] developed low-temperature glassy carbon films for SPME using a sol-gel process. The films were prepared by coating silica particles with a diethynyl oligomer precursor and heat curing at 300–1000°C. Using a sol-gel process, the coated particles were then immobilized into stainless-steel fibers, which were subsequently used for HS and liquid extractions. Aromatic hydrocarbons, geosmin, 2-methylisoborneol, and 2,4,6-trichloroanisole were efficiently extracted from water samples using these films.[65]

Teng and Chen[68] coupled SPME with sol-gel-assisted laser desorption/ionization time-of-flight MS. An optical fiber with a sol-gel-derived 2,5-dihydroxybenzoic acid film was used to extract trace organics from aqueous samples. The fiber was then inserted into the mass spectrometer for the analysis of the extracted compounds. This method easily coupled SPME with laser desorption/ionization MS.[68]

Azenha and coworkers[69] developed a sol-gel ultrathin phenyl-functionalized fiber coating for SPME. These coatings were thermally stable at 320°C and were able to withstand 150 extraction cycles involving solvent and thermal treatments. The sol-gel coatings were most efficient in the extraction of long-chain and aromatic compounds. The most successful sol-gel coating was the one prepared using a sol solution characterized by an MTMS : phenyltrimethoxysilane (PTMOS) ratio of 4:1, a water : precursor ratio of 2:1, and NaOH as the catalyst. This sol-gel coating was able to extract the compounds in amounts similar to, or higher than, a commercial PDMS fiber. This sol-gel coating was also able to extract long aliphatic chain compounds better than a commercial CW/DVB fiber, but the extraction of polar and aromatic compounds was better with the CW/DVB fiber. Observed run-to-run peak area RSD values were between 25% and 30%.[69]

Lee and coworkers[70] synthesized hydrophilic oligomers and developed sol-gel oligomer coatings for SPME coupled to GC-MS analysis. For organochlorine pesticides, triazine herbicides, estrogens, alkylphenols, and bisphenyl A, this coating was found to have similar or considerably better extraction efficiencies compared with commercial SPME fibers. The extraction efficiencies for triazine herbicides were significantly better on this coating than on commercial coatings. Detection limits between 0.001 and 0.005 µg/L were observed, with run-to-run peak area RSD values between 5.0% and 11.0%. Natural water from a reservoir was also used as a matrix for the triazine herbicides. Extraction recoveries from the natural water were higher than 84% with run-to-run peak area RSD values less than 11%. The efficiency of extraction was not significantly altered by matrix effects.[70]

Wang[71] developed sol-gel SE-54 fibers for SPME. These fibers demonstrated temperature stability to 310°C and run-to-run peak area RSD values less than 2.7% in GC analysis.[71]

Recently, sol-gel carbon nanotube-containing SPME fibers were developed.[72] A stainless-steel fiber was coated with a sol solution, which contained carbon nanotubes. These fibers demonstrated high-temperature resistance, mechanical strength, and a long service life.[72]

22.4.6.4. Other nonsilica-based non-polysiloxane-based sol-gel fiber coatings. Wei et al.[73] developed a titania-based sol-gel dimethyl-3,7-diaminobenzothiophene-5,5-dioxide-3,3,4,4,-diphenylsulfone tetracarboxylic dianhydride fiber for the efficient determination of BTEXs and low-polar halocarbons using SPME-GC-FID.[64] Recently, a sol-gel titania PEG coating was applied to anodized aluminum fibers and used for HS-SPME of aromatic hydrocarbons from water samples.[73]

Xu and Lee[75] developed a zirconia hollow fiber and applied it to SPME. A polypropylene hollow fiber was used as the template. It was ultrasonically cleaned in acetone for 5 min. The fiber was removed from acetone and air-dried. The sol solution was prepared by dissolving 6 g of $ZrOCl_2{:}8H_2O$ into 30 mL of ethanol : water (5:3, v/v). The mixture was sonicated for 30 min. The fiber was immersed into the sol solution for 1 h, then dried at 393 K for 2 h. This immersion and drying process was repeated several times. Finally, the fiber was heated from room temperature to 873 K at 5 K/min and maintained at the highest temperature for 4 h. This removed the template and crystallized the zirconia. The fiber was not installed into an SPME syringe.[75] A cross-sectional SEM image of the sol-gel zirconia fiber is shown in Figure 22.4.[75]

HPLC with MS was used in the analysis of pinacolyl methylphosphonic acid, a nerve agent degredation product of a chemical warfare agent. The pH of the sample solution was adjusted to 6.5. The extraction was performed by immersing the zirconia fiber into 10 mL of sample solution in a 12-mL vial. A shaker was used to facilitate the extraction process. The extractions were performed for 30 min, even though this time was not enough to reach equilibrium. After extraction, the fiber was removed and placed in ultrapure water to remove any surface contaminants. The fiber

Figure 22.4. A cross-sectional scanning electron microscope image of a zirconia hollow fiber. Reprinted from reference 75 with permission from the American Chemical Society. Copyright 2007.

was dried and placed in 200 µL of 1% ammonia solution for desorption, which took 30 min. A 20-µL volume of this solution was directly injected into the LC-MS system for analysis. After each use, the zirconia fiber was washed with concentrated ammonia solution for 10 min to remove any possible residual analyte. The fiber can be used again with no carryover effects. This fiber was capable of extracting pinacolyl methylphosphonic acid with a detection limit of 0.07 ng/mL and a run-to-run peak area RSD value of 3.7%. A lake water sample was also analyzed, and the recovery of pinacolyl methylphosphonic acid was 94.2%.[75] This sol-gel zirconia fiber represents a new configuration in microextraction and may be suitable for the determination of chemical warfare agent residues in environmental samples, as well as phosphorylated peptides and proteins in biological samples.[75]

22.5. SOL-GEL MATERIALS IN IN-TUBE SPME (CME)

22.5.1. Introduction

In sol-gel SPME, the sample capacity is still limited by the relatively short segment of the fiber that is coated with the sol-gel sorbent. This limits the sensitivity of sol-gel SPME. Also, since sol-gel SPME coatings are coated on the outside of the fibers, mechanical damage is still possible. Furthermore, it is difficult to hyphenate sol-gel SPME to HPLC and CE applications, as a special desorption device is needed,[51] which results in sample loss.

Sol-gel CME was developed in order to address these problems.[76,77] In sol-gel CME, the sol-gel coating is chemically bonded to the inner surface of a fused-silica capillary. Because of this, sol-gel CME offers increased solvent and thermal stability when compared with traditional in-tube SPME. Since a longer segment of capillary can be coated, this increases the sample capacity and sensitivity.[76] The coating is also protected from mechanical damage and can easily be hyphenated to HPLC[79,90,91,93] and CE.[81,83]

22.5.2. Capillary Pretreatment Procedures

The inner surface of a fused-silica capillary contains a small concentration of silanol groups. The chemical bonding of a sol-gel coating to the inner surface of a fused-silica capillary is accomplished through condensation reactions between silanol, alkoxy, and other sol-gel active groups in the sol-gel network and silanol groups on the inner surface of the fused-silica capillary. Therefore, like in sol-gel fiber SPME, it would be advantageous to have more silanol groups on the fused-silica surface to allow for more chemical bonding between the sol-gel coating and the surface of the capillary. This is usually accomplished through hydrothermal pretreatment. This pretreatment is conducted by first cleaning the fused-silica capillary by rinsing sequentially with methylene chloride, methanol, and deionized water. Next, the ends are typically sealed with an oxyacetylene torch and the capillary is placed in a GC oven and heated between 250 and 350°C for approximately 2h. The ends are then cut and the capillary is ready to be coated.[76,81,83,85,87,91,93,94,96] In some cases,[89,95] the formation of more silanol groups on the fused-silica surface has also been accomplished by rinsing sequentially with 1M NaOH for 2h, water for 30min, 1M HCl for 2h, and water for 30min. After cleaning, the capillary is ready to be coated with the sol-gel material.

The capillary is usually coated by filling it with the sol solution using either a gas pressure-operated filling system[97] or a syringe. The sol solution is allowed to stand in the capillary for an appropriate amount of time (typically 20–30min), followed by purging. The coated capillary is subsequently conditioned in a GC oven under helium flow[76,81,83,85,87,91,93,94,96] or in a muffle furnace[89,95] by heating to a specific temperature for a certain amount of time to facilitate completion of the sol-gel reactions. The coated capillary may be cleaned or conditioned again prior to each use.

22.5.3. Silica-Based Sol-Gel Coatings in CME

22.5.3.1. Polysiloxane-based monofunctional sol-gel capillary coatings. Malik and coworkers[76,77] introduced sol-gel CME as a solventless extraction technique for the preconcentration of trace analytes. This was the first report on the use of sol-gel-coated capillaries in analytical microextraction. A sol-gel PDMS-coated capillary was prepared in this study. The sol-gel PDMS solution used to make the capillary coating was prepared in the same manner as the sol-gel fiber PDMS coating.[2] A 250-μm internal diameter (i.d.) fused-silica capillary was previously cleaned and hydrothermally treated, then filled with the sol solution using a homemade helium pressure-operated filling/purging device. A sol-gel network evolved in the sol solution within the fused-silica capillary and a thin layer of the sol-gel material chemically bonded to the inner all of the capillary through condensation with the silanol groups on the capillary walls. After this time period, the remaining sol-gel solution was expelled from the capillary under helium pressure. Helium was purged through the capillary for an additional hour. The capillary was then placed in a GC oven and heated from 40 to 350°C at a rate of 1°C/min, holding it at the final temperature for 5h under helium purge. The final conditioning temperature was determined by the thermal stability of each sol-gel coating. The capillary was subsequently rinsed with methylene chloride and methanol and reheated with the same temperature program in a GC, except that it was held at the final temperature only for 30min.[76]

Extraction was carried out using gravity-fed sample dispensers (Fig. 22.5)[76] made from modification of a Chromaflex AQ column in which the glass was deactivated by rinsing with 5% hexamethyldisilazane (HMDS) in methylene chloride and heating to 250°C. For extraction, a previously conditioned sol-gel capillary was attached to the bottom end of the sample dispenser. The sample to be analyzed (25mL) was added to the dispenser and allowed to drip through the capillary for 30min, the time required to reach equilibrium. After extraction, the capillary was installed inside the GC injector with its lower end connected to the inlet of the GC column via a two-way press-fit quartz connector. For this, the quartz wool was removed from the glass insert to accom-

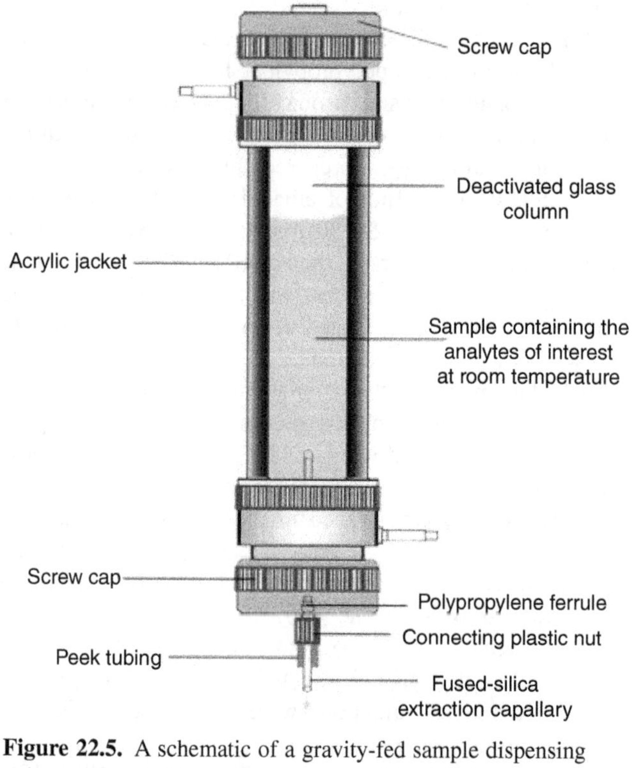

Figure 22.5. A schematic of a gravity-fed sample dispensing unit used for capillary microextraction. Reprinted from reference 94 with permission from Elsevier. Copyright 2005.

modate the deactivated two-way fused-silica connector within the insert. Before installing the microextraction capillary, the GC injector was first cooled down to room temperature. To connect the capillary with the GC column, the column nut at the bottom of the injector was loosened and the column was slid up through the top of the injector. The extraction capillary and the GC column were press-fitted into the fused-silica connector from the top and bottom, respectively. The column was pulled back down until the microextraction capillary disappeared below the septum support of the injector. Finally, the septum was reinstalled and the injector nut and capillary column nut were tightened. The extracted analytes were then thermally desorbed through heating the injection port from 30 to 330°C at a 100°C/min ramp rate for 5 min. The column temperature was held at 30°C to allow for analyte focusing. The GC analysis was performed with the split remaining closed, using helium flow, and flame ionization detection.[76]

PAHs, aldehydes, and ketones were extracted using the sol-gel PDMS capillary. PAHs demonstrated excellent detection limits of 0.31 ng/L for naphthalene to 0.94 ng/L for phenanthrene. Aldehydes demonstrated detection limits between 28.36 ng/L for *n*-decylaldehyde and 103.20 ng/L for benzaldehyde. Ketones demonstrated detection limits between 32.67 ng/L for anthraquinone and 215.7 ng/L for valerophenone. The compounds tested had excellent run-to-run peak area RSD values between 2.5% and 6%.[76]

Sol-gel CME offers the advantage of creating thicker coatings than those used in in-tube SPME where a short segment of GC capillary column is used for extraction. These thicker coatings allow for increased sensitivity and detection limits in the picogram per liter range. Sol-gel coatings are also chemically bonded to the inner walls of the capillary, which allows for greater thermal stability and solvent resistance. Because of their excellent solvent resistance, sol-gel capillary coatings are suitable for use in HPLC. Also, coating the inner surface of the capillary allows for a more uniform coating compared with sol-gel-coated fibers. The development of sol-gel CME promoted the creation of many different types of sol-gel-coated capillaries, with different extraction capabilities and applications, many of which will be discussed in this chapter.[76]

22.5.3.2. Polysiloxane-based multifunctional sol-gel coating in CME. Malik and coworkers[78] developed a sol-gel-immobilized cyano-PDMS coating for CME of aqueous trace analytes ranging from PAHs to free fatty acids using GC-FID analysis. The detection limits ranged from 2.9 to 4.1 ng/L for PAHs, 12.0 to 22.4 ng/L for aldehydes, 2.3 to 7.0 ng/L for ketones, 10.4 and 114.9 ng/L for aromatic amines, 39.1 and 161.5 ng/L for phenols, 1.4 and 60.3 ng/L for alcohols, and 4.4% and 5.9% for free fatty acids, with run-to-run peak area RSD values between 1.3% and 5.3% and capillary-to-capillary RSD values (for peak area) between 1.8% and 8.7% for all analytes. The sol-gel cyano-PDMS coating demonstrated excellent repeatability and low nanogram per liter detection limits for both polar and nonpolar analytes from aqueous media without the need for derivatization, pH adjustment, or salting out procedures. This coating was found to be highly resistant to solvents and thermally stabile at 330°C.[78]

Recently, Segro and Malik[79] developed a sol-gel polydimethyldiphenylsiloxane (PDMDPS) coating for online hyphenation of CME with HPLC using UV detection. A 40-cm-long segment of 0.25-mm i.d. fused-silica capillary was coated with the sol-gel PDMDPS coating, provided with the appropriate fittings, and installed as an external sampling loop in an HPLC injection port. The online CME-HPLC setup is presented in Figure 22.6.[79] The bottom of a vertically held gravity-fed sample dispenser was fitted with a deactivated fused-silica capillary (520-μm i.d.) using appropriate ferrules and connections. The other end of this capillary was fitted with a syringe needle. The gravity-fed dispenser was filled with the sample solution containing trace amounts of PAHs, aromatic compounds, ketones, and aldehydes. The HPLC system was placed in the load position, and the needle was inserted into the HPLC injection port and the sample was allowed to pass through the sol-gel capillary for the time required to reach equilibrium between the solution and the sol-gel coating. The samples were desorbed from the sol-gel capillary by putting the HPLC injection valve into the inject position. The injected analytes were then separated on a commercial C_{18} column using isocratic or gradient elution with acetonitrile/water mobile phases. This sol-gel coating was capable of efficiently extracting these analytes with low nanogram per liter detection limits, good run-to-run reproducibility, and capillary-to-capillary reproducibility. The test results also indicate that the sol-gel PDMDPS-coated capillary is resistant to high-temperature solvents. This makes the sol-gel PDMDPS-coated capillary suitable for applications in high-temperature HPLC.[79]

22.5.3.3. Sol-gel coating with cavity ligands for CME. Feng and coworkers[80] determined nonsteroidal anti-inflammatory drugs in urine samples using a sol-gel β-cyclodextrin-coated capillary for in-tube SPME coupled to HPLC. A sample loop was replaced with the sol-gel-coated extraction capillary. Before extraction, the analytical HPLC column was equilibrated with the mobile phase using one pump. The mobile phase consisted of 70% methanol and 30% acetate buffer solution at pH 5.0. Urine samples were collected and centrifuged at 4500 rpm for 10 min to remove precipitates. The supernatant of the urine was diluted 10 times with double-distilled water and spiked with analytes. The sample solution (250 μL) was pumped through the capillary at a flow rate of 300 μL/min. This method required less than 1 min for the extraction process. After extraction, desorption was performed by allowing the mobile phase to pass through the capillary. The desorbed analytes were transferred to the inlet of an analytical column for separation followed by UV detection at 223 nm. Ketoprofen (KEP),

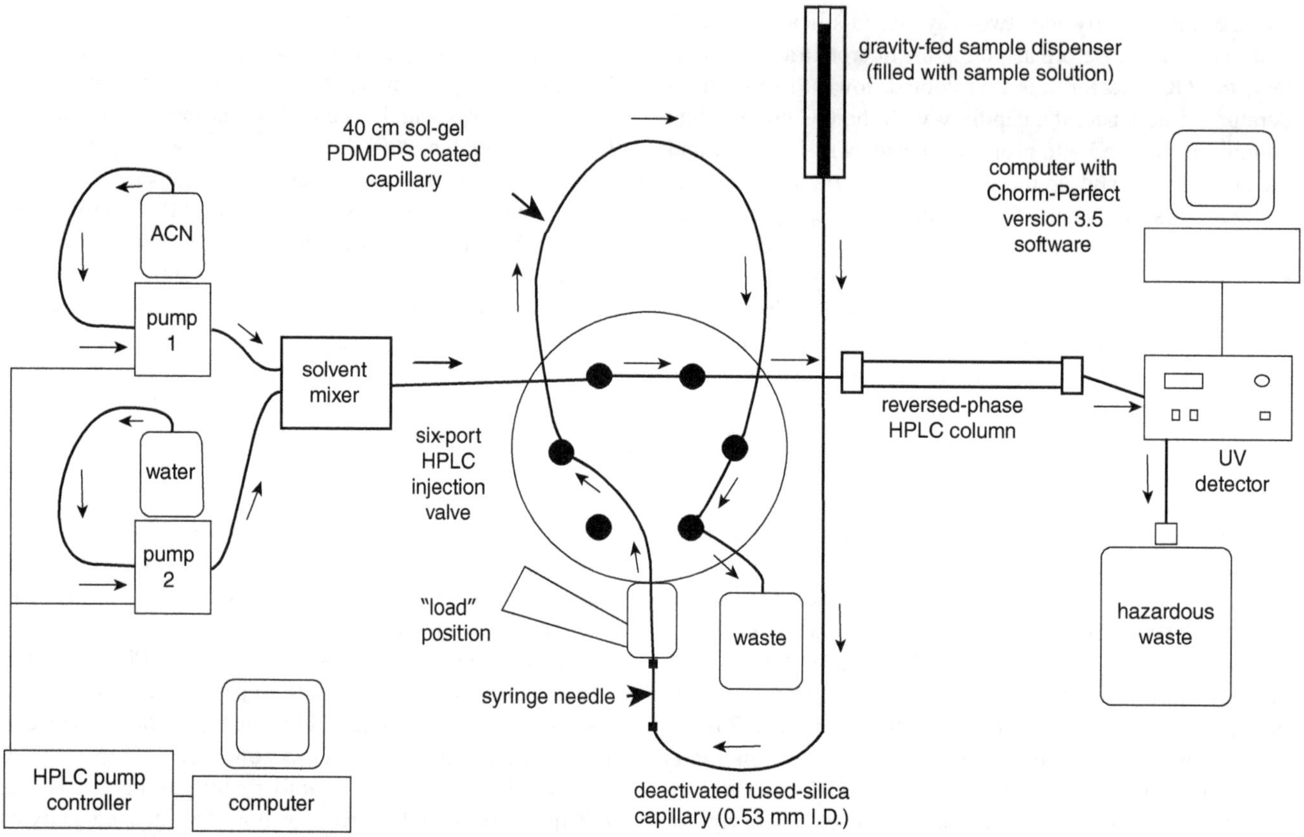

Figure 22.6. A schematic diagram of the setup used for on-line capillary microextraction on-line hyphenated with HPLC analysis. ACN, acetonitrile. Reprinted from reference 79 with permission from Elsevier. Copyright 2008.

fenbufen (FEP), and ibuprofen (IBP) were extracted and analyzed with detection limits of 38, 18, and 28 ng/mL, respectively. The intraday and interday peak area RSD values were less than 4.9% and 6.9%, respectively. This sol-gel capillary was used in over 250 extractions without a decrease in extraction efficiency. This sol-gel capillary and method were convenient for common laboratory use. This capillary can also be used in the in-tube SPME of other biological samples and environmental pollutants.[80]

22.5.3.4. Electrically charged sol-gel coatings for CME. Malik and coworkers[81,82] developed positively charged sol-gel coatings for online preconcentration of amino acids in CE. The capillary was first sealed on one end using an oxyacetylene torch. It was then filled with the sol-gel coating solution from the open end using a homemade helium pressure-operated filling device. Gas was trapped at the sealed end of the capillary, keeping the surface near the sealed end untouched by the sol solution. The sol solution was kept in the capillary for 20 min, then released from the capillary by the compressed gas pocket when the external pressure was removed. The open end was sealed and the capillary was conditioned at 150°C for 2 h in a GC oven. The sealed ends were then cut open, and the capillary was purged with helium for 30 min, rinsed with 100% acetonitrile, deionized water, and the appropriate running buffer. The polyimide coating was burned off the outside of the capillary in the uncoated section to create an optical window for in-line UV detection.[81]

In-line extraction and preconcentration was conducted in a capillary electrophoretic setup. This was accomplished by installing the sol-gel-coated capillary on the CE system and filling it with running buffer. The pH of the sample solution was maintained above the pI of the amino acid in the sample to impart a negative charge on the amino acid. Next, the inlet end of the column was inserted into the sample vial, and the sample was hydrodynamically injected for 3 min at 100 psi. This is an extended injection time, which allowed for more than one column volume of sample to pass through the electrically charged sol-gel capillary, which increased the extraction sensitivity. Electrostatic interaction between the positively charged sol-gel coating and the negatively charged amino acid molecules allowed the extraction to occur. The sample solution was then removed from the column by purging with deionized water or by reversed electroosmotic flow. The use of reversed electroosmotic flow resulted in lower detection limits. The inlet of the capillary was returned to the buffer solution, and a high electric field was applied to perform the CE analysis.[81]

This was the first report of an on-column extraction and preconcentration effect in CZE using a positively charged sol-gel column. The on-line extraction and preconcentration capabilities of the sol-gel-coated column dramatically improved the detection limits of the amino acids using CE. The observed detection limits were 139 nM for alanine, 98 nM for asparagine, 141 nM for phenylalanine, and 115 nM for tryptophan. A 150,000-fold enrichment effect for alanine was obtained using this sol-gel column and method. This method does not limit the injection volume or require modifications of the CE system.[81]

Malik and coworkers[82,83] also developed a negatively charged sol-gel-coated capillary with stable electroosmotic flow for online preconcentration of zwitterionic biomolecules in capillary electromigration separations. Myoglobin and asparagine were preconcentrated using the negatively charged sol-gel column and analyzed using CE with UV detection at 214 nm. For myoglobin, the 74% mercaptopropyltrimethoxysilane (MPTMS) column demonstrated the best sensitivity enhancement factor of 973 by peak height and 3104 by corrected peak area. For asparagine, the 60% MPTMS column demonstrated a sensitivity enhancement factor of 4450 by peak height and 7335 by corrected peak area.[83]

22.5.3.5. Non-polysiloxane-based sol-gel coatings for CME. Malik and coworkers[76] developed a sol-gel PEG coating for CME-GC-FID analysis. Phenols, alcohols, and amines were extracted using the sol-gel PEG capillary. Phenols demonstrated detection limits between 6.0 ng/L for 2,3-dimethylphenol and 16.1 ng/L for 2,6-dimethylphenol. Alcohols and amines provided detection limits between 2.0 ng/L for myristyl alcohol and 6.0 ng/L for benzanilide.

Recently, Malik and coworkers[84] developed a sol-gel-immobilized short-chain PEG coating for CME of underivatized polar analytes. GC-FID with helium carrier gas was used for the separation of extracted analytes. An in-house prepared sol-gel PDMS capillary GC column was used in the separation of the extracted analytes. The sol-gel short-chain PEG coating was used to extract polar analytes, including aldehydes, ketones, aromatic amines, phenols, alcohols, and acids, for which detection limits were between 11.8 and 20.4 ng/L, 7.2 and 119 ng/L, 5.2 and 398.2 ng/L, 23.3 and 158.6 ng/L, 8.1 and 133.8 ng/L, and 19.7 and 67.8 ng/L, respectively. For the same classes of analytes, the run-to-run peak area RSD values were between 3.0% and 4.1%, 2.9% and 4.1%, 2.7% and 4.9%, 2.6% to 4.3%, 2.5% to 3.9%, and 1.9% to 3.7%, respectively. This coating was thermally stable to 340°C. The performance of the sol-gel PEG-coated capillary in CME remained practically unchanged after rinsing it with 50 mL of dichloromethane : methanol mixture (1:1, v/v) over a 24-h period. This was the first report on the use of low-molecular-weight PEG to prepare sol-gel coatings for microextraction capillaries.[84]

Malik and coworkers[85,86] developed a sol-gel dendrimer coating for CME (Fig. 22.7). In-house prepared sol-gel PDMS and sol-gel PEG capillary GC columns were used in the separation of the extracted analytes. The sol-gel dendrimer coating was used to extract analytes belonging to various chemical classes with detection limits between 2.1 and 19.4 ng/L for PAHs, 3.3 and 44.3 ng/L for aldehydes, 1.9 and 15.2 ng/L for ketones, 9.88 and 16.03 ng/L for alcohols, and 10.53 and 12.33 ppt for phenols. PAHs, aldehydes, ketones, alcohols, and phenols demonstrated run-to-run peak area RSD values between 2.04% and 9.20%, 3.70% and 8.97%, 2.08% and 6.45%, 2.53% and 7.59%, and 2.50% and 6.18%, respectively. This sol-gel dendrimer capillary was capable of low nanogram per liter detection limits for both polar and nonpolar analytes. This coating demonstrated

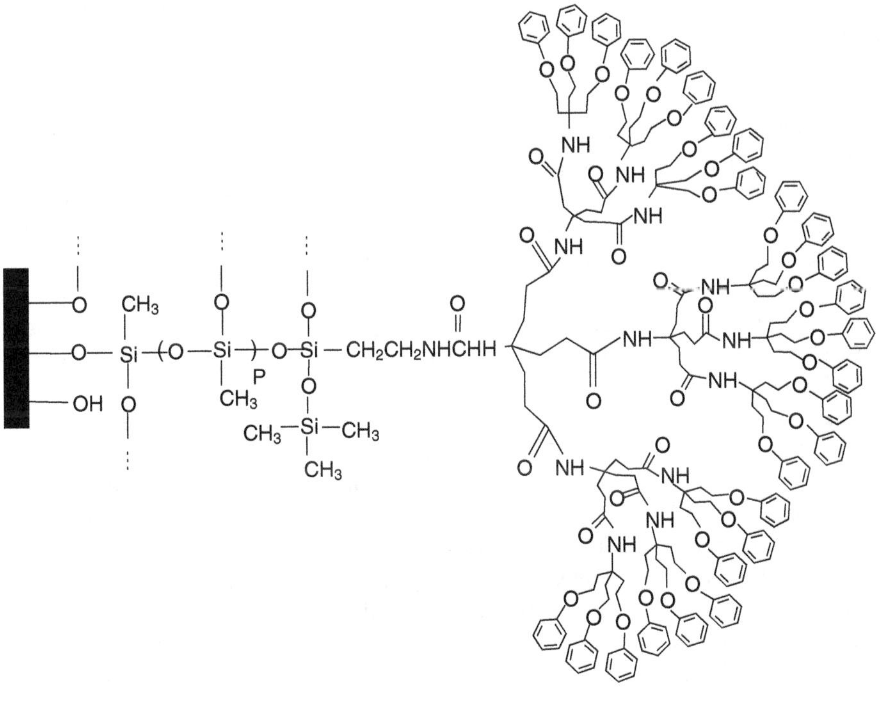

Figure 22.7. The structure of the surface-bonded sol-gel dendrimer coating. Reprinted from reference 85 with permission from Elsevier. Copyright 2004.

thermal and solvent stability, making it suitable for GC and HPLC analysis.[85]

Malik and coworkers[87,88] developed a sol-gel poly(tetrahydrofuran) (poly-THF) coating for high-sensitivity sample preconcentration using CME. This coating was capable of extracting PAHs, aldehydes, ketones, phenols, and alcohols, for which the detection limits were between 0.26 and 0.75 ng/L, 0.625 and 1 ng/L, 0.34 and 1 ng/L, 18 and 150 ng/L, and 0.59 and 13 ng/L, respectively. The capillary-to-capillary peak area RSD values for PAHs, aldehydes, ketones, phenols, and alcohols were between 2.13% and 6.45%, 4.01% and 10.31%, 3.02% and 8.01%, 3.72% and 8.74%, and 1.21% and 11.75%, respectively. For the same classes of analytes of analytes, the run-to-run peak area RSD values were between 1.07% and 5.05%, 2.19% and 7.48%, 2.03% and 8.36%, 2.21% and 7.32%, and 2.21% and 6.78%, respectively. This coating was very successful in the extraction of highly polar, moderately polar, and nonpolar compounds. It can even be used in the simultaneous extraction of polar, moderately polar, and nonpolar compounds. This coating is thermally stable to 320°C and resistant to solvents. It could be successfully coupled to HPLC and capillary electrochromatography (CEC) analysis.[87]

Zeng and Hu[89] developed a sol-gel 3-MPTMS-modified silica coating for CME online hyphenated with inductively coupled plasma atomic emission spectrometry for the determination of Cu, Hg, and Pb in biological samples. Inductively coupled plasma MS was used to perform the analysis in hyphenation with CME. The sol-gel MPTMS capillary was connected to the system. The flow rate used was 0.25 mL/min. The aqueous sample solution containing Cu, Hg, and Pb (with pH previously adjusted to 6.0 with 0.01 M NaAc and 0.1 M HCl) was passed through the sol-gel capillary and into a waste container. The analytes were adsorbed into the sol-gel coating during this step. For the elution, a 0.1 M HCl with 4% thiourea (m/v) solution was pumped through the capillary with a flow rate of 0.2 mL/min. The analytes in the eluent were detected by ICP-MS.[89]

This sol-gel capillary demonstrated detection limits of 0.17 ng/mL for Cu, 0.22 ng/mL for Hg, and 0.52 ng/mL for Pb. The RSDs were 4.2%, 2.6%, and 3.8% ($C = 4$ ng/mL, $n = 7$, sample volume = 1 mL) for Cu, Hg, and Pb, respectively. This sol-gel capillary and method was also applied to the determination of Cu, Hg, and Pb in human serum, human urine, and preserved egg. The analytical results were within the range of reported values for these reference materials. This was the first report of sol-gel MPTMS-silica-coated CME coupled to ICP-MS for trace element analysis. This coating demonstrated excellent sensitivity with little solvent and sample consumption.[89]

Recently, Segro and Malik[90] developed a sol-gel methyl coating in CME hyphenated online with HPLC. In the preparation of this capillary, MTMS was used as the sol-gel precursor.[90] The capillary was coated and installed as an external sampling loop, as described previously.[79] The methyl groups on the sol-gel precursor molecules ultimately turned into pendant groups within the sol-gel network and were primarily responsible for the extraction of nonpolar analytes, including PAHs and ketones, where low and sub-nanogram per liter detection limits were observed.[90] This sol-gel methyl coating also demonstrated a counterintuitive extraction capability for polar analytes, including aromatic phenols, alcohols, and amines. This extraction capability can be attributed to a symbiotic extraction between methyl groups and silanol groups within the sol-gel coating and the analyte molecules. This study demonstrates that sol-gel sorbents with short alkyl side chains have the potential to offer a polymer-free alternative to traditional sol-gel coatings, which are typically prepared using polymers in the sol solution. The elimination of polymers is conducive to improved thermal stability and solvent tolerance in the created sol-gel extracting phase and also makes the preparation process more facile and cost-effective.[90]

22.5.4. Nonsilica-Based Sol-Gel Coatings in CME

22.5.4.1. Titania-based sol-gel coatings in CME. Malik and coworkers[91,92] developed a sol-gel titania hybrid organic–inorganic coating for CME coupled to HPLC with UV detection. A capillary was coated and installed as an external sampling loop, as described previously.[79] This sol-gel capillary was successful in the extraction of PAHs, ketones, and alkylbenzenes. The detection limits were between 200 and 3100 ng/L for PAHs, 2500 and 11600 ng/L for ketones, and 700 and 5500 ng/L for alkylbenzenes. This sol-gel coating was resistant to solvents and was successfully used in CME coupled to HPLC. It was also thermally stable at 320°C, making it suitable for use in GC analysis. This titania-based sol-gel coating demonstrated impressive resistance to high pH. It was rinsed with 0.1 M NaOH for 12 h with no significant change in performance. This is very significant, as silica-based coatings are notoriously unstable in extreme pH conditions.[91]

Recently, Malik and coworkers[93] developed a sol-gel titania poly-THF coating for use in CME-HPLC with UV detection. This coating offered part per trillion detection limits for PAHs, ketones, phenols, alcohols, and amines. It was even capable of extracting underivatized acids. This coating demonstrated incredible pH resistance, surviving 18-h exposures to 1.0 M NaOH (pH ≈ 14.0) and 1.0 M HCl (pH ≈ 0.0). The sol-gel titania poly-THF-coated capillaries were applied to the extraction of polypeptides at pH values near their respective isoelectric points.[93]

22.5.4.2. Zirconia-based sol-gel coatings in CME. Malik and coworkers[94] also developed a sol-gel organic–inorganic hybrid zirconia coating for CME coupled to GC-FID analysis. This coating successfully extracted PAHs, aldehydes, and ketones. The observed detection limits were between 0.03 and 0.57 ng/mL for PAHs, 0.05 and 0.33 ng/mL for aldehydes, and 0.02 and 0.92 ng/mL for ketones. The run-to-run peak RSD values were between

2.45% and 7.25% for PAHs, 1.29% and 5.45% for aldehydes, and 1.24% and 5.57% for ketones. The capillary-to-capillary RSD values for GC peak areas were 4.60% for undecanal, 1.61% for hexanophenone, 5.40% for fluorene, and 4.91% for phenanthrene. This coating also demonstrated high pH stability. It was rinsed with 0.1 M NaOH (pH 13) for 24 h. It survived this high pH treatment with only a 2.15–5.64% change in peak area for PAHs. Nanogram per liter detection limits were achieved for both polar and nonpolar analytes using this sol-gel coating.[94]

Hu and coworkers[95] developed a sol-gel zirconia coating for CME online hyphenated with ICP-MS for the determination of Cr, Cu, Cd, and Pb in biological samples. The sol-gel zirconia capillary was connected to the system. The flow rate used was 0.25 mL/min. In the prefill stage, pump 1 with the injector valve in the fill position and pump 2 with the injector valve in the injection position were activated so that all of the tubing would be filled with solution. In step 1, pump 1 was still in the fill position, and the capillary was conditioned with 0.01 M NH_4OH solution. In step 2, the valve was still in the fill position, but pump 1 was stopped and pump 2 was activated. The aqueous sample solution containing Cr, Cu, Cd, and Pb (with pH previously adjusted to 8.0 with 0.01 M NH_4OH and 0.01 M HNO_3) was passed through the sol-gel capillary and into a waste container. The analytes were adsorbed into the sol-gel coating during this step. In step 3, the valve was still in the fill position, but pump 2 was stopped and pump 3 was activated. High-purity deionized water was sucked through the sol-gel capillary to remove the residual matrix. In step 4, the valve was turned to the inject position, and 0.5 M HNO_3 solution was pumped to elute the sorbed analytes. The analytes in the eluent were detected by ICP-MS. The required desorption time was 60 s. This analysis required a short analysis time of only 450 s. The capillary was used repeatedly after regeneration with high-purity deionized water and 0.01 M NH_4OH solutions.[95]

This sol-gel capillary demonstrated detection limits of 6.1 pg/mL for Cr, 13.6 pg/mL for Cu, 1.4 pg/mL for Cd, and 1.6 pg/mL for Pb. The precisions for nine replicate measurements of 1 ng/mL Cr, Cu, Cd, and Pb were 4.9%, 2.2%, 2.0%, and 3.2% (RSD), respectively. This sol-gel capillary and method was used in the analysis of human urine samples. The samples were spiked with the metals, and the coating and method demonstrated extraction recoveries between 89.2% and 101.8%. A NIES No. 10-a Rice Flour-Unpolished certified reference material and a BCR No. 184 Bovine Muscle certified reference material were also analyzed using this sol-gel capillary and method. The analytical results were within the range of reported values for these reference materials. This was the first report of sol-gel CME coupled to ICP-MS for trace element analysis. Coupling of the sol-gel zirconia-coated capillary to ICP-MS demonstrated high sensitivity and selectivity with low solvent consumption in the trace/ultratrace elemental analysis of various samples in complicated matrices.[95]

22.5.4.3. Germania-based sol-gel coatings in CME. Recently, Malik and coworkers[96] developed a germania-based sol-gel hybrid organic–inorganic coating for CME with GC analysis. Sol-gel germania PDMS-coated capillaries also proved to be useful as GC columns.[96] The sol-gel germania PDMS-coated microextraction capillaries demonstrated detection limits between 11.4 and 83.8 ng/L for PAHs, 89.7 and 139.7 ng/L for aldehydes, and 30.6 and 92.3 ng/L for ketones. The observed run-to-run peak area RSD values were between 2.33% and 5.05% for PAHs, 1.32% and 6.46% for aldehydes, and 2.38% and 7.18% for ketones. The capillary-to-capillary RSD values for GC peak area were between 5.4% and 5.9%. The sol-gel germania PDMS capillary was able to withstand extreme pH values between 1.5 and 13.0. It was separately rinsed with 0.05 M HCl and 0.1 M NaOH for 24 h. The rinsing actually improved the extraction characteristics of the capillary, as demonstrated by the chromatograms in Figure 22.8.[96]

The sol-gel germania PDMDPS capillary was also capable of extracting chlorophenols. This was the first report on the use of sol-gel germania coatings in microextraction.[96]

Malik and coworkers[96] also developed a sol-gel germania capillary using 3-aminopropyltrimethoxysilane. This sol-gel germania 3-aminopropyltrimethoxysilane capillary was capable of extracting highly polar compounds, such as free fatty acids.[96]

22.6. SOL-GEL MATERIALS IN STIR-BAR SORPTIVE EXTRACTION

22.6.1. Sol-Gel PDMS-Based Coated Stir Bars

Guan and coworkers[98] developed stir bars for sorptive extraction using sol-gel technology. A bare glass bar (30 × 1.8 mm outer diameter [o.d.]) was used for the preparation of the sorptive stir bars. The sorptive stir bar contained an iron bar inside the glass tube, while the ultrasonic sorptive bar contained no iron bar. These bars were sequentially cleaned with water and methylene chloride, then sequentially soaked in 1 M NaOH and 0.01 M HCl for 8–12 h. The bars were washed again with water and purged with nitrogen flow at 120°C in a 40 × 10 mm i.d. stainless-steel tube. The bars were coated by immersing them in the sol solution. After coating, the bars were placed into a vacuum desiccator for 8 h to allow coating gelation to occur. The bars were then heated again at 120°C and purged with nitrogen flow in a 40 × 10 mm i.d. stainless-steel tube. This tube was then put into a GC oven and heated under temperature programming from 40 to 120·C at 1°C/min, holding at 120°C for 180 min, then heating at 240°C at the same rate, holding another 180 min, and finally heating at 300°C at the same rate, this time holding for 240 min. The bars were extracted by boiling methylene chloride for 4 h. Before extraction, the bars were purged under nitrogen at 300°C for 2 h.[98]

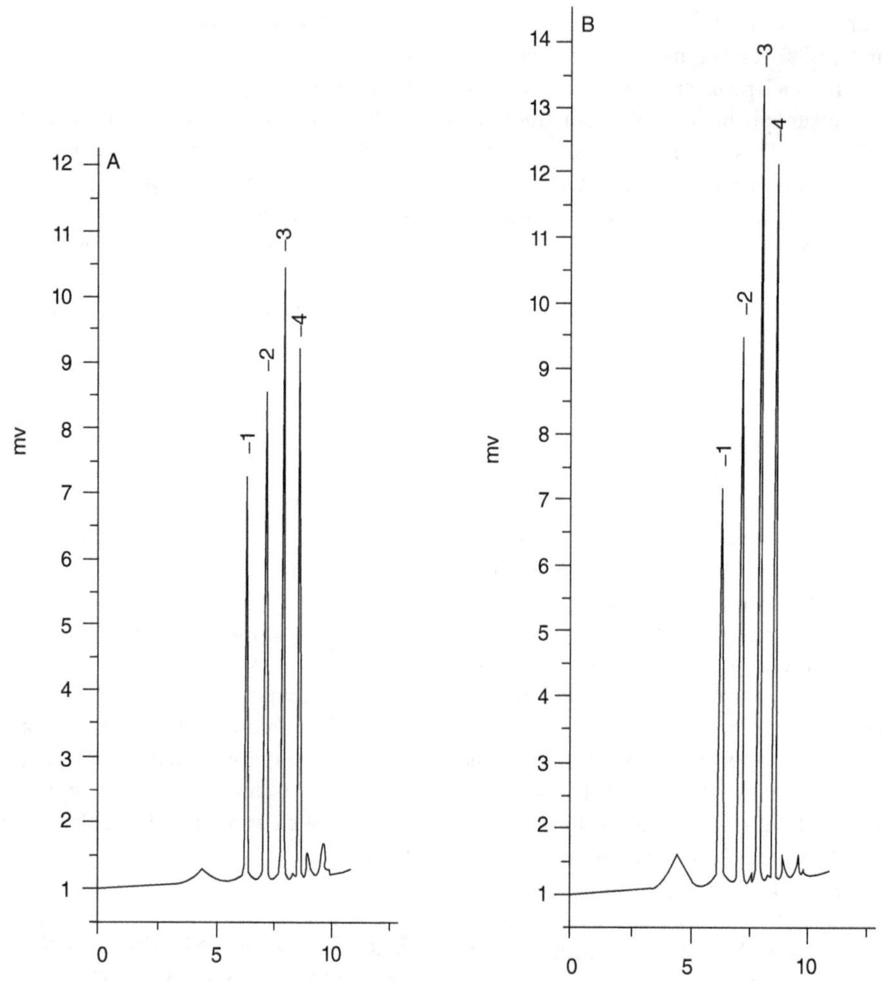

Figure 22.8. Excellent stability of sol-gel germania-PDMS coating under highly acidic conditions demonstrated through CME-GC analysis of aldehydes using a germania-PDMS-coated microextraction capillary before (A) and after (B) continuously rinsing the capillary with a 0.05 M HCl (pH 1.3) solution for 24 h. Extraction conditions: 10 cm × 0.25 mm i.d. microextraction capillary and extraction time, 40 min (gravity fed at room temperature). Other conditions: 5 m × 0.25 mm i.d. sol-gel PDMS-coated GC capillary column; splitless desorption splitless analyte desorption by rapidly increasing the GC injector temperature from 30 to 300°C, 5 min; GC oven temperature programmed from 30 to 300°C at a rate of 20°C/min; helium carrier gas; FID 350°C. GC peaks: (1) nonanal, (2) decanal, (3) undecanal, and (4) dodecanal. Reprinted from reference 96 with permission from the American Chemical Society. Copyright 2007.

GC-FID was used for the analysis of *n*-alkanes and PAHs. Thermionic-specified detection was used for the analysis of organophosphorus pesticides. For stir-bar sorptive extraction, the stir bar was placed in a 50-mL sample vial at room temperature. For ultrasonic sorptive extraction, the bar was placed into a 50-mL sample vial, which was put into an ultrasonic bath. Using stir-bar sorptive extraction with this coating, 90 min of extraction time was required to reach equilibrium. Using ultrasonic sorptive extraction, only 10 min of extraction time was required to reach equilibrium. For organophosphorus pesticides, 120 min of stir-bar extraction was required. A special apparatus is required for thermal desorption, and it was made in the laboratory. It consists of a quartz liner and heaters, two gas flow controllers, temperature control units, and a sample transfer line. A 90-mm-long stainless-steel cylinder with an i.d. of 6.7 mm and an external diameter of 12 mm was used to hold the desorption liner and transfer line. A gas flow controller provided purge gas for the desorption from the top of the liner, and another provided sweep gas from the bottom of the liner to prevent diffusion and adsorption of the sample outside of the quartz liner. The transfer line was inserted 20 cm deep through the injector septum into the GC precolumn. Thermal desorption of the analytes was accomplished by drying the stir bars with tissue paper and placing them into the desorption apparatus for 2 min, the time required to reach the desorption temperature (260–280°C), and then holding for an additional 5 min.[98]

This stir-bar sorptive sol-gel coating provided detection limits between 0.74 and 20.0 pg/mL for *n*-alkanes, 0.18 and 2.76 pg/mL for PAHs, and 0.3 and 8.0 pg/mL for organophosphorus pesticides. This coating did not exhibit carryover problems when desorption temperatures between 260 and 280°C were used. This sol-gel bar coating was also solvent resistant and thermally stable at 300°C.[98]

Guan and coworkers[99] used this same sol-gel-coated stir bar for the determination of organophosophorus pesticides in cucumber and potato samples. Extraction recovery values for the organophosphorus compounds in the extracts were between 93% and 105%. The observed detection limits were between 0.004 and 0.15 ng/g in cucumber samples and between 0.0012 and 0.098 ng/g in potato samples. The sol-gel-coated stir bar was also applied to the determination of dimethoate and parathion-methyl in real cucumbers and potatoes. Dimethoate and parathion-methyl were detected in cucumbers, and parathion-methyl was detected in potatoes.[99]

22.6.2. Sol-Gel Multifunctional Stir-Bar Coatings

Yu and Hu[100] used stir-bar sorptive extraction coupled with ultrasonic-assisted extraction for the determination of brominated flame retardants (BFRs) in environmental samples using HPLC. The stir bar was coated with sol-gel PDMS-β-cyclodextrin. Stir-bar sorptive extraction with HPLC-UV was performed to analyze aqueous samples containing BFRs. The detection limits ranged from 2.9 to 4.2 µg/L for the BFRs. The bar-to-bar peak area RSD ranged from 1.3% to 15.7%. This sol-gel-coated stir bar was also applied to the extraction of BFRs from soil and dust samples. One soil sample and one dust sample were used in a recovery test. In the determination of BFRs in these samples, the method provided extraction recoveries between 56% and 118%.[100] The same research group recently developed sol-gel PDMS/poly(vinyl alcohol)[101] and sol-gel PEG/PDMS/poly(vinyl alcohol)[102] coated stir bars for sorptive extraction.

Li and coworkers[103] used a sol-gel PDMS/β-cyclodextrin stationary phase for stir-bar sorptive extraction. This stir-bar sorptive extraction was coupled with HPLC analysis. The observed detection limits were between 0.04 and 0.11 µg/L for estrogens using UV detection and 8 ng/L for bisphenol A (BPA) using fluorescence detection.[103]

22.7. SOL-GEL MONOLITHS IN ANALYTICAL MICROEXTRACTION

The focus of this chapter to this point has been sol-gel coatings in fiber SPME and in-tube SPME, or CME. Sol-gel coatings have been used extensively in the extraction of many types of compounds. The majority of the sol-gel coatings were used in SPME for GC analysis of the extracted samples, but some SPME coatings have been used in conjunction with HPLC and CE as well. Continuous, porous sol-gel materials are known as sol-gel monoliths. Sol-gel monoliths completely fill fused-silica capillaries, and liquids can pass through the pores of the sol-gel monolith. Sol-gel monoliths have been used successfully as stationary phases in HPLC and CE. While there have not been as many publications on sol-gel monoliths in analytical microextraction, they have also been used successfully for this purpose. In this section, they will be discussed in chronological order. Future developments will likely lead to the more frequent use of sol-gel monoliths in analytical microextraction.

Malik and coworkers[104] reported the preparation of porous sol-gel monolithic beds for use in analytical microextraction. Like sol-gel coatings, these monoliths are also chemically bonded to the silica substrate. A sol-gel octadecylsilane (ODS) monolithic bed was prepared inside a fused-silica capillary using the sol-gel precursors tetramethoxysilane (TMOS) and *N*-octadecyl-dimenthyl [3-(trimethoxy-silyl) propyl] ammonium chloride. Phenyldimethylsilane was used as the deactivation reagent and TFA was used as the catalyst. A sol-gel coating was also prepared using the same reagents for comparison. An aqueous sample containing a PAH and an alcohol was extracted under identical conditions using both the monolithic capillary and the coated capillary and analyzed using GC-FID. The results indicate that the sample capacity for the monolithic capillary is at least two orders of magnitude higher than that of the sol-gel-coated capillary. Chromatograms comparing the extraction sensitivity in CME using a sol-gel monolithic capillary and a sol-gel-coated capillary are depicted in Figure 22.9.[105]

Zare and coworkers[106,107] developed a sol-gel monolithic stationary phase for CEC and noticed that it was capable of preconcentrating a variety of neutral and charged analytes.[106,107] They used a porous photopolymerized sol-gel monolith with solvent gradient and sample stacking for online preconcentration in CEC.[108] Through combining the preconcentration effects of the sol-gel monolith with solvent gradient and sample stacking, it was possible to preconcentrate eight alkyl phenyl ketones and four PAHs. UV detection was used at wavelengths of 214 and 254 nm. Electric field-enhanced sample injection yielded 1000-fold improvement in detection sensitivity for five peptides.[108]

Takeuchi and coworkers[109] used monolithic precolumns for sample enrichment in microcolumn LC. The samples were loaded into the precolumn using a handmade rubber band-driven pumping device. The precolumn allowed for the enrichment of aqueous samples with volumes up to 1 mL. The desorption solvent was then allowed to flow through the precolumn and into the separation column. Phthalates were detected using a UV detector at 204 nm. The run-to-run peak area RSD values were between 0.92% and 5.5% for the capillaries. The detection limits of phthalates in HPLC-UV with the sol-gel monolithic precolumns were between 0.21 and 0.87 ng/mL. A 150-µm i.d. sol-gel monolithic precolumn was applied to the trace analysis of phthalates in laboratory distilled water and tap water samples.[109]

Toyo'oka and coworkers[110] developed a sol-gel monolith for the preconcentration of biogenic amines prior to their on-column derivatization for subsequent analysis by CEC. Both UV and fluorescence detectors were used in this analysis. Derivatization was accomplished by first passing the derivatizing agent (*o*-phthalaldehyde [OPA], 2-mercaptoethanol [2ME]) through the monolith in buffer solution, then passing buffer solution without the derivatizing agent through the monolith. Online preconcentration was performed by passing the amine sample solution in 9% sodium chloride through the monolith for up to 600 s. The derivatization produced an OPA-thiolamine derivative, which can be detected at 340 nm using a UV/visible detector

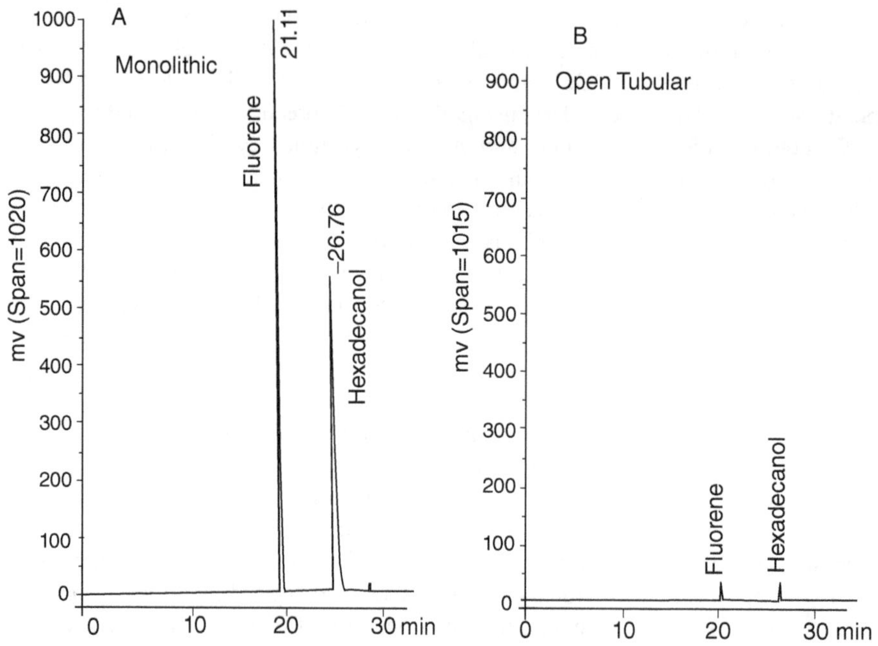

Figure 22.9. A sample of a PAH and an alcohol extracted on (A) a sol-gel ODS monolith and (B) a sol-gel ODS-coated capillary. Reprinted from reference 105 with permission from Elsevier. Copyright 2002.

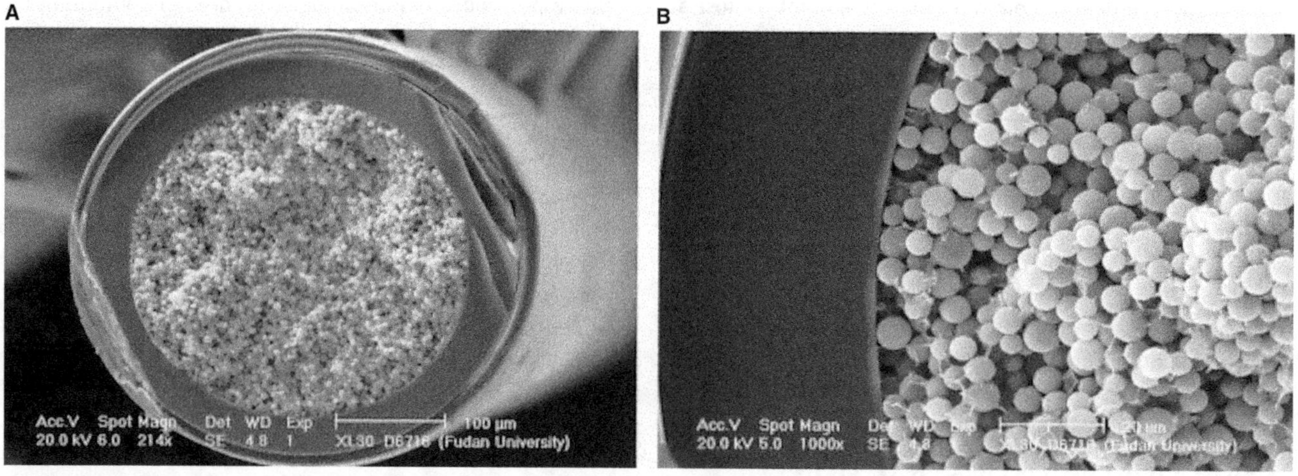

Figure 22.10. Scanning electron micrographs of the 320-μm i.d. sol-gel monolithic precolumn (A = 240×; B = 1000×). Reprinted from reference 111 with permission from Elsevier. Copyright 2005.

and at 480 nm using a fluorescence detector. Using the fluorescence detector with the preconcentration provided detection limits of 0.1 μM, resulting in a 1000-fold increase in the detection sensitivity compared with standard CEC with UV/visible detection.[110]

Zhang and coworkers[111] developed a monolithic precolumn for online trapping and preconcentration of peptides in a multidimensional LC system for proteome analysis. Both 320-μm i.d. and 530 μm i.d. precolumns were prepared. On-column frits were fabricated using Zorbax BP-SIL particles. The capillaries were packed with 5-μm Hypersil C_{18} particles. Figure 22.10 is a SEM of the monolithic precolumn.[111]

These sol-gel monoliths were mechanically stable under the pressures encountered in HPLC. BSA tryptic digest, lecucine enkaphalin, and oxytocin were preconcentrated using the sol-gel monolithic precolumns in this study. High concentrations of salt buffers did not present a noticeable effect on the preconcentration abilities of the precolumns. The detection limits for lecucine enkephalin and oxytocin were 0.87 and 0.53 ng/μL, which is equivalent to a 70-fold preconcentration factor. The precolumns were capable of recovering over 90% of the BSA tryptic digest (25 μg/μL), 98% of the lecucine enkaphalin (20 μg/μL), and 99% of the oxytocin (20 μg/μL). The day-to-day RSD values for recoveries of BSA peptides on a single precolumn ranged from 4.66% to 7.56%, while the column-to-column RSD values were 3.51–6.13% for recoveries of BSA peptides. The precolumns were resistant to continuous flushing with acidic loading buffer and had good reproducibility for back pressure, with back pressure %RSD values between 2.68% and 3.05% and column-to-column back pressure RSD values between 1.22% and 1.26%. The precolumns were used in over 150 preconcentration/desorption cycles with no significant changes in performance. These sol-gel monolithic precolumns appear to be suitable for the preconcentration of dilute peptides.[111]

Li and coworkers[112] developed a sol-gel PDMS/β-cyclodextrin membrane for SPME. Extraction was accomplished through a membrane extraction procedure. For the extraction, a 1×0.2 cm membrane was dipped into the sample solution with magnetic stirring at 1080 rpm for the time required for equilibrium to be established. After extracting, the membrane was removed with tweezers and dried carefully with filter paper. The membrane was cut into small pieces and transferred into a small glass vial with 200 μL of acetonitrile. The vial was agitated ultrasonically for desorption of the analytes from the membrane into acetonitrile. Desorption took about 5 min. The acetonitrile extracts were then directly injected into a GC-MS for analysis.[112]

This sol-gel membrane was capable of extracting PAHs with detection limits between 0.01 and 0.2 μg/L. The run-to-run peak area RSD values in the extraction of PAHs were between 5.4% and 9.3%. This sol-gel membrane was also used to determine PAHs in a real river water sample. This membrane was capable of extracting phenols with detection limits between 0.02 and 1.5 μg/L. The run-to-run peak area RSD values observed in the extraction of the phenols ranged from 4.9% to 13.1%. This membrane is stable in methanol, acetone, and acetonitrile solutions. It does not present carryover problems but can be disposable because of its low cost. The membrane was many times less efficient in the extraction of PAHs and phenols than sol-gel fiber SPME coatings. This is most likely due to the incomplete injection of the extracted compounds since direct injection of a small amount of desorbed sample was used. The detection limits could be improved with complete injection of the extracted samples through large volume sample injection or thermal desorption into the GC. The membrane has a large surface area to extraction phase volume ratio.[112]

Xue and coworkers[113] used sol-gel monoliths for optical determination of Cr(VI). Optically transparent monoliths were obtained and stored in water before use. In order to perform the extraction of Cr(VI), each monolith was placed in a solution with a known concentration of Cr(VI) in 0.1 M HCl. The monolith was exposed to the solution for about 300 min, then removed from solution, rinsed thoroughly with deionized water, and analyzed by UV-VIS spectroscopy. The detection limits could be decreased by placing the monolith in diphenylcarbazide solution prior to UV-VIS spectroscopy analysis. The monolith was regenerated by soaking in 6.0 M HCl after soaking in diphenylcarbazide. To perform the UV-VIS spectroscopy, absorbance at 350 nm was measured. At milligram per liter concentrations of Cr(VI), the absorbance at 350 nm could be used to detect the Cr(VI). At microgram per liter concentrations of Cr(VI), there is not enough color change to detect the Cr(VI) at 350 nm. After immersing the monolith in diphenylcarbazide solution, a magenta color forms, which can be detected at 540 nm. Using the monolith to extract the Cr(VI) and subsequently immersing it in diphenylcarbazide solution resulted in the detection of Cr(VI) at concentrations as low as 10 μg/L. This sol-gel monolith was capable of removing 77.3–81.7% of Cr(VI) from solution. A sample of lake water was spiked with 85 μg/L Cr(VI), and the pH was adjusted to 1. The monolith was exposed to 20 mL of this solution for 6 h, rinsed with deionized water, exposed to diphenylcarbazide solution for 1 h, and analyzed using UV-VIS spectrometry. The concentration of Cr(VI) in the lake water was accurately determined. The matrix of the lake water did not present a problem in this analysis, making this sol-gel monolith and method suitable for environmental analysis.[113]

Bergquist and coworkers[114] used a sol-gel monolith column for online biological sample cleanup for electrospray MS. Electrospray ionization with time-of-flight MS was used to perform the analysis of peptides in this study. The exit end of the monolith was directly connected to the mass spectrometer to effectively hyphenate the electrophoretic separation with electrospray MS. Peptide stock solutions were prepared in water at concentrations of 15 μg/mL. The running buffer used consisted of 5 mM ammonium acetate (pH 6.8) in 40% water and 60% acetonitrile. A washing buffer of 5 mM ammonium acetate (pH 6.8) in water was used. Prior to analysis, urine samples were filtered and stored at −20°C. A freshly thawed sample was used each day of analysis. The sol-gel monolith column was conditioned with washing buffer for 5 min using a slight overpressure of 1 bar. Urine samples spiked with 15 μg/mL concentrations of neurotensin, oxytocin, angiotensin II, leucine-enkephalin, and leutenizing hormone-releasing hormone were injected (1.4 μL) into the column at 1 bar pressure for 5 min. Washing buffer was then allowed to flow through the column for 15 min to remove salts and electrolytes. The inlet of the column was then placed into a vial containing running buffer, and a separation voltage of +20 kV was applied. A 70-fold preconcentration was achieved for all of the peptides in the sample. This column had a lifetime of about 30 runs, due to the analysis of raw urine samples. This monolithic column was capable of selectively preconcentrating the hydrophobic peptides while the hydrophilic species, such as urea and salt, passed through the column. The peak area RSD values were about 4% on the same day and 5% between days. This sol-gel monolith was capable of both online desalting and preconcentration of peptide samples.[114]

Johannesson and Bergquist[115] used a similarly prepared[114] sol-gel monolith column for the online extraction and quantification of escitalopram from urine.[115] Electrospray ionization with time-of-flight MS detection was used to perform the analysis in this study. The observed detection limit for ascitalopram was 10 pg/mL. The intraday RSD value was less than 6.3%. This sol-gel monolith was also used in the determination of the concentration of ascitalopram in a human urine sample.[115]

Feng and coworkers[116] developed a sol-gel hybrid organic–inorganic octyl monolithic column for in-tube SPME coupled to capillary HPLC with UV detection. Two microflow pumps (pump A and pump B) were used. Pump A was combined with valve A and a 1-mL stainless-steel sampling loop. Two segments were connected using polyether ether ketone (PEEK) tubing between valves A and B. The sample loop of valve B was replaced with the monolithic capillary column. Before extraction, valve A was switched to the load position, and the carrier solution was driven by pump A to flow through the monolithic capillary for conditioning at 0.05 mL/min. The stainless-steel sampling loop was filled with sample solution using a syringe. Valve A was rotated to the inject position for extraction. Extraction efficiency was increased by adding NaCl to the 40% methanol in the sample solution. Extraction efficiency improved with increasing extraction volume, but a volume of over 1 mL was undesirable since a longer extraction time was required at a flow rate of 0.05 mL/min. After extraction, the valve was returned to the load position. Desorption was performed by switching valve B from load to inject, and 90 µL of mobile phase (85% methanol, 15% water) was driven through the monolith by pump B. The desorbed analytes were transferred to the inlet of an analytical column for separation followed by UV detection.[116]

This sol-gel monolith was capable of extracting PAHs.[116] The observed detection limits were 6.5 ng/mL for biphenyl, 7.1 ng/mL for fluorene, 2.4 ng/mL for phenanthrene, and 8.1 ng/mL for fluoranthene. The intra- and interday recovery RSD values were less than 7.4% and 8.1%, respectively. Standard solutions of 2 mg/L concentrations of PAHs were extracted using this monolith, and the observed recoveries ranged from 83.2% to 110.2%.[116]

Recently, a sol-gel tetraethoxysilane-based monolith was used as an in-line concentrator for the determination of methionine enkephalin in cerebral spinal fluid using CE hyphenated with MS.[117] A 40-fold preconcentration was demonstrated. The observed detection limit in cerebral spinal fluid was approximately 1 ng/mL.[117]

22.8. CONCLUSION

Sol-gel materials have been very successful in the area of analytical sample preparation. The preceding examples are indicative of this. This chapter should be consulted as a reference material for anyone interested in doing further research using sol-gel coatings, monoliths, or particles for any sample preparation purposes. Sol-gel materials are versatile and capable of a wide range of extraction capabilities for numerous types of analytes. They are generally very resistant to solvents and high temperatures and demonstrate long lifetimes and can even handle difficult matrices. This chapter should promote awareness of the different sol-gel materials that have been developed for analytical sample preparation and facilitate future research and development in this area.

REFERENCES

1. Brinker, C.J.; Scherer, G.W. *Sol-Gel Science*. New York: Academic Press; 1990.

2. Chong, S.L.; Wang, D.; Hayes, J.D.; Wilhite, B.W.; Malik, A. Sol-gel coating technology for the preparation of solid-phase microextraction fibers of enhanced thermal stability. *Anal. Chem.* **1997**, *69*, 3889–3898.

3. Pawliszyn, J. *Applications of Solid Phase Microextraction*. Cambridge, UK: The Royal Society of Chemistry; 1999.

4. Malik, A.; Wang, D. Solid-phase microextraction fiber structure and method of making. U.S. Patent 6759126 (2004).

5. Vonderheice, A.P.; Montes-Bayon, M.; Caruso, J.A. Solid-phase microextraction as a sample preparation strategy for the analysis of seleno amino acids by gas chromatography-inductively coupled plasma mass spectrometry. *Analyst* **2002**, *127*, 49–53.

6. Bagheri, H.; Es-haghi, A.; Rouini, M.R. Sol-gel-based solid-phase microextraction and gas chromatography-mass spectrometry determination of dextromethorphan and dextrorphan in human plasma. *J. Chromatogr. B* **2005**, *818*, 147–157.

7. Bagheri, H.; Es-haghi, A.; Khalilian, F.; Rouini, M.R. Determination of fentanyl in human plasma by head-space solid-phase microextraction and gas chromatography-mass spectrometry. *J. Pharm. Biomed. Anal.* **2007**, *43*, 1763–1768.

8. Fernandes de Oliveira, A.; Berta da Silveira, C.; Denafre de Campos, S.; Antonio de Campos, E.; Carasek, E. The use of a thin glass-ceramic rod as a surface for sol-gel coating in the preparation of SPME fibers. *Chromatographia* **2005**, *61*, 277–283.

9. Liu, W.; Hu, Y.; Zhao, J.; Xu, Y.; Guan, Y. Physically incorporated extraction phase of solid-phase microextraction by sol-gel technology. *J. Chromatogr. A* **2006**, *1102*, 37–43.

10. Guan, Y.; Liu, W.; Zhao, J.; Xu, Y. Method for preparation of coating for solid phase microextraction. Faming Zhuanli Shenqing Gongkai Shuomingshu; 2007.

11. Azenha, M.A.; Nogueira, P.J.; Silva, A.F. Unbreakable solid-phase microextraction fibers obtained by sol-gel deposition on titanium wire. *Anal. Chem.* **2006**, *78*, 2071–2074.

12. Azenha, M.; Nogueira, P.; Fernando-Silva, A. An improved bonded-polydimethylsiloxane solid-phase microextraction fiber obtained by a sol-gel/silica particle blend. *Anal. Chim. Acta* **2008**, *610*, 205–210.

13. Yang, C.G.; Ji, X.J.; Lan, X.Z. Preparation of PDMS-coated microspheres by sol-gel method for sorptive extraction of PAHs. *Chin. Chem. Lett.* **2008**, *19*, 996–999.

14. Liu, M.; Liu, Y.; Zeng, Z.; Peng, T. Preparation and characteristics of high pH-resistant sol-gel alumina-based hybrid organic-inorganic coating for solid-phase microextraction of polar compounds. *J. Chromatogr. A* **2006**, *1108*, 149–157.

15. Zeng, Z.; Liu, M. Manufacture of solid phase microextraction fiber coated with alumina and hydroxy silicone oil. Faming Zhuanli Shenqing Gongkai Shuomingshu; 2006.

16. Li, X.; Gao, J.; Zeng, Z. pH-resistant titania hybrid organic-inorganic sol-gel coating for solid-phase microextraction of polar compounds. *Anal. Chim. Acta* **2007**, *590*, 26–33.

17. Lopes, A.L.; Augusto, F. Preparation and characterization of polydimethylsiloxane/poly(vinylalcohol) coated solid phase microextraction fibers using sol-gel technology. *J. Chromatogr. A* **2004**, *1056*, 13–19.

18. Zuin, V.G.; Lopes, A.L.; Yariwake, J.H.; Augusto, F. Application of a novel sol-gel polydimethylsiloxane-poly(vinyl alcohol) solid-phase microextraction fiber for gas chromatographic determination of pesticide residues in herbal infusions. *J. Chromatogr. A* **2004**, *1056*, 21–26.

19. da Silva, R.C.; Zuin, V.G.; Yariwake, J.H.; Eberlin, M.N.; Augusto, F. Fiber introduction mass spectrometry: Determination of pesticides in herbal infusions using a novel sol-gel PDMS/PVA fiber for solid-phase microextraction. *J. Mass Spectrom.* **2007**, *42*, 825–829.

20. da Silva, G.A.; Augusto, F.; Poppi, R.J. Simultaneous optimization by neuro-genetic approach of a multiresidue method for determination of pesticides in *Passiflora alata* infuses using headspace solid phase microextraction and gas chromatography. *J. Chromatogr. A* **2007**, *1138*, 251–261.

21. Yang, M.; Zeng, Z.R.; Qiu, W.L.; Wang, Y.L. Preparation and investigation of polymethylphenylvinylsiloxane-coated solid-phase microextraction fibers using sol-gel technology. *Chromatographia* **2002**, *56*, 73–80.

22. Yang, M.; Qiu, W.; Zeng, Z.; Niu, X. Determination of toluene and xylenes in wastewater samples by headspace solid-phase microextraction using a new fiber. *Fenxi Shiyanshi* **2002**, *21*, 40–42.

23. Wang, Y.; Zeng, Z.; Wang, C. Determination of methylphosphonates and phosphates in air samples using solid-phase microextraction coupled with gas chromatography. *Jilin Daxue Xuebao, Lixueban* **2007**, *45*, 661–664.

24. Dong, C.; Zeng, Z.; Yang, M. Determination of organochlorine pesticides and their derivations in water after HS-SPME using polymethylphenylvinylsiloxane-coated fiber by GC-ECD. *Water Res.* **2005**, *39*, 4204–4210.

25. Wang, Y.; Yang, M.; Dong, C.; Song, F. Headspace solid-phase microextraction coupled to gas chromatography analysis of five organophosphorus pesticides in water. *Huazhong Shifan Daxue Xuebao, Ziran Kexueban* **2006**, *40*, 374–378.

26. Dong, C.; Wang, W. Determination of organochlorine pesticides and the similar in water by headspace solid-phase microextraction and gas chromatography. *Huaxue Fenxi Jiliang* **2006**, *15*, 27–29.

27. Cai, L.; Xing, J.; Dong, L.; Wu, C. Application of polyphenylmethylsiloxane coated fiber for solid-phase microextraction combined with microwave-assisted extraction for the determination of organochlorine pesticides in Chinese teas. *J. Chromatogr. A* **2003**, *1015*, 11–21.

28. Jinno, K.; Taniguchi, M.; Sawada, H.; Hayashida, M. Microcolumn liquid chromatography coupled with solid phase microextraction (SPME/micro-LC) for the analysis of benzodiazepines in human urine. *Analusis* **1998**, *26*, M27–M30.

29. Wang, Y.; Zeng, Z.; Yang, M.; Dong, C.; Wang, Z. Determination of organophosphorus pesticides by solid-phase microextraction combining with gas chromatography. *Wuhan Ligong Daxue Xuebao* **2005**, *27*, 37–39.

30. Hu, Y.; Fu, Y.; Li, G. Preparation of anilinemethyltriethoxysilane/polydimethylsiloxane sol-gel coatings for solid-phase microextraction of aromatic compounds. *Anal. Chim. Acta* **2006**, *567*, 211–217.

31. Liu, M.M.; Zeng, Z.R.; Wang, C.L.; Tan, Y.J.; Liu, H. Solid-phase microextraction of phosphate and methylphosphonate using novel fibers coated with a sol-gel-derived silicone-divinylbenzene co-polymer. *Chromatographia* **2003**, *58*, 597–605.

32. Biajoli, A.F.P.; Augusto, F. Solid phase microextraction fibers coated with sol-gel aminopropylsilica/polydimethylsiloxane: Development and its application to screening of beer headspace. *Anal. Sci.* **2008**, *24*, 1141–1146.

33. Bianchi, F.; Bisceglie, F.; Careri, M.; Di Berardino, S.; Mangia, A.; Musci, M. Innovative sol-gel coatings for solid-phase microextraction. *J. Chromatogr. A* **2008**, *1196*, 15–22.

34. Bagheri, H.; Babanezhad, E.; Khalilian, F. A novel sol-gel based amino-functionalized fiber for headspace solid-phase microextraction of phenol and chlorophenols from environmental samples. *Anal. Chim. Acta* **2008**, *616*, 49–55.

35. Zeng, Z.; Qiu, W.; Huang, Z. Solid-phase microextraction using fused-silica fibers coated with sol-gel-derived hydroxy-crown ether. *Anal. Chem.* **2001**, *73*, 2429–2436.

36. Zeng, Z.; Qiu, W.; Yang, M.; Wei, X.; Huang, Z.; Li, F. Solid-phase microextraction of monocyclic aromatic amines using novel fibers coated with crown ether. *J. Chromatogr. A* **2001**, *934*, 51–57.

37. Yun, L. High extraction efficiency solid-phase microextraction fibers coated with open crown ether stationary phase using sol-gel technique. *Anal. Chim. Acta* **2003**, *486*, 63–72.

38. Cai, L.; Zhao, Y.; Gong, S.; Dong, L.; Wu, C. Use of a novel sol-gel dibenzo-18-crown-6 solid-phase microextraction fiber and a new derivatizing reagent for determination of aliphatic amines in lake water and human urine. *Chromatographia* **2003**, *58*, 615–621.

39. Wang, D.; Xing, J.; Peng, J.; Wu, C. Novel benzo-15-crown-5 sol-gel coating for solid-phase microextraction. *J. Chromatogr. A* **2003**, *1005*, 1–12.

40. Wang, D.; Peng, J.; Xing, J.; Wu, C.; Xu, Y. Solid-phase microextraction and gas chromatography-electron capture detection analysis of trace organochlorine pesticides in water using novel benzo-15-crown-5 sol-gel coating. *J. Chromatogr. Sci.* **2004**, *42*, 57–61.

41. Yu, J.; Wu, C.; Xing, J. Development of new solid-phase microextraction fibers by sol-gel technology for the determination of organophosphorus pesticide multiresidues in food. *J. Chromatogr. A* **2004**, *1036*, 101–111.

42. Cai, L.; Gong, S.; Chen, M.; Wu, C. Vinyl crown ether as a novel radical crosslinked sol-gel SPME fiber for determination of organophosphorus pesticides in food samples. *Anal. Chim. Acta* **2006**, *559*, 89–96.

43. Li, X.; Zeng, Z.; Gao, S.; Li, H. Preparation and characteristics of sol-gel-coated calix[4]arene fiber for solid-phase microextraction. *J. Chromatogr. A* **2004**, *1023*, 15–25.

44. Li, X.; Zeng, Z.; Zhou, J. High thermal-stable sol-gel-coated calix[4]arene fiber for solid-phase microextraction of chlorophenols. *Anal. Chim. Acta* **2004**, *509*, 27–37.

45. Li, X.; Zeng, Z.; Zhou, J.; Gong, S.; Wang, W.; Chen, Y. Novel fiber coated with amide bridged-calix[4]arene used for solid-phase microextraction of aliphatic amines. *J. Chromatogr. A* **2004**, *1041*, 1–9.

46. Dong, C.; Zeng, Z.; Li, X. Determination of organochlorine pesticides and their metabolites in radish after headspace solid-phase microextraction using calix[4]arene fiber. *Talanta* **2005**, *66*, 721–727.

47. Liu, M.; Zeng, Z.; Lei, Y.; Li, H. High extraction efficiency fiber coated with calix[4] open-chain crown ether for solid-phase microextraction of polar aromatic and aliphatic compounds. *J. Sep. Sci.* **2005**, *28*, 2306–2318.

48. Zeng, Z.; Liu, M. Extraction head coated with C[4] open-chain crown ether for solid phase microextraction and its manufacture. Faming Zhuanli Shenqing Gongkai Shuomingshu; 2006.

49. Li, X.; Gang, S.; Zeng, Z. Development of a sol-gel procedure for preparation of a diglycidyloxycalix[4]arene solid-phase microextraction fiber with enhanced extraction efficiency. *Chromatographia* **2005**, *62*, 519–525.

50. Li, X.; Zeng, Z.; Hu, M.; Mao, M. High operationally stable sol-gel diglycidyloxycalix[4]arene fiber for solid-phase microextraction of propranolol in human urine. *J. Sep. Sci.* **2005**, *28*, 2489–2500.

51. Zhou, X.; Li, X.; Zeng, Z. Solid-phase microextraction coupled with capillary electrophoresis for the determination of propranolol enantiomers in urine using a sol-gel derived calix[4]arene fiber. *J. Chromatogr. A* **2006**, *1104*, 359–365.

52. Li, X.; Zeng, Z.; Xu, Y. A solid-phase microextraction fiber coated with diglycidyloxycalix[4]arene yields very high extraction selectivity and sensitivity during the analysis of chlorobenzenes in soil. *Anal. Bioanal. Chem.* **2006**, *384*, 1428–1437.

53. Zhou, J.; Zeng, Z. Novel fiber coated with β-cyclodextrin derivatives used for headspace solid-phase microextraction of ephedrine and

methamphetamine in human urine. *Anal. Chim. Acta* **2006**, *556*, 400–406.

54. Yu, J.; Dong, L.; Wu, C.; Wu, L.; Xing, J. Hydroxyfullerene as a novel coating for solid-phase microextraction fiber with sol-gel technology. *J. Chromatogr. A* **2002**, *978*, 37–48.

55. Yang, Z.; Zhang, W.; Wang, C.; Xu, D.; Pan, K.; Wu, C. Determination of the leaching quantity of phthalic diesters in polyvinyl chloride (PVC) toys dipped in simulated saliva. *Chin. J. Chromatogr.* **2003**, *21*, 617–620.

56. Bianchi, F.; Mattarozzi, M.; Betti, P.; Bisceglie, F.; Careri, M.; Mangia, A.; Sidisky, L.; Ongarato, S.; Dalcanale, E. Innovative cavitand-based sol-gel coatings for the environmental monitoring of benzene and chlorobenzenes via solid-phase microextraction. *Anal. Chem.* **2008**, *80*, 6423–6430.

57. Wang, Z.; Xiao, C.; Wu, C.; Han, H. High-performance polyethylene glycol-coated solid-phase microextraction fibers using sol-gel technology. *J. Chromatogr. A* **2000**, *893*, 157–168.

58. Gomes da Costa Silva, R.; Augusto, F. Highly porous solid-phase microextraction fiber coating based on poly(ethylene glycol)-modified ormosils synthesized by sol-gel technology. *J. Chromatogr. A* **2005**, *1072*, 7–12.

59. Liu, M.; Zeng, Z.; Xiong, B. Preparation of novel solid-phase microextraction fibers by sol-gel technology for headspace solid-phase microextraction-gas chromatographic analysis of aroma compounds in beer. *J. Chromatogr. A* **2005**, *1065*, 287–299.

60. Liu, M.; Zeng, Z.; Tian, Y. Analysis of fatty acids in lung tissues using gas chromatography-mass spectrometry preceded by derivatization-solid-phase microextraction with a novel fiber. *Anal. Chim. Acta* **2005**, *540*, 341–353.

61. Liu, M.; Zeng, Z.; Fang, H. Preparation and application of the sol-gel-derived acrylate/silicone co-polymer coatings for headspace solid-phase microextraction of 2-chloroethyl ethyl sulfide in soil. *J. Chromatogr. A* **2005**, *1076*, 16–26.

62. Fang, H.; Liu, M.; Zeng, Z. Solid-phase microextraction coupled with capillary electrophoresis to determine ephedrine derivatives in water and urine using a sol-gel derived butyl methacrylate/silicone fiber. *Talanta* **2006**, *68*, 979–986.

63. Cha, D.; Liu, M.; Zeng, Z. Analysis of fatty acids in lung tissues using gas chromatography-mass spectrometry preceded by derivatization-solid-phase microextraction with a novel fiber. *Anal. Chim. Acta* **2006**, *572*, 47–54.

64. Gbatu, T.P.; Sutton, K.L.; Caruso, J.A. Development of new SPME fibers by sol-gel technology for SPME-HPLC determination of organometals. *Anal. Chim. Acta* **1999**, *402*, 67–79.

65. Giardina, M.; Olesik, S.V. Application of low-temperature glassy carbon films in solid-phase microextraction. *Anal. Chem.* **2001**, *73*, 5841–5851.

66. Giardina, M.; Olesik, S.V. Application of low-temperature glassy carbon-coated macrofibers for solid-phase microextraction analysis of simulated breath volatiles. *Anal. Chem.* **2003**, *75*, 1604–1614.

67. Giardina, M.; Ding, L.; Olesik, S.V. Development of fluorinated low temperature glassy carbon films for solid-phase microextraction. *J. Chromatogr. A* **2004**, *1060*, 215–224.

68. Teng, C.; Chen, Y. Fiber introduction mass spectrometry: Coupling solid-phase microextraction with sol-gel-assisted laser desorption/ionization time-of-flight mass spectrometry. *Rapid Commun. Mass Spectrom.* **2003**, *17*, 1092–1094.

69. Azenha, M.; Malheiro, C.; Silva, A.F. Ultrathin phenyl-functionalized solid phase microextraction fiber coating developed by sol-gel deposition. *J. Chromatogr. A* **2005**, *1069*, 163–172.

70. Basheer, C.; Jegadesan, S.; Valiyaveettil, S.; Lee, H.K. Sol-gel-coated oligomers as novel stationary phases for solid-phase microextraction. *J. Chromatogr. A* **2005**, *1087*, 252–258.

71. Wang, D. Preparation of SE-54 fibers for solid-phase microextraction with sol-gel method. *Fenxi Ceshi Xuebao* **2006**, *25*, 11–15.

72. Ouyang, G.; Jiang, R.; Zhu, F.; Luan, T. Carbon nanotube-containing solid phase micro extraction fiber and its preparation method. Faming Zhuanli Shenqing Gongkai Shuomingshu; 2008.

73. Wei, L.M.; Ou, Q.Y.; Li, J.B. Preparation of a new fiber by sol-gel technology in solid-phase microextraction (SPME). *Chin. Chem. Lett.* **2004**, *15*, 1127–1130.

74. Farhadi, K.; Tahmaebi, R.; Maleki, R. Preparation and application of the titania sol-gel coated anodized aluminum fibers for headspace solid phase microextraction of aromatic hydrocarbons from water samples. *Talanta* **2009**, *77*, 1285–1289.

75. Xu, L.; Lee, H.K. Zirconia hollow fiber: Preparation, characterization, and microextraction application. *Anal. Chem.* **2007**, *79*, 5241–5248.

76. Bigham, S.; Medlar, J.; Kabir, A.; Shende, C.; Alli, A.; Malik, A. Sol-gel capillary microextraction. *Anal. Chem.* **2002**, *74*, 752–761.

77. Malik, A. Sample preconcentration tubes with sol-gel surface coatings and/or sol-gel monolithic beds. U.S. Patent 6783680 (2004).

78. Kulkarni, S.; Fang, L.; Alhooshani, K.; Malik, A. Sol-gel immobilized cyano-polydimethylsiloxane coating for capillary microextraction of aqueous trace analytes ranging from polycyclic aromatic hydrocarbons to free fatty acids. *J. Chromatogr. A* **2006**, *1124*, 205–216.

79. Segro, S.S.; Malik, A. Solvent-resistant sol-gel polydimethyldiphenylsiloxane coating for on-line hyphenation of capillary microextraction with high-performance liquid chromatography. *J. Chromatogr. A* **2008**, *1205*, 26–35.

80. Fan, Y.; Feng, Y.; Da, S.; Wang, Z. In-tube solid phase microextraction using a β-cyclodextrin coated capillary coupled to high performance liquid chromatography for determination of non-steroidal anti-inflammatory drugs in urine samples. *Talanta* **2005**, *65*, 111–117.

81. Li, W.; Fries, D.; Alli, A.; Malik, A. Positively charged sol-gel coatings for on-line preconcentration of amino acids in capillary electrophoresis. *Anal. Chem.* **2004**, *76*, 218–227.

82. Malik, A.; Hayes, J.D. Sol-gel open tubular ODS columns with charged inner surface for capillary electrochromatography. U.S. Patent 6998040 B2 (2006).

83. Li, W.; Fries, D.; Malik, A. Negatively charged sol-gel column with stable electroosmotic flow for online preconcentration of zwitterionic biomolecules in capillary electromigration separations. *J. Sep. Sci.* **2005**, *28*, 2153–2164.

84. Kulkarni, S.; Shearrow, A.M.; Malik, A. Sol-gel immobilized short-chain poly(ethylene glycol) coating for capillary microextraction of underivatized polar analytes. *J. Chromatogr. A* **2007**, *1174*, 50–62.

85. Kabir, A.; Hamlet, C.; Yoo, K.S.; Newkome, G.R.; Malik, A. Capillary microextraction on sol-gel dendrimer coatings. *J. Chromatogr. A* **2004**, *1034*, 1–11.

86. Malik, A.; Kabir, A.; Newkome, G.R.; Yoo, K.S. Sol-gel dendron separation and extraction capillary column. *U.S. Pat. Appl. Publ.* No. US 2005/0106068 B1 (2005).

87. Kabir, A.; Hamlet, C.; Malik, A. Parts per quadrillion level ultra-trace determination of polar and nonpolar compounds via solvent-free capillary microextraction on surface-bonded sol-gel polytetrahydrofuran coating and gas chromatography-flame ionization detection. *J. Chromatogr. A* **2004**, *1047*, 1–13.

88. Malik, A.; Kabir, A. Sol-gel polytetrahydrofuran-based coating for capillary microextraction. *U.S. Pat. Appl. Publ.* No. 2006/0013981 A1 (2006).

89. Zeng, F.; Hu, B. MPTS-silica coated capillary microextraction on line hyphenated with inductively coupled plasma atomic emission spectrometry for the determination of Cu, Hg and Pb in biological samples. *Talanta* **2007**, *73*, 372–379.

90. Segro, S.S.; Malik, A. Sol-gel methyl coating in capillary microextraction on-line with high-performance liquid chromatography. *J. Chromatogr. A* **2008**, *1200*, 62–71.

91. Kim, T.; Alhooshani, K.; Kabir, A.; Fries, D.P.; Malik, A. High pH-resistant, surface-bonded sol-gel titania hybrid organic-inorganic coating for effective on-line hyphenation of capillary microextraction (in-tube solid-phase microextraction) with high-performance liquid chromatography. *J. Chromatogr. A* **2004**, *1047*, 165–174.

92. Malik, A.; Kim, T. Titania-based coating for capillary microextraction. United States Patent No. US 7,622,191 B2 (2009).

93. Segro, S.S.; Cabezas, Y.; Malik, A. Ultra-high-stability, pH resistant sol-gel titania poly(tetrahydrofuran) coating for capillary microextraction on-line coupled to high-performance liquid chromatography. *J. Chromatogr. A* **2009**, *1216*, 4329–4338.

94. Alhooshani, K.; Kim, T.; Kabir, A.; Malik, A. Sol-gel approach to in situ creation of high pH-resistant surface-bonded organic-inorganic hybrid zirconia coating for capillary microextraction (in-tube SPME). *J. Chromatogr. A* **2005**, *1062*, 1–14.

95. Wu, Y.; Hu, B.; Jiang, Z.; Feng, Y.; Lu, P.; Li, B. Sol-gel zirconia coating capillary microextraction on-line hyphenated with inductively coupled plasma mass spectrometry for the determination of Cr, Cu, Cd and Pb in biological samples. *Rapid Commun. Mass Spectrom.* **2006**, *20*, 3527–3534.

96. Fang, L.; Kulkarni, S.; Alhooshani, K.; Malik, A. Germania-based, sol-gel hybrid organic-inorganic coatings for capillary microextraction and gas chromatography. *Anal. Chem.* **2007**, *79*, 9441–9451.

97. Hayes, A.; Malik, A. Sol-gel chemistry-based Ucon-coated columns for capillary electrophoresis. *J. Chromatogr. B* **1997**, *695*, 3–13.

98. Liu, W.; Wang, H.; Guan, Y. Preparation of stir bars for sorptive extraction using sol-gel technology. *J. Chromatogr. A* **2004**, *1045*, 15–22.

99. Liu, W.; Hu, Y.; Zhao, J.; Xu, Y.; Guan, Y. Determination of organophosphorus pesticides in cucumber and potato by stir bar sorptive extraction. *J. Chromatogr. A* **2005**, *1095*, 1–7.

100. Yu, C.; Hu, B. Novel combined stir bar sorptive extraction coupled with ultrasonic assisted extraction for the determination of brominated flame retardants in environmental samples using high performance liquid chromatography. *J. Chromatogr. A* **2007**, *1160*, 71–80.

101. Yu, C.; Hu, B. Sol-gel polydimethylsiloxane/poly(vinylalcohol)-coated stir bar sorptive extraction of organophosphorus pesticides in honey and their determination by large volume injection GC. *J. Sep. Sci.* **2009**, *32*, 147–153.

102. Yu, C.; Li, X.; Hu, B. Preparation of sol-gel polyethylene glycol-polydimethylsiloxane-poly(vinyl alcohol)-coated sorptive bar for the determination of organic sulfur compounds in water. *J. Chromatogr. A* **2008**, *1202*, 102–106.

103. Hu, Y.; Zheng, Y.; Zhu, F.; Li, G. Sol-gel coated polydimethylsiloxane/β-cyclodextrin as novel stationary phase for stir bar sorptive extraction and its application to analysis of estrogens and bisphenol A. *J. Chromatogr. A* **2007**, *1148*, 16–22.

104. Malik, A.; Bigham, S.; Medlar, J.; Ashford, C.; Kabir, A. Sol-gel approach to in situ creation of surface coatings and monolithic beds for analytical microextraction. *Extech 2000 Advances in Extraction Technologies*. Department of Chemistry, University of Waterloo, Waterloo, Ontario, Canada; May 1–3, 2000.

105. Malik, A. New polymeric extraction materials. *Compr. Anal. Chem.* **2002**, *37*, 1023–1071.

106. Dulay, M.T.; Quirino, J.P.; Bennett, B.D.; Kato, M.; Zare, R.N. Photopolymerized sol-gel monoliths for capillary electrochromatography. *Anal. Chem.* **2001**, *73*, 3921–3926.

107. Quirino, J.P.; Dulay, M.T.; Bennett, B.D.; Zare, R.N. Strategy for on-line preconcentration in chromatographic separations. *Anal. Chem.* **2001**, *73*, 5539–5543.

108. Quirino, J.P.; Dulay, M.T.; Zare, R.N. On-line preconcentration in capillary electrochromatography using a porous monolith together with solvent gradient and sample stacking. *Anal. Chem.* **2001**, *73*, 5557–5563.

109. Lim, L.W.; Hirose, K.; Tatsumi, S.; Uzu, H.; Mizukami, M.; Takeuchi, T. Sample enrichment by using monolithic precolumns in microcolumn liquid chromatography. *J. Chromatogr. A* **2004**, *1033*, 205–212.

110. Oguri, S.; Tanagaki, H.; Hamaya, M.; Kato, M.; Toyo'oka, T. On-line preconcentration prior to on-column derivatization monolith octadecasiloxane capillary electrochromatography for the determination of biogenic amines. *Anal. Chem.* **2003**, *75*, 5240–5245.

111. Gu, X.; Wang, Y.; Zhang, Z. Large-bore particle-entrapped monolithic precolumns prepared by a sol-gel method for on-line peptides trapping and preconcentration in multidimensional liquid chromatography system for proteome analysis. *J. Chromatogr. A* **2005**, *1072*, 223–232.

112. Hu, Y.; Yang, Y.; Huang, J.; Li, G. Preparation and application of poly(dimethylsiloxane)/β-cyclodextrin solid-phase microextraction membrane. *Anal. Chim. Acta* **2005**, *543*, 17–24.

113. Carrington, N.A.; Thomas, G.H.; Rodman, D.L.; Beach, D.B.; Xue, Z. Optical determination of Cr(VI) using regenerable, functionalized sol-gel monoliths. *Anal. Chim. Acta* **2007**, *581*, 232–240.

114. Johannesson, N.; Pearce, E.; Dulay, M.; Zare, R.N.; Bergquist, J.; Markides, K.E. On-line biological sample cleanup for electrospray mass spectrometry using sol-gel columns. *J. Chromatogr. B* **2006**, *842*, 70–74.

115. Johannesson, N.; Bergquist, J. Rapid on-line extraction and quantification of escitalopram from urine using sol-gel columns and mass spectrometric detection. *J. Pharm. Biomed. Anal.* **2007**, *43*, 1045–1048.

116. Zheng, M.; Lin, B.; Feng, Y. Hybrid organic-inorganic octyl monolithic column for in-tube solid-phase microextraction coupled to capillary high-performance liquid chromatography. *J. Chromatogr. A* **2007**, *1164*, 48–55.

117. Ramautar, R.; Ratnayake, C.K.; Somsen, G.W.; de Jong, G.J. Capillary electrophoresis-mass spectrometry using an in-line sol-gel concentrator for the determination of methionine enkephalin in cerebral spinal fluid. *Talanta* **2009**, *78*, 638–642.

Chapter 23

The Use of Molecularly Imprinted Polymers for Sampling and Sample Preparation

Börje Sellergren and Antonio Martin Esteban

23.1. INTRODUCTION

The detection and quantification of high-priority analytical targets typically rely on labor-intensive methods, which often fail to satisfy the required method criteria in terms of speed, robustness, and sensitivity. The bottleneck and the source of the poor performance of these methods are commonly found in the sample pretreatment steps, which are often lengthy and laborious. This illustrates the need for developing improved sample extraction techniques capable of cleaning up and enriching samples to easily measurable levels. One of the most powerful and reproducible ways of doing this is by molecularly imprinted solid-phase extraction (MISPE).[1–7]

Molecular imprinting allows the molding of complementary binding sites usually for small lipophilic molecules into synthetic polymers. The technique, as outlined in Figure 23.1, entails copolymerization of mono- and polyfunctional monomers in the presence of a template (Schemes 23.1 and 23.2), which is thereafter removed to leave sites that can be reoccupied by the template or closely related compounds (Scheme 23.3). This leaves a highly cross-linked and durable molecularly imprinted polymer (MIP) containing templated sites or imprints of the template added in the polymerization step.[8] Vis-á-vis biological receptors, MIPs are distinguished by their robustness and ease of synthesis, which has led to their assessment in a range of molecular recognition-based applications targeting small molecules.

The recognition properties displayed by MIPs can be on a par with their biological counterparts, and they have therefore gained ground as a robust selector phase in solid-phase extraction (SPE). Numerous examples now demonstrate their superior performance over more generic phases in delivering clean extracts and reproducible recoveries of mainly low-molecular-weight analytes.[9] The analyst may gain by using these phases in obtaining methods that are faster, more sensitive, and selective compared with those relying on generic sample pretreatment methodology. This review will explain the basic principles of the MISPE technology, briefly outlining the synthetic procedure used to prepare MIPs, summarize the recent contributions encompassing analytical chemistry, and, finally, provide a guide on how to use the phases in method development.

23.2. AFFINITY-BASED SPE

The complex sample matrices in combination with the need for lower detection limits have increased the demand for selectivity in the sample pretreatment step. Biological molecular recognition elements (e.g., antibodies) can be used for this purpose. For instance, immunoaffinity (IA)-based enrichment techniques can be used for specific determination of a particular analyte or a class of analytes.[10,11] As with most affinity techniques relying on biological recognition elements, method robustness and reproducibility, receptor immobilization technique, and the time and cost involved in providing the receptor for a new target analyte are factors that have hampered this technology.

MIPs provide a low-cost solution to this problem in combining high affinity and molecular discrimination capability with physical robustness.[12–14] Moreover, these phases are not matrix dependent in the sense that analytes can be effectively enriched from both aqueous as well as apolar sample matrices. This opens up a range of cleanup applications for analytes present in an apolar matrix or in methods involving sample preparation by liquid–liquid extraction.

23.3. SYNTHETIC PROCEDURES TO PRODUCE MIPs

The most versatile approach to the synthesis of molecularly imprinted sorbents is based on the self-assembly of the

Handbook of Sample Preparation, Edited by Janusz Pawliszyn and Heather L. Lord

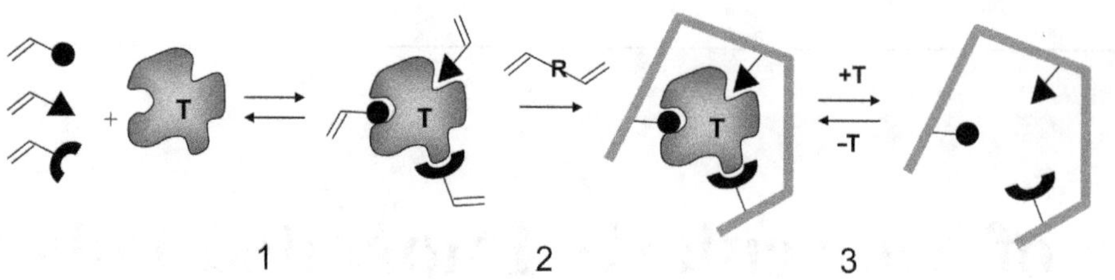

Figure 23.1. Principle of molecular imprinting relying on self-assembly of functional monomers and a template (1), polymerization (2), and (3) removal of the template providing empty cavities.

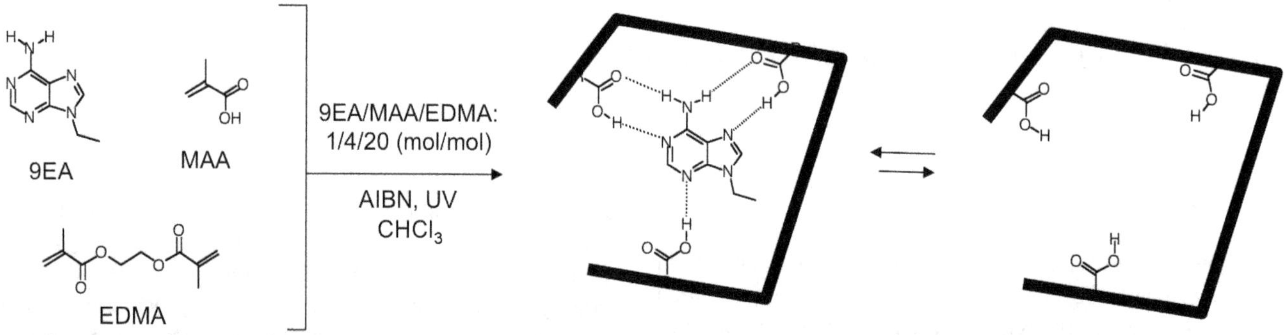

Figure 23.2. Noncovalent imprinting of 9-ethyladenine (9EA) leading to highly cross-linked monoliths from which particles are obtained by repetitive crushing and sieving cycles.

template and a complementary functional monomer (e.g., methacrylic acid [MAA]) prior to polymerization (Fig. 23.2).[15] Thus, the template remains associated with the growing polymer during synthesis, and by adding a large portion of a cross-linking monomer (e.g., ethylenglycoldimethacrylate [EDMA]), sites complementary to the template are formed that remain stable after template removal.

Although the preparation of an MIP by this method is technically simple, it relies on the success of the stabilization of the relatively weak interactions between the template and the functional monomers. The use of solvents of low polarity and the addition of an excess of functional monomer are the common means ensuring that the template molecule is complexed to a maximal degree. However, a large proportion of the functional monomer is not involved in complexation of the template and is instead distributed randomly throughout the polymer matrix during the polymerization.[16] This is a major cause for the binding site heterogeneity observed in these materials. For instance, using 9-ethyladenine as template and MAA as functional monomer, MIPs were obtained, which displayed a heterogenous binding site population for the template in chloroform with one class of high-energy binding sites ($K = 77{,}000\,M^{-1}$, $Q = 20\,\mu mol/g$) and one class of lower energy sites ($K = 2400\,M^{-1}$, $Q = 86\,\mu mol/g$).[17] It could be shown that the low energy site class represented nontemplated sites contributing to the nonspecific binding of the MIP.

The polymerization is performed in the presence of a pore-forming solvent called a porogen. In order to stabilize the electrostatic interactions between the functional monomers and the template, the porogen is often chosen to be aprotic and of low to moderate polarity. Obviously, it must also solubilize all components in the prepolymerization mixture. This restricts the choice of porogen and limits the ability to fully control the morphology of the materials. One difficulty in the process is thus to find a compromise between these two aspects: materials morphology and molecular recognition. Since water disrupts the weak electrostatic interactions present during imprinting, the use of suspension and emulsion polymerization procedures is usually not possible, and polymerization is most commonly carried out in one phase leading to polymer monoliths, which need to be crushed and sieved in order to provide particles suitable as packing material.[18] This tedious processing can be circumvented by precipitation polymerization yielding, in favorable cases, monodisperse micron-sized beads.[19] Alternatively, polymer beads may, in some cases, be produced by various bead polymerization techniques[18] or preformed support beads may be used to produce imprinted composites.[20] After washing and conditioning, the so obtained sorbent is then ready for testing or direct use in SPE.

MAA is the most widely used functional monomer in imprinting, resulting in MIPs showing high selectivity for a large number of target molecules (Fig. 23.2).[16] However, it does not provide a universal solution to imprinting problems, and for several targets, it fails to produce sufficient binding affinity and selectivity. In this case, another monomer or combinations of monomers need to be found. Another fundamental problem in imprinting concerns the selection or design of a template that is likely to produce a sufficient cross-reactivity for the desired target.[21] Apart from functional and shape complementarity with the target, the template needs to be stable and soluble in the prepolymerization mixture and during polymerization. Commonly, the polymer matrix of the MIP needs to be optimized in order to minimize nonspecific binding of the target or matrix components in the sample matrix for which they are intended. This requires fine-tuning of the type and amount of cross-linking monomer as well as the porogenic solvent.

23.3.1. Factors to Consider when Choosing the Template

The majority of analytes targeted so far exhibit moderate to high solubility in the corresponding polymerization medium, and they can therefore be used directly in the conventional procedure. However, for several reasons, a structural analog to the target molecule is commonly preferred as the template:

1. The target is unstable under the polymerization conditions or inhibits the polymerization.[22]
2. The target is not available in sufficient quantities to make imprinting worthwhile.[23]
3. To avoid template bleeding, that is, the bleeding of low levels of template from the MIP, thus precluding SPE applications in trace analysis.[23–28]
4. The target is insoluble or poorly soluble in the prepolymerization mixture.[29]
5. The target is toxic or hazardous in other ways, for example, explosive.[30]

Therefore, a close target analog is often chosen as template, leading to binding sites that can also capture the target analyte.[23,25–29] In addition to being a closely related analog to the target, the template should possess functionalities that interact with the functional monomer, it should be stable and inert during the polymerization, and it should not interfere during the subsequent quantification of the analyte.

23.3.2. MIP Cross-Reactivity and Group Selective MIPs

In light of the previous arguments, the engineering of MIPs capable of not only recognizing its template but also another target compound or a group of structurally related compounds is a key activity when developing a new MIP.[9] Although MIPs typically exhibit a very high selectivity and affinity for their template, not unlike the way antibodies recognize their antigens, they can also cross-react with structurally similar compounds.

For instance, the 9-ethyladenine (9EA)-imprinted polymer prepared as shown in Figure 23.2 does not totally exclude other nucleic acid bases but instead exhibits enhanced retention for all compounds containing a free adenine substructure (Fig. 23.3).[17]

This type of cross-reactivity can be easily understood since all retained analogs do contain the adenine substructure, and the principle has been exploited in the development of broad spectrum MISPE applications for classes of related compounds (e.g., nitrophenols, nitroimidazoles, benzodiazepines, triazines, amino alcohols [e.g., β-receptor agonists or antagonists], β-lactam antibiotics, fluoroquinolones, endocrine-disrupting compounds) (see Table 23.1). One of the most widely used MISPE sorbents target the general class of 1,2-amino alcohols (Scheme 23.1).

Thanks to a careful template selection (a dummy template such as Scheme 23.3 of minor analytical relevance), the sorbents can cross-react with the majority of β-receptor agonists and antagonists, such as the high-priority growth promoter clenbuterol (Scheme 23.2), making this a sorbent of choice for food quality control,[31,32] environmental analysis,[33] doping,[34] and pharmaceutical analysis.[9]

MIPs targeting single high-priority analytes may also find broad usage, and the role of the template is here to create a site capable of accommodating only one compound; thus, it can exhibit a closer structural resemblance with the target. For instance, the nitrosylated nicotine metabolite 4-(methylnitrosamino)-1-(3-pyridyl)-1-butanol (NNAL) (Scheme 23.4), a biomarker for nicotine smoke exposure, could be selectively retained by an MIP prepared using the isosteric enamine (Scheme 23.5) as template.

In another recent example, covalent boronate-mediated imprinting of 5-O-TBDMS-protected adenosine (Scheme 23.6) led to an MIP capable of recognizing adenosine monophosphate (Scheme 23.7).[35]

When allowing a library of structurally diverse compounds to contact an MIP, it is often found that the MIP retains a few compounds that may appear structurally unrelated to the template. This nonobvious cross-reactivity has been exploited by the MIP manufacturer MIP Technologies in the Explorasep concept, which in brief consists in screening of 96-well plate MIP libraries grouped according to the interacting functional group of the polymer (i.e., acid, base, neutral, etc.).[36] For example, it was found that an MAA-based MIP imprinted with nicotinamide (Scheme 23.8) also showed selectivity for theophylline (Scheme 23.9).

The strongest interactions between the functional monomer and the template are here expected to involve the pyridine nitrogen and the amide group. If theophylline is superimposed on nicotinamide, good hydrogen-bond acceptors in theophylline match, as highlighted in gray, the hydrogen-bond acceptors in nicotinamide. In other words, nicotinamide should provide a good template for theophylline. One key advantage of the Explorasep approach is that template analogs found by screening will typically be different from the target molecules and thus unlikely to interfere with the analysis.

Unlike this random screening approach, it may pay off to design somewhat smaller template libraries directed toward the functionalities of a given target. Complex natural products containing a limited number of functional groups may,

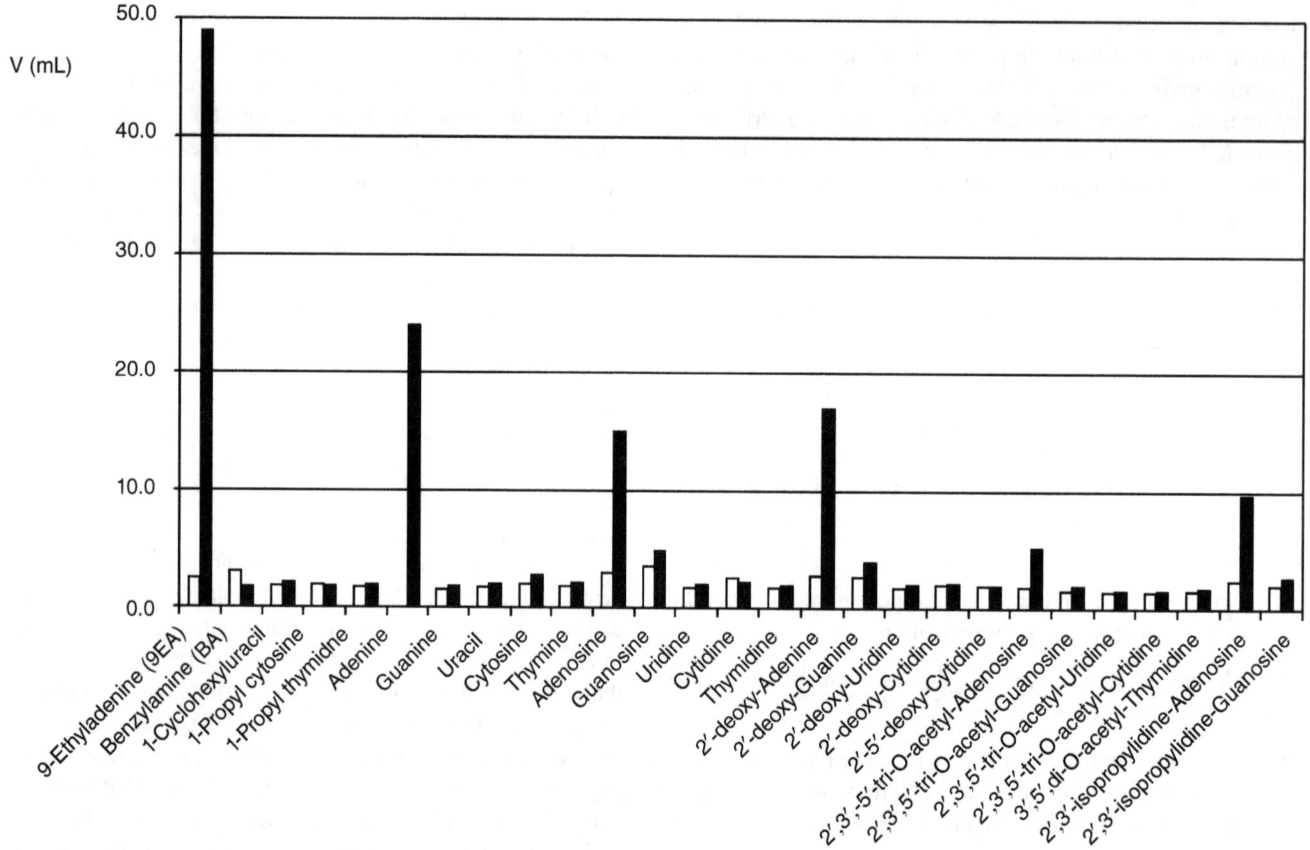

Figure 23.3. Elution volumes of the DNA bases, nucleosides, and their derivatives from chromatographic runs using columns packed with a polymer imprinted with 9EA (black bars) or a control polymer imprinted with benzylamine (white bars). The mobile phase was acetonitrile : acetic acid : water: 92.5:5:2.5 (v/v/v). Data from reference 17.

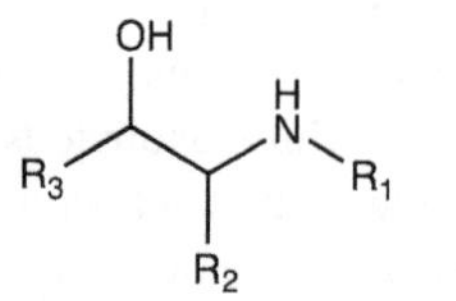

Scheme 23.1

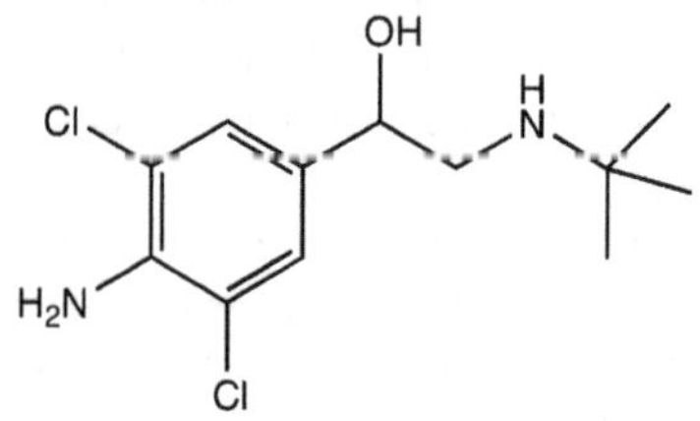

Scheme 23.2

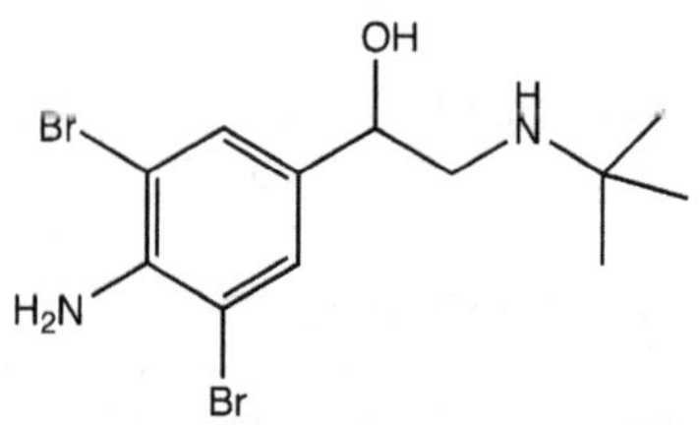

Scheme 23.3

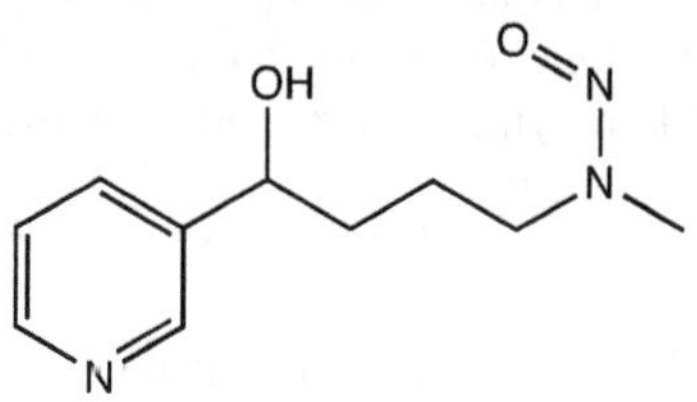

Scheme 23.4

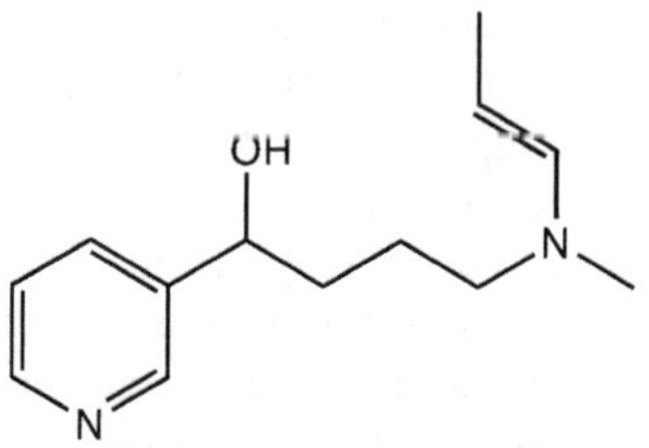

Scheme 23.5

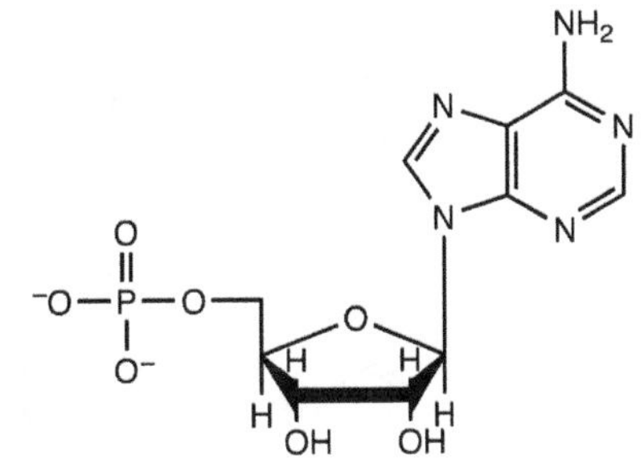

Scheme 23.6

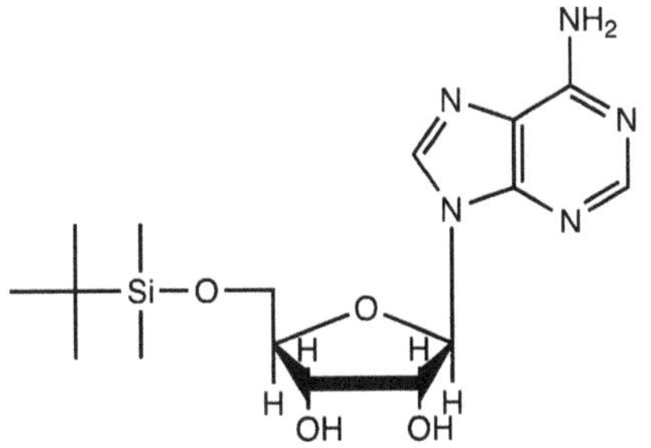

Scheme 23.7

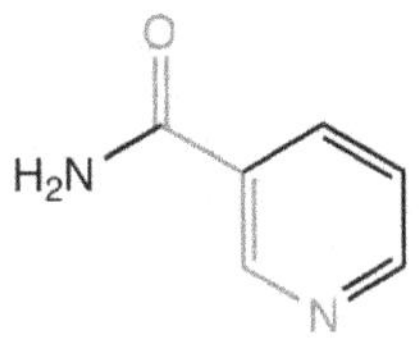

Scheme 23.8

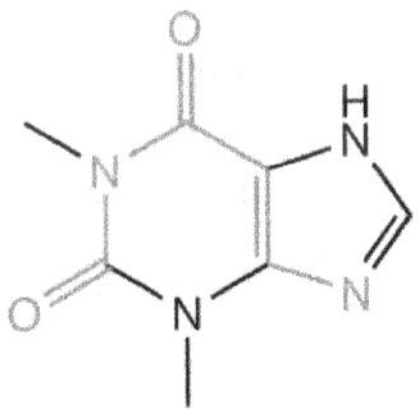

Scheme 23.9

for instance, bind to MIPs made using a considerably simpler template exhibiting part or all of these functional groups placed at geometrically complementary positions to the target. This principle was used by Nemoto et al. in targeting the fish poison domoic acid (Scheme 23.10) by MIPs prepared using simple commodity di- or tri-acids as templates.[37] It was found here that the tri-acid pentane-1,3,5-tricarboxylic acid (Scheme 23.11) led to an MIP displaying sufficient retentitivity for the target to allow efficient cleanup of the target from fish.

23.3.3. Fragment Imprinting

Traditional imprinting fails when the targets are complex, poorly soluble, or simply unavailable in sufficient quantities for imprinting. Most biomolecules fall under this category and especially the macromolecular ones, that is, proteins,

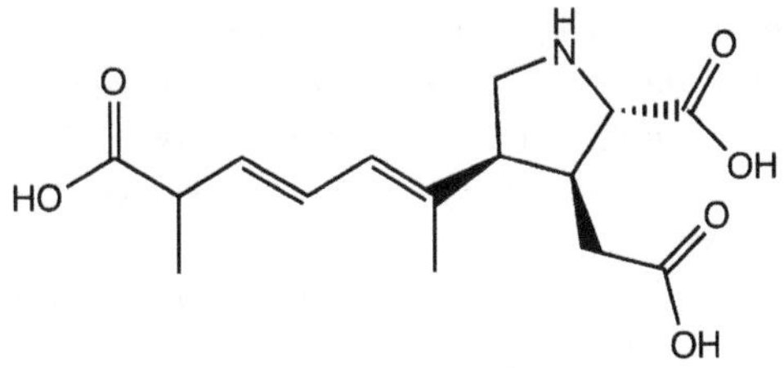

Scheme 23.10

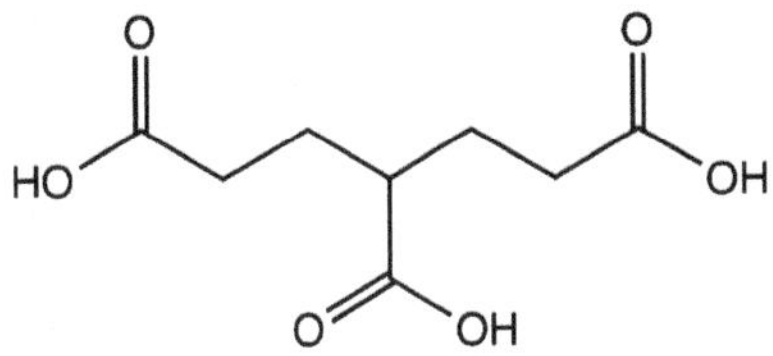

Scheme 23.11

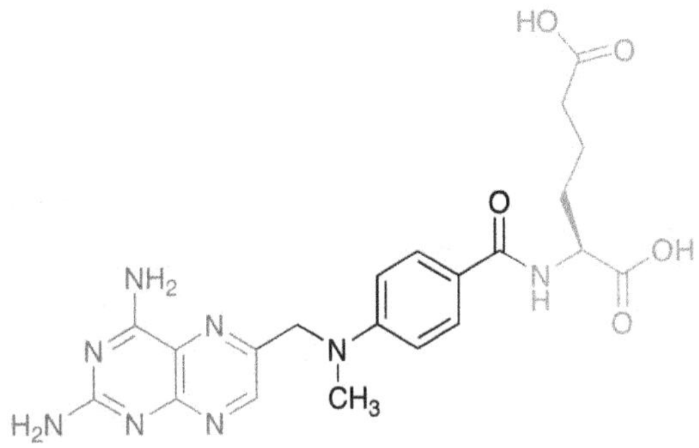

Scheme 23.12

nucleic acids, and polysaccharides. In these cases, the polymer may need to be synthesized in an aqueous medium to solublize the target and to stabilize it in a conformation close to its native low energy conformation. However, in water, the strong hydration forces and the polar environment commonly prevent stable electrostatic interactions between the target and the functional monomers that are commonly used in molecular imprinting. Moreover, such targets are difficult, if at all possible, to come across in sufficient quantities, precluding the use of the target as template.

In order to overcome this problem, fragments complementary to substructures of the target may be used as templates.[22,38] Although the resulting sites are by consequence less specific for the target, this strategy uses simple available templates compatible with the imprinting conditions. This may result in sites complementary to the substructure that still bind the larger biomolecules with significant affinities. This concept was first demonstrated for peptides[38] and simple vitamins.[22] Thus, the cytostatic drug methotrexate (Scheme 23.12) is unsuited as a template since it fails on all selection criteria in Section 23.3.1.

Therefore, substructures (highlighted in gray) corresponding to the pteridine part (left substructure) or glutamate part (right substructure) were chosen as templates and MIPs were subsequently prepared by conventional imprinting. Both substructure MIPs resulted in sorbents capable of effectively retaining the target as well as other folic acid reductase inhibitors and the native substrate folic acid.[39] Combining the two complementary MIPs either in one column or sequentially would be one strategy to enhance the selectivity further.

Following an epitope imprinting approach, short peptides were used as templates for binding a larger target containing the epitope sequence as its terminal part.[38] This concept has subsequently been combined with surface imprinting techniques to produce MIPs exhibiting cross-reactivities with peptides[40] or more impressively with proteins[41] based on just a short peptide complementary terminal sequence.

Combining epitope imprinting with the use of strongly interacting host monomers, substructures can be targeted more effectively. Thus, 1,3-diaryl urea host monomers were used to imprint phosphorylated amino acids in order to create binding sites for binding phosphorylated peptides. In this way, MIPs were obtained, which effectively discriminated between the different amino acid side chains of phosphorylated peptides from tryptic digests.[42]

23.3.4. Factors to Consider when Choosing the Functional Monomer

The guiding principle in the choice of functional monomer is functional group complementarity.[43] Thus, for templates containing acidic groups, Brönsted basic functional monomers (e.g., 2- or 4-vinylpyridine [VPy], diethylaminoethylmethacrylate [DEAEMA]) are preferably chosen, whereas acidic functional monomers (e.g., MAA, trifluoromethylacrylic acid [TFM], itaconic acid [ITA]) are used to target Brönsted bases. For carboxylic acids and amides, high selectivities have also been observed with primary amide-containing monomers (e.g., methacrylamide [MAAM]). Other neutral solvating monomers that commonly enhance the imprinting effect are *N*-vinylpyrrolidone (NVP) and hydroxyethylmethacrylate (HEMA). Scheme 23.13 shows the structures for these functional monomers.

In the case of weakly polar to apolar templates, with few polar interaction sites, it may instead be beneficial to use amphiphilic monomers, stabilizing the monomer–template assemblies by hydrophobic and van der Waals forces or, in the case of extended π-systems, through charge transfer stabilization. Similarly to biological systems, a large number of complementary interactions are expected to increase the strength and fidelity of recognition. Thus, the use of comonomers, which may orthogonally target different subunits of a complex template, is often a successful strategy.[44,45] This leads to terpolymers or higher polymers. The screening and optimization task is here considerably more complex and is best carried out using high-throughput polymer synthesis techniques, experimental design, and parallel techniques to analyze binding data.[21] As discussed below, comonomers can also be chosen in order to impart compatibility of the material with a certain solvent, that is, to reduce nonspecific binding of target analogs or matrix components.[29,46]

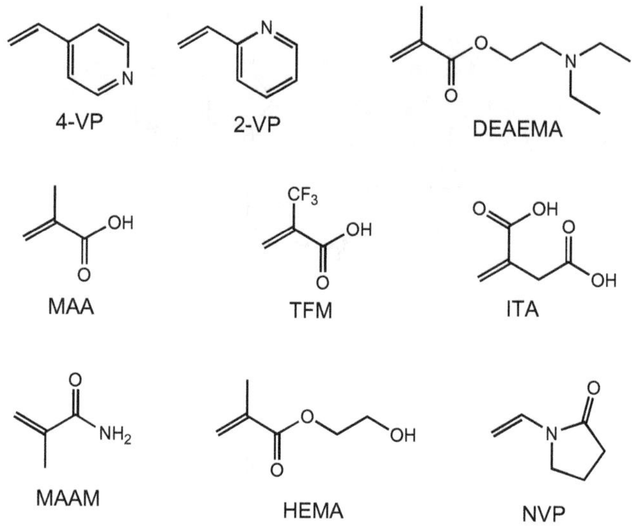

Scheme 23.13

The selection of suitable monomers may also be aided by molecular modeling and computational methods.[47] A virtual library of functional monomers is created and screened for all possible interactions the individual monomers may engage in with the template. Monomers with the highest binding scores are subsequently selected to produce full-scale MIPs with hopefully superior recognition properties.

Finally, the search for the optimal structural motif, which can complement the template functionality, may also be guided by results from the areas of host–guest chemistry. Polymerizable host monomers designed to strongly complex a given guest in solution have been used to successfully imprint similar guests.[48,49] In some cases, this has resulted in a quantitative yield of high-energy binding sites (stoichiometric imprinting), which can recognize the guest also under aqueous conditions.

23.4. MISPE FORMATS

Several examples of the successful application of MIPs in SPE have been described in the literature (Tables 23.1–23.3), and the field is presently experiencing a rapid growth with over 50 papers published in 2008. The main reason for the success of MISPE is that they offer an affordable and robust mode of affinity SPE programmable to capture single analytes or groups of analytes from complex samples. Thus, clean extracts can be obtained with minimal sample pretreatment translating into significant time savings, lower detection limits, and rugged methods. A notable positive contribution to the enhanced method ruggedness is the reduced ion suppression experienced using mass spectrometry (MS) detectors post-MIP extractions (see Section 23.5.1).[33,50] MISPE has been demonstrated using several operational modes including online SPE, conventional offline SPE where the MIP is packed into cartridges, and batch SPE where the MIP is incubated with the sample.

Table 23.1. Examples of MISPE in Pharmaceutical Analysis and Bioanalysis

Analytes	Sample and Matrix	Note	Reference
Pharmaceutical drugs			
Pentamidine	Urine	In-line	51
Sameridine	Plasma	Batch	25
7-Hydroxycoumarine	Aqueous samples	Offline	52
Theophylline	Serum	In-line, PE	53
Theophylline	Plasma	In-line, DPE	54
2-Aminopyridine	Serum	In-line, DPE, V	55
Phenytoin	Plasma	Offline, V	56
Drug lead compounds	Serum	Offline, V	23
Metformin	Plasma, urine	In-line, DPE, V	57,58
Verapamil and its metabolites	Urine, plasma, and cell cultures	Online (RAM-MIP), V	59
Propranolol and β-blockers	Plasma	Offline, V	60
Scopolamine	Calf urine and serum	Offline	61
Clenbuterol	Urine	Offline	62
Phenobarbital	Urine	Offline	63
Cephalexin	Serum	In-line, DPE	64
Antioxidants	Food matrix	Offline	65
Naproxen	Urine	Offline, V	66
Ropivacaine	Plasma	Offline, V, C	67
Ropivacaine	Plasma	Online, V	68
Ropivacaine and bupivacaine	Plasma	Online, V, C	69
β-Agonists	Urine	Offline, V	50
Trimethoprim	Urine	Offline, V	70
Diazepam and metabolites	Hair	Offline, V, C	71
Benzodiazepines	Hair	Offline, V, C	72
Methylxanthines	Urine	Offline, V	73
Cephalexin and amoxicillin	Urine	Offline, V	74
Tamoxifen	Urine	Offline	75
Tamoxifen and metabolites	Urine	Offline, V	76
Carbamezapine	Urine and wastewater	Offline, V	77
Cefathiamidine	Plasma	Offline	78
Mycophenolic acid	Plasma of transplant patients	Offline, V	79
Alfuzosin	Plasma and soil	Offline	80
Sildenafil and desmethyl metabolite	Plasma	Offline, V, C	81
Ibuprofen	Aqueous samples	Offline	82
Reserpine		Offline	83
Pazufloxacin mesilate	Urine	Offline, V	84
Lamotrigine	Serum	Offline, V, C	85
β-Blockers	Plasma	Offline, V	86
Diclofenac	Urine	Offline	87
Methotrexate	Serum, urine	Offline, V	88
Phenobarbital	Plasma	Offline, V	89
Piroxicam	Capsule	Offline, V	90
β-Blockers and β-agonists	Various	Offline, V, C	91
Ciprofloxacin	Urine	Offline, V	92
Biomarkers, endogenous compounds, natural products			
NNAL (tobacco-specific nitrosamine)	Urine	Offline, V, C	93
Cotinine	Smoker's urine	Offline, V	94
Nicotine	Hair	Offline, V	95
Pyrimidine nucleosides	Urine	Offline	96
Cocaine metabolites	Aqueous samples	Offline	97
Uric acid	Urine	SPE sensor, V	98

Table 23.1. (*Continued*)

Analytes	Sample and Matrix	Note	Reference
Homocysteine	Plasma	Offline	99
Creatine	Plasma	Offline, V	100,101
Dopamine	Water	Offline	102
Lipomonosaccharide	Water	Offline	103
Tocopherol	Leaves	Offline	104
Diphenylphosphate	Flame retardants in urine	Offline, V	105
Nerve agents	Serum	Offline, V	106
Quercitin	Plasma	Offline, V	107
Diniconazole	Biological fluids	Offline	108
Inhibitors of epidermal growth factor receptor	Plant extract	Offline	109
Sinomenine	Herb and plasma	Offline, V	110
Transreservatrol and emodin	*Polygonum cuspidatum*	Offline	111
Ephedrine	Herbal extracts	Offline, V	112
Benzoic acid derivatives	Plant extracts: *Melissa officinalis*	Offline, MSPD	113
Triterpenes	Plant extracts	Offline	114
Ursolic acid	Herb	Offline	115

V, method validated; C, method compared with reference method; MSPD, matrix solid-phase dispersion; PE, pulsed elution; DPE, differential pulsed elution.

Table 23.2. Examples of MISPE in Environmental Analysis

Analytes	Sample	MISPE Mode	Reference
Pesticides			
Triazines	Water	Offline	116
Triazines	Groundwater and sediments	Offline, V, C	117
Triazines	Water (industrial effluent)	Offline, V, C	118
Triazines	Water	Online, V	119
Triazines	Tap water, groundwater, soil, and corn	Online, V	120
Triazines	Soil and grape juice	Offline, V	121
Sulfonylureas	Water and soil	Offline, V	122
Monosulfuron	Soil	Offline, V	123
Triazines	River water	Online, V, C	27
Chlorinated phenoxyacids	River water	Offline, V	124,125
Triazines	Water	Offline (SPE sensor)	126
Pentachlorphenol	Lake, waste, and river water	Online, V	127
Organophosphorus pesticides	Soil	Offline, V	128
Methylthiotriazines	River water	Online, V	129
Monocrotophos	Water samples	Stir bar	130
Monocrotophos	Water samples	Membrane	131
Monocrotophos	Soil	Offline, MSPD, V	132
Sulfamerazine	Fish and pond water	Offline, V	133
Triazines	River water	Offline, V	134
Triclosan and triclocarban	Water, soil, and biosolids	Offline, V, C	135
Antibiotics and drugs			
Penicillin G and derivatives	Tap and river water	Offline, V, C	136
Fluoroquinolones	Soil	In-line, V	137
Fluoroquinolones	River water	Offline, V	138
β-Blockers	Wastewater	Offline, V, C	33
Diclofenac	Surface and wastewater	Offline, V	139
Indomethacin	Tap and river water	Offline, V	140
NSAIDs and clofibric acid	Wastewater	Offline, V, C	141

Table 23.2. (*Continued*)

Analytes	Sample	MISPE Mode	Reference
Metals			
Tributyltin	Muscle and oyster tissue, marine sediments	Offline	142
Lanthanides	Water	Offline	143
Nickel	Water	Offline	144,145
Thorium	Water	Offline	146
Yttrium-90	Milk ash	Offline	147
Copper	Water	Offline	148,149
Yttrium and europium	Nuclear power plant monitoring	Offline	150
Erbium III	Water	Offline	151
Cadmium	Water	Offline, V	152
Iron (III)	Plants and water	Offline, V	153
Zirconium (IV)	Alloys, tap, and wastewater samples	Offline, V	154
Nickel	Seawater	Offline, V	155
Samarium (III)	Water	Offline	156
Phenols and endocrine disruptors			
4-Nitrophenol	River water	Online, C	157
2,4-Dinitrophenol	Water	Offline, V, C	158
Catechol	River water, paper industry effluents	Offline	159
Diethylstilbestrol	River and tap water	Offline	160
Bisphenol A	River water	Online, V	161
Bisphenol A	River water	Offline, V	162
Bisphenol A	Tap and lake water	In-line, V	163
Bisphenol A	Water	Offline	164
Nonylphenol	Sludge, sediments	Offline	165
Nonylphenol	Water	Offline	166
Estrone	River, lake, and tap water	Offline, V	167
Others			
Esculetin	Ash bark	Offline	168
Microcystein LR	Drinking water	Offline	169
Phosphate	River water	Offline	170
Nerve agents	Soil	Offline	171

V, method validated; C, method compared with reference method; MSPD, matrix solid-phase dispersion; NSAIDs, nonsteroidal anti-inflammatory drugs.

Table 23.3. Examples of MISPE in Food Analysis

Analytes	Sample	Note	Reference
Pesticides and fungicides			
Atrazine	Beef liver	Offline	172
Thiabendazole	Orange, lemon, grape, and strawberry	In-line, V	173
Triazines	Corn	Offline, V	120
Triazines	Grape juice	Offline, V	121
Triazines	Apple	Online (C_{18} + MIP)	174
Triazines	Corn, pea, and potato	Offline (NIP-MIP), V	175
Triazines	Corn and potato	Offline, V	176
Fenuron	Wheat, barley, and carrot	Offline	177
Phenylureas	Potato, corn, pea, and carrot	Offline (NIP + MIP)	178
Phenylureas	Potato, pea, and corn	In-line	179
Pyrimethanil	Wine	Offline	180

Table 23.3. *(Continued)*

Analytes	Sample	Note	Reference
Tebuconazole	Cabbage, pannage, shrimp, orange juice, and tap water	Offline, V	181
Sulfonamides	Pork and chicken	Offline, V	182
Chlorpromazine	Pig urine	Offline, V	183
Sulfonylureas	Soybean	Offline	184
Malachite green	Fish and fish feed	Offline, V	185
Sudan	Food matrices	Offline	186
Antibiotics and veterinary drugs			
Fluoroquinolones	Urine and pig liver	Offline (SPE-MIP), V	187
Fluoroquinolones	Eggs and tissue	Offline, V	188
Chloramphenicol	Raw milk, skimmed milk, and milk powder	Offline, V, C	189
Chloramphenicol	Honey, milk, plasma, and urine	Offline, V	190
Tetracyclines	Fish	Offline, V	191
Erythromycin	Pig muscle	Offline	192
Chloramphenicol	Shrimp	Offline, V, C	193
Enrofloxacin and ciprofloxacin	Chicken muscle	Offline, V	194
Enrofloxacin and ciprofloxacin	Milk	Offline, V	195
Ofloxacin	Chicken	Offline, V	196
5-Nitroimidazoles	Egg-based samples	Offline, V, C	197
Clenbuterol	Animal feeds	Offline	31
Clenbuterol	Bovine liver	MSPD, V	198
Clenbuterol	Food	Offline, V	199
β-Agonists	Bovine muscle	Offline, V	200
Mycotoxins			
Zearalenone and its metabolite	Swine feed, wheat, corn, barley, rye, and rice	Offline, V	201
Ocharatoxin A	Wheat	In-line, PE, V	202
Ocharatoxin A	Wine	Offline, V	203
Deoxynivalenol	Pasta	Offline, V	204
Others			
Nicotine and oxidation products	Chewing gum extracts	In-line, offline, V, C	205
Bisphenol A	Shrimp	Offline, V	206
Diethylstilbestrol	Fish	Offline	207
Benzo[a]pyrene	Coffee	Offline	208
Caffeine	Green tea	Offline	209
Cholesterol	Cow milk, yolk, shrimp, pork, and beef	Offline, C	210
Cholesterol	Cheese	Offline	211
Epicatechin	Tea beverage	Offline	212
Flavonoid antioxidants	Wine, orange juice, tea	Offline	213
Caffeine, catechin	Tea	Offline, C	214
Zink (II)	Celery, *Apium graveolens*	Offline, V	215
Polyphenols	Olive oil	Offline	216
Ceramides	Yeast extracts	Offline	217

V, method validated; C, method compared with reference method; MSPD, matrix solid-phase dispersion; PE, pulsed elution.

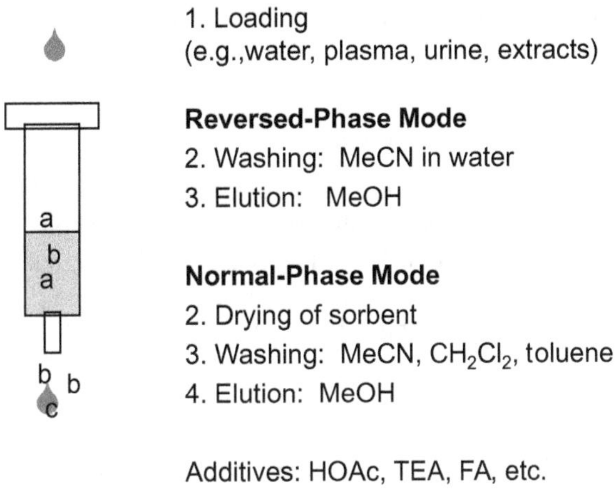

Figure 23.4. Operation modes in offline MISPE showing examples of wash and elution solvents.

MISPE can be operated in the reversed-phase or normal-phase mode (Fig. 23.4).[1] The former refers to the extraction of hydrophobic analytes from aqueous media, for example, biological or environmental samples, where the loading step occurs in the liquid chromatography (LC) reversed-phase mode of operation and is usually nonselective. Here, the selectivity is introduced by selective wash procedures, which serve to remove nonselectively bound matrix components. The analyte is then recovered by eluting with solvents of increasing eluotropic strength. Alternatively, a so-called solvent switch is performed commonly employing an exhaustive drying step followed by a switch to a water-immiscible washing solvent. This essentially means that the remaining washing and elution steps are carried out in a normal-phase mode. Otherwise, normal-phase mode of operation requires that the analyte is transferred to an organic solvent of lower polarity prior to the MISPE. This may be the case for analytes that are not sufficiently retained in aqueous mobile phases. The adsorption then occurs in a low-polarity solvent, since in such media, specific interactions are maximized, allowing high selectivity to be achieved not necessarily involving washing steps.

23.4.1. Offline Protocols

Offline is the most common mode of imprint-based SPE, probably due to its simplicity and flexibility with respect to experimental variations. Offline MISPE protocols do not differ in principle from other SPE procedures. Typically, depending on the sample volume and the required saturation capacity, an amount of 15–500 mg of imprinted polymer is packed into polyethylene cartridges. Then, after the conditioning, loading, and washing steps, analytes are eluted, ideally free of coextractives, and the elution extract is further analyzed by LC, gas chromatography (GC), or capillary electrophoresis.

Recent years have seen a rapid increase in the number of reported offline MISPE methods for the determination of a great variety of analytes in environmental samples (river water, groundwater, wastewater, seawater, and soil extracts), biomatrices (urine, plasma, serum and blood, hair), and food samples. Some relevant contributions will be discussed in Section 23.5. In general, sample is loaded into the MIP cartridge, and, after a washing step for the removal of compounds nonspecifically bound to the polymeric matrix, analytes are eluted with a solvent able to disrupt the typical noncovalent interactions that take place between the analyte and the imprinted polymer.

Preceeding the offline applications of MIPs, however, was the use of MIP in a batch format relying on equilibrium partitioning.[25] This was reported in one of the earlier descriptions of MISPE for the determination of sameridine in human plasma samples. An imprinted polymer prepared using a sameridine analog as template was incubated for 1 hour with spiked human plasma. After the corresponding washing step, sameridine was eluted and further determined by GC. The performance of the developed method was found to be better than that of the standard method based on liquid–liquid extraction thanks to the high selectivity provided by the MIP leading to cleaner chromatographic traces. Apart from demonstrating the general benefits of MISPE, it also highlighted the importance of using a close structural analog of the analyte as template in order to circumvent the template bleeding problem (see above).[24] Solid-phase matrix dispersion (SPMD) is a related method for food, plant, or soil extractions, which has been successfully combined with MISPE. Thus, Crescenzi et al.[198] demonstrated the feasibility of combining these two techniques for analyzing clenbuterol in liver. Bovine liver samples were first homogenized with a C18 sorbent; the mixture was packed into an empty cartridge and then placed on top of the MISPE cartridge (Fig. 23.5). Subsequently, clenbuterol was eluted from the matrix solid-phase dispersion (MSPD) cartridge onto the MIP cartridge by acetonitrile containing 1% acetic acid. Under these conditions, the MIP was able to retain the target analyte, and, after a washing step with the same solvent, clenbuterol could be eluted with acidified methanol for further LC analysis. The complete extraction procedure was rapid and simple, and provided quantitative recoveries allowing a detection limit below 0.1 µg/kg. There are numerous more recent examples of MIP-based MSPD (see Table 23.1) including examples where the MIP is used in the first extraction step.

MIPs have also been adapted to the solid-phase microextraction (SPME) format relying on equilibrium partitioning between the sample and an MIP-coated fiber designed for direct injection followed by LC or GC analysis.[6] This is a less investigated format for MIP-based sample pretreatment probably due to difficulties in reproducibly preparing high-density coatings and/or the lack of selectivity in the aqueous extraction step requiring the use of additional washing steps in order to boost extraction selectivity. Some recent contributions describing robust techniques for fiber production[218–220] and high-performance methods[221] may

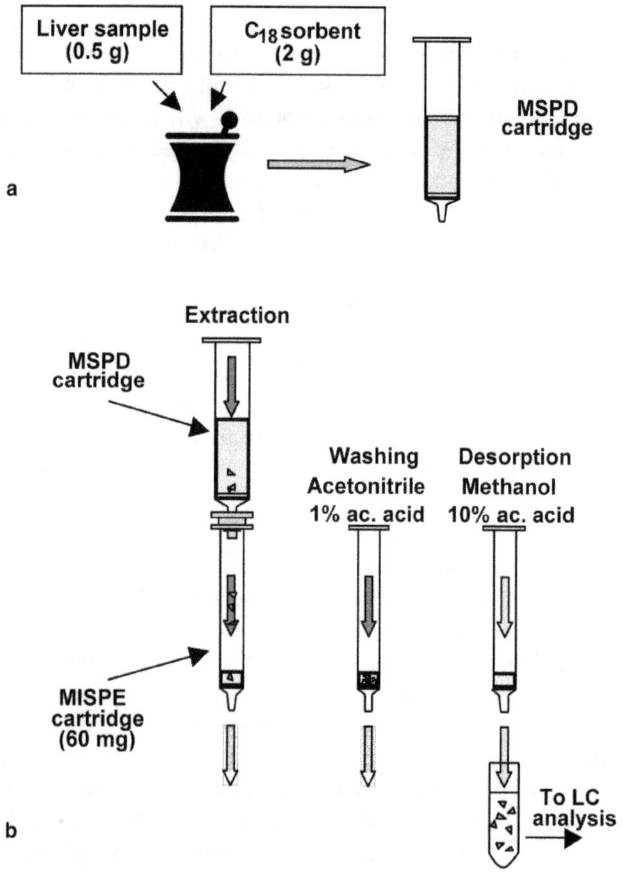

Figure 23.5. Sample preparation used by Crescenzi et al.[198] (a) MSPD cartridge preparation; (b) tandem extraction by MISPE. Reprinted from reference 198 with permission from the American Chemical Society. Copyright 2001.

indicate that the format will become more important in the future.

23.4.2. Online Protocols

Online extraction, on the other hand, may offer benefits for routine analysis since it allows complete automation of the extraction–separation–detection sequence. In the simplest format, a small precolumn packed with the imprinted polymer (typically about 50 mg) is placed in the loop of a six-port injection valve. After loading the sample and washing out interfering compounds, the analytes are eluted by the mobile phase and then separated in the analytical column.[157,174,222] One of the more convincing demonstrations of this format was reported by Boos et al.[27] This integrated three columns online for the analysis of environmental or biological samples. The sample was first allowed to pass a restricted access material (RAM) column, which retained low-molecular-weight hydrophobic substances but excluded high-molecular-weight matrix compounds such as humic acids and proteins. Thereafter, the analytes were transferred to the MIP column where the extract was polished further for final separation and quantification on an analytical column. This resulted in a highly rugged and relatively fast analytical method for the group of triazines.

In analogy with this method, Mullett et al.[59] developed an online two-step SPE-LC-MS procedure for the determination of verapamil and its metabolites in urine, plasma, and cell cultures using a RAM cartridge and a verapamil-imprinted polymer (Fig. 23.6). Using this approach, verapamil and its metabolites were easily quantified by LC-MS in less than 35 min.

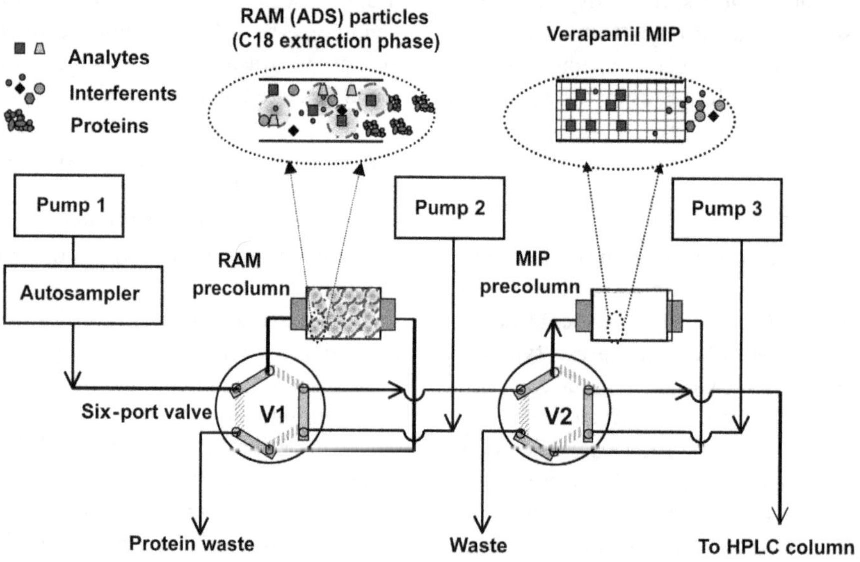

Figure 23.6. Experimental configuration of the online two-step SPE-LC procedure for the determination of verapamil and its metabolites using a restricted access material (RAM) cartridge and a verapamil-imprinted polymer. Reprinted from reference 59 with permission from Elsevier. Copyright 2004.

In spite of the clear advantages of online protocols, less than 20 papers have been published on online MISPE studies. One common problem with online formats is related to the mobile-phase incompatibility, that is, the usual lack of compatibility of the elution solvent necessary to desorb analytes from the MIP precolumn and the mobile phase required to perform the separation on the analytical column. This problem can be overcome by directing the eluent from the MIP precolumn to the injection loop and subsequently injecting on the chromatographic system,[223] or by mixing the organic elution solvent with an aqueous-rich solvent upstream from the analytical column. However, although both methods were successfully applied, extra instrumentation (i.e., additional pumps, multiposition valves) for the automation of the whole system is required, increasing the complexity and cost of the analysis.

Another problem is related to reusability. The extensive analysis times using the two-column format reflect difficulties in reconditioning MIPs exposed to biological samples where the hydrophobic sorbent surface is rapidly fouled by sticky plasma proteins. This problem has been successfully addressed by combining the RAM and the MIP properties into one single material, thus omitting the need for the RAM or C18 precolumn fractionation step. Two approaches were reported to achieve this. Haginaka et al. reported that hydrophilized MIPs prepared by postgrafting of a hydrophilic polymer on the surface of a preformed MIP, allowed the repeated direct injection of plasma samples with minor fouling by sticky proteins.[224] Dirion et al. instead focused on the direct inclusion of a hydrophilic comonomer in the imprinting polymerization (Fig. 23.7).[46] It was anticipated that this could contribute to defining the imprinted site in addition to creating a more hydrophilic polymer matrix. In order to screen monomer combinations for such synergistic effects, combinatorial and high-throughput synthesis techniques were deemed necessary.[46]

The starting point for the search was the well-characterized MIP consisting of poly(MAA-*co*-EDMA) imprinted with bupivacaine.[225] This polymer was chosen as the reference for subsequent comparisons. The library was constructed by slightly modifying the procedure used to make the reference polymer. The modifications comprised (1) the use of the hydrophilic comonomer 2-HEMA, known to impart water compatibility in a number of unrelated systems, (2) the use of four porogens (dichloromethane [DCM], 1,1,1-trichloroethane [TCE], toluene, and methyl *tert*-butylether [MTBE]), and (3) the relative ratios of (A) HEMA/MAA and (B) (HEMA + MAA)/EDMA. Polymers synthesized using low cross-linking levels (50%) and high HEMA concentrations, in combination with the use of two of the porogens (toluene or TCE), showed high imprinting factors. These were high mainly as a result of low nonspecific binding.

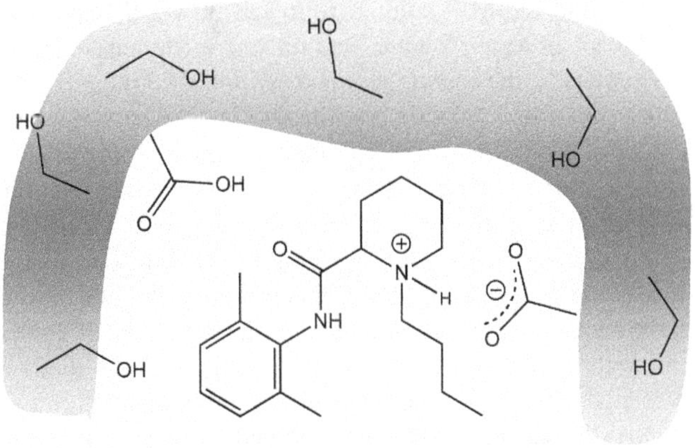

Figure 23.7. Plausible binding site in a water-compatible MIP for bupivacaine prepared using MAA as functional monomer and HEMA as hydrophilic comonomer.

The best performing polymers were upscaled for assessment as sorbents for extractions of bupivacaine from blood plasma samples.[69] As can be seen in Table 23.4, an SPE procedure involving aqueous washes containing maximum 10% organic solvent led to effective removal of bound bupivacaine from the water-compatible non-imprinted polymer (NIP) contrasting with the quantitative recovery obtained using the corresponding MIP. Meanwhile, the recoveries measured using the reference hydrophobic MIP and NIP were quantitative, indicating a pronounced nonspecific binding in this case. Protein binding, measured by breakthrough experiments, to the optimized polymers was also lower compared with that for the reference polymer. Thus, implementing MIPs in online methods may be very rewarding.

23.4.3. In-Line Protocols

Thanks to the high selectivity provided by MIPs, it is possible to directly couple an MIP column in-line with the detection system. In this manner, extraction, enrichment, separation, and determination of target analytes can be achieved in one single step. MISPE with direct in-line UV detection was first described by Sellergren for the determination of pentamidine in urine.[51] The strong interactions between the amidine groups of the template and the acid groups of the functional monomer (MAA) allowed the polymerization in polar porogens and resulted in essentially stoichiometric imprinting of the template. Thus, the resulting MIP could be used for direct high affinity extraction from aqueous samples. After loading 100 mL of diluted

Table 23.4. Recoveries (%) of Bupivacaine from Imprinted (MIP) and Non-Imprinted (NIP) Water-Compatible Polymer (AquaMIP) and Reference Polymers (Reference) in SPE Using 10% Ethanol in Water as Washing Solvent

Polymer	MIP	NIP
Reference	105	105
AquaMIP	104	4

urine sample, the column was washed with 100 mL of a buffer solution (pH 9) in order to remove sorbed basic compounds, and pentamidine eluted in ~1.5 mL of buffer at pH 3. The selective elution performed allowed the successful detection of pentamidine free of coextractives by UV at 270 nm. Another in-line protocol was developed by the same group for the analysis of nicotine and its decomposition products in chewing gums.[205] Due to the apolar matrix (heptane), the MIP could be used in the normal-phase mode for direct selective enrichment of the nicotine-related analytes from the remaining matrix components. Elution with a polar solvent led to a sharp elution peak containing the nicotine-related analytes. Thus, the recognition capabilities of MIPs are often enhanced in organic media. In this regard, Turiel et al.[173] developed an analytical method for the determination of the fungicide thiabendazole (TBZ) in fruit sample extracts in acetonitrile. Organic sample extracts (50 μL) were injected onto a TBZ-imprinted polymer column using a 100% MeOH mobile phase. Then, after 2.4 min, the mobile phase was switched to methanol : acetic acid (80:20, v/v) in 0.1 min, keeping these conditions constant for 5 min before returning to the initial conditions. Following this methodology, the high selectivity of the MIP column permitted the target analyte to be retained on the column while the interferences were rapidly eluted, and thus, TBZ was unambiguously detected and quantified in less than 15 min.

MISPE with pulsed elution (MISPE-PE), proposed by Mullett and Lai,[53] represents an alternative in which a small volume of elution solvent, instead of a complete solvent switch, is used. In this first work, theophylline in chloroform-diluted serum samples was injected onto a theophylline-imprinted polymer packed into a stainless-steel column using chloroform as mobile phase. Subsequently, after interfering compounds had passed through the column, 20 μL methanol was injected, allowing theophylline to be eluted free of coextractives and to be directly determined spectrophotometrically at 270 nm. MISPE-PE has been subsequently improved by the application of successive 20-μL pulses of different solvents in a procedure known as MISPE with differential pulsed elution (MISPE-DPE).[54] This approach is more efficient in removing both remaining interfering compounds and the nonspecifically retained analyte fraction. Both MISPE-PE and MISPE-DPE allow analyte enrichment through injection of large volumes of sample and provide a high analytical sensitivity, thanks to the narrow band obtained by pulsed elution.

23.5. ANALYTICAL METHODS USING MISPE

Since the first reported example of MISPE by Sellergren for the determination of pentamidine in urine samples,[51] the use of MIPs in sample preparation has grown in importance, and routine methods are now starting to emerge. In this section, some selected examples of MISPE describing the extraction of analytes from biological, environmental, and food samples will show that the technique is now mature enough to offer the analyst fast access to a highly selective sample pretreatment, which will pay off in terms of lower detection limits, faster analysis times, and a more rugged method. For an extensive coverage of each respective field, the reader is referred to some recent general reviews on MISPE[6,226,227] and MISPE of bio- and environmental samples[228,229] and food samples.[4,230]

23.5.1. Pharmaceutical Analysis and Bioanalysis

As shown by the numerous recent examples (Table 23.1), the quantitative analysis of pharmaceutical drugs, drugs of abuse, or endogenous biomarkers in various biomatrices (urine, serum, plasma, hair, tissue samples) can strongly profit from MISPE.[2,7] Again, the main arguments for using this form of sample pretreatment are the cleaner extracts that can be obtained by this method and the added benefits in terms of lower detection limits, enhanced method ruggedness, and higher sample throughput that it allows. Using MS detectors, higher sensitivity can also be expected due to reduced influence of ion suppression.[62] Ion suppression is commonly observed when one or more interfering and coeluting species affect droplet formation and ionization in the MS detector. This results in poor reproducibility, accuracy, and sensitivity of the method, which is particularly common in bioanalysis with its highly complex sample matrices. One example of how MISPE can overcome this problem was shown in a recent report describing the analysis of the tobacco-specific biomarker NNAL (Scheme 23.4) in urine samples.[93] This lung carcinogen is an indicator for tobacco exposure in smokers as well as in nonsmokers exposed to secondhand smoke. The low concentrations found particularly in the latter group require highly sensitive and selective analytical methods. Thus, an MIP targeting NNAL was developed (see Section 23.3.2) and implemented in an offline SPE method. Urine samples (5 mL) containing internal standard were diluted with phosphate buffer, and the glucoronate metabolite of NNAL was either enzymatically hydrolyzed back to NNAL (total NNAL) or left nonhydrolyzed (free NNAL). After hydrolysis, the solution was loaded into the MIP column followed by wash steps with 1 mL each of water, toluene, and toluene/methylene chloride (9:1). Finally, the analyte was eluted with 3 × 1 mL of methylene chloride. After reconstitution in water, the clean samples were analyzed by tandem mass spectrometry (LC-MS/MS) using atmospheric pressure ionization. Overall, the resulting method was sufficiently sensitive (limit of detection [LOD] = 1.7 pg/mL, mainly due to the low ion suppression of <4%), precise, accurate, and fast to be implemented in large scale analysis in support of epidemiological investigations. The time savings of the method was considerable, on the order of several hours.

MISPE has also broken new ground in the area of forensic analysis.[71,72] The group of Miller and Ariffin compared MISPE with mixed-mode SPE for the analysis of benzodiazepines (Scheme 23.14) in postmortem hair samples.

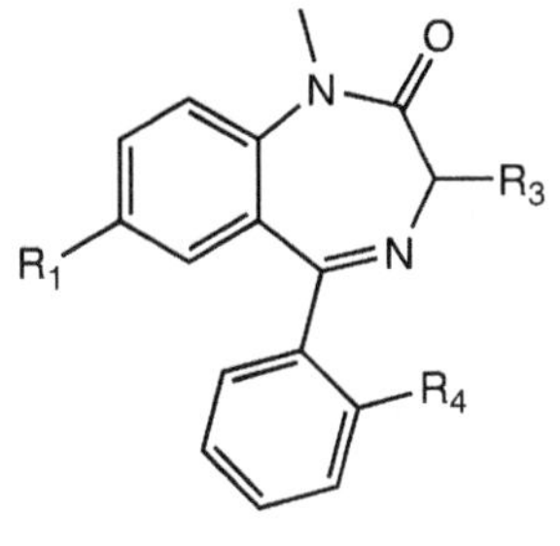

Scheme 23.14

This class of drugs includes the most frequently prescribed and abused tranquilizers, and they are involved in most cases of drug-related deaths in the United Kingdom.[71] After washing, the hair samples were incubated in MeOH/25% ammonium hydroxide solution (20:1) overnight, and the extract was subsequently reconstituted in buffer (SPE) or toluene (MISPE). In the MISPE, sample was added to the cartridges (preconditioned in toluene) and allowed to drip through without vacuum. This was followed by washing with toluene and elution using 15% acetic acid in acetonitrile. After reconstitution of the samples in mobile phase, they were analyzed by LC-MS/MS. Both the MISPE and the SPE methods were validated for nine benzodiazepines. The efficiency of the MISPE method became very evident when diazepam was detected in four samples where the SPE method detected none. This was due to the higher extraction recovery and lower LOD using the MISPE method. The authors conclude that the MISPE method appeared to be significantly more streamlined than the SPE method, providing a simpler and more time-efficient procedure.

23.5.2. Environmental Analysis

The need for reliable methods allowing monitoring of harmful substances in the environment is rapidly increasing.[229] Synthetic organic compounds, such as pesticides, pharmaceutically active compounds, and endocrine-disrupting compounds, are important classes of such contaminants, which are relatively easy to target with MIPs. Environmental samples (mainly environmental waters, sediments, and soils) have hence been subjected to MISPE for the extraction of pesticides, pharmaceuticals, and phenolic compounds including endocrine-disrupting compounds, among others (Table 23.2).[229] The first offline MISPE work describing an environmental application was reported by Ferrer et al.[117] In this study, an MIP prepared using terbuthylazine as the template molecule was used to enrich six chlorotriazines from groundwater and sediment samples. After a cleanup step with DCM, the matrix components were removed, whereas target analytes were selectively retained. Obtained recoveries were higher than 80% (except for propazine, with a recovery of 53%) after percolating 100 mL of water sample. An in-depth study of this system by Horvai et al. has shed important light on the key parameters to consider in the method development.[119] Column drying is a

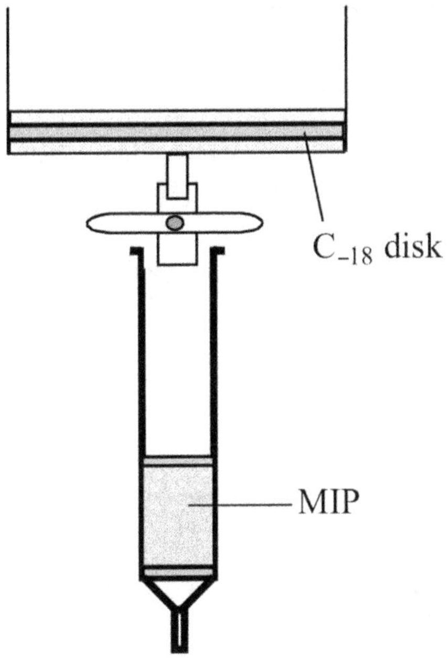

Figure 23.8. Coupling of a C_{18} disk with MISPE for processing of large sample volumes. Reprinted from reference 119 with permission from Elsevier. Copyright 2002.

very critical step in order to achieve high recoveries with MISPE using non-water-miscible solvents. This is because residual water on the MIP may shield the sorbent surface and reduce the efficiency of the washings leading to irreproducible and low recoveries. Extensive drying was found to partially provide a solution to this problem. In addition, matrix effects due to the presence of divalent cations in real water (such as calcium ions) have been observed to reduce the recoveries of that target analytes.[118] An ion exchange interaction between these cations and the acidic protons of the MIP occurs, thus preventing subsequent interactions (by H bonding) between the triazines and the MIP. However, this matrix effect can be easily avoided by introducing an acidic washing step after the percolation of the water sample. Alternatively, as first reported by Horvai et al., it is possible to perform a two-step SPE procedure (Fig. 23.8). Water samples are first loaded into conventional sorbents (e.g., C_{18} disks), allowing the preconcentration of large sample volumes, and then retained analytes are eluted using a suitable organic solvent and loaded on the MIP.[119,120]

Although to a lesser extent, MISPE has also been successfully applied to the cleanup of soil sample extracts. The method has been applied for the determination of triazines,[120,121] monosulfuron,[123] and fluoroquinolones.[137] As an example of the suitability of MISPE, Figure 23.9 shows the chromatograms obtained by high-performance liquid chromatography (HPLC)-UV in the analysis of soil samples spiked with 0.1 μg/g of five fluoroquinolones with and without performing the extraction onto the ciprofloxacin-imprinted polymer cartridges.[137]

It is clear that the detection and identification of the analytes without the MISPE process would be impossible due to the high amount of matrix interference compounds that

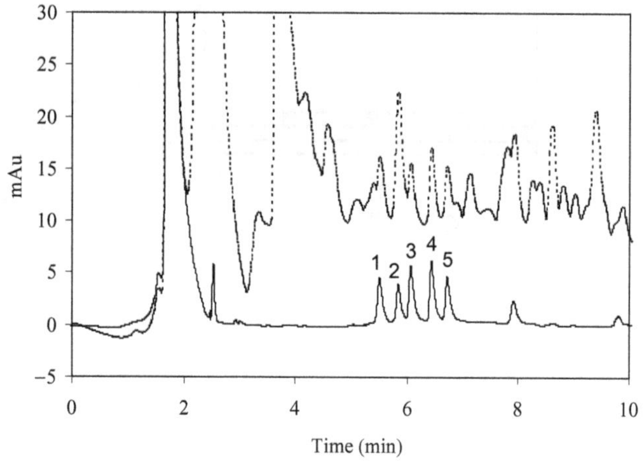

Figure 23.9. Chromatograms obtained in the analysis of a soil sample extract spiked with 0.1 µg/g of the selected fluoroquinolones with (upper profile) and without (lower profile) performing MISPE into a ciprofloxacin-imprinted polymer cartridge. Peak assignment: 1, enoxacin; 2, norfloxacin; 3, ciprofloxacin; 4, danofloxacin; 5, enrofloxacin. Reprinted from reference 137 with permission from Elsevier. Copyright 2007.

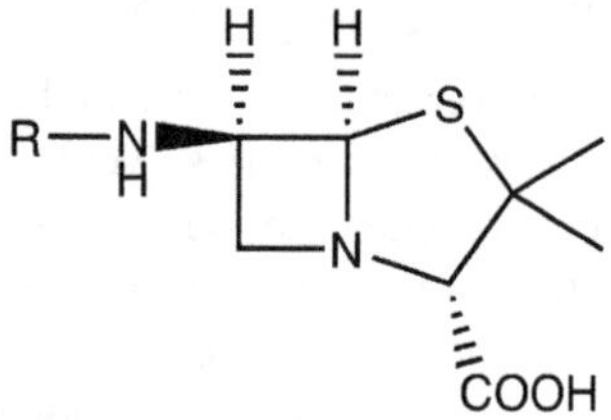

Scheme 23.15

were coextracted. However, by incorporating a cleanup step using the MIP, detection and quantification of target analytes were easily accomplished with recoveries for the tested fluoroquinolones higher than 80%.[137]

The development of water-compatible MIPs relying on a direct selective extraction upon sample loading has proven to be rewarding in some cases.[69,136,137] This will reduce the solvent consumption and simplify the SPE procedure leading to faster and greener methods. The design of the MIP phases poses different challenges depending on the hydrophobicity and functionality of the target analyte. Thus, for lipophilic targets where good imprinting is seen in organic solvents, the key challenge is to passivate or hydrophilize the surface of the sorbent in order to reduce the stickiness of the analytes. However, for more hydrophilic analytes, the binding site design is the key challenge since such compounds exhibit limited solubility in the imprinting solvents, and binding in water may not be hydrophobically assisted as for lipophilic targets. Examples of the latter are the recent reports on MISPE of β-lactam antibiotics (BLAs) (Scheme 23.15) and fluoroquinolones (Scheme 23.16) by stoichiometric imprinting.

The contributions from the group of Moreno-Bondi are particularly noteworthy.[136,138] Stoichiometrically imprinted polymers showing good target binding in aqueous media

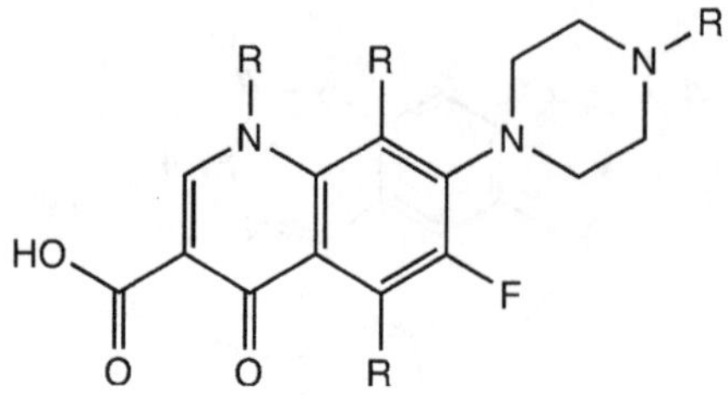

Scheme 23.16

were prepared using penicillin G (PenG) procaine salt or enrofloxacin (ENR) as template molecules and a stoichiometric quantity of a recently developed urea-based functional monomer to target the single oxyanionic species in the template molecule. The optimized MISPE/HPLC procedures allowed direct extraction of BLAs on the PenG MIP and fluoroquinolones on the ENR MIP from river water samples with minor use of organic solvents. For instance, in the latter case, the selective washing was achieved using acetonitrile: HEPES (N-2-hydroxyethylpiperazine-N′-2-ethanesulfonic acid) buffer pH 7.5 (10: 90, v/v) and elution with 2% trifluoroacetic acid (TFA) in methanol. Good recoveries and precision were obtained for river water samples for all the antibiotics tested resulting in limits of detection ranging from 0.01 to 0.30 µg/L when 100-mL water samples were processed.

Another important contribution came from the group of Barcelo who reported on trace analysis of β-blockers in wastewaters.[33] β-Blockers are an important class of antihypertensive drugs, which are widely used to treat numerous disorders. Sewage water treatment plants contribute to their presence in the environment where they can reach surface as well as ground and drinking water resulting in potentially serious ecotoxicological effects. In order to boost the sensitivity of the existing methods, the authors turned to MISPE based on a commercially available β-receptor MIP and validated the method against a method relying on a generic mixed-mode SPE phase. Although recoveries for the group of β-blockers with both materials were similar, the main advantage of MISPE versus Oasis HLB was that good recoveries and sensitivity were achieved also for lower concentration factors, that is, requiring only 25-mL sample volumes versus 100–200 mL for Oasis. Nevertheless, the detection limits obtained for MISPE were in the same range as the ones achieved using Oasis HLB and in some cases lower. An important consequence of the lower sample volumes is the time saving this implies and the reduced matrix effects on the separation and detection. External calibration curves prepared with wastewater extracts were compared with those prepared in solvent. From this comparison, it was found that the MIPs resulted in minimal signal suppression because both curves overlapped, whereas Oasis HLB showed significant suppression of the analyte signal (Fig. 23.10). Combining MISPE with quadropole linear ion trap MS offered a powerful method for monitoring this group of analytes according to the European Union (EU) set limits at the low nanogram per liter range.

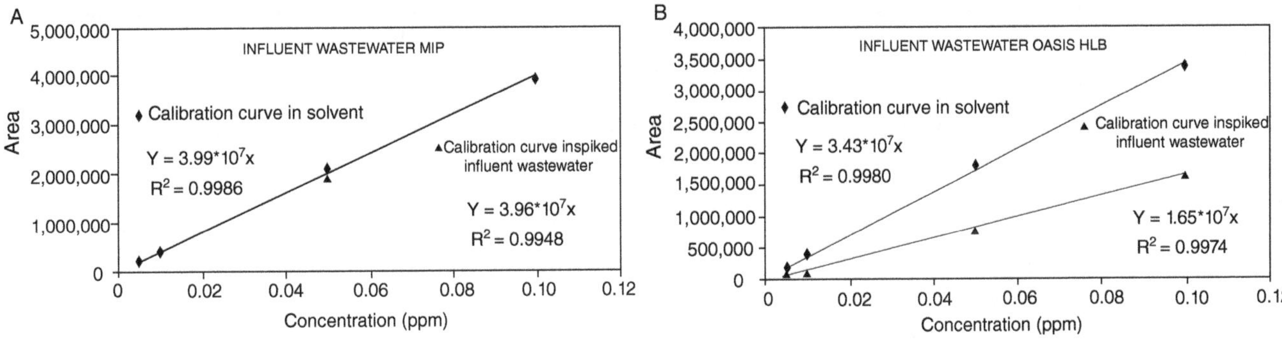

Figure 23.10. Calibration curves in solvent versus those prepared in spiked wastewater influent extracts for the β-blocker atenolol in MIPs (A) and Oasis HLB (B) cartridges. From Reference 33.

Table 23.5. Detection Limits (ng/g) of Triazines in Vegetable Samples Obtained Using the Direct and Two-Step MISPE Procedures Using a Propazine-Imprinted Polymer

	Peas		Potatoes		Corn	
	Direct	Two-Step	Direct	Two-Step	Direct	Two-Step
DIA	0.6	0.7	n.d.	0.9	n.d.	0.7
DEA	0.5	0.3	1.7	0.8	n.d.	0.9
Simazine	0.5	0.5	1.4	0.6	1.6	0.8
Atrazine	0.6	0.4	1.6	0.7	2.1	0.7
Propazine	0.5	0.3	1.6	0.7	n.d.	0.7

Adapted from reference 175 with kind permission of Springer Science and Business Media. Copyright 2003.
n.d., not detected; DIA, desisopropylatrazine; DEA, desethylatrazine.

23.5.3. Food Analysis

Pesticides, antibiotics, and endocrine disrupters are well-known toxic contaminants, which have been detected in foods due to the enormous increase in industrial and agricultural activities during recent years. In addition to the synthetic chemicals, naturally occurring toxins (mycotoxins) have been detected in some food products, mainly of plant origin.[231] Accordingly, national and international authorities have established maximum residue levels (MRLs) for some of these compounds in food. Thus, it is necessary to develop analytical methods that will allow the determination of contaminant levels, in a precise and accurate manner. MISPE can, in this context, contribute to method performance particularly for contaminants where no safe MRL can be set due to uncertainty about chronic toxicity. Although at present, the application of MISPE to food and beverage analysis represents a small fraction of published papers, the fraction is rapidly growing and the technique will most likely impact method development in this field in the future (Table 23.3). Pesticides, antibiotics, and mycotoxins are the analytes most commonly targeted by this technique.

For instance, MISPE of triazines (Scheme 23.17) has been applied to the cleanup of fruits, vegetables and juices.[120,121,174–176]

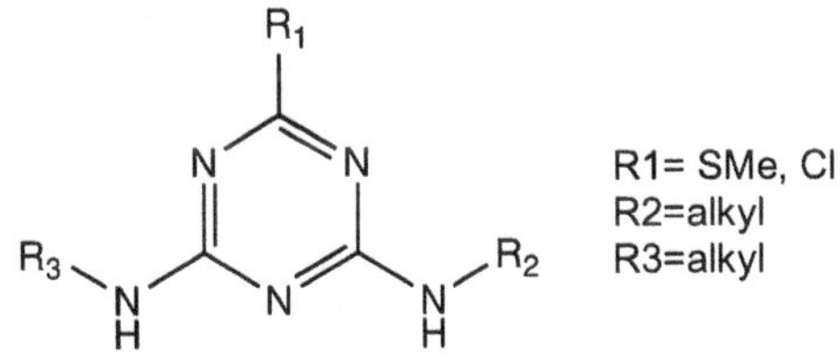

Scheme 23.17

In general, a high degree of selectivity was achieved, but, in some cases, the complexity of the samples under study lead to the coextraction of some matrix components, which complicated the chromatographic determination of target compounds after MISPE. This problem has been circumvented by introducing a pre-cleanup step using an NIP.[175,177,178] In this approach, the sample is first loaded onto the corresponding NIP, and thus, analytes and matrix components are nonspecifically retained. Then, target analytes (and a small amount of interfering compounds) are eluted during the washing steps, whereas the majority of matrix compounds remain bound to the NIP, thus achieving a first cleanup of the sample extracts. Subsequently, the obtained washing solution is loaded onto the corresponding MIP for further cleanup. Table 23.5 shows a comparison of the detection limits obtained in the determination of several triazines in vegetables after direct and two-step MISPE.[175] As shown by the reported data, inserting an NIP-based pre-cleanup step improves the SPE procedure in cases where matrix components are strongly bound to the polymeric matrix.

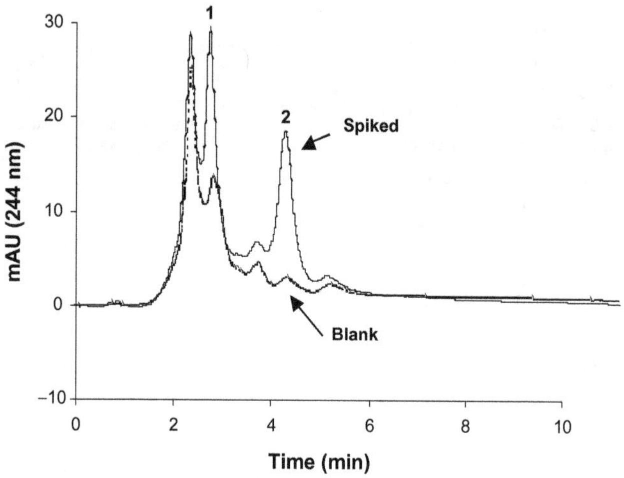

Figure 23.11. Chromatograms obtained after the injection of a blank and spiked potato sample extracts directly onto a linuron-imprinted column. Injection volume was 20 µL, mobile phase 100% acetonitrile at a flow rate of 1 mL/min. Peak numbers: (1) linuron and (2) isoproturon. Reprinted from reference 232 with permission from Elsevier. Copyright 2005.

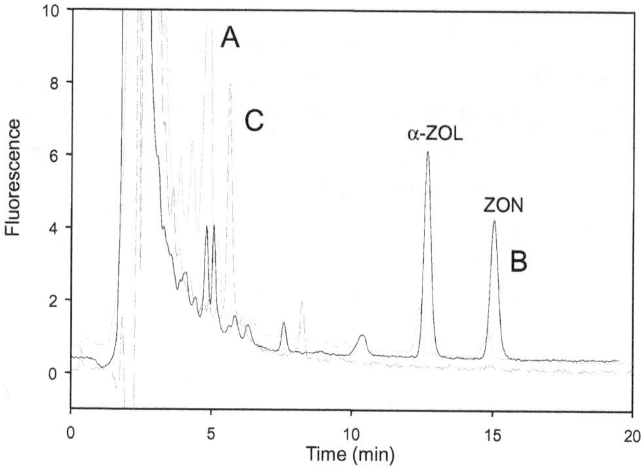

Figure 23.12. Chromatograms of a wheat extract (5-g sample spiked with 100 ng/g of ZON and α-ZOL): (A) without and (B) with cleanup on the imprinted polymer. (C) Chromatogram resulting from cleanup on NIP. Reproduced from reference 201 with kind permission of Springer Science and Business Media. Copyright 2006.

An example exploiting MISPE in order to obtain a higher sample throughput and performance for pesticide analysis was reported by Tamayo and Martin-Esteban.[232] A direct coupling of the MIP column in-line to the detection system allowed here fast and selective screening of phenylurea herbicides in vegetable samples. Thus, a linuron-imprinted polymer was packed in an HPLC column and directly connected to a UV spectrophotometer. Sample extracts were injected, resulting in separation of the target analytes from matrix-interfering compounds as well as into two different groups of phenylureas depending on the presence or absence of a methoxy group in its chemical structure. Figure 23.11 shows the chromatograms obtained for blank and spiked (100 ng/g concentration level of linuron and isoproturon) potato sample extracts. As it can be observed, both analytes were separated from each other and also from interferences, allowing the screening of these analytes at concentration level required by current legislation in less than 10 min.

The application of MISPE to the extraction and cleanup of mycotoxins from food samples has also been demonstrated as superior in analytical performance compared with traditional methods.[231] For instance, MISPE has been applied to the determination of zearalenone (ZON) and its metabolite *trans*-α-zearalenol (α-ZOL) in cereal and swine feed samples.[201] In this work, after the extraction of target analytes by pressurized liquid extraction, sample extracts in acetonitrile were loaded into a imprinted polymer. Then, a simple washing step using 1 mL of water was performed before elution with 3 mL of methanol : phosphoric acid (95:5, v/v). This fraction was evaporated to dryness and dissolved in 1 mL of methanol for final analysis by HPLC with fluorescence detection. Figure 23.12 shows the chromatograms obtained for the analysis of wheat extracts spiked with 100 ng/g of ZON and α-ZOL without and with cleanup by MISPE and that obtained using an NIP. The amount of matrix interference was much lower after MISPE, allowing the determination of target analytes at trace concentration levels. The detection limits obtained were, however, in this case comparable to those obtained using other cleanup procedures (i.e., IA columns) or techniques (i.e., LC-MS).

As was highlighted in the previous section, a more effective cleanup procedure developed using MISPE may result in considerable savings in time and cost of the analysis. This was convincingly demonstrated by the group of Guy[189] for the determination of chloramphenicol in milk (raw milk, skimmed milk, and milk powder). In their work, the advantages of MISPE using a commercial MIP were assessed by comparing the data generated from a reference method relying on standard generic SPE. The MIP-based procedure only required a prior centrifugation step, whereas the traditional approach involved a protein precipitation step, an SPE cleanup, and three liquid–liquid extraction steps. Apart from the expected cleaner chromatograms obtained following the MISPE procedure, a considerable increase in sample throughput was achieved (18 samples processed within 3 h compared with 8 h with the classical approach), demonstrating convincingly the suitability of MISPE in food analysis. Another aspect of the improved sample cleanup is seen in the required instrumentation. Whereas the reference method relying on generic SPE requires LC–electrospray ionization (ESI)–MS/MS to achieve the necessary sensitivity and selectivity, acceptable LODs using the MISPE procedure may be achieved using more available LC-UV or fluorescence detectors.

23.5.4. Peptide and Protein Analysis

The MISPE examples reported thus far in the literature concern methods for low molecular target analytes. However, given the recent advances in peptide and protein imprinting, MISPE may also come to play a role in proteomics-related sample pretreatment strategies. This area faces the challenge of having to separate some 10,000 different protein species present in a single cell with relative abundances differing by 5–10 orders of magnitude. The use of affinity reagents for protein fractionation or target identification is undeniably of central importance in the endeavor. In this area, MIPs could play an important role, complementing currently used immunological and chemical methods. One first proof of concept for MISPE in proteomics was reported by Emgenbroich et al.[42] They used an epitope imprinting strategy combined with urea-based host monomers to prepare phosphotyrosine (pTyr)-imprinted polymers capable of selective enrichment of pTyr-containing peptides (Fig. 23.13). The polymers exhibited binding affinities versus the pTyr template in the range of pTyr antibodies and furthermore displayed a pronounced selectivity for pTyr over phosphoserine (pSer). As sorbents in MISPE, the polymers offered high enrichment factors for phosphopeptides (Table 23.6) also in complex samples such as tryptic digests, with pTyr versus pSer selectivity, solvent compatibility, and high robustness. Using a model system consisting of phosphorylated and nonphosphorylated angiotensin, it is seen in Table 23.6 that the phosphorylated peptide is retained much stronger on the MIP and can be eluted in a clean and enriched form.

23.6. THE DEVELOPMENT OF NEW MISPE PROTOCOLS

Three different steps are involved in the development of a new SPE protocol based on molecular imprint: the synthesis of the material, the assessment of its recognition properties, and the development of extraction protocol and the final validation of the protocol.

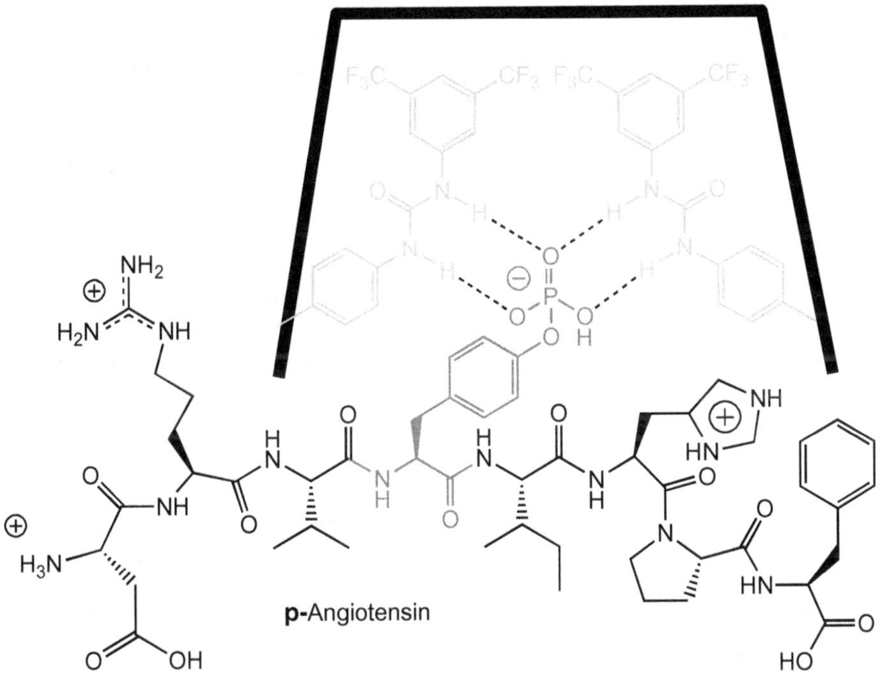

Figure 23.13. An MIP prepared by urea-mediated imprinting of N, O-protected phosphotyrosine (pTyr) for selective phosphopeptide extractions. From Reference 42.

Table 23.6. Peak Intensities of Peptides Identified by Matrix-Assisted Laser Desorption Ionization (MALDI)-MS Analysis of Fractions Collected during Solid-Phase Extraction Experiments Performed Using a Phosphotyrosine MIP and NIP

	MIP				NIP			
Peptide	Load 1	Load 2	Elute 1	Elute 2	Load 1	Load 2	Elute 1	Elute 2
Angiotensin	1632	0	0	0	919	0	0	0
p-Angiotensin	66	31	38	25	55	63	0	0
Ser-436/ZAP70	1065	0	0	0	512	0	0	0

The fractions were collected with 5-min intervals after injection (10 μL) of a model peptide mixture on MIP or NIP using a loading mobile phase A (Load) and after switch to an eluting mobile phase B (Elute). Mobile phase: 1–10 min: A = MeCN/water: 95/5 (v/v) (0.1% TFA); 10–20 min: B = MeOH (0.1% TFA).

The peptide mixture consisted of nine peptides each at a concentration of 11 mg/mL except for pAng and pZAP70, which were present at a concentration of 0.11 μg/mL in water.

First of all, a suitable combination of template, monomers, and a porogenic solvent has to be found where the template can be successfully imprinted.[43] The solubility and thermodynamic stability of the monomer–template assemblies is the key factor in this step. Guidelines for choosing or designing the template and monomers were provide in Section 23.3. In brief, one or more functionalized vinyl monomers are sought, which are capable of forming strong noncovalent interactions with the functional groups of the template. The monomer–template assemblies need to be stable enough for them to be transformed to specific binding sites during the polymerization. The structure and morphology of the polymer matrix are strongly influenced by the type and amount of cross-linker and porogen.[233] These factors affect, in turn, the sorption–desorption kinetics and the stability of the material and, therefore, require careful optimization as well.

Once a suitable material has been found, a chromatographic characterization step is required where the recognition properties of the sorbent are evaluated in different mobile phases. In order to predict suitable extraction protocols, the sorbents should thus be characterized in terms of affinity, selectivity, and sample load capacity (breakthrough volumes) for the analytes. Except for establishing recognition based on the retention times of the template and analogs at one or two different concentrations, frontal analysis or batch rebinding at several concentrations yield thermodynamic information in terms of binding energy and mass transfer kinetics.[234]

The search for suitable wash and elution conditions in the SPE procedure starts by considering the nature of the analyte in terms of solubility, hydrophobicity, functional groups, and their pK_a values.[15] This will determine whether offline MISPE can be operated in a normal-phase and reversed-phase modes. For low to moderately polar templates, good recognition is generally seen using organic wash solvents, where the template interacts mainly electrostatically with the binding sites. In fact, most rebinding experiments to MIPs demonstrating high affinity and selectivity have been performed using organic solvents and, for most templates, the first chromatographic rebinding evaluation is most often carried out in organic mobile-phase systems.[43] Suitable wash solvent may be found here by loading the analyte on the column in such a solvent followed by stepwise elution with the solvent containing increasing amounts of additive (e.g., acetic acid) or a second solvent (e.g., water) (Fig. 23.4).

The sample pretreatment strategy for triazines described by Pap et al. can serve as a good example for how such a method development can be performed (Fig. 23.14).[119] A cartridge containing 50 mg of MIP for terbuthylazine was used in the study. Procedure 1 involved direct application of an aqueous solution of the analytes on the column. This results in complete retention of the relatively hydrophobic analytes on both the MIP and NIP columns. A suitable wash solvent should now remove the nonspecifically bound matrix components while leaving the analyte on the column. Figure

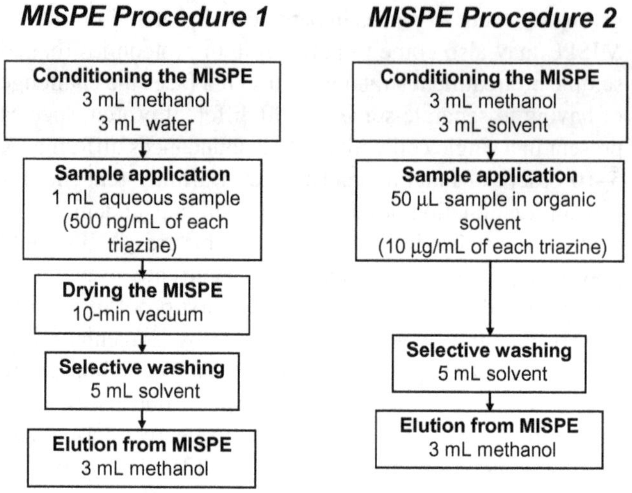

Figure 23.14. SPE procedures assessed by Pap et al.[119]

23.15 shows the cumulative recoveries of the triazines in consecutive 0.5-mL fractions of acetonitrile containing different amounts of water. Using pure acetonitrile, the elution curves show that both chlorotriazines are completely retained, whereas the s-triazines are released.

This agrees with previous observations showing that polymers imprinted with chlorotriazines preferentially retain chlorotriazines, whereas conversely s-triazine MIPs prefer s-triazines. Now by adding 2% water to the wash solvent, the MIP starts to discriminate the non-imprinted from the imprinted chlorotriazine, and using 10% water in acetonitrile, all triazines apart from the template are eventually eluted in the fourth 0.5-mL fraction. A different scenario emerges when washing is performed using 50% water in acetonitrile. Now, presumably due to its higher log P_{ow} value, terbuthylazine is completely retained, whereas all nontemplate triazines are fully eluted. Thus, the latter wash solvent appears to be a good choice if the goal is to obtain a method for terbuthylazine alone. The recognition in the aqueous system is due to the hydrophobic effect.[235,236] Obviously, depending on the hydrophobicity of the template, the hydrophobic effect can either enhance or attenuate the selectivity in the molecular recognition of imprinted polymers. This effect is analogous to the hydrophobic binding exhibited by enzymes and antibodies.[237]

Procedure 2 in Figure 23.14 was then used to assess the influence of different wash solvents on the cumulative recovery of the different triazines (Fig. 23.16). Among the four different solvents tested, toluene and DCM gave the most promising imprinting effects. Using the least polar solvent toluene, significant retention of all triazines was observed on both MIP and NIP, and excessive wash volumes are needed to elute the triazines from the NIP whereas the MIP retains them all. In the more polar solvent DCM, however, the analytes elute from the NIP in the first fraction, whereas they are retained on the MIP. Interestingly, using this solvent, a fractionation between the chlorotriazines (template-related: terbuthylazine) and s-triazines (ametryn,

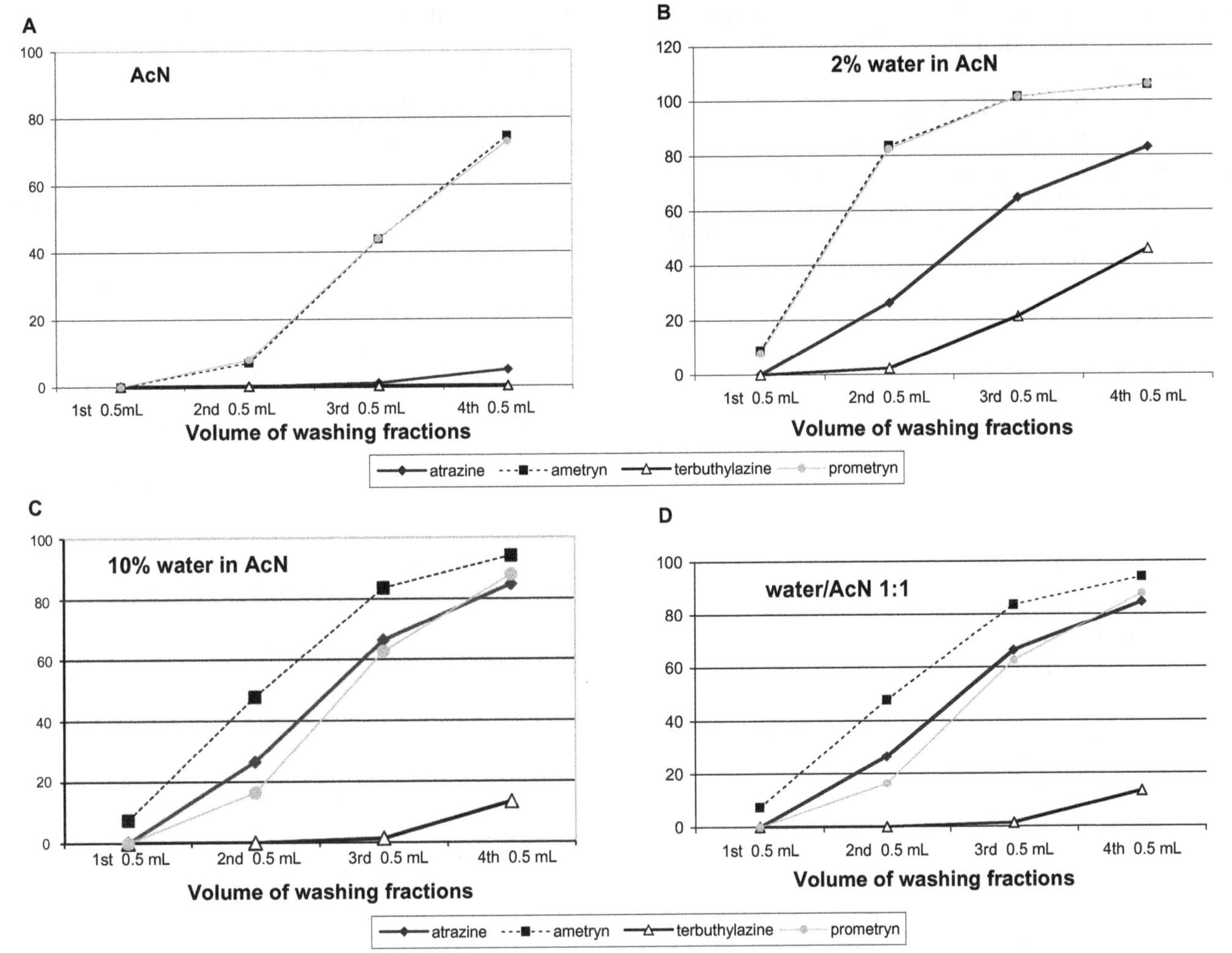

Figure 23.15. Cumulative recovery of triazines on an MISPE in acetonitrile (A), in acetonitrile with 2% water (B), in acetonitrile with 10% water (C), and in acetonitrile with 50% water (D). From Reference 119.

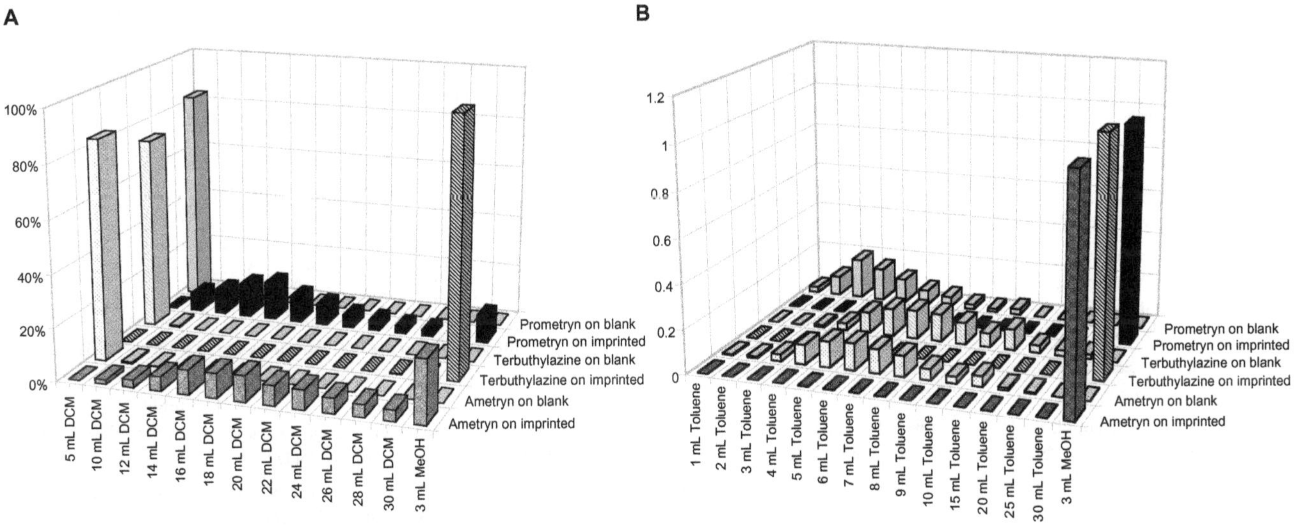

Figure 23.16. Elution profile of triazines on NIP and MIP using different organic media as selective washing solvent: dichloromethane (A) and toluene (B). From Reference 119.

prometryn) is possible. This shows the importance of carefully optimizing this factor as well in the MISPE method development.

An appropriate pH adjustment is critical in MISPE targeting analytes with protolytic functional groups, that is, Brönsted basic or acidic groups. In the case where, respectively, an acidic (i.e., MAA) or basic (i.e., VPy) functional monomer are used in the preparation of the corresponding MIPs, the polymeric phases operate as weak ion exchangers, and the retention is governed by an electrostatic model.[238] For MAA MIPs retaining an amine template, the selectivity is typically high at low pHs, whereas it drops when pH exceeded the pK_a value of the solute. Moreover, the retention displays a maximum when $pH = pK_a$ of the solute, suggesting that the retention is controlled by a simple ion exchange process.

23.7. CONCLUSIONS

MIPs programmed to recognize a specific analyte or a group of analytes are compatible with currrent sample pretreatment procedures. In contrast to generic SPE sorbents, these dedicated sorbents are capable of cleaning up complex samples to unprecedented levels from small sample volumes. Overall, this translates into fast and robust analytical methods with unprecedentedly low detection limits satisfying regulatory requirements. Although today's advanced high-resolution MS detectors can sometimes provide sufficient sensitivities with minimal sample pretreatment, the method robustness gain with MISPE, and the clean extracts obviate the need for these still expensive detectors. Allowing interdisciplinary research to further refine these molecular recognition elements may pave the way for MISPE as one of the breakthrough technologies in twenty-first century separation science.

REFERENCES

1. Lanza, F.; Sellergren, B. The application of molecular imprinting technology to solid phase extraction. *Chromatographia* **2001**, *53*, 599–611.

2. Theodoridis, G.A.; Papadoyannis, I.N. Novel advanced approaches in sample preparation and analyte detection for bioanalysis. *Curr. Pharm. Anal.* **2006**, *2*(4), 385–404.

3. Baggiani, C.; Anfossi, L.; Giovannoli, C. Molecular imprinted polymers: Useful tools for pharmaceutical analysis. *Curr. Pharm. Anal.* **2006**, *2*(3), 219–247.

4. Baggiani, C.; Anfossi, L.; Giovannoli, C. Solid phase extraction of food contaminants using molecular imprinted polymers. *Anal. Chim. Acta* **2007**, *591*(1), 29–39.

5. Pichon, V. Selective sample treatment using molecularly imprinted polymers. *J. Chromatogr. A* **2007**, *1152*(1–2), 41–53.

6. Tamayo, F.G.; Turiel, E.; Martin-Esteban, A. Molecularly imprinted polymers for solid-phase extraction and solid-phase microextraction: Recent developments and future trends. *J. Chromatogr. A* **2007**, *1152*(1–2), 32–40.

7. Mullett, W.M. Determination of drugs in biological fluid by direct injection of samples for liquid-chromatographic analysis. *J. Biochem. Biophys. Methods* **2007**, *70*(2), 263–273.

8. Sellergren, B. *Molecularly Imprinted Polymers. Man-Made Mimics of Antibodies and Their Applications in Analytical Chemistry*, Vol. *23*. Amsterdam: Elsevier Science B.V.; 2001.

9. Widstrand, C.; Boyd, B.; Billing, J.; Rees, A. Efficient extraction of toxic compounds from complex matrices using molecularly imprinted polymers. *Am. Lab.* **2007**, *39*(17), 23–24.

10. Hennion, M.-C. Solid-phase extraction: Method development, sorbents, and coupling with liquid chromatography. *J. Chromatogr. A* **1999**, *856*, 3–54.

11. Nelson, J.O.; Karu, A.E.; Wong, R.B. *Immunoanalysis of Agrochemicals*, Vol. *586*. Washington, DC: American Chemical Society; 1995.

12. Bartsch, R.A.; Maeda, M. *Molecular and Ionic Recognition with Imprinted Polymers*. Washington, DC: Oxford University Press; 1998.

13. Kriz, D.; Ramstrom, O.; Mosbach, K., Molecular imprinting. New possibilities for molecular imprinting. *Anal. Chem.* **1997**, *69*(11), 345A–349A.

14. Barceló, D. *Special Issue on Molecular Imprints and Related Approaches for Solid-Phase Extraction and Sensors in Chemical Analysis*, Vol. *18*. 1999.

15. Sellergren, B. The non-covalent approach to molecular imprinting. In *Molecularly Imprinted Polymers: Man-Made Mimics of Antibodies and Their Applications in Analytical Chemistry*, Sellergren, B., Ed., Vol. *23*. Amsterdam: Elsevier Science B.V.; 2001; pp. 113–184.

16. Sellergren, B.; Lepistoe, M.; Mosbach, K. Highly enantioselective and substrate-selective polymers obtained by interactions. NMR and chromatographic studies on the nature of recognition. *J. Am. Chem. Soc.* **1988**, *110*(17), 5853–5860.

17. Shea, K.J.; Spivak, D.A.; Sellergren, B. Polymeric complements for adenine prepared by molecular imprinting. *J. Am. Chem. Soc.* **1993**, *115*, 3368–3369.

18. Mayes, A.G. Polymerisation techniques for the formation of imprinted beads. In *Molecularly Imprinted Polymers: Man-Made Mimics of Antibodies and Their Applications in Analytical Chemistry*, Sellergren, B., Ed., Vol. *23*. Amsterdam: Elsevier Science B.V.; 2001; pp. 305–324.

19. Ye, L.; Cormack, P.A.G.; Mosbach, K. Molecular imprinting on microgel spheres. *Anal. Chim. Acta* **2001**, *435*(1), 187–196.

20. Sulitzky, C.; Rückert, B.; Hall, A.J.; Lanza, F.; Unger, K.; Sellergren, B. Grafting of molecularly imprinted polymer films on silica supports containing surface-bound free radical initiators. *Macromolecules* **2002**, *35*, 79–91.

21. Lanza, F.; Sellergren, B. Molecularly imprinted polymers via high-throughput and combinatorial techniques. *Macromol. Rapid Commun.* **2004**, *25*(1), 59–68.

22. Quaglia, M.; Chenon, K.; Hall, A.J.; De Lorenzi, E.; Sellergren, B. Target analogue imprinted polymers with affinity for folic acid and related compounds. *J. Am. Chem. Soc.* **2001**, *123*(10), 2146–2154.

23. Dirion, B.; Lanza, F.; Sellergren, B.; Chassaing, C.; Venn, R.; Berggren, C. Selective solid phase extraction of a drug lead compound using molecularly imprinted polymers prepared by the target analogue approach. *Chromatographia* **2002**, *56*, 237–241.

24. Ellwanger, A.; Karlsson, L.; Owens, P.K.; Berggren, C.; Crecenzi, C.; Ensing, K.; Bayoudh, S.; Cormack, P.; Sherrington, D.; Sellergren, B. Evaluation of methods aimed at complete removal of template from molecularly imprinted polymers. *Analyst* **2001**, *126*(6), 784–792.

25. Andersson, L.I.; Paprica, A.; Arvidsson, T. A highly selective solid-phase extraction sorbent for preconcentration of sameridine made by molecular. *Chromatographia* **1997**, *46*(1/2), 57–62.

26. Dzygiel, P.; O'Donnell, E.; Fraier, D.; Chassaing, C.; Cormack, P.A.G. Evaluation of water-compatible molecularly imprinted polymers as solid-phase extraction sorbents for the selective extraction of sildenafil and its desmethyl metabolite from biological samples. *J. Chromatogr. B* **2007**, *853*, 346–353.

27. Köber, R.; Fleischer, C.T.; Lanza, F.; Boos, K.-S.; Sellergren, B.; Barcelo, D. Evaluation of a multidimensional solid phase extraction platform (six-SPE) for highly selective on-line clean-up and high throughput LC-MS analysis of triazines in river water samples using molecularly imprinted polymers (MIPs). *Anal. Chem.* **2001**, *73*, 2437–2444.

28. Karlsson, J.G.; Andersson, L.I.; Nicholls, I.A. Probing the molecular basis for ligand-selective recognition in molecularly imprinted polymers selective for the local anesthetic bupivacaine. *Anal. Chim. Acta* **2001**, *435*(1), 57–64.

29. Manesiotis, P.; Hall, A.J.; Courtois, J.; Irgum, K.; Sellergren, B. An artificial riboflavin receptor prepared by a template analogue imprinting strategy. *Angew. Chem. Int. Ed. Engl.* **2005**, *44*(25), 3902–3906.

30. Hall, A.J.; Emgenbroich, M.; Sellergren, B. Imprinted polymers. In *Templates in Chemistry II*, Schalley, C.A.; Vögtle, F.; Dötz, K.H., Eds., Vol. 249. Berlin: Springer-Verlag; 2005; pp. 317–349.

31. Brambilla, G.; Fiori, M.; Rizzo, B.; Crescenzi, V.; Masci, G. Use of molecularly imprinted polymers in the solid-phase extraction of clenbuterol from animal feeds and biological matrices. *J. Chromatogr. B Biomed. Sci. Appl.* **2001**, *759*(1), 27–32.

32. Fiori, M.; Civitareale, C.; Mirante, S.; Magaro, E.; Brambilla, G. Evaluation of two different clean-up steps, to minimise ion suppression phenomena in ion trap liquid chromatography-tandem mass spectrometry for the multi-residue analysis of beta agonists in calves urine. *Anal. Chim. Acta* **2005**, *529*(1–2), 207–210.

33. Gros, M.; Pizzolato, T.-M.; Petrovic, M.; Lopez de Alda, M.J.; Barcelo, D. Trace level determination of beta-blockers in waste waters by highly selective molecularly imprinted polymers extraction followed by liquid chromatography-quadrupole-linear ion trap mass spectrometry. *J. Chromatogr. A* **2008**, *1189*(1–2), 374–384.

34. Percival, C.; Harrison, T.; Shallcross, D. Catching the cheats: Detecting drugs in sport. *Chem. Rev. (Deddington, United Kingdom)* **2006**, *15*(4), 24–27.

35. Breton, F.; Delepee, R.; Jegourel, D.; Deville-Bonne, D.; Agrofoglio, L.A. Selective adenosine-5′-monophosphate uptake by water-compatible molecularly imprinted polymer. *Anal. Chim. Acta* **2008**, *616*(2), 222–229.

36. Widstrand, C.; Yilmaz, E.; Boyd, B.; Billing, J.; Rees, A. Molecularly imprinted polymers: A new generation of affinity matrices. *Am. Lab. (Shelton, CT, United States)* **2006**, *38*(19), 12–14.

37. Nemoto, K.; Kubo, T.; Nomachi, M.; Sano, T.; Matsumoto, T.; Hosoya, K.; Hattori, T.; Kaya, K. Simple and effective 3D recognition of domoic acid using a molecularly imprinted polymer. *J. Am. Chem. Soc.* **2007**, *129*(44), 13626–13632.

38. Rachkov, A.; Minoura, N. Towards molecularly imprinted polymers selective to peptides and proteins. The epitope approach. *Biochim. Biophys. Acta* **2001**, *1544*(1–2), 255–266.

39. Hall, A.J.; Quaglia, M.; Manesiotis, P.; De Lorenzi, E.; Sellergren, B. Polymeric receptors for the recognition of folic acid and related compounds via substructure imprinting (IV). *Anal. Chem.* **2006**, *78*, 8362–8367.

40. Titirici, M.M.; Hall, A.J.; Sellergren, B. Hierarchical imprinting using crude solid phase peptide synthesis products as templates. *Chem. Mater.* **2003**, *15*(4), 822–824.

41. Nishino, H.; Huang, C.-S.; Shea, K.J. Selective protein capture by epitope imprinting. *Angew. Chem. Int. Ed. Engl.* **2006**, *45*(15), 2392–2396.

42. Emgenbroich, M.; Borrelli, C.; Shinde, S.; Lazraq, I.; Vilela, F.; Hall, A.J.; Oxelbark, J.; De Lorenzi, E.; Courtois, J.; Simanova, A.; Verhage, J.; Irgum, K.; Karim, K.; Sellergren, B. A phosphotyrosine-imprinted polymer receptor for the recognition of tyrosine phosphorylated peptides. *Chem. Eur. J.* **2008**, *14*, 9516–9529.

43. Sellergren, B. Polymer- and template-related factors influencing the efficiency in molecularly imprinted solid-phase extractions. *Trends Anal. Chem.* **1999**, *18*(3), 164–174.

44. Ramström, O.; Andersson, L.I.; Mosbach, K. Recognition sites incorporating both pyridinyl and carboxy functionalities prepared by molecular imprinting. *J. Org. Chem.* **1993**, *58*(26), 7562–7564.

45. Lübke, C.; Lübke, M.; Whitcombe, M.J.; Vulfson, E.N. Imprinted polymers prepared with stoichiometric template-monomer complexes: Efficient binding of ampicillin from aqueous solutions. *Macromolecules* **2000**, *33*, 5098–5105.

46. Dirion, B.; Cobb, Z.; Schillinger, E.; Andersson, L.I.; Sellergren, B. Water-compatible molecularly imprinted polymers obtained via high-throughput synthesis and experimental design. *J. Am. Chem. Soc.* **2003**, *125*(49), 15101–15109.

47. Piletsky, S.A.; Karim, K.; Piletska, E.V.; Day, C.J.; Freebairn, D.W.; Legge, C.; Turner, A.P.F. Recognition of ephedrine enantiomers by molecularly imprinted polymers designed using a computational approach. *Analyst* **2001**, *126*, 1826–1830.

48. Wulff, G.; Knorr, K. Stoichiometric noncovalent interaction in molecular imprinting. *Bioseparation* **2002**, *10*, 257–276.

49. Hall, A.J.; Manesiotis, P.; Emgenbroich, M.; Quaglia, M.; De Lorenzi, E.; Sellergren, B. Urea host monomers for stoichiometric molecular imprinting of oxyanions. *J. Org. Chem.* **2005**, *70*(5), 1732–1736.

50. Van Hoof, N.; Courtheyn, D.; Antignac, J.-P.; Van de Wiele, M.; Poelmans, S.; Noppe, H.; De Brabander, H. Multi-residue liquid chromatography/tandem mass spectrometric analysis of beta-agonists in urine using molecular imprinted polymers. *Rapid Commun. Mass Spectrom.* **2005**, *19*(19), 2801–2808.

51. Sellergren, B. Direct drug determination by selective sample enrichment on an imprinted polymer. *Anal. Chem.* **1994**, *66*, 1578.

52. Walshe, M.; Howarth, J.; Kelly, M.T.; O'Kennedy, R.; Smyth, M.R. The preparation of a molecular imprinted polymer to 7-hydroxycoumarin and its use as a solid-phase extraction material. *J. Pharm. Biomed. Anal.* **1997**, *16*(2), 319–325.

53. Mullett, W.M.; Lai, E.P.C. Determination of theophylline in serum by molecularly imprinted solid-phase extraction with pulsed elution. *Anal. Chem.* **1998**, *70*(17), 3636–3641.

54. Mullett, W.M.; Lai, E.P.C. Molecularly imprinted solid phase extraction micro-column with differential pulsed elution for theophylline determination. *Microchem. J.* **1999**, *61*(2), 143–155.

55. Mullett, W.M.; Dirie, M.F.; Lai, E.P.C.; Guo, H.; He, X. A 2-aminopyridine molecularly imprinted polymer surrogate micro-column for selective solid phase extraction and determination of 4-aminopyridine. *Anal. Chim. Acta* **2000**, *414*(1–2), 123–131.

56. Bereczki, A.; Tolokán, A.; Horvai, G.; Horváth, V.; Lanza, F.; Hall, A.J.; Sellergren, B. Determination of phenytoin in plasma by molecularly imprinted solid phase extraction. *J. Chromatogr.* **2001**, *930*, 31–38.

57. Lai, E.P.C.; Feng, S.Y. Molecularly imprinted solid phase extraction for rapid screening of metformin. *Microchem. J.* **2003**, *75*(3), 159–168.

58. Feng, S.Y.; Lai, E.P.C.; Dabek-Zlotorzynska, E.; Sadeghi, S. Molecularly imprinted solid-phase extraction for the screening of antihyperglycemic biguanides. *J. Chromatogr. A* **2004**, *1027*(1–2), 155–160.

59. Mullett, W.M.; Walles, M.; Levsen, K.; Borlak, J.; Pawliszyn, J. Multidimensional on-line sample preparation of verapamil and its metabolites by a molecularly imprinted polymer coupled to liquid chromatography-mass spectrometry. *J. Chromatogr. B Anal. Technol. Biomed. Life Sci.* **2004**, *801*(2), 297–306.

60. Martin, P.D.; Jones, G.R.; Stringer, F.; Wilson, I.D. Comparison of normal and reversed-phase solid phase extraction methods for extraction of beta-blockers from plasma using molecularly imprinted polymers. *Analyst (Cambridge, United Kingdom)* **2003**, *128*(4), 345–350.

61. Theodoridis, G.; Kantifes, A.; Manesiotis, P.; Raikos, N.; Tsoukali-Papadopoulou, H. Preparation of a molecularly imprinted polymer for the solid-phase extraction of scopolamine with hyoscyamine as a dummy template molecule. *J. Chromatogr. A* **2003**, *987*(1–2), 103–109.

62. van Hout, M.W.J.; Niederlander, H.A.G.; de Zeeuw, R.A.; de Jong, G.J. Ion suppression in the determination of clenbuterol in urine by solid-phase extraction atmospheric pressure chemical ionisation ion-trap mass spectrometry. *Rapid Commun. Mass Spectrom.* **2003**, *17*(3), 245–250.

63. Hu, S.-G.; Wang, S.-W.; He, X.-W. An amobarbital molecularly imprinted microsphere for selective solid-phase extraction of phenobarbital from human urine and medicines and their determination by high-performance liquid chromatography. *Analyst (Cambridge, United Kingdom)* **2003**, *128*(12), 1485–1489.

64. Lai, E.P.C.; Wu, S.G. Molecularly imprinted solid phase extraction for rapid screening of cephalexin in human plasma and serum. *Anal. Chim. Acta* **2003**, *481*(2), 165–174.

65. Bruggemann, O.; Visnjevski, A.; Burch, R.; Patel, P. Selective extraction of antioxidants with molecularly imprinted polymers. *Anal. Chim. Acta* **2004**, *504*(1), 81–88.

66. Caro, E.; Marcé, R.M.; Cormack, P.A.G.; Sherrington, D.C.; Borrull, F. Naproxen bioanalysis. *J. Chromatogr. B* **2004**, *813*, 137.

67. Cobb, Z.; Andersson, L.I. Determination of ropivacaine in human plasma using highly selective molecular imprint-based solid phase extraction and fast LC-MS analysis. *Anal. Bioanal. Chem.* **2005**, *383*(4), 645–650.

68. Abdel-Rehim, M.; Andersson, L.I.; Altun, Z.; Blomberg, L.G. Microextraction in packed syringe online with liquid chromatography-tandem mass spectrometry: Molecularly imprinted polymer as packing material for MEPS in selective extraction of ropivacaine from plasma. *J. Liq. Chromatogr. Relat. Technol.* **2006**, *29*(12), 1725–1736.

69. Cobb, Z.; Sellergren, B.; Andersson, L.I. Water-compatible molecularly imprinted polymers for efficient direct injection on-line solid-phase extraction of ropivacaine and bupivacaine from human plasma. *Analyst* **2007**, *132*(12), 1262–1271.

70. Hu, S.-G.; Li, L.; He, X.-W. Comparison of trimethoprim molecularly imprinted polymers in bulk and in sphere as the sorbent for solid-phase extraction and extraction of trimethoprim from human urine and pharmaceutical tablet and their determination by high-performance liquid chromatography. *Anal. Chim. Acta* **2005**, *537*(1–2), 215–222.

71. Ariffin, M.M.; Miller, E.I.; Cormack, P.A.G.; Anderson, R.A. Molecularly imprinted solid-phase extraction of diazepam and its metabolites from hair samples. *Anal. Chem.* **2007**, *79*(1), 256–262.

72. Anderson, R.A.; Ariffin, M.M.; Cormack, P.A.G.; Miller, E.I. Comparison of molecularly imprinted solid-phase extraction (MISPE) with classical solid-phase extraction (SPE) for the detection of benzodiazepines in post-mortem hair samples. *Forensic Sci. Int.* **2008**, *174*(1), 40–46.

73. Silva, R.G.D.C.; Augusto, F. Sol-gel molecular imprinted ormosil for solid-phase extraction of methylxanthines. *J. Chromatogr. A* **2006**, *1114*(2), 216–223.

74. Beltran, A.; Marce, R.M.; Cormack, P.A.G.; Sherrington, D.C.; Borrull, F. Selective solid-phase extraction of amoxicillin and cephalexin from urine samples using a molecularly imprinted polymer. *J. Sep. Sci.* **2008**, *31*(15), 2868–2874.

75. Rashid, B.A.; Briggs, R.J.; Hay, J.N.; Stevenson, D. MISPE of tamoxifen. *Anal. Commun.* **1997**, *34*, 303–305.

76. Claude, B.; Morin, P.; Bayoudh, S.; de Ceaurriz, J. Interest of molecularly imprinted polymers in the fight against doping. Extraction of tamoxifen and its main metabolite from urine followed by high-performance liquid chromatography with UV detection. *J. Chromatogr. A* **2008**, *1196–1197*, 81–88.

77. Beltran, A.; Caro, E.; Marce, R.M.; Cormack, P.A.G., Sherrington, D.C.; Borrull, F. Synthesis and application of a carbamazepine-imprinted polymer for solid-phase extraction from urine and wastewater. *Anal. Chim. Acta* **2007**, *597*(1), 6–11.

78. Tang, Y.; Huang, Z.; Yang, T.; Hu, X.; Jiang, X. The characteristic and application of molecularly imprinted polymer: Efficient sample preconcentration of antibiotic cefathiamidine from human plasma and serum by solid phase extraction. *Anal. Lett.* **2005**, *38*(2), 219–226.

79. Yin, J.; Wang, S.; Yang, G.; Yang, G.; Chen, Y. Molecularly imprinted solid-phase extraction for rapid screening of mycophenolic acid in human plasma. *J. Chromatogr. B Anal. Technol. Biomed. Life Sci.* **2006**, *844*(1), 142–147.

80. Chapuis, F.; Mullot, J.-U.; Pichon, V.; Tuffal, G.; Hennion, M.-C. Molecularly imprinted polymers for the clean-up of a basic drug from environmental and biological samples. *J. Chromatogr. A* **2006**, *1135*(2), 127–134.

81. Dzygiel, P.; O'Donnell, E.; Fraier, D.; Chassaing, C.; Cormack, P.A.G. Evaluation of water-compatible molecularly imprinted polymers as solid-phase extraction sorbents for the selective extraction of sildenafil and its desmethyl metabolite from plasma samples. *J. Chromatogr. B* **2007**, *853*(1–2), 346–353.

82. Farrington, K.; Regan, F. Investigation of the nature of MIP recognition: The development and characterization of a MIP for ibuprofen. *Biosens. Bioelectron.* **2007**, *22*(6), 1138–1146.

83. Shi, X.; Wu, A.; Qu, G.; Li, R.; Zhang, D. Development and characterisation of molecularly imprinted polymers based on methacrylic acid for selective recognition of drugs. *Biomaterials* **2007**, *28*(25), 3741–3749.

84. Yang, C.; Zhang, Z.; Chen, S.; Yang, F. Molecularly imprinted on-line solid-phase extraction combined with chemiluminescence for the determination of pazufloxacin mesilate. *Microchim. Acta* **2007**, *159*(3–4), 299–304.

85. Mohajeri Seyed, A.; Ebrahimi Soltan, A. Preparation and characterization of a lamotrigine imprinted polymer and its application for drug assay in human serum. *J. Sep. Sci.* **2008**, *31*(20), 3595–3602.

86. Aurand, C.; Santasania, C.T.; Shollenberger, D. Extraction of beta adrenergic blocking agents by molecularly imprinted solid phase extraction. *Abstracts, 39th Middle Atlantic Regional Meeting of the American Chemical Society*. Collegeville, PA; May 16–18, 2007; MARM-264.

87. Fernandez-Llano, L.; Blanco-Lopez, M.C.; Lobo-Castanon, M.J.; Miranda-Ordieres, A.J.; Tunon-Blanco, P. Determination of diclofenac in urine samples by molecularly-imprinted solid-phase extraction and adsorptive differential pulse voltammetry. *Electroanalysis* **2007**, *19*(15), 1555–1561.

88. Chen, S.; Zhang, Z. Molecularly imprinted solid-phase extraction combined with electrochemical oxidation fluorimetry for the determination of methotrexate in human serum and urine. *Spectrochim. Acta A* **2008**, *70A*(1), 36–41.

89. Pilau, E.J.; Silva, R.G.C.; Jardim, I.C.F.S.; Augusto, F. Molecularly imprinted sol-gel silica for solid phase extraction of phenobarbital. *J. Braz. Chem. Soc.* **2008**, *19*(6), 1136–1143.

90. Rezaei, B.; Mallakpour, S.; Rahmanian, O. A selective solid-phase extraction and preconcentration method with using molecularly imprinted polymer for piroxicam in pharmaceutical sample. *Anal. Lett.* **2008**, *41*(10), 1818–1831.

91. Widstrand, C.; Kronauer, S.; Bjoerk, H.; Wihlborg, A.-K.; Trinh, A. The simultaneous class-selective extraction and analysis of beta-blockers and beta-agonists using molecularly imprinted polymer SPE. *Am. Lab.* **2008**, *40*(9), 6–7.

92. Yan, H.; Row, K.H. Novel molecularly imprinted monolithic column for selective on-line extraction of ciprofloxacin from human urine. *Biomed. Chromatogr.* **2008**, *22*(5), 487–493.

93. Xia, Y.; McGuffey, J.E.; Bhattacharyya, S.; Sellergren, B.; Yilmaz, E.; Wang, L.; Bernert, J.T. Analysis of the tobacco-specific nitrosamine 4-(methylnitrosamino)-1-(3-pyridyl)-1-butanol in urine by extraction on a molecularly imprinted polymer column and liquid chromatography/atmospheric pressure ionization tandem mass spectrometry. *Anal. Chem.* **2005**, *77*(23), 7639–7645.

94. Yang, J.; Hu, Y.; Cai, J.-B.; Zhu, X.-L.; Su, Q.-D. A new molecularly imprinted polymer for selective extraction of cotinine from urine samples by solid-phase extraction. *Anal. Bioanal. Chem.* **2006**, *384*(3), 761–768.

95. Yang, J.; Hu, Y.; Cai, J.B.; Zhu, X.L.; Su, Q.D.; Hu, Y.Q.; Liang, F.X. Selective hair analysis of nicotine by molecular imprinted solid-phase extraction: An application for evaluating tobacco smoke exposure. *Food Chem. Toxicol.* **2007**, *45*(6), 896–903.

96. Jegourel, D.; Delepee, R.; Breton, F.; Rolland, A.; Vidal, R.; Agrofoglio, L.A. Molecularly imprinted polymer of 5-methyluridine for solid-phase extraction of pyrimidine nucleoside cancer markers in urine. *Bioorg. Med. Chem.* **2008**, *16*(19), 8932–8939.

97. Zurutuza, A.; Bayoudh, S.; Cormack, P.A.G.; Dambies, L.; Deere, J.; Bischoff, R.; Sherrington, D.C. Molecularly imprinted solid-phase extraction of cocaine metabolites from aqueous samples. *Anal. Chim. Acta* **2005**, *542*(1), 14–19.

98. Prasad, B.B.; Sharma, P.S.; Lakshmi, D. Molecularly imprinted polymer-based solid-phase extraction combined with molecularly imprinted polymer-based sensor for detection of uric acid. *J. Chromatogr. A* **2007**, *1173*(1–2), 18–26.

99. Puoci, F.; Garreffa, C.; Curcio, M.; Iemma, F.; Bloise, E.; Picci, N. Determination using molecularly imprinted solid phase extraction (MISPE) of derivatized homocysteine. *Macromolecules (Rajkot, India)* **2007**, *3*(1), 31–36.

100. Sharma, P.S.; Lakshmi, D.; Prasad, B.B. Molecularly imprinted solid-phase extraction combined with molecularly imprinted polymer-sensor: A diagnostic tool applicable to creatine deficiency syndrome. *Biomed. Chromatogr.* **2007**, *21*(9), 976–986.

101. Sharma, P.S.; Lakshmi, D.; Prasad, B.B. Highly sensitive and selective detection of creatinine by combined use of MISPE and a complementary MIP-sensor. *Chromatographia* **2007**, *65*(7/8), 419–427.

102. Lulinski, P.; Maciejewska, D.; Bamburowicz-Klimkowska, M.; Szutowski, M. Dopamine-imprinted polymers: Template-monomer interactions, analysis of template removal and application to solid phase extraction. *Molecules* **2007**, *12*(11), 2434–2449.

103. Maury, D.; Couderc, F.; Garrigues, J.C.; Poinsot, V. Synthesis and evaluation of new lipomonosaccharide-imprinted polymers as MISPE supports. *Talanta* **2007**, *73*(2), 340–345.

104. Puoci, F.; Cirillo, G.; Curcio, M.; Iemma, F.; Spizzirri, U.G.; Picci, N. Molecularly imprinted solid phase extraction for the selective HPLC determination of alpha-tocopherol in bay leaves. *Anal. Chim. Acta* **2007**, *593*(2), 164–170.

105. Moeller, K.; Crescenzi, C.; Nilsson, U. Determination of a flame retardant hydrolysis product in human urine by SPE and LC-MS. Comparison of molecularly imprinted solid-phase extraction with a mixed-mode anion exchanger. *Anal. Bioanal. Chem.* **2004**, *378*(1), 197–204.

106. Zi-Hui, M.; Qin, L. Determination of degradation products of nerve agents in human solid phase extraction using molecularly imprinted polymer. *Anal. Chim. Acta* **2001**, *435*(1), 121–127.

107. Xie, J.; Chen, L.; Li, C.; Xu, X. Selective extraction of functional components derived from herb in plasma by using a molecularly imprinted polymer based on 2,2-bis(hydroxymethyl)butanol trimethacrylate. *J. Chromatogr. B* **2003**, *788*(2), 233–242.

108. Yang, G.; Liu, H.; Wang, M.; Liu, S.; Chen, Y. Chromatographic characterization and solid-phase extraction on diniconazole-imprinted polymers stationary phase. *React. Funct. Polym.* **2006**, *66*(5), 579–583.

109. Zhu, L.; Xu, X. Selective separation of active inhibitors of epidermal growth factor receptor from *Caragana jubata* by molecularly imprinted solid-phase extraction. *J. Chromatogr. A* **2003**, *991*(2), 151–158.

110. Lin, L.-Q.; Zhang, J.; Fu, Q.; He, L.-C.; Li, Y.-C. Concentration and extraction of sinomenine from herb and plasma using a molecularly imprinted polymer as the stationary phase. *Anal. Chim. Acta* **2006**, *561*(1–2), 178–182.

111. Zhuang, X.; Dong, X.; Ma, S.; Zhang, T. Selective on-line extraction of trans-resveratrol and emodin from *Polygonum cuspidatum* using molecularly imprinted polymer. *J. Chromatogr. Sci.* **2008**, *46*(8), 739–742.

112. Dong, X.; Wang, W.; Ma, S.; Sun, H.; Li, Y.; Guo, J. Molecularly imprinted solid-phase extraction of (-)-ephedrine from Chinese Ephedra. *J. Chromatogr. A* **2005**, *1070*(1–2), 125–130.

113. Karasova, G.; Lehotay, J.; Klodzinska, E.; Gadzala-Kopciuch, R.; Buszewski, B. Comparison of several extraction methods for the isolation of benzoic acid derivatives from *Melissa officinalis*. *J. Liq. Chromatogr. Relat. Technol.* **2006**, *29*(11), 1633–1644.

114. Claude, B.; Morin, P.; Lafosse, M.; Belmont, A.-S.; Haupt, K., Selective solid-phase extraction of a triterpene acid from a plant extract by molecularly imprinted polymer. *Talanta* **2008**, *75*(2), 344–350.

115. Liu, H.; Liu, C.; Yang, X.; Zeng, S.; Xiong, Y.; Xu, W. Solid-phase extraction of ursolic acid from herb using beta-cyclodextrin-based molecularly imprinted microspheres. *J. Sep. Sci.* **2008**, *31*(20), 3573–3580.

116. Matsui, J.; Okada, M.; Tsuruoka, M.; Takeuchi, T. Solid-phase extraction of a triazine herbicide using a molecularly imprinted synthetic receptor. *Anal. Commun.* **1997**, *34*(3), 85–87.

117. Ferrer, I.; Lanza, F.; Tolokan, A.; Horvath, V.; Sellergren, B.; Horvai, G.; Barcelo, D. Selective trace enrichment of chlorotriazine pesticides from waters and sediment samples using terbuthylazine. *Anal. Chem.* **2000**, *72*(16), 3934–3941.

118. Chapuis, F.; Pichon, V.; Lanza, F.; Sellergren, S.; Hennion, M.C. Optimization of the class-selective extraction of triazines from aqueous samples using a molecularly imprinted polymer by a comprehensive approach of the retention mechanism. *J. Chromatogr. A* **2003**, *999*(1–2), 23–33.

119. Pap, T.; Horvath, V.; Tolokan, A.; Horvai, G.; Sellergren, B. Effect of solvents on the selectivity of terbutylazine imprinted polymer sorbents used in solid-phase extraction. *J. Chromatogr. A* **2002**, *973*(1–2), 1–12.

120. Turiel, E.; Martin-Esteban, A.; Fernandez, P.; Perez-Conde, C.; Camara, C. Molecular recognition in a propazine-imprinted polymer and its application to the determination of triazines in environmental samples. *Anal. Chem.* **2001**, *73*(21), 5133–5141.

121. Chapuis, F.; Pichon, V.; Lanza, F.; Sellergren, B.; Hennion, M.C. Retention mechanism of analytes in the solid-phase extraction process using molecularly imprinted polymers. Application to the extraction of triazines from complex matrices. *J. Chromatogr. B* **2004**, *804*, 93–101.

122. Zhu, Q.-Z.; Degelmann, P.; Niessner, R.; Knopp, D., Selective trace analysis of sulfonylurea herbicides in water and soil samples based on solid-phase extraction using a molecularly imprinted polymer. *Environ. Sci. Technol.* **2002**, *36*(24), 5411–5420.

123. Dong, X.; Wang, N.; Wang, S.; Zhang, X.; Fan, Z. Synthesis and application of molecularly imprinted polymer on selective solid-phase extraction for the determination of monosulfuron residue in soil. *J. Chromatogr. A* **2004**, *1057*, 13–19.

124. Baggiani, C.; Giovannoli, C.; Anfossi, L.; Tozzi, C. MISPE chlorinated phenoxyacids. *J. Chromatogr. A* **2001**, *938*, 35.

125. Baggiani, C.; Giovannoli, C. Molecular imprinted solid-phase extraction for cleanup of chlorinated phenoxyacids from aqueous samples. *Methods Biotechnol.* **2006**, *19*, (Pesticide Protocols), 421–433.

126. Breton, F.; Euzet, P.; Piletsky, S.A.; Giardi, M.T.; Rouillon, R. Integration of photosynthetic biosensor with molecularly imprinted polymer-based solid phase extraction cartridge. *Anal. Chim. Acta* **2006**, *569*(1–2), 50–57.

127. Han, D.-M.; Fang, G.-Z.; Yan, X.-P. Preparation and evaluation of a molecularly imprinted sol-gel material for on-line solid-phase extraction coupled with high performance liquid chromatography for the determination of trace pentachlorophenol in water samples. *J. Chromatogr. A* **2005**, *1100*(2), 131–136.

128. Zhu, X.; Yang, J.; Su, Q.; Cai, J.; Gao, Y., Selective solid-phase extraction using molecularly imprinted polymer for the analysis of polar organophosphorus pesticides in water and soil samples. *J. Chromatogr. A* **2005**, *1092*(2), 161–169.

129. Sambe, H.; Hoshina, K.; Haginaka, J. Molecularly imprinted polymers for triazine herbicides prepared by multi-step swelling and polymerization method. Their application to the determination of methylthiotriazine herbicides in river water. *J. Chromatogr. A* **2007**, *1152*(1–2), 130–137.

130. Zhu, X.; Cai, J.; Yang, J.; Su, Q.; Gao, Y. Films coated with molecular imprinted polymers for the selective stir bar sorption extraction of monocrotophos. *J. Chromatogr. A* **2006**, *1131*(1–2), 37–44.

131. Zhu, X.; Su, Q.; Cai, J.; Yang, J.; Gao, Y. Molecularly imprinted polymer membranes for substance-selective solid-phase extraction from aqueous solutions. *J. Appl. Polym. Sci.* **2006**, *101*(6), 4468–4473.

132. Liang, F.; Zhu, X.; Yang, J.; Su, Q. Accelerated solvent extraction and matrix solid phase dispersion using molecularly imprinted polymer for the analysis of monocrotophos in soil. *Asian J. Chem.* **2008**, *20*(5), 3954–3960.

133. Guo, L.; Jiang, X.; Yang, C.; Zhang, H. Analysis of sulfamerazine in pond water and several fishes by high-performance liquid chromatography using molecularly imprinted solid-phase extraction. *Anal. Bioanal. Chem.* **2008**, *391*(6), 2291–2298.

134. Guzzella, L.; Pozzoni, F.; Baggiani, C. Synthesis and characterization of a propazine imprinted polymer for the extraction of triazines herbicides. *Water Sci. Technol.* **2008**, *57*(1), 139–144.

135. Verma, K.S.; Xia, K. Application of molecularly imprinted polymers to selective extraction of triclosan and triclocarban from water, soil, and biosolids for analysis on HPLC/UV. *Abstracts, 60th Southeast Regional Meeting of the American Chemical Society*. Nashville, TN; November 12–15, 2008; SERM-660.

136. Urraca, J.L.; Moreno-Bondi, M.C.; Hall, A.J.; Sellergren, B. Direct extraction of penicillin G and derivatives from aqueous samples using a stoichiometrically imprinted polymer. *Anal. Chem.* **2007**, *79*(2), 695–701.

137. Turiel, E.; Martin-Esteban, A.; Tadeo Jose, L. Molecular imprinting-based separation methods for selective analysis of fluoroquinolones in soils. *J. Chromatogr. A* **2007**, *1172*(2), 97–104.

138. Benito-Pena, E.; Urraca, J.L.; Sellergren, B.; Moreno-Bondi, M.C. Solid-phase extraction of fluoroquinolones from aqueous samples using a water-compatible stoichiometrically imprinted polymer. *J. Chromatogr. A* **2008**, *1208*(1–2), 62–70.

139. Sun, Z.; Schuessler, W.; Sengl, M.; Niessner, R.; Knopp, D. Selective trace analysis of diclofenac in surface and wastewater samples using solid-phase extraction with a new molecularly imprinted polymer. *Anal. Chim. Acta* **2008**, *620*(1–2), 73–81.

140. Yang, T.; Li, Y.-H.; Wei, S.; Li, Y.; Deng, A. Development of a selective molecularly imprinted polymer-based solid-phase extraction for indomethacin from water samples. *Anal. Bioanal. Chem.* **2008**, *391*(8), 2905–2914.

141. Zorita, S.; Boyd, B.; Joensson, S.; Yilmaz, E.; Svensson, C.; Mathiasson, L.; Bergstroem, S. Selective determination of acidic pharmaceuticals in wastewater using molecularly imprinted solid-phase extraction. *Anal. Chim. Acta* **2008**, *626*(2), 147–154.

142. Gallego-Gallegos, M.; Liva, M.; Olivas, R.M.; Camara, C. Focused ultrasound and molecularly imprinted polymers: A new approach to organotin analysis in environmental samples. *J. Chromatogr. A* **2006**, *1114*(1), 82–88.

143. Garcia, R.; Vigneau, O.; Pinel, C.; Lemaire, M. Solid-liquid lanthanide extraction with ionic-imprinted polymers. *Sep. Sci. Technol.* **2002**, *37*(12), 2839–2857.

144. Ersoz, A.; Say, R.; Denizli, A. Ni(II) ion-imprinted solid-phase extraction and preconcentration in aqueous solutions by packed bed columns. *Anal. Chim. Acta* **2004**, *502*(1), 91–97.

145. Jiang, N.; Chang, X.; Zheng, H.; He, Q.; Hu, Z. Selective solid-phase extraction of nickel(II) using a surface-imprinted silica gel sorbent. *Anal. Chim. Acta* **2006**, *577*(2), 225–231.

146. Birlik, E.; Buyuktiryaki, S.; Ersoz, A.; Denizli, A.; Say, R. Selective separation of thorium using ion imprinted chitosan-phthalate particles via solid phase extraction. *Sep. Sci. Technol.* **2006**, *41*(14), 3109–3121.

147. Chauvin, A.-S.; Bunzli, J.-C.G.; Bochud, F.; Scopelliti, R.; Froidevaux, P. Use of dipicolinate-based complexes for producing ion-imprinted polystyrene resins for the extraction of yttrium-90 and heavy lanthanide cations. *Chem. Eur. J.* **2006**, *12*(26), 6852–6864.

148. Birlik, E.; Ersoez, A.; Denizli, A.; Say, R. Preconcentration of copper using double-imprinted polymer via solid phase extraction. *Anal. Chim. Acta* **2006**, *565*(2), 145–151.

149. Wang, S.; Zhang, R. Selective solid-phase extraction of trace copper ions in aqueous solution with a Cu(II)-imprinted interpenetrating polymer network gel prepared by ionic imprinted polymer (IIP) technique. *Microchim. Acta* **2006**, *154*(1–2), 73–80.

150. Froidevaux, P.; Happel, S.; Chauvin, A.-S. Ion-imprinted polymer concept for selective extraction of 90Y and 152Eu for medical applications and nuclear power plant monitoring. *Chimia* **2006**, *60*(4), 203–206.

151. Ramakrishnan, K.; Prasada Rao, T. Ion imprinted polymer solid phase extraction (IIP-SPE) for preconcentrative separation of erbium(III) from adjacent lanthanides and yttrium. *Sep. Sci. Technol.* **2006**, *41*(2), 233–246.

152. Fang, G.-Z.; Tan, J.; Yan, X.-P. An ion-imprinted functionalized silica gel sorbent prepared by a surface imprinting technique combined with a sol-gel process for selective solid-phase extraction of cadmium(II). *Anal. Chem.* **2005**, *77*(6), 1734–1739.

153. Chang, X.; Jiang, N.; Zheng, H.; He, Q.; Hu, Z.; Zhai, Y.; Cui, Y. Solid-phase extraction of iron(III) with an ion-imprinted functionalized silica gel sorbent prepared by a surface imprinting technique. *Talanta* **2007**, *71*(1), 38–43.

154. Chang, X.; Wang, X.; Jiang, N.; He, Q.; Zhai, Y.; Zhu, X.; Hu, Z. Silica gel surface-imprinted solid-phase extraction of Zr(IV) from aqueous solutions. *Microchim. Acta* **2008**, *162*(1–2), 113–119.

155. Otero-Romani, J.; Moreda-Pineiro, A.; Bermejo-Barrera, P.; Martin-Esteban, A. Synthesis, characterization and evaluation of ionic-imprinted polymers for solid-phase extraction of nickel from seawater. *Anal. Chim. Acta* **2008**, *630*(1), 1–9.

156. Shirvani-Arani, S.; Ahmadi, S.J.; Bahrami-Samani, A.; Ghannadi-Maragheh, M. Synthesis of nano-pore samarium (III)-imprinted polymer for preconcentrative separation of samarium ions from other lanthanide ions via solid phase extraction. *Anal. Chim. Acta* **2008**, *623*(1), 82–88.

157. Masqué, N.; Marcé, R.M.; Borrull, F.; Cormack, P.A.G.; Sherrington, D.C. *Anal. Chem.* **2000**, *72*, 4122.

158. Luo, W.; Zhu, L.; Yu, C.; Tang, H.; Yu, H.; Li, X.; Zhang, X. Synthesis of surface molecularly imprinted silica micro-particles in aqueous solution and the usage for selective off-line solid-phase extraction of 2,4-dinitrophenol from water matrixes. *Anal. Chim. Acta* **2008**, *618*(2), 147–156.

159. Tarley, C.R.T.; Kubota, L.T. Molecularly-imprinted solid phase extraction of catechol from aqueous effluents for its selective determination by differential pulse voltammetry. *Anal. Chim. Acta* **2005**, *548*(1–2), 11–19.

160. Bravo, J.C.; Garcinuno, R.M.; Fernandez, P.; Durand, J.S. A new molecularly imprinted polymer for the on-column solid-phase extraction of diethylstilbestrol from aqueous samples. *Anal. Bioanal. Chem.* **2007**, *388*(5–6), 1039–1045.

161. Sanbe, H.; Hosoya, K.; Haginaka, J. Preparation of uniformly sized molecularly imprinted polymers for phenolic compounds and their application to the assay of bisphenol A in river water. *Anal. Sci.* **2003**, *19*(5), 715–719.

162. Kawaguchi, M.; Hayatsu, Y.; Nakata, H.; Ishii, Y.; Ito, R.; Saito, K.; Nakazawa, H. Molecularly imprinted solid phase extraction using stable isotope labeled compounds as template and liquid chromatography-mass spectrometry for trace analysis of bisphenol A in water sample. *Anal. Chim. Acta* **2005**, *539*(1–2), 83–89.

163. Jiang, M.; Zhang, J.-H.; Mei, S.-R.; Shi, Y.; Zou, L.-J.; Zhu, Y.-X.; Dai, K.; Lu, B. Direct enrichment and high performance liquid chromatography analysis of ultra-trace bisphenol A in water samples with narrowly dispersible bisphenol A imprinted polymeric microspheres column. *J. Chromatogr. A* **2006**, *1110*(1–2), 27–34.

164. Martin-Esteban, A.; Tadeo, J.L. Selective molecularly imprinted polymer obtained from a combinatorial library for the extraction of bisphenol A. *Comb. Chem. High Throughput Screen.* **2006**, *9*(10), 747–751.

165. Nunez, L.; Turiel, E.; Martin-Esteban, A.; Tadeo, J.L. Molecularly imprinted polymer for selective extraction of endocrine disrupters nonylphenol and its ethoxylated derivates from environmental solids. *J. Sep. Sci.* **2008**, *31*(13), 2492–2499.

166. Guerreiro, A.; Soares, A.; Piletska, E.; Mattiasson, B.; Piletsky, S. Preliminary evaluation of new polymer matrix for solid-phase extraction of nonylphenol from water samples. *Anal. Chim. Acta* **2008**, *612*(1), 99–104.

167. Wang, S.; Xu, Z.; Fang, G.; Zhang, Y.; He, J. Separation and determination of estrone in environmental and drinking water using molecularly imprinted solid phase extraction coupled with HPLC. *J. Sep. Sci.* **2008**, *31*(6–7), 1181–1188.

168. Hu, S.-G.; Li, L.; He, X.-W. Solid-phase extraction of esculetin from the ash bark of Chinese traditional medicine by using molecularly imprinted polymers. *J. Chromatogr. A* **2005**, *1062*(1), 31–37.

169. Chianella, I.; Piletsky, S.A.; Tothill, I.E.; Chen, B.; Turner, A.P.F. MIP-based solid phase extraction cartridges combined with MIP-based sensors for the detection of microcystin-LR. *Biosens. Bioelectron.* **2003**, *18*(2–3), 119–127.

170. Kugimiya, A.; Takei, H. Selective recovery of phosphate from river water using molecularly imprinted polymers. *Anal. Lett.* **2008**, *41*(2), 302–311.

171. Le Moullec, S.; Begos, A.; Pichon, V.; Bellier, B. Selective extraction of organophosphorus nerve agent degradation products by molecularly imprinted solid-phase extraction. *J. Chromatogr. A* **2006**, *1108*(1), 7–13.

172. Muldoon, M.T.; Stanker, L.H. Molecularly imprinted solid phase extraction of atrazine from beef liver extracts. *Anal. Chem.* **1997**, *69*(5), 803–808.

173. Turiel, E.; Tadeo, J.L.; Cormack, P.A.G.; Martin-Esteban, A. HPLC imprinted-stationary phase prepared by precipitation polymerisation for the determination of thiabendazole in fruit. *Analyst* **2005**, *130*(12), 1601–1607.

174. Bjarnason, B.; Chimuka, L.; Ramstroem, O. Online solid-phase extraction of triazine herbicides using a molecularly-imprinted polymer for selective sample enrichment. *Anal. Chem.* **1999**, *71*(11), 2152–2156.

175. Cacho, C.; Turiel, E.; Martin-Esteban, A.; Perez-Conde, C.; Camara, C. Clean-up of triazines in vegetable extracts by molecularly-imprinted solid-phase extraction using a propazine-imprinted polymer. *Anal. Bioanal. Chem.* **2003**, *376*(4), 491–496.

176. Cacho, C.; Turiel, E.; Martin-Esteban, A.; Ayala, D.; Perez-Conde, C. Semi-covalent imprinted polymer using propazine methacrylate as template molecule for the clean-up of triazines in soil and vegetable samples. *J. Chromatogr. A* **2006**, *1114*(2), 255–262.

177. Tamayo, F.G.; Casillas, J.L.; Martin-Esteban, A. Highly selective fenuron-imprinted polymer with a homogeneous binding site distribution prepared by precipitation polymerisation and its application to the clean-up of fenuron in plant samples. *Anal. Chim. Acta* **2003**, *482*(2), 165–173.

178. Tamayo, F.G.; Casillas, J.L.; Martin-Esteban, A. Clean up of phenylurea herbicides in plant sample extracts using molecularly imprinted polymers. *Anal. Bioanal. Chem.* **2005**, *381*(6), 1234–1240.

179. Tamayo, F.G.; Casillas, J.L.; Martin-Esteban, A. Evaluation of new selective molecularly imprinted polymers prepared by precipitation polymerization for the extraction of phenylurea herbicides. *J. Chromatogr. A* **2005**, *1069*(2), 173–181.

180. Baggiani, C.; Baravalle, P.; Giraudi, G.; Tozzi, C. Molecularly imprinted solid-phase extraction method for the high-performance liquid chromatographic analysis of fungicide pyrimethanil in wine. *J. Chromatogr. A* **2007**, *1141*(2), 158–164.

181. Hu, M.-L.; Jiang, M.; Wang, P.; Mei, S.-R.; Lin, Y.-F.; Hu, X.-Z.; Shi, Y.; Lu, B.; Dai, K. Selective solid-phase extraction of tebuconazole in biological and environmental samples using molecularly imprinted polymers. *Anal. Bioanal. Chem.* **2007**, *387*(3), 1007–1016.

182. He, J.; Wang, S.; Fang, G.; Zhu, H.; Zhang, Y. Molecularly imprinted polymer online solid-phase extraction coupled with high-performance liquid chromatography-UV for the determination of three sulfonamides in pork and chicken. *J. Agric. Food Chem.* **2008**, *56*(9), 2919–2925.

183. Song, S.; Shi, X.; Li, R.; Lin, Z.; Wu, A.; Zhang, D. Extraction of chlorpromazine with a new molecularly imprinted polymer from pig urine. *Process Biochem.* **2008**, *43*(11), 1209–1214.

184. Tang, K.; Chen, S.; Gu, X.; Wang, H.; Dai, J.; Tang, J. Preparation of molecularly imprinted solid phase extraction using bensulfuron-methyl imprinted polymer and clean-up for the sulfonylurea-herbicides in soybean. *Anal. Chim. Acta* **2008**, *614*(1), 112–118.

185. Li, Y.-H.; Yang, T.; Qi, X.-L.; Qiao, Y.-W.; Deng, A.-P. Development of a group selective molecularly imprinted polymers based solid phase extraction of malachite green from fish water and fish feed samples. *Anal. Chim. Acta* **2008**, *624*(2), 317–325.

186. Puoci, F.; Garreffa, C.; Iemma, F.; Muzzalupo, R.; Spizzirri, U.G.; Picci, N. Molecularly imprinted solid phase extraction for detection of sudan I in food matrices. *Food Chem.* **2005**, *93*(2), 349–353.

187. Caro, E.; Marce, R.M.; Cormack, P.A.G.; Sherrington, D.C.; Borrull, F. Novel enrofloxacin imprinted polymer applied to the solid-phase extraction of fluorinated quinolones from urine and tissue samples. *Anal. Chim. Acta* **2006**, *562*(2), 145–151.

188. Yan, H.; Qiao, F.; Row Kyung, H. Molecularly imprinted-matrix solid-phase dispersion for selective extraction of five fluoroquinolones in eggs and tissue. *Anal. Chem.* **2007**, *79*(21), 8242–8248.

189. Mohamed, R.; Richoz-Payot, J.; Gremaud, E.; Mottier, P.; Yilmaz, E.; Tabet, J.-C.; Guy Philippe, A. Advantages of molecularly imprinted polymers LC-ESI-MS/MS for the selective extraction and quantification of chloramphenicol in milk-based matrixes. Comparison with a classical sample preparation. *Anal. Chem.* **2007**, *79*(24), 9557–9565.

190. Boyd, B.; Bjoerk, H.; Billing, J.; Shimelis, O.; Axelsson, S.; Leonora, M.; Yilmaz, E. Development of an improved method for trace analysis of chloramphenicol using molecularly imprinted polymers. *J. Chromatogr. A* **2007**, *1174*(1–2), 63–71.

191. Xiong, Y.; Zhou, H.; Zhang, Z.; He, D.; He, C. Molecularly imprinted on-line solid-phase extraction combined with flow-injection chemiluminescence for the determination of tetracycline. *Analyst* **2006**, *131*(7), 829–834.

192. Song, S.; Wu, A.; Shi, X.; Li, R.; Lin, Z.; Zhang, D. Development and application of molecularly imprinted polymers as solid-phase sorbents for erythromycin extraction. *Anal. Bioanal. Chem.* **2008**, *390*(8), 2141–2150.

193. Wihlborg, A.-K.; Boyd, B.; Kronauer, S.; Widstrand, C.; Trinh, A. The highly selective extraction of chloramphenicol from shrimp using molecularly imprinted polymer solid-phase extraction. *Am. Lab.* **2008**, *40*(13), 6–7.

194. Yan, H.; Row, K.H. Molecularly imprinted solid-phase extraction for determination of enrofloxacin and ciprofloxacin in chicken muscle. *Bull. Korean Chem. Soc.* **2008**, *29*(6), 1173–1178.

195. Yan, H.; Tian, M.; Row, K.H. Determination of enrofloxacin and ciprofloxacin in milk using molecularly imprinted solid-phase extraction. *J. Sep. Sci.* **2008**, *31*(16–17), 3015–3020.

196. Lu, Y.; Zhao, N.; Qin, X.; Liu, Y.; Lu, G. Preparation and application of the sacrificial mesoporous silica imprinted polymers for the selective solid-phase extraction of ofloxacin residues in chicken. *Chem. J. Internet* **2008**, *10*(6).

197. Mohamed, R.; Mottier, P.; Treguier, L.; Richoz-Payot, J.; Yilmaz, E.; Tabet, J.-C.; Guy, P.A. Use of molecularly imprinted solid-phase extraction sorbent for the determination of four 5-nitroimidazoles and three of their metabolites from egg-based samples before tandem LC-ESIMS/MS analysis. *J. Agric. Food Chem.* **2008**, *56*(10), 3500–3508.

198. Crescenzi, C.; Bayoudh, S.; Cormack, P.A.G.; Klein, T.; Ensing, K. Determination of clenbuterol in bovine liver by combining matrix solid-

phase dispersion and molecularly imprinted solid-phase extraction followed by liquid chromatography/electrospray ion trap multiple-stage mass spectrometry. *Anal. Chem.* **2001**, *73*(10), 2171–2177.

199. Shimelis, O.; Aurand, C.; Trinh, A. The selective extraction of clenbuterol using molecularly imprinted polymer SPE. *LC-GC North Am.* **2007**, (Suppl.), 19.

200. Kootstra, P.R.; Kuijpers, C.J.P.F.; Wubs, K.L.; van Doorn, D.; Sterk, S.S.; van Ginkel, L.A.; Stephany, R.W. The analysis of beta-agonists in bovine muscle using molecular imprinted polymers with ion trap LCMS screening. *Anal. Chim. Acta* **2005**, *529*(1–2), 75–81.

201. Urraca Javier, L.; Marazuela Maria, D.; Moreno-Bondi Maria, C. Molecularly imprinted polymers applied to the clean-up of zearalenone and alpha-zearalenol from cereal and swine feed sample extracts. *Anal. Bioanal. Chem.* **2006**, *385*(7), 1155–1161.

202. Zhou, S.N.; Lai, E.P.C.; Miller, J.D. Ochratoxin by MISPE PE. *Anal. Bioanal. Chem.* **2004**, *378*, 1903.

203. Maier, N.M.; Buttinger, G.; Welhartizki, S.; Gavioli, E.; Linder, W. Ochratoxin in wine by MISPE. *J. Chromatogr. B* **2004**, *804*, 104.

204. Pascale, M.; De Girolamo, A.; Visconti, A.; Magan, N.; Chianella, I.; Piletska, E.V.; Piletsky, S.A. Use of itaconic acid-based polymers for solid-phase extraction of deoxynivalenol and application to pasta analysis. *Anal. Chim. Acta* **2008**, *609*(2), 131–138.

205. Zander, A.; Findlay, P.; Renner, T.; Sellergren, B.; Swietlow, A. Analysis of nicotine and its oxidation products in nicotine chewing gum by a molecularly imprinted solid-phase extraction. *Anal. Chem.* **1998**, *70*(15), 3304–3314.

206. Zhang, J.-H.; Jiang, M.; Zou, L.; Shi, D.; Mei, S.-R.; Zhu, Y.-X.; Shi, Y.; Dai, K.; Lu, B. Selective solid-phase extraction of bisphenol A using molecularly imprinted polymers and its application to biological and environmental samples. *Anal. Bioanal. Chem.* **2006**, *385*(4), 780–786.

207. Jiang, X.; Zhao, C.; Jiang, N.; Zhang, H.; Liu, M. Selective solid-phase extraction using molecular imprinted polymer for the analysis of diethylstilbestrol. *Food Chem.* **2008**, *108*(3), 1061–1067.

208. Lai, J.P.; Niessner, R.; Knopp, D. Benzapyrene. *Anal. Chim. Acta* **2004**, *522*, 137.

209. Jin, Y.; Row, K.H. Molecularly imprinted solid-phase extraction of caffeine from green tea. *J. Ind. Eng. Chem. (Seoul, Republic of Korea)* **2006**, *12*(3), 494–499.

210. Shi, Y.; Zhang, J.-H.; Shi, D.; Jiang, M.; Zhu, Y.-X.; Mei, S.-R.; Zhou, Y.-K.; Dai, K.; Lu, B. Selective solid-phase extraction of cholesterol using molecularly imprinted polymers and its application in different biological samples. *J. Pharm. Biomed. Anal.* **2006**, *42*(5), 549–555.

211. Puoci, F.; Curcio, M.; Cirillo, G.; Iemma, F.; Spizzirri, U.G.; Picci, N. Molecularly imprinted solid-phase extraction for cholesterol determination in cheese products. *Food Chem.* **2007**, *106*(2), 836–842.

212. Ding, L.; Li, H.; Tang, F.; Yao, S. Molecularly imprinted solid-phase extraction of epicatechin from tea beverage. *Anal. Lett.* **2006**, *39*(12), 2373–2385.

213. Theodoridis, G.; Lasakova, M.; Skerikova, V.; Tegou, A.; Giantsiou, N.; Jandera, P. Molecular imprinting of natural flavonoid antioxidants: Application in solid-phase extraction for the sample pretreatment of natural products prior to HPLC analysis. *J. Sep. Sci.* **2006**, *29*(15), 2310–2321.

214. Jin, Y.; Row, K.H. Solid-phase extraction of caffeine and catechin compounds from green tea by caffeine molecular imprinted polymer. *Bull. Korean Chem. Soc.* **2007**, *28*(2), 276–280.

215. Zhao, J.; Han, B.; Zhang, Y.; Wang, D. Synthesis of Zn(II) ion-imprinted solid-phase extraction material and its analytical application. *Anal. Chim. Acta* **2007**, *603*(1), 87–92.

216. Michailof, C.; Manesiotis, P.; Panayiotou, C. Synthesis of caffeic acid and p-hydroxybenzoic acid molecularly imprinted polymers and their application for the selective extraction of polyphenols from olive mill waste waters. *J. Chromatogr. A* **2008**, *1182*(1), 25–33.

217. Zhang, M.; Xie, J.; Zhou, Q.; Chen, G.; Li, Z. On-line solid-phase extraction of ceramides from yeast with ceramide III imprinted monolith. *J. Chromatogr. A* **2003**, *984*(2), 173–183.

218. Turiel, E.; Tadeo, J.L.; Martin-Esteban, A. Molecularly imprinted polymeric fibers for solid-phase microextraction. *Anal. Chem.* **2007**, *79*(8), 3099–3104.

219. Hu, X.; Hu, Y.; Li, G. Development of novel molecularly imprinted solid-phase microextraction fiber and its application for the determination of triazines in complicated samples coupled with high-performance liquid chromatography. *J. Chromatogr. A* **2007**, *1147*(1), 1–9.

220. Djozan, D.; Ebrahimi, B. Preparation of new solid phase micro extraction fiber on the basis of atrazine-molecular imprinted polymer: Application for GC and GC/MS screening of triazine herbicides in water, rice and onion. *Anal. Chim. Acta* **2008**, *616*(2), 152–159.

221. Hu, X.; Pan, J.; Hu, Y.; Huo, Y.; Li, G. Preparation and evaluation of solid-phase microextraction fiber based on molecularly imprinted polymers for trace analysis of tetracyclines in complicated samples. *J. Chromatogr. A* **2008**, *1188*(2), 97–107.

222. Boos, K.-S.; Fleischer, C.T. Multidimensional on-line solid-phase extraction (SPE) using restricted access materials (RAM) in combination with molecular imprinted polymers (MIP). *Fresenius J. Anal. Chem.* **2001**, *371*(1), 16–20.

223. Theodoridis, G.; Zacharis, C.K.; Tzanavaras, P.D.; Themelis, D.G.; Economou, A. Automated sample preparation based on the sequential injection principle solid-phase extraction on a molecularly imprinted polymer coupled on-line to high-performance liquid chromatography. *J. Chromatogr. A* **2004**, *1030*(1–2), 69–76.

224. Haginaka, J.; Takehira, H.; Hosoya, K.; Tanaka, N. Uniform-sized molecularly imprinted polymer for (S)-naproxen selectively modified with hydrophilic external layer. *J. Chromatogr.* **1999**, *849*, 331–339.

225. Andersson, L.I. Efficient sample pre-concentration of bupivacaine from human plasma solid-phase extraction on molecularly imprinted polymers. *Analyst* **2000**, *125*(9), 1515–1517.

226. Lanza, F.; Sellergren, B. Molecularly imprinted extraction materials for highly selective sample cleanup and analyte enrichment. *Adv. Chromatogr.* **2001**, *41*, 137–173.

227. Masque, N.; Marce, R.M.; Borrull, F. Molecularly imprinted polymers: New tailor-made materials for selective solid-phase extraction. *Trends Anal. Chem.* **2001**, *20*(9), 477–486.

228. Caro, E.; Marce, R.M.; Borrull, F.; Cormack, P.A.G.; Sherrington, D.C. Application of molecularly imprinted polymers to solid-phase extraction of compounds from environmental and biological samples. *Trends Anal. Chem.* **2006**, *25*(2), 143–154.

229. Pichon, V.; Chapuis-Hugon, F. Role of molecularly imprinted polymers for selective determination of environmental pollutants—A review. *Anal. Chim. Acta* **2008**, *622*(1–2), 48–61.

230. Appell, M.; Maragos, C.M.; Kendra, D.F. Molecularly imprinted polymers for mycotoxins. *ACS Symp. Ser.* 2008, *1001* (Food Contaminants: Mycotoxins and Food Allergens), 152–169.

231. Baggiani, C.; Anfossi, L.; Giovannoli, C. Molecular imprinted polymers as synthetic receptors for the analysis of myco- and phyco-toxins. *Analyst* **2008**, *133*(6), 719–730.

232. Tamayo, F.G.; Martin-Esteban, A. Selective high performance liquid chromatography imprinted-stationary phases for the screening of phenylurea herbicides in vegetable samples. *J. Chromatogr. A* **2005**, *1098*(1–2), 116–122.

233. Sellergren, B. Molecular imprinting by noncovalent. *Makromol. Chem.* **1989**, *190*(11), 2703–2711.

234. Shimizu, K.D. Binding isotherms. In *Molecularly Imprinted Materials: Science and Technology*, Yan, M.; Ranström, O., Eds., New York: Marcel Dekker, 2005; pp. 419–434.

235. Dauwe, C.; Sellergren, B. Influence of template basicity and hydrophobicity on the molecular recognition properties of molecularly imprinted polymers. *J. Chromatogr. A* **1996**, *753*(2), 191–200.

236. Nicholls, I.A.; Ramström, O.; Mosbach, K. Insights into the role of the hydrogen bond and hydrophobic effect on recognition in molecularly imprinted polymer synthetic peptide receptor mimics. *J. Chromatogr. A* **1995**, *691*(1–2), 349–353.

237. Fersht, A. *Enzyme Structure and Mechanism*, 2nd ed. New York: W.H. Freeman and Company; 1985.

238. Sellergren, B.; Shea, K.J. Chiral ion-exchange chromatography. Correlation between solute retention and a theoretical ion-exchange model using imprinted polymers. *J. Chromatogr. A* **1993**, *654*, 17–28.

Index

Handbook of Sample Preparation, Edited by Janusz Pawliszyn and Heather L. Lord

www.ingramcontent.com/pod-product-compliance
Lightning Source LLC
LaVergne TN
LVHW062137200926
848484LV00007B/1824